Visuomotor Coordination

Amphibians, Comparisons, Models, and Robots

Visuomotor Coordination

Amphibians, Comparisons, Models, and Robots

Edited by

Jörg-Peter Ewert

University of Kassel
Kassel, Federal Republic of Germany

and

Michael A. Arbib

University of Southern California
Los Angeles, California

Plenum Press • New York and London

Library of Congress Cataloging in Publication Data

International Workshop on Visuomotor Coordination in Amphibians: Experiments, Comparisons, Models, and Robots (1987: Kassel, Germany)
Visuomotor coordination: amphibians, comparisons, and models, and robots / edited by Jörg-Peter Ewert and Michael A. Arbib.
p. cm.
"Proceedings of an International Workshop on Visuomotor Coordination in Amphibians: Experiments, Comparisons, Models, and Robots, held August 25-27, 1987, in Kassel, Federal Republic of Germany"—T.p. verso.
Bibliography: p.
Includes index.
ISBN 0-306-43230-7
1. Amphibians—Physiology—Congresses. 2. Vision—Congresses. 3. Sensorimotor integration—Congresses. I. Ewert, Jörg-Peter, 1938- II. Arbib, Michael A. III. Title.
QL669.2.I59 1987 89-8467
597.6′041—dc20 CIP

Proceedings of an International Workshop on Visuomotor Coordination in Amphibians: Experiments, Comparisons, Models, and Robots, held August 25-27, 1987, in Kassel, Federal Republic of Germany

Printed in the United States of America

This book is dedicated to the memory of our friends

Hans-Wilhelm Borchers, Rolando Lara, and Elena Sandoval

killed in a car crash close to Cuernavaca in Mexico on January 19, 1985.
We remember their significant contributions
to the understanding of the amphibian visual system
as we mourn their loss.
A sampling of their last papers is included in the present volume.

Contributors

P. ARAIZA, Instituto de Fisiologia Celular, Universidad Nacional Autónoma de México, 04510 México, D.F., *México*

MICHAEL A. ARBIB, Departments of Computer Science, Neurobiology, Physiology, Biomedical Engineering, Electrical Engineering, and Psychology, University of Southern California, Los Angeles, CA 90089-0782, *USA*

RONALD C. ARKIN, Department of Information and Computer Science, Georgia Institute of Technology, Atlanta, Georgia 30332, *USA*

ULRICH BÄSSLER, Fachbereich Biologie, Universität Kaiserslautern, D-6750 Kaiserslautern, *FR Germany*

THOMAS W. BENEKE, Abteilung Neuroethologie, Fachbereich Biologie/Chemie, Universität Kassel, D-3500 Kassel, *FR Germany*

M. B. BERKINBLIT, Institute for Information Transmission Problems, Academy of Sciences and Moscow State University, Moscow 101447, *USSR*

BILL BETTS, Program in Neural, Informational, and Behavioral Sciences, University of Southern California, Los Angeles, CA 90089-0782, *USA*

HANS-WILHELM BORCHERS[+], Abteilung Neuroethologie, Fachbereich Biologie/Chemie, Universität Kassel, D-3500 Kassel, *FR Germany*

FRANCISCO CERVANTES-PEREZ, Departamento de Neurosciencias, Instituto de Fisiologia Celular, Universidad Nacional Autónoma de México, 04510 México, D.F., *México*

THOMAS EGGERT, Max-Planck-Institut für Verhaltensphysiologie Seewiesen, D-8130 Seewiesen/Starnberg, *FR Germany*

JÖRG-PETER EWERT, Abteilung Neuroethologie, Fachbereich Biologie/Chemie, Universität Kassel, D-3500 Kassel, *FR Germany*

A. G. FELDMAN, Institute for Information Transmission Problems, Academy of Sciences and Moscow State University, Moscow 101447, *USSR*

J. FERNANDEZ, Instituto de Fisiologia Celular, Universidad Nacional Autónoma de México, 04510 México, D.F., *México*

THOMAS FINKENSTÄDT, Abteilung Neuroethologie, Fachbereich Biologie/ Chemie, Universität Kassel, D-3500 Kassel, *FR Germany*

EDDA M. FRAMING, Abteilung Neuroethologie, Fachbereich Biologie/Chemie, Universität Kassel, D-3500 Kassel, *FR Germany*

O. I. FUKSON, Institute for Information Transmission Problems, Academy of Sciences and Moscow State University, Moscow 101447, *USSR*

FRANCISCO GONZALEZ-LIMA, Department of Anatomy, College of Medicine, Texas A&M University, College Station, TX 77843, *USA*

PAUL GROBSTEIN, Department of Biology, Bryn Mawr College, Bryn Mawr, Pennsylvania 19010, *USA*

EDWARD R. GRUBERG, Biology Department, Temple University, Philadelphia, PA 19122, *USA*

UWE AN DER HEIDEN, Fakultät für Naturwissenschaften, Universität Witten/ Herdecke, D-5810 Witten, *FR Germany*

BARRY HORWITZ, Laboratory of Neurosciences, National Institute on Aging, National Institutes of Health, Bldg. 10, RM. 12S-207, Bethesda, MD 20892, *USA*

REINHARD KOY-OBERTHÜR, Institut für Elektrische Informationstechnik, Technische Universität Clausthal, D-3392 Clausthal-Zellerfeld, *FR Germany*

PETER R. LAMING, Department of Biology, Queen's University of Belfast, Belfast BT7 1NN, Northern Ireland, *UK*

ROLANDO LARA[+], Centro de Investigaciones en Fisiologia Celular, Universidad Nacional Autónoma de México, 04510 México D.F., *México*

GYULA LAZAR, Department of Anatomy, University Medical School, Pecs, H-7643, *Hungary*

HANSPETER A. MALLOT, Institut für Zoologie III (Biophysik), Johannes Gutenberg-Universität, D-6500 Mainz, *FR Germany*

CHRISTOPH VON DER MALSBURG, Department of Computer Science and Neurobiology, University of Southern California, Los Angeles, CA 90089-0782, *USA*

GERHARD MANTEUFFEL, Fachbereich Biologie, Universität Bremen, D-2800 Bremen, *FR Germany*

L. MASSIEU, Instituto de Fisiologia Celular, Universidad Nacional Autónoma de México, 04510 México, D.F., *México*

NOBUYOSHI MATSUMOTO, Department of Biophysical Engineering, Faculty of Engineering Science, Osaka University, Toyonaka, Osaka 560, *Japan*

HORST MITTELSTAEDT, Max-Planck-Institut für Verhaltensphysiologie Seewiesen, D-8130 Seewiesen/Starnberg, *FR Germany*

CHRISTOPH PINKWART, Hölderlinstr. 10, D-8500 Nürnberg 20, *FR Germany*

HERBERT J. REITBOECK, Angewandte Physik und Biophysik, Philipps-Universität Marburg, D-3550 Marburg, *FR Germany*

GERHARD ROTH, Fachbereich 2/Biologie, Universität Bremen, D-2800 Bremen 33, *FR Germany*

M. E. SANDOVAL[+], Instituto de Fisiologia Celular, Universidad Nacional Autónoma de México, 04510 México, D.F., *México*

EVELYN SCHÜRG-PFEIFFER, Abteilung Neuroethologie, Fachbereich Biologie/Chemie, Universität Kassel, D-3500 Kassel, *FR Germany*

WOLFGANG W. SCHWIPPERT, Abteilung Neuroethologie, Fachbereich Biologie/Chemie, Universität Kassel, D-3500 Kassel, *FR Germany*

WERNER VON SEELEN, Institut für Zoologie III (Biophysik), Johannes Gutenberg-Universität, D-6500 Mainz, *FR Germany*

CARME TORRAS, Institut de Cibernètica (CSIC-UPC), Diagonal 647, 08028-Barcelona, *Spain*

ANANDA WEERASURIYA, Department of Physiology, Faculty of Medicine, University of Colombo, Colombo-8, *Sri Lanka*

Preface

Various brain areas of mammals can phyletically be traced back to homologous structures in amphibians. The amphibian brain may thus be regarded as a kind of "microcosm" of the highly complex primate brain, as far as certain homologous structures, sensory functions, and assigned ballistic (pre-planned and pre-programmed) motor and behavioral processes are concerned. A variety of fundamental operations that underlie perception, cognition, sensorimotor transformation and its modulation appear to proceed in primate's brain in a way understandable in terms of basic principles which can be investigated more easily by experiments in amphibians. We have learned that progress in the quantitative description and evaluation of these principles can be obtained with guidance from theory. Modeling - supported by simulation - is a process of transforming abstract theory derived from data into testable structures. Where empirical data are lacking or are difficult to obtain because of structural constraints, the modeler makes assumptions and approximations that, by themselves, are a source of hypotheses. If a neural model is then tied to empirical data, it can be used to predict results and hence again to become subject to experimental tests whose resulting data in turn will lead to further improvements of the model. By means of our present models of visuomotor coordination and its modulation by state-dependent inputs, we are just beginning to simulate and analyze how external information is represented within different brain structures and how these structures use these operations to control adaptive behavior.

Visuomotor Coordination: Amphibians, Comparisons, Models, and Robots presents for a larger audience the proceedings of the third international workshop on this topic, organized by Peter Ewert at the University of Kassel in August 25-27, 1987. The interdisciplinary workshops on visuomotor coordination in frogs and toads have a successful tradition. The first meeting was organized by Michael Arbib at the University of Massachusetts at Amherst in 1981 and the second by Rolando Lara at the University of Mexico in 1982. Meanwhile a wealth of new experimental data on visuomotor functions in target-oriented behaviors of amphibians - in terms of ethology, physiology, anatomy, and pharmacology - has accumulated. Our third international workshop proceeds in the tradition of furthering the dialog between experimentalists and theorists.

Because there are no boundaries between artificial and natural sensorimotor systems in cybernetics, cross-disciplinary studies are particularly fruitful. Implicit in the goals of our interdisciplinary research is another branch touching on interests of applied science. The study of biological information processing systems holds promise of indicating elegant solutions to practical problems in robots. Natural pattern recognition systems, for example, successfully deal with different invariance conditions in visual sensorimotor tasks. Comparable properties have proven difficult to implement in optoelectronic devices or machines (e.g., position/orientation/velocity invariance in the recognition of work-pieces on assembly lines). Computer simulations represent operating algorithms derived from the basic properties of the model which can be implemented in robotic systems. Hence, with the aid of models constructed from data on biological mechanisms of visuomotor coordination a first step in the development and the testing of algorithms for performance and tuning of robotic devices may be taken. Moreover, the interplay of empirical and theoretical analyses will actually increase the pace toward our understanding of the properties of information processing carried out by central nervous systems.

PART I opens the workshop proceedings with a general discussion on experimentation and modeling in neuroscience, transcribed from tape recordings at the meeting. Main topics of general interest concern the dialog between model and experiment, the structure/function problem, the modulation of sensorimotor function, and methodological interpretational problems.

PART II offers two overviews of the theme of this volume. The neuroethological approach (**Ewert**) develops the neurophysiological equivalent of the ethological Lorenzian concept of an "innate releasing schema" (also called releasing mechanism after Tinbergen) as a "sensorimotor code" constituted by different combinations of neurons, each of these expressing information processing in a functional unit of interacting cells and specified to encode significant coincidences of stimulus cues that refer to an object, e.g. a visual releaser. The corresponding command releasing systems result from parallel distributed processing in a neural macro-network that evaluates visual input under various aspects, such as features, background, space, and past experience. The systems theoretical treatment (**Arbib**) uses perceptual and motor schemas to provide models of amphibian visuomotor coordination bridging behavior and neural circuits, and allowing approaches to perceptual robotics, both in the control of dexterous robot hands and the design of machine vision systems.

PART III reports new results on the cellular architecture and connectivity of the anuran's optic tectum and pretectum (**Lázár**) and the response properties of labeled neurons (**Matsumoto**). We are faced here with an enormous richness in cytological features of local circuits and of connecting and projecting neurons whose bifurcating axons might provide a substrate for parallel processing. Response properties raise the question of isomorphisms between the geometry of dendritic trees, axonal projections, and the function of neurons. Application of the pulse-triggered averaging technique provides insight into retino-tectal synaptic connectivity. Biochemical investigations are becoming of increasing importance to understand interneuronal communication and its control. Using isolated nerve endings or brain slices from the optic tectum, various neurochemical criteria - sodium dependent uptake and calcium dependent release - are applied to identify amino acid and amine neurotransmitters (**Sandoval, Massieu, Araiza & Fernández**). A model of retino-tectal information processing drawing on Gaussian receptive fields and layered *on-* and *off*-units captures many neuronal response properties (**an der Heiden & Roth**), while a mathematical dynamic model of a functional unit in the tectum suggests that subclasses of tectal cells can be interpreted as more or less stable states of a thalamic-pretectal controlled tectal system responsible for object recognition and capable of adaptability and changeability (**Betts**).

PART IV considers in particular the role of so-called visual centers and their interactions. The notion is put forward that various functions related to feature filtering, generating nystagmus, and estimating depth in amphibians take advantage of the same basic mechanism in which perception of coherent motion on the one hand and form discrimination and object localization on the other hand are ascribed to integrative, intrinsic, and interactive operations in tectal and pretectal structures (**Manteuffel**). The nucleus isthmi, a post-tectal feedback system associated with the retino-tectal pathway, seems to directly modify retinal input (**Gruberg**), which is not necessarily incompatible with its role in prey-localization as suggested previously. These more global treatments of visual structures challenge the question "why cortices?" It can be shown that neuronal networks of cortical organization responsible for visual information processing involve certain structural principles (average anatomy, retinotopy, patchy columnar organization) whose significance for neural computation and technical considerations is discussed mathematically (**Mallot & von Seelen**). Furthermore, we learn that our present models of invariance generation in visual pattern recognition in mammals have many attractive features whose main function is to point to specific problems, to develop concepts, to stimulate new experiments, and to give a theoretical background for machine vision (**Reitboeck**). This theme leads to the development of a schema-based "hybrid" contour perception system in sensory substitution for the blind, in which reduced pictorial information is transformed via a somatosensory channel, and in which human intelligence is used for pattern recognition (**Koy-Oberthür**).

PART V draws on the various questions related to sensorimotor interfacing, starting with the presentation of Schema Theory as an interscientific bridge that provides us, on the one hand, with a global language to explain cognition at a level that can be used and understood within all cognitive sciences and, on the other hand, with a key methodology for "top-down" studies (**Cervantes-Pérez**). Neurophysiological equivalents of perceptual and motor schemas, e.g. configuration-selective tecto-bulbar/spinal projecting cells studied during the toad's behavior show recognition, premotor, motor feedback, and modulatory properties (**Schürg-Pfeiffer**). Intracellular recording and labeling studies in toads suggest that various types of sensory information converge on highly integrative interacting bulbar cells to yield frequency (burst) coded pre-motor signals (**Schwippert, Beneke & Framing**). It is suggested that directed movements depend on "activity gated divergence," a property of neural networks (involving tectal, tegmental, and bulbar structures) which allows them to associate with a given topographically specified sensory input any of a number of outputs, depending on the character of other relevant neuronal signals (**Grobstein**). This calls for a formulation of principles for transforming map-ordered spatial information into motor commands: The approach to the positive phototaxis of an idealized "mathematical planaria" offers the concept of "map-weighting" - an intermediate information processing system between eyes and motor apparatus - which can be applied to sensorimotor transformation in general (**Mittelstaedt & Eggert**).

PART VI focusses on motor control, starting with an anatomical study in search of the motor pattern generator for snapping in toads. This approach suggests that a vital component for tongue/jaw coordination consists of two unilateral subunits located in the bulbar medial reticular formation that receive visual tectal, tactual toral, and visceral solitary inputs, and these subunits have largely independent access to the relevant motoneuronal pools (**Weerasuriya**). Investigations on the various forms of frog's wiping patterns in response to cutaneous stimulation corroborate the hypothesis that motor behaviors are organized of components and segments that can be combined in different ways to form new movement patterns, and this is consistent with the concept of motor schema (**Berkinblit, Feldman & Fukson**). Furthermore, these results suggest that every limb joint is subserved by a set of individual control systems which interact in the process of solving a common motor task, similar to motor coordination in locomotion. Neurobiological investigations on the variable walking patterns in stick insects point to a large amount of democracy among the motor pattern generating subunits under the influence of sense organs, a democracy which does not exclude the possibility that higher centers command or, rather, recommend to locomote (**Bässler**). The two last chapters of PART VI refer to robot motor control. A mobile robot path navigation system has been developed that - traversing open spaces and corridors - strongly correlates with a schema-based model for toad's detour behavior whereby multiple concurrent schemas contribute to the determination of the overall path of the vehicle (**Arkin**). Perceptual robotics hence prove the relevance of Arbib's *Rana computatrix*, the computational frog. In robotic assembly systems, sensorimotor integration can either take the form of a two-stage process accomplished by interleaving sensation and action or arise from the simultaneous operation of sensors and actuators. Here the so-called on-line specialists encompass the parametric code for the sensory strategies available to corresponding off-line specialists, where the former can be seen as complex motor schemas that have perceptual schemas embedded into them (**Torras**).

PART VII reviews new results on brain structures and suggested neural mechanisms responsible for arousal, habituation, and learning. A concept of arousal explaines how processes related to high frequency waveforms in the electro-encephalogram and to sustained potential shifts increase neuronal responsiveness, partly in connection with glial activity. It is suggested that studies on neuronal sensitization may also reveal the causes of neuronal hyperactivity in clinical disorders like epilepsy (**Laming**). Investigations applying the deoxyglucose technique show that non-associative (habituation) and associative (classical conditioning) learning in toads involve prosencephalic circuits, extrinsic to the retinotectal one, which share particular structures (**Finkenstädt**) that are homologous to those in mammals for comparable functions (**Gonzalez-Lima**). An interesting outcome of these studies is that arousal and habituation are accompanied by opposite patterns of activity in certain brain structures whose prediction has a long tradition in neuroscience. The dialog between experiment and theory again leads to construction of models that on the one hand allow us to understand processes involved in stimulus-specific habituation and associative learning and on the other hand show us how to design new experiments (**Lara**). New experimental perspectives for the study of single neurons during arousal and habituation in behavioral sequences are feasible by means of a three-function telemetric system designed for single cell recording, stimulating, and marking (**Borchers & Pinkwart**). A novel approach to the analysis of functional neural systems by use of interregional correlation of energy metabolism is presented and its application to studies in humans and animals is discussed (**Horwitz**). At the end of PART VII a retrospective view stresses some goals, problems, and perspectives involved in modeling so-called simple behaviors with respect to the various levels of data structure (**von der Malsburg**).

The workshop was sponsored by Hessischer Minister für Wissenschaft und Kunst, Präsident der Universität Kassel, Kanzler der Universität Kassel, Bayer Pharma AG, Hoffmann-La Roche AG, National Institutes of Health NIH under Grant 1 R01 NS 24926 from NINCDS, and in part by a travel grant from the Program in Neural, Informational, and Behavioral Sciences of the the University of Southern California.

We thank Priv. Doz. Dr. Thomas Finkenstädt, Dr. Evelyn Schürg-Pfeiffer, Dr. Wolfgang Schwippert, Dipl. Biol. Thomas Beneke, Dipl. Biol. Claudia Merkel-Harff, Mrs. Gisela Burggraf, Mrs. Gudrun Frühauf, and Mrs. Gerda Kruk for their assistance in the local organization of the meeting. We gratefully acknowledge the generous help of Dr. Ananda Weerasuriya in reworking the general discussions from the tape recordings. The collaboration of all participants contributed greatly toward a successful workshop and a reasonably fast publication of the proceedings.

We express our special gratitude to Mrs. Ursula Reichert (Abteilung Neuroethologie, Universität Kassel) for text processing, type setting, and lay-out.

Kassel, Summer 1988

Jörg-Peter Ewert
Michael A. Arbib

Contents

PART I: Introduction

Experimentation and Modeling: An Introductory Discussion
M.A. ARBIB, U. BÄSSLER, J.-P. EWERT, T. FINKENSTÄDT, F. GONZALEZ-LIMA, P. GROBSTEIN, E.R. GRUBERG, U. AN DER HEIDEN, B. HORWITZ, P.R. LAMING, G. LAZAR, H.A. MALLOT, C. VON DER MALSBURG, H. MITTELSTAEDT, G. MOORE, E. MÜHLENFELD, J.I. NELSON, and H.J. REITBOECK

First Session: Experimenters, Modelers, and Robotics 3
Second Session: Neurons, Networks, and Building Blocks 16
Third Session: Distributed Properties, Modulation, and Memory 27

PART II: An Opening Perspective

The Release of Visual Behavior in Toads: Stages of Parallel/Hierarchical Information Processing
J.-P. EWERT
1 Introduction 39
2 Action Patterns and Their Releasing Schemas/Mechanisms 43
2.1 Action Patterns 43
2.2 The Conditions of Sequential Release 45
3 Configural Properties of Sign Stimuli 45
3.1 Natural Prey 46
3.2 Prey/Background Relationships 47
3.3 Innate Correspondence Between Stimulus Features and RMs 47
3.4 Learned Correspondence Between Stimulus Features and RMs 54

3.5 How Are Moving Objects Sorted? 56
4 Approach Toward Behaviorally Relevant Brain Structures 63
4.1 Regional Distribution of Brain Activity 64
4.2 Functional Relevance of Visual Maps and Related Structures 65
4.3 Properties of a Macro-Network 68
5 Feature-Analyzing Neurons and Integrative Functional Units 69
5.1 Information Processing in the Retina 70
5.2 Feature-Analyzing Tectal and Pretectal Neurons 72
5.3 Toward a Feature Analyzing Neuronal Circuit 79
5.4 Comparisons Across Species 84
5.5 How are Moving Objects Represented in Anuran's Visual System? 86
5.6 Evaluation of T5.2 Neurons 89
6 Sensorimotor Interfacing: The Command System Approach 92
6.1 Specialized Neurons 92
6.2 Tecto-Bulbar/Spinal Output 93
6.3 The "Command Releasing System" Concept 95
6.4 Modulatory Influences 98
6.5 What General Properties Do CRSs Have? 102
7 Concluding Remarks 104
7.1 Stages of Parallel/Hierarchical Processing 104
7.2 Comparisons 106
7.3 From Instinct to Cognition 107
7.4 Modeling and Engineering 108
8 References 109

Visuomotor Coordination: Neural Models and Perceptual Robotics
M. A. ARBIB
1 What is a Schema? 121
2 Schemas for Rana computatrix 126
2.1 Schemas for Prey Recognition 126
2.2 Schemas for Detour Behavior 128
2.3 Schemas for Prey-Acquisition 133
3 Tectal Columns 137
3.1 Facilitation of Prey Catching Behavior 137
3.2 An Array of Tectal Columns that Models Pattern Recognition 140
4 Depth Perception 146
4.1 The Dev Model 146
4.2 The Cue Interaction Model 147
4.3 The Prey Localization Model 149
5 Path-Planning and Detours 152
6 Schemas for Hand Control 159
7 Schemas for Vision 163
8 Challenges for Cooperation 166
8.1 Cooperation Between Neuroscience and Computer Science 167
8.2 Cooperation in the Study of the Brain 167
9 References 169

PART III: Cellular Mechanisms in Tectum and Pretectum

Cellular Architecture and Connectivity of the Frog's Optic Tectum and Pretectum
G. LAZAR
1 Introduction 175
2 Optic Tectum 176
2.1 General Organization 176
2.2 Columnar Organization of the Optic Tecum 177
2.3 Tectal Neurons 179
2.4 Non-Optic Afferent Pathways of the Optic Tectum 187
3 Pretectum 187
4 Discussion 191
4.1 Projection Neurons, Local Circuit Neurons 191
4.2 Columnar Organization of the Tectum 192
4.3 Pretectum 193
4.4 General Organization of Supraspinal Control System 195
5 References 195

Morphological and Physiological Studies of Tectal and Pretectal Neurons in the Frog
N. MATSUMOTO
1 Introduction 201
2 Receptive Fields and Response Properties of Identified Tectal Neurons 202
2.1 Responses to Electrical Stimulation of the Optic Tract 202
2.2 Responses to Changes of Diffuse Light 204
2.3 Responses to Moving Configurational Stimuli 206
2.4 Axonal Projections of Large Ganglionic Neurons 209
3 Morphological Identification of Prectectal Neurons 209
4 Synaptic Retino-Tectal Connections 213
5 Discussion 218
5.1 Response Properties and Cellular Morphology 218
5.2 Projection of Tectal and Pretectal Neurons 218
5.3 Neuronal Circuit 219
6 References 220

Toward an Identification of Neurotransmitters in the Frog's Optic Tectum
M. E. SANDOVAL+, L. MASSIEU, P. ARAIZA, and J. FERNANDEZ
1 Introduction 223
2 Methodology 225
2.1 General Considerations 225
2.2 Uptake Experiments 227
2.3 Release Experiments 227

2.4 Enucleation Experiments 228
2.5 Enzymatic Quantification of Glutamic Acid 228
3 Uptake Studies 229
3.1 Amino Acid Neurotransmitters 229
3.2 Amine Neurotransmitters 230
4 Release Studies 230
4.1 Glutamic Acid Release 230
4.2 GABA Release 233
4.3 Glycine Release 235
4.4 Aspartic Acid Release 235
5 Discussion 236
5.1 Sodium-Dependent Uptake 236
5.2 Calcium-Dependent Release 237
6 References 238

Retina and Optic Tectum in Amphibians:
A Mathematical Model and Simulation Studies
U. AN DER HEIDEN and G. ROTH
1 Introduction 243
2 Response Properties of Tectal Neurons 245
3 Mathematical Model of Retinal Ganglion Cells 248
4 Comparison of Real and Simulated Responses of the Retinal Network 251
5 Mathematical Model for the Constitution of Some Tectal Cell Types 257
5.1 Basic Model: Summatory Processes in the Tectum 257
5.2 Extended Model: Lateral Inhibition in the Tectum 261
6 Discussion 262
7 References 265

The T5 Base Modulator Hypothesis:
A Dynamic Model of T5 Neuron Function in Toads
B. BETTS
1 Introduction 269
1.1 Behavioral Flexibility and Neuronal Tuning 270
1.2 Toward a "T5 Base Modulator" 271
1.3 Other Models in Perspective 275
2 A Model of T5 Neuron Function 276
2.1 General Model Anatomy 276
2.2 A Mechanism for Enhancing Worm Discrimination 277
2.3 Antiworm Preference Generated by Retinal Ganglion Cell Inputs 279
2.4 Antiworm Preference Generated by Stimulus Configuration 281
2.5 Antiworm Preference Generated by Tectal Interneurons 285
2.6 Long-Term Changes in T5 Neurons 287
2.7 The Model 289
3 Discussion 290
3.1 The T5 Base Modulator Hypothesis 290
3.2 T5 Subclasses and the T5 Base Modulator 293

3.3 Retinal Inputs and the T5 Base Modulator 294
3.4 Tectal Interneurons 294
3.5 Ethological Implications 295
3.6 Comparison with Other Models 297
3.7 Directional Sensitivity 298
3.8 Conclusion 300
4 Appendix: Mathematical Description of the Model 301
4.1 Retinal Inputs: R2, R3, and R4 301
4.2 Tectal and Thalamic Neurons 302
5 References 305

PART IV: The Role of Visual Centers

Compensation of Visual Background Motion in Salamanders
G. MANTEUFFEL
1 Introduction 311
2 Behavioral Analysis of the OKR 314
3 Neural Circuits Underlying the Control of OKR 317
4 Behavioral Analysis of the Vestibulo-Collic Reflex 324
5 Visual-Vestibular Interactions in Gaze Stabilization 325
6 Modeling of Gaze Stabilization in Salamanders 327
7 The Intrinsic Self-Motion Computer 329
8 Discussion and Conclusions 332
8.1 Consequences of the Model for Internal Self-Motion Computation 334
8.2 Implications for the Function of Visual Systems 335
8.3 Can the Salamander Help Engineers to Improve Technical Image Analysis? 336
9 References 336

Nucleus Isthmi and Optic Tectum in Frogs
E. R. GRUBERG
1 Introduction 341
2 The Nucleus Isthmi 342
3 The Behavioral Deficits 343
3.1 Scotomas 344
3.2 Permanence of the Deficit 344
3.3 Size of Scotoma vs. Size of Lesion 345
3.4 Ibotenic Acid Application 345
3.5 Bilateral N. Isthmi Lesions 346
3.6 Conflicting Behavioral Data 346
4 Anatomy 347
5 Acetylcholine as the Isthmo-Tectal Neurotransmitter 350
6 What Tectal Elements Do N. Isthmi Fibers Affect? 351

7 What Does N. Isthmi Do? 351
8 References 355

Why Cortices? Neural Networks for Visual Information Processing
H. A. MALLOT and W. VON SEELEN
1 Introduction 357
1.1 Computational Theory and Structured Neural Networks 357
1.2 Structural Principles of Neural Networks Subserving Vision 358
2 Lamination and Intrinsic Feedback 359
2.1 Average Anatomy 359
2.2 Positive Feedback and Quadratic Regularization 363
2.3 Summary 364
3 Retinotopic Mapping 365
3.1 Formal Description of Retinotopic Maps 365
3.2 Inverse Perspective Mapping for Optical Flow Computation 369
3.3 Mapped Filters 372
3.4 Summary 376
4 Non-Topographic Mapping 377
4.1 Patchy Connectivity 377
4.2 Columnar Organization 377
4.3 Computational Maps 378
4.4 Summary 379
5 Conclusions 379
6 References 380

Invariances in Pattern Recognition
H. J. REITBOECK
1 Introduction 383
2 Shift-Invariance 384
2.1 Generation of Shift-Invariance via the Fourier Transform 385
2.2 R-Transforms as a Model of Shift-Invariance in the Visual System 387
3 Size- and Rotation-Invariance 390
4 Invariance Generation in the Visual System via Global Transforms 391
4.1 Problems Due to the Cyclic Periodicity of the Transforms 391
4.2 Object Separation 392
5 Conclusion 393
6 References 393

Perception by Sensorimotor Coordination in Sensory Substitution for the Blind
R. KOY-OBERTHÜR
1 Introduction 397
2 General Description of the Contour-Perception System 399
3 The Sensory Substitution Channel 402
4 A Model of Sensorimotor Coordination 404
5 Preliminary Experimental Results 411
6 Discussion and Conclusions 414

6.1 What Has Been Shown in this Approach? 414
6.2 What Can Schema Theory Perform in Principle? 414
6.3 Situations Described by Schemas 415
6.4 Further Investigations 416
7 References 417

PART V: The Visuomotor Interface

Schema Theory as a Common Language to Study Sensorimotor Coordination
F. CERVANTES-PEREZ
1 Introduction 421
2 The Cognitive Science Tri-Cycle 423
2.1 The Initial Cognitive Sciences 425
2.2 The Theory-Experiment Cycle 425
2.3 Intelligent Machines: Robotics and Artificial Intelligence 425
2.4 Perceptual Robotics 427
2.5 Philosophy and Cybernetics 427
2.6 A Multiple Interaction: The Tri-Cycle 428
3 Schema Theory as the Bridge to Study Sensorimotor Coordination 428
4 Schema-Theoretic Models of Visuomotor Coordination 432
4.1 The Frogs and Toads Model 432
4.2 The Praying Mantis Model 435
5 Schema Theory for Top-Down Studies 436
6 Conclusions 447
7 References 448

Behavior-Correlated Properties of Tectal Neurons in Freely Moving Toads
E. SCHÜRG-PFEIFFER
1 Introduction 451
2 Methodological Remarks 452
2.1 Terminology 452
2.2 Requirements 453
3 Responses of Retinal Ganglion Cells 458
3.1 Are Barlow's "Fly Detectors" Correlated with Prey Capture? 458
3.2 Are Lettvin's "Bug Perceivers" Correlated with Prey Capture? 459
4 Responses of Feature Analyzing Tectal Cells 460
4.1 Properties of Feature-Sensitive T5.1, T5.3, and T5.4 Neurons 460
4.2 Properties of Prey-Selective T5.2 Neurons 461
4.3 Properties of T5.2 Neurons Before and After a Pretectal Lesion 465
4.4 Evaluation of Temporal Discharge Patterns 467
4.5 Recording and Electrical Stimulation with Discharge Patterns 467
5 Spontaneously Active Tectal Neurons 468
5.1 Class T8.1 Neurons 469

5.2 Class T8.2 Neurons 469
6 Tectal Wide Field Neurons 470
7 Discussion and Conclusions 472
7.1 Tectal Command Functions 472
7.2 Command Elements for Toad's "Visual Grasp Reflexes" 472
7.3 Command vs. Motor Functions 475
7.4 Comparisons with Command Functions in Other Vertebrates 475
8 References 476

Visual Integration in Bulbar Structures of Toads: Intra/Extra-Cellular Recording and Labeling Studies
W. W. SCHWIPPERT, T. W. BENEKE, and E. M. FRAMING
1 Introduction 481
2 Extracellular Approach toward Properties of Bulbar Neurons 487
2.1 "Monocular Small Field" Neurons 489
2.2 "Binocular Small Field" Neurons 492
2.3 "Frontal Wide Field" Neurons 492
2.4 "Horizontal Wide Field" Neurons 493
2.5 "Dorsal Wide Field" Neurons 493
2.6 "Total Field" Neurons 493
2.7 Spontaneously Active Neurons 496
2.8 Reverberation Displaying Neurons 496
2.9 Cyclic Bursting Neurons 497
2.10 Projective Bulbar/Spinal Neurons 500
2.11 Multiple Property Displaying Neurons 501
3 Description of Bulbar Neurons by Intracellular Recording and Labeling 501
3.1 Survey of Intracellular Response Properties of Bulbar Neurons 503
3.2 Morphological Properties of Rostral Bulbar Neurons 504
3.3 Intracellular Activities of Visual Sensitive Bulbar Neurons 508
4 Discussion 520
4.1 General Properties of Bulbar Neurons in Toads 520
4.2 Problems with the M-Classification 524
4.3 Comparisons 524
5 References 529

Organization in the Sensorimotor Interface: A Case Study with Increased Resolution
P. GROBSTEIN
1 Introduction 537
2 Activity-Gated Divergence: A New Concept for the Sensorimotor Interface 538
3 Abstract Spatial Representation: A New Concept of Processing Within the Sensorimotor Interface 543
4 Discovery of New Pathways and Structures in the Sensorimotor Interface 551
5 A New Provisional Map of the Anatomical and Information Processing Organization of the Sensorimotor Interface 555
6 General Concepts of Sensorimotor Processing: What's New ? 560

7 References 565

How to Transform Topographically Ordered Spatial Information into Motor Commands
H. MITTELSTAEDT and T. EGGERT
1 Introduction 569
2 Fundamentals of Orientation 570
2.1 Understanding Loeb's Phototaxis 571
2.2 Performance 573
3 Generation of Motor Commands by Map-Weighting 575
3.1 Weighting and Necessary Conditions 575
3.2 Resolving vs. Tuning Power 577
4 Mathematical Appendix 578
4.1 Definitions 578
4.2 Comparison of the Result of "Map-Weighting" with the Statistical Optimum 581
5 Conclusion 583
6 References 584

PART VI: Motor Control

In Search of the Motor Pattern Generator for Snapping in Toads
A. WEERASURIYA
1 Introduction 589
2 Why Analyse Snapping? 591
3 Topography of Motoneurons 592
3.1 Tongue Muscle Motoneurons 592
3.2 Jaw Muscle Motoneurons 593
4 Tecto-Bulbar Outflow 593
5 Toward the Internuncial Circuitry 597
5.1 Spatial Summation 597
5.2 Specification of Afferents to Bulbar Nuclei 597
6 Discussion, Conclusions, and Speculations 604
6.1 What is the Evidence for a Snapping MPG? 605
6.2 How Many Interneurons are Traversed from Tectal Efferents to Motoneurons? 605
6.3 Which Kinds of Cells are Probably Involved in Snapping? 606
6.4 Functional Aspects of Snapping Pattern Generation 606
6.5 Predicted Locations and Interconnections of the Interneurons 607
6.6 Complexity and Variability of Snapping 609
6.7 Are Interneurons of the MPG "Monopolized" for Snapping? 610
7 References 611

Wiping Reflex in the Frog: Movement Patterns, Receptive Fields, and Blends
M. B. BERKINBLIT, A. G. FELDMAN, and O. I. FUKSON
1 Introduction 615
2 Methodological Comments 616
3 The Wiping Reflex 618
4 Wiping Forms and Their Receptive Fields 619
5 Hindlimb Movement Patterns 621
6 Hybrid Form of Wiping 623
7 Discussion and Conclusions 623
7.1 Movement Strategies 623
7.2 Signal Integration 624
7.3 Specification of Wiping by Command Neurons 624
7.4 The Problem of Choice 625
7.5 Hybrid Wiping 626
7.6 Sets of Collectively Working Control Systems 626
7.7 Levels of Sensorimotor Processing 627
8 References 628

Pattern Generation for Walking Movements
U. BÄSSLER
1 Introduction 631
2 Movement of All Legs 632
3 Control of a Single Leg 636
4 The "Active Reaction" 639
5 Discussion 646
6 References 647

Neuroscience in Motion: The Application of Schema Theory to Mobile Robotics
R. C. ARKIN
1 Introduction 649
2 Intention 650
2.1 Path Planning and Execution 650
2.2 Schemas 652
3 Methodology of Schema-Based Navigation within AuRA 653
4 Simulation 659
5 Experimental Results 661
5.1 Avoidance 661
5.2 Exploration 661
5.3 Hall Following 663
5.4 Navigation in the Presence of Obstacles 663
5.5 Single Wall Following 663
5.6 Impatient Waiting 663
5.7 Follow-The-Leader 670
5.8 Motor Schema Experiment Summary 670
6 Summary and Conclusions 670
7 References 671

Sensorimotor Integration in Robots
C. TORRAS
1 Introduction 673
2 Types of Sensors Used in Robotics 674
3 Placing Sensorimotor Integration in Context 676
4 An Illustrative Project 677
5 Degrees of Sensorimotor Integration 683
6 Some Sample Tasks 684
7 Discussion and Conclusions 686
8 References 687

PART VII: Arousal, Habituation, Conditioning

Central Representation of Arousal
P. R. LAMING
1 The Orientation Reaction: Arousal and Attention 693
2 Peripheral Physiological Correlates 694
3 Adaptive Nature of Responses to Novelty 695
4 Changes in Sensory and Motor Performance During Arousal and Attention 696
5 Changes in EEG During Behavioral Arousal 698
6 Oscillating Membrane Potentials: The Basis of the Synchronized EEG 700
6.1 Origin of EEG Waveforms 700
6.2 Frequency Increase in Membrane Potential Oscillations 702
6.3 Amplitude Increases in Oscillations 703
6.4 Changes in Synchrony of Oscillations 703
7 Nature and Origin of the SPS 707
7.1 The SPS as a Feature of Brain Activation 707
7.2 Involvement of Neuroglia 708
7.3 Implication of the Cytoarchitecture 709
7.4 Correlation with Sensory Experience 710
8 Abnormal Sensitization, Arousal and Glia 712
8.1 Epileptogenesis 712
8.2 Arousal and Seizure Incidence 713
8.3 Histopathology 715
8.4 Seizure Electrophysiology 716
8.5 Changes in Potassium Dynamics 717
9 Summary and Conclusions 718
10 References 719

Functional Brain Circuitry Related to Arousal and Learning in Rats
F. GONZALEZ-LIMA
1 Introduction 729
2 Methodological Comments 730

3 Thalamo-Cortical Involvement in Arousal 734
4 Arousal Effects on Limbic System 738
5 Basal Forebrain Involvement in Arousal 739
6 Motor and Autonomic Correlates of Arousal 740
7 Arousal Effects on the Auditory System 742
8 Learning Effects on the Auditory System 745
9 Nonauditory Forebrain Structures Involved in Associative Learning 750
10 Implications for the Neuroanatomical Organization of Auditory Learning 753
11 Summary and Conclusions 757
12 Abbreviations of Brain Structures 759
13 References 760

Stimulus-Specific Habituation in Toads: 2DG Studies and Lesion Experiments
T. FINKENSTÄDT
1 Introduction 767
2 Methodological Comments 769
3 The Experimental Paradigm: "Habituation Group" and "Naive Group" 769
4 Comparison of 2DG-Uptake Across Brain Structures 772
5 Brain Lesions 777
6 Electrical Stimulation and 2DG-Uptake 778
7 Narcosis and 2DG-Uptake 779
8 Discussion 783
8.1 Concepts of Stimulus-Response Mediating and Modulating Systems 783
8.2 Sokolov's Hypothesis of Habituation 784
8.3 The Lara & Arbib-Model of Prey Habituation in Toads 784
8.4 Brain Structures Involved in Prey Habituation 785
8.5 Hypothesis for Processes Underlying Prey Habituation 786
8.6 Habituation of Other Behaviors in Toads 788
8.7 Comparisons with Other Vertebrates 788
8.8 Habituation and Arousal 789
8.9 Consideration of Other Brain Structures 790
8.10 Critical Evaluation of 2DG-Uptake in "Habituation" and "Naive" Groups 791
9 Conclusions 792
10 Abbreviations of Brain Structures 792
11 References 793

Visual Associative Learning:
Searching for Behaviorally Relevant Brain Structures in Toads
T. FINKENSTÄDT
1 Introduction 799
2 The Training Paradigm: "Training Groups" and "Naive Groups" 801
2.1 Training Groups 801
2.2 Naive Group 803
3 Mapping of Brain Activity with 2DG 804
3.1 Technical Comments:
"Interhemispherical" and "Brain-to-Brain" Comparisons 804

3.2 Survey of Metabolic Activity Across Brain Structures 804
4 Brain Lesions 808
5 Olfactory Learning in Toads 811
6 Discussion 814
6.1 Associative Learning 814
6.2 General Comments on Learning-Correlated Activity in Brain Metabolism 815
6.3 Changes of Metabolic Activity in the Central Visual Pathway 816
6.4 Changes of Metabolic Activity in Non-Visual Forebrain Areas 819
6.5 Changes of Metabolic Activity in the Cerebellum 824
6.6 Olfactory Learning 824
7 Conclusions 826
8 Abbreviations of Brain Structures 827
9 References 828

Learning and Memory in the Toad's Prey/Predator Recognition System: A Neural Model
R. LARA+
1 Introduction 833
2 Basic Structure of the Model 835
3 General Behavior of the Model 838
4 Mathematical Definition of the Model 840
5 Computer Simulation 841
5.1 Long-Term Habituation 841
5.2 Learning to Avoid Bees 843
5.3 Learning to Attack a Predator-Like Stimulus 843
5.4 Learning to Discriminate Between Palatable and Unpalatable Stimuli 843
6 Discussion 848
7 Appendix 851
7.1 Description of the Basic Structure of the Model 851
7.2 Threshold Functions 852
7.3 Membrane Constants 853
7.4 Weights 853
7.5 Constants 853
8 References 853

Telemetric Transmission System for Single Cell Studies in Behaving Toads
H.-W. BORCHERS+ and C. PINKWART
1 Introduction 857
2 Telemetric Recording System 858
2.1 The Radio Transmitter 858
2.2 The VHF Receiver 862
3 Telestimulation System 863
3.1 The Inductive Energizer 863
3.2 The Telestimulator 864
4 Telemarking System 868
5 Subsystems in Independent Operation 868

6 Conclusions 868
7 References 870

Functional Neural Systems Analyzed by Use of Interregional Correlations of Glucose Metabolism
B. HORWITZ
1 Introduction 873
2 Simultaneous Recordings of Electrical Activity 874
2.1 Multiunit Microelectrode Recording 874
2.2 Optical Recording of Voltage Sensitive Dyes 874
2.3 Scalp Recorded Electrical Activity 875
3 Metabolic Measures of Functional Activity 875
3.1 Principles and Techniques 875
3.2 Problems in Interpretation 877
4 Correlation Matrix Method 879
4.1 Correlation Method 879
4.2 Application to Alzheimer's Disease 883
4.3 Application to Deoxyglucose Data from Rats 886
5 Discussion and Conclusions 887
6 References 890

Neural Models, Rana and Robots
C. VON DER MALSBURG
1 Introduction 893
2 Goals and Problems 893
3 Perspective 895
4 References 895

Subject Index 897

Visuomotor Coordination

Amphibians, Comparisons, Models, and Robots

PART I
Introduction

Experimentation and Modeling:

An Introductory Discussion

M.A. ARBIB, U. BÄSSLER, J.-P. EWERT, T. FINKENSTÄDT,
F. GONZALEZ-LIMA, P. GROBSTEIN, E.R. GRUBERG,
U. AN DER HEIDEN, B. HORWITZ, P.R. LAMING, G. LAZAR,
H.A. MALLOT, C. VON DER MALSBURG, H. MITTELSTAEDT,
G. MOORE, E. MÜHLENFELD, J.I. NELSON, and H.J. REITBOECK

Discussants in the general discussion sessions at the
Third International Workshop on Visuomotor Coordination in Amphibians
University of Kassel, August 25-27, 1987, D-3500 Kassel, FR Germany

Abstract. On the occasion of the lectures given at this interdisciplinary workshop, various topics of common interest were discussed by neurophysiologists, neuroanatomists, neuroethologists, systems theorists, physicists, and computational neuroscientists. The dialogs presented in this chapter are from tape recordings of three general discussion sessions. After introductory remarks the following issues were treated: (1) Promotion of the interaction between model and experiment; the advantage of modeling for neuroscience; the question of the adequate model; responsible *vs.* romantic modeling; problems with "detailism," "reductionism," and "reality"; the heuristic value of modeling; application of brain models to artificial intelligence and engineering. (2) The structure/function problem; synaptic, connectional, inferential, and intermediate levels of approach; the question of isomorphism between dendritic geometries and neuronal response properties; the notion of parametric *vs.* topographic representational maps; implicit, explicit, and cooperative processing; internal patterns of selforganization. (3) Modulation of sensorimotor functions; evaluation of distributed brain activities by monitoring energy metabolism; parallel distributed processing; coding problems; controversial experimental approaches and modeling; methodological and interpretational problems.

First Session: Experimenters, Modelers, and Robotics

MICHAEL ARBIB:

I would like to start a discussion going on what we can do over the next few years to catalyze a much stronger integration of theory and experiment. We already have a number of good examples, but I don't think they are enough. Let me pick a few points. We have models of the retina and the optic tectum. What we need to do next between us is to do two things. One is to provide the integrated software environment in which it is very easy to play with models and explore where they differ. Then in terms of that experience, we need to have our experimentalist colleagues engage in trying to really pin down experimentally whether the differences that seem to be in the models are indeed such that we can decide between models or design the next ones. That leads firstly into the criticism of theorists. We have discovered recently that models developed by our own group, as we try to reimplement them at the University of Southern California, and models in the literature by other groups are not specified well enough that we could reimplement them without having to guess many details. In other words the papers by the modelers often are enough to give you the basic structure of the model, but it turns out that there are a number of little missing details that are crucial before the thing will actually run on a computer. So, I am interested in the notion that in the next few years we can converge on some conventions for presenting models and get to the point where enough of us share computers that can run the same software, so that it becomes quite common for people who have developed a model to send it to others who can play with the model and test it. Conversely, one would hope then that the same computer systems can become available in the laboratory, so that more and more of the data from experiments is not simply that you have 6,000 slides from your work and you pick 7 of them which you then publish in the literature but that somehow this wealth of data is available in some kind of form which will allow one to then run much more thorough tests. So, I think there is an interesting sociological problem before us over the next few years.

In my own group now we have chosen some work stations using the C language using Unix operating systems. In the process of taking all the different models our group has developed and several other models and implementing them, we have come up with a common language. So, I would like to anounce here and now that we would be very happy to export that software as we develop it to any one of you and I invite you to start talking to me now or correspond with me in the future. Because I think, as I say, if we can develop common tools for modeling and then begin to have a critique of them, it can finally get to the point where there are sufficiently good packages that experimentalists who might want to get excited about the model can at least play with it. I have talked to others who have presented papers here today who say that they are very excited about interaction with modelers but they themselves are not modelers. They don't quite know how to do this. So, I think there is a tremendous problem for us somehow to get a good interaction going.

A comment I want to make on the paper of Francisco Cervantes-Pérez. He pointed out that he has used a model of a schema in which the notion of the goal, the classic cybernetic idea, is quite explicitly set forth, and Horst Mittelstaedt then argued that in his model of spider navigation the explicit representation of the goal was not necessary. What I want to say is that I think the crucial point is that as we come to analyze complex behaviors, we can't model everything at the lowest level. It was bad enough when the lowest level was the neuron. Now it is the transmitter-modulated membrane patch. It becomes hopeless if we have to model everything at that level without bridging constructs, whether they are quite real or just our ways of imposing order on complexity. We have to have bridging constructs. Now it seems to me that the ethologists presented the classic language and Horst Mittelstaedt gave a beautiful example of how to take complex behaviors and tease them apart into pieces. I would not like to think that we leave it with, here are the schemas that were presented by one group, here is the type of modeling that comes out, and that somehow the question of goals means that there is an unbridgeable gap. I think what we really have to do is get to work at the higher level, the coarser level of analysis, to try to understand what would be an adequate theoretical language for that level of analysis such that the sort of models we made of preying toads or praying mantises and the models we have of walking stick insects and spider navigation can be done. In other words, schema theory is not a final language that exists like "here it is, take it or leave it." It's part of a properly humble attempt to find a good level for describing complex behaviors, which could then be used to guide, organize, connect different models at the neural level.

Let me close by again noticing how confused we are at the neural level now because we have been learning so many new subtle properties of neurons that to try and model individual neurons in a way that takes into account everything that we know about the neuron can itself be overwhelming. So, again we need some way of understanding collective properties of neurons and then how to modulate them without necessarily understanding the subtlety of neurons. What I am really offering for discussion is my fervent hope that much of our discussion in the rest of the meeting is not modelers talking to each other about their computer models and experimentalists talking to each other about the latest stains, but that there will also be a lot of cross talk even if we don't have each other's vocabulary well enough in hand to begin to make a model. Professor Lázár has this wonderful anatomical data and it's so much richer than anything we put in the sort of connectivity we have in our models. Bill Betts will present a first attempt to take one of the cells Professor Matsumoto and Peter Ewert are talking about and respond to some of the richness of cellular structure that we glimpsed. Let's see a lot of discussion of how we can begin to build a research community where the line between modeling and experimentation doesn't really exist.

PETE LAMING:

Michael, as an experimenter, could I ask you as a modeler what you regard as an adequate model. Is it the model that simply represents the behavioral phenomenon as the biological system performs it, or will it also represent all the anatomical pieces of that biological system. You have talked about retinal-tectal models of prey-catching behavior in amphibia, a couple of them at least, but we are fairly sure now that the system that controls visually guided behavior in anurans includes, for instance, the nucleus isthmi, pretectum certainly, the telencephalon, reticular formation, and the motor systems. Isn't an attempt being made therefore, by modelers, to incorporate all these structures into their model systems? I have looked at the toad prey-catching system, superficially compared to what Peter Ewert and others have done; nevertheless there are several things that are easily recognized. You quickly realize that in the experimental situation there are changes related to the state of the organism. You will see that there are perhaps some days you cannot record, you cannot find a particular unit at all. You are looking at the basic system -

the retina, the tectum and pretectum - but there are other systems modulating it or each other somehow. As an experimenter it would be wonderful if a modeler could come and join in examining this situation.

MICHAEL ARBIB:

I want to answer in two different ways. There is what I call responsible modeling and romantic modeling, and I think romantic modeling is just as good as responsible modeling. For example, Christoph von der Malsburg is in the class of romantics rather than responsible modelers because he doesn't say "look at all the detailed information about cortex and memory, I must try and find some way of juggling all those pieces of the jigsaw into the real pattern." He says "we already have a class of models which uses one type of synaptic connectivity and here is a critique with respect to some general remarks about pattern recognition." And he then introduces a new type of connection and shows that this expands the space. So, I think there is one type of modeling which is simply developing new classes of models, so that we have a richer vocabulary. It is not concerned whether any animal really does it this way. It is a comparative study into think space rather than into biological space. For example, there are different models of depth perception, just to understand what would be the different ways of getting a network of components to do effective depth perception, so that when you now come to approach a particular animal system, you have a vocabulary with which you can make a particular statement. Is it a cooperative phenomenon, does it use different levels of resolution, is there sensory fusion involved? So, I think there is one direction of the modeler who will make contact with people in artificial intelligence and robotics in building this space of modeling concepts. Now we finally come to responsible modeling where we say we want to look at a particular animal and try to explain what is going on in that animal. But again the point I am making is that if we were to try immediately to say every little twist and turn of a "Matsumoto dendrite" had to be included in the model, we will not succeed. In the old sense we would not see the forest for the trees. Or another one would be dealing with a full scale map. You have to cover the territory before you can read the map, because its full scale doesn't help you very much.

PETER EWERT:

Michael, let me elaborate a bit on that what Pete said. Experimentation in neuroethology is introduced by certain questions about structural and functional properties of a system that controls a behavior pattern. The emerging experimental data allow the experimenter to draw a global working model which leads the subsequent experimental steps in that predictions can be made and tested. These models are not quantitative ones. They simply help us thinking in terms of conceptual boxes. These boxes have suggested functions and they are connected by arrows that specify the assumed influences. Such box-models are rather efficient as long as you count on a system's stable property determined by a couple of parameters. When in the course of experimentation further variables are tested, the number of boxes and arrows increases and you will recognize that some connections between boxes turn out to be rubberband-like. Dynamics, plasticity, variability, and adaptability invade the experimenter's box-model making it to some extent unpredictable, as the living organism. I guess, here in particular is work for the modeler because thinking in terms of little boxes is not feasible anymore. Thinking in terms of dynamic neural structures calls for modeling. Can you give us examples where, in this context, quantitative models have opened new realistic perspectives, have pointed to unexpected phenomena, and thus have stimulated experimental research?

CHRISTOPH VON DER MALSBURG:

I can give you some. There was for a very long time the issue of how retina maps onto tectum in an ordered fashion. This is a very important issue because it promises a paradigm for the way in which the brain is wired up. There have been ten or fifteen models, each of these suggesting its own set of experiments. There has been total disagreement. But over the last 5 to 7 years one model has gained a fairly wide ranged acceptance. Let me briefly summarize the point which I am referring to. The idea is that there are mechanisms which are important for the guidance of fibers from the retina to the target point in the tectum - one of them a positioning mechanism, the other a sorting mechanism. The positioning mechanism seems to involve, as everyone believes, chemical markers or gradients, although nobody has shown the existence of these markers. Many people are trying to search for this mechanism. The other mechanism, the sorting mechanism, is activity dependent. It depends on correlations between neighbouring fibers coming from the retina. This has led to experiments. People have poisoned activity in retina and found a reduction in map precision. The fiber sorting model has strengthened the belief that maps can shift. I am proud to have taken part in this modeling effort. Such models have changed the way people think about experiments and have suggested new experiments.

PAUL GROBSTEIN

I would like to amplify on this point. Certainly, in the case of retino-tectal connections there are multicomponent models that have popularized the notion that what we are dealing with in pattern formation is not a single mechanism but an interaction of several distinguishable mechanisms. By the time the models appeared in that field, however, almost all experimentalists knew already, at least at a gut level, that the problem could in principle be solved that way. There was a more interesting model published in connection with that whole history, the so-called arrow model. What was special about it was that it showed that a previously made experimental observation believed to prove definitively that one of the processes existed didn't actually do so. It had not occurred to anybody before, none of the earlier experimentalists, that there was an additional way to describe mapping rules. My point, I guess, is to emphasize the importance of what Michael has called "romantic modeling," modeling which suggests new alternatives rather than verifying the adequacy of known ones.

CHRISTOPH VON DER MALSBURG

As a modeler I am used to working with experimentalists and to design a system that mimics effects. Models of that kind, on the whole, are in fact very interesting to the experimentalist because they are describing in a formal form what now could be done in one form or another. I would rather put the notion that modeling of the romantic kind brings to the experimentalist's attention things that would not have occurred to them before, of a kind that was suggested by Professor Reitboeck in his contribution.

HORST MITTELSTAEDT:

Our models Michael has been referring to were not yet complete, that is, they were not specified down to the neuronal level. Nevertheless they were rigorously formulated and hence could be crucially tested. I wonder whether, at the stage we are in now, it is useful to demand models that are exhaustively explicated. Rather I think before we go onto the neuronal level we need an intermediate step.

PAUL GROBSTEIN:

I agree that we need work at intermediate levels. Let me try and make that more concrete. It does not seem to me at this point that we know enough about the information processing characteristics of the tectum to make it likely that putting together models based on the known connectivity of the tectum will give us useful new information. We need first to know more about the role of the tectum in the broader information processing schemes of which it is a part.

UWE AN DER HEIDEN:

One could have the impression that there are two separate worlds. One world where the experimentalists are and another world where the modelers are. But I do not think it would be correct to speak of a world of abstract models and a world of experimental results. The reason is that behind and at the foundation of each model there are certain ideas of how the system could work, and likewise each experiment is founded on and guided by some ideas of how the system under investigation could work. No experimentalist can work without such ideas. Otherwise he would behave absolutely at random which would be chaotic and lead to nothing. Moreover, he would not be able to interpret the experimental data. So, there is a common world of ideas in the background of abstract models and of experiments. It would be optimal if the models and the experiments would be made by the same person. Then there would be generally no conflict. Unfortunately, the mathematical and experimental techniques are so intricate that this personal union cannot be achieved generally. This is also the reason for many misunderstandings. The ideas in the background can be called qualitative models. They are what Peter Ewert called global working models. The task and the relevance of mathematical modeling is to develop quantitative models from the qualitative models and to draw conclusions from the quantitative models either by mathematical analysis or by computer simulations. Only the quantitative models are really elaborated models. It was the mistake of early philosophers that they only formed concepts and did nothing else but thinking about these concepts. We know about the mysteries which arose from these works. The quantitative elaboration and analysis of a working model reveals whether the ideas in the background are not in conflict to each other and are indeed sufficient to explain the experimental data. Thus, in a strict sense an experimentalist cannot, if he wants to complete his work, get along without such quantitative models. On the other hand, at least the responsive modeler cannot live without reference to experimental findings.

Otherwise he would produce constructions without connection to reality.

PETE LAMING:

I would like to agree basically with what's just been said. We all have models, also the experimenter has a global model of what he thinks is happening. The difference is he uses the animal to test out the model. He may have to do it several thousand times to get to the point where he believes he is right. The modeler, I suspect, is taking the data he has and then constructs some mathematical computer-based model that would predict what would be the outcome. And so, I think, both experimentalists and modelers are basically doing the same thing. One is using a biological system to test, it might not be a nice test as in many ways it may not come out with such a precise answer as in a test the modeler has. But basically the computer modeler is doing the same. The experimentalist tests the biological system and elucidates structure/function relationships. The modeler uses these correlations to test his model for simulation and prediction.

UWE AN DER HEIDEN:

A mathematical model can also help to elucidate structure/function relationships. An essential difference between the model and the animal is that in case of the model all components of the system are completely described and made explicit and you have complete insight and control, such that you can have confidence that nothing else will interfere and the conditions are in fact sufficient. With the animals you will never have complete control or insight.

PETE LAMING:

Precisely, that's what Peter and I mean, biological systems are state-dependent, they include features that are not, as yet, predictable.

CHRISTOPH VON DER MALSBURG:

I want to move for a short moment to another field in which models play a role, namely physics. There is a very important dividing line between biological and physical thinking. Physicists have attentively been developing simple abstract models like the two body motion or the harmonic oscillator. They are ready to abstract these models to the extreme, to cut away all unessential detail, and they are ready to work for decades on the formulation of these models, so that they can be stated in a more precise way. That's the big dividing line between biology and physics. Physicists are ready to accept that the world, which is infinitely more complicated, is nothing but an inessential complication of their models. Biologists on the other hand think the only sensible thing to talk about is reality in all its complexity. If you take away bits from this reality, in drawing a model, you have taken away reality. That's a big difference in thinking, and I think that's a very important point of misunderstanding. I am, as Michael said, on the side of romantic modelers. I am a physicist, trying to purge away inessential details in my model in order to convey an idea. That, of course, creates a difficulty in talking to biologists.

PETE LAMING:

I beg to differ, because I think biologists, too, try to reduce and to control the variables.

CHRISTOPH VON DER MALSBURG:

Often I had heard that a model or theory wasn't talking about reality because nerve cells did this or that. They were simplified logical entities. Biologists are discontent with this type of simplification. Of course, they would like to describe their animals, and this with all their variability down to the cells, to the molecules and so on. Maybe they don't have a choice.

JERRY NELSON:

My work is on the neurophysiological basis of orientation: orientation, contrast, stereoscopic depth perception in globality. Michael Arbib has mentioned two kinds of models, romantic and responsible ones. I would like to define two classes of problems to which we can apply models. We can call these the regularity problem and the final functionality problem. Let me give you some examples of these kinds of problems. In the regularity problem, we are trying to understand how the nervous system manages to display or to discover a certain regularity. One example would be the display of columnar architecture in the cortex. This is the architecture we see in either orientation iso-slabs, or in orientation sequence slabs. In trying to display this kind of regularity, the genetic code is overwhelmed. It is the great contribution of Christoph von der Malsburg's work to show how very few, very simple principles can generate this regularity. In globality, the visual system is overwhelmed with local detail. In a random dot stereogram, models such as those of Marr, Poggio, Dev and myself show how certain very simple neurophysiological principles, such as mutual facilitation and mutual inhibition, can surmount the flood of local information and discover the regularity which lies in the disparity cues of a stereogram. Interestingly enough, these successful applications of models to regularity problems lie in the field of so-called romantic models. Now turning to final function, here, we seek to know some of the ultimate answers. How do we read? How do we navigate a room, build an egocentric or an allocentric map of the world around us? Here I have to make a short digression. It seems to me that biology is inherently different from the physical sciences. The nature of explanation is different, because in biology we have a much deeper hierarchy extending from the interesting problems all the way down past membrane properties, all the way into, ultimately, biophysics and physical sciences. But in this very deep hierarchy, the vertical relationships are weaker since they are more diffuse than in physics. We have another problem. As we ascend these levels of complexity, we are often unpleasently surprised by unexpected emergent phenomena, and that is, I think, the key to what is needed for the successful application of models to final functionality problems. One has to identify their emergent phenomena or the abstract function that is being performed. The job of the modeler is then to remove those unpleasant surprises for us and explain how they arise from underlying mechanisms. Those models may well be of the responsible kind. But they will depart from what Michael Arbib calls full scale models. By concentrating on the emergent phenomena we explain their role for the experimentalist, who is interested not in the regularity problem of morphogenesis or globality but in functional questions. We also identify the abstract functions which most need explanation. I hope that's of some help.

CHRISTOPH VON DER MALSBURG:

You refer to differences between physical sciences and biology. I think it is also true in the physical sciences that new emergent properties are discovered experimentally, that only after the facts comes the theory to explain them. Think of superconductivity. They have a very hard time understanding it. Now this new step of "high temperature" superconductors discovered recently was a complete surprise to theory. They have a hard time over that, too.

THOMAS FINKENSTÄDT:

I wanted to comment on the interaction between experimentalists and theorists. As an experimenter I am quite reluctant to trust too much in models, because I know that in the present tectal models much data is not involved, such as the telencephalic and hypothalamic structures and related functions. So, we are looking through a narrow window. Now that we have started at our university a little computer group, I know how intense the exchange of data must be, so that we can all gain from each other. First of all we had to learn the same vocabulary. In the beginning this seemed to be almost impossible, since we have used the same words but actually with completely different meanings. So, very intense exchange was necessary. The best, therefore, in my opinion is a small group in which computer scientists doing the modeling are just next door the experimentalists collecting the data. It creates various problems if an experimentalist will become a modeler and *vice versa*.

MICHAEL ARBIB:

The question is how you create social structures in which you are addressing what you can learn from the theorists as well as what you can learn from the experimentalists, as we are doing in USC, to build groups in which theory and experiments come together. In fact we have in our group competing models of retinal and tectal interactions. I don't say waiting until we have understood the telencephalon is going in any way to contribute to the issues between the models at the moment. One can already ask, for example, how important is the inhibition coming from the pretectum in a certain set of behaviors. I think we can add to that list the role of the nucleus isthmi. There are now specific models and they can be addressed in the unfolding of the research that Dr. Wang is doing in Beijing and Dr. Gruberg is doing in Philadelphia. So, I would really hope that, quite apart from the difference in style, level of romance and so on, that we can really begin to have experimentalists give theorists some clues on assumptions of models that are waiting for test, or in which interesting experiments can be done.

GEORGE MOORE:

Despite the fact that experimenters and modelers are present here in the same room, much of the discussion suggests that they are, in fact, quite a great distance apart. The experimenters will have to move several levels of generality or abstraction higher, and the theoreticians several levels of concreteness lower.

CHRISTOPH VON DER MALSBURG:

There is this famous idea that our civilization is cut into two subcivilizations, the scientists and the humanists. There were people who objected to that by saying that there are engineers who play the piano. That objection is not valid because the cut may go through the mind of that engineer. I think we have a similar phenomenon here. People are doing experiments and are doing theory and they have a hard time uniting the two types of activity in their brain.

EDWARD GRUBERG:

Also isn't there a stylistic problem? We have talked about the issue of having romantic modelers and responsible modelers. We are left with the default notion that there is one kind of experimentalist. In a certain way, I think, there are also equivalent romantic biologists and responsible biologists. A responsible modeler would probably want to have a responsible biologist working on the same system. He would want to say just do these kinds of experiments, generate this kind of data so I can build my responsible model. The romantic biologist is more inspirational and harder to pin down. He doesn't necessarily gather data day after day in a well defined system.

PAUL GROBSTEIN:

Michael and Christoph have in fact adopted very different postures with regard to modeling, and I presume they regard that heterogeneity as valuable within their modeling community. We, the experimentalists, are entitled to the same appreciation of our heterogeneity, as Ed suggested. I am sorry that Christoph's experience has been that some biologists don't read his models because they are too simple. Peter and I go looking for modelers who will make us realize, at some higher level, something that never occurred to us. Michael, on the other hand, bitches at me because I won't give him the particular data that he can plug into his model. So, as an experimentalist, I am getting hit on both sides. There are those of us who would like to see some data incorporated in models for the purpose of testing our intuitions, or to have modelers test theirs. But I would like again to put in a plug what seems to me a role for modeling at least as significant as verification of intuitions. Modeling has the capability of opening new horizons, creating new ways of thinking about things, and that ought to be a major thrust of the modeling community.

BARRY HORWITZ:

To elaborate on Paul's point, I agree that it is very frustrating for modelers to propose a new idea and not to have it examined by the experimental community. However, there are some modelers who work closely with an experimental group; it is almost as if the experimental group has adopted the modeler. I work very closely with an experimental laboratory at the NIH, and I find it very valuable, indeed much more valuable than when I worked alone. But I also would like to make another point. I think there is another aspect of modeling that must be understood and appreciated by the experimental community. This feature of modeling has occurred in physics in numerous ways, but it has not been mentioned here as such. There are a number of characteristics of the nervous system that are not directly accessible to the experimentalists, at least in the near future. For example, one cannot record the electrical activity from a dendritic spine. On the other hand, the nervous system also is associated with phenomena which have such a diffuse character that, at least at present, it is impossible for experimentalists to investigate them directly. Both of these situations can be approached by the use of theories, namely models, so long as these theories embody ideas that are testable experimentally. If the testable consequences of such a model are borne out by experiment, then the features of the model which are not accessible experimentally have some chance of being accepted as correct. I think that this kind of model is much more valuable than those that just explain what can be tested, although both types have their place in neuroscience.

ULRICH BÄSSLER:

I think we have discussed more or less only models as a heuristic way of helping the experimenter to find new ideas. So, the modeler's task with a behavior is to model this behavior and find out the principles of how the system could work. But I think we have also a large number of systems which are now quite well understood on the basis of connectivity between cells or parts of the central nervous system. But if that is becoming too complex, we have no idea of what the system characteristics are in this system. Here, the modeler can help very much by trying to find out the characteristics of the system. I want to give you an example. Working on the walking of insects, we found some coordinating pathways that exist between the legs of an insect. We found their properties. An insect has six legs and there are many of these pathways. Then we tried to model it. We found that the model produced the proper coordination. It can also do so when we make some ablations. Afterwards we found that there are other coordinating pathways as well. Then we could make a model on that basis which also produced the correct coordination, and so on. By this we found that the whole system is redundant. It uses much more information than is necessary. I think that is another type of model which uses facts found by the experimenter and comes to find out the characteristics of the whole system. I would like to adopt this as a task for modelers.

HERBERT REITBOECK:

I want to come back to a question Michael Arbib would put into the romantic model category: What can we learn from computer pattern recognition, particularly with respect to models of the visual system? Should a computer program be viewed as possibly containing transcriptions of operational principles of the visual system if that program has been shown to succeed in a specific pattern recognition task? Let me give you an example: Consider a program that generates a syntactic scene description based on line analysis and on features made up of characteristic line intersections. Programs like that can recognize simple 3-dimensional objects and they can even "understand" object relations in more complex structures. Such programs are presently used in industry; they are fast and reliable, and they perform beautifully in an artificial environment of objects defined by straight lines that are seen against a uniform background. There can even be shadows in the scenes. But one cannot extend that concept to deal with the specific problems that arise in a natural scene with irregularely shaped objects seen against a patterned background. It's the old curse of the "initially successful approach," like the ladder that gets you closer to the moon, but adding more steps won't bring you there. It's the same in pattern recognition: There are techniques available that are very successful in a particular task but they have certain limitations - and when one tries to overcome those limitations, the whole approach breaks down. As long as we recognize the limitations, we can speculate whether some aspects of an algorithm could be realized in the nervous system. Even "romantic" models that are biologically implausible can stimulate important questions, because they can make us think about what alternative approaches might have evolved in biological systems in order to solve a particular problem. The least what we can learn from many of the present technical pattern recognition systems is how the visual system is not doing it. And this I think is important, too.

PETER EWERT:

This stimulates me touching on another topic with reference to the impressive talks by Torras, Arkin, and Koy-Oberthür. In a reminiscence to the slogan "my brain is my best computer," which so far as I know goes back to Jerry Lettvin in the sixties, there is today an increasing interest in neuro-computers, transputers, and other kinds of developments. Does the study of biological information processing systems really hold promise of solutions to practical problems in robots? Can we say already that algorithms derived from the basic properties of a visual system model are suitable to be implemented by optoelectronic devices in robotic systems?

PETE LAMING:

My question is in the same direction. Animals have hierarchical systems. They have spinal reflexes which control tension in muscles and so forth, they have higher-order reflexes in the medulla that tell that set of reflexes basically what to do and these in turn are controlled by distributed cerebral structures. Has anyone designed in robotics a system similar to that, having a control system somewhere and another combination control system that tells the whole what to do?

MICHAEL ARBIB:

Carme Torras and Ron Arkin are addressing similar issues to robotics. They distinguish between off-line and on-line learning, whether it's finding the right way to bring something or whether to combine the pieces to assemble, or whether it's a mobile robot trying to move in the countryside where there is the notion of having an overall path in place that must be modulated according to circumstances, disturbances, or obstacles it encounters. Therefore, the hierarchical level of planning and the ability of the more detailed plans to be affected by more and more of run-time data is a very important part of robot design. Going back to your question, Peter, in the design of low level visual systems, the parallelism which is obvious from the work of Lettvin, Hubel, and Wiesel, has been very important in the machine community in helping them avoid the dominance of other parts of technology by the serial paradigm. Vision was the one stronghold of parallel processing. Now, I think, there is going to be an increasing amount of interchange. Ron Arkin talks about the detour behavior of the frog and toad and uses it to get run-time corrections in the path of a mobile robot. Psychological and neurophysiological models of optic flow have added much to the world of machine vision. Seeing how the brain could do it is suggesting new ideas to the person building machines. And then the person building machines giving us his richer vocabulary which poses other experimental questions is a theme that I hope will be at the center of this technology for the next 20 years.

EIKE MÜHLENFELD:

Well, I am the sort of parasite you are talking about. I come from the engineering field with the intent to build visual system automata that are better than what's already known and who joins workshops like this occassionally to see whether there are improvements in models. We don't expect complete models describing entire systems. What I find here are ideas or principles which might govern the processing of artificial intelligent systems which we have in our hands, and this might help us to create ideas for technical solutions. We cannot expect to find a solution we can copy and enter into a technical realization. It is always better to keep yourself open for ideas that come out of the field of biocybernetics.

CHRISTOPH VON DER MALSBURG:

May I enlarge on that. Three years or so ago we tried in Germany to put together neurobiologists and engineers in a conference. I think you were among the engineers. This was to find out whether there was any kind of common interest. But it was a complete failure. Now three years later, the atmosphere has completely changed. Artificial intelligence people and robotics people are realizing their ideas are in deep trouble. In order to improve the flexibility of their robots they need more volume in the computer, volume in terms of knowledge, space in terms of operations per second. They will only fill that volume if they go parallel, if they have thousands of little processors in one box. In addition there is another problem, that is the software explosion. In order to add more flexibility to robots, you have to add even more software. By adding more and more subroutines, you have the problem of integrating them into a flexible functioning system. There is a huge problem there. Many of these people are realizing that they are in trouble, that they need new concepts. They are looking to the brain where all these problems are solved by very slow processors that work in parallel in huge numbers. The brain is able to adapt to new situations, completely new situations, which could not have been forseen by evolution in detail, to adapt to deal with a situation very smoothly. So, many of those engineers are looking over the fence expecting news from the neurobiology community.

EIKE MÜHLENFELD:

I do not quite agree with the key of artificial intelligence as a consensus one. It may be this might suggest that all problems are solved as soon as we get parallel computers, and I don't think so. It is a structural problem. Artificial intelligence is based on building graphic nets on artificial grammars, and these grammars must allow algorithms. This can only be done in a subset of these grammars as long as they are context-free. But to build models of perception and models of visuomotor coordination, I am sure, and artificial intelligence people have found out, that you can't get very much further with context-free grammars, and context sensitive grammars are not solvable yet. So many disappointed people from artificial intelligence try to find solutions by studying what is done by biological systems. I must stress again that it might be dangerous to put all hopes on parallelism. This might be a solution that is very good for biological systems. We might have the same for technical systems, but we haven't found it yet.

CHRISTOPH VON DER MALSBURG:
I don't see a disagreement. There is this quantitative problem: In order to use knowledge you have to get it into search routines in a matter of fractions of seconds, and you can't do it with your CRAY computer, that is much too slow. Acknowledging this quantitative problem you immediatedly get into a qualitative problem. You can't put an AI-system into a parallel machine. That's a severe problem. You have to organize it and control it in a good algorithmic way. So, I didn't make the point that the problem is only quantitative.

EIKE MÜHLENFELD:
I want to say that if we solve the quantitative problem we haven't solved all the problems.

Second Session: Neurons, Networks, and Building Blocks

GYULA LAZAR:
From the lectures given by Ewert, Manteuffel, and Betts we have learned that visual information is processed in separate brain structures. More precisely, visual pattern recognition, localizing of the prey, and avoidance behavior are processed through the retino-tectal channel with the involvement of the pretectum, while whole-field pattern movements are mediated by the pretectum and by the accessory optic system which in the amphibians is only a tract and a basal nucleus in the tegmentum. One of the most striking observations shown by Ewert was that almost the whole brain is involved in processing visual information. I think that it would be a nice project to study which parts of the brain are not involved in processing visual information. These new discoveries create further difficulties in understanding visual functions. We have been rather brave until now in saying that we are studying a simple visual system in amphibians, but in the future we have to think in a similar scale as scientists who work on higher vertebrates are supposed to think even now. Manteuffel's talk also pointed towards complexity. Betts seems to be moving in the other direction to make his model as simple as possible. If you consider the whole range of anatomical, electrophysiological, and behavioral studies, and modeling, we may recognize several gaps which could not be bridged until now. Let us see first the anatomy. We identified several types of cells, analyzed the distribution of cells and the different kinds of afferent and efferent systems, but synaptic connections of identified neurons have never been studied anatomicially in amphibians. Further steps in analyzing structure cannot be made without synaptological studies. The next gap is between the morphological and physiological properties of identified neurons. Matsumoto's fine work narrowed the gap, but data are too few to find correlation between morphological features of the cell and its physiological properties. One

would expect that a neuron having certain morphological characteristics may have at first a function which is characteristic only to this cell. We do not know which are the parameters - size of dendrites, the shape of dendritic tree, or location of cell or the transmitter it uses - which determine physiological properties. Finally, a gap can be detected between physiological properties of an individual cell and the behavior of the animal. We have seen good examples in Ewert's talk, how the width of this gap can be reduced. In connection with modeling I should like to say that working on a model which contains verified synaptic connections would be more promising than the models which could be generated until now on the basis of poor anatomical data.

BARRY HORWITZ:

I would like to comment on one thing that Dr. Lázár said about hoping to find some relative isomorphisms between the geometry of dendritic trees and function. I do not think there is very much evidence that one can really do this in a general way. People working in other preparations who have studied dendritic morphology of different cell types have found that there is very little correlation between dendritic geometry and neuronal functioning. I have done modeling of changes in dendritic geometry, and the resultant changes in the spread of postsynaptic potentials. Essentially, there is an interrelationship among the location of the synapse, the geometrical structure of the dendritic tree, and the amplitude and time course of postsynaptic potentials at some location such as the cell soma. You can take a different morphology and just rearrange the locations of the synapses and get the same kind of neuronal functioning. This is not to suggest that one might not find some approximate correlation between dendritic geometry and neuronal functioning, but I think the problems may be too great for these approximate correlations to be of much use immediately. Once parameters are obtained for some kinds of models - perhaps of the sort presented by Bill Betts - by fitting the model to experimental data, then I think it would be a reasonable goal for people who do dendritic analysis to try to determine how an arrangement of synapses can result in a particular neuron having a particular property, such as directional selectivity. However, going in the other direction is harder: to predict from the dendritic morphology what the cell is doing physiologically and behaviorally.

GYULA LAZAR:

I think there are some systems in which there is a correlation between dendritic morphology and the function. George Székely's group showed that there are many kinds of motoneurons in the frog spinal cord. At least two main groups can be distinguished. One has a crossing dendritic branch and may be stimulated on the contralateral side. There are also dendrites that project dorsally, so that they can be contacted directly by the primary afferents. There are some other dendritic processes which can be contacted by the descending pathways. In the spinal cord, you can have special structures where the wiring of the descending and ascending pathways are stereotyped or well-defined. In these systems, the dendritic morphology may be of some functional significance. In the tectum, I can imagine that a few cells which have restricted narrow dendritic trees are able to make synaptic contact with certain kinds of afferents, and maybe this defines the function of the cell. However, although a few hundred cells were recorded and filled with cobalt, I think it is not enough if you take into account that there are half a million cells in the tectum. At present I do not see any reliable clues for a correlation between tectal dendritic morphology and function.

ULRICH BÄSSLER:

Working on invertebrates I see there a long tradition to fill cells after you have recorded from them, and the main purpose of doing this is to identify the cell as a unit or a member of a group subserving the same function. Another purpose is to look where the inputs are coming from and where the output is and so to determine whether it is a local circuit neuron or a descending or an ascending neuron. But I think still the main purpose is to identify the neuron.

PETE LAMING:

So, the consensus that seem to be developing is that there is not necessarily a relationship between dendritic or cellular morphology and function.

ULRICH BÄSSLER:

If you know a certain animal well, your long-lasting experience may in certain cases allow you to say a certain interneuron with a certain dendritic tree subserves a certain function.

BARRY HORWITZ:

I can tell you about an interesting study of dendritic variability in invertebrates. In Al Selverston's lab, an investigation was undertaken examining an identifiable neuron in the lobster stomatogastric ganglion. A graduate student, Sheryl Glasser, injected Procion yellow, a fluorescent dye, into the same identifiable neuron in seven lobsters. She found that in the same way that weeping willow trees, let us say, look like one another, the dentritic trees of these neurons resembled one another. Then she did a computer reconstruction of the dendrites of these neurons. She measured the diameters and lengths of all the branches and found a 30% variance in the values of these geometric parameters. She then did a compartmental analysis to determine the voltage attenuation characteristics of these neurons. Interestingly, there was only about a 10% variance in the electrical properties of these cells. Obviously, these cells did the same thing in every one of the lobsters. So, the results of the electrical analysis were good. What is bad, for neuroscientists, was the fact that apparently during development, dendritic geometry in one part of the dendritic tress can change to compensate for geometrical changes somewhere else in the cell, and this produces a system that functions invariantly. This is depressing for neuroscientists who hope to find a correlation between dendritic morphology and physiology. It seems that dissimilar dendritic geometries may have the same function, and conversely, similar geometries may have quite different functions.

ULRICH BÄSSLER:

I certainly agree if you have identified neurons from animal to animal and then you see that they change from animal to animal.

PAUL GROBSTEIN:

Let me generalize. There are a number of published reports of variability in dendritic morphology of invertebrate neurons. There is no case that I know of in which it has been determined whether that variability in dendritic morphology has a behaviorally meaningful accompanying variability in function. It seems clear to me that the variability is there, and that dendritic morphology is not necessarily irrelevant. I find myself occupying a somewhat awkward position here because normally I would not argue that the best way to understand the nervous system is by looking at the dendritic morphology. What I am trying to get across is that dendritic morphology poses a number of very interesting questions, which may or may not ultimately relate to understanding how the brain works at a more global level. In some cases, at least, it does. The classic structure and function correlation was made on dendrites in the frog retina, and still holds as far as I know. So, I guess, I like the posture that almost anything is worth looking at if you are inclined to do that particular sort of looking. One can almost invariably detect an interesting pattern of some sort, and such patterns will sometimes even prove significant for understanding higher order properties.

PETER EWERT:

I also want to address a warning for predicting response properties from dendritic morphology. On the one hand, we cannot deny that there are many cases in which so-called identified neurons have been described based on features like afferent and efferent branches, cytoarchitecture, pharmacology, electrophysiology, stimulus-response properties and so on. Paul mentioned the retina, other examples are the granular, basket, stellate, and Purkinje neurons in mammalian cerebellum, the Mauthner neurons in tailed aquatic amphibians, the giant electromotor neurons in electric fish, the Rohon-Beard cells in amphibian embryos, the omega neuron in crickets. Although there is correlation between cell type and function, this does not exclude the possibility that all these neurons are not exclusively dedicated, and the function of some after their selective elimination may be taken over by other neurons as Eaton's investigations on Mauthner neurons and recent experiments by Atkin on cricket's identified neurons have shown.

However, on the other hand what I want to emphasize is that I am tempted to warn everybody who tries to guess physiological properties of a neuron just from the spread and shape of dendritic trees relative to afferent fiber terminals. I'll give you some examples from Dr. Matsumoto's intracellular recording/labeling studies in our lab, in which we tried to find correlations between visual response properties and morphological features of tectal neurons. We have found many cells whose dendrites extended through all the afferent laminae and which seemed to arborize mainly, if not exclusively, in the superficial laminae where class R2 optic nerve fibers terminate. But in most of these neurons you will find that they resemble class R3 *on-off* properties which indicates substantial input from R3. Let me select another example. There are tectal cells with huge dendritic arborizations that appear to spread across all laminae within almost one third of the tectum. You might suggest that such a neuron is highly integrative, which may well be the case, but the visual receptive field of these neurons was often tiny. In another neuron, whose dendritic tree was extremely narrow we anticipated to discover a tiny-field-neuron, however, its receptive field could encompass nearly the entire field of vision, suggesting that the widefield character was determined by integrative processes taking place in other brain structures, such as pretectum or the medulla oblongata. The problem of isomorphism between morphology and function is particularly obvious in bulbar structures, as the papers by Ananda Weerasuriya and Wolfgang Schwippert show. In summary then, there are in neurobiology well investigated examples of identified neurons and I agree with Paul that looking at the site of the soma and its dendritic and axonal features is instructive for various reasons, but we cannot necessarily derive response properties of neurons just from the dendritic morphology.

PAUL GROBSTEIN:

A classic example of that is in the mammalian spinal cord. In the IA reflex pathway, dendrites of motoneurons going to antagonistic muscles are literally twisted around one another, and yet physiologically there is clearly a specificity displayed by IA inputs onto those motoneurons.

HANSPETER MALLOT:

I think the relationship between dendritic morphology and function can be somewhat more subtle if the entire neural network is considered. Our model studies on visual cortical functions have shown that each dendrite contributes to the network in that the function recorded from a unit is somewhat determined by the special geometry of its dendrites. Maybe, one could try to find at a higher-level description of dendritic trees, such as average density of synapses, things that do in fact have an influence on the function of neurons or even the entire network.

GEORGE MOORE:

Let me introduce a new issue with reference to Paul Grobstein's and Horst Mittelstaedt's work. I would like to express my admiration to these papers since they exhibit a style of thinking in neurobiology in which inference places a major part. They both use inference to report the existence of something they had not seen or injected or recorded from. Paul announced a function that must exist in the nervous system. He hadn't seen it but by inference he did use the existence of it. Inference allowed Horst Mittelstaedt to connect weight with a corresponding inference that he had made. My observation is that because they both moved into a different level than where they had started, the theoretical level moved down to something and caused experimental data. To move up to something abstract, it provided the singular instance of a connection between theory and experiment which is exactly the characteristic that Christoph's example had of the interaction between theory and experiment. The distance between single unit stimulation, recording, filling, and modeling may be in certain cases too large. The experimenters are then going to have to conceive their experiments at a higher inferential level in order that the connection between theory and experiment be made.

PETER EWERT:

I think this touches on a general question. How shall we proceed in neuroethology? Can we understand behavior in terms of neurons, their connections and their interactions, or do we have to deal with larger building blocks? In the latter case, we must distinguish between neuronal building blocks and those blocks or boxes that exist in our conceptual world. By the way, interestingly it is cell recording, stimulation, labeling, tracing, and lesioning that allows us to describe these blocks, for instance showing that their components may be spatially distributed and multiply installed. I wonder whether the neuronal and the building block approaches, in an essence, are really that different.

CHRISTOPH VON DER MALSBURG:

Maybe I have part of an answer. Listening to Horst Mittelstaedt's stories, what strikes me is if you need sine functions implemented in neural hardware, you absolutely need mechanisms which regulate details, so that they conform to these precise values. This is made evident in lesion experiments where, for instance, you take out one vestibular system and the animal is in a very poor condition for a while and then recovers to a certain extent, showing that indeed there are these mechanisms regulating detail, so that the whole works in some useful way. So, one way of responding to Dr. Ewert's question is to look for regulatory mechanisms, for organizational mechanisms, which are behind the details. This seems the kind which we have to look for, instead of details of an axon or a dendritic arbor or the distribution of individual synapses and so on, which anyway would regulate if we impaired something, would go to another state to reinstall function.

PETE LAMING:

All experimenters, certainly in neuroethology, have concepts and models of what must be happening in the animal, before they do an experiment where they look for it and how grossly they look for it. How they look for it in the first instance depends on how much is known, this being true for most of our work. It is certainly true of the work of Peter's and my own labs. We start out with behavior and have innovative ideas about what must be happening and where it might be. Then by a combination of techniques like lesions or ablations, gross recording or evoked potentials the more important structures and mechanisms are recognized. Then you get the finer and finer work, recording from single units and then recording intracellularly and marking individual cells to determine their morphology. In this way the individual components of the neuronal systems that ultimately evoke a complex behavior can be determined. You will find some answers to the original questions but also see principles which were not obvious when you started with a set of behaviors and a concept but otherwise very little to go on. So I am not sure that I accept George Moore's criticisms that experimenters do not conceive their experiments at a sufficiently high level. In neuroethology the tendency is to try to cover the whole span of understanding from behavior to neuron. As George says most people try to model the biological system, the real actual existing system. Similarly, in neuroethology the conceptual understanding of a behavior requires a mental effort as well as designing the simple experiment and plotting from *a* to *b* to *c*. Most of us wouldn't do that. We would, if possible, try to go from *a* to *d*.

PAUL GROBSTEIN:

I guess I have a feeling that there is a level of organization problem as Peter defines it. If one actually sits and looks at the history of invertebrate neurobiology, neuroethology, and neurobiology generally over the last twenty years, it can be instructive. The field got started with an enormous enthusiasm that there was going to be, that there were, meaningful isomorphisms between elements of behavioral descriptions. It seems to me that invertebrate neurophysiologists, who were at the forefront pushing that notion, and subsequently visual system people, have out of their own experience begun to realize that promise is not going to hold up. The isomorphisms that were originally seen, or expected, have begun to fall apart: if you track the visual pathway you do not actually see the progressive abstraction of features one might have hoped would be there. Pattern generating circuitry has proven remarkably more complex than the simple neuronal circuits which motivated research in that area. My sense is that there is a need for an intermediate level of approach which serves the function of linking aspects of behavior and of neuronal organization more directly than it is likely ever to prove possible to do at the single neuron level. A central pattern generator bears a close isomorphism to an element of behavior. Its relation to a map - another intermediate level concept - can be explored. It is these intermediate blocks, it seems to me, that one needs to understand nervous system organization in relation to behavior. They in turn define further questions which can be explored at the cellular level by people so inclined.

PETER EWERT:

I see the weights in Paul's historical review somewhat differently. After the initial disappointments with the detector concept realizing, for example, that no cells in frog's retina are specifically tuned to bugs and that in frog's auditory nerve there are no specific mating call detecting fibers, neurophysiologists in the sixties denied that such behavior relevant specificity, at single cell level, exists anywhere in sensory systems. So it was claimed that the behavioral specificity is a matter of certain population effects of neurons that are more or less feature sensitive. In the last decade, I think, neuroethology opened a new chapter. Not that the population and ensemble coding concepts are out of date, but many studies in central sensory systems provided evidence of highly selective neurons, such as sign-selective neurons in electric fish, prey-selective neurons in toads, mating call selective neurons in frogs, song-specific neurons in birds, face-selective neurons in monkey. Depending on the system and its function, the neuronal operations for stimulus discrimination and object recognition may range - within a continuum - from single cell coding to ensemble coding, as Horace Barlow has recently pointed out in the 1985 Bartlett Lecture on "The role of single neurons in the psychology of perception." So, in this field I cannot see much frustration. Nevertheless, there are various problems the CNS has solved that the experimenter cannot yet grasp. One problem is that the high feature specificity at cellular level requires convergence obviously at the expenditure of localization properties as regards the increases of the receptive fields of visual neurons. So, I agree with Paul that when tracking the visual pathway, people interested in localization properties become frustrated because topographic representational maps appear to fade, whereas people interested in feature analysis become excited about the emerging parametric, computational maps, given they ask behavior-relevant questions. While we are going to understand the latter, the question arises how information on space, that seems to get lost in parametric maps, is so-to-speak reinstalled, for instance by directional asymmetries, lateral inhibition, vector operation, or by parallel processing. The paper by Hanspeter Mallot is particularly instructive in this context. Furthermore, there is controversial discussion about how strongly the specialized neurons are specified for perceptual tasks and how this specification is reached, for instance by interaction, iterative processes, or sensorimotor codes.

ULRICH BÄSSLER:

I want to add another issue, that we ultimately all want to make some general statements about pattern generation, about how walking is organized, about general behavior, given that there are common kinds of properties in all animals. In a large number of animal species then perhaps these generalizations may be made on an intermediate level, and not on the identified cell level. I am absolutely certain that we use other neurons for walking than the stick insect, but the general principles of control, at a middle level, seem to be the same. Nevertheless, I am going to single neurons because I want to find out how certain principles of coordination are realized with neurons or neuronal networks.

MICHAEL ARBIB:

Aplysia has become popular in the neuroscience community with the notion that certain basic learning principles can be tested at synaptic level in invertebrates. But I think there is an alternative in the computer frog *Rana computatrix* because in terms of that there are certain organizational principles that cannot be easily seen in invertebrates which have a very different plan of organization from the vertebrates. But, moreover, these principles can't be seen at the moment in experimental preparations where we can analyze only one or a few cells at a time. We need to understand coordinated patterns of activity. I think what has been exciting about today is that we have seen a network of insights at a number of different levels. We are finally getting to the point where the pattern generation for walking can be placed down at the single cell level and there we had a very nice demonstration of entrainment of oscillators from Dr. Bässler. Having that range of examples, we begin to feel that we can understand locomotion in a rather deep sense. In probing the interaction between the sorts of general mapping in visuomotor coordination, Paul Grobstein gives us the domain looking for detailed synaptic connectivity. We have to have this notion of islands of reliability. Each part is a context for another. If we try to stick with one level, it wouldn't work. I would like to hear from Horst Mittelstaedt a little more, given his confession that one of his more complicated slides was his first attempt towards a neural model of the transformation of topographically ordered spatial information into motor commands.

HORST MITTELSTAEDT:

Well, I'll give you another example of the effect of this kind of top-down modeling on neurophysiology and *vice versa*. Typically, the first step was deductive: in creatures with a mobile neck, vestibular information about head tilt must be complemented by information about the angle between head and trunk if it is to be used for orientating the trunk to the vertical. In my first model of how this could be done the utricular output is added to a respective output of neck proprioceptors. Properly calibrated the system would nicely respond to trunk tilt, whereas the two summands would cancel if only the head is tilted. Recently, Pompeiano and Wilson and their co-workers found neurons in the vestibular nuclei of the cat's brainstem which showed the postulated summation and cancellation very neatly indeed. But in the meantime, I had begun to test an inevitable consequence of the model: as a matter of fact, the cancellation would miserably and dramatically fail if the trunk is turned upside-down, or if the otoliths are removed, or if the g-load is increased on the centrifuge, or in weigthlessness: strong tonic neck reflexes on the legs would occur upon head tilt with respect to the trunk. By the way, as far as I know none of the Russian or NASA authorities had thought of such a catastrophe. Yet in fact nothing of the sort happened in either of the four cases. The saving grace in my dilemma turned out to be that much-neglected organ: the saccule. All extant facts could be accounted for, if its output were multiplied with a sine component of the neck proprioceptors and utricular output with the cosine of the neck tilt, respectively. Subsequently, the two products would be summed to yield the cancellation. Consequently, if model Number 2 is correct, these multiplications must occur *upstream* of those summating brainstem neurons of the cat. We're now all very curious, if it's done, where it's done and how it's done.

HERBERT REITBOECK:

It seems that most of us, or at least the biologists among us, prefer models that are based on anatomical and neurophysiological evidence. However, as long as our knowledge of essential functions of the nervous system still is as fragmentary as it is, even experimental findings can be misleading in their functional interpretation. Let me give you an example: When Hubel and Wiesel made their discoveries of cells that respond to oriented bars and edges, their findings stimulated a particular approach in computer image processing: It was assumed to be proven that primary processing operations in the visual system involve the detection of lines and edges, and that subsequent cortical processing is likely to be based on line analysis too. It can be argued, therefore, that technical pattern recognition systems should use the same approach. I have mentioned before that computer pattern recognition algorithms based on line analysis can perform very well in an artificial object recognition task, as long as there is a plain background. But when there is a patterned background - and this is generally the case in a natural viewing situation - line analysis at early processing stages easily brings a program into non-terminal loops: as long as object boundaries are not defined, a line analyzing algorithm will combine object lines with intersecting background structures, ending up with astronomical numbers of nonsense objects that have to be processed and compared with data in memory. But how can the object boundaries be known if the line analyzing algorithm itself is supposed to define the object, and its borders? We arrive, therefore, at a situation of an infinite regress. In order to recognize an object in a patterned environment, we generally need *a priori* knowledge about the object that is to be discovered, as we have seen in

Michael Arbib's beautiful examples. So, the start with line analysis too early in the processing scheme, before region definition and figure/ground separation have been achieved, is an inefficient and dangerous way to go. An important processing operation for region definition is texture analysis. There are, indeed, more recent neurophysiological findings that show that the receptive fields of visual cortical cells have internal structures and dynamics. Not only are the classical receptive fields well suited for texture analysis, even better suited are such structured receptive fields. It is thus most likely that cells in the primary visual areas predominantly function as texture analyzers. Neurophysiological results, therefore, can be seen in many functional contexts, and some hypotheses can be misleading.

HORST MITTELSTAEDT:

As far as pattern recognition is concerned, what I need for implementing my model is something much more primitive. All I need are functions, trigonometric functions, which are proportional to the orientation of the entire picture with respect to the vertical meridian of the retina. The simple cells would do marvellously because they could be used to extract all those second and fourth harmonics. So, my question is addressed really to Peter but related to all those people who are either modeling or are using toad's tectal T5.2 neurons, or are working with them. It concerns something I've asked in the discussion of Bill Betts' paper, namely how it happens that these neurons operate in a very precise and configuration-selective manner for about half a year. What happens to them the rest of the year? Do they change their configurational selectivity to have another function? How is selection made and is there a clue from Evelyn's work in freely moving animals when she did the pretectal lesions and the receptive field of T5.2 neurons increased by a factor of two or three? I am just asking Peter what he thinks about all that.

PETER EWERT:

I think that toad's configuration selective T5.2 neurons are of a general interest because they show how cells with radially symmetric receptive fields can select asymmetric dynamic stimuli, such as differently oriented bars that move in the same direction. The discriminative ability is based on a simple, efficient principle. This principle takes advantage of two things, namely the asymmetry in the time domain of the spatiotemporal stimulus pattern, and the discrete parallel weighting of the spatial cues and the patiotemporal cues by means of different network operations. It is then the subtractive interaction between these networks that determines the high degree of configural selectivity as well as its velocity and direction invariance. This property is expressed in the toad's prey selecting behavior, and I think Evelyn Schürg-Pfeiffer's chronic recordings convincingly demonstrate what happens with neuronal and behavioral selectivities if one network operation, the pretectal one, is impaired. To understand the selective principle of T5.2 neurons, you should ask an engineer to construct a T5.2-like operating processor. He would implement this function, explicitly, by means of asymmetric processors and he would have to connect a huge number of different processors in order to gain direction invariance. In toads, parallel distributed processing and subtractive interaction allow two types of isotropic network structures to solve this problem, so-to-speak implicitly. I think it's the simple efficient discrimination principle that's of general interest here.

What are prey selective neurons doing in the mating season, or what are mating call selective neurons doing in the hunting season? We don't know. Hormonal controlled threshold setting of these cells or adjustments in the network that determines these neuron's selectivity are discussed. Bill Betts' "T5-base modulator model" shows us how the latter possibility might be implemented neuronally.

CHRISTOPH VON DER MALSBURG:

I would like to give here a little follow-up to the remark I made before. We are used to thinking in terms of computer analogies or digital machine analogies. Looking into neural hardware, we often have the tendency to believe that every little detail there has been put together according to a plan, a plan which can only work if all the details are precisely as they ought to be. If you change one little connection in a computer then you can throw it away. There might be another view of nervous structures which is of an entirely different philosophy which might help in finding relevant aspects. I would like to remind you of this instructive experiment by Martha Constantine-Paton. She put a third eye into a frog's head, and this extra eye projected to one of the two tecta. This was for the first time in the phylogeny of this animal that two eyes projected onto one tectum. The marvellous thing Martha found was that the fibers from the two eyes sorted out into stripe-like patterns. Now if you look at ocularity stripes which had been found in the monkey visual cortex, if you are honest, you always had the impression that genetic programming had created this pattern of stripes with great labour, and that there must be a purpose for having these ocularity stripes. Now comes this experiment which is a complete freak. So, there must be some inborn tendency for fibers to sort out in this order. They obviously like this order. Maybe the nervous system is like a crystal. There are very simple mechanisms which simply like simple patterns. So, maybe what we should look for in the brain is repeating patterns, internal patterns which are put together in the style of crystals, for instance. Maybe we should look for the plans, for the principles and not so much for details. This could focus the matter enormously.

PAUL GROBSTEIN:

I would like to draw attention to one generality which came out across all the discussions. I was trained in the notion of the wiring diagram, the notion that if one knew the connectivity of the nervous system, one would thereby understand its function. It seems to me that among the clearest things to come out of a number of the papers here is that one could know the entire structure of a given nervous system and still not know how that nervous system is going to respond to a given stimulus. The network structure is not enough, one has to know in addition the activity pattern in the network structure at the time the stimulus occurs. The nearly exclusive focus on network structure was too optimistic, and that is in part what has caused trouble over the past ten or fifteen years.

GEORGE MOORE:

So, Paul, to paraphrase what you said, you can have a complete, detailed wiring diagram and still not know how it works.

PAUL GROBSTEIN:

Because part of its organization reflects the fact that it is designed to work using activity patterns as significant information parameters.

Third Session: Distributed Properties, Modulation, and Memory

EDWARD GRUBERG:

I would like to make a few general comments given that we are in the last round of our discussions. First an anecdote. Forty years ago there was an American couple who decided to see if they could teach a chimpanzee to speak English. They reasoned that the chimpanzee has a similar brain structure to ours and is able to vocalize. They obtained an infant and raised it in their house for about seven years. At the end of the seven years, the chimpanzee could fluently say only four words: "mama," "papa," "up," and "cup." That is a hell of an experiment to spend seven years on. The question it raises for us here is how do we know that we are not doing that kind of experiment? A lot of work for little return. Obviously we don't know. One can't predict that one's approach is a productive one. By the way, the couple got divorced soon after and the major bone of contention in the divorce proceedings was who was going to get custody of the chimpanzee.

I would like to make a small defense of anatomy, probably self-serving and unneccessary. It was interesting to see at this meeting how many of the experimental papers depend on crucial new findings in anatomy. There is still much to figure out about the details of connectivity. Learning more anatomy is a useful exercise and has helped light the way for physiological studies. I certainly agree with Paul Grobstein and others who have said that just because we know the whole circuit diagram does not mean that we know how the brain works. In Peter Ewert's and his collegues studies we may have the first approximation of the anatomy of the sensorimotor system that they have been studying. It really is impressive to see a wide range of techniques being used so beautifully, HRP, intracellular cobalt-lysine, 2DG, and the recordings from freely moving animals. I think it's very impressive to see that being done and to find out what is correlative. It's an excellent piece of work.

Next, a comment on information theory. The word "information" has been used here a lot. I was trained to use information in the narrow Shannon sense. The amount of information needed to specify a particular thing of a set of things according to Shannon's definition is log to the base two of the number of things in the set, assuming the things in the set are equally likely to appear. So information is related to choosing from a set of things. When we talk about information in the nervous system what are we talking about? When we listen to a neuron is it saying something informationally complicated or something less interesting, "it is warm," "it is less warm" and that's all? There are examples in the literature of what are called codes. The power output per time can be very similar for different messages. Should neurons be assumed to have very simple codes? Are they multiplexing devices? Should we disregard the apparent complexity that we see in neurons?

Thomas Finkenstädt using 2DG compared naive preresponding toads with habituated ones and showed that the intrinsic pathways for prey-catching were reduced in habituated animals while the extrinsic systems were enhanced. It is an interesting result and is a clear indication of the kind of problem we have in trying to analyze distributed systems. What are the elements that comprise the system? How do we deal with the problem? It is easy to raise the question but hard to answer. It is shown very nicely in Bill Betts' model how to modify other systems. Michael Arbib and Francisco Cervantes have demonstrated the elements that go into modeling habituation and obviously the work of Thomas Finkenstädt is related to that. Sometimes, I have the feeling that experimenters should stop for three years for the modelers to get the models all setup. I guess that's hard to do. Francisco Gonzalez-Lima demonstrated that the functional brain circuitries related to arousal and learning are sensitive to very different kinds of patterns in both those systems. The studies do provide us with a very rich amount of information that is hard to digest.

PETE LAMING:

I was very impressed with the 2-deoxyglucose works. One thing that intrigues me about it in relation especially to arousal is that one of the features of arousal, certainly in mammals and possibly, we don't know, in lower vertebrates, is that blood distribution in the brain changes. That may be a factor we have to take into consideration when we look at 2DG distribution or metabolism distribution. One other point which relates to the paper I presented is that glia also change their metabolism in almost all situations that neurons do. Thus, our interpretation that changes in 2-deoxyglucose uptake indicate only neuronal activity may be wrong. There is evidence that glia and neurons may respond differently during behavior and thus we may be measuring the metabolism of glia or neurons or both. Indeed, as I have suggested glia may be part of the system which modulates neuronal activity during arousal.

FRANCISCO GONZALEZ-LIMA:

I would like to review the issues from two different points of view, one will consider some of the conceptual issues already raised, the other will be a somewhat methodological view. First, I would like to make three points regarding concepts. The first issue has to do with modulation that we heard from Thomas Finkenstädt's and Pete Laming's stories in lower vertebrates, of how modulation affects the sensorimotor system. We saw two examples of these general kinds of modulation, arousal and habituation. I think it was very interesting to compare the effects of habituation, for example reported by Finkenstädt, and some of the effects related to arousal that Laming has worked out, because these two systems are operating in a way that can be contrasted, such as when you examine a number of structures in the brain that show increases in activity due to habituation. And the same structure, such as the ventral medial pallium, is suppressed by various kinds of arousing stimuli. Interestingly, the same thing happens, for example, in the hippocampal formation in the mammal. Therefore, there are a number of brain structures whose activity level is consistently modulated by arousal level. These structures show opposite changes to increases in arousal and during habituation.

The second point deals with the meaning of habituation. We have seen two processes in toads, one that can be referred to as short-term habituation which takes place during a session when you are presenting a repetitive stimulus to the animal and there is a decrease in the response. But there is another type we can refer to as long-term, the one that takes habituation from one session through the next days. We know in mammals these two processes have different neural substrates. The long-term one involves substrates responsible for the conditioning type of changes. The short-term one does not seem to involve such memory related functions. For example, we know that mammals with a section behind the midbrain can perform perfect short-term habituation responses but these decerebrate animals lose completely the capacity for long-term habituation. Since there is a lot of confusion in the literature by using the word habituation to refer to fundamentally different processes, I think this is to be cleared up from the beginning when models are considered.

My next point refers to conditioning. The question of coding in the "conditioned stimulus channel" is related to the general question of how sensory information is coded in sensory maps. There is, for instance, the primary paradoxical question of whether sensory maps are coding the significance of signals, since how can a sensory map represent the significance of the stimulus and at the same time represent the physical properties of the stimulus. Here seems to be one of the advantages of the 2DG technique. For example, if you have an area in which the neuronal map representing the stimulus is defined to a particular neuronal space like tonotopy and retinotopy, this representation is the same during learning. In other words, you have the same spatial pattern of representation, but the activity we see in the neuronal space will change to be able to code the significance of that stimulus, a precondition for discriminating this stimulus from other ones. So, I think this needs not necessarily be a paradox that sensory systems can represent physical stimuli and perhaps at the same time code the significance of the stimulus within the same neuronal space. This goes along with the Hebbian idea of how associations are made. But outside sensory maps, the extrinsic system seems to operate in different roles. In the extrinsic system we do not see the overlapping effect in terms of neuronal representations of the stimuli that are involved in sensory structures during conditioning. My experiences in several conditioning paradigms and in habituation are mainly supporting the notion that sensory structures are involved in learning.

Regarding the methodological points, I would like briefly to put forward for discussion the 2DG method. The first point is regarding the functional mapping. It should be understood that this technique is useful for spatial resolution but it has the limitation of the temporal dimension. We can only see the cumulative metabolic changes on a high resolution scale. The second point concerns the question what is the correlation between electrical and metabolic activity and other activity related to this. What are the cellular mechanisms involved in these metabolic changes? Are they involving neurons or glia as Pete Laming asks? In fact I believe there is very good correlation between electrical and metabolic activity. For example, it is possible to penetrate an electrode in a structure where you know there are tonotopically organized laminae. Penetrating an electrode and recording the discharge frequencies of the neurons along this track, you can define the best frequency of these neurons to different tones. Then, in the same animal, you can apply the 2DG-technique and presenting the same stimulation you will see a tonotopic band mapping exactly the position of the best frequency of the neurons that you recorded from this area. You can do experiments like this for different positions and you will find that the position of the best frequency obtained by recording corresponds with the 2DG metabolic map. So, in sensory systems, and I believe independent of the system you are working with, you should be able to demonstrate the coupling between electrical and metabolic activity, at least to some extent. In fact when we go to the question of cellular mechanisms, it seems to be that the neuropil is the one showing the greatest change in metabolic activity. One point we should keep in mind is that the 2DG uptake represents the entire sum, the net metabolic activity of a particular area. So it includes of course the glia, without any doubt. If we move outside of the sensory maps, for example to areas that belong to the reticular activating system, the so-called integrating regions, I have found a very good correlation between what you may call long-lasting unit responses and metabolic activity, net metabolic activity to be precise. Neurophysiologists record in these regions only for a few milliseconds, so that they see in the reticular activating system, for example, a decrease in activity for a few milliseconds and then an increase in firing rate that lasts for several seconds. The way it comes out in terms of 2DG uptake is that you have a net increase in metabolic activity. So, if you correlate changes that take place over more than one second, you actually see very good correlation between unit activity and the cumulative net metabolic activity. This is found in virtually every region of the brain that involves the reticular activating system. In most of these areas which showed biphasic responses, the dominant response is the one that is shown in the 2DG uptake. The evidences against this view are based on very short range responses like in the case of the hippocampus. I leave these points for the discussion.

PAUL GROBSTEIN:

I would like to follow up on a methodological point that puzzles me. I can certainly understand that if you are in a situation where the net activity increases then there is more uptake and hence more labeling. I am interested in the concern, the reservation, reflected in the fact that metabolic activity registered by the 2DG method is neuropil metabolic activity related to the secretion of transmitters. It ought then to be the case that in a particular structure which is receiving large amounts of inhibitory synaptic input would be expected to show heavy 2DG labeling, just like one receiving large amounts of excitatory input. Do you disagree with that point?

FRANCISCO GONZALEZ-LIMA:

If we look at inhibition, for example, in the ventromedial region of the basal forebrain just beneath the ventromedial globus pallidus, you will see inhibition of unit activity and metabolic activity as a result of reticular stimulation. This area shows a very large suppression in terms of 2DG uptake. So, if the unit responses are long lasting and you don't have biphasic unit responses, where it changes from one to the other, you can clearly see correlation between unit activity and net metabolic activity.

PAUL GROBSTEIN:

How is that consistent with the evidence that it is primarily neuropil metabolic activity which determines the labeling pattern? You should show high neuropil metabolic activity in that case.

FRANCISCO GONZALEZ-LIMA:

Okay, that is true, the neuropil is the dominant effect, simply because there is a large volume fraction of membrane in neuropil as opposed to cell body. With the 2DG technique, since glucose is used for energy metabolism, what we are really seeing at the molecular level is the sodium/potassium pump. Its activity is larger in neuropil. If you look at structures that are very well laminated, for example the hippocampal formation, it is clear that the molecular neuropil layers are the ones that show the largest 2DG uptake. But we have to realize that I cannot define whether this reflects only presynaptic or postsynaptic events. I believe it is both. If there is integration, then the 2DG uptake maps the net result of both. So, if there is presynaptic inhibition, this will increase the depolarization of the membrane of the axonal bouton leading to a decrease in transmitter secretion and, postsynaptically, to a reduced depolarization, so that the net metabolic effect is smaller than for an excitatory synapse. Postsynaptic inhibition involves the sodium/potassium pump less than postsynaptic excitation, both in the neuropil and in the cell bodies.

BARRY HORWITZ:

I would like to continue with the point I was trying to make in my presentation. It was not that an increase in activity necessarily implied excitation, or inhibition, or suppression of inhibition. The point I was trying to emphasize was that the kind of change in 2DG uptake that one sees in a region of interest depends very much on the particular system one is examining. In one system I talked about, Ackermann's model where you could have long duration inhibition, in that particular case the axonal firing in the region more than overcame the decline in activity in the cell body. Thus, you found an increase in metabolism. However, this might not hold in another system where the geometry might be arranged such that you have slightly less neuropil, and slightly more active cell bodies. The point I am trying to make is simply that it is very complicated and you cannot infer necessarily that a suppression in activity means inhibition. Conversely, you cannot necessarily rule out that either. The significant point is that knowledge about the 2DG pattern must be coupled with knowledge about the postsynaptic potentials, that means whether they are excitatory or inhibitory.

THOMAS FINKENSTÄDT:

Michael, do you think we should include the insights derived from 2DG studies into *Rana computatrix*? What we actually need furthermore are data from experiments that are very hard or almost not possible to do, for example, multi-electrode recordings from the behaving animal in the course of conditioning, since conditioning takes many days to weeks. So, we encounter technical limitations. Would your model, taking our 2DG data into account, allow us to economize, or even to abandon experimentation in certain cases?

MICHAEL ARBIB:

It is an interesting issue in terms of what we have been hearing to ask whether we should expand the basic model of the neuron we use, so that it includes a model of energy metabolism, so that in fact we could start doing simulations of 2DG uptake, correlated with our simulations of the function of a network and carry out comparisons there. That leads into the more general issue of hoping that we can evolve a little bit more of a social mechanism for experimentalists to adopt theorists and start building a relationship with them where they are not in a position of only seeing your polished paper when it is in print or as a preprint, but rather being able to see raw data, so that it becomes possible to start testing a model not against a view of graphs published in a paper, but rather to test it with far more intimacy against the body of experimental findings. This is an implicit invitation to each experimentalist who thought that a theorist was saying something interesting and challenging.

BARRY HORWITZ:

Neuroanatomically-based models of metabolism, as far as I know, have never been done. Such models would ask the question: if you have this kind of electrical activity, either in a neuron or in a small group of neurons, what would be a metabolism (2DG uptake) measured over some time interval? I think that would be a very interesting theoretical question to address. Very little has been done on it, and nobody knows the relationship. But no one has really tried to work out the details. Although it is complicated, I think it would be a reasonable area within which a modeler could work.

HERBERT REITBOECK:

Let me introduce another issue. Ed Gruberg suggests that we should think again about coding in the nervous system. Up to now, concepts taken from information and coding theory have been applied mainly to signals in single neural channels in analogy to the technical transmission channels these theories were developed for. The single neuron, however, is not a decoder for sequential signals in a signal channel, but it responds to coincidences in parallel channels. In order to be better suited for the particular problems in neural systems, coding and information theoretical concepts must be extended to deal with multi-channel codes. I think this is an important point that has not yet attracted sufficient attention. A related point is the old serial *vs.* parallel processing dichotomy. Many computer scientists still emphasize that anything that can be done in parallel can be done sequentially, too. The Turing machine, the most universal computer, is a sequential machine. But many important operations in image processing, ranging from local analysis to global filtering, can be faster by orders of magnitude if done in parallel. The present excitement about recent results in neural network simulations and neurocomputing has stimulated renewed interest in parallel processing. Although neural network algorithms usually are simulated on sequential machines, the power of these algorithms can only be realized with special parallel computer architectures.

PETER EWERT:

I would like to emphasize the advantage of the 2DG technique which allows us to monitor concurrent activities in brain structures indicative for parallel distributed information processing. Rumelhart and McClelland's book on parallel distributed processing, PDP, is cited in many places of brain science. Michael, is the PDP enterprise really a new break-through? How do you think about this?

MICHAEL ARBIB:

I am reminded of the character in Molière who was amazed to discover that he had been talking prose all his life. And so I am amazed to discover that I have been doing parallel distributed processing all my life! Brain theorists have always been committed to the idea that there are simultaneously active regions and each region is constituted of a population of neurons. The slogan parallel distributed processing only carries information to people in the field of artificial intelligence and cognitive psychology who have been dominated by serial processing models. It is really saying to them that even if you don't want to be a responsible neural modeler worrying about the hard data of neurophysiology, at least pay attention to that general style of the brain of distributed subsystems, with parallel activity in each. I think that all the models that have been talked about here, because they are addressed to the brain, are distributed models, also the Mittelstaedt models where they do not get into the neuronal details. Nevertheless, PDP has attracted a lot of workers, and exciting progress is beeing made in the field of multi-layer adaptive networks.

PETER EWERT:

Well, in this sense our 1968 tectal/pretectal box model of toad's innate prey-predator discrimination was already an example of PDP. Recently, Finkenstädt's 2DG and lesion studies have shown that learning to select visual stimuli involves other distributed properties which allow the brain to store information about the learned significance of stimulus cues and to compare this with information related to a present stimulus. Michael, do your computer simulations provide new insights into this function?

MICHAEL ARBIB:

I think the answer is that I don't know how it is done in the brain. The state of the art in modeling tends to be to look at a system that is learning a few associations or learning a few reflexes. There is usually a model of a single network, perhaps a synaptic matrix, and then the latest version of some learning rule is used to put that in a network with a small number of layers. It is trained with a certain input-output response and the hidden units are trained to yield that relationship. What I think was significant today was that we were all reminded very strongly that any model which simply says here is the place where the association is learned or this is the response that is acquired, is not good enough. We have some discussion of this in the modeling literature where we talk about the credit assignment problem. The original motivation was in game playing machines where you play a game of checkers and go on for many moves before you finally either win or lose. Now, how do you decide that your 37th move was a good one or a bad one? How do you propagate the final judgement back? This is the temporal credit assignment problem, of assigning credit or blame for this final result back in time. Here, I think we are talking not so much about that temporal assignment problem but the structural assignment problem. You have got a whole lot of subsystems, but only one of them is producing the output that will be rewarded. How is it that effects are distributed through a network? The answer is at the moment that we have begun to understand the importance of this, and hopefully the stimulus for the sort of experiments we heard about will help to push us a lot further.

PETER EWERT:

Let me select another point regarding distributed processing of visual information. I was very impressed by Ed Gruberg's new data on the nucleus isthmi, since these tell me that we may have to regard the function of this structure from different points of view. Ed, which is, or rather are isthmic functions and, Michael, how can these be incorporated in *Rana computatrix*?

EDWARD GRUBERG:

The nucleus isthmi is a small nucleus. It has only about 8000 cells compared to the million or so cells in the tectum. So it looks like it is not doing fine analysis. It could be carrying out some kind of weighting function. But this is hard to test in ablation experiments because you lose the nucleus. It requires doing experiments with intact animals. We have been doing the physiology and have not technically solved the problem yet. Assuming the nucleus is carrying out a weighting function does it affect the tectal output or input by modifying, informationally, what is encoded or, is it a power switch turning something on while suppressing other things. It is still not clear.

MICHAEL ARBIB:

The nucleus isthmi has an interesting story in the relationship between experimentalists and modelers. Donald House and I originally had a model of depth perception which was essentially a phenomenological model responding to the observation that first Ingle and then Collett found that the animal could use monocular cues. But if stereoscopic disparity cues were available, they seemed to be more important. So we came up with a design of a depth perception system that would be able to use either cue alone but if both cues were available, it had to integrate the cues. In doing so, it would solve one of the crucial problems, the ghost suppression problem or the correspondence problem, making sure that the corresponding two points on the two retinae were matched up to correctly locate things in the world. Then along came Tom Collett and Sue Udin with an experiment in which they ablated nucleus isthmi and the animal could still snap accurately for depth, but seemed to be snapping at ghosts more often. So we came up with a rather new model. It was interesting in that we incorporated things we already knew how to do. We already had a vocabulary in place which let us come with a new model rather quickly and was able to explain this phenomenon. Then Ed Gruberg comes along and says it is not that way at all, you cut the isthmi and you get scotomas. Now there is a very interesting question. What do you do at this stage? Here you are a poor innocent modeler with two people you love and respect coming up with totally discordant experimental data. One thing is just to stop modeling for five years in the hope that eventually there will be a consensus among experimenters as to who was right. Well, the other is to keep going.

Some years ago, I was at a meeting on the cerebellum. One of my students presented our model of the cerebellum and immediately an experimentalist said "Don't you realize the main data you have been using in your model are wrong? We have shown in our lab that's not the way it is at all, and that proves my point, it is premature to do any modeling." Of course my response was "If anything, the evidence you have given us is that experimentation is premature!" To make some sense out of this, what I want to say is that I don't think it is useful for the modeler to wait till all the information is in. I think rather what the modeler can do is to say here are two models at this particular point. Here is a model which we already have, which is correct for Collett, and here is the model we don't have yet but we should develop it now to address the data of Gruberg. By looking at those models together, we can think of new experiments that would advance the field. So I think about that notion of having a range of models to which the experimentalist can respond and challenge rather than having the one true model coupled to the one true set of data.

PAUL GROBSTEIN:

The story is an interesting one in connection with our earlier discussion of the level at which one ought to model, and one suggestion that perhaps the level where the patterns in experimental observations are the most clear. What I am suggesting is that attempting to incorporate nucleus isthmi into a model related to behavior may in fact be premature. There are certain facts about the nucleus isthmi which are quite clear in the sense that a number of labs have reported similar phenomena. If one wants to model the nucleus isthmi, it would seem better to start with the consistent evidence that it is a highly topographic system which links various points in and between the tectal lobes in particular ways. I would be inclined to work with the clear pattern and see what emerges from that as possible functional roles, rather than presuming a function where the pattern of evidence is less clear and trying to incorporate that into a more behaviorally oriented model.

MICHAEL ARBIB:

The basic feature of the model is that it has all those properties that you say are agreed upon yet still leaves mysterious why its removal should, as Ed tells us, create such problems. But I think there is something else to be said about why we need both romantic and responsible models. Perhaps we need a way to read the sort of models that I already have in a different way. For certain functions, they provide architectures that achieve them. It may not matter very much in terms of getting the functional ideas whether its isthmi or some other part of the brain that's doing this particular job. For example, our early models of prey-selection were all stuck in the tectum. Now, thanks to Peter's influence, we have considered pretectal and tectal circuits and have added new functions to understand the discrimination of moving configural stimuli. So I think there may be something to be said just for having the models as objects, where one round of debate and investigation tells us whether this is a good way of achieving a certain function, and a separate debate tells us to what extent those particular functions are installed in the brain.

HERBERT REITBOECK:

That brings us to the end of our discussion sessions. This meeting covered an unusually wide span of topics, ranging from visuomotor coordination to robotics and from information processing mechanisms in amphibians and mammals to a glimpse at the future sixth generation of computers. It's that multi-facetted view of biological and technical information processing systems that made this meeting so stimulating. It was a lively forum for an interdisciplinary dialog, and sometimes it generated sparks of an excitement reminiscent of what must have been at the group around Warren McCulloch when they first tried to model neural functions. Tremendous progress has been made since that time in both the neurosciences and in computer science. But as the fields became more and more specialized they also drifted apart. It is only recently that new - and very successful - attempts are made on a larger scale to bring the two fields together again. This meeting is an excellent example for that trend. I think to speak in the name of all participants when I thank Professor Ewert and Professor Arbib for having organized this most interesting and successful conference.

Arbib, M. A., Computer Science Department, Salvatori Computer Science Center, University of Southern California, University Park, Los Angeles CA 90089-0782, USA. - **Bässler, U.**, Fachbereich Biologie, Universität Kaiserslautern, Pfaffenbergstraße, D-6750 Kaiserslautern, FR Germany. - **Ewert, J.-P.**, Abteilung Neuroethologie, Fachbereich Biologie/Chemie, Universität Kassel (GhK), Heinrich-Plett-Str. 40, D-3500 Kassel, FR Germany. - **Finkenstädt, T.**, Abteilung Neuroethologie, Fachbereich Biologie/Chemie, Universität Kassel (GhK), Heinrich-Plett-Str. 40, D-3500 Kassel, FR Germany. - **Gonzalez-Lima, F.**, Department of Anatomy, College of Medicine, Texas A & M University, College Station, Texas 77843, USA. - **Grobstein, P.**, Department of Biology, Bryn Mawr College, Bryn Mawr, Pennsylvania 19010, USA. - **Gruberg, E.R.**, College of Arts and Sciences, Department of Biology, Temple University, Philadelphia, Pennsylvania 19122, USA. - **an der Heiden, U.**, Institut für Naturwissenschaften/Mathematik & Theorie Komplexer Systeme, Stockumer Str. 10, D-5810 Witten-Annen, FR Germany. - **Horwitz, B.**, Laboratory of Neurosciences, National Institute of Aging, National Institute of Health N.I.H., Bethesda, MD 20892, USA. - **Laming, P.R.**, Department of Zoology, The Queens University of Belfast, Belfast BT7 1NN, Northern Ireland. - **Lázár, G.**, Department of Anatomy, University Medical School, Szigeti ut 12., 7643 Pecs, Hungary. - **Mallot, H.A.**, Institut für Zoologie III, Fachbereich Biologie, Universität Mainz, D-6500 Mainz, FR Germany. - **von der Malsburg, C.**, Department of Computer Science and Neurobiology, Salvatori Computer Science Center, University of Southern California, University Park, Los Angeles CA 90089-0782, USA. - **Mittelstaedt, H.**, Max-Planck-Institut für Verhaltensphysiologie, Abteilung Mittelstaedt, D-8131 Seewiesen, FR Germany. - **Moore, G.**, Computer Science Department, Salvatori Computer Science Center, University of Southern California, University Park, Los Angeles CA 90089-0782, USA. - **Mühlenfeld, E.**, Technische Universität Clausthal, Institut für Elektrische Informationstechnik, Leibnitzstr. 28, D-3392 Clausthal-Zellerfeld, FR Germany. - **Nelson, J.I.**, Arbeitsgruppe Angewandte Physik und Biophysik, Philipps-Universität Marburg, Renthof 7, D-3550 Marburg, FR Germany. - **Reitboeck, H.J.**, Arbeitsgruppe Angewandte Physik und Biophysik, Philipps-Universität Marburg, Renthof 7, D-3550 Marburg, FR Germany.

PART II

An Opening Perspective

The Release of Visual Behavior in Toads: Stages of Parallel/Hierarchical Information Processing

JÖRG-PETER EWERT

Abteilung Neuroethologie, Fachbereich Biologie/Chemie, Universität Kassel
D-3500 Kassel, FR Germany

Abstract. The release of toad's prey-catching behavior (orienting, approaching, fixating and snapping) is mediated by *releasing schemas,* also called releasing mechanisms. The present neuroethological investigation suggests the neurophysiological equivalent of a schema as a combination of neurons specified to encode significant coincidences of stimulus cues in space and time. A correspondingly coded *command releasing system* results from parallel/hierarchical processing in a neural macro-network whose interacting components evaluate visual input under various aspects such as features, background, space, motivation, and past experience. Essential components of prey-related *sensorimotor codes* are T5.2 neurons. These neurons are tuned to the spatiotemporal properties of moving objects, in such a manner that elongated worm-like moving shapes are preferred and shapes elongated perpendicular to the direction of movement are neglected. Unlike cortical "orientation detectors," the function of these tectal "configuration discriminators" does not depend on an anisotropic structure; the *anisotropy* required for configural discrimination lies in the time domain of the spatiotemporal properties of the moving object itself. Evidence is provided that the direction- and velocity-invariant discriminatory properties of T5.2 neurons are brought about by discrete, parallel evaluations of the spatial and the spatiotemporal stimulus parameters by two-filter operations in pretectal and tectal structures, respectively, and through their subtractive interaction. In amphibians, with motionless eyes, the movement-specific analysis of moving objects seems to take advantage of an economical system that has sidestepped the need for a sophisticated anisotropic cortex-like structure known to involve a huge number of asymmetric processors in order to meet direction-invariant discriminatory operations.

1 Introduction

A stimulus is referred to as a "sign stimulus" if it represents a behaviorally relevant object by a few characteristic features adequate to release a specific pattern of behavioral response whenever the animal's motivation is appropriate (Tinbergen

1951). In classical ethological terms such responses are called "instinctive actions." *Instinct* has been defined by Tinbergen (1951) as "a hierarchically organized nervous mechanism which reacts to priming, releasing, and directing stimuli of either endogeneous or exogeneous character. The reaction consists of a coordinated series of movements that contributes to the preservation of the individual and species." Instinct refers to a species' knowledge about its environment allowing an animal in response to predictable situations adequate behavior often consisting of directed *appetitive reaction(s)* and the *consummatory act*. A bird, for example, uses knowledge about nest building by putting pieces of straw into the form of a nest according to an inborn "nest schema." A toad uses knowledge about prey by comparing visual cues with an inborn "prey schema" that makes prey-catching successful. Releasing these actions presupposes analyses of the significant coincidence of certain external stimulus features (sign stimulus) and space parameters in, and only in, the presence of appropriate internal drive (motivation).

The neural operation responsible for recognition and localization of the sign stimulus has been termed the *innate releasing schema* (v. Uexküll 1909; Lorenz 1935, 1943) or the *innate releasing mechanism* IRM (Tinbergen 1948, 1951). Although the neural mechanisms underlying an IRM were unknown when the concept of IRM was developed, a relatively "hardwired" process was presumed because of a certain degree of stereotypy in stimulus-response relationships. Current neuroethological concepts of IRM suggest a complex, to some extent adaptive neural sensory/associative/motivational/motor interface (Ewert 1988). The motorside of this face has access to motor pattern generating systems which - according to a pattern of excitation and inhibition (the motor pattern) - activates respective motoneurons to coordinate muscle contractions for the appropriate *fixed action pattern* (FAP). The question of how "fixed" FAPs are and to what extent they are "innate," has been debated ever since Lorenz and Tinbergen (e.g., see Hoyle 1984). Despite redefinitions, uncertainty about these adjectives often remaines. An evolving concept of RM refers to the predispositions operating in the context of developmental experience and learning (Fentress 1983, 1987). This offers neuroethology an advantage of studying relatively "hardwired," dedicated processes in connection with "softwired" operations, whereby an analysis of (I)RM may actually guide this approach (Baerends 1987; Fernald 1987). There may be fundamental neural processes that *instinct* and *cognition* share (Ewert 1987c, 1988). We will return to this issue at the end of the chapter.

What neural operations connect a sign stimulus to a corresponding action pattern? Advances in the neuroethological investigation of sign stimulus recognizing and behavior generating systems have been made in various domains, such as auditory (e.g, in crickets), electrosensory (in weakly electric fish), and visual (e.g., in toads) (summarized in Ewert et al. 1983b; Huber & Markl 1983; Guthrie 1987). Among terrestrial vertebrates, amphibians of the genera *Rana* (frogs) and *Bufo* (toads) are especially appropriate experimental animals for the following reasons: (i) They have a limited behavioral repertoire that includes FAPs; (ii) in response to relatively simple sign stimuli, they consistently exhibit predictable behaviors whenever their motivation is appropriate; (iii) stimulus-response relationships can be analyzed quantitatively by changing certain stimulus parameters in simulated natural objects or dummies, thereby identifying both releasing and ineffective features; (iv) the neuroanatomy is reasonably well understood (e.g., see Lázár and Gruberg, this volume), a prerequisite for neurobiological recording, labeling, stimulation, and lesion experiments.

The visually released prey-catching of common toads *Bufo bufo* is particularly suitable for a neuroethological analysis of visuomotor function (Eibl-Eibesfeldt 1951; Schneider 1954; Ewert 1968, 1984a; Ingle 1968, 1983). Toads respond to prey with distinct behavioral patterns: Orienting, approaching, fixating, and snapping, which often yields a sequence. Prey-catching thus calls for certain perceptual operations by RMs concerned with stimulus recognition, stimulus localization, sensorimotor transformation, motor pattern generation, and state-dependent modulation. The investigation of such processes touches on questions of general significance to scientists working in the fields of *neurobiology* (see Bässler, Berkinblit, Borchers, Finkenstädt, Gonzalez-Lima, Grobstein, Gruberg, Horwitz, Laming, Lázár, Manteuffel, Matsumoto, Sandoval, Schürg-Pfeiffer, Schwippert, and Weerasuriya, this volume), *systems neurosciences* (see Arbib, Betts, Cervantes-Pérez, an der Heiden, Lara, Mallot, Mittelstaedt, and Reitboeck, this volume), and *artificial intelligence* and *engineering* (see Arbib, Arkin, Koy-Oberthür, v. d. Malsburg, and Torras, this volume). Such questions concern:

(1) *Perception-Action Cycles*. We note that *behavioral sequences* are often hierarchically organized (Tinbergen 1951). So we will ask whether in a toad's prey-catching one particular action must first be accomplished before another one can begin. Is such an action necessary or even sufficient to trigger a subsequent response? Or does the release of each action pattern depend on the ongoing stimulus situation, drawing on operations related to both stimulus recognition and stimulus localization? The answers to these questions are important for understanding the toad's prey recognition system.

(2) *Concept of Sign Stimulus*. Given that the visual environment of the toad is structured, the basic question arises as to how *signals* (prey) are distinguished from *noise* (background). What are the releasing (visual) features of prey? Is there in the visual domain a unique prey sign stimulus, or do the features that release prey-catching refer to a certain range within a stimulus continuum?

(3) *Parallel vs. Serial Processing*. We ask whether the perceptual operations required for toad's prey-selection are performed by a feature-detector (Fig. 1A), which has often been proposed as a necessary component of innate RMs (for a discussion see Capranica 1983 and Scheich 1983), or by a feature analyzing network (Fig. 1B). The *feature-detector* hypothesis (e.g., see Hubel & Wiesel 1962, 1965; cf. also Konorski 1967; Barlow 1985a) usually postulates distinct "hierarchical" feature extracting processes: The higher the level of convergence on feature-sensitive neurons (Fig. 1A), the higher the degree of feature selectivity. At one extreme, highly selective gnostic neurons (say, a "grandmother cell" in a human brain) would only respond if a complex signal occurred (say, a specific set of individual facial features). In such a serial processing system, information flow would be from a level of lesser selectivity to one of greater selectivity. The *network analyzer* hypothesis, on the other hand, involves "parallel" information processing in a neuronal network: Here, an ensemble of different classes of interconnected feature-analyzing neurons would be responsible for the recognition of a visual object (Fig. 1B). Unlike in a feature detection system, the feature selective properties of neurons in such a network would be determined by

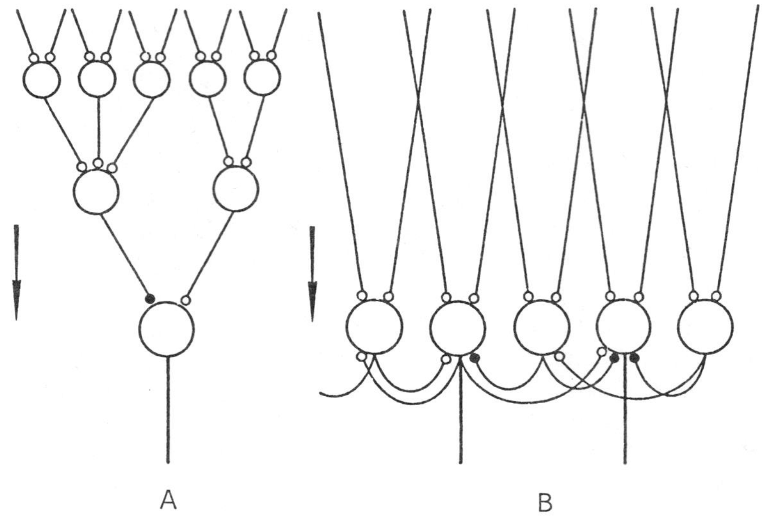

Figure 1. Principles of serial and parallel information processing by interconnected neurons. Neuronal wiring principles underlying a "feature-detector" (A) and a "feature analyzing network" (B). Large circles refer to neurons; black and white small circles refer to different types of synapses, here arbitrarily chosen.

mutual interactions. In addition, a class of neurons, through interconnections, could take part in different ensembles (Fig. 1B), thus contributing to the recognition of different objects. Here, the same basic network, but with distinct spatiotemporal activity patterns, could take part in the analysis of different visual objects (e.g., see Hebb 1949, 1959; Mountcastle 1957; John & Schwartz 1978; Creutzfeldt in Szentágothai & Arbib 1974). We will ask whether the "detector" and "network analyzer" concepts are mutually exclusive, or whether the toad's brain might be taking advantage of both parallel and hierarchical principles of information processing.

(4) *Sensorimotor Interfacing.* Here we will ask how the signal-related information is *coded* in the brain, transferred to the appropriate motor systems and finally translated into goal-related *motor commands*.

(5) *Modulation of Stimulus-Response Mediation.* The question arises as to how RM circuitry mediating stimulus-response relations is influenced by modulatory circuits signaling motivational states and learning (as defined by Kandel et al. 1979). In this context it is important to ask whether both circuits work independently, except at their point of convergence, or whether there is substantial overlap, so that certain neurons may be used for both *mediation* (coupling an appropriate behavioral response with a given stimulus) and *modulation* (influencing the mediation).

The answers to these questions require a multidisciplinary *top-down* approach from toad's behavior to neuron, involving: (i) Behavioral studies on innate and learned aspects of prey stimuli; (ii) neurobiological investigations of brain structures and their interactions in prey-catching and avoidance behaviors; (iii) the evaluation of feature-analyzing and space-monitoring neurons as putative components of RMs; (iv) the study of sensorimotor links between visual pattern analyzing and motor pattern generating systems; and (v) the investigation of neural loops in the modulation of RMs as a function of motivation, habituation, and learning.

We begin with a global black box study in a behavioral paradigm and then analyze functionally related brain structures and interacting subsystems down to single cell responses and underlying changes in membrane potential. We then propose a conceptual synthesis that leads to integrative and interacting functional units (cell assemblies) as putative components of cooperative releasing systems, the central core of the behavioral black box. Since our view in the description and evaluation of a functional structure is restricted by limitations arising from the technical instrument (analogous to "keyhole peeping"), the application of ethological, anatomical, and neurophysiological techniques will allow us to tackle a problem from many different points of view (comparable to "multihole peeping").

2 Action Patterns and Their Releasing Schemas/Mechanisms

2.1 Action Patterns

Postmetamorphic prey-naive common toads (the tadpoles are vegetarian) raised in the laboratory from the egg exhibit prey-catching patterns after their transition to terrestrial life, and these are subject to maturation (Traud 1983; Ewert 1985). The prey-directed visual behavior consists then of two kinds of action pattern (Fig. 2): The target-oriented appetitive responses, orienting toward prey (*o*), approaching (*a*), and fixating (*f*), and the consummatory act, snapping (*s*). Whereas the former (*o, a, f*) can vary in shape and order, e.g. depending on prey's behavior in space, the latter (*s*) is relatively rigid; both depend on motivation (for definitions see Tinbergen 1951; Manning 1979). The activation of the components of prey-catching is mediated by different *releasing schemas*, also called *releasing mechanisms*, RMs, which use comparable prey recognition processes but are distinguishable by the operations that locate the visual stimulus in space (Fig. 2) (Ewert et al. 1983a).

The notion that visual perception of "what" and "where" precedes prey-catching behavior in toads and frogs (Ewert 1974) has to be evaluated with reference to the ballistic, programmed character of the action patterns emphasized by various authors. Once triggered by an appropriate stimulus in the peripheral visual field, orienting toward the prey proceeds to completion - in conjunction with proprioceptive inputs (Comer et al. 1985) - without feedback from the target during the turning movement (Grobstein et al. 1983). Toads also program the route of the approach toward prey to some extent before they start to move (Collett 1983). If the prey disappears during a critical phase of fixation, toads snap accurately at the empty stimulus site (Hinsche 1935). It is noteworthy that the subsequent ingestive behavioral patterns (gulping and mouth wiping) may also occur, despite the absence of the prey. This suggests that certain motor programs are somehow connected in the brain (Ewert 1967b). Frogs and toads also possess various "compound motor coordinations," e.g., for jumping/snapping. These too are ballistic, but may involve feedback-guided correctional manoeuvres (Gans 1961).

a

b

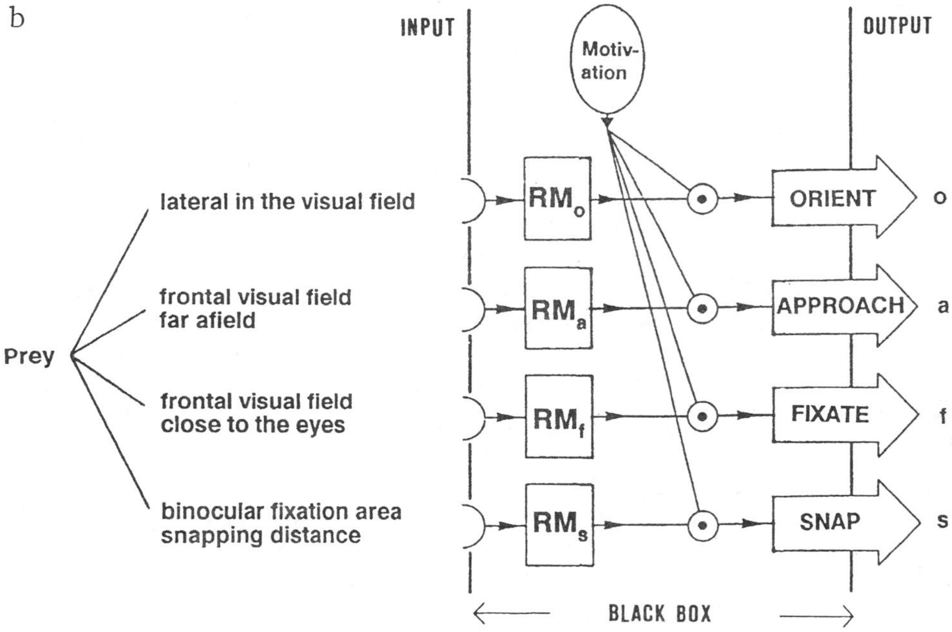
INPUT
Motiv-
ation
OUTPUT
lateral in the visual field
frontal visual field
far afield
Prey
frontal visual field
close to the eyes
binocular fixation area
snapping distance
RM_o
RM_a
RM_f
RM_s
ORIENT
APPROACH
FIXATE
SNAP
o
a
f
s
BLACK BOX

2.2 The Conditions of Sequential Release

Hierarchical influences among RMs or motor pattern generating structures (Tinbergen 1951; Baerends 1976) are not essential for the sequential activation of prey-catching patterns in common toads (Ewert et al. 1983a). It is not the previous action but the ongoing stimulus situation "prey sign stimulus and its locus in space" that determines the subsequent response. If the distance between prey and toad is short, prey-catching consists of *o,f,s*; if the prey suddenly appears close to the animal, only *o,s*, or *f,s*, or just *s* is elicited; if the prey flees, the toad's appetitive pursuit responses occur - depending on the prey's behavior - in variable succession, such as *o,o,o,a,o,a,f,a,f,o,f,s*; if, in an experimental paradigm (Ewert 1969a), a prey object is moved around the toad at constant distance from the animal, then *o,o,o,o,o,o*.... Studies in which frogs readily snap toward prey even though their orienting is blocked by selective transection of the tecto-spinal pathways show that orienting is not a prerequisite for eliciting snapping (Ingle 1983; Kostyk & Grobstein 1982).

In summary, the release of each action pattern requires stimulus analysis regarding *prey features*; the choice of action - the capture strategy - further demands an appropriate evaluation of the *local sign* in visual field coordinates. Furthermore, we note that sequential activation of action patterns is determined by *significant coincidence* of sensory information (Fig. 2b): Sign stimulus and its locus in space. All this favors a sensory recognition concept in contrast to a "motor concept" (Grüsser & Grüsser-Cornehls 1970, 1976) according to which object recognition should not precede a motor response, but motor feedback from the various action patterns during the behavioral sequence in toto - *o,a,f* and *s* - was believed to be necessary for prey recognition. It can be shown in toads that the probability of occurrence of a single object-oriented action pattern *o,a,f* or *s* in an object-motivated animal is positively correlated with the object-defining features.

3 Configural Properties of Sign Stimuli

The ethological analysis of prey recognition is concerned with quantitative measurements of the prey orienting and/or snapping activities, R [responses/t* for t* =30 or 60 s] in response to black two-dimensional moving stimuli with variable configural features (Ewert 1968, 1969a). "Configuration" refers to the combination of characteristic features of an object and to the mode of their arrangement; its holistic property is determined by particular inherent relationships (Tinbergen 1951; see also Section 3.5.2).

←

Figure 2. Toad *Bufo bufo spinosus* (a) in its environment fixating an earthworm (photo by H. Burghagen). (b) Releasing conditions for prey-catching action patterns; depending on prey and its location in visual space, the various action patterns orienting (*o*), approaching (*a*), fixating (*f*), and snapping (*s*) are activated by appropriate releasing mechanisms (RMs), so-called releasing schemas. (Simple flow-chart according to a scheme by B. Hassenstein in Czihak et al. 1981).

3.1 Natural Prey

Before we begin to analyze and describe features of visual prey quantitatively, some remarks on natural prey are necessary. Metamorphosed common toads are carnivorous. From investigations of stomach contents it is known that anurans feed on earthworms, slugs, millipedes, beetles, ants, and other small invertebrates (Porter 1972). All these have more or less elongated shapes and their longer body axis is oriented parallel to the direction of movement. Objects must usually move in order to be categorized as prey: A non-moving retinal image fails to elicit prey-catching. In the analysis of prey recognition we will therefore focus on *moving* stimuli that contrast with their background, investigating features in relation to the direction of movement.

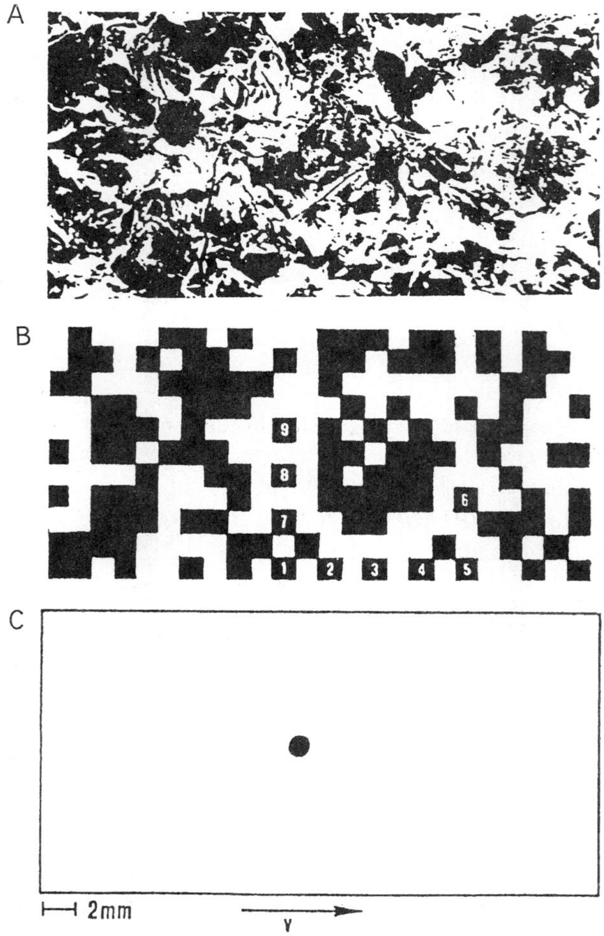

Figure 3. Visual environments: (A) Ground in a forest; (B) artificial texture; (C) small stationary object against a white background. Arrow indicates movement direction of retinal image; for explanation see Sections 3.2 and 3.3. (Ewert et al. 1970, 1982, 1983a; Burghagen & Ewert 1983).

We will ignore color, since common toads are active during twilight and nighttime during their hunting season. (For color vision see, e.g., Kondrashev 1987).

3.2 Prey/Background Relationships

The visual environment of the toad includes things like leaves, branches, and stones. Among these there are also objects of prey size. Why does the toad ignore any moving retinal image as potential prey when the source is a stationary texture in front of which he moves his head (Fig. 3A or B), which makes the retinal images move? Two explanations come to mind: First, the many moving retinal images may suppress the capture response because of "surround inhibition" through inhibitory interactions in the central visual system. Second, the brain of the moving toad may anticipate information about corresponding moving retinal images, which is cancelled by means of the "reafference principle" (v. Holst & Mittelstaedt 1950). The observation that for a moving toad the self-induced motion of the retinal image of a single stationary stimulus against a contrasting homogeneous white background (Fig. 3C) readily elicits prey-catching (orienting and snapping) favors the inhibitory surround hypothesis (Burghagen & Ewert 1983; cf. also Frost 1982).

The "swarm phenomenon" is an ethological example of inhibitory surround effects: A toad's prey-catching is generally inhibited when there are many simultaneously moving visual stimuli, originating from the presence of multiple prey (Schneider 1954). This phenomenon has been investigated quantitatively with simulated prey objects (Ewert & Härter 1968, 1969). The inhibition of the toad's attack under these conditions is not necessarily caused by "interocular competition" or "confusion" owing to too many visual releasers: A double stimulus presented in the animal's monocular visual field can be inhibitory even if one of the two stimuli is subliminal for the release of prey capture. Detailed quantitative experimental studies show that these inhibitory effects depend on the spatial arrangement of images on the retina in relation to their direction of movement (Ewert et al. 1970). This effect may be the result of certain properties of interacting central nervous processes, which leads to the "interaction concept."

3.3 Innate Correspondence Between Stimulus Features and RMs

3.3.1 The "Interaction Concept"

Suppose a toad is sitting in front of the stationary texture shown in Fig. 3B. How could a signal with features of prey emerge? It must be different from the rest of the texture in certain spatiotemporal properties: If just one small stimulus object S1 is moving, while all others are stationary, the toad responds with prey-catching; the probability of responding increases - within limits - with the number of such stimuli, provided they move at the same speed and are aligned parallel to the direction of movement at a short interstimulus distance: S1-S2-S3-S4-S5. But if two stimuli move simultaneously and are aligned perpendicular to the direction of movement, S1-S7, the toad's prey-catching activity is lower than with a singly moving stimulus; it decreases further the greater the extent of the pattern perpendicular to the direction of movement: S1-S7-S8-S9. If a further stimulus, S6, is added to the optimal prey stimulus pattern S1-S2-S3-S4-S5 (Fig. 3B), the entire configuration's capacity to release prey-catching is markedly reduced and may even become zero. (For quantitative details see Ewert et al. 1970, 1982). Thus, to distinguish prey from nonprey, RMs may take advantage of certain interacting central processes:

- o many simultaneously moving, randomly distributed retinal images of small visual objects have masking effects on the perception of a single image because of inhibitory interactions in the central visual system;
- o several small moving images aligned at small distances parallel to the direction of movement are interpreted as prey because of excitatory interactions in the central visual system;
- o the arrangement of small images perpendicular to the direction of movement signals nonprey because of inhibitory interactions in the visual system.

The first property allows toads to select prey from textured backgrounds and to discriminate object motion from self-induced motion; the latter properties applying to patterned as well as to continuous shapes determine the configural features of prey. (For details on prey/background relationship see Burghagen & Ewert 1983; see also Manteuffel, this volume). We shall return to the "interaction concept" in Sections 4.3, 5.3.4, and 5.5.3.

Figure 4. Capability of common toads to discriminate between moving configural objects. Top: During the experiment the animal sat within a cylindrical glass vessel; the stimulus object, a black two-dimensional stimulus, was moved around the vessel at a constant angular velocity (v) and a constant distance (d) against a white background. The toad's orienting activity, measured as the number of successive turning movements per time interval $t^* = 30$ (or 60 s), correlates with the degree of resemblance of that object to prey (in this procedure, the subsequent action patterns of prey-catching failed to occur because of the lack of appropriate stimulus conditions according to Fig. 2, provided that d was relatively long; for details see Ewert 1969a). Stimulus objects with different configural features were investigated: (a) Rectangular stripes of 2.5 mm width whose longer axis xl(1) is oriented ("worm-like") parallel to the direction of movement, and (b) stripes of 2.5 mm width whose longer axis xl(2) is oriented ("antiworm-like") perpendicular to the direction of movement, whereby the lengths xl(1) and xl(2) were variable, respectively; (c) squares of comparable edge length xl(1,2). In these studies $l(1)=l(2)=2.5$ mm; $x=1,2,4,8$, and 16; $v=10°/s$; $d=70$ mm. - (A) Average prey-catching orienting activity in normal toads (from Ewert et al. 1983b; corresponding results for $v=23°/s$ see Ewert 1969a). (B) Comparable experiments in toads following a pretectal lesion (from Ewert & v.Wietersheim 1974b; see also Ewert 1968; averages from $n=20$ animals each group). (C) Maturation of worm(*W*)/antiworm(*A*) discrimination calculated by the discrimination values $D_{W,A}$ in young postmetamorphic common toads during the first (1) and the seventh (7) day of their terrestrial life; for definition of $D_{W,A}$ see text. Thick lines: animals were caught in a pond as they approached land; thin lines: control animals raised in the laboratory singly from eggs in homogeneous white environment. Averages from $n=12$ each group. These animals were not fed after metamorphosis. Test stimuli: stripes of 0.7 mm width and variable length xl for $l(1)=l(2)=0.7$ mm; $x=1,2,4$, and 8; $d=20$ mm. (From Traud 1983).

3.3.2 Configural Features

It is obviously the geometry of a visual pattern in relation to its direction of movement that plays a critical role in prey selection, as expressed by different levels of prey-catching activity. Systematically varying the area of a two-dimensional rectangular 2.5 mm x 2.5 mm object in different directions (cf. Figs. 4Aa-c and 5) shows that the relationship between the edge lengths parallel, xl(1), and perpendicular, xl(2), to the direction of movement is critical (Ewert 1968, 1969a): Successively extending xl(1) [for xl(1):xl(2) > 1 and xl(2) = constant] increases within

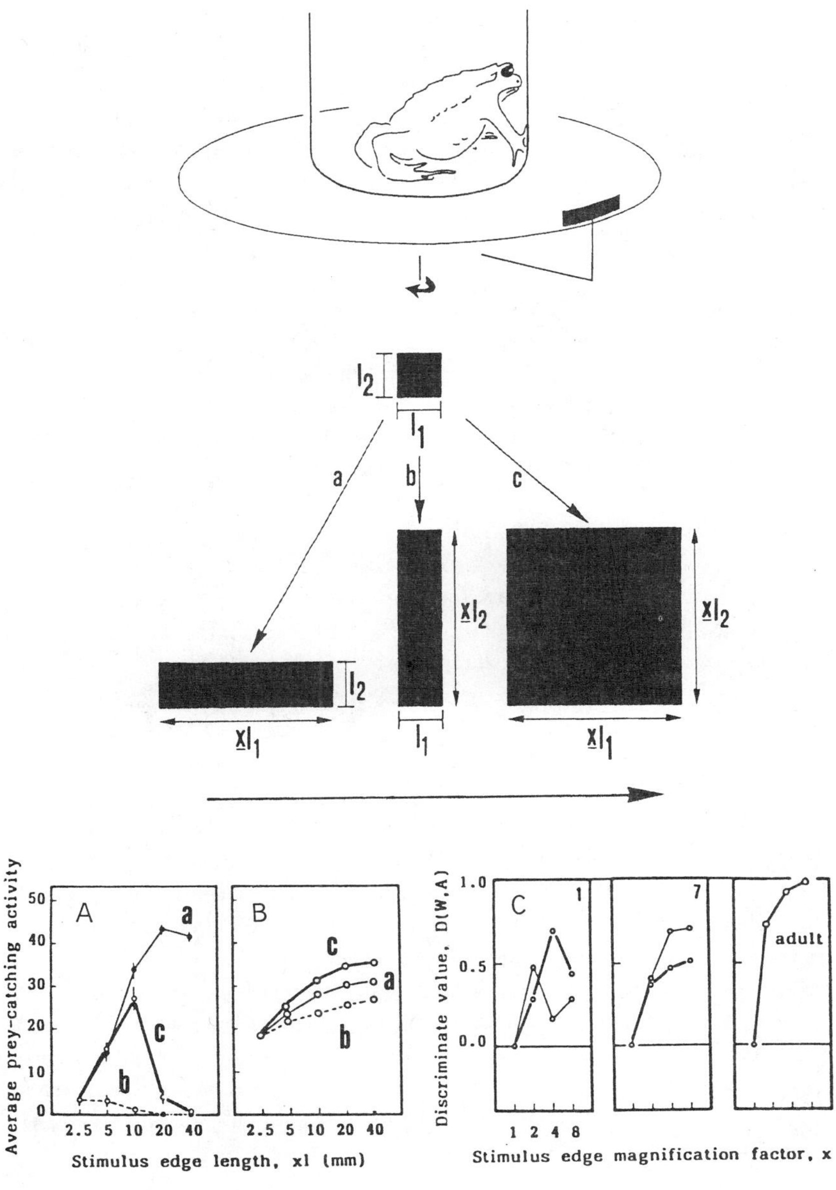

limits the releasing value of prey (Fig. 4Aa), whereas extending of xl(2) [for xl(1): xl(2) < 1 and xl(1) = constant] progressively decreases this value (Fig. 4Ab). To evaluate these feature effects it is important to note that for certain ranges of xl(1) or xl(2), respectively, the toad's prey-catching orienting activity R [orienting responses/t*] varies roughly according

$$R = \begin{cases} \beta_1 \log xl(1) + k_1 & \text{for } 2.5<xl(1)<20 \text{ with } xl(2)=2.5 \text{ [mm]} \\ \beta_2 \log xl(2) + k_2 & \text{for } xl(2)>2.5 \text{ with } xl(1)=2.5 \text{ [mm]} \end{cases} \quad (1)$$

with constants $\beta_{1,2}$ and $k_{1,2}$ of which the latter, in particular, depend on various stimulus parameters and the animal's motivation. Configural selection is determined by $\{\beta_1 > 0; \beta_2 < 0\}$. The configural properties of a stimulus [expressed by xl(1) : xl(2)] are more important than the area [expressed by xl(1) *x* xl(2)] for a relatively wide range, so long as the area is varied within behaviorally relevant limits (Ewert 1968). For example, a 2 x 20 mm^2 and an 8 x 80 mm^2 black stripe are equally effective, whereas a 16 x 160 mm^2 stripe is much weaker, provided each one moves along its longer axis at a given speed.

However, if stimuli are "configuration-neutral" in the present respect, xl(1) : xl(2) = 1, as with squares (Fig. 4Ac) or disc shaped objects of different sizes, prey-catching exhibits a nonlinear interaction between the influences of both configural features xl(1) and xl(2) expressed by Eq. (1). The area then becomes important for discriminating prey from nonprey, which requires estimating absolute size (Ingle 1968; Ewert & Gebauer 1973). When the area of such a moving object exceeds a critical range, it is categorized as a predator (Ewert & Rehn 1969).

3.3.3 The "Dynamic Gestalt" Approach

A simple way of fundamentally altering the "dynamic Gestalt" (configuration) of a visual stimulus (while preserving its shape and area) is to move a small 2.5 x 30 mm^2 stripe either parallel or perpendicular to its longer axis (Ewert 1968). The two configurations, have different releasing values. Whereas the former (called the "worm-configuration", W) provides an optimal prey signal, the latter ("antiworm-

---

Figure 5. Prey sign stimulus for common toads emerging from a stimulus continuum of two-dimensional dynamic configurations according to Ewert (1969a). Bottom: Stimulus continuum of two-dimensional rectangular black objects (5 by 5), each moving against a white background at constant velocity of v = 10°/s; xl(1) is the length of the edge which is oriented parallel and xl(2) that of the edge which is oriented perpendicular to the direction of movement; l(1) = l(2) = 2.5 mm; magnification factors x = 1,2,4,8, and 16; arrow indicates the direction of movement (of a single object). Above: Corresponding grid consisting of a 5 by 5 array serving as a coordinate frame; the part of the stimulus continuum that refers to the 25 objects shown below is surrounded by thick lines. (A) Orienting activity R_o in response to moving single objects for d = 70 mm as shown below; computer-processed data of Fig. 4A; intermediate values are approximated. (B) Snapping activity R_s; data for d = 20 mm from Herbst (1981). Scales on the ordinate refer to 5 responses per t* = 60 s (R_o) or 30 s (R_s).

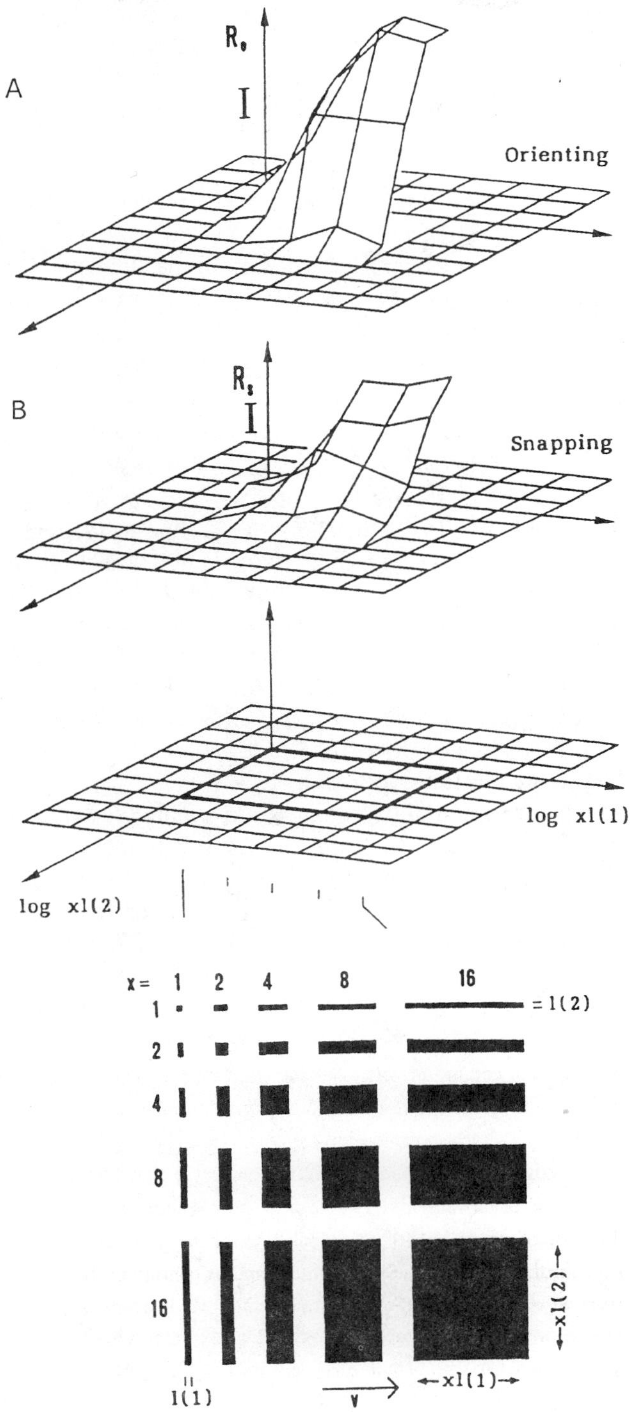
A
R_o
I
Orienting
B
R_s
I
Snapping
log xl(1)
log xl(2)
x = 1 2 4 8 16
1
2
4
8
16
= l(2)
l(1)
v
←xl(1)→
←xl(2)→

configuration", A) is treated as nonprey and may even be regarded as a threat, since toads sometimes freeze in their postures or avoid such a stimulus. The selective response to a stripe moving in a W- or an A-configuration can be quantified by the expression $D_{W,A} = (R_W - R_A)(R_W + R_A)^{-1}$, in which R_W and R_A are the prey-catching activities in response to worm and antiworm, respectively (Fig. 4C, adult). The discrimination value is positive and increases within limits as a function of stripe length magnification x

$$D_{W,A} = f(x) \tag{2}$$

(Ewert 1968; Ewert et al. 1978). That is, for positive values of $D_{W,A}$, stripes are preferred to the same ones presented in antiworm configuration. The important point here is that two identical stripes are discriminated by the toad depending on the orientation of their longer axes relative to the direction of movement.

Since the efficacy of a stimulus as a prey signal depends on its intensity (e.g., velocity and contrast), the question of invariance in worm *vs.* antiworm preference has to be investigated. If the speed of movement (Fig. 6B) (Burghagen 1979; Burghagen & Ewert 1983; Ewert et al. 1983a), the movement pattern (Borchers et al. 1978), the direction of movement (Fig. 6A) (Beck & Ewert 1979; Ewert et al. 1979a), or the contrast between stimulus and background (Ewert et al. 1979a) are changed equally in both W- and A- configurations, the $D_{W,A}$ values remain positive, although their magnitudes may vary; that is, the "basic patterns" of the stimulus-response relationships shown in Fig. 4Aa-c are not altered. However, if such a stimulus parameter is changed differently in W- and A-configuration (e.g., the contrast direction with the background "white/black" *vs.* "black/white"), the $D_{W,A}$ value can be positive and significantly different from zero only up to a certain ratio xl(1) : xl(2). Invariance in configuration preference thus requires a "minimal ratio" of the configural features.

In summary, a comparison between prey-catching activities (R) in responses to squares (S) and stripes moving in W- or A-configuration yields

$$R_W > R_S > R_A \quad \text{for } xl(1,2) \geq 10 \text{ mm}$$
$$R_W \geq R_S > R_A \quad \text{for } xl(1,2) < 10 \text{ mm}$$

The "worm" preference here becomes particularly obvious when the stimulus efficacies E [prey-catching responses/mm^2 stimulus area] are compared; then $E_W > E_S > E_A$ for xl > 2.5 (e.g., 1 : 0.50 : 0.13 > for xl = 5 mm).

3.3.4 Some Remarks

What does "worm" mean? This laboratory jargon refers to the W *vs.* A configural comparison of stripes of equal length and whose width is in an "edible" range (Ewert 1968). Interestingly, the maximal value $D_{W,A} = 1$ is almost reached when relatively short stripes of xl = 10 mm are presented, so that "W" stands for any invertebrate which moves in the direction of its long body axis. Consider that the invariance in worm preference referring to W *vs.* A comparison is not trivial: (i) The directional invariance shown in Fig. 6A cannot be predicted; (ii) in certain species of urodeles - but not in any investigated anuran amphibians - the sign of $D_{W,A}$ may depend (Roth 1987) or may not depend (Himstedt 1983) on the velocity of movement of the stimulus.

It is important to note that the toad's discrimination pattern for W-, S-, and A-stimuli is independent of whether the respective retinal images are produced by a moving visual stimulus, or if they occur in a situation where the toad is passively moved (continuously, stepwise, or with varying acceleration) in front of a stationary stimulus (for details see Burghagen & Ewert 1983; Ewert et al. 1983a).

The results reported refer to black stimuli moving against a white background. Burghagen & Ewert (1982) have shown for this contrast direction that common toads orient toward and snap at the leading edge of a worm-like stripe, whereas they prefer the trailing edge if the stimulus/background contrast is reversed, i.e., white stripe *vs.* black background (see also Ewert & Institut für den Wissenschaftlichen Film 1982a). These results on edge preference are clearly not in conflict with the concept of configuration that refers to the visual analysis of a figure in relation to its background. We learn that the toad's visual pattern recognition system is particularly sensitive to *off*-effects, hence allowing sharp configural discrimination for objects darker than the background (see also Tsai & Ewert 1987). For objects with reversed contrast there may be some confusion between figure and ground resulting in a less sharp discrimination.

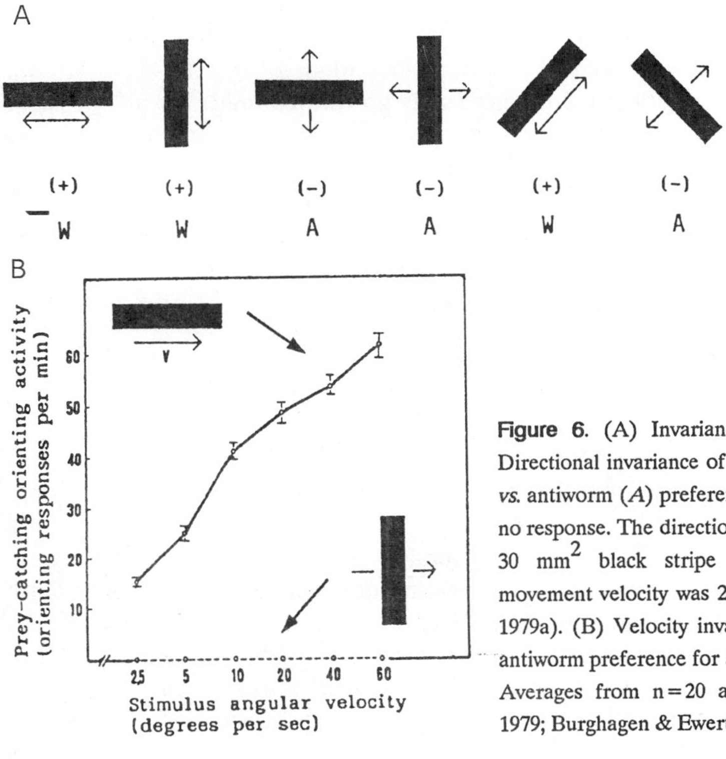

Figure 6. (A) Invariance in Gestalt perception. Directional invariance of common toad's worm (*W*) *vs.* antiworm (*A*) preference. (+), prey-capture; (-), no response. The direction of movement of the 2.5 x 30 mm^2 black stripe is indicated by arrows; movement velocity was 25 mm/s (from Ewert et al. 1979a). (B) Velocity invariance of toad's worm *vs.* antiworm preference for a 2.5 x 40 mm^2 black stripe. Averages from n=20 animals. (From Burghagen 1979; Burghagen & Ewert 1983).

Moderate changes in prey motivation (e.g., those related to hunger or to the season) emerge in the general level of prey-catching activity expressed mainly by changes in the constants $k_{1,2}$ of Eq. (1); the basic stimulus-response relationships of Fig. 4Aa-c are maintained (Ewert 1984a). For example, the $D_{W,A}$ values decrease with increasing prey-catching motivation, but they remain positive.

3.3.5 Ontogentic and Phylogenetic Considerations

The W *vs.* A preference is innate in common toads. At the completion of metamorphosis, immediately after the transition to terrestrial life, the RMs of the prey-naive toads are already functional, although the acuity of configural stimulus discrimination (Fig. 4C), distance estimation, and snapping performance improve with maturation (Ewert & Burghagen 1979a; Traud 1983; for results in salamanders see Himstedt et al. 1976). It has been further shown that $D_{W,A} > 0$ irrespective of whether an animal was raised from the egg in a homogeneous white environment or in differently structured environments consisting of stationary or moving worm-like stripes or moving antiworm-like ones (Traud 1983). In the course of the first postmetamorphic year both sensory and motor performances improve (Ewert & Burghagen 1979a) in parallel with special neural differentiations in the central visual system (Clairambault 1976; Ewert 1984b; see also Section 5.4). In adult toads, configural discrimination is rather robust: The stimulus-response relationships in test animals freshly caught in the forest display the same basic pattern (Fig. 4Aa-c) as in those kept for several months or years in the laboratory and fed exclusively on mealworms (Herbst 1981).

In the subspecies *Bufo bufo bufo* there are no obvious differences in configural prey discrimination if the responses of small males and big females are compared. Other subspecies (*B.b.spinosus*), species (*B.asper*, *B.viridis*), or genera (*Hyla arborea*, *Eleutherodactylus coqui*, *Rana temporaria*) show some variations with respect to optimal sizes of worm and square shaped objects, but the basic stimulus response pattern according to Fig. 4Aa-c is comparable, hence suggesting a common fundamental algorithm (Burghagen 1979; Ewert & Burghagen 1979b; Finkenstädt unpubl.; see also Himstedt 1982 and Finkenstädt & Ewert 1983a for results in urodeles). Burghagen (1979) has measured in seven investigated anuran species (*B. bufo, R. temporaria, R. esculenta, H. arborea, H. cinerea, Alytes obstetricans, Bombina variegata)* a positive correlation between the width (j) of an animal's jaw (for $5 \leq j \leq 25$ mm) and the edge length (xl) of a preferred square prey dummy, yielding xl = 0.43 j [mm].

3.4 Learned Correspondence Between Stimulus Features and RMs

Various visual features of natural prey objects allow toads to link these with individual experience and to store this information in order to recall it when faced with the corresponding stimulus (e.g., see Cott 1936; Freisling 1948; Eibl-Eibesfeldt 1951; Eikmanns 1955; Birukow & Meng 1955; Brower et al. 1960). Depending on the stimulus events, these experiences may be concerned with stimulus-specific habituation or associative conditioning.

3.4.1 Successive Stimulus Events

With the repeated release of the prey-catching orienting response by the same prey object during a longterm series of presentations, the toad habituates specifically to certain visual cues varied within worm-like dimensions (Birukow & Meng 1955; Ewert & Kehl 1978). For example, after the habituation of prey orienting in response to a small rectangular triangle with the small cathetus (vertex) leading in the direction of movement, the toad immediately responds to the mirror image of the triangle, i.e., when the object is moved with its sharp tip leading in the same direction (see also Finkenstädt, this volume). Since the releasing values of both stimulus objects - measured separately in terms of the initial response rates per min - are almost the same, it can be concluded that a stimulus-specific property of habituation keeps the prey-catching RMs responsive to "novel" prey stimuli. Motor pattern generating components of prey-catching are not subject to habituation.

3.4.2 Concurrent Stimulus Events

Associative conditioning in toads may involve different processes that (i) add new stimulus selecting properties, or (ii) permanently modify the innate ones. As an example of addition (i), toads for whom the smell of mealworms has been associated with prey respond either to the prey stimulus or to the combination "prey odor" and "any visual moving stimulus" (Ewert 1968; Shinn & Dole 1978). In the presence of known prey odor, the $D_{W,A}$ values become smaller. Furthermore, surround inhibition is decreased: For example, toads may respond with prey-catching to components of the kind of moving textured background shown in Fig. 3B. In the absence of a known prey odor, surround inhibition is effective and visual stimuli are sharply classified according to the RM's original properties. Other studies in which toads combined negative experience with particular cues associated with specific prey have been reported elsewhere (e.g., see "bee avoidance," Cott 1936).

As an example of modification (ii), toads who were fed several times on mealworms presented by hand finally respond to the hand alone, suggesting an association between food and hand, first described by Brzoska & Schneider (1978). This phenomenon is not "hand specific," since Burghagen (1979) has shown that it involves the extension of the entire spectrum of potential prey to include even large black square objects which normally elicit escape (for a demonstration see Ewert & Institut für den Wissenschaftlichen Film 1982b; see also Finkenstädt, this volume). The $D_{W,A}$ values are decreased but positive, and surround inhibition is attenuated. This modification in prey recognition often persists for several months. If in a comparable experiment a mealworm was presented several times together with an antiworm-like moving holder, toads become less cautious with the A-stimulus and finally respond to it alone with prey-catching (Päselt & Ewert, unpubl. data). Here too, the $D_{W,A}$ values are significantly less than in unconditioned controls, but they remain positive, which means the preference is not inverted in favour of the A-configuration.

Recalling the "interaction concept" in configural selection and prey/background discrimination, we conclude that the underlying neural mechanisms can be modulated by special inputs related to motivation and learning which are expressed by changes of both $\beta_{1,2}$ and $k_{1,2}$ in Eq. (1).

3.5 How Are Moving Objects Sorted?

First of all we must accept the guidance from the quantitative behavioral data showing that prey recognition precedes each of the toad's ballistic prey-catching action patterns, orienting as well as snapping. The subsequent study of the neuronal basis of stimulus discrimination (Section 5) can thus be focussed on the analysis of processes that translate visual input into target-oriented motor output (see also Schürg-Pfeiffer, this volume).

3.5.1 What Do We Measure?

We must take into account the fact that the "peripheral" visual system of frogs or toads, compared to higher organized vertebrates, displays a variety of peculiarities: They have no retinal foveal pit (Gordon & Hood 1976), no mammalian-type involuntary saccadic eye movements (Autrum 1959), nor any voluntary scanning or overt eye movements. This does not mean that the anuran eye is completely motionless, but they only have optokinetic (and vestibular-coupled) nystagmus, and movements concerned with binocular adjustments or monitoring the stationary environment (e.g., for such things as obstacles; see Grüsser & Grüsser-Cornehls 1976 and Dieringer 1986 for reviews). The peculiarities appear to determine certain properties of toad's visual behavior: (i) Prey-catching or escape behavior is usually released by moving visual objects, (ii) prey recognition is feasible in any part of the large visual field, and (iii) perceptual decisions related to prey or nonprey precede the corresponding ballistic behavior patterns (Ewert et al. 1983a). If a visual stimulus has no prey features, the goal-oriented turning in a prey-motivated toad fails to occur or, rather, the probability of its occurrence is extremely low. Hence, the frequency of orienting and/or snapping in response to a visual stimulus can be taken as a measure of its resemblance to prey. When an experimentally simulated prey-object circles at a constant distance and a constant angular velocity around the toad sitting in a glass vessel, the effective path in the visual field, through which the stimulus is displaced to elicit a prey-oriented turning movement, decreases as the resemblance of the dummy to prey increases (Ewert 1969a). Thus, in this procedure smaller values for the "effective displacement" are concerned with greater orienting activity [responses/t*] and *vice versa*. More specifically, the effective displacement depends on the response latency which is inversely correlated with prey features. That is, the less the stimulus resembles prey, the longer the animal takes to decide to turn toward a prey object (which follows the "all-or-none principle"). Since the stimulus-response relationships measured for orienting to prey (Fig. 4A) are maintained for approaching, fixating and snapping at it (e.g., cf. Fig. 5A and B; see also Herbst 1981; Ewert et al. 1983a), we conclude that prey recognition processes are basically the same in the corresponding RMs (Fig. 2).

3.5.2 What is Prey in Visual Configural Terms?

Before we can discuss visual features, some remarks on the term recognition are necessary. According to Zusne (1970), "recognition" refers to an ability to decide whether forms are different from those previously seen. In the present study, the term recognition (comparison with stored image) will refer to a genetically determined information store: If the vegetarian tadpoles of common toads are raised from the egg in natural or artificial differently textured (or homogeneous) visual environments, at the completion of metamorphosis with transition to terrestrial life the predatory juvenile animals respond - largely independent of these environments - to certain moving visual objects with action patterns related to prey-catching; we hence infer that this is based on comparing the incoming signal with a "prey-image" (*schema*) that has emerged and been established in the toad's visual system during its evolutionary history (Traud 1983; Ewert 1985). Schemas as internal abstraction of external reality are the means by which animals (and humans) in terms of relationships between features (*configuration*) extract a symbolic picture of their physical environments. Schemas do not copy but contain (or gain) knowledge of the outside world through combining certain pieces of sensory information into coherent perceptual wholes. (For definitions of "schema" see also the contributions by Arbib, Arkin, Cervantes-Pérez, and Koy-Oberthür, this volume).

More specifically, pattern recognition is defined here as the assignment of space-time dependent contrast distributions (features) from the visual environment to innate and/or learned classes (categories) that have behavioral significance (for details of this concept see Section 5.6). How organisms sort objects in their environment into different categories according to invariant features is one of the most basic questions of cognitive science:

> "The problem is very general, for an 'object' can be any recurring class of experience... and 'sorting' can be any differential response to the object category... Categorization hence plays a critical role in perception, thinking, and language" (Harnad 1987a).

What defines categories? In toads, the stimulus parameters of contrast and movement are prerequisites for categorizing the retinal image of a visual object as prey or predator. Within wide ranges of variation of these parameters, prey is distinguished from nonprey by size and configural features. According to Tinbergen (1951), configural properties of a visual sign stimulus are determined by particular inherent spatial or spatiotemporal relationships between specific features. The recognition of such a holistic stimulus configuration ("configural stimulus") is invariant under changes of certain stimulus parameters. Depending on the way features are related to each other, Tinbergen (1951) distinguishes between a "static configuration" (disregarding object motion) (Fig. 7A) and a "dynamic configuration" (including object motion) (Fig. 7B,C). Configuration (in the sense of Gestalt) thus refers to the arrangement (combination) of characteristic parts (features) of an object, whereas shape refers more to the form of its contour. That is two distinct dynamic configurations (e.g., W- and A-configuration) could, in principle, have identical shapes (rectangular stripe).

To evaluate configural recognition it is not sufficient to investigate some releasing features of a stimulus, rather it is also important to consider parameters that extinguish releasing values. The fact that a stripe is differently categorized by the toad depending on the orientation of its longer axis relative to the direction of movement is a very convincing example of "dynamic configural" selection (Fig. 7Ca). Another

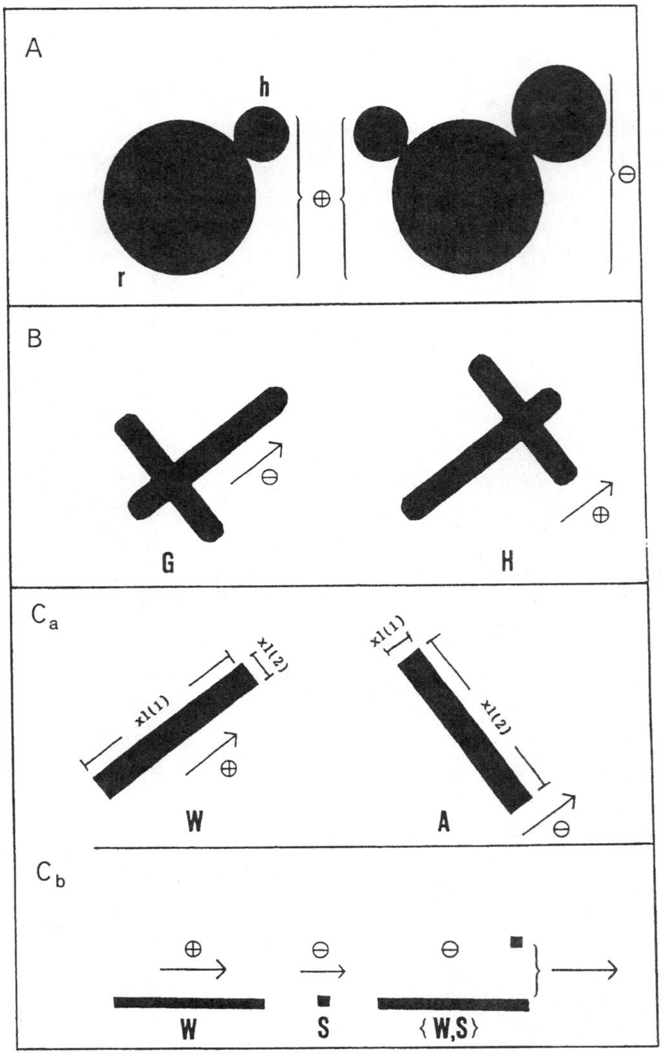

impressive example is that an optimal worm-like stripe loses its efficacy as prey if a small spot (itself subliminal for the release of prey-catching) is arranged just above the stripe (Fig. 7Cb).

A stimulus object must also be evaluated in comparison with the set of stimuli it is assigned to, by variation of its features along a physical continuum (Gibson 1950, 1951; Garner 1966). Studying the effect of variation of configural features xl(1) and xl(2) (Fig. 5 bottom) on prey-catching (Fig. 5 top) allows us to describe the range of the prey category in the stimulus continuum (Fig. 8A). From this approach we learn that toad's prey-selection takes advantage of the spatiotemporal *relationships between features*, namely the area of extension x of an object parallel to, xl(1), and perpendicular to, xl(2), the direction of movement (Eqs. (1) and (2)). More

Figure 7. Examples of "static" (A) and "dynamic" (B,C) configural properties of sign stimuli. (A) The head(h)-rump(r) model of a parent bird consisting of two adjacent solid discs of different diameters: The model elicits gaping responses (+) in blackbird nestlings if the ratio of head to rump length is 1:3, but the model remains unresponded to (-) if the difference in size between h and r is less, which can be tested by means of a "rump/two-head" model (Tinbergen & Kuenen 1939). Here r and h are features; their combination (spatial arrangement) yields a static configuration. A characteristic property of a configuration is the non-additive nature of its features: The algebraic sum of the nestling's responses (R) to either feature alone ($R_h + R_r$) is much less than the response to the combination of both features: $R_{h,r} > (R_h + R_r)$. The recognition is invariant under changes of other stimulus parameters such as size (within limits) provided the h:r=1:3 ratio is maintained. (B) Dynamic configurations involve movement and, therefore, take into account the spatiotemporal relationships between features, as examplified by the "goose(*G*)/hawk(*H*)" phenomenon (Tinbergen 1951): In domestic fowls or ducks the model of an airborne bird elicits no escape (-) if the long neck leads goose-like in the direction of movement (indicated by an arrow), but the same model releases escape (+) if the perpendicularly oriented wings lead, hawk-like. (Ca) A comparable example of dynamic configural properties is provided by the "worm(*W*)/antiworm(*A*)" phenomenon in common toads (Ewert 1968): A stripe moving worm-like in the direction of its longer axis elicits prey-catching (+); but the same stripe remains unresponded to or is associated with threat (-) if its longer axis is oriented perpendicular to the direction of movement, antiworm-like. The sign stimulus for prey is determined by the spatiotemporal relationships between two features, namely the area of extension of an object parallel to [xl(1)] and perpendicular to [xl(2)] the direction of movement (see arrow), within a behaviorally relevant range of area: xl(1) > xl(2) signalling "prey" and xl(2) > xl(1) signalling "nonprey." This structure of feature relationships is invariant under changes of other stimulus parameters, such as speed and direction of movement in the x-y coordinates of the visual field. (Cb) Illustration of the non-additive nature of features in a dynamic configuration: Optimal worm-like prey-stimulus (*W*); small square stimulus subliminal for prey-catching (*S*); stimulus configuration (*W,S*) signalling nonprey or threat: $R_W > R_{W,S} = R_S =$ zero. (Ewert et al. 1970).

specifically, it is the proportion xl(1) : xl(2) of an object - also taking into account its area xl(1) *x* xl(2) - that determines the degree of resemblance between that object and prey. In other words prey is not represented by a specific feature but by an invariant structure of feature relationships (see also Section 5.6). This implies that prey categorization is approximate and that prey-catching involves a *schema* which includes various prey objects within the limits of the defining spatiotemporal features. The boundaries of the prey category within the stimulus continuum, as well as the acuity of stimulus discrimination and the accuracy of recognition, are influenced by motivation and learning.

3.5.3 Information Bearing Parameters

What do "worm" and "antiworm" mean? Worm and antiworm can be interpreted as invariant configurations reflecting the extremes of a two-dimensional stimulus continuum (Figs. 4 and 8A). However, neither the "worm" nor the "antiworm" appears to exist in the toad's visual world, that is, they are probably not explicitly represented in the brain by specific detectors. But "worm" and "antiworm" denote behaviorally

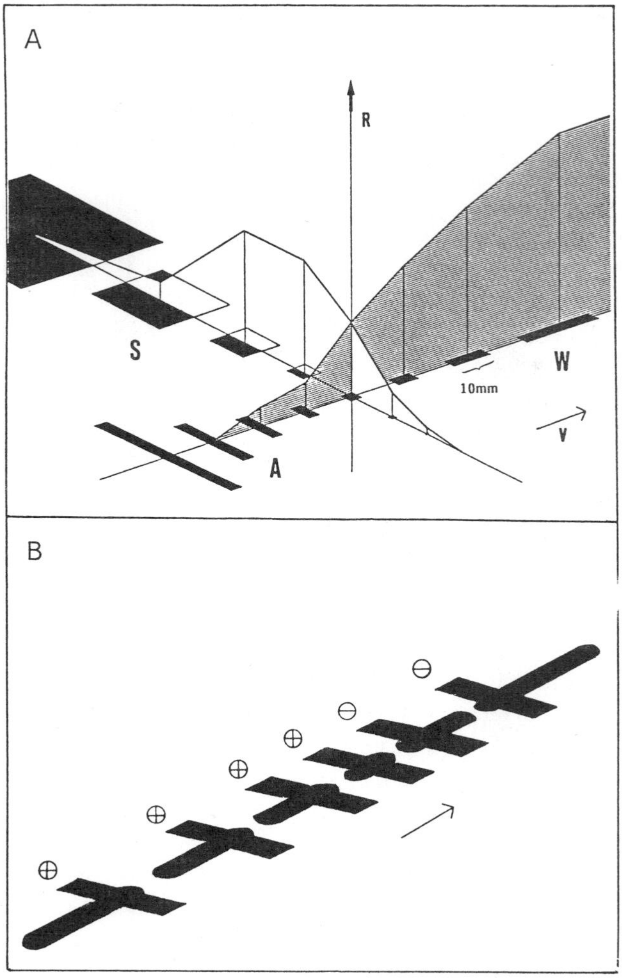

Figure 8. Stimulus continuum: Variation of moving objects in their dynamic configural (spatiotemporal) properties by the change of feature-relationships. (A) Stimulus continuum corresponding to changes of two-dimensional black forms in the side parallel [xl(1)] and/or perpendicular to [xl(2)] the direction of movement for v=23°/s (see arrow): worm-like *W*, antiworm-like *A*, and square *S* objects. In this continuum the degree of resemblance of an object to prey is described by the toad's prey-catching orienting activity, R (according to data from Ewert 1969a). (B) Stimulus continuum corresponding to changes of configural features in models of airborne birds that release escape (+) or remain unresponded (-) by domestic fowls (according to Tinbergen 1951; for a discussion see also Schleidt 1961). Arrow indicates the direction of movement of a single object.

significant features of moving objects, besides other possible features not considered here (e.g., see Kondrashev 1987). In the terminology of Suga (1984), the corresponding variables xl(1) and xl(2) are "information-bearing parameters" essential for determining the range in a stimulus continuum which carries the visual information that is important for a species in service of category formation. "Worm" [parameter xl(1)] and "antiworm" [parameter xl(2)] are sufficiently generalized to "describe" different categories of significant forms without being too specific. Worm preference is thus not limited to worms, but includes a variety of other invertebrates, such as beetles, ants, woodlice, and millipedes. The parameter xl(2) is most decisive in service of category formation: It accentuates a single object's surface extension perpendicular to the direction of movement, which obviously means "be cautious" to the toad. Parameter xl(1) in combination with an appropriate size of the side of the stimulus perpendicular to the direction of movement, xl(2), fits the category "predator", e.g., moving shadows simulating airborn predators such as owls, or large looming objects simulating ground predators such as hedgehogs (Ewert & Traud 1979). The image of a snake - the "arch-enemy" of toads - includes many such antiworm-like features, e.g., coils of the body; in addition, the raised head and "neck" often assume a position perpendicular to the movement of the body (Ewert & Traud 1979). The notion that antiworm-like features in "threat postures" may serve as vital signals for intra- and interspecific communication in the animal kingdom (e.g., see Tinbergen 1951; Eibl-Eibesfeldt 1979) suggests that this concept may have a broader application for configuration-analyzing systems in different perceptual tasks.

3.5.4 Approximateness *vs.* Exactness

We generally note that sign stimulus is not comparable to a highly specific "key," but refers to a *structure of feature relationships*. For common toads it is the proportion xl(1) : xl(2) of a moving object - taking also into account its area xl(1) *x* xl(2) - which determines the configural resemblance between that object and prey. For gaping black-bird nestlings, the feeding parent is characterized by a 1:3 ratio of a head-rump schema (Fig.7A) (Tinbergen 1951). For a female treefrog, to mention an example in the auditory domain, the calling male is characterized by certain low- and high-frequency tone components of its mating call which, in order to be optimal, must occur in a particular ratio (Gerhardt 1981; Capranica & Moffat 1983).

From a teleological point of view, a highly specific representation of a behaviorally meaningful object by a "hardwired" innate recognition system would have an adverse impact on an animal's survival ability. Representation by a "schema" allows the recognition of a variety of objects belonging to a category by their shared defining features. In anurans, with a "prey schema" (flexible within a limited range) the probability of correct decisions is high enough to guarantee behavioral success, and there is also room for adaptations based on individual experience (see Section 3.4). In songbirds it has been suggested that an innate "song template" only crudely defines the species song, leaving considerable freedom for imitation, improvisation, and invention (Konishi 1985). The mechanisms underlying face recognition in primates probably involve "broad tuning" which has the advantage of allowing interference between different memories stored in the neural network (in contrast to very fine tuning which has the disadvantage of losing plasticity) (Rolls 1987). In human speech, correctness of categorization also turns out to be an approximate rather than an exact matter, which is regarded as an advantage rather than a handicap: Approximateness gives the verbal system the potential for the universal expressive power of natural language and guarantees that meaning revision will always converge (see Harnad 1987b).

3.5.5 Innateness

A few remarks about "innateness" are in order at this point. The controversy surrounding this term is enormous; currently at least four meanings are associated with it: (a) something that is present at birth, (b) an ability that is unlearned, (c) a property that is phylogenetically adapted, and (d) a disposition arising from genetic differences. In the case of the toad's configural prey recognition, meaning (c) seems the appropriate one, although many authorities in developmental ethology would dispute whether that means "genetically determined," since genes alone may not be sufficient for the expression of a property, that is, gene products depend upon developmental contexts (Fentress 1987). In our opinion, the prey-configuration-selecting behavior of an adult toad has been established as a result of genetic

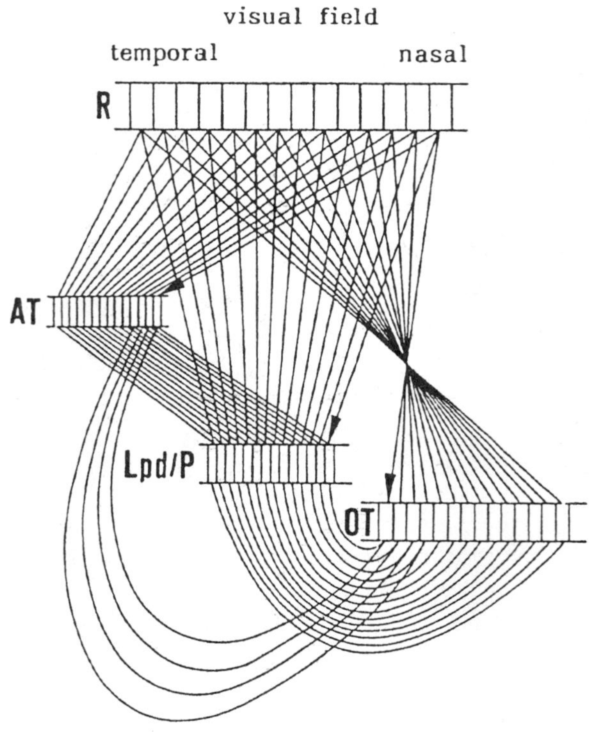

Figure 9. Central visual maps in anurans. Schematic illustration showing central projections from the contralateral eye and the ipsilateral interactions among projection fields in the central visual system. These relationships are not precisely "point-to-point" as regards convergence and divergence (e.g., see Fig. 12C). R, retina; AT, anterior thalamus; OT, optic tectum; Lpd/P, pretectal nuclei; "nasal" and "temporal" refers to visual field positions. (According to Gaze 1958; Fite & Scalia 1976; Lázár 1971; Ewert et al. 1974; Wilczynski & Northcutt 1977; Neary & Northcutt 1983; Weerasuriya & Ewert 1983).

dispositions (designed neural network properties) involving maturational processes (the programmed development of neural connectivity) and behavioral practice (perceptual sharpening). The term "established" may be preferable to "innate" in this context as it leaves open a variety of factors that might be involved in RM (see also Fentress 1987).

4 Approach Toward Behaviorally Relevant Brain Structures

The visual field is topographically represented by a multiplicity of maps in different laminae of the optic tectum and in various structures of the thalamus (Fig. 9). Hence there are different "visual channels," but these are not isolated. Central visual maps are (mutually) connected, so that they can communicate with and influence each other. The initial neuroethological analysis of behaviorally relevant brain areas is pursued using two experimental methods: (1) The study of regional glucose utilization in the brain of the behaving toad by applying the [^{14}C]-2-deoxy-D-glucose (2DG) method; (2) functional evaluation of these structures by means of stimulation and lesion experiments.

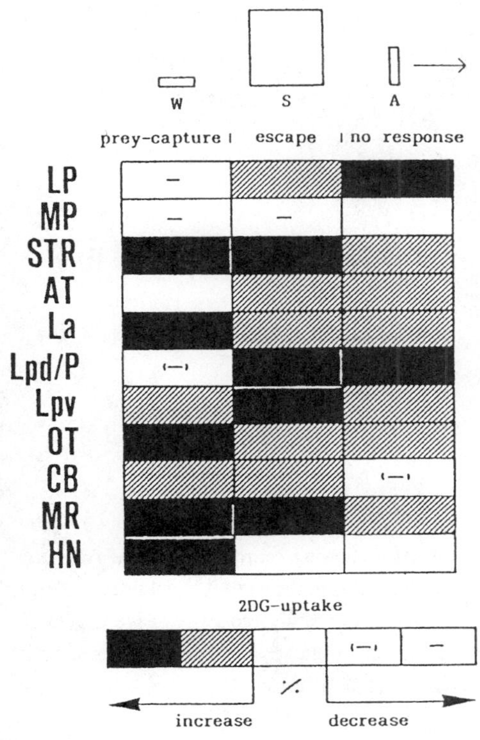

Figure 10. Regional patterns of brain activity measured with the 2DG autoradiographic technique in common toads which were repetitively orienting toward a moving prey object (4 x 28 mm^2 worm-like stripe, *W*), avoiding a moving predator object (84 x 84 mm^2 square object, *S*), or sitting motionless before a moving nonprey object (4 x 28 mm^2 antiworm-like stripe, *A*; the black objects were moved against a white background). The patches indicate relative uptake of 2DG in various brain structures in comparison with reference structures and controls. AT, anterior thalamus; CB, cerebellum; HN, hypoglossal nucleus; La, lateral anterior thalamic nucleus; Lpd, lateral posterodorsal thalamic nucleus; Lpv, lateral posteroventral thalamic nucleus; LP, lateral pallium (posterior part); MP, medial pallium (posteroventral division); MR, medial reticular formation of the medulla oblongata; OT, optic tectum; P, posterior thalamic nucleus (lateral part); STR, striatum (ventral division). Lpd, Lpv, and P are thalamic pretectal nuclei; nomenclature of diencephalic structures according to Neary & Northcutt 1983. (For details see Finkenstädt et al. 1985, 1986; Finkenstädt & Ewert 1988).

4.1 Regional Distribution of Brain Activity

The question of whether central visual maps (Fig. 9) and related structures are involved in various kinds of visual behavior can be investigated by means of the 2DG method. Figure 10 shows that prey-catching in response to a worm-like moving stripe, W, predator avoidance in response to a moving large square, S, and no motor response to an antiworm-like moving stripe, A, are associated with different patterns of increase and decrease in 2DG-uptake in various brain structures. In view of the distinct releasing properties of the W-, S-, and A-stimuli, the following suggestions can be made (Figure 10): If the uptake of ^{14}C labeled 2DG in a given brain area is different in S- and W-experiments ($^{14}C_S \neq {}^{14}C_W$), a correlation with prey-catching or predator-avoidance, respectively, can be inferred (e.g., see thalamic pretectal Lpv nucleus); if in addition $^{14}C_A \neq {}^{14}C_W$, the structure under investigation may also be involved in stimulus-feature analysis (e.g., see optic tectum OT, thalamic pretectal Lpd nucleus). Cases in which $^{14}C_S = {}^{14}C_W$ indicate a correlation with prey-catching and avoidance behavior (e.g., striatum STR, cerebellum CB). (For details see Finkenstädt 1987; Finkenstädt et al. 1985, 1986; see also Finkenstädt, this volume).

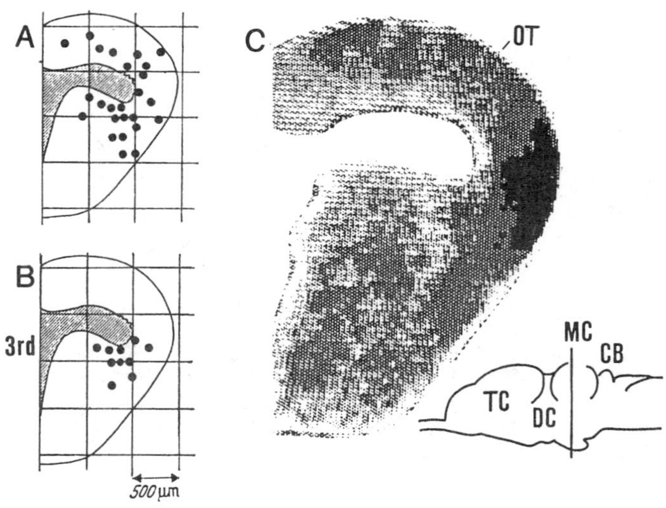

Figure 11. Areas relevant to snapping behavior in the common toad's midbrain shown in transverse section. (A,B) Reconstruction of sites which, when electrically stimulated, elicited snapping (B) and/or orienting (A) (from Ewert 1967b). (C) "Snap evoking area" labeled by means of the 2DG method during repetitive release of snapping with a worm-like visual object moving to and fro close to the toad in the frontal visual field. Note that heaviest 2DG-uptake (indicated by black label in the computer-processed autoradiographic image) converges from superficial to central tectal layers. Furthermore, note that structures involved in the visual release of snapping (C) do not include also subtectal (toral) structures which are responsible for tactually elicited snapping and are thus effective too when electrically stimulated (B). CB, cerebellum; DC, diencephalon; MC, mesencephalon; OT, optic tectum; TC, telencephalon; 3rd, third ventricle. (From Finkenstädt et al. 1985).

4.2 Functional Relevance of Visual Maps and Related Structures

4.2.1 Behaviorally Effective Brain Areas

Results from brain stimulation experiments with chronically implanted electrodes in freely moving toads suggest that (1) central programs for prey-catching and avoidance behaviors exist; (2) the former can be triggered by tectal stimulation and the latter mainly by posterior thalamic (pretectal) stimulation. The optic tectum is involved in "visual grasping" (Akert 1949; Hess et al. 1946), whereas thalamic nuclei participate in different types of "avoidance" related to visual information from a predator (escape), from a stationary obstacle (detour), or to vestibular information from body tilt (maintenance of posture).

There are topographic relationships between brain stimulation sites and behavioral patterns. Depending on the position of the electrode in the tectal visual map, orienting responses are directed toward appropriate loci of the visual field of the contralateral eye (Ewert 1967b,c; 1974; see also Schürg-Pfeiffer, this volume). The tectal area from which snapping can be elicited is smaller (Fig. 11A,B; cf. also C) and represents the animal's visual snapping region as determined in behavioral studies (Ingle 1976). Electrical stimulation of the caudal thalamus elicits avoidance behaviors such as ducking, ipsiversive turning, moving backward, jumping, running, "freezing" in body posture, puffing up and assuming a stiff-legged avoidance posture accompanied by secretion of the skin glands, sidestepping, tilting the body along its sagittal or transverse axes, or a tendency to rotate along the sagittal body axis (Rehn 1977; cf. also Ewert 1984a; for comparable results in salamanders see Finkenstädt & Ewert 1983b, and in cats see Hess 1949, 1968). In the toad's posterior thalamus, stimulation sites for changes in body posture and detour-like movements are distributed in median regions, while those for directed escape are found more laterally.

4.2.2 Thalamic/Tectal Interactions

The effects of various tectal lesions show that the optic tectum participates in both prey-catching and predator avoidance behaviors (Bechterev 1884; Rehn 1977; Ingle 1983; Ewert et al. 1983a). The coordinated control of both types of behavior, however, requires thalamic/tectal interactions (Fig. 12B).

Functional thalamo-tectal connections are suggested from electrical stimulation experiments: Repetitive prolonged stimulation with trains of impulses in an area between the optic tectum and the ipsilateral pretectal thalamic neuropil - close to the rostral tectum - initially elicits escape, then with an increase of the escape threshold it leads to nonresponsiveness and may finally release orienting and snapping (Ewert 1968). Interestingly, after repetitive pretectal electrical stimulation toads often respond to large moving visual stimuli with prey-catching as well: Tetanization of posterior thalamic (pretectal) neurons has presumably reduced their excitability, thus facilitating the visual response of tectal neurons in the ipsilateral hemisphere. Rehn (1977) has supported this hypothesis with experiments using two stimulation electrodes: The threshold for electrically elicited prey-catching in the tectum decreased when tectal stimulation was preceded by longterm repetitive stimulation of the ipsilateral posterior dorsal thalamus. These experiments suggest inhibitory connections from caudal dorsal thalamus to tectum.

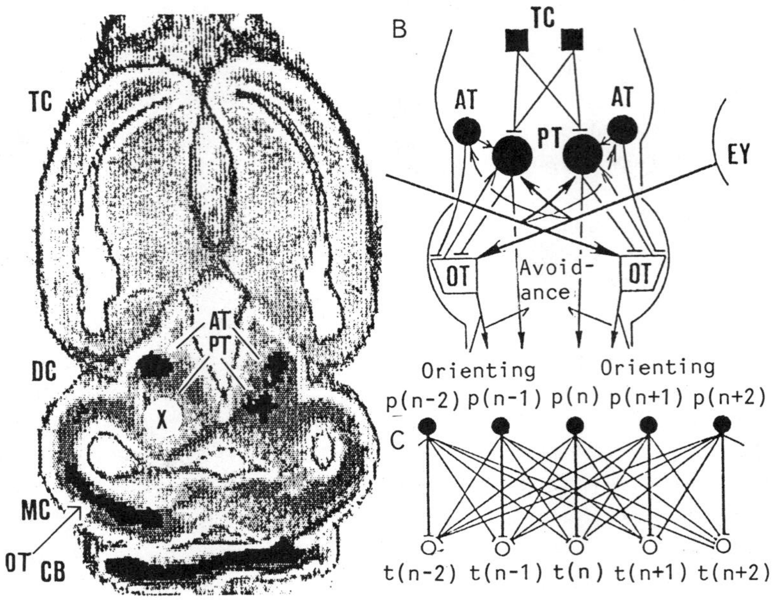

Figure 12. Central interactions in the control of prey-catching and predator-avoidance behaviors in common toads. (A) Glucose utilization in the brain of a unilateral pretectally lesioned toad, measured with the 2DG method during repetitive release of "disinhibited" prey-catching orienting in response to an antiworm-like stripe moving in the horizontal visual field of either eye. The 2DG-uptake shown in the autoradiographic image of a horizontal brain section was much stronger in the left mediocaudal tectal lobe (see arrow) ipsilateral to the pretectal lesion (see "x") than in the tectal lobe of the right hemisphere in which the pretectal region (PT), known to inhibit the tectum, displayed considerable 2DG-uptake. Note that the anterior thalamus (AT), which is thought to inhibit the rostral tectum, was strongly labeled bilaterally, which explains the weak 2DG-uptake in the rostral tectum. (For connectivity cf. Figs. A and B) (Ewert 1985). (B) Integrative and interactive aspects of information processing related to visually guided prey-catching and predator-avoidance, derived from brain stimulation and lesion studies (Ewert 1967a,b; Ewert et al. 1983a). Arrows indicate putative excitatory influences and lines with cross bars inhibitory influences. AT, anterior thalamus; ON, optic nerve; OT, optic tectum; PT, pretectal Lpd/P nuclei; TC, telencephalon (striatum). (C) Proposed principle of central lateral inhibition; p, pretectal neurons; t, tectal neurons; lines with cross bars: putative inhibitory connections; retinal inputs to p and t are not shown. (Adapted from Ewert & Härter 1968 and Ewert 1981).

Evidence regarding the behavioral significance of thalamo-tectal interactions is provided by lesion studies (Ewert 1968; Ewert et al. 1974). After unilateral lesions in the posterior dorsal thalamus, toads respond to any stimulus moving in the visual field of their contralateral eye with prey-catching. This effect resembles a sort of "agnosia" whereby the retino-tectal pathways are intact but visual pattern recognition is disturbed. This "disinhibition syndrome" is characterized by:

(a) Loss of configural discrimination, so that $D_{W,A} = 0$ and $ß_2 > 0$ (Fig. 4B);
(b) failure of prey/predator discrimination, e.g., increase of the prey-catching activity with successive elongation of the edge of a moving square stimulus (Fig. 4Bc);
(c) strong attenuation of surround inhibition, e.g., prey-catching in response to moving background structures;
(d) eventual lack of longterm stimulus-specific habituation;
(e) eventual occurrence of sustained post-stimulus activities in orienting behavior;
(f) loss of discrimination between object- and self-induced motion;
(g) failure of visual escape, weak response to painful stimuli.
(Ewert 1967a, 1968, 1969b; Ewert et al. 1970; Ewert & v.Wietersheim 1974b; for comparable results in frogs see Kondrashev & Dimentman 1978; for results in the fire salamander see Finkenstädt 1981; Finkenstädt & Ewert 1983b).

The experimental data allow us to draw three main conclusions: First, the optic tectum includes an intrinsic network characterized by mutual excitatory, reverberatory connectivity. Second, this network is under the inhibitory control of extrinsic prosencephalic (thalamic, pretectal, telencephalic) inputs. Third, thalamic pretectal nuclei provide essential elements for "discrimination filters" that determine important properties of visual RMs. Fourth, the agnostic and hyperexcitatory effects of caudal thalamic lesions are related to disorders in the central visual system, since the threshold for tactually elicited turning or snapping (Grobstein et al. 1983) is not reduced in thalamic lesioned toads with subsequent bilateral optic nerve transection (Weerasuriya, unpublished data). Fifth, the source of the putative inhibitory thalamo-tectal influences (Fig. 12B) is the anterior dorsal thalamus (AT), the pretectal (PT) lateral posterodorsal thalamic nucleus, Lpd, and the lateral portion of the posterior thalamic nucleus, P, respectively, but not the median part of P. This has been investigated with various lesion techniques using either sagittal microknife cuts, electro-coagulation, or the axon-sparing neurotoxins kainic acid or ibotenic acid (Ewert et al. 1983a; Finkenstädt & Ewert 1983b; Ewert 1984a; Schürg-Pfeiffer, this volume). Median transections of the posterior or tectal commissures are not effective in this context. Since it has not yet been possible to distinguish between effects of lateral P and those of Lpd, in the following we will be referring to "pretectal" (Lpd/P) lesions.

It is important to note that the above mentioned "agnostic" phenomena produced by unilateral anterior thalamic lesions are always restricted to the frontal visual field, whereas those after pretectal lesions can be related to any portion of the entire visual field of the contralateral eye with an emphasis on extrafrontal regions (Ewert et al. 1983a). The effects evoked by a small unilateral Lpd/P-lesion correspond to a small circumscribed region in the visual field resembling the reverse properties of a scotoma (Ewert et al. 1974). Evidence of ipsilateral inhibitory pretecto-tectal influences is provided by 2DG studies showing strong glucose utilization in the tectum ipsilateral to a unilateral pretectal lesion (see "x" in Fig.12A). Here, the size and position of the "agnostic visual field" coincides with increased 2DG-uptake in a topographically corresponding area of the tectum (Finkenstädt & Ewert, unpubl.). Anatomical connections between anterior thalamus, pretectum, and tectum (Fig. 12B) have been described in frogs (Trachtenberg & Ingle 1974;

Wilczynski & Northcutt 1977; Neary & Northcutt 1983; Lázár 1984; this volume) and in toads (Weerasuriya & Ewert 1983) suggesting that structures of the rostral thalamus project to the rostral third of the tectum, whereas the pretectal projections include the mediocaudal tectal regions.

Telencephalic influences on the release of prey-catching are inferred from brain stimulation, lesion, and 2DG studies (cf. Diebschlag 1935; Ewert 1967a; Finkenstädt et al. 1986; Finkenstädt & Ewert 1986). The results suggest that striatal and medial pallial structures via thalamic/pretectal nuclei determine and modulate tecto-bulbar/spinal output (Fig. 12B) (see also Section 6.4.5). After removal of both telencephalic hemispheres there is visual neglect of prey and a decrease in the visual threshold for escape. One explanation for this result is that the putative inhibitory pretecto-tectal influences override the tectal efferents related to prey-catching. This idea is supported by large bilateral pretectal lesions in telencephalic ablated toads: visual prey-catching is hyperexcited ("disinhibited") whereas predator-avoidance fails to occur (Ewert 1967a).

4.3 Properties of a Macro-Network

The experiments obtained in amphibians suggest that release of prey-catching results from parallel distributed processing of visual input involving various brain structures (Ewert 1967, 1968). This implies that different visually guided behaviors are regulated by the same "macro-network" with different spatiotemporal patterns of excitation and inhibition (Figs. 9, 10 and 12B). We therefore hasten to admit that it would be dangerous to interpret the release of an action pattern in terms of information processing in one structure, such as the optic tectum, only (see also Kondrashev 1987). We learn that the so-called visual centers, maps, and channels do not operate privately to the exclusion of other. Prey-catching is controlled by the optic tectum depending on inputs from prosencephalic pretectal, thalamic, and telencephalic nuclei (Fig.12B; Ewert 1968; 1974; 1984a), by isthmic nuclei (Gruberg, this volume), and by tegmental structures (Grobstein, this volume). Thalamic pretectal nuclei are also involved in the control of predator avoidance (Fig. 12B; Ewert 1968; Ewert & v.Wietersheim 1974a), barrier avoidance (Ingle 1971, 1977, 1980, 1983; Ewert 1971), and optokinetic nystagmus (Lázár 1973; Katte & Hoffmann 1980; Fite et al. 1983; see also Manteuffel, this volume). The optic tectum is involved in the control of other behaviors than prey-catching, such as predator avoidance and escape (Ingle 1983; Ewert et al. 1983).

Drawing on the experimental data and recalling the "interaction concept" of visual prey recognition, we suggest that the toad's configural stimulus discrimination can be traced back to a property inherent in interacting pretectal/tectal networks (Ewert 1967a; 1968; see also Ewert et al. 1970). In this context, inhibitory pretecto-tectal connections play a prominent role in the control of excitation in the tectum and probably act in a network of lateral inhibition as suggested by Ewert & Härter (1969; see also Ewert 1981). There is indirect evidence to show that pretectal inhibition is not only related to tectal regions topographically but also spreads with decreasing strength over large tectal areas (Fig. 12C; Ewert 1968; Ewert & Härter 1968; 1969; Ewert et al. 1970). According to this hypothesis, pretecto-tectal inhibition interferes with lateral excitation in the tectum (Ewert et al. 1970; Székely 1973; Ingle 1975), and this inhibition determines the structure of configural feature relationships in the stimulus continuum from which prey emerges (Fig. 5):

- o a moving retinal image of a small object weakly activates pretectal structures, thus producing less inhibition to corresponding tectal structure, compared to several randomly distributed images which would strongly activate pretectum;
- o several small retinal images that are narrowed (or a corresponding compact image extended) parallel to the direction of movement are less effective in pretectal structures, but they are optimal to activate the excitatory mechanisms in the tectum which is expressed by the positive values of $ß_1$ in Eq. (1):
- o the same small retinal images aligned (or an appropriate compact extended) perpendicular to the direction of movement, strongly activates pretectal structures so that pretecto-tectal inhibition will override the comparatively weak tectal excitatory processes which determines the negative values of $ß_2$ in Eq. (1);
- o in response to overall distributed images (or a corresponding overall extended compact image) tectal and pretectal effects interact appropriately.

These perceptual operations take the size of the stimulus area into consideration in relation to the direction of movement; the processing principle, however, is invariant under changes of the direction in which the stimulus moves (Fig. 6). According to this concept, intrinsic tectal excitatory processes, under the influence of extrinsic thalamic pretectal inhibitory inputs, determine the functional properties of RM's required for (a) extraction of prey from the background, (b) selection between prey and non-prey based on configural stimulus cues, (c) discrimination between prey and predator, (d) surround inhibition, and (e) discrimination between moving retinal images and self-induced moving images. Superposition of corresponding anterior thalamic input on the pretecto-tectal inhibitory processes presumably influences visual discrimination with special reference to the frontal visual field, thus providing a sort of "central fovea." Modulatory inputs to thalamic/pretectal structures arriving from the telencephalon (e.g., medial pallium, striatum) may determine adaptive properties in their interaction. Such influences could, for example, attenuate anterior thalamic and pretectal effects, thus allowing surround inhibition to decrease and discriminatory processes of RMs to adapt to appropriate internal or external signals (Fig. 12B) in relation to motivational states and learning (Section 3.4.2).

5 Feature-Analyzing Neurons and Integrative Functional Units

The cellular approach to the analysis of configural features is guided by (i) the "landmarks" obtained from 2DG, brain stimululation and lesion studies (Section 4), (ii) neuroanatomical investigations of the cytoarchitecture of retinal, pretectal, and tectal structures (see also Lázár, this volume), and (iii) extracellular microelectrode recordings from single neurons (partly combined with electrical stimulation and electrolytic lesioning) in paralyzed or freely moving animals (see also Schürg-Pfeiffer, this volume). Intracellular recording and iontophoretic staining allows us to identify neurons and to evaluate their responses in terms of changes in membrane potential (see Matsumoto, this volume). It can be shown that selectivity for configural stimulus properties is accomplished by feature-analyzing neurons integrated in interacting pretectal/tectal networks which obtain certain combinations of retinal input.

5.1 Information Processing in the Retina

The question of what the frog's retina tells its central projection fields was - given Ramon y Cajal's (1896) cytological treatise on the retina and the pioneering physiological studies by Hartline (1940) and Barlow (1953) - systematically examined by Lettvin et al. (1959), and quantitatively investigated by Grüsser & Grüsser-Cornehls (1968, 1976). (For comparative studies in toads see Ewert & Hock 1972).

In the frog's and toad's retina there are at least four different classes of ganglion cell (R1-R4) that mediate the output of functionally connected retinal neurons. These ganglion cells send their axons to the superficial layers of the optic tectum (classes R1-R4) (Gaze 1958; Grüsser & Grüsser-Cornehls 1976), to the pretectal neuropil (probably classes R1, R3, R4) (Ewert 1971; Ewert et al. 1974), and to other

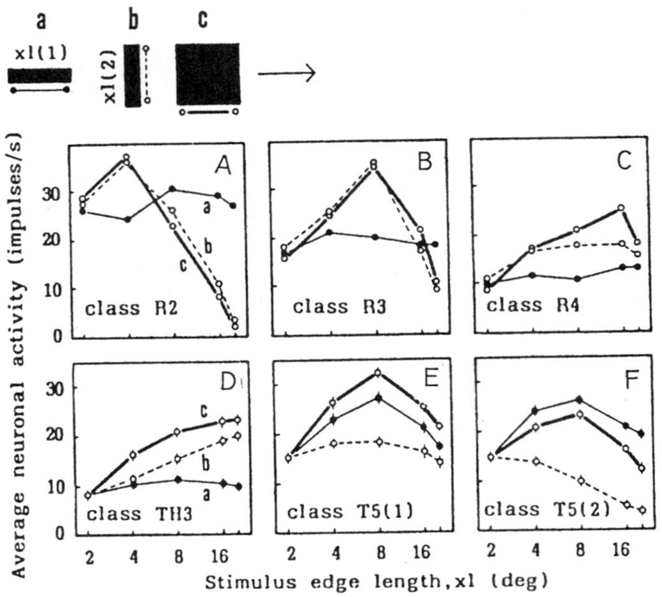

Figure 13. Response properties of retinal ganglion cells (A-C) and central visual neurons (D-F). Neurons were recorded extracellularly in paralyzed common toads. Neuronal activities [impulses/s] were calculated in response to black worm-like (a), antiworm-like (b), and square (c) stimuli each traversing the centers of the excitatory receptive fields of single neurons at v=7.6°/s against a white background; l(1)=l(2)=2°; x=1,2,4,8, and 10. Averages from n=10 (A-C) and n=20 (D-F), respectively. (From Ewert & Hock 1972; Ewert & v.Wietersheim 1974a).

diencephalic targets (Lázár 1971; Lázár, this volume). The various ganglion cell classes exhibit different sensitivities to the stimulus parameters of movement and contrast. Furthermore, as a result of different angular size (diameter) of the excitatory visual receptive field (ERF) and the strength of the adjoining inhibitory receptive field (IRF), these neurons display various preferences for the area of a moving stimulus: Class R1 and R2 neurons (ERF diam. = 4°) are activated maximally by relatively small compact moving objects of about 4° diameter, class R3 neurons (ERF diam. = 8°) by such objects about double in size, and class R4 neurons (ERF diam. = 12-16°) by large dark objects against a bright background (Fig. 13A-C,c).

In the analysis of retinal ganglion cell responses of common toads we have taken advantage of testing behaviorally relevant configural stimuli known from ethological studies (Ewert & Hock 1972; Ewert et al. 1978). The results of these (Figs. 13A-C and 14, R2-R4) and other investigations, together with the arguments presented by Grüsser & Grüsser-Cornehls (1970) and Hood & Gordon (1981), invalidate retinal recognition concepts (Barlow 1953; Lettvin et al. 1959) according to which a particular class of retinal ganglion cell should exert the function of a "prey detector." Here are our main arguments:

(i) The behavioral (Fig. 4A) and neuronal stimulus-response relationships do not correlate well for any class of retinal ganglion cell (Fig. 13A-C).

(ii) The $D_{W,A}$ values are negative for stripes with an edge length of xl = ERF

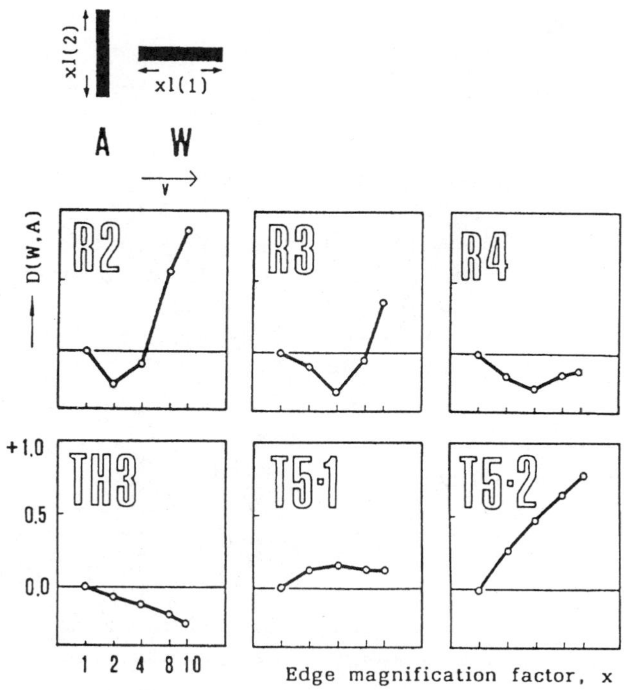

Figure 14. Discrimination values $D_{W,A}$ for stripes moving in *W*- or *A*-configuration as a function of x for retinal (R2, R3, R4), pretectal (TH3), and tectal (T5.1, T5.2) neuron classes in common toads; l(1) = l(2) = 2°. (From Ewert et al. 1978).

diam., both in R2 and R3 neurons (cf. Figs. 14 and 4C adult).

(iii) Mismatch between activation of R2 and R3 ganglion cells and prey-catching in response to W- and A-stimuli has been demonstrated by chronic recordings in behaving toads (see Schürg-Pfeiffer, this volume).

(iv) Although with a moving background structure class R2 neurons are less activated than class R3 and R4 (Tsai & Ewert 1988), the behavioral phenomenon of "surround inhibition" cannot be explained solely by inhibitory receptive fields of R2 cells. Behavioral inhibition is strongly reduced in toads stimulated with the odor of prey (Section 3.4.2) and is almost abolished in toads with pretectal lesions (Ewert 1968; Ewert et al. 1970; cf. Section 4.2.2), while retinal IRF properties are unaffected (Ewert & v. Wietersheim 1974b).

What is the function of the various retinal ganglion cell classes? They obviously transmit to the optic tectum and to other central visual structures information regarding different stimulus parameters (e.g., contrast, movement, surface), with each cell class emphasizing various aspects of these. For example, to a 3° x 30° black stripe traversing the ERF center against a white background in W-configuration, the *off*-dominated class R3 neurons fire strongest toward the *off*-producing leading edge entering the ERF and less, or not at all, toward the *on*-producing trailing edge leaving the ERF; with a white stripe moving against a black background, the *off*-producing trailing edge is most effective. Since class R2 neuron's discrimination of leading *vs.* trailing edges is not statistically significant, prey-catching obviously takes advantage of information provided by contrast border "*off* position" indication by R3 neurons (see Section 3.3.4; see also Barlow 1953; for quantitative details see Tsai & Ewert 1987). Class R2 neurons are particularly sensitive to slow movements of small objects, whereas class R4 prefer fast moving large dark objects.

5.2 Feature-Analyzing Tectal and Pretectal Neurons

The quantitative behavioral studies in intact and brain-lesioned toads suggest that tectal and pretectal structures are responsible for prey recognition. Single cell recordings from retinal projection fields in the optic tectum and pretectum allow us to describe classes of neurons that exhibit different kinds and degrees of sensitivity to variation in the configural features xl(1) and xl(2).

5.2.1 Morphological and Physiological Survey

The optic tectum consists of a laminated structure (Fig. 15Aa). Based on Golgi studies, Székely & Lázár (1976) have described morphologically distinct types of tectal neurons which are suitable for collecting retinal inputs, processing these by intrinsic links, and mediating the output of the tectum, e.g., to bulbar/spinal structures (Fig. 15Ab). In the pretectal Lpd/P region (Fig. 15Bb), neurons pick up retinal input from the adjacent posterior thalamic (pretectal) neuropil and others provide connections between Lpd and P nuclei (Lázár 1979). Investigations of thalamic/pretectal/tectal connections have shown that cells from Lpd/P nuclei project to the tectum (Wilczynski & Northcutt 1977; Weerasuriya & Ewert 1983; Neary & Northcutt 1983), while some tectal cells of layer 6 (top) and layer 8 send axons to pretectal and rostral thalamic nuclei (Lázár et al. 1983; Bieger & Neuman 1984; Antal et al. 1986;

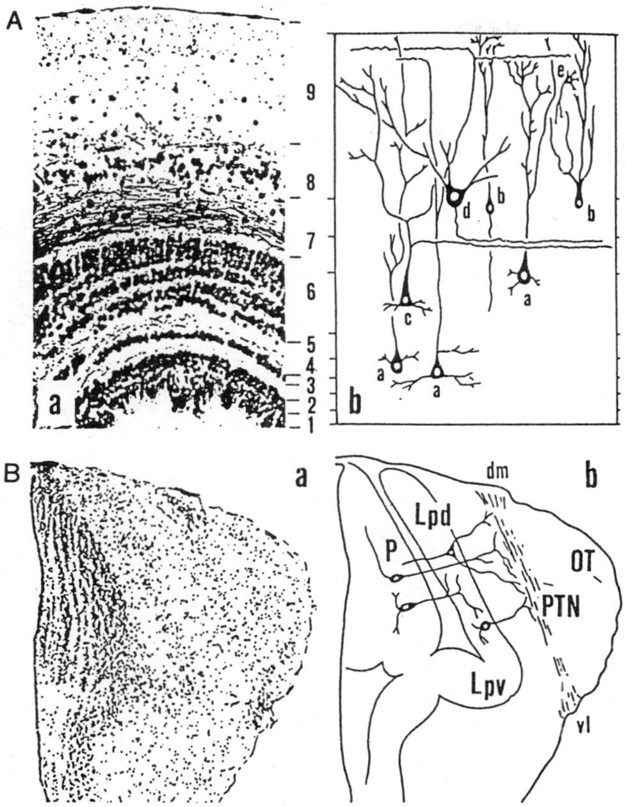

Figure 15. (A) Cytoarchitecture of the anuran optic tectum (cut-out). (a) Klüver-Barrera stain of a transverse section of the optic tectum showing the laminar structure, layers 1-9 (Ewert et al. 1983b); (b) neuronal morphologies from Golgi studies; a,b, pear-shaped neurons; c, pyramidal neuron; d, large ganglionic neuron; e, stellate neuron (drawing after Székely & Lázár 1976). (B) Cytoarchitecture of the anuran pretectal region. (a) Nissl stain of a transverse section (Ewert et al. 1983a); (b) some morphological types of neurons from Golgi studies (drawing after Lázár 1979); dm,vl, dorsomedial and ventrolateral optic tract, respectively; PTN, posterior thalamic (pretectal) neuropil; for other abbreviations see Fig. 10.

see also contributions by Lázár and Matsumoto, this volume).

Applying various physiological criteria, extracellular microelectrode recording studies in awake paralyzed toads and frogs have led to a classification of posterior thalamic pretectal (TH1-TH10) neurons (by Ewert 1971) and tectal (T1-T9) neurons (the classification T1 to T7 by Grüsser & Grüsser-Cornehls 1970 draws on size and location of the ERF of a neuron and the presence of monocular or binocular inputs; our classification of T5-type "small field" T5.1-T5.4 neurons takes into account their different sensitivities to changes of dynamic configural stimulus features; for reviews see Grüsser & Grüsser-Cornehls 1976; Ewert et al. 1983a; Ewert 1984a; 1987a).

Among pretectal neurons (Fig. 16A-C), many types are optimally activated in visual stimulus situations suitable for releasing avoidance behaviors: Turning away from a large moving object (e.g., class TH3) or a looming approaching object (class TH6), or detouring around a stationary obstacle (e.g., classes TH10.1 and TH10.2). The responses of other pretectal neurons display wide fields (e.g., class TH4), they are flexible (e.g., class TH5.2), and some cells even seem to have memory (class TH9). Class TH1 and TH2 neurons are non-visual, class TH7 display tonic sensitivity to brightness and darkness, and TH8 change their tonic activity in response to object motion through large ERF/IRF sections.

Among tectal neurons (Fig. 17A-D) different receptive field properties might be utilized to obtain information from large sections of the contralateral visual field such as the frontal (class T2.1), the inferior (class T2.2), the superior (class T6), or the entire field (class T4). So-called small-field neurons may serve to monitor moving objects of different configurations in the x-y coordinates (T5-type neurons) or on the z-axis (classes T1 and T3). Responses of the T7-type, referring to tiny ERFs of less than ten degrees diameter, have been interpreted as dendritic spikes (Grüsser & Grüsser-Cornehls 1976). Class T8 and T9 neurons (Ewert 1984a) have no defined receptive fields (see also Schürg-Pfeiffer this volume).

5.2.2 Responses to Moving Configural Stimuli

Since monocular input is sufficient for prey recognition, we have focussed our quantitative investigations on monocularly driven tectal and pretectal "small field" neurons. The relationship between recording sites and the approximately circular excitatory receptive fields (ERF) of these neurons corresponds roughly with retinotopy in the tectum and pretectum, respectively (Ewert et al. 1974). Regarding the sensitivity to configural features of moving visual stimuli, five principal response types can be distinguished: T5.1, T5.2, T5.3, T5.4, and TH3 neurons. (For results in toads see: Ewert 1971, 1974; Ewert & v. Wietersheim 1974a; v. Wietersheim & Ewert 1978; Ewert et al. 1979b; Satou & Ewert 1985; for comparable data in frogs from extracellular studies see Schürg-Pfeiffer & Ewert 1981, and from intracellular recording and labeling see Matsumoto et al. 1986; Matsumoto, this volume). In the following, we present a brief description of the properties of these neurons in response to W-, A-, and S-stimuli of corresponding maximal edge length varied in the range $2° \leq xl \leq 20°$.

Figure 16 (Left column). Response properties of thalamic pretectal neurons in the toad *Bufo americanus*; extracellular responses in paralyzed animals. (A) Selective sensitivity of a TH6 neuron (ERF=180°) to a large object approaching the toad's contralateral eye from the dorsal (a) or lateral (b) visual field; the neuron was weakly activated (c) if the object moved around the toad in the horizontal plane at constant distance. (B) Selective sensitivity of a TH3 neuron (ERF=45°) to different configural stimuli (a-c) moving through the ERF center of the contralateral eye. (C) Sensitivity of TH10-type neurons (ERF=30-90°) to a stationary obstacle (SO); (a,b) records of TH10.1 neurons with spontaneous activity and (c) without; the response to the obstacle ceased after the room lights were turned *off* and was regained when they were turned *on*; (d) a TH10.2 neuron responding to a stationary obstacle (so) or to offset of the room lights. (From Ewert 1971).

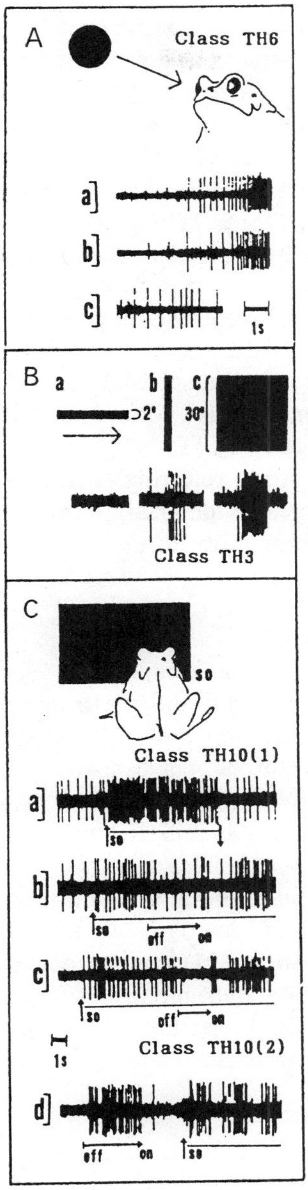

Fig. 16

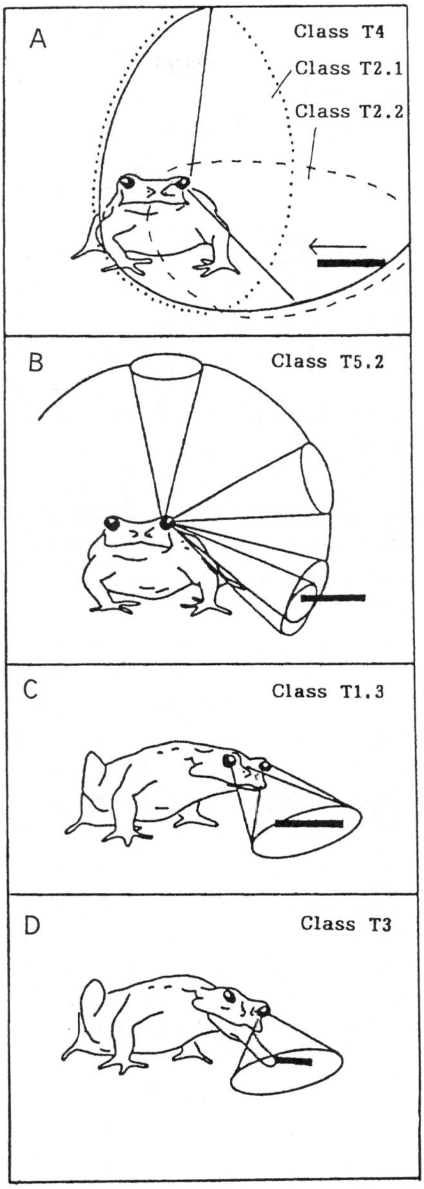

Fig. 17

Figure 17 (Right column). Receptive fields of tectal neurons in frogs and toads, suitable for discovering (A) and localizing (B-D) visual objects such as prey. (A) Class T2.1, T2.2, and T4 monocular wide field neurons whose ERFs have different sizes and arrangements, the ERF of T4 neurons encompassing the entire visual field of the contralateral eye; (B) several class T5.2 monocular small field neurons (ERF = 27°); (C) class T1.3 binocular neuron (ERF = 25°); (D) class T3 approach-sensitive neuron (ERF = 25°) (according to Lettvin et al. 1961; Grüsser & Grüsser-Cornehls 1970; Ewert & Borchers 1971; Ewert 1974). Note that the release of each prey-catching action pattern probably requires cooperation (combination) of different classes of tectal neurons, in which prey-feature selective T5.2 cells are involved (see Fig. 24).

Class TH3 neurons (ERF diam. = 45°), recorded from Lpd and lateral P, are best activated by large moving square (S) objects. When responses to configural features are tested with stripes of 2° width and varying length xl, the discharge frequency *R* to the W-configuration averages less than that to the A-configuration:

$$R_S > R_A > R_W$$

(Figs. 13D and 18A). These neurons are weakly activated by a small 2° square object but respond well to several randomly distributed, simultaneously moving stimuli that fill a large portion of the visual field, such as a moving background texture (Tsai & Ewert 1988).

Class T5.3 neurons (ERF = 25-35°), recorded mainly from tectal layer 8, are similar to TH3 neurons as far as responses to configural stimuli and moving textured backgrounds are concerned.

Class T5.1 neurons (ERF = 27°), recorded frequently from layer 8 and sometimes at the border of layers 6/7, respond strongly to square stimuli in a limited size range; configurationally, however, they are more responsive to the W- than to the A-configuration:

$$R_S > R_W > R_A$$

(Figs. 13E and 18B). These neurons cannot be activated by a moving textured background.

Class T5.2 neurons (ERF = 27°), recorded at the border of layers 6/7, exhibit a selective property with respect to the W- and A-configuration of a moving stripe in that

$$R_W > R_S > R_A$$

(Figs. 13F and 18C), and they fail to respond to a moving background texture (Tsai & Ewert 1988).

Class T5.4 neurons are selective for compact large moving objects (such as squares or circular shapes bigger than 4°) and do not respond to small worm-like or antiworm-like stripes:

$$R_S > R_W = R_A = \text{zero.}$$

If the correlation coefficients [rA;rW] for comparison between neuronal discharge frequency and prey-catching activity in response to black stripes in W and A configuration of varying length are calculated (Borchers & Ewert 1979), class T5.2 neurons show a better correlation [rA;rW = +0.9; +0.7] than class T5.1 neurons [rA;rW = 0.0; +0.8] or class TH3 neurons [rA;rW = -0.9; +0.9]. These evaluations agree with comparisons of the discrimination values for $D_{W,A} = f(x)$ (Fig. 14). The distinction between T5.2 and T5.1 neurons is also demonstrated by calculating β_2 from the equation $R_A = \beta_2\ xl(2) + k_2$ [spikes/s]. It can be shown that the inhibitory effect of xl(2), which determines the configuration selective acuity, is about 10 times stronger in class T5.2 than in T5.1 neurons:

$$^{T5.2}\text{-}\beta_2 = 9.6\ [^{T5.1}\text{-}\beta_2]$$

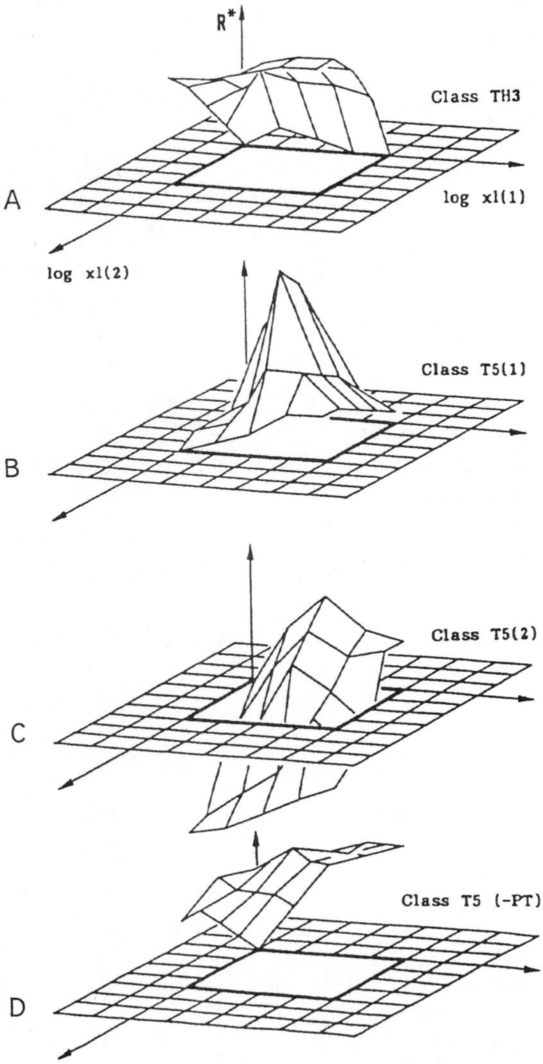

Figure 18. Tuning ranges of common toad's central visual neurons (A-D) across a 5 x 5 array of the stimulus continuum related to dynamic configuration shown in Fig. 5 (bottom): class TH3, T5.1, T5.2 neurons, and T5 neurons after ipsilateral pretectal (-PT) lesions; computer-processed data from Ewert & v.Wietersheim 1974a,b; intermediate values approximated. Ordinate: Average neuronal discharge frequency to stimuli (as shown in Fig. 5) minus the response to a 2° x 2° square stimulus serving as reference.

(v. Wietersheim & Ewert 1978; cf. also curves b in Fig. 13E and F). All this supports the conclusion that the activity of class T5.2 neurons in response to moving stimuli is positively correlated with the probability that an object's configuration fits the prey category:

- The algorithm for configural stimulus selection implemented in prey-catching according to Eq. (1) is provided by the output of T5.2 neurons.

In a manner comparable to prey-catching behavior, the $D_{W,A}$ values of T5.2 neurons are always positive, independent of the direction of movement (Fig. 19A) (Ewert et al. 1979b) and the angular velocity (Ewert et al. 1978). The latter result has been investigated in detail by Roth & Jordan (1982) who demonstrated that the configural preferences of neurons corresponding to classes T5.1, T5.2, and T5.3 were velocity-invariant in the range tested $2 \leq v \leq 20°/s$ (see an der Heiden & Roth, this volume). In common toads, of the entire sample of tectal "smallfield" neurons examined in this study, only 4% indicated a velocity-dependent preference inversion with W- and A-stimuli; 12% showed a velocity-dependent inversion with W- and S-stimuli of corresponding edge lengths.

To summarize, among feature-analyzing tectal and pretectal neurons (Fig.18A-C) there are (i) feature-sensitive neurons (classes T5.1, T5.3, TH3) that are optimally tuned to sets of configural features [xl(1) ; xl(2)] assigned to different, partly overlapping ranges in a continuum, and (ii) feature-selective neurons that display selective sensitivity to feature combinations, in either lumping together xl(1) and xl(2) (class T5.4), or differentiating between xl(1) and xl(2) (class T5.2).

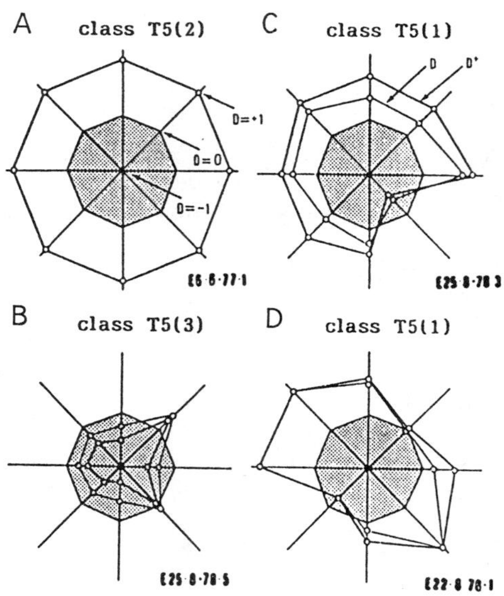

Figure 19. Discrimination values $D_{W,A}$ plotted in polar coordinates for worm-like and antiworm-like 2° x 8° stripes that traversed the centers of the approximately radially symmetrical receptive fields of tectal T5-type neurons in different directions of the visual field x-y coordinates at 7.6°/s. Values inside the "zero-circle" are negative, indicating antiworm preference, and those outside are positive, indicating worm preference. D and D* correspond to different data processing of extracellular records in common toads. (For details see Ewert et al. 1979b).

5.3 Toward a Feature Analyzing Neuronal Circuit

The results regarding an interaction between tectum and pretectum (Fig. 12B,C) in conjunction with the presence of tectal and pretectal feature-sensitive/selective neurons in laminated structures (Figs. 9 and 15) raise the question of circuitry. There are various hints about neuronal connectivity patterns.

5.3.1 Functional Considerations

Recent data from intracellular recording and iontophoretic labeling of the grass frog's tectal neurons confirm the properties of T5-type neurons and further characterize their postsynaptic activities in response to (i) electrical stimulation of the optic tract, (ii) moving configural stimuli, and (iii) brisk changes of diffuse illumination of the entire visual field at *off* and *on* (Ewert et al. 1985; Matsumoto et al. 1986; see also Matsumoto, this volume). Four main points are of interest here:

First, the responses of T5.1 and T5.2 cells to (i)-(iii) are characterized by the occurrence of excitatory and inhibitory potentials (EPSPs, spikes, IPSPs), whereas T5.3 cells show mainly EPSPs and spikes. Class T5.1 neurons are not homogeneous with respect to their response properties and morphological characteristics.

Second, the various combinations of excitatory and inhibitory postsynaptic potentials (EPSP/IPSP) of T5-type cells in response to brisk changes of diffuse light compared with the *on/off* spike activities of retinal ganglion cell fibers make it possible to estimate the combination and proportion of direct retinal input and indirect visual inputs from tectal and pretectal cells.

Third, postsynaptic reverberatory activities in some T5.1 cells indicate mutual excitatory intratectal connections.

Fourth, the fact that altering the configuration of a moving stripe from W to A is responded to by a T5.2 cell with a change from EPSPs to IPSPs (or a strong decrease of any PSPs) provides evidence of "configuration selective" excitatory and inhibitory inputs to this neuron. The inhibitory inputs may come from neurons (TH3 and/or T5.3) that respond well to the A-configuration. (For details see Matsumoto, this volume).

5.3.2 Pharmacological Considerations

The lack of "flat vesicles" in retino-tectal axon terminals (Lázár pers. comm. to N. Matsumoto, Matsumoto et al. 1986) in conjunction with the notion that the retino-tectal transmission in anurans is mainly nicotinic cholinergic (Freeman & Norden 1984) suggests that the IPSPs recorded from tectal neurons arise primarily from intrinsic tectal and/or extrinsic (e.g., pretectal) connections. Milson & Mitchell (1977) have reported that GABA in the frog tectum exerts a widespread inhibitory influence on field potentials which is antagonized by picrotoxin. We have observed that topical application of picrotoxin to the tectal surface of toads leads to visual disinhibition of prey-catching, which resembles the "agnostic" phenomena observed with pretectal thalamic lesions (Schürg-Pfeiffer & Ewert, unpubl. data). Similar results have been reported for application of strychnine to the toad's tectum (Ewert et al. 1974; for a study on amino acid and amine neurotransmitters in the frog's tectum see Sandoval et al., this volume).

Pretectal inhibitory influences on the tectum might be concerned with at least two types of action: One depends on visual input and exerts brief influences by "conventional" presynaptic or postsynaptic (e.g., GABAergic) transmission to determine the configural response properties of certain tectal neurons; the other depends on motivational states and induces longterm adjustments by peptidergic neuroregulators (e.g., enkephalin). The latter idea arises from results of immunocytochemical investigations (Reiner pers. comm.; Reiner et al. 1982, Kuljis & Karten 1982; see also Lázár, this volume; for a discussion on this subject see Section 5.3.5; Doty 1987; Ewert 1987b).

5.3.3 Cytological Considerations

Combined intracellular recording and iontophoretic labeling techniques have described T5-type neurons morphologically (Matsumoto & Antal 1984; Matsumoto et al. 1986). If we consider both the site of the somata and the extent of the dendritic trees, pear-shaped T5.1 neurons in layer 8 appear to be suitable for integrating retinal input, whereas large ganglionic T5.1 and T5.3 neurons of the same layer may be particularly appropriate for mediating local interactions. T5.2 (pyramidal) and T5.1 (large ganglionic) cells at the border of layers 6/7 are tecto-bulbar/spinal projection neurons (Satou & Ewert 1985) and/or tecto-thalamic connecting neurons (Antal et al. 1986; for details see Matsumoto, this volume).

5.3.4 The Proposed Circuitry

Given the strong inhibitory pretecto-tectal influences (Section 4.3), we suggest that the configuration-selectivity of T5.2 neurons is determined both by excitatory input from tectal T5.1 neurons [sensitive to extension xl(1) and area] and by direct and/or indirect inhibitory input from pretectal TH3 neurons [sensitive to xl(2) and

---

Figure 20. Neuronal correlates of the "interaction concept" for configural prey-selection in anurans. Visual information is processed in a network consisting of retinal (R), thalamic pretectal (TH), and tectal (T) neurons. Note the stages of parallel/hierarchical processing, involving distributed functions as regards different processing steps by TH- and T-neurons in response to certain combinations of R-type neurons. Modulatory systems are not considered here (see Figs. 24, 25).The network pieces shown in (A) and (B) are hypothesized to be parts of an integrated whole. Nomenclature: (a) *Functional unit of neurons* as proposed by Ewert (1974) and illustratively described as window discriminators, which broadly corresponds to the "column concept" described anatomically by Székely & Lázár (1976) and analyzed mathematically by Lara et al. (1982); (b) *neuronal network*; (c) *neuronal population*. Arrows indicate excitatory inputs and lines with cross bars inhibitory inputs; the lateral excitatory and inhibitory connections are not restricted to immediately adjacent neurons (see also Fig. 12C). (C) Concept of "goal-related interacting units." Note that the projection neurons T5.2 and T5.1' in this scheme are each expressing the result of information processing that takes place in assemblies (a_1 and a_2, respectively) of interacting neurons integrated in basically the same network; a_1 corresponds to the functional unit shown in (Aa). For further explanations see Section 5.3. The network shown in A) underlies the initial systems theoretical model of Ewert & v. Seelen 1974 . For advanced models, that consider the tectal cytoarchitecture, see Arbib and Cervantes-Pérez, this volume.

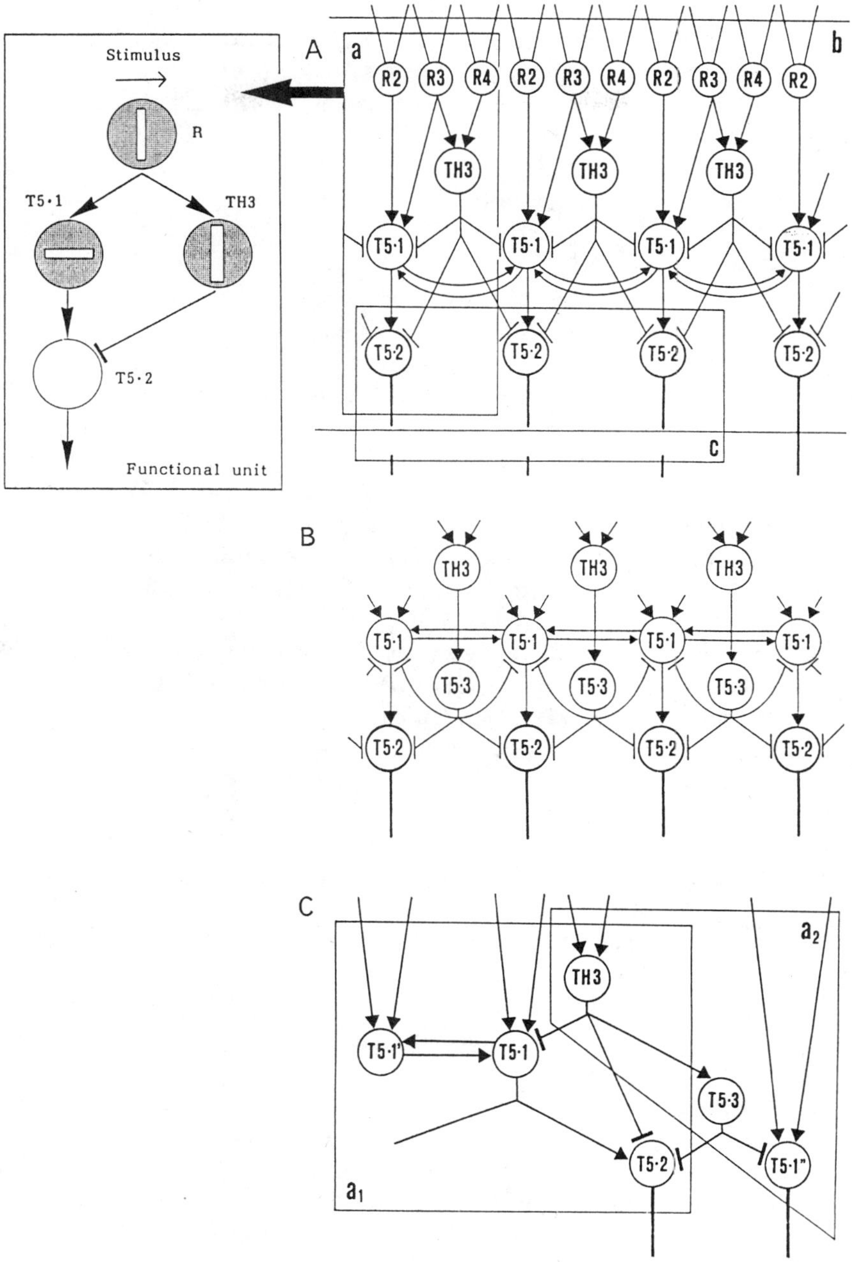

area] which in turn receive different combinations of retinal input (Fig. 20Aa and B). By its firing rate a class T5.2 neuron would thus encode the outcome of information processing that takes place in a functional unit (assembly) of connected cells (Fig. 20Aa) integrated in a network (Fig. 20Ab). Since a prey stimulus activates several adjacent assemblies of the same type, it is reasonable to suggest that configuration related information is transmitted to bulbar/spinal systems by a small population of T5.2 neurons (Fig. 20Ac). The proposed network (Fig. 20A,B) is characterized mainly by: (i) Combinations of certain retinal inputs to tectal T5.1 cells in layer 8 and to pretectal TH3 cells, (ii) lateral excitation presumably via other types of T5.1 cells of layer 8, and (iii) lateral inhibition probably mediated by TH3 and T5.3 neurons (Ewert 1974, 1981; Ewert & v. Seelen 1974; Matsumoto et al. 1986). This network is the neurophysiological equivalent of the "interaction concept" introduced in Sections 3.3 and 4.3:

o if small retinal images are narrowed (or a corresponding compact image is extended) parallel to the direction of movement [xl(1) > xl(2)], class T5.1 neurons of layer 8 are strongly activated due to mutual excitatory connections, whereas TH3 cells respond only weakly, so that T5.2 neurons convey worm-configuration related strong activity to the motor systems;
o if the same images are aligned (or an appropriate compact image is extended) perpendicular to the direction of movement [xl(1) < xl(2)], the TH3 neurons will fire more strongly than T5.1, so that T5.2 neurons are inhibited;
o in response to overall extended images [xl(1) = xl(2)] the respective excitatory inputs from T5.1 and inhibitory inputs from TH3 to T5.2 neurons will interact.

The basic property of this configuration selective system can be tested. If the responses of the same T5.2 neuron are recorded extracellularly before (Fig. 21A) and after a direct-current (Fig. 21Ba) or kainic-acid (Fig. 21Bb) induced ipsilateral pretectal lesion in an immobilized animal, a dramatic change in the response characteristics of the T5.2 neuron occurs, comparable to the "disinhibited prey-catching behavior" (Section 4.2.2):
(a) Loss of configural selectivity, yielding $D_{W,A} = 0$ and $ß_2 > 0$;
(b) strong increase in the discharge rate with the elongation of the edge of a moving square stimulus yielding no optimal size;
(c) strong decrease of surround inhibition, i.e., responses to a moving textured background;
(d) lack of neuronal adaptation in response to repetitive visual stimulation;
(e) increased probability of self-sustained activities and after-discharges;
(f) enlargement of the functional excitatory receptive field;
(Laming & Ewert 1983; Ewert 1985; Schürg-Pfeiffer et al. 1988). Deficits in stimulus discrimination (Ewert & v. Wietersheim 1974b) and habituation (Ingle 1973) of T5 neurons were also obtained in immobilized toads and frogs whose lesion-induced "disinhibited" prey-catching behaviors were determined before recording (see also Fig. 18D). Correlation between a pretectal lesion-induced change of the configuration selective property in a T5.2 neuron and in the toad's prey-catching behavior has been investigated in freely moving preparations by Schürg-Pfeiffer (this volume).

The observation that pretectal lesion-induced changes - particularly with respect to (b) and (f) - also occur in T5.1 neurons, supports the idea that inhibitory pretectal influences arrive at different tectal layers (Fig. 20A,B) to attenuate mutually excitatory intratectal processes. Thus, the "original" properties of T5.1 and T5.2

neurons - arising from retinal input and intrinsic tectal connectivity - may actually emerge after pretectal lesions (see Fig. 18D) and might be termed T5.0.

5.3.5 The "TP-Lesion Syndrome"

Regarding the behavioral and neurophysiological changes in configural selection after thalamic pretectal (TP) lesions, a few remarks on the underlying mechanisms

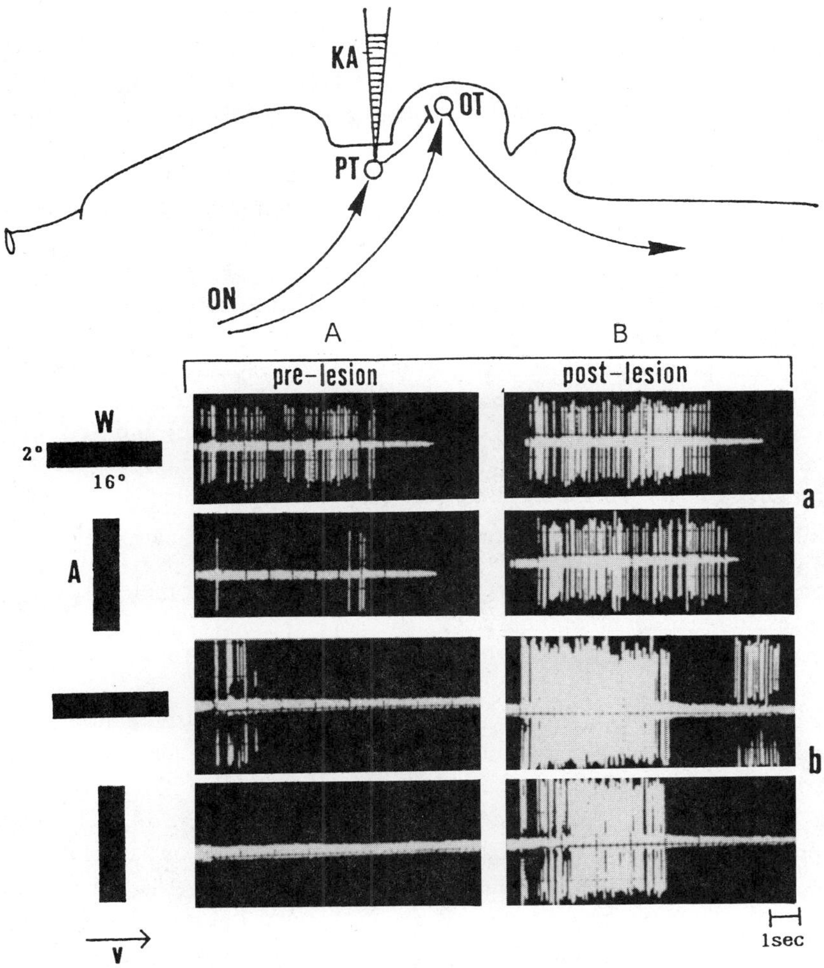

Figure 21. Evidence of pretecto-tectal inhibitory connections. (A) Examples of extracellular records from single class T5.2 neurons in a paralyzed common toad (a) and a grass frog (b) in response to a 2° x 16° black stripe traversing the center of the excitatory receptive field in a worm-like *W* or an antiworm-like *A* fashion at 7.6°/s. (B) Changes of the response of the same T5.2 neuron following an ipsilateral pretectal lesion induced by direct-current delivered by means of an electrode (a), or by Kainic acid (KA) solution applied through a micro-pipette (b); ON, optic nerve; OT, optic tectum; PT, pretectal Lpd/P region; rec, recording electrode. (From Schürg-Pfeiffer et al. 1989).

are in order (Doty 1987; Ewert 1987b). Two distinct pretecto-tectal inhibitory functions are possible.

Hypothesis (1): Configural selectivity in tectal neurons in response to retinal inputs is achieved by inhibitory influences of configuration sensitive pretectal neurons. These inhibitory pretectal influences are effective only *in response* to visual stimulation.

Hypothesis (2): Configural selectivity in tectal neurons is achieved alone by tectal intrinsic processes in response to retinal inputs (see an der Heiden & Roth, this volume). In this case, stimulus non-specific tonic inhibitory pretectal influences spread across the tectum, and this influences the overall tectal excitability.

We prefer the first hypothesis for the following main reasons based on experimental results: (i) In the pretectal Lpd/P region there is a retinotopic map (Lázár 1971; Fite & Scalia 1976) whose TH3 neurons display configuration sensitivity *in response* to moving visual objects (Ewert et al. 1974). These neurons show no spontaneous activity. (ii) There are projections from Lpd/P to tectum (e.g., see Wilczynski & Northcutt 1977; see also contributions by Lázár and Matsumoto, this volume). Lpd/P lesions abolish configural T5.2 and prey-catching selectivity by turning $\beta_2 < 0$ to $\beta_2 > 0$; see Eq. (1) in Sections 3.3.2 and 5.2.2 and Fig. 4B. If the second hypothesis were true, a pretectal lesion should strongly raise $k_{1,2}$ but not β_2 in Eq. (1). (iii) Functional recovery of configurational discrimination following pretectal lesions depends on the size of the lesion. After relatively large lesions, in the course of several postoperative weeks (or months) configural selection in prey-catching is only roughly "repaired" suggesting a kind of compensation under the influence of (lateral spread from) intact pretectal structures and intrinsic tectal connectivity. The stimulus-response pattern in compensated toads is much different from intact animals. The sharp configural discrimination of intact toads is never regained (see Ewert et al. 1983a).

Although the experimental evidence supports the first hypothesis, we infer at least two types of modulatory pretecto-tectal actions, a selective specifying one and a general adjusting one (Section 5.3.2), which would actually acknowledge features of both hypotheses (see also Cervantes-Pérez, this volume).

5.4 Comparisons Across Species

Complementary studies in different amphibians, *Bufo bufo, Rana temporaria, Salamandra salamandra, Hydromantes italicus,* show partly comparable T5-type neurons. Furthermore, the configural discrimination of moving visual stimuli implemented in prey-catching is reflected in the response characteristic of certain tectal neurons (cf. Ewert 1974, 1984a; Schürg-Pfeiffer 1979; Finkenstädt 1981; Himstedt 1982; Roth 1982; an der Heiden & Roth, this volume).

For the evaluation of species differences in configural selection, it is necessary to compare the effects of changes of the information bearing parameters [xl(1); xl(2)] across the stimulus continuum (as shown in Fig. 8A). Our comparative studies of T5.2 neurons in *B. bufo* (Fig. 22A) and *R. temporaria* (Fig. 22B) show that (i) a common fundamental algorithm according to Eq. (1) is implemented in both species; (ii) species-different variations refer to certain ranges of the stimulus continuum.

If we compare the configural selective abilities between *S. salamandra* (Finkenstädt & Ewert 1983a) and *B. bufo* (Ewert & v. Wietersheim 1974a) across the stimulus continuum of Fig. 5, it is interesting to note that $ß_2$ in Eq. (1), indicative for the configuration selective acuity, is significantly different between fire salamander and common toad both for prey-catching and tectal neurons. In the fire salamander only $ß_2$-values were found, largely comparable to those of toad's T5.1 cells (Section 5.2.2). Fire salamanders hence are lacking a clear dissociation in T5.1 and T5.2 neurons. We admit that this difference between toad and fire salamander cannot be seen if only three W-, S-, and A-stimuli of a certain size are tested (see an der Heiden & Roth, this volume). If we accept the hypothesis that in *B. bufo* the strong configural acuity is achieved by "two-stage" pretecto-tectal inhibitory inputs (see Fig. 20A,B), yielding T5.1 and T5.2 properties, then the "second stage" obviously fails to occur in *S. salamandra*. In fact, a comparison of pretectal thalamic nuclei reveals differences: In anurans the dorsal posterior thalamus is parcellated into a lateral (Lpd) and a central

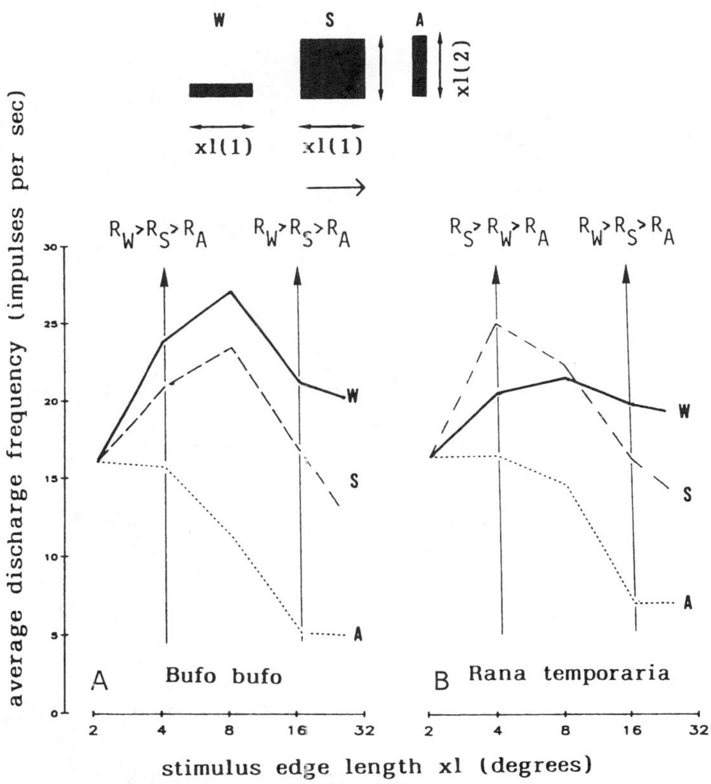

Figure 22. Species differences in prey-selective T5.2 neurons. Comparisons between *Bufo bufo* (A) and *Rana temporaria* (B) class T5.2 neurons responses across three segments of a stimulus continuum: Extracellularly recorded spike activity, R, in response to W-, S-, and A-stimuli of varying edge length (xl) moving at 7.6°/s. Note that in toad T5.2 neurons the relation $R_W > R_S > R_A$ is valied for $2° < xl \leq 32°$; in frog T5.2 neurons the relation $R_S > R_W > R_A$ is valied for $2° < xl \leq 8°$ and $R_W > R_S > R_A$ for $8° < xl \leq 32°$. (From Ewert & v. Wietersheim 1974a (A); Schürg-Pfeiffer & Ewert 1981 (B)).

(P) nucleus, whereas in urodeles a rather unique posterocentral cell group is identifiable (Clairambault 1976). Interestingly, young postmetamorphic anurans exhibit a posterior thalamic organization similar to the one urodeles show as adults. The parcellation of anuran's thalamus is completed 6-12 months after metamorphosis (Clairambault 1976), which is paralleled with remarkable improvements of configural selection, size constancy, and depth estimation. These results are discussed in the context of Ebbesson's (1984) *parcellation theory* of brain evolution (Ewert 1984b; Ebbesson 1987). It is interesting to note that neuroanatomical and immunocytochemical investigations reveal distinct projections from different pretectal nuclei to various tectal layers (Reiner pers. communication; Kuljis & Karten 1982; Reiner et al. 1984; Lázár, this volume).

Evidence that the properties of salamander's "T5.1" neurons are determined by prosencephalic inputs is provided by lesion studies. After anterior thalamic or pretectal lesions in the fire salamander, changes in the configural sensitivity of "T5.1" neurons and prey-catching behavior (Finkenstädt & Ewert 1983b) were partly comparable to those seen in pretectally lesioned toads and frogs (Ewert & v. Wietersheim 1974b).

5.5 How are Moving Objects Represented in Anuran's Visual System?

We now return to the introductory questions of Section 1 and to some conclusions drawn in Section 3.5. From a historical point of view, there are three concepts of prey-recognition: The "retina concept," the "motor concept," and the "interaction concept."

5.5.1 Retina Concept

The *retina concept*, initiated by Barlow (1953) and put forward by Lettvin et al. (1959), restricted the filter functions essential for categorization (e.g., detection of prey) to a class of retinal ganglion cells. Thus, Barlow (1953) favored the frog's R3 neurons when he said:

> "In fact, 'on-off' units [class R3] seem to possess the whole of the discriminatory mechanism needed to account for this rather simple behaviour [prey capture], ... and it is difficult to avoid the conclusion that the 'on- off' units are matched to this stimulus (image of a fly at 2 in. distance) and act as 'fly detectors' (p.86)."

Lettvin et al. (1959) preferred R2 neurons

> "We have been tempted, for example, to call the convexity detectors [class R2] 'bug perceiver' ... Could one better describe a system for detecting an accessible bug? (p.1951) ... The operations found in the frog [retina] make unlikely later processes in his system ... (p.1950)."

Such processes were not excluded in the discussion by Maturana et al. (1960):

> "The [retinal] ganglion cells can be considered as emitting parcels of information about particular useful aspects of the visual image, which aspects depend on the class to which the cell belongs (p.161) ... In a system like this, a unique combination of the four qualitative contexts [provided by the

> operations mediated by four retinal ganglion cell classes] in a certain spatial relation may define a class of objects ... We do not yet know how the recognition of such unique combinations of contexts is made at the tectum, but this can be studied experimentally (p.164)."

Although there is no specific retinal prey detector like the one originally suggested, the classes of retinal ganglion cells continue to have the fascinating property that they do say something perceptually important to the central visual system (see also Barlow 1985a).

5.5.2 Motor Concept

The *motor concept* was developed by Grüsser & Grüsser-Cornehls (1970, 1976; see also Grüsser-Cornehls 1984). Both authors claimed that the motor acts (i.e., the accomplishments of the prey-catching action patterns) are essential for prey recognition, emphasizing that

> "the construction of the invariants predator, prey, hiding place etc. apparently does not precede the motor response; instead, the entire neuronal reaction chain that leads to the observable behavioral sequence *in toto* constitutes the invariant" (Grüsser & Grüsser-Cornehls 1970, Engl. transl. of p.215: "die Invariantenbildung Feind, Beute, Versteck usw. vor der motorischen Reaktion nicht erfolgt, sondern die gesamte neuronale Reaktionsfolge, die zu der beobachtbaren Verhaltensfolge führt, *in toto* die Invariantenbildung darstellt")

and re-emphasizing that it is

> "the sequence of motor responses which we interpret as 'recognition' of universals" (Grüsser & Grüsser-Cornehls 1976, p. 374), i.e. that "the 'recognition' of the behavioral universal 'prey' or 'enemy' ... is the result of a network for which the motor acts play an essential role" (Grüsser-Cornehls 1984, p. 238).

This hypothesis derived from a problem of how one can know what one sees before one responds, was both attractive and wrong, as far as the toad's ballistic action patterns are concerned. (However, recognition by sensorimotor coordination can play an important role, for example, in sensory substitution in the blind; see Koy-Oberthür, this volume). The argument that toad's prey-recognition does not precede a prey-catching response - that the frog or toad must instead run through the behavioral sequence starting with orienting and terminating with snapping in order to recognize prey - was invalidated by the animal's behavior: Orienting toward prey presupposes prey recognition (Figs. 4A and 5A,B; see also Section 2.2). To evaluate the "motor concept" it is important to know that in their own investigations Grüsser & Grüsser-Cornehls (1970, 1976) did not distinguish neurons with T5.1, T5.2, T5.3, and T5.4 configural response characteristics. Both authors initially even disputed the existence and significance of tectal neurons displaying configuration-selective T5.2 properties that were first described by Ewert (1974) as "small field type II neuron" responses, quantitatively studied by Ewert & v. Wietersheim (1974a; see also v. Wietersheim & Ewert 1978), and confirmed by Roth & Jordan (1982; see also an der Heiden & Roth, this volume).

5.5.3 Interaction Concept

Our neuroethological investigations have led to the *interaction concept* of configural stimulus selection (Section 3.3.1; Ewert et al. 1970; Ewert 1968, 1974; for systems theoretical analyses and computer simulations see Ewert & v. Seelen 1974; Arbib 1982; Lara et al. 1982). The main components of our hypothesis are:

(1) The analysis of configural features of a moving visual stimulus [xl(1); xl(2)] is carried out by functional units (Fig. 20Aa) of connected feature-analyzing neurons (T5.1, T5.2, T5.3, TH3) inherent in a pretectal/tectal network (Fig. 20A,B) related to retinal projection fields (Fig. 9). The idea is that an extensive continuous network (Fig. 20Ab) is subdivided into "integrative interacting units" (Szentágothai 1967) of minimal size (Fig. 20Aa) capable of the same basic information processing as the corresponding network (consisting of these units, also called "columns" by Arbib, Lara, and by Lázár, this volume). The retinal input to such an assembly is mediated by particular combinations of ganglion cells (e.g., R2, R3, R4). Feature-analyzing tectal and pretectal neurons are distinguished by the degree and quality of integration and interaction. Neurons of adjacent functional units are partially coupled through mutual excitation and lateral inhibition. Outputs of these units are mediated, e.g., by T5.2 configuration-selective tecto-bulbar/spinal projective neurons (Fig. 20Ac). (For feature analysis in central visual maps with a cortical columnar structure see Mallot & v. Seelen, this volume).

(2) Configural stimulus selection is determined by the properties of feature-sensitive neurons (e.g., T5.1, T5.3, TH3) which are tuned to various, partly overlapping ranges in a stimulus continuum (Fig. 18A,B). Moving visual objects that are different in their configural characteristics thus elicit corresponding patterns of activity in these neurons. Specific neuronal interactions determine the range in the stimulus continuum that defines the prey category. In other words, this connectivity provides the configural prey schema with which visual input is compared: Configural cues are extracted by T5.2 prey-selective cells through specific excitatory and inhibitory synaptic contacts with feature-sensitive neurons. Class T5.2 neurons are optimally tuned to the sets of configural features that characterize prey (cf. Figs. 18C and 5). The result of this analysis in adjacent functional units, expressed in the spike frequency of adjacent T5.2 neurons, is transmitted to the appropriate bulbar/spinal motor systems.

(3) Feature-sensitive neurons of classes T5.1, T5.3, T5.4, and TH3 are probably not monopolized for a specific behavioral goal. Certain classes of neurons of assemblies related to prey-feature analysis (Fig. 20C,a1) may take part in other goal-related assemblies (Fig. 20C,a2) contributing in that case to the analysis of other objects. The extension of the idea described in (1) is that different types of functional units (goal-related assemblies) inherent in the same network are defined by the specific integrative properties of particular bulbar/spinal projective neurons (e.g., see T5.2 and T5.1" in Fig. 20C). Pattern recognition in the toad's visual system thus involves *parallel-hierarchical* information processing.

(4) The feature-analyzing system is relatively stable, given constant inputs from the internal state of the organism related to circadian determinants, nutrition, hormonal balance and so on.

(5) The feature-analyzing system is to some extent modifiable through modulatory prosencephalic influences depending on internal and external signals that may change the properties of either the whole network or parts of it (see Section 6.4).

5.6 Evaluation of T5.2 Neurons

5.6.1 Filter Operations

Class T5.2 neurons can be judged from different points of view. If one focuses just on the selectivity in worm/antiworm discrimination arising from convergent excitatory and inhibitory inputs by neurons that are sensitive to worm and antiworm configuration, respectively, one might be tempted to call T5.2 cells "worm detector," in accordance with a wiring principle illustrated in Fig. 1A. This concept is misleading for reasons mentioned in Section 3.5.

If we consider that prey category is determined by a structure of feature relationships assigned to a range within a configural stimulus continuum (Fig. 5), we can regard (small populations of) T5.2 neurons whose activities are correlated with changes in these prey-defining feature relationships (Fig. 18C) as "prey filters." Such a filter compares information related to configural (spatiotemporal) features of visual stimuli with stored information here provided by neural circuitry. The filter actively picks out information related to prey and simultaneously suppresses information related to non-prey.

This calls for a specification of the terms *category* (class) and *comparison* (classification). According to v. Seelen (1970) pattern recognition is defined as the assignment of two-dimensional, space-time dependent contrast distributions from the environment to classes of functional significance. The recognition system's operation proceeds in two steps: First, the extraction of significant features ${}^{\nu}x_i$ from the pattern to be perceived; second, the separation of the feature vectors x by means of a discriminating function U_D in such a way that upon presentation of a pattern of the νth class of significance:

$$U_D({}^{\nu}x) > U_D({}^{\mu}x) \text{ for } \nu,\mu = 1, 2, \ldots M \text{ and } \nu \neq \mu$$

where M is the number of possible classes to be discriminated, in the present study M=2 referring to prey or non-prey (Ewert & v. Seelen 1974). This separation procedure involving weighted neuronal threshold-value operations, is called *classification*. Configural selection can be implemented in such a way that the T5.2 prey filters *compare* - cross-correlate according to Eq. (3) - the incoming visual information with their filter property: If y(r,s,t) is the space(r,s)- and time (t)-dependent input signal to a neural model network, H(r,s,t) the coupling function of neurons involving lateral inhibition and lateral excitation, and z(r,s,t) the system's output, then

$$z(r,s,t) = \int_{0}^{t}\int_{-\infty}^{+\infty}\int_{-\infty}^{+\infty} H(r-r', s-s', t-\tau)\, y(r',s',\tau)\, dr'ds'd\tau \qquad (3)$$

The strength of the system's output depends on the number of times the threshold is crossed with a latency depending on the neuronal impulse frequency. With the aid of Fourier-transformations the convolution integral according to Eq. (3) becomes

$$\mathbf{F}_z = \mathbf{F}_H \cdot \mathbf{F}_y. \quad (4)$$

Given the operations of feature extraction are implemented by a retino-tectal filter, described by H_1, and a retino-pretectal filter, described by H_2, then the space-dependent course of their subtractive interaction reproduces the function of U_D (for details see Ewert & v. Seelen 1974). According to this model, specific inhibitory pretecto-tectal interactions establish a structure of feature relationships that can be modulated by prosencephalic inputs. These interactions determine the range in the stimulus continuum [xl(1); xl(2)] that defines prey. Species-specific differences in configural selection can be explained by variations in coupling functions H_1 and H_2 and by the weights of their interaction which specifies $ß_1$ and $ß_2$ in Eq. (1). The size constancy effect (Section 3.3.2) in addition may be achieved by appropriate tuning in relation to depth information provided by the lens accommodation mechanism (Collett 1977; for models of depth perception see Arbib, this volume). Note that more general models of prey recognition, such as those of Arbib, Lara, and Cervantes-Pérez, involve non-linear operations (Lara et al. 1982). One of the interesting results of these approaches is that a preference inversion with respect to W- and S-stimuli in prey-catching (according to our and Roth's experimental results) can be computer-simulated by means of tectal model neurons under certain conditions of parameter-setting for pretecto-tectal inhibition, whereas a velocity-dependent preference inversion regarding W- and A-stimuli (incompatible with our and Himstedt's results; Section 3.3.4) has not yet been simulated (see Cervantes-Pérez 1985).

5.6.2 Representational *vs*. Computational Mapping

An important outcome of this study is the notion of topographic *representational* and parametric *computational* mapping. The former refers to a *representation* of the x-y coordinates of the retina in pretectal (Lázár 1971) and tectal (Gaze 1958) laminated structures; such maps subserve orientation in space. The latter refer to topographic functional units in which relations between significant features of a visual object are *computed*, probably for its recognition. Such feature maps (e.g., displayed by T5.2 cells) are inherent in representational maps, and this seems to be a general property of laminated sensory structures (see Mallot & v. Seelen, this volume). The relatively high degree of integration required for the computational properties is achieved at the expense of spatial resolution (e.g., note the relatively large receptive fields of T5.2 cells, ERF=27°). The latter can be maintained via parallel processing visual input in tegmental and bulbar structures (see Grobstein, this volume), and/or regained through interaction of axon terminals of adjacent T5.2 neurons in pre-motor structures, for example (see also Mittelstaedt, this volume).

5.6.3 Implicit *vs*. Explicit Computing

Both mammal's cortical Area 17 "orientation detectors" and the anuran's tectal T5.2 "configuration discriminators" are able to distinguish two stripes whose long axes deviate by 90° from each other. The networks that underlie these functions, however, seem to be entirely different. The Area 17 neurons compute the two stripe orientations *explicitly* with asymmetric processors organized in cortical orientation columns involving an anisotropic cytological structure; for the various possible orientations of the stripes, a huge number of asymmetric processors is required. The T5.2 neurons, however, only need one type of processor, provided each one of the two stripes moves at a given speed in the same direction. This *implicit* computing (Stevens 1987) is feasible with an approximately isotropic histological structure, while the asymmetry required for discrimination lies in the time domain of the spatiotemporal stimulus parameters itself: Configural discrimination takes advantage of the asymmetry in the spatiotemporal properties of the stimulus. This discrimination principle is invariant under the direction and the velocity of stimulus movement. These invariances are brought about through discrete evaluations of spatial and spatiotemporal stimulus features by pretectal and tectal filter operations and their subtractive interactions (Ewert 1987d).

We infer that to analyze visual moving objects (prey, predator, mate), amphibians with motionless eyes have evolved a highly economic system that so-to-speak has "circumvented" a redundant, sophisticated, anisotropic cortex-like structure.

5.6.4 Network Properties

Do all T5-type neurons contribute to prey feature analysis? They might in so far as they were components of a *feature analyzing network* (Fig. 20A-C). However, even if all monocular small- and widefield neurons T2.1, T2.2, T4, T5.1, T5.2, T5.3, and T5.4 (and others) were *parallel processing* components of a network continuum, whatever this means with regard to a distribution over tectal laminae, we cannot deny that neurons with distinct response characteristics exist, that these can be recorded separately at a given time, that these project to bulbar/spinal motor systems, that it would make no sense if all types of these projection neurons had excitatory connections to any motor pattern generating circuitry (Fig. 24), and that the message these neurons convey must be *read* in some way by respective pre-motor networks. Information necessary to characterize different behaviorally relevant objects (e.g., prey, mate, predator) can be mediated by feature analyzing projection neurons (e.g., classes T5.2, T5.1, T5.4) of basically the same network.

The information provided by (a small population of neurons of) such a class could be *essential*, but must not necessarily be *sufficient* (see Section 6.3). The qualifying notion "essential" is used here with respect to a given functional principle [F]. If an integral component (f_i) were eliminated, the brain would not just be dealing with [F-(f_i)], but rather with some different self-organizing structure [F^*]. Given the multi-connectivity in the pretectal/tectal network - of which Fig. 20A,B suggests only some main connections with respect to the function under investigation - one should not be surprised if the role of prey-selective T5.2 neurons (after their selective elimination, if this were possible) would be executed by other cells, e.g., modified T5.1 projective neurons (Fig. 20C). Long-term functional recovery after pretectal lesions is an example of "network compensation" (Ewert et al. 1983a) suggesting modified prey-selective cells. The dynamic model of "T5 base modulator" developed by Betts (this volume) shows how a given T5 unit can be tuned to different configural objects.

6 Sensorimotor Interfacing: The Command System Approach

The prey-catching action patterns *o,a,f,* and *s* are activated by different releasing mechanisms (RMs) which, using comparable feature-analyzing processes, are distinguishable by operations that locate the visual stimulus in space. The various stimulus situations can be encoded by particular combinations of "specialized" feature selective and space monitoring neurons that are connected to the corresponding motor systems, thus representing a "sensorimotor interface." Modulatory inputs - related to arousal, motivation, and learning - make such releasing systems adaptive to changes in internal state and external conditions. The experimental approach is concerned with neuronal recording from and electrical stimulation of behaving and paralyzed toads, and with retrograde filling and antidromic firing of tectal cells via axons traveling to bulbar/spinal systems.

6.1 Specialized Neurons

Compared to other tectal neurons, class T5.2 display a specialization in the way they discriminate configural features of moving ojects, a property which allows the toad to recognize prey (Ewert 1974; v. Wietersheim & Ewert 1978). When in the following description of neurons the adjective "specialized" is used, this refers to a selectional property (not necessarily excluding other properties). Specialist properties of neurons are suggested to be long-term adjusted, and these may change under certain conditions (e.g., see Section 6.4.4 and the contribution by Betts, this volume).

Class T2.1, T2.2, T4, and TH4 wide field neurons are specialized to obtain visual information from large distinctly arranged receptive fields (Fig. 17A), e.g., suitable for arousing or sensitizing a RM (Lettvin et al. 1961; Grüsser & Grüsser-Cornehls 1970; Ewert & Borchers 1971; Ewert 1971; Brown & Ingle 1973; Gaillard & Galand 1979).

The locus of a stimulus in the x-y coordinates of the visual field can be determined by information from a number of T5.2 or T5.1 neurons with partly overlapping receptive fields (Fig. 17B). Furthermore, it is possible that the required space and head position signals (Rocoux & Crommelinck 1987) are processed and evaluated by tegmental structures feeding into bulbar/spinal circuits (see Grobstein, this volume). The tegmental neurons suggested to take part in space-coding remain to be discovered. In this context, it must be checked whether T7 neurons, whose tiny ERFs of 4 to 6° diameter would be suitable for accurate prey localization, feed into tegmental structures, since antidromic stimulation studies failed to determine T7 mediated tecto-bulbar/spinal projections (Satou & Ewert 1985). Interestingly, Class T7 neurons display the same *off*-related edge positioning properties as R3 and T5.2 neurons (Tsai & Ewert 1987).

Depth estimation along the z-axis may involve binocular class T1.3 neurons specialized to receive concurrent inputs from "disparate" retinal areas (Fig.17C) (e.g., see Grüsser & Grüsser-Cornehls 1976), a property resulting from special neuronal circuitry (e.g., see Gruberg & Udin 1978; Gruberg & Lettvin 1980; cf. also Collett & Udin 1983; House 1984; Arbib, this volume). Monocular depth estimation responsible for size constancy (Section 3.3.2) may involve pretecto-tectal inhibitory circuitry in conjunction with information from the lens accommodation mechanism (Collett 1977). The monocular class T3 neurons (Fig.17D) are specialized in that the approach of a small object along the z-axis is required to drive the cell maximally, a characteristic that can also be used in judging object distance (Grüsser & Grüsser-Cornehls 1976).

6.2 Tecto-Bulbar/Spinal Output

6.2.1 Physiological and Anatomical Evidence

Questions related to access of "specialized neurons" to motor systems have been investigated in paralyzed toads by extracellularly recording their antidromic activation in response to an electrical stimulus applied to the tecto-bulbar/spinal tracts in the medulla oblongata (Satou & Ewert 1985). Class T1.3, T2.1, T3, T4, T5.1, T5.2, and T5.4 neurons were activated antidromically by contralateral bulbar stimulation (e.g., see Fig. 23). Among T2.1, T4, and T5.1, however, not all recorded neurons could be backfired - and T7 not at all - suggesting that within a class there is a distinction between "locally interacting" (processing) and tecto-bulbar/spinal descending (processing and projecting) neurons (see also Lázár and Matsumoto, this volume). It

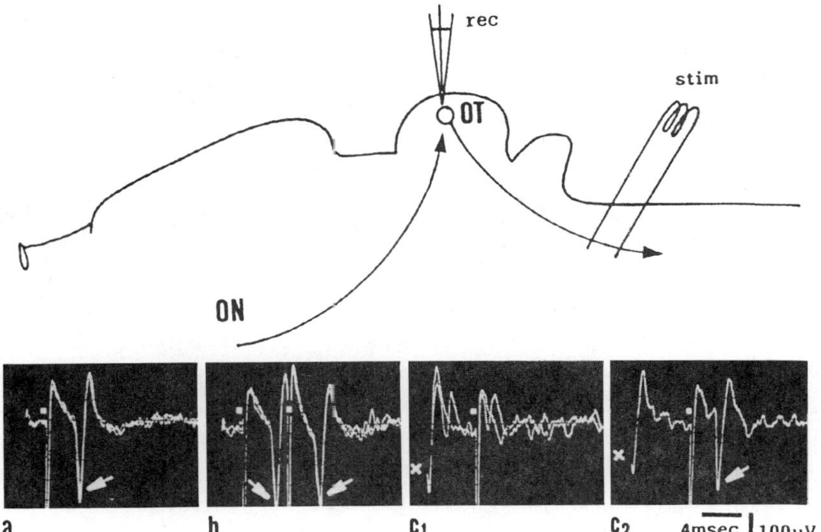

Figure 23. Neurophysiological specification of tecto-bulbar/spinal projections in paralyzed toads. Example of antidromic activation ("backfiring") of a physiologically identified tectal class T5.2 projection neuron in response to a 0.1 ms electrical stimulus applied through bipolar electrodes to the tractus tecto-bulbaris et spinalis cruciatus in the contralateral caudal medulla oblongata. Criteria: (a) *constant latency response* (several superimposed traces); the onset of electrical stimulation is marked by a small spot (immediately followed by the stimulus artifact), and the antidromic spike is indicated by an arrow; (b) *following ability* to electrical double stimuli at intervals down to less than 4 ms; (c) *collision* between a visual spike (indicated by a cross) and an electrically evoked antidromic spike at a critical spike-stimulus delay of 5.1 ms (c1); no collision (see arrow) at 5.9 ms (c2). (Extracellular records from Satou & Ewert 1985).

is worth mentioning that all investigated T5.2 neurons from frontal and extrafrontal sites of the tectal visual map could be backfired (Satou & Ewert 1985; Buxbaum-Conradi & Ewert, unpubl. data). This indicates that the results of prey feature analysis from various parts of the visual field are transmitted by T5.2 neurons to bulbar/spinal motor systems.

The morphological types of tecto-bulbar/spinal projection neurons have been investigated by injecting horseradish peroxidase (HRP) unilaterally into the medulla oblongata (Weerasuriya & Ewert 1981; for results in frogs see Ingle 1983). In toads various types of retrogradely filled cells were found on top of layer 6 or at the border of layers 6/7 in both tecta - particularly dense in the contralateral tectal hemisphere and especially in the lateral corner - from which neurons with prey-selective and localization properties can be recorded (see Weerasuriya, this volume). Furthermore, labeling studies with HRP and cobalt techniques have provided evidence of pretecto-bulbar/spinal projection neurons (Weerasuriya & Ewert 1981; Ingle 1983, Lázár 1984; Tóth et al. 1985; see also Lázár and Matsumoto, this volume).

6.2.2 Question of Behavior-Related Efferent Channels

Ingle (1983) has shown in frogs that transection of the tractus tecto-bulbaris et spinalis cruciatus (tr.tbs.c.) impairs orienting toward prey, but not snapping for adequate stimulus positions, while lesioning the tractus tecto-bulbaris directus (tr.tb.d.) abolishes snapping. Therefore, it seems likely that some tectal efferents running in tr.tbs.c. are essential for turning, whereas some tectal efferents running in tr.tb.d. are critically involved in snapping. Ingle (1983) claims that T5.2 axons descend only in tr.tb.d. and that they are monopolized for snapping, whereas the efferent tectal cells descending in tr.tbs.c. are responsible for stimulus localization and triggering prey-orienting. According to Ingle, the latter should be driven (almost) exclusively by R2 neurons. Ingle estimates the weights of R2 fiber terminals from the adjacent extent of dendritic arborizations of tectal cells backfilled with HRP applied to the tr.tbs.c., concluding that: "most (if not all) of the retinal input to cells projecting into the tectospinal tract receive input from class-2 [R2] fiber terminals" (Ingle 1983, p.207); "efferent cells which participate in orienting to prey have heavy (if not exclusive) inputs from the class-2 ('bug detecting') retinal fibers" (Ingle & Quinn 1982, p.406). This hypothesis is in conflict with various facts, such as: (i) Prey orienting and snapping, respectively, are tuned to the same ranges of the configural stimulus continuum (Fig. 5), hence suggesting that release of either action pattern requires basically the same stimulus recognizing process drawing on the properties of T5.2 neurons. (ii) If the tectal efferents projecting in the tr.tbs.c. were responsible for prey-catching orientation and were activated exclusively by retinal R2 (so-called bug-detecting) input, it is difficult to explain that, disconnected from pretectal influences, the same efferents mediate orienting toward large predator-like objects (see Schürg-Pfeiffer, this volume). (iii) If *the* tecto-spinal efferents projecting in tr.tbs.c. were R2-driven and monopolized for prey-orienting, then we must ask for tectal efferents activating orientational turning toward other goals, such as a sexual partner (Kondrashev 1987), not attractive to "bug detectors." Even if (i) were specific to *Bufo*, the arguments (ii) and (iii) also hold for *Rana*.

We suggest that the prey-catching motor pattern generating circuits for *o,a,f* and *s* receive tectal T5.2 input mediated via tr.tbs.c. and tr.tb.d. of which the former tract is essential for orienting, and it is possible that the turn generator obtains also parallel inputs from "space-coding" putative tegmental neurons (see Grobstein, this volume). T5.2 cell recordings, electrical stimulation, and marking via the same microelectrode in the visual tectal map of freely moving preparations show that orienting is approximately in register with the position of the center of the neuron's receptive field as measured before the animal started to turn. For correlations between visual stimulation, T5.2 neuron firing, and prey-catching behavior see Schürg-Pfeiffer (this volume).

6.3 The "Command Releasing System" Concept

6.3.1 Sensorimotor Codes

It must be borne in mind that the release of each prey-catching action pattern (*o,a,f,s*) requires basically the same prey feature analyzing processes but different localization mechanisms (Fig. 2b), that tectal neurons with feature-analyzing and space-monitoring properties feed into bulbar/spinal motor systems, and that electrical stimulation in the close vicinity of prey-feature analyzing neurons elicits prey-catching. It is reasonable to suggest that the activation of the corresponding motor pattern generators - in a manner similar to an AND-gate - requires simultaneous activation of appropriate combinations of such specialized neurons. Generalizing Kupfermann & Weiss's (1978) definition of "command systems," we call such a combination a "command releasing system" (Fig. 24). Its components (specialized neurons) - the "command elements" - operate between sensory and motor systems and are thus part of a sensorimotor interface. Their simultaneous activation represents "prey" and its "location in space," and this produces the command to activate the related behavioral response. The idea is that different stimulus situations related to a prey object can be monitored by activity in certain combinations of different types of neurons specialized for feature-analysis (class T5.2), localization (e.g., T1.3), and arousal (e.g., T4), and that T5.2 neurons with prey-feature selecting properties are shared in each combination (Ewert 1980), e.g.:

[(T5.2)(T4)]	[(T5.2')(T2.1)(T1.1)]	[(T5.2')(T1.2)]	[(T5.2")(T1;3)(T3)]
o	*a*	*f*	*s*

( ) command element
[ ] command releasing system, CRS

According to this hypothesis, CRSs provide *sensorimotor codes* (Fig. 24) and are neuronal correlates of RMs, the "releasing schemas." Command elements may be compared with a key-board (to use an analogy of Hoyle's 1984) on which certain combinations elicit specific responses (see also Fig. 2): "stimulus moving somewhere in the visual field (T4)" AND "stimulus recognized as prey, n degrees outside the fixation area (T5.2)" yield the command to orient (*o*); "stimulus moving in the frontal visual field (T2.1)" AND "stimulus recognized as prey" (T5.2')" and "stimulus far afield (T1.1)" yield the command to approach (*a*); "stimulus recognized as prey near the fixation area (T5.2')" AND "stimulus close to the toad (T1.2)" yield the command to fixate (*f*); "stimulus recognized as prey inside the fixation area (T5.2")" AND "both retinae adequately stimulated (T1.3)" AND "stimulus within snapping distance (T3)" yield the command to snap (*s*).

In these systems, configuration-selective neurons of the type T5.2 are essential and decisive for prey-catching. We infer that no prey-catching command by any CRS can be executed if these neurons are not appropriately activated. If these neurons lose their prey-selective properties (after pretectal lesions), any moving visual object sufficiently activates the prey-catching CRSs (see Schürg-Pfeiffer, this volume; for modulatory processes see Section 6.4.4.).

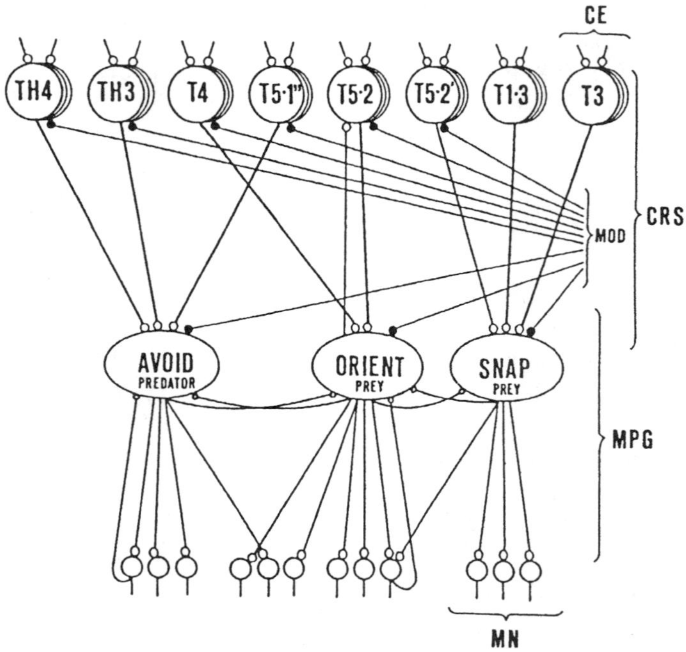

Figure 24. Concept of sensorimotor codes in command releasing systems. Triggering a motor pattern generating circuit (MPG) for an action pattern (e.g., orient-prey, snap-prey, or avoid-predator) requires inputs of a certain combination of so-called specialized neurons functioning as command elements (CE). Such a combination provides the sensorimotor code for the appropriate command releasing system (CRS). Modulatory influences (MOD) probably enter a CRS at the level of particular command elements and the MPG. Upon activation by the CRS, a specific "routine" is called into play in the appropriate MPG to produce a spatiotemporal pattern of excitation and inhibition to coordinate the activities in the corresponding motoneuronal pools (MN). Among the various CRSs, and MPGs, respectively, certain neurons may be shared. A circle stands for a set of neurons belonging to the same class; an oval symbol stands for a circuit; small circles indicate excitatory or inhibitory inputs which are not differentiated here. Compare with Figs. 2 and 20. (After Ewert 1980, 1985).

Additional or alternative elements of CRSs remain to be discovered. So we will have to consider as yet unidentified "space-coding" tegmental (see Grobstein, this volume) and "color-coding" thalamic (see Kondrashev 1987) neurons. As an example of alternative (or substitute) command elements, it is suggested by Grobstein et al. (1983) that following bilateral tectal ablation, neurons of the lateral torus semicircularis mediate snapping in response to tactile stimulation. We conjecture that other types of behavior require different *sensorimotor codes* and are thus released by distinct combinations of command elements, in which certain elements of prey-catching CRSs may be shared: For example, a predator traversing the visual field might yield the command [(TH3)(T5.1)(TH4)] to avoid; a startle response to a looming large object could be triggered by [(T5.4)(TH4)] or [(T5.4) (T4)]. Whereas orienting toward "prey" may be released by [(T5.2)(T4)], the orienting command for a different goal - e.g., toward "mate" during spring time - could be issued by [(T5.1)(T4)], which involves modulatory inputs from motivational systems.

6.3.2 "Sensorimotor" Terminology

Finally, some comments on "sensorimotor" are in order. *Sensorimotor functions* involve neural mechanisms that link afferent sensory processes to motor pattern generation. *Sensorimotor transformation* refers to the operations that translate a specific pattern of sensory input into an appropriate spatiotemporal pattern of excitation and inhibition in motoneurons (the motor pattern) necessary to activate and coordinate the muscle contractions for the corresponding action pattern. The term *sensorimotor coordination* has a broader applicability in that it also considers sensory feedback loops necessary for the coordination of actions (for details see Bässler and Berkinblit et al., this volume). The "specialized" neurons which encode the outcome of feature/space-related sensory information processing and transmit the resulting informations to the "motor pattern generating circuits" (MPGs) are part of the *sensorimotor interface* (see also Scheich 1983). *Sensorimotor integration* points to the modes by which processed sensory information from different sources about the same target - mediated by specialized (projective) neurons - is integrated by neurons of the MPG, for example, in a manner of an AND- or an OR-gate. An MPG is triggered (like a safe is unlocked) according to a *sensorimotor code* (Fig. 24), requiring concurrent activities of specialized neurons in a specific combination, cooperating for a common purpose. With respect to their efferent effects, the specialized neurons of such a combination are acting as "command elements" and together form a "command releasing system" CRS which is the central core of the "releasing mechanism" RM, also called the "releasing schema." The prefix "command" points to the initiating, primarily ballistic function of the system and "releasing" refers to its coded property (concept of "sensorimotor code").

6.3.3 "Schema" Terminology

Expressing our concepts in the language of perceptual and motor schemas (see Arbib 1987a; Arbib and Cervantes-Pérez, this volume), we may say that Fig. 2 introduces four motor schemas of prey-catching. In Section 2 we show that these need not occur in this generic order - rather, each proeeds to completion (of the control program rather than of the motor activity it elicits) to be followed by perceptual schemas that will determine which motor schema is to be executed next. Motor schemas may take the form of "compound motor coordinations" (such as a programmed detour route to prey, or the snap-gulp-mouthwipe sequence) which comprise a set that will proceed to completion without intervening perceptual tests in such a manner that: (i) schema *A* proceeds to completion, (ii) completion of schema *A* triggers initiation of schema *B*, or (iii) schema *A* passes a parameter x to schema *B*.

The "sensorimotor code" of a CRS (Fig. 24) embodies a perceptual schema that exists for only one purpose, to determine the conditions for the activation of a specific MPG embodying a motor schema, and the CRS must also ensure that the resultant movement is directed toward the prey. A schema and its instantiation usually appear to be coextensive, i.e., instantiation of a schema appears to be identifiable with "appropriate" activity in certain populations of neurons of the toad's brain (Fig. 24) whereby each schema may involve several cell types/brain regions, while a given cell type/brain region may be involved in several schemas. This raises various interesting questions for experimental research: (1) How could one detect that schema *A* had been instantiated in network N when schema *A* can have multiple instantiations? (2) Since each cell of Fig. 24 corresponds to an array of cells in the toad's brain and since cells in each array have overlapping receptive fields, the question arises whether a weak distributed pattern of activity proves as effective a trigger as a stronger localized activity, which indeed seems to be the case (cf. "effective displacement" in Section 3.5.1; Ewert 1969a).

6.4 Modulatory Influences

Prey-catching behavior in toads is influenced by (i) motivation (changes attributable to an alteration in the internal, e.g. homeostatic, environment of the animal), (ii) attention (changes attributable to processes responsible for alertness and arousal), and (iii) learning (changes attributable to the history of previous exposure to stimuli). Although motivation is not just a matter of homeostasis, the latter as the dominant concept in regulatory biology has loomed large in theoretical concepts of motivation (e.g., see Stellar 1954; Zucker 1983). According to Fentress (1983) motivation (a) allows animals to link physically different classes of action into functionally coherent ensembles, (b) integrates the same class of actions into different functional contexts, and thus (c) frees animals from enslavements of simple stimulus-response pairings. Modulation of stimulus-response relationships in prey-catching could be brought about through modulatory circuits acting on a stimulus-response mediating CRS.

6.4.1 Motivation

Prey-catching commands cannot be executed if the toad's motivation is not appropriate. These modulatory long-term adjustments may be acting on the CRS by influencing the T5.2 prey-selective command elements (Fig. 24). In summer, during the hunting season, responses of T5.2 neurons to prey cues are very weak in satiated animals, suggesting that these neurons receive inputs related to the nutritional state. In winter, during the hibernation period, both prey-catching motivation and the activity of class T5.2 neurons are extremely low in laboratory animals (Ewert 1984a), suggesting hypothalamo-hypophysial influences (Hanke 1976; Dierickx 1969) known to project to deep tectal layers (Wilczynski & Northcutt 1977). In spring, during the mating season, the threshold for visually released escape is high and prey-catching appears to be turned off, whereas visually guided appetitive orientation and approach behaviors by the mating-motivated male toward the female are turned on (Heusser 1959, 1960; Kondrashev 1976, 1987). Presumably, the T5.2 prey-selective command elements in the CRSs for prey-related orienting and approaching are functionally replaced by T5.1 projective neurons of layer 6/7, tuned in this context for partner detection. (For an alternative explanation see Betts, this volume). It has not yet been possible to identify responses of class T5.2 neurons in male or female toads during the mating period (Ewert 1984a), which suggests that these neurons receive information on changes in the balance of sex hormones (Guha et al. 1980).

6.4.2 Attention

Prey-catching action patterns are also subject to short-term attentional shifts which are reflected in the activity of T5.2 neurons (Schürg-Pfeiffer, this volume). According to a definition used by Laming (this volume) *attention* refers to a specific registration response towards the stimulus source. In such a context we may say that T5.2 neurons deliver commands related to goal oriented, prey-selective attention. (For details see Schürg-Pfeiffer, this volume).

6.4.3 Visual Habituation

Longterm repetitive presentation of the same visual prey object leads to stimulus-specific after-effects (Ewert 1967a; Finkenstädt & Ewert 1988). These are presumably associated with a "negative value" by telencephalic "habituating neurons" and translated into a build-up of inhibition on tectal T5.2 neurons which is mediated via thalamic relay neurons. Computer simulations with model networks suggest that in the course of such a hypothesized process, information related to an "image" of stimulus cues is stored in prosencephalic structures that can be compared with the present stimulus (Lara & Arbib 1985; see also Lara, this volume). 2DG studies and lesion experiments show that the telencephalic posteroventral medial pallium (vMP), for example, plays an important role in long-term habituation which can be regarded as non-associative learning (see Finkenstädt, this volume).

6.4.4 Associative Learning

Prey-selecting properties can be modified or extended through conditioning (cf. Section 3.4.2). There are different ways of explaining these behavioral changes in terms of circuit modulations: (i) During "visual/visual" (worm/hand) conditioning in the course of hand-feeding (Fig. 25A) there could be long-term modifications of the properties of T5.2 prey-selective neurons if hypothetical "associative *As* neurons," receiving repeated concurrent inputs from T5.2 (responsive to worm) and TH3 (responsive to hand), developed a sensitivity to TH3 input and then, by recurrent inhibition, attenuated the inhibitory influence of TH3 on T5.2. This might explain why $D_{W,A}$ values are decreased and why large nonprey objects (e.g., hand) elicit prey-catching. (ii) The properties of T5.2 neurons could be changed temporarily by hypothetical "associative *As'* neurons" (e.g., receiving excitatory visual T5.2 and olfactory input) if, and only if, the conditioned stimulus (prey odor) were present (Fig. 25B). This might explain why configural prey discrimination is impaired in the presence of a familiar prey odor. (iii) The existing prey recognition properties could be extended by an "associative As'/mT5 unit" of which hypothetical modulatory mT5 command elements were activated only in the presence of the associated (e.g., olfactory) stimulus and, while inhibiting T5.2, they might take over their function (Fig. 25C). It is important to note from 2DG and lesion studies that the medial pallium and the lateral amygdala are significantly involved in associative learning. (For neurophysiological details see Finkenstädt, for a "T5-base modulator" hypothesis see Betts, and for a learning model see Lara, this volume).

6.4.5 Internal Loop Operations

What are the potential routes through which modulatory signals could reach the tectum? Various diencephalic structures are involved in circuits (Fig. 26A) by which striatal and pallial telencephalic influences could determine or modulate feature-selective properties of tectal CRSs (Ewert 1967a, 1980, 1987a; Kicliter & Ebbesson 1976; Wilczynski & Northcutt 1977, 1983a,b; Northcutt & Kicliter 1980; Neary & Northcutt 1983; Reiner et al. 1984). One such internal loop (1) is concerned with the

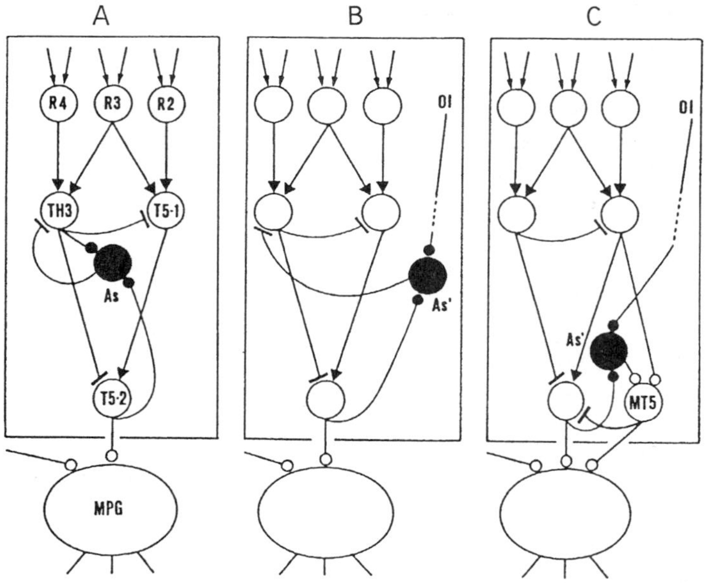

Figure 25. Speculative approach to the adaptive aspects of feature-selective properties in CRSs. Associative conditioning might act on a CRS either by influencing the property of T5.2 command elements in changing the network characteristics of the corresponding prey-feature analyzing neuronal assembly (A,B) (cf. Figs. 20Aa) or by functionally replacing T5.2 in activating substitute modulatory command elements mT5 (C). As and As', hypothetical associative neurons; Ol, olfactory input; mT5, hypothetical command element mediating "associative" messages; arrows: excitatory inputs; small open circles: excitatory inputs to an AND gate; small solid circles: associative excitatory inputs; lines with cross bars: inhibitory inputs. For explanation see text. (From Ewert 1987a).

striatum ("paleo-striatum") and involves the lateral forebrain bundles: The retina projects via the optic tectum (OT) to the lateral anterior thalamic nucleus (La) and from there to the striatum (STR), which in turn feeds back to the optic tectum via the pretectal Lpd/P nuclei or tegmental cell groups. Another loop (2) is concerned with the medial pallium ("primordium hippocampi" of Herrick 1933) and involves the medial forebrain bundles: The retina projects directly (or via the optic tectum and the pretectal P nucleus) to the anterior thalamus (AT) and from there to the medial pallium (MP), which in turn feeds back to the tectum, either in the same way (loop 2a), or via the preoptic area (PO) and the hypothalamus (HY), sending axons to deep tectal layers (loop 2b).

Loop operations might be concerned with: (a) the integration of external and internal excitatory inputs (e.g., "priming" a CRS, Fig. 26Ba), (b) the build-up of stimulus-specific inhibition (e.g., "screening" a CRS, Fig. 26Bb), (c) polysynaptic disinhibition (e.g. "arousing" a CRS, Fig. 26Bc), (d) disynaptic disinhibition (e.g., "gating" a CRS, Fig. 26Bd), (e) associative disinhibition (e.g., "modifying" a CRS, Fig. 26Be), and (f) associative inhibition (e.g., "extending" a CRS by replacing an innate feature-selective command element, Fig. 26Bf). We suggest that the innate prey selecting properties are mediated by loop 1 (as illustrated in the wiring shown in Fig. 26Bd), whereas the modulating influences are integrated through loops 2a and b (e.g.,

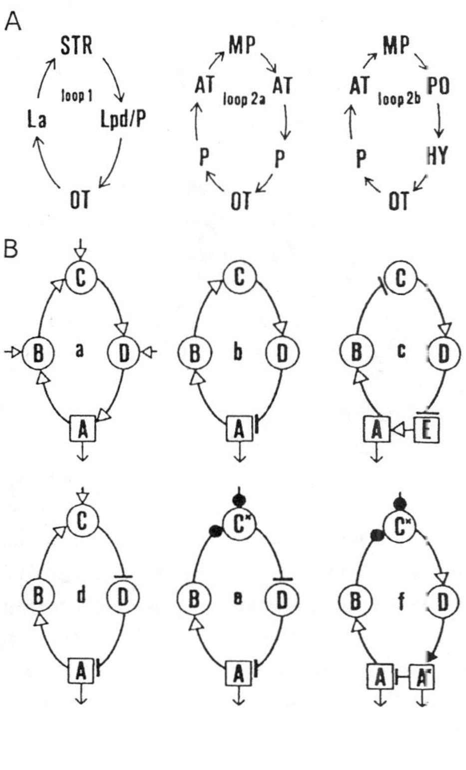

Figure 26. Concept of "internal loops" by which prosencephalic structures might determine the properties of tectal neurons and modulate their respective activities. (A) Anatomical substrates (according to data by Wilczynski & Northcutt 1977; 1983a,b; Neary & Northcutt 1983). HY, hypothalamus; PO, preoptic area; for other abbreviations see Fig. 10. Arrows indicate projecting pathways. (B) Simple diagrams illustrating some possible loop-operations in hypothetic neural structures *B,C,D,E* exerting effects on *A*: (a) integration ("priming"), (b) build-up of inhibition ("screening"), (c) polysynaptic disinhibition ("arousing"), (d) disynaptic disinhibition ("gating"), (e) associative disinhibition ("modifying"), or (f) associative inhibition ("replacing/extending"). Arrows: excitatory influences; small solid circles: associative concurrent excitatory influences; lines with cross bars: inhibitory influences. *A* may stand here for tectal T5.2 command elements and, *A** for a substitute mT5 command element; *C* refers to telencephalic, *B* and *D* to diencephalic, and *E* to tegmental reticular structures; retinal inputs are not shown.

according to 26Bb,c,e, or f). Needless to say, the diagrams shown in Fig. 26B are rather simplistic representations of complex processes, but they are nevertheless useful conceptual guides for experimental approaches (e.g., see Fig. 25; Finkenstädt & Ewert 1985, 1986, 1987; Finkenstädt et al. 1986; Finkenstädt, this volume).

6.5 What General Properties Do CRSs Have?

CRSs are the neurophysiological equivalents of RMs. They address sensory processes to motor functions in that (i) they combine stimulus feature analysis and localization, (ii) they link sensory analysis and command functions by activating the motor pattern generating system in response to specific sensory input, (iii) they may participate in the temporal sequence of certain motor patterns through feedback of the motor pattern generating network, and (iv) they integrate the input of modulatory circuits. Properties (ii)-(iv) have been analyzed in great detail for command functions in invertebrates (e.g., see Davis & Gillette 1978; Gillette et al. 1978; Davis & Kovac 1981).

The concept of "command neuron" (Wiersma & Ikeda 1964) has received support from intracellular recording and stimulation showing that a single identified cell triggers a fixed action pattern (Hoyle 1977; Davis & Kovac 1981; Eaton 1983). Kupfermann & Weiss (1978) proposed the concept of "command system" (CS) for the release of those action patterns in which concurrent activations of several neurons of the same type (command elements) was necessary and sufficient to activate the respective motor pattern generating circuit. There are problems with the experimental usefulness of this concept because it is not yet technically possible to depolarize, hyperpolarize, or eliminate all the appropriate cells of a CS simultaneously in order to check its conformity with the "necessity"- and "sufficiency"-criteria. The present concept of "command releasing system" CRS (Ewert 1980) considers "sensorimotor codes" and thus refers to a system composed of differently specified command elements each corresponding to a small population of neurons. The CRS concept proposed here does not rely rigorously on the necessity- and sufficiency-definition in that it includes dynamic properties (state-dependence), alternatives (co-existing circuits), and modifications (adaptive mechanisms). A comparable concept is discussed by DiDomenico & Eaton (1987). CRSs have several important features:

(1) CRSs are economical, since certain command elements can be shared by different systems (Fig. 24). Feature-selective command elements (e.g., T5.2 neurons) participate in the different CRSs of goal-related action patterns (of prey-catching) and are thus essential and decisive for a behavioral sequence. State-dependent inputs to these neurons would be powerful tools to switch a behavior (e.g., prey-catching) *on* or *off*, while orienting toward and approaching other goals (e.g., a sexual partner) can be released by the same space-monitoring neurons in combination with other appropriate feature-selecting command elements.

(2) A command element does not necessarily correspond to a single neuron; it can be a small population of adjacent neurons of the same class which have similar thresholds and whose receptive fields partly overlap. This assumption has two

desirable features: First, it is extremely unlikely that a prey object activates just one T5.2 neuron. Second, stimulus localization by a T5-type neuron with an ERF of 20-30° would be inaccurate compared with the information provided by cooperating adjacent T5-type neurons whose receptive fields include a small common area.

(3) Command elements may have disproportionate distributions in visual feature maps. The frontal visual field requires more discrete perceptual operations than the extrafrontal field. This may explain why the frontal visual field in the lateral tectum displays a greater density and variety of tecto-bulbar/spinal projective neurons (Weerasuriya & Ewert 1981; Satou & Ewert 1985); this would also agree with the idea of a "central fovea."

(4) The outputs of a distinct combination of command elements "converging" to meet the criterion for a CRS need not impinge on one cell, but may do so on different cells of the motor pattern generating network. Concurrence of appropriate activities in a CRS is required to bring the central pattern generator into action (Schwippert et al., this volume).

(5) Estimating absolute object size is an important function of the CRS involving different depth cues in the visual field. This information comes from the lens accommodation mechanism and/or binocular cues (Arbib, this volume).

(6) Different CRSs interact if toads have to reach the prey by making detours around stationary obstacles, a problem that has been investigated in detail by Ingle (1971, 1977, 1980, 1983) and Collett (1983; see also Arbib and Arkin, this volume).

(7) CRSs are to some extent plastic. Alternative or substitute command elements - mediating the output of co-existing circuits (functional units) - will make a system resistant to disturbances.

(8) CRSs are adaptive. Influences from modulatory circuits involving diencephalic and telencephalic structures make a CRS adaptive to internal and external signals. Modulatory inputs, for example, may act directly on feature-selecting command elements or on certain neurons of the circuitry of the corresponding functional units that determine the feature-selective property (e.g., see Fig. 25A). It is likely that certain modulatory influences (e.g., those related to circadian rhythms) also act on the premotor network and/or the motor neurons themselves (Parent 1973; see also Weerasuriya, this volume).

Finally, some general comments on the link between "sensory space" and "motor space" and the corresponding "motor routines." The networks responsible for motor pattern generation are situated in the medial reticular formation of the medulla and the central grey of the spinal cord, both operating in connection with rostral (e.g., tegmental) brainstem structures (Nieuwenhuys & Opdam 1976; Székely & Czeh 1976; Weerasuriya 1983; Masino & Grobstein 1985; Comer et al. 1985; Grobstein, this volume). The different programs for appropriate spatiotemporal coordination of the motoneurons are partly built in these networks (Székely & Czeh 1976). If adequate signals from a CRS concurrently reach certain elements of the corresponding circuitry, a specific routine is called into play (see contributions by Schwippert et al. and Weerasuriya, this volume). Problems arise with the question of how sensory space is translated into appropriate motor space (Roucoux & Crommelinck 1987; see contributions by Grobstein and Mittelstaedt, this volume). Other problems are concerned with locomotion routines (Bässler, this volume). There are still many open questions about coordinating principles serving to link motor pattern generating circuits (see contributions by Bässler and Berkinblit et al., this volume).

7 Concluding Remarks

7.1 Stages of Parallel/Hierarchical Processing

The release of visual behavior in toads involves various stages of parallel/hierarchical information processing. The neuroanatomical substrates for parallel distributed processing are central visual structures that evaluate a stimulus under different aspects (Fig. 27). At an initial stage, retinal R2 and R3 inputs to tectal intrinsic excitatory circuits are evaluated in terms of the area of a moving object and its extension in the direction of movement. In *parallel* to this, retinal R3 and R4 inputs to pretectal circuits are evaluated in terms of texture, area and an object's extension perpendicular to the direction of movement (Fig. 20). *Convergent* inhibitory pretectal and excitatory tectal inputs to neurons in central tectal layers determine the property of T5.2 neurons specified (i) to discriminate dynamic configural features, (ii) to select object from background texture, and (iii) to distinguish between object motion and self-induced motion in a textured environment. *Parallel* processing of visual input in basically the same network expressed by other neurons (e.g., classes T5.1, T5.3, T5.4) may characterize (or contribute to the characterization of) other objects. In *parallel* visual input computed in an anteriorthalamic relay may amplify these feature analyzing processes with special reference to the frontal visual field. In *parallel* to this, visual input is processed in tectal, tegmental, and isthmic structures in conjunction with other information (e.g., from lens accommodation, triangulation, proprioception) to judge objects in space. In *parallel*, visual input is evaluated in various diencephalic/telencephalic structures under the aspects of motivation, novelty, and past experience.

The concept of coded commands in a CRS stresses that firing of various classes of efferent neurons in certain combination characterizes objects in space, and that these neurons as command elements cooperatively select the appropriate goal-directed action pattern. The same class of command elements in different *combinations* with other command elements may take part in CRSs for different action patterns; the same action pattern can be triggered by CRSs distinguished by the combination of their command elements (cf. "multiple instantiations of perceptual schemas" in Arbib, this volume). An advantage of the combinatorial nature of codes could be a relatively high potential of information content with respect to "neural parsimony." Commanding behavior according to a code is at another stage a property of the decoding premotor circuitry (Schwippert et al., this volume). For a discussion of the concept of command function see also Comer (1987) and DiDomenico & Eaton (1987).

In this multi-stage parallel/hierarchically organized macro-network (Figs. 20, 24-27) information flow does not proceed just from a level of lesser to a level of greater selectivity and specialty. Various *feedback loops* within and among these stages determine *modulatory* (e.g., tecto-prosencephalo-tectal loops), *warming-up* (e.g., tecto-reticularbulbo-tectal loops), and *spatiotemporal* (e.g., tecto-tegmental/bulbo-tectal loops) properties (see also Finkenstädt, Gonzalez-Lima, Grobstein, Lázár, and Schwippert et al., this volume). Feature selective T5.2 neurons, integrated in this macro-network, exhibit most of these characteristics (see Schürg-Pfeiffer, this volume). An important property of this macro-network is a degree of plasticity so that, for example, modulatory influences can modify object recognition in adaptation to internal (motivation) and external (conditioning) signals. A toad may thus snap

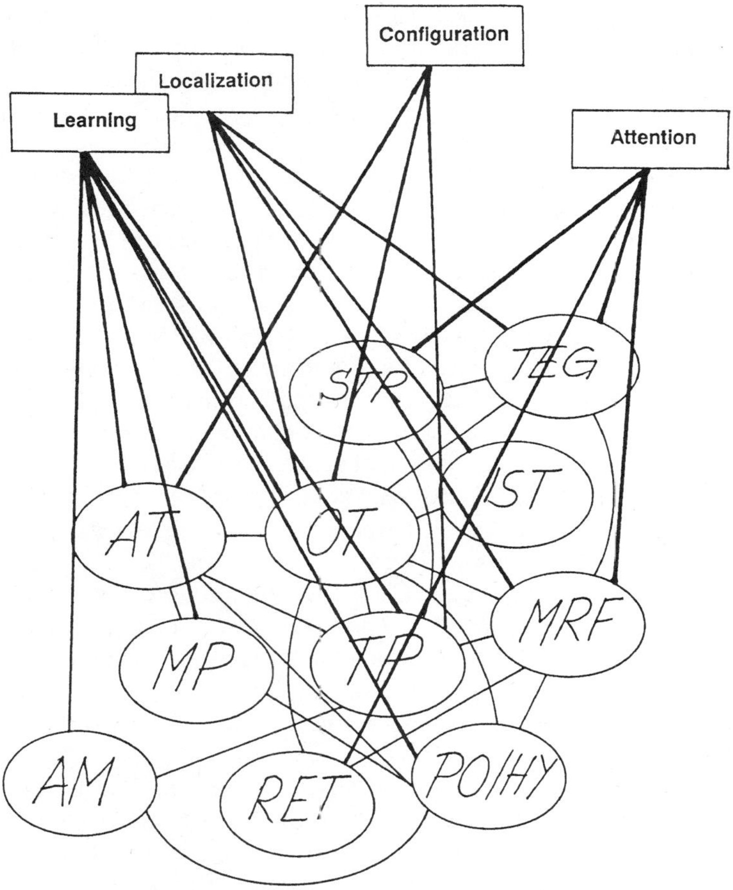

Figure 27. Concept of distributed properties in a macro-network. Visual information is evaluated under various aspects (see boxes) involving different, partly interacting brain structures (see oval symbols). AM, nucleus amygdalae; AT, anterior thalamus; IST, nucleus isthmi; MP, medial pallium; MRF, bulbar medial reticular formation; OT, optic tectum; PO/HY, hypothalamic/preoptic areas; RET, tegmental reticular formation; STR, striatum; TEG, tegmentum; TP, thalamic pretectal region. This simplified diagram emphasizes some of the connectivity under investigation in this paper, but it does not acknowledge all assignments (see thick lines), all structures (oval symbols), and all interactions (thin lines).

even toward a swarm of prey, or a big moving object if the motivation is appropriate. We are reminded of Bullock's (1961) concept of *multiple-input metastable feedback loop* as a functional unit responsible for recognition in an analyzer made of neurons. For a dynamic model see Betts, this volume.

7.2 Comparisons

In all vertebrates so far investigated, the optic tectum (or superior colliculus, respectively) integrates inputs from various sources in order to link sensory, motivational, associative, and motor functions for orientational behavior. This structure processes information by intrinsic circuits which are subject to shaping, gating, and modulatory influences. These extrinsic influences arrive via di- or polysynaptic pathways from striatal (basal ganglionic) and limbic (medial pallial, hippocampal) nuclei thus forming tecto-prosencephalo-tectal circuits.

7.2.1 Goal Directed Attention

It is suggested that certain *striatal* structures are involved in property-shaping and response-gating effects on certain tecto-bulbar/spinal output neurons to determine whether or not a behaviorally meaningful stimulus can be taken as a target for an orienting movement, referring to goal directed attention (e.g., see Ewert 1967a, 1984a, 1987e; Chevalier et al. 1984; Finkenstädt et al. 1986). In anurans, the putative anatomic substrate is provided by a *striato-pretecto-tectal* and a small *striato-tegmento-tectal* pathway (Fig. 26A, loop 1; Wilczynski & Northcutt 1983a,b; Neary & Northcutt 1983). In reptiles and birds, the former (striato/pallido-pretecto-tectal) is prominent while another pathway takes the form of a small *striato-nigro-tectal* route (Reiner et al. 1984). According to Wilczynski & Northcutt (1983), such a "dual-route" system is the ancestral tetrapod condition. In mammals, the comparable pathway mediated by the substantia nigra is a prominent circuit in this context (besides the main basalganglio-ventrothalamo-motocortical pathway), whereas the one via the pretectum may have been "lost" (Wilczynski & Northcutt 1983, p. 342). Despite the absence of homology between the anuran striato-pretecto-tectal pathway and the mammalian striato-nigro-tectal pathway, the disinhibitory properties of both (as illustrated in the wiring shown in Fig. 26Bd) exhibit some interesting similarities which suggests comparable operational principles in subserving the animal's specific behavioral requirements (see Ewert 1987e for a review).

From a phylogenetic point of view, comparative immunocytochemical studies of pretectal nuclei are promising since they show that bird's lateral spiriform nucleus SpL and reptile's dorsal nucleus of the posterior commissure nDCP (equivalent to the anuran's pretectal P nucleus) harbor enkephalin (*ENK*) containing neurons terminating in deep tectal layers close to the tecto-bulbar/spinal projection neurons (Reiner et al. 1980, 1982; Reiner pers. communication 1984; Karten et al. 1982). The bird's nucleus pretectalis PT and the reptile's dorsal pretectal nucleus NPd (equivalent to the anuran's pretectal Lpd nucleus), on the other hand, harbor neurons that produce an avian pancreatic polypeptide (*APP*)-like substance giving rise to scattered *APP*ergic terminals in more superficial retinorecipient tectal layers. Presumably the frog's pretectal Lpd is the source of *APP*ergic terminals which have been identified by Kuljis & Karten (1982) in frog's superficial tectal layers, whereas pretectal P is the source of *ENK*ergic terminals found in frog's deep tectal layers. Influences from the striatum via lateral forebrain bundles to the P nucleus (cf. Fig. 26A, loop 1) involve *ENK*ergic transmission (Merchenthaler in Lázár, this volume). Since pretectal nuclei and tectal projections are also present in mammals, a comparative approach to pretectal functions is challenging. Reiner et al. (1984), however, infer that the pretectal SpL/nDCP system was greatly de-emphasized in the line of evolution leading to modern mammals.

7.2.2 Arousal, Motivation, and Learning

In anurans, the sensory/associative/motivational/motor interface is mediated by *medialpallio-hypothalamo-tectal* pathways (Fig. 26A, loops 2a,b) that are connected to other homologous structures of the "limbic system" (e.g., nucleus amygdalae). The posterior medial pallium is significantly involved in arousal, stimulus-specific habituation, and associative conditioning (see Finkenstädt & Ewert 1985, 1987, 1988; Finkenstädt, this volume). Herrick's (1933) term "primordium hippocampi" for this brain structure is thus justified not only from anatomical homology but also with respect to functional properties. In mammals, too, the hippocampus together with its limbic connections influences (modifies) stimulus-response mediation at the collicular level:

> "Hippocampus is not concerned with the inital stages of involuntary orienting to particular spatial locations, but rather with more subtle aspects of stimulus selection and scanning involving much preprocessed information... probably relating to subtle stimulus characteristics - stimulus ambiguity, reward significance, and perhaps spatial position"... [hippocampus also functions to] "detect novelty and, as appropriate, to 'block' or 'deblock' the arousal system and so to alter motor behaviour" (Foreman & Stevens 1986, p. 108, 109).

In mammals, the prominent sensory/associative/motivational interface is represented by *cortico-limbo-cortical* circuits (e.g., see Feigenbaum et al. 1986; Rolls 1987a,b). For a discussion on evolutionary aspects in this context see Ehret (1987), and the contributions by Gonzalez-Lima and Finkenstädt (this volume).

7.3 From Instinct to Cognition

The operations between sensory analysis and motor response - represented by sensorimotor codes - are of general importance, since these are in a broad sense related to *cognition* (Ewert 1987c, 1988). Asking for the phyletic roots of cognitive processes at *algorithmic* levels, it is likely that the rules according to which significant coincidences of events determine a toad's behavioral strategy (Fig. 2) take advantage of *sensorimotor codes* (Fig. 24) and the decoding premotor circuitry. These can be interpreted as "species-specific cognition" (= *instinct*). The information processing provided by modulatory circuits enabling CRSs to behave adaptively in relation to past and present events includes further algorithmic levels. Paillard (1987) suggests that the tuning of *sensorimotor instruments* can be regarded as a primordial form of an organism's adaptation to environmental constraints, and that this has prepared the way for the emergence of *cognitive instruments* dealing with the control of perception and action. The way in which fundamental algorithms of cognition (related to perception, memory, retention, revival, retrieval) are implemented in the human brain may therefore become to some extent understandable in terms of the evolution of related structures and functions of other vertebrate animals' brains.

Principles of cognition rest upon and express relations of the universal elements in conscious experience, viz., space and time. Cognition as the process by which knowledge is gained about the world refers to internal activities in response to antecedent (internal or external) signals which they classify, transform, and coordinate prior to initiating actions (at least potentially), processes that are influenced by memories of past experiences and expectations of future ones. Context dependent "ensemble coding" may underlie formation of abstract *concepts* and combination of *ideas* which characterizes human cognition. We are thus reminded of David Hume who has argued that a theory of knowledge - according to which perception and judgement of relations in approximate exactness were essential for cognition - will acquire additional authority if it explains related phenomena in animals - thus implying comparable basic neural mechanisms and suggesting that human's superiority is in some respect a matter of degree (Hume 1902). The discontinuity obviously results from language which involves special tuning of neural systems that allows stages of intelligence only humans can reach. Anything depending on language, its symbolic structure, and its logic may not be available to non-verbal species, hence suggesting perceptual processes built up by linguistic competence and experience. The question is open as to whether these processes are added to, or develop from, or take advantage of neural organizations and representations which arise generally in the brains of vertebrates.

7.4 Modeling and Engineering

Configural stimulus discrimination in toads (Ewert 1968, 1969a) was subject to various model approaches using *systems-theory* in homogeneous two-dimensional neural networks (Ewert & v. Seelen 1974), *computer simulation* in three-dimensional networks considering neuroarchitecture (see Arbib, Betts, and Cervantes-Pérez, this volume), and *Gaussian functions* with special reference to on/off channels (an der Heiden, this volume). Speculating on a technological application of principles that underlie toad's pattern recognition it may be possible, for example, to improve the codes and the decoding systems used for identification of industrial products. The present *bar codes* (Fig. 28Aa) take advantage of scanning the parameter $x1_1$ (cf. Section 3.3.2.) of sequential bars. A *figure code* (Fig. 28Ab), we are suggesting here, would consider $x1_1$ also in relation to $x1_2$ and thus

(a) involve a higher information potential,

(b) permit a fast evaluation of the sequential dynamic configurations by parallel processing $x1_1$ and $x1_2$,

(c) include the possibility of modification by changing the $x1_1/x1_2$ algorithm, and

(d) operate in correspondence with symmetric decoding processors. (For a mathematical treatment of this problem see Ewert & v. Seelen 1974).

In particular the last aspect (d) is of general technological interest. We illustrate this with a special example. Suppose an engineer was asked to design a processor that

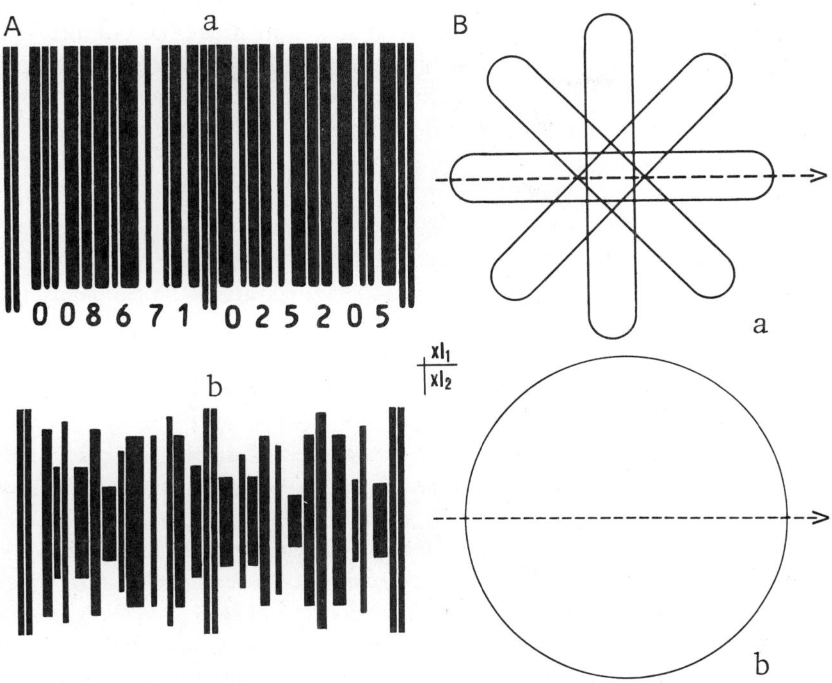

Figure 28. (A) Object identifying codes: *bar code* [a], *figure code* [b]. (B) Decoding "asymmetric" [a] and "symmetric" [b] processors for various scanning directions in the orientation of the bar/figure sequence. (Note that the principle of dynamic figure code could also be applied to industrial robots for the recognition of objects moving on assembly lines).

selects elongated shapes traveling in the orientation of their longer axes [W-configuration] from the same shapes whose longer axes are oriented perpendicular to the direction of movement [A-configuration] (cf. Fig. 6A). The first design that comes to mind is *explicit* processing of stimulus orientation with the aid of "asymmetric processors." To guarantee movement-direction invariance (Fig. 6A), however, this would require an enormous amount of processing steps presuming a hardware comparable to mammalian cortical "orientation columns." In amphibians evolution obviously sidestepped the need for explicit processing in that a relatively simple economic principle permits *implicit* computation (cf. p.90) of the spatiotemporal stimulus features by means of a few interacting "radially symmetric processors." This principle is implemented by parallel computing an object's edge length in and perpendicular to the direction of movement. We take the lesson that the toad's brain has solved a problem of configural selection in a manner which an engineer might not have thought of by intuition, since the axes along which toads sort information are different from the ones that we might first imagine (Stevens 1987).

Acknowledgements. Substantial portions of this paper have appeared in somewhat different form in *The Behavioral and Brain Sciences* 10: 337-405 (1987) and are used here with kind permission of the editor Dr. Stevan Harnad. I am indebted to Mrs. Ursula Reichert for text processing, and Mrs. Gisela Burggraf, Mrs. Gudrun Frühauf, and Mrs. Gerda Kruk for technical assistance. The author's research was supported by grants of the Deutsche Forschungsgemeinschaft, the Fritz Thyssen-Stiftung, the Foundations' Fund for Research in Psychiatry, and the University of Kassel.

8 References

Akert K (1949) Der visuelle Greifreflex. *Helv Physiol Pharmacol Acta* 7: 112-134

Antal M, Matsumoto N, Székely G (1986) Tectal neurons of the frog: intracellular recording and labeling with cobalt electrodes. *J Comp Neurol* 246: 238-253

Arbib MA (1982) Modelling neural mechanisms of visuomotor coordination in frog and toad. In: Amari S, Arbib MA (eds) *Competition and cooperation in neural nets.* Springer-Verlag, Berlin Heidelberg New York, pp 342-370

Arbib MA (1987a) Levels modeling of mechanisms of visually guided behavior. *Behav Brain Sci* 10:407-465

Arbib MA (1987b) *Brains, machines, and mathematics.* Springer-Verlag, New York Berlin Heidelberg London Paris Tokyo

Autrum H (1959) Das Fehlen unwillkürlicher Augenbewegungen beim Frosch. *Naturwiss.* 46: 436

Baerends GP (1976) The functional organization of behaviour. *Anim Behav* 24: 726-738

Baerends GP (1987) Ethology and physiology: a happy mariage. *Behav Brain Sci* 10: 369-370

Barlow HB (1953) Summation and inhibition in the frog's retina. *J Physiol (Lond)* 173: 377-407

Barlow HB (1985a) The twelfth Barlett memorial lecture: The role of single neurons in the psychology of perception. *Quart J Exp Biol* 37(A): 121-145

Barlow HB (1985b) The role of nature, nurture, and intelligence in pattern recognition. Pontificiae Academiae Scientiarum Scripta Varia 54. *Ex Aedibus Academicis in Civitate Vaticana Roma* 1985

Bechterev W (1884) Über die Funktion der Vierhügel. *Pflügers Arch* 73: 501-534

Beck A, Ewert J-P (1979) Prey selection by toads (*Bufo bufo* L) in response to configurational stimuli moving in the visual field z,y-coordinates. *J Comp Physiol* 129: 207-209

Bieger D, Neuman RS (1984) Selective accumulation of hydroxytryptamines by frogs tectal neurons. *Neurosci* 12: 1167-1177

Birukow G, Meng M (1955) Eine neue Methode zur Prüfung des Gesichtssinnes bei Amphibien. *Naturwissenschaften* 42: 652-653

Borchers H-W (1982) Correlation between behavior patterns and single unit responses from the optic tectum in the freely moving toad (*Bufo bufo*). In: Trappl R, Ricciardi L, Pask G (eds) *Progress in biocybernetics and systems research Vol 9.* Hemisphere Publication Corp/McGraw-Hill, Washington London, pp 109-117

Borchers H-W, Ewert J-P (1979) Correlation between behavioral and neuronal activities of toads *Bufo bufo* (L) in response to moving configurational prey stimuli. *Behav Processes* 4: 99-106

Borchers H-W, Pinkwart C (1983) A telemetry system for single unit recording in the freely moving toad (*Bufo bufo* L) In: Ewert J-P, Capranica RR, Ingle DJ (eds) *Advances in vertebrate neuroethology.* Plenum Press, New York, pp 603-612

Borchers H-W, Burghagen H, Ewert J-P (1978) Key stimuli of prey for toads (*Bufo bufo* L): configuration and movement patterns. *J Comp Physiol* 128: 189-192

Borchers H-W, Schürg-Pfeiffer E, Megela AL, Ewert J-P (1983) Single neuron activity in the optic tectum of intact and thalamic-pretectal (TP)-lesioned behaving toads. *Neurosci Lett Suppl* 14: 36

Brower LP, Brower JVZ, Westcott PW (1960) Experimental studies of mimicry. 5: The reaction of toads

(*Bufo terrestris*) to bumblebees (*Bombus americanorum*) and their robberfly mimics (*Mallophora bomboides*), with a discussion of aggressive mimicry. *Amer Naturalist* 93: 343-356

Brown WT, Ingle D (1973) Receptive field changes produced in frog thalamic units by lesions of the optic tectum. *Brain Res* 59: 405-409

Brzoska J, Schneider H (1978) Modification of prey-catching behavior by learning in the common toad (*Bufo b bufo* L, Anura, Amphibia): changes in response to visual objects and effects of auditory stimuli. *Behav Processes* 3: 125-136

Bullock TH (1961) The problem of recognition in an analyzer made of neurons. In: Rosenblith WA (ed) *Sensory communication.* MIT Press, Cambridge Mass, pp 717-724

Bullock TH (1983) Implications for neuroethology from comparative neurophysiology. In: Ewert J-P, Capranica RR, Ingle DJ (eds) *Advances in vertebrate neuroethology.* Plenum Press, New York

Burghagen H (1979) *Der Einfluß von figuralen, visuellen Mustern auf das Beutefangverhalten verschiedener Anuren.* PhD Thesis, Univ of Kassel

Burghagen H, Ewert J-P (1982) Question of "head preference" in response to worm-like dummies during prey capture of toads *Bufo bufo. Behav Processes* 7: 295-306

Burghagen H, Ewert J-P (1983) Influence of the background for discriminating object motion from self-induced motion in toads *Bufo bufo* (L). *J Comp Physiol* 152: 241-249

Capranica RR (1983) Sensory processing of key stimuli. In: Ewert J-P, Capranica RR, Ingle DJ (eds) *Advances of vertebrate neuroethology.* Plenum Press, New York, pp 3-6

Capranica RR, Moffat AJM (1983) Neurobiological correlates of sound communication in anurans. In: Ewert J-P, Capranica RR, Ingle DJ (eds) *Advances of vertebrate neuroethology.* Plenum Press, New York, pp 701-730

Cervantes-Pérez F (1985) Modelling and analysis of neural networks in the visuomotor system of anuran amphibians. *COINS Tech Rep* 85-27. Univ of Mass, Amherst MA

Chevalier G, Vacher S, Deniau JM (1984) Inhibitory nigral influence on tecto-spinal neurons, a possible implication of basal ganglia in orienting behavior. *Exp Brain Res* 53: 320-326

Clairambault P (1976) Development of the prosencephalon. In: Llinas R, Precht W (ed) *Frog neurobiology.* Springer-Verlag, Berlin Heidelberg New York, pp 924-945

Collett TS (1977) Stereopsis in toads. *Nature* 267: 349-351

Collett TS (1983) Picking a route: do toads follow rules or make plans? In: Ewert J-P, Capranica RR, Ingle DJ (eds) *Advances in vertebrate neuroethology.* Plenum Press, New York, pp 321-330

Collett TS, Udin SB (1983) The role of the toad's nucleus isthmi in prey-catching behavior. In: Lara R, Arbib MA (eds) *Proceedings of the 2nd workshop on visuomotor coordination in frog and toad: models and experiments.* COINS-Techn Rep 83-19, Univ of Massachusetts, Amherst

Comer CM (1987) Sensorimotor functions: what is a command, that a code may yield it? A commentary. *Behav Brain Sci* 10:372

Comer C, Schotland J, Grobstein P (1985) Short and longterm effects of unilateral vestibular lesions on posture and orienting movements in the frog. *Soc Neurosci Abstr* 11: 289

Cott HB (1936) The effectiveness of protective adaptations in the hive-bee, illustrated by experiments on the feeding reactions, habit formation and memory of the common toad (*Bufo bufo bufo*). *Proc Zool Soc (London)* 1: 113-133

Czihak G, Langer H, Ziegler H (1981) *Biologie.* Springer-Verlag, Berlin Heidelberg New York

Davis WJ, Gillette R (1978) Neural correlate of behavioral plasticity in command neurons of *Pleurobranchaea. Science* 199: 801-803

Davis WJ, Kovac MP (1981) The command neuron and the organization of movement. *TINS* 4: 73-76

DiDomenico R, Eaton RC (1987) Toward a reformulation of the command concept. A commentary. *Behav*

Brain Sci 10: 374-375

Diebschlag E (1935) Zur Kenntnis der Großhirnfunktionen einiger Urodelen und Anuren. *Z Vergl Physiol* 21: 343-394

Dierickx K (1969) Hypothalamo-hypophysial regulation of food intake in *Rana temporaria*. *Gen Comp Endocrinol* 13: 361-366

Dieringer N (1986) Image fading - a problem for frogs? *Naturwissenschaften* 73: 330

Doty RW (1976) The concept of neural centers. In: Fentress JC (ed) *Simpler networks and behavior*. Sinauer Ass, Sunderland Mass

Doty RW (1987) Has the gready toad lost its soul, and if so, what was it? *Behav Brain Sci* 10: 375

Eaton RC (1983) Is the Mauthner cell a vertebrate command neuron? A neuroethological perspective on an evolving concept. In: Ewert J-P, Capranica RR, Ingle DJ (eds) *Advances in vertebrate neuroethology*. Plenum Press, New York, pp 629-636

Ebbesson SOE (1984) Evolution and ontogeny of neural circuits. *Behav Brain Sci* 7: 321-366

Ebbesson SOE (1987) Prey-catching in toads: an exceptional neuroethological model. A commentary. *Behav Brain Sci* 10:375-376

Eibl-Eibesfeldt I (1951) Nahrungserwerb und Beuteschema der Erdkröte (*Bufo bufo* L). *Behaviour* 4: 1-35

Eibl-Eibesfeldt I (1979) *Ethology, the biology of behavior*. Holt, Rinehart and Winston Inc, New York

Eikmanns K-H (1955) Verhaltensphysiologische Untersuchungen über den Beutefang und das Bewegungssehen der Erdkröte (*Bufo bufo* L). *Z Tierpsychol* 12: 229-253

Ewert J-P (1967a) Untersuchungen über die Anteile zentralnervöser Aktionen an der taxisspezifischen Ermüdung beim Beutefang der Erdkröte (*Bufo bufo* L). *Z Vergl Physiol* 57: 263-298

Ewert J-P (1967b) Aktivierung der Verhaltensfolge beim Beutefang der Erdkröte (*Bufo bufo* L) durch elektrische Mittelhirnreizung. *Z Vergl Physiol* 54: 455-481

Ewert J-P (1967c) Elektrische Reizung des retinalen Projektionsfeldes im Mittelhirn der Erdkröte (*Bufo bufo* L). *Pflügers Arch* 295: 90-98

Ewert J-P (1968) Der Einfluß von Zwischenhirndefekten auf die Visuomotorik im Beute- und Fluchtverhalten der Erdkröte (*Bufo bufo* L). *Z Vergl Physiol* 61: 41-70

Ewert J-P (1969a) Quantitative Analyse von Reiz-Reaktions-Beziehungen bei visuellem Auslösen der Beutefang-Wendereaktion der Erdkröte (*Bufo bufo* L). *Pflügers Arch* 308: 225-243

Ewert J-P (1969b) Das Beutefangverhalten zwischenhirndefekter Erdkröten (*Bufo bufo* L) gegenüber bewegten und ruhenden visuellen Mustern. *Pflügers Arch* 306: 210-218

Ewert J-P (1971) Single unit response of the toad (*Bufo americanus*) caudal thalamus to visual objects. *Z Vergl Physiol* 74: 81-102

Ewert J-P (1974) The neural basis of visually guided behavior. *Sci Amer* 230: 34-42

Ewert J-P (1980) *Neuroethology. An Introduction to the neurophysiological fundamentals of behavior*. Springer-Verlag, Berlin Heidelberg New York

Ewert J-P (1984a) Tectal mechanisms that underlie prey-catching and avoidance behaviors in toads. In: Vanegas H (ed) *Comparative neurology of the optic tectum*. Plenum Press, New York, pp 247-416

Ewert J-P (1984b) Behavioral selectivity based on thalamo-tectal interactions: ontogenetic and phylogenetic aspects in amphibians. A commentary. *Behav Brain Sci* 7: 337-338

Ewert J-P (1985) The Nico Tinbergen lecture 1983: concepts in vertebrate neuroethology. *Anim Behav* 33: 1-29.

Ewert J-P (1987a) Neuroethology of releasing mechanisms: prey-catching in toads. *Behav Brain Sci* 10: 337-405

Ewert J-P (1987b) Toad's prey-catching: a complex system with heuristic value. *Behav Brain Sci* 10: 389-398

Ewert J-P (1987c) The evolutionary aspect of cognitive functions. A commentary to the target article by JR

Anderson. *Behav Brain Sci* 10(3)

Ewert J-P (1987d) Advantage of modeling in neuroscience. A commentary to the target article by MA Arbib. *Behav Brain Sci* 10(3)

Ewert J-P (1987e) Neuroethology: toward a functional analysis of stimulus-response mediating and modulating neural circuitries. In: Ellen P, Thinus-Blanc C (eds) *Cognitive processes and spatial orientation in animal and man, Vol I.* Martinus Nijhoff, Dordrecht, pp 177-200

Ewert J-P (1988) Instinct, current concepts. In: Adelman G (ed) *Neuroscience year 1988: The yearbook of the enzyclopedia of neuroscience.* Birkhäuser, Boston

Ewert J-P, Borchers H-W (1971) Reaktionscharakteristik von Neuronen aus dem Tectum opticum und Subtectum der Erdkröte *Bufo bufo* (L). *Z Vergl Physiol* 71: 165-189

Ewert J-P, Burghagen H (1979a) Ontogenetic aspects on visual 'size-constancy' phenomena in the midwife toad *Alytes obstetricans* (Laur). *Brain Behav Evol* 16: 99-112

Ewert J-P, Burghagen H (1976b) Configurational prey selection by *Bufo, Alytes, Bombina,* and *Hyla. Brain Behav Evol* 16: 157-175

Ewert J-P, Finkenstädt T (1986) Modulation of tectal functions by prosencephalic loops in amphibians. A commentary. *Behav Brain Sci* 10: 122-123

Ewert J-P, Gebauer L (1973) Größenkonstanzphänomene im Beutefangverhalten der Erdkröte (*Bufo bufo* L). *J Comp Physiol* 85: 303-315

Ewert J-P, Härter H-A (1968) Inhibitionsphänomene im visuellen System der Erdkröte. *Naturwissenschaften* 55: 237

Ewert J-P, Härter H-A (1969) Der hemmende Einfluß gleichzeitig bewegter Beuteattrappen auf das Beutefangverhalten der Erdkröte. *Z Vergl Physiol* 64: 135-153

Ewert J-P, Hock FJ (1972) Movement sensitive neurones in the toad's retina. *Exp Brain Res* 16: 41-59

Ewert J-P, Kehl W (1978) Configurational prey-selection by individual experience in the toad *Bufo bufo. J Comp Physiol* 126: 105-114

Ewert J-P, Rehn B (1969) Quantitative Analyse der Reiz-Reaktionsbeziehungen bei visuellem Auslösen des Fluchtverhaltens der Wechselkröte (*Bufo viridis* Laur). *Behaviour* 35: 212-234

Ewert J-P, Seelen W v (1974) Neurobiologie und System-Theorie eines visuellen Muster-Erkennungsmechanismus bei Kröten. *Kybernetik* 14: 167-183

Ewert J-P, Traud R (1979) Releasing stimuli for antipredator behaviour in the common toad *Bufo bufo* (L). *Behaviour* 68: 170-180

Ewert J-P, Wietersheim A v (1974a) Musterauswertung durch tectale und thalamus/praetectal Nervennetze im visuellen System der Kröte (*Bufo bufo* L). *J Comp Physiol* 92: 131-148

Ewert J-P, Wietersheim A v (1974b) Der Einfluß von Thalamus/Praetectum-Defekten auf die Antwort von Tectum-Neuronen gegenüber bewegten visuellen Mustern bei der Kröte (*Bufo bufo* L). *J Comp Physiol* 92: 149-160

Ewert J-P, Speckhardt I, Amelang W (1970) Visuelle Inhibition and Exzitation im Beutefangverhalten der Erdkröte (*Bufo bufo* L). *Z vergl Physiol* 68: 84-110

Ewert J-P, Hock FJ, Wietersheim A v (1974) Thalamus/Praetectum/Tectum: retinale Topographie und physiologische Interaktionen bei der Kröte (*Bufo bufo* L). *J Comp Physiol* 92: 343-356

Ewert J-P, Borchers H-W, Wietersheim A v (1978) Question of prey feature detectors in the toad's *Bufo bufo* (L) visual system: a correlation analysis. *J Comp Physiol* 126: 43-47

Ewert J-P, Arend B, Becker V, Borchers H-W (1979a) Invariants in configurational prey selection by *Bufo bufo* (L). *Brain Behav Evol* 16: 38-51

Ewert J-P, Borchers H-W, Wietersheim A v (1979b) Directional sensitivity, invariance, and variability of tectal T5 neurons in response to moving configurational stimuli in the toad *Bufo bufo* (L). *J Comp*

Physiol 132: 191-201

Ewert J-P, Burghagen H, Albrecht L, Kepper J (1982) Effects of background structure on the discrimination of configurational moving prey dummies by toads *Bufo bufo* (L). *J Comp Physiol* 147: 179-187

Ewert J-P, Burghagen H, Schürg-Pfeiffer E (1983a) Neuroethological analysis of the innate releasing mechanism for prey-catching behavior in toads. In: Ewert J-P, Capranica RR, Ingle DJ (eds) *Advances in vertebrate neuroethology.* Plenum Press, New York, pp 413-475

Ewert J-P, Capranica RR, Ingle DJ (1983b) (eds) *Advances in vertebrate neuroethology.* Plenum Press, New York

Ewert J-P, Schürg-Pfeiffer E, Weerasuriya A (1984) Neurophysiological data regarding motor pattern generation in the medulla oblongata of toads. *Naturwissenschaften* 71: 590-591

Ewert J-P, Matsumoto N, Schwippert WW (1985) Morphological identification of prey-selective neurons in the grass frog's optic tectum. *Naturwissenschaften* 72: 661-662

Feigenbaum J, Cahusac P, Rolls ET, Miyashita Y, Niki H (1986) Neuronal activity in the parahippocampal gyrus of the behaving primate. *Int Congr of Neuroethology Abstr* 1: 83

Fentress JC (1983) The analysis of behavioral networks. In: Ewert J-P, Capranica RR, Ingle DJ (eds) *Advances in vertebrate neuroethology.* Plenum Press, New York, pp 939-968

Fentress JC (1987) Ethological invariants: boxes, rubberbands, and biological processes. A commentary. *Behav Brain Sci* 10: 377-378

Fernald RD (1987) More than meets the eye. A commentary. *Behav Brain Sci* 10: 378-379

Finkenstädt T (1981) *Der Einfluß des Tectum opticum und Vorderhirns auf die Steuerung des Beutefang- und Fluchtverhaltens beim Salamander Salamandra salamandra (L): eine quantitative Untersuchung mit Hilfe von Ableitungs-, Läsions- und Hirnreizungstechniken.* PhD Thesis, Univ of Kassel

Finkenstädt T (1987) *Verschaltung, Interaktion und Funktion visuell beeinflußbarer Hirngebiete bei Amphibien.* Habil Thesis, Univ Kassel

Finkenstädt T, Ewert J-P (1983a) Processing of area dimensions of visual key stimuli by tectal neurons in *Salamandra salamandra. J Comp Physiol* 153: 85-98

Finkenstädt T, Ewert J-P (1983b) Visual pattern discrimination through interactions of neural networks: a combined electrical brain stimulation, brain lesion, and extracellular recording study in *Salamandra salamandra. J Comp Physiol* 153: 99-110

Finkenstädt T, Ewert J-P (1985) Glucose utilization in the toad's brain during anesthesia and stimulation of the ascending reticular arousal system: a ^{14}C-2- deoxyglucose study. *Naturwiss.* 72: 161-162

Finkenstädt T, Ewert J-P (1987) Visual associative conditioning in the toad: a 2-deoxyglucose study. In: Elsner N, Creutzfeldt O (eds) *New frontiers in brain research.* G Thieme-Verlag, Stuttgart

Finkenstädt T, Ewert J-P (1988) Stimulus-specific long-term habituation of visually guided orienting behavior toward prey in toads: a ^{14}C-2DG study. *J Comp Physiol* 163: 1-11

Finkenstädt T, Adler NT, Allen TO, Ebbesson SOE, Ewert J-P (1985) Mapping of brain activity in mesencephalic and diencephalic structures of toads during presentation of visual key stimuli: a computer assisted analysis of ^{14}C-2DG autoradiographs. *J Comp Physiol* 156: 433-445

Finkenstädt T, Adler NT, Allen TO, Ewert J-P (1986) Regional distribution of glucose utilization in the telencephalon of toads in response to configurational visual stimuli: a ^{14}C-2DG study. *J Comp Physiol* 158: 457-467

Fite KV, Scalia F (1976) Central visual pathways in the frog. In: Fite KV (ed) *The amphibian visual system: a multidisciplinary approach.* Academic Press, New York San Francisco London, pp 87-118

Fite KV, Hayden D, Montgomery N, Bengston L (1983) Visual orienting responses following selective reduction of optic input to AOS and/or optic tectum in frog. *Soc Neurosci Abstr* 9: 818

Foreman N, Stevens R (1987) Relationships between the superior colliculus and hippocampus: neural and behavioral considerations. *Behav Brains Sci* 10: 101-151

Freeman JA, Norden JJ (1984) Neurotransmitters in the optic tectum of non mammalians. In: Vanegas H (ed) *Comparative neurology of the optic tectum.* Plenum Press, New York, pp 469-546

Freisling J (1948) Studien zur Biologie und Psychologie der Wechselkröte (*Bufo viridis* Laur). *Österreichische Zool Z (Wien)* 1: 383-440

Frost BJ (1982) Mechanisms for discriminating object motion from self-induced motion in the pigeon. In: Ingle DJ, Goodale MA, Mansfield RJW (eds) *Analysis of visual behavior.* The MIT Press, Cambridge Mass, pp 177-196

Gaillard F, Galand G (1979) Diencephalic binocular wide field neurons in the frog. *Exp Brain Res* 34: 511-520

Gans C (1961) A bullfrog and its prey. *Nat Hist* 70: 26-37

Gans C, Gorniak GC (1982) Functional morphology of lingual protrusion in marine toads (*Bufo marinus*). *Amer J Anat* 163: 195-222

Garner WR (1966) To perceive is to know. *Amer Psychol* 21: 11-19

Gaze RM (1958) The representation of the retina on the optic lobe of the frog. *Quart J Exp Physiol* 43: 209-314

Gerhardt HC (1981) Mating call recognition in the green tree frog (*Hyla cinerea*): importance of two frequency bands as a function of sound pressure level. *J Comp Physiol* 144: 9-16

Gibson JJ (1950) *The perception of the visual world.* Houghton, Boston Mass

Gibson JJ (1951) What is form? *Psychol Rev* 58: 403-412

Gillette R, Kovac MP, Davis WJ (1978) Command neurons in *Pleurobranchaea* receive synaptic feedback from the motor network they excite. *Science* 199: 798-801

Gordon J, Hood DC (1976) Anatomy and physiology of the frog retina. In: Fite KV (ed) *The amphibian visual system: a multidisciplinary approach.* Plenum Press, New York, pp 29-86

Grobstein P, Comer C, Kostyk SK (1983) Frog prey capture behavior: between sensory maps and directed motor output. In: Ewert J-P, Capranica RR, Ingle DJ (eds) *Advances in vertebrate neuroethology.* Plenum Press, New York, pp 331-347

Gruberg ER, Lettvin JY (1980) Anatomy and physiology of a binocular system in the frog *Rana pipiens.* *Brain Res* 192: 313-325

Gruberg ER, Udin SB (1978) Topographic projections between the nucleus isthmi and the tectum of the frog *Rana pipiens.* *J Comp Neurol* 179: 487-500

Grüsser-Cornehls U (1984) The neurophysiology of the amphibian optic tectum. In: Vanegas H (ed) *Comparative neurology of the optic tectum.* Plenum Press, New York, pp 211-245

Grüsser O-J, Grüsser-Cornehls U (1968) Neurophysiologische Grundlagen visueller angeborener Auslösemechanismen beim Frosch. *Z Vergl Physiol* 59: 1-24

Grüsser O-J, Grüsser-Cornehls U (1970) Die Neurophysiologie visuell gesteuerter Verhaltensweisen bei Anuren. *Verh Deutsch Zool Ges Köln* 64: 201-218

Grüsser O-J, Grüsser-Cornehls U (1976) Neurophysiology of the anuran visual system. In: Llinas R, Precht W (eds) *Frog neurobiology.* Springer-Verlag, Berlin Heidelberg New York, pp 298-385

Guha K, Jorgensen CB, Larsen LO (1980) Relationship between nutritional state and testes function, together with the observations on patterns of feeding, in the toad. *J Zool (London)* 192: 147-155

Guthrie DM (1987) (ed) *Aims and methods in neuroethology.* Manchester Univ Press, Manchester

Hanke W (1976) Neuroendocrinology. In: Llinas R, Precht W (eds) *Frog neurobiology.* Springer-Verlag, Berlin Heidelberg New York, pp 975-1020

Harnad S (1987a) Psychophysical and cognitive aspects of categorical perception: a critical overview. In:

Harnad S (ed) *Categorical perception.* Cambridge University Press, Cambridge Mass

Harnad S (1987b) Category induction and representation. In: Harnad S (ed) *Categorical perception.* Cambridge University Press, Cambridge Mass

Hartline HK (1940) The receptive fields of optic nerve fibers. *Amer J Physiol* 130: 690-699

Hebb DO (1949) *The organization of behavior.* Wiley, New York

Hebb DO (1959) A neuropsychological theory. In: Koch S (ed) *Psychology: a study of a science Vol I,* McGraw-Hill, New York

Herbst G (1981) Quantitative Untersuchungen zur Frage der Invariantenbildung beim Beuteerkennen verschiedener Subspecies der Erdkröte (*Bufo bufo bufo* und *Bufo bufo spinosus*). Staatsexam Thesis, Univ Kassel

Herrick CJ (1933) The amphibian forebrain. VIII: Cerebral hemispheres and pallial primordia. *J Comp Neurol* 58: 737-759

Heusser H (1959) Die Lebensweise der Erdkröte *Bufo bufo* (L); Nahrungsaufnahme und Pigmentierung der Daumenschwielen im Jahresverlauf. *Biol Zbl* 88: 457-467

Heusser H (1960) Instinkterscheinungen an Kröten unter besonderer Berücksichtigung des Fortpflanzungsinstinktes der Erdkröte (*Bufo bufo* L). *Z Tierpsychol* 17: 67-81

Himstedt W (1982) Prey selection in salamanders. In: Ingle DJ, Goodale MA, Mansfield RJW (eds) *Analysis of visual behavior.* The MIT Press, Cambridge Mass, pp 47-66

Himstedt W, Freidank U, Singer E (1976) Die Veränderung eines Auslösemechanismus im Beutefangverhalten während der Entwicklung von *Salamandra salamandra* (L). *Z Tierpsychol* 41: 235-243

Hinde RA (1954) Changes in responsiveness to a constant stimulus. *Behaviour* 2: 41-54

Hinsche G (1935) Ein Schnappreflex nach "Nichts" bei Anuren. *Zool Anz* 111: 113-122

Holst E v, Mittelstaedt H (1950) Das Reafferenzprinzip. *Naturwissenschaften* 37: 464-476

Hood DC, Gordon J (1981) The frog ganglion cell: not a feature detector and not a monkey cortical cell. *Perception* 10: 421-422

House D (1984) *Neural models of depth perception in frog and toad.* PhD Thesis, Univ Mass Amherst

Hoyle G (ed) (1977) Identified neurons and behavior of arthropods. Plenum Press, New York:

Hoyle G (1984) The scope of neuroethology. *Behav Brain Sci* 7: 367-412

Hubel DH, Wiesel TN (1962) Receptive fields, binocular interaction and functional architecture in the cat's visual cortex. *J Physiol* 160: 106-154

Hubel DH, Wiesel TN (1965) Receptive fields and functional architecture in two non-striate visual areas (18 and 19) of the cat. *J Neurophysiol* 28: 229-289

Huber F, Markl H (1983) (eds) *Neuroethology and behavioral physiology.* Springer-Verlag, Berlin

Hume D (1902) *Enquiries concerning the human understanding and concerning the principles of morals* (Selby-Bigge LA, ed). Clarendon Press, Oxford

Ingle D (1968) Visual releasers of prey-catching behavior in frogs and toads. *Brain Behav Evol* 1: 500-518

Ingle D (1971) Prey-catching behavior of anurans toward moving and stationary objects. *Vision Res Suppl* 3: 447-456

Ingle D (1973) Disinhibition of tectal neurons by pretectal lesions in the frog. *Science* 180: 422-424

Ingle D (1975) Selective visual attention in frogs. *Science* 188: 1033-1035

Ingle D (1976) Spatial vision in anurans. In: Fite KV (ed) *The amphibian visual system: a multidisciplinary approach.* Academic Press, New York, pp 119-140

Ingle D (1977) Detection of stationary objects by frogs (*Rana pipiens*) after ablation of optic tectum. *J Comp Physiol Psychol* 91: 1359-1364

Ingle D (1980) Some effects of pretectum lesions on the frog's detection of stationary objects. *Behav Brain Res* 1: 139-163

Ingle DJ (1983) Brain mechanisms of visual localization by frogs and toads. In: Ewert J-P, Capranica RR, Ingle DJ (eds) *Advances in vertebrate neuroethology.* Plenum Press, New York, pp 177-226

Ingle DJ, Quinn S (1982) Retrograde labelling of neurons of known behavioral function in frog tectum. *Soc Neurosci Abstr* 8: 406

John ER, Schwartz EL (1978) The neurophysiology of information processing and cognition. *Ann Rev Psychol* 29: 1-29

Kandel ER, Krasne FB, Strumwasser F, Truman JW (eds) (1979) *Cellular mechanisms in the selection and modulation of behavior.* Neurosci Research Program Bulletin, Vol 17. MIT Press, Cambridge Mass

Karten HJ, Reiner A, Brecha N (1982) Laminar organization and origins of neuropeptides in the avian retina and optic tectum. In: Liss AR (ed) *Cytological methods in neuroanatomy*

Katte O, Hoffmann K-P (1980) Direction specific neurons in the pretectum of the frog (*Rana esculenta*). *J Comp Physiol* 140: 53-57

Kicliter E, Ebbesson SOE (1976) Organization of the 'non olfactory' telencephalon. In: Llinas R, Precht W (eds) *Frog neurobiology.* Springer-Verlag, Berlin Heidelberg New York, pp 946-972

Kondrashev SL (1976) Influence of the visual stimulus size on the breeding behavior of anuran males. *Akademija Nank Zoologiceskij Jurnal* 55: 1576-1579

Kondrashev SL (1987) Neuroethology and color vision in amphibians. *Behav Brain Sci* 10: 385

Kondrashev SL, Dimentman AM (1978) Role of dorsal thalamus in the organization of visually-mediated behavior in amphibians. In: Orlov OY (ed) *Mechanisms of vision of animals.* Nauka, Moscow

Konishi M (1985) Birdsong: from behavior to neuron. *Ann Rev Neurosci* 8: 125-170

Konorski J (1967) *Integrative activity of the brain.* University of Chicago Press, Chicago Ill

Kostyk SK, Grobstein P (1982) Visual orienting deficits in frogs with various unilateral lesions. *Behav Brain Res* 6: 379-388

Kuljis RO, Karten HJ (1982) Laminar organization of peptide-like immunoreactivity in the anuran optic tectum. *J Comp Neurol* 212: 188-201

Kupfermann I, Weiss KR (1978) The command neuron concept. *Behav Brain Sci* 1: 3-39

Laming PR, Ewert J-P (1983) The effects of pretectal lesions on neuronal, sustained potential shift and electroencephalographic responses of the toad tectum to presentation of a visual stimulus. *Comp Biochem Physiol* 76: 247-252

Lara R, Arbib MA (1985) A model of the neural mechanisms responsible for pattern recognition and stimulus specific habituation in toads. *Biol Cybern* 51: 223-237

Lara R, Cervantes F, Arbib MA (1982) Two-dimensional model of retinal-tectal-pretectal interactions for the control of prey-predator recognition and size preference in amphibia. In: Amari S, Arbib MA (eds) *Competition and cooperation in neural nets.* Springer-Verlag, Berlin Heidelberg New York, pp 371-393

Lázár G (1969) Efferent pathways of the optic tectum in the frog. *Acta Biol Acad Sci Hung* 20: 171-183

Lázár G (1971) The projection of the retinal quadrants on the optic centers in the frog: a terminal degeneration study. *Acta Morph Acad Sci Hung* 19: 325-334

Lázár G (1973) Role of accessory optic system in the optokinetic nystagmus of the frog. *Brain, Behav Evol* 5: 443-460

Lázár G (1979) Organization of the frog visual system. In: Lissak R (ed) *Recent development of biology in Hungary Vol 8.* Akademiai Kiado, Budapest

Lázár G (1984) Structure and connections of the frog optic tectum. In: Vanegas H (ed) *Comparative neurology of the optic tectum.* Plenum Press, New York, pp 185-210

Lázár G, Tóth P, Csank G, Kicliter E (1983) Morphology and location of tectal projection neurons in frogs. A study with HRP and cobalt-filling. *J Comp Neurol* 215: 108-120

Lettvin JY, Maturana HR, McCulloch WS, Pitts WH (1959) What the frog's eye tells the frog's brain. *Proc Inst Radio Engin* 47: 1940-1951

Lettvin JY, Maturana HR, McCulloch WS, Pitts WH (1961) Two remarks on the visual system of the frog. In: Rosenblith WA (ed) *Sensory communication.* MIT Press, Cambridge Mass, pp 757-776

Lorenz K (1935) Der Kumpan in der Umwelt des Vogels. Der Artgenosse als auslösendes Moment sozialer Verhaltensweisen. *J Ornithol* 83: 137-213, 289-413

Lorenz K (1943) Die angeborenen Formen möglicher Erfahrung. *Z Tierpsychol* 5: 235-409

Manning A (1979) *An introduction to animal behaviour.* 3rd edn. Springer-Verlag, Berlin Heidelberg

Masino T, Grobstein P (1985) The organization of tectal projections to the ventral midbrain in *Rana pipiens. Soc Neurosci Abstr* 11: 289

Matsumoto N, Antal M (1984) Physiological and morphological study of prey and predator detecting neurons in the frog tectum. *Proc 6th Ann Conf Japan Soc Gen Comp Physiol Abstr* 6: 169

Matsumoto N, Schwippert WW, Ewert J-P (1986) Intracellular activity of morphologically identified neurons of the grass frog's optic tectum in response to moving configurational visual stimuli. *J Comp Physiol* 159: 721-739

Maturana HR, Lettvin JY, McCulloch WS, Pitts WH (1960) Anatomy and physiology of vision in the frog (*Rana pipiens*). *J Gen Physiol* 43: 129-176

Megela AL, Borchers H-W, Ewert J-P (1983) Relation between activity of tectal neurons and prey-catching behavior in toads *Bufo bufo. Naturwissenschaften* 70: 100-101

Milson JA, Mitchell JF (1977) The action of amino acids on evoked responses in the frog optic tectum. *Br J Pharmacol* 59: 484P

Mountcastle VB (1957) Modality and topographic properties of single neurons of cat's somatic sensory cortex. *J Neurophysiol* 20: 403-434

Neary TJ, Northcutt RG (1983) Nuclear organization of the bullfrog diencephalon. *J Comp Neurol* 213: 262-278

Northcutt RG, Kicliter E (1980) Organization of the amphibian telencephalon. In: Ebbesson SOE (ed) *Comparative neurology of the telencephalon.* Plenum Press, New York, pp 203-255

Paillard J (1987) Cognitive versus sensorimotor encoding of spatial information. In: Ellen P, Thinus-Blanc C (eds) *Cognitive processes and spatial orientation in animal and man Vol II.* Martinus Nijhoff Publ, Dordrecht

Parent A (1973) Distribution of monoamine-containing neurons in the brainstem of the frog, *Rana temporaria. J Morphol* 139: 67-78

Porter KR (1972) *Herpetology.* WB Saunders Comp, Philadelphia London Toronto

Ramon y Cajal S (1896) Die Retina der Wirbeltiere (transl by R Greef). JF Bergmann Wiesbaden. [Engl transl by Rodieck RW (1973) *The vertebrate retina.* WH Freeman & Comp, pp 773-904]

Rehn B (1977) *Cerebrale Repräsentation des Fluchtverhaltens der Erdkröte (Bufo bufo L).* PhD Thesis, Technical Univ Darmstadt

Reiner A, Brauth SE, Karten HJ (1980) Basal ganglionic pathways to the tectum: studies in reptiles. *J Comp Neurol* 193: 565-589

Reiner A, Karten HJ, Brecha NC (1982) Enkephalin-mediated basal ganglia influences over the optic tectum: immunohistochemistry of the tectum and the lateral spirifrom nucleus in pigeon. *J Comp Neurol* 208: 37-53

Reiner A, Brauth SE, Karten HJ (1984) Evolution of the amniote basal ganglia. *TINS* 7: 320-325

Rolls ET (1987a) A neurophysiological systems approach to neuroethology. In: Guthrie DM (ed) *Aims and methods in neuroethology.* Manchester Univ Press, Manchester, pp 231-259

Rolls ET (1987b) Neuronal activity underlying perception and learning in the primate. In: Elsner N,

Creutzfeldt O (eds) *New frontiers in brain research.* Thieme Verlag, Stuttgart New York, pp 49-51

Roth G (1982) Responses in the optic tectum of the salamander *Hydromantes italicus* to moving prey stimuli. *Exp Brain Res* 45: 386-392

Roth G (1987) *Visual behavior in salamanders.* Springer-Verlag, Berlin Heidelberg New York

Roth G, Jordan M (1982) Response characteristics and stratification of tectal neurons in the toad *Bufo bufo* (L). *Exp Brain Res* 45: 393-398

Roucoux A, Crommelinck M (1987) Sensorimotor maps in the tectum. *Behav Brain Sci* 10: 386-387

Rubinson E (1968) Projections of the tectum opticum of the frog. *Brain Behav Evol* 1: 529-561

Satou M, Ewert J-P (1985) The antidromic activation of tectal neurons by electrical stimuli applied to the caudal medulla oblongata in the toad, *Bufo bufo* (L). *J Comp Physiol* 157: 739-748

Satou M, Matsushima T, Ueda K (1984) Neuronal pathways from the tectal "snapping-evoking area" to the tongue muscle controlling motoneurons in the Japanese toad: evidence of the intervention of excitatory interneurons. *Zool Sci* 1: 829-832

Scalia F (1976) The optic pathway of the frog: nuclear organization and connections. In: Llinas R, Precht W (eds) *Frog neurobiolog.* Springer-Verlag, Berlin Heidelberg New York, pp 386-406

Scheich H (1983) Sensorimotor interfacing. In: Ewert J-P, Capranica RR, Ingle DJ (eds) *Advances in vertebrate neuroethology.* Academic Press, New York, pp 7-14

Schleidt WM (1961) Reaktionen von Truthühnern auf fliegende Raubvögel und Versuche zur Analyse ihrer AAMs. *Z Tierpsychol* 18: 534-560

Schleidt WM (1962) Die historische Entwicklung der Begriffe "Angeborenes auslösendes Schema" und "Angeborener Auslösemechanismus" in der Ethologie. *Z Tierpsychol* 19: 697-722

Schneider D (1954) Beitrag zu einer Analyse des Beute- und Fluchtverhaltens einheimischer Anuren. *Biol Zbl* 73: 225-282

Schneider GE (1969) Two visual systems. *Science* 163: 895-920

Schürg-Pfeiffer E (1979) *Quantitative neurophysiologische Untersuchungen zur Frage nach Gestaltdetektoren im visuellen System des Frosches Rana temporaria.* PhD Thesis, Univ of Kassel

Schürg-Pfeiffer E, Ewert J-P (1981) Investigation of neurons involved in the analysis of Gestalt prey features in the frog *Rana temporaria. J Comp Physiol* 141: 139-152

Schürg-Pfeiffer E, Ewert J-P (1988) Correlation between responses of prey-selective T5.2 neurons and prey-catching in freely moving toads (submitted)

Schürg-Pfeiffer E, Finkenstädt T, Cromarty A, Ewert J-P (1989) Change of response properties of a T5.2 neuron in the course of pretectal lesions in the toad (submitted)

Seelen W v (1970) Zur Informationsverarbeitung im visuellen System der Wirbeltiere I/II. *Kybernetik* 7: 43-60; 89-106

Shinn EA, Dole JW (1978) Evidence for a role for olfactory cues in the feeding response of leopard frogs *Rana pipiens. Herpetologia* 34: 167-172

Stellar E (1954) The physiology of motivation. *Psychol Rev* 61: 5-22

Stevens KH (1987) Implicit versus explicit computation. A commentary. *Behav Brain Sci* 10: 387-388

Suga N (1984) The extent to which biosonar information is represented in the bat auditory cortex. In: Edelman GM, Gall WE, Cowan WM (eds) *Dynamic aspects of neocortical function.* Wiley & Sons, New York

Székely G (1973) Anatomy and synaptology of the optic tectum. In: Jung R (ed) *Handbook of sensory physiology Vol VII/3B.* Springer-Verlag, Berlin Heidelberg New York, pp 1-26

Székely G, Czeh G (1976) Organization of locomotion. In: Llinás R, Precht W (eds) *Frog neurobiology.* Springer-Verlag, Berlin Heidelberg New York, pp 765-792

Székely G, Lázár G (1976) Cellular and synaptic architecture of the optic tectum. In: Llinás R, Precht W (eds) *Frog neurobiology.* Springer-Verlag, Berlin Heidelberg New York, pp 407-434

Szentágothai J (1967) The anatomy of complex integration units in the nervous system. In: Lissak K (ed) *Recent developments of neurobiology in Hungary. Vol 1: Results in neuroanatomy, neurochemistry, neuropharmacology and neurophysiology.* Akademiai Kiado, Budapest

Szentágothai J, Arbib MA (1974) *Conceptual models of neural organization.* Neurosciences Research Program Bulletin Vol 12. MIT Press, Cambridge Mass

Tinbergen N (1948) Social releasers and the experimental method required for their study. *Wilson Bulletin* 60: 6-52

Tinbergen N (1951) *The study of instinct.* Clarendon Press, Oxford

Tinbergen N, Kuenen DJ (1939) Über die auslösenden und richtungsgebenden Reizsituationen der Sperrbewegung von jungen Drosseln (*Turdus merula merula* L und *Turdus ericetorum ericetorum* Turdon). *Z Tierpsychol* 3: 37-60

Tóth P, Csank G, Lázár G (1985) Morphology of the cells of origin of descending pathways to the spinal cord in *Rana esculenta.* A tracing study using cobalt-lysine complex. *J Hirnforsch* 26: 365-383

Trachtenberg MC, Ingle D (1974) Thalamo-tectal projections in the frog. *Brain Research* 79: 419-430

Traud R (1983) *Einfluß von visuellen Reizmustern auf die juvenile Erdkröte (Bufo bufo L).* PhD Thesis, Univ of Kassel

Tsai H-J, Ewert J-P (1987) Edge preference of retinal and tectal neurons in common toads (*Bufo bufo*) in response to worm-like moving stripes: the question of behaviorally relevant "position indicators." *J Comp Physiol* 161: 295-304

Tsai H-J, Ewert J-P (1988) Influence of stationary and moving textured backgrounds on the response of visual neurons in toads (*Bufo bufo* L). *Brain Behav Evol* 32: 27-38

Uexküll J v (1909) *Umwelt und Innenwelt der Tiere.* Springer-Verlag, Berlin

Weerasuriya A (1983) Snapping in toads: some aspects of sensorimotor interfacing and motor pattern generation. In: Ewert J-P, Capranica RR, Ingle DJ (eds) *Advances in vertebrate neuroethology.* Plenum Press, New York, pp 613-627

Weerasuriya A, Ewert J-P (1981) Prey-selective neurons in the toad's optic tectum and sensori-motor interfacing: HRP studies and recording experiments. *J Comp Physiol* 144: 429-434

Weerasuriya A, Ewert J-P (1983) Afferents of some dorsal retino-recipient areas of the brain of *Bufo bufo. Soc Neurosci Abstr* 9: 536

Wiersma CAG, Ikeda K (1964) Interneurons commanding swimmeret movements in the crayfish, *Procambarus clarkii* (Girard). *Comp Biochem Physiol* 12: 509-525

Wietersheim A v, Ewert J-P (1978) Neurons of the toad's (*Bufo bufo* L) visual system sensitive to moving configurational stimuli: a statistical analysis. *J Comp Physiol* 126: 35-42

Wilczynski W, Northcutt RG (1977) Afferents to the optic tectum of the leopard frog: an HRP study. *J Comp Neurol* 173: 219-229

Wilczynski W, Northcutt RG (1983a) Connections of the bullfrog striatum: efferent projections. *J Comp Neurol* 214: 333-343

Wilczynski W, Northcutt RG (1983b) Connections of the bullfrog striatum: afferent organization. *J Comp Neurol* 214: 321-332

Wurtz RH (1979) Modulation of the primate visual system by attention and readiness to respond. In: Kandel ER, Krasne FB, Strumwasser F, Truman JW (eds) *Cellular mechanisms in the selection and modulation of behavior.* Neurosci Research Program Bulletin Vol 17. MIT Press, Cambridge Mass

Zucker I (1983) Motivation, biological clocks, and temporal organization of behavior. In: Satinoff E, Teitelbaum P (eds) *Handbook of behavioral neurobiology, Vol 6.* Plenum Press, New York

Zusne L (1970) *Visual perception of form.* Academic Press, New York London

Visuomotor Coordination: Neural Models and Perceptual Robotics

MICHAEL A. ARBIB

Departments of Computer Science, Neurobiology, Physiology,
Biomedical Engineering, Electrical Engineering, and Psychology
University of Southern California
Los Angeles, CA 90089-0782, USA

Abstract. This paper addresses two complementary questions: What is the appropriate set of tools for the study of the networks of animal and human brains, and what are the strategies for building computers with "intelligence"? We argue that there are overall architectural principles which unite both sides of this study, namely, that a computer no longer be thought of as a unitary system but rather as a network of more specialized devices, and that many of these devices be structured as highly parallel arrays of interacting neuron-like components. We illustrate this with a discussion of the architecture of the frog's brain as revealed in studies of the mechanisms of visuomotor coordination, and of the design of vision and motor conrol systems for robots. The following topics are treated: (1) What is a schema? (2) Schemas for *Rana computatrix*. (3) Tectal columns. (4) Depth perception. (5) Path-planning and detours. (6) Schemas for hand control. (7) Schemas for vision. (8) Challenges for cooperation.

1 What is a Schema?

In this paper we introduce the complementary, yet overlapping, fields of *Neural Computing* - where the aim is to develop ideas from neuroscience into a form that is promising for the development of the next generation of computers and programs - and *Computational Neurobiology* - in which models are developed which stand or fall in relation to the data of experimental study of the brain. The metaphor "the brain is a computer" must not be read as reducing the brain to the level of serial computers, but rather must expand our concepts of computation to embrace the style of the brain, depending on the constant interaction of concurrently active systems, many of which express their activity in the interplay of spatiotemporal patterns in manifold layers of neurons. The last few years have seen a dramatic re-awakening of interest in these approaches, with a major surge of activity in neural computing - the give-and-take between the analysis of brain mechanisms and the design of parallel and distributed computers and robotic systems. Workers in machine vision have contributed much to the understanding of the importance of parallelism in the "low-level" processing of images, and of distributed problem-solving in the "high-level" visual problem of interpreting the objects in a scene and the relationships between them (cf. Arbib & Hanson 1987). Meanwhile, the development of VLSI and computer networking has enabled many computer scientists to abandon the serial computer paradigm, and seek to understand how problem-solving can be distributed across a network of interacting highly sophisticated, concurrently active processors.

In discussing modeling style, we use the term *top-down* to refer to a functional analysis which starts from some overall behavior and seeks to break it down into the function of interacting subsystems, whereas *bottom-up* analysis starts with a set of components and seeks to infer the properties of systems constructed therefrom. By contrast, in models of perception (cf. Section 6), *bottom-up* perception is data-driven proceeding "up" from the raw sensory data, while *top-down* perception is hypothesis-driven, proceeding "down" from some model of the world. In general, successful modeling and successful perception both require an interleaving of top-down and bottom-up processes.

A schema-theoretic approach (Arbib 1981) to modeling the CNS offers a framework for performing a top-down analysis of perceptual and motor behavior in both hierachical and distributed terms. *Perceptual* and *motor schemas* are defined as units of perceptual analysis and of motor control, respectively, to account for the ability of the system to serve these requirements. Multiple brain regions contribute to a given schema, and it is in the interaction of these schemas that sensory guidance of motor behavior emerges. The task, then, for the brain modeler is to not only identify potential schemas which represent the motor behavior, but to also show how these schemas map into the known neural circuitry. Turning from computational neuroscience to neural computing, we offer schemas as providing programs for both perceptual analysis and motor control in *perceptual robotics* - an emerging field of robotics in which, far from executing repetitive and stereotyped movements, the robot integrates perceptual data into the control of a flexible repertoire of behaviors adapted to current circumstances.

In this Section, and in Section 2, I shall mainly speak of schemas as functional units for the analysis of animal brains, leaving it in most cases to the reader to draw the parallel implications for neural computing and perceptual robotics. By contrast, in Sections 6 and 7 the emphasis will be on computer systems, but with rich implications to be drawn for neuroethology and cognitive science.

A set of *basic motor schemas* is hypothesized to provide sensorimotor neural activity corresponding to simple, prototypical patterns of movement. These elementary "building blocks" of motor behavior combine with *perceptual schemas* to form *coordinated control programs* which interweave their activations in accordance with the current task and sensory environment. Thus, motor schemas in general may be either basic, or built up from other schemas as coordinated control programs. While basic motor schema activations may be sensory-driven, schema activations in general are largely task-driven, reflecting the intention-related goals of the individual and the physical and functional requirements of the task.

For a motivating example, consider that human grasping movements may be analyzed in terms of the temporal phases involved in the reaching and grasping

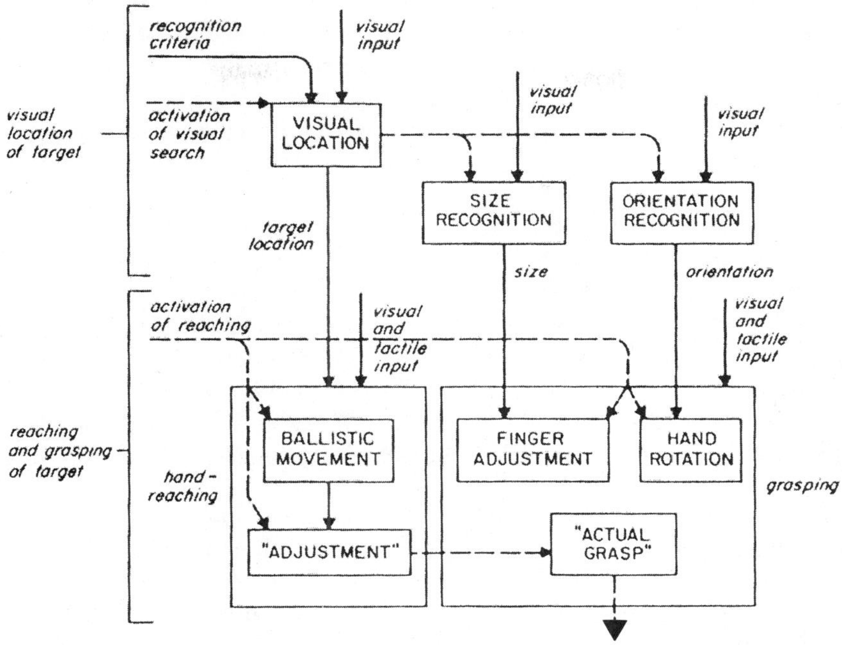

Figure 1. Control of the movements for a prehensile task can be described by a coordinated control program of interacting perceptual and motor schemas (Arbib 1981). Visual location and parameterization of the target is performed by three perceptual schemas. Control of the arm movements is done by two motor schemas (on the bottom left) and control of the finger and wrist movements is done by three motor schemas (bottom right).

components, as well as in terms of the basic schemas involved. Our coordinated control program for reaching and grasping (Fig. 1; Arbib 1981) illustrates the high-level channeling of pertinent sensory information to the control structures responsible for different submovements. Subschemas for finger adjustment and hand rotation, movements which precede the actual grasping of the object, receive visual and proprioceptive information necessary for their respective sensorimotor transformation (more of this in Section 6).

More formally, schemas are programs developed to satisfy the following criteria:
(a) They serve to represent, at least, perceptual structures and distributed motor control.
(b) Schemas may be instantiated. For example, given a schema that represents generic knowledge about some domain of interaction, we may need several active copies of the schema - which we call schema *instances* - each suitably tuned, to subserve our perception of several instances of that domain.
(c) The programs are concurrent. We postulate that unlike serial computers, the brain can support concurrent activity of many schemas for the recognition of different objects, and the planning and control of different activities.

In summary, then, the analysis of interacting computing agents called *schema instances* can serve as a valuable intermediary between overall specification of some animal, human, or robot behavior and the neural networks (or other low-level machine instructions) that subserve it. We add that an object may have different representations appropriate to different tasks, and, within such a representation, different coordinate systems may be set up for different parts of the object and/or different stages of task execution.

We have asserted that, in a particular application domain, we may work with a fixed set of basic schemas, modeling a complex behavior in terms of interactions among more simple behaviors represented by such schemas. As Grobstein's paper in this volume demonstrates, such a decomposition is succesful to the extent that it passes the test of, e.g., brain lesion studies. However, I do not know of any methodology *guaranteed* to yield the "right" decomposition. In a general setting, moreover, there is no fixed repertoire of basic schemas. Rather, I see new schemas being formed as assemblages of old schemas; but once formed a schema may be tuned (by something akin to an adaptive network mechanism), much as a skill is honed into a unified whole from constituent pieces (cf. the last Section of Arbib 1981, on "Program Synthesis and Visuomotor Coordination"). It is the tunability of schema-assemblages that allows them to start as composite but emerge as primitive. It is also the reason why one must realize that a model expressed in a schema-level formalism may only approximate the behavior of a model expressed in a neural net formalism. (In analyzing perceptual and motor development, one also needs to study the *differentiation* of schemas into assemblies of more refined schemas. But this is beyond the scope of the present paper.)

It is an improtant challenge to explore the "neural reality" of schemas. A variety of methodologies are required. This exploration can be constrained by noting that claims about the localization of schemas in the brain may be tested by lesion experiments. Then, once a schema-theoretic model of some animal behavior has been refined to the point of hypotheses about the localization of schemas, we may model a brain region by seeing if its known neural circuitry can indeed be shown to implement the posited schema. With this, we emphasize the following points:

(1) A given schema, defined functionally, may be distributed across more than one brain region; conversely, a given brain region may be involved in many schemas. A top-down analysis may advance specific hypotheses about the localization of (sub)schemas in the brain, and these may be tested by lesion experiments, for example.

(2) In some cases, we will then proceed to model each schema by interacting layers of neuron-like elements, or by nets of "intermediate-level" units. Even if the nets are little constrained by anatomy or physiology, such studies can be valuable in extending our understanding of the properties of neural networks, better preparing us to handle new data as they become available.

(3) However, once a schema-theoretic model of some animal behavior has been refined to the point of hypotheses about the localization of schemas, we may then model a brain region by seeing if its known neural circuitry can indeed be shown to implement the posited schema. In some cases the model will suggest properties of the circuitry that have not yet been tested, thus laying the ground for new experiments.

(4) If a number of models have been established, further modeling should - to the *extent possible* - be incremental, in that new models should refine, modify and build upon prior models, rather than being constructed ab *initio*.

Starting with the Didday model of prey-selection (Didday 1970, 1976), my colleagues and I have analyzed frog and toad visuomotor coordination in terms of models based on schemas, on interacting layers of "neuron-like" units, and on neural networks closely coupled to details from neuroanatomy and neurophysiology. These studies are all part of the evolution of *Rana computatrix*, the "computational frog," and suggest the excitement of "evolving" an integrated account of a single animal, seeing the challenges posed by integrating different aspects of vision with mechanisms for the control of an expanding repertoire of behavior. We shall study *Rana computatrix* at the level of schemas and in terms of neural networks in the next four Sections. The work will have implications for both computational neuroscience - the use of computational techniques to analyze information processing in the nervous system - by giving case studies of how models evolve in interaction with experimental data; and neural computing - the use of concepts gleaned from study of the brain to catalyze the development of new technological devices - by studying frog and toad as "biological robots" to gain general insights into the role of neural computing in perceptual robotics in general.

In fact, the study of *Rana computatrix* has implications that go far beyond the study of frogs and toads, *per se*. It is not many years ago that the study of invertebrates was regarded as being, at best, peripheral by the majority of neuroscientists. Yet the increasing attention to mechanisms of neural function has made *Aplysia* and other invertebrates invaluable in the study of basic cellular mechanisms of facilitation, rhythm generation, and habituation, for example. However, drastic differences in organizational principles separate the primate brain from the *Aplysia* nerve net. I thus argue that if creatures evolved by chance can provide insight into cellular mechanisms, then "creatures" which "evolve in the computer" can provide opportunities for understanding organizational principles; and that these organizational principles are not to be sought solely in terms of cellular mechanisms but in terms of structural constructs (layers and modules), functional constructs (schemas), and computational strategies (cooperative computation in neural nets, etc.).

The schemas of *Rana computatrix* which we introduce in Sections 2 and 4 - various functional subsystems underlying visuomotor coordination in the frog and toad - have the very important but special property that each schema is implemented within *fixed* neural ciruitry. (This does not preclude the use of a given region of the brain by more than one schema.) In contrast with this situation, when we turn in Sections 6 and 7 to the study of perceptual and motor schemas in perceptual robotics, we will consider systems in which it is not so plausible that the same circuitry is used every time a copy of the schema is activated. The analysis of a schema-based system for high-level vision (Section 7) will present processes whereby the "model of the environment" can be formed as an *assemblage of instances of perceptual schemas*. While this is a system for high-level *computer* vision, I will present it as also being a hypothesis about the way in which scene understanding may be implemented in the human brain.

This raises the problem of *schema instantiation* in neural nets which is, I claim, one of the most important open problems to be faced in bridging between cognitive science and neuroscience. How does the brain mobilize the appropriate assemblies of neurons when many copies of the schema appear to be active at the same time? How economically can inferential and other mechanisms respond to recruited - and so in some sense unpredictable - assemblies? How are recruited assemblies demobilized? This is especially difficult when an assembly shares many neurons with other assemblies. Von der Malsburg & Schneider (1986) offer a solution to some of these problems with the dynamic link architecture (see also v. d. Malsburg, this volume). This architecture uses synaptic change on a long time scale for long-term memory; fast modulation of synaptic weights about the long-term values then provides the dynamic relationships within and across assemblies. This architecture appears to be promising for neural instantiation of schemas.

2 Schemas for Rana computatrix

Lettvin, Maturana, McCulloch & Pitts (1959) initiated the behaviorally oriented study of the frog visual system with their classification of retinal ganglion cells into four classes each projecting to a retinotopic map at a different depth in the optic tectum, the four maps in register. We view the analysis of such interactions between layers of neurons as a major approach to modeling "the style of the brain." This Section, and the three which follow, will develop various aspects of *Rana computatrix* to exemplify the general view of cooperative computation at the level of schemas, between neurons within a layer, and between layers in specific models exhibiting cooperative computation.

2.1 Schemas for Prey Recognition

Lettvin et al. (1959) found that group 2 retinal cells responded best to the movement of a small object within the receptive field and group 4 cells responded best to the passage of a large object across the receptive field. It became common to speak of R2 cells as "bug detectors" (Barlow 1953 referred to R3 ganglion cells as "fly detectors") and of R4 cells as "enemy detectors," although subsequent studies make it clear that a given frog or toad behavior will depend on far more than the activity of a single class of retinal ganglion cells (Ewert 1976).

Given the mapping of retinal "feature detectors" to the tectum and the fact that tectal stimulation could elicit a snapping response, it became commonplace to view one task of the tectum to be directing the snapping of the animal at small moving objects - it being known that the frog would jump away from large moving objects and would not respond with prey-catching or escape when there were only stationary objects. This might suggest that the animal is controlled by, *inter alia*, two schemas, one for prey-catching, which is triggered by the recognition of small moving objects, and one for predator avoidance, which is triggered by large moving objects. However, Ewert (1968) has observed that animals with lesions of the pretectum will snap at large moving objects that a normal toad will avoid. This suggests a new analysis in terms of a prey-selection schema activated by moving objects of any size, and a predator-recognition schema that serves not only to activate avoidance bahavior but also to inhibit prey acquisition behavior. Thus, even gross lesion studies can distinguish between alternative top-down analyses of a given behavior.

Of course, such an analysis can be refined by more detailed behavioral and neurobiological studies which let us determine what features of a moving object serve to elicit one form of behavior or another. Ewert placed a toad in a perspex cylinder from which it could see a stimulus object being rotated around it (Fig. 2). He then observed how often the animal would respond with an orienting movement (this

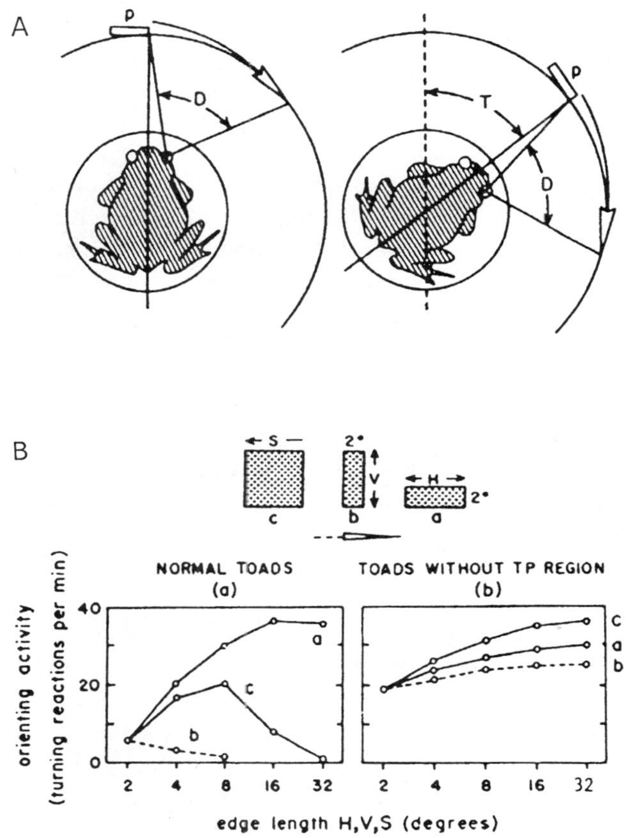

Figure 2. Prey-catching orienting behavior of the toad in response to different configurations of the moving stimulus. (A) Turning reaction to the stimulus presentation. Note that *D*, the effective angular displacement of the stimulus, differs from *T*, the angle of turning movements. (B) Orienting activity to three stimulus configurations, horizontal ("worm": type a) and perpendicular ("anti-worm": type b) rectangles, and squares (type c). Ba) Normal animal's response becomes more frequent when we increase the dimension (H) of a stimulus of type a, whereas response frequency rapidly drops to zero when we increase the dimension (V) of a type b stimulus, and a sort of summation of these two responses is obtained when we increase both dimensions of stimulus type c. Bb) This discrimination is lost in toads with pretectal lesions. (From Ewert 1976).

frequency being his measure of how "prey-like" the object was) for different stimulus objects. As we can see from Fig. 2a, the worm-like stimulus (rectangle moved in the direction of its long axis) proved increasingly effective with increasing length; whereas for 8° or more extension on its long axis, the antiworm stimulus (rectangle moved in the direction orthogonal to its long axis) proved ineffective in releasing orienting behavior. The square elicited an intermediate behavior, the response rising to a maximum at 8°, but being extinguished by 32°.

With such quantitative data to hand, Ewert & v. Seelen (1974) suggested a systems-theoretical model (Fig. 3A shows a version of the model due to Ewert 1974). In this model, retinal output is passed in parallel to a tectal "worm filter" (type I) sensitive to area extension in the direction of movement, and to a thalamic pretectal "antiworm filter" sensitive to area extension perpendicular to the movement direction; the output of the latter then serves to inhibit tectal type II activity excited by type I. A worm-like stimulus (Fig. 3B) would then tend to yield much excitation of the worm filter whose influence would be little inhibited by the thalamic antiworm response, thus yielding a vigorous type II output; the antiworm (Fig. 3D) would yield weak tectal type I activity, strong thalamic activity, and resultant weak tectal type II output. The square (Fig. 3C) would yield intermediate behavior. Ewert & v. Seelen were able to adjust the parameters in this model to fit the data over a linear subrange of the results. Note that the filters need not respond exclusively to worm and antiworm stimuli, respectively, for the model to yield the appropriate response functions. Ewert & v. Seelen stress that the shorthand use of the terms "worm filter" and "antiworm filter" is not meant to imply that there are neurons which serve as specific worm or antiworm detectors in the toad's brain (see also Ewert 1987).

While the average rate of response of the output correlates well with the average turning rate of the toad, the model of Fig. 2 can neither explain the spatial locus at which the toad snaps nor the time at which it snaps. We shall thus turn in Section 3 to a model which can indeed explain the spatiotemporal distribution of the animal's behavior.

2.2 Schemas for Detour Behavior

To see how we may use schemas to move beyond models of the recognition of visual patterns that serve to trigger stereotyped, though appropriately spatially directed, responses, we now consider situations in which the animal exhibits behavior which takes account of a complex spatial context. Specifically, we shall start from data on a toad viewing a vertical paling fence behind which there is a worm. It has been shown that the animal may either snap directly at the worm, or may detour around the barrier. However, it will not go around the barrier if there is no worm behind it. Thus, we may still see the worm as triggering the animal's response, but we now see a complex trajectory dependent upon the relative spatial position of worm and barrier.

A first view of these data is given in Fig. 4 from Collett (1982). The row of dots indicates a paling fence. The two circles indicate two alternative placements of worms which are to attract a toad's attention, while the T indicates an opaque barrier which prevents the toad from seeing the worms after it starts to move. We represent the toad with a dot for its head and a line for its orientation. The sequence of such

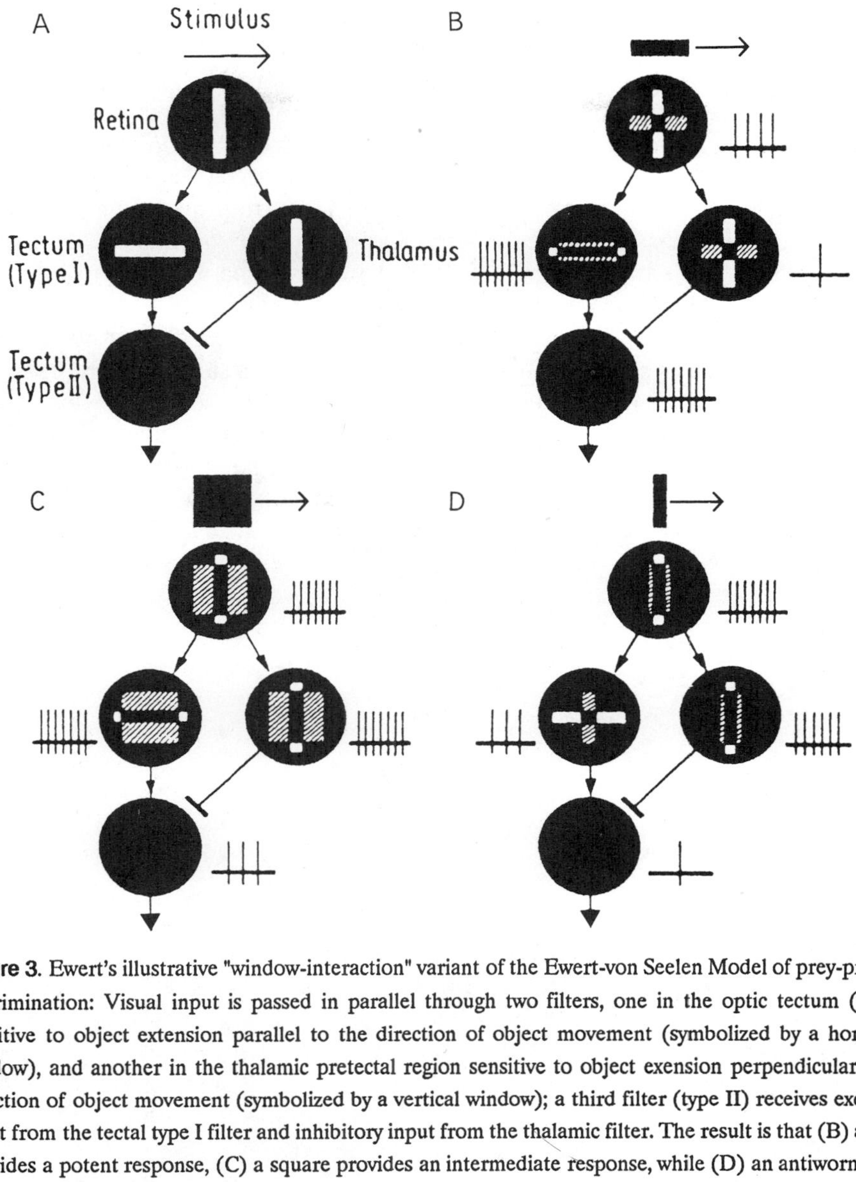

Figure 3. Ewert's illustrative "window-interaction" variant of the Ewert-von Seelen Model of prey-predator discrimination: Visual input is passed in parallel through two filters, one in the optic tectum (type I) sensitive to object extension parallel to the direction of object movement (symbolized by a horizontal window), and another in the thalamic pretectal region sensitive to object exension perpendicular to the direction of object movement (symbolized by a vertical window); a third filter (type II) receives excitatory input from the tectal type I filter and inhibitory input from the thalamic filter. The result is that (B) a worm provides a potent response, (C) a square provides an intermediate response, while (D) an antiworm yields little or no response. (From Ewert 1974).

"arrows" on the right-hand side of the figure indicates successive positions taken by the toad in a single approach to the prey. The animal sidesteps around the barrier, pauses for several seconds, and then continues to a position at which it stops, pointing in approximately the direction of the worm - but note that, due to the opaque barrier, the worm is no longer visible. On the left-hand side of the figure, we indicate the position of the toad on a number of different occasions, at the pause. The solid arrows show the pause for the nearer position of the worms, the dashed arrows give the position of the pause for the further position. What is of interest is that even though the worms are no longer visible to the toad at the time of the pause, the orientation of the animal correlates well with the position of the target. Thus, we must not only explain how it is that the animal chooses whether to proceed directly toward the prey or to sidestep around the barrier, but also come to understand how the position of the target can be encoded in such a way as to be available to guide the animal's behavior even if the target does not continue to be visible.

Didday (1976) gave a simple model of the tectum as a row of neurons selecting its maximal stimulus. Epstein (1979) adapted this model by positing a more structured form of input (Fig. 5). Each visible prey-stimulus provides a tectal input with a sharp peak at the tectal location corresponding retinotopically to the position of the stimulus in the visual field, with an exponential decay away from the peak. Each barrier-stimulus provides a trough of inhibition whose tectal extent is slightly greater, retinotopically, than the extent of the barrier in the visual field. When the model

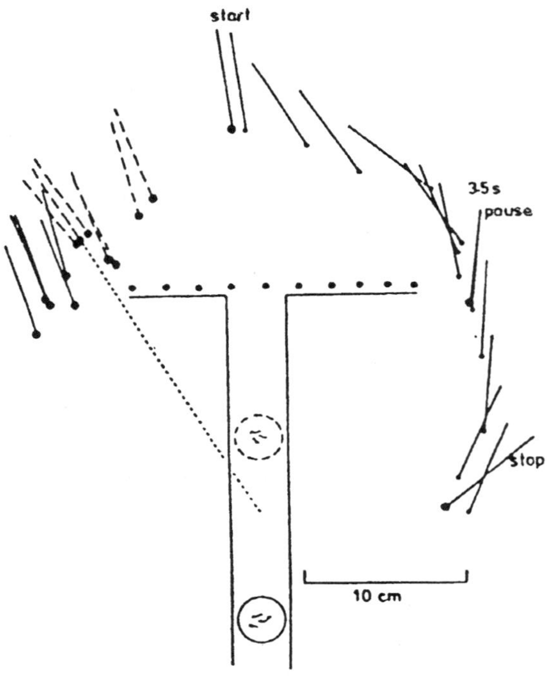

Figure 4. The right hand side of the figure shows the trajectory of a toad which has sighted a number of mealworms behind a paling fence (the row of dots) and then detours around the fence to approach the prey. Note that when it stops (the prey no longer being in view) its position shows that it has retained a representation of prey position, which is relatively accurate despite the intervening movement of the toad. (From Collett 1982).

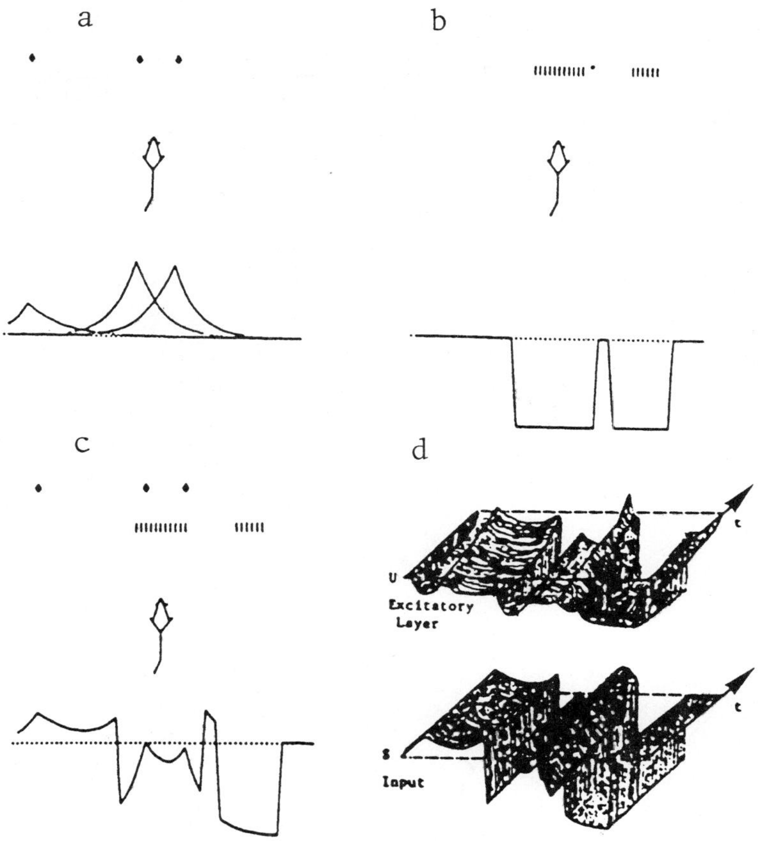

Figure 5. The Epstein Model of detour behavior: Each prey-object provides a broad peak of excitation to the tectum, each barrier a narrow trough of inhibition (Epstein 1979). The tectum then acts as a maximum selector to choose a prey or barrier-edge as the target of the first move. In (a-c) the central object shows the position and orientation of a gerbil, the diamond-shaped icons represent food pellets, and vertically hatched regions represent barriers. The curves below the gerbil show spatially distributed patterns of excitation (regions above the dotted line), and inhibition (below the dotted line) elicited in the model by the configuration of pellets and barriers. In (a) the presence of three prey objects results in an overlapping pattern of excitation, whereas in (b) the barriers result in a trough of inhibition extending a small distance beyond each barrier end. The net effect of summing excitation due to pellets and inhibition due to barriers is shown in (c). The presence of inhibition leaves the maximally excited position to the right of the left fence. The curves traced in (d) show the time-course of the model in response to the stimulus pattern in (c). The time dimension t is drawn going into the paper, the horizontal axis represents the spatial dimension and the vertical axis the level of excitation. The curve of input vector S (bottom) simply shows the stimulus pattern of (c) held constant as time advances. The curve of excitation level U (top) shows that the model eventually converges with all orientations suppressed except for the one corresponding with the initial maximal input. According to the assumptions of the model, this triggers a turn to the right edge of the left barrier.

tectum acts upon the resultant input pattern, it will choose either a prey-locus or the edge of a barrier as the maximum. Thus, Epstein's model can exhibit choice of the direction of the prey or the barrier edge, but not the spatial structure of the behavior.

Given that the behavior of the toad - whether to approach the prey directly, or to detour around the barrier - depends upon the distance at which the worms are behind the barrier, a full model of this behavior must incorporate an analysis of the

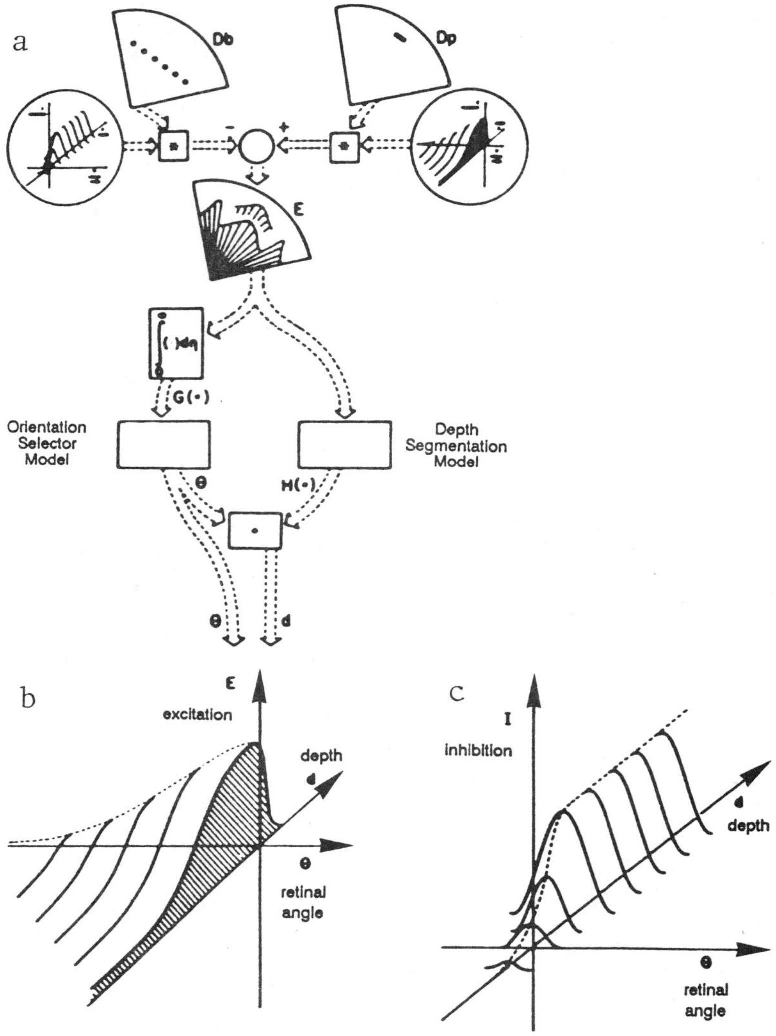

Figure 6. The Orientation Model of toad detour behavior: Assumptions made in this diagram are (i) that separate depth maps are maintained for prey and barrier stimuli, (ii) that direction for an orientation turn is obtained by combining information from these two depth maps, and (iii) that information on preferred orientation and depth of prey and barriers is available simultaneously to motor schemas. These schemas are capable of integrating this information to produce a coordinated motor-output. (From Arbib & House 1987).

animal's perception of depth. To address this, Arbib & House (1987) gave two models for detour behavior which make use of separate depth maps for prey and barriers. In the first, the Orientation Model (Fig. 6) the retinal output of both eyes is processed for "barrier" and "worm" recognition to provide separate depth mappings for "barrier" and "worm." We suggest that the animal's behavior reflects the combined effects of prey "attraction" and barrier "repulsion." Formally, generalizing Epstein's model, the barrier map B is convolved with a mask I which provides a (position-dependent) inhibitory effect for each fencepost; while the worm depth map W is convolved with a mask E which provides an excitatory effect for each worm. The resultant total map

$$T = B * I + W * E$$

is then subject to further processing which will determine the chosen target. E is an excitatory mask which projects broadly laterally, and somewhat less broadly towards the animal; I is an inhibitory mask such that there is a short distance behind the edge in which there is little inhibition (to model the fact that the toad does snap at a worm close behind a fence), after which inhibition is equally strong at all distances. The total excitation T is summed in each direction, and then a maximum selector network chooses that direction with maximal activity. If this corresponds to the prey, the animal will approach and snap, otherwise, further processing is required.

We may note, with Collett (1982) and Ingle (1983), that the full detour behavior exhibited here is quite complex: The animal does not simply orient towards the prey or the end of the barrier; rather, if it does not proceed directly toward the prey, it sidesteps around the barrier orienting in a way that depends upon the position of the target and the length of the sidestep. We postulate that each component of the behavior (sidestepping, orienting, snapping, etc.) is governed by a specific *motor schema*. We then see detour behavior as an example of the coordination of motor schemas (Arbib 1981), where the sidestepping schema acts to modulate the orienting schema. Ingle (1983) has observed that a lesion of the crossed-tectofugal pathway will remove orienting; lesion of the crossed-pretectofugal pathway will block sidestepping; while lesion of the uncrossed-tectofugal pathway will block snapping.

In their second model, the Path-Planning Model, which we will present in Section 5, Arbib & House (1987) associate with each point of the depth map a two-dimensional vector. Instead of a single scalar indicating a measure of confidence that there is a target for the first move at the corresponding position in the visual field, the vector is to indicate the preferred direction of the animal were it to follow a path through that corresponding position. The model specifies how this vector field (a so-called *potential field*) is generated, and begins to specify how the vector field is processed to determine the appropriate parameters for the coordinated activation of motor schemas.

2.3 Schemas for Prey-Acquisition

Lara et al. (1984) offer an alternative model of detour behavior in the presence of barriers in which recognition of gaps is an explicit step in detour computation.

(Consider a human walking through a doorway: The analogous claim is that we recognize the opening we can walk through, rather than avoiding the door-frame.) It is thus a challenge to experimentalists to design ways to discriminate between the hypotheses that "the brain 'recognizes' barriers as inhibitory" and that "the brain 'recognizes' gaps as excitatory" in the cooperative computation of behavior. The same paper also reports models - at the level of interacting schemas, rather than layers of neuron-like elements - for prey-acquisition in environments containing chasms as well as barriers, and for predator-avoidance. Future research will test these models by developing their possible neural instantiations. We now turn to a presentation of the schemas for prey-acquisition by toads posited in this approach.

Figure 7 presents the prey acquisition schema. It is activated by the instantiation of the perceptual schema for *Prey*, but the actual motor schemas executed will depend on whether further perceptual processing activates the *Free-Prey* perceptual schema (*FP*) representing a situation in which no obstacles intervene between the animal and its prey, perceptual schema *Barr* which represents a situation in which a barrier intervenes, or perceptual schema *Chasm* which represents an intervening barrier. The

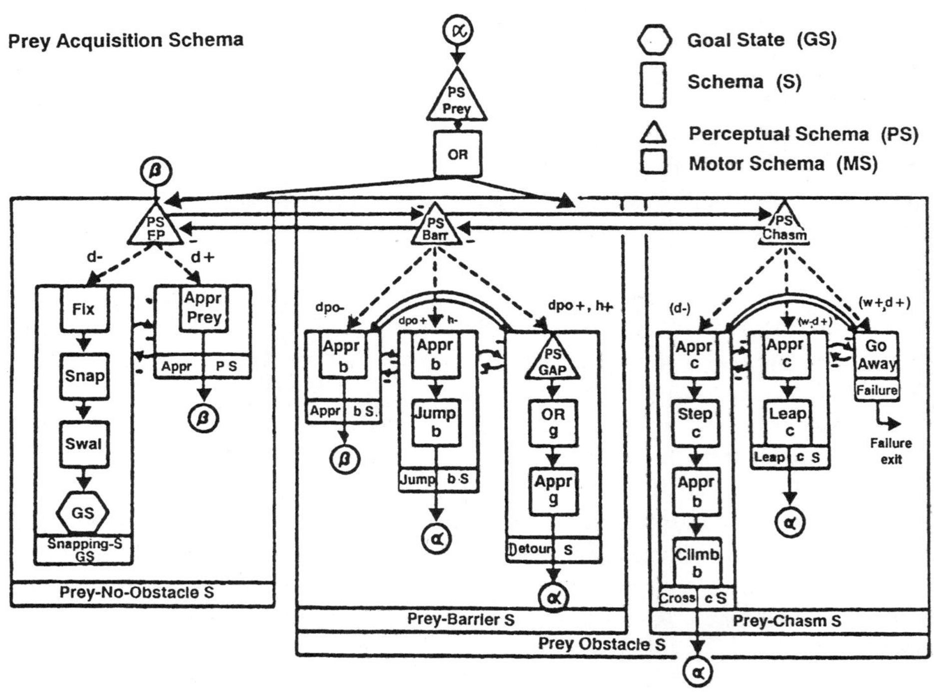

Figure 7. If prey is in the visual field, the orient motor schema *OR* is activated. The consequent unfolding of action depends on whether a barrier (activating perceptual schema *Barr*) or a chasm (activating perceptual schema *Chasm*) or "free space" (activating perceptual schema *FP*) is perceived. See the text for further details. (From Lara et al. 1984).

activation of the corresponding coordinated control program - *Prey-No-Obstacle*, *Prey-Barrier* or *Prey-Chasm*, respectively - is represented in Fig. 7 as the outcome of competitive interaction between the three perceptual schemas:

(1) If perceptual schema *FP* is instantiated, it generates a parameter *d* for the distance of the prey. If *d* is small (d-) then the animal fixates, snaps and swallows the prey to achieve the goal state. If *d* is larger (d+) then the animal approaches the prey. The arrow to *ß* indicates that the animal will return to the circle marked *ß* above: so long as the prey remains visible, the animal will continue to approach it until *d* is small enough for it to snap and swallow the prey. It might be argued that both of the *ß*-returns in Fig. 7 should be replaced by returns to *a* since a barrier might be interposed or become visible after the animal has executed an approach motor schema. In any case, such returns (to be discussed further below) represent the behavioral fact that the toad seems to execute a single behavior (which may itself be composite, directed by a coordinated control program, as in fixate → snap → swallow) to completion, with the passage to the next requiring a fresh perceptual trigger. By contrast, humans formulate some overall plan to be pursued and modulate it in the light of changing circumstances as when, going to get some object, we initiate a search strategy should the object not be in its expected place.

(2) If perceptual schema *Barr* is instantiated, it makes available two parameters, *dpo*, the distance of the prey behind the barrier, and *h*, the height of the barrier. If *dpo* is small or negative (dpo-), the toad approaches the barrier and proceeds as if no barrier were present (transfer to *ß*). If the prey is further behind the barrier and the barrier is low (dpo+, h-) the animal will jump over the barrier, whereas if the barrier is higher (dpo+, h+), the animal will detour. More specifically, instantiation of the perceptual schema for a gap will elicit the motor schema for orienting to (*Or g*) and approaching (*Appr g*) that gap. It seems necessary to include the "gap" at the end of a barrier as also activating the gap perceptual schema. In either dpo+ case, control is returned to ß after execution of the specified motor schema.

(3) Finally, if perceptual schema *Chasm* is activated, it makes available parameters representing the depth *d* and width *w* of the chasm. If the depth is small (d-) the toad will walk across the floor of the chasm; if the chasm is deep but not wide (w-, d+), the toad will leap the chasm; but if the chasm is wide and deep (w+, d+), the animal will not approach the chasm.

However, general considerations (Arbib 1981) suggest that the animal's behavior should depend on the representation of the environment by an assemblage of schemas embodying the spatial relations between multiple objects in the environment, rather than sequential activation of a perceptual schema for one prey and then at most one obstacle. In fact, in the description Lara et al. (1984) give of the actual implementation of gap approach, they do use a mode of competitive interaction which is only hinted at in the Fig. 7 by the various arrows labeled "-" to express competitive interactions. More generally, the idea seems to be that the *Gap* schema is instantiated for each gap in the environment, each schema is given an activity level based on its position relative to toad, prey and other gaps, and it is the most active *Gap* schema which provides the parameters for the next motor schema activation. Once it has executed this motor schema, the animal's behavior is determined *ab initio* by the activation of its perceptual schemas in its new situation.

Note that this explicit view of motor schema activation by the "winner" (cf. the Didday Model of prey-selection) of perceptual schema competition obviates the α-β question raised above, since we simply postulate that, once triggered, a motor schema executes to completion with the parameters supplied by the perceptual schema, whereupon activation of perceptual schemas is re-initiated to trigger the next motor schema. As noted above, this seems an appropriate model for frog and toad - the schema assemblage is completely "refreshed" after each integral action. This is in contrast to the action-perception cycle stressed in, e.g., human behavior, in which activation of perceptual schemas serves to update an extant schema assemblage and coordinated control program, rather than necessarily to create new ones (see Koy-Oberthür, this volume).

An example of two trajectories predicted by the model of Lara et al. (1984) is presented in Fig. 8. In (a), the toad chooses the gap nearer the prey, but in (b), the presence of the wide gap in the further fence biases the animal's "choice" to favor the other gap in the nearer fence. Clearly, such predictions can be used to test the model, and data culled from observations of animal behavior in this fashion can be used to refine the algorithmic specification of the constituent schemas, just as psychophysical data can be used to tune schema-based models of visual function in humans. Such specifications can serve as the endpoint of analysis for the ethologist uninterested in neural mechanisms; but for the neuroethologist, they can provide the formal "specification" of the task whose implementation in neural circuitry is to be analyzed.

a

b

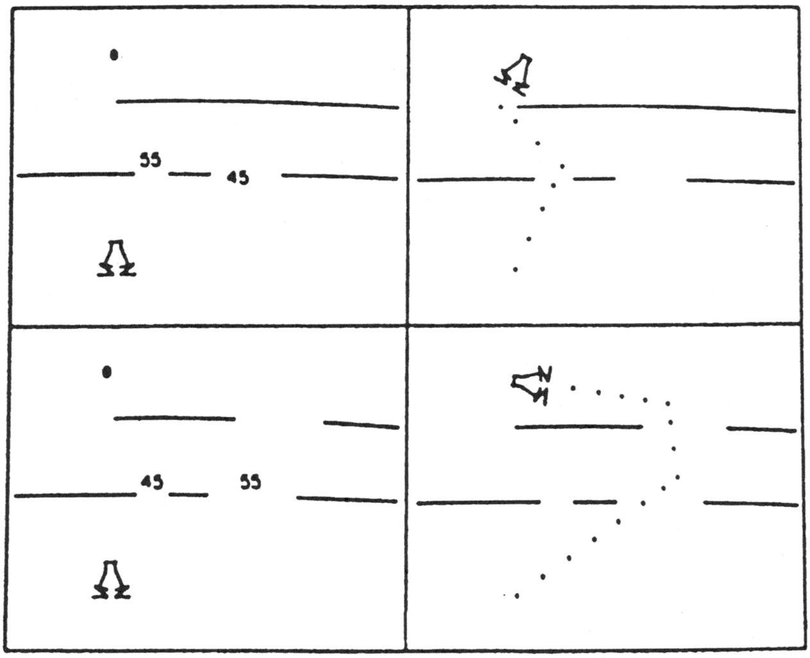

Figure 8. Computer simulation, using a program elaborating the schemas of Fig. 7, of a toad's behavior in response to a worm in the presence of barriers. In (a) the toad always prefers the closer gap, in (b) the further gap is broader and is the one chosen. (From Lara et al. 1984).

3 Tectal Columns

We now turn to models of actual neural circuitry in anuran tectum. The first model addresses mechanisms for facilitation of prey-catching behavior found at a single locus of tectum. We will then see how interactions among a number of such loci ("tectal columns") "unlump" the Ewert-v.Seelen Model of prey-recognition of Section 2.1.

3.1 Facilitation of Prey Catching Behavior

Presenting a worm to a frog for 0.3 s may yield no response, whereas orientation is likely to result from a 0.6 s presentation. Ingle (1975) observed a facilitation effect: If a worm is presented initially for 0.3 s, then removed, and then restored for only 0.3 s, the second presentation suffices to elicit a response as long as the intervening delay is at most a few seconds. Ingle observed tectal cells whose time course of firing accorded well with this facilitation effect, leading Lara et al. (1982) to a model in which the "short-term memory" is encoded as reverberatory activity in a neural circuit. Their tectal column model is abstracted somewhat crudely from the anatomy of Székely & Lázár (1976) - and thus is currently being updated in the light of new and detailed anatomy and neurophysiology (see the contributions by Lázár and Matsumoto in this volume for examples of the data, and the paper by Betts for a first step on the new modeling).

Each column (Fig. 9) comprises one pyramidal cell (PY) as sole output cell, one large pear-shaped cell (LP), one small pear-shaped cell (SP), and one stellate interneuron (SN). All cells are modeled as excitatory, save for the stellates. The retinal input to the model is a lumped "foodness" measure, and activates the column through glomeruli with the dendrites of the LP cell. LP axons return to the glomerulus, providing a positive feedback loop. A branch of LP axons also goes to the SN cell. There is thus competition between "runaway positive feedback" and the stellate inhibition. Glomerular activity also excites SP which likewise sends its axon back to the glomerulus. SP also excites LP to recruit the activity of the column. PY is excited by both SP and LP. The cells in our model interact only via their "firing rates," while the firing rate of a cell depends on its own membrane potential. Thus, in computer simulation of this model we do not need separate variables to hold "old" and "new" values, but instead cycle through two steps:

Step 1: Combine current values of firing rates to compute new values of membrane potentials.

Step 2: Using new values of membrane potentials, form the new values of the corresponding firing rates.

We shall use the following notation: We use the same symbol to denote the cell and its firing rate, while we use lower case to indicate the corresponding membrane potential. For example, at time t cell PY has firing rate PY(t) and membrane potential py(t).

Step 1. Updating the Membrane Potentals

The membrane potential of each cell is described by a differential equation of the form

$$\tau_m \, dm(t)/dt = - m(t) + S_m(t).$$

Here the subscript m indicates the specific cell type, while m(t) denotes the membrane potential of that cell at time t. τ_m is the time constant for the rate of change dm(t)/dt of this potential. $S_m(t)$ represents the total input that the cell of type m receives from other cells. If this input is 0, the above equation reduces to $\tau_m dm(t)/dt = -m(t)$ whose solution is exponential decay toward 0 with time constant τ_m. If the total input $S_m(t)$ is positive, the differential equation will drive m(t) toward a higher value; if negative, m(t) will tend toward a lower value.

In our computer simulations, we have taken the simplest possible approach to solving the above differential equation. We choose some small time step Δt (say 20 ms of "toad time") and then approximate dm(t)/dt by the ratio $(m(t+\Delta t) - m(t))/\Delta t$. We thus replace our differential equation by the difference equation

$$\tau_m[(m(t+\Delta t) - m(t))/\Delta t] = - m(t) + S_m(t)$$

which, on rearranging terms, yields

$$m(t+ \Delta t) = (1- \Delta t/ \tau_m)m(t) + (\Delta t/ \tau_m)S_m(t). \quad (1)$$

Step 2. Updating the Firing Rates

A more detailed model than ours would explicitly model spike generation in response to changes in membrane potential. However, we use instead a coarse approximation, choosing one of three simple functions which express the intuition that "the higher the membrane potential, the higher the firing rate." The "firing rate" of a cell is obtained by passing the membrane potential through a "thresholding function" of one of the types below:

(i) $f(x) = 1$ if $x \geq 0$; whereas $f(x) = 0$ for $x < 0$.

In this case, 0 is the threshold value for the membrane potential x at which the cell makes the transition from "no firing" to "maximal firing," which we standardize to the value 1.

(ii) $g(x) = x$ if $x \geq 0$; whereas $g(x) = 0$ for $x < 0$.

Again, 0 is the threshold, but now the firing rate increases proportionately as the membrane potential x increases above threshold.

(iii) $s(x, A, B) = 0$ if $x \leq A$

thus A is the firing threshold;

$$s(x, A, B) = (x-A)/(A-B) \text{ for } A \leq x \leq B$$

firing then increases linearly until x reaches the "saturation level" B; while

$$s(x, A, B) = 1 \text{ for } x \geq B$$

the cell fires maximally for any value of the membrane potential above B.

The updating of the firing rate is then accomplished simply by applying a thresholding function to the corresponding membrane potential. Recalling our convention that upper case denotes firing rate, while lower case indicates the corresponding membrane potential, we may succinctly summarize our choice of

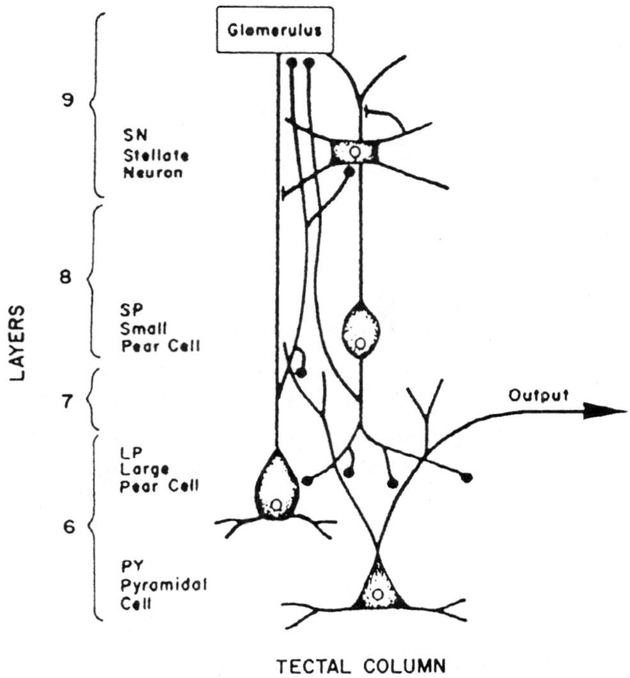

Figure 9. The tectal column. The glomerulus is constituted by the LP and SP dendrites and recurrent axons as well as by optic and diencephalic terminals. The LP excites the PY, the SN, and the GL, and is inhibited by the SN. The SP excites the LP and PY cells, and it sends recurrent axons to the glomerulus; it is inhibited by the SN. The SN is excited by LP neurons and diencephalic fibers and it inhibits the LP and SP cells. The PY is activated by the LP, SP, and optic fibres, and is the efferent neuron of the tectum. (Lara Arbib 1982).

thresholding functions as follows:

```
Glomerulus                GL = gl
Large Pear-Shaped Cell    LP = f(lp - 1.0)
Small Pear-Shaped Cell    SP = f(sp - 2.0)
Stellate Neuron           SN = g(sn - 0.2)
Pyramidal Neuron          PY = s(py, 2.3, 5.0)
Pretectal Cell            TP = g(tp - 3.8).
```

In the glomerulus (Fig. 9), we simply take GL=gl, but we still need here two distinct values because of our updating convention. Thus, LP and SP provide the maximal firing rate (which we have standardized to 1) when their membrane potential passes the threshold values of 1.0 and 2.0, respectively. SN and TP provide a graded firing rate above thresholds 0.2 and 3.8 respectively. PY fires above threshold 2.3, reaching maximal firing rate at saturation level 5.0.

Recall that $S_m(t)$ represents the sum of all the excitation and inhibition received by the cell of type m. Thus, the overall dynamics determined by Eq.(1) will depend upon the actual choice of excitatory and inhibitory weights in each $S_m(t)$, as well as the choice of membrane time constants τ_m. Appropriate choices ensure that excitation of the input does not lead to runaway reverberation between the LP and its glomerulus as well as ensuring that this activity is "chopped" by stellate inhibition to yield a period of alternating LP and SN activity. SP has a longer time constant and is recruited only if this alternating activity continues long enough. In one simulation experiment, we graphed the activity of cells as a function of how long a single stimulus is applied (Lara et al. 1982). There is, as in the experimental data, a critical presentation length below which there is no pyramidal response. Input activity activates LP, which re-excites the glomerulus but also excites the SN, which reduces LP activity. But if input continues, it builds on a larger base of glomerular activity, and so over time there is a build-up of LP-SN alternating firing. If the input is removed too soon, the reverberation will die out without activating SP enough for its activity to combine with LP activity and trigger the pyramidal output. If input is maintained long enough, the reverberation may continue, though not at a high enough level to trigger output. However, re-introducing input shortly after this "subthreshold" input can indeed "ride upon" the residual activity to build up to pyramidal output after a presentation time too short to yield output with an initial presentation.

It required considerable computer experimentation to find weights that yield the neural patterns discussed below. More recently, Cervantes-Pérez (1985) has given a mathematical analysis of how weighting patterns affect the model's overall behavior. We hope our hypotheses concerning the ranges of the parameters involved in the model will stimulate more detailed neurobiological studies of tectal activity.

3.2 An Array of Tectal Columns that Models Pattern Recognition

To exemplify the "evolutionary" development of *Rana computatrix*, we now show how an array of tectal columns may be connected to model how the tectum may interact with pretectum in subserving "prey schemas" - recall Ewert's (1974, 1976) data summarized in Fig. 2. We now further note Ewert's recordings of the activity of different types of neurons in toads. Pretectal thalamic TH3 neurons have a response that is uniform for worms, increases with increasing length for antiworms, but is greatest for squares, increasing with their size. The activity of tectum type I neurons (T5.1) seems to correlate fairly well with increases in the length of the stimulus in the direction of motion (rather than to the length perpendicular to the direction of movement), although it is somewhat higher for squares than for worms. It is the tectum type II neuron (T5.2) whose overall rate of *neural* response seems to best match the overall frequency of the *behavioral* response. However, the discrimination between worm, antiworm, and square is essentially abolished by pretectal lesion. Our task now is to give a neural network model.

The structure of our model is shown in Fig. 10 (Cervantes-Pérez et al. 1985). Rather than using a single column to represent the tectum, we now use a 8 x 8 array of columns, and we provide the column not only with the R2 input of our initial study of facilitation, but also with R3 and R4 input. The connections of R3 and R4 to TH3 are tuned to yield the observed TH3 responses.

The retinal input to the model is based on overall ganglion cell response curves, in which only the average rate of firing of a cell is given for each stimulus, rather than the temporal pattern of that response. Since the response of tectum and pretectum must depend on the spatiotemporal pattern of retinal ganglion cell firing, current work (Teeters 1988) in my laboratory (extending the work of Lee 1986) is aimed at more detailed modeling of the response of frog retina to varied activity.

The tectum-pretectum model comprises six layers, each of which is characterized by two arrays: The first, labeled with a lower case letter, represents membrane potentials, while the second, bearing the corresponding upper case label, represents the firing rates. There are five layers in the tectum - glomeruli (gl, GL), large pear-shaped cells (lp, LP), small pear-shaped cells (sp, SP), stellate neurons (sn, SN), and pyramidal cells (py, PY) - while there is just one layer (tp, TP) of cells in the current model of the pretectum. Recalling Eq. (1) for updating the membrane potential of each cell, we must now specify how each term $S_m(t)$ incorporates the input from all the other cells to which the given cell is connected. The influence of one layer, a, on another, b, will be represented by a matrix $W_{a.b}$. In the present model, each matrix is 3 x 3, and is constant save for its central element. In what follows we shall use the abbreviation W{x,y} for the matrix

$$\begin{bmatrix} y & y & y \\ y & x & y \\ y & y & y \end{bmatrix}$$

In what follows, the notation B=W*A denotes that the array of activity B is obtained by "convolution" connecting array A to array B according to the pattern of connection weights W, and then summing up the firing rates accordingly. Thus the activity of the (i, j) element of the B array satisfies the formula

$$B(i,j) = \sum_{k,l} W(k,l)A(i+k,\ j+l)$$

where the summation runs over the indices (k,l) of the matrix W of "synaptic weights."

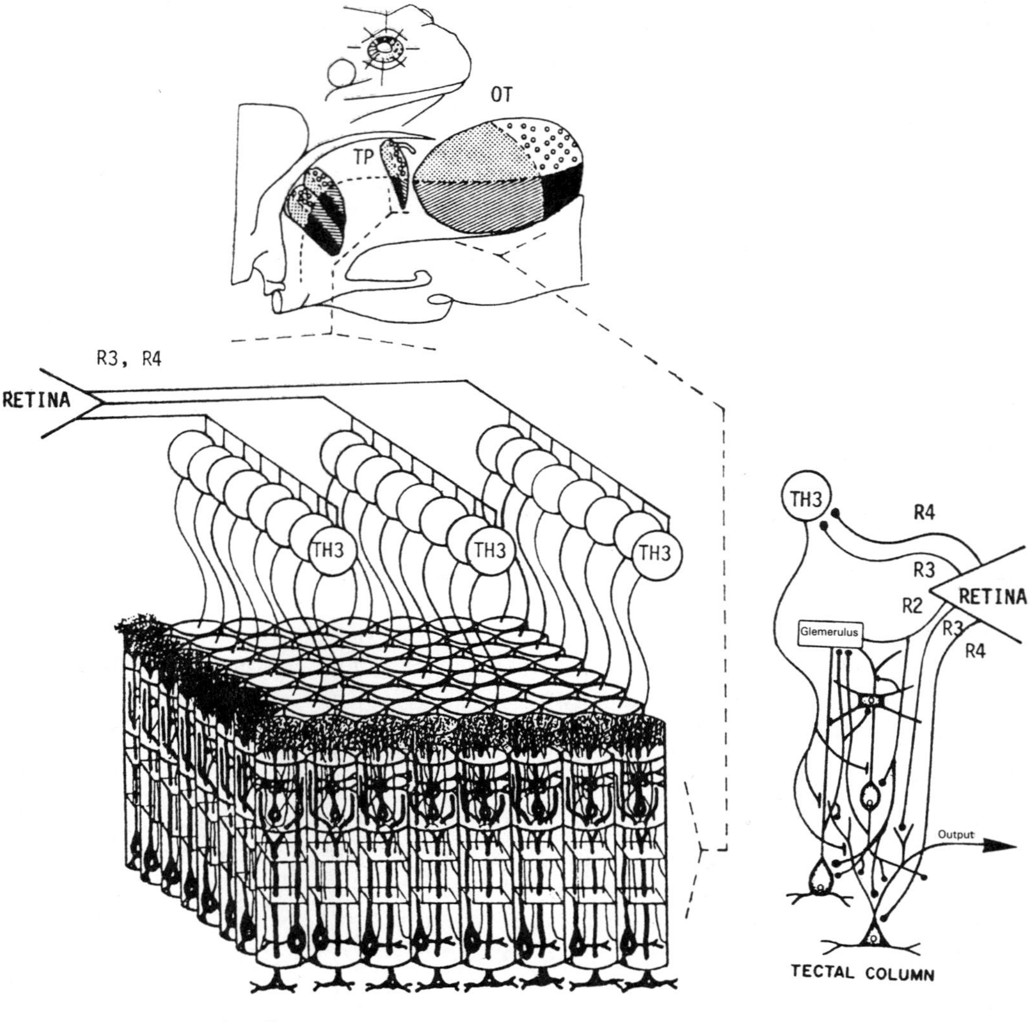

Figure 10. Interactions among retina, optic tectum and pretectum in the model of Cervantes-Pérez, Lara & Arbib 1985. The retina sends fibers in a retinotopical fashion to both optic tectum (classes R2, R3 and R4) and pretectum (classes R3 and R4). (a) TH3 neurons also project retinotopically to the optic tectum. For simplicity we only show the projection of three rows of TH3 cells projecting upon the tectal columns. (b) A closer look of the interactions among retinal, tectal, and pretectal cells. The TH3 cell of the pretectal column inhibits LP, SP, and PY of the tectal column corresponding to its retinotopic projection.

Glomerulus: Here $\tau_{gl} = 2.3$ while

$$S_{gl} = W_{r2.gl}*R2 + W_{lp.gl}*LP + W_{sp.gl}*SP$$

where $W_{r2.gl} = \{6.7, 0\}$, $W_{lp.gl} = W\{8, 5.3\}$ and $W_{sp.gl} = W\{0.7, 0.7\}$.

The form of the equation simply expresses the fact that, in our model, the glomeruli receive inputs from the R2 retinal array, as well as from the arrays of large and small pear cells.

Large Pear-Shaped Cell: Here $\tau_{lp} = 0.3$ while

$$S_{lp} = W_{gl.lp}*gl + W_{r2.lp}*R2 + W_{sp.lp}*SP - W_{tp.lp}*TP - W_{sn.lp}*SN$$

with $W_{gl.lp} = W_{r2.lp} = W\{1,0\}$, $W_{sp.lp} = W\{0.8,0.6\}$, $W_{tp.lp} = W\{0.1,0\}$, and

$W_{sn.lp} = W\{8.0,8.2\}$

Small Pear-Shaped Cell: Here $\tau_{sp} = 0.9$ while

$$S_{sp} = W_{r2.sp}*R2 + W_{gl.sp}*GL - W_{tp.sp}*TP - W_{sn.sp}*SP$$

with $W_{r2.sp} = W\{1,0\}$, $W_{gl.sp} = W\{1.0,0.5\}$, $W_{tp.sp} = W\{0.1,0\}$, and $W_{sn.sp} = W\{20.0,0\}$.

Stellate Neuron: Here $\tau_{sn} = 1.6$ while

$$S_{sn} = W_{lp.sn}*LP$$

with $W_{lp.sn} = W\{5.2,5.2\}$.

Pyramidal Neuron: Here $\tau_{py} = 0.12$ while

$$S_{py} = W_{r2.py}*R2 + W_{r3.py}*R3 + W_{r4.py}*R4 + W_{sp.py}*SP + W_{lp.py}*LP - W_{tp.py}*TP$$

where $W_{r2.py} = W\{3.5,0\}$, $W_{r3.py} = W\{0.3,0\}$, $W_{r4.py} = W\{7.0,0\}$, $W_{sp.py} = W\{2.0,0\}$,

$W_{lp.py} = W\{0.7,0.56\}$, and $W_{tp.py} = W\{0.9,0\}$.

Pretectal Cell: Here $\tau_{tp} = 0.02$ while

$$S_{tp} = W_{r3.tp}*R3 + W_{r4.tp}*R4$$

where $W_{r3.tp} = W\{0.3,0\}$ and $W_{r4.tp} = W\{5.0,0\}$.

With the above setting of parameters, the model does indeed exhibit in computer simulation responses to moving stimuli of different types that match well

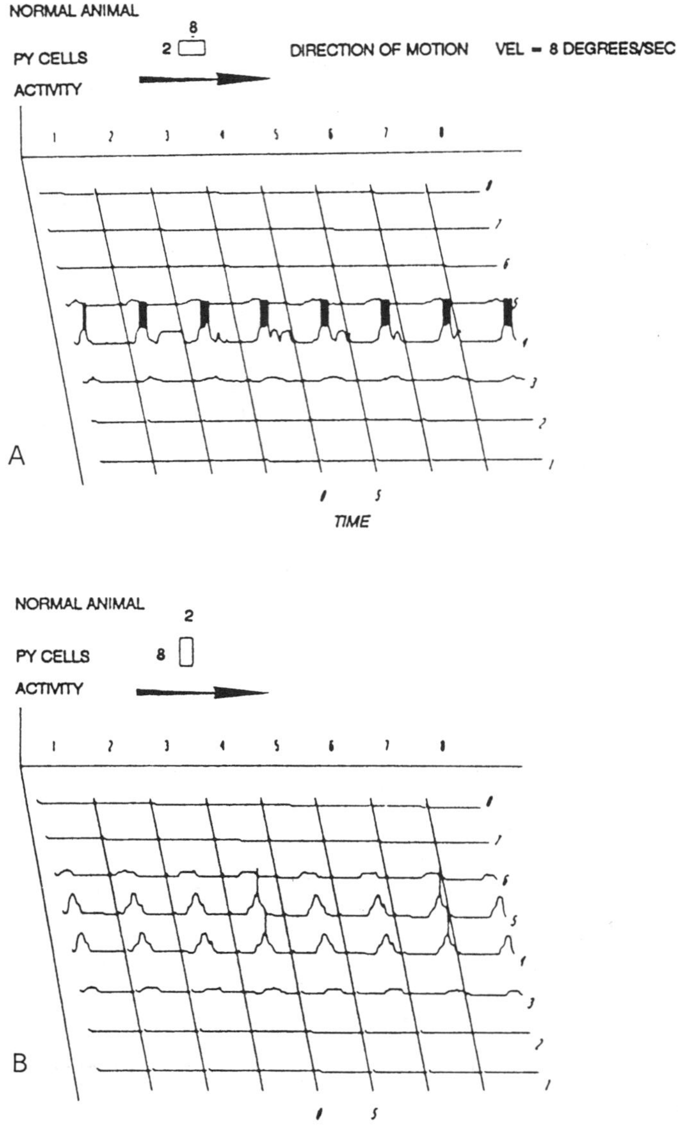

the neural data. However, the model is only approximate at a quantitative level, so that - if our goal is prediction of detailed neural firing rather than just a general understanding of pattern recognition networks - further work must be done on tuning the model parameters. Figure 11 shows the response, for a brief period of time, of the 8 x 8 array of tectal columns, subject to pretectal inhibition, to a moving worm (Fig. 11A), antiworm (Fig. 11B), and square (Fig. 11C). Each figure represents an 8 x 8 array of graphs of neural activity corresponding to the pyramidal cell activity of the column, with time measured from left to right in each subgraph. We thus see that the

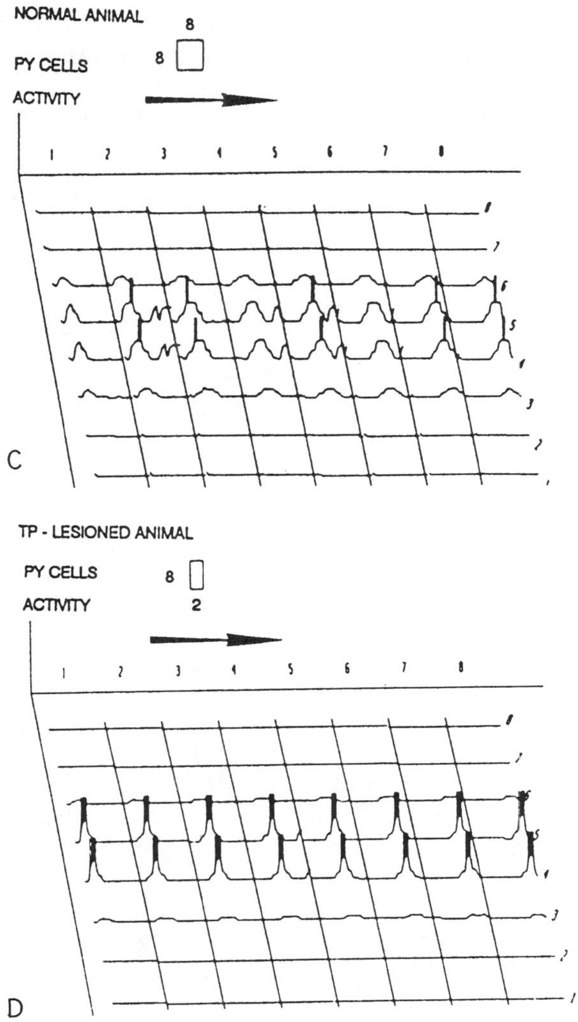

Figure 11. A computer simulation of toad's PY cell response of the 64 columns of the model of the interactions among retina, optic tectum, and pretectum. All stimuli are moved from left to right with a speed of 8 degrees/s. The figure shows four-dimensional graphs, where the x and y axis are used to represent the spatial localization of the (i, j) column. The y axis of this plane is also used to show the time scale for the response of every column's PY cell, while the vertical axis (z axis) represents its local membrane potential. In the graph when the PY local potential is about the threshold, this is indicated by spikes. (A) Response to a "worm-like" stimulus of 2 x 8 degrees (type a). (B) Response to an "antiworm-like" stimulus of 8 x 2 degrees (type b). (C) Response to a square stimulus of 8 x 8 degrees (type c). The response of the PY cells is strongest to stimulus type a, then to type c, and finally to type b. It can be seen that the response to type c stimulus, although wider spread, is weaker than that to type a, whereas the weakest response is again to stimulus type b. PY cells - like T5.2 neurons - are most sensitive to "worm-like" stimuli, rather than to squares or to "antiworm-like" stimuli. (D) Response of a TP-lesioned animal to stimulus type b. (From Cervantes-Pérez et al. 1985).

response to a worm moving from left to right is delayed later and later as we look at tectal columns arrayed from left to right in the path corresponding retinotopically to the path of the worm. Here we see that there is a vigorous response to the worm, virtually no response to the antiworm, and only a weak response to the square. For further details of the model, the results of further simulation, a demonstration of directional invariance of response, and a discussion of motivation, the reader is referred to Cervantes-Pérez et al. (1985). Extensions of the model are provided by the contributions of Lara, Betts and Cervantes-Pérez in this volume.

4 Depth Perception

4.1 The Dev Model

The problem for many models of binocular perception is to suppress ghost targets. The essence of the Dev 1975 scheme (Fig. 12) is to have those neurons which represent similar features at nearby visual directions and approximately equal depths excite each other, whereas those neurons which correspond to the same visual direction but different depths are (via interneurons) mutually inhibitory (Arbib 1987a offers a brief review of related models).

Amari & Arbib (1977) have formalized the Dev Model as follows. It receives input from two one-dimensional retinas, each with coordinate axis x. It comprises a two-dimensional "excitatory field" where the membrane potential $u_d(x,t)$ of the cell at position (x,d) represents the confidence level at time t that there is an object in direction x on the left retina whose disparity on the right retina is d, and a one-dimensional inhibitory field with v(x,t) the activity of the cell at position x at time t. We think of d as taking a set of discrete values. The output of these cells are given as

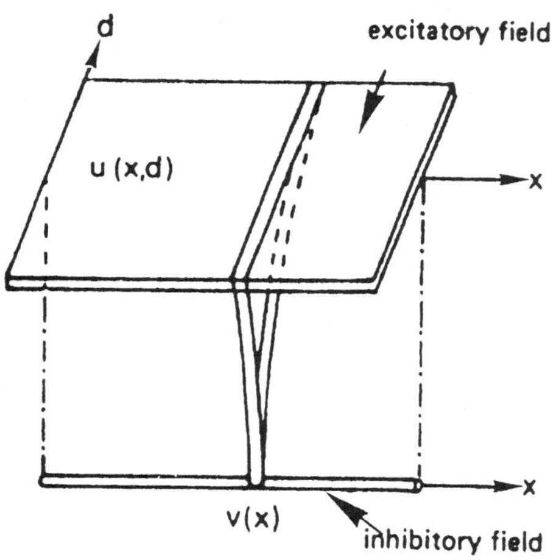

Figure 12. The Dev Model for stereopsis involves competition along the disparity axis (constant x) and cooperation along the direction axis (constant d).

$f(u_d(x,t))$ and $g(v(x,t))$, respectively, where (recalling the thresholding functions of Section 3.1)

$f(u) = 1$ if $u > 0$, and is otherwise 0, while
$g(v) = v$ if $v > 0$, and is otherwise 0.

Employing the B= W*A convention of the previous Section, the model can then be expressed mathematically as:

$$\tau_u \partial u/\partial t = -u + w_1 * f(u) - w_2 * g(v) - h_u + s$$

$$\tau_u \partial v/\partial t = -v + \Sigma_d f(u_d) - h_v,$$

where the first differential equation describes the dynamics of the 2-dimensional array u and the second that of the one-dimensional array v. Here h_u and h_v are threshold levels (constants) and s_d is the input array. We set

$R_L(x) = 1$ if some object projects to point x on the Left retina

and otherwise set it to 0, and similarly for R_R. We then define the stereo input to the model as

$$s_d(x) = R_L(x)\ R_R(x+d)$$

which is 1 only if there is an object at position x on the left retina as well as at x plus *disparity* d on the right retina. A more refined model would make $s_d(x)$ a measure of the correlation of the features of the pattern near position x on the left retina and those centered at x+d on the right retina.

In this way, neurons which could represent elements of a surface in space will cooperate, whereas those which would represent paradoxical surfaces at the same depth will compete. The result is that, in many cases, the system will converge to an adequate depth segmentation of the image. However, as we shall now see, such a system may need extra cues. For example, in looking at a paling fence, if several fenceposts are matched with their neighbors on the other eye in a systematic fashion, the cooperative effect can swamp out the correct pairing and lead to the perception of the fence at an incorrect depth. In animals with frontal facing eyes such ambiguity can be reduced by the use of vergence information to drive the system with an initial depth estimate. For example, Hirai & Fukushima (1978) make explicit use of cues from vergence, favoring solutions with lower parallax. Another method is to use accommodation information to provide the initial bias for a depth perception system; this is more appropriate to the amphibian, with its lateral-facing eyes. It is the latter possibility we pursue here, as part of our concern with *Rana computatrix*.

4.2 The Cue Interaction Model

Ingle (1976) had observed that a monocular frog can snap fairly accurately at prey presented within its monocular field, implying that depth can be estimated from monocular cues. Collett (1977) used experiments with prisms and lenses placed in fornt of the eyes of the toad to show that, in its binocular field, the toad relied mainly on stereopsis, but that the monocular toad did make depth judgements based on accommodation cues in the monocular animal, but which would nonetheless be most dependent upon binocular cues within the binocular field. This evidence for the use of multiple depth cues led House (1982, 1984) to develop the Cue Interaction Model which extends the Amari & Arbib version of the Dev Model (see Chipalkatti & Arbib 1988 for a mathematical analysis). It uses both accommodative and disparity-matching cues to guide a cooperative-competitive process that segments a visual scene into depth regions. Further, model parameters are so tuned that depth cues from binocularity dominate, while monocular accommodative cues remain sufficient to determine depth in the absence of binocular cues.

The model produces a complete depth map of the visual field, and so is appropriate for building a representation of barriers for use in navigation. The model is shown in Fig. 13. At the top we see an accommodation-driven field, M, which receives information about accommodation and which - left to its own devices - would

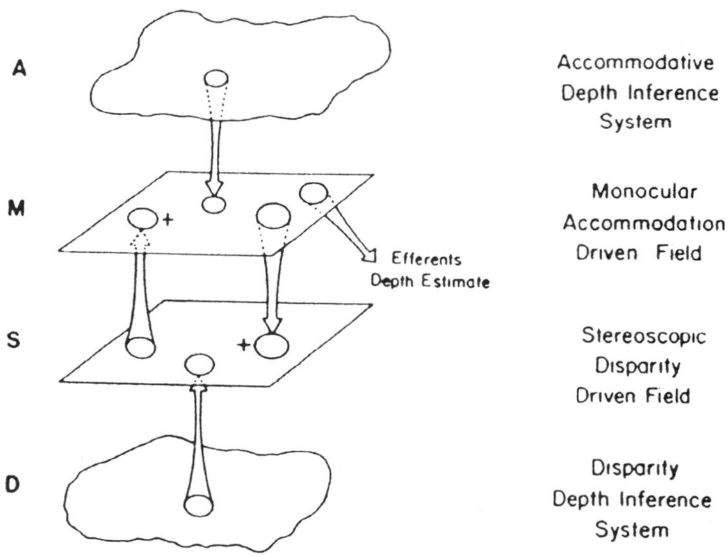

Figure 13. The Cue Interaction Model. Layer A represents an inference system that provides monocular depth cues from lens accommodation, and layer D represents an inference system that provides binocular cues from disparity matching. Layers M and S represent spatially organized fields over which two depth-mapping processes operate. Arrows from layers A and D to these fields indicate that field M receives only monocular depth cues, and field S receives only binocular cues. The ovals and arrows between fields M and S indicate mutual excitatory interconnections that map each local region (oval) of one field onto the corresponding local region of the other field. By means of these interconnections, points of high excitation in one field provide additional excitation to corresponding points in the other field. Competition among depth estimates within each field assures that points excited in only one field will have little chance to sustain this excitation when there are other points receiving stimulation in both fields. (From House 1982).

sharpen up that information to yield relatively refined depth estimates. Below, we see the type of system, S, posited by Dev to use disparity information and to suppress ghost targets. However, the systems are so intercoupled that a point in the accommodation field M will excite the corresponding point in the disparity field S, and *vice versa*. Thus, a high confidence in a particular (direction, depth) coordinate in one layer will bias activity in the other layer accordingly. The result is that the system will converge to a state affected by both types of information - although the monocular system can, by itself, yield depth estimates.

Figure 14 shows stages in the processing by this model of a scene comprising a fence and two worms. The left-hand column of each Fig. 14b shows the accommodation (above) and the disparity field (below) for the fence information. In the top image of Fig. 14b, we see the initial state of the accommodation field. The information is blurred, representing the lack of fine tuning offered by accommodation. Below, we see the initial state of the stereopsis field. The targets are better tuned, but they offer ghost images in addition to the correct images. Figure 14f shows the outcome of such interaction. We see that virtually all the ghost fence targets have also been suppressed. In addition, we see that the accommodation information has been sharpened considerably. The information is now precise and unambiguous, and thus can be used to guide the further behavior of the animal.

4.3 The Prey Localization Model

Collett & Udin (1983) used lesion experiments to demonstrate that, at least for the more limited problem of unobstructed prey-catching, toads are able to make accurate binocularly-based depth estimates even after the major cross-brain binocular relay (nucleus isthmi) has been severed. To explain their results, Collett & Udin postulated that the toad may use a neurally-implemented triangulation process to localize prey, rather than a process based on disparity matching. This postulate suggests a depth resolving system that does not attempt to produce a complete depth map but instead selects a single point and locates that point in space (i.e., the position of a prey).

Our most recent depth modeling efforts have been directed at exploring the postulate of Collett & Udin (1983) , while still incorporating the use of multiple depth cues. These efforts lead to the development of the Prey Localization Model (House 1984; for a mathematical analysis of this system, the reader is referred to Chipalkatti & Arbib 1987). This model is summarized in Fig. 15. It uses pattern recognizers and prey selectors on each side of the brain to select and identify the position of a prey target on each eye. The right and left eye positions are used to compute a depth which is used, in turn, to adjust lens focus. If the selected points on the two eyes correspond with the same point in space then the depth estimate will be accurate and the object will be brought into clearer focus. This means that the images of the points will be selected by the left and right prey selectors with even greater confidence, ensuring that the system will indeed choose that object. If the points do not correspond then the computed depth will be incorrect, and the resulting lens adjustment will usually serve to bring one of the objects into better focus while

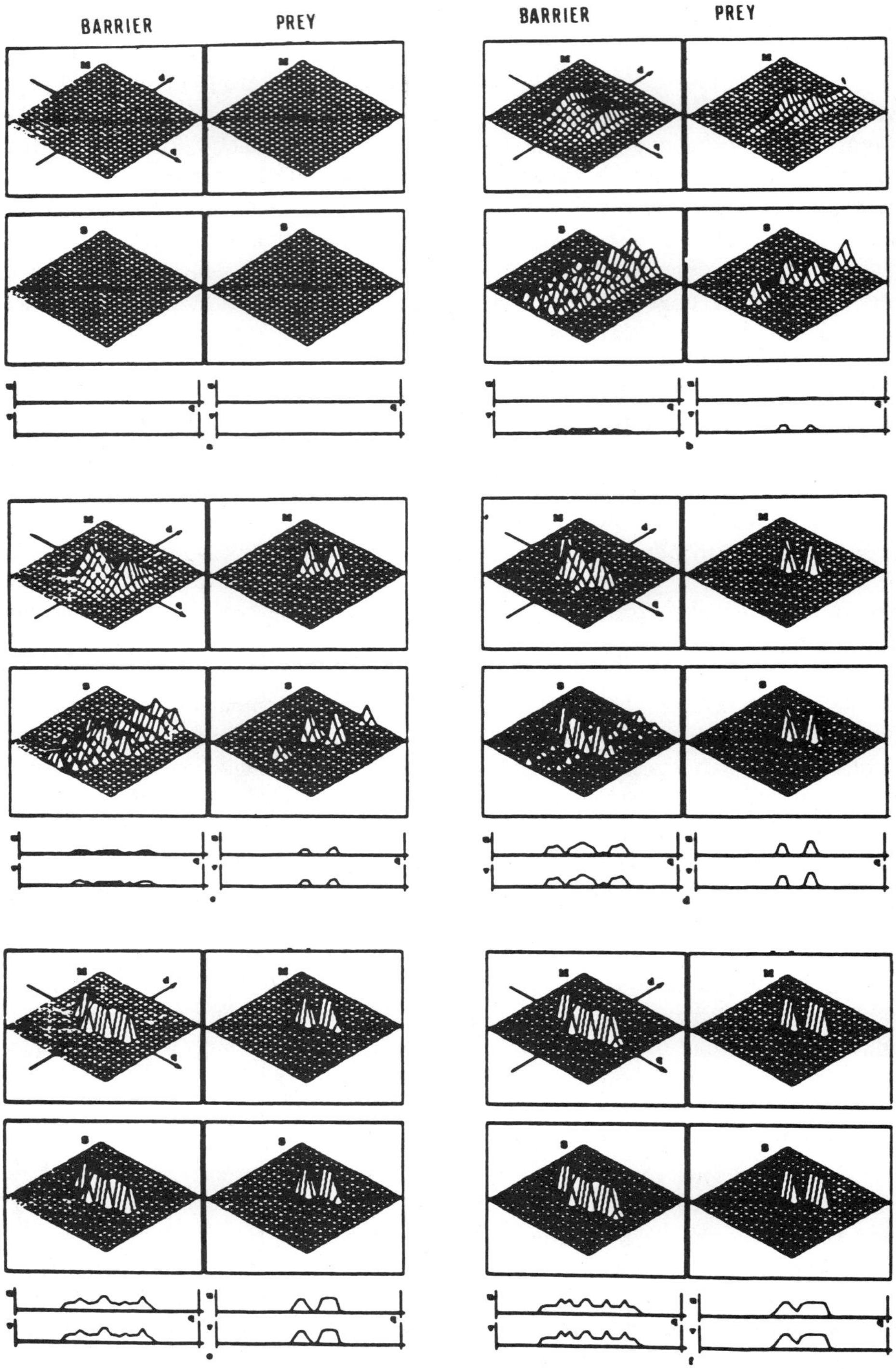
BARRIER
PREY
BARRIER
PREY

simultaneously degrading the focus of the other object. Consequently, the system will tend to choose one of the objects over the other and will eventually correctly locate this object.

Neither of our depth models can, by itself, fully explain the complete range of data on the depth resolving system of toads. The Cue Interaction Model successfully integrates binocular and accommodation cues in a way which allows it to replicate behavioral data. However, it relies on a neural connectivity that does not appear to be necessary for binocular depth perception. The Prey Localization Model successfully addresses the Collett-Udin data, and integrates both depth cue sources (but see the discussion by Gruberg, this volume). However, it is not capable of operating in a purely monocular mode. Further, this model does not allow us to address as broad a range of visual data as does the first model. In particular, it locates only a single point in space and is, therefore, not well suited to locating barrier-like objects. We argue that, instead of there being a single general depth-perception mechanism, there are various neural strategies functioning either cooperatively or alternatively to cope with the vast array of visuomotor tasks required for the freely functioning animal. House (1984) describes one way in which both of the processes outlined in our models might be integrated in the brain of the toad.

Unfortunately, the results reported by Caine & Gruberg (1985), contradict those of Collett & Udin, posing a clear challenge for both future theory and future experiment (see also Gruberg, this volume). They found that frogs with lesions of nucleus isthmi failed to respond to either threat or prey stimuli presented in the corresponding region of the visual field (contra Collett & Udin), while they exhibited normal barrier-avoidance and optokinetic nystagmus in all areas of the visual field (consistent with the scheme of Fig. 9). Caine & Gruberg conclude that the behavioral deficit seems to be "visual" and not "motor," and appears to be identical to that induced by unilateral tectal aspiration.

Lara & Arbib (1982) study an 8 x 1 array of tectal columns and show how this system can reproduce Didday's result of exhibiting a distributed network in which outputs are highly selective, giving a high weight to points receiving the maximum input stimulation and suppressing all others - but here the inhibition is provided by the pretectum, rather than intratectal connections. However, in the prey selection model, House (1984, see also 1988) hypothesizes that it is ipsilaterally projecting isthmi fibers doing the suppression. Gruberg (personal communication) suggests that his results are consonant with our thinking that the nucleus isthmi is a global inhibitory circuit. My impression is that this is not so - that removal of competition

Figure 14. The time course of the Cue Interaction Model is shown from its initially-inert state (a) and immediate response to input (b) to a satisfactory depth segmentation (f) here. All figures are in the retinal angle *vs.* disparity coordinate system. Successive figures are temporally spaced 1.4 field time-constants apart. In each subfigure, the left column shows the barrier fields and the right the prey fields; in each case, the upper two-dimensional grid shows the level of excitation of the accommodation field, and the lower the disparity field. The line-graphs under the grids indicate the intensity and localization on the retinal angle axis of excitation in the inhibitory pools. (From House 1982).

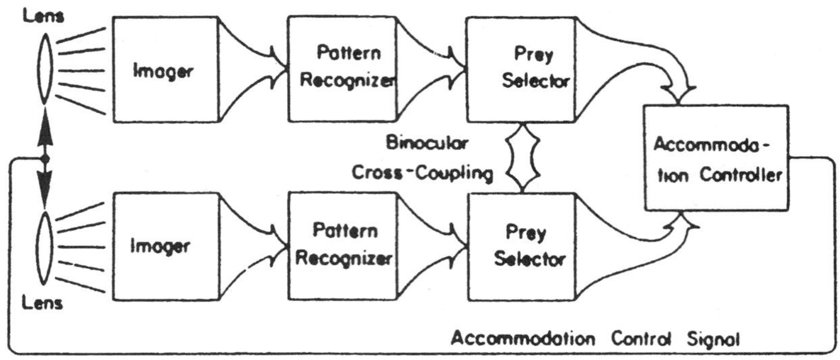

Figure 15. The Prey Localization Model. Lenses are coupled, so that they are accommodated to the same depth. Images produce visuotopically mapped output signals whose intensity is proportional to the crispness-of-focus at each image point. Pattern recognizers produce visuotopically mapped output signals that are strongest wherever the image most closely corresponds to the recognizer's matching criteria. Prey selectors take their input from the recognizers and from cross-coupled connections with each other. Their outputs are highly selective, giving a high weight to points receiving the maximum input stimulation and suppressing all others. Since the prey selectors take their input both from the pattern recognizers and from cross-coupling, they are biased in favor of points receiving strong stimulation on both sides of the binocular system. The accommodation controller converts the weighted image coordinates from both prey selectors into an estimate of depth. It then uses this calculated depth to adjust the lenses. (From House 1988).

would yield snapping at "average flies," rather than visual scotomas - but it is certainly a shortcoming that we have not conducted a simulation of the effects of ablating isthimi. However, such a simulation would be incomplete without a specific account of what patterns of tectal activity constitute the "trigger" for a motor act. If indeed the motor apparatus needs an unequivocal peak of tectal activity to direct a snap at the corresponding position, then a diffusion of tectal activity in the absence of isthmi might yield a consequent appearance of a motor scotoma. To resolve such questions calls for a delicate interplay between theory and experiment.

5 Path-Planning and Detours

Returning to the phenomena of detour behavior introduced in Section 2.2, we now present the Path-Planning Model. It is our intention that, by considering a variety of models of detour behavior, we can create a space of alternatives in which the design of a rich set of neuroethological and neurophysiological experiments will be possible. Both the Epstein Model and the Orientation Model of Section 2.2 serve simply to choose the direction and, in the latter model, the depth of the target for the animal's first move. In the present Section, then, we suggest how the brain plans overall paths of action which would require the coordination of several motor schemas. We present both the Orientation Model and the Path-Planning Model because we believe that at this preliminary stage of the search for the neural substrates of detour behavior, it is premature to focus on a single model. It is hoped that the contrast between these models will serve to stimulate the design of new behavioral and physiological experiments. We also stress that the models are not tightly constrained, in that they do not attempt to specify what particular neurons are doing in the posited behaviors. Rather, they represent processing schemes which could plausibly be carried out in neural structures, and thus represent postulates that there are populations of neurons which carry out the indicated operations. We expect that the refinement of our models will go hand in hand with the development of further data of this kind, and that theory and experiment will provide each other with important stimulation.

We saw at the end of the last section that there is some consensus that we have separate schemas for mapping the position of barriers and for localizing prey, but that there is no consensus as to the neural substrates of these schemas. However, the power of our schema methodology is that it allows us to model one structure or function within the context of a schematic description of its neural environs without requiring a fully articulated neural description of how those environs operate. All that matters is that the information flow between the specified subsystem and its environs be adequately represented. Thus, in developing our barrier negotiation models, we have made the assumption that toads are able to infer the depth of barriers and prey simultaneously and that this information is either determined by different neural substrates or is at least separable by object category - but our use of these "depth schemas" does *not* depend on any specific choice of neural mechanism for their implementation.

The Path-Planning Model is described in terms of an array of elements which are somewhat more abstract than individual neurons. It remains an open question as to how best to map this model onto a neural network, and whether this can be done in a manner consistent with available physiology. We associate with each position in the ground plane a vector to indicate the preferred direction of motion of the animal were it to follow a path through the corresponding point. For conceptual simplicity, the coordinate system used in this model is the Cartesian (x,y) system. Our task with this model will be twofold: To specify how the vector field is generated, and to specify how the vector field is processed to determine the appropriate parameters for the coordinated activation of motor schemas. In the technical jargon of differential geometry , then, the neural surface corresponds to a manifold representing space in some internal coordinate system, while the firing of a group of neurons associated with a particular coordinate is to represent the vectors of a tangent field, or flow. The question is how those local vectors are to be integrated to determine an overall trajectory for the animal.

Our provisional choice is shown in Fig. 16. A single prey (Fig. 16a) will set up an attractant field, in which from every point in the animal's representation of space there is an arrow suggesting a choice of movement toward the prey, with the length of the vector (the strength of choice for a movement in the given direction) being the greater, the closer is the point to the prey. We have associated a repellant field with a single fencepost (Fig. 16b), with the strength of the field contributing mostly to the determination of a lateral movement relative to the position of the fencepost from the viewpoint of the animal. Finally, the animal's representation of itself (Fig. 16c) within

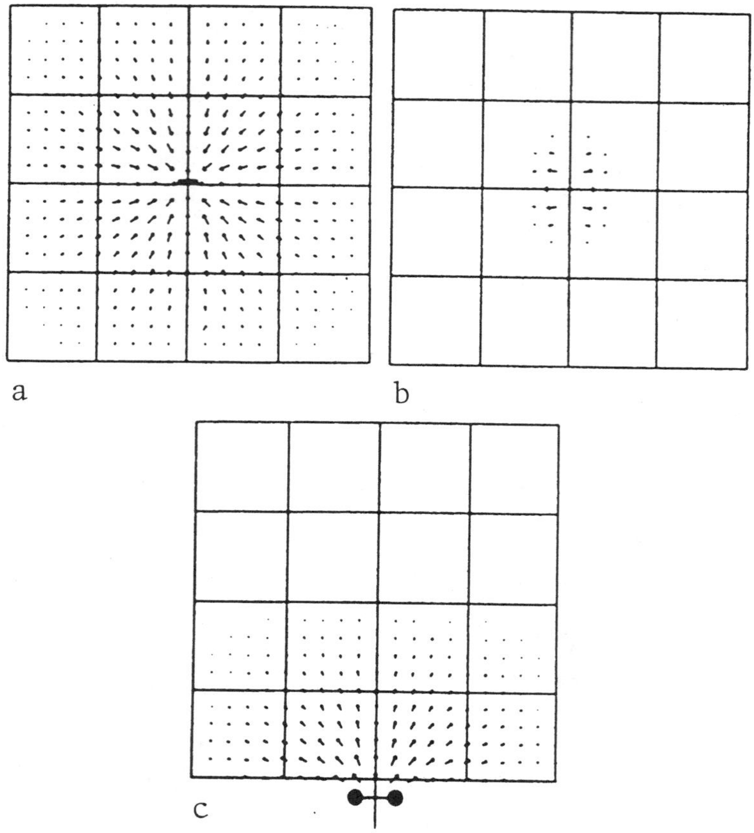

Figure 16. The Path-Planning (vector field) Model envisions objects in the animal's visual world as determining a space of potential motor activities. The potential fields depicted here represent an exploratory attempt at defining a set of primitive fields which will interact in interpreting a more complex scene. (a) A single prey object sets up a radially symmetric attractant field whose strength decays gradually with distance from the prey. (b) A single barrier object sets up a repellant field whose effect is more localized to its point of origin than is that of the prey field. The barrier field is not radially symmetric but has a lateral component that is stronger but decays more rapidly with distance than does its opposing component. (c) The vector field model also contains a representation of the animal itself. This representation is simply the converse of the prey representation, i.e., it is radially symmetric but diverges from its point of origin. (From Arbib & House 1987).

this field consists of a set of vectors radiating out in all directions from the animal's current position with a decay similar to that for the prey field. Figure 17 shows the various effects obtained by summing the vectors for each point of the manifold: The bug attractant field in interaction with the animal's self representation (Fig. 17a); the summed effect of the fenceposts yields a field which repels from the fence as a whole but with especially strong lateral flow at the edges of the barrier (Fig. 17b); and the

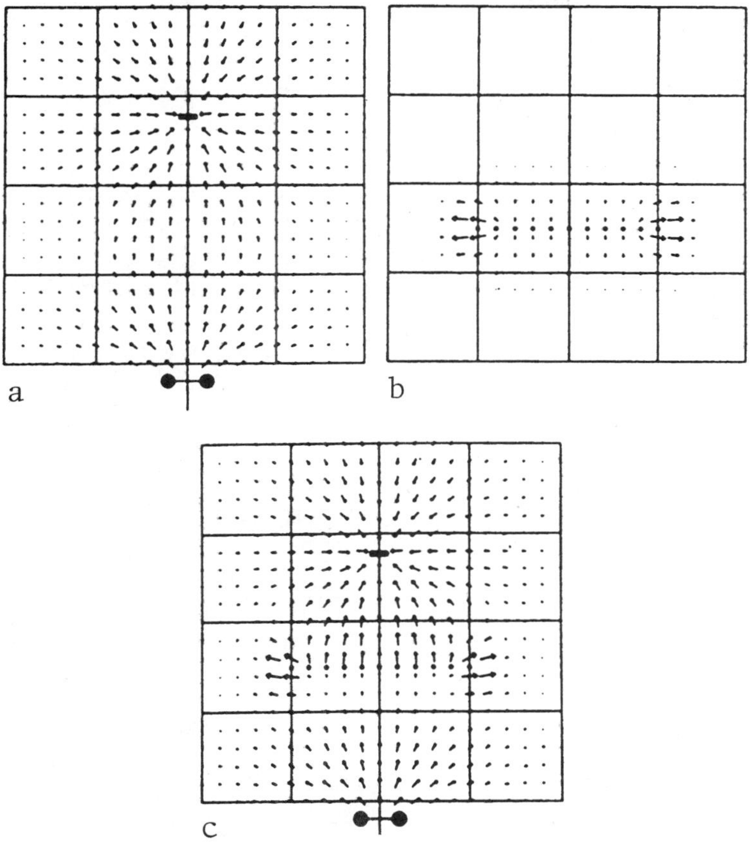

Figure 17. The three types of primitive field of Fig. 16 are shown here in interaction. (a) The prey attractant field in interaction with the animal's self-representation produces a field suggesting various curved paths terminating at the position of the prey. (b) The effect of the interaction of the fields from several barrier objects arranged to form a fence is to provide a strong lateral thrust at the fence ends. The lateral components produced by the interior posts is effectively cancelled by neighboring posts. (c) The net field produced by the interaction of all of the elements of the configuration can be thought of as tracing out a set of paths, most of which are diverted around the fence ends. (From Arbib & House 1987).

overall field set up by a scene containing a single worm behind a fence (Fig. 17c).

The total field may be interpreted as representing the "net motor effect" of the scene upon the animal, whether the animal is an essentially ballistic creature like a frog or a toad, or a more "tracking" creature like a gerbil. In the case of the gerbil we would postulate that the vector field is integrated to yield a variety of trajectories, with a weight factor for each trajectory. We would then see that this field has two "bundles" of trajectories receiving high weight, that bundle which goes round the left end of the barrier to approach the worm, and that which goes around the right end of the barrier to approach the worm. Thus, if we change "worm" to "sunflower seed," we would posit that the gerbil actually builds within its brain a representation of the entire path, one of the paths is selected, and this path regulates the pattern of footfalls that will move the animal along this trajectory. In yet more sophisticated models, we could see the path not as being generated once and for all, but rather as being dynamically updated on the basis of optic flow as the animal proceeds along a chosen direction.

In toad, however, we postulate that the vector field is processed not to yield a continuous trajectory - or a bundle of continuous trajectories of which one is to be chosen - but rather serves to generate a map of motor targets, appropriately labeled as to type. The divergence operator is a likely candidate for this form of processing. Once a suitably constructed representation of a vector field is set up, the computation of divergence is a simple local process which may be carried out in the parallel distributed fashion associated with neural mechanisms. Further, the divergence of a vector field is a scalar field. The negative of the divergence will contain peaks where the flow lines in the field tend to converge and valleys where they tend to diverge. Figure 18 is a display of the negative of the divergence of the net field of Fig. 17c. This contour indicates a trough of high divergence in front of the fence, peaks of convergence corresponding to the two edges of the fence, and a third peak corresponding to the worm. We would postulate that a scalar field of this sort could be used by motor schemas, so that one of the fence ends is chosen as the motor target for a sidestep. However, the relative position of the worm is simultaneously available to determine the coordinates for the orienting schema, and for the subsequent second leg of the motor sequence that we saw to follow the pause in the trajectory shown in the right-hand side of Fig. 4.

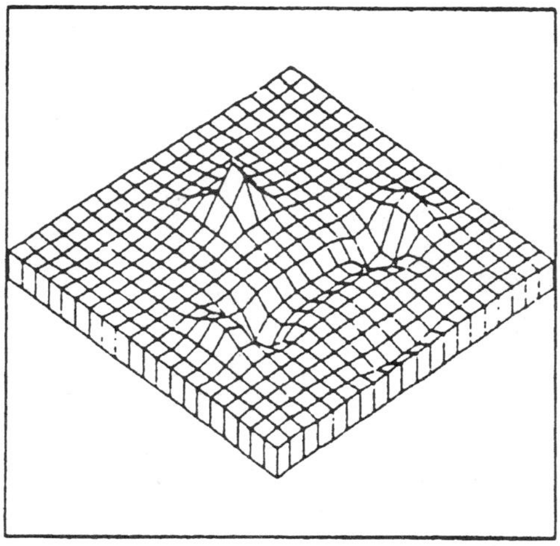

Figure 18. The negative of the divergence of the net field of Fig. 17 is shown here as a three-dimensional plot. Peaks on this plot represent regions of strong path convergence or bundling of paths, whereas valleys represent a strong divergence. The two peaks separated from each other by a trough are centered on the two fence ends. The large rear peak is centered on the prey. There is a trough of divergence in front of the fence. (From Arbib & House 1987).

In Fig. 19, we offer corresponding analyses for the cases of a fence at the position of the fence in Fig. 17c but with a central gap (Fig. 19a), a solid fence near to the prey (Fig. 19b), a solid fence behind a fence with a gap (Fig. 19c), and a cage (Fig. 19d). Our research has not yet resulted in isolating the most suitable algorithm for extracting a path from data of this sort. However, these preliminary results are suggestive of a strong agreement with the behavioral data. In particular, the powerfully attractive quality of fence gaps noted by Collett (1982) is especially apparent.

The vector model differs significantly from the simple orientation model of Section 2. The primary source of the difference is that here visual stimuli are not seen as setting up a simple decision surface which can be processed to select among several optional actions. Rather, what is set up is a spatially encoded map of potential motor activity which in some sense is the net result of the interaction of all of the pertinent visual stimuli. Although in the simple Cartesian representation used in this paper the vectors are described in terms of components of forward and lateral motion, there is no reason to expect that the nervous system would encode vector quantities in this way. What is more likely is that they would be encoded in terms of the various types of schematized motor patterns available to the animal. For instance, a particular vector could be envisioned as having components governing side stepping, turning, and snapping. The coordinate system for such a vector field would, most appropriately, be body centered rather than eye centered.

Ingle (1982) also suggests a model for detour behavior based upon principles similar to those employed in our simple orientation model. In this model he envisions wide-field tectal neurons driven by retinal R2 ganglion cells as providing the kind of spread effect we hypothesised in our model. He proposes that if inhibition from pretectal cells driven by barrier detectors is sufficient to suppress excitation in narrow-field tectal neurons, the effect of the wide-field neurons will be to provide a lateral shift of the locus of tectal excitation. This shift in locus would then be translated into a corresponding shift in orientation turning angle. However, his idea is quite different in character from either of the models which we propose. He has shown that pretectum is involved in governing side stepping, whereas tectum governs orientation turning and snapping. This has led him to suggest that the signals to the motor area from tectum and pretectum are of the nature of commands which are intended to be executed in a coordinated way by the motor area. In our models, however, we see these signals as describing a field of potential motor activity which contains within it a spatial model of the animal's visual world. We would agree with Ingle's placement of the barrier related field in pretectum and that associated with prey in tectum. Further, we agree that final processing and decision making based upon this kind of scheme must take place in the motor area. However, we differ in that we do not see the nature of this process being to select among and coordinate independent motor commands. Rather, we see it as being a process of deriving appropriate motor responses based upon a set of spatially distributed motor cues which already contain within themselves the results of an interaction among the various visual cues. This seems to be in harmony with the notion of activity-gated divergence developed by Grobstein (this volume). It seems clear that the full experimental analysis of these phenomena will require a much richer understanding

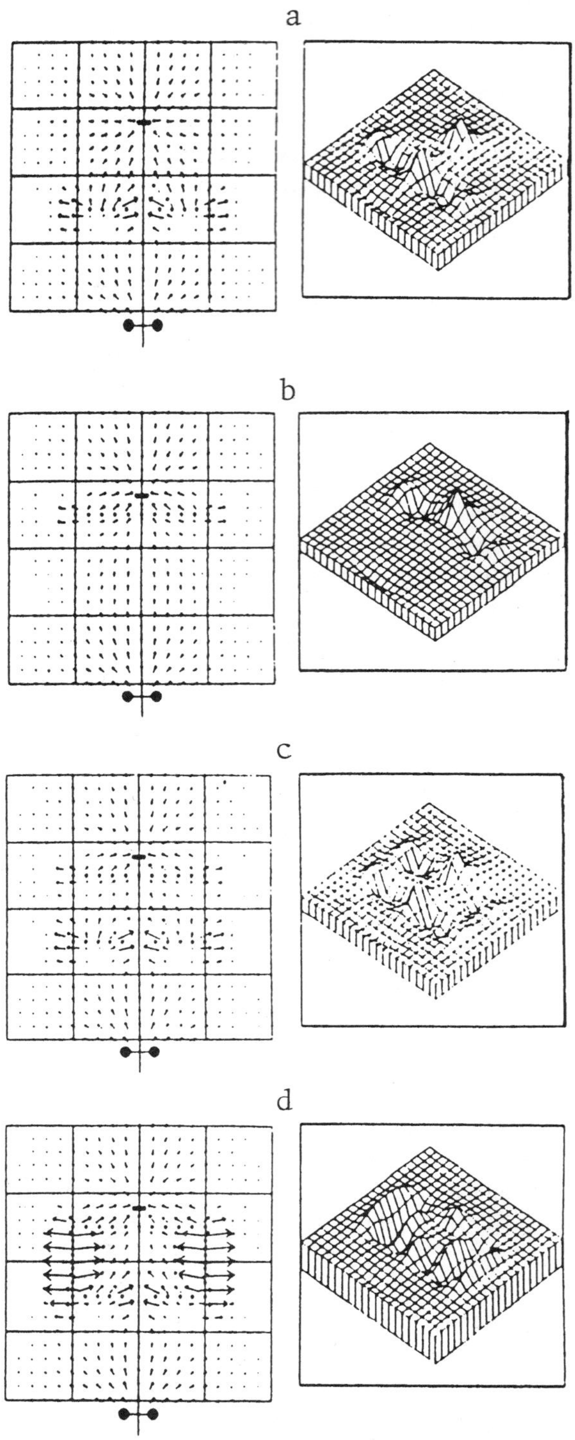
a
b
c
d

of pre-motor and motor structures. The contributions to this volume by Bässler, Berkinblit, Grobstein, Schwippert and Weerasuriya constitute an encouraging start in this direction.

We close this Section by pointing out the relevance of this model of detour behavior for technological applications. Arkin (this volume) discusses AURA, an AUtonomous Robot Architecture which provides the control system for a mobile robot. A subsystem called the Navigator conducts off-line path planning on the basis of a terrain map using relatively conventional AI (Artificial Intelligence) techniques. Another subsystem, called the Pilot, extracts more specific features, e.g., landmarks, to elaborate the plan into motor schemas, with perceptual schemas embedded within the motor schemas. Such schemas include *Stay-On-Path, Move-To-Goal, Move-Ahead, Find-Landmark* and *Avoid-Obstacles*. What is significant here is that a potential field is given for each motor schema. In this way, high-level path planning is model-driven (at a level "above" that posited in *Rana computatrix* to date), but execution is data-driven (though in the "trajectory" mode of the gerbil, rather than the "ballistic" mode of the toad). This system has been used with a real mobile robot traversing the open spaces and corridors of the University of Massachusetts at Amherst, proving the relevance of *Rana computatrix* for perceptual robotics. And with this we turn to look at perceptual robotics in more detail in the following two Sections.

6 Schemas for Hand Control

In the present Section, our concern is with neural networks controlling primate hands, but elsewhere we have been concerned with robotics, seeking to exploit these principles for the control of dextrous hands in distributed control of robot assembly tasks. We first recall our coordinated control program for grasping (Fig. 1), where we noted that subschemas for finger adjustment and hand rotation, movements which precede the actual grasping of the object, receive visual and proprioceptive information necessary for their respective sensorimotor transformation. We hypothesize that this information is integrated into functional groupings called *virtual fingers*. Arbib et al. (1985) introduced this concept by considering the example of lifting a mug. Depending on the size of the handle, one, two, three or even four fingers will be inserted through the handle to provide the force to lift the mug; the thumb will be placed atop the handle to oppose this force, while the remaining fingers will be placed beneath the handle, possibly to stabilize the grasp. It is claimed that a perceptual schema analyzes the handle size to then group the physical fingers into three virtual fingers, *VF1*, *VF2*, and *VF3* - where *VF1* is always the thumb, VF2 comprises the next j fingers ($1 \leq j \leq 4$), while VF3 comprises any fingers that remain (and will thus be empty if $j=4$) - and that the coordinated control program thereafter

Figure 19. Although an analysis has not yet been completed to identify a specific means for deriving motor activity from the vector fields, these figures indicate that the relevant information is efficiently encoded by the vector model. (a) Fence with gap. (b) Fence near to prey. (c) Fence with gap in front of a solid fence. (d) Cage. The model results are consistent with behavioral results. (From Arbib & House 1987).

works in terms of virtual fingers, with physical fingers inheriting their commands from the virtual finger to which they are assigned.

This still leaves open the question of how the movements of virtual fingers are to be precisely specified. It was the discovery of Iberall et al. (1986) that a large variety of movements could be described using the oppositions shown in Fig. 20. What is stressed is not only the nature of the opposition but the associated axis - or *opposition space* - along which the opposing elements of the virtual fingers are to move. In pad opposition, the pad of the thumb (*VF1*) is opposed to the pads of *VF2*. In palm opposition, a group of fingers excluding the thumb forms *VF2*, which is opposed to the palm as *VF1* (showing that a virtual finger need not comprise physical fingers). Palm opposition sacrifices flexibility in favor of stability. The object is fixed in the hand, along an axis roughly normal to the palm, with greater stability due to the larger size of *VF1*, and the higher strength of the more proximal finger areas. Finally, in side opposition, the pad of the thumb as *VF1* is opposed to the side of a group of other fingers as *VF2*. It is *not* claimed that all manual tasks are captured by this vocabulary - consider stroking, sign language or piano playing, for example. But it is claimed that this vocabulary can reduce many complex grasps to simply parameterized forms, especially when it is realized that *composites* may be used. For example, the grip on a screwdriver might combine a pad opposition to grasp the handle with a side opposition to direct the shaft.

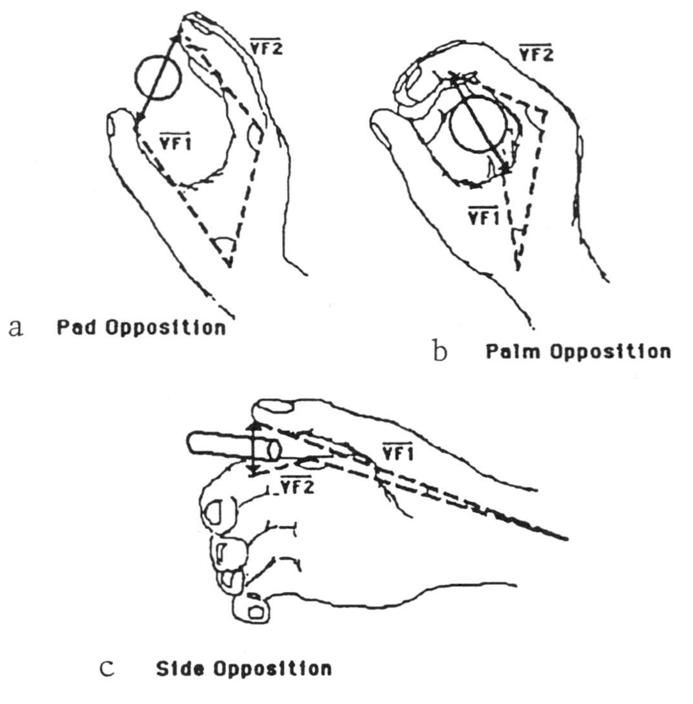

Figure 20. Three basic oppositions for human prehension (a) Pad opposition, (b) palm opposition, and (c) side opposition. Solid lines show axis of opposition, dashed lines show virtual finger vectors *VF1* and *VF2*. (From Iberall 1987).

Of course, when an object is grasped, the virtual fingers moving along the opposition axis may come to rest their opposing surfaces on the object between them, rather than making direct contact with each other. This makes possible a theory of preshaping. One task of vision is to determine, from the retinal input, an opposition space embedded in the object which is to be the target for the positioning of the appropriate opposition space of the hand. As we indicate in Fig. 21, there is a saftey margin extending the opposition space beyond the boundaries of the object. Preshaping then forms the hand so that the opposing surfaces of the virtual fingers will be correspondingly separated. An approach vector, between the origin of opposition space and the center of the opposition vector, distinguishes the orientation and distance of the hand relative to the object. We show in Fig. 21 how to characterize this. A virtual finger configuration must be moved from its current state to a desired state, under the control of a motor schema. As the hand preshapes to meet this specification (the *Preshape* schema), the arm transports it (*Move Arm* schema) and the wrist rotates it (the *Orient* schema) to approximately the right position. Note that the target position of the wrist depends on the proposed preshape of the hand. Thereafter, the position of the hand is adjusted to align the two opposition spaces (that in the hand with that in the object) and the virtual fingers then close along the aligned axes to firmly grasp the object under tactile control (the *Enclose* schema).

Iberall & Arbib (1988) have postulated that various of these activities may be distributed in the primate brain as shown in Fig. 22:

(i) Area 7 sends task features to Area 5, cues the *Approach-Vector-Selector* perceptual schema (hypothesized to be in the frontal eye field area FEF) and cues the *Orient-Trunk* and *Move-Arm* schemas (hypothesized to be in the premotor area PM).

(ii) The frontal eye fields FEF (Areas 8 and 9) are posited not only to monitor eye movements but also to collect visual information about objects from peristriate cortex and Area 7. They seem able to suppress reaching movements until the trunk is

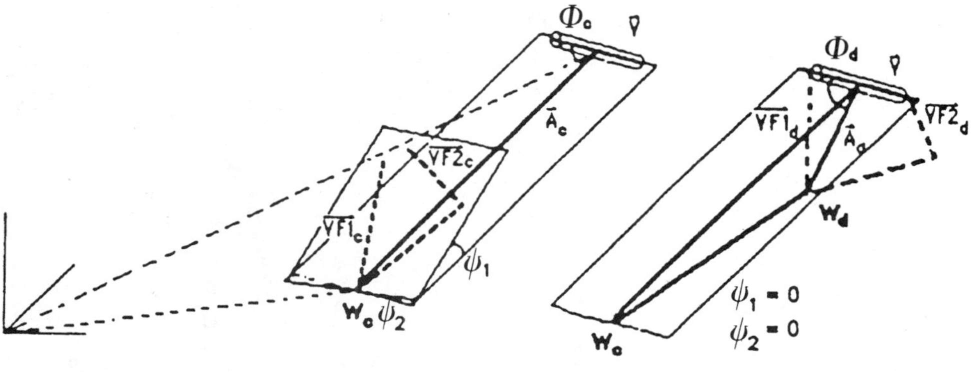

Figure 21. The opposition axis embedded in the object sets goals for the preshape of the hand. After the opposition axis of hand and object are aligned, enclosing can take place. (From Iberall 1987).

oriented correctly. We hypothesize that FEF serves as the approach vector selector, with Area 7 as requestor. The output of FEF would be the desired wrist location to be sent to PM for orienting arm and trunk. To do this, FEF would maintain a map of object locations and suppress reaching until the body is oriented; at which time it would trigger the *Orient-Trunk* and *Move-Arm* schemas (PM).

(iii) Area 6 comprises PM and the supplementary motor area, SMA. PM seems to be involved in the transport component of reaching using FEF to direct proximal areas of the arm. A large hand region in PM known as the APA, the arcuate premotor area, is involved in the grasping component. Lesion of SMA yields tactile following, an inability to let go of objects. We thus posit that SMA inhibits enclosing as part of the *Preshape* schema, while SMA, APA, and Area 5 work as a unit for

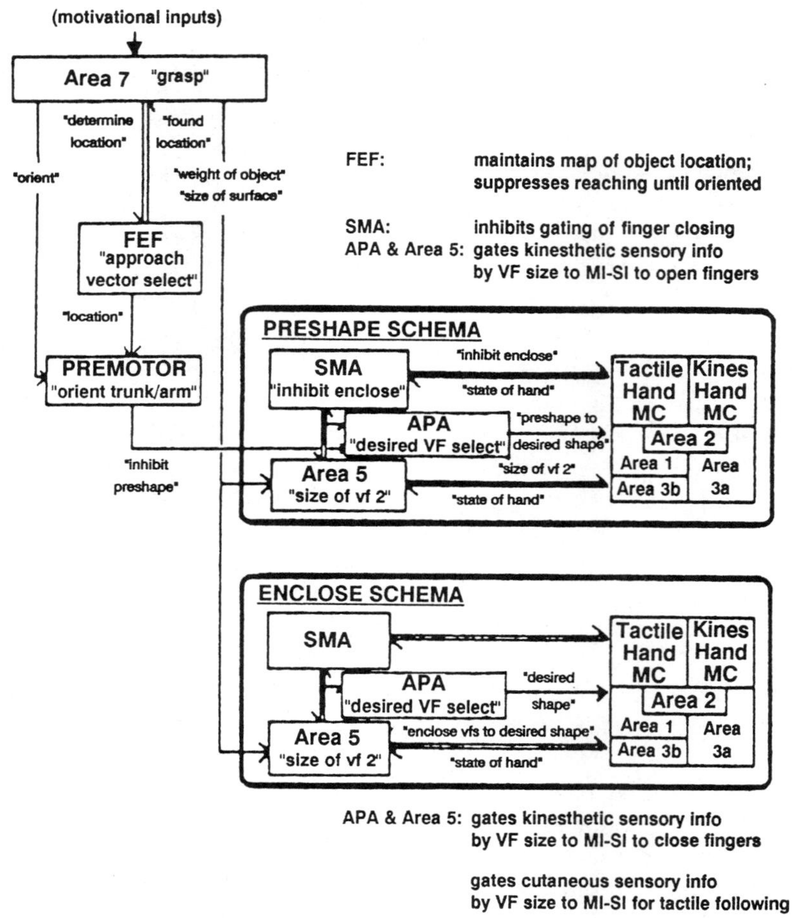

Figure 22. Hypothetical mapping of Coordinated Control Program onto CNS. Pad opposition is a subset of the prehensile repertoire of the CNS. We identify the roles various cortical areas might have in prehensile movements, but especially in pad opposition. Subcortical contributions are not included. (From Iberall 1987).

gating tactile and kinesthetic information through the transcortical somatosensory cortex/motor cortex loop.

(iv) We suggest that Area 5 with the APA can gate kinesthetic information from somatosensory cortex to motor cortex to flex the fingers during the closing of the hand. After contact is made Area 5 and the APA can then gate the cutaneous information to motor cortex for tactile following of the object.

(v) We postulate that APA, SMA, and Area 5 contribute to both the *Preshape* and *Enclose* schemas. APA facilitates preshaping, using virtual finger size information form Area 5 to gate desired *VF* configurations to motor cortex. We thus posit that APA contains knowledge about opposition space, using desired wrist location to compute the desired preshape state based on *VF* information from Area 5. SMA ensures success of the preshape by inhibiting gating of finger flexion information to cortex until the wrist is correctly located. When the wrist is correctly located, inhibition is released and the *Preshape* schema closes the fingers using *VF* information from Area 5.

7 Schemas for Vision

The present Section outlines the approach to schemas for high-level vision adopted by Weymouth (1986; see also Riseman & Hanson 1987), but so presented as to be seen as a model of schema interaction within the brain which extends that given in Arbib (1981). The system is to automatically go from an image to an interpretation network describing the scene, with pointers from object-nodes to the corresponding region of the image.

It is impossible to interpret Fig. 23a in isolation; but if I tell you that it is a sample of the image in Fig. 23b near the dividing line between the two walls of the house, then there is no problem recognizing it as a bush. This seems to pose a chicken and egg problem. You have no trouble recognizing the bush because you recognize the house ... but how did you recognize the context (the house)? Perhaps to recognize the context, you had to recognize the bush which was the context for the context! How do we get the recognition process going? The answer is to find pieces of the image that we can interpret with sufficient confidence to provide an "island of reliability" to get further processes going.

The general strategy is then shown in Fig. 24. On the left, we see the processes of "low-level vision" which take the image and break it up into regions which are candidates for interpretation. The right hand side employs various schemas that can guide the interpretation of the various regions of the image. These schemas are then arranged into a network which allows context to play its role in the process of interpretation. However, rather than go through all the details of how the system really works, let me just run through one example. Fig. 25a shows what happens when the *Sky* schema runs on the image after it has been segmented and divided into different regions. The *Sky* schema finds regions reasonably high in the image and with the right color for sky. In this case it recognizes the top regions satisfactorily. The roof region also is interpreted as sky but since it is lower in the image and the color is not quite sky, the confidence level is much lower. But meanwhile another schema for roof is running and it is very excited about that region because it has just the right

Figure 23. It is impossible to recognize the pattern of (a) without placing it in the context provided by (b). (Courtesy of A. Hanson and E. Riseman).

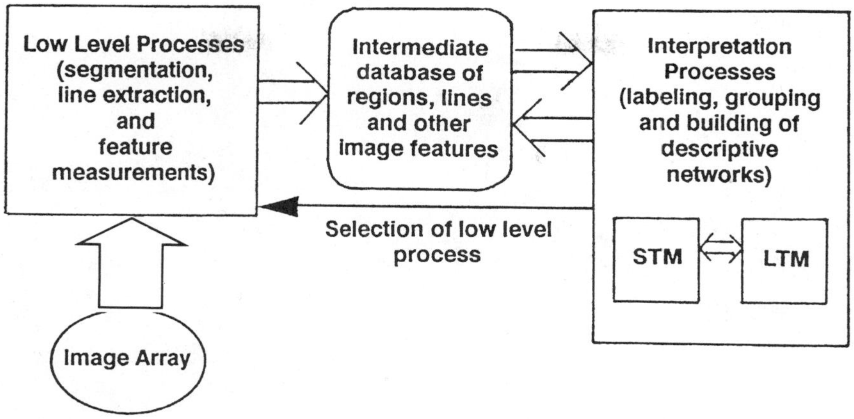

Figure 24. The overall architecture of the VISIONS machine vision system. Low-level processes extract edges and regions from the image which are candidates for interpretation by the schemas of high-level vision. (From Weymouth 1986).

geometrical characteristics and is in the right position of the image. Thus, the system is very quickly confident that the roof really is a roof. As a result, the low confidence hypothesis that it might be sky does not play any further part in the computation.

However, returning to Fig. 25a, we see that something unfortunate happened. The segmentation left out a crucial edge, and so the "sky" in fact includes one of the walls of the house. And the *Wall* schema has also been running in this distributed system. It got turned on when the *Roof* schema sought confirming context, invoking the *Wall* schema to search underneath to see if the criteria for a wall were met, as indeed they were. There is a major problem. Because of the missing edge there is one big region which is interpreted both as wall and sky. We saw in the case of the roof that if one hypothesis were much stronger than another, the system will tend to ignore the low confidence hypothesis. But here the two hypotheses are both too strong to be ignored. This particular implementation is told to go back and look for more edges in the image, and so it takes the offending region and goes back with a lower threshold to look for the weak edges that it ignored before; coming up with the segmentation shown in Fig. 25b. The system then has simply to decide for each little region there whether to call it wall or sky. Now of course the whole sky region can only be sky because it does not pass the test of being underneath the roof. The regions which are wall could still be sky but they are lower in the image, so that the confidence that they are sky is lowered, whereas the confidence that they are wall becomes raised. And so the system confidently moves on to the interpretation of Fig. 25c. As can be seen, the system "knows" about walls, shutters, roof, sky, and grass, but does not know about people, roads, or windows. The knowledge engineering required to extend the system's abilities is an exciting and ongoing problem for research.

Figure 25. Because of a missing edge, the region of (a) is rated with high confidence as both sky and wall. In response to this conflict, the system performs a finer segmentation of the region, coming up with the edges shown in (b). This allows the system to correctly distinguish walls from sky and to move on to the interpretation shown in (c). (From Riseman & Hanson 1987).

8 Challenges for Cooperation

In this concluding Section, I briefly address two different challenges for cooperation: That between neuroscientist and computer scientist as we seek to garner information-processing principles that both enrich our understanding of brains and extend our design techniques for the next generation of computers, and that between experimentalist and theorist in neuroethology as we try to synthesize our insights from many different levels of analysis.

8.1 Cooperation Between Neuroscience and Computer Science

Inspiration from the brain leads away from an emphasis on a single "universal" machine and suggests that the computer of the future will be a device composed of different structures, just as the brain may be divided into hippocampus, motor cortex, cerebellum, etc. Thus, we can expect to contribute to neural computing as we come to better chart the special power of each structure. Each "sixth generation computer" will, I predict, be a network of subsystems, including general-purpose engines and special-purpose machines some of which (such as the front ends for perceptual processors, and devices for matrix manipulation) will be highly parallel machines. Some subsystems will use optical computing (lenses for Fourier transforms; holograms for high-density connectivity, etc.); some will be designed using the emerging principles of neural computing, while, on a longer time-scale, some will employ biomaterials. Just as the interface to computers has progressed from bits to symbols to interactive graphics, the next generation will be more "action oriented" with computers including robotic actuators and multi-modal intelligent interfaces among their subsystems (see Torras, this volume). The design of these perceptual robotic systems will be heavily influenced by the analysis of neural mechanisms for perception and the control of movement. Finally, sixth generation computers will be machines that learn, with principles of adaptive programming that incorporate lessons from brain theory and connectionism (see Arbib 1987b, Chapters 4 and 5).

Just as it still takes much effort to program today's computers for novel tasks, so may we expect neural computing to develop many special techniques adapted to specific classes of problems. Within current computer science, there are such different research areas as hardware, software, and AI, with AI itself fractionated into specific subfields like expert systems, natural language processing, vision, robotics, intelligent tutoring systems, and adaptive database techniques. As we develop neural computing, we can learn much from neuroscience but, continuing the above theme, much of it will be focussed on specific domains. Moreover, there is no ready-made body of wisdom in neuroscience waiting to be applied. Thus, we cannot simply follow the "neural blueprint" to build a better computer. Rather, fundamental research in neuroscience must continue, but - in this context of neural computing - structured to extract key principles of information processing that can enter into the design of the next generation of computers.

8.2 Cooperation in the Study of the Brain

Models evolve in complexity in (at least) two ways: (a) Through replacing units by more detailed networks, and (b) through integrating more and more schemas into explanations of an increased range of behaviors and/or brain regions. A problem with (a) is that the simulation of a large array of units may become infeasible if each unit is described in the most refined way currently available. A problem with (b) is that the analysis of one "macro-schema" may itself be a full research task, yet the interfaces with other schemas may introduce constraints that simply cannot be ignored. The resolution will require much greater "cooperative computation" between research groups, and the development of standards for interfacing their work. Here are three scenarios:

1) We discussed the "evolution" of two different, but overlapping, subsystems of *Rana computatrix*: Tectal column models (Section 3), and models for depth perception and detour behavior (Sections 2, 4, and 5). We must now synthesize these into an overall system. In the process, we may better understand some of the complexity of tectal circuitry as we probe its computations as subserving several rather different schemas.

2) Much of our modeling addresses the ways in which retinotopically structured subsystems may interact to determine the location of objects or task-related features so as to commit the organism to some spatially structured course of action. In the meantime, a number of workers (e.g., Robinson 1981) have used control theory to explicate how the (numerically, rather than spatially, encoded) position of a target can be transformed to commands for muscle contractions to bring the eye to bear on the target. This raises fascinating problems in the interfacing of such efforts:

(i) What control system for, say, snapping at prey, would be analogous to Robinson's systems for eye movements?

(ii) What constraints on such a control system are imposed by making it accept retinotopic, rather than numerical, inputs?

(iii) How can a better appreciation of variables involved in muscle control change our conception of the proper output for the "more or less visual" portions of the visuomotor control system?

To the extent that different groups work on these problems, it would be most helpful to have an agreed on "interface language" to express the flow of information between subsystems. This should be so designed that researchers could couple programs developed at different laboratories, and use the experience to suggest changed input/output specifications to constrain model development at each site. My group is currently developing a new programming language NSL (Neuron/Schema Language: pronounced "nissl") with this end in mind.

3) Our versions of *Rana computatrix* to date have used "black-box" models of the retina and so, as already mentioned in Section 3.2, we now seek more insight into the details of spatiotemporal patterns of retinal output, simulating a 10,000 neuron network on a CRAY supercomputer (Teeters 1988). Meanwhile, other workers find it necessary to exploit the full power of a LISP machine to explore the detailed synaptic and membrane interactions within a single neuron. Here, the suitable interface seems to be a description of the cell's input/output relations at an intermediate level of complexity. At any time, the task of one group would be to test how well a network of such model neurons would explain overall retinal response, while another group would explore how adequately such a representation corresponds to the fine details of neuronal form and function.

As a corollary to these methods should come far better tools for the exchange of data between experimentalist and theorist. Published data often lack details crucial for model testing and development. In the future, we should see increased availability of large data sets recorded in forms which will allow easy processing for detailed comparison of model and experiment. The flow will also go the other way, as new concepts developed to model the brain come to increasingly enter the language of the experimentalist, so that models become an integral part of the design of experiments. As a result, the encouraging, but nonetheless still limited, interactions between computational and experimental neuroethology recorded in this volume will grow into a powerful methodology to let us chart neural systems of a complexity far beyond our current reach.

Acknowledgements. Preparation of this paper was supported in part by NIH under grant 1 R01 NS24926 from NINCDS. Portions of the paper have appeared in somewhat different form in *The Behavioral and Brain Sciences* (Arbib 1987a) and are used here with kind permission of the Editor. My grateful thanks to the colleagues with whom much of the research was conducted and to Terry Weymouth for permission to reproduce his figures in this article. Much of the argument here is developed at greater length and in richer detail in the forthcoming volume *The Metaphorical Brain 2: Neural Networks and Beyond* to be published by Wiley-Interscience.

9 References

Amari S-I, Arbib MA (1977) Competition and cooperation in neural nets. In: Metzler J (ed) *Systems neuroscience.* Academic Press London New York, pp 119-165

Arbib MA (1975) Artificial intelligence and brain theory: unities and diversities. *Ann Biomed Eng* 3: 238-274

Arbib MA (1981) Perceptual structures and distributed motor control. In: Brooks VB (ed) *Handbook of physiology - The nervous system II. Motor control.* American Physiological Society Bethesda MD, pp 1449-1480

Arbib MA (1987a) Levels of modeling of mechanisms of visually guided behavior. *Behav Brain Sci* 10: 407-465

Arbib MA (1987b) *Brains, machines, and mathematics,* 2nd Edn, Springer-Verlag, New York

Arbib MA (1989) *The metaphorical brain 2: neural networks and beyond.* Wiley Interscience, New York (in press)

Arbib MA, Hanson AR (eds) (1987) *Vision, brain, and cooperative computation.* A Bradford Book/MIT Press Cambridge MA

Arbib MA, House DH (1987) Depth and detours: an essay on visually-guided behavior. In: Arbib MA, Hanson AR (eds) *Vision, brain, and cooperative computation.* A Bradford Book/MIT Press Cambridge MA, pp 129-163

Arbib MA, Iberall T, Lyons D (1985) Coordinated control programs for movements of the hand. In: Goodman AW, Darian-Smith I (eds) *Hand function and the neocortex.* Springer-Verlag, New York, pp 135-170

Arbib MA, Overton KJ, Lawton DT (1984) Perceptual systems for robots. *Interdiscipl Sci Rev* 9(1): 31-46

Barlow HB (1953) Summation and inhibition in the frog's retina. *J Physiol (Lond)* 119: 69-88

Caine HS, Gruberg ER (1985) Ablation of nucleus isthmi leads to loss of specific visually guided behavior in the frog *Rana pipiens. Neurosci Lett* 54: 307-312

Cervantes-Pérez F (1985) *Modelling and analysis of neural networks in the visuomotor system of anuran amphibia.* PhD Thesis and COINS Techn Rep 85-27, Computer and Information Science Department, Univ of Massachusetts at Amherst MA

Cervantes-Pérez F, Lara R, Arbib MA (1985) A neural model of interactions subserving prey-predator discrimination and size preference in anuran amphibia. *J Theor Biol* 113: 117-152

Chipalkatti R, Arbib MA (1987) The prey localization model: a stability analysis. *Biol Cybern* 57: 287-300

Chipalkatti R, Arbib MA (1988) The cue interaction model of depth perception: a stability analysis. *J Math Biol* 26: 235-262

Collett T (1982) Do toads plan routes? A study of the detour behaviour of *Bufo viridis. J Comp Physiol* 146: 261-271

Collett T, Udin S (1983) The role of the toad's nucleus isthmi in prey-catching behaviour. In: Lara R, Arbib MA (eds) *Proceedings of the second workshop on visuomotor coordination in frog and toad: models and experiments.* COINS Techn Rep 83-19, Univ of Massachusetts at Amherst MA

Dev P (1975) Perception of depth surfaces in random-dot stereograms: a neural model. *Int J Man-Machine Studies* 7: 511-528

Didday RL (1970) *The simulation and modelling of distributed information processing in the frog visual system.* PhD Thesis Stanford University

Didday R (1976) A model of visuomotor mechanisms in the frog optic tectum. *Math Biosci* 30: 169-180

Epstein S (1979) *Vermin users manual.* Unpublished MS Thesis, Department of Computer and Information Science, Univ of Massachusetts at Amherst MA

Ewert J-P (1968) Der Einfluß von Zwischenhirndefekten auf die Visuomotorik im Beute- und Fluchtverhalten der Erdkröte (*Bufo bufo* L). *Z Vergl Physiol* 61: 41-70

Ewert J-P (1974) The neural basis of visually guided behavior. *Sci Amer* 230: 34-42

Ewert J-P (1976) The visual system of the toad: behavioural and physiological studies on a pattern recognition system. In: Fite KV (ed) *The amphibian visual system: a multidisciplinary approach.* Academic Press, New York, pp 142-202

Ewert J-P (1987) Neuroethology of releasing mechanisms: prey-catching in toads. *Behav Brain Sci* 10: 337-405

Ewert J-P, Seelen W v (1974) Neurobiologie und Systemtheorie eines visuellen Muster-Erkennungsmechanismus bei Kröten. *Kybernetik* 14: 167-183

Hanson AR, Riseman EM (1978) VISIONS: a computer system for interpreting scenes. In: Hanson AR, Riseman EM (eds) *Computer vision systems,* Academic Press, New York, pp 129-163

Hirai Y, Fukushima K (1978) An inference upon the neural network finding binocular correspondence. *Biol Cybern* 31: 209-217

House DH (1982) The frog/toad depth perception system - A cooperative/competitive model. In: Arbib MA (ed) *Proceedings of the workshop on visuomotor coordination in frog and toad: models and experiments.* COINS Techn Rep 82-16, Univ of Massachusetts at Amherst MA

House DH (1984) *Neural models of depth perception in frog and toad.* PhD Thesis, Department of Computer and Information Science, Univ of Massachusetts at Amherst MA

House D (1988) A model of the visual localization of prey by frog and toad. *Biol Cybern* 58: 173-192

Iberall T (1987) *A neural model of human prehension.* PhD Thesis, Department of Computer and Information Science, Univ of Massuchusetts at Amherst MA

Iberall T, Arbib MA (1988) Schemas for the control of hand movements: an essay on cortical localization. In: Goodale M (ed) *Vision and action: the control of grasping* (in press)

Iberall T, Bingham G, Arbib MA (1986) Opposition space as a structuring concept for the analysis of skilled hand movements. *Exp Brain Res Series* 15: 158-173

Ingle D (1975) Selective visual attention in frogs. *Science* 188: 1033-1035

Ingle D (1976) Spatial vision in anurans. In: Fite KV (ed) *The amphibian visual system: a multidisciplinary approach*. Academic Press, New York, pp 119-140

Ingle DJ (1982) Organization of visuomotor behaviors in vertebrates. In: Ingle DJ, Mansfield RJW, Goodale MA (eds) *Advances in the analysis of visual behavior*. The MIT Press Cambridge MA, pp 67-109

Ingle DJ (1983) Brain mechanisms of visual localization by frogs and toads. In: Ewert J-P, Capranica RR, Ingle DJ (eds) *Advances in vertebrate neuroethology*. Plenum Press, New York, pp 177-226

Jeannerod M, Biguer B (1982) Visuomotor mechanisms in reaching within extrapersonal space. In: Ingle DJ, Mansfield RJW, Goodale MA (eds) *Advances in the analysis of visual behavior*. The MIT Press Cambridge MA, pp 387-409

Jeannerod M, Michel F, Prablanc C (1984) The control of hand movements in a case of hemianeaesthesia following a parietal lesion. *Brain* 107: 899-920

Lara R, Arbib MA (1982) A neural model of interaction between tectum and pretectum in prey selection. *Cognition and Brain Theory* 5: 149-171

Lara R, Arbib MA, Cromarty AS (1982) The role of the tectal column in facilitation of amphibian prey-catching behaviour: a neural model. *J Neurosci* 2: 521-530

Lara R, Carmona M, Daza F, Cruz A (1984) A global model of the neural mechanisms responsible for visuomotor coordination in toads. *J Theor Biol* 110: 587-618

Lee Y (1986) A neural network model of frog retina: a discrete time-space approach. Computer Science Department, Univ of Southern Califonia, Los Angeles CA. *Techn Report* TR-86-219

Lettvin JY, Maturana H, McCulloch WS, Pitts WH (1959) What the frog's eye tells the frog's brain. *Proc IRE* 47: 1940-1951

Lyons DM, Arbib MA (1988) A formal model of distributed computation for schema-based robot control. *IEEE J Robotics and Automation* (in press)

Malsburg C vd, Schneider W (1986) A neural cocktail-party processor. *Biol Cybern* 54: 29-40

Riseman EM, Hanson AR (1987) A methodology for the development of general knowledge-based vision systems. In: Arbib MA, Hanson AR (eds) *Vision, brain and cooperative computation*. A Bradford Book/ The MIT Press Cambridge MA, pp 285-328

Robinson DA (1981) The use of control systems analysis in the neruophysiology of eye movements. *Ann Rev Neurosci* 4: 463-503

Székely G, Lázár G (1976) Cellular and synaptic architecture of the optic tectum. In: Llinás R, Precht W (eds) *Frog neurobiology*. Springer-Verlag, Berlin Heidelberg New York, pp 407-434

Teeters J (1988) A simulation system for the amphibian retina. Techn Rep, Department of Computer Science, University of Southern California

Weymouth TE (1986) *Using object descriptions in a schema network for machine vision*. PhD Thesis and COINS Techn Rep 86-24, Department of Computer and Information Science, Univ of Massachusetts at Amherst MA

PART III

Cellular Mechanisms in Tectum and Pretectum

Cellular Architecture and Connectivity of the Frog's Optic Tectum and Pretectum

GYULA LAZAR

Department of Anatomy, University Medical School
Pecs, H-7643, Hungary

Abstract. The result of studies, in which horseradish peroxidase or cobalt was used to label neurons and fiber tracts in the frog's brain are summarized in this paper. Focal cobalt injections into various layers of the *optic tectum* label small groups of neurons which appear as columns formed by dendrites oriented perpendicular to the surface. This arrangement of neurons resembles that of the cortex of mammals. In frogs tectal neurons projecting to the medulla oblongata and the spinal cord were identified as large ganglionic, pyramidal, and large piriform cells with widely arborizing dendritic trees. Large piriform neurons with narrow dendritic trees and small piriform neurons project to the isthmic nucleus. All tectal cell types which project to the medulla, are also efferents for the *pretectum* and other diencephalic nuclei, but the majority of neurons projecting rostrally are small piriform cells of layer 8. Afferents to the tectum originate bilaterally from the retina, pretectum, and the isthmic nucleus, and ipsilaterally from certain thalamic nuclei, the nucleus profundus lateralis, the anterodorsal and posterodorsal tegmental nuclei, and from the contralateral optic tectum. Retinal and isthmic afferents terminate in the superficial strata of the tectum (layers 8 and 9) and non-retinal afferents distribute in the deep layers (layers 2 to 6). Two nuclei receive retinal afferents in the pretectal area: the nucleus lentiformis mesencephali and the posterior thalamic nucleus. Both nuclei project to the ipsilateral medulla and to both tecta. Non-retinal afferents arrive at the pretectum from the ipsilateral striatum, anterior and lateral thalamic nuclei, pretectal grey, optic tectum and the medullary reticular formation, and from the contralateral superior vestibular nucleus and posterior thalamic nucleus.

1 Introduction

In lower vertebrates the optic tectum is the most preferred brain area for studying its morphology, physiology, chemical architecture, and its role in the behavior of animals. The 19th century knowledge on the organization and connections of the frog's tectum was collected by Gaupp (1899) in his great book. At the beginning of this century the cytoarchitecture of the tectum was described in detail by

Cajal (1911). In the subsequent 40 years there was not too much added to that body of knowledge. During this period of time Röthig's (1927) work on tectal connections and Ströer's (1940) studies on the topography of retinal projections have to be mentioned. This first period of tectal research was closed by two significant Golgi studies (Cajal 1946; Kiro 1948), and the first chapter of research on the brain of urodeles was finished by Herrick's book on the brain of the tiger salamander (Herrick 1948).

Sperry's experimental works (Sperry 1943; 1944; 1945) mark the beginning of the second period in tectal research. In this era scientists concentrated their attention primarily on the function of the amphibian visual system. Gaze (1958) verified the topographical organization of the retinotectal projection by recording electric activity of the axon terminals of retinal ganglion cells in the tectum. Lettvin et al. (1959) and Maturana et al. (1960) showed that different types of retinal ganglion cells are tuned to various features of objects in the visual field, and that the projections of these ganglion cells are separated within the tectum. Their work is the first example of a complex morphological-electrophysiological approach in the study of the retino-tectal system. A series of morphological, physiological, and behavioral investigations followed these works which covered other visual brain areas as well, including the pretectum. The results of this period was summarized in the textbooks edited by Llinás & Precht (1976) and by Fite (1976).

Since the early seventies, techniques based on the axoplasmic transport of radioactive substances, horseradish peroxidase (HRP) or cobalt were used in studies on the amphibian visual system, and further details of its organization were published (for references see: Lázár 1978; Ingle 1983; Lázár et al. 1983b; Weerasuriya 1983). The chemoarchitecture of the tectum was also described in this period (Northcutt 1974; Contestabile 1976; Sood 1978). The results of recent immunohistological studies (Inagaki et al. 1981; Kuljis & Karten 1982, 1983; Takagi et al. 1987) and functional mapping of the amphibian brain with (^{14}C)-2-deoxyglucose (Finkenstädt et al. 1985; Finkenstädt, this volume) will force us to speculate about tectal structure and function on a much broader scale than before. Matsumoto's work with intracellular labeling of tectal cells opened new dimensions for studying relationships between structure and function (Matsumoto & Bando 1980; Antal et al. 1986; Matsumoto et al. 1986; Matsumoto, this volume).

The description of the organization of tectum and pretectum is based on the results of studies performed on *Rana pipiens* in American laboratories, *Rana esculenta* in my laboratory, and *Bufo bufo* in Ewert's laboratory. The greater part of the following account is based on works, in which HRP and cobalt labeling techniques were used to show neurons and fiber pathways. The nomenclature introduced by Potter (1965) and Neary & Northcutt (1983) will be used.

2 Optic Tectum

2.1 General Organization

The optic tectum is a laminated, cortex-like structure. Its innermost layer is formed by ependymo-glial cells. The others are alternating cellular and fibrous layers

(Figs. 1 and 2). These layers were numbered from 1 to 9 by Gaupp (1899) (Fig. 2), and conventionally this numbering is used for orientation in the tectum. Most of the cells are located in the inner two fifth of the tectum surrounding the optic ventricle (Fig. 1). The outer two fifth (layer 9) is a neuropil which, among others, receives visual information, as the main input to the tectum. In layer 9 only 6% of all the tectal cells can be found.

The optic fibers are distributed in well-defined sublayers or laminae. Myelinated retinal fibers occupy four laminae, labeled B, D, F, and G, while unmyelinated fibers terminate in three other laminae (A, C, and E; Potter 1969) (Fig. 2). Lamina G is located just below layer 8. In reduced silver preparations this lamina cannot be distinguished from the fibers of layer 7 which collects mainly the efferent axons. Layer 7, lamina G, and layer 8 occupy the middle one fifth of the thickness of the tectum. Myelinated retinal fibers form flat, terminal arborizations. It was found in Golgi and HRP studies that one type of retinal terminals measures 250-300 μm in length, 30-40 μm in thickness and 100-150 μm in width. Another type is 100-200 μm in length and width, and 30-40 μm in thickness (Cajal 1946; Lázár & Székely 1967; Potter 1972; Constantine-Paton et al. 1983; Lázár 1984). The largest terminals occur in lamina G. There is a considerable overlap between neighbouring terminals. In laminae C and E, the profusely arborizing unmyelinated fibers form dense networks, and the shape of individual terminals cannot be recognized.

A well-defined stratification could also be observed in the tectum, when a number of peptides were discovered here (Kuljis & Karten 1982). Some of them were related to retinal fibers, others belonged to tectal neurons or non-optic afferents (Kuljis & Karten 1983).

2.2 Columnar Organization of the Optic Tecum

In contrast with the above described pattern of afferents, the dendrites of most tectal cells and the processes of ependymo-glial cells, are oriented perpendicular to the layers of incoming optic fibers and the surface of the tectum (Fig. 3). This perpendicular arrangement of dendrites resembles that seen in the cortex of higher vertebrates. In comparison with the cortex (see Mallot & von Seelen, this volume) it was proposed that the frog's tectum had also *columnar organization* and that tectal columns might be functional units (Lázár & Székely 1967).

Recently, I have attempted to study the organization of suggested tectal columns with the aid of anatomical techniques. I used two approaches. In the first approach a piece of filter paper, soaked with cobaltous lysine complex was placed on the surface of the tectum. The result was heavy labeling of a large number of tectal cells. Most of the cobalt-filled dendrites were oriented toward the surface of the tectum and the borders of the labeled spot were sharp in contrast to the unlabeled tectal tissue (Fig. 5). The number of cobalt-filled axons and their terminals was surprisingly low in the vicinity of the cobalt labeled area (Fig. 4).

In the second approach cobalt was iontophoretically injected into various layers of the tectum. A focal cobalt injection into layer 9 resulted in heavy labeling of a cell column in the deep tectal layers, and there was a rather broad area with lower intensity in layer 9 (Fig. 7). Here, primarily retinal fibers were filled. Cobalt injected

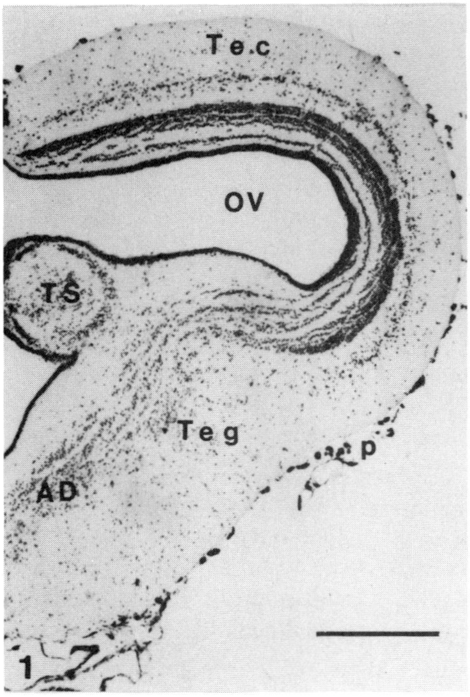
Te.c
OV
TS
Teg
AD
p
1

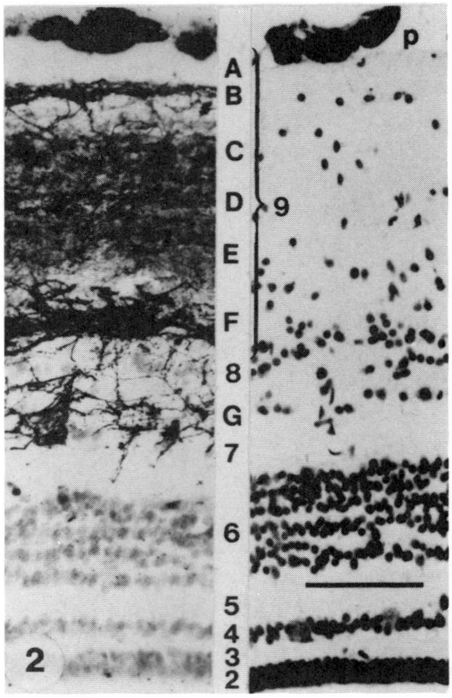
p
A
B
C
D
9
E
F
8
G
7
6
5
4
3
2

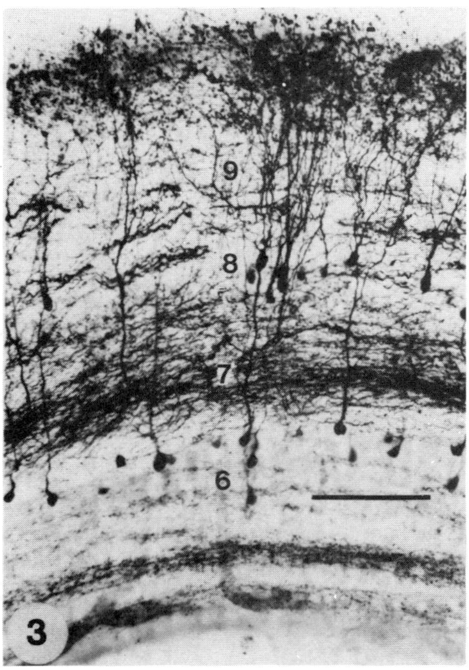
9
8
7
6
3

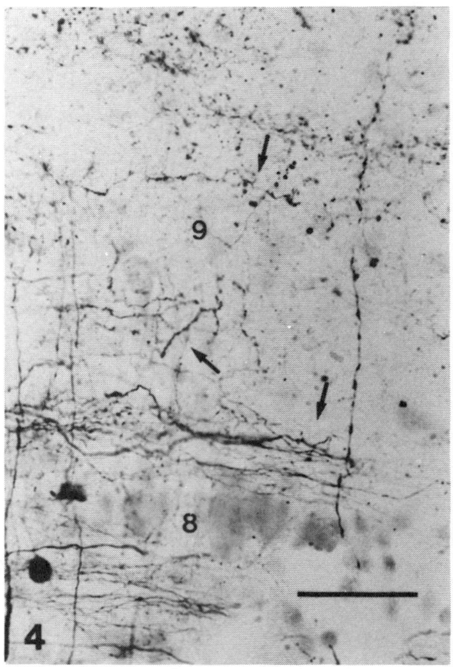
9
8
4

into layer 7 and 6 labeled a dendritic column peripheral to the injection site, and perikarya in the deep layers. Labeled axons could be followed for a longer distance in layer 7, but in layer 9 the cobalt containing area was clearly separated from the rest of tectal tissue (Fig. 6). A few axon terminals could be identified at 300-350 μm distance from the periphery of the labeled dendritic column in layer 9.

A cobalt focus located in layers 1 to 5 resulted in heavy labeling of layer 5. A dendritic column with well-defined borders could be recognized in layers 6 to 9 (Fig. 8.). Several fine, beaded axons, and probably basal dendrites of the neurons of the deep tectal layers, were filled with cobalt within a few hundreds μm distance from the site of cobalt injection (Fig. 9). Terminal branches were rare, but the large number of beads suggests the existence of abundant synaptic connections.

Within the dendritic column, fine, beaded axons form loose networks indicating that most of the connections are formed within the column. The low number of labeled axons outside the column shows that lateral connections are very limited.

2.3 Tectal Neurons

Several types of tectal cells were distinguished by different authors in Golgi preparations (Cajal 1911; Kiro 1948; Potter 1969). In our Golgi studies (Lázár & Székely 1967; Lázár 1984) we classified the neurons as small and large piriform or pear-shaped cells, pyramidal cells, large ganglionic cells, and different small cells in layer 9. It was inferred from Golgi studies that pyramidal cells and large ganglionic cells may be *projection neurons*, and the others may be *local circuit neurons*.

2.3.1 Projection Neurons and the Termination of Their Axons

Studies with retrograde HRP filling of tectal cells from the isthmic nucleus (IN) led to the conclusion that almost all known cell types may be projection neurons (Gruberg & Lettvin 1980). These results, however, should be evaluated with caution. The tectum and the IN are located close to each other and the HRP injected into the

Figures 1-4. Cytology of the frog's optic tectum. **Fig. 1.** Transverse section of the frog's mesencephalon showing the optic tectum (Tec), optic ventricle (OV), the most anterior part of the torus semicircularis (TS) and the mesencephalic tegmentum (Teg). Note stratification of the nuclei of cells in the tectum. AD, anterodorsal tegmental nucleus; p, pigment granules. Nissl staining. Scale bar: 500 μm. **Fig. 2.** Stratification of the optic tectum. The left half of the figure shows cobalt-filled retinal fibers, the right half shows the cell layers of the tectum in a Nissl preparation. Transverse section. Capital letters label laminae of optic terminals, Arabic numerals indicate layers, formed by perikarya of neurons; p, pigment granules. These labels will be used also in the subsequent pictures. Scale bar: 100 μm. **Fig. 3.** Cobalt preparation showing that the majority of dendrites are oriented perpendicular to the surface of the tectum. A large amount of cobalt was injected into the diencephalon, just in front of the tectum. Scale bar: 100 μm. **Fig. 4.** Collaterals (arrows) of the axons of tectal cells which were filled with cobalt through their apical dendrites. Note the low number of labeled axon terminals. Scale bar: 50 μm.

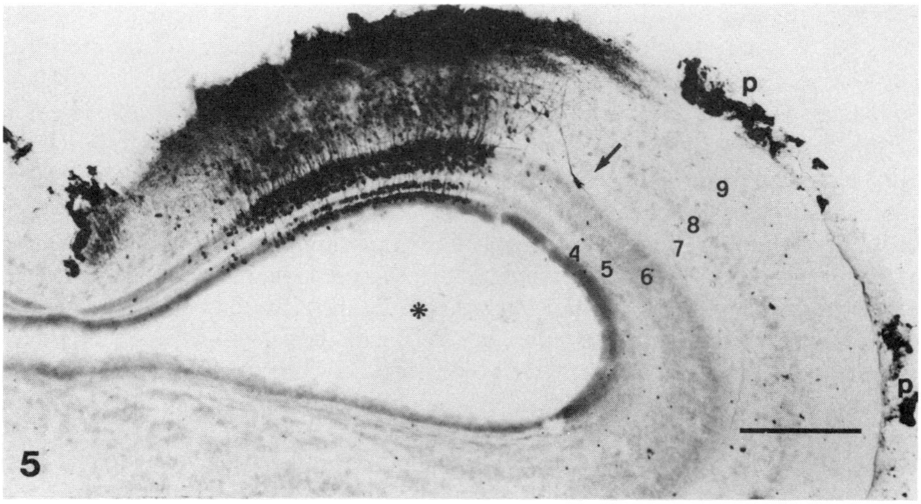

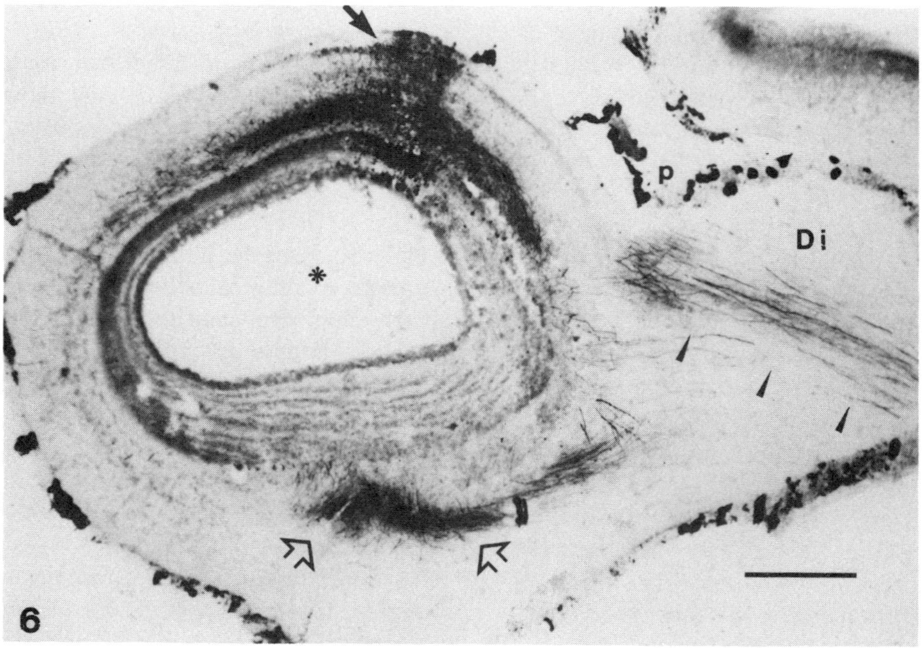

Figures 5 and 6. Labeling of tectal cells by cobalt lysine. **Fig. 5**. Cobalt-labeled cells in the optic tectum in a transverse section. Cobaltous lysine complex soaked into filter paper was placed on the surface of the tectum. Arrow points to a large ganglionic cell. Note the sharp boundaries of the cobalt-filled tectal area. Asterisk here and in the subsequent figures indicate the optic ventricle. Scale bar: 250 μm. **Fig. 6**. A cobalt injection site (arrow) in layers 6 and 7 of the optic tectum in a sagittal section. Open arrows show the caudally (left) and rostrally (right) turning axons of cobalt-filled cells. Arrow heads point to fibers in the tectothalamic tract. More dorsally retinal fibers are filled in the marginal optic tract. Di, diencephalon. Scale bar: 250 μm.

IN may diffuse into the tectum. This may result in direct labeling of tectal cells through their perikarya or dendrites.

2.3.2 The Cells of Origin of the Tecto-Bulbo-Spinal Tract

With the aid of retrograde cobalt-filling of the tecto-bulbo-spinal tract we could prove that pyramidal cells were really projection neurons, and among the large piriform cells there were also many with descending axons (Lázár et al. 1983b). Large ganglionic cells at the surface of layer 6, were regularly labeled on the side contralateral to the cobalt injection (Fig. 10). Labeled large ganglionic cells occurred in layer 7 ipsilaterally. Camera lucida drawings show that most of the dendrites of layer 6 ganglion cells project mainly to lamina B (Fig. 11), while dendrites of layer 7 ganglion cells terminate mainly in laminae F, G, and layer 8 (Fig. 12). This tendency of dendritic orientation was also shown by intracellular cobalt filling of large ganglionic cells. The number of labeled dendritic branches was, however, considerably higher in these preparations, than after retrograde filling (Antal et al. 1986; Matsumoto, this volume). In contrast to this pattern a few layer 6 ganglion cells have dendrites which are oriented parallel to the surface of layer 6, and their terminal branches do not reach the superficial laminae of the tectum. Also a few layer 7 ganglion cells send dendrites up to lamina A. Weerasuriya & Ewert (1981), Ingle (1983), Kostyk & Grobstein (1987a,b) found similar distribution of the cells of origin of the tecto-bulbo-spinal tract (see also Weerasuriya, this volume). Based on these observations it was suggested that contralaterally projecting neurons might be involved in the orientation movements towards the prey, while the ipsilateral cells might be responsible for turning away threatening objects (Ingle 1983; Lázár et al. 1983b) (for a discussion see Grobstein, this volume).

The descending axons of tectal cells are joined by axons of multipolar cells at the tecto-tegmental border. Caudally projecting axons give several collaterals to cells in the nucleus profundus lateralis and the anterodorsal tegmental nucleus. In the medulla, collaterals can be observed mainly in the superior reticular nucleus (Fig. 13). Most of the synapses probably are formed on dendrites, however, beaded, very fine fibers are also closely related to perikarya, which suggests the presence of axosomatic synapses (Fig. 14). The tecto-bulbo-spinal pathway has a crossed and an uncrossed component. Both give a few collaterals that cross to the other side.

2.3.3 Neurons Projecting to the Isthmic Nucleus

The connections between the optic tectum and the IN are topographically organized and reciprocal. The IN also projects to the contralateral optic tectum (Gruberg & Udin 1978; Grobstein et al. 1978; see also Gruberg, this volume). We identified the cells of origin of the tecto-isthmic pathway as large piriform neurons with a narrow dendritic tree, piriform neurons of layer 8, and some multipolar neurons of layer 9 (Lázár, unpublished). After tectal cobalt injections, very sharply delineated spots formed by cobalt-filled axon terminals could be visualized in the IN (Fig. 15). Within such a spot, thin (0.3-0.5 µm diam.) tectal fibers formed broom-like terminal arbors which intermingled with the dense dendritic arborization of isthmic cells (Fig. 16).

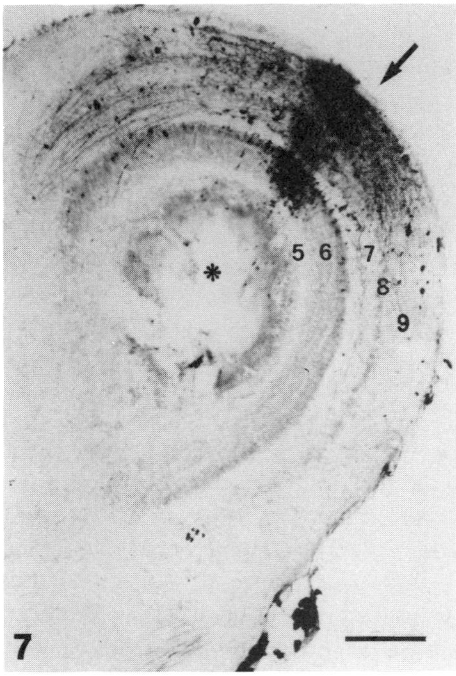
5
6
7
8
9
7

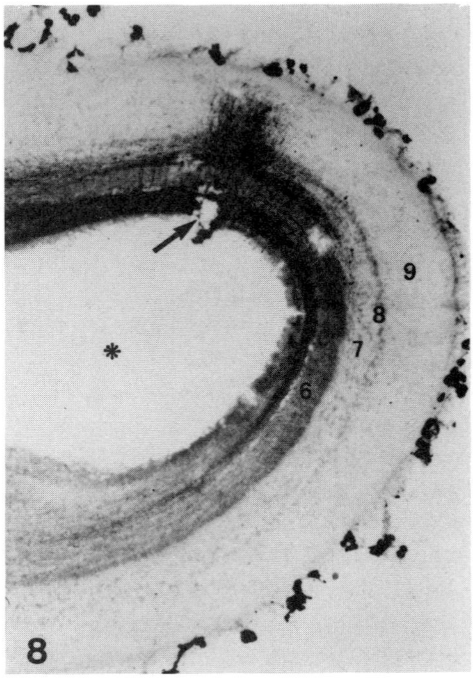
9
8
7
6
8

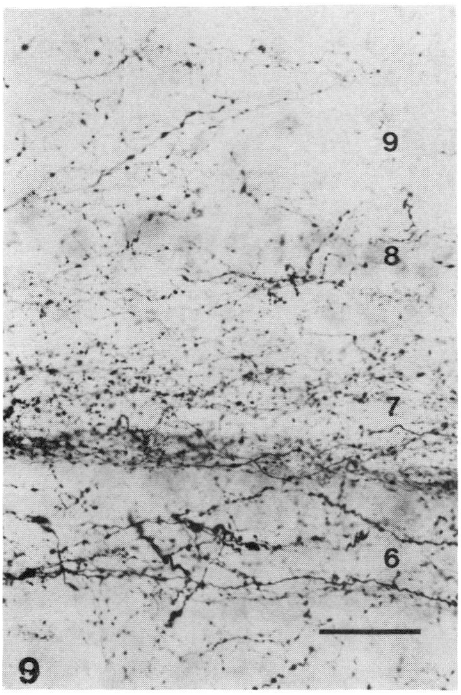
9
8
7
6
9

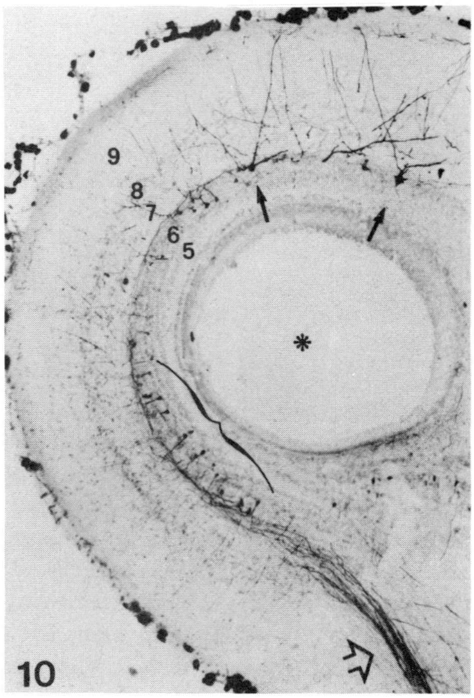
9
8
7
6
5
10

Many of the large piriform cells with narrow dendritic trees in layer 6 contain enkephalins (Kuljis & Karten 1982). We obtained similar results (Merchenthaler et al. 1987, 1988) and also found intense enkephalin-like immunoreactivity in fiber terminals in the core of the IN. They may belong to enkephalin containing tectal cells. The possible function of the IN is discussed in detail by Gruberg (this volume).

2.3.4 Neurons Projecting to the Thalamus

Another important target of tectal efferent axons is the thalamus. Layer 7 ganglionic cells, large piriform, and pyramidal cells send efferent axons to the ipsilateral thalamus. The most frequently labeled cell types after cobalt injections into the tecto-thalamic tract were small piriform neurons in layer 8. Their axons left the tectum via laminae F and D and finally layer 7 (for details see Lázár et al. 1983b).

2.3.5 Local Circuit Neurons

The identification of these neurons is difficult. In our present studies, in which cobalt was injected into the tectum, some of the cells not labeled in cases of extratectal cobalt injections were filled in the vicinity or in a few hundreds of µm distance from the cobalt focus. Figure 17 shows such a cell which has a fusiform perikaryon in layer 6, and the dendrites arborize in layers 6 and 7. Similar cells were intracellularly filled by Antal et al. (1986). In one preparation, a small cell, similar to certain large ganglionic cells, was labeled in layer 8 in the vicinity of the site of cobalt injection (Fig. 18). A cell with round perikaryon was labeled in a distance of 500 µm also in layer 8 in another frog (Fig. 19). The size and shape of this neuron suggest that it may be a local circuit neuron. The site of injection, however, was close to the caudal edge of the tectum and involved, in part, layer 7, so that this neuron may (also) be a projection neuron.

Figures 7-10. Cobalt studies of tectal columns and tectofugal projections. **Fig. 7.** The site of cobalt injection (arrow) in layer 9 in a transverse section of the optic tectum. A column formed by the dendrites of cobalt-labeled cells is clearly visible in the deep layers. The scale bar is 250 µm. **Fig. 8.** The site of cobalt injection (arrow) in the deep tectal layers in a horizontal section. Note the cobalt-filled dendritic column in layer 9 and heavy labeling of fibers in layer 5 (black stripe). Scale bar: 250 µm. **Fig. 9.** Cobalt-filled fine, beaded fibers, in a distance of more than 1 mm from the cobalt focus shown in Fig. 8. Most of the labeled fibers are probably non-retinal afferents, others are basal dendrites. Scale bar: 50 µm. **Fig. 10.** Cobalt-filled large ganglionic cells (arrow) and piriform neurons (bracket) in a transverse section of the optic tectum. Cobaltic lysine complex was injected into the tecto-bulbar/spinal tract (open arrow) in the mesencephalic tegmentum. Scale bar: 250 µm.

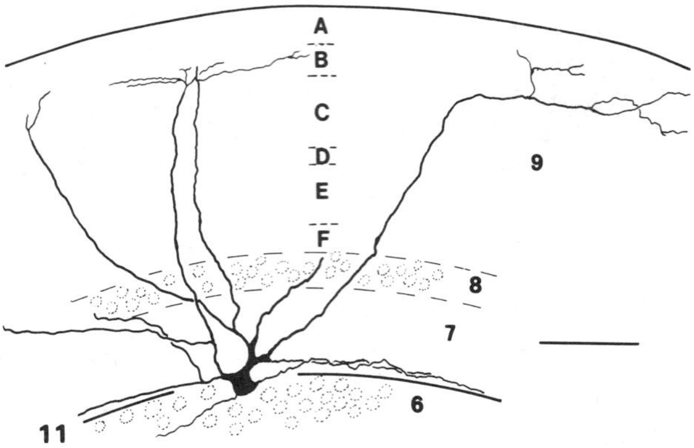

11

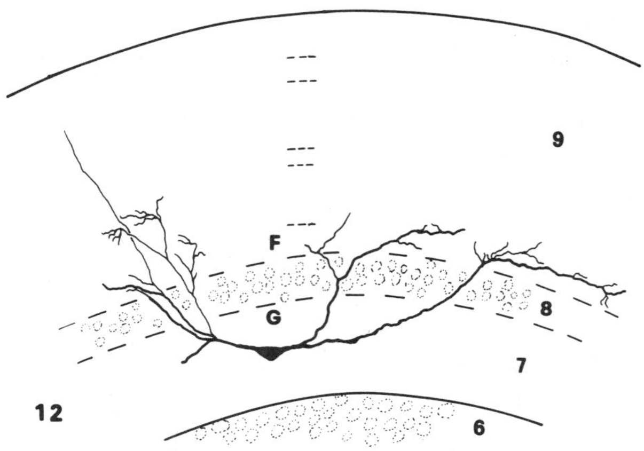

12

Figures 11 and 12. Tecto-bulbar/spinal projecting large ganglionic cells. **Fig. 11.** Camera lucida drawing of a layer 6 large ganglionic cell retrogradely filled by way of the contralateral tecto-bulbar/spinal tract. Note that the majority of dendrites terminate in the outermost part (lamina B) of layer 9. Scale bar: 100 µm. **Fig. 12.** Camera lucida drawing of a layer 7 large ganglionic cell which was retrogradely filled through the ipsilateral tecto-bulbar/spinal tract. Note that the majority of dendrites terminate in the vicinity of layer 8. Scale bar: 100 µm.

Figure 13-16. Tecto-reticular and tecto-isthmic projections. **Fig. 13.** Terminals of tectal fibers (arrows) in the superior reticular nucleus ipsilateral to a cobalt injection into the tectum. Horizontal section. Scale bar: 50 µm. **Fig. 14.** Fine, beaded terminals (arrows) of ipsilateral tectal axons around a perikaryon in the superior reticular nucleus after cobalt injection into the ipsilateral tectum. Horizontal section. Scale bar: 50 µm. **Fig. 15.** Terminals of tectal fibers in the isthmic nucleus (IN). Cobalt was injected into the ipsilateral tectum. Cobalt-filled neurons are visible in the nucleus profundus lateralis (open arrows). Arrows point to the tecto-isthmic tract. Horizontal section. Medial is to the right, caudal is upward. Scale bar: 100 µm. **Fig. 16.** Retrogradely filled cells and anterogradely filled tectal axons in the isthmic nucleus. Cobalt was injected into the ipsilateral tectum. Arrow points toward the hilus of the IN. Horizontal section. Scale bar: 50 µm.

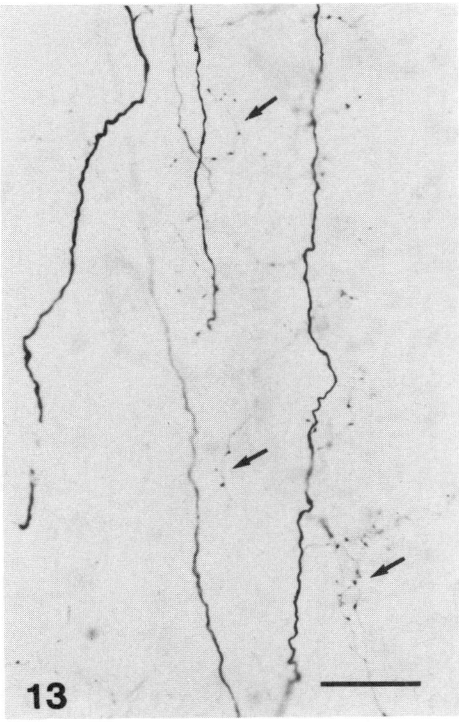
13

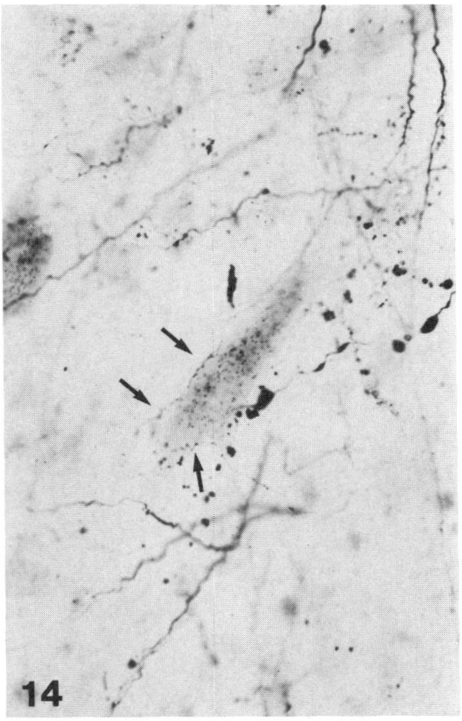
14

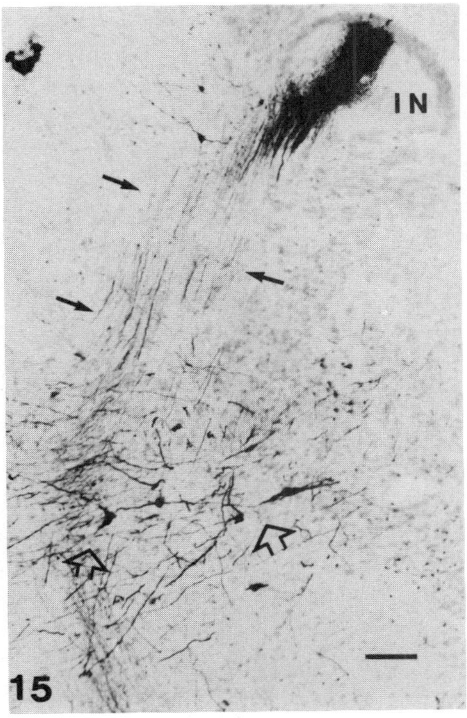
IN
15

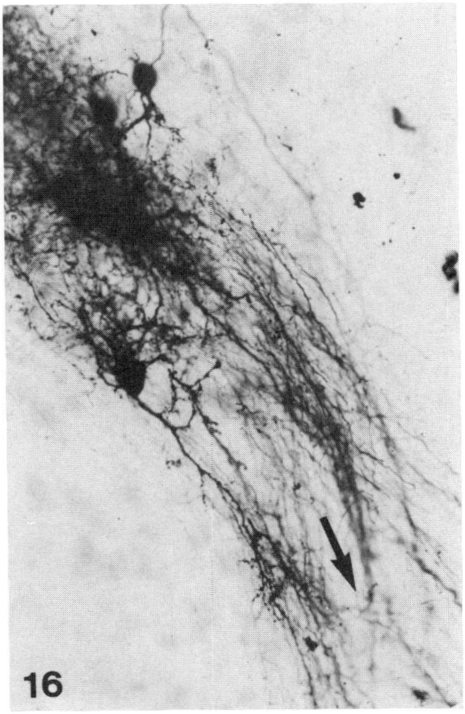
16

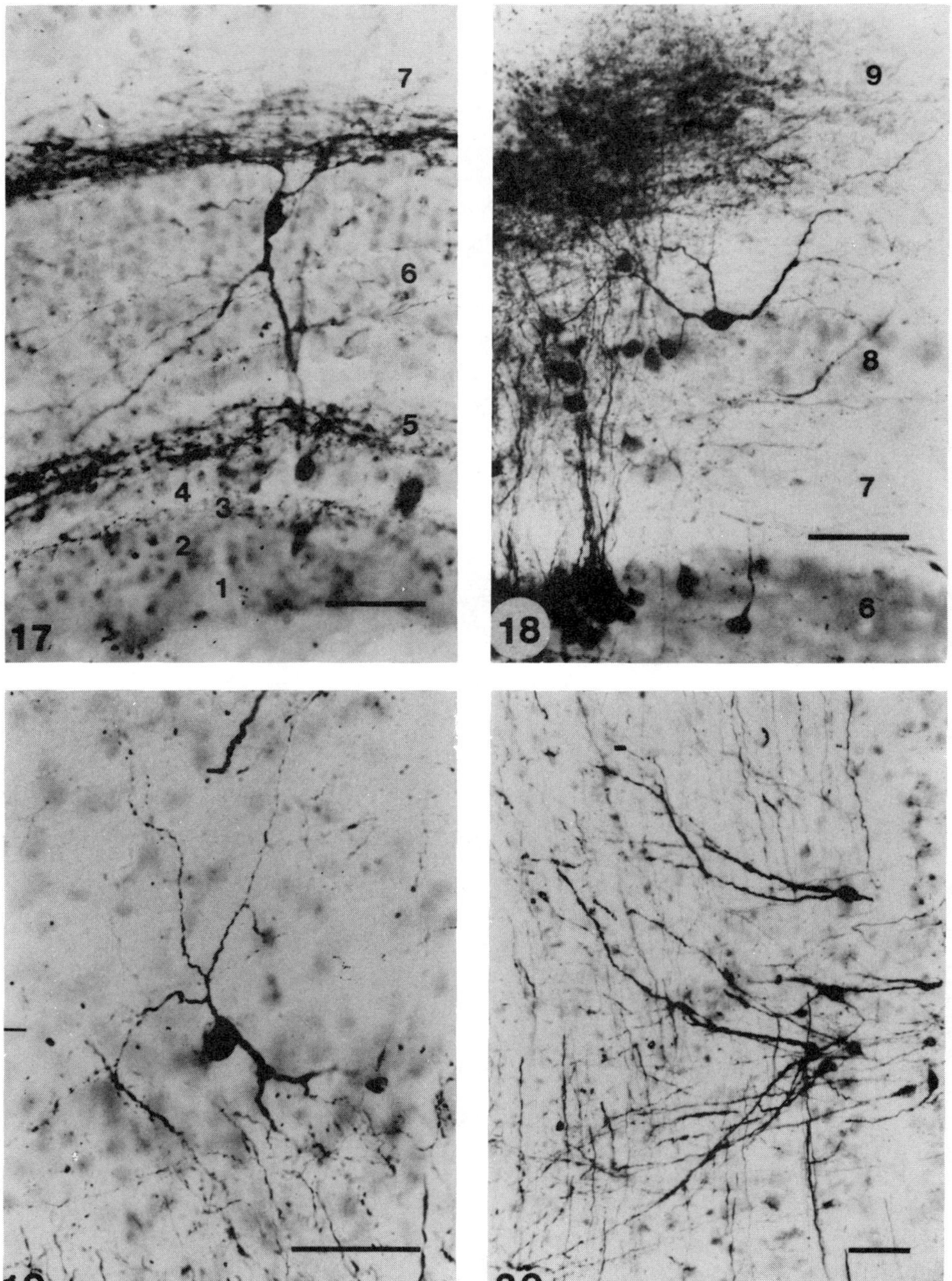
7
6
5
4
3
2
1
17
9
8
7
6
18
19
20

2.4 Non-Optic Afferent Pathways of the Optic Tectum

In earlier degeneration studies (Trachtenberg & Ingle 1974) it was shown that afferents arrived to the tectum from the thalamus, pretectal area, mesencephalic tegmentum, and from the medulla. In their HRP studies Wilczynski & Northcutt (1977) showed that non-optic afferent pathways to the tectum originate from the ipsilateral posterior thalamic nucleus, nucleus lentiformis mesencephali (NLM), anterodorsal, posterodorsal and posteroventral tegmental nuclei, IN, and the dorsal horn of the cervical spinal cord. Few fibers arrive from the contralateral tectum and pretectum and bilaterally from a suprapeduncular cell group.

In our recent cobalt-filling studies (Lázár, unpublished) we found that a large number of afferents entered the tectum alongside its lateral border through layers 3, 5, and 7, and distributed among the neurons of ipsilateral deep tectal layers. Retrograde filling of the cells of origin of these afferents showed that they were located in the same tegmental nuclei as reported by Wilczynski & Northcutt (1977). Their perikarya were piriform, triangular, or fusiform. One or two long dendrites projected laterally and ventrolaterally. Thus, the dendritic arborization of these cells was in the horizontal plane, chiefly in the white matter of the tegmentum (Fig. 20). In this part of the mesencephalon many fibers, which arrive here via the lateral forebrain bundle, terminate (Halpern 1972; Kicliter 1979; Northcutt & Kicliter 1980; Vesselkin et al. 1980; Lázár & Kozicz 1988). Afferents from the brain stem also terminate here and descending pathways originate (Tóth et al. 1985).

3 Pretectum

This region of the frog brain is better known from physiological and behavioral studies rather than from morphological works. The following structures may be included in the pretectal region: pretectal neuropil area, uncinate field, NLM, lateral thalamic nucleus, posterodorsal and posteroventral divisions, central thalamic nucleus, and the pretectal grey. Many scholars described retinal projections to this region (for references see Scalia & Fite 1974), and it was shown that retinal fibers terminated in the large lateral neuropil area and the smaller medial uncinate field.

←

Figure 17-20. Various types of tectal neurons labeled by cobalt lysine. **Fig. 17.** Cobalt-filled fusiform cell in layer 6 of the tectum in horizontal section. Cobalt was iontophoretically injected into the deep tectal layers close to this cell. The dendrites do not reach the retinal recipient laminae. Scale bar: 50 µm. **Fig. 18.** Labeled cells at the periphery of a cobalt injection site in the tectum. Arrow points to a small multipolar cell (similar to large ganglionic cells) in layer 8. Transverse section. Scale bar: 50 µm. **Fig. 19.** Small neuron with round perikaryon and radially oriented dendrites in layer 8, in a horizontal section. Cobalt was injected close to the caudal border of the tectum into layer 6 and part of layer 7. Scale bar: 50 µm. **Fig. 20.** Cobalt-filled piriform neurons in the anterodorsal tegmental nucleus in a horizontal section. The same animal as in Fig. 19. Medial is to the right. Scale bar: 50 µm.

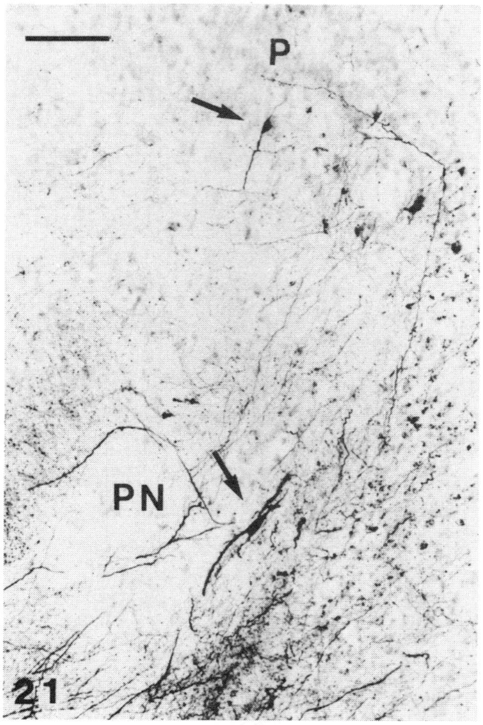
P
PN
21

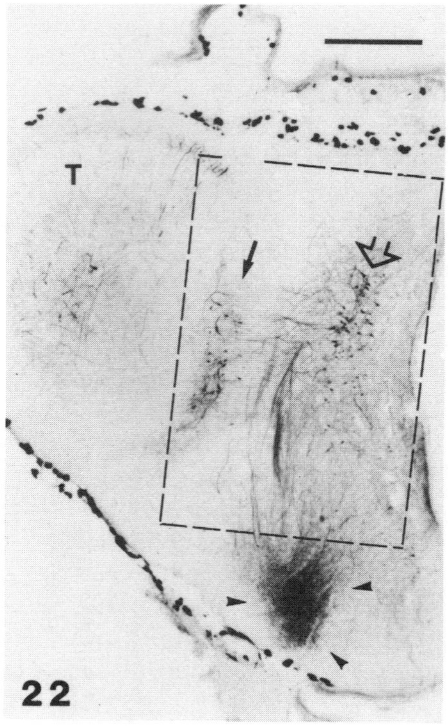
T
22

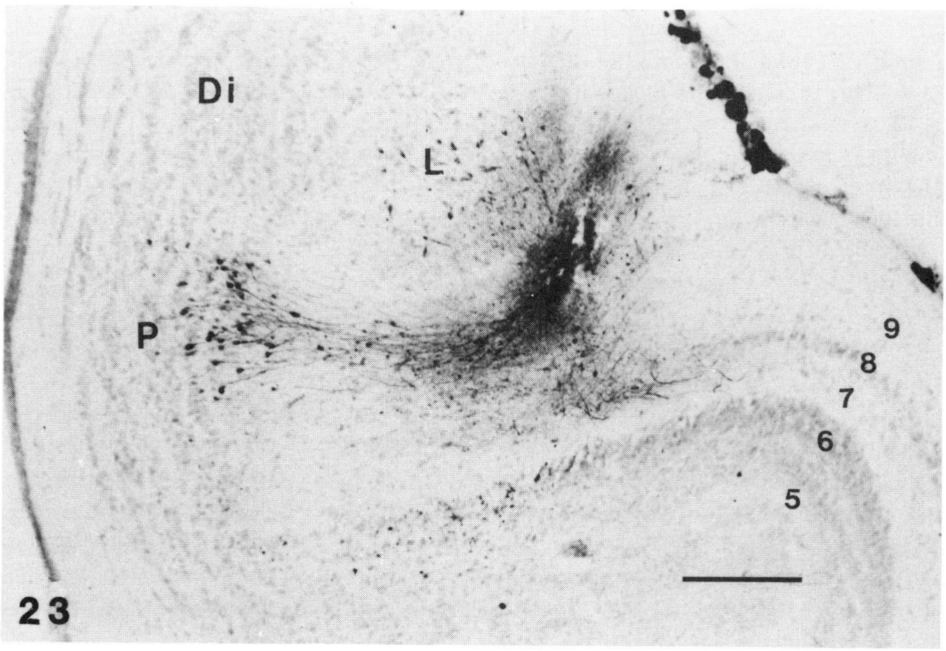
Di
L
P
9
8
7
6
5
23

Two cell goups are in close relationship with the terminal fields of retinal fibers. One of them is the magnocellular pretectal nucleus which was termed by Neary & Northcutt (1983) the nucleus lentiformis mesencephali. Montgomery et al. (1985) could label cells in the NLM with HRP and described three trypes of cells: large neurons, fusiform, and stellate neurons. The other cell group is the posterior thalamic nucleus whose neurons send long dendrites to the optic neuropils (Scalia & Gregory 1970; Lázár 1979). When we injected cobalt into the lateral neuropil, piriform cells were filled through their dendrites in this nucleus (Fig. 23).

Retinal terminals form dense plexuses among the cells of the NLM suggesting extensive synaptic coupling, probably even with perikarya. Most of the retinal fibers arrive here through the axial optic tract (Wlassak 1893; Lázár 1978). Montgomery et al. (1985) reported that contralateral retinal fibers form a dense core in the central portion of the NLM, while ipsilateral fibers terminate throughout the nucleus, except the dense core. Dendrites from the posterior thalamic nucleus arborize within that part of the pretectal neuropil, where thinner retinal fibers terminate. This peculiar anatomical arrangement suggests that the two pretectal nuclei may be functionally different. Montgomery et al. (1985) described reciprocal connections between the NLM and the basal optic nucleus. They also reported that non-optic afferents arrive to the NLM from the mesencephalic pretectal grey, optic tectum, anterior thalamus, and the nucleus of the fasciculus longitudinalis medialis. We can confirm these results with our cobalt labeling data (Lázár unpublished). In a recent cobalt labeling study we identified some of the non-retinal afferent sources of the posterior thalamic nucleus (Lázár & Tompos 1988). This nucleus receives afferents from the ipsilateral striatum, the contralateral superior vestibular nucleus, and is reciprocally interconnected with the contralateral posterior thalamic nucleus, the ipsilateral superior reticular nucleus, tectum, pretectal grey, and ventrolateral thalamic nucleus.

We also studied projection neurons of the pretectal area. Cobalt injected into the tectum retrogradely labeled piriform neurons in the posterior thalamic nucleus and multipolar cells in the NLM (Fig. 21). When the labeling substance was injected close to the pretectal region, into fiber bundles at the level of the basal optic nucleus, two cell groups were filled in the pretectal region. One of these corresponded to the

Figure 21-23. Cytology of pretectal nuclei. **Fig. 21.** Cobalt-labeled cells (arrows) in the pretectal region in a horizontal section. Cobalt was injected into the rostral pole of the ipsilateral tectum. Medial is upward, rostral is to the left. P, posterior thalamic nucleus; PN, pretectal neuropil. Scale bar: 100 μm. **Fig. 22.** Retrogradely filled neurons in the nucleus lentiformis mesencephali (arrow) and the posterior thalamic nucleus (open arrow). Cobalt was injected into descending pretectal fibers lateral to the basal optic nucleus (arrow-heads). Transverse section; T, optic tectum. The enframed area is shown with higher magnification in Fig. 24. Scale bar: 500 μm. **Fig. 23.** A cobalt injection site in the anterolateral part of the pretectal neuropil area in horizontal section. Labeled piriform cells in the posterior thalamic nucleus (P) were filled by way of their dendrites. A few labeled cells are also visible in the lateral thalamic nucleus (L). Cobalt-filled axons enter the tectum via layer 7. Scale bar: 250 μm.

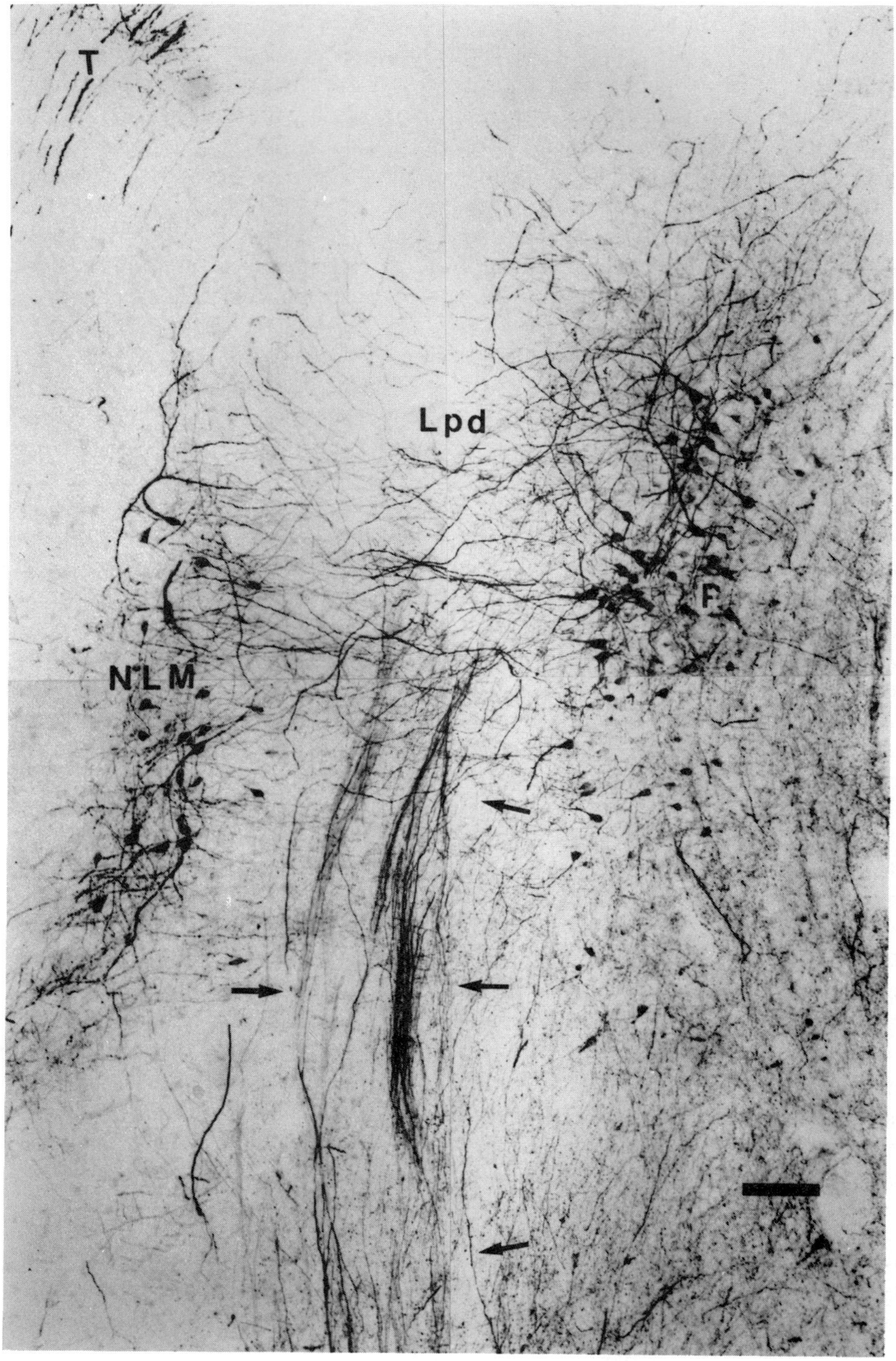
T
Lpd
P
NLM

NLM, and the other to the posterior thalamic nucleus (Fig. 22). In the NLM mainly fusiform cells were labeled, and the main dendrites were oriented dorsoventrally (Fig. 24). Seemingly the dendrites did not leave the area occupied by perikarya, so the NLM can be regarded as a closed nucleus. Most of the labeled cells in the posterior thalamic nucleus have piriform perikarya, but some multipolar cells also occur (Fig. 24). Using the HRP technique, Ingle (1983) made similar observations regarding the location of cells which project toward the medulla oblongata (see also Matsumoto, this volume).

Efferent fibers from the posterior thalamic nucleus could be followed ipsilaterally to the anterodorsal and anteroventral tegmental nuclei and to the NLM, bilaterally to the basal optic nucleus and to the contralateral pretectal grey.

4 Discussion

4.1 Projection Neurons, Local Circuit Neurons

We can conclude from the results of anatomical and behavioral studies that large ganglionic cells and pyramidal cells may be regarded as *projection neurons* for different targets. Most of them send axons either rostralward or caudalward, but there are some large ganglionic cells which have descending axons and ascending collaterals, although the termination site has not been determined yet (Antal et al. 1986). Retrograde double labeling studies are required for further analysis of these neurons.

Based on retrograde filling studies, it seems reasonable to conclude that the majority of the small cells described in Golgi studies (stellate, fusiform, and amacrine cells) in layer 9, are *local circuit neurons*, because they were only occasionally labeled when neurons were back-filled with tracer substances (Gruberg & Lettvin 1980; Ingle 1983). It was proposed by Setalo & Székely (1967) that those cells which have presynaptic perikarya might be inhibitory interneurons. An inhibitory neurotransmitter, however, has not been demonstrated in tectal cells to date.

Cells with widely arborizing dendritic trees in layer 8 (shown by Antal et al. 1986) may also be local circuit neurons because of reasons mentioned above. Many cells in the deep tectal layers which contain different peptides (Kuljis & Karten 1982) may also be interneurons. Morphologically they are large piriform neurons with narrow dendritic trees. Many of them, however, may also be projection neurons as shown by Gruberg & Lettvin (1980). Immunocytochemical studies of neuropeptides and transmitters, combined with classical neuroanatomical lesion experiments, may reveal, what kind of cells can be classified as local circuit neurons in the tectum.

Figure 24. Pretectal projection neurons in the nucleus lentiformis mesencephali (NLM) and the posterior thalamic nucleus (P). Enlarged from Fig. 22. Arrows point to descending axons originating from these nuclei. Lpd, lateral thalamic nucleus posterodorsal division; T, optic tectum. Scale bar: 100 μm.

At the present stage of tectal research a substantial amount of information is available on the morphology of tectal cells. In physiological experiments, using extracellular recordings, several aspects of the functions of tectal cells were learned (see Grüsser & Grüsser-Cornehls 1976; Ewert et al. 1983; Ewert 1984, and this volume). In these experiments morphologically classified tectal cell types could not be associated with physiologically classified cells. By introducing intracellular recording and labeling techniques, the gap between cell morphology and physiological properties of neurons, may be bridged. Attempts to do that have already been made (Antal et al. 1986; Matsumoto et al. 1986; Matsumoto, this volume) and the results are promising.

4.2 Columnar Organization of the Tectum

Our present studies with focal cobalt injections suggest that anatomically the tectum may have columnar organization. However, columns cannot be visualized in normal animals. Experimentally, a columnar (or stripped) arrangement of retinal afferents can be induced in the tectum (Constantine-Paton & Law, 1978; Straznicky et al. 1980; Law & Constantine-Paton 1981; Ide et al. 1983). The width of such columns matches well with the width of most of the terminal arborizations of myelinated retinal axons. Furthermore, the dendritic spread of pyramidal cells and large piriform neurons with widely arborizing dendritic tree are also in the same range. Cell columns may be aligned in the direction of retinal axons entering the tectum, and form strips in double innervated tecta. Tectal columns may be *functional units*, tuned to processing various aspects of visual information (for conceptions see also Arbib, Ewert, Mallot & v. Seelen, and Koy-Oberthür, this volume).

It is not known yet, what kind of neurons establish synaptic contacts within the supposed column and what the synaptic relations are between afferent fibers and tectal cells. The only identified afferents in the tectum, at the ultrastructural level, are the optic terminals (Székely et al. 1973; Matsumoto & Scalia 1981). Synaptological studies are needed to analyze tectal circuitry.

Due to the lack of detailed knowledge regarding intercellular connections in the tectum, we can only speculate as to what tectal columns look like. In columns projecting to the medulla, large ganglionic cells, large pear-shaped, and pyramidal cells should be the output elements. I suppose that in this pathway at least two columns may exist; a large one and a smaller one. The output elements of a large column may be the large ganglionic cells. The dendritic spread of these cells reaches 1 mm in diameter and is always larger than 300 μm. The output of contralaterally projecting columns may trigger turning towards the prey, and ipsilaterally projecting columns may initiate avoidance movements by the animal.

Pyramidal or large piriform cells may be the output elements of the smaller columns. Such a column, 150-250 μm in diameter, may control fine adjustment of ballistic motor activity in prey-catching. The smallest column may belong to projection neurons in layer 8. Schematic drawings of such hypothetical tectal columns were published earlier (Székely & Lázár 1976; Lázár 1984).

The concept of tectal columnar organization was discussed in many theoretical papers, and models were proposed based upon relatively poor anatomical background (Arbib 1982; Arbib & Lara 1982; Lara et al. 1982; Arbib, this volume). The establishment of the proposed neuronal circuitry requires combined electrophysiological and morphological studies with multiple labeling and electron microscopy.

It is generally accepted now that the tectum is the place in the frog's brain, where the commands for motor activities necessary for prey-catching (detection of movements of small objects, the localization of the prey in the visual field, tracking the movements of a potential prey, orientation towards the prey, and the initiation of prey-catching movements) are elaborated. Commands toward spinal motor neurons are mediated via crossed and uncrossed tectal descending pathways. Only few fibers reach the spinal cord directly (Rubinson 1968; Lázár 1969). The majority of fibers terminate in the medullary reticular formation (see also Weerasuriya, this volume). Ingle (1983) demonstrated crossing collaterals from the ipsilateral tecto-bulbar tract in the superior reticular nucleus. We found crossing fibers here originating both from the crossed and the uncrossed descending tectal pathways (Lázár, unpublished observation). The reticulo-spinal tracts also have crossed and uncrossed components (TenDonkelaar et al. 1981; Tóth et al. 1985). This multiple crossing of descending pathways enables the animal to respond with symmetric movements when the stimulus appears in the visual field of one eye (e.g., forward movement toward a prey after orientation). The tectum is also involved in avoidance behavior, in which ipsilaterally descending fibers play an important role. All these functions were discussed in detail (for references see Ewert 1980, 1984; Ingle 1983; Kostyk & Grobstein 1987a,b; Grobstein, this volume).

4.3 Pretectum

The structure of pretectal nuclei which belong to the visual system seem to be more simple than that of the tectum. The number of cell types is rather limited morphologically. In electrophysiological experiments pretectal cells proved to be functionally more variable (Ewert 1971). As it was shown above, the anatomical connections of pretectal nuclei are surprisingly rich, indicating that the pretectum may be involved in various functions. In the posterior thalamic nucleus, direction selective units, sensitive dominantly to temporonasal movement were detected (Cochran et al. 1984). Large pretectal lesions dramatically reduced the frequency of the horizontal optokinetic nystagmus (Montgomery et al. 1982; Lázár et al. 1983a). Pretectal lesions also altered the size of the receptive fields of tectal cells, "disinhibited" tectal neurons, and changed their selective properties (Ewert 1984). The pretectum plays an important role in (i) the discrimination of configural moving visual objects, (ii) the discrimination between object and background (see contributions by Ewert and Manteuffel, this volume), and (iii) the avoidance of barriers and of large stationary objects (Ingle 1976, 1983).

Recently, it was found that pretectal structures possessed intense immunoreactivity with glutamic acid decarboxylase (GAD) antiserum (Yucel et al. 1988). Immunopositive perikarya were located in the NLM and the posterior thalamic

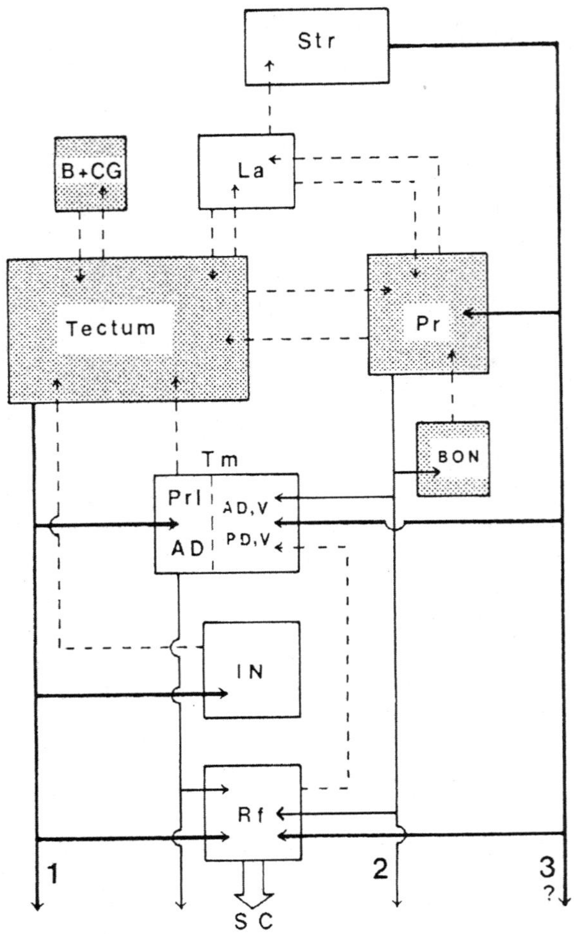

Figure 25. Simplified schema summarizing current knowledge regarding anuran premotor control systems related to the visual system. Dotted areas receive direct retinal afferents. Solid lines ending in arrows show descending pathways and their collaterals. Dashed lines indicate internuclear connections. AD,V, anterodorsal and anteroventral tegmental nuclei; B+CG, lateral geniculate complex; BON, basal optic nucleus; IN, isthmic nucleus; La, lateral thalamic nucleus; PD,V, posterodorsal and posteroventral tegmental nuclei; Prl, nucleus profundus lateralis; Pr, pretectal region; Rf, reticular formation; SC, spinal cord; Str, striatum; Tm, mesencephalic tegmentum.

nucleus, while the optic neuropil areas contained GAD positive "puncta." It was proposed by Yucel et al. (1988) that GABAergic mechanisms might be responsible for directional asymmetry of the horizontal optokinetic nystagmus in monocular frogs. We found enkephalin-containing perikarya in the caudal part of the pretectal area (Merchenthaler et al. 1988). These neurons may also be involved in pretectal/tectal inhibitory circuits. The connections with the superior vestibular nucleus may mean that the posterior thalamic nucleus is also involved in a relay station for ascending vestibular information or it is part of a feed-back loop in the vestibulo-ocular reflex arch.

4.4 General Organization of Supraspinal Control System

The connections of brain regions which are probably involved in processing visual information are summarized in Fig. 25. This summary is rather schematic and many details are omitted for the sake of clarity. Descending tectal pathways are labeled with 1, the pretectal ones with 2. The information descending through these two channels converges in the mesencephalic tegmentum and the medullary reticular formation via collaterals. These two areas project to the spinal cord and are reciprocally interconnected. With the involvement of pathway 1, a tecto-tegmental (see also Grobstein, this volume) and tecto-isthmic (Gruberg, this volume) feed-back loop can be clearly recognized, and the existence of a tecto-reticulo-tegmento-tectal loop can be supposed (see also Schwippert et al., this volume). Pathway 2 may be involved in a similarly organized loop which is closed by tecto-pretectal connections. Pathway 3 descends from the striatum and may directly influence pretectal activity and indirectly tectal activity (see also Ewert, this volume). Finally most of the fibers of this pathway terminate in the anterodorsal and posterodorsal tegmental nuclei which also receive afferents from pathways 1 and 2. The rest of striatal fibers ends in the reticular formation. Pathway 3 is part of the lateral forebrain bundle and seems to be enkephalinergic (Merchenthaler et al. 1988). Ascending tectal outflow mediated by the lateral thalamic nucleus may influence the activity of the striatum. Thus, descending striatal fibers may take part in the formation of a long striato-tecto-striatal feed-back loop.

Even this simplified scheme shows that a considerable part of the brain can be associated with the visual system anatomically. Functionally, the system is even more complex as it is shown by Ewert in this volume.

5 References

Antal M, Matsumoto N, Székely G (1986) Tectal neurons of the frog: intracellular recording and labeling with cobalt electrodes. *J Comp Neurol* 246: 238-253

Arbib MA (1982) Modelling neural mechanisms of visuomotor coordination in frog and toad. In: Amari S, Arbib MA (eds) *Competition and cooperation in neural nets.* Springer-Verlag, Berlin Heidelberg New York, pp 342-370

Arbib MA, Lara R (1982) A neural model of the interaction of tectal columns in prey-catching behavior. *Biol Cybern* 44: 185-196

Cajal PR y (1946) El cerebros de los batracios. *Trab Inst Cajal Invest Biol* 38:41-111

Cajal SR y (1911) *Histologie due systeme nerveux de l'homme et des vertebres.* Maloine, Paris

Cochran SL, Dieringer N, Precht W (1984) Basic optokinetic-ocular reflex pathways in the frog. *J Neurosci* 4: 43-57

Constantine-Paton M, Law MI (1978) Eye-specific termination bands in tecta of three-eyed frogs. *Science* 202: 639-641

Constantine-Paton M, Pitts EC, Reh TA (1983) The relationship between retinal axon ingrowth, terminal morphology, and terminal patterning in the optic tectum of the frog. *J Comp Neurol* 218: 297-313

Contestabile A (1976) Comparative survey on enzyme localization, ultrastructural arrangement and functional organization in the optic tectum of nonmammalian vertebrates. *Experientia (Basel)* 32: 1223-1229

Ewert J-P (1971) Single unit responses of the toad's (*Bufo americanus*) caudal thalamus to visual objects. *Z Vergl Physiol* 74: 81-102

Ewert J-P (1980) *Neuroethology. An introduction to the neurophysiological fundamentals of behavior.* Springer-Verlag, Berlin Heidelberg New York

Ewert J-P, Burghagen H, Schürg-Pfeiffer E (1983) Neuroethological analysis of the innate releasing mechanism for prey-catching behavior in toads. In: Ewert J-P, Capranica RR, Ingle DJ (eds) *Advances in vertebrate neuroethology.* Plenum Press, New York, pp 413-475

Ewert J-P (1984) Tectal mechanisms that underlie prey-catching and avoidance behaviors in toads. In: Vanegas H (ed) *Comparative neurology of the optic tectum.* Plenum Press, New York, pp 247-399

Finkenstädt T, Adler NT, Allen TO, Ebbesson SOE, Ewert J-P (1985) Mapping of brain activity in mesencephalic and diencephalic structures of toads during presentation of visual key stimuli: a computer assisted analysis of (^{14}C)2DG autoradiographs. *J Comp Physiol* 156: 433-445

Fite KV (1976) (ed) *The amphibian visual system: a multidisciplinary approach.* Academic Press, New York

Gaupp E (1899) *Anatomie des Frosches.* Vieweg, Braunschweig

Gaze RM (1958) The representation of the retina on the optic lobe of the frog. *Quart J Exp Physiol* 43: 209-214

Grobstein P, Comer C, Hollyday M, Archer SM (1978) A crossed isthmotectal projection in *Rana pipiens* and its involvement in the ipsilateral visuotectal projection. *Brain Res* 156: 117-123

Gruberg ER, Lettvin JY (1980) Anatomy and physiology of a binocular system in the frog *Rana pipiens. Brain Res* 192: 313-325

Gruberg ER, Udin SB (1978) Topographic projections between the nucleus isthmi and the tectum of the frog *Rana pipiens. J Comp Neurol* 179: 487-500

Grüsser O-J, Grüsser-Cornehls U (1976) Neurophysiology of the anuran visual system. In: Llinás R, Precht W (eds) *Frog neurobiology.* Springer-Verlag, Berlin Heidelberg New York, pp 297-385

Halpern M (1972) Some connections of the telencephalon of the frog, *Rana pipiens.* An experimental study. *Brain Behav Evol* 6: 42-68

Herrick CJ (1948) *The brain of the tiger salamander.* University of Chicago Press, Chicago

Ide CF, Fraser SE, Meyer RL (1983) Eye dominance columns from an isogenic double-nasal frog eye. *Science* 221: 293-295

Inagaki S, Senba E, Shiosaka S, Takagi H, Kawai Y, Takatsuki K, Sakanaka M, Matsuzaki T, Tohyama M (1981) Regional distribution of substance P-like immunoreactivity in the frog brain and spinal cord: immunohistochemical analysis. *J Comp Neurol* 201: 243-254

Ingle DJ (1976) Behavioral correlates of central visual function in anurans. In: Llinás R, Precht W (eds) *Frog neurobiology.* Springer-Verlag, Berlin Heidelberg New York, pp 435-451

Ingle DJ (1983) Brain mechanisms of visual localization by frogs and toads. In: Ewert J-P, Capranica RR, Ingle DJ (eds) *Advances in vertebrate neuroethology.* Plenum Press, New York, pp 177-226

Kicliter E (1979) Some telencephalic connections in the frog, *Rana pipiens. J Comp Neurol* 185: 75-86

Kiro CM (1948) A comparative histology of the midbrain of amphibians. In: *Collection in memory of AA Zavarzin.* USSR Academy of Sciences Press, Moscow, pp 54-80 (in Russian)

Kostyk SK, Grobstein P (1987a) Neuronal organization underlying visually elicited prey orienting in the frog. II: Anatomical studies on the laterality of central projections. *Neurosci* 21: 57-82

Kostyk SK, Grobstein P (1987b) Neuronal organization underlying visually elicited prey orienting in the frog. III: Evidence for the existence of an uncrossed descending tectofugal pathway. *Neurosci* 21: 83-96

Kuljis RO, Karten HJ (1982) Laminar organization of peptide-like immunoreactivity in the anuran optic tectum. *J Comp Neurol* 212: 188-201

Kuljis RO, Karten HJ (1983) Modifications in the laminar organization of peptide-like immunoreactivity in the anuran optic tectum following retinal deafferentation. *J Comp Neurol* 217: 239-251

Law MI, Constantine-Paton M (1981) Anatomy and physiology of experimentally produced striped tecta. *J Neurosci* 1: 741-759

Lara R, Arbib MA, Cromarty AS (1982) The role of the tectal column in facilitation of amphibian prey-catching behavior; a neural model. *J Neurosci* 2: 521-530

Lázár Gy (1969) Efferent pathways of the optic tectum in the frog. *Acta Biol Acad Sci Hung* 20: 171-183

Lázár Gy (1978) Application of cobalt-filling technique to show retinal projections in the frog. *Neurosci* 3: 725-736

Lázár Gy (1979) Organization of the frog visual system. In: Lissak K (ed) *Recent developments of neurobiology in Hungary.* Akademiai Kiado, Budapest, pp 9-50

Lázár Gy (1984) Structure and connections of the optic tectum. In: Vanegas H (ed) *Comparative neurology of the optic tectum.* Plenum Press, New York, pp 185-210

Lázár Gy, Kozicz T (1988) Neuronal pathways of the lateral forebrain bundle in the frog, *Rana esculenta* (in preparation)

Lázár Gy, Székely Gy (1967) Golgi studies on the optic center of the frog. *J Hirnforsch* 9: 329-344

Lázár Gy, Tompos G (1988) Neuronal connectivity and cell morphology of the frog's diencephalon. A study with cobalt labeling (in preparation)

Lázár Gy, Alkonyi B, Tóth P (1983a) Reinvestigation of the role of the accessory optic system and the pretectum in the horizontal head nystagmus of the frog. Lesion experiments. *Acta Biol Acad Sci Hung* 34: 385-393

Lázár Gy, Tóth P, Csank Gy, Kicliter E (1983b) Morphology and location of tectal projection neurons in frogs: a study with HRP and cobalt filling. *J Comp Neurol* 215: 108-120

Lettvin JW, Maturana HR, McCulloch WS, Pitts WH (1959) What the frog's eye tells the frog's brain? *Proc IRE* 47: 1940-1951

Llinás R, Precht W (eds) (1976) *Frog neurobiology.* Springer-Verlag, Berlin Heidelberg New York

Matsumoto N, Bando T, (1980) Excitatory synaptic potentials and morphological classification of tectal neurons of the frog. *Brain Res* 192: 39-48

Matsumoto DE, Scalia F (1981) Long-term survival of centrally projecting axons in the optic nerve of the frog following destruction of the retina. *J Comp Neurol* 20: 135-155

Matsumoto N, Schwippert WW, Ewert J-P (1986) Intracellular activity of morphologically identified neurons of the grass frog's optic tectum in response to moving configurational visual stimuli. *J Comp Physiol* 159: 721-739

Maturana HR, Lettvin JY, McCulloch WS, Pitts WH (1960) Anatomy and physiology of vision in the frog (*Rana pipiens*). *J Gen Physiol* 43: 129-176

Merchenthaler I, Lázár Gy, Maderdrut JL (1988) Distribution of proenkephalin-peptides in the brain of *Rana esculenta* (in preparation)

Merchenthaler I, Maderdrut JL, Lázár Gy, Gulyas I, Petrusz P (1987) Immunocytochemical analysis of proenkephalin-derived peptides in the amphibian hypothalamus and optic tectum. *Brain Res* 416: 219-227

Montgomery NM, Fite KV, Grigonis AM (1985) The pretectal nucleus lentiformis mesencephali of *Rana pipiens. J Comp Neurol* 234: 264-275

Montgomery N, Fite KV, Taylor M, Bengston L (1982) Neural correlates of optokinetic nystagmus in the mesencephalon of *Rana pipiens*: functional analysis. *Brain Behav Evol* 21: 137-150

Neary TJ, Northcutt RG (1983) Nuclear organization of the bullfrog diencephalon. *J Comp Neurol* 213: 262-278

Northcutt RG (1974) Some histochemical observations on the telencephalon of the bullfrog, *Rana catesbeiana. J Comp Neurol* 157: 379-390

Northcutt RG, Kicliter E (1980) Organization of the amphibian telencephalon. In: Ebbesson SOE (ed) *Comparative neurology of the telencephalon.* Plenum Press, New York, pp 203-256

Potter HD (1965) Mesencephalic auditory region of the bullfrog. *J Neurophysiol* 28: 1132-1154

Potter HD (1969) Structural characteristics of cell and fiber population in the optic tectum of the frog (*Rana catesbiana*). *J Comp Neurol* 136: 203-232

Potter HD (1972) Terminal arborizations of retinotectal axons in the bullfrog. *J Comp Neurol* 144: 269-284

Röthig P (1927) Beiträge zum Studium des Zentralnervensystems der Wirbeltiere. 11. Über die Faserzüge in Mittelhirn, Kleinhirn, und Medulla oblongata der Urodelen und Anuren. *Z Mikr Anat Forsch* 10: 381-472

Rubinson K (1968) Projections of the tectum opticum of the frog. *Brain Behav Evol* 1: 529-561

Scalia F, Fite KV (1974) A retinotopic analysis of the central connections of the optic nerve in the frog. *J Comp Neurol* 158: 455-478

Scalia F, Gregory K (1970) Retinofugal projections in the frog, location of the postsynaptic neurons. Brain Behav Evol 3: 16-29

Setalo G, Székely G (1967) The presence of membrane specializations indicative of somato-dendritic synaptic junctions in the optic tectum of the frog. *Exp Brain Res* 4: 237-242

Sood PP (1978) Chemoarchitectonics of the optic tectum of frogs (*Rana tigrina*). *Cell Mol Biol* 23: 195-205

Sperry RW (1943) Effect of 180° degree rotation of the retinal field on visuomotor coordination. *J Exp Zool* 92: 263-279

Sperry RW (1944) Optic nerve regeneration with return of vision in anurans. *J Neurophysiol* 7: 57-71

Sperry RW (1945) Restortion of vision after crossing the optic nerves and after contralateral transplantation of eye. *J Neurophysiol* 8: 15-29

Straznicky C, Tay D, Hiscock J (1980) Segregation of optic fibre projections into eye-specific bands in dually innervated tecta in *Xenopus. Neurosci Lett* 19: 131-136

Ströer WFH (1940) Zur vergleichenden Anatomie des primären optischen System bei Wirbeltieren. *Z Anat Entwickl Gesch* 110: 301-321

Székely G, Lázár Gy (1976) Cellular and synaptic architecture of the optic tectum. In: Llinás R, Precht W (eds) *Frog neurobiology.* Springer-Verlag, Berlin Heidelberg New York, pp 407-434

Székely G, Setalo G, Lázár Gy (1973) Fine structure of the frog's optic tectum: optic fibre termination layers. *J Hirnforsch* 14: 189-225

Takagi S, Tsuji T, Amagai T, Takamatsu T, Fujisawa H (1987) Specific cell surface labels in the visual centers of *Xenopus laevis* tadpole identified using monoclonal antibodies. *Develop Biol* 122: 90-100

TenDonkelaar HJ, de Boer-van Huizen R, Schouten FTM, Eggen SJH (1981) Cells of origin of descending pathways to the spinal cord in the clawed toad (*Xenopus laevis*). *Neurosci* 6: 2297-2312

Tóth P, Csank Gy, Lázár Gy (1985) Morphology of the cells of origin of descending pathways to the spinal cord in *Rana esculenta*. A tracing study using cobaltic-lysine complex. *J Hirnforsch* 26: 365-383

Trachtenberg MC, Ingle DJ (1974) Thalamo-tectal projection in the frog. *Brain Res* 79: 419-430

Vesselkin NP, Ermakova TV, Kenigfest B, Goikovic M (1980) The striatal connections in frog *Rana temporaria*: an HRP study. *J Hirnforsch* 21: 381-390

Weerasuriya A (1983) Snapping in toads: some aspects of sensory-motor interfacing and motor pattern generation. In: Ewert J-P, Capranica RR, Ingle DJ (eds) *Advances in vertebrate neuroethology*. Plenum Press, New York, pp 613-626

Weerasuriya A, Ewert J-P (1981) Prey-selective neurons in the toad's optic tectum and sensorimotor interfacing: HRP studies and recording experiments. *J Comp Physiol* 144: 429-434

Wilczinsky W, Northcutt RG (1977) Afferents to the optic tectum of the leopard frog: an HRP study. *J Comp Neurol* 173: 219-230

Wlassak R (1893) Die optischen Leitungsbahnen des Frosches. *Arch Anat Physiol (Physiol Abtlg)* Suppl: 1-28

Yucel YH, Hindelang C, Stoeckel ME, Bonaventure N (1988) GAD immunoreactivity in pretectal and accessory optic nuclei of the frog mesencephalon. *Neurosci Lett* 84: 1-6

Morphological and Physiological Studies of Tectal and Pretectal Neurons in the Frog

NOBUYOSHI MATSUMOTO

Department of Biophysical Engineering
Faculty of Engineering Science, Osaka University
Toyonaka, Osaka 560, Japan

Abstract. Intracellular recording and staining techniques were applied to visual neurons in frog's optic tectum and pretectum to study receptive fields, response characteristics, cellular morphologies, axonal projections, and synaptic organizations. In the first experiment, response properties of labeled neurons, classified according to extracellular studies, could be correlated with morphology. *Class T5.1 neurons* were identified as large ganglionic neurons in layer 8 or 7, and pear-shaped neurons in layer 8 or the top of 6. No obvious relationship was found between physiological and morphological properties. *Class T5.2 neurons* were located on top of layer 6 and resembled the pyramidal type; an axon could be traced to the medulla oblongata. *Class T5.3 neurons* were identified as large ganglionic neurons in layer 8. Neurons responding predominantly by IPSPs could be identified as pear-shaped neurons in layers 2, 4, or 6. In the second experiment, axonal projections of neurons in the pretectal caudal thalamus were examined. Some neurons from the Lpd (lateral posterodorsal) nucleus projected bilaterally to both optic lobes or to the contralateral tectum; from P (posterocentral) nucleus neurons projected toward the medulla oblongata. To electrical stimulation of the optic tract, the former neurons exhibited an EPSP and the latter an EPSP followed by IPSP. In the third experiment, a pulse-triggered averaging technique was applied to investigate retino-tectal connections. Synaptic organizations between retinal *on-off* fibers and tectal E-E type neurons (EPSP at *on* and at *off* of diffuse light) and EI-EI type neurons (EPSP and IPSP at *on* and at *off* of diffuse light), respectively, were examined. It was found that tectal neurons of these types are monosynaptically excited by *on-off* retinal fibers.

1 Introduction

Since Lettvin et al. (1959) had classified the responsiveness of retinal ganglion cells of the frog into 4 classes, many investigators have performed physiological experiments to understand the mechanism of information processing in the visual centers of frogs and toads. Another important approach to understand the brain is provided by anatomical studies such as the examination of cellular morphology, axonal projection, and fine structure of the neuronal connections (Székely & Lázár 1976; Lázár, this volume).

Physiological and anatomical studies of the frog's brain have been performed independently because of limitations of experimental techniques. Recently, physiological and morphological approaches were combined utilizing intracellular recording and labeling techniques.

This paper deals with three different but closely related experimental subjects. The first part presents a brief description on the correlation between cellular morphology and receptive field properties of tectal neurons. In the second part preliminary results on morphological identification and the axonal projections of pretectal neurons are described. The third part deals with synaptic organization of retino-tectal connections studied by simultaneous recording from both retinal fiber terminals and tectal neurons.

2 Receptive Fields and Response Properties of Identified Tectal Neurons

Grüsser & Grüsser-Cornehls (1976) have divided frog's tectal neurons into seven classes (T1-T7), according to shape, size, and location of ERF (excitatory receptive field) and binocular or monocular properties. These classes can be grouped into "small field neurons" (T1, T3, T5, and T7) and "large field neurons" (T2, T4, and T6). Ewert and coworkers considered the sensitivity to changes of configuration parameters of moving stimuli, showing that among T5 neurons the T5.1, T5.2, T5.3, and T5.4 types are distinguished (for reviews see Ewert 1984, and this volume). Based on investigations of cellular morphology and synaptic organization of tectal neurons Székely & Lázár (1976) proposed a column model for the neuro-architecture underlying visual information processing (see also Lázár, this volume).

Matsumoto & Bando (1980) first applied intracellular recording and Procion Yellow staining techniques to tectal neurons of the frog, which made it possible to combine morphological and physiological studies. Using potassium-citrate/Co^{3+}-lysine filled micropipettes, Antal et al. (1986) and Matsumoto et al. (1986) studied response properties and cellular morphology in more detail (for staining technique see Görcs et al. 1979; Urban & Székely 1983). In the following, the results of the above mentioned experiments will be reviewed.

2.1 Responses to Electrical Stimulation of the Optic Tract

Synaptic organization in the retino-tectal connection of the frog was initially studied at intracellular level by applying electrical stimuli to the optic nerve or to the dorsal optic tract. There are several response types as shown in Fig. 1. The response with only a single EPSP was named E type, and that with only a single IPSP, I type. Other responses were composed of combinations of an EPSP (E) and an IPSP (I), yielding EE, II, EI, and EEI types. The EI type was most frequently observed (about 50% of the cases). When the stimulus intensity was raised, the EPSP grew up until it generated an action potential; in some cases, however, the following inhibitory response was too large, so that generation of an action potential failed to occur. The

E type (about 25% of the cases), I type (about 20% of the cases), EEI type, and EE type were less frequently recorded. The number of synapses involved was analyzed by measuring the latencies of EPSPs showing that some large ganglionic neurons of the EEI type were monosynaptically excited by myelinated optic fibers (Matsumoto & Bando 1978, 1980).

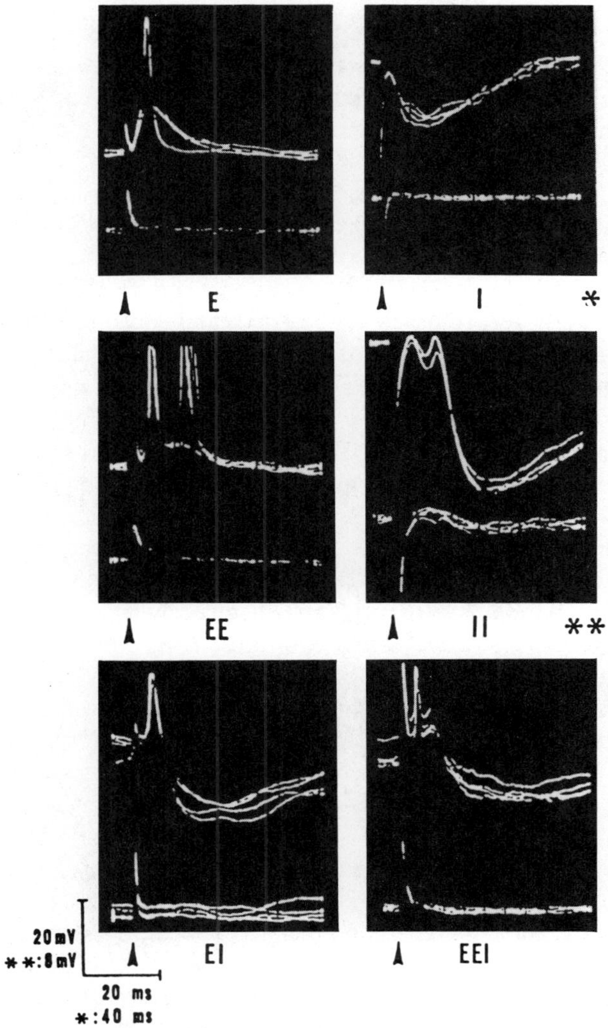

Figure 1. Intracellular responses of tectal neurons of the grass frog, *Rana temporaria*, in response to electrical stimulation of the optic tract; E: a single EPSP; I: a single IPSP; EE: two EPSPs; II: two IPSPs; EI: an EPSP followed by an IPSP; EEI: two EPSPs followed by an IPSP. The lower traces indicate the field potentials recorded just outside the neurons. Arrows indicate the onset of electrical stimulation (0.1 ms in duration) applied to the optic tract. (From Matsumoto et al. 1986)

2.2 Responses to Changes of Diffuse Light

Switching the diffuse light on and off evoked various kinds of responses, the main types of which are shown in Fig. 2A. They are E-E type (EPSPs at *on* and *off*), I-I type (IPSPs at *on* and *off*), IE-IE type (EPSP and IPSP at *on* and *off*) and IE-IE type (IPSP and EPSP/rebound excitation at *on* and *off*). These neuron types exhibited "symmetrical" responses for *on* and *off*; however, "asymmetrical" types, such as I-EI, were only seldom found in some of the neurons. A moving small visual object elicited EPSPs (Fig. 2B1) or IPSPs (Fig. 2B2) or a combination of EPSP and IPSP (Fig. 2B3). As shown in Fig. 6B(E), the moving edge of a *large* black object against a white background typically evoked a transient EPSP or IPSP when entering and when leaving neuron's receptive field.

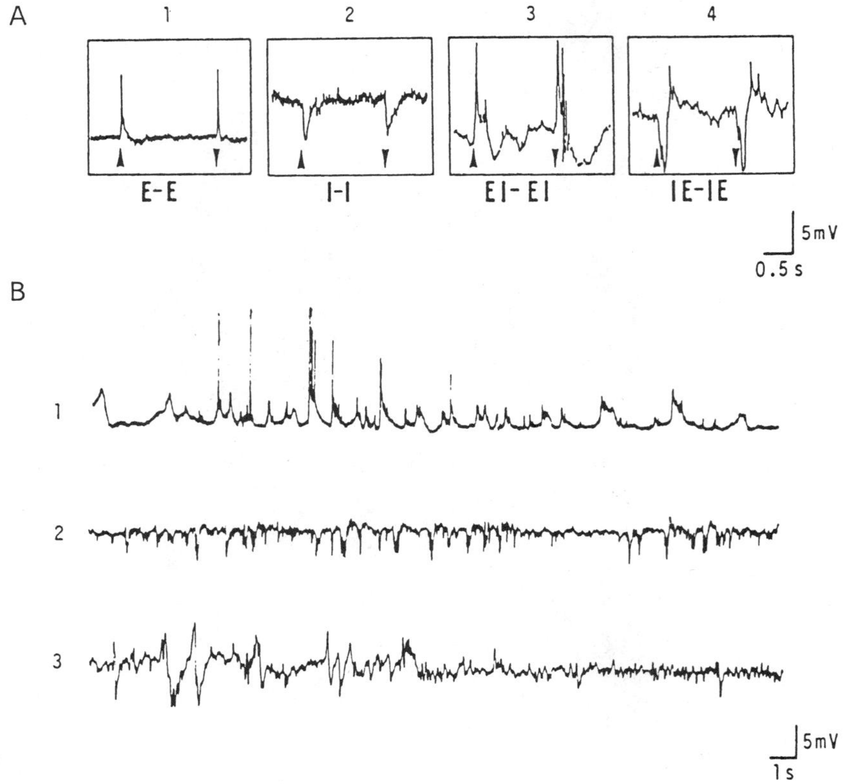

Figure 2. Intracellular responses of tectal neurons of the grass frog to visual stimuli. (A) Responses to changes of diffuse light at *on* and *off*. A1: E-E type (EPSP at *on* and *off*); A2: I-I type (IPSP at *on* and *off*); A3: EI-EI type (EPSP-IPSP at *on* and *off*); A4: IE-IE type (IPSP-EPSP/rebound excitation at *on* and *off*). Upward and downward arrows indicate *on* and *off* of light, respectively. (B) Response to a moving stimulus; B1: excitatory type; B2: inhibitory type; B3: mixture of excitatory and inhibitory responses. (Modified from Matsumoto et al. 1986, and Antal et al. 1986).

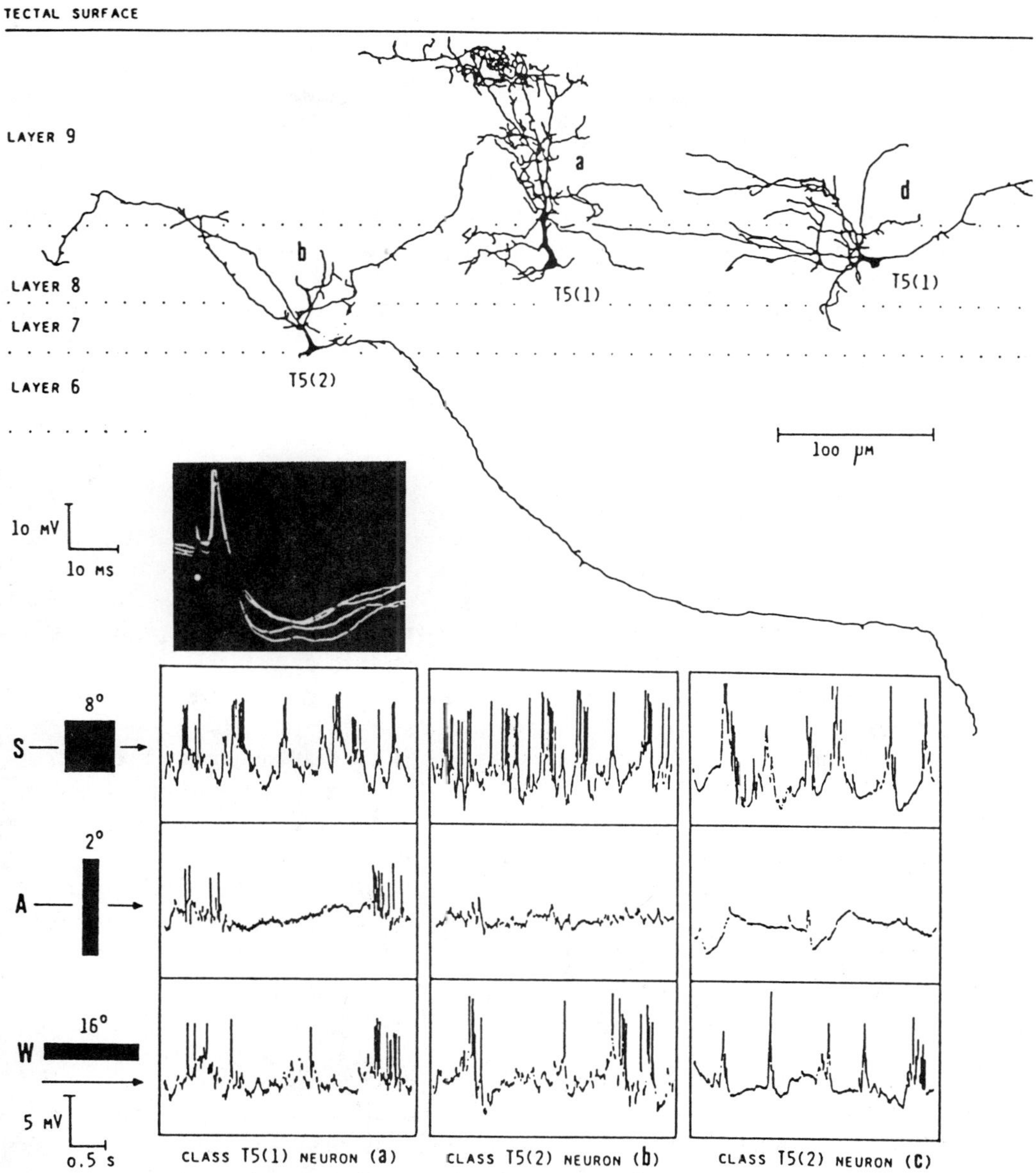

Figure 3. Cellular morphology and intracellular response properties of T5.1 and T5.2 neurons in the grass frog *Rana temporaria.* Neurons were labeled following intracellular iontophoretic application of Co^{3+}-lysine. Top: Camera lucida drawings of two T5.1 neurons (a and d) and one T5.2 neuron (b); note that the axon of the T5.2 cell, which could be followed down to the rostral medulla oblongata, is distorted due to the pictorial representation. Middle: Postsynaptic potentials of EI type (an EPSP followed by an IPSP) evoked by electrical stimulation of the optic tract; white dot indicates the onset of electrical stimulation 0.1 ms in duration. Below: Examples of intracellular records of a T5.1 neuron (a) and two T5.2 neurons (b and c) in response to configurational moving stimuli (S, square; W, worm-like stripe; A, antiworm-like stripe); a and b refer to labeled neurons shown above. (From Ewert et al. 1985).

2.3 Responses to Moving Configurational Stimuli

Antal et al. (1986) examined properties of morphologically identified tectal neurons in response to diffuse light changes, to a small moving spot, and to a moving large edge, respectively. The precise receptive field properties of the neurons, however, were not studied in detail. Ewert et al. (1985) and Matsumoto et al. (1986) performed a similar experiment, and they successfully determined the receptive field properties. Visual stimuli used by Matsumoto et al. (1986) were changes of diffuse light and configurational stimuli consisting of a black 8° x 8° square (S), a 2° x 16° worm-like (W) stripe, or a 16° x 2° antiworm-like (A) stripe each one traversing a neuron's ERF center at 25°/s against a white background. Intracellularly recorded responses were assigned to types of neurons by comparison with results obtained from extracellular studies. The properties of small field neurons, especially T5-type neurons, are described below.

Class T5.1 Neurons

Excitatory responses (R) of class T5.1 neurons were strongest to S, weaker to W, and somewhat less than (or equal) to A, yielding the relation: $R_S > R_W \geq R_A$. The size of the RF (receptive field) was about 33° in diameter. Electrical stimulation of the optic tract indicated EI type responses most frequently, and diffuse light changes yielded almost all kinds of response types, such as E-E, I-I, EI-EI, IE-IE. Class T5.1 neurons could be identified as large ganglionic neurons whose cell bodies were located in layer 8 (Fig. 3d) or layer 7 (not shown in the figure), and as pear-shaped neurons with cell bodies in layer 8 (Fig. 3a) and at the top of layer 6 (not shown in the figure). In these studies no obvious correlations were found between morphological (neuron type, location of the soma, dendritic arborization) and physiological (ERF size, response properties to S, W, and A) properties.

It was interesting to note that some of the T5.1 large ganglionic neurons in layer 8 showed repetitive and sustained EPSPs to moving stimuli (Fig. 4). The sustained responses were strongest when a W-stimulus was moved into the receptive field and then stopped within the field center.

Class T5.2 Neurons

This class is characterized by $R_S > R_W > R_A = 0$. The RF size was about 34° in diameter. As shown in Fig. 3, the 8° x 8° S-stimulus was a stronger releaser for neurons of this class than the 2° x 16° W-stimulus. (For species differences of frog's and toad's class T5.2 neurons see Ewert, this volume). When the 2° x 16° stripe was moved in A-configuration, some neurons showed only small EPSPs but no action potentials at all (Fig. 3, response b), and some other neurons exhibited IPSPs mainly (Fig. 3, response c). Changes in diffuse light yielded EI-EI or I-I type responses, respectively, and electrical optic tract stimulation assigned all these neurons to the EI type. Out of five physiologically classified T5.2 neurons, one neuron was successfully stained and identified as a pyramidal neuron with the soma at the top of layer 6. Its axon could be followed caudally up to the level of the medial reticular formation in the medulla oblongata. Although the stimulus patterns were slighly different from the present ones, Antal et al. (1986) recorded similar responses from a morphologically identified pyramidal neuron, i.e., a small moving spot evoked EPSPs and action potentials, whereas a large edge (black/white border oriented perpendicular to the direction of movement) elicited no obvious responses.

The intracellular data from T5.2 cells are limited, and therefore it cannot be concluded that T5.2 properties are restricted exclusively to pyramidal neurons. However, it must be emphasized that from the neurons stained in this study, no pyramidal neuron was recorded with different response properties. An explanation for the small number of pyramidal neurons (see also Antal et al. 1986) sampled in the present study could be due to the "pyramidal" shape of the cell, which may make the membrane difficult to penetrate. Retrograde labeling studies in frogs (Lázár et al. 1983) and toads (Weerasuriya & Ewert 1981; Weerasuriya, this volume) have indicated a much greater probability of finding pyramidal cells at the top of layer 6. In extracellular recordings in frogs (Schürg-Pfeiffer & Ewert 1981) the sample ratio of T5.1 to T5.2 was 1:1 (for common toads 1:3, cf. Satou & Ewert 1985), compared with 6:1 in the present intracellular study.

Class T5.3 Neurons

Neurons of this class responded to S, W, and A according to $R_S > R_A > R_W$. The RF size of this class was about 29° in diameter. Electrical stimulation of the optic tract evoked E type response in all investigated T5.3 neurons, and changes in diffuse light most frequently yielded the E-E type, suggesting that T5.3 neurons are predominantly excited by class R3 retinal fibers. Some neurons displayed prolonged

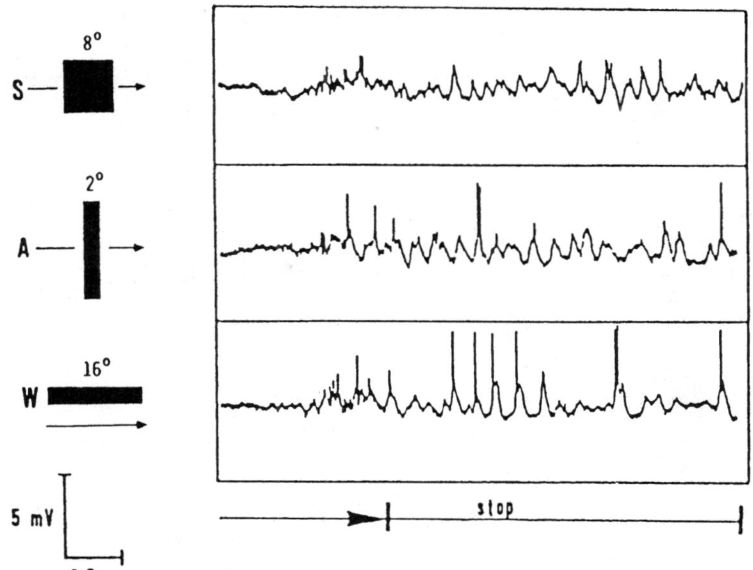

Figure 4. Sustained excitation in a T5.1 large ganglionic neuron. When the S, A, or W stimulus was moved into the receptive field and stopped there, postsynaptic potentials sustained in a manner depending on the stimulus configuration: EPSPs and IPSPs for S or A; EPSPs mainly for W. (From Matsumoto et al. 1986).

spike activities after turning the light *off*, which suggests the contribution of class R4 retinal inputs. All of the three successfully stained neurons of this class could be identified as large ganglionic neurons whose somata were located in layer 8.

Other Small Field Neurons

Neurons with T5.4 property ($R_S > R_A = R_W = 0$), and T1.1 property (distance dependent activity) were found, but no clear relationship between responses and cellular morphology could be obtained. Class T3 and T7 neurons could not be recorded in this study. Three neurons responded to moving stimuli almost exclusively with IPSPs. The receptive field size of these neurons was 20° to 30° in diameter. Inhibitory responses were strongest to S and weaker to W and A (Fig. 5).

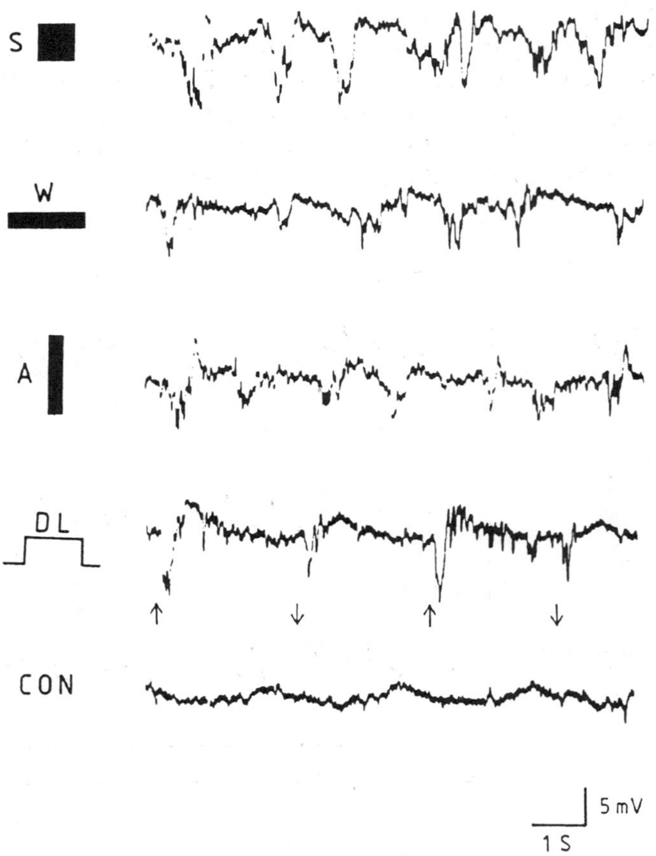

Figure 5. Responses of an "IPSP neuron" to diffuse light and to moving stimuli; 8° x 8° square (S); 2° x 16° worm-like stripe (W); 16° x 2° antiworm-like stripe (A). DL, diffuse light; CON, control indicating the potential recorded without visual stimuli. (Modified from Matsumoto et al. 1986).

These neurons evoked I type response to electrical optic tract stimulation, and I-I response to changes in diffuse light, which suggests inhibitory inputs due to information mediated by class R3 retinal fibers. Successful labeling identified these neurons as pear-shaped neurons in layers 8, 6, or 4. Some neurons recorded from deep tectal layers displayed lasting spontaneous EPSPs and/or IPSPs.

2.4 Axonal Projections of Large Ganglionic Neurons

The somata of large ganglionic neurons were found in layers 8, 7, and at the top of layer 6. A variety of responses was recorded from large ganglionic neurons; but it was not possible to correlate these with the neuron's morphology.

Axons were identified in most neurons examined, and they all entered in layer 7. In some cases, axons could be followed for quite a long distance. Figure 6A shows a schematic drawing of axonal projections of four large ganglionic neurons. In some neurons (cf. neurons *c* and *d* in Fig. 6A) axons bifurcate: One branch projects rostrally through the dorsal part of the brain; the branch of neuron *d* also sends collaterals to terminate in the pretectal region of both sides. The second branches of these two neurons *c* and *d* run around the caudal margin of the optic tectum, and then turn back rostrally toward the ventrolateral diencephalon. The axon of neuron *b* also bifurcates: Its rostral branch joins the axons of neurons *c* and *d*, and the caudal branch together with the axon of neuron *a* runs ipsilaterally, hence joining the tecto-bulbar tract.

3 Morphological Identification of Prectectal Neurons

Functional properties of the pretectal region have been studied electro-physiologically as well as behaviorally. Electrical stimulation of pretectal structures can elicit various patterns of avoidance behavior in a freely moving toad. If pretectal region was lesioned, the animal showed prey-catching behavior to large visual objects which normally release avoidance responses (Ewert 1968). These and other results suggest that the pretectal region of amphibians harbors neurons that are necessary for the commands of avoidance behaviors and which inhibit tectal T5.2 prey-selective neurons. (For pretectal functions related to optokinetic nystagmus see Manteuffel, this volume).

Neuronal responses to sensory inputs in the caudal thalamus have been examined at the extracellular single unit level (Ewert 1971), and they were divided into ten classes (TH1-TH10). Class TH3 neurons have receptive fields of 30° to 45° in diameter and are best activated by moving objects that fill the ERF or by moving large textures (Tsai & Ewert 1988). Neurons of this class can be recorded from the pretectal Lpd (lateral posterodorsal) and the lateral P (posterocentral) regions of the caudal dorsal thalamus. Class TH3 neurons are suggested to inhibit tectal T5.2 neurons (Ewert 1984).

Some anatomical experiments have been performed to demonstrate the connection between pretectum and optic tectum. Using the HRP method, Wilczynski & Northcutt (1977) and Weerasuriya & Ewert (1983) described reciprocal connections between ipsilateral Lpd nucleus and the optic tectum. Following lesions

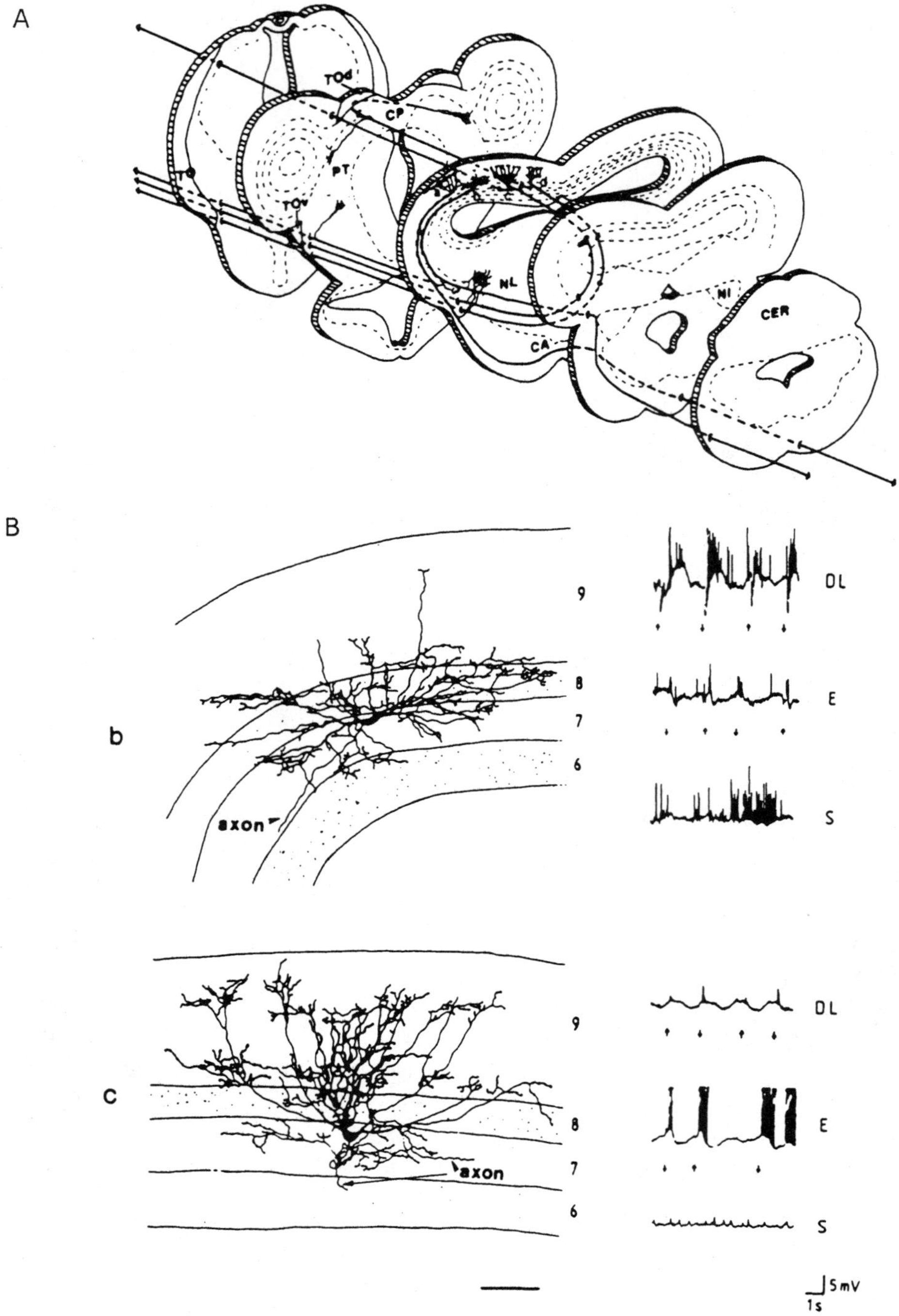
A
TOd
CP
PT
TOv
NL
NI
CER
CA
B
b
c
9
8
7
6
axon
axon
DL
E
S
5mV
1s

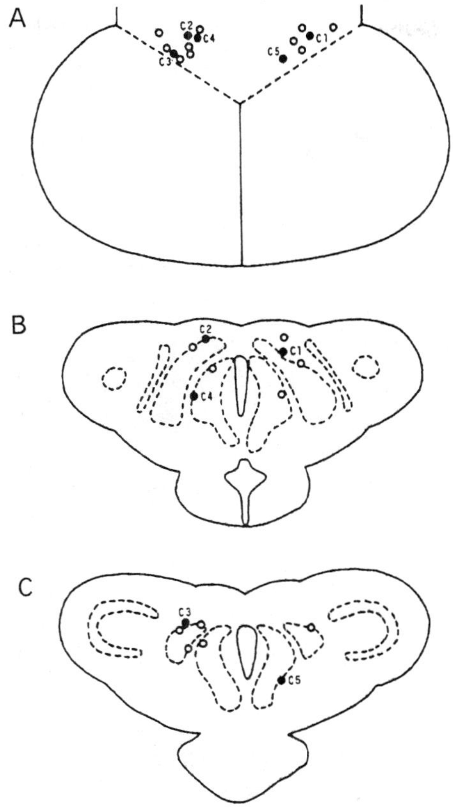

Figure 7. Recording sites of neurons in the pretectal Lpd (lateral posterodorsal) and P (posterocentral) nuclei of the caudal dorsal thalamus. (A) Dorsal view of the optic tectum and the caudal thalamus. (B) and (C) Transverse sections indicating locations of the cell bodies of neurons (see filled circles) whose axonal projections could be examined. Neurons c1 and c2 project bilaterally to both optic lobes, whereas c3 sends its axon to the contralateral optic tectum; neurons c4 and c5 project ipsilaterally toward the medulla oblongata.

Figure 6. Cellular morphology, axonal projections, and response properties of large ganglionic tectal neurons. (A) Axonal projections of four large ganglionic neurons. (B) Left: Drawings of neurons "b" and "c" shown in A; right: responses of these neurons to diffuse light (DL), to the edge of a black large object moving against a white background (E), and to a moving spot (S); arrows in (DL) refer to *on*-set (upward arrows) and *off*-set (downward arrows) of diffuse light and in (E) to the leading (downward arrows) and to the trailing (upward arrows) contrast border of the large object during traverse of the neuron's ERF. Calibration bar: 100 μm. (Modified from Antal et al. 1986).

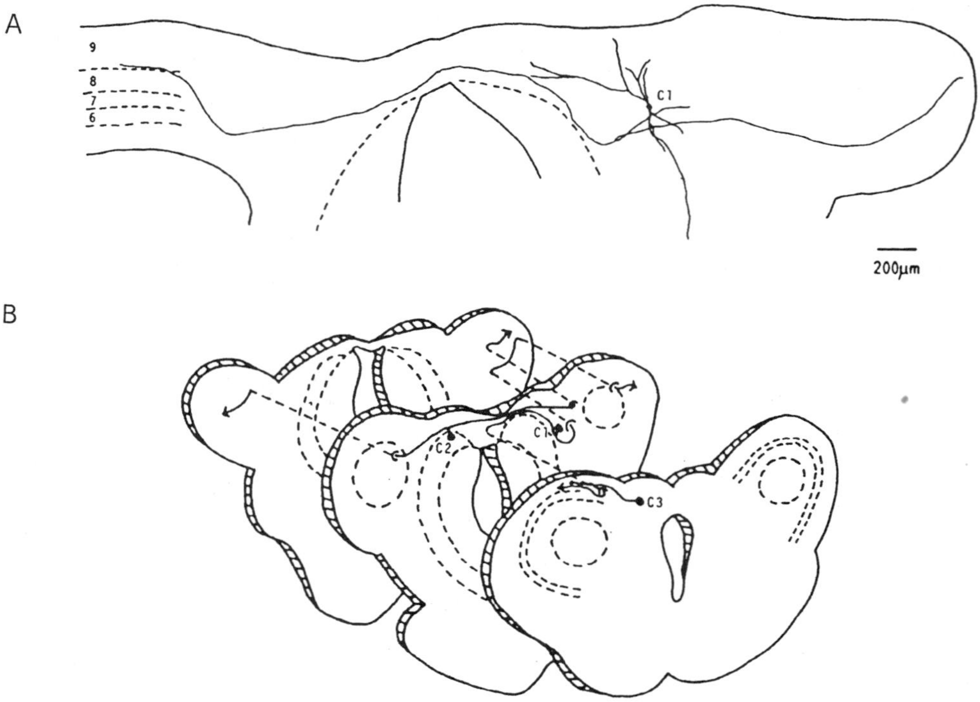

Figure 8. Reconstruction of neurons of the pretectal Lpd nucleus and schematic drawings of the axonal processes. (A) Reconstruction of the neuron c1 which projects to both optic lobes. (B) Schematic drawing of three pretectal neurons, c1 and c2 projecting bilaterally, and c3 projecting to the contralateral optic tectum.

in the optic tectum, degenerated axon terminals were found in the ipsilateral and contralateral optic tectum (Rubinson 1968; Lázár 1969). These experiments, however, could not clearly identify the neuronal types in both regions. Using an intracellular marking technique, the purpose of the present experiment (Kawakatsu & Matsumoto, unpublished) was to examine the morphology and axonal projection of thalamic-pretectal neurons.

Fifteen neurons were successfully stained with lucifer yellow, and the positions of those neurons are shown in Fig. 7. In five cases, axonal projections could be traced. Two cells (c1 and c2) projected bilaterally to the optic tectum, and one cell (c3) projected only to the contralateral optic tecum. All these three neurons were located in the Lpd region. A reconstruction of c1 is shown in Fig. 8A, and axonal projections of cells c1 to c3 are depicted in Fig. 8B. Two other cells (c4 and c5) were located in the P region, and they sent its axon to the medulla oblongata. A reconstruction of cell c4 is shown in Fig. 9A, and its axonal projection depicted in Fig. 9B. Responses of the neurons to electrical stimulation of the optic nerve were also examined. All neurons

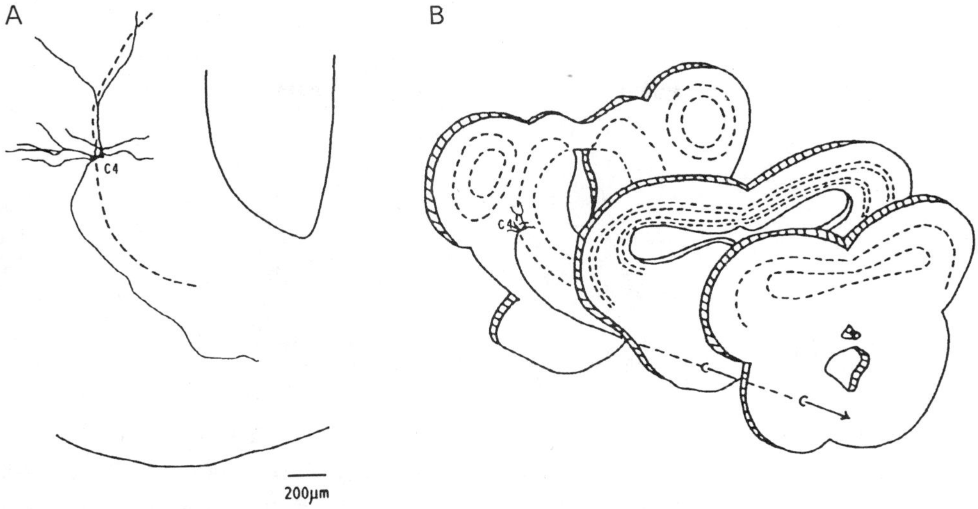

Figure 9. Reconstruction and schematic drawing of the axonal process of a neuron of the pretectal P nucleus. (A) Reconstruction of a caudally projecting neuron c4. (B) Axonal process of this neuron.

(c1 to c3) which projected to the optic tectum responded with an EPSP, and those neurons (c4 and c5) which projected toward the descending pathway responded with an EPSP followed by IPSP.

4 Synaptic Retino-Tectal Connections

In the preceeding experiments we could successfully examine the receptive field and stimulus-response properties of tectal neurons. We have thus accumulated substantial data forming the foundation for understanding (i) the synaptic organization between retinal fibers and tectal neurons and (ii) the modes of intratectal circuits. To date, however, it is unknown which class of retinal fibers makes synaptic connections with which type of tectal neurons. Since from a total of 103 intracellularly recorded tectal neurons about 94% showed *on-off* responses to brisk changes of diffuse illumination, a strong contribution of direct, or indirect input of retinal R3 *on-off* fibers can be suggested. For example, when we record a tectal E-E type response, we may suppose that this neuron receives excitatory input from *on-off* retinal fibers, however, we cannot be sure whether it is activated monosynaptically or polysynaptically. To answer this question, simultaneous recordings from retinal terminals (extracellular) and tectal neurons (intracellular) was carried out, and postsynaptic potentials were analyzed by applying the "pulse-analogue correlation (pulse-triggered averaging) method" (Nagano et al. 1988). In this study, we describe E-E and EI-EI type responses generated by class R3 retinal fibers. Figure 10 shows a

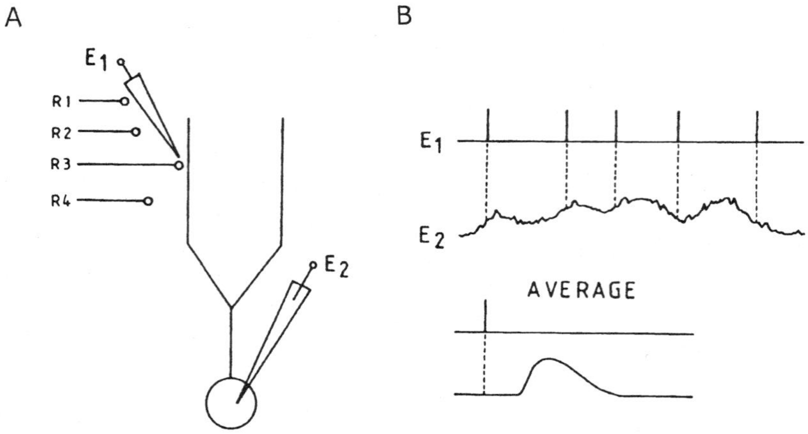

Figure 10. Schematic diagram of the experimental procedure of the pulse-analogue correlation method. (A) Action potentials of class R3 retinal ganglion cell axon terminals were recorded extracellularly with a tungsten electrode (E1); intracellular responses of a neighbouring tectal neuron were recorded with a glass microelectrode (E2). (B) Postsynaptic potentials recorded with electrode (E2) were averaged by triggering with the impulses of a single class R3 fiber recorded by electrode (E1).

schematic diagram of the experimental procedure. A tungsten electrode was first inserted into the superficial layers of the optic tectum to isolate a single retinal R3 fiber. A glass microelectrode was then used to penetrate a tectal neuron in the vicinity of the tungsten electrode. Postsynaptic potentials were averaged by triggering with the impulses of an isolated presynaptic single unit.

It is generally believed that low resistance metal electrodes pick up activities of terminals of retinal ganglion cell axons (Witpaard & ter Keurs 1975) rather than activities of tectal neurons. To ascertain that the *on-off* activities recorded with a metal electrode (E1) are of retinal origin, the following criteria were tested: (i) small size of receptive field (about 8° diam.); (ii) short and constant latency (<1.4 ms) of the spike in response to electrical stimulation of the optic tract; (iii) very little or no adaptation; (iv) coincidence of the recording position with established retino-tectal map. Only those units were kept for further analysis, which fulfilled the physiological criteria as retino-tectal afferents.

The *pulse-analogue correlation* method can be used only when there is neither autocorrelation nor cross-correlation in presynaptic impulses. Figure 11 shows the autocorrelogram and the cross-correlogram of presynaptic impulses. The maximum number in one bin is 4 for autocorrelogram and 7 for cross-correlogram. These numbers are small enough compared to the total number of impulses examined ($n=388$ for autocorrelogram and $n_1=388$ and $n_2=754$ for cross-correlogram).

Figure 12 shows an example of correlation between a class R3 retinal unit and an E-E type tectal response. A multi unit *on-off* retinal response and a simultaneously recorded intracellular response of a tectal neuron are shown in Fig. 12A. After identifying the response types with a diffuse light stimulation, a moving stimulus was

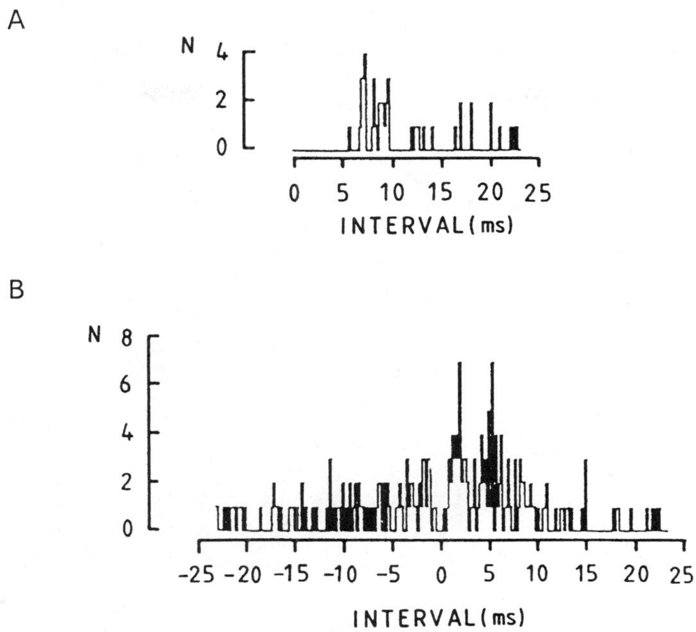

Figure 11. Autocorrelogram and cross-correlogram of impulses of retinal fibers. (A) Autocorrelogram of n=388 impulses. A single unit (a fiber's response) was chosen from a multi-unit response. (B) Cross-correlogram of impulses of two units (n_1 = 388, n_2 = 754). These two units were selected from the same multi-unit as in A. There is neither significant autocorrelation nor cross-correlation. bin: 180 us. (From Nagano et al. 1988).

presented (Fig. 13B). Small EPSPs seem to occur coincidentally with the arrival of presynaptic impulses. Some of those EPSPs would be related to recorded impulses, but others not. The top and the second traces of Fig. 13C show an averaged presynaptic impulse and the related postsynaptic potential (PSP), respectively. The third trace is the field potential recorded just outside the recorded cell. It is clear that the field potential does not contribute to the averaged EPSP.

It is known that the averaged response represents the sum (gross response) of (i) the neurally-related response (response correlated only with the impulses used for triggering) and (ii) the stimulus-related response (background response due to other inputs). To obtain the neurally-related response, the stimulus-related response should be subtracted from the gross response. Stimulus-related response can be obtained by averaging the shuffled data. (For theoretical details see Perkel et al. 1967; for results see Toyama et al. 1981). The fourth trace of Fig. 13C is the stimulus-related response thus obtained. Obviously, the stimulus-related response does not contribute to the gross response (second trace). This neuron was stained with ethidium bromide (Aghajanian & Vandermaelen 1982) and identified as a large ganglionic neuron in layer 8 (Fig. 12D); its dendrites spread horizontally in lamina F where class R3 *on-off* retinal fibers terminate.

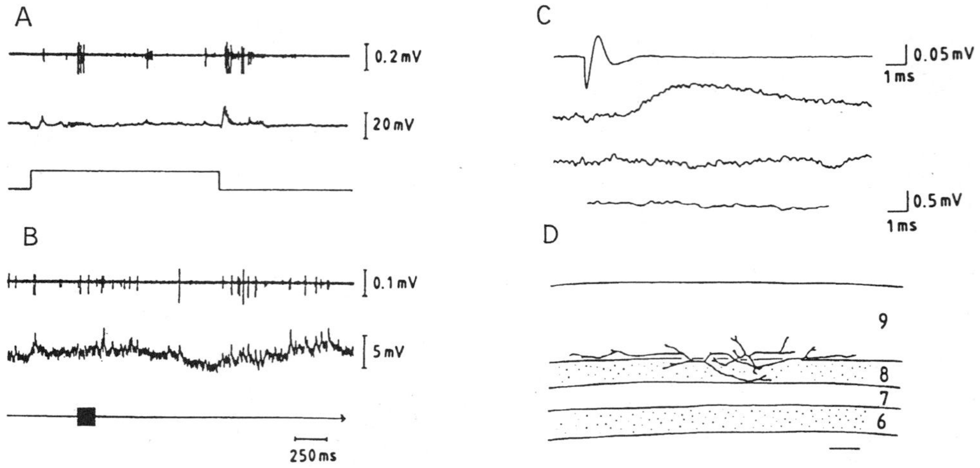

Figure 12. An example of correlation between a class R3 retinal *on-off* unit and an E-E type tectal response. (A) Multi-unit *on-off* response of presynaptic fibers (top trace) and a simultaneously recorded intracellular response of a tectal neuron (middle trace). Upward deflection of the bottom trace indicates the duration of diffuse light illumination. (B) Responses of the same pre- and postsynaptic pair to a moving 2° x 2° square. (C) The top trace is an averaged presynaptic impulse, and the second trace an averaged EPSP; the third trace is an averaged field potential recorded just outside the neuron; the bottom trace shows the stimulus-related response obtained by averaging after shuffling the data. These results were obtained from the same unit as in A and B. (D) Ethidium bromide was injected into the tectal neuron examined in A-D; it was identified as a large ganglionic neuron in layer 8. Calibration bar: 100 μm. (Modified from Nagano et al. 1988).

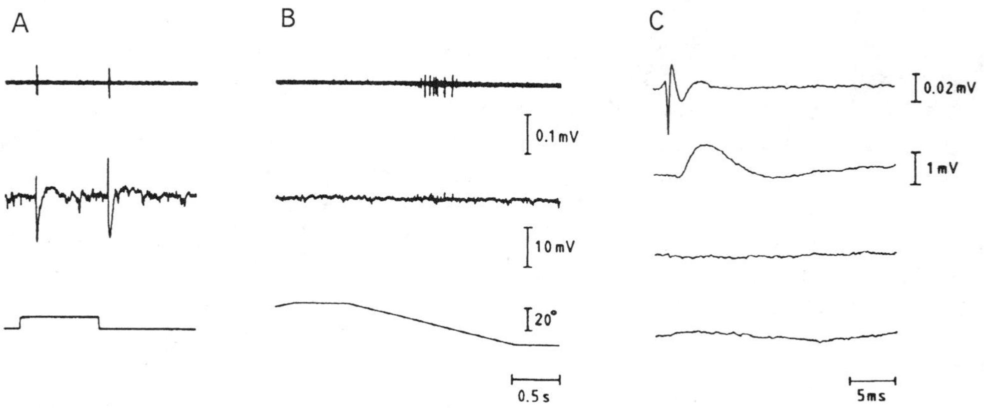

Figure 13. Correlation between class R3 retinal *on-off* single unit and tectal EI-EI type responses. (A) Simultaneous representation of presynaptic (top trace) and postsynaptic (middle trace) responses. Duration of illumination is indicated by upward deflection in the bottom trace. (B) Responses of the same pre- and postsynaptic pair to a moving 2° x 2° square. Movement of the stimulus is indicated at the bottom. (C) From top to bottom: traces indicate averaged presynaptic response, averaged EPSP, field potential, and stimulus-related response, respectively. (From Nagano et al. 1988).

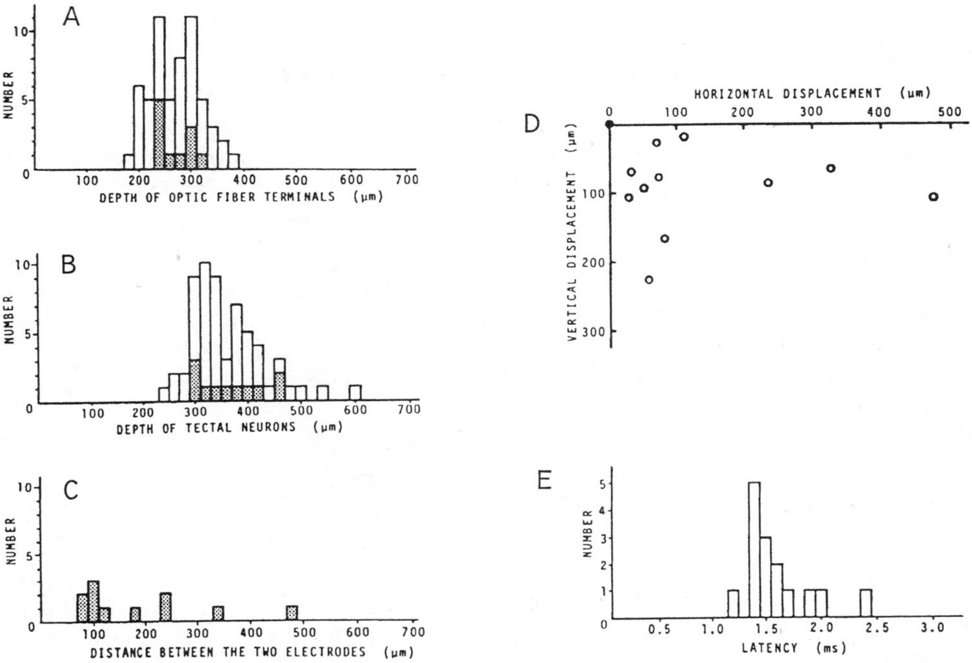

Figure 14. Sample distributions. (A) Distribution of recording depth of optic fiber terminals. (B) Distribution of recording depth of tectal neurons. (C) Distribution of distance between the two electrode tips. (D) Horizontal and vertical displacements between the two recording sites. (E) Frequency histograms of the latency of averaged EPSPs. Shadowed columns indicate the element for which positive correlation was found.

An example of correlation between *on-off* retinal fibers and an EI-EI type response is depicted in Fig. 13A. The large IPSP follows after an EPSP in response to diffuse light changes at *on* and *off*; however, the IPSP was not found in the averaged response, probably due to the polysynaptic nature.

Figure 14A shows the distribution of depth of recording sites of *on-off* retinal units. Shadowed bars indicate the unit from which we could record an EPSP after averaging. Figure 14B shows the depth distribution of tectal neurons. Distances between the two recording sites in successful cases are illustrated in Fig. 14C and D.

Latencies of averaged EPSPs were measured and the histograms are shown in Fig. 14E; the peak occurs at 1.4 ms. This value is comparable to the synaptic delay time in the central nervous system of the frog (Brookhart & Fadiga 1960; Matsumoto & Bando 1980), and it was thus concluded that averaged EPSPs in Fig. 13 and 14 were monosynaptic. Histograms in Fig. 14E indicate that most of the EPSPs were monosynaptic. It could be shown that a synapse about 500 μm apart from the soma of a tectal neuron could generate a detectable monosynaptic EPSP in the soma.

5 Discussion

5.1 Response Properties and Cellular Morphology

The investigations by Antal et al. (1986) were mainly morphologically oriented and the ones by Matsumoto et al. (1986) more physiologically. Both studies revealed that large ganglionic neurons display much more varieties in morphology and physiology than we had expected. Although all investigated T5.3 neurons resembled morphological features of large ganglionic neurons whose somata were located in layer 8, no further relationship could be found between responses and dendritic field sizes or lamination of dendritic arborizations. Mismatch between response properties and neuronal morphology was particularly obvious in T5.1 neurons which resembled large ganglionic or pear-shaped neurons whose somata could be located in various tectal layers including the output layer 7 and border 6/7.

The identification of the successfully stained T5.2 prey-selective neuron as a pyramidal neuron was a clear result. The axon of this neuron could be followed to the medulla oblongata. It is important to note that no pyramidal neuron with properties different from T5.2 has as yet been recorded. In an extracellular recording study, T5.2 cells could be antidromically activated if the medulla oblongata was electrically stimulated (Satou & Ewert 1985). These findings confirm the idea that T5.2 neurons send "prey-related information" to motor system (see also Schürg-Pfeiffer, this volume).

With respect to intracellular responses it is important to note that T5.1 and T5.2 cells are characterized by the occurrence of excitatory and inhibitory potentials (EPSPs, spikes, IPSPs), whereas T5.3 cells show mainly EPSPs and spikes. Neurons with T5.1 properties are heterogeneous in this context, too.

It was also interesting to note that among pear-shaped neurons inhibitiory responses were found in cells with beaded dendrites, and excitatory responses in cells with smooth dendrites. Bipolar cells in layer 6 did not respond to any visual stimulus, although they were activated by electrical stimulation. Some large ganglionic neurons with huge dendritic fields displayed very small synaptic potentials. There are various possibilities to explain this result. One is that the stimuli used were not adequate, and another that habituation was too fast. We hence address a warning toward attempts trying to guess neuronal functional properties only drawn from morphological features. For example, neurons with large dendritic fields that extend through wide regions across various tectal layers can have very small RFs (Matsumoto et al. 1986).

5.2 Projection of Tectal and Pretectal Neurons

As already suggested (Lázár 1969; Lázár et al. 1983, and this volume), large ganglionic tectal neurons were found to project to many regions: Rostrally (diencephalic targets), caudally (bulbar targets), or within the optic tectum (Fig. 6). One of the interesting findings here was that the axons of some large ganglionic neurons bifurcate and thus may project rostrally and caudally. The fact that a brain region receives the axons of large ganglionic neurons of various sources, and that a single large ganglionic neuron can project to various sites suggests the physiological and morphological complexities of this neuron type as significant features of macro-network structure.

The farthest brain region in which axonal terminals of tectal cells were identified was the bulbar medial reticular formation in the medulla oblongata. This region is considered to relay information from the optic tectum to the motor systems. Schwippert et al. (this volume) recorded the intracellular responses of neurons in the rostral medulla oblongata and identified those neurons morphologically. Some cells in this region exhibited response characteristics of tectal T2.1, T2.2, T4, T5.1, T5.2, T5.4 neurons or pretectal TH3 neurons. Receptive fields were large and the response properties suggested convergence of different classes of tectal neurons onto bulbar neurons.

Some other large ganglionic neurons are projecting to pretectal regions. Projections of pretectal Lpd neurons to the optic tectum were found as well (Fig. 6). These results support the idea of mutual connections between optic tectum and pretectum as has been suggested from physiological studies (Ewert et al. 1974). Intracellular response characteristics of these projections, however, remained as yet known. Neurons of the P nucleus projected to the medulla oblongata. It seems likely that these neurons mediate information regarding large objects, obstacles, and textured backgrounds (cf. contributions by Ewert and Manteuffel, this volume). The presence of two different projection systems in pretectal neurons is consistent with the results reported by Lázár (this volume).

5.3 Neuronal Circuit

With reference to the site of the somata and the extent of the dendritic trees it is reasonable to suggest that T5.1 pear-shaped neurons of layer 8 are suitable to integrate retinal input, whereas T5.1 and T5.3 large ganglionic neurons (whose somata are located in the same layer) may be particularly suitable to mediate local interactions (Fig. 3). Dendro-dendritic or axo-dendritic reverberatory circuits were proposed by Székely & Lázár (1976) in their anatomical studies. Some physiological evidences which might support their idea were obtained in a series of intracellular studies. Matsumoto & Bando (1980) found that some large ganglionic neurons evoke a monosynaptic EPSP followed by a disynaptic EPSP and an IPSP. Since the conduction velocities responsible for the two EPSPs were similar, this response type could be explained by the anatomically introduced positive feedback circuit. When electrical stimulation was used, no sustained excitation could be observed, because it activates all optic fibers and produces strong lateral inhibition. Some of the T5.1 large ganglionic neurons displayed reverberatory responses when moving visual stimuli were used. The degree of excitation was the greatest when a W-stimulus was presented, where the effect of lateral inhibition was the smallest (Fig. 4). Loops of mutual excitation may be responsible for the sensitivity of tectal T5.1 and T5.2 neurons in response to a small stimulus that is extended parallel to the direction of movement (see Arbib and Ewert, this volume).

About 75% of the tectal neurons examined showed IPSPs to electrical stimulation (I, EI and EEI types). There are two possibilities for the origin of these IPSPs; they are produced (i) through a direct retino-tectal pathway, and/or (ii) by a pathway via the pretectum. The latter pathway is considered to act for the discrimination of moving configurational objects. Altering the configuration of a moving stripe from W to A is responded to by a T5.2 cell with a change from EPSPs to IPSPs, or a remarkable decrease of PSPs (Fig. 3), and this provides evidence of "configuration selective" excitatory and inhibitory inputs to this neuron. The source of the latter might be TH3 and/or T5.3 which are tuned to the A-configuration. Class T5.1 neurons, to mention another example, that display I-I responses to changes of diffuse light, can be explained by direct excitatory input from retinal R2 neurons (which show no *on-off* responses) and inhibitory input from tectal T5.3 and/or pretectal TH3 neurons mainly driven by retinal class R3 *on-off* neurons. An inhibitory transmitter has not been demonstrated in frog's tectal cells to date (see Lázár, this volume). It remained unknown, however, which pathway, (i) or (ii), was responsible for the IPSPs recorded. Pure IPSPs (I type) were found in about 20% of the neurons examined. The latency of some I type neurons was very short, but the number of synapses involved could not be determined. In the pigeon's optic tectum, however, monosynaptic inhibition of tectal neurons by optic nerve fibers has been suggested by Hardy et al. (1985). An assumption of direct inhibitory inputs from retinal R3 fibers to tectal neurons in the frog's study would be in conflict with data showing that after optic tract stimulation in no tectal cell of the 314 tested was an EPSP preceded by an IPSP. (Note that the conduction velocity in R2 fibers is less than in R3.)

To analyze synaptic organization between visual cortical neurons of the cat, Toyama et al. (1981) have applied the pulse-correlation method. Using the same technique, Tanaka (1983) investigated geniculo-cortical connections. In the present study, intracellular recordings were made from the postsynaptic neurons, and the pulse-triggered averaging technique was successfully applied to analysis of retino-tectal connections. The results showed that EPSPs of E-E and EI-EI types are generated monosynaptically by class R3 retinal *on-off* fibers. For measurements of synaptic delay times, this technique is more direct than other methods, and it has an advantage when applied to synaptic connections, in which slow conduction fibers are involved, such as class R1 or R2 unmyelinated retinal fibers.

Arbib (1982) and Lara et al. (1982) have constructed models of neural circuits underlying visually guided behaviors (see contributions by Arbib, Betts, and Cervantes-Pérez, this volume). The model circuit of the optic tectum considered the anatomical results by Székely & Lázár (1976), and the activities of pyramidal and other tectal neurons involved were calculated, appropriately. In connection with pretectal circuits, this model successfully simulated the physiological data concerning prey and predator selection. It should be kept in mind, however, that large ganglionic neurons are not sufficiently considered in the model because very little anatomical information was available. To build "*Rana computatrix*," large ganglionic neurons should be considered in the model as elements of the hardware (Lázár 1984).

6 References

Aghajanian AK, Vandermaelen CP (1982) Intracellular identification of central noradrenergic and serotonergic neurons by a new double labeling procedure. *J Neurosci* 12: 1786-1792

Antal M, Matsumoto N, Székely G (1986) Tectal neurons of the frog: intracellular recording and labeling with cobalt electrodes. *J Comp Neurol* 246: 238-253

Arbib MA (1982) Modelling neuronal mechanisms of visuomotor coordination in frog and toad. In: Amari S, Arbib MA (eds) *Competition and cooperation in neuronal nets*. Springer-Verlag, Berlin Heidelberg New York, pp 342-370

Brookhart JM, Fadiga E (1960) Potential field initiated during monosynaptic activation of frog motoneurons. *J Physiol* 150: 633-655

Ewert J-P (1968) Der Einfluß von Zwischenhirndefekten auf die Visuomotorik im Beute- und Fluchtverhalten der Erdkröte (*Bufo bufo* L). *Z Vergl Physiol* 61: 41-70

Ewert J-P (1971) Single unit response of the toad (*Bufo americanus*) caudal thalamus to visual objects. *Z Verg Physiol* 74: 81-102

Ewert J-P (1984) Tectal mechanisms that underlie prey-catching and avoidance behaviors in toads. In: Vanegas H (ed) *Comparative neurology of the optic tectum*. Plenum Press, New York London, pp 247-399

Ewert J-P, Hock FJ, Wietersheim A v (1974) Thalamus/Praetectum/Tectum: retinale Topographie und physiologische Interaktionen bei der Kröte (*Bufo bufo*). *J Comp Physiol* 92: 343-356

Ewert J-P, Matsumoto N, Schwippert WW (1985) Morphological identification of prey-selective neurons in the grass frog's optic tectum. *Naturwissenschaften* 72: 661-662

Görcs T, Antal M, Olah E, Székely G (1979) An improved cobalt labelling technique with complex compounds. *Acta Biol Acad Sci Hung* 30: 79-86

Grüsser O-J, Grüsser-Cornehls U (1976) Neurophysiology of the anuran visual system. In: Llinás R, Precht W (eds) *Frog neurobiology*. Springer-Verlag, Berlin Heidelberg New York, pp 298-385

Hardy O, Leresche N, Jassik-Geschenfeld (1985) Morphology and laminar distribution of electrophysiologically identified cells in the pigeon's optic tectum: an intracellular study. *J Comp Neurol* 233: 390-404

Lara R, Cervantes F, Arbib MA (1982) Two-dimensional model of retinal-tectal-pretectal interactions for the control of prey-predator recognition and size preference in amphibia. In: Amari S, Arbib MA (eds) *Competition and cooperation in neural nets*. Springer-Verlag, Berlin Heidelberg New York, pp 371-393

Lázár G (1969) Efferent pathways of the optic tectum in the frog. *Acta Biol Acad Sci Hung* 19: 325-334

Lázár G (1984) Structure and connection of the frog optic tectum. In: Vanegas H (ed) *Comparative neurology of the optic tectum*. Plenum Press, New York London, pp 185-210

Lázár G, Tóth P, Csank G, Kicliter E (1983) Morphology and location of the tectal projection neurons in frogs: a study with HRP and cobalt-filling. *J Comp Neurol* 215: 108-120

Lettvin JY, Maturana HR, McCulloch WS, Pitts WH (1959) What the frog's eye tells the frog's brain. *Proc Inst Radio Engin* 47: 1940-1951

Matsumoto N, Bando T (1978) Intracellular recordings of tectal cells of the frog. *Proc Jap Acad* 54B: 386-390

Matsumoto N, Bando T (1980) Excitatory synaptic potentials and morphological classification of tectal neurons of the frog. *Brain Res* 192: 39-48

Matsumoto N, Schwippert WW, Ewert J-P (1986) Intracellular activity of morphologically identified neurons of the grass frog's optic tectum in response to moving configurational visual stimuli. *J Comp Physiol* 159: 721-739

Nagano K, Li QL, Tamada A, Matsumoto N (1988) An analysis of postsynaptic potentials of tectal neurons of the frog: correlation with impulses recorded from the terminals of retino-tectal afferents. *Exp Brain Res* 70: 429-432

Perkel DH, Gerstein GL, Moor GP (1967) Neuronal spike trains. II: Simultaneous spike trains. *Biophysical J* 7: 419-440

Rubinson K (1968) Projections of the tectum opticum of the frog. *Brain Behav Evol* 1: 529-561

Satou M, Ewert J-P (1985) The antidromic activation of tectal neurons by electrical stimuli applied to the caudal medulla oblongata in the toad *Bufo bufo* (L). *J Comp Physiol* 157: 739-748

Schürg-Pfeiffer E, Ewert J-P (1981) Investigation of neurons involved in the analysis of Gestalt prey features in the frog *Rana temporaria*. *J Comp Physiol* 141: 139-158

Székely G, Lázár G (1976) Cellular and synaptic architecture of the optic tectum. In: Llinás R, Precht W (eds) *Frog neurobiology*. Springer-Verlag, Berlin Heidelberg New York, pp 407-437

Tanaka K (1983) Cross-correlation analysis of geniculostriate neuronal relationships in cats. *J Neurophysiol* 49: 1303-1318

Toyama K, Kimura M, Tanaka K (1981) Cross-correlation analysis of interneuronal connectivity in cat visual cortex. *J Neurophysiol* 46: 191-201

Tsai HJ, Ewert J-P (1988) Influence of stationary and moving background textures of the response of visual neurons in toads (*Bufo bufo*). *Brain Behav Evol* (in press)

Urban L, Székely G (1983) Intracellular staining of motoneurons with cobalt complex compounds in the frog. *J Neurobiol* 14: 157-161

Weerasuriya A, Ewert J-P (1981) Prey-selective neurons in the toad's optic tectum and sensorimotor interfacing: HRP studies and recording experiments. *J Comp Physiol* 144: 429-434

Weerasuriya A, Ewert J-P (1983) Afferents of some dorsal retino-recipient areas of the brain of *Bufo bufo*. *Soc Neurosci Abstr* 9: 536

Wilczynski W, Northcutt R (1977) Afferents to the optic tectum of the leopard frog: an HRP study. *J Comp Neurol* 173: 219-229

Witpaard J, ter Keurs HEDL (1975) A reclassification of retinal ganglion cells in the frog, based upon tectal endings and response properties. *Vision Res* 15: 1933-1938

Toward an Identification of Neurotransmitters in the Frog's Optic Tectum

M. E. SANDOVAL *), L. MASSIEU, P. ARAIZA, and J. FERNANDEZ

Instituto de Fisiologia Celular
Universidad Nacional Autónoma de México
04510 México, D.F., México

Abstract. The amphibian optic tectum offers an excellent model to study neurotransmission. Using partially isolated nerve endings and slices from frog's optic tectum, various neurochemical criteria (Na^+-dependent transmitter uptake and Ca^{2+}-dependent release, respectively) have been applied to identify neurotransmitters. We looked for the amino acids glutamic acid, aspartic acid, gamma-aminobutyric acid (GABA), and glycine, and for the amines dopamine, norepinephrine, and 5-hydroxytryptamine (serotonin).

1 Introduction

Since the optic tectum of amphibia has been studied from anatomical and physiological points of view and its role in the control of the visuomotor coordination is well known (Ewert 1976, 1984; Ewert et al. 1983; Grüsser & Grüsser-Cornehls 1976; Ingle 1976, 1983), it offers an excellent model to investigate the participation of neurotransmitters in the control of interneuronal communcation. Our present knowledge of synapses arises mainly from anatomical and neurophysiological studies. The biochemical approach has to be incorporated.

Still relatively little is known on the neurotransmitters acting in the visuomotor centers of amphibians (see also Lázár and Gruberg, this volume). There is some evidence that glutamic acid, gamma-aminobutyric acid (GABA), and glycine may serve as neurotransmitters in the optic tectum of frogs and toads. A Na^+-dependent high-affinity transport system for glutamic acid has been described in frog tectal synaptosomes (Roberts & Yates 1976; Oswald & Freeman 1977). Unilateral enucleation in the frog resulted in a significant decrease of glutamic acid levels in the contralateral optic tectum, whereas the high-affinity uptake did not change (Yates &

*) This paper was completed by the late Elena Sandoval and her co-workers in 1984. We publish it here in a copy-edited form. The Editors.

Roberts 1974; Roberts & Yates 1976). However, there are no studies on the effect of enucleation upon release of glutamic acid in frog's tectum. GABA exerts a widespread inhibitory action on field potential of tectal cells after electrical stimulation of the optic nerve (Milson & Mitchell 1977). Both glutamic acid decarboxylase (GAD) and high-affinity uptake of GABA have been reported in this tissue (Yates & Taberner 1975; Roberts & Yates 1976) and after retinal ablation, a progressive decrease of both parameters has been observed (Roberts & Yates 1976). Glycine may also act as a tectal inhibitory transmitter as indicated by its strychnine-sensitive action on the depression of the first wave of the field potential of tectal cells following electrical stimulation of the optic nerve (Milson & Mitchell 1977). Furthermore, the presence of both high-affinity specific binding of strychnine and a Na^+-dependent high-affinity transport system for glycine has been reported in frog tectal synaptosomes (Davidoff & Adair 1976; Müller & Snyder 1978).

The amphibian brain contains relatively high concentrations of the catecholamines epinephrine (less norepinephrine) and dopamine and in particular the indoleamine 5-hydroxytryptamine (5-HT or serotonin). Serotonin-containing cells are mainly located in tegmental and raphé regions whose projections to the optic tectum have been described (for discussion see Freeman & Norden 1984; Weerasuriya, this volume). (For a review on neurotransmitters in the optic tectum of non-mammalians see Freeman & Norden 1984; see also Kuljis & Karten 1982; Reiner et al. 1984).

The aim of the present work is to apply the most crucial neurochemical criteria in order to identify neurotransmitters in the frog's optic tectum. We use the criteria of the presence of a high-affinity transport system and the calcium-dependent release from nerve endings (Rubin 1970). The putative neurotransmitters that we looked for in the tectal tissue were glutamic acid, aspartic acid, gamma-aminobutyric acid, glycine, among the amino acids, and dopamine, norepinephrine, and 5-hydroxytryptamine for amines. With respect to glutamic acid, as already mentioned, there is some evidence of its action in the optic tectum of frogs and toads (Johnson 1972; Yates & Roberts 1974; Roberts & Yates 1976; Oswald & Freeman 1977). Since this substance has been suggested as excitatory transmitter in the retino-tectal pathway of the pigeon (Bondy & Purdy 1977; Fonnum & Henke 1982), we decided to further evaluate its role in the frog's optic tectum measuring the levels of glutamic acid and its calcium-dependent release in normal and enucleated tecta.

In this study, we used partially isolated nerve endings from the optic tectum. It has been demonstrated that isolated nerve endings maintain the basic physiological features of the nerve terminals *in situ*, such as glucose utilization, respiration, ATP synthesis, as well as those processes related to synaptic transmission, namely synthesis and the storage of neurotransmitters, membrane potential, neurotransmitter accumulation, and calcium-dependent release under depolarizing conditions (Whittaker 1969; Bradford et al. 1975). Procedures for isolating crude preparations of nerve endings from specific areas and whole brain have been reported by Douglas (1968) and Whittaker (1969). In the uptake and release experiments, we used the methodology that has been applied to similar studies in the hippocampus and cerebellum (Sandoval et al. 1978; Sandoval & Cotman 1978). For release experiments, we also used tectal slices and studied the role of calcium in amino acids' release when slices were depolarized by potassium or veratrine.

2 Methodology

2.1 General Considerations

In the vertebrate's nervous system, the transfer of information in most of the synapses is given by chemical signals mediated by neurotransmitters. It is known that at the nerve endings exists the biochemical machinery for the synthesis, storage, and release of neurotransmitters. Studies on neurotransmitters have shown that they may be of different chemical nature. There is increasing evidence that *amino acids* (such as glutamic acid, aspartic acid, glycine, gamma-aminobutyric acid GABA) and *amines* (norepinephrine, dopamine, serotonin, and acetylcholine) are neurotransmitters in the central nervous system across all vertebrates (Freeman & Norden 1984).

When an action potential arrives at the nerve terminal, specific voltage-sensitive calcium channels are opened; calcium ions enter the nerve ending, and the increasing internal calcium concentration then triggers neurotransmitter release through a largely unknown mechanism (DeRobertis et al. 1962; Baker et al. 1973; Miledi 1973). In the synaptic cleft the neurotransmitter molecules diffuse towards the postsynaptic membrane where they bind to a specific receptor. The interaction of the neurotransmitter with its receptor leads to a change in the membrane permeability to either sodium, potassium, or chloride ions changing the polarity of the membrane. The neurotransmitter inactivation can take place in two ways, an enzymatic modification of the molecule in the synaptic cleft, or transmitter uptake by the nerve ending.

For the identification of neurotransmitters, two classical approaches can be applied, the physiological and the neurochemical one (Werman 1966; Orrego 1979; Sandoval & Lara 1984).

Physiological criteria are:
(i) *Changes in the neuronal excitability*. A putative neurotransmitter must modify the neuronal excitability. This may be due to an increase or a decrease in the neuronal firing or to slow changes in the membrane potential.
(ii) *Identity of action.* When a putative neurotransmitter is applied to a neuronal membrane, it must mimic the response of the neuron to the stimulation of the respective pathway. That is, it should lead to a specific postsynaptic response similar to that elicited by the release of the natural neurotransmitter.
(iii) *Specificity of the postsynaptic receptor.* The action of the putative neurotransmitter on the postsynaptic membrane must be enhanced by agonists of the natural transmitter or inhibited by antagonists.

Neurochemical criteria are:
(i) *Localization at the nerve ending.* The postulated neurotransmitter must be naturally present at the nerve ending.
(ii) *Synthesis and storage at the nerve endings*. The synaptic terminal must have the biochemical machinery to synthesize and store the neurotransmitter. It is known for well established neurotransmitters that they can be synthesized either in the nerve ending or in the cell body and then transported to the nerve terminal through

axoplasmic flow. At the nerve ending the neurotransmitter may be stored in the synaptic vesicles or in a soluble pool (Snyder et al. 1970; Israel et al. 1979).

(iii) *Calcium-dependent release from the nerve ending*. The postulated neurotransmitter must be released in a calcium-dependent manner from nerve endings when a stimulus that mimics the action potential is given (Fig. 1). This criterion is perhaps one of the most critical to postulate a substance as a neurotransmitter (Fagg & Lane 1979; Orrego 1979).

(iv) *Specific postsynaptic receptor*. The interaction of the neurotransmitter with its postsynaptic receptor must be specific. The agonist and antagonist agents of the natural transmitter must displace or inhibit the binding of the putative neurotransmitter to its receptor, respectively (Snyder 1975; Enna & Snyder 1975).

(v) *Neurotransmitter inactivation*. A specific mechanism to terminate the action of the neurotransmitter must exist at the synapse. Two systems for neurotransmitter inactivation have been described. The first is related to an enzymatic system in the synaptic cleft that changes the chemical nature of the neurotransmitter to make it ineffective. The second and more common system for neurotransmitter removal is a specific high-affinity transport system in the membrane of the nerve ending that is able to accumulate neurotransmitter from the environment (Fig. 1). This transport system requires sodium ions, and it has also been demonstrated in glial cells (Iversen 1971; Henke et al. 1976; Fagg & Lane 1979; Orrego 1979).

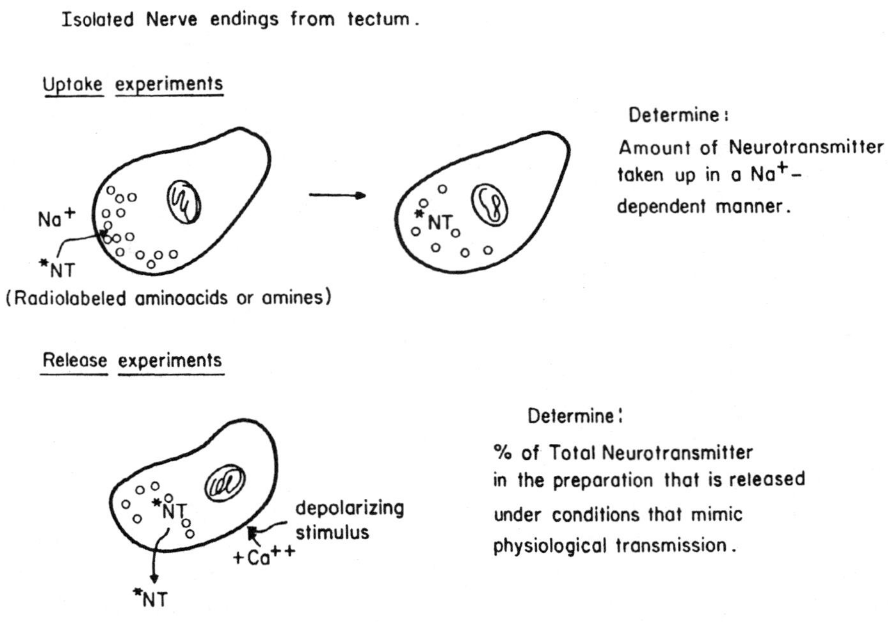

Figure 1. General procedure for "uptake" and "release experiments" to identify neurotransmitters in isolated nerve endings of the frog's optic tectum.

Regarding the criteria mentioned above, a large number of molecules can be listed as neurotransmitters in the vertebrate's nervous system. However, there is little information on the identification of neurotransmitters at specific nerve endings. Glutamic acid, for example, has been proposed as the neurotransmitter of the granule cells in the cerebellum (Chujo et al. 1975; Sandoval & Cotman 1978; Foster & Roberts 1980) in the lateral olfactory tract of the olfactory bulb (Harvey et al. 1975; Yamamoto & Matsui 1976) and at synapses between the enthorinal cortex and the pyramidal cells of the hippocampus (Nadler et al. 1976; Sandoval et al. 1978). GABA has been described as the inhibitory transmitter for some interneurons and the Purkinje cells of the cerebellar cortex (Storm-Mathisen 1976; Woodward et al. 1976). But no attempt has been made for relating the control of synaptic transmission with specific functions of the nervous system.

The release and uptake experiments in our study were performed according to Sandoval et al. (1978) and Sandoval & Cotman (1978):

2.2 Uptake Experiments

Adult frogs *Rana berlandieri forreri* were used throughout the experiments. Synaptosomal fractions (P_2) were prepared from optic tecta of decapitated frogs. Tecta were homogenized in 0.32 M sucrose and homogenates centrifuged at 23000 x g for 10 min; the resulting pellet was resuspended in 0.32 M sucrose and recentrifuged at 23000 x g for 10 min. The resulting P_2 fraction was resuspended in a Na^+-free medium and aliquots (about 150 ug protein/ml) preincubated at 21°C in either two of the following media: (i) a Krebs-bicarbonate medium containing (mM): NaCl, 115; KCl, 3 ; $CaCl_2$, 2; NaH_2PO_4 1.2; $MgSO_4$, 1.2; $NaHCO_3$, 25; glucose, 10; pH=7.4 maintained with CO_2/O_2 mix (5%/95%); (ii) choline chloride, 137; KH_2PO_4, 1.2; $MgSO_4$, 1.2; glucose, 10; HEPES (N-2-hydroxyethylpeperazine-N'-2-ethanesulfonic acid), 10; pH=7.4 with tris. After 3 min of preincubation, aliquots of the radiolabeled aminoacids were added: [^{14}C]glutamic acid, 0.37 µCi (0.85 µM); [^{14}C]aspartic acid, 0.37 µCi (1.1 µM); [^{3}H]GABA, 0.33 µCi (1.03 µM), and [^{14}C]glycine, 0.30 µCi (30 µM). Incubation followed for another 10 min. Aliquots of the incubation medium were filtered under vacuum on unipore-disc prefilters, and tissue samples were washed at room temperature with 10 ml of Na^+-free medium. Radioactivity in the filters was counted in tritosol after tissue digestion in 2% sodium dodecyl sulphate, 20 mM EDTA (ethylenediaminetetraacetic acid); pH=8.0. Na^+-dependent uptake was taken as the radioactivity accumulated by the tissue samples in Krebs-bicarbonate medium after subtracting the radioactivity obtained in the Na^+-free medium (see Table 1).

2.3 Release Experiments

Slices of frog optic tectum were used for release experiments. Coronal sections (0.5 mm) of tecta were obtained in a McIlwain tissue chopper. The slices were preloaded with the radiolabeled amino acids as described for uptake experiments and washed twice by filtration on millipore discs with 5 ml of Krebs-bicarbonate medium but containing EGTA (ethyleneglycol-bis [gamma-amino-ethylether] N,N'tetra-acetic acid) 1 mM and no calcium (calcium-free medium). The slices were transferred to a continuous superfusion system connected to a Gilson Miniplus 2 peristaltic pump where eight tissue samples can be simultaneously studied. Each tissue sample consisted of all the slices obtained from both tecta of each frog. In order to obtain a constant baseline of radioactivity efflux, all samples were perfused for 8 min with the calcium-free medium at room temperature at an efflux rate of 1 ml/min and the perfusates media collected in scintillation vials. The samples were then perfused for eigth min with one of the following media: (i) calcium-free medium, (ii) calcium-free medium containing either 53 mM KCl (NaCl reduced to 53 mM) or veratrine 60 µg/ml, or (iii) Krebs-bicarbonate medium containing 2 mM $CaCl_2$ and 53 mM KCl or veratrine 60 µg/ml. From these experimental conditions we could determine the baseline efflux of radioactivity, the amino acid efflux induced by depolarization in the absence of calcium, and the calcium-dependent efflux induced by depolarization as reported by Sandoval et al. (1984); furthermore, the role of Ca^{2+} was studied by adding Mg^{2+} ions to the perfusion medium (cf. solutions a-c described in Fig. 3). Radioactivity in the collected media was determined in tritosol while in the tissue was determined in tritosol after digestion in 2% sodium dodecyl sulphate. Results are expressed as the ratio of radioactivity released after each minute of perfusion to the total radioactivity in the tissue before superfusion according to Cotman et al. (1976).

2.4 Enucleation Experiments

The right eye of adult frogs was removed under ether anesthesia, and after 10 to 17 postoperative days tecta were dissected and processed exactly as described for release experiments except that the slices were stimulated with 53 mM KCl only in the presence or in the absence of 2 mM $CaCl_2$. Throughout these experiments [^{3}H]glutamate instead of [^{14}C]glutamate was used (0.5 µCi, 1 µM).

2.5 Enzymatic Quantification of Glutamic Acid

Glutamic acid levels were enzymatically determined according to the method described by Hughey et al. (1980). After 10 postoperative days, the tecta contralateral and ipsilateral to the enuclation site were separately homogenized in 1 ml of 0.6% perchloric acid and centrifuged at 12000 x g for 10 min. The resulting supernatant was neutralized with 170 µl 2.5 M $KHCO_3$. The neutralized supernatant was recentrifuged at 12000 x g for 10 min and 200 µl aliquotes were added to a mix containing (mM): PO_4^{3-}, 0.05; tris acetate, 0.05; hydrazine, 0.1; ADP, 2.5 and 400 µl aqua dest. The optic density of the samples was determined at 340 nM; after a stable reading was obtained, the reaction was started adding 1 µl aliquote of glutamic acid dehydrogenase (GDH). The optic density was determined and the concentration of glutamate in the tissue was obtained in terms of the reduced NAD by the GDH according to:

$$(D_o/C_{em}) \times (V_t/V_s) \times (1/\text{prot})$$

where D_o is the optic density increment in the sample after the addition of GDH,

C_{em} the glutamic acid molar extinction coefficient, V_t the total volume, and V_s the sample volume. Protein (prot) was determined by means of the method of Lowry et al. (1951).

3 Uptake Studies

3.1 Amino Acid Neurotransmitters

Regarding amino acid neurotransmitters, we found that GABA and glycine, both inhibitory transmitters in the central nervous system of vertebrates, are rapidly accumulated by tectal nerve endings (Table 1). However, it was observed that GABA was mostly taken up through a sodium-dependent transport system, whereas glycine was accumulated by two different mechanisms, one sodium-dependent and the other sodium-independent.

Table 1. Uptake of amino acid neurotransmitters by tectal nerve endings in frogs (in nmoles/mg protein /5min).

	Total Activity	Activity in Na^+-free medium	Na^+-dependent uptake (% total)
GABA	5.98 ± 0.57 (9)	0.78 ± 0.07 (9)	87.0
Glycine	5.35 ± 0.61 (7)	2.02 ± 0.33 (7)	62.3
Glutamic acid	0.94 ± 0.07 (9)	0.15 ± 0.01 (9)	84.0
Aspartic acid	1.26 ± 0.15 (3)	0.58 ± 0.08 (3)	54.0

Table 2. Uptake of amine neurotransmitters by tectal nerve endings in frogs (in pmoles/mg protein/5min).

	Total Activity	Activity in Na^+-free medium	Na^+-dependent uptake (% total)
Norepinephrine	21.50 ± 4.8 (5)	9.48 ± 1.34 (5)	55.9
Dopamine	2.61 ± 0.32 (8)	1.55 ± 0.26 (7)	40.6
Serotonin	2580 ± 460 (5)	1200 ± 330 (5)	53.5

In contrast to GABA and glycine, aspartic and glutamic acids (postulated as excitatory transmitters in the vertebrate's central nervous system) were less accumulated by tectal nerve endings. However, it is important to note that glutamic acid uptake was mostly due to a sodium-dependent mechanism, whereas the sodium dependency for aspartic acid accumulation was unclear.

3.2 Amine Neurotransmitters

We looked for a high-affinity transport system for dopamine, norepinephrine, and serotonin (Table 2), known to be inhibitory amines in the central nervous system of vertebrates. In contrast to amino acids (Table 1), the total activity of the amines studied was relatively small, except in the case of serotonin. The sodium dependency of the amines' uptake by isolated nerve endings from tectal tissue was in most cases less than that of the amino acids. This is particularly obvious when dopamine and GABA (or glutamic acid) are compared.

4 Release Studies

4.1 Glutamic Acid Release

Glutamic acid was released from isolated nerve endings of the tectal tissue in a calcium-dependent manner. Under unstimulated conditions, the transmitter

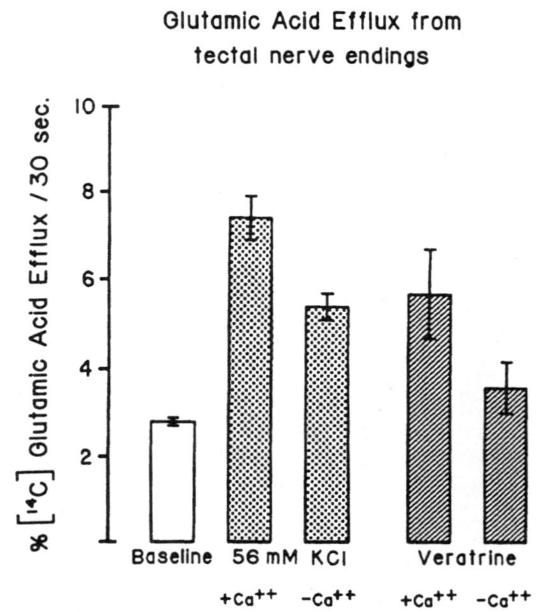

Figure 2. [^{14}C]glutamic acid efflux from frog's tectal nerve endings. Baseline efflux represents the spontaneous release of the transmitter under nonstimulated conditions.

substances are released spontaneously from the nerve ending. Figure 2 shows the spontaneous or baseline efflux of [^{14}C]glutamic acid from tectal nerve endings. When nerve endings were stimulated with depolarizing agents that mimic the action potential (56 mM KCl or veratrine) in the presence of calcium ions, an about twofold increase in efflux was obtained. In addition, when calcium ions were removed from the incubation medium, glutamic acid efflux decreased remarkably.

As in nerve endings, glutamic acid was released from tectal slices in a calcium-dependent manner. Adding 53 mM KCl in the presence of 2 mM $CaCl_2$ to the medium resulted in about threefold increase in glutamic acid efflux (Fig. 3Aa), and this stimulation was significantly reduced when calcium was omitted (Fig. 3Ab) or 15 mM $MgCl_2$ was added (Fig. 3Ac). Moreover, when veratrine was used as depolarizing agent, a clear calcium-dependent (Fig. 3Ba) magnesium-sensitive (Fig. 3Bc) release of glutamic acid was observed.

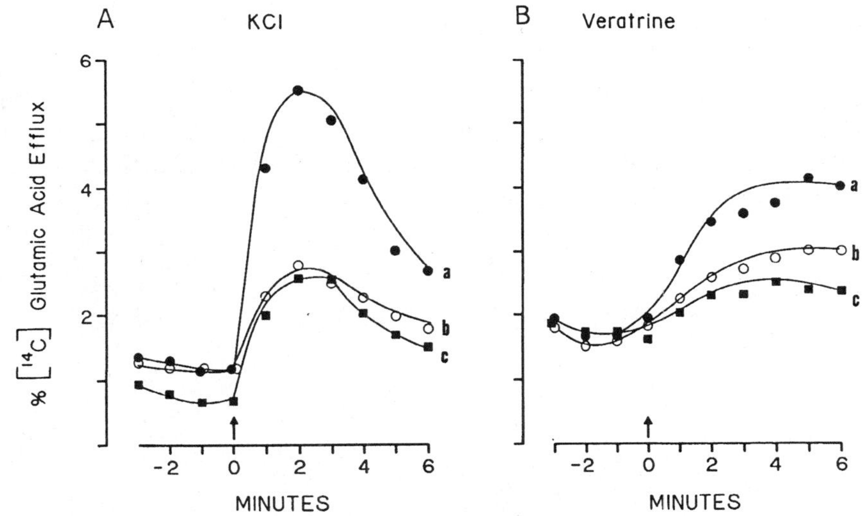

Figure 3. [^{14}C]glutamic acid efflux from frog tectal slices. Effect of 53 mM KCl (A) or veratrine (B). Slices were washed by filtration with Ca^{2+}-free Krebs-bicarbonate medium containing 1 mM EGTA and were transferred to a continuous superfusion system at room temperature (about 20°C). The slices were perfused for 8 min at a superfusion rate of 1 ml/min with the same medium in order to obtain a constant radioactivity efflux. The slices were then superfused for 6 min (see arrow) with one of the following experimental solutions: (a) *calcium-containing medium* (cf. filled circular symbols): Krebs-bicarbonate medium containing 2 mM $CaCl_2$ and either 53 mM KCl (NaCl reduced to 53 mM) or veratrine 60 ug/ml; (b) *calcium-free medium* (cf. open circles): Krebs-bicarbonate medium containing 1 mM EGTA and either 53 mM KCl or veratrine; (c) *magnesium-containing medium* (cf. filled squares): Krebs-bicarbonate medium containing 15 mM $MgCl_2$, 2 mM $CaCl_2$, and either 53 mM KCl or veratrine. This scheme of treatment is maintained in Figs. 7-9. The radioactivity in the collected media and in the tissue slices was determined as described in Section 2. Data represent the mean of 5 to 9 different experiments. In all cases the standard errors were under the 10% of the mean values.

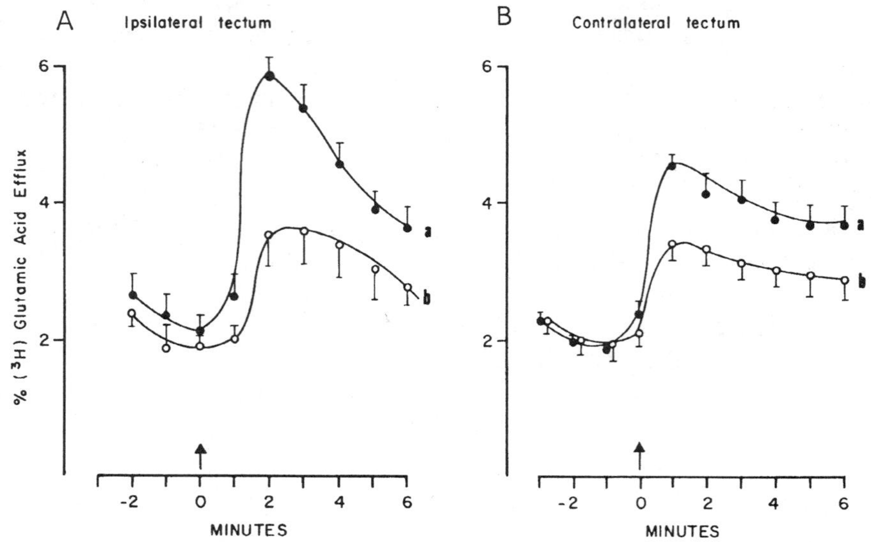

Figure 4. [^{3}H]glutamic acid efflux from tectal slices after 10 days of unilateral eye removal. Effect of 53 mM KCl on ipsilateral (A) and contralateral (B) optic tecta. After loaded with [^{3}H]glutamic acid, slices were treated as described in Fig. 3. (a) $CaCl_2$ and KCl; (b) EGTA and KCl. Standard error represents about 10% of the mean value.

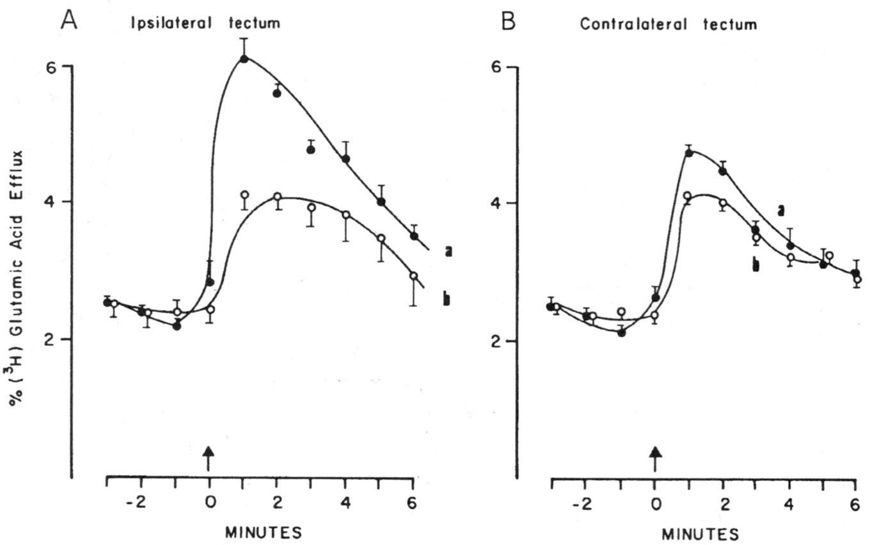

Figure 5. [^{3}H]glutamic acid efflux from tectal slices after 17 days of unilateral eye removal. Effect of 53 mM KCl on ipsilateral (A) and contralateral (B) optic tecta. After loaded with [^{3}H]glutamic acid, slices were treated as described in Fig. 3. (a) $CaCl_2$ and KCl; (b) EGTA and KCl. Standard error represents about 10% of the mean value.

Furthermore, the enucleation experiments (Figs. 4 and 5) showed a strong decrease in the contralateral tectum 10 days (Fig. 4) or 17 days (Fig. 5) after enucleation. Moreover, the endogenous glutamic acid levels diminished significantly ten days after the contralateral eye was removed (32.5 $\pm$ 0.7 SEM nm/mg protein and 19.49 $\pm$ 0.95 SEM nm/mg protein in control and enucleated tecta, respectively, for n=6).

4.2 GABA Release

In contrast to glutamic acid, GABA was released from tectal nerve endings in a calcium-independent manner. The efflux of GABA was increased twice in relation to the baseline efflux after stimulation with 56 mM KCl. However, this efflux was unchanged after the removal of calcium ions from the incubation medium (Fig. 6).

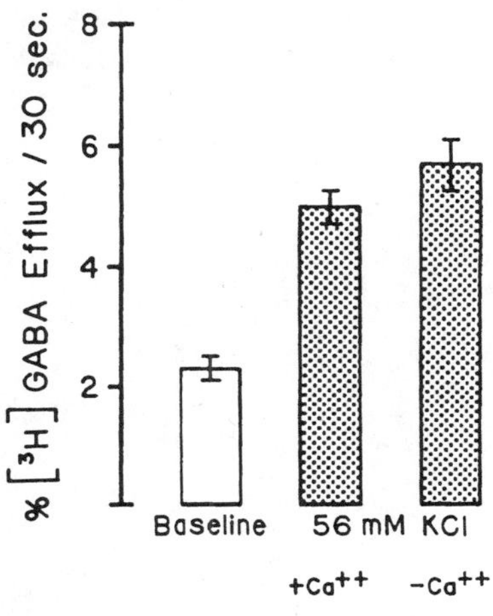

Figure 6. [^{3}H]GABA efflux from frog's tectal nerve endings.

When the efflux of recently accumulated [^{3}H]GABA from tectal slices was studied, we also found a calcium-independent release of GABA when the slices were stimulated by KCl (Fig. 7Aa,b); this effect was partially reduced by magnesium (Fig. 7Ac). However, when the effect of veratrine was studied, a clear calcium-dependent release was observed (Fig. 7Ba), and it was inhibited both when calcium was removed (Fig. 7Bb) or when magnesium was added to the medium (Fig. 7Bc).

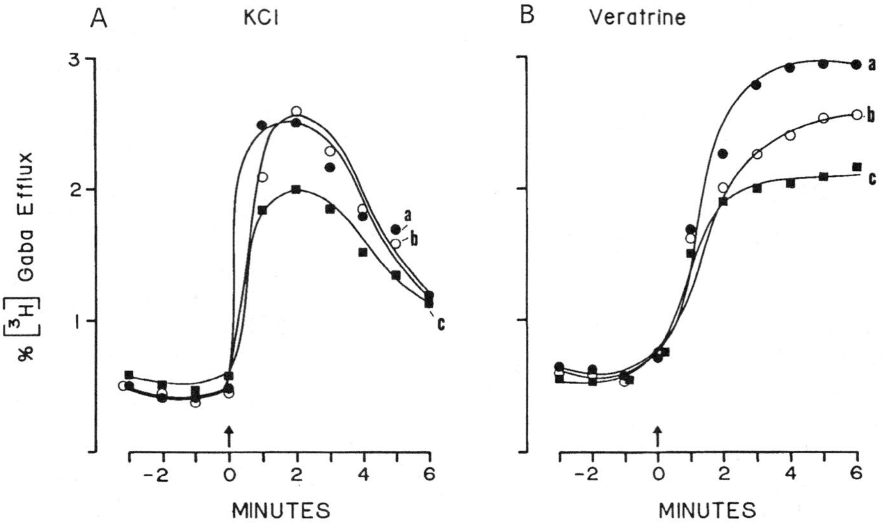

Figure 7. [^{3}H]GABA efflux from tectal slices. Effect of 53 mM KCl (A) or veratrine (B). After loaded with [^{3}H]GABA, slices were treated as described in Fig. 3. (a) $CaCl_2$ with KCl or veratrine; (b) EGTA with KCl or veratrine; (c) $MgCl_2$ and $CaCl_2$ with KCl or veratrine. Data represent the mean of 5 to 10 experiments. In all cases standard errors were under the 10% of the mean value.

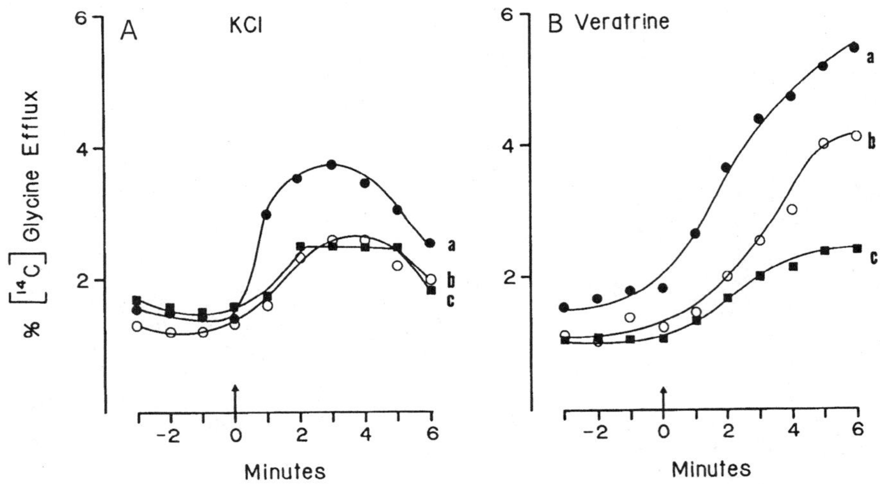

Figure 8. [^{14}C]glycine efflux from frog tectal slices. Effect of 53 mM KCl (A) or veratrine (B). After loaded with [^{14}C]glycine, slices were treated as described in Fig. 3. (a) $CaCl_2$ with KCl or veratrine; (b) EGTA with KCl or veratrine; (c) $MgCl_2$ and $CaCl_2$ with KCl or veratrine. Data represent the mean of 4 to 6 different experiments. In all cases standard errors were under the 10% of the mean values.

4.3 Glycine Release

A calcium-dependent release system was observed for recently accumulated [^{14}C]glycine in tectal slices. Although the rate of efflux induced by 53 mM KCl or veratrine (Fig. 8) was lower than that found for glutamic acid (Fig. 3), the addition of KCl in the presence of calcium ions (Fig. 8Aa) clearly stimulated the baseline efflux, and it was inhibited both by removal of calcium ions (Fig. 8Ab) or in the presence magnesium (Fig. 8Ac). When the effect of veratrine was studied (Fig. 8B), we found that it rapidly induces a calcium-dependent magnesium-sensitive glycine efflux. A second component of glycine efflux inhibited by magnesium, but independent of calcium, was evident after 5 min of veratrine depolarization.

4.4 Aspartic Acid Release

Compared to glutamic acid (Fig. 3) and glycine (Fig. 8), the efflux of aspartic acid (Fig. 9) from tectal slices was independent of the presence of calcium. As shown in Fig. 9Aa, the addition of KCl and 2 mM $CaCl_2$ induced an increase in the baseline efflux of about 2.5 times. However, the omission of calcium resulted in only a 20% inhibition (Fig. 9Ab), whereas $MgCl_2$ reduced the stimulated efflux (Fig. 9Ac). When the effect of veratrine - a more selective depolarizing agent - was tested, no calcium-dependent release was detected (Fig. 9Ba-c).

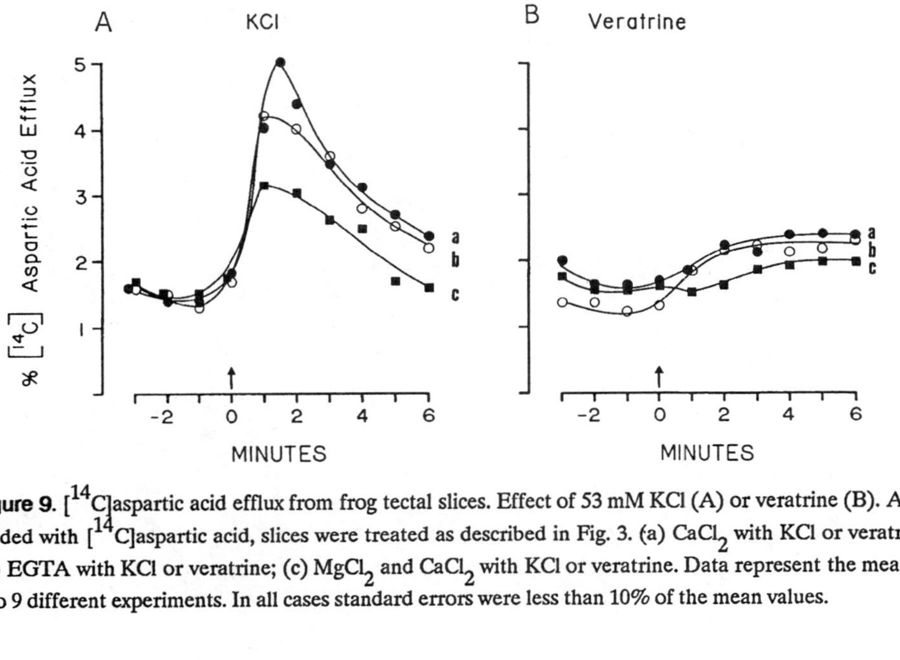

Figure 9. [^{14}C]aspartic acid efflux from frog tectal slices. Effect of 53 mM KCl (A) or veratrine (B). After loaded with [^{14}C]aspartic acid, slices were treated as described in Fig. 3. (a) $CaCl_2$ with KCl or veratrine; (b) EGTA with KCl or veratrine; (c) $MgCl_2$ and $CaCl_2$ with KCl or veratrine. Data represent the mean of 6 to 9 different experiments. In all cases standard errors were less than 10% of the mean values.

5 Discussion

5.1 Sodium-Dependent Uptake

The rates of amino acid neurotransmitters' efflux observed in the isolated nerve endings from frog tectal tissue are similar to those reported for these amino acids in other areas of the nervous system in similar preparations and to those observed in the pigeon's tectum (Snyder et al. 1970; Sandoval & Cotman 1978; Sandoval et al. 1978; Cuenod & Henke 1978; Hertz 1979). It seems that the high-affinity transport systems for glutamic acid, GABA, and glycine observed in the frog's optic tectum have the same characteristics as those reported for well established neurotransmitters: the requirement of sodium ions and the high transport activity at very low amino acid concentration (Snyder et al. 1970; Sandoval & Cotman 1978; Cuenod & Henke 1978; Fagg & Lane 1979; Hertz 1979). Thus, it may be suggested that the transport systems for glutamic acid, GABA, and glycine are related to the removal of neurotransmitter from the synaptic clefts in the tectum (Iversen 1971).

The presence of a high-affinity transport system has been considered as one of the criteria to determine a substance as a neurotransmitter, since it is related to the inactivation of the transmitter from the synaptic cleft after its release from presynaptic nerve terminals. This process has been reported both for neurons and glial cells, and it is dependent on energy metabolism and temperature, as well as on the presence of sodium ions in the incubation medium (Fagg & Lane 1979).

Our results show that the amino acids studied were accumulated by P_2 fractions of frog's optic tectum through a sodium-dependent high-affinity process. GABA and glutamic acid in particular were incorporated almost in a totally Na^+-dependent manner. The high-affinity uptake found for these amino acids was temperature-dependent, since the activiy was considerably reduced at 4°C. However, the presence of such an uptake is not a definitive criterion for assessing a substance as a neurotransmitter, since it has been shown that glial cells actively accumulate amino acids (Fagg & Lane 1979). The results thus may suggest the presence of an uptake system of neurotransmitters by glial elements surrounding nerve terminals. Freeman & Norden (1984) point out that denervation induces a number of profound metabolic changes in the optic tectum often resulting in an extensive reactive gliosis, and that these effects have not yet been satisfactorily accounted for in studies dealing with high-affinity uptake: "The high-affinity uptake data require a quantitative estimation of the amount of gliosis, sprouting, and transneuronal effects and a knowledge of how these affect the density and turnover of remaining high-affinity uptake sites" (Freeman & Norden 1984, p. 476).

The role of amines in the synaptic transmission of the pigeon's optic tectum has been poorly studied. Both dopamine and serotonin are accumulated by isolated nerve endings. Serotonin is taken up by a high-affinity system, whereas dopamine is accumulated by a low-affinity transport mechanism. A transport system for norepinephrine has not been observed in the pigeon's tectum (Cuenod & Henke 1978; Hertz 1979). For evidence of catecholamines in frog's tectal pear-shaped cells see Bieger & Neuman (1984). From the results of our investigations we suggest that serotonin plays a prominent role in the frog's tectum. However, it is necessary to study also other parameters in order to evaluate the contribution of the various amines to the tectal intrinsic neurotransmission.

5.2 Calcium-Dependent Release

Depolarization induced calcium-dependent release is considered as one of the most critical tests to determine a substance as a neurotransmitter candidate. Here, we have shown that recently accumulated glutamic acid (Fig. 3) and glycine (Fig. 8) are released from tectal slices after depolarization with either KCl or veratrine, and that this release is inhibited both by the removal of calcium or in the presence of magnesium, a well known inhibitor of the calcium transport induced by depolarization (Miledi 1973). A clear calcium-dependent release was also demonstrated for glutamic acid in isolated nerve terminals (Fig. 2).

Glutamate is likely to be a non-specific neuroexcitant in the frog tectum or a neuromodulatory agent in the bird tectum (Freeman & Norden 1984). This substance has been suggested as a neurotransmitter candidate in the retino-tectal pathway of the pigeon, since high-affinity uptake and glutamate levels were decreased in the contralateral tectum after unilateral enucleation (Bondy & Purdy 1977; Fonnum & Henke 1982). Moreover, calcium-dependent release of glutamic acid has been described (Bondy & Harrington 1979). In the frog, glutamate levels decreased in the contralateral tectum after unilateral enucleation, whereas the total glutamic acid high-affinity uptake remained unchanged, which suggested that uptake in the frog's optic tectum may be a result of glial activity (Yates & Roberts 1974; Roberts & Yates 1976). A role of glutamate in the transfer between tectal neurons and adjacent glial cells responsible for sensitization processes underlying arousal is discussed by Laming (this volume). The calcium-dependent release of glutamic acid observerd in our studies strengthens the suggestion that this amino acid acts as a transmitter in the frog's tectum. Our enucleation experiments showed a clear decrease in both tectal endogenous glutamate levels and a calcium-dependent release induced by high potassium depolarization from contralateral tectal slices after 10 and 17 of eye removal. Lázár (1980) found a substantial reduction in the number of myelinated axons from the retina in superficial layers of the contralateral optic tectum after 15 days of enucleation, while unmyelinated fibers did not degenerate until fifty days after eye removal. Matsumoto & Scalia (1981) obtained similar results in *Rana pipiens*. These data may suggest that the decrease in calcium-dependent release of glutamic acid is due to a reduction of retinal terminals in tectal superficial layers which could be glutamatergic in nature.

Glycine as a tectal transmitter in the pigeon is well documented. A high-affinity transport system and a calcium-dependent release process as well as a high-affinity strychnine binding to tectal membranes have been described (Reubi & Cuenod 1976; Henke & Cuenod 1978; Lefort et al. 1978). In amphibians, glycine seems to play a more important role as neurotransmitter than in mammals, since both high-affinity transport system and strychnine binding are clearly present in the frog's brain and spinal cord (Davidoff & Adair 1976; Müller & Snyder 1978). Our results show that recently accumulated glycine is released from tectal slices of the frog's tectum in a calcium-dependent manner and thus suggest a transmitter role in this tissue.

Although aspartic acid has been suggested as a transmitter in the frog's optic tectum (Milson & Mitchell 1977), our present results rule out this possibility, since preloaded aspartic acid was released from tectal slices in a calcium-independent manner after depolarization with KCl. Moreover, veratrine failed to stimulate the aspartic acid efflux in the presence of calcium. These results suggest either that in frog's tectum exogenous aspartic acid is taken up by the high-affinity transport system for glutamic acid, but is not recognized by the release processes, or that this amino acid is mainly accumulated by glial cells that release it in a calcium-independent manner after depolarization with KCl. Such a mechanism has been postulated for glial cells of the mammalian CNS (Levi et al. 1982).

In contrast to glutamic acid and glycine, from our results, the role of GABA as a tectal transmitter in the frog is still unclear. We showed that GABA is accumulated by synaptosomes through a Na^+-dependent, high-affinity transport system at a greater extent than the other amino acids studied. Moreover, calcium-dependent release of GABA was demonstrated after depolarization by veratrine. However, a clear calcium-independent release of GABA induced by KCl was also observed. Depolarizing KCl concentrations enhanced the baseline efflux of GABA, but this release was unaffected by the presence of calcium. However, it was considerably inhibited in the presence of magnesium. Similar effects have been described for mammals (Fagg & Lane 1979). There is some evidence in the frog's spinal cord that Mg^{2+} may reduce both the high potassium-depolarization induced release and the electrically induced release of transmitters by means of a reduction in the membrane sensitivity to changes in potassium concentrations, or by a decrement in the conductance to potassium ions (Sykova & Vyklicky 1977; Shefner & Levy 1981).

The fact that veratrine induces release by opening the voltage-sensitive sodium channels in neuronal membranes, suggests correlations with neuronal activity. It has been shown that depolarization of rat brain synaptosomes produces an increase of GABA efflux with both calcium-dependent and calcium-independent components (Simon et al. 1974; Raitieri et al. 1975). A calcium-independent release of GABA from horizontal cells in the toad's retina has been described by Schwartz (1982).

There is further evidence to show that GABA plays a role as a neurotransmitter in the optic tectum. In the pigeon's tectum, GABA is mainly accumulated by nerve endings. In the goldfish tectum, exogenous GABA is accumulated by neurons' synaptic terminals and dendrites (Villani et al. 1982). In the frog's tectum, Roberts & Yates (1976) and Milson & Mitchell (1977) have demonstrated that GAD, an enzyme considered as a neuronal marker for GABAergic terminals (Fonnum et al. 1970; Pérez de la Mora et al. 1981), is present in relatively high concentration. From all these results we conclude that further studies are needed in order to understand the role of GABA in the synaptic processes of the frog's optic tectum.

6 References

Baker PF, Meves H, Ridgway EB (1973) Calcium entry in response to maintained depolarization of squid axons. *J Physiol* 231: 527-548

Bieger D, Neuman RS (1984) Selective accumulation of hydroxytryptamines by frog tectal neurons.

Neurosci 12: 1167-1177

Bondy SC, Harrington ME (1979) Calcium-dependent release of putative neurotransmitters in the chick visual system. *Neurosci* 4: 1521-1527

Bondy SC, Purdy JL (1977) Putative neurotransmitters of the avian visual pathway. *Brain Res* 119: 417-126

Bradford HF, Jones DG, Ward HK, Booher J (1975) Biochemical and morphological studies of the short and long term survival of isolated nerve endings. *Brain Res* 90: 245-259

Cotman CW, Haycock JW, White WF (1976) Stimulus-secretion coupling processes in brain: analysis of noradrenaline and gamma-aminobutyric acid release. *J Physiol* 254: 475-505

Cuenod M, Henke H (1978) Neurotransmitters in the avian visual system. In: Fonnum F (ed) *Amino acids as chemical transmitters*. Plenum Press, New York, pp 221-239

Chujo T, Yamada Y, Yamamoto C (1975) Sensitivity of Purkinje cell dendrites to glutamic acid. *Exp Brain Res* 23: 293-300

Davidoff RA, Adair R (1976) GABA and glycine transport in frog CNS: high-affinity uptake and potassium-evoked release in vitro. *Brain Res* 118: 40-415

Davidson N (1976) *Neurotransmitter amino acids*. Academic Press, New York

Douglas WW (1968) Stimulus-secretion coupling: the concepts and clues from chromaffin and other cells. *J Pharmacol* 34: 451-474

De Robertis E, Pellegrino de Iraldi A, Rodriguez de Lores Arnaiz G, Salganicoff L (1962) Cholinergic and non-cholinergic nerve endings in rat brain. I: Isolation and subcellular distribution of acetylcholine and acetylcholinesterase. *J Neurochem* 9: 23-35

Enna SJ, Snyder SH (1975) Properties of gamma-aminobutyric acid (GABA) receptor binding in rat brain synaptic membrane fraction. *Brain Res* 100: 81-97

Ewert J-P (1976) The visual system of the toad. Behavioral and physiological studies on a pattern recognition system. In: Fite KV (ed) *The amphibian visual system: a multidisciplinary approach*. Academic Press, New York San Francisco London, pp 141-202

Ewert J-P (1984) Tectal mechanisms that underlie prey-catching and avoidance behavior in toads. In: Vanegas H (ed) *Comparative neurology of the optic tectum*. Plenum Press, New York London, pp 247-416

Ewert J-P, Burghagen H, Schürg-Pfeiffer E (1983) Neuroethological analysis of the innate releasing mechanism for prey-catching behavior in toads. In: Ewert J-P, Capranica RR, Ingle DJ (eds) *Advances in vertebrate neuroethology*. Plenum Press, New York London, pp 413-475

Fagg GE, Lane JD (1979) The uptake and release of putative amino acid neurotransmitters. *Neurosci* 4: 1015-1036

Fonnum F, Henke H (1982) The topographical distribution of alanine, aspartate, aminobutyric acid, glutamate, glutamine and glycine in the pigeon optic tectum and the effect of retinal ablation. *J Neurochem* 38: 1130-1134

Fonnum F, Storm-Mathisen J, Walberg F (1970) Glutamate decarboxylase in inhibitory neurons. A study of the enzyme in Purkinje cell axons and boutons in the cat. *Brain Res* 20: 259-275

Foster AC, Roberts PJ (1980) Endogenous amino acid release from rat cerebellum in vitro. *J Neurochem* 35: 517-519

Freeman JA, Norden JJ (1984) Neurotransmitters in the optic tectum of non mammalians. In: Vanegas H (ed) *Comparative neurology of the optic tectum*. Plenum Press, New York, pp 469-546

Fricke U (1975) Tritosol: a new scintillation cocktail based on triton x 100. *Analyt Biochem* 63: 555-558

Grüsser O-J, Grüsser-Cornehls U (1976) Neurophysiology of the anuran visual system. In: Llinás R, Precht W (eds) *Frog neurobiology*. Springer-Verlag, Berlin Heidelberg New York, pp 297-385

Harvey JA, Schofield CN, Graham LT, Aprison MH (1975) Putative transmitters in deprivated olfactory cortex. *J Neurochem* 24: 445-449

Henke H, Cuenod M (1978) Uptake of l-alanine, glycine and l-serine in the pigeon central nervous system. *Brain Res* 152: 105-119

Henke H, Schenker TM, Cuenod M (1976) Uptake of neurotransmitter candidates by pigeon tectum. *J Neurochem* 26: 125-130

Hertz L (1979) Functional interactions between neurons and astrocytes. I: Turnover and metabolism of putative amino acid neurotransmitters. *Prog Neurobiol* 13: 277-323

Hughey PR, Rankin BB, Curthoys NP (1980) Accute acidosis and renal arteriovenous differences of glutamine in normal and adrenalectomized cats. *Amer Physiol Soc* 238: 199-204

Ingle DJ (1976) Spatial vision in anurans. In: Fite KV (ed) *The amphibian visual system: a multidisciplinary approach.* Academic Press, New York, pp 119-140

Ingle DJ (1983) Brain mechanisms of visual localization by frogs and toads. In: Ewert J-P, Capranica RR, Ingle DJ (eds) *Advances in vertebrate neuroethology.* Plenum Press, New York London, pp 177-226

Israel M, Dunant Y, Manarauche R (1979) The present status of the vesicular hypothesis. *Prog Neurobiol* 13: 571-591

Iversen LL (1971) Role of transmitter uptake mechanisms in synaptic transmission. *Br J Pharmacol Chem* 41: 571-591

Johnson JL (1972) Glutamic acid as a synaptic transmitter in the nervous system. A review. *Brain Res* 37: 1-7

Kuljis RO, Karten HJ (1982) Laminar organization of peptide-like immunoreactivity in the anuran optic tectum. *J Comp Neurol* 212: 188-201

Lázár G (1980) Long term persistance after eye-removal of unmyelinated fibres in the frog visual pathway. *Brain Res* 199: 219-224

Lefort B, Henke H, Cuenod M (1978) Glycine specific ^{3}H-strychnine binding in the pigeon CNS. *J Neurochem* 30: 1897-1291

Levi G, Gordon RD, Gallo V, Wilkin GP, Balaz R (1982) Putative acidic amino acid transmitters in the cerebellum. *Brain Res* 239: 425-445

Lowry OH, Rosebrough NJ, Fair AL, Randall RJ (1951) Protein measurement with the folin phenol agent. *J Biol Chem* 193: 265-275

Matsumoto DE, Scalia F (1981) Long-term survival of centrally projecting axons in the optic nerve of the toad (*Bufo bufo*). *J Anat* 92: 21-27

Miledi R (1973) Transmitter release induced by injection of calcium ions into nerve terminals. *Proc R Soc (B)* 183: 421-425

Milson JA, Mitchell JF (1977) The action of amino acids on evoked responses in the frog optic tectum. *Br J Pharmacol* 59: 484

Müller WE, Snyder SH (1978) Glycine high-affinity uptake and strychnine binding associated with glycine receptors in frog central nervous system. *Brain Res* 143: 487-498

Nadler JV, Vaca KW, White WF, Lynch GS, Cotman CW (1976) Aspartate and glutamate as possible transmitters of excitatory hippocampal afferents. *Nature (Lond)* 260: 538-540

Orrego F (1979) Criteria for the identification of central neurotransmitters and their application to studies with some nerve tissue preparations *in vitro. Neurosci* 4: 1037-1057

Oswald RE, Freeman JA (1977) Amphibian optic nerve transmitters: Ach, yes; GABA and glutamate, no. *Soc Neurosci Abstr* 3: 411

Pérez de la Mora M, Possani LD, Tapia R, Teran L, Palacios R, Fuxe K, Horkfelt T, Ljungdahl A (1981) Demonstration of central gamma-aminobutyric containing nerve terminals by means of antibodies against glutamate decarboxylase. *Neurosci* 6: 875-895

Raitieri M, Federico R, Coletti A, Levi G (1975) Release and exchange studies relating to the synaptosomal uptake of GABA. *J Neurochem* 24: 1243-1250

Reiner A, Brauth SE, Karten HJ (1984) Evolution of the amniote basal ganglia. *TINS* 7: 320-325

Reubi JC, Cuenod M (1976) Release of exogenous glycine in the pigeon optic tectum during stimulation of a midbrain nucleus. *Brain Res* 112: 347-361

Roberts PJ, Yates RA (1976) Tectal deafferentation in the frog: selective loss of L-glutamate and gamma-aminobutyric acid. *Neurosci* 1: 371-374

Rubin RP (1970) The role of calcium in the release of neurotransmitter substances and hormones. *Pharmacol Res* 22: 389-428

Sandoval ME, Cotman CW (1978) Evaluation of glutamate as neurotransmitter of cerebellar parallel fibers. *Neurosci* 3: 199-206

Sandoval ME, Lara R (1984) Caracteristicas de la transmision sinaptica quimica. In: Pasantes-Morales H, Arechiga H (eds) *Aminoacidos y peptidos neuroactivos.* Universidad National Autonoma de Mexico, Mexico

Sandoval ME, Horch P, Cotman CW (1978) Evaluation of glutamate as a hippocampal neurotransmitter: glutamate uptake and release from synaptosomes. *Brain Res* 142: 285-299

Sandoval ME, Torner CA, Medrano L (1984) High-affinity uptake and Ca-dependent release of glutamic acid in the developing cerebellum. *Neurosci* 11: 867-875

Schwartz EA (1982) Calcium-independent release of GABA from isolated horizontal cells of toad retina. *J Physiol* 323: 211-227

Shefner SA, Levy RA (1981) The contribution of increases in extracellular potassium to primary afferent depolarization in the bullfrog spinal cord. *Brain Res* 205: 321-335

Simon JR, Martin DL, Kroll M (1974) Sodium-dependent efflux and exchange of GABA in synaptosomes. *J Neurochem* 23: 981-991

Snyder SH, Kuhar MJ, Green AI, Coyle JT, Shoskan EG (1970) Uptake and subcellular localization of neurotransmitters in the brain. *Int Rev Neurobiol* 13: 127-158

Snyder SH (1975) The glycine synaptic receptor in the mammalian central nervous system. *Br J Pharmacol* 53: 473-484

Storm-Mathisen J (1976) Distribution of the components of the GABA system in neuronal tissue: cerebellum and hippocampus-effects of axotomy. In: Roberts E, Chase TU, Tower DB (eds) *GABA in nervous system function.* Raven Press, New York, pp 149-168

Sykova E, Vyklicky L (1977) Changes of extracellular potassium activity in isolated spinal cord of frog under high Mg concentrations. *Neurosci Lett* 4: 161-165

Villani L, Contestabile A, Niso A (1982) Electron microscopic demonstation of neurons and synaptic terminals selectively accumulating ^{3}H-GABA in goldfish optic tectum. *Bas Appl Histochem* 26: 185-191

Werman R (1966) Criteria for identification of a central nervous system transmitter. *Comp Biochem Physiol* 18: 745-766

Whittaker UP (1969) The synaptosome. In: Lajtha A (ed) *Handbook of neurochemistry Vol II.* Plenum Press, New York, pp 327-364

Woodward DJ, Hoffer RJ, Siggins GR, Oliver AP (1976) Inhibition of Purkinje cells in the frog cerebellum. II. Evidence for GABA as the inhibitory transmitter. *Brain Res* 33: 91-100

Yamamoto C, Matsui S (1976) Effect of stimulation of an excitatory nerve tract on release of glutamic acid from olfactory cortex slices *in vitro. J Neurochem* 26: 487-491

Yates RA, Roberts PJ (1974) Effects of enucleation and intraocular colchicine on amino acids of frog optic tectum. *J Neurochem* 233: 891-894

Yates RA, Taberner PV (1975) Glutamic acid, GABA and their metabolizing enzymes in the frog central nervous system. *Brain Res* 84: 397-404

Retina and Optic Tectum in Amphibians: A Mathematical Model and Simulation Studies

UWE AN DER HEIDEN and GERHARD ROTH

Fakultät für Naturwissenschaften, Universität Witten/Herdecke
D-5810 Witten, FR Germany
Fachbereich 2/Biologie, Universität Bremen
D-2800 Bremen 33, FR Germany

Abstract. The processing of visual information within the retino-tectal system of amphibians is decomposed into five major operational stages, three of them taking place in the retina and two in the optic tectum. The stages in the retina involve (i) a spatially local high-pass filtering in connection to the perception of moving objects, (ii) separation of the receptor activity into *on*- and *off*-channels regarding the distinction of objects moving against light or dark backgrounds, (iii) spatial integration via near excitation and far-reaching inhibition. Variation of the spatial range of excitation and inhibition allows to account for typical activities observed in a variety of classes of retinal ganglion cells. A mathematical description of the operations in the optic tectum include (a) spatial summation of retinal output (mainly of class R2 and class R3 retinal ganglion cells), and (b) lateral inhibition between tectal cells. In the computer simulation, first the output of the mathematical retina model is computed which, then, is used as the input to the tectum model. The full spatiotemporal dynamics is taken into account. The simulations show that different combinations of strength of lateral inhibition on the one side and the response properties of the retinal ganglion cells on the other side determine the response properties of tectal cell types T5.1, T5.2, and T5.3 involved in object recognition.

1 Introduction

For prey-catching behavior of most anuran and urodele amphibians (e.g., frogs, toads, salamanders), vision plays a dominant role. In these animals, the visual system consists of the retina and of retino-recipient structures such as (i) the optic tectum as the dorsal part of the midbrain (mesencephalon), (ii) various dorsal thalamic nuclei as portions of the diencephalon, (iii) the pretectum including several nuclei of the posterior thalamus, and (iv) the basal optic nucleus at the junction of the caudal hypothalamus and rostral mesencephalic tegmentum. Whereas these structures receive direct retinal input, there are various other nuclei involved in processing visual information, such as the mesencephalic nucleus isthmi and the telencephalic striatum. The optic tectum can be regarded as a center for localization and identification of prey as well as predator stimuli. Thalamic structures seem to be involved in barrier detection (Ingle 1983), whereas certain pretectal nuclei coordinate optomotor behavior (Manteuffel 1985; Roth 1987). It is suggested that thalamic pretectal structures also determine prey-selective properties of tectal neurons (Ewert 1968, 1974, 1984; Ewert, this volume). The midbrain tegmentum and the medulla oblongata are important for the integration of these functions and for their translation into motor programs (Herrick 1948; see also contributions by Grobstein, Schwippert, and Weerasuriya, this volume).

The optic tectum receives the majority of primary visual afferent fibers constituted by the axons of the retinal ganglion cells (RGC). In all amphibians studied so far, there seem to be three to four major functional types of RGC: a first type (R1/R2) with small excitatory receptive fields (ERF=2-6° in anurans, 4-10° in salamanders), strong inhibitory surround and responding best to small moving objects; in frogs, a R1 type can be distinguished with a longer time constant of decrease in neural activity after stimulus position change (Grüsser & Grüsser-Cornehls 1976); in toads, there is only one small-field type of R2 (Ewert & Hock 1972); a second type, R3, with ERFs of intermediate size (6-10° in anurans, 8-16° in salamanders), weaker inhibitory surround and responding preferentially to objects of intermediate size; a third type, R4, with large ERFs (10-16° in anurans, 10-20° in salamanders), with only weak or no inhibitory surround and responding best to large objects or to changes in ambient illumination (for reviews see Grüsser & Grüsser-

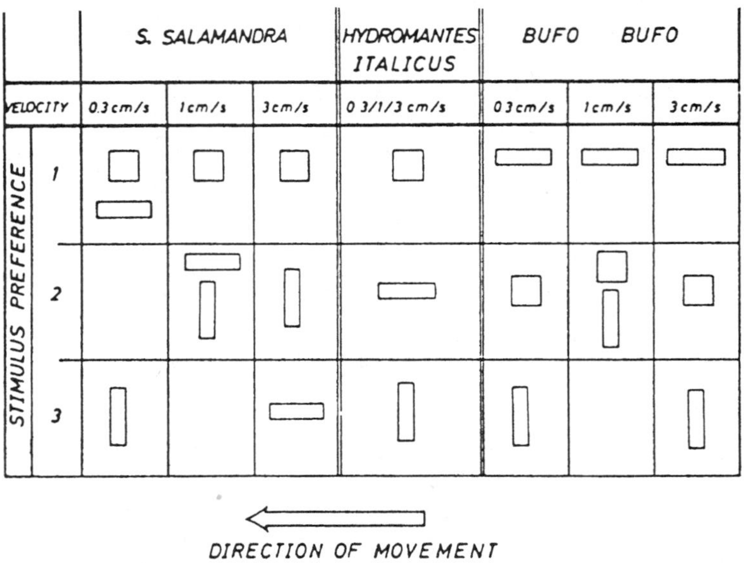

Figure 1. Schematic representation of stimulus preference ranking from (1 to 3) in prey-catching behavior of *Salamandra salamandra*, *Hydromantes italicus*, and *Bufo bufo* with respect to squares, horizontal and vertical rectangles moving at three different speeds.

Cornehls 1976; Ewert 1984, 1987; Roth 1987). Whereas the first and the second type seem to be involved primarily in the localization and identification of prey objects, the third type seems to play a major role in ambient vision. The fibers of these RGC terminate in the superficial part of the optic tectum where they form distinct horizontal layers in which they make contacts with the dendrites of tectal neurons.

Whereas the response properties of RGC and tectal neurons have been studied in greater detail in a number of amphibian groups, little is known about the actual connectivity between the RGC and tectal cells on the one hand, and between tectal cells on the other (Matsumoto, this volume). Both types of connectivity mainly constitute the response properties of tectal cells. However, afferents from other parts of the visual system, e.g., the pretectum, also seem to play an important role (Ewert 1984; Roth 1987; Manteuffel, this volume).

Most studies of the visual system of amphibians have concentrated on the guidance of prey-catching behavior. This type of behavior can easily be elicited by rather simple prey dummies, such as pieces of cardboard moved in front of the animals. By using a set of prey dummies differing in size, configuration, velocity etc., a variety of amphibian groups have been studied with regard to their prey preferences. Figure 1 shows the prey-stimulus preferences of the fire salamander *Salamandra salamandra*, the common toad *Bufo bufo*, and the tongue-projecting salamander *Hydromantes italicus*. All of these respond quite well to square-shaped stimuli (S); *Salamandra* and *Bufo* respond more or less equally well to elongate stimuli moved *parallel* to their longitudinal axis (H) (Ewert 1968, 1969; Finkenstädt & Ewert 1983a) whereas in *Hydromantes* the response to this stimulus is much weaker than to S (Roth 1987), and this seems to be true for most frogs, too. In all amphibian groups mentioned there is always a rather weak response to elongated stimuli moved *perpendicular* to their longitudinal axis (V) (H refers to "worm-configuration" *W*, and V to "antiworm-configuration" *A* of Ewert's terminology). The direction of movement itself has no significant influence on the strength of the behavioral response.

How are the animals able to perceive the differences of S-, H-, and V-shaped moving objects? The answer to this question is the subject of this paper.

2 Response Properties of Tectal Neurons

In the optic tectum a variety of types of neurons can be found, some of which are universally present whereas others are found only within certain groups of amphibians. In Figs. 2 and 3 the most common types are illustrated. Figure 2 shows the results of recordings from the tectum of the toad *Bufo bufo* (Ewert & v. Wietersheim 1974a). In these experiments prey stimuli were used that differed in size, configuration, and orientation with regard to the direction of movement according to Ewert (1969). Here, a series of different stimuli was constructed by increasing a small square of 2° edge length geometrically either with regard to its horizontal or to its vertical edge or to both edges, so that horizontally (H) or vertically (V) oriented rectangles of increasing length, and squares (S) of increasing size were obtained. Each stimulus was moved through the visual field of the immobilized subject at a constant angular velocity of 7.6°/s in horizontal direction.

Figure 2A shows tectal type I neurons, recorded by Ewert & v. Wietersheim (1974a), which regarding their relatively small monocular receptive fields are T5 neurons according to Grüsser & Grüsser-Cornehls (1976) and with respect to their configural response properties are called T5.1 by Ewert et al. (1978). This type characteristically responds more strongly to squares than to "horizontal rectangles," and weakest to "vertical rectangles" (S > H > V); a type II, called T5.2, is shown in Fig. 2B in which the sequence of preference is H > S > V, and a type III, shown in Fig. 2C, is called T5.3 and has the preference S > V > H, provided that the stimuli are moved in horizontal direction (for a review see Ewert 1984); interestingly, tectal response types that prefer V to S and H are very rarely found (Roth 1987). This corresponds well to the above mentioned fact that in behavioral experiments vertically oriented bars which are moved horizontally only rarely elicit prey-catching responses.

The response characteristics of some tectal cells become more complicated if stimulus velocity is varied. Figure 3 shows the results of experiments, again in *Bufo bufo*, in which three types of stimuli were used, a square measuring 8° x 8° as well as rectangles measuring 2° x 8° oriented either horizontally or vertically and moved in horizontal direction at velocities of 2, 6, and 20°/s (Roth & Jordan 1982). As can be seen, certain cell types do not change their stimulus preference with changing velocity, and they correspond to the above described types T5.1, T5.2 and T5.3 (Fig. 3A-C) (see also Ewert, this volume). On the other hand, Fig. 3D and E represent two types in which stimulus preference (S *vs.* H) changed with changing velocity: One of them (Fig. 3D) exhibits the preference H > S > V at low velocity (2°/s) and the preference S > H > V at high velocity (20°/s); in another type (Fig. 3E) S is equally

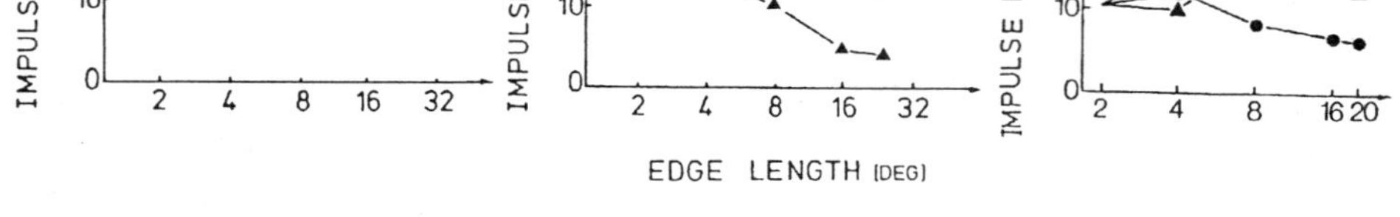

Figure 2. Responses of three cell types (A,B,C) in the optic tectum of the toad *Bufo bufo* (L) to moving stimuli (v = 7.6°/s) of different size and orientation with respect to the direction of movement. Square symbols refer to the average impulse frequencies in response to squares *S* with edge length given on the abscissa (measured in degrees of visual angle); dot symbols refer to responses to rectangles with shorter edge measuring 2° in all cases and length of the longer edge oriented *parallel, H,* to the direction of movement given on the abscissa; triangle symbols refer to stripes of 2° width whose longer edge is oriented *vertical, V,* to the horizontal direction of movement. (A) Cell type T5.1; (B) cell type T5.2; (C) cell type T5.3. (After Ewert & v. Wietersheim 1974 (A, B) and Schürg-Pfeiffer & Ewert 1981).

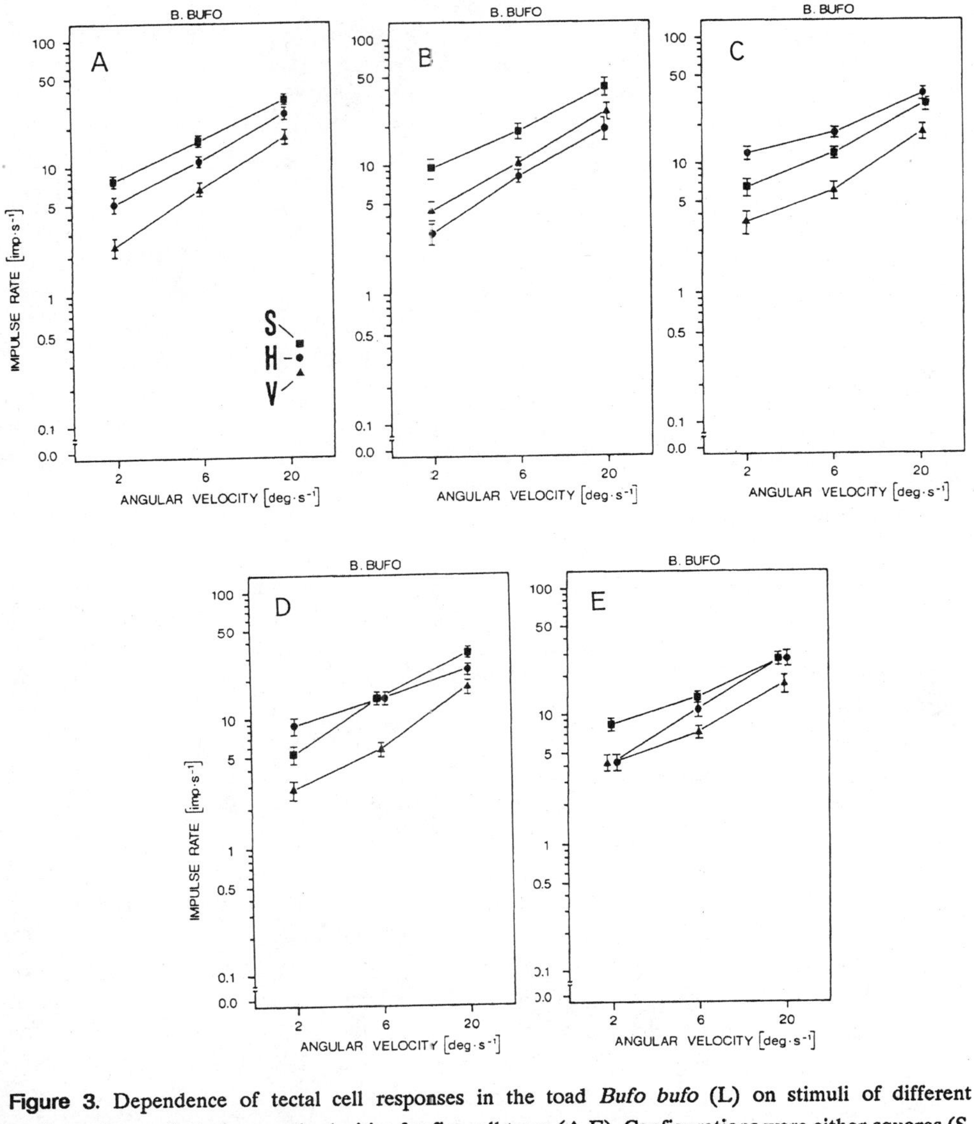

Figure 3. Dependence of tectal cell responses in the toad *Bufo bufo* (L) on stimuli of different configurations, orientations, and velocities for five cell types (A-E). Configurations were either squares (S, see square symbols) measuring 8°x 8° or rectangles measuring 2°x 8°. The rectangles were oriented either parallel (H, see dot symbols) or vertical (V, see triangle symbols) to the horizontal direction of movement. Magnitudes of responses depend on the absolute value of the velocity (given on the abscissa) but not on the direction of movement. In neurons of the types shown in (A)-(C) the preferences S>H>V according to T5.1 (A), S>V>H according to T5.3 (B), and H>S>V according to T5.2 (C) are independent of the velocity of movement. Neurons of the type shown in (D) exhibit a phenomenon of preference *inversion* and in (E) a sort of *convergence* with respect to S *vs.* H: In (D) the preference is H>S>V for low velocities and S>H>V for high velocities. In (E) the preference S>H=V for low velocities and the preference S=H>V for high velocities are observed. (After Roth & Jordan 1982).

Table 1. Distribution of neuronal response types in the optic tectum of four different amphibian species in response to 8°x 8°(S), 2°x 8°(H), and 8°x 2°(V) objects, each one moved in horizontal direction; "x" indicates cases in which the preference depends on the stimulus angular velocity; *) calculated on the basis of data from Schürg-Pfeiffer & Ewert 1981; if 8° is replaced by 16° then H>S>V in *Rana temporaria*; ?= not measured in *Rana*. In *B.bufo*, 17% of the neurons displayed a property shown in Fig. 3E.

	Cell Types					
Species	S>H>V T5.1	S>V>H T5.3	H>S>V T5.2	S=H>V T5.2*)	S>VxH	HxS>V
Bufo bufo	26%	10%	19%	-	4%	12%
Salamandra salamandra	18%	26%	15%	-	6%	9%
Hydromantes italicus	15%	32%	-	-	20%	-
Rana temporaria	23%*)	13%*)	-	21%	?	?

preferred to both H and V at low velocity, whereas at high velocity, H has the same efficacy as S.

In Table 1 the occurrence of tectal cell types in the tecta of different groups of amphibians is listed. Whereas in the optic tectum of *Bufo* and *Salamandra* all response types mentioned occur and type T5.2 (which prefers elongate, "wormlike" stimuli) is well presented, in the tectum of *Hydromantes* this type is completely absent and in the tectum of *Rana* only an intermediate type between T5.1 and T5.2 is found but no clear T5.2 type (cf. Roth 1986, 1987; for a discussion see Ewert, this volume). These findings correspond closely with the fact that in their prey-catching behavior both *Bufo* and *Salamandra* respond well to wormlike objects, whereas *Hydromantes* and *Rana* prefer small compact prey objects. (If the response properties for stimuli of variable edge length are measured it seems, however, that the algorithms for configuration selection are less different between *Bufo* and *Rana* than between *Bufo* and *Salamandra*; for a discussion see Ewert, this volume).

3 Mathematical Model of Retinal Ganglion Cells

Having demonstrated the relevance of tectal activity for prey recognition and prey-catching activity, we now turn to the question of how the variety of different response types in the optic tectum is constituted by the interaction of various elements in the visual system of amphibians.

Since the tectum receives direct input from the retina, this question is coupled to the problem concerning the form in which the information about the stimuli is conveyed from the retina to the tectum. A systems-theoretic mathematical description of the tectum presupposes a corresponding description of the operations of the retina. In the early model of Ewert & v. Seelen (1974) the operations of the retina and of the tectum are lumped together into a single step. Previous models of Arbib, Cervantes-Pérez, and Lara for tectal activity did not actually use the output of a retina model as an input to a tectum model (see Lara et al. 1982; Cervantes-Pérez et al. 1985; Lara & Arbib 1985). The latter is, however, necessary in order to verify that the operations of the tectum are adequately conceived. In this section, therefore, a mathematical model is developed and analyzed which represents the essential operations performed by the retinal network on optic patterns in the visual field.

We disregard here the problem of color coding. Accordingly, the visual stimulus may be represented by a function x(s,t) giving the light intensity x at the point s in the visual field at time t. The spatial coordinates $s=(s_1,s_2)$ of the two-dimensional visual field are measured in degrees of visual angle (rad).

For the understanding of the visual system, it is important to realize that the retina is by no means a device for a point-wise transduction of external light signals to the brain. The retinal ganglion cells whose axons project to the tectum form the output of a fairly complex neural network within the retina itself. Besides the receptors (which are also made up of several classes of cells) the retinal network comprises cell types such as horizontal, bipolar, and amacrine cells. The present model is not aimed at providing a detailed description of the functional properties of all of these elements, but is more a "lumped circuit" description intended to adequately produce the activity at the output level (the RGC) of the retina given any external light distribution input x(s,t).

When the excitatory receptive field (ERF) of a RGC is illuminated with small spots of light (e.g., with a diameter of 0.2°) for a brief period (1.3 s) then, depending on the position of the light spot within the ERF, the recorded cell responds with a

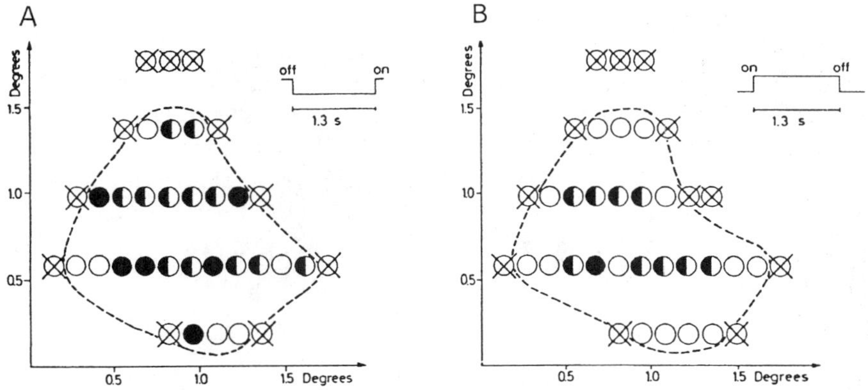

Figure 4. *On-*, *on-off* and *off*-zones in the receptive field of retinal class-1 (R1) neurons of the toad *Hyla septendrionalis.* Black circles: *off*-activation; white circles: *on*-activation, crosses: no response. The retina was stimulated by 0.2° light spots which were either turned off for 1.3 s (A) or turned on for 1.3 s (B). These experimental findings form one of the motivations for including local *on-* and *off*-channels into the retina model. (Reproduced from Grüsser et al. 1976 by permission).

short activity burst after the light is turned on (*on*-response) or after it is turned off (*off*-response) (Fig. 4).

There is evidence (Levine & Shefner 1977) that these local *on*- and *off*-responses of RGC are mediated by separate *on*- and *off*-channels leading from the receptor layer to the ganglion cells. The activity of these channels is probably a result of complex interactions between receptors, horizontal and bipolar cells. Here, according to a suggestion by Varjú (1978), we simply assume that the *on*-response of an *on*-channel is the positive part of a high-pass filter response and that the *off*-response of an *off*-channel corresponds to the negative part of a high-pass filter response. In mathematical terms the high-pass filter operation is described by the time integral

$$y(s,t) = \int_{-\infty}^{t} x(s,t') \, [\delta(t-t') - T^{-1} e^{-(t-t')/T}] \, dt' \tag{1}$$

where T is the time constant of the filter and the expression in brackets represents the impulse response of a high-pass filter (δ= Dirac's function). The positive part of y is

$$z_{on}(s,t) = \max(0, y(s,t)) \tag{2}$$

and the negative part is

$$z_{off}(s,t) = \max(0, -y(s,t)) \tag{3}$$

where the function *max* means selecting the maximum of the two values given in parentheses (neurophysiologically, *max* could be realized by a threshold and the minus sign in Eq. (3) by inhibitory synapses).

One of the most important features of a retinal ganglion cell is spatial integration. The RGC is not only excited by a single local *on*- or *off*-channel but from a multitude of channels receiving receptor input from a rather large area of the visual field. This area is generally called the "excitatory receptive field" (ERF) of the RGC. Of course, the boundaries of the ERF are not sharply defined. Instead, it is assumed that the influence of the channels on a certain RGC decreases continuously with the distance of the channel (more precisely, of the central location of the channel's visual input) from the center s_0 of the ERF. Mathematically, this is expressed by the two-dimensional spatial integration

$$v_e(s_0,t) = \iint_F [(b_{1e} \, z_{on}(s,t) + b_{2e} \, z_{off}(s,t)] \, exp\,(-k_e|s-s_0|) \, ds. \tag{4}$$

The integration extends across the whole visual field F. The constant factors b_{1e} and b_{2e} give the relative contribution of the *on*- and *off*-channels to the excitation of the RGC. This contribution specifically decays with a Gaussian rate (rate constant k_e) of the eccentricity $|s-s_0|$ of the channel position (s) from the center (s_0) of the ERF.

In the next step it has to be taken into account that an RGC is not only excited but also inhibited by other neurons in the retina. The strength of inhibition also decays from the center of the receptive field of the RGC to the periphery of the visual field. Therefore, inhibition is decribed essentially in the same way as excitation:

$$v_i(s_{0,t}) = \iint_F [b_{1i}\, z_{on}(s,t) + b_{2i}\, z_{off}(s,t)]\ exp\ (-k_i|s-s_0|)\ ds. \tag{5}$$

Detailed measurements (Grüsser & Grüsser-Cornehls 1976) show that excitation exceeds inhibition in a (nearly circular) area around the center of the ERF, but inhibition strength exceeds excitation in the periphery. This relation between excitation and inhibition requires

$$k_i < k_e \quad \text{and} \quad b_{1e}+b_{2e} > b_{1i}+b_{2i}. \tag{6}$$

Assuming a low-pass filter characteristic for the RGC's membrane potential w with a membrane time constant $\overline{T}$, the effectiveness of the netto excitation v_e-v_i is given by

$$\overline{T}\ dw(s_0,t)/dt = v_e(s_0,t) - v_i(s_0,t) - w(s_0,t) \tag{7}$$

(note that the RGC is spatially identified by the center s_0 of its receptive field). The functional model of the retinal output cells is completed by an expression for the transduction of membrane potential w into a firing frequency f of RGC axons:

$$f(s_0,t) = \max(0,w(s_0,t)) \tag{8}$$

meaning that f is proportional to w if w exceeds the constant threshold value, here normalized to 0. This is one of the simplest assumptions which can be made about the transformation of membrane potentials into nerve impulses.

4 Comparison of Real and Simulated Responses of the Retinal Network

Does this model for the input-output relationship of the retina fit the actually observed reactions of RGC in response to moving rectangular objects?

As already mentioned, there are three to four physiological classes of retinal ganglion cells (R1/R2, R3, and R4). Figure 5 shows recordings from an R1-neuron (Schürg-Pfeiffer & Ewert 1981). The left column shows responses to squares (S) with edge lengths increasing geometrically from top to bottom, the middle column shows responses to vertical rectangles (V), and the right column to horizontal rectangles (H) of geometrically increasing length (all stimuli moved horizontically at a velocity of 7.6°/s).

Figure 6A-C shows simulated responses of the membrane potentials of the model to the same set of stimuli as in the recording experiments. These curves have been obtained by choosing suitable values of the parameters (T, $\overline{T}$, k_e, k_i, b_{1e}, etc.) as given in the legend of Fig. 6 and solving the Eqs. (1)-(7) numerically on a digital computer. According to Eq. (8), only values of the membrane potential exceeding a certain threshold value contribute to the formation of impulses. Assuming a threshold value equal to or slightly above 0, the reader may convince himself that there is good qualitative agreement between the activity of the real RGC of class R1 and the model RGC.

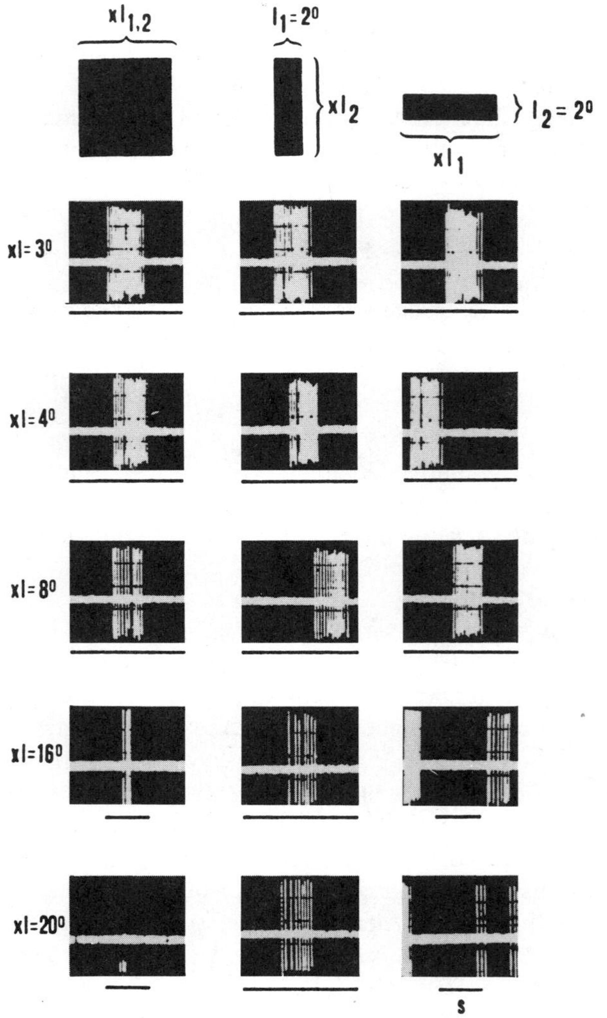

Figure 5. Original registrations from retinal ganglion cells of class R1. Stimuli were black rectangles moved at a velocity of v=7.6°/s in front of a white background in horizontal direction through the visual field of the paralyzed animal (*Rana temporaria*). Left vertical column: responses to squares (S) with edge length indicated beside the records. Middle vertical columns: responses to rectangles (V) oriented perpendicular to the direction of movement, shorter axis of 2° length, the length of the longer axis given in the columns. Right vertical column: responses to rectangles (H) oriented parallel to the direction of movement, shorter axis 2°, lenght of longer axis given in the columns. (Reproduced from Schürg-Pfeiffer & Ewert 1981).

It is possible to discuss and to understand the temporal development of the simulated RGC membrane potential on the basis of the model assumptions. Negative potentials (hyperpolarization) result if inhibitory influences outweigh excitatory influences. Since, in traversing the receptive field centrally, the stimulus first enters and passes with its leading edge the periphery of the receptive field, where inhibition prevails, the cell is at first hyperpolarized. Afterwards, a depolarization occurs resulting from the excitation of the central region of the receptive field by the off-response producing leading edge of the stimulus. This depolarization is followed by a second hyperpolarization when the leading edge of the stimulus crosses again the periphery of the field.

Depending on the length of the stimulus in the direction of movement, more or less prominent further excitatory and inhibitory wave components appear originating from the *on*-responses caused by the trailing edge of the stimulus. (Note that the moving stimuli in all these examples are black against a white background). In this way the two bursts in the response to very long horizontal bars may be explained which occur both in reality and in the model (see also Tsai & Ewert 1987).

Because of the high-pass filter components in the model, non-moving stimuli under stationary illumination would not elicit any response of the model RGC. This is in accordance with real RGC. However, a response occurs both in the model and the real RGC when a light source is turned on and off and the stimulus is not moving.

If we plot the responses of an RGC to the standard set of stimuli in the format of Fig. 2 (average impulse frequency *vs*. edge length), then Fig. 7A is obtained for a real R2-neuron and Fig. 7B for the model neuron. (Average imulse frequency is calculated in the model as the mean of the membrane potential above threshold.) As may be seen from Figs. 7A and 8A, cells of classes R2 and R3 differ significantly in their responses to the set of stimuli (moving at $v=7.6°/s$). R2-cells respond maximally to squares and vertical bars with major edge length measuring 4°, whereas R3-cells have their maximum at 8°. We explain the difference between the responses of these two types as caused by stronger weights and spatial extension of excitatory influences of *on*-and *off*-channels in the case of R3-cells. Accordingly, our model for R3-cells coincides with that for the R2-cells in all respects, with the exception that the parameters b_{2i}, k_e and k_i are chosen differently. More specifically, these parameters are chosen to satisfy the condition that the strength of excitatory synapses is equal to the strength of inhibitory synapses on a circle (around the center of the receptive field of the RGC) measuring 8° in diameter for R2-cells and 16° for R3-cells, i. e.

$$(b_{1e}+b_{2e})\ exp\ (-k_e s^2) - (b_{1i}+b_{2i})\ exp\ (-k_i s^2) = 0 \tag{9}$$

for $s=8°$ in case of an R2-cell, and for $s=16°$ in case of R3-cell. Obviously this choice of parameters is based on the observation that real RGCs show maximal responses to squares of corresponding edge lengths. Figure 8B shows the result of computer simulations using this R3-neuron model. Again the behavior of the model neurons satisfactorily coincides with that of the real neurons. In our model we did not consider the retinal class R4 because it is most probably not directly involved in prey recognition.

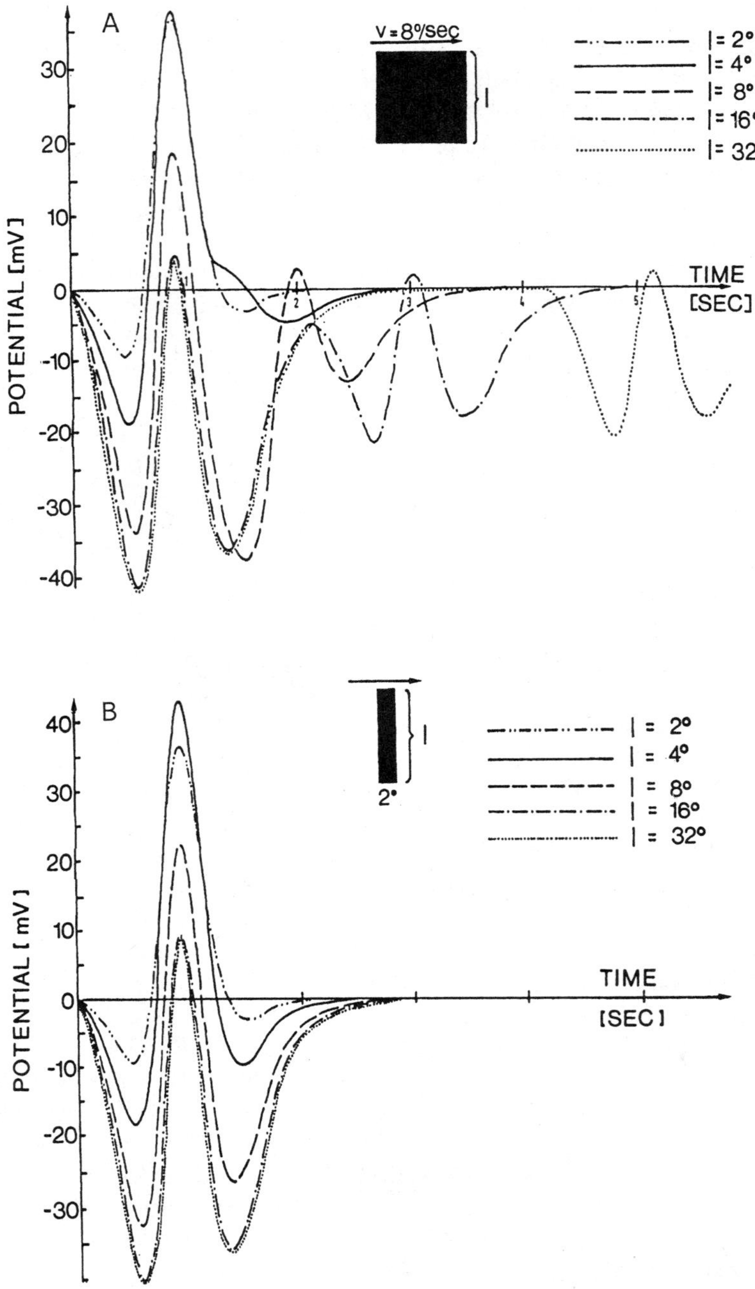
A
v = 8°/sec
l
l = 2°
l = 4°
l = 8°
l = 16°
l = 32°
POTENTIAL [mV]
30
20
10
0
-10
-20
-30
-40
TIME
[SEC]
B
l
2°
l = 2°
l = 4°
l = 8°
l = 16°
l = 32°
POTENTIAL [mV]
40
30
20
10
0
-10
-20
-30
TIME
[SEC]

Responses of RGC strongly depend on stimulus velocity. In a first test, we computed the responses of our model RGC to squares measuring 2° x 2° moving at different velocities. The results are shown in Fig. 9A for both R2 and R3. It may be observed in our simulation that at all velocities the responses of R2-cells are stronger than those of R3-cells. This agrees well with the experimental data, as can be seen in Fig. 9B. The model shows a clear dependence of the activity on the velocity of moving prey objects. Since the first stage of our model consists of a layer of high-pass filters, it was to be expected that stationary objects would not elicit any response in the model RGC.

Summarizing the last two sections, a mathematical model for the activity of the major class of retinal ganglion cells projecting to the optic tectum has been developed and numerically analyzed. It had the form of a combination of integral and differential equations with space and time as independent variables, and it has the capacity to fully generate the spatiotemporal dynamics of these ganglion cells under arbitrary conditions of non-colored visual input to the retina.

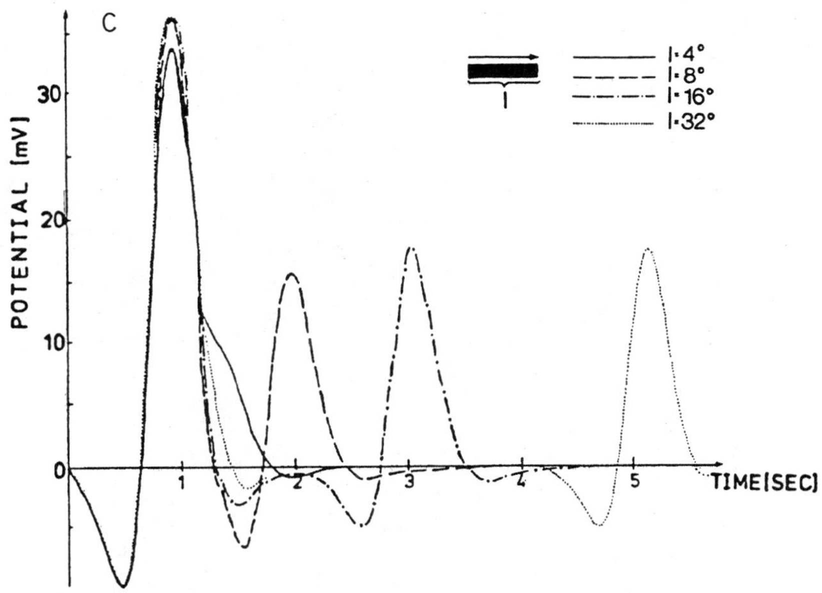

Figure 6. (A-C) Time courses of membrane potentials in the model retinal ganglion cell of class R1 in response to stimuli described in the previous figures. (A) Responses to five different squares with edge length given in the figure. (B,C): Responses to four rectangles with longer axis (length given in the figure) oriented vertical to the direction of movement (B) or parallel to the direction of movement (C). Shorter axis measured 2°, v=7.6°/s throughout. Curves were obtained by solving Eqs. (1)-(7) numerically on a digital computer. Parameters: $T=0.3$ s, $b_{2e}=50$, $b_{2i}=14.7$, $b_{1e}=20$, $b_{1i}=8.3$, $k_e=0.37$, $k_i=0.06$, $\bar{T}=5$ ms. Note that, according to Eq. (8), the time course of the impulse frequency of this model cell is simply given by those parts of the membrane potential curves lying above the time axis (t). Note that the model correctly predicts the two bursts (see Fig. 5, lower part of right column) occurring in the presence of long "horizontal" stimuli (H), the *off*-burst in response to the leading edge being stronger than the *on*-burst to the trailing edge of the stripe.

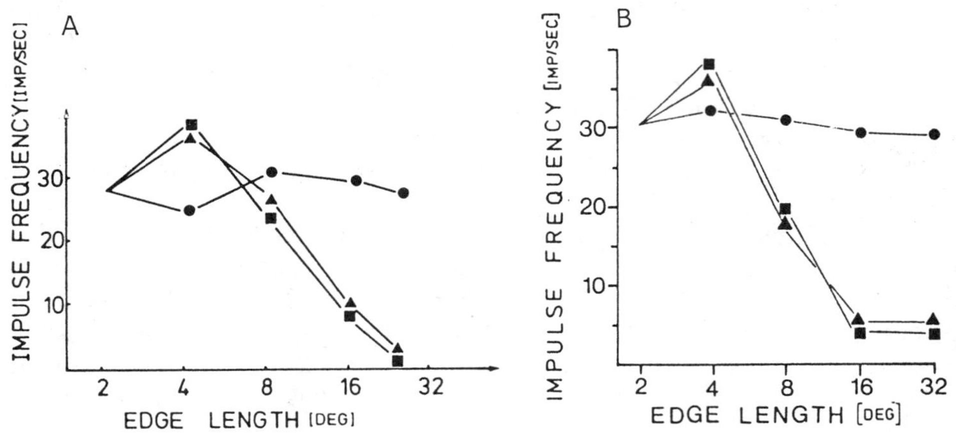

Figure 7. Characteristics of retinal ganglion cells of class R2 according to experiments (A) and model predictions (B). Stimulus configurations as described in Fig. 3. Here and in the following figures curves with square symbols show responses to squares, curves with dots (triangles) refer to elongated rectangles with longer axis oriented parallel (perpendicular) to the direction of movement. Experimental data were obtained from measurements in toads (*Bufo bufo*) by Ewert & Hock (1972). Parameters in the model described by Eqs. (1)-(7) are as in Fig. 6, with the exception of $b_{1e} = 0$. In (B) model predictions appear essentially correct, revealing, however, that normal experimental estimates of the size of the inhibitory receptive field on these cells (diam. about 16°) are too small (showing up at 16° in (B)).

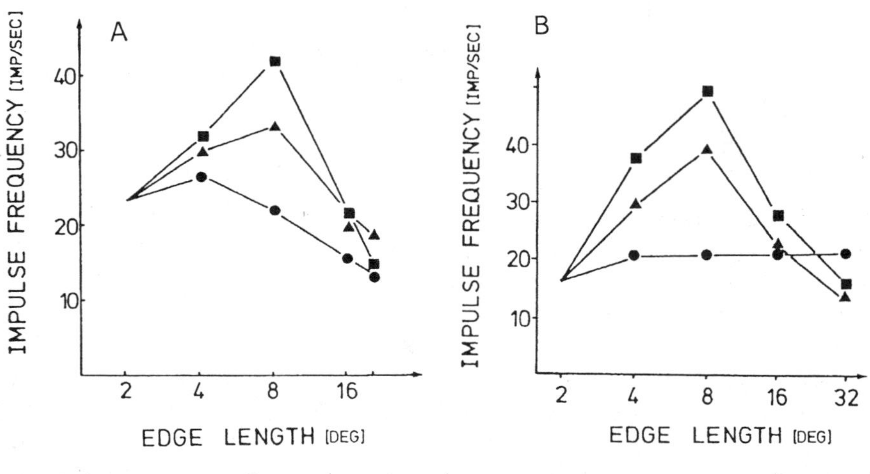

Figure 8. Characteristics of retinal ganglion cells of class R3 in experiment (A) and model (B). Same stimuli as in Figs. 3 and 7. Parameters in Eqs. (1)-(7) as given in Fig. 6, with the exception of $b_{2i} = 23.5$, $k_e = 0.07$, $k_i = 0.023$.

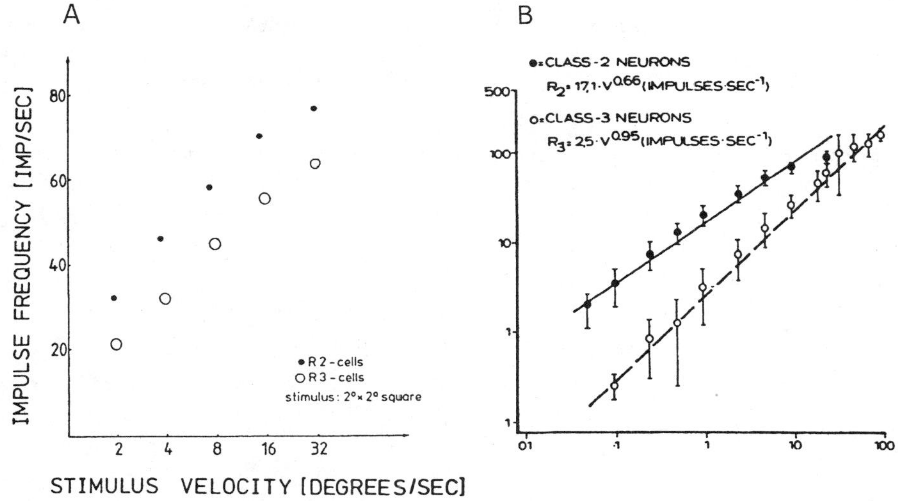

Figure 9. Dependence of the activity of retinal ganglion cells on the velocity of the stimuli. (A) Model cells. (B) Real cells (reproduced from Grüsser & Grüsser-Cornehls 1976, by permission). In (A) the stimulus is a square of 2°x2°. In (B) the stimulus is a circle of 1.2° diameter for class-2 (R2) neurons, and of 2.7° diameter for class-3 (R3) neurons. Black stimuli against a white background throughout.

5 Mathematical Model for the Constitution of some Tectal Cell Types

5.1 Basic Model: Summatory Processes in the Tectum

Generally, a tectal cell receives input not only from a single RGC but from a multitude of retinal cells. The receptive fields of those RGC projecting to a single tectal cell generally overlap with each other, and together they form the receptive field of this tectal cell. The diameters of receptive fields of tectal cells are larger than those of RGCs and most often exceed 15° to 30° (there are tectal cells whose ERFs are larger than 200°).

When a rectangular stimulus passes centrally through the receptive field of a certain tectal cell, then in most cases it does not move centrally through the receptive fields of those RGC projecting to this cell but more or less shifted toward the periphery. To our knowledge no model approach in relevant literature has adequately treated this problem before. Our model for retinal cell activity allows us to calculate the responses to stimuli moving through any region of their receptive fields. At this point it becomes quite clear why a model of the tectum presupposes a model of the retina (at least of its output).

In trying to describe the physiological properties of tectal cells mathematically one is faced with the fact that little is known about the actual circuitry of neurons in this region of the brain. For this reason we tried to build our model on as few and simple assumptions as possible. The most elementary and certainly realistic assumption is that the majority of tectal cells integrate (summate) in some form the activity of RGC projecting onto them. Contrary to retinal cells, tectal cells lack an inhibitory component (surround) in their receptive field. Therefore, a purely summatory effect on their retinal input is well-grounded. The summation of a tectal cell concerns the output of those RGC which share their receptive field with the receptive field of this cell and are connected by their axons to this tectal cell.

The number of RGC of a certain type (e.g., R2 or R3) projecting to a tectal cell, depends on the degree of overlap between the excitatory receptive fields of RGC. We define the degree of overlap to be the inverse of the average distance (measured in spatial units, here degrees of visual angle) between the centers of neighboring receptive fields. Note that this definition is independent of the size of receptive fields, which is important since this size may vary under different physiological conditions.

If Δ denotes the degree of overlap and d the diameter of the receptive field of the tectal cell, then a rule of thumb for the relation between the number N of RGC projecting to this cell and d is given by

$$N = \pi (d \cdot \Delta)^2/4 \tag{10}$$

under the assumption, which is most often observed, of a circular-shaped receptive field of the tectal cell.

Given the number N of RGC of a certain type (R2 or R3) projecting to a certain tectal cell, the summation process of this cell is defined precisely as follows. The centers of the receptive field of these RGC are assumed to be distributed in the receptive field of the tectal cell according to a uniform random distribution. The excitation e(t) of the tectal cell is given by

$$e(t) = a \sum_{i=1}^{N} w^i_+(t) \; exp\,(-kd_i^2)_i. \tag{11}$$

Here, a is a proportionality factor, w^i_+ the output of the *ith* RGC (arbitrarily numbered), and d_i measures the distance (in units of the visual field) of the center of the *ith* RGC to the center of the receptive field of the tectum cell (see Fig. 10). The factor $exp(-k\, d_i^2)$ specifies the assumption that the synaptic strength per unit area decays according to a Gaussian distribution with constant rate k from the center to the periphery of the tectal cell's receptive field. This assumption presupposes the well-established hypothesis that there is a nearly topographic projection from the retina to the tectum; only then it is possible to use the spatial coordinates of the visual field also in the tectum and to define the quantities d_i.

The low-pass filter dynamics of the tectum-cell membrane potential u(t) transforms the input e(t) according to the differential equation

$$T^* \, du(t)/dt = e(t) - u(t) \tag{12}$$

where T* is the membrane time constant. Our basic model for tectal cells is completed by an expression for the transformation of the membrane potential u into the impulse frequency u_+ analogous to the situation of RGC:

$$u_+(t) = a^* \max(0, u) \tag{13}$$

with a constant factor a* and a threshold value arbitrarily set to 0.

Let us now consider the behavior of this elementary model for tectal cells under the influence of the set of stimuli moving through the visual field. For this purpose the model equations (i.e., the combined retinal and tectal model system) were solved numerically on a digital computer.

In the first series of simulations only the output of R2 retinal cells was used as input to the model tectum cell. To study the effect of the degree of overlap between receptive fields of RGC projecting to this cell, we applied small and large values of Δ corresponding to a small or large number N of R2-cells, respectively (Eq. 10), connected to this tectal cell (a fixed value $d = 27°$ was used throughout).

Figure 11A shows the case of high overlap. The membrane potentials in response to differently large square stimuli rise smoothly to a unique maximum and afterwards decay to the resting potential. If the degree of overlap is low, then - as may be seen from Fig. 11B - several more or less sharply separated bursts of activity may occur. In fact, such a type of interrupted neural activity was often observed in recorded neurons. Our model suggests a rather simple explanation for this phenomenon.

However, in all of the following numerical simulations Δ was chosen large enough to produce only one single peak of activity for each stimulus. Figure 12A shows the average impulse frequencies plotted as a function of stimulus extensions under the condition that the tectal cell only sums up the output of R2 cells. Obviously,

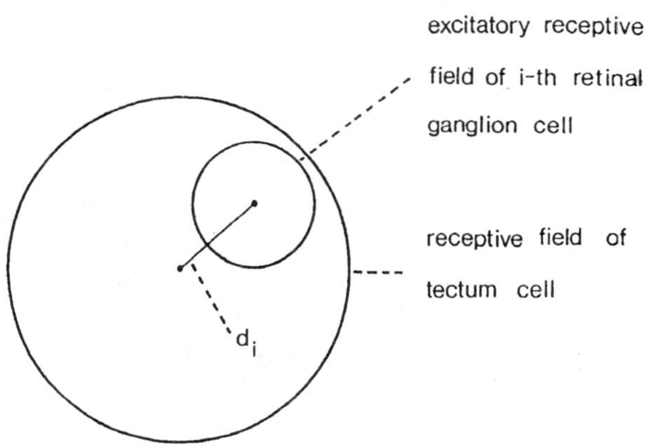

Figure 10. Determination of the distance d_i between the centers of the receptive fields of a tectal cell (large circle) and of the *ith* retinal ganglion cell (small circle) projecting to this tectal cell. The index i runs from 1 to N = number of retinal cells projecting to the tectal cell. The centers of the receptive fields of these retinal ganglion cells are assumed to be homogeneously distributed across the receptive field of the tectal cell.

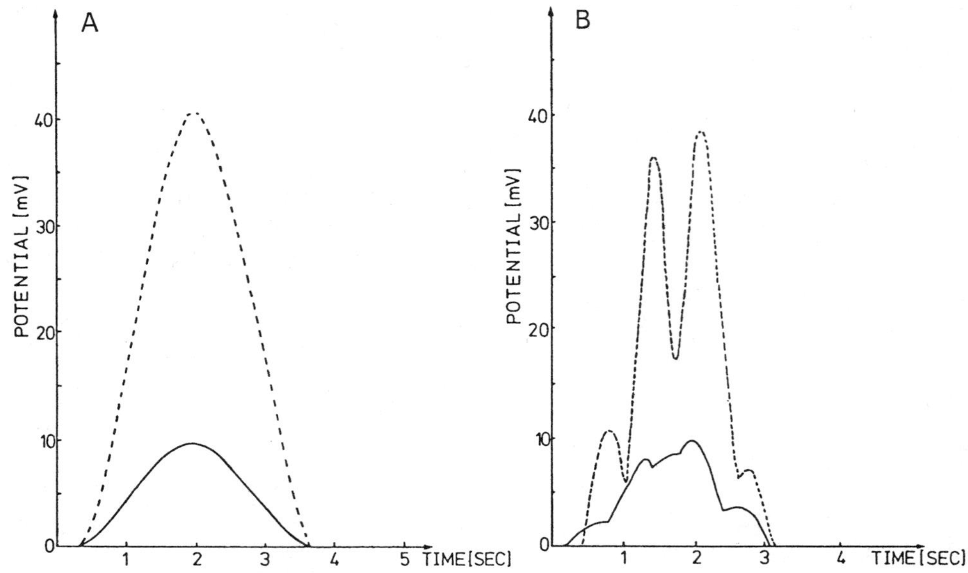

Figure 11. Simulated membrane potentials of model tectal cells described by Eqs. (11) and (12), with input from model retinal ganglion cells of class R2. Visual input to model R2-cells were dark squares with edge length 2° (continuous line) and 8° (dotted line), moving at v = 7.6°/s centrally through the receptive field of the model tectal cell. Parameters in Eqs. (11) and (12): a = 1, T* = 5 ms, d = 27°. Figures (A) and (B) differ in the degree (Δ) of overlap between receptive fields of neighbouring retinal ganglion cells projecting to the tectal cell. In (A) the degree of overlap was taken to be Δ = 1/1°, resulting in N ≈ 547 (see Eq. 10) retinal ganglion cells projecting to this model cell. In (B) the degree of overlap is Δ = 1/5°, resulting in N ≈ 22 retinal ganglion cells projecting to this model cell.

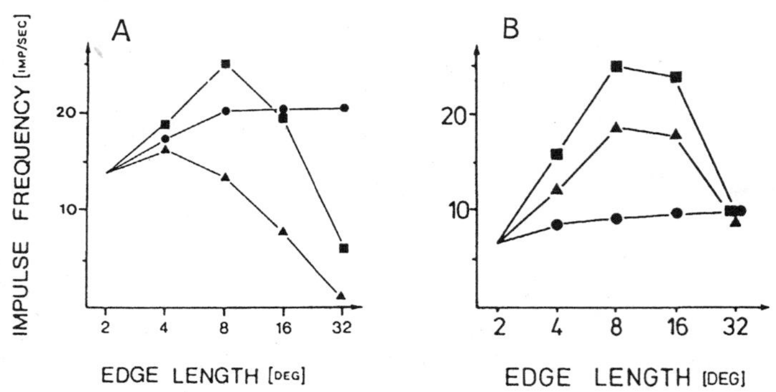

Figure 12. Results of computer simulations of the combined retina-tectum model, described by Eqs. (1)-(13). No lateral inhibition between tectal cells is assumed. (A) Output of model tectal cells receiving input from R2 retinal model cells. Same stimulus parameters as in Figs. 7 and 10A. Under these conditions essentially the tectal cell type T5.1 (S > H > V) is obtained. (B) Output of model tectal cells if input from R3 retinal model cells is used (parameters as in Fig. 8). In this way the tectal cell type T5.3 (S > V > H) is constituted.

the responses are rather similar to those observed with real T5.1 cells exhibiting the preference type S > H > V (at least for stimulus angles less than 16°). Note that the maximal response is obtained for squares with 8° edge length which coincides neither with the diameter (27°) of the receptive field of the model tectal cell nor with the maximum (4°) of the R2 cells.

We now assume that only R3 cells have contact to the tectal cell. Then the response characteristic of Fig. 12B is obtained in the computer simulations. In this case, the type T5.3 with preference S > V > H was successfully reconstructed.

By taking suitable mixtures of inputs both from R2 and R3 cells it is clearly possible to obtain intermediate configurations between Fig. 12A and B. Since the squares are the preferred stimuli both in T5.1 and T5.3 cells, in all these intermediate cases the squares lead to the largest impulse frequencies. In fact, in all our computer simulations of this basic tectal cell model we never obtained the sequence H > S > V characteristic for T5.2 cells. Therefore, it appears necessary to extend or modify this model in some way.

5.2 Extended Model: Lateral Inhibition in the Tectum

The optic tectum contains many interneurons which are generally believed to act inhibitory on neighboring cells, one candidate being the stellate cells. It seems reasonable to postulate some intratectal inhibition. Although such an assumption should not preclude inhibitory influences arriving from other brain regions, e.g., from the pretectum (Ewert 1968; 1984), these will be neglected here. We included inhibitory interactions among tectal cells as an extension of the basic model, leaving all its other properties unmodified. Recurrent lateral inhibition is assumed to act either directly between tectal cells or across mediating interneurons. Formally these

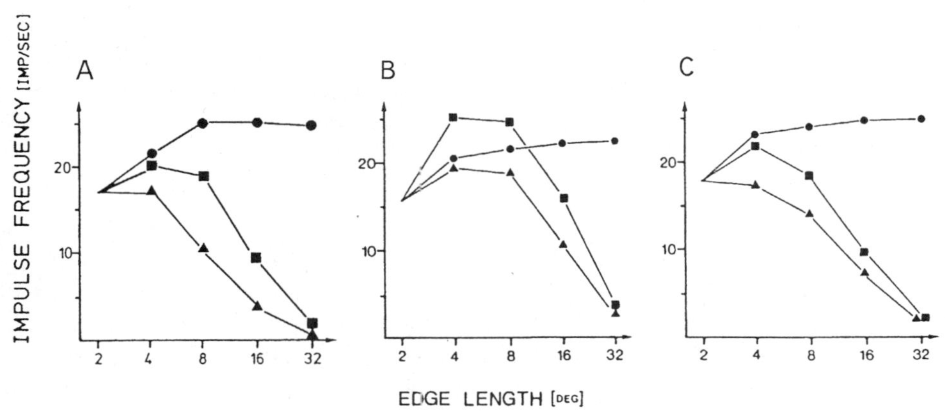

Figure 13. Results of computer simulations if lateral inhibition between tectal cells is additionally assumed. (A) Same conditions as in Fig. 11A with the exception that the inhibitory coefficient b (see Eq. 14) is sufficiently different from zero, here b = 0.04. This network structure produces the tectal cell type T5.2 (H > S > V). (B) Same conditions as in Fig. 11B, but now b = 0.16. Essentially type T5.1 (S > H > V) is obtained. If inhibition is sufficiently strong (e.g., b = 2) then the type T5.2 (H > S > V) results again (C).

two cases are not distinguished in our model. The extended model coincides with the basic model, with the exception of Eq. (12) which is to be replaced by the more general equation

$$T_3\, du_i(t)/dt = e_i(t) - u_i(t) - b \sum_{j=1}^{L} u_{j+}(t), \quad i=1,2,3,\ldots,L \tag{14}$$

Here the tectal cells involved in the inhibitory interactions are enumerated by the index i running from 1 to L. Besides the excitatory input e_i, the *ith* cell receives an inhibitory input which is the sum of the firing rates u_{i+}, of the L coupled tectal cells, multiplied by an inhibitory coupling coefficient b. Altogether Eq.(14) is a system of L coupled nonlinear differential equations describing recurrent lateral inhibition under non-autonomous conditions.

We solved the total system including Eq.(14) numerically. The same parameters as with the basic model were used, thus all parametric variations were restricted to the inhibitory coefficient b. If new phenomena appeared, they could only be due to the effect of lateral inhibition. Of course, if the inhibitory coefficient is very small, then the extended model behaves very similarly to the basic model (note that the basic model is obtained by choosing $b=0$); hence, in this case the same two cell types, T5.1 and T5.3, are produced as with the basic model.

However, the situation changes with larger values of b. In the first alternative, where the model tectal cells receive input only from model retinal cells of R2, it turned out that for $b=0.04$ a clear case of T5.2 response characteristic with preference type $H>S>V$ is generated. Numerical results are shown in Fig. 13A.

In the second alternative, all involved tectal cells receiving input only from R3 cells, the following result was obtained. For values of b not too large (e.g., for $b=0.16$) the model cells showed the response characteristic $S>H>V$ at least for stimulus extensions smaller than 16°; see Fig. 13B. This is a second possibility for the constitution of the type T5.1. For larger values of b (e.g., for $b=2$), again type T5.2 resulted (see Fig. 13C) which thus may be constituted in two different ways. A schematic representation of coupling types and the related response types is given in Fig. 14.

6 Discussion

We could show that it is possible to construct the basic response types observed in the optic tectum and believed to underlie the capability of amphibians to distingsuish between moving objects differing in configuration. The generation of these cell types can be understood on the basis of simple integrative neural operations and inhibitory interactions. In the case of tectal model cells receiving only input from R3 cells, variation of a single parameter, the inhibitory coupling coefficient b, led to the constitution of three different cell types.

Our model shows that a set of rather complex response properties of visual tectal cells can be reduced to very few parameters, namely to the properties of retinal input to tectal cells and to tectal cell coupling. Thus, a huge variety of response characteristics can be understood on the basis of a variation of these few parameters.

The presented model calls for a consideration of the so-called TP-phenomenon. This phenomenon was first observed by Ewert (1968) in *Bufo bufo*: A lesion in the thalamic pretectal (TP-)region led to a loss of "worm/antiworm" (H *vs.* V) discrimination, i.e. visual objects of any configuration elicited prey-catching with increasing size of a moving object. After such lesions, T5.2 characteristics in the toad's tectum transformed into a response property similar to the one obtained in TP-lesioned toad's prey-catching (Ewert & v. Wietersheim 1974b; see also Ewert and Schürg-Pfeiffer, this volume). Additionally, the receptive field sizes were enlarged from about 27° to more than 40° (Ingle 1983). The same "TP-phenomenon" was also described by Finkenstädt & Ewert (1983b) in *S. salamandra*. This phenomenon vanishes partly after several weeks and toads return to some "semi-normal" prey-selecting behavior (Ewert et al. 1983).

In their original model of prey recognition in the toad, Ewert & v. Seelen (1974) assume that the T5.2 properties are due to a subtractive inhibitory interaction

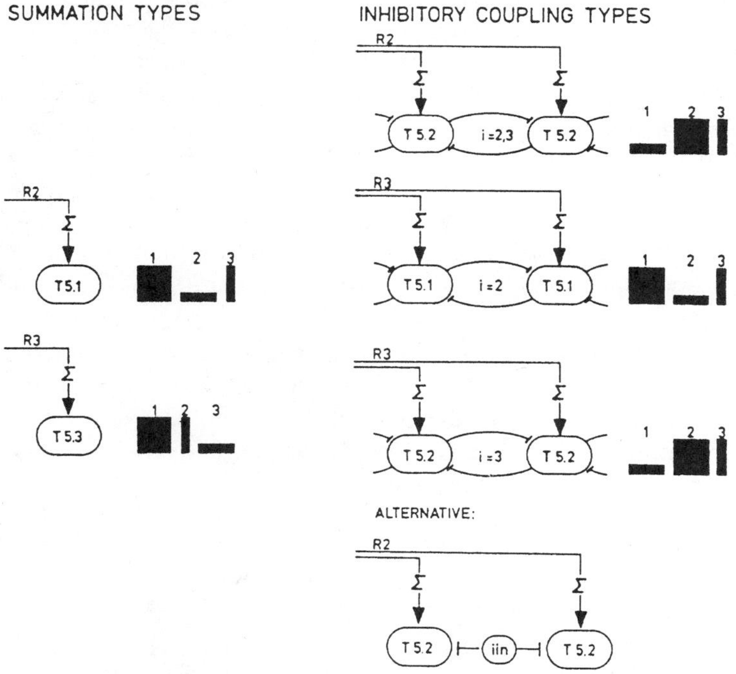

Figure 14. Graphical summary of the network architecture of retina-tectum model and resulting constitution of various preference types. On the left hand side response types constituted by mere summation of activity of retinal ganglion cells (RGC) are given, on the right hand side response types constituted by summation of RGC-output plus lateral recurrent inhibition between tectal cells are shown; (i) indicates the strength of inhibitory coupling expressed in arbitrary units. Row of black stimuli indicates the respective stimulus preference ranking from (1) to (3). R2 = RGC class-2, R3 = RGC class-3.

between tectal (T5.1) characteristics and pretectal (TH3) characteristics (Ewert & v. Seelen 1974). If this alone were the case then a destruction of the thalamic/pretectal region should turn T5.2 characteristics back into T5.1 ones. This, however, is not observed during recording from "disinhibited" T5.2 cells whose discharge rates increase with the area of a moving stimulus. The model by Lara et al. (1982), also based on Ewert's proposed pretecto-tectal inhibition, solves this problem with the consideration of tectal intrinsic reverberatory loops which, released from pretectal inhibition by a TP lesion, indeed can explain the neuroethological results (see also Arbib and Cervantes-Pérez, this volume).

Our model shows that T5.2 properties can also be explained by intrinsic tectal inhibition, i.e. through recurrent inhibition between T5.1 cells, without pretectal influence. We assume that the extrinsic inhibitory influence on which the "TP-phenomenon" is based is a non-specific one and hence exclude the possibility that the inhibitory interneurons are themselves influenced by feature sensitive thalamic pretectal cells. We propose that intrinsic tectal inhibition is mediated by stellate cells (Székely & Lázár 1976) and that this process is modulated by stimulus non-specific tonic input from pretectal neurons. Skékely & Lázár claim that such an inhibitory process is involved in the restriction of lateral spread of local activation over the whole tectum. It may well be that the extrinsic non-specific inhibitory influence consists of controlling and restricting the lateral spread of activity and along with this the receptive field size of tectal cells, whereas the intratectal inhibitory stellate cells produce the specific inhibitory effect among T5 cells. For a discussion on this topic see Ewert (this volume). Our model does not account for the TP-phenomenon seen after thalamic pretectal lesions. In order to check the specificity of extrinsic tectal inputs, a continuation of our model would have to include a pretectal modulatory mechanism, e.g., as proposed by Manteuffel (this volume). A model of a TP-modulated T5-base unit is presented by Betts (this volume).

Let us finally and briefly consider the relationship of the mathematical model of retina and optic tectum to other models of the amphibian retina and/or tectum in the literature. Some elements of our retina model may be found in the models of Grüsser et al. (1967), Ewert & v. Seelen (1974), and Varjú (1978). We consider our model as a continuation of these descriptions. Realism, simplicity, and flexibility were the guiding principles. We could not convince ourselves to simply take one of the earlier retina models to use it as input to the tectum model to be developed here. To our knowledge, we have the first mathematical theory coupling models for the retina and for the optic tectum as interacting unities.

Acknowledgement. This work was supported by the Deutsche Forschungsgemeinschaft. The paper has been published by the authors in a similar form in *Acta Biotheoretica* 36: 179-212 (1987) and is presented here with permission by Martinus Nijhoff/ DrW Junk Publishers.

7 References

An der Heiden U (1980) *Analysis of neural networks.* Springer-Verlag, Berlin Heidelberg New York

An der Heiden U, Roth G (1983) Cooperative neural processes in amphibian visual prey recognition. In: Basar E, Flohr H, Haken H, Mandell AJ (eds) *Synergetics of the brain.* Springer-Verlag, Berlin Heidelberg New York Tokyo

An der Heiden U, Roth G (1987) Mathematical model and simulation of retina and tectum opticum of lower vertebrates. *Acta Biotheoretica* 36: 179-212

Cervantes-Pérez F, Lara R, Arbib MA (1985) A neural model of interactions subserving prey-predator discrimination and size preference in anuran amphibia. *J Theor Biol* 113: 117-152

Ewert J-P (1968) Der Einfluß von Zwischenhirndefekten auf die Visuomotorik im Beute- und Fluchtverhalten der Erdkröte (*Bufo bufo* L) *Z Vergl Physiol* 61: 41-70

Ewert J-P (1969) Quantitative Analyse von Reiz-Reaktionsbeziehungen bei visuellem Auslösen der Beutefang-Wendereaktion der Erdkröte (*Bufo bufo* L). *Pflügers Arch* 308: 225-243

Ewert J-P (1974) The neural basis of visually guided behavior. *Sci Amer* 230: 34-42

Ewert J-P (1976) The visual system of the toad: behavioral and physiological studies on a pattern recognition system. In: Fite KV (ed) *The amphibian visual system: a multidisciplinary approach.* Academic Press, New York San Francisco London, pp 141-202

Ewert J-P (1984) Tectal mechanisms that underlie prey-catching and avoidance behaviors in toads. In: Vanegas H (ed) *Comparative neurology of the optic tectum.* Plenum Press, New York London, pp 247-416

Ewert J-P (1987) Neuroethology of releasing mechanisms: prey-catching in toads. *Behav Brain Sci* 10: 337-405

Ewert J-P, Hock FJ (1972) Movement sensitive neurons in the toad's retina. *Exp Brain Res* 16: 41-59

Ewert J-P, Seelen W v (1974) Neurobiologie und Systemtheorie eines visuellen Muster-Erkennungsmechanismus bei Kröten. *Kybernetik* 14: 167-183

Ewert J-P, Wietersheim A v (1974a) Musterauswertung durch Tectum- und Thalamus/Praetectum-Neurone im visuellen System der Kröte (*Bufo bufo* L). *J Comp Physiol* 92: 131-148

Ewert J-P, Wietersheim A v (1974b) Der Einfluß von Thalamus/Praetectum-Defekten auf die Antwort von Tectum-Neuronen gegenüber bewegten visuellen Mustern bei der Kröte (*Bufo bufo* L). *J Comp Physiol* 92: 149-160

Ewert J-P, Borchers H-W, Wietersheim A v (1978) Question of prey feature detectors in the toad's *Bufo bufo* (L) visual system: a correlation analysis. *J Comp Physiol* 126: 43-47

Ewert J-P, Burghagen H, Schürg-Pfeiffer E (1983) Neuroethological analysis of the innate releasing mechanism for prey-catching in toads. In: Ewert J-P, Capranica RR, Ingle DJ (eds) *Advances in vertebrate neuroethology.* Plenum Press, New York London, pp 413-475

Finkenstädt T, Ewert J-P (1983a) Processing of area dimensions of visual key stimuli by tectal neurons in *Salamandra salamandra. J Comp Physiol* 153: 85-98

Finkenstädt T, Ewert J-P (1983b) Visual pattern discrimination through interactions of neural networks: a combined electrical brain stimulation, brain lesion, and extracellular recording study in *Salamandra salamandra. J Comp Physiol* 153: 99-110

Grüsser O-J, Finkelstein D (1967) Analyse eines auf "Bewegungswahrnehmung" spezialisierten Neuronensystems in der Froschnetzhaut. In: Kroebel W (ed) *Fortschritte der Kybernetik.* Oldenbourg-Verlag, München, pp 833-96

Grüsser O-J, Grüsser-Cornehls U (1968) Neurophysiologische Grundlagen visueller angeborener Auslösemechanismen beim Frosch. *Z Vergl Physiol* 59: 1-24

Grüsser O-J, Grüsser-Cornehls U (1970) Die Neurophysiologie visuell gesteuerter Verhaltensweisen bei Anuren. *Verh Dtsch Zool Ges* 64: 201-218

Grüsser O-J, Grüsser-Cornehls U (1976) Neurophysiology of the anuran visual system. In: Llinas R, Precht W (eds) *Frog neurobiology*. Springer-Verlag, Berlin Heidelberg New York, pp 297-385

Grüsser O-J, Grüsser-Cornehls U, Licker MD (1968) Further studies on the velocity function of movement-detecting class-2 neurons in the frog retina. *Vision Res* 8: 1173-1185

Grüsser-Cornehls U (1984) The neurophysiology of the amphibian optic tectum. In: Vanegas H (ed) *Comparative neurology of the optic tectum*. Plenum Press, New York London, pp 211-245

Grüsser-Cornehls U, Himstedt W (1973) Responses of retinal and tectal neurons of the salamander (*Salamandra salamandra* L.) to moving visual stimuli. *Brain Behav Evol* 7: 145-168

Grüsser-Cornehls U, Himstedt W (1976) The urodele visual system. In: Fite KV (ed) *The amphibian visual system: a multidisciplinary approach*. Academic Press, London New York, pp 203-266

Herrick CJ (1948) *The brain of the tiger salamander Ambystoma tigrinum*. Univ Chicago Press

Himstedt W, Roth G (1980) Neuronal responses in the tectum opticum of *Salamandra* to visual prey stimuli. *J Comp Physiol* 135: 251-257

Ingle DJ (1973) Two visual systems in the frog. *Science* 181: 1053-1055

Ingle DJ (1980) Some effects of pretectum lesions on the frog's detection of stationary objects. *Behav Brain Res* 1: 139-163

Ingle DJ (1983) Brain mechanisms of visual localization by frogs and toads. In: Ewert J-P, Capranica RR, Ingle DJ (eds) *Advances in vertebrate neuroethology*. Plenum Press, New York London, pp 177-226

Lara R, Arbib MA (1985) A model of the neural mechanisms responsible for pattern recognition and stimulus specific habituation in toads. *Biol Cybern* 51: 223-237

Lara R, Cervantes F, Arbib M (1982) Two-dimensional model of retinal-tectal-pretectal interactions for the control of prey-predator recognition and size preference in amphibia. In: Amari S, Arbib MA (eds) *Competition and cooperation in neural nets*. Springer-Verlag, Berlin Heidelberg New York, pp 371-393

Levine MW, Shefner IM (1977) Variability in ganglion cell firing patterns: implications for separate "on" and "off" processes. *Vision Res* 17: 765-776

Manteuffel G (1985) Monocular and binocular optic inputs to salamander pretectal neurons: an intracellular recording and HRP-labelling study. *Brain Behav Evol* 27: 1-10

Pfeiffer E (1975) Musterauswertung durch retinale Ganglienzellen beim Frosch (*Rana esculenta* L). Diploma Thesis, Tech Univ Darmstadt

Roth G (1978) The role of stimulus movement patterns in the prey catching behavior of *Hydromantes genei* (Amphibia, Plethodontidae). *J Comp Physiol* 123: 261-264

Roth G (1982a) Responses in the optic tectum of the salamander *Hydromantes italicus* to moving prey stimuli. *Exp Brain Res* 45: 386-392

Roth G (1982b) Beuteerkennungsmechanismen im Tectum opticum von Amphibien - eine vergleichende Untersuchung. *Funkt Biol Med* 1: 90-98

Roth G (1986) Neural mechanisms of prey recognition: an example in amphibians. In: Feder ME, Lauder GV (eds) *Predator-prey relationships*. Univ Chicago Press, Chicago London, pp 42-68

Roth G (1987) *Visual behavior in salamanders*. Springer-Verlag, Berlin Heidelberg New York Tokyo

Roth G, Jordan M (1982) Response characteristics and stratification of tectal neurons in the toad *Bufo bufo* (L). *Exp Brain Res* 45: 393-398

Schürg-Pfeiffer E, Ewert J-P (1981) Investigations of neurons involved in the analysis of gestalt prey features in the frog *Rana temporaria*. *J Comp Physiol* 141: 139-152

Székely G, Lázár G (1976) Cellular and synaptic architeture of the optic tectum. In: Llinás R, Precht W (eds) *Frog neurobiology*. Springer-Verlag, Berlin Heidelberg New York, pp 407-434

Tsai H-J, Ewert J-P (1987) Edge preference of retinal and tectal neurons in common toads (*Bufo bufo*) in response to worm-like moving stripes: the question of behaviorally relevant "position indicators." *J Comp Physiol* 161: 295-304

Varjú D (1978) Excitation and inhibitory processes giving rise to the delayed response in the retinal ganglion cell of the frog. In: Heim R, Palm G (eds) *Theoretical approaches to complex systems*. Springer-Verlag, Berlin Heidelberg New York

The T5 Base Modulator Hypothesis: A Dynamic Model of T5 Neuron Function in Toads

BILL BETTS

Program in Neural, Informational, and Behavioral Sciences
University of Southern California
Los Angeles, CA 90089-0782, USA

Abstract. Prey-catching behavior in toads exhibits flexibility. Toads can be conditioned to regard objects which normally elicit escape behavior as prey. The range of acceptable sizes for prey varies with motivation. In spring, prey recognition is "exchanged" for female recognition introducing mating behavior. To account for this behavioral flexibility within a single dynamic model, T5 neuron function is modeled as a *T5 base modulator* consisting of two components. The first, called the *T5 base*, consists of a T5 unit and all of the influences upon it except for *thalamic pretectal (TP) inhibition* which constitutes the second component. The T5 base modulator is a TP modulated tectal structure in which the T5 base provides a basis for T5 neuron function while TP inhibition modulates that basis to achieve the desired characteristics (subclasses T5.1, T5.2, or T5.3) for an individual T5 neuron at a given time. A T5 neuron thus has the potential, e.g., to be either a prey or mate feature analyzer. In the absence of TP modulation, the T5 base is optimized for recognizing objects that are among the least prey-like objects that T5 neurons analyze. Three mechanisms for achieving this are explored. The T5 base modulator hypothesis does not argue against the division of T5 neurons into subclasses, but instead suggests that these subclasses can be regarded as (more or less) stable states of a modulated system capable of adaptability and changeability.

1 Introduction

Early attempts to determine the mechanism by which frogs and toads recognize prey focused on retinal ganglion cells as "bug detectors." Thus, Barlow (1953) suggested that R3 cells act as fly detectors while Lettvin et. al. (1959) suggested that R2 cells fulfill such a role. More recently, investigators have emphasized tectal integrative (Grüsser & Grüsser-Cornehls 1968, 1976) and interactive processes involving the optic tectum *and* thalamic pretectal (TP) region (Ewert 1968, 1987), noting that anuran prey-catching behavior is too complex to be explained by simple retinal bug detectors.

1.1 Behavioral Flexibility and Neuronal Tuning

Prey-catching in toads is flexible. For example, toads which are allowed to eat mealworms out of an experimenter's hand can be conditioned to respond to the moving-hand alone (Brzoska & Schneider 1978; Burghagen 1979; Finkenstädt, this volume). Heatwole & Heatwole (1968) have shown that the upper size threshold for acceptable prey increases with long-term food deprivation while the lower size threshold remains constant. In spring, prey-catching behavior decreases or fails to occur. Moving females release orienting, approaching, and clasping behavior in the male (Heusser 1960; Kondrashev 1976, 1987). A unique neural filter with this flexibility could have adaptive advantage over a system in which prey and mate are coded by distinct filter systems (Ewert 1987).

In considering the neural implementation of such a system, we note that ablation of the TP region results in "disinhibition" of prey-catching with animals snapping at objects much larger than normal prey (Ewert 1968; Ewert et al. 1970). The fact that these animals respond to an experimenter's hand and to other toads suggest parallels between the effects of TP ablation and the conditioning results and changes that occur during mating season described above. The fact that, in many species, males will attempt to clasp practically any moving object including other males during mating season (Wells 1977) reinforces these parallels.

Tectal T5 neurons have relatively small excitatory receptive fields (8-30°), are located in tectal layers 6 to 8, and are activated by moving configurational stimuli with objects about 8° in size being optimal (Ewert 1984). They have been divided into at least three subclasses in *Bufo bufo* (Ewert & v.Wietersheim 1974; Ewert et al. 1979a) and *Rana temporaria* (Schürg-Pfeiffer & Ewert 1981). The T5.1 and T5.2 subclasses are comparable to types I and II in *B. bufo*, first described by Ewert (1974), while T5.3 is comparable to T5 *delta*-type cells (Ewert et al. 1979a). T5.1 are sensitive to the entire area of a moving stimulus and, with respect to configurational cues, mainly to stimulus expansion in the direction of movement. This subclass may play a role in mate recognition; T5.2 are influenced by stimulus expansion both parallel and perpendicular to the direction of movement and function as prey selectors (Ewert 1987); T5.3 show response characteristics similar to pretectal TH3 neurons (Ewert 1971), i.e. they are sensitive to the entire area of a moving object and, regarding configural cues, mainly to stimulus expansion perpendicular to the direction of movement.

The disinhibition effect observed after TP ablation is reflected in the activity of T5 neurons (Ewert & v. Wietersheim 1974; see also Schürg-Pfeiffer, this volume). Ewert suggests that the characteristics of the disinhibited state represent general properties of T5 neurons:

> "Since after TP lesions T5 neurons with only "disinhibited" response characteristics are recorded from the optic tectum, it seems likely that these are general properties of T5 neurons and, hence, we might designate them T5(0). It is possible that in the intact brain different neuronal populations in the optic tectum are "produced" by differentially weighted inputs to these neurons from the thalamic-pretectal region. One population would have the response

characteristics of class T5(1), another of T5(2) neurons, and there may be still others that remain to be discovered." (Ewert 1984, p. 363).

If TP influence can "produce" a T5.1 or T5.2 neuron from one with the characteristics of T5.0, then it might be able to modify a T5.1 neuron, once produced, to be a T5.2 neuron. If a T5 neuron is flexible enough that it can be conditioned to respond to an object the size of an experimenter's hand, then that same neuron might be able to respond to a female toad particularly given the lack of specificity exhibited by males in detecting females. By "combining", e.g., prey feature and female feature analyses, the toad could make more efficient use of a population of neurons. Given that prey-catching decreases during mating season, that is, males in pursuit of females are not simultaneously interested in prey, this combination would result in no loss of functionality. Note that such a neuron, while optimized for prey selection, is no longer dedicated solely to that function and, therefore, might be more aptly described as an "approachable object selector", i.e., a "selector" required for the release of orienting and approaching responses by the toad toward a target.

1.2 Toward a "T5 Base Modulator"

We hypothesize that T5 neurons in the toad tectum are such "approachable object selectors," so far as they can assume T5.1 and T5.2 properties. In modeling T5 neurons, we start with an assembly of tectal neurons which we will refer to as the *T5 base*. The T5 base consists of a T5 unit and all of the influences upon it except for those from the TP region. The TP region modulates the T5 base to produce the characteristics of a T5.1, T5.2, or T5.3 neuron desired at a given time, thus allowing various kinds of feature analysis with one basic network. We hypothesize that there is a single type of *T5 base modulator* for all T5 neurons consisting of a T5 base and the modulation through TP (mediated) inhibition.

In considering the implementation of a system which only recognizes prey, one should focus on the structures and functions that allow it to recognize prey. With an "approachable object selector," however, the T5 base itself - in the absence of TP modulation - might be generalized for recognition of prey, females, and even abstract objects which the animal will never encounter in its environment, but which exist in the broad spatiotemporal domain to which T5 neurons are tuned. The important consideration is that, whatever the starting point provided by the T5 base, the TP region must be capable of modulating the T5 base to detect the object type desired at a given time. We hypothesize that, for the T5 base, the starting point is not prey at all but is somewhere in the continuum in which prey objects are included.

In this paper, we begin modeling T5 neurons in terms of the T5 base modulator by looking at the study of Ewert et al. (1979a) on directional sensitivity in T5 neurons. We will not explicitly model the role of T5 neurons in mate detection. Instead we focus on the worm-antiworm continuum of Ewert and show that the full range of T5 neuron responses to worm and antiworm stimuli can be accounted for in a single dynamic model of T5 neuron function.

Ewert et al. (1979a) recorded the response of T5 neurons to worm and antiworm stimuli traversing the center of their receptive fields at constant angular velocity in different directions. Worm and antiworm objects are constructed of

Table 1. Average discrimination values $D_{W,A}$ for each of four tectal neuron subclasses to worm and antiworm stimuli moved horizontally and vertically across the receptive field. (From Ewert et al. 1979a).

Response Type of T5 Neurons (n=42)	Horizontal Stimulus Movement [degrees] 0	180	Vertical Stimulus Movement [degrees] 90	270
alpha (n=11)	0.71	0.62	0.59	0.58
beta (n=13)	0.32	0.25	0.20	0.13
gamma (n=11)	0.44	0.34	0.03	-0.13
delta (n=7)	-0.24	0.08	-0.18	-0.38

cardboard; worm stimuli are rectangles elongated in the direction of movement and elicit prey-catching behavior while antiworm stimuli are rectangles elongated perpendicular to the direction of movement. Individual stimuli vary in size. The stimuli used in the above mentioned recordings were 2° x 8° and 2° x 20°. As a measure of the toad's ability to discriminate between worm and antiworm stimuli, the authors computed a discrimination value:

$$D_{W,A} = \frac{R_W - R_A}{R_W + R_A}$$

where R_W and R_A are the average discharge frequency measured in response to worm and antiworm stimuli, respectively. In Fig. 1, $D_{W,A}$ was plotted for T5 neurons as a function of the direction that the stimulus traversed the receptive field. Values plotted outside the octagons represent worm preference while those inside represent antiworm preference. Four types of T5 neurons were distinguished:

alpha-types preferred worm stimuli equally in all directions;

beta-types are similar to alpha with one sector in which there was no discrimination or antiworm was preferred;

gamma-types preferred worm stimuli in a given direction and the opposite direction and had antiworm or no preference in perpendicular directions;

delta-types preferred worm stimuli in only one direction and had antiworm or no preference in all other directions.

The *alpha*-types correspond to class T5.2 while *beta* and *gamma* correspond to T5.1, and *delta* match T5.3 (Ewert et al. 1979a; Ewert 1984).

The average discrimination values for each of the four classes for stimuli entering the receptive field of the T5 neuron from 0, 90, 180, and 270 degrees were computed and are shown in Table 1: *Alpha*-types have the greatest average worm preference followed by *beta*-types while *delta*-types have the greatest average antiworm preference. *Gamma*-types have an average worm preference between that of *alpha* and *beta* for horizontally traversing stimuli while they have a slight antiworm preference on average lying between that of *beta* and *delta* types for vertically traversing stimuli.

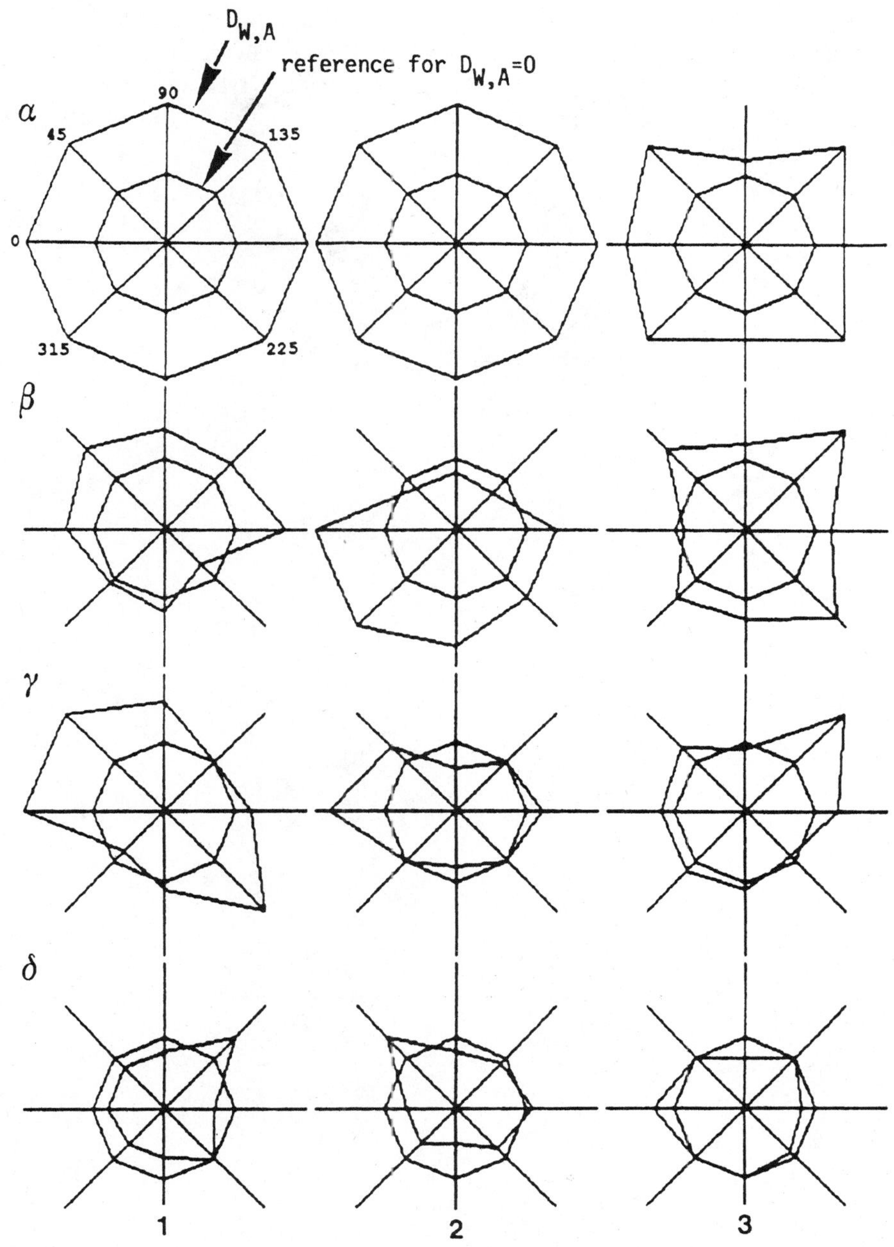

Figure 1. Discrimination values for black worm and antiworm stimuli traversing the receptive field of T5 neurons in various directions against a white background at constant visual angular velocity of 7.6°/s. The figure shows the response of four types of T5 neurons: *alpha*, *beta*, *gamma*, and *delta*. Radial axes run from a $D_{W,A}$ of -1 at the center of the plot through 0 at the octagon to 1 at the outer edge. Values outside the octagon indicate worm preference while those inside indicate antiworm preference. Directions are clockwise starting with 0 degrees at the left. All T5 neurons in the figure were recorded in response to 2° x 8° stimuli except *alpha*2, *alpha*3, and *beta*3. (From Ewert et al. 1979a).

Some cells were recorded over a long period of time (4-5 hours) as depicted in Fig. 2. It was shown that some of these cells vary in their worm/antiworm discriminative ability even to the extent of changing type according to the classification above. Thus, one cell changed from *beta* to *gamma* (and possibly even to

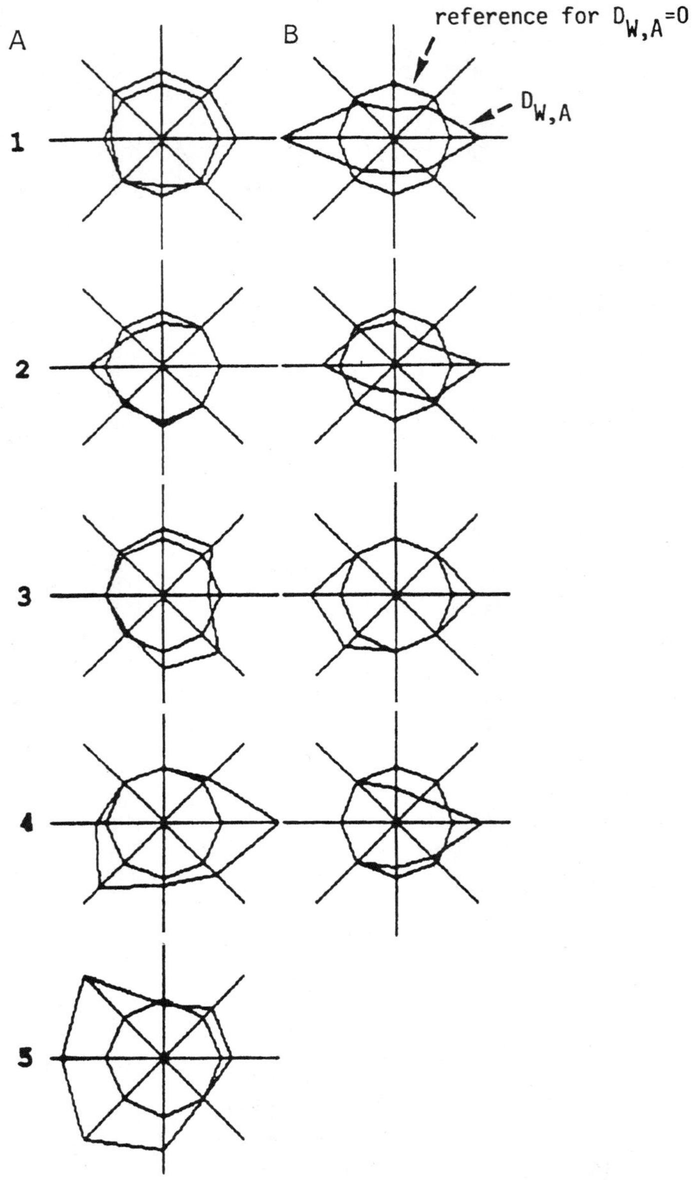

Figure 2. Variation in discrimination values $D_{W,A}$ over time during long-term recordings from two T5 neurons. See Fig. 1 for an explanation of the plots. For each cell, plots are separated by one hour. (A) was recorded in response to 2° x 20° stimuli while (B) was recorded in response to 2° x 8°. A1 and A2 are *beta*, A3 is *gamma*, A4 and A5 are *beta*. B1, B2, and B3 are *gamma* while B4 is *delta*. (From Ewert et al. 1979a).

alpha) while another cell changed from *gamma* to *delta*. The changes over time were seen primarily in the antiworm response while the worm response remained relatively stable. Ewert et al. (1979a) suggest that these changes were due to modulation by thalamic inhibition.

From a modeling perspective, these data can be divided into three aspects. First, there is the discriminative ability of a T5 neuron without reference to the direction of movement. For example, Fig. 1 *alpha*1 exhibits perfect worm preference while, in general, Fig. 1 *delta*3 exhibits slight antiworm preference. We refer to *discrimination irrespective of direction*. Second, there is variation of *discrimination with direction*. For example, Fig. 1 *gamma*2 exhibits greater worm preference for horizontally traversing stimuli than for vertically traversing stimuli. We shall refer to this aspect of the discrimination data as "symmetric directional sensitivity." Finally, there is variation in discrimination for a given direction and the opposite direction. For example, Fig. 1 *beta*2 exhibits antiworm preference for 90 degrees and worm preference for 270 degrees. We will refer to this as "asymmetric directional sensitivity." Asymmetric directional sensitivity is singled out because modeling it involves a temporal component in which excitatory and inhibitory influences are skewed with respect to time. Examples of such models include Barlow & Levick (1965) and Poggio & Koch (1987). Although the division of T5 neurons into classes depends on all three of these characteristics, the mechanisms underlying them are modeled separately. We will model all three aspects of discrimination. However, the focus of the model is on the first aspect: discrimination irrespective of direction.

In terms of the T5 base modulator hypothesis, it is shown that these data can be explained by an inhibitory mechanism that enhances worm discrimination by reducing antiworm response. It is suggested that the thalamus provides this inhibitory mechanism in agreement with Ewert et al. (1979a). This implies the need for the T5 base, an assembly of cells serving as input to a T5 neuron which generate maximum antiworm response. Three mechanisms for generating maximum antiworm response are modeled. The T5 base is then modulated by the thalamus to achieve the desired discrimination. The long-term recordings of T5 neurons suggest that a single type of T5 base may serve as the basis for all T5 neurons. Directional sensitivity in T5 neurons is modeled in terms of directional sensitivity in R3 neurons and by providing weights for thalamic neurons which vary with direction. These directional weights are meant to indicate that the thalamus supports a mechanism for directional sensitivity which may result from retinal inputs to the thalamus or from other sources. Evidence from the long-term recordings is discussed which supports this role.

1.3 Other Models in Perspective

An der Heiden & Roth (1983) presented a model of retinal ganglion cell and tectal neuron function. Retinal function is divided into three stages: A spatially local high pass filter, separate *on*- and *off*-channels for detecting objects on light and dark backgrounds, and spatial integration via near excitation and far inhibition. The model of tectal function consists of spatial summation of retinal inputs and lateral inhibition between tectal cells largely resembling the model of Ewert & v. Seelen (1974) with the main difference that An der Heiden & Roth (see this volume) do not consider extrinsic prosencephalic determinants, which in reality exist. Thus, the An der Heiden & Roth model does not include the influence of the TP region on T5 function and, in general, lacks the flexibility to account for the full range of T5 functionality. Lara et al. (1982) developed a neural model for the tectal column similar to the model discussed in this chapter. With this model, they were able to simulate prey selection from among several stimuli (Lara & Arbib 1982), stimulus-specific habituation (Lara & Arbib 1985; Lara, this volume), facilitation of prey-catching by short duration exposures to worm, antiworm, and square stimuli responses as a function of edge length, the directional invariance of prey-catching behavior (Lara et al. 1982), and size preference and response latency changes as a function of motivational state (Cervantes-Pérez et al. 1985). However, the behaviors modeled were restricted to prey-catching behavior and the model has not been used to simulate a full range of T5 responses (e.g., mating). The emphasis in this chapter is in building a single model which provides a flexible, dynamic framework in which to model all T5 functions rather than prey recognition *per se*.

2 A Model of T5 Neuron Function

2.1 General Model Anatomy

The model of T5 neuron function is presented as a set of mechanisms which account for various aspects of the data. Figure 3 shows the general anatomy for the model. Székely & Lázár (1976) suggest that optic inputs to the tectum terminate

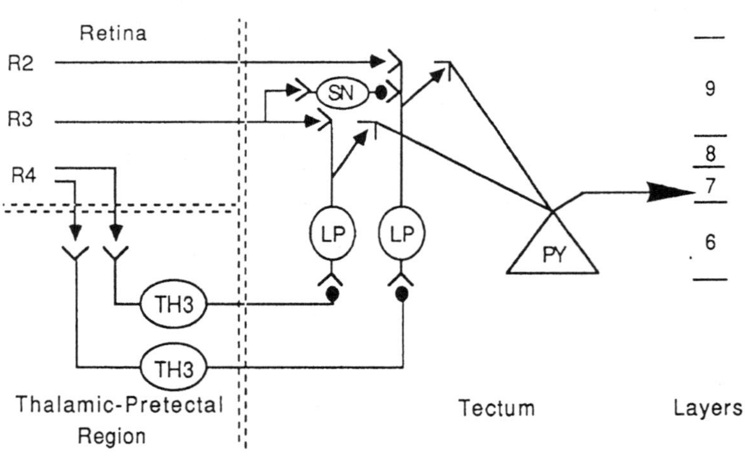

Figure 3. The general anatomy of the T5 neuron model. The dashed lines divide the diagram into retinal, tectal, and thalamic pretectal sections with the numbers at the right indicating tectal layers. The pyramidal (PY) cell serves as the T5 neuron and generates the output of the model which is used to calculate discrimination values. It receives input from many large pear (LP) cells which, in turn, receive excitatory input from R2 or R3 retinal ganglion cells and inhibitory input from stellate (SN) neurons and thalamic pretectal (TH3) neurons. The TH3 cells receive excitatory input from R4 retinal ganglion cells.

primarily on large pear (LP) cells. We model R2 and R3 retinal ganglion cells as terminating exclusively on LP cells. We assume that R2 and R3 cells are the primary visual input to T5 neurons as do Ewert et al. (1979a) and have not included R4 neurons in the tectal model. We model a one to one correspondence of R2 or R3 to LP cells and a density of R2 or R3 inputs of 0.5 cell per degree (see Section 4.1). R2 and R3 inputs are generated by a black box model of the retina which considers firing rate only and not the time course of firing. Worm and antiworm stimuli are modeled with a mask matrix containing ones for the appropriate edges of the stimulus (see Table 8, Appendix).

The LP cells receive lateral inhibition from stellate (SN) cells which receive R3 input. LP cells are also inhibited by TH3 cells from the TP region which are, in turn, excited by R4 retinal ganglion cells. Figure 3 shows TH3 cells inhibiting LP cells through basal dendrites although they may inhibit them through apical dendrites as well. All LP cells terminate on a single pyramidal (PY) cell. The PY cell is the T5 neuron and generates the output from which $D_{W,A}$ values are computed. The $D_{W,A}$ values are computed using values for worm and antiworm response averaged over the duration of stimulus exposure. The average diameter of the excitatory receptive field for T5 neurons was ERF = 26° (Ewert et al. 1979a) and an R2 or R3 input density of 0.5 cell per degree, therefore, results in approximately 169 R2 or R3 inputs to the PY cell.

2.2 A Mechanism for Enhancing Worm Discrimination

Ewert et al. (1979a) measured the discrimination value $D_{W,A}$ of T5 neurons to worm *vs.* antiworm stimuli. In the discussion that follows, we refer to the characteristics of the neurons displayed in that paper. Forty-two neurons were recorded of which the 15 displayed are representative of the rest of the neurons. As illustrated in Fig. 1, the maximum $D_{W,A}$ measured in these neurons is 1.0 (see Fig. 1 *alpha*1) which indicates perfect discrimination for worm stimuli, i.e., perfect worm preference. Perfect discrimination can only be achieved when there is either no response to the opposite stimulus (antiworm) in the input to the T5 neuron or that response is inhibited. The visual input to the tectum comes from retinal ganglion cells. Retinal ganglion cells respond to both worm and antiworm stimuli for stimulus sizes of 2° x 8° and 2° x 20°, the stimulus sizes used in Fig. 1. Therefore, the fact that some T5 neurons achieve perfect worm discrimination suggests *an inhibitory mechanism to enhance worm discrimination by reducing antiworm response*.

While the maximum discrimination value, both theoretically and experimentally, is 1.0, the minimum experimental value only reaches -0.5 (see Fig. 1 *delta*2, 270 degrees) in contrast to the theoretical minimum of -1.0. We can compute the antiworm response of a T5 neuron from the worm response and the discrimination value with the equation:

$$R_A = R_W \frac{1 - D_{W,A}}{1 + D_{W,A}}$$

The term $(1 - D_{W,A})/(1 + D_{W,A})$ is plotted in Fig. 4. This curve indicates that, as $D_{W,A}$ decreases, it becomes increasingly hard to achieve the necessary antiworm response given a reasonable worm response. For example, suppose the worm response is 6 impulses/s; then, in order to achieve $D_{W,A}$ values of -0.4, -0.5, and -0.6, the antiworm response must be 14, 18, and 24 impulses/s, respectively. If T5 cells fire in the range of 0 to 20 impulses/s, a $D_{W,A}$ value of -0.5 is the maximum that can be achieved.

The curve showing R_W as a function of R_A and $D_{W,A}$ is a mirror image of the curve in Fig. 4 reflected about $D_{W,A}=0$. Clearly, perfect worm preference cannot be achieved by increasing R_W in the presence of R_A. Even a $D_{W,A}$ of 0.8 would require R_W to be 9 times R_A. Therefore, whereas the maximum worm preference implies an inhibitory mechanism, the maximum antiworm preference is probably a natural result of the responses of retinal ganglion cells and the firing ranges of T5 neurons and no special influences are needed. T5 neurons are "asymmetric" in having a mechanism for increasing worm preference with no analogous mechanism for increasing antiworm preference.

We will assume that the inhibitory mechanism which is necessary to generate maximum worm preference acts only by inhibiting the antiworm response of a T5 neuron although it could have an effect on worm response as well. If we isolate this mechanism, we are left with the other factors that influence a T5 neuron including retinal ganglion cell inputs, tectal interneurons, and extratectal influences. We will refer to the T5 neuron and all the influences acting upon it except for the inhibitory mechanism discussed above as the T5 base. This *T5 base* has some inherent discrimination value of its own which can be changed by the influence of the inhibitory mechanism to increase worm discrimination. Since the inhibitory mechanism can only increase worm discrimination, this inherent discrimination must provide as great or greater antiworm preference (less worm preference) than the discrimination value it will have under the influence of the inhibitory mechanism. If the inhibitory mechanism is capable of modulating the discrimination of the T5 base

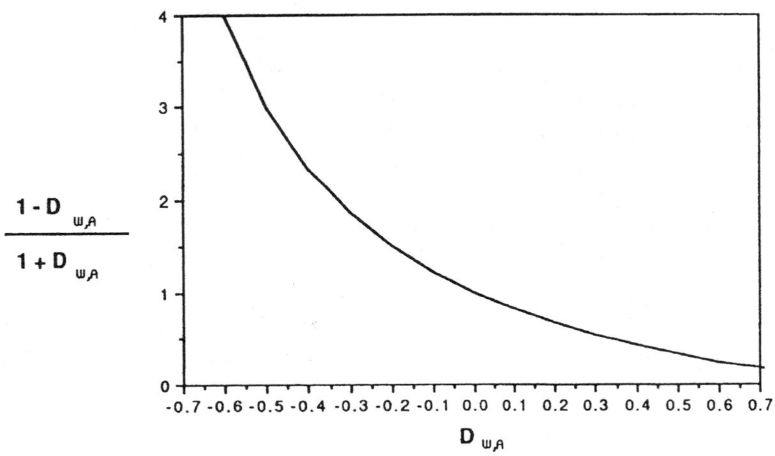

Figure 4. The curve plots the term $(1 - D_{W,A})/(1 + D_{W,A})$ which indicates the ratio of antiworm to worm response necessary to achieve a given $D_{W,A}$. The lowest $D_{W,A}$ values observed for T5 neurons were -0.5 indicating a R_A/R_W ratio of 3.

over time then, *the T5 base must, by itself, generate the maximum antiworm preference that it is capable of obtaining.*

Ewert et al. (1979a) showed that certain T5 neurons can change their discrimination over time even to the extent of changing types. Thus, one cell changed from *beta* to *gamma* (and possibly to *alpha*) while another changed from *gamma* to *delta*. They observed that the changes were primarily in antiworm rather than worm response and suggested that these changes over time were mediated by thalamic inputs. This suggestion was supported by observations that configurational selectivity in prey-catching behavior (Ewert 1968) and the response of T5 neurons (Ewert & v. Wietersheim 1974; Ewert et al. 1974) is decreased by lesions to the caudal dorsal thalamic pretectal region (see also Schürg-Pffeifer, this volume).

The fact that the change in T5 discrimination over time seen by Ewert et al. (1979a) was primarily in antiworm response supports our earlier assumption that the inhibitory mechanism inhibits antiworm response only and has little or no effect on worm response. We model this inhibitory mechanism with the TH3 cell which receives exclusively R4 input and have not considered the R3 input known to influence the TP region (see Section 3.6).

If, in the dorsal thalamus, there is an inhibitory mechanism which can enhance worm preference by inhibiting antiworm response and this mechanism can change T5 neuron discrimination over time even to the extent of changing types, then the hypothesis that single T5 bases generate the maximum antiworm response for individual T5 neurons may be extended across T5 neuron types. *There may be a single type of T5 base for all T5 neurons.* In other words, T5 bases for all T5 neurons may generate the maximum antiworm discrimination (approximately $D_{W,A} = -0.5$) seen in any T5 neuron and these bases are then modulated by the thalamus to achieve the discrimination values seen in individual neurons. We will model T5 neurons assuming a single type of T5 base for all T5 neurons.

We have now separated the influences contributing to the discrimination characteristics of a T5 neuron into a T5 base and an inhibitory mechanism. We have suggested that the inhibitory mechanism may be in the thalamus and that it acts to inhibit antiworm response only. We have also suggested that T5 bases must therefore achieve not only the maximum antiworm discrimination for the T5 neuron with which the base is associated but the maximum antiworm discrimination for all T5 neurons.

The next question is how does the T5 base achieve this? Although there are numerous possibilities, three will be explored. It is possible that the maximum antiworm preference is generated by

(i) the retinal ganglion cell inputs,
(ii) the stimulus configuration, and/or
(iii) the influence of tectal interneurons.

2.3 Antiworm Preference Generated by Retinal Ganglion Cell Inputs

Firing rates for R2 and R3 neurons in response to worm and antiworm stimuli are shown in Table 2 while discrimination values are shown in Table 3. The basic firing rates for R2 and R3 neurons were taken from Ewert et al. (1978). Ewert et al. (1979b) observed that R3 neurons had a greater response to 2° x 6° antiworm stimuli

Table 2. Firing rates for R2, R3, and R4 neurons as a function of edge length in worm- or antiworm-like stripes of 2° width. All data are from Ewert et al. (1978) except the ones for R3 indicated by *) which were calculated as explained in the Appendix (Section 4); 0 degree refers to horizontally and 90 degrees to vertically traversing stimuli.

	Stimulus Edge Length [°]			
	6	8	16	20
R2-Worm at 0 degrees	29	32	30	28
R2-Antiworm at 0 degrees	33	26	8	3
R3-Worm at 0 degrees	21	20	18	19
R3-Antiworm at 0 degrees	32	37	18	10
R3-Antiworm at 0 degrees	25	29	14	8 *)
R3-Antiworm at 90 degrees	34	39	19	11 *)
R4-Worm at 0 degrees	11	11	14	14
R4-Antiworm at 0 degrees	18	18	18	17

traversing the receptive field vertically than to those traversing horizontally (see Table 2). It is assumed that a similar difference would be seen for other stimulus sizes and the R_A and $D_{W,A}$ values of R3 for 2° x 8°, 2° x 16°, and 2° x 20° stimuli are adjusted accordingly (see Section 4.1). Note that this is one mechanism by which the model implements symmetric directional sensitivity.

Ewert et al. (1979a) recorded from T5 neurons in response to 2° x 8° and 2° x 20° stimuli. For both of these stimulus sizes, R3 show greater antiworm preference than R2 (see Table 3). Therefore, in order to achieve maximum antiworm preference in terms of retinal ganglion cell input, we postulate that the primary functional input to a T5 base is R3 cells.

Ewert et al. (1979a) computed average $D_{W,A}$ values for each of the four T5 cell types (see Table 1). It is interesting to note that the maximum antiworm preference in response to horizontally and vertically traversing stimuli is $D_{W,A}$ = -0.24 and -0.38, respectively, in *delta*-types and these values correspond closely with those for R3 neurons in response to 2° x 8° stimuli: -0.18 and -0.32 (see Table 3). Since none of the *gamma*- or *delta*-types shown in Ewert et al. (1979a) were recorded in response to 2° x 20° stimuli (see Figs. 1 and 2) except for one cell during one hour of a long-term recording (see Fig. 2A3), it is suggestive that all the *delta*-types were recorded in response to 2° x 8° stimuli, which indeed was the case (Ewert, pers. communication). On the other hand, the neurons with maximum antiworm response to 2° x 20° stimuli were *beta*-types (see Fig. 1 *beta*3) and their average discrimination values in response to horizontally and vertically traversing stimuli of 0.25 to 0.32 and 0.13 to 0.20, respectively, correspond somewhat with the values for R3 neurons in response to 2° x 20° stimuli: 0.41 and 0.27. The correspondence is not as close for these neurons as it is

Table 3. $D_{W,A}$ values of R2 and R3 ganglion cells to worm and antiworm stimuli based on responses in Table 2.

	Stimulus Edge Length [°]			
	6	8	16	20
R2 at 0 degrees	-0.06	0.08	0.58	0.81
R3 at 0 degrees	-0.21	-0.30	0.00	0.31
R3 at 0 degrees	-0.09	-0.18	0.12	0.41 *)
R3 at 90 degrees	-0.24	-0.32	-0.03	0.27 *)

for *delta*-types in response to 2° x 8° stimuli, but this may be due to the fact that *beta*-types in response to 2° x 8° and 2° x 20° stimuli were mixed when computing the average discrimination values in Table 1. If *gamma*- and *delta*-types are not recorded in response to 2° x 20° stimuli, these observations strengthen the argument for R3 cells as the primary functional input to a T5 base.

2.4 Antiworm Preference Generated by Stimulus Configuration

Like R3 neurons, spatiotemporal properties of stimulus configuration will tend to favor antiworm stimuli. However, while R3 neurons exhibit greater antiworm preference in response to 2° x 8° than to 2° x 20° stimuli, the opposite is true for the effect of stimulus configuration.

As mentioned above, T5 neurons have an average receptive field diameter of ERF = 26° and are likely to receive hundreds of retinal inputs via LP cells. The total visual input to a T5 neuron is a function both of the response of individual retinal ganglion cells and the number of cells firing in response to the stimulus. Retinal ganglion cells respond to edges moving through their receptive fields which are oriented perpendicular to the direction of movement. Thus, when a worm stimulus traverses the center of the receptive field, only the leading and trailing edges which are 2° each generate a response. However, an antiworm stimulus of the same size generates a response along its leading edge which is more than 2°. (In this model, we assume that, due to neuronal adaptation, the trailing edge of antiworm stimuli does not generate responses. Note that, if it did, it would strengthen the argument being made.) For example, for 2° x 8° stimuli, the worm configuration generates response in retinal ganglion cells along 4° of edge (2° leading and 2° trailing) while the antiworm configuration generates response along 8° of leading edge (see Table 8, Appendix in Section 4). For 2° x 20° stimuli, this difference is even greater. This will result in the cumulative retinal input to a T5 neuron being biased toward antiworm preference relative to the response of individual retinal ganglion cells and this bias will be greater for 2° x 20° than 2° x 8° stimuli. Note that a more sophisticated retinal model (e.g., An der Heiden & Roth 1983) would result in variation of the retinal response along the leading edge of antiworm stimuli. However, the response would tend to increase at the ends of the stimulus while the responses in Table 8 are measured with the stimulus centered with respect to the receptive field. Thus, a more sophisticated retinal model would probably further bias the model in favor of antiworm preference.

The combined effect of R3 input and stimulus configuration on discrimination in T5 neurons is modeled in Fig. 5. Following the convention of Lara et al. (1982), the state of tectal neurons is defined by a membrane potential (in lower case) and a firing rate (in upper case). Model neurons interact only via firing rates while the firing rate of a neuron depends on its membrane potential. The fundamental equation describing membrane potential dynamics is:

$$\tau \frac{dm(t)}{dt} = -m(t) + S_m(t)$$

Where τ is the membrane time constant, $m(t)$ is the membrane potential, and $S_m(t)$ is the weighted sum of excitatory and inhibitory inputs. A weight $w_{a.b}$ exists on a connection from a neuron of type a to one of type b.

As stated above, each R3 cell excites a LP cell. The following equations describe membrane potential dynamics and firing rate for the LP cell:

$$\tau_{lp} \frac{dlp_{i,j}(t)}{dt} = -lp_{i,j}(t) + w_{r3.lp} R3_{i,j}(t)$$

$$LP_{i,j}(t) = ramp(lp_{i,j}(t))$$

where $w_{r3,lp}$=0.2 for all model simulations, τ_{lp}=0.1, and $ramp(x)$=MAX(0,x). In general, τ is 0.1 for excitatory neurons and 0.01 for inhibitory neurons. These values were chosen so that the model would respond quickly to inhibitory influences.

LP cells, in turn, excite a single PY cell which is the T5 neuron:

$$\tau_{py} \frac{dpy(t)}{dt} = -py(t) + \sum_{i=1}^{M} \sum_{j=1}^{M} w_{lp.py} LP_{i,j}(t)$$

$$PY(t) = ramp(py(t))$$

where M is the number of LP inputs to the PY neuron in the X or Y dimension. Thus, M^2 is the total number of LP inputs. In this model, M=13 and M^2=169 as mentioned above. $w_{lp.py}$=1.8 for all model simulations while τ_{py}=0.1.

Figure 5A models the response to 2° x 8° stimuli while Fig. 5B models the response to 2° x 20° stimuli. The discrimination values are similar for both stimulus sizes due to the competing effects of R3 input and stimulus configuration and the fact that averaging PY cell response over the duration of stimulus exposure decreases R_W for 2° x 20° stimuli relative to 2° x 8° stimuli.

As discussed in the last section, the T5 data may support greater antiworm preference for 2° x 8° than 2° x 20° stimuli thus favoring the role of R3 $D_{W,A}$ in determining T5 response. However, the competing effect of stimulus configuration must be taken into account. In order to model the effect of R3 $D_{W,A}$ alone on T5 discrimination, we postulate a lateral inhibitory effect. SN neurons receive input from R3 cells:

$$\tau_{sn} \frac{dsn_{i,j}(t)}{dt} = -sn_{i,j}(t) + w_{r3.sn} R3_{i,j}(t)$$

$$SN_{i,j}(t) = ramp(sn_{i,j}(t))$$

where $w_{r3.sn} = 0.2$ for all model simulations and $\tau_{sn} = 0.01$.

SN cells, in turn, inhibit adjacent LP cells and the modified LP cell membrane potential equation is:

$$\tau_{lp} \frac{dlp_{i,j}(t)}{dt} = -lp_{i,j}(t) + w_{r3.lp} R3_{i,j}(t) - \sum_{k=\pm 1} \sum_{l=\pm 1} w_{sn.lp} SN_{i+k,j+l}(t)$$

Figures 5C and D model the effect of SN cell inhibition in response to 2° x 8° and 2° x 20° stimuli, respectively. Figure 5C is analogous to Fig. 5A while Fig. 5D is analogous to Fig. 5B. $w_{sn.lp}$ weights of 0.3 and 0.42 were chosen in Figs. 5C and D, respectively, so that the $D_{W,A}$ values match the R3 $D_{W,A}$ to 2° x 8° and 2° x 20°

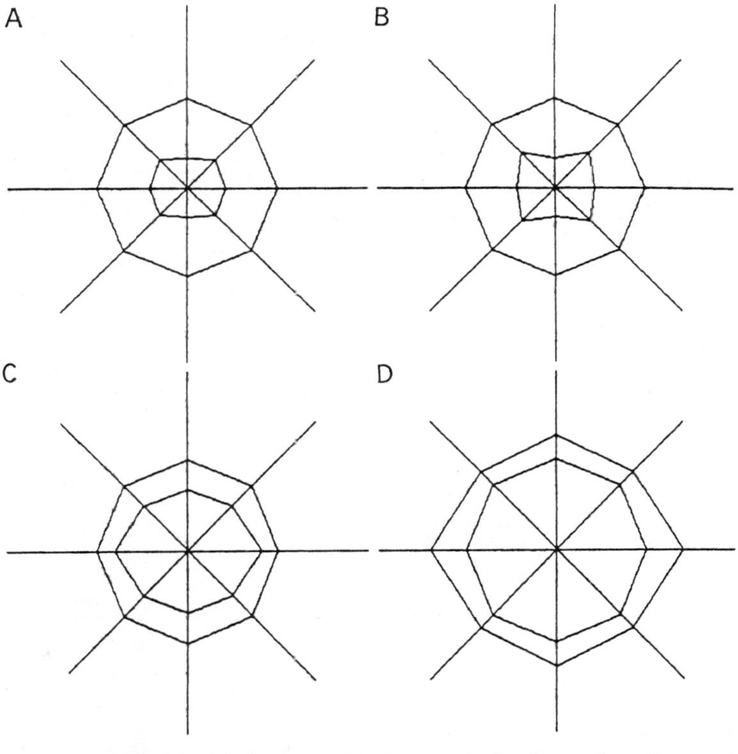

Figure 5. Model results demonstrating the effect of R3 input and stimulus configuration on the T5 base. (A) and (B) model T5 bases in response to 2° x 8° and 2° x 20° stimuli, respectively. In (C) and (D), stellate cell inhibition is added resulting in the discrimination values that reflect discrimination values of R3 neurons for 2° x 8° and 2° x 20° stimuli, respectively.

stimuli. We will use the model simulated in Figs. 5C and D to represent the T5 base in the absence of thalamic pretectal (TP) inhibition throughout the rest of the paper. In Fig. 6, thalamic pretectal inhibition is added to the models simulated in Figs. 5C and D. The thalamic TH3 cell receives excitatory input from R4 cells:

$$\tau_{th3} \frac{dth3_{i,j}(t)}{dt} = -th3_{i,j}(t) + w_{r4.th3} R4_{i,j}(t)$$

$$TH3_{i,j}(t) = binary(th3_{i,j}(t) - 15.0)$$

where $w_{r4.th3} = 1.0$ for all model simulations, $\tau_{th3} = 0.01$, and $binary(x) = 1$ if $x > 0$ and 0 otherwise.

R4 cells have greater response to antiworm than to worm stimuli for all stimulus sizes greater than 2°. The threshold of the TH3 cell (15.0) is set between the worm and antiworm input levels from the R4 cell such that the TH3 cell will fire in response to antiworm and not to worm stimuli. $w_{r4.th3}$ is related to the TH3 cell threshold and was chosen to be 1.0 to clarify the choice of threshold in terms of the R4 firing rates

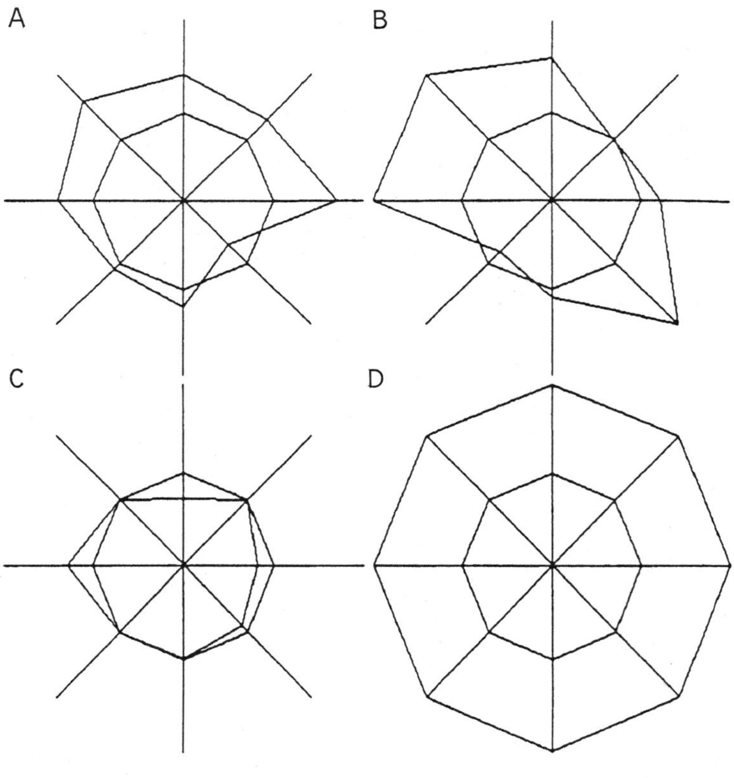

Figure 6. Thalamic inhibition is added to the models simulated in Figs. 5C and D in order to simulate neurons in Fig. 1. A-D are models of *beta1*, *gamma1*, *delta3*, and *alpha2* respectively. (A), (B), and (C) simulate response to 2° x 8° stimuli processed by the T5 base in Fig. 5C while (D) simulates response to 2° x 20° stimuli processed by the T5 base in Fig. 5D.

Table 4. TH3 weights ($w_{th3.lp}$) for neurons (A-D) shown in Fig. 6 A-D as a function of movement direction in degrees.

	Direction of Stimulus Movement [degrees]							
	0	45	90	135	180	225	270	315
A	2.0	3.8	3.2	2.8	2.9	0.1	2.5	2.0
B	4.1	5.5	3.8	1.6	1.5	5.5	2.1	0.8
C	1.7	1.65	0.33	1.65	0.0	1.3	1.7	1.65
D	1.0	2.0	2.0	2.0	1.0	2.0	2.0	2.0

in Table 2. The $w_{r3.lp}$ and $w_{r3.sn}$ were set at 0.2, as stated earlier, in order to balance TH3 inhibition.

The LP cell membrane potential equation is modified to receive inhibitory input from TH3 cells:

$$\tau_{lp}\frac{dlp_{i,j}(t)}{dt} = -lp_{i,j}(t) + w_{r3.lp}R3_{i,j}(t) - w_{th3.lp}(\phi)TH3_{i,j}(t) - \sum_{k=\pm 1}\sum_{l=\pm 1} w_{sn.lp}SN_{i+k,j+l}(t)$$

where Ø is the direction of the stimulus in degrees. Variation of $w_{th3.lp}$ by direction is the mechanism by which the model implements *asymmetric directional sensitivity*. It is also, along with R3 firing rates (see Table 2), a mechanism for *symmetric directional sensitivity*. The fact that this weight varies with direction is meant to indicate that the TP region supports a mechanism for directional sensitivity and not that the TH3 cell itself implements this mechanism. Evidence supporting this role is discussed in Section 3.7.

With thalamic inhibition, the T5 base can be modulated to achieve any worm preference greater than that shown in Figs. 5C and D. However, no analogous mechanism exists for increasing antiworm preference. Therefore, the neurons simulated in Fig. 6 were chosen from Fig. 1 so that the antiworm preference never exceeded the antiworm preference in Figs. 5C and D. Figure 6A models Fig. 1 *beta*1, Fig. 6B models Fig. 1 *gamma*1, and Fig. 6C models Fig. 1 *delta*3 in response to 2° x 8° stimuli while Fig. 6D models Fig. 1 *alpha*2 in response to 2° x 20° stimuli. The weights for TH3 cells used in Fig. 6 are shown in Table 4.

2.5 Antiworm Preference Generated by Tectal Interneurons

Tectal interneurons may also contribute to the antiworm preference of the T5 base. For example, antiworm preference could be increased by setting up a competition between R2 and R3 inputs. R2 cells have a greater response to worm stimuli than to antiworm stimuli for sizes equal to or greater than 8° (see Table 2). If

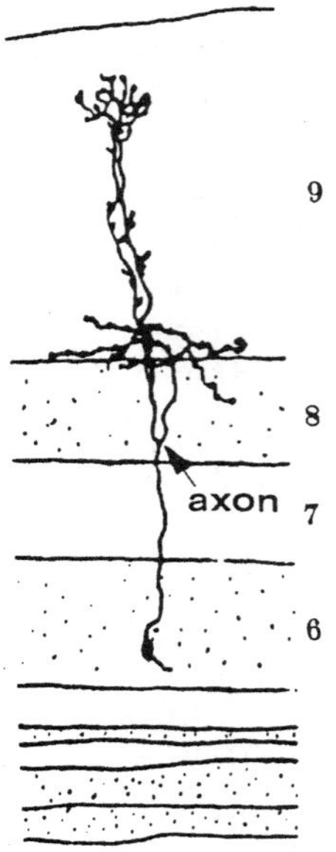

Figure 7. Tectal layer 6 pear-shaped cell which receives input from tectal lamina B and, therefore, possibly from R2 retinal ganglion cells. Its axon terminates in lamina F among R3 retinal ganglion cell terminals. If this cell were inhibitory, it could set up a competition between R2 and R3 cells which would have the effect of enhancing the antiworm discrimination. (From Antal et al. 1986).

R2 inputs inhibited R3 responses to worm stimuli to a greater extent than antiworm responses, they would cause the $D_{W,A}$ to drop.

Since R2 input is excitatory, this competition could only be set up by acting through a tectal inhibitory interneuron. Antal et al. (1986) stained the layer 6 pear cell shown in Fig. 7. This cell appears to receive input from tectal lamina B which contains R2 optic terminals and terminates in lamina F which contains R3 optic terminals. If this cell were inhibitory, it might have the effect of setting up the competition between R2 and R3 inputs suggested above. We will refer to cells of this type as inhibitory large pear (ILP) cells in the model. ILP cells are modeled as receiving R2 input:

$$\tau_{ilp} \frac{dilp_{i,j}(t)}{dt} = -ilp_{i,j}(t) + w_{r2.ilp} R2_{i,j}(t)$$

$$ILP_{i,j}(t) = binary(ilp_{i,j}(t) - 27.0)$$

where $w_{r2.ilp} = 1.0$ to provide clarity in choosing a threshold as with TH3 cells and $\tau_{ilp} = 0.01$.

The ILP cell threshold (27.0) is set so that ILP cells respond only to worm stimuli. The PY cell membrane potential equation is modified to receive input from all ILP cells:

$$\tau_{py} \frac{dpy(t)}{dt} = -py(t) + \sum_{i=1}^{M} \sum_{j=1}^{M} w_{lp.py} LP_{i,j}(t) - \sum_{i=1}^{M} \sum_{j=1}^{M} w_{ilp.py}(i,j) ILP_{i,j}(t)$$

where M has the same meaning as before. Note that, whereas there is a single weight for all LP cells, each ILP cell has its own weight and, if this weight is set to 0, the ILP cell is effectively disconnected. As mentioned earlier, $w_{lp.py}=1.8$ and this value was chosen to balance the effect of ILP cells.

In Fig. 8, simulations using the model with the addition of ILP cells are shown. Again, the T5 base model is the one simulated in Figs. 5C and D with the addition of ILP cells. Unlike the simulations in Fig. 6, neurons were chosen for Fig. 8 in which the $D_{W,A}$ values in some directions were less than those shown in Figs. 5C and D and, therefore, requiring the antiworm enhancing effect of ILP cells. Figure 8A models Fig. 1 *delta*1 in which the $D_{W,A}$ is less than Fig. 5C for 180 (-0.3 *vs*. -0.19) degrees. In Fig. 8B, Fig. 1 *delta*2 is modeled with $D_{W,A}$ less than Fig. 5C for 0 (-0.3 *vs*. -0.19) and 270 (-0.5 *vs*. -0.33) degrees. Figure 8C models Fig. 1 *beta*3 for 2° x 20° stimuli with $D_{W,A}$ less than Fig. 5D for 0 (-0.1 *vs*. 0.41), 90 (0.2 *vs*. 0.27), and 180 (0.2 *vs*. 0.41) degrees. Finally, Fig. 8D models Fig. 1 *alpha*3 with a $D_{W,A}$ less than Fig. 5D for 90 (0.2 *vs*. 0.27) degrees. The TH3 weights used in this Figure are shown in Table 5. A matrix of weights for individual ILP cells is shown in Table 6 for each of the T5 neurons modeled in Fig. 8 reflecting the antiworm enhancements indicated above.

2.6 Long-Term Changes in T5 Neurons

Ewert et al. (1979a) recorded some T5 neurons for periods of 4 to 5 hours. While the *alpha*-type neurons (resembling class T5.2) - so far investigated - were stable in their worm/antiworm preference, among *beta* and *gamma* types (resembling class T5.1) response properties could change over time. Two of the latter long-term recordings are depicted in Fig. 2. The authors show that changes in $D_{W,A}$ over time were due primarily to changes in the antiworm response. In the current model, TP inhibition of the T5 base is exclusively inhibition of antiworm response. Therefore, in order to model the changing properties of T5 neurons during long-term recordings, we must start with a T5 base that exhibits the maximum antiworm response in each direction observed during the entire recording. Note that the worm response of the T5 neuron will not change during a long-term recording and, therefore, the worm response reflects the response of the T5 base in isolation from TP inhibition. For the long-term recording in Fig. 2B, the worm and antiworm responses from neuron E24-8-78-2 (see Fig. 3D in Ewert et al. 1979a) have been reproduced in Fig. 9A. The model must start with a T5 base that approximates the worm response in Fig. 9A1 while achieving the maximum antiworm response seen during the recording in all directions. Starting with the T5 base in Fig. 5C, it was necessary to increase antiworm preference for 90, 135, 270, and 315 degrees with the addition of ILP cells. The worm response of the resulting T5 base is shown in Fig. 9B1 while the antiworm response is shown in Fig. 9B2 and the $D_{W,A}$ is shown in Fig. 9B3. The ILP weights for the T5 base in Fig. 9B are shown in Table 7.

Figure 9C presents the antiworm response resulting from thalamic inhibition of the T5 base in Fig. 9B for each of the four hour recordings shown in Fig. 2B. Figure 10 shows the resulting $D_{W,A}$ values. The TH3 weights for Figs. 9C and 10 are shown in Table 7.

Table 5. TH3 weights ($w_{th3.lp}$) for neurons (A-D) shown in Fig. 8 A-D as a function of movement direction in degrees.

	Direction of Stimulus Movement [degrees]							
	0	45	90	135	180	225	270	315
A	0.8	0.9	1.0	3.3	0.5	1.8	0.3	0.3
B	0.0	3.3	1.8	1.5	1.5	0.9	0.0	0.3
C	0.19	0.7	0.12	2.0	0.41	1.0	0.2	0.4
D	0.3	2.0	0.05	2.0	0.0	2.0	0.25	2.0

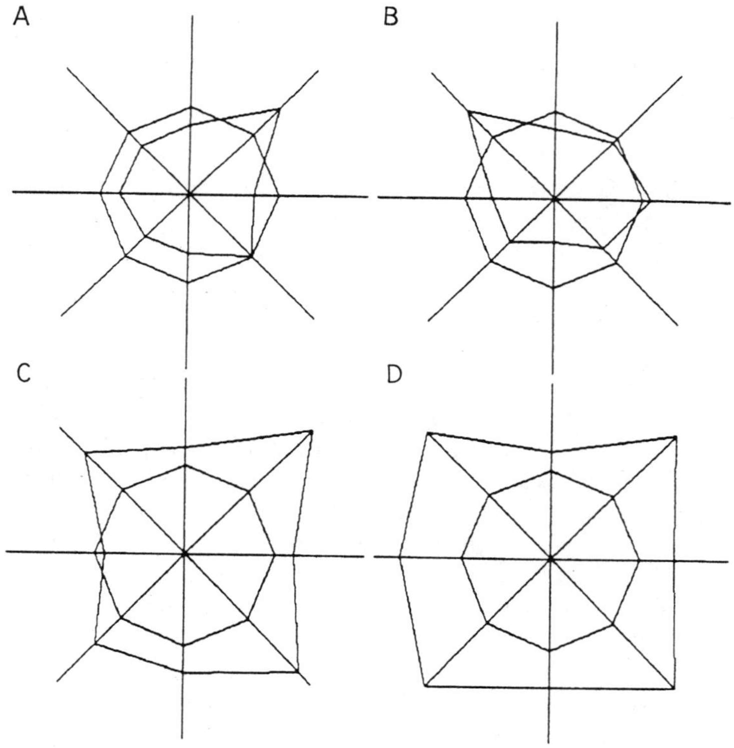

Figure 8. Model simulation with ILP cells to enhance antiworm preference in the T5 base modulator. (A) and (B) model *delta1* and *delta2* of Fig. 1, respectively, in response to 2° x 8° stimuli based on the T5 base in Fig. 5C, while (C) and (D) model *beta3* and *alpha3* of Fig. 1, respectively, in response to 2° x 20° stimuli based on the T5 base in Fig. 5D.

Table 6. ILP weights ($w_{ilp.py}$) for neurons shown in Fig. 8. Stimulus movement directions are clockwise starting with the left as 0 degrees.

A:

0	0	0	0	0	0	0	0	0	0	0	0	0
0	0	0	0	0	0	0	0	0	0	0	0	0
0	0	0	0	0	0	0	0	0	0	0	0	0
0	0	0	0	0	0	0	0	0	0	0	0	0
0	0	0	0	0	0	0	0	0	0	0	0	0
2	2	2	2	2	2	2	2	2	2	2	2	2
2	2	2	2	2	2	2	2	2	2	2	2	2
2	2	2	2	2	2	2	2	2	2	2	2	2
0	0	0	0	0	0	0	0	0	0	0	0	0
0	0	0	0	0	0	0	0	0	0	0	0	0
0	0	0	0	0	0	0	0	0	0	0	0	0
0	0	0	0	0	0	0	0	0	0	0	0	0
0	0	0	0	0	0	0	0	0	0	0	0	0

B:

0	0	0	0	0	2	2	2	0	0	0	0	0
0	0	0	0	0	2	2	2	0	0	0	0	0
0	0	0	0	0	2	2	2	0	0	0	0	0
0	0	0	0	0	2	2	2	0	0	0	0	0
0	0	0	0	0	2	2	2	0	0	0	0	0
1	1	1	1	1	2	2	2	1	1	1	1	1
1	1	1	1	1	2	2	2	1	1	1	1	1
1	1	1	1	1	2	2	2	1	1	1	1	1
0	0	0	0	0	2	2	2	0	0	0	0	0
0	0	0	0	0	2	2	2	0	0	0	0	0
0	0	0	0	0	2	2	2	0	0	0	0	0
0	0	0	0	0	2	2	2	0	0	0	0	0
0	0	0	0	0	2	2	2	0	0	0	0	0

C:

0	0	0	0	0	1	1	1	0	0	0	0	0
0	0	0	0	0	1	1	1	0	0	0	0	0
0	0	0	0	0	1	1	1	0	0	0	0	0
0	0	0	0	0	1	1	1	0	0	0	0	0
0	0	0	0	0	1	1	1	0	0	0	0	0
4	4	4	4	4	4	4	4	4	4	4	4	4
4	4	4	4	4	4	4	4	4	4	4	4	4
4	4	4	4	4	4	4	4	4	4	4	4	4
0	0	0	0	0	1	1	1	0	0	0	0	0
0	0	0	0	0	1	1	1	0	0	0	0	0
0	0	0	0	0	1	1	1	0	0	0	0	0
0	0	0	0	0	1	1	1	0	0	0	0	0
0	0	0	0	0	1	1	1	0	0	0	0	0

D:

0	0	0	0	0	1	1	1	0	0	0	0	0
0	0	0	0	0	1	1	1	0	0	0	0	0
0	0	0	0	0	1	1	1	0	0	0	0	0
0	0	0	0	0	1	1	1	0	0	0	0	0
0	0	0	0	0	1	1	1	0	0	0	0	0
0	0	0	0	0	1	1	1	0	0	0	0	0
0	0	0	0	0	1	1	1	0	0	0	0	0
0	0	0	0	0	1	1	1	0	0	0	0	0
0	0	0	0	0	1	1	1	0	0	0	0	0
0	0	0	0	0	1	1	1	0	0	0	0	0
0	0	0	0	0	1	1	1	0	0	0	0	0
0	0	0	0	0	1	1	1	0	0	0	0	0
0	0	0	0	0	1	1	1	0	0	0	0	0

2.7 The Model

In summary, a PY cell, which is the T5 neuron, receives R3 input via LP cells. R3 and LP cells have a density of 0.5 cell per degree resulting in 169 R3 inputs to the PY cell based on the 26° ERF of T5 neurons. R3 inputs provide the maximum antiworm preference, in general, that is seen in T5 neurons. The effect of stimulus configuration is normalized by SN cell lateral inhibition. SN cells receive R3 input and inhibit LP cells. Finally, competition between R2 and R3 inputs mediated by ILP cells serves to increase antiworm preference in some cases. ILP cells receive R2 input and inhibit the PY cell directly. These cells form the T5 base. The T5 base is modulated by TP inhibition from TH3 cells. TH3 cells receive R4 input and terminate

Table 7. TH3 ($w_{th3.lp}$) weights for neurons in Figs. 9C and 10 and ILP ($w_{ilp.py}$) weights for the T5 base shown in Fig. 9B.

	Direction of Stimulus Movement [degrees]							
	0	45	90	135	180	225	270	315
1	4.5	1.5	0.1	1.1	2.5	1.1	0.8	1.1
2	1.9	1.5	1.7	0.0	2.5	1.5	0.1	0.0
3	2.5	1.9	2.5	1.9	2.2	1.9	2.5	2.9
4	1.0	1.9	0.9	0.0	2.4	1.6	1.8	1.9

0	0	0	0	0	2	2	2	0	0	0	0	0
0	0	0	0	0	2	2	2	0	0	0	0	0
0	0	0	0	0	2	2	2	0	0	0	0	0
0	0	0	0	0	2	2	2	0	0	0	0	0
0	0	0	0	0	2	2	2	0	0	0	0	0
0	0	0	0	0	2	2	2	0	0	0	0	0
0	0	0	0	0	2	2	2	0	0	0	0	0
0	0	0	0	0	2	2	2	0	0	0	0	0
0	0	0	0	0	2	2	2	0	0	0	0	0
0	0	0	0	0	2	2	2	0	0	0	0	0
0	0	0	0	0	2	2	2	0	0	0	0	0
0	0	0	0	0	2	2	2	0	0	0	0	0
0	0	0	0	0	2	2	2	0	0	0	0	0

on LP cells. They inhibit the antiworm response of the T5 base. The equations for the model are summarized in the Appendix in Section 4.

3 Discussion

3.1 The T5 Base Modulator Hypothesis

The *T5 base modulator* hypothesis states that most T5 neuron functions can be accounted for with a single dynamic model. Three subhypotheses were presented. First, it was postulated that T5 neurons can be functionally divided into a T5 base which provides a basis for T5 response and a TP inhibitory influence that modulates the T5 base to generate the desired response characteristics at a given time. Second, we hypothesized that there is a single type of T5 base for all T5 neurons and that, therefore, any individual T5 neuron should be capable of a range of prey and mate recognition depending on TP influences. Third, the hypothesis was made that the T5 base, in the absence of TP inhibition, maximizes antiworm preference.

The first hypothesis resulted from attempting to account for the range of $D_{W,A}$ values recorded by Ewert et al. (1979a). Whereas the maximum antiworm response was regarded as not requiring special influences, the fact that T5 neurons achieve perfect worm preference indicates an inhibitory influence. Alternative possibilities exist. T5 neurons that achieve perfect worm preference represent only a portion of all T5 neurons. It may be that the retinal inputs to these neurons do not respond to

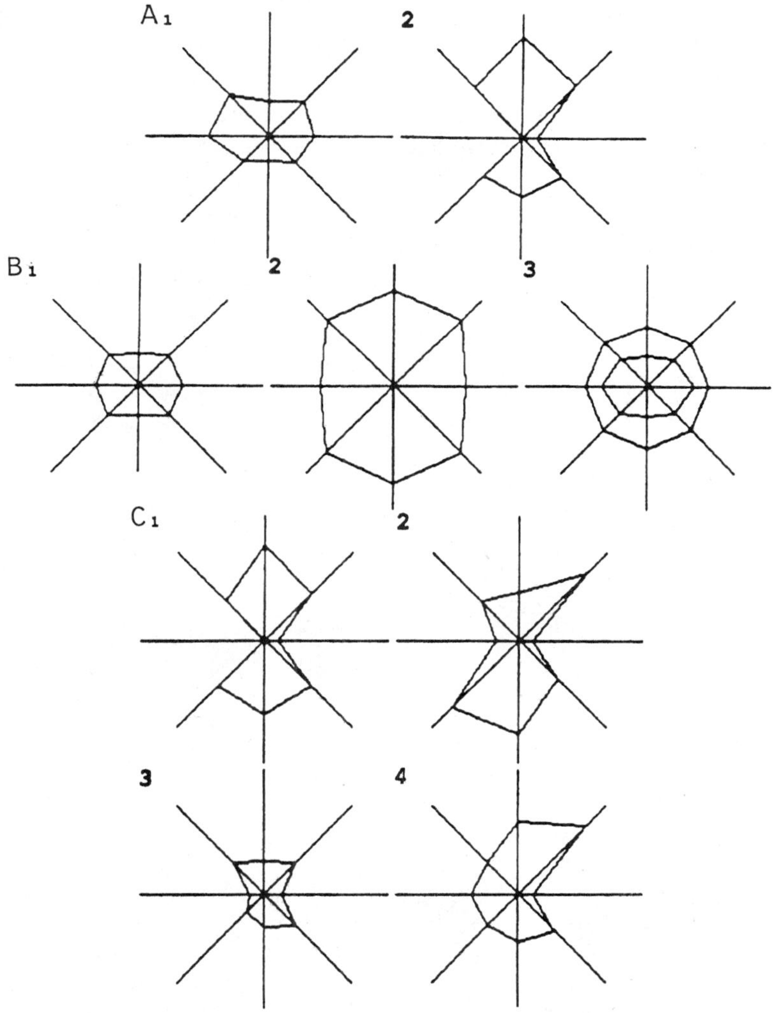

Figure 9. Model simulations of long-term changes in configural sensitivity from the T5 neuron in Fig. 2B. (A) shows the response to worm (A1) and antiworm (A2) recorded by Ewert et al. (1979a) which generate the $D_{W,A}$ values seen in Fig. 2B1. (B) shows simulations of the response to worm (B1) and antiworm (B2), and $D_{W,A}$ (B3) for the T5 base used to simulate these long-term changes. (C) shows the response to antiworm during the four hours shown in Figs. 2B1-4, respectively. Response to worm remains as shown in B1. (A is from Ewert et al. 1979a).

antiworm stimuli even though the average antiworm response of both R2 and R3 cells is greater than 0 for stimulus sizes of 2° x 8° and 2° x 20°. However, we have argued that T5 neurons are likely to receive hundreds of retinal inputs (see Section 4.1) and it is not likely that all of them fail to respond to antiworm stimuli. Even if those neurons exhibiting perfect worm preference represent an aberrant subclass, many T5 neurons that do not exhibit perfect worm preference also require an inhibitory influence. For example, the discrimination values for R2 and R3 cells in response to 2° x 8° stimuli are 0.08 and -0.30, respectively. Many of the T5 neurons recorded in response to 2° x 8° stimuli have discrimination values greater than this. Finally, it is possible that T5 neurons have a threshold that would discriminate the input generated by worm *vs*. antiworm stimuli. However, it is difficult to explain how such a mechanism would work since, for example, R3 cells prefer antiworms for 2° x 8° stimuli and worms for 2° x 20° stimuli while T5 neurons achieve perfect worm discrimination for both stimulus sizes (see Figs. 1 *alpha*1 and 1 *alpha*2).

Regarding the TP modulatory influence, note that two quite different influences have been discussed: one which modifies a T5 base modulator with T5.2 prey recognition characteristics to assume T5.1 mate recognition characteristics (see

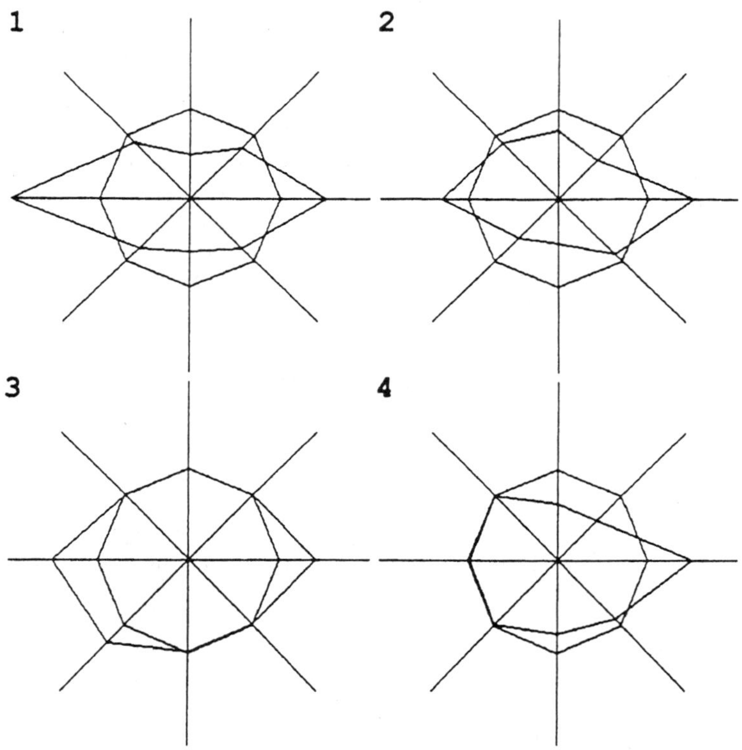

Figure 10. Model discrimination values produced from the response to worm in Fig. 9B and response to antiworm in Figs. 9C1-4 during a simulation of the long-term recording. These discrimination values simulate those in Figs. 2B1-4.

Introduction) and another which modulates worm/antiworm discrimination by inhibiting antiworm response. The first is long-term remaining relatively stable for several weeks while the second can change over a period of hours as evidenced by long-term recordings of T5 neurons. Anatomically, it has been shown that the tectum receives diverse thalamic inputs from hypothalamic/preoptic, anterior thalamic, posterior thalamic/pretectal (posterocentral, P; lateral posterodorsal, Lpd), and large-celled pretectal nuclei (see contributions by Lázár and Finkenstädt, this volume). These inputs enter several tectal layers (3, 5, and 9; Székely & Lázár 1976). It might be appropriate for those inputs entering layer 9 to mediate long-term stable modulation since they could prevent certain visual inputs from entering tectal circuitry at all. Short-term modulation might be mediated through layer 3 and 5 inputs since these inputs could modulate information that had already entered tectal circuitry.

Since the T5 base modulator hypothesis infers various modes of modulation, further research, particularly on the neurotransmitters found in these cells, will help to clarify their role in pretecto-tectal function. With respect to transmitters, it is known that sauropside homologues of P and Lpd nuclei contain different types of peptidergic neuroregulators (Reiner et al. 1982) which in the anuran's tectum probably give rise to corresponding laminated "peptidergic bands" (Kuljis & Karten 1982). To date at least three types of neurotransmitters are discussed for mediating thalamic/pretecto-tectal inhibition: enkephalin (Reiner et al. 1982; Lázár, this volume), GABA (Ewert 1987), and acetylcholine (Stevens 1973; Ewert et al. 1974).

3.2 T5 Subclasses and the T5 Base Modulator

The hypothesis that there is a single type of T5 base for all classes of T5 neurons is supported, in part, by recordings of T5 neurons over time (Ewert et al. 1979a). Figure 2B shows an example of transition over time from *gamma* (T5.1) to *delta* (T5.3). Although an example of a transition from *beta* or *gamma* (T5.1) to *alpha* (T5.2) has not (yet) been observed, T5 neurons did achieve perfect worm preference in individual directions during long-term recordings and one cell achieved perfect worm discrimination in all directions, if worm and antiworm responses were measured as the product of firing frequency and the number of spikes (see Ewert et al. 1979a, Fig. 5A5). In fact, one cell changed from a $D_{W,A}$ of -0.2 to 1.0 in one direction in one hour (Fig. 2A3-4, 180 degrees), thus covering almost the entire range of all neurons recorded (-0.5 to 1.0).

It should be emphasized that the hypothesis presented in this chapter does not argue against the division of T5 neurons into subclasses. Instead it suggests that the subclasses should be regarded as more or less stable states of a modulated system rather than in terms of hardwired connections. During long-term recordings (Ewert et al. 1979a), T5.2 neurons appeared resistant to change suggesting that this subclass represented a more stable state of the T5 system than, for example, T5.1. This makes sense, in terms of the T5 base modulator hypothesis, since T5.2 neurons represent maximal TP inhibition of the T5 base and this maximal inhibition is likely to produce a stable state.

One way to differentiate between T5 subclasses as independent populations and as different states of the same system would be in terms of anatomy. If they represent separate populations, it might be possible to distinguish them either by cell type or the layer they are located in the tectum. Matsumoto et al. (1986) recorded T5-type neurons in the tectum and then injected them with cobalt-lysine (see also Matsumoto, this volume). Fourteen T5.1 neurons were either pear-shaped or large ganglionic with pear cells located in layers 6 and 8 and large ganglionic cells located in layers 7 and 8. A single T5.2 neuron was found to be a pyramidal neuron located in layer 6. Three T5.3 neurons were large ganglionic cells in layer 8. Finally, two T5.4 cells were recorded from layer 8; one was a pear cell and the other was a large ganglionic neuron. Although results are preliminary, further research of this type would contribute to evaluating the T5 base modulator hypothesis. Even if class T5.2 neurons were pyramidal cells, their adaptive recognition properties could take advantage of the dynamic nature of a T5 base modulator.

3.3 Retinal Inputs and the T5 Base Modulator

The third hypothesis presented in this chapter is that the T5 base, in the absence of TP inhibition, maximizes antiworm preference. In terms of retinal input, we hypothesized that this can best be modeled by pure R3 input. The ability to generate worm preference was modeled purely in terms of TP modulation without considering R2 inputs since compositions of R2 and R3 inputs could account for some of the discrimination values observed but could not account for the maximum worm preference. However, the analysis of stimuli could potentially be refined if TP inhibition was capable of not only inhibiting a T5 base but changing the composition of R2 and R3 inputs to it. In Betts (1988), we model selective extratectal modulation of R2 and R3 inputs to a T5 base and show that, by utilizing the different response characteristics of R2 *vs*. R3 cells, this selective modulation can modify a T5 neuron from a specific prey selector to a selector of less specific stimuli, including perhaps mates.

Tsai & Ewert (1987) showed that the preference of T5.2 cells for the leading and trailing edges of worm stimuli corresponds most closely to the edge preference of R3 cells. They conclude that *off*-dominated R3 cells are responsible for the edge preference while R2 cells are responsible for area evaluation in T5 neurons. This suggests a mechanism in which R2 cells can influence the magnitude of T5 response without influencing its time course. Mechanisms similar to that provided by the ILP cell might provide such an influence particularly if ILP cells terminate presynaptically on R3 terminals rather than directly on the PY cell.

3.4 Tectal Interneurons

We have postulated a competition between R2 and R3 inputs to T5 neurons caused by ILP cells. Currently there is no evidence to confirm or deny the existence of inhibitory pear cells. The anatomy of the pear cell in Fig. 7 suggests that it can modulate R3 inputs in terms of R2. If the cell is excitatory, that would result in an R2/R3 cooperation while, if it is inhibitory, an R2/R3 competition would result as postulated in this chapter. Further research on the neurotransmitters found in tectal interneurons will help to clarify their role in tectal function.

3.5 Ethological Implications

It was argued that, if a T5 neuron can be modulated to respond to an experimenter's hand which presumably results from changes in size preference of single neurons (see also Ewert and Finkenstädt, this volume) rather than from switching between cell populations, then it should be able to respond to a female toad. Wells (1977) divides the reproductive behavior of anurans into two basic patterns: explosive breeding and prolonged breeding. Examples of species with explosive breeding patterns include *R. pipiens, R. temporaria, R. esculenta, R. japonica, B. americanus,* and *B. marinus* while species with prolonged breeding patterns include *R. catesbeiana, R. clamitans,* and *B. calamita*. *Bufo bufo*, the species studied by Ewert et al. (1979a), are categorized as explosive breeders. For explosive breeders, the breeding period is a few days to a few weeks while, for prolonged breeders, it is more than a month. Explosive breeders breed in dense aggregations which may include hundreds of individuals. Males search and compete for females at the breeding site. Males will approach and attempt to clasp almost any moving object within a range of absolute size (see Kondrashev 1976, 1987). They apparently bearly discriminate visually between males and females. When they clasp other males, the male being clasped gives a stereotyped release call.

In contrast, for prolonged breeders, males call from stationary positions and vocalizations help to maintain intermale spacing. Vocalizations play an important role in attracting females as indicated by experiments in which females approach loudspeakers playing conspecific calls. There are no examples of active searching among prolonged breeders. Males do not rush forward and clasp every frog that moves nearby and the notion that males attempt to clasp any moving object of appropriate size does not hold for many prolonged breeders. In several species, a male will only clasp a female when she initiates physical contact, in some cases by jumping on the male.

Wells (1977) suggests that it is more important for explosive breeders to be quick than efficient while the reverse is true for prolonged breeders. For example, *B. bufo* return to traditional breeding grounds every year and have a breeding period of 6 to 14 days. Once males clasp females, they may remain in amplexus, the mating embrace, for 1 to 14 days. Therefore, once a female is in amplexus with a male, she may remain unavailable for the entire breeding season. This suggests that a male, arriving at the breeding site, must be concerned with rapidly acquiring a mate before available females are taken. Efficiency is not an important consideration particularly given that a male, once clasped, will give a release call (i.e., mistakes are likely to be temporary ones). Given the density of toads, the likelihood of clasping a moving object that is not a toad is relatively low. Due to the short breeding period, it is not necessary for these toads to eat while breeding and, therefore, concern about prey selection is eliminated. For prolonged breeders, on the other hand, the arrival of a receptive female is spread over a long period of time. There is little or no competition from other males. Males may have to be more concerned about food during a longer breeding season. Therefore, Wells suggests that the emphasis, among prolonged breeders, is on efficiency rather than speed.

These considerations suggest that the visual Gestalt for females in *B. bufo* is rather vague. It is sufficient for the male to discriminate large moving objects (Kondrashev 1976). Given the lack of concern with feeding during a short breeding season, it may be reasonable for prey and mate recognition to represent the specific and nonspecific ranges of the same continuum and for the tectum to implement both mechanisms with the same population of neurons. In prolonged breeders, females initiate mating. The combination of specific vocalizations from the male and specific tactile interaction may, once again, result in the visual Gestalt necessary for mate recognition by females being vague. However, feeding requirements during the mating season may differ from the requirements for explosive breeders. Perhaps subtle differences in the visual processing of explosive and prolonged breeders will be found which relate to these considerations.

Ewert (1987) pointed out that it is not possible to record T5.2 prey-selective neurons during mating season. He suggests that this is because T5.2 neurons are inhibited during mating season while T5.1 tectal efferent neurons may act as mate recognizers. The present model suggests that T5.2 neurons can be modulated by the TP region to become T5.1 neurons which results in a specific prediction. Since, in explosive breeders, female toads do not require female toad recognizers, Ewert's suggestion might result in the expectation that female toads lack a population of T5 neurons found in the male toad. The present model predicts that both sexes have a single population of "approachable object selectors," although active visual approach toward "mates" in the breeding period is observed in males only in explosive breeders.

Kondrashev (1987) emphasizes points of similarity between prey-catching (orienting, approaching, fixating, and snapping) and mating (orienting, approaching, mounting, and clasping). For the release of both types of behavior, he suggests similar functional properties of command elements, since as mating motivation fades, within a short period the same prey-sized visual stimulus can alternatively elicit prey-catching or mating in the same male grass frog. In the context of our hypothesis, this could be explained by a T5 base modulator in transition from T5.1 to T5.2 response characteristics. Orienting and approaching motor pattern generating circuitry may be shared by both behaviors while the circuitry responsible for switching between snapping *vs.* clasping may be in transition under the same modulatory influence as that modulating the T5 base.

Ewert (1984, p. 388) reports another observation which seems to be important in this context. After conditioning toads in the terrarium to large moving objects in the presence of known prey odor, an excited toad (exposed to the odor of mealworms) showed orienting toward a (untrained) conspecific and snapping at its moving head or eyes; when the conspecific was walking, the excited toad followed and clasped it - similar to the behavior seen during mating, although this observation was made in summer outside the breeding season. With reference to our hypothesis, this behavior can be explained by the fact that the T5.2-tuned T5 base modulator, due to conditioning in the presence of known prey odor, changes its configural selective properties and accidentally channels tectal efferents to mating motor pattern generating circuitry, thus eliciting mating behavior.

Other important tasks for a T5 base modulator are adaptative modification of prey-selection due to individual experience (e.g., the "hand-feeding" phenomenon) and ontogenetic sharpening of prey-selection. Transition from T5.1-like to T5.2-like tuned prey-selecting behavior (Traud 1983) proceeds within the first year after metamorphosis and is paralleled by a morphological parcellation of the posterior dorsal thalamus into P and Lpd nuclei (Ebbesson 1987).

3.6 Comparison with Other Models

An der Heiden & Roth (1983) simulate the response of T5 neurons to worm, antiworm, and square stimuli with varying edge length using a model which does not include TP inhibition. They suggest that the "disinhibition" effect seen after TP lesion is due to a general increase in the receptive field size of T5 neurons and, although not excluding the influence of the thalamic pretectal (TP) region in producing this effect, do not model it explicitly. They cite the fact that disinhibition effects caused by TP lesions are reduced after several weeks post-lesion (Ewert et al. 1983) as evidence against the involvement of the TP region in determining specific T5 response characteristics.

Ewert et al. (1983) present a number of observations that suggest that the effect of TP inhibition is specific rather than general. Furthermore, although the effect of TP lesion is reduced after several weeks, prey-selection does not return to normal levels. The plot of prey-catching activity 30 days post-lesion (see Fig. 15 in Ewert et al. 1983) resembles the response of a combination of R2 and R3 cells with antiworm response peaking at 4 degrees and square response peaking at 8 degrees more than it does normal prey-catching behavior. Parallel changes of T5.2 pre-motor and the toad's prey-catching activity after TP-lesion during chronic recordings presently most convincingly demonstrate the role of the TP region in T5 function (Schürg-Pffeifer, this volume).

One might imagine that the response characteristics of T5 neurons could be generated without the need for TP influence. It is, however, much harder to imagine how modulation of these responses due to motivation, habituation, and conditioning could occur without extratectal influences (see Finkenstädt, this volume). The hypothesis presented in this chapter that mate recognition results from modulation of T5 neurons otherwise used for prey-recognition is not possible without extratectal influences. It is reasonable that those extratectal influences are mediated by the TP region. Resolving whether the changes seen in T5 neurons during breeding season are the result of modulation of single T5 neurons or switching between populations of cells may contribute to our understanding of TP involvement in T5 function.

The tectal column model of Lara et al. (1982) includes the inhibitory effects of TH3 neurons from the TP region and models variability in prey recognition capabilities with changes in motivational state (Cervantes-Pérez et al. 1985). In general, the model presented in this chapter differs from the tectal column model in stressing a single dynamic model to explain all T5 function rather than in explaining the prey-selecting responses of T5.2 neurons. This resulted in an emphasis on modeling the role of the T5 base in providing a basis for T5 function. Significant differences between the models include:

(i) The present model's stress on R3 input to the T5 neuron: The tectal column model emphasizes R2 input, although modeling R3 and R4 input to the T5 neuron itself. R2 inputs are appropriate for prey analysis but are less likely to provide a basis for a dynamic T5 model which accounts for both prey and mate recognition, and modulation of prey-selection.

(ii) The greater density of retinal inputs to the T5 neuron in the present model: This feature of the model arose from consideration of the density of ganglion cells in the retina (Carey 1975; see Section 4.1).

The tectal column model considers the role of recurrent excitation in the facilitation of prey-catching behavior.The present model is a feedforward network but neglects the role of positive feedback (intrinsic to T5 and in lateral interactions between T5) only to simplify analysis for the behaviors modeled. In addition, the tectal column model considers the role of R3 and R4 cells as input to the TH3 neurons while this model only considers R4. The role played by R4 input in our model is to discriminate antiworm from worm stimuli, so that TH3 inhibition occurs only during antiworm stimulus presentation. R3 input would enforce this discrimination around 8 degrees, but abolishes it for larger edge lengths. However, R4 cells provide only slightly greater response to antiworm than to worm stimuli and future models should include a more refined model for the TP region which will probably incorporate both R3 and R4 input.

3.7 Directional Sensitivity

As mentioned in the Introduction, the data of Ewert et al. (1979a) can be divided into three aspects: discriminative ability irrespective of the direction of movement, symmetric directional sensitivity, and asymmetric directional sensitivity. Directional invariance is of particular importance for a pattern recognition system that judges objects by their extension relative to the direction of object movement, as required for configural prey-selection. This section discusses the potential contribution of the retina, tectum, and thalamus to discrimination and directional sensitivity.

A number of investigators have shown directional sensitivity in retinal ganglion cells (Andrew 1955; Lettvin et al. 1959; Maturana et al. 1960; Reuter & Virtanen 1976; Backstrom et al. 1978; Ewert et al. 1979b; Grüsser-Cornehls & Saunders 1981; and Grüsser-Cornehls & Langeveld 1985). Maturana et al. (1960) demonstrated directional sensitivity in "sustained" and "convex edge detectors" (classes R1 and R2) with preferred directions that were not correlated to any specific axis of the eye or body. They suggested that they were related to elongated receptive fields. Reuter & Virtanen (1976) recorded R2 cells which responded to stimuli in one direction and not at all or very little in the opposite direction. The preferred directions were randomly distributed. The directional sensitivity disappeared more or less completely when a spot was moved to and fro within the excitatory receptive field suggesting that it may be due to asymmetries in the inhibitory receptive field. Backstrom et al. (1978) also found R2 cells that were directionally sensitive with preferred directions typically covering a sector of 180 degrees. However, while Maturana et al. (1960) frequently encountered directionally sensitive cells and Reuter & Virtanen (1976) found directional sensitivity in 8 out of 20 class R2 cells recorded, Backstrom et al. (1978) found only 29 out of 171 class R2 cells to be directionally sensitive.

On the other hand, Grüsser-Cornehls & Langeveld (1985) demonstrated chromatic directional sensitivity in R3 cells. Class R3 cells are sensitive to light at 450, 500, and 580 nm. Directional sensitivities were found at all three wavelengths but primarily at 450 and 500 nm. In all but one case the directional sensitivities at all three wavelengths preferred stimuli that moved in a temporonasal direction. These authors suggest that the fast response of green rod photoreceptors, which are responsible for ganglion cell sensitivity at 450 nm, could contribute to the worm-antiworm differential sensitivity of T5 neurons to grey-scale stimuli since they would recover fast enough to respond to both the leading and trailing edge of worm stimuli while only being able to respond to the leading edge of antiworm stimuli. This coupled with the temporonasal directional preference of the green rods would result in directional sensitivity in R3 ganglion cells to grey-scale worm-antiworm stimuli. Grüsser-Cornehls & Langeveld suggest that these chromatic directional sensitivities could help explain the directional sensitivity in T5 neurons. However, the fact that all but one cell preferred temporonasal movement would limit the contribution of this mechanism to the directional sensitivity seen in T5 neurons.

Due to the variety of directional sensitivities reported by different investigators for retinal ganglion cells, only those reported by Ewert et al. (1979b) were incorporated in the present model (see Tables 2 and 3). It is interesting to note, however, that the absolute difference in mean directional sensitivities between 0 and 180 degrees and between 90 and 270 degrees for types of T5 neurons (see Table 1) increases from type *alpha* to *delta*. For horizontal movements, the difference in $D_{W,A}$ values increases from 0.09 for *alpha*-type to 0.32 for *delta*; while, for vertical movements, the difference increases from 0.01 for *alpha*-type to 0.2 for *delta*. We have hypothesized that *delta*-types correspond most closely to the T5 base in the absence of TP inhibition and that *delta*-types receive primarily R3 input. It may be that the $D_{W,A}$ differences in *delta*-types reflect contributions from R3 neurons.

The ILP cell is a postulated tectal mechanism for symmetric directional sensitivity but would also contribute to asymmetric directional sensitivity if the dendritic and axonal terminations were skewed with respect to a given location in the retinotopic representation of the visual field in tectal lamina B and F and the ILP cell terminated presynaptically on the R3 optic terminals rather than on the PY cell directly. For example, if the output of an ILP cell is nasal to the input, then, for a worm stimulus moving temporonasally, R2 inputs would excite the ILP cell which would block R3 input to the PY cell while, for nasotemporal movements, the blocking effect would come after the R3 inputs had already excited the PY cell. Thus, antiworm preference would be higher for nasotemporal movement than for temporonasal. This mechanism is similar to the one proposed by Barlow & Levick (1965) for the rabbit retina.

We have modeled symmetric directional sensitivity in R3 cells (see Figs. 5C and D) and both symmetric and asymmetric directional sensitivity with directionally sensitive weights on TH3 neurons. Since, in the present model, only R3 cells influence the T5 base, it exhibits only symmetric directional sensitivity while, in combination with TP inhibition, the T5 neuron exhibits both types of directional sensitivity. Does the T5 base itself exhibit asymmetric directional sensitivity? If TP inhibition acts only on antiworm response as postulated in this chapter, then the worm response of a T5 neuron offers a rare glimpse of the T5 base in isolation. In Ewert et al. (1979a, Fig. 3), examples of T5 neuron responses to worm stimuli are shown (see Fig. 9A for one example). Although the response to worm exhibits primarily symmetric directional sensitivity in these examples, there is some evidence for asymmetric directional sensitivity. This asymmetric directional sensitivity may be due to retinal ganglion cells or mechanisms such as the ILP cell described in the last paragraph.

Asymmetric directional sensitivities change over time. For example, between Figs. 2A4 and 2A5, the T5 neuron increases its worm preference for 0 degrees while decreasing this discrimination for 180 degrees. This suggests that TP inhibition determines not only symmetric but asymmetric directional sensitivity. Although TP inhibition may have resulted in symmetric directional sensitivity changes to which the tectum added an asymmetric component, we have modeled asymmetric directional sensitivity in terms of TP inhibition by modeling TH3 cells with weights that vary with direction. Our intention is not to suggest that the TH3 cells themselves have directional weights, but to suggest that some mechanism influencing TH3 neurons has directional sensitivity. This may be provided by retinal ganglion cells or thalamic interneurons. Note that the tectal ILP cell mechanism for asymmetric directional sensitivity discussed above cannot account for changes in directional sensitivity over time because it operates by modulating the response to worm and the present model follows the observation of Ewert et al. (1979a) in modeling directional sensitivity changes over time purely in terms of response to antiworm.

Finally, we have discussed directional sensitivity as though it was an intentional mechanism. It is possible that asymmetric directional sensitivity, in particular, is random. The fact that the plots of worm and antiworm response shown in Ewert et al. (1979a) are primarily symmetric would argue in favor of this idea. The fact that the neuron in Fig. 2B changes from $R_A = 0$ and $D_{W,A} = 1.0$ in the first hour (Fig. 2B1) to $R_A = 9.9$ and $D_{W,A} = 0.0$ in the fourth hour (Fig. 2B4) for 0 degrees while the $D_{W,A}$ and, therefore, the R_A remain unchanged over four hours for 180 degrees argues for the intentionality of asymmetric directional sensitivity.

3.8 Conclusion

The T5 base modulator hypothesis presents the view that most variations in the response characteristics of T5 neurons are due to a single dynamic assembly of cells consisting of a T5 base and inhibitory neurons in the TP region. This hypothesis is in contrast to hypotheses suggesting that variations in T5 response are due to switching between populations of T5 neurons with different response characteristics. The truth is likely to lie somewhere on the continuum which exists between these positions. Further research is necessary to clarify the extent to which anurans take advantage of modulating a single dynamic system or switch between populations of less flexible systems in order to analyze objects in their environment.

4 Appendix: Mathematical Description of the Model

In the model of T5 worm/antiworm discrimination discussed in this chapter, matrices representing retinal ganglion cell firing rates serve as input to thalamic-pretectal (TH3), tectal large pear (LP and ILP), and tectal stellate (SN) neurons. TH3 and SN neurons, in turn, inhibit LP neurons. Finally, LP and ILP neurons serve as input to a single pyramidal (PY) neuron which is the output of the model used for calculating $D_{W,A}$ values. This appendix discusses the retinal black box model and the mathematical description of tectal and thalamic pretectal neurons.

4.1 Retinal Inputs: R2, R3, and R4

Carey (1975) determined the density of ganglion cells in the retinas of several species of frogs and toads. He found that *B. americanus* has an average density of 10-40 ganglion cells per visual degree^2 depending on location in the retina while *B. marinus* has an average density of 15-80 ganglion cells per visual degree^2. These figures correspond to a density of 3.2-6.3 cells per visual degree for *B. americanus* and 3.9-8.9 cells per visual degree for *B.marinus*. Given that these densities are irrespective of physiological cell type, it might be expected that densities of R2, R3, and R4 cells are closer to 1-3 cells per visual degree. In order to simplify the model presented in this chapter and reduce its computational cost, we model R2, R3, and R4 cells as having a density of 0.5 cell per visual degree.

Retinal firing rates are taken from Table 2 in response to a given type (worm or antiworm) and size of stimulus. The time course of retinal firing is not modeled. A mask matrix representing the stimulus position multiplied by the appropriate firing rate and transferred to the appropriate section of a retinal firing rate matrix represents the stimulus. Mask matrices for 2° x 8° stimuli are shown in Table 8. The stimulus, centered in the receptive field, traverses the receptive field of the T5 neuron with a velocity of 7.6°/s. Membrane potentials and firing rates for tectal and thalamic neurons are updated every 10 ms. However, retinal firing rates are only updated when the stimulus has moved 2 degrees.

R3 firing rates (Table 2) based on data from Ewert et al. (1978) were adjusted to reflect the difference in firing rates in response to horizontally and vertically traversing 2° x 6° stimuli found in Ewert et al. (1979b). We will use the superscripts 78 and 79 to refer to stimulus responses from each paper, respectively. The response to worm measured from Ewert et al. (1979b) (R_W^{79}) is 29 while the antiworm response to horizontally (H) traversing stimuli (R_A^{H79}) is 35 and to vertically (V) traversing stimuli (R_A^{V79}) is 47. All other firing rates in the calculations below are from Table 2. To adjust the antiworm response of R3 cells to account for the difference in antiworm response to horizontally and vertically traversing stimuli seen with 2° x 6° stimuli, the R_W values were first normalized to the R_W in Ewert et al. (1978) by calculating a correction factor:

$$C_W = R_W^{78}/R_W^{79} = 21/29 = 0.72.$$

It was then applied to the horizontal and vertical R_As from the 79 paper:

$$R_A^{H79} = C_W R_A^{H79} = 0.72 \times 35 = 25, \text{ and } R_A^{V79} = C_W R_A^{V79} = 0.72 \times 47 = 34.$$

Correction factors for R_A values were then computed:

$$C_A^H = R_A^{H79}/R_A^{78} = 25/32 = 0.78, \text{ and } C_A^V = R_A^{V79}/R_A^{78} = 34/32 = 1.06.$$

These correction factors were applied to the R_A for other stimulus sizes. For example, for 2° x 8° stimuli:

$$R_A^H = C_A^H R_A^{78} = 0.78 \times 37 = 29, \text{ and } R_A^V = C_A^V R_A^{78} = 1.06 \times 37 = 39.$$

The $D_{W,A}$ values in Table 3 were calculated from the adjusted responses in Table 2.

4.2 Tectal and Thalamic Neurons

Following the convention of Lara et al. (1982), the state of tectal and thalamic neurons is defined by two quantities: A membrane potential (in lower case) and a firing rate (in upper case). Model neurons interact only via firing rates while the firing rate of a neuron depends on its membrane potential. Thus, the computation for a layer of neurons involves (i) computing the membrane potential of the current layer from the firing rates of its inputs and (ii) computing the firing rate of the current layer from the membrane potential.

The fundamental equation describing membrane potential dynamics is:

$$\tau_m \frac{dm(t)}{dt} = -m(t) + S_m(t)$$

where τ_m is the membrane time constant, $m(t)$ is the membrane potential, and $S_m(t)$ is the weighted sum of excitatory and inhibitory inputs. This equation is replaced by the difference equation:

$$\tau_m \frac{m(t+\Delta t) - m(t)}{\Delta t} = -m(t) + S_m(t)$$

which yields:

$$m(t + \Delta t) = (1 - \frac{\Delta t}{\tau_m})\, m(t) + \frac{\Delta t}{\tau_m} S_m(t)$$

Excitatory and inhibitory inputs are multiplied by a weighting factor, w_{ab}, on a connection from a neuron of type a to one of type b.

The firing rate is related to the membrane potential by a transfer function, F(m - θ) where θ is a threshold and F is one of:

$$ramp(x) = \begin{cases} x \text{ if } x > 0 \\ 0 \text{ otherwise} \end{cases} \qquad binary(x) = \begin{cases} 1 \text{ if } x > 0 \\ 0 \text{ otherwise} \end{cases}$$

The membrane time constants, threshold functions, and thresholds for tectal and thalamic neurons are shown in Table 9. For each neuron, we compute the membrane potential using the difference equation, so, with the membrane time constants in Table 9, it remains to specify the inputs S_m. These are as follows for each cell type:

Inhibitory Large Pear cell (ILP):

$$S_{ILP}(i,j) = w_{r2.ilp} R2_{i,j}(t)$$

Stellate cell (SN):

$$S_{SN}(i,j) = w_{r3.sn} R3_{i,j}(t)$$

Pretectal cell (TH3):

$$S_{TH3}(i,j) = w_{r4.th3} R4_{i,j}(t)$$

Large Pear cell (LP):

$$S_{LP}(i,j) = w_{r3.lp} R3_{i,j}(t) - w_{th3.lp}(\phi) TH3_{i,j}(t) - \sum_{k=\pm 1} \sum_{l=\pm 1} w_{sn.lp} SN_{i+k,j+l}(t)$$

where Ø is the direction of the stimulus in degrees.

Pyramidal cell (PY):

$$S_{PY} = \sum_{i=1}^{M} \sum_{j=1}^{M} w_{lp.py} LP_{i,j}(t) - \sum_{i=1}^{M} \sum_{j=1}^{M} w_{ilp.py}(i,j) ILP_{i,j}(t)$$

where M is the number of LP or ILP inputs to the PY cell in the X or Y dimension. M^2 is the total number of LP or ILP inputs to the PY cell. In this chapter, M = 13.

All weights except those for connections from TH3 to LP cells ($w_{th3.lp}$) and from ILP cells to the PY cell ($w_{ilp.py}$) are shown in Table 10. In general, weights were chosen to be in a reasonable range (0.0 to 5.5). Specific weights were chosen to provide balance between various inputs to the model. Cells which receive retinal input and whose firing rate is determined with the ramp function have a weight of 0.2 on inputs. For inhibitory neurons receiving retinal inputs whose firing rates are binary, the weight on the inputs is related to the threshold. For clarity, the input

Table 8. Retinal input masks: Columns refer to the type of stimulus while rows refer to the direction that the stimulus, centered in the receptive field, traverses the receptive field. The masks are for 2° x 8° stimuli but each mask covers an area of 10 degrees^2 with each number representing 2 degrees^2. For worm stimuli, the leading and trailing edge are "on" (1.0) while, for antiworm stimuli only the leading edge is on assuming that the trailing edge does not respond due to neuronal adaptation. Analogous masks in 11 x 11 matrices exist for 2° x 20° stimuli.

	Worm	Antiworm
	0.0 0.0 0.0 0.0 0.0	0.0 0.0 1.0 0.0 0.0
	0.0 0.0 0.0 0.0 0.0	0.0 0.0 1.0 0.0 0.0
Horizontal	1.0 0.0 0.0 0.0 1.0	0.0 0.0 1.0 0.0 0.0
	0.0 0.0 0.0 0.0 0.0	0.0 0.0 1.0 0.0 0.0
	0.0 0.0 0.0 0.0 0.0	0.0 0.0 0.0 0.0 0.0
	0.0 0.0 1.0 0.0 0.0	0.0 0.0 0.0 0.0 0.0
	0.0 0.0 0.0 0.0 0.0	0.0 0.0 0.0 0.0 0.0
Vertical	0.0 0.0 0.0 0.0 0.0	1.0 1.0 1.0 1.0 0.0
	0.0 0.0 0.0 0.0 0.0	0.0 0.0 0.0 0.0 0.0
	0.0 0.0 1.0 0.0 0.0	0.0 0.0 0.0 0.0 0.0
	1.0 0.0 0.0 0.0 0.0	0.0 0.0 0.0 0.0 1.0
	0.0 0.0 0.0 0.0 0.0	0.0 0.0 0.0 1.0 0.0
Left Diagonal	0.0 0.0 0.0 0.0 0.0	0.0 0.0 1.0 0.0 0.0
	0.0 0.0 0.0 0.0 0.0	0.0 1.0 0.0 0.0 0.0
	0.0 0.0 0.0 0.0 1.0	0.0 0.0 0.0 0.0 0.0
	0.0 0.0 0.0 0.0 1.0	1.0 0.0 0.0 0.0 0.0
	0.0 0.0 0.0 0.0 0.0	0.0 1.0 0.0 0.0 0.0
Right Diagonal	0.0 0.0 0.0 0.0 0.0	0.0 0.0 1.0 0.0 0.0
	0.0 0.0 0.0 0.0 0.0	0.0 0.0 0.0 1.0 0.0
	1.0 0.0 0.0 0.0 0.0	0.0 0.0 0.0 0.0 0.0

weights were chosen to be 1.0, so that the threshold could be determined from retinal firing rates (see Table 2) in a straightforward manner $w_{sn.lp}$ was determined to be 0.3 for 2° x 8° stimuli and 0.42 for 2° x 20° stimuli $w_{lp.py}$ was chosen to be 1.8, so that LP input balanced ILP input to the PY cell in the models simulated in Figs. 8 and 9. Membrane time constants were 0.1 for excitatory cells and 0.01 for inhibitory cells. These settings insured that the model responded rapidly to inhibitory inputs.

Acknowledgements. I would like to thank Dr. Michael Arbib and Dr. Jörg-Peter Ewert for their generous support of this research and for careful reviews of manuscripts which played a crucial role in shaping the ideas expressed in this chapter. The research described in this chapter was supported in part by grant no. 1RO1 NS 24926 from the National Institutes of Health (M.A. Arbib, Principal Investigator).

Table 9. Time constants, threshold functions, and thresholds for tectal and thalamic pretectal neurons.

Cell Type	Time Constant	Threshold Function	Threshold
ILP	0.01	binary	27.0
SN	0.01	ramp	0.0
TH3	0.01	binary	15.0
LP	0.1	ramp	0.0
PY	0.1	ramp	0.0

Table 10. Weights for all model simulations. $w_{th3.lp}$ and $w_{ilp.py}$ are not included and are listed separately for each simulation. For $w_{sn.lp}$ the values in parentheses are the stimulus size.

$W_{r2.ilp}$	1.0
$W_{r3.lp}$	0.2
$W_{r3.sn}$	0.2
$W_{r4.th3}$	1.0
$W_{sn.lp}$(2°x 8°)	0.3
$W_{sn.lp}$(2°x 20°)	0.42
$W_{lp.py}$	1.8

5 References

An der Heiden U, Roth G (1983) A mathematical network model for retino-tectal prey recognition in amphibians. In: Lara R, Arbib MA (eds) *Proc of the 2nd workshop on visuomotor coordination in frog and toad: models and experiments.* Univ of Massachusetts, COINS Technical Report 83-19, Amherst, MA

Andrew AM (1955) Action potentials from the frog colliculus. *J Physiol (Lond)* 130: 25

Antal M, Matsumoto N, Székely G (1986) Tectal neurons of the frog: Intracellular recording and labeling with cobalt electrodes. *J Comp Neurol* 246: 238-253

Backstrom A-C, Hemila S, Reuter T (1978) Directional selectivity and colour coding in the frog retina. *Med Biol* 56: 72-83

Barlow HB (1953) Summation and inhibition in the frog's retina. *J Physiol (Lond)* 119: 69-88

Barlow HB, Levick WR (1965) The mechanism of directionally selective units in rabbit's retina. *J Physiol* 178: 477-504

Betts B (1988) The toad optic tectum as a recurrent on-center off-surround neural net with quenching threshold. In: *Proc of IEEE International Conference on Neural Networks II*, pp 47-54

Brzoska J, Schneider H (1978) Modification of prey-catching behavior by learning in the common toad (*Bufo b bufo* L, Anura, Amphibia): changes in response to visual objects and effects of auditory stimuli. *Behav Process* 3: 125-136

Burghagen H (1979) *Der Einfluß von figuralen, visuellen Mustern auf das Beutefangverhalten verschiedener Anuren.* PhD Thesis, Univ of Kassel

Carey RG (1975) *A quantitative analysis of the distribution of the retinal elements in frogs and toads with special emphasis on the areae retinalis.* Master's Thesis, Univ of Massachusetts, Amherst, MA

Cervantes-Pérez F, Lara R, Arbib M (1985) A neural model of interactions subserving prey-predator discrimination and size preference in anuran amphibians. *J Theor Biol* 113: 117-152

Ebbesson SOE (1987) Prey-catching in toads: an exceptional neuroethological model. A commentary. *Behav Brain Sci* 10: 375-376

Ewert J-P (1968) Der Einfluß von Zwischenhirndefekten auf die Visuomotorik im Beute- und Fluchtverhalten der Erdkröte (*Bufo bufo* L). *Z Vergl Physiol* 61: 41-70

Ewert J-P (1971) Single unit response of the toad (*Bufo americanus*) caudal thalamus to visual objects. *Z Vergl Physiol* 74: 81-102

Ewert J-P (1974) The neural basis of visually guided behavior. In: Held R (ed) *Recent Progress in Perception.* WH Freeman, San Francisco, pp 96-104

Ewert J-P (1984) Tectal mechanisms that underlie prey-catching and avoidance behaviors in toads. In: Vanegas H (ed) *Comparative neurology of the optic tectum.* Plenum, New York, pp 247-416

Ewert J-P (1987) Neuroethology of releasing mechanisms: prey-catching in toads. *Behav Brain Sci* 10: 337-405

Ewert J-P, Seelen W v (1974) Neurobiologie und Systemtheorie eines visuellen Muster-Erkennungsmechanismus bei Kröten. *Kybernetik* 14: 167-183

Ewert J-P, Wietersheim A v (1974) Einfluß von Thalamus/Praetectum-Defekten auf die Antwort von Tectum-Neuronen gegenüber bewegten visuellen Mustern bei der Kröte *Bufo bufo* (L). *J Comp Physiol* 92: 149-160

Ewert J-P, Speckhardt I, Amelang W (1970) Visuelle Inhibition und Excitation im Beutefangverhalten der Erdkröte (*Bufo bufo* L). *Z Vergl Physiol* 68: 84-110

Ewert J-P, Hock FJ, Wietersheim A v (1974) Thalamus/praetectum/tectum: retinale topographie und physiologische Interaktionen bei der Kröte *Bufo bufo* L. *J Comp Physiol* 92: 343-356

Ewert J-P, Borchers H-W, Wietersheim A v (1978) Question of prey feature detectors in the toad's *Bufo bufo* (L) visual system: a correlation analysis. *J Comp Physiol* 126: 43-47

Ewert J-P, Borchers H-W, Wietersheim A v (1979a) Directional sensitivity, invariance and variability of tectal T5 neurons in response to moving configurational stimuli in the toad *Bufo bufo* (L). *J Comp Physiol* 132: 191-201

Ewert J-P, Krug H, Schönitz G (1979b) Activity of retinal class R3 ganglion cells in the toad *Bufo bufo* (L) in response to moving configurational stimuli: influence of the movement direction. *J Comp Physiol* 129: 211-215

Ewert J-P, Burghagen H, Schürg-Pffeifer E (1983) Neuroethological analysis of the innate releasing mechanisms for prey-catching behavior in toads. In: Ewert J-P, Capranica RR, Ingle DJ (eds) *Advances in vertebrate neuroethology.* Plenum, New York, pp 413-475

Grüsser O-J, Grüsser-Cornehls U (1968) Neurophysiologische Grundlagen visueller angeborener Auslösemechanismen beim Frosch. *Z Vergl Physiol* 59: 1-24

Grüsser O-J, Grüsser-Cornehls U (1976) Neurophysiology of the anuran visual system. In: Llinás R, Precht W (eds) *Frog neurobiology.* Springer-Verlag, Berlin, Heidelberg, New York, pp 298-385

Grüsser-Cornehls U, Langeveld S (1985) Velocity sensitivity and directional selectivity of frog retinal ganglion cells depend on chromaticity of moving stimuli. *Brain Behav Evol* 27: 165-185

Grüsser-Cornehls U, Saunders RMcD (1981) Response of frog retina ganglion cells to monochromatic spots under photopic conditions. *Vision Res* 21: 1617-1620

Heatwole H, Heatwole A (1968) Motivational aspects of feeding behavior in toads. *Copeia* 4: 692-698

Heusser H (1960) Instinkterscheinungen an Kröten unter besonderer Berücksichtigung des Fortpflanzungsinstinktes der Erdkröte (*Bufo bufo* L). *Z Tierpsychol* 17: 67-81

Kondrashev SL (1976) Influence of the visual stimulus size on the breeding behavior of anuran males. *Akad Nayk Zool J* 55: 1576-1579

Kondrashev SL (1987) Neuroethology and color vision in amphibians. A commentary. *Behav Brain Sci* 10: 385

Kuljis RO, Karten HJ (1982) Laminar organization of peptide-like immunoreactivity in the anuran optic tectum. *J Comp Neurol* 212: 188-201

Lara R, Arbib MA (1982) A neural model of interaction between pretectum and tectum in prey selection. *Cognition and Brain Theory* 5(2): 149-171

Lara R, Arbib MA (1985) A model of the neural mechanisms responsible for pattern recognition and stimulus specific habituation in toads. *Biol Cybern* 51: 223-237

Lara R, Arbib MA, Cromarty AS (1982) The role of the tectal column in facilitation of amphibian prey-catching behavior: a neural model. *J Neurosci* 2(4): 521-530

Lettvin JY, Maturana HR, McCulloch WS, Pitts WH (1959) What the frog's eye tells the frog's brain. *Proc Inst Radio Eng* 47: 1940-1951

Maturana HR, Lettvin JY, McCulloch WS, Pitts WH (1960) Anatomy and physiology of vision in the frog. *J Gen Physiol* 43: 129-175

Matsumoto N, Schwippert WW, Ewert J-P (1986) Intracellular activity of morphologically identified neurons of the grass frog's optic tectum in response to moving configurational visual stimuli. *J Comp Physiol* 159: 721-739

Poggio T, Koch C (1987) Synapses that compute motion. *Sci Amer* 256(5): 46-52

Reiner A, Karten HJ, Brecha NC (1982) Enkephalin-mediated basal ganglia influences over the optic tectum: immuno-histochemistry of the tectum and the lateral spiriform nucleus in pigeon. *J Comp Neurol* 208: 37-53

Reuter T, Virtanen K (1976) Color discrimination mechanisms in the retina of the toad (*Bufo bufo*). *J Comp Physiol* 109: 337-343

Schürg-Pfeiffer E, Ewert J-P (1981) Investigation of neurons involved in the analysis of Gestalt prey features in the frog *Rana temporaria*. *J Comp Physiol* 141: 139-158

Stevens RJ (1973) A cholinergic inhibitory system in the frog optic tectum: its role in visual electrical responses and feeding behavior. *Brain Res* 49: 309-321

Székely G, Lázár G (1976) Cellular and synaptic architecture of the optic tectum. In: Llinás R, Precht W (eds) *Frog neurobiology*, Springer-Verlag, Berlin Heidelberg New York, pp 407-434

Traud R (1983) *Einfluß von visuellen Reizmustern auf die juvenile Erdkröte (Bufo bufo* L). PhD Thesis, Univ of Kassel

Tsai HJ, Ewert J-P (1987) Edge preference of retinal and tectal neurons in common toads (*Bufo bufo*) in response to worm-like moving stripes: the question of behaviorally relevant 'position indicators'. *J Comp Physiol* 161: 295-304

Wells KD (1977) The social behavior of anuran amphibians. *Anim Behav* 25: 666-693

PART IV

The Role of Visual Centers

Compensation of Visual Background Motion in Salamanders

GERHARD MANTEUFFEL

Fachbereich Biologie, Universität Bremen
D-2800 Bremen, FR Germany

Abstract. Self-motion produces a shift of the whole visual environment across the retinae. Any object thus may fuse with the background or at least may be incorrectly localized. Compensation of visual shifts caused by self-motion can be achieved by different strategies: (i) Head and/or eyes are stabilized with respect to the environment during locomotion; (ii) proprioceptive, labyrinthine and visual background motion signals are computed together with object signals in the localization systems of the CNS. Like in other vertebrates, gaze stabilization is present in salamanders and is mediated by the (visual) optokinetik reflex and the vestibulo-collic reflex. These reflexes, however, relayed in the pretectum and the vestibular nuclei, are able to compensate only 50-80% of the shift velocity in salamanders. As a consequence, high retinal slip velocities may persist during locomotion. It is assumed that this residual retinal shift is directly computed in the optic tectum. Pretectal and vestibular nuclei were found to project to the optic tectum. Whereas the signal for retinal shift velocity from the pretectum is sent directly and bilaterally to the tectum, the vestibular nucleus projects to the contralateral dorsal tegmentum, and cells in this area send head velocity related signals to both tecta. The hypothesis is proposed that these synergistic inputs establish a directional selectivity in tectal columns which suppresses responses that would be elicited by objects moving in the same direction and speed as the background.

1 Introduction

In a natural environment a seeing animal is faced with an abundant number of visual objects. Not all of these cues are stationary, but many, under certain circumstances all of them, may move relative to the retina. Particularly when the animal is moving actively, it is confronted with the problem that the whole visual world becomes blurred since the entire picture shifts over the retina and so any visual object would be wiped out. The problem becomes even more complicated if a moving animal needs to localize moving objects. In this case, different movements are imaged simultaneously on the retina and have to be distinguished from each other. Thus, the visual system is concerned not only with the problem of discriminating different features of the environment but also with the much more difficult problem of evaluating the site of an object, its relative and absolute velocity and its complete shape, which may be obscured during a movement.

As pointed out, all objects, whether their perceived movement is a result of an object- or a self-motion, constitute a retinal image with different local intensity and velocity features. The animal can solve one part of this complex problem if it is able to distinguish between those retinal image shifts produced by self-motion and those resulting from active object movements. Under these circumstances, the animal can eliminate all patterns of stimulation attributable only to its own movements. All objects which pass this "self-motion filter" are of potential interest since their motion across the retina is not, or not completely, the result of the subject's motion. If, in addition, an animal is able to eliminate all other general motions in the picture (e.g., grass moved by the wind) then it has solved a great deal of the problem of extracting a relevant object from the visual environment.

Let us now consider a typical but simple terrestrial vertebrate, such as a salamander, which moves at moderate speed continuously rather than "jerkily" like a frog. We will observe several components of motion vectors which will unify at the two-dimensional layer of the retina into a resulting image shift. A forward motion of the salamander will produce a naso-temporal image shift and the characteristic locomotion of a tailed lower vertebrate will produce trunk oscillations which are transferred to the head. As a result, the whole visual environment constituting the visual background undergoes a shift with a translatory and a rotatory component. The latter is frequently changing its direction at a frequency that roughly corresponds to the straight component velocity. (Note that the retinal image shift is always reversed with regard to the direction in the environment due to the reversing properties of the lens; but for the sake of clarity in the following I will describe the situation in terms of the perceived motion).

Obviously, many salamanders can compensate the image shifts produced by self-motion. They are able, for example, to correctly localize objects while walking and to hit them if they are prey. This can be observed in fire salamanders, but it is even more impressive in some plethodontid salamanders which very precisely hit the prey with their projectile tongues.

Theoretically, several strategies exist for compensating image shifts related to self-motion. Figure 1 depicts a simplified model of a moving salamander and the most important mechanisms are added in a rough scheme of functional central nervous units and connections:

(i) The *gaze stabilization reflex*. Body undulations are compensated by antagonistic head and/or eye movements: e.g. when the anterior part of the trunk moves to the right, the head and the eyes are moved to the left, preserving a straight orientation of gaze direction. The gaze stabilization system is under the sensory control of the visual background system and of the vestibular system.

(ii) The *intrinsic self-motion computer*. The information on visual shifts induced by self-motion is fed into the discrimination and localization systems, where it is computed in order to distinguish a particular moving object from the shifting environment. The source of sensory information can be the visual background system,

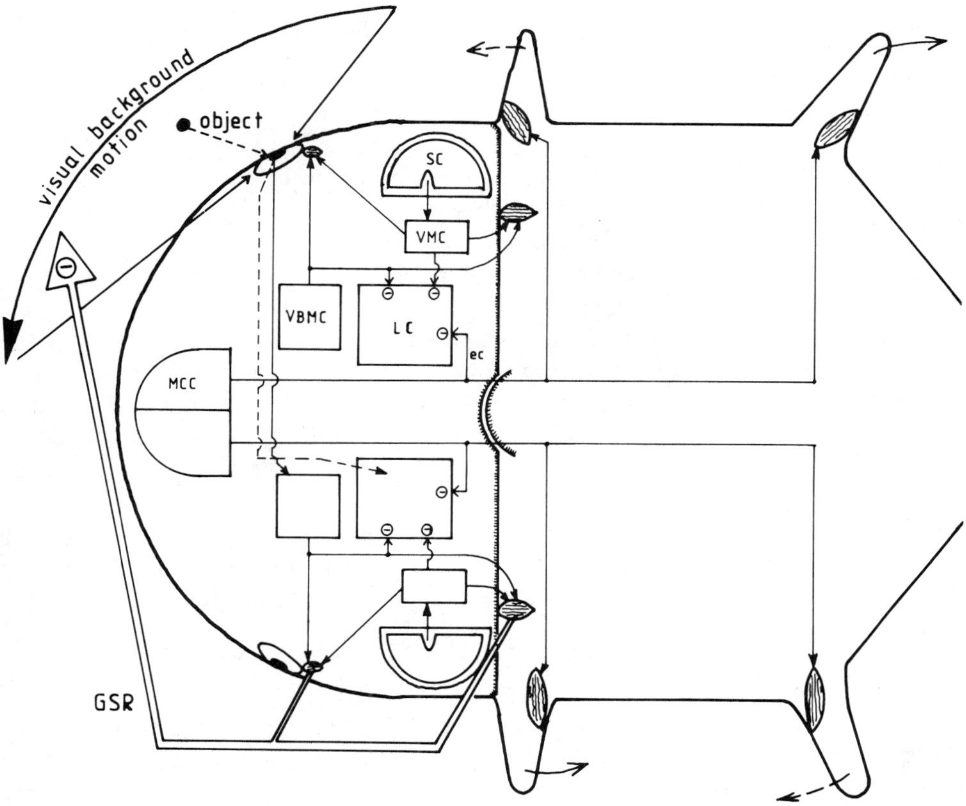

Figure 1. Schematic diagram of a walking salamander. The motor control center (MCC) induces walking which is accompanied by undulating trunk movements due to the diagonally linked power strokes of the legs (arrows). The trunk movement produces a stimulation of the (horizontal) semicircular canals (SC) and induces movements of the visual world relative to the head. The motion of the entire visual scene (the "background") is perceived by the visual background-motion center (VBMC) and the vestibular stimulation by the vestibular motion center (VMC). These two motion centers control compensatory head and eye movements, that are effected by head and eye muscles. Head and eye movements contribute to the gaze stabilizing reflex (GSR) that reduces the image shift. A distinct object is perceived in the localization center (LC). Object perception that is invariant of self-motion can be achieved by subtracting signals on self-motion originating from VMC and VBMC. An efference copy (ec) of the motor command could, in principle, also provide information on self-motion.

the vestibular system, or an efference copy from the motor control system. However, the latter case of a corollary discharge from the motor centers requires an additional adjustment via reafference from the muscle and body position sensors (von Holst & Mittelstaedt 1950) since the mere motor command is not a sufficient measurement for real motion (see also Burghagen & Ewert 1983; Ewert, this volume). Such proprioceptive information is relayed in the cerebellum which is not well developed in amphibians and apparently does not project to visual centers.

As mentioned, gaze stabilization is one of the mechanisms by which a stable image of the visual environment can be preserved during self-motion. Other mechanisms exist, however, without gaze stabilization the retinal shift velocities could become too large for the working range of visual units. Therefore, gaze stabilization reflexes (GSRs) are present in all animals with a well developed visual system. Since GSRs constitute a major part of background motion compensation, I will first describe these reflexes in salamanders and point out what is known about their central control. Then a preliminary description of central nervous circuits underlying the intrinsic self-motion computer is given.

In general, the GSRs of vertebrates are mediated by two sensory systems. The first is the vestibular semicircular canal system which mediates the vestibular gaze stabilization reflex. In animals with well developed eye movements it is mainly displayed as the vestibulo-ocular reflex (VOR). In amphibians, however, eye movements play only a minor role in gaze stabilization (Dieringer 1987), and the reflex is mainly performed by head movements (vestibulo-collic reflex, VCR). The second major sensory system contributing to GSRs is the visual system which mediates the optokinetic reflex (OKR). Again, it is to a large extent effected by eye movements in many mammals and by head movements in amphibians.

2 Behavioral Analysis of the OKR

The pure OKR of a salamander can be examined in an optomotor drum. In this set-up the animal sits in the center of a horizontally revolving cylinder whose inner surface is covered entirely with vertical stripes or random dots. Thus, a moving surround is generated and the animal displays the OKR which is measured by means of a search coil attached to the head surface (for details see Kopp & Manteuffel 1984).

When the stimulus consists of a prolonged turning of the drum, the head of the subject follows the pattern for a certain time, then it resets to approximately the initial position and again starts to pursue the stimulus. It thus displays the typical optokinetic nystagmus. If the stimulus movement is presented only for a short time, reset saccades often fail to occur (Fig. 2). Sinusoidal movements of the drum elicit sinusoidal head movements interrupted by some saccades, especially at lower frequencies. At higher frequencies, the OKR becomes nonlinear, as shown by a distorted shape of the head movement, and the response amplitudes decrease significantly (Fig. 2) indicating a low-pass characteristic of the OKR.

When the stimulation is stopped after a while by switching off the drum illumination, the response velocity decreases to zero with an apparent time constant of up to 5 s in the fire salamander and up to 6 s in the plethodontid salamander *Hydromantes italicus* (Manteuffel et al. 1986). This optokinetic afternystagmus indicates a considerable velocity storage in the OKR of some salamanders which significantly increases the overall capacity of the system, since it maintains the output although the input decreases as a result of the compensatory movement. However,

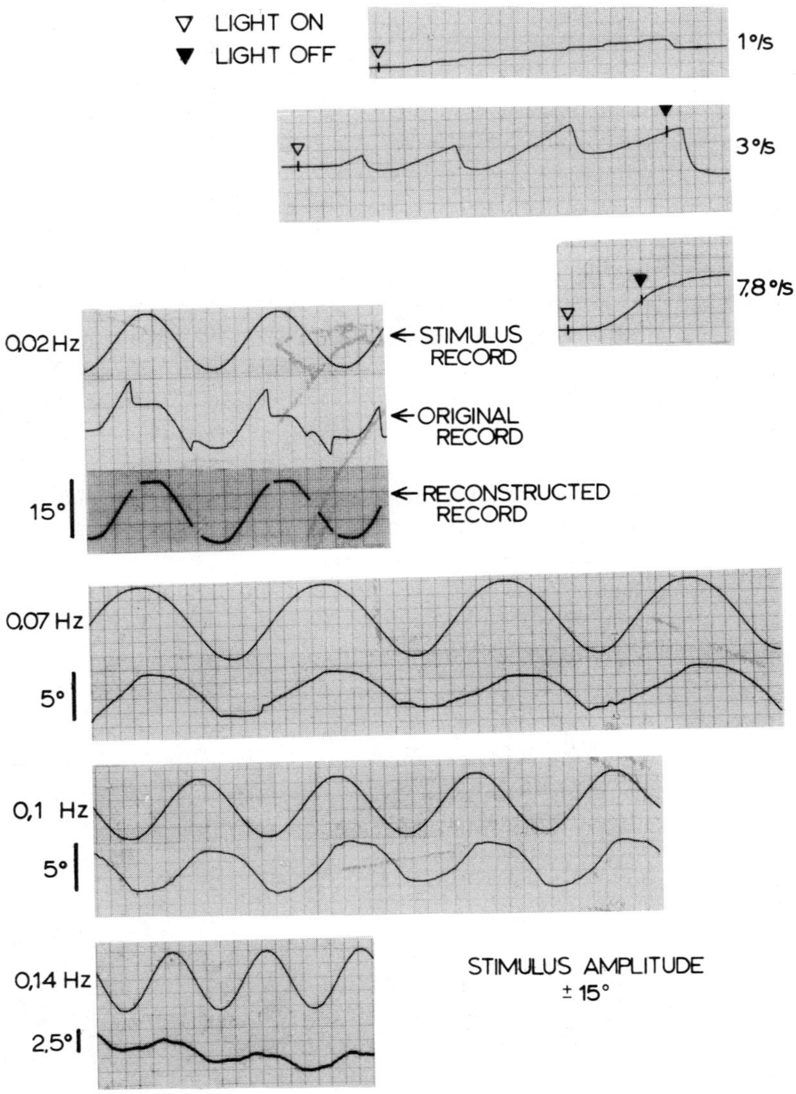

Figure 2. Optokinetic responses of a fire salamander using various kinds of stimulus movements of an optokinetic drum. The position of the head under constant velocity stimulation is depicted in the upper right; note the after-reaction at 7.8°/s stimulus velocity. The lower left shows the responses to various stimulus frequencies of sinusoidal drum movements (position records).

the occurrence of response oscillations at very low stimulus velocities (Fig. 2) and the failure of a fairly complete stimulus compensation (see below) demonstrate that the velocity storage system in salamanders may differ from that of mammals.

The capacity to compensate visual shifts, as measured in terms of the gain (response-velocity/stimulus-velocity), was 0.6-0.7 in the fire salamander and most other amphibians (Manteuffel et al. 1986) and reached 0.8 under optimal conditions (optimum stimulus velocity combined with full alertness of the animal). The optimum stimulus angular velocity was in the range of 3-5°/s and proved to be invariant to the pattern used. At higher stimulus velocities the stimulus pattern becomes more critical because higher spatial frequencies may then result in temporal frequencies of retinal stimulation that exceed the optimal response range of retinal units (Fig. 3).

The gain with sinusoidal stimulations corresponds to the gain with continuous stimulus movements. It is slightly amplitude dependent (Fig. 3). Lower stimulus movement amplitudes result in less prominent gain drops at higher frequencies, probably as an effect of a greater content of optimal velocities.

The constant delay in the OKR of approximately 1 s, i.e., the latency between stimulus onset and the start of head movement, produces an increasing phase lag of the response. This phase shift reaches values up to 70°/cycle and is largely amplitude invariant (Kopp & Manteuffel 1984).

In salamanders, the OKR is generated only when a large part of the visual field is stimulated by a moving textured pattern. The stimulus area may be smallest (60°) in the rostral part of the visual field which thus seems to be most effective in generating the OKR. This is apparently not attributable to binocular vision because the same holds for monocular animals (Manteuffel et al. 1983). In contrast to the situation in the rostral visual field, the caudal visual field seems to be completely ineffective in eliciting the OKR.

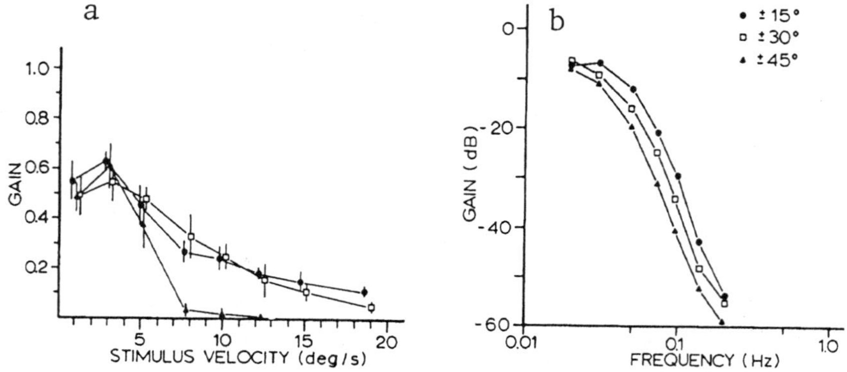

Figure 3. (a) Gain (ratio between head velocity and stimulus velocity) as a function of stimulus velocity averaged over 5 fire salamanders. Three different stimulus patterns were used: a striped fence with a period of 3.5° (closed triangles), or 7.6° (open squares), and a random-dot pattern with an average period of 6° (s.d. 4.8°) (closed circles). (b) Position gains with sinusoidal stimulation at various amplitudes of drum movement (see inset legend).

Under monocular stimulus conditions, salamanders, like most other vertebrates with panoramic vision (e.g., see Birukow 1938; Tauber & Atkin 1968), generate nystagmus only if the pattern moves from temporal to nasal (T-N direction) with respect to the stimulated eye; N-T stimulations fail to elicit a response.

3 Neural Circuits Underlying the Control of OKR

Pretectal nuclei have been shown to be the first relay station in the neural circuitry mediating visual gaze stabilization in all vertebrates examined to date (e.g., Collewijn 1980; Montgomery et al. 1982; Gioanni et al. 1983; Hoffmann & Huber 1983; Kato et al. 1986), including salamanders (Manteuffel et al. 1983). The influence of the accessory optic system (AOS) on horizontal OKR, first claimed by Lázár (1973), is questionable because more restricted lesions in the AOS-region have failed to corroborate the earlier findings (see Montgomery et al. 1982; Manteuffel et al. 1983). It is now usually believed that the AOS which has a pronounced vertical direction selectivity (Simpson & Soodak 1978; Manteuffel 1982; Gioanni et al. 1984; Grasse & Cynader 1984), is mainly involved in the *vertical* gaze stabilization, whereas the pretectum mediates the *horizontal* OKR (Fite et al. 1980; Lázár & Tóth 1983; Manteuffel et al. 1983). Both systems interact via reciprocal connections (Montgomery et al. 1981; Gruberg & Grasse 1984; Naujoks-Manteuffel & Manteuffel 1986), probably ensuring compensation of environmental shifts in oblique directions (Simpson 1984). In accordance with this hypothesis, Lázár (1983) reported the loss of vertical OKR in frogs after restricted AOS-lesions. The finding that horizontal OKR remains normal after disruption of the retinal input to the AOS (the basal optic tract), however, is abolished by lesions in the accessory optic nucleus (Manteuffel et al. 1983), is also consistent with this view, since lesions to the basal optic nucleus (nBON) most probably disrupt descending pretectal fibers. These fibers run in the vicinity of, or even through, the accessory optic nucleus (Manteuffel 1984a, 1985).

In salamanders, as in other vertebrates, the pretectum covers the most dorsal and anterior part of the synencephalon, a brain area at the meso-diencephalic junction. The synencephalon typically consists of a group of adjacent nuclei with the nucleus praetectalis (nPT) in the most antero-dorsal position. Further nuclei, in a ventro-caudal sequence, are the nucleus of Darkschewitsch (nD), the nucleus of the medial longitudinal fasciculus (nflm) and the nucleus of the basal optic neuropil (nBON) of the accessory optic system. In salamanders, these nuclei, with the exception of the nBON, are located within the periventricular cell mass and do not exhibit the differentiated morphological pattern found in most other vertebrates, including frogs and toads (Fig. 4).

The salamander's synencephalon receives direct visual input via two retino-recipient neuropils: the posterior thalamic (pretectal) neuropil (PTN) and the basal optic neuropil (BON). Whereas large parts of the PTN contain terminals from both eyes, the BON is supplied almost exclusively from the contralateral eye (Fritzsch 1980; Rettig & Roth 1986). The retinal terminals in the PTN contact mainly cells located in the nPT, but some dendrites of nD-cells and of nflm-cells also reach the PTN (Manteuffel 1985; Naujoks-Manteuffel & Manteuffel 1986). In the BON, retinal fibers contact the cells of the nBON and of the nflm.

Recordings in the PTN revealed a pronounced sensitivity of retinal terminals to large moving patterns (Sperl & Manteuffel 1987). These units are directionally selective, mostly for T-N stimulations. This holds for afferents from both eyes. However, when compared to the afferents from the contralateral eye, ipsilateral afferents more often prefer directions other than T-N. Furthermore, they frequently

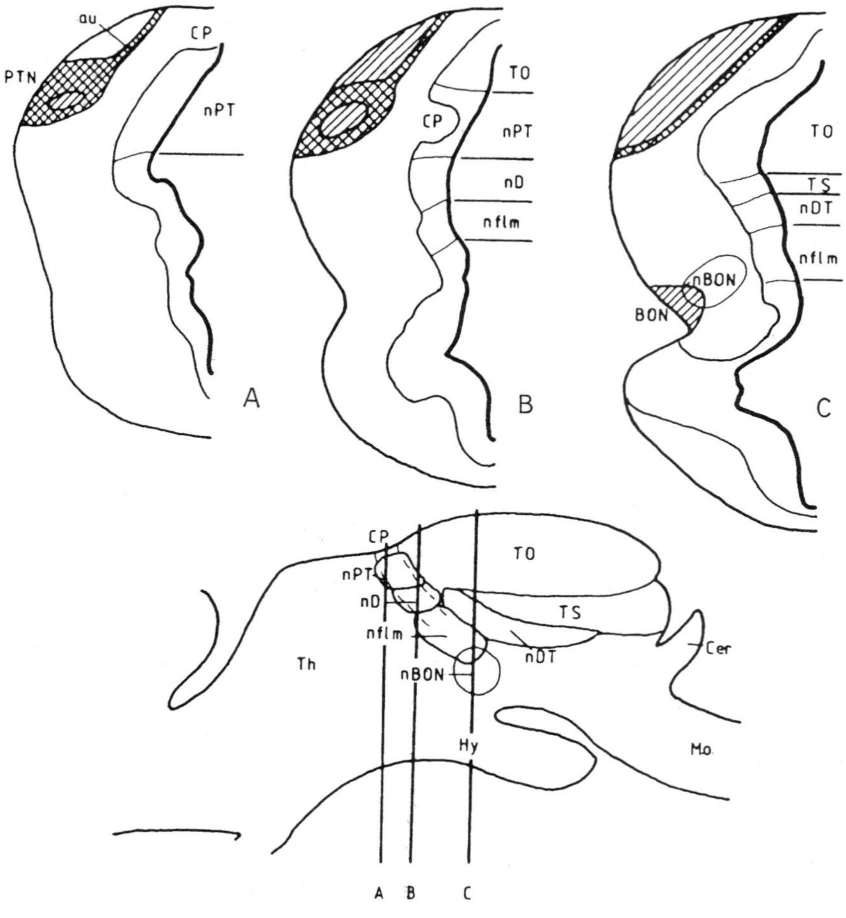

Figure 4. Nuclear organization of the salamander synencephalon as revealed by tracing studies. Sites of termination of retinal afferents are hatched; right-up hatching: terminals from the contralateral eye; right-down hatching: terminals from the ipsilateral eye; cross-hatching: bilateral afferents. Abbreviations: au, area uncinata; BON, basal optic neuropil of the accessory optic system; Cer, cerebellum; CP, posterior commissure; Hy, hypothalamus; Mo, medulla oblongata; nBON, nucleus of the basal optic neuropil; nD, nucleus of Darkschewitsch; nDT, dorsal tegmental nucleus; nflm, nucleus of the medial longitudinal fasciculus; nPT, pretectal nucleus; PTN, pretectal neuropil; Th, thalamus; TO, optic tectum; TS, torus semicircularis.

display less pronounced directional selectivities (Fig. 5). Most retinal afferents are optimally stimulated by low pattern velocities, however, the directional preferences usually do not change with varying stimulus velocities (Fig. 6).

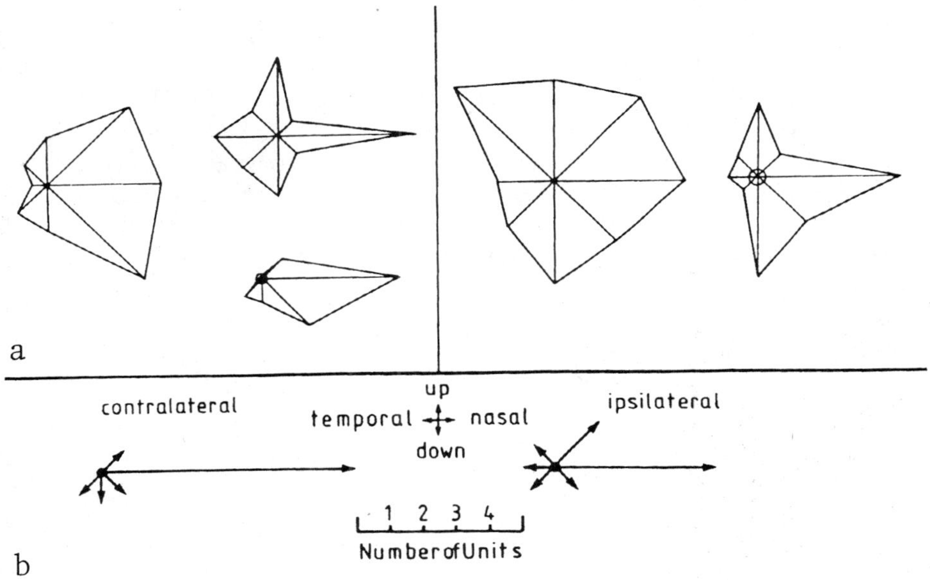

Figure 5. (a) Directional sensitivities of retinal terminals originating from the contralateral eye (left) and the ipsilateral eye (right), in the pretectum of the fire salamander. (b) Number of afferents in the pretectal neuropil and their optimum directions.

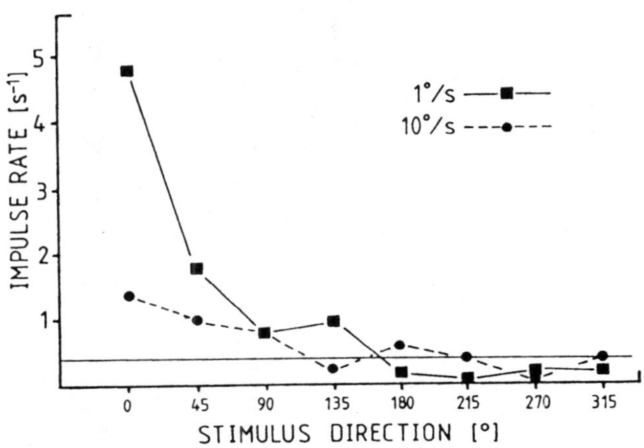

Figure 6. Directional tuning curves of a contralateral retinal terminal in the pretectum for two stimulus velocities. The continuous horizontal line indicates the level of spontaneous activity (0°: nasal, 90°: up).

Basically, the same type of retinal afferent characteristics were found in the BON. The main exceptions are the strictly contralateral afferents and the preferred directional selectivities for vertical or oblique stimulus movements (Manteuffel 1982).

Cells responsive to movements of the whole visual environment were recorded throughout the synencephalic region of the salamander (Manteuffel 1984a) (Fig. 7); they may be interpreted as background motion detectors. However, depending on their location in different nuclei, these cells displayed different characteristics concerning the ocular inputs and the preferred movement directions. Neurons in the nPT and in the nD showed a preference for horizontal movements whereas units in the nflm preferred vertical directions. Binocular cells were restricted to the nPT. Most of the cells in the nPT and the nD preferred a stimulus movement in the T-N direction with respect to the contralateral eye, i.e., ipsiversive turnings of the visual environment (Manteuffel 1984a). This observation is in accordance with the results of lesion studies in which an unilateral pretectal lesion was found to abolish the OKR only in the ipsiversive direction, whilst contraversive stimulation still elicited a nystagmus.

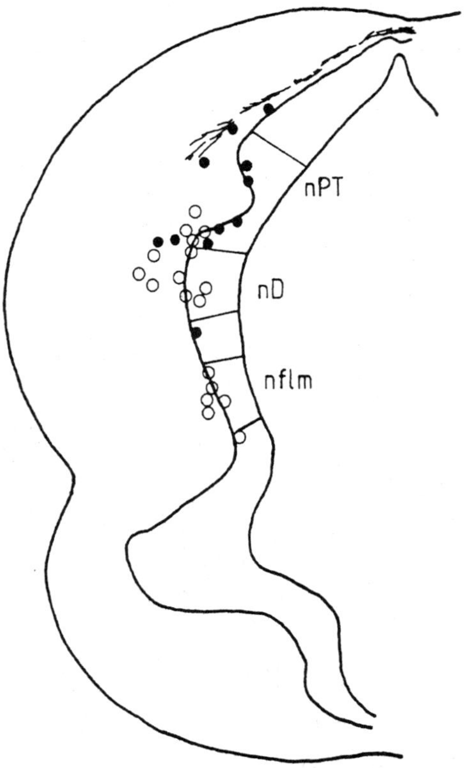

Figure 7. Location of synencephalic cells sensitive to full-field stimulation in the optokinetic drum. Binocular units, closed circles; merely contralateral units, open circles. For abbreviations see Fig. 4.

All of the units displayed comparatively low velocity optima when stimulated with a whole-field pattern (vertical stripes, period 3.3°); the optimum velocities were below 5°/s in 80% of the units and no unit preferred velocities above 20°/s. However, higher stimulus velocities can activate pretectal units when the stimulus pattern consists of a large (15° x 15°) square. Under these conditions, responses can be evoked even in the range of 100°/s (Himstedt et al. 1987). Thus, pretectal cells are best activated by large patterns consisting of a number of small elements moving coherently, or by a single large element. Similar results have been obtained in toad's thalamic pretectal class TH3 neurons (Tsai & Ewert 1988; see also Ewert, this volume).

Most of the cells showed a tonic discharge to moving background patterns as long as the motion persisted, sometimes with a short burst at the onset of the movement. Some units displayed a more complex behavior, however. One group of such neurons were active after a movement had stopped. Since their response rate correlated well with the foregoing stimulus velocity, it seems likely that such units are subject to inhibition from movement detectors.

As mentioned above, the pretectal nucleus is the only nucleus in the synencephalon of the fire salamander receiving binocular afferents. Since lesion experiments have demonstrated that the nPT plays a major role in the control of horizontal OKR (Manteuffel et al. 1983), it was not unexpected that the monocular OKR-gain is weaker than the binocular gain even in the preferred T-N direction. This decrease in gain may be as large as 0.3 in individual animals (Kopp 1987a).

In most binocular pretectal neurons, the impulse rate was highest when the stimulus was perceived binocularly while moving in the T-N direction relative to the contralateral eye and in the N-T direction relative to the ipsilateral eye, but it was depressed during a simultaneous contralateral and ipsilateral T-N movement. Mere contralateral stimulation activated the cells only moderately (Fig. 8). These subtractive cells (contralateral T-N minus ipsilateral T-N) which constituted 57 % of the binocular pretectal cells, are not activated by ipsilateral N-T stimulations (Manteuffel 1987).

Only a small group of cells (14%) were best activated by contralateral T-N and ipsilateral N-T movements and were also responsive to mere contralateral T-N and mere ipsilateral N-T movements. They showed no depression during parallel

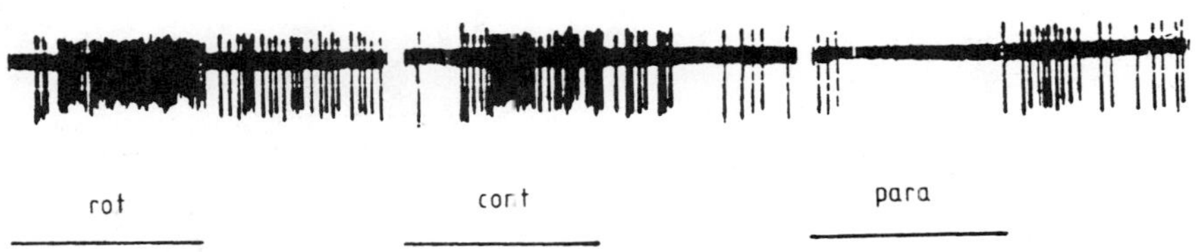

Figure 8. Responses of a subtractive binocular pretectal cell to rotatory (contralateral T-N, ipsilateral N-T), monocular contralateral T-N, and parallel (contralateral and ipsilateral T-N) stimulation. In each case, the stimulus was presented for 10 s (indicated by the bars) at a velocity of 5°/s.

(contralateral and ipsilateral T-N) stimulation. In some of these cells, the proposed inhibitory effect of the accessory optic system upon pretectal neurons was likewise tested. A frequency reduction occurred with vertical stimulus movements presented to the contralateral eye which was slightly more pronounced in the down-direction. Since direct contralateral visual afferents to the pretectum are only slightly sensitive to vertical stimulus movements (Sperl & Manteuffel 1987; Fig. 5), this depression was probably mediated via an inhibition originating from the AOS which is selectively sensitive to vertical directions.

Another group (29%) of binocular pretectal cells was activated *only* when the stimulus was presented binocularly and moved in the ipsiversive direction.

Monocular cells were not frequently found in the salamander's pretectum using extracellular techniques. However, when intracellular recording techniques were applied, 60 % of the pretectal cells were monocular and were excited by contralateral optic nerve stimulation. These cells displayed only graded potentials (mostly EPSPs), i.e., they were apparently unable to generate action potentials. When labeled with

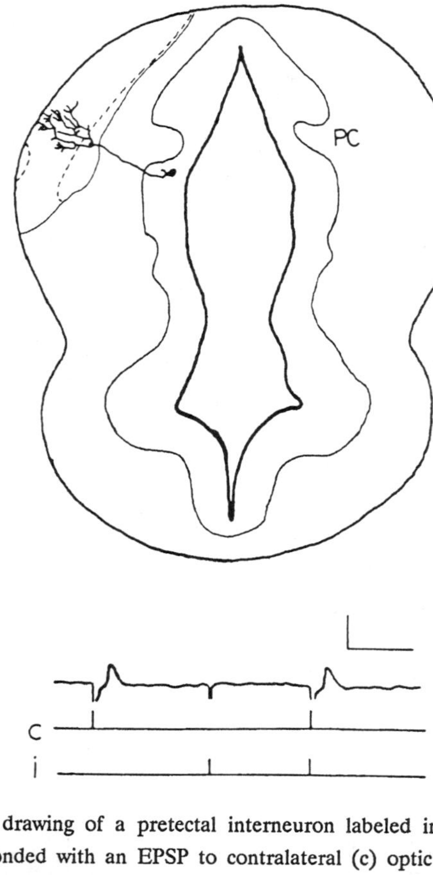

Figure 9. Camera lucida drawing of a pretectal interneuron labeled intracellularly with horseradish peroxidase. The cell responded with an EPSP to contralateral (c) optic nerve stimulation but not to ipsilateral (i) stimulation. Calibration: 10 mV vertical, 100 ms horizontal.

horseradish peroxidase, these non-spiking cells showed the morphological features of interneurons: their apical dendrits were comparatively narrow and they seemed to lack an axon (Fig. 9).

In contrast to monocular cells, most of the binocular cells recorded intracellularly generated action potentials (Manteuffel 1985). These cells usually had large dendritic trees, sometimes covering nearly the entire PTN. Most of them were excited by contralateral and inhibited by ipsilateral optic nerve stimulation (Fig. 10). This type included more than 50% of the population of binocular units. The remaining binocular cells displayed various kinds of retinal input couplings with the additive group (excited by both optic nerves) being most frequently represented (37%).

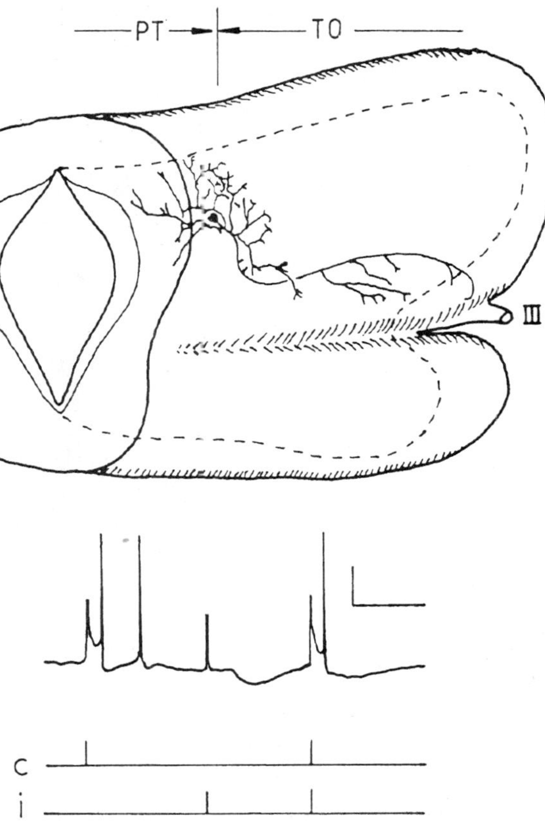

Figure 10. Reconstruction of a binocular pretectal projection neuron after intracellular horseradish peroxidase labeling. The cell projected to the anterior dorsal tegmentum, the anterior optic tectum and the BON. The descending axon decussated at the level of the oculomotor nerve exit (III). It responded with excitation (two spikes) to contralateral (c) optic nerve stimulation and with hyperpolarization to ipsilateral (i) optic nerve stimulation. When both nerves were stimulated simultaneously, the IPSP suppressed the second spike. Calibration: 10 mV vertical, 100 ms horizontal.

A disadvantage of electrical stimulation studies is that cells cannot be clearly categorized with respect to their visual properties. It is nevertheless tempting to speculate that those optokinetic pretectal cells which are responsive to rotatory stimulation (contralateral T-N, ipsilateral N-T) and are silent during parallel stimulation are identical to those cells which are excited by contralateral and inhibited by ipsilateral stimulation. If this were the case, the T-N sensitive pretectal afferents, which make up the majority of retinal inputs from both eyes, would be excitatory from the contralateral eye and inhibitory from the ipsilateral eye. As a result, the pretectal cells would be most responsive when the ipsilateral inhibitory input were itself inhibited, because a disinhibitory enhancement could then take place. Some other cells may be excited by contralateral T-N sensitive inputs and by ipsilateral N-T inputs. The latter input type occurs in a small number (Sperl & Manteuffel 1987).

To summarize, it is obvious that in the salamander the vast majority of pretectal cells are sensitive to rotatory background movements in the horizontal plane, and that some, if not all, of them are inhibited by vertical movements. In the adjacent nD, cells are not rotation specific, because they receive only contralateral ocular input. They may respond to translatory background movements as well (i.e., when both eyes are simultaneously stimulated in the T-N or the N-T direction).

4 Behavioral Analysis of the Vestibulo-Collic Reflex (VCR)

The vestibulo-collic reflex (VCR) is a compensatory head movement which occurs when the body is turned around a vertical axis. Hence, if the body is turned to the left, the head turns to the right roughly maintaining the gaze direction. In an experimental situation, the VCR is evoked by a passive body movement on a horizontal turntable in the dark, thus achieving pure vestibular stimulation.

The compensatory head movement reduces (but does not abolish) the vestibular stimulation induced by the turntable movement. However, this is not crucial because of the dynamic time constant of the primary afferents, which is approximately 3 s in frogs (Blanks & Precht 1976). Therefore, the initial stimulation governs the output of the system for a sufficiently long time.

In contrast to amniotes (Robinson 1976; Büttner et al. 1978; Dieringer et al. 1983) and teleost fishes (Hartmann & Klinke 1975), in which the VOR or VCR time constant is significantly longer than the primary afferent time constant, the amphibian VCR-time constant is approximately equal to that of the primary afferents (Fig. 11) (see also Dieringer et al. 1983). This points to a lack of an additional central velocity storage in the vestibular system of amphibians.

The VCR gain in the fire salamander (head-velocity/turntable-velocity after acceleration) was 0.7 within a turntable velocity range between 10°/s and 60°/s (Kopp 1987a). The gain is not affected by different amounts of acceleration of the turntable; i.e., the final head velocity is the same regardless of whether it is reached by a modest acceleration over a long period or by a strong acceleration over a short period. This indicates an integration of the acceleration in the time domain. When the animal is moved sinusoidally on the turntable, instead of a peak acceleration with subsequent constant velocity, the head movement begins to compensate the stimulus movement above a stimulus frequency of 0.02 Hz (30° stimulus amplitude). In this situation, the VCR reaches its optimal compensation at frequencies above 0.05 Hz (Fig. 12).

5 Visual-Vestibular Interactions in Gaze Stabilization

In most amphibians the visual-vestibular interactions are very similar to those found in other vertebrates (for review see Collewijn 1981). Some amphibians have a comparatively well developed optokinetic afternystagmus (OKAN) (Manteuffel et al. 1986). This visual reaction, which prolongs the compensatory head movement for a short time after the end of stimulation, suppresses the vestibular postrotatory nystagmus that is opposite to the vestibular compensation movement during stimulation. This can be demonstrated in an experimental situation by moving an animal passively on a turntable and allowing it to look at an earth-fixed optomotor drum. When the table stops and the illumination is simultaneously switched off,

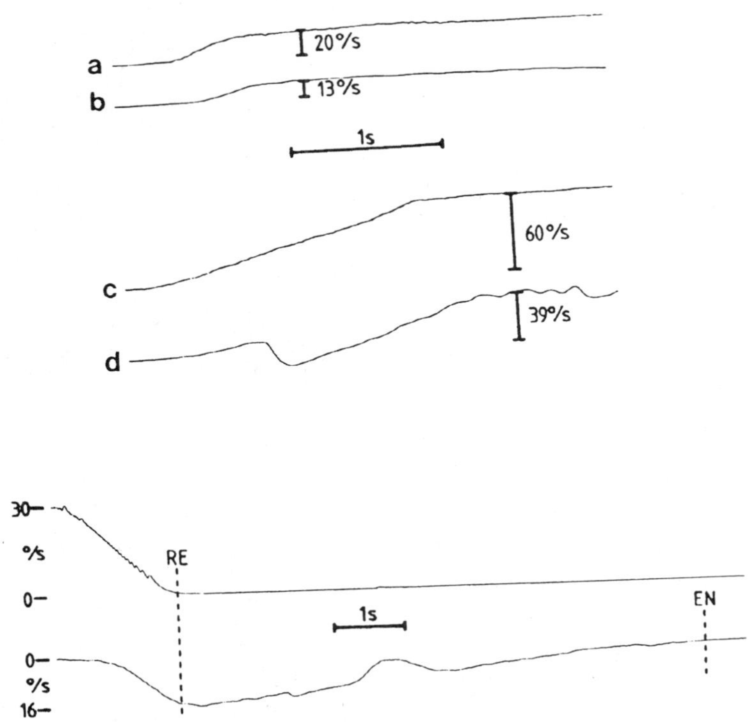

Figure 11. Top: Velocity registrations of the VCR (traces b and d) in the fire salamander at two different terminal velocities of the stimulus (traces a and c). Bottom: Postrotatory response (lower trace) after a deceleration of the turntable (upper trace). The apparent time constant of the decelerating head-velocity is 4 s. RE, end of deceleration; EN, end of head movement. (From Kopp 1987a).

OKAN abolishes the vestibular postrotatory reaction and the head movement immediately ceases. In species without OKAN, the postrotatory nystagmus takes place in this situation resulting in a head movement opposite to the perrotatory one (Dieringer et al. 1983; Manteuffel et al. 1986).

As in other vertebrates, the amphibian OKAN depends on an intact vestibular system, being abolished after bilateral labyrinthectomy (Manteuffel et al. 1986). Moreover, the entire performance of the OKR is weakened or even abolished at higher stimulus velocities if both labyrinths are destroyed (Kopp 1987b).

Since the VCR compensates well at higher stimulus frequencies and the OKR acts mainly in the range of the low frequencies, both together ensure gaze stabilization from nearly 0 Hz to comparatively high frequencies (Fig 12). A combined VCR/OKR stimulation can be achieved by rotating the animal in the light on a turntable surrounded by a stationary optokinetic drum. When the animal is turned to the left, the visual environment shifts to the right. The VCR, which acts opposite to the movement of the turntable, and the OKR acting in the direction of the perceived visual shift, work together synergistically. The respective head velocities and phases relative to the stimulus are summated to provide the overall capacity of the gaze stabilizing reflexes. This form of stimulation resembles the natural condition

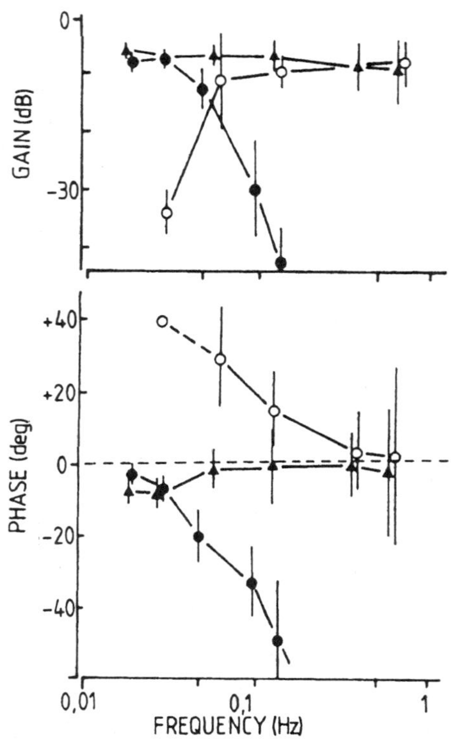

Figure 12. Sinusoidal gain and phase relationships of the OKR (closed circles), the VCR (open circles) and a combined stimulation (closed triangles) in the fire salamander. (From Kopp 1987b).

during walking, whereby the animal stimulates its semicircular canals by undulating body movements whilst viewing its visual environment. By virtue of the different working ranges of the vestibular and optic systems, gaze stabilization is thus performed predominantly by the OKR during slow self-motion and mainly by the VCR during fast movement. However, the gaze stabilizing system is able to compensate for only 50 to 80% of the evoked gaze shifts (Fig. 12). In amphibians, this rather imperfect compensation by reflexive head movements is not significantly improved by additive eye movements (Dieringer 1987).

6 Modeling of Gaze Stabilization in Salamanders

Gaze stabilization can be basically modeled as a negative feedback loop in which the compensatory movement is subtracted from the stimulating background movement. The result of this subtraction is the system's input.

During self-movement the vestibular signal is added to the system's output. A fast system switches the output off if the stimulus direction has changed its sign (Fig. 13). This mechanism has to be introduced because in a sinusoidal stimulation the new head movement direction is immediately displayed (with only the internal delay) and is not superimposed by the afternystagmus which would prolong movement in the original direction. This applies only to binocular stimulation, however. When

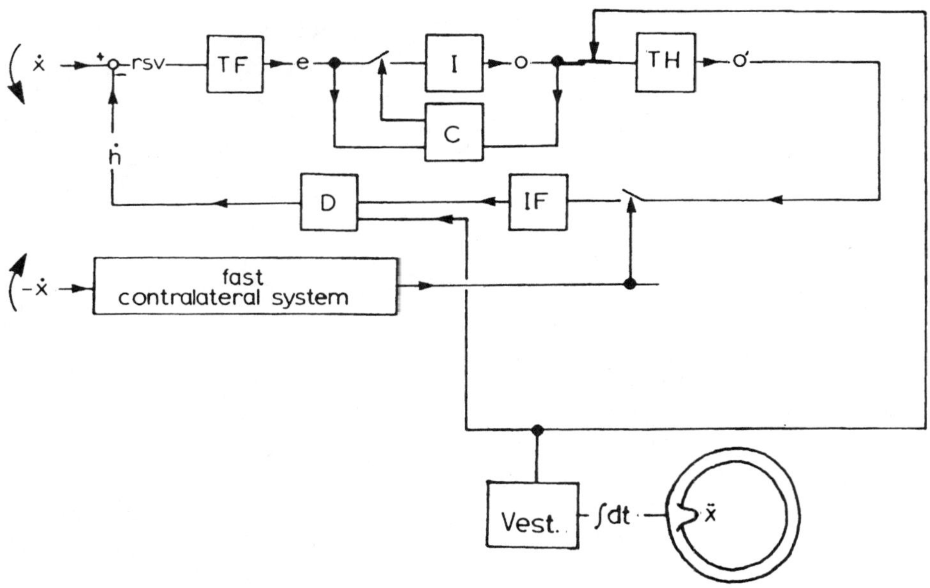

Figure 13. Model of the vestibular and visual gaze-stabilization system in the fire salamander. C, comparator; D, delay; e, neural excitation of the primary and secondary visual relays (0...1); $\dot{h}$, head velocity; I, velocity storage integrator; IF, internal neuro-motor transformation factor; o, output of the integrator (0...1); rsv, retinal slip velocity; TF, transfer function of the primary and secondary visual relays; TH, threshold; Vest, vestibular system nuclei.

monocular salamanders are stimulated sinusoidally in an optokinetic drum, the reaction does in fact persist for the length of the afternystagmus after the stimulus has changed its direction (remember that monocular salamanders perform the OKR only when the seeing eye is stimulated in the T-N direction). It seems, therefore, that the switching-off of the formerly active system is relayed by the other eye which is activated after the stimulus has changed its sign (Kopp 1987a). Candidates for mediating this process in the central nervous system may be the pretectal neurons that are activated when the stimulus stops (see above).

The input stimulation, the actual retinal slip velocity, is transferred to a neural signal by the retinal units and subsequently by the pretectal cells according to a velocity-dependent transfer function. Having passed all internal thresholds, the signal can be transformed into a motor signal according to a neuro-motoric algorithm. This process implies the transformation of a velocity-coded neuronal signal into a position-coded signal for the motoneurons. The final result, however, will be the compensatory head velocity.

As the computation is time-consuming (apparently 1 s in the fire salamander) the compensatory head movement subtracts from the stimulus movement with a delay. Internally, the loop seems to be controlled by the vestibular system in a non-specific, all-or-none mode. The loop is thus closed when the vestibular system is active on an arbitrary level. In contrast, the intact visual input is not necessary for the animal to perform the VCR, so that a simple "and"-gate cannot be used for modelling visual-vestibular interactions.

The model system, as presented so far, is unable to describe the real behavior of a salamander, i.e., there is no realistic possibility of adjusting the parameters in such a way that the animal's behavior is simulated over the entire working range and in all stimulus situations.

If a velocity-storage integrator with an appropriate time constant is introduced, the system suddenly displays a mammalian-like OKR: it saturates at very low retinal slip velocities. To weaken this effect to a degree found in the fire salamander, a comparator element has to be introduced which enables the integrator to load only if its input is higher than its actual loading condition. This comparator element can be realized by a simple interneuron network which subtracts the integrator's afferent signal from its efferent signal by algebraic excitation and inhibition. The interneuron, if excited, may then shunt the input channel of the velocity-storage network.

This model (Manteuffel 1984b) reproduces all the behavioral data found in the animal including the correct numerical gain values and the non-linearities in all stimulus situations tested. The transfer function for retinal slip velocity to neural excitation transformation was directly derived from the velocity functions measured in the pretectal neurons (Manteuffel 1984a). The integrator's time constants for loading and unloading were set to 2.5 s and to 4.2 s, respectively; the threshold was assigned a value of 0.2 and the internal neuro-motor transformation factor was taken as 2.75.

The model predicts several effects of manipulations in the CNS. If, for example, the threshold is lowered (which may be achieved by an application of facilitating neuro-modulators), the gain in the low velocity range would drop as a result of vigorous oscillations in the response, and the gain in the high velocity range would be increased. It has been shown by Dieringer (1987) that eye movements of low amplitude are able to compensate the non-linearity distortions of the head movements in the OKR of frogs by virtue of their own phase shifted non-linearities. These eye movements can be produced by a very similar model if the internal transformation factor is set to a lower value and the delay is shortened.

7 The Intrinsic Self-Motion Computer

As stated above, gaze stabilizing reflexes make an important contribution to image stabilization during self-motion. However, they are by no means able to produce a completely stable image. When the animal moves actively, 20 to 50% of the image shift caused by trunk undulations remains uncompensated and translatory shifts remain completely uncompensated. As a result, the system for visual detection of environmental shifts (the pretectum and the nucleus of Darkschewitsch in the case of horizontal movements) and the semicircular canals continue to be stimulated even when the gaze stabilizing reflexes are in action. Their output signals are proportional to the remaining image shifts and can therefore be used to inform the localization and identification center about expected image shifts resulting from a self-motion.

In all vertebrates, the optic tectum is necessary for visual object localization (cf. Bürgi 1957; Ingle 1970; Ewert 1973; Springer et al. 1977; Stein & Gaither 1981; Kostyk & Grobstein 1987). In amphibians, it is involved to a major extent in visual shape recognition as well (cf. Ewert 1974, 1984; Himstedt & Roth 1980; Lara et al. 1982; Roth 1982, 1987; Ewert, this volume). According to the introductory remarks, the tectum is assumed to receive afferents from both the visual and the vestibular background motion systems. Indeed, a well developed bilateral pretectal projection to the optic tectum has been demonstrated in amphibians (Wilczynski & Northcutt 1977; Finkenstädt et al. 1983; Montgomery et al. 1985). Furthermore, a caudally adjacent synencephalic nucleus projects ipsilaterally to the optic tectum in many vertebrates (Finkenstädt et al. 1983; Dacey & Ulinski 1986; Taylor et al. 1986). In the salamander, this nucleus has been termed the nucleus of Darkschewitsch by Herrick (1948), although it may not be homologous to its mammalian homonym. (For pretecto-tectal projections in frogs see contributions by Lázár and Matsumoto, this volume). The nPT and the nD might carry visual information about rotatory (nPT) and translatory (nD) background shifts to the optic tectum.

A direct vestibular projection to the optic tectum has not been found in amphibians. Ascending projections from the vestibular nuclei reach the oculomotor complex and, more importantly in this context, the intermediate zone of the mesencephalon via the lateral lemniscus (Wilczynski 1981; Naujoks-Manteuffel & Manteuffel 1987). Cells which are postsynaptic to this tract are located in the torus semicircularis and in the dorsal tegmental nucleus (nDT). These anatomical results are consistent with the electrophysiological finding that neurons in this region respond well to acoustic, vibratory and vestibular stimulation (Manteuffel & Naujoks-Manteuffel 1987). However, cells responsive to vestibular stimulation on a turntable are located only in the nDT (Fig. 14). All of them displayed a strict selectivity for contraversive rotations, i.e., they responded to that part of the sinusoidal cycle in which the animal was moved toward the side opposite the recorded nucleus (Fig. 15). The responses were roughly in phase with the stimulus velocity.

Since the nDT, like the nPT, projects bilaterally to the optic tectum (Fig. 16), it could well carry vestibular signals to this target. In this respect it is interesting to note that the nuclei which are probably involved in gaze stabilizing reflexes *and* internal computation of self-motion project not only to the optic tectum but also to the medulla oblongata (Naujoks-Manteuffel et al. 1986; Naujoks-Manteuffel & Manteuffel 1988). Therefore, the same nuclei may mediate both gaze stabilization and internal self-motion computation. Furthermore, the nuclei involved in the compensation of rotatory components of self-motion, the visual nPT and the vestibular nDT, are functionally lateralized in the same way. When the head is turned to the right, for example, the visual world shifts to the left. The rightward turning activates the left nDT and the visual shift activates the left nPT.

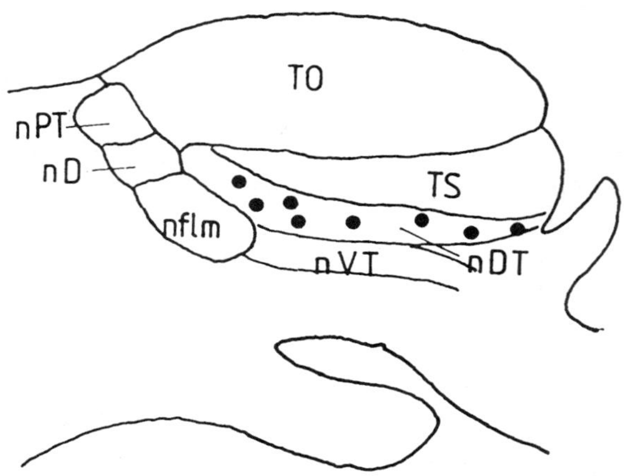

Figure 14. Recording sites of neurons sensitive to contraversive rotation on a turntable in the dark (closed circles) as projected onto a parasagittal plane of the salamander midbrain. For abbreviations see Fig. 4.

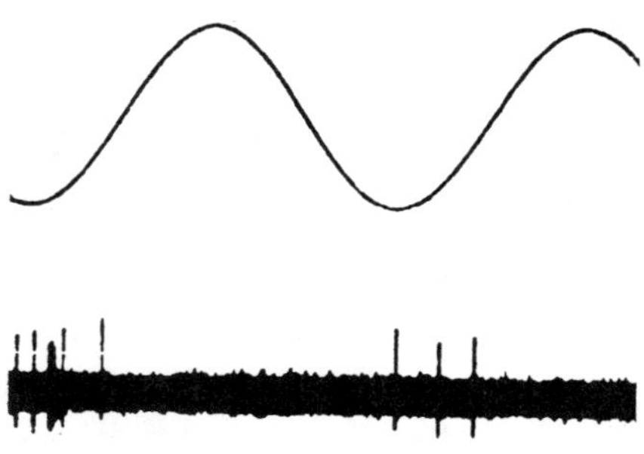

Figure 15. Response of an nDT neuron to sinusoidal stimulation on a turntable in the dark. Stimulus frequency, 0.1 Hz; stimulus amplitude, 20°. The contraversive component of the turntable movement corresponds to the negative part of the stimulus velocity recording (upper trace).

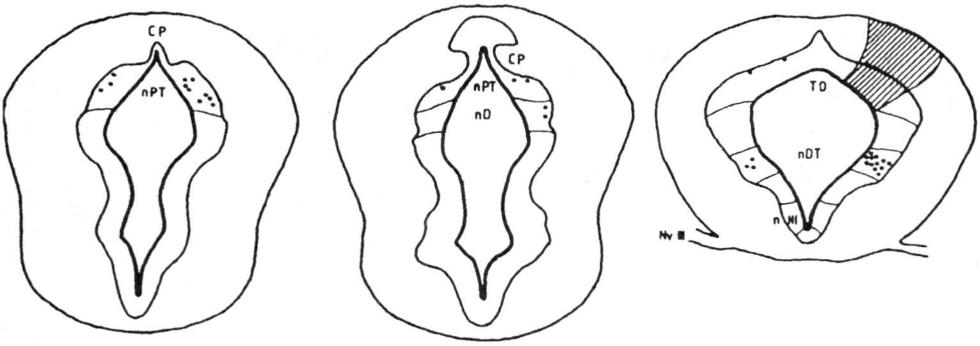

Figure 16. Cells afferent to the optic tectum as revealed by retrograde labeling with horseradish peroxidase injected into the area indicated by hatching. After Finkenstädt et al. (1983), Rettig (1984), and own unpublished results.

Summarizing the anatomical and electrophysiological data presented so far, all prerequisites for an effective self-motion computation are present in the localization and identification systems of the optic tectum. The tectum receives afferents from the visual background motion centers (nPT, nD) working in the low velocity range, and from the vestibular motion center, working in the range of high velocities. These inputs are able to predict the tectal shift of the center of excitation caused by visual objects moving as a result of self-motion. All objects that do *not* move in the predicted way are of potential interest, because they move independent of the subject. By this procedure, a non-topographic ("parametric") map (see also Mallot & von Seelen, this volume) is superimposed onto the topographically organized array of tectal columns.

According to the proposed model (Fig. 17), this prediction is performed by temporally establishing a directional selectivity among functional columns in the tectum that is oriented opposite to the direction of the shift of the visual environment. If we make the reasonable assumption that lateral inhibition exists between tectal columns, then the afferents from the nPT and nDT may enhance this inhibition asymmetrically. For afferents signaling rotatory movements, this operation has to happen in both tectal hemispheres, but in different directions. The reasons are the following: If an object is in a more or less nasal position in front of the animal and if the animal's head turns to the right, then the object will apparently shift to the left. In the tectum the point of excitation moves from a caudal to a rostral position in the left hemisphere and from a rostral to a caudal position in the right tectal hemisphere, due to the properties of the retino-tectal topography. The head turning to the right will also stimulate the left nPT and nDT (and only the left ones) which, in turn, suppress the responses to caudo-rostral excitation movements in the left tectum and to rostro-caudal movements in the right tectum. In fact, the nuclei sensitive for rotatory movements project bilaterally to the optic tectum (Fig. 16). In contrast to the "rotation nuclei", a nucleus sensitive to translatory movements (i.e., the nD) projects

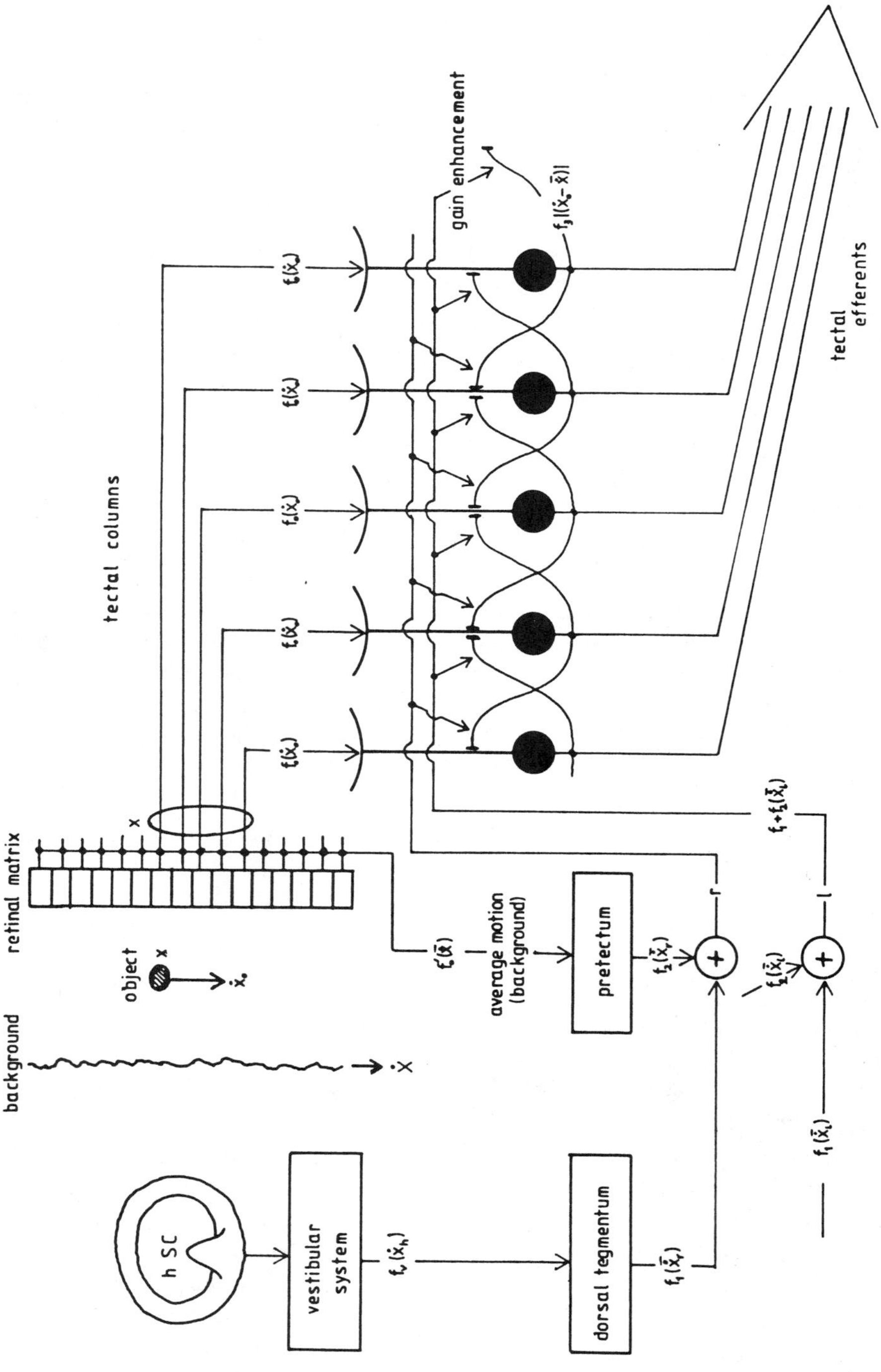

tectal columns
retinal matrix
background
object
average motion (background)
pretectum
h SC
vestibular system
dorsal tegmentum
gain enhancement
tectal efferents

ipsilaterally to the tectum, since the same happens on both sides during forward or backward movement.

The major consequence of this model is that the optic tectum is re-adjusted by suppressing all excitations that are equal in velocity and direction to the background shift. The tectum will therefore be stimulated only by those objects which differ from the background movements in one or both of these parameters. The tectum can then calculate the "real" movement of the object (i.e., the object's motion with respect to the earth's coordinates), although the retinal shift of the object's image remains within the animal's coordinate system. This can be easily demonstrated in the following situation: Imagine that an object moves to the right at a velocity of 5°/s and the animal turns its head to the right at a velocity of 10°/s. The background shift over the retina would then amount to 10°/s to the left and the object's retinal image velocity would be 5°/s to the left. The model predicts a perceived velocity in the optic tectum of 5°/s to the right, which is the object's real velocity. However, the site of excitation in the optic tectum remains within subject-dependent coordinates, ensuring the execution of correct saccades.

8 Discussion and Conclusions

It was shown that in salamanders extraction of objects from the background during self-motion is probably achieved by two cooperative mechanisms. Firstly, the gaze-stabilizing reflexes produce a fairly stable image by preserving a sufficiently straight gaze direction even when the body undulates during walking. Thus, the retinal units are enabled to remain within their low-velocity working range. Secondly, the internal self-motion computer calculates the remaining uncompensated image shifts. It produces a "neuronal gaze stabilization" that enables the animal to perceive the relevant visual world as being stable. Both systems apparently use the same sensory inputs (i.e., visual and vestibular) that divide into a descending component, mediating the motor commands for the gaze stabilization reflexes, and a tectal

Figure 17. Model of the intrinsic system of self-motion computation in salamanders. An object with a velocity of $\dot{x}_o$ at position x is moving across a background which is itself moving at a velocity $\dot{x}$. The object velocity is encoded by the firing frequency of the retinal units according to the function $f_o(\dot{x}_o)$. The object position is encoded by the retinal matrix and preserved by the topic projections onto the tectum, whereby the continuously changing position of the object on the retina successively stimulates adjacent tectal columns. Simultaneously, the responses to all objects constituting the background are averaged and transferred to the pretectum by specialized retinal ganglion cells on the basis of their particular velocity functions. If background movement occurs as a result of self-motion, the vestibular system will send a corresponding signal to the dorsal tegmentum. Efferent signals from the pretectum and the dorsal tegmentum are summated (probably in the tectum itself). Output from the nPT or nDT is dependent on movement in a particular direction (right, r; left, l). Such output can then enhance the lateral inhibition of the tectal columns asymmetrically, according to the direction coded by a particular nucleus. The efferent tectal signal will then be proportional to the difference between the velocities of object and background, thus encoding relative movement. For further explanations see text.

afferent component establishing the intrinsic visual shift computation. Since the retinal slip velocity and the actual head movement velocity represent the inputs not only for gaze stabilization but also for intrinsic self-motion computation, the models described for each of the two mechanisms, can be easily combined at the levels of the pretectal and vestibular outputs.

8.1 Consequences of the Model for Internal Self-Motion Computation

Since the proposed model of internal self-motion computation introduces the pretectum as a structure sensitive for (background-) movements, it would seem necessary to consider whether the model is consistent with previous behavioral findings following pretectal lesions. The model is readily compatible with that proposed by an der Heiden & Roth (1983) for the optic tectum (an der Heiden & Roth, this volume), since in their model the filter properties of tectal units are intrinsic and mediated by intra-tectal lateral inhibition and various types of coupling of retinal units. Ewert & v. Wietersheim (1974) and Finkenstädt & Ewert (1983), however, have reported a disinhibition of prey-catching behavior after pretectal lesions (see also Schürg-Pfeiffer, this volume). According to this result, pretecto-tectal inhibitory interactions are involved in configurational analysis, and this idea is used in Arbib's (1982) model of "Rana computatrix" (see Arbib, this volume).

What would happen according to the model of pretecto-tectal interaction presented here, if an amphibian were faced with a non-prey object (a large square or a rectangle elongated perpendicular to its movement)? Such stimuli would, in fact, activate pretectal cells (Ewert 1971; Himstedt et al. 1987). Any pretectal activation, however, would lead to an appropriate tectal inhibition in the direction of stimulus movement, and the tectal cells would show little or no response to these stimuli. Furthermore, it has been demonstrated that two objects which evoke prey capture when presented separately are ineffective when moved coherently (Ewert et al. 1970). This is easily explained by the model. Since multiple objects moving coherently activate the "summing-up units" in the pretectum, the tectum will be effectively inhibited by such objects. The model predicts, however, that this inhibition would not occur if the two objects move incoherently, e.g., in different directions. Since it is reasonable to assume that single pretectal cells do not project to the entire tectum, a restricted pretectal lesion would produce only a focal disinhibition in the tectum (see Ewert, this volume) although each pretectal cell sums up the average motion over nearly the whole visual field. In this respect it is also important to remember that pretectal lesions lead to a significant enlargement of tectal neuron's receptive fields (Ingle 1973; Finkenstädt & Ewert 1983). This is likely to be a result of the lost sustained enhancement of lateral inhibition from spontaneous activity of the pretectum.

The effect of pretectal lesions on depth perception in amphibians would seem, at first glance, to be less amenable to interpretation within the framework of the proposed pretecto-tectal model. Frogs and toads are able to estimate depth and to avoid obstacles on their way toward a prey object (Ingle 1976; Lock & Collett 1979). These animals, therefore, seem to construct a representation of the three-dimensional world in their visual systems. This capability is lost substantially after pretectal lesions

(Ingle 1980). If we consider that an emergent property of the proposed model is to extract any kind of motion that is incoherent with the average background motion, then this lesion-induced effect can also be explained. Despite the existence of several other mechanisms for estimating depth, such as binocular disparity and accommodation, a major depth-perception mechanism seems to be motion parallax, since it works under monocular conditions as well. Imagine an animal that moves its head either intentionally or during walking. In either case, an object that is closer to the subject than the background will *apparently* move relative to the background. This relative motion will be perceived by the tectum and will be the more effective the closer the object is, since the amplitude of the apparent movement increases with decreasing distance between subject and object. This dependency of the degree of activation on the distance of an object thus provides the animal with an additional mechanism for depth perception. The capability to perceive relative apparent motion, however, depends on an intact pretectum.

Instead of teleological conceptions such as "the pretectum is made for feature filtering" or is made "for generating nystagmus" or is made "to perceive depth", one should rather accept that the pretectum has evolved in such a way that all of the above statements are correct but none of them represents the complete truth. All the performances that are seemingly located within the pretectum have been studied as single phenomena during the last years. The model presented here, does not contradict earlier theories on pretecto-tectal function but tries to show that all of the observations gained in the past are consequences of one and the same mechanism in which perception of coherent motion, on the one hand, and form identification and object localization, on the other, are ascribed to two central structures, the pretectum (and probably the nDT) and the optic tectum, respectively.

8.2 Implications for the Function of Visual Systems

It has been shown in a variety of vertebrates that feature-extracting neurons in the visual system are sensitive to relative movements (for review see Allman et al. 1985). This finding has recently been corroborated in amphibians (Tsai & Ewert 1987). The model presented here proposes that the underlying neuronal mechanism is based upon parallel distributed processing of motion and form parameters. Motion processing can thereby take advantage of a number of sensory systems.

The computation of relative rather than absolute movement enables the visual system to readily achieve an object-background discrimination, which is one of the basic requirements in vision. In the cat, Hoffmann & Schoppmann (1981) have demonstrated that visual pretectal neurons are inhibited in their responses to large area patterns if parts of the pattern are moved in different directions through the receptive fields. This can be interpreted as a disrupted movement coherence in a part of the visual environment evoked by a separately moving object. If this result is applied to the model proposed here, then any object would be able to locally disinhibit the tectum and could thus be interpreted as a cohesive form even if it were not surrounded by solid lines.

Recently, data have accumulated which show that in advanced mammals a similar mechanism seems to be established by interactions of the suprasylvian cortex (area MT in monkeys) and the striate visual cortex (cf. Ferrera & Wilson 1987; Hamada 1987; Orban et al. 1987). Since the extrastriate area MT has many functional similarities with the pretectum, and the striate cortex is generally accepted as a feature filter, the interactions of both can perform what has been proposed for salamanders and toads using pretecto-tectal interactions.

8.3 Can the Salamander Help Engineers to Improve Technical Image Analysis?

As shown above, motion can provide the information about which parts of a visual scene constitute an object. Motion is by no means a disturbing parameter but may be helpful in order to solve some of the problems associated with artificial image processing (e.g., see Reitboeck, this volume).

The visual systems of animals have evolved in a natural three-dimensional environment, and probably took advantage of the depth of each scene and of different kinds of motion at various locations. Motion can help to re-establish the three-dimensional world within the visual system, even with just one eye (or camera). Reliance on motion cues would seem to be an even simpler mechanism than creating depth using binocular disparity, whereby each point in one image has to be unequivocally attributed to the corresponding point in the other image.

By contrast, coherent movement links all points belonging to one object. Under the assumption that an object is rigid, coherent movement itself (or the steady change in movement vectors) will determine surface and borders of the object. I have tried to demonstrate how, in a simple animal, nature may have constructed such a mechanism. It is possible that object constitution, object-background discrimination, and depth perception may be achieved in a relatively simple manner by using motion as a cue.

Acknowledgements. I am indepted to all co-workers who, during the years, helped to gain insight into the salamander brain. Helpful remarks on the manuscript by C. Naujoks-Manteuffel and G. Roth are gratefully acknowledged. I thank A. Burt for improving the English style.

9 References

Allman J, Miezin F, McGuiness E (1985) Stimulus specific responses from beyond the classical receptive field: neurophysiological mechanisms for local-global comparisons in visual neurons. *Ann Rev Neurosci* 8: 407-430

Arbib MA (1982) Modelling neural mechanisms of visuomotor coordination in frog and toad. In: Amari S, Arbib MA (eds) *Competition and cooperation in neural nets. Lecture notes in biomathematics, Vol 45.* Springer-Verlag, Berlin Heidelberg New York, pp 342-370

Birukow G (1938) Untersuchungen über den optischen Drehnystagmus und über die Sehschärfe des Grasfrosches (*Rana temporaria*). *Z Vergl Physiol* 25: 92-142

Blanks RHI, Precht W (1976) Functional characterization of primary vestibular afferents in the frog. *Exp Brain Res* 25: 369-390

Burghagen H, Ewert J-P (1983) Influence of the background for discriminating object motion from self-induced motion in toads *Bufo bufo* (L). *J Comp Physiol* 152: 241-249

Bürgi S (1957) Das Tectum opticum. Seine Verbindungen bei der Katze und seine Bedeutung beim Menschen. *Dtsch Zeitschr Nervenheilk* 176: 701-729

Büttner U, Waespe W, Miles TS (1978) Transfer characteristics of the vestibular system determined from nystagmus and neuronal activity in the alert monkey. In: Butenandt E, Hauske G (eds) *Kybernetik '77*. Oldenbourg-Verlag, pp 126-136

Collewijn H (1980) Sensory control of optokinetic nystagmus in the rabbit. *TINS* 11: 277-280

Collewijn H (1981) The oculomotor system of the rabbit and its plasticity. In: Braitenberg V (ed) *Studies of brain function, Vol 5*. Springer Verlag, Berlin Heidelberg New York

Dacey DM, Ulinski PS (1986) Optic tectum of the eastern garter snake, *Thamnophis sirtalis*. V: Morphology of brainstem afferents and general discussion. *J Comp Neurol* 245: 423-453

Dieringer N (1987) The role of compensatory eye and head movements for gaze stabilization in the unrestrained frog. *Brain Res* 404: 33-38

Dieringer N, Cochran SL, Precht W (1983) Differences in the central organization of gaze stabilizing reflexes between frog and turtle. *J Comp Physiol* 153: 495-508

Ewert J-P (1971) Single unit response of the toad's (*Bufo americanus*) caudal thalamus to visual objects. *Z Vergl Physiol* 74: 81-102

Ewert J-P (1973) Lokalisation und Identifikation im visuellen System der Wirbeltiere. *Fortschr Zoologie* 21: 307-337

Ewert J-P (1974) The neural basis of visually guided behavior. *Sci Amer* 230:34-42

Ewert J-P (1984) Tectal mechanisms that underlie prey-catching and avoidance behaviors in toads. In: Vanegas H (ed) *Comparative neurology of the optic tectum*. Plenum Press, London New York, pp 247-416

Ewert J-P, Wietersheim A v (1974) Einfluß von Thalamus/Praetecttum-Defekten auf die Antworten von Tectum-Neuronen gegenüber bewegten visuellen Mustern bei der Kröte *Bufo bufo* (L). *J Comp Physiol* 92: 149-160

Ewert J-P, Speckhardt I, Amelang W (1970) Visuelle Inhibition und Exzitation im Beutefangverhalten der Erdkröte (*Bufo bufo* L). *Z Vergl Physiol* 68: 84-110

Ferrera VP, Wilson HR (1987) Direction specific masking and the analysis of motion in two dimensions. *Vision Res* 27: 1783-1796

Finkenstädt T, Ewert J-P (1983) Visual pattern discrimination through interactions of neural networks: a combined electrical brain stimulation, brain lesion, and extracellular recording study in *Salamandra salamandra*. *J Comp Physiol* 153: 99-110

Finkenstädt T, Ebbesson SOE, Ewert J-P (1983) Projections to the midbrain tectum in *Salamandra salamandra* L. *Cell Tiss Res* 234: 39-55

Fite KV, Montgomery N, Wojcicki C, Bengston L (1980) Visuomotor correlates of the anuran accessory optic system. *Neurosci Abstr* 6: 839

Fritzsch B (1980) Retinal projections in European salamandridae. *Cell Tiss Res* 213: 325-341

Gioanni H, Rey J, Villalobos J, Richard D, Dalbera A (1983) Optokinetic nystagmus in the pigeon (*Columba livia*). II: Role of the pretectal nucleus of the accessory optic system (AOS). *Exp Brain Res* 50: 237-247

Gioanni H, Rey J, Villalobos J, Dalbera A (1984) Single unit activity in the nucleus of the basal optic root (nBOR) during optokinetic, vestibular and visuo-vestibular stimulations in the alert pigeon (*Columba livia*). *Exp Brain Res* 57: 49-60

Grasse KL, Cynader MS (1984) Electrophysiology of lateral and dorsal terminal nuclei of the cat accessory optic system. *J Neurophysiol* 51: 276-293

Gruberg ER, Grasse KL (1984) Basal optic complex in the frog (*Rana pipiens*): a physiological and HRP study. *J Neurophysiol* 51: 98-110

Hamada T (1987) Neural response to the motion of textures in the lateral suprasylvian area of cats. *Behav Brain Res* 25: 175-185

Hartmann R, Klinke R (1975) System analysis properties of primary vestibular fibers. *Exp Brain Res Suppl* 23: 85

Heiden U an der, Roth G (1983) Cooperative neural processes in amphibian visual prey recognition. In: Basar E, Flohr H, Haken H, Mandell AJ (eds) *Synergetics of the brain.* Springer-Verlag, Berlin Heidelberg New York, pp 299-310

Herrick CJ (1948) The brain of the tiger salamander *Ambystoma tigrinum.* Univ Chicago Press, Chicago

Himstedt W, Roth G (1980) Neuronal responses in the tectum opticum of *Salamandra* to visual prey stimuli. *J Comp Physiol* 135: 251-257

Himstedt W, Heller K, Manteuffel G (1987) Neuronal responses to moving visual stimuli in different thalamic and midbrain centers of *Salamandra salamandra* (L). *Zool Jb Physiol* 91: 243-256

Hoffmann K-P, Schoppmann A (1981) A quantitative analysis of direction-specific responses of neurons in the cat's nucleus of the optic tract. *Exp Brain Res* 42: 146-157

Hoffmann K-P, Huber HP (1983) Responses to visual stimulation in single cells in the nucleus of the optic tract (NOT) during optokinetic nystagmus (OKN) in the awake cat. *Neurosci Abstr* 9: 1048

Holst E v, Mittelstaedt H (1950) Das Reafferenzprinzip (Wechselwirkungen zwischen Zentralnervensystem und Peripherie). *Naturwissenschaften* 37: 464-476

Ingle DJ (1970) Visuomotor functions of the frog optic tectum. *Brain Behav Evol* 3: 57-71

Ingle DJ (1973) Two visual systems in the frog. *Science* 181: 1053-1055

Ingle DJ (1976) Spatial vision in anurans. In: Fite KV (ed) *The amphibian visual system, a multidisciplinary approach.* Academic Press, New York, pp 119-140

Ingle DJ (1980) Some effects of pretectum lesions on the frog's detection of stationary objects. *Behav Brain Res* 1: 139-163

Kato I, Harada K, Hasegawa T, Igarashi T, Koike Y, Kawasaki T (1986) Role of the nucleus of the optic tract in monkeys in relation to optokinetic nystagmus. *Brain Res* 364: 12-22

Kopp J (1987a) *Quantitative Analyse des optokinetischen und vestibulären Nystagmus bei Amphibien.* PhD Thesis, Tech Univ Darmstadt

Kopp J (1987b) Vestibulo-collic reflex (VCR) and optokinetic reflex (OKR) interaction in the fire salamander (*Salamandra salamandra*). In: Elsner N, Creutzfeldt O (eds) *New frontiers in brain research.* Thieme-Verlag, Stuttgart, p 154

Kopp J, Manteuffel G (1984) Quantitative analysis of salamander horizontal head nystagmus. *Brain Behav Evol* 25: 187-196

Kostyk SK, Grobstein P (1987) Neuronal organization underlying visually elicited prey orienting in the frog. I: Effects of various unilateral lesions. *Neuroscience* 21: 41-55

Lara R, Arbib MA, Cromarty AS (1982) The role of the tectal column in facilitation of amphibian prey-catching behavior: a neural model. *J Neurosci* 2: 521-530

Lázár G (1973) Role of the accessory optic system in the optokinetic nystagmus of the frog. *Brain Behav Evol* 5: 443-460

Lázár G (1983) Transection of the basal optic root in the frog abolishes vertical optokinetic head-nystagmus. *Neurosci Lett* 43: 7-11

Lázár G, Tóth P (1983) Optokinetic horizontal head nystagmus in the frog after regeneration of transected retinal fibers. *Acta Biol Acad Sci Hung* 34: 371-383

Lock A, Collett T (1979) A toad's devious approach to its prey: a study of some complex uses of depth vision. *J Comp Physiol* 131: 179-189

Manteuffel G (1982) The accessory optic system in the newt, *Triturus cristatus*: unitary response properties from the basal optic neuropil. *Brain Behav Evol* 21: 175-184

Manteuffel G (1984a) Electrophysiology and anatomy of direction-specific pretectal units in *Salamandra salamandra*. *Exp Brain Res* 54: 415-425

Manteuffel G (1984b) A "physiological" model for the salamander horizontal head nystagmus. *Brain Behav Evol* 25: 197-205

Manteuffel G (1985) Monocular and binocular optic inputs to salamander pretectal neurons: intracellular recording and HRP labelling study. *Brain Behav Evol* 27: 1-10

Manteuffel G (1987) Binocular afferents to the salamander pretectum mediate rotation sensitivity of cells selective for visual background motions. *Brain Res* 422: 381-383

Manteuffel G, Naujoks-Manteuffel C (1987) Synergistic visual and vestibular self motion related inputs to the optic tectum of salamanders. *Neuroscience Suppl* 22: 737

Manteuffel G, Petersen J, Himstedt W (1983) Optic nystagmus and nystagmogen centers in the European fire salamander (*Salamandra salamandra*). *Zool Jb Physiol* 87: 113-125

Manteuffel G, Kopp J, Himstedt W (1986) Amphibian optokinetic afternystagmus: properties and comparative analysis in various species. *Brain Behav Evol* 28: 186-197

Montgomery N, Fite KV, Bengston L (1981) The accessory optic system of *Rana pipiens*: neuroanatomical connections and intrinsic organization. *J Comp Neurol* 203: 595-612

Montgomery N, Fite KV, Taylor M, Bengston L (1982) Neural correlates of optokinetic nystagmus in the mesencephalon of *Rana pipiens*: a functional analysis. *Brain Behav Evol* 21: 137-150

Montgomery N, Fite KV, Grigonis AM (1985) The pretectal nucleus (lentiformis mesencephali) of *Rana pipiens*. *J Comp Neurol* 234: 264-275

Naujoks-Manteuffel C, Manteuffel G (1986) Internuclear connections between the pretectum and the accessory optic system in *Salamandra salamandra*. *Cell Tiss Res* 243: 595-602

Naujoks-Manteuffel C, Manteuffel G (1987) Afferents to the torus semicircularis in *Salamandra salamandra* (Amphibia, Urodela). *Neurosci Suppl* 22: 773

Naujoks-Manteuffel C, Manteuffel G (1988) The origins of descending projections to the medulla oblongata and rostral medulla spinalis in the urodele *Salamandra salamandra* (Amphibia). *J Comp Neurol* 273: 187-206

Naujoks-Manteuffel C, Manteuffel G, Himstedt W (1986) Deszendierende Bahnen aus dem Di-und Mesencephalon des Feuersalamanders (*Salamandra salamandra* L). *Verh Dtsch Zool Ges* 79: 373

Orban GA, Gulyas B, Vogels R (1987) Influence of a moving textured background on direction selectivity of cat striate neurons. *J Neurophysiol* 57: 1792-1812

Rettig G (1984) *Neuroanatomische Untersuchungen der visuellen Projektionen bei Salamandern (Ordnung Caudata)*. PhD Thesis, Univ Bremen

Rettig G, Roth G (1986) Retinofugal projections in salamanders of the family Plethodontidae. *Cell Tiss Res* 243: 385-396

Robinson DA (1976) Adaptive gain control of vestibulo-ocular reflex by the cerebellum. *J Neurophysiol* 39: 954-969

Roth G (1982) Beuteerkennungsmechanismen im Tectum opticum von Amphibien - eine vergleichende Untersuchung. *Funkt Biol Med* 1: 90-98

Roth G (1987) Visual behavior in salamanders. In Braitenberg V (ed) *Studies of brain function, Vol 14*. Springer Verlag, Berlin Heidelberg New York

Simpson JI (1984) The accessory optic system. *Ann Rev Neurosci* 7: 13-41

Simpson JI, Soodak RE (1978) The accessory optic system: a visual system in vestibular coordinates. *Soc*

Neurosci Abstr 4: 645

Sperl M, Manteuffel G (1987) Directional selectivities of visual afferents to the pretectal neuropil in the fire salamander. *Brain Res* 404: 332-334

Springer AD, Easter SS, Agranoff BW (1977) The role of the optic tectum in various visually mediated behaviors of goldfish. *Brain Res* 128: 393-404

Stein BE, Gaither NS (1981) Sensory representation in reptilian optic tectum: some comparisons with mammals. *J Comp Neurol* 202: 69-87

Tauber ES, Atkin A (1968) Optomotor responses to monocular stimulation: relation to visual system organization. *Science* 160: 1365-1367

Taylor AM, Jeffery G, Lieberman AR (1986) Subcortical afferent and efferent connections of the superior colliculus in the rat and comparisons between albino and pigmented strains. *Exp Brain Res* 62: 131-142

Tsai HJ, Ewert J-P (1987) Edge preference of retinal and tectal neurons in common toads (*Bufo bufo*) in response to worm-like moving stripes: the question of behaviorally relevant "position-indicators". *J Comp Physiol* 161: 295-304

Tsai HJ, Ewert J-P (1988) Influence of stationary and moving background structures on the response of visual neurons in toads (*Bufo bufo*). *Brain Behav Evol* (in press)

Wilczynski W (1981) Afferents to the midbrain auditory center in the bullfrog, *Rana catesbeiana*. *J Comp Neurol* 198: 421-433

Wilczynski W, Northcutt RG (1977) Afferents to the optic tectum of the leopard frog: an HRP study. *J Comp Neurol* 173: 219- 230

Nucleus Isthmi and Optic Tectum in Frogs

EDWARD R. GRUBERG

Biology Department, Temple University
Philadelphia, PA 19122, USA

Abstract. Our laboratory has been investigating the role of nucleus isthmi in tectally mediated visual information processing. We have found that unilateral ablation of n. isthmi results in a contralateral monocular scotoma. Within the scotoma the animal does not respond to visually presented prey or threat stimuli. The size of the scotoma is dependent on the volume of n. isthmi ablated. When the entire nucleus is ablated there is no recovery of function within the scotoma for up to 11 months. Unilateral injection into n. isthmi of the excitatory neurotoxin ibotenic acid results in a temporary loss of responses to visually presented prey in the contralateral monocular field. Normal responses return within 1 to 2 weeks. Bilateral ablation of n. isthmi results in loss of responsiveness to visually presented prey and threat stimuli in the entire visual field. After 4 to 5 weeks there is a return of responsiveness in the binocular field. Nucleus isthmi fibers are the predominant non-retinal extrinsic input to the superficial tectal layers. The contralaterally projecting isthmo-tectal fibers cross in the posterior part of the optic chiasm. When this part is transected, the animal no longer responds to visually presented threat stimuli but responds normally to visually presented prey stimuli. Nucleus isthmi is virtually the only cholinergic input to the tectum. There is evidence from other laboratories that some retino-tectal fibers contain membrane bound acetylcholine receptors. Thus, n. isthmi might directly modify retinal input. Our evidence is compatible with the model of Arbib & House with n. isthmi playing a role in prey localization.

1 Introduction

There is a flow of information in the brain from the sensory side to the motor side. There are also feedback pathways at different levels of sensory/motor processing. I will describe a feedback system centered on the frog's nucleus isthmi that is associated with the retino-tectal system. It has become of interest to my colleagues and me because disruption of n. isthmi affects the responsiveness of frogs to certain important visual stimuli. The work I will discuss is based on collaborative efforts with Mark Wallace, Robert Waldeck, Weslia Patterson, Hanan Caine, Alex Ricciuti, Paul Desan, Jerome Lettvin, and Susan Udin.

It has been long known that the optic tectum mediates visually significant forms of information. Ingle (1973), for example, showed that unilateral removal of the optic tectum produced a contralateral monocular field scotoma to visually presented prey and threat stimuli. We have found that unilateral ablation of the nucleus isthmi leads to a similar deficit (Caine & Gruberg 1985). Nucleus isthmi receives its principal input from the ipsilateral optic tectum and projects bilaterally to the optic tectum (Gruberg & Udin 1978). Mammals have an equivalent structure called the parabigeminal nucleus. It has similar connections with the superior colliculus. In addition the parabigeminal nucleus projects to the lateral geniculate nucleus (Graybiel 1978; Sherk 1979).

In this paper I will begin with a brief description of n. isthmi and then review some of our findings concerning the behavioral consequences of disrupting the nucleus. I will then show what we know of the anatomical substrate on which the behavior rests. I will also outline how we determined the likely neurotransmitter for isthmo-tectal fibers. Finally, I will speculate about a role for the n. isthmi feedback loop.

2 The Nucleus Isthmi

The nucleus isthmi of frogs is located in the posterior part of the midbrain tegmentum below the posterior pole of the optic tectum (Fig. 1). In the adult leopard frog, *Rana pipiens*, it is well developed and contains about 8,000 cells (Fig. 2). By comparison, each retina contains approximately 500,000 ganglion cells and each tectal lobe has about 1 million cells.

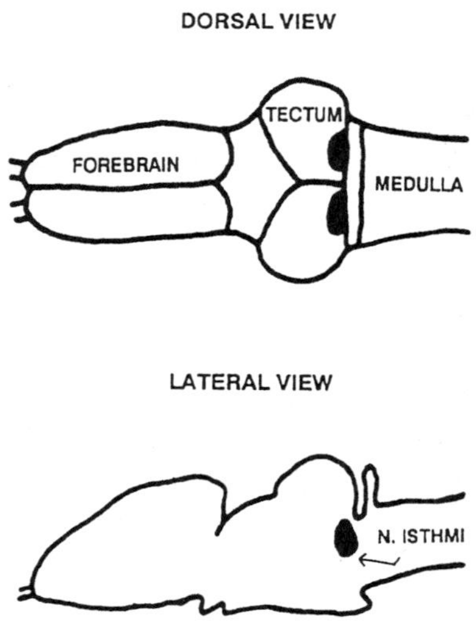

Figure 1. Location of n. isthmi (dark area) in frog brain. (From Caine & Gruberg 1985).

In curarized animals the nucleus can be monitored using a microelectrode that is passed through the posterior pole of the tectum. At the dorsal border of the nucleus are visually responsive units whose receptive fields are centered in the contralateral temporal field. As the electrode is advanced in 50 µm ventral steps there is a progressive nasal shift in the location of the receptive fields (Gruberg & Lettvin 1980). At the ventral boundary of the nucleus the receptive fields are centered in the ipsilateral nasal field. Our principal method of ablating n. isthmi is to pass current through a recording electrode placed in the nucleus. After the animal recovers overnight we start behavioral testing.

3 The Behavioral Deficits

Animals with unilateral n. isthmi lesions no longer respond to prey or threat stimuli in the contralateral monocular field. For prey stimulus a live cricket is tethered on a string that is attached to a rod. The cricket is placed at locations around the animal. For threat stimulus an 8 cm x 10 cm black object fixed at the end of a rod is moved toward the animal from different directions.

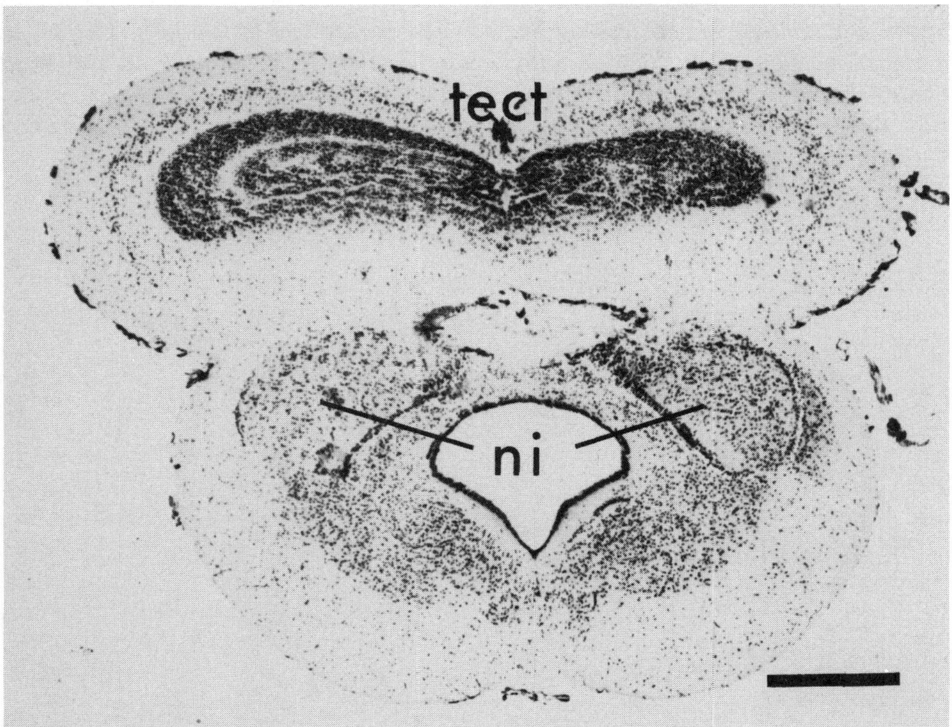

Figure 2. Transverse section of posterior midbrain. Nucleus isthmi (ni) below caudal pole of tectum (tect) Scale 500 µm.

3.1 Scotomas

We have mostly studied animals with unilateral n. isthmi lesions. We can compare the effects of stimuli inside and outside the scotoma. Thus, we can distinguish between circumscribed visual deficits and systemic effects. Depending on the extent of the n. isthmi lesion, the scotomas range from 30 degrees to 120 degrees. Within the scotomas we find no responses to prey or threat stimuli. The scotomas generally are stable in size and location. Usually there is a sharp border to the scotoma, and beyond its boundary the animal responds normally (Fig. 3). Outside the scotoma, if the prey is far away, the frog will orient, approach, and snap. If the prey is closer, the frog will orient and snap. Almost all scotomas include the extreme posterior field. Larger scotomas have more of the nasal area included. For threat stimuli outsided the scotoma the animal will turn and jump away from the stimulus. A threat approaching from a position opposite the scotoma will cause the animal to turn towards the scotoma. Thus, the frog has the ability to turn in any direction. Nucleus isthmi ablation does not appear to compromise turning abilities. As with normal frogs, after a series of threat stimuli outside the scotoma the animal tends to freeze to further stimuli.

3.2 Permanence of the Deficit

We have maintained animals with unilateral scotomas that extend over the entire monocular field for up to 11 months. We find no obvious return of responsiveness within the scotoma. Animals with initial scotomas of 90 degrees or less show some recovery of function within the scotoma. We are currently exploring how much recovery occurs.

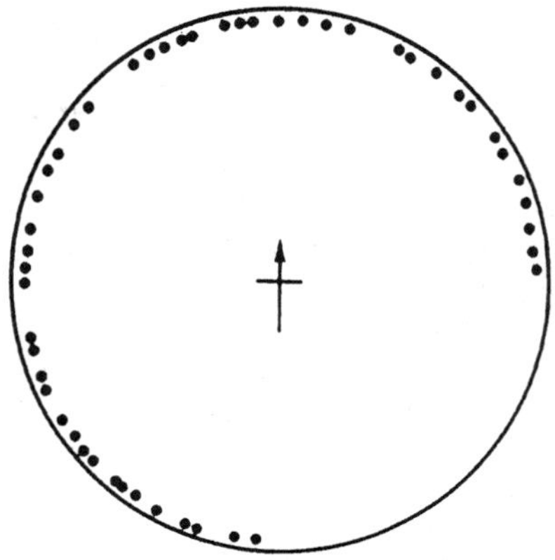

Figure 3. Angular locations of prey (filled circles) which elicited a turning response of a frog after a left n. isthmi lesion. Combined records of 3 test sessions in the second week after n. isthmi lesion. Arrow points in anterior direction. Line perpendicular to arrow is the axis of the frog's eyes.

3.3 Size of Scotoma *vs*. Size of Lesion

Throughout n. isthmi there is very high acetylcholinesterase (AChE) activity. AChE, which hydrolyzes acetylcholine, is a robust enzyme and activity remains after mild fixation of the brain. Staining for AChE is easy and reliable. We determine the extent of n. isthmi lesions by making serial sections through n. isthmi, staining for AChE and comparing the volume of the normal nucleus with whatever is left of the ablated nucleus. We find that there is a rough correspondence between size of scotoma and amount of isthmi ablated (Fig. 4).

3.4 Ibotenic Acid Application

In several animals we have injected ibotenic acid, an excitatory neurotoxin into n. isthmi. In such animals there is a temporary scotoma in the contralateral

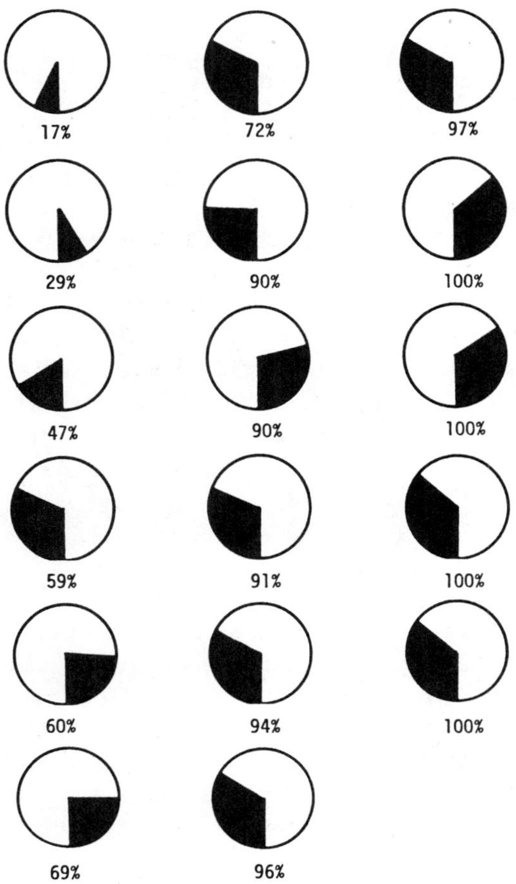

Figure 4. Size of scotoma (black area) compared to percentage of n. isthmi unilaterally ablated. Each case represents an animal studied for a month post-lesion before histological preparation.

monocular field. Within the scotoma there are no responses to visually presented prey stimuli. However, threat responses are normal everywhere. Our working assumption is that we have selectively injured a certain class of cells in n. isthmi. After 1 to 2 weeks the animal becomes behaviorally normal. In subsequent histology n. isthmi appears normal.

3.5 Bilateral N. Isthmi Lesions

Each tectal lobe receives retinal input from the contralateral retina. Yet when one records in the superficial tectum it is easy to record units that are driven by the ipsilateral retina. Nucleus isthmi is the link that makes each tectal lobe binocular. After n. isthmi ablation, visual units in the contralateral tectal lobe are driven only by the contralateral eye (Glasser & Ingle 1978; Grobstein et al. 1978). We have found that after bilateral n. isthmi lesions the animal is unresponsive to prey and threat stimuli everywhere in the visual field. However, after 4 to 5 weeks responses to prey and threat return in the binocular part of the visual field.

3.6 Conflicting Behavioral Data

So far only our laboratory has found the behavioral deficits described above. At the previous workshop Collett & Udin (1983) reported on the behavioral effects of ablating both n. isthmi. Their aim was to study animals with no ipsilateral retinal input to the tectum in order to assess the ability of the frogs to carry out accurate estimates of depth. They found that these frogs could still accurately assess depth. Collett & Udin (1983) proposed that such frogs used a method of triangulation which did not require binocular cues in each tectal lobe. As described above, in our hands, bilateral ablation of n. isthmi blinds animals to visually presented prey and threat stimuli throughout their visual field for at least a month. We thus have a clear discrepancy in results. We cannot easily account for this difference. Despite our histological assessment perhaps we have damaged other structures adjacent to n. isthmi which could compromise the frogs behavior. Conversely, Collett & Udin (1983) could have incompletely ablated n. isthmi.

Kostyk & Grobstein (1987) have hemisected the brain at the midbrain/hindbrain border in order to assess the behavioral effects caused by cutting descending tectal output. Such hemisected animals respond everywhere to visually presented prey, but they cannot turn past the midline on one side (they jump forward when the prey is in the hemifield). Our laboratory has confirmed these results. What happens when the hemisection is slightly further forward, so that n. isthmi is ablated? Kostyk & Grobstein (1987) have found no behavioral differences between animals where n. isthmi is ablated and where n. isthmi is spared. However, we find that when we hemisect and also ablate n. isthmi, we have additive behavioral defects. The animal has a turning disorder in the hemifield and a contralateral monocular scotoma to prey and threat stimuli. We cannot yet explain these contradictory results. As with the previous disagreement, it is possible that we were too vigorous in our ablation and so damaged structures adjacent to n. isthmi or, Grobstein's group had partially spared n. isthmi. It is also possible that we test the animals differently. Perhaps with our testing protocols there is a scotoma. With others doing the testing the animals are responsive. We intend to try to resolve these differences in the future.

4 Anatomy

The dramatic change in visually guided behavior after ablation of n. isthmi suggests that n. isthmi fibers are strategically located. We have studied the anatomy of the isthmo-tectal projection using autoradiographic techniques after injection of tritiated proline into n. isthmi. We have also orthogradely labeled isthmo-tectal fibers with horseradish peroxidase (HRP) and in the same animals injected both retinas with tritiated proline. Susan Udin and I showed that ipsilateral n. isthmi projects to most of the superficial layers of the tectum (Gruberg & Udin 1978). Retinal fibers project to much of the same layers that receive the ipsilaterally projecting n. isthmi input (Fig. 5). The contralaterally projecting n. isthmi fibers terminate in the most superficial layer of the tectum and a deeper (but still superficial) layer. The pathway taken by contralaterally projecting n. isthmi fibers is impressively indirect (Fig. 6). After leaving the nucleus, the fibers course laterally and slightly ventrally to the lateral surface of the midbrain tegmentum. The fibers then are directed forward along the surface through the entire midbrain below the lateral branch of the retino-tectal projection. They remain at the surface of the diencephalon below the principal optic tract until they reach the level of the posterior part of the optic chiasm. They then cross in the dorsal, posterior part of the chiasm admixed with retinal fibers (and not outside the chiasm as we earlier thought). From the chiasm to the tectum the n. isthmi fibers continue to be admixed with retinal fibers.

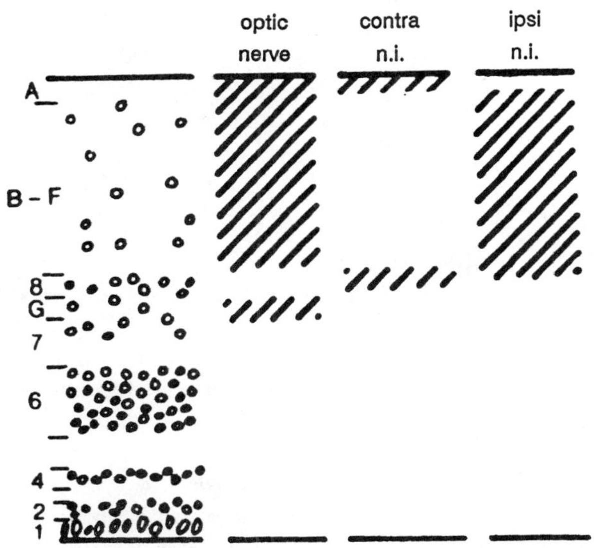

Figure 5. Approximate distribution of retinal and n. isthmi terminals in frog tectum. "A" is most superficial layer; "1" is layer adjacent to the ventricle. (From Gruberg & Lettvin 1980).

In order to see if other fiber systems also cross in the chiasm, we have cut the posterior part of the optic chiasm including the contralaterally projecting isthmi fibers and applied HRP to the cut. We find that only retina and n. isthmi are retrogradely labeled with HRP. An unexpected behavioral change occurs after cutting the back half of the chiasm. The post-lesion animals appear to be normal in their ability to carry out visually guided prey-catching. However, they no longer respond to visually presented threat stimuli anywhere in the visual field! So far we have seen this effect in

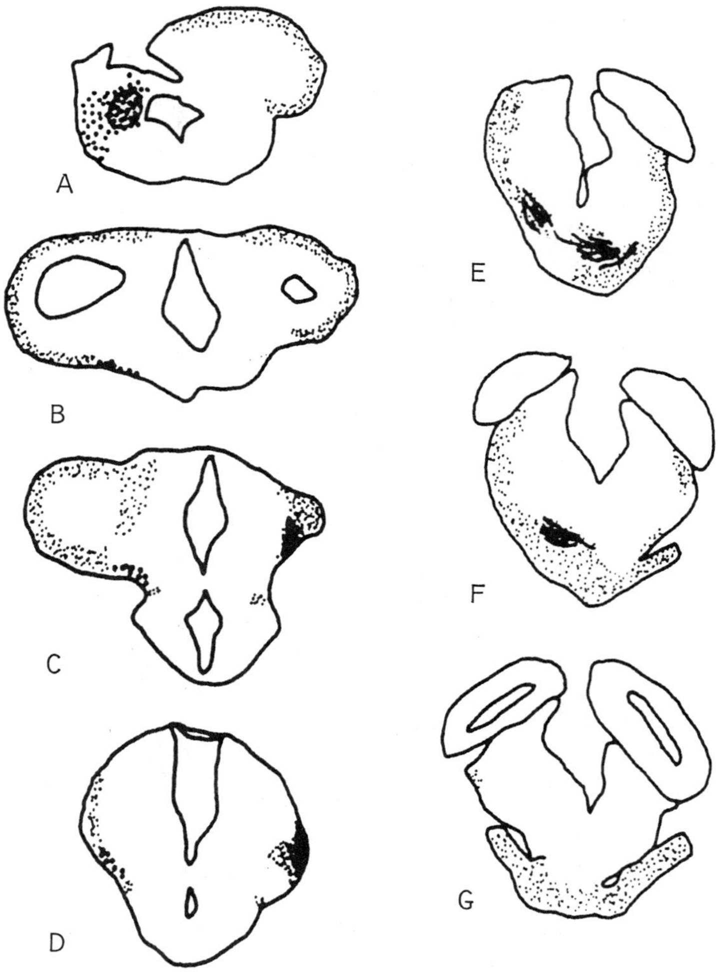

Figure 6. Camera lucida drawings of transverse sections of double labeled brain. HRP (large dots) was applied to n. isthmi (section A). Tritiated proline (fine dots) was injected into both retinas. Heavy lines and black areas indicate locations of overlap of retinofugal fibers and contralaterally projecting isthmo-tectal fibers. Section A is most posterior. Section G is most anterior section and is at the level of the anterior part of the optic chiasm.

seven animals. The effect lasts for several months. We are currently following one animal 5 months after such a cut and some threat avoidance behavior is now occasionally elicited. We cannot yet ascribe the behavioral deficit to either retinal fibers or n. isthmi fibers (or both).

We have also investigated whether other fiber systems end in the same superficial layers of the tectum as isthmo-tectal fibers and retino-tectal fibers. In some animals we have injected HRP superficially and in other animals more deeply into the tectum. After a survival period of several days we determine which structures backfill with HRP. We have also applied HRP to the tectal surface. The results are fairly consistent with the notion that n. isthmi and retina are the dominant extrinsic

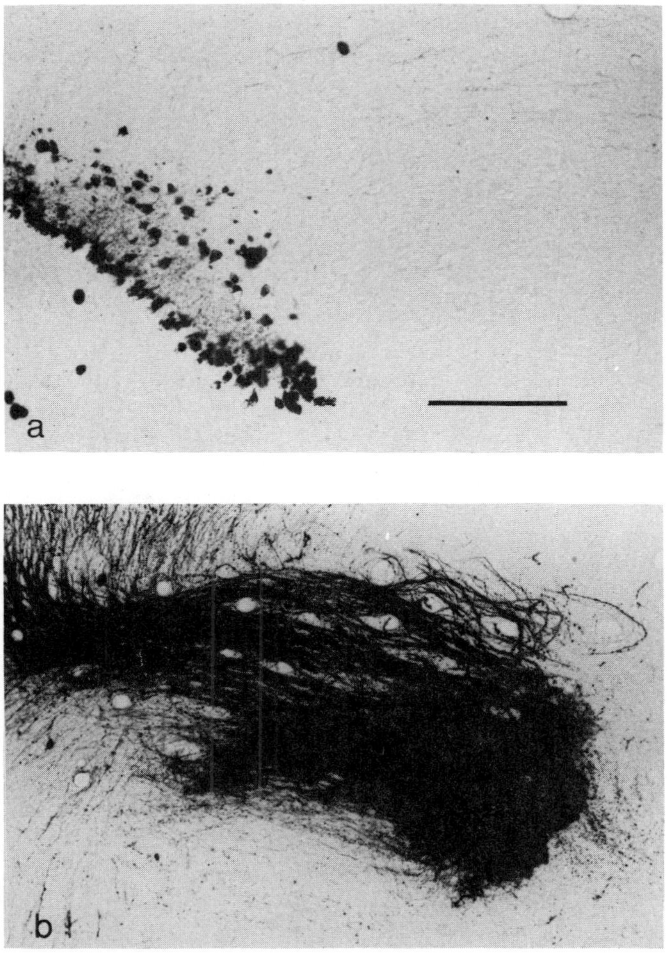

Figure 7. Photomicrographs from an animal where HRP was applied to the tectal surface. (a) Nucleus isthmi contralateral to application site showing retrograde labeling of cells; (b) Nucleus isthmi contralateral to application site showing orthogradely labeled tecto-isthmal fibers, but no retrogradely stained cells. Scale 100 μm.

inputs in the most superficial tectal layers. When we inject small volumes of HRP solutions into the superficial layers of the tectum, we find retrogradely labeled cells in n. isthmi bilaterally and retina contralaterally. We also find some stained cells in posterior thalamus and midbrain tegmentum. In deeper injections the midbrain tegmentum and posterior thalamus are more heavily represented.

There is however an anomaly. In other animals we have applied blotting paper containing dried HRP and lysolecithin (which aids in HRP diffusion past the pia) to the surface of the tectum. In the majority of cases only retinal cells and contralateral n. isthmi cells backfill with HRP (Fig. 7). In ipsilateral n. isthmi there are well stained orthogradely labeled tecto-isthmal fibers. In other cases we find retina and both n. isthmi backfilled. However, recently we have found other cases of HRP application where we backfill retinal cells, contralateral n. isthmi and cells in midbrain tegmentum and posterior thalamus but not ipsilateral n. isthmi. One possibility is that in certain areas of the tectum ipsi n. isthmi cells terminate near the tectal surface while in other places tegmental and pretectal terminals are near the surface. However, with our HRP injections we have yet to backfill tegmental and pretectal targets without our backfilling ipsilateral n. isthmi. On the other hand with superficial HRP injections we have backfilled n. isthmi bilaterally without backfilling tegmental and pretectal areas. We need more data.

5 Acetylcholine as the Isthmo-Tectal Neurotransmitter

The superficial layers of the tectum contain high activity of the enzyme AChE (Shen et al. 1955). These layers correspond to retinal recipient layers. It was originally supposed that the retino-tectal fibers were cholinergic. AChE is not an entirely reliable marker for cholinergic fibers. There can be false positives: some non-neural structures have high AChE levels. However, the tectum also contains significant activity of choline acetyltransferase (ChAT), the enzyme that synthesizes ACh. Generally, there is a close correspondence between cholinergic fibers and the presence of ChAT. We have found that unilateral eye enucleation or cutting the optic nerve has no detectable influence on ChAT activity in the tectum (Ricciuti & Gruberg 1985). However, when n. isthmi is unilaterally ablated, the ChAT activity in the ipsilateral tectum is approximately 1/3 the ChAT activity in the contralateral tectum. When bilateral n. isthmi lesions are made, the loss of ChAT activity is directly related to the amount of n. isthmi ablated. With regression analysis n. isthmi accounts for over 90% of the tectal ChAT activity. We have also carried out an immunocytochemical study using a polyclonal antibody to pig ChAT that cross-reacts with frog ChAT (Desan et al. 1987). In normal animals virtually all n. isthmi cells stain positively for ChAT (the enzyme is synthesized in cell bodies and is transported to the terminals (Fig. 8). The most superficial tectal layer has fairly dense staining for ChAT. The next layer is mostly free of staining. The deeper superficial layers contain a rete of stained fibers (Fig. 9). When n. isthmi is ablated unilaterally we find no staining of the most superficial layer of the contralateral tectal lobe and no staining of the rete of fibers of the ipsilateral tectal lobe (Fig. 9). Thus n. isthmi can account for all the tectal ChAT immunostaining.

6 What Tectal Elements Do N. Isthmi Fibers Affect?

The scotoma that occurs after n. isthmi ablation might suggest that n. isthmi processes are interposed between tectal and retinal elements. However, ultrastructural evidence (Székely et al. 1973) suggests that retinal fibers terminate primarily onto tectal dendritic processes. Nucleus isthmi could affect tectal function by influencing tectal cells. However, it has been found in goldfish that acetylcholine receptors (AChRs) are synthesized by retinal ganglion cells and transported to the optic tectum (Henley et al. 1986). Peter Sargent (pers. comm.) has found AChRs in the frog tectum, and they are also associated with retino-tectal fibers. An implication of these receptor studies is that cholinergic n. isthmi fibers could directly control cholinoceptive retino-tectal input. Possible arrangements of retinal, tectal, and isthmal elements are illustrated in Fig. 10. In order to determine the actual arrangement, further ultrastructural studies are needed.

7 What Does N. Isthmi Do?

The change in visually guided behavior following n. isthmi ablation is both dramatic and unrevealing about function. If there had been a subtler change in behavior, we might reasonably suggest a role for n. isthmi. House (1984) and Arbib & House (1985) suggest that the ipsilaterally projecting isthmo-tectal fibers could be used as a global tectal inhibitory circuit suppressing tectal output from everywhere

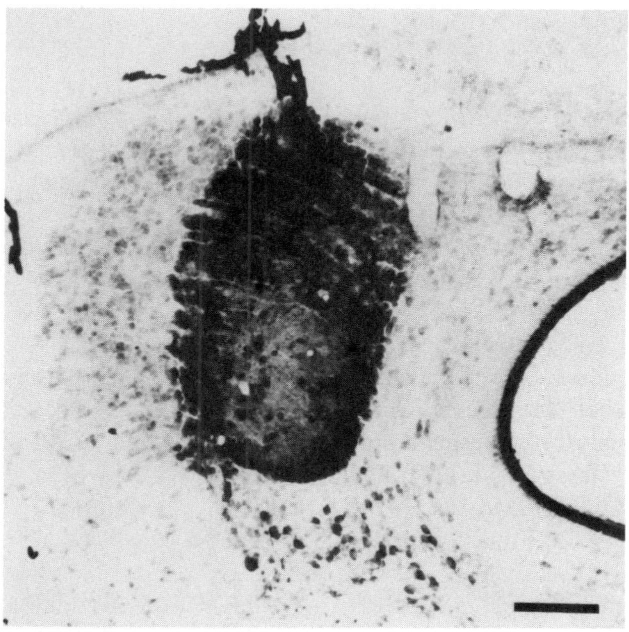

Figure 8. Transverse section through n. isthmi. Tissue reacted for ChAT immunostaining. Both cells and neuropil are darkly stained. Scale 100 μm. (From Desan et al. 1987).

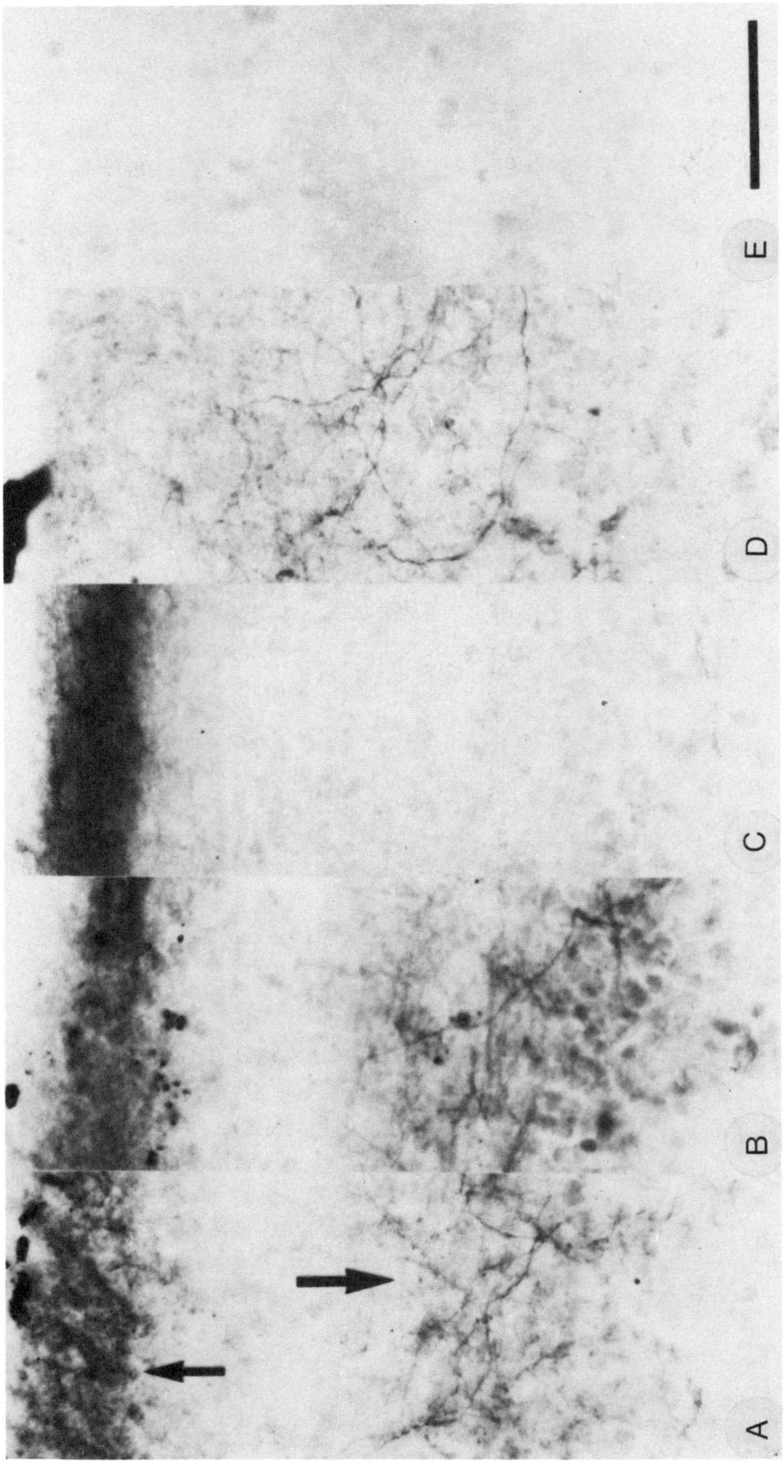
A
B
C
D
E

but at a circumscribed region. This would enable the animal to attend to a particular prey object in the visual field. Koch & Ullman (1985) have developed a generalized scheme for selective attention. They assume that a part of the brain is made up of a set of independent topographic "feature maps." Each feature map analyzes an "elementary" feature. These maps project topographically to another area, a "saliency map," where the features are in alignment and can be combined. The saliency map in

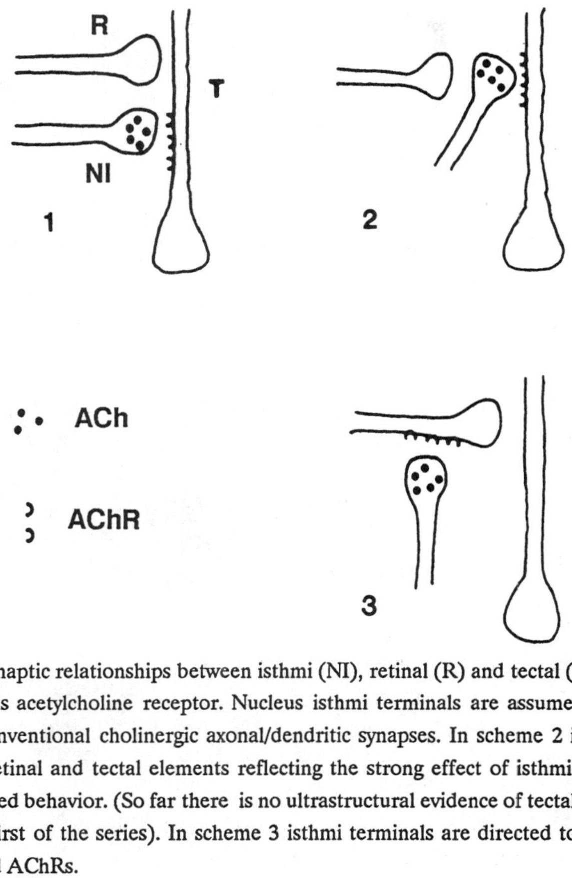

Figure 10. Possible synaptic relationships between isthmi (NI), retinal (R) and tectal (T) elements. ACh is acetylcholine. AChR is acetylcholine receptor. Nucleus isthmi terminals are assumed to be cholinergic. Scheme 1 assumes conventional cholinergic axonal/dendritic synapses. In scheme 2 isthmi terminals are interposed between retinal and tectal elements reflecting the strong effect of isthmi removal on tectally mediated visually guided behavior. (So far there is no ultrastructural evidence of tectal serial synapses with retinal elements the first of the series). In scheme 3 isthmi terminals are directed towards retinal fibers with membrane-bound AChRs.

Figure 9. Superficial layers of tecta reacted for ChAT immunostaining. (A) Normal animal with dense staining of superficial layer (above upper arrow) and more diffuse staining of deeper superficial layers (below lower arrow). (B) Eye enucleated animal showing no obvious changes in ChAT staining pattern. (C) Animal with unilateral n. isthmi ablation showing ipsilateral tectum. Note absence of deeper staining. (D) Same animal as C but showing contralateral tectum. Note absence of superficial staining. (E) Animal incubated with non-immunized serum showing only background staining. Scale 50 μm. (Desan et al. 1987).

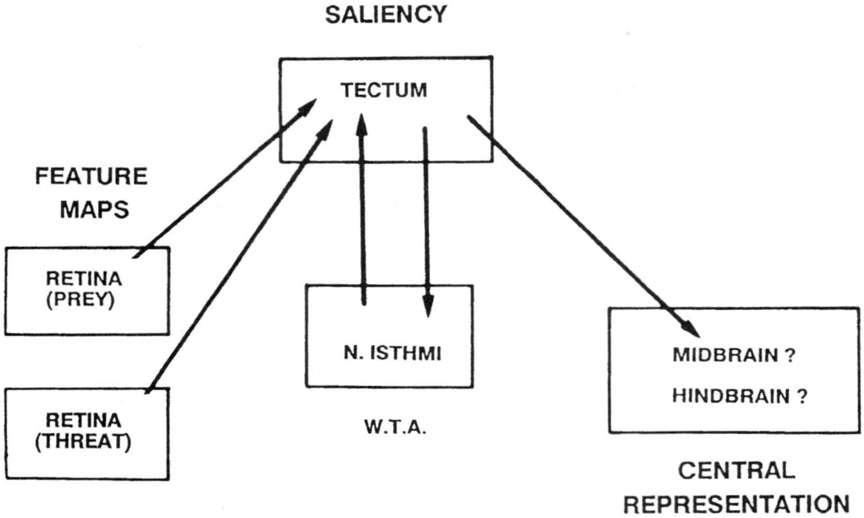

Figure 11. Possible relation between selective attention model and the retino-tectal-isthmal system.

turn projects to two areas: 1) an area called "winner take all" (WTA), that is reciprocally and topographically connected to the saliency map and which performs a weighting function across the map and suppresses all but a single local area. That local area in turn is the only area to influence 2) the "central representation," which is no longer strictly topographic.

We can make a correspondence between the retino/tectal/isthmal system and such a model (Fig. 11):

(a) There are "feature maps" present in the frog retina. Our behavioral results suggest that there might be separate prey and threat channels since we can create conditions where frogs can respond to prey but not threat (by cutting the posterior part of the optic chiasm) and, conversely, frogs can respond to threat and not to prey (by ibotenic acid injection into n. isthmi).
(b) The saliency map could correspond to the tectum. It has long been known that physiologically different types of visual units are encountered in different strata in the tectum.
(c) Part of n. isthmi makes reciprocal projections with the ipsilateral tectum, and this part could play the role of a WTA. There are only 8,000 cells in each n. isthmi so fine detailed analysis of the visual scene is probably limited especially since the 1 million cell tectum can do much of that processing.
(d) The location of a central representation is still under investigation. Evidence at this conference points to both the midbrain tegmentum and the medullary reticular formation. It is likely that the post-tectal structure (or structures) is not organized along strict retinotopic lines.

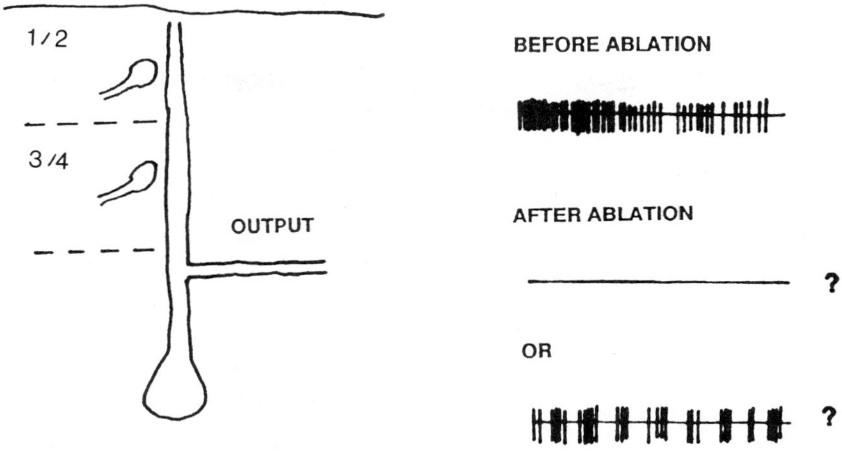

Figure 12. Two alternative possibilities for the effect of isthmi ablation on tectal output. Either tectal output is absent (a power gate) or, tectal output is informationally aberrant and cannot be read by target cells.

It has puzzled us that ablation of so small a structure as n. isthmi has such a profound effect on behavior. Both the retinal ganglion cell layer and the tectum have approximately two orders of magnitude more cells than n. isthmi. After n. isthmi ablation we do not see any obvious signs of tectal deterioration. Most of the visual processing apparatus is still intact. One possibility is that in the absence of isthmo-tectal input tectal output is lost. Another possibility is that without isthmo-tectal input the tectal output is coded in a way that cannot be read unambiguously by the central representation (Fig. 12). If n. isthmi performs the WTA role, then there should be ways of testing how tectal output is controlled by n. isthmi. This remains for the future.

Acknowledgement. Much of this work was supported by NIH grant EY04366.

8 References

Arbib MA, House DH (1985) Depth and detours: an essay on visually guided behavior. *COINS Tech Rep* 85-28

Caine HS, Gruberg ER (1985) Ablation of nucleus isthmi leads to loss of specific visually elicited behaviors in the frog *Rana pipiens. Neurosci Lett* 54: 307-312

Collett TS, Udin SB (1983) The role of the toad's nucleus isthmi in prey-catching behavior. In: Lara R, Arbib MA (eds) *Proceedings of the second workshop on visuomotor coordination in frog and toads: models and experiments.* University of Massachusetts, Amherst, *COINS Tech Rep* 83-19

Desan PH, Gruberg ER, Grewell KM, Eckenstein F (1987) Cholinergic innervation of the optic tectum in the frog *Rana pipiens. Brain Res* 413: 344-349

Glasser S, Ingle D (1978) The nucleus isthmus as a relay station in the ipsilateral visual projection to the frog's optic tectum. *Brain Res* 157: 214-218

Graybiel AM (1978) A satellite system of the superior colliculus: the parabigeminal nucleus and its projections to the superficial collicular layers. *Brain Res* 145: 365-374

Grobstein P, Comer C, Hollyday M, Archer SM (1978) A crossed isthmo-tectal projection in *Rana pipiens* and its involvement in the ipsilateral visuotectal projection. *Brain Res* 156: 117-123

Gruberg ER, Lettvin JY (1980) Anatomy and physiology of a binocular system in the frog *Rana pipiens. Brain Res* 192: 313-325

Gruberg ER, Udin SB (1978) Topographic projections between the nucleus isthmi and the tectum of the frog *Rana pipiens. J Comp Neurol* 179: 487-500

Henley JM, Lindstom JM, Oswald RE (1986) Acetylcholine receptor synthesis in retina and transport to optic tectum in goldfish. *Science* 232: 1627-1629

House DH (1984) Neural models of depth perception in frogs and toads. *COINS Tech Rep* 84-16

Ingle D (1973) Two visual systems in the frog. *Science* 181: 1053-1055

Koch C, Ullman S (1985) Shifts in selective visual attention: towards the underlying neural circuitry. *Human Neurobiol* 4: 219-227

Kostyk S, Grobstein P (1987) Neuronal organization underlying visually elicited prey orienting in the frog. I Effects of various unilateral lesions. *Neurosci* 21: 41-55

Shen SC, Greenfield P, Boell EJ (1955) The distribution of cholinesterase in the frog brain. *J Comp Neurol* 102: 717-743

Sherk H (1979) Connections and visual-field mapping in cat's tectoparabigeminal circuit. *J Neurophysiol* 42: 1656-1668

Székely G, Setalo G, Lázár Gy (1973) Fine structure of the frog's optic tectum: optic fibre termination layers. *J Hirnforsch* 14: 189-225

Why Cortices? Neural Networks for Visual Information Processing

HANSPETER A. MALLOT and WERNER VON SEELEN

Institut für Zoologie III (Biophysik), Johannes Gutenberg-Universität
D-6500 Mainz, FR Germany

Abstract. Neural networks for the processing of sensory information show remarkable similarities between different species and across different sensory modalities. As an example, *cortical* organization found in the mamalian neopallium and in the optic tecta of most vertebrates appears to be equally appropriate as a substrate for visual, auditory, and somatosensory information processing. In this paper, we formulate three structural principles of the vertebrate visual cortex that allow to analyze structure and function of these neural networks on an intermediate level of complexity. Computational applications are taken from the field of early vision. The proposed principles are: (a) *Average anatomy*, i.e., lamination, average axonal and dendritic domains, and intrinsic feedback, determines the spatiotemporal interactions in cortical processing. Possible applications of the resulting filters include continuous motion perception and the direct measurement of high-level parameters of image flow. (b) *Retinotopic mapping* is an emergent property of massively parallel connections. With a local intrinsic operation in the target area, mapping combines to a space-variant image processing system as would be useful in the analysis of optical flow. (c) Further space-variance is brought about by both, discrete or *patchy connections* between areas and periodic or *columnar arrangement* of specialized neurons within the areas. We present preliminary results on the significance of these principles for neural computation.

1 Introduction

1.1 Computational Theory and Structured Neural Networks

A basic idea in the investigation of both biological and technical information processing is David Marr's (1982) distinction of a *computational*, an *algorithmic*, and a *hardware* level of description. By separating functional aspects from the structure that they are built on, this heuristic distinction triggered a wealth of research in what is now generally called the computational theory of vision. The algorithms as well as the hardware used for their implementation are considered more or less independent from the basic computations required for a given information processing task. However, as more dedicated or special-purpose processors became available, the problem of a mutual adaptation of computation (i.e., function) and hardware (i.e., structure) has gained new interest. We feel that the investigation of biological information processing can contribute to this problem of hardware adaptation; some principles of information processing "wetware" with respect to fault tolerant computing have been formulated in a recent paper (v. Seelen & Mallot 1988). In addition, computational theory itself can receive new impulses from structural principles found in information processing "wetware." In this paper, we suggest some new solutions to old problems which are in fact triggered by anatomical studies.

Neural networks, the substrate of biological information processing, have become an intensively studied topic over the past five years. Many of the so-called "connectionist" studies focus on the cooperative capabilities of large unstructured networks of relatively simple units. Realistic neural networks, however, exhibit a wealth of structural differentiations which occur typically on the level of global architecture. This structure is characteristic for the task the network has to deal with. In accordance with a basic idea of biological science, i.e., the correlation of structure and function, a correspondence may be expected between the structure of a neural network and the information processing strategy it pursues (e.g., see also contributions by Arbib, Ewert, Lázár, and Matsumoto, this volume).

In the study of this adaptation of structure and function, it is of particular interest to investigate structures that are not phylogenetically homologous but have evolved similar specializations to meet similar functional requirements. Once structures of this type are identified, one can separate common features of their organization and try to understand their functional implications. An analysis on the level of these features is likely to be the most adequate approach to the function in question, since it was this function itself that induced the analogies in the first place. In this paper, we discuss the neural networks subserving vision as an example for similar adaptations of structure and function evolved in independent parts of the brain.

1.2 Structural Principles of Neural Networks Subserving Vision

In both the mammalian visual cortex and the optic tectum of the other vertebrates, the grey matter is organized as a cortex, i.e., in two-dimensional layers of neurons with strong vertical connectivity. Over the past decade, many details of cortical organization in several sensory and motor systems have been elucidated. In an earlier paper (v. Seelen et al. 1987), we formulated a number of *structural principles*, three of which will be discussed here in more detail:

1) *Average Anatomy*: The fiber anatomy of cortical neurons is basically uniform within a cortical area. Average intrinsic coupling is characterized by layering, strong vertical connectivity, and positive feeback via the pyramidal cell axon collaterals.
2) *Retinotopic Mapping*: Parallel projections between different regions of the brain usually preserve continuity but cause systematic distortions. By altering the effect of

the uniform intrinsic operation in a systematic way, these distortions act as a preprocessing applied to the input.

3) *Patchy Connectivity and Columnar Organization*: Projections of different areas to a common target area segregate into distinct patches or stripes (e.g., ocularity stripes in area V1). This too, leads to new neighborhoods for intrinsic processing. Columnar or periodic organization of the intrinsic operation adds to the space-variance produced by different types of mapping.

Information processing in two-dimensional sheets of neurons is natural in the visual system where two-dimensional images are the input. Cortical organization seems, however, to be a useful strategy in other sensory systems as well. As an example, consider the auditory system, where a stimulus which by its physical nature is purely sequential, is transformed into a spatiotemporal distribution of exitation by the basilar membrane. Most of the above features can be found in the auditory system as well.

2 Lamination and Intrinsic Feedback

Cortices are two-dimensionally extended, layered structures with strong vertical connectivity. As for their functional properties, cortical areas may be considered as multi-input multi-output systems composed of a small number of layers with characteristic connectivity. In this section, we discuss a number of computational implications of the fiber anatomy of cortical tissue.

2.1 Average Anatomy

Despite the multiplicity of shapes, cortical neurons can be classified into a few cell classes, the most frequent one being the pyramidal cell (for review see Peters & Jones 1984). The dendritic trees of these neurons mostly reach from the layer where the cell body is situated up to layer 1. This geometrical design suggests that information processing consists in combining information from different layers in a systematic way: The deeper a neuron is located, the more layers supply it with input. The axonal arborizations extend over several layers as well, thus distributing output in a feedback loop. Since these intrinsic connections outnumber the sensory afferents by two to three orders of magnitude (for the number of input fibers, see Braitenberg 1985), we consider the combination of the momentary input with a large amount of partially evaluated data a basic strategy of cortical information processing. Pyramidal cells, which comprise about three quarters of the cortical neurons, have exitatory connections. Inhibitory influences are mediated by a second cell type, the nonspiny stellate cells whose axonal and dendritic trees are situated in the proximity of the cell bodies.

We define the "average anatomy" of a type of neuron by the mean densities of its pre- and postsynaptic sites, or its dendritic and axonal arborizations, the mean being taken over a cortical area. The usefulness of this average anatomy for an overall model rests on two assumptions, first, that the coupling of cells is largely determined by the overlap of their dendritic and axonal arborizations and second, that an average system exhibits average behavior, at least if linearity is assumed.

a

b

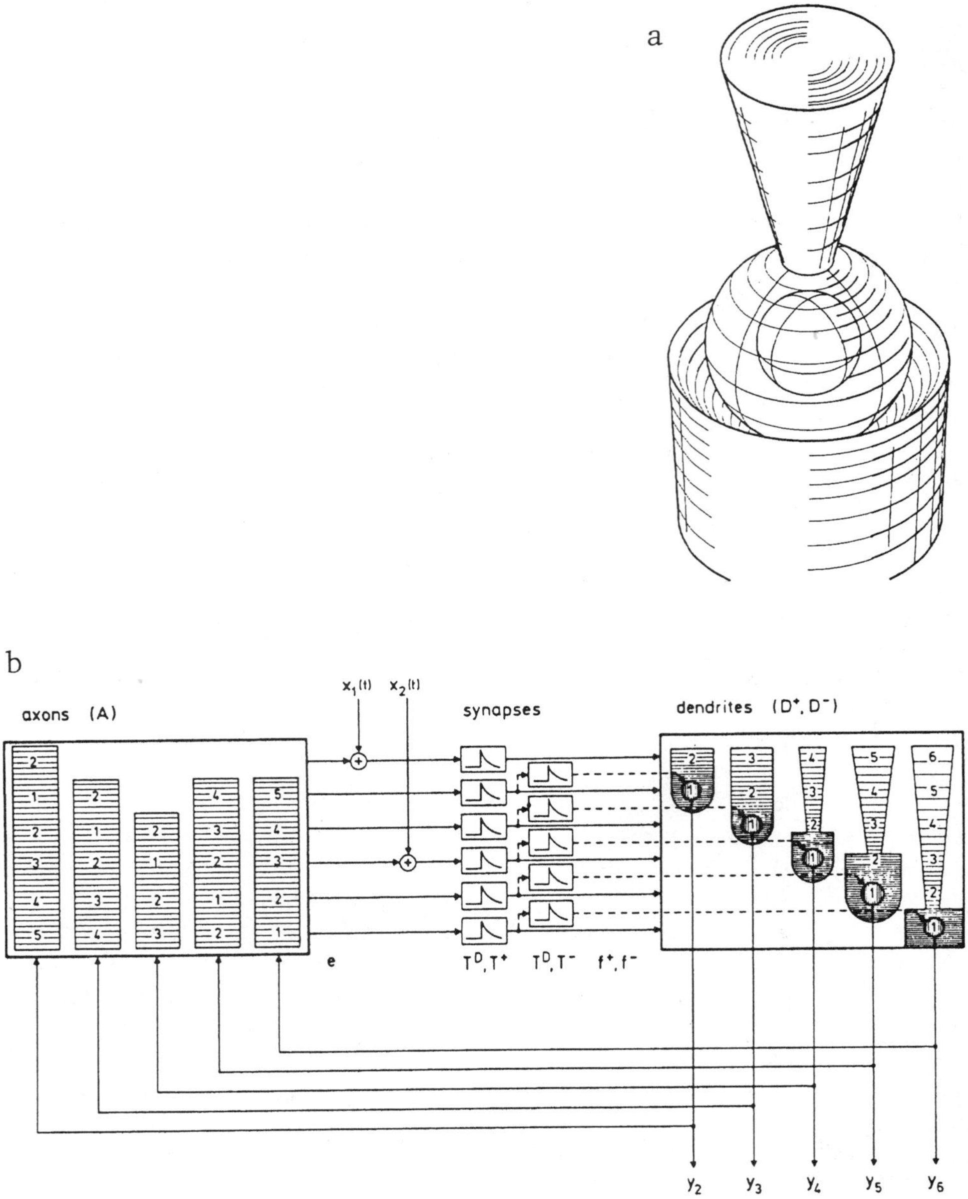

Figure 1. Network model of intrinsic cortical organization based on the average anatomy. (a) General shape of pre- and postsynapic connectivity domains of a pyramidal cell. The large sphere and the cone represent basal and apical dendrites, respectively, i. e. regions of excitatory inputs. The inner sphere represents inhibitory inputs mediated by basket cells. The excitatory output fibers of the cell sparsely fill the lower cylinder. (b) Block diagramm of the complete model. The arrows represent two-dimensional distributions of excitation (solid lines) or inhibition (dashed lines). The spatial behavior is determined by the connectivity domains discussed above. The temporal behavior is due to synaptic delays and time constants as well as to propagation times.

We designed a continuous model of the network of pyramidal and stellate cells on the basis of average dendritic and axonal domains (Fig. 1). Due to the overall uniformity of the visual cortex, we can thus study the effects of dense two-dimensional connectivity and feedback. For a detailed description of the model and the properties of the linear version, see Krone et al. (1986). First results for a non-linear version (neurons with threshold and saturation) were presented by v. Seelen et al. (1987).

The main results of the linear version of the network concern the spatiotemporal behavior induced by feedback (Fig. 2). In the open loop, spatial filtering is due to the axonal and dendritic domains depicted in Fig 1a. Temporal characteristics included in the model are synaptic delays and time constants as well as very short propagation times between the layers. In the closed loop, spatiotemporal behavior is no longer separable into two independent parts. In addition, if non-separability is due to feedback (v. Seelen et al. 1987) it can not be described by a small number of different spatial "mechanisms" each with separable but different temporal behavior. Such "piecewise" separability has been found in retinal ganglion cells (Enroth-Cugell et al. 1983). In a forthcoming paper (Dinse et al., in prep.), we present neurophysiological evidence for the complete non-separability of spatiotemporal receptive field profiles in the cat's visual cortex. This study also confirms the damped oscillations of the response predicted by the model.

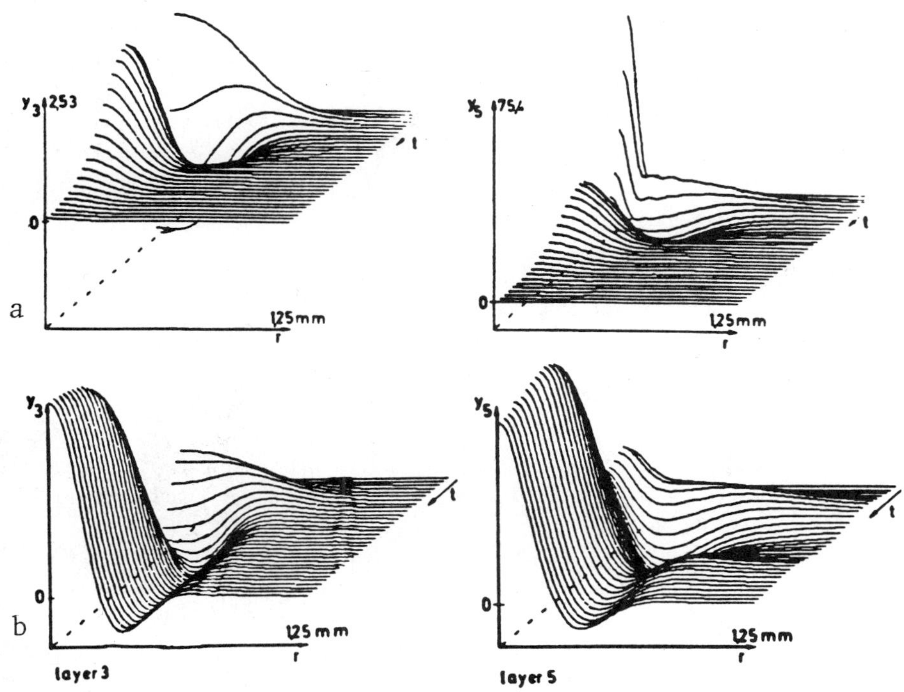

Figure 2. Spatiotemporal activity profiles for layers 3 and 5 of the cortex model of Fig. 1 following a point stimulation in layer 4. (a) Flashed (transient) stimulus. (b) Sustained stimulus.

The non-separable receptive field profiles found in our simulations have a number of computationally interesting properties: First, the spatiotemporal distribution of excitation emerging in most simulations can be described as a damped oscillation in time modulating a spatial spread or contraction (Fig. 2a). More specifically, the impulse responses can be approximated by functions of the type:

$$f(r,t) = \exp[-r^2/(B_0 e^{-t/T})^2] e^{-t/S} \cos(wt), \; t>0. \tag{1}$$

Here, the spatial spread is modeled by a Gaussian function with a temporally modulated width described by the parameters B_0 and T, whereas the damped temporal oscillation is represented as a cosine function with frequency w and an exponential envelope with time constant S. Functions of this type resemble the matched filters for a "zoomed" stimulus as it occurs when the observer approaches a viewed object. They may thus play a role in the construction of "looming detectors" in a certain spatiotemporal frequency channel (i.e., detectors for the divergence of a vector field) as have been proposed by Koenderink & Van Doorn (1976) for the analysis of optical flow. Psychophysical evidence for the existance of looming detectors in the human visual system has been presented by Regan & Beverly (1978).

Second, functions of type (1) could be used in motion analysis instead of the spatiotemporal Gabor-functions proposed by Heeger (1987). As compared to those, they have the advantage of being causal. Spatiotemporal non-separability (which is not used in Heeger's model) is advantageous for all types of motion perception, since the optimal filters for these tasks are spatiotemporally oriented (Korn & v. Seelen 1972, Adelson & Bergen 1985). However, without distributed feedback, usefull non-separabilities are difficult to obtain.

Finally, in terms of pattern recognition or feature extraction, spatiotemporal filters should not simply be considered edge-detectors of a special kind. As is

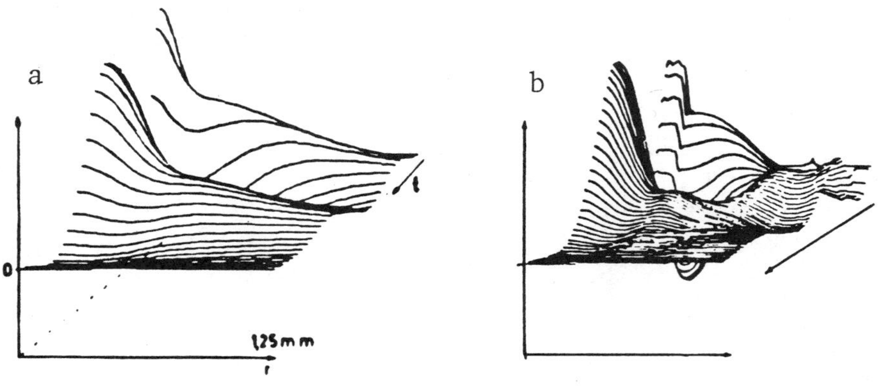

Figure 3. Spatiotemporal activity profiles for layer 6 of the cortex model of Fig. 1 following a brief point stimulation in layer 4. (a) Lateral connectivity weighted by a Gaussian function. (b) Lateral connectivity constant over an equivalent range. Initial differences in the impulse responses vanish as the eigenfunction of the network becomes dominant due to feedback.

illustrated in Fig. 2b, temporal modulation of a stimulus critically influences the spatial features extracted. In other words, the primitives used by the visual system are elementary events rather than elements of a static sketch of the scene. In terms of motion detection, spatiotemporal filters can directly use the continuity of motion that is always present in natural scenes. They thus help overcome the correspondance problem which arises when motion detection is based on a sequence of instantaneous frames (sampled or apparent motion).

Another interesting property of distributed feedback is illustrated in Fig. 3. Suppose that the synapse density functions describing the average anatomy had small systematic deviations from the normal shapes. In Fig. 3, simulation results for a Gaussian weighting function (Fig. 3a) and for constant weights in a corresponding range (Fig. 3b) are compared. The difference is well visible in the early part of the response, but vanishes as feedback and the overall network properties become important, i.e., this "error" can be suppressed over a period of time. Thus, the reliability of the system and therefore the quality of a parameter estimation, say, increases with time. Computation time is the prize to be paid for an increased reliability and accuracy.

2.2 Positive Feedback and Quadratic Regularization

As a further feature of intracortical feedback relevant to neural computation, the interpretation as an iterative algorithm should be mentioned. This interpretation is possible if the feedback loop containes a time lag such as synaptic delays. In this case, stable feedback with a symmetric spatial coupling minimizes a quadratic functional (v. Seelen et al. 1987). In the computational theory of vision, regularizations of this type play an important role. As Poggio et al. (1985) pointed out, the problem of reconstructing a scene description from images can be interpreted as "inverse optics" and is generally ill-posed. Additional regularization constraints which may be characterized by quadratic as well as by non-convex cost-functions (Marroquin et al. 1987) are used to obtain unique and stable solutions.

Let us characterize the recursion implemented by delayed feedback by:

$$z_0 := 0$$

$$z_t := Lz_{t-1} + My, \tag{2}$$

where z_t denotes the distribution of exitation over a neural layer at time step t, y is the external input which is taken to be sustained, and L and M are linear operators describing the intrinsic coupling and the weighting function for the input, respectively. We interpret **y** as the data and **z** as the regularized solution of some ill-posed problem. Suppose we want to minimize the following cost function (Poggio et al. 1985):

$$F(y) := \| Az-y \|^2 + \beta \| Pz \|^2 , \quad \beta>0. \tag{3}$$

For the sake of simplicity, we treat **y** and **z** as (finite-dimensional) vectors and L,

M, A, and P as matrices; // . // denotes the Euclidian norm and ß is the regularization parameter; A describes the relation between object and image that we want to invert and P is a smoothness constraint. The cost function thus describes a trade-off between the correctness of the solution (first term) and its smoothness (second term) balanced by the regularization parameter ß.

An iterative algorithm for the minimization of Eq. (3) is the well-known steepest descent. At any given state $\mathbf{z}_t$, we proceed in the direction of the gradient and eventually approach the unique minimum of the convex cost-function. A little computation yields:

$$\text{grad } F(z) = 2\,A^T A z - 2\,A^T \mathbf{y} + 2\beta\, P^T P z. \tag{4}$$

To formulate the recursive algorithm, let q denote the amount by which $\mathbf{z}$ is altered in each step, $q > 0$. We have:

$$z_0 = 0$$

$$z_t = z_{t-1} - q \text{ grad } F(z_{t-1})$$

$$= (I - 2q\,A^T A - 2\beta q\, P^T P) z_{t-1} + 2qA^T \mathbf{y}, \tag{5}$$

where I is the identity; a comparison with Eq. (2) yields:

$$L = I - 2q\,A^T A - 2\beta q\, P^T P \tag{6a}$$

$$M = 2q\,A^T. \tag{6b}$$

Eq. (6) has the following interesting properties:
1) The constant q can always be chosen in such a way that the algorithm converges. Since F is quadratic, the convergence approaches the global minimum.
2) The matrix L in Eq. (6a) is always symmetric. Therefore, feedback with a nonsymmetric coupling does not minimize a functional of this type.
3) Positive feedback via certain spatial lowpasses, as was present in the average anatomy model (Section 2.1), minimizes a cost function where A is the identity and P gives a spatial derivative (i.e., $(Pz)_i = (z_{i+1} - z_{i-1})/2$) or a measure of curvature (i.e., $(Pz)_i = z_i - (z_{i-1} + z_{i+1})/2$). This is a spatial smoothing or interpolation. In the second case (minimal curvature), it implements an interpolation by cubic splines which is an important tool in surface interpolation (Grimson 1981).

2.3 Summary

In this section, we have shown how anatomic knowledge on the organization of cortical tissue can be utilized in computational considerations. The main results are:

1) Cortical processing is characterized by complicated spatiotemporal interrelations. As an example, non-separable spatiotemporal filters that could be used in motion or looming detection, are obtained by distributed feedback. Filters of this

type are difficult to realize by other means. Even in the processing of stationary data, part of the evaluation is transferred to the dynamics of the system.

2) An analytic approach to feedback systems with delays was sketched in Section 2.2. If delayed feedback is treated as an iterative algorithm, it can implement a steepest descent minimization. The minimized functional can be determined from the spatial filter in the feedback loop. It turns out that the filters required for typical applications of optimization in vision are excitatory low-passes as are actually present in the visual cortex.

3 Retinotopic Mapping

Receptotopic mapping is a general feature of almost all higher sensory centers. The generation of new spatial neighborhoods by coordinate transform and the representation of all kinds of information in a spatial data format are extremely useful in an information processing system that relies on local interactions of its elements. We shall discuss the detailed structure of topographic maps in the cat visual cortex and present examples for how information can be processed via mapping.

3.1 Formal Description of Retinotopic Maps

In Fig. 4, topographic maps from the visual system of the cat are shown together with a model that describes all three maps by minor variations of the same mapping function. For the approximation of a cortical mapping function, a model should be built not by the summation of terms (e.g., a power series expansion), but by composing steps in a sequence. Such sequences might then lead to models of the maps found at different stages in the visual pathways. Here, a composition of four steps was found:

$$R\colon \mathbb{R}^2 \longrightarrow \mathbb{R}^2,\ R = R_2 \circ L_2 \circ R_1 \circ L_1. \tag{7}$$

The mappings R_2 and R_1 combine to the complex power function, i.e., in real polar coordinates:

$$R_1(r,\beta) = (r^p,\beta) \tag{8a}$$

$$R_2(r,\beta) = (r,p\beta); \qquad p>0,\ r>0,\ -\pi<\beta<\pi. \tag{8b}$$

The mappings L_1 and L_2 are affine (shifted linear) transformations. If L_2 contains a mirroring at the vertical meridian, an area 18 type map with a branched representation of the horizontal meridian results. The initial steps, $R_1 \circ L_1$ are roughly equal for all three areas and combine to a reasonable model of the mapping in the lateral geniculate body. The single steps of the overall maps are depicted in Fig. 5.

Unlike the mapping in area V1 in the monkey, topographic maps in the cat visual cortex are not conformal, nor can they be approximated by the complex

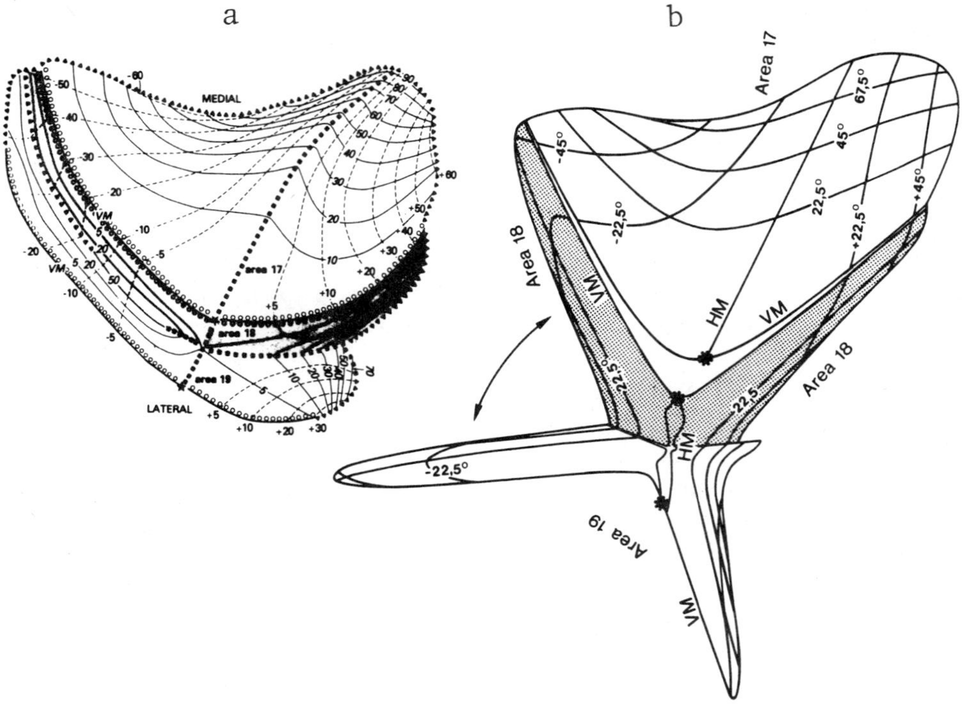

Figure 4. (a) Multiple representation of the visual field in the cat's cortical areas 17, 18, and 19 (Tusa et al. 1979). (b) Formalization by mathematical functions, in this case by a composition of a *complex power function* with certain affine transformations. The functions for all three areas are closely related (Mallot, 1985). The model fits the overall shape of the areas, the differences between upper and lower visual field in area 17, the magnification factors, the continuous bordering between areas, and the branching of the horizontal meridian in areas 18 and 19. The gap between the lower peripheries in areas 18 and 19 (see arrow) can be closed in a three-dimensional representation. The resulting fold corresponds to the lateral sulcus of the cat brain.

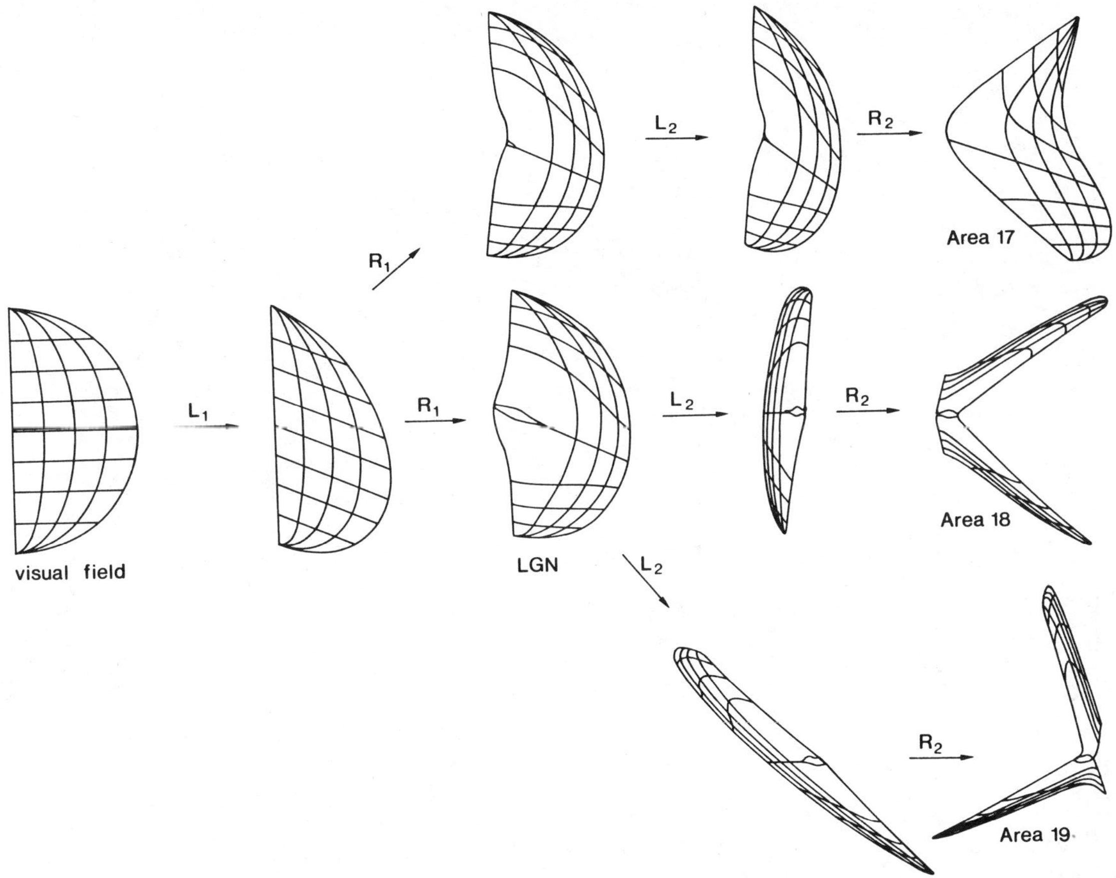

Figure 5. Construction of retinotopic maps as a composition of steps. The visual field is represented by a grid of spherical polar coordinates. The first two steps are basically the same for all three mappings and give a reasonable approximation of the LGN map. For details see text and Mallot (1985).

logarithm as proposed by Fischer (1973) and Schwartz (1980). This is important since most applications of coordinate transforms in computational problems that have been proposed so far rely heavily on the special properties of the logarithmic map (e.g., Sawchuk 1974; Reitboeck & Altmann 1984; Jain et al. 1987). This runs counter to the fact that in all investigated mammals, multiple retinotopic representations of the visual field with different maps are found. For example, in macaque monkeys where some 15 retinotopically organized areas can be distinguished (Van Essen 1985) it is only the area V1 map that approximates the complex logarithm.

We present two examples for the computational advantages of coordinate tranforms that are not restricted to the complex logarithmic mapping. The general idea of our approach is that mapping prepares data from distant parts of the visual field for subsequent processing by the same space-invariant local operation.

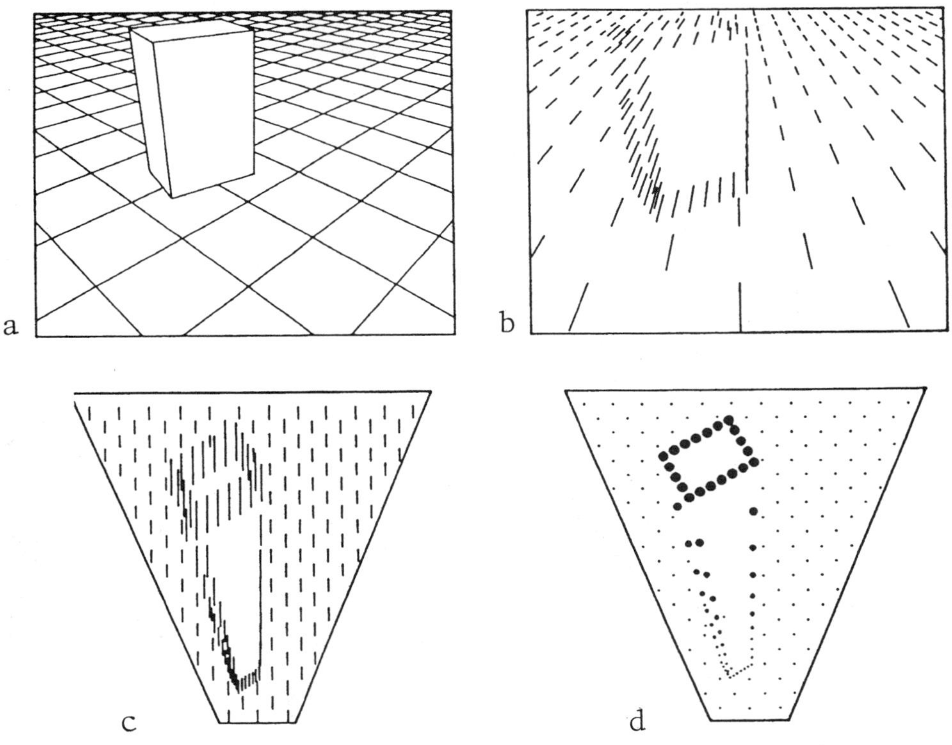

Figure 6. Optical flow analysis by retinotopic mapping. By inverting the perspective projection of the floor unto the image plane, the velocity vectors of elevated points can be detected. The method is robust against small changes in elevation (pot-holes or bumps), since the output scales with elevation. (a) Scene with obstacle. (b) Needle-plot of the velocity field. (c) Mapped velocity field. (d) Detected obstacle.

3.2 Inverse Perspective Mapping for Optical Flow Computation

Consider the optical flow generated by an observer moving between stationary obstacles in a plane (Fig 6a). It is convenient to compute the perspective projection for a coordinate system centered at the object nodal point of the eye (Fig. 7). (In the following discussion, vectors are printed in bold face, while scalar quantities appear in plain print.) Let *x,y* span the image plane while *z* points in the direction of the optical axis of the eye. Let further denote f the focal length of the lens. The projection formula for the camera-centered coordinate system is (cf. Horn 1986):

$$P: R^3 \longrightarrow R^2, \ (x,y,z) \longrightarrow (x',y') := f/z\ (x,y) \tag{9}$$

where $\mathbf{x} = (x, y, z)$ is a point in three-dimensional space and x', y' are the coordinates of the projection of $\mathbf{x}$ into the image plane. For a purely translatory egomotion $\mathbf{m}$, we have to substitute $\mathbf{x} = \mathbf{x}_0 - \mathbf{m}t$ in Eq. (9). For a point $\mathbf{x}$ on a visible surface, a motion vector $\mathbf{m}_I$

$$\mathbf{m}_I := -J_P(\mathbf{x})\ \mathbf{m} \tag{10a}$$

$$= -f/z\ (m_1 - xm_3/z\ ,\ m_2 - ym_3/z) \tag{10b}$$

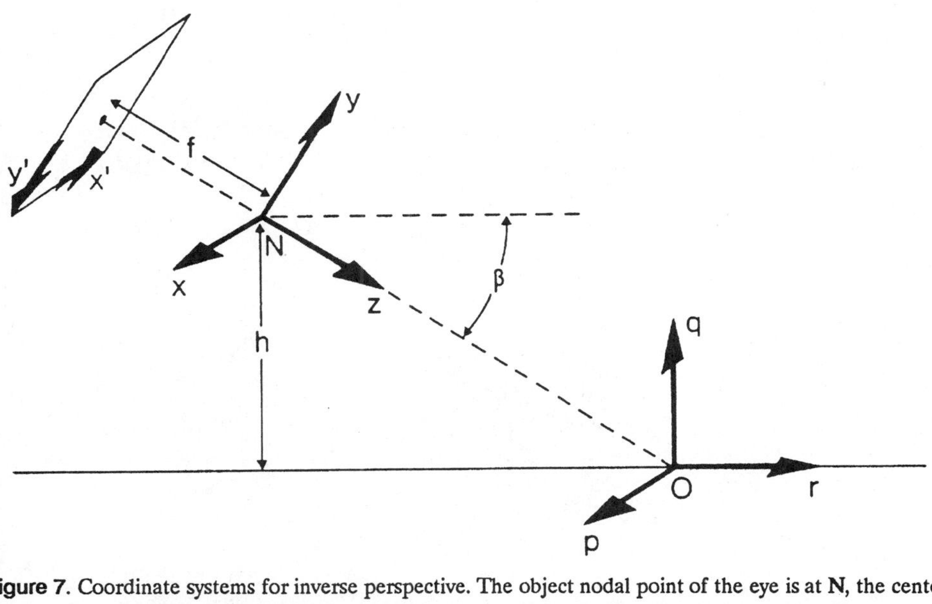

Figure 7. Coordinate systems for inverse perspective. The object nodal point of the eye is at **N**, the center of the eye-centered coordinate system (**x, y, z**). The image plane has the coordinates x' and y', its distance from the nodal point is f, the focal length. The eye is at elevation h above the horizontal plane, its optical axis (z-direction) is inclined towards that plane by the (positive) angle ß; **O** is the intersection of the optical axis with the horizontal plane and the center of the world-coordinate system (**p, q, r**). For details see text.

is obtained in the image plain. m_1, m_2, and m_3 denote the components of the egomotion vector **m** and J_P is the Jacobian matrix of P. Note that Eq. (10a) holds for any differentiable mapping. In order to compute the image flow, one has to trace the ray through the nodal point and an image-point (x', y') towards the visible surface. The relative motion of the intersection point **x** projected according to Eq. (10) is the image-flow at (x', y'). Note that the ray-tracing may be considered the inversion of the perspectivic projection which can only be well-defined if **x** is constraint to a two-dimensional manifold, i.e., a surface. The result is depicted in Fig. 6b. For a given direction of view, the stimulus velocity scales with the distance z of the object from the observer or, equivalently, its elevation above the ground.

The problem is now to determine from the image-flow of Fig. 6b, whether further movement is possible in the same direction **m**. Clearly, if the flow field was that of a horizontal plane, the observer may continue without concern. All deviations from this motion field, on the other hand, must be considered obstacles. We introduce a new coordinate system to describe the horizontal plane or, strictly speaking, the plane on which the observer moves (Fig. 7). We assume that the eye does not make roll movements around its optical axis which differ from roll movements of the observer as a whole and that the height h of the eye above the ground is constant while the optical axis points downwards with an inclination angle ß. We define:

$$\mathbf{0} := (0,\ 0,\ h/\sin \beta);$$

$$\mathbf{p} := (1,\ 0,\ 0); \qquad \mathbf{q} := (0,\ \cos \beta,\ -\sin \beta)$$

$$\mathbf{r} := (0,\ \sin \beta,\ \cos \beta). \tag{11}$$

To keep definitions straight, we chose **p, q, r** such that for small ß, they roughly equal the eye-centered coordinate system **x, y, z**. The origin of the new frame is the intersection of the optical axis with the horizontal plane, i.e., by Eq. (9), it will be projected to the point (0,0) in the image plane. We denote the coefficients of **p, q, r** by p, q, r and obtain the points $\mathbf{x}_h$ in the horizontal plane by:

$$\mathbf{x}_h(p,r) := \mathbf{0} + p\,\mathbf{p} + r\,\mathbf{r}. \tag{12}$$

We can now explicitly perform the ray-tracing for points in the horizontal plane or, in other words, define an *inverse perspective mapping* for optical flow analysis by:

$$\mathrm{IP}: \mathbb{R}^2 \longrightarrow \mathbb{R}^2,$$

$$(x',y') \longrightarrow (p,r) := h/(f \sin \beta - y' \cos \beta)\ (x',\ y'/\sin \beta). \tag{13}$$

From Eq. (10) it follows that the "save" motion field is given by

$$\mathbf{m}_I(x',y') = -J_P\{\mathbf{x}_h[(\mathrm{IP}(x',y')]\}\ \mathbf{m}. \tag{14}$$

If all we want to detect are the deviations of the perceived image flow from the field given in Eq. (14), which in fact would be sufficient for obstacle avoidance, it is advantageous to apply the inverse perspective mapping defined in Eq. (13). By this mapping (Fig. 6c), the velocity vectors assigned to points in the plane become equal and by means of a suitable filter and threshold, elevated points can be detected (Fig. 6d). Egomotion is the smallest velocity vector occurring in the mapped images (Fig. 6c). To show this, we apply the composition of IP with P (Eqs. 9,13), to an arbitrary point (p, q, r) in world coordinates. The deviation of a point from its image under IP o P is a measure for its deviation from the expected position in the horizontal plane;

$$IP \circ P: R^3 \longrightarrow R^2,$$

$$(p,q,r) \longrightarrow (p',r') = h/(h-q)\ (p\ ,\ r + q \cot \beta). \qquad (15)$$

For points in the plane (q = 0), IP o P is of course the identity. From Eq. 10a, we can now compute the motion field induced in the inversely mapped image. Egomotion is restricted to the horizontal plane, i.e., in world coordinates: $\mathbf{m} = (m_p, 0, m_r)$. Some computation yields:

$$\mathbf{m}_I = -J_{IP \circ P}(p,q,r)\ (m_p,\ 0,\ m_r)^T$$

$$= -h/(h-q)\ (m_p,\ m_r), \text{ and finally}$$

$$//\ \mathbf{m}_I\ // = h/(h-q)\ //\ m\ //\ \text{for } q<h \qquad (16)$$

This result clearly shows the advantages of inverse perspective mapping for optical flow analysis. The vectors become parallel, their length scales with the ratio of the elevations of the observer and possible obstacles. Thus, small bumbs or pot-holes produce less output than tall objects. What is more important, the opical axis needs not to be aligned with the direction of motion nor point towards the focus of expansion since Eq. (16) does not change for rotations of **m** in the horizontal plane. On the contrary, in complex logarithmic mapping, the vector field is parallelized only if the focus of expansion is found and fixated. Finally, at least for stationary obstacles, the aperture problem in avoided, since the detection of one component of the motion vector given in Eq. (16) would already be sufficient. In ongoing work, we combine this mapping principle with an image flow algorithm proposed recently by Little et al. (1987) and test this scheme with a mobile robot.

Inverse perspective mapping, although very useful for visual information processing, has not been demonstrated in electrophysiological studies. However, Epstein (1984), investigating the deviation of the cat's area 17 map from a conformal mapping argued that inverse perspective does account for that deviation. In monkeys, who do not live on a plane, this compensation of perspective has not been found and would in fact be useless.

3.3 Mapped Filters

In Fig. 8, the effect of retinotopic mapping in the cat's visual cortex is shown for a natural scene. It is difficult to see, what the computational advantages of these distortions might be. We therefore took a more systematic approach to computational interpretations of retinotopic mapping. Consider an intrinsic operation performed in a cortical area such as the spatiotemporal filtering described in Sect. 2. Retinotopic mapping changes the effective surrounds of the neurons in the target area and thereby influences the meaning of local intrinsic computations with respect to the original image. Consider a simple space-invariant DOG- (Difference-of-Gaussians) filter in a topographically mapped area (Fig 9a). Let R denote the mapping function, u,v the cortical and x,y the retinal coordinates, i.e., R(x,y) = (u,v). Let further denote k(u,v) the kernel of the intrinsic convolution, i.e., the DOG function in Fig. 9a. The composition of mapping and convolution transforms a stimulus distribution s(x,y) into a distribution of excitation e(u,v) by the following equation:

$$e(u,v) = \iint s[R^{-1}(u',v')]\; k[(u',v')-(u,v)]\; du'\, dv' \quad (17a)$$

$$= \iint s(x,y)\; k\{[R(x,y)-(u,v)]\}\; |J_R(x,y)|\; dx\, dy, \quad (17b)$$

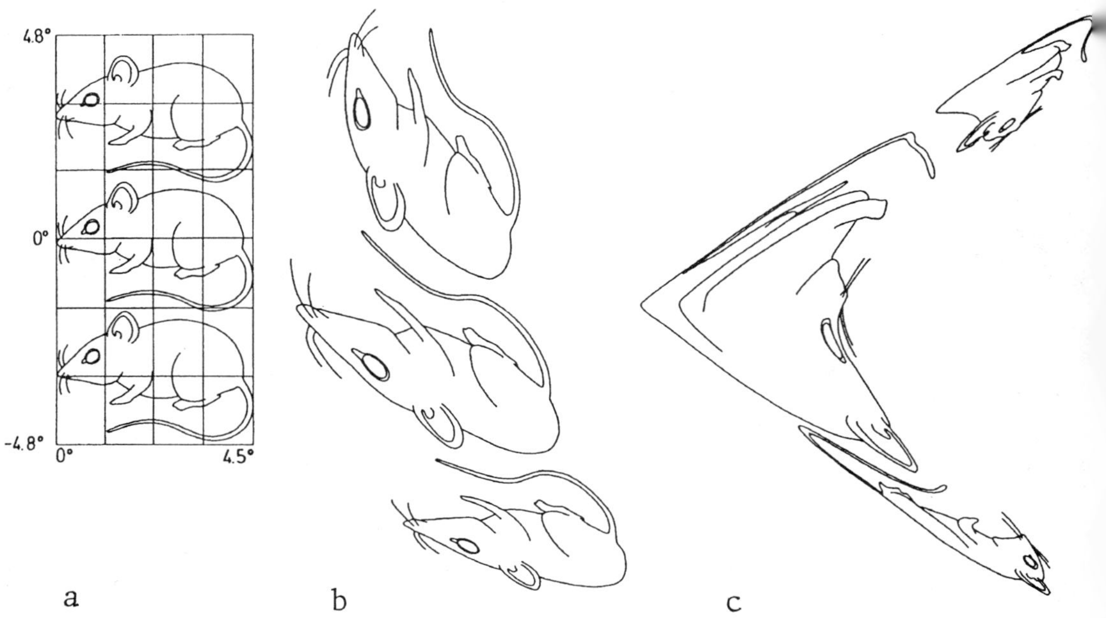

a b c

Figure 8. Effect of retinotopic mapping on realistic images. (a) Scene from the natural environment of the cat; (b) distorted version in area 17; (c) distorted version in area 18. The distortions were computed with the mapping function depicted in Fig. 3. It is difficult to see functional advantages of these mappings.

(Mallot 1985, 1987). Here, J_R denotes the (determinant of the) Jacobian of R. The resulting receptive field, i.e., the kernel of the linear operator composed of mapping and convolution (*"mapped filter"*, Eq. 17b), is depicted for two different maps in Fig. 9b,c. For an affine transformation (Fig. 9b), a simple distortion of the mask's range occurs. In general, however, the mask becomes asymmetric in a space-variant way. This is shown for the complex logarithm in Fig. 9c.

In Fig. 10, the space-variance of the resulting operation is shown for the simulated maps of the cat's areas 17, 18, and 19 by suitable contour lines of the corresponding kernels. In these cases, radial orientations are imposed on isotropic cortical operators by retinotopic mapping. This is in spite of the fact that the complex power function used in this example does not map radial directions onto parallel ones.

Mapped filters are a special class of space-variant linear operators and one might be interested to know, how large that class is, or what general space-variances can be generated by the cascade of mapping and convolution or a combination of such cascades. It is easy to derive a necessary condition for an operator to be a composition of mapped filters: Consider the zeroth moment of the kernel of the integral equation (Eq. 17b). Obviously, this does not depend on the retinal

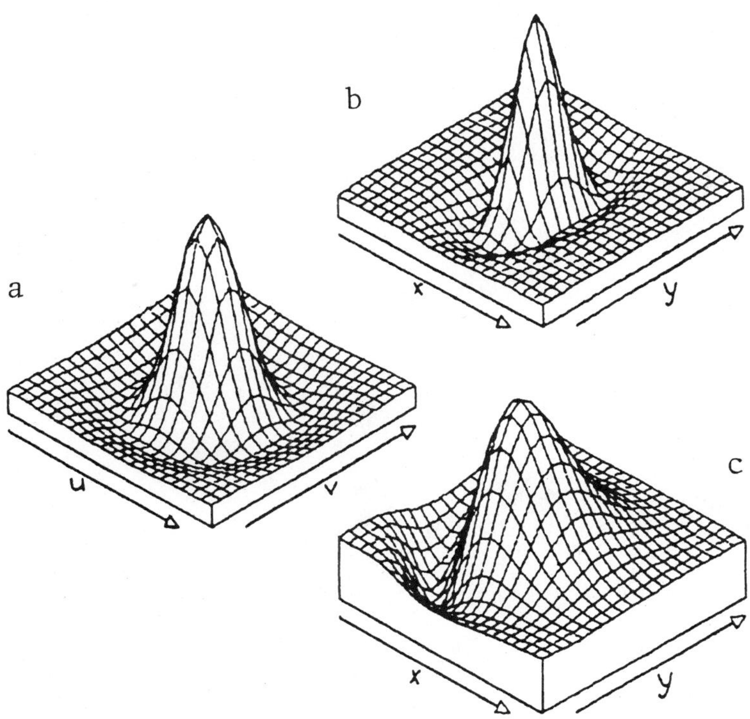

Figure 9. Effect of retinotopic mapping on an intrinsic kernel (Eq. 17b). (a) DOG-function; (b) resultant kernel for an affine coordinate transformation; (c) resultant kernel for the complex logarithmic mapping.

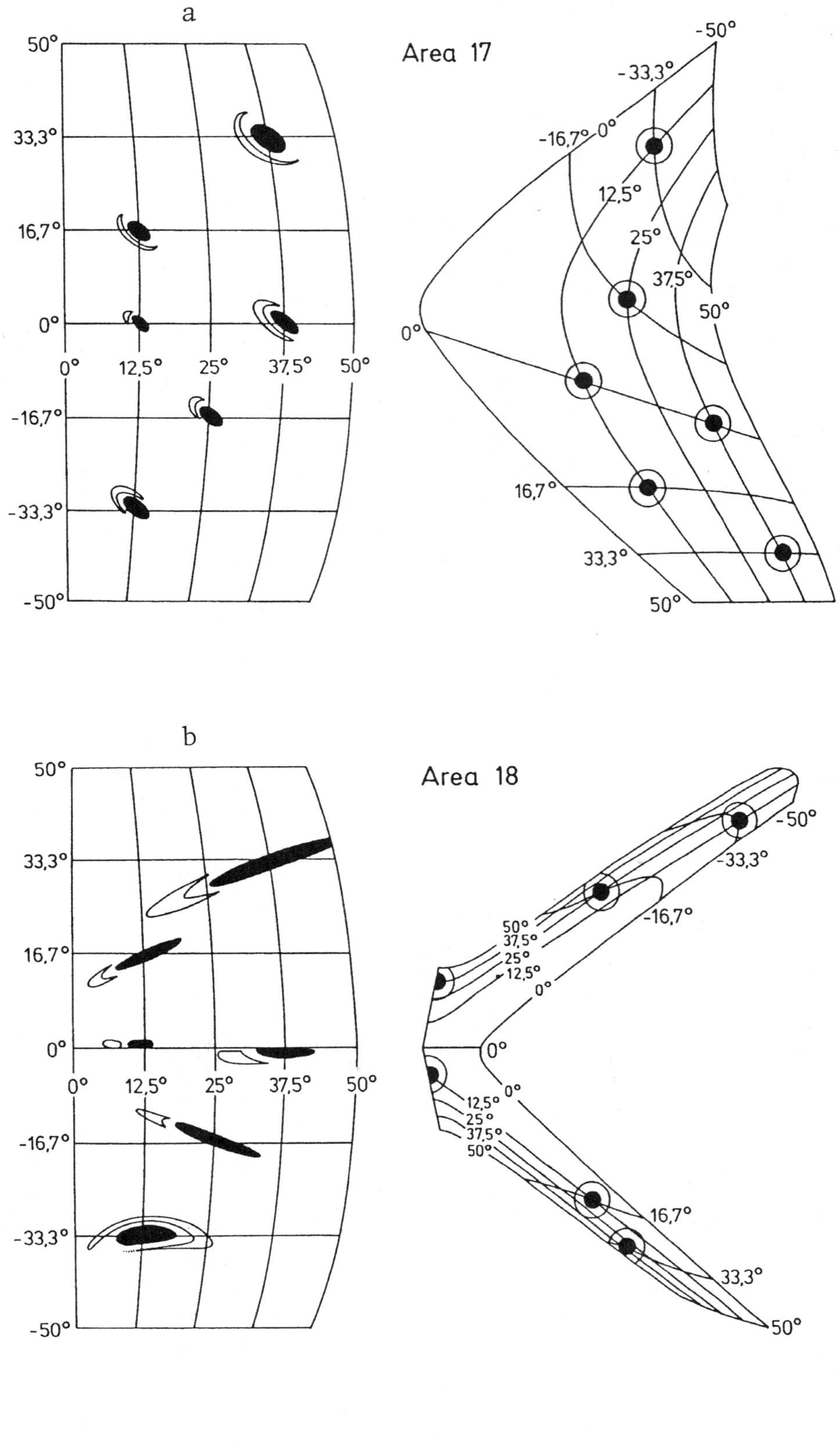
a
Area 17
50°
33,3°
16,7°
0°
-16,7°
-33,3°
-50°
0° 12,5° 25° 37,5° 50°
-50°
-33,3°
-16,7° 0°
12,5°
25°
37,5°
50°
0°
16,7°
33,3°
50°
b
Area 18
-50°
-33,3°
-16,7°
50°
37,5°
25°
12,5°
0°
0°
0°
12,5°
25°
37,5°
50°
16,7°
33,3°
50°

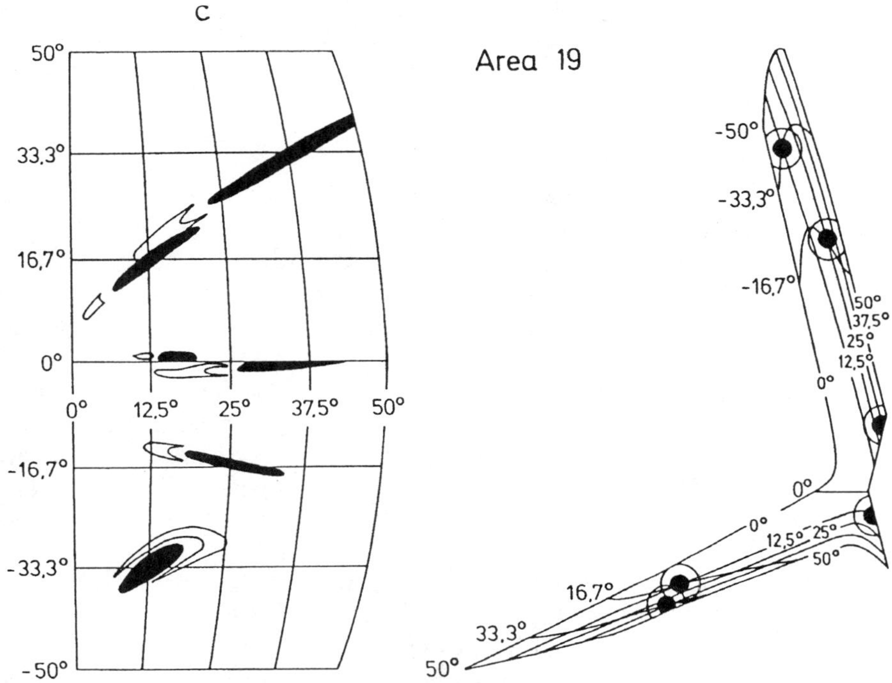

Figure 10. (a-c) Space-variance of the resultant kernel for the simulated maps of area 17, 18, and 19 (Fig. 4). Contour lines for 20% excitation and 80% inhibition. Kernel like in Fig. 9a. Although the cortical coupling functions are rotationally symmetric (right parts), a systematic asymmetry is induced by topographic mapping in the retinal receptive fields (left parts).

coordinates (x,y), i.e., the operator is "*space-invariant in the mean*" and the same holds for compositions of operators of this type. We can therefore conclude that operators which are not space-invariant in the mean, can not be approximated by mapped filters. As an example, consider the space-variant filtering performed by the retinal ganglion cells. Braccini et al. (1981) modelled this filtering by an operator of the type

$$e(x,y) = \iint s(x',y')\ k\{(x-x',y-y')\ (x^2 + y^2)^{-0.5}\}\ dx'dy'. \tag{18}$$

The width of this kernel increases with eccentricity without a decrease in amplitude. Therefore, this operator is not space-invariant in the mean and cannot be modeled by mapped filters. Interestingly, electrophysiological studies of retinal ganglion cells show that the product of receptive field size and sensitivity does not vary with eccentricity (Fischer & May 1970; Creutzfeldt et al. 1970) , i.e., the zeroth moment of the corresponding kernel is constant.

In ongoing work, we are investigating mappings and kernels that produce certain desired space-variances, e.g., for applications in optical flow analysis. A more technical discussion of this topic will be published elsewhere (Mallot et al., in prep.).

3.4 Summary

Topographic maps are a general principle of cortical organization. In this section, we presented a model of the mapping of the areas 17, 18, and 19 of the cat's visual cortex. This model includes the so-called "second-order transformation" of areas 18 and 19, which is a weak discontinuity well-known in the theory of complex variables. It also provides a coarse model of the mapping in the lateral geniculate body.

The computational advantages of topographic mapping were discussed in an example taken from the analysis of optical flow. Inverse perspectivic mapping is a powerful tool for a fast evaluation of special data relevant to obstacle avoidance. Most previous examples of computational advantages of topographic mapping have relied on the complex logarithm which, however, is found only in area V1 of certain monkeys.

The space-variant filters obtained from the composition of topographic mapping and uniform intrinsic processing were studied in Section 3.3. It turns out that even for topographic maps, computational differences are induced in the overall processing of information. Retinotopic mapping appears as a tool to make the most out of the

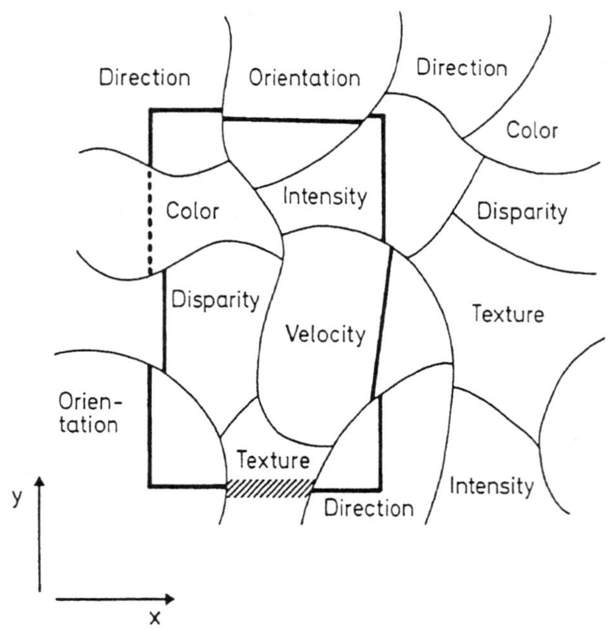

Figure 11. Integration of image segmentation data from different modules in a patchy representation. For details see text.

parallelism available. In modular architectures, just one convolution type operation may subserve many different tasks if the corresponding input is preprocessed by appropriate mappings.

4 Non-Topographic Mapping

In the previous sections, we discussed the average intrinsic structure of cortical areas and the space-variance added to the information processing by retinotopic mapping. There are further sources of space-variance both in the interconnections of areas and in their intrinsic organization that can be summarized as non-topographic maps. One has to bear in mind, however, that some of these space-variances are localized in the inter-area connectivity (e.g., ocularity stripes) while others are due to intrinsic organization (columns).

4.1 Patchy Connectivity

Parallel connections from two areas to a common target are typically segregated into distinct regions such as the ocularity stripes in area 17. In other cases, overlapping patches are found (Sherk 1986). As in the case of topographic mapping, this results in the generation of new neighborhoods on which local neural computations are performed. Also, if multi-dimensional feature vectors are to be represented on a two-dimensional sheet, patchy connections are optimal in the sense that continuity is maximized (Mitchison & Durbin 1986; Shen et al. 1987).

As an example of how to use this principle in the integration of different modules of early vision, we study the following strategy for image segmentation, i.e., the definition of regions in an image based on discontinuities (edges) in different parameter maps. Suppose in a topographically mapped area neurons that are tuned to different stimulus parameters occur in clusters as indicated schematically in Fig. 11. In general, edges will not occur in all modules simultaneously nor at exactly corresponding positions. If a patchy map is used to interlace the different parameter maps, the interaction problem reduces to the straightening and combination of edge elements found in the different patches. As indicated in the patch labeled "color," this might also provide a model of subjective contour.

4.2 Columnar Organization

Periodic or patchy variations in the intrinsic organization of the visual cortex have been known for a long time; for recent developments on the issue, see Löwel et al. (1986), Swindale et al. (1987) and Götz (1988). The question how the orderly layout of orientation-tuned cells in the visual cortex can be brought about in development has triggered a large amount of research. In this paper, we want to focus on functional implications of this layout, taking the organization ifself for granted. Oriented (or for that matter, directed) quantities in visual information processing occur for instance in the analysis of motion (velocity vectors) or surface orientation (surface normals). Two-dimensional vector fields are therefore a commonly used data

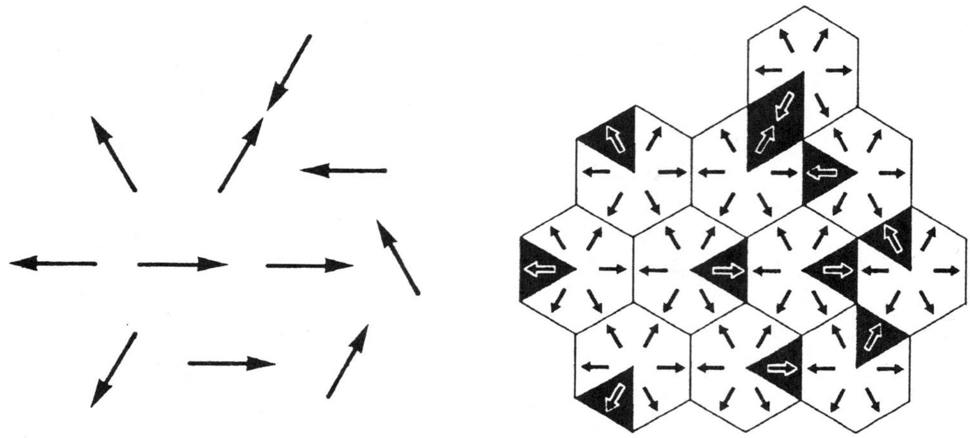

Figure 12. Representation of a vector field (left) in a two-dimensional grid of "hypercolumns" (right). Active regions in the hypercolumns are colored in black. Small shifts of excitation implement a regularization. A low-pass filtering would result in a divergence (looming) map.

type in computer vision problems such as optical flow analysis or surface representation. The grouping of orientation tuned cells into hypercolumns can be considered as a means to represent a vector-field in a two-dimensional sheet which otherwise could only represent scalar quantities. The coarse position of an excitation codes for retinal position, whereas the vector is represented by the fine position within the hypercolumn (Fig. 12). Note that this representation lents itself very easily to the computation of regularizations: Small shifts as produced by lateral inhibition will tend to equalize the distances between the peaks of excitation which, when the represented vector-field is considered, corresponds to a smoothing operation. In computational vision, constrained regularizations, either by means of the calculus of variations (Poggio et al. 1985) or by statistical methods (Marroquin et al. 1987), are considered a general tool for many different information processing tasks.

4.3 Computational Maps

The term non-topographic mapping has been used by Barlow (1986) for maps plotting two arbitrary parameters. Although the computational advantages of this strategy are obvious from his paper, it is difficult to find biological examples. We therefore favor an alternative view of non-topographic mapping that is based on receptive fields rather than on mere coordinate transforms. (In fact, the discussion in Section 3.3 clearly shows that pure coordinate transforms without receptive field variations are not possible.) In an ideal topographic mapping, position is the only receptive field parameter that varies as the recording electrode proceeds through the cortex. In general, however, a continuous but steady change of any receptive field property or a combination thereof can be considered a mapping of this property to the cortical surface. Mappings of this type are common in electrophysiology, for review see Knudsen et al. (1987). The patchy and columnar organizations presented in Sections 4.1 and 4.2 illustrate the computational capabilities of these types of representation.

4.4 Summary

Topographic mapping is an idealized concept that, strictly speaking, is not found in the visual cortex. For more realistic approximations, mappings have to be described in terms of their space-variant receptive fields rather than coordinate transforms. We presented two examples for actual computations than could be facilitated by representation of visual data in two-dimensional maps.

5 Conclusions

Neural networks show obvious anatomical differences that can be studied directly. Especially in the field of neural computation, the direct investigation of existing structures may be more rewarding than the search for emerging structure in randomly wired nets.

We propose a basic cortex-like neural network for visual information processing which contains the following elements: First, a space-invariant intrinsic operation with non-separable spatiotemporal interactions and, second, a number of mappings and other connectivity schemes (such as patches) which recombine information into new neighborhoods suited for being processed by the intrinsic operation. Although the examples presented here were taken from early vision, it seems likely that middle and high level vision can be dealt with in patchy and non-topographically mapped networks.

The structural principles reported in this chapter imply a level of description for visual information processing that we believe will be fruitful for technical considerations as well. The distinction of the *computational*, the *algorithmic*, and the *hardware* level of information processing, althought extremely successful in the construction of seeing machines, is probably not observed in the evolutionary "construction" of biological vision systems. The interactions between the function and the structure of a biological organism, examplified in the question why cortices are particularly suited for vision, allow insight into the information processing strategies that should not be neglected. In future research, we attempt to validate the cortex-like network defined above in technical vision applications.

6 Synopsis

In the previous Sections of this chapter, we have presented an approach to neural information processing that may be characterized as functional anatomy; it is guided by the structural adaptions of an important neural network, the visual cortex.

The goal that might eventually be achieved along this line is to "explain" features of cortical anatomy in terms of the computational operations they subserve. Unfortunately, it is far from obvious what exactly is meant by the term "computational operations" in a living animal. One possibility is to assume that the brain is a computer of some kind or, put more cautiously, that the computational operations used by the brain are of the same type as those studied in computer science. Here, we list some aspects of neural information processing that conflict with this view:

The Purpose of Neural Information Processing. Brains are no universal computers nor do they run programs. Rather, brains built working models that support the animal's actions in the world (cf. Barlow 1985). These models and the corresponding patterns of behavior are hierarchically built on one another. The hierarchy is a result of the evolution of informed behavior.

Isomorphic Processing. There is an isomorphism between information processing tasks and the neural processes representing these tasks in the brain. Therefore, small errors or deviations in processing correspond to small errors in the resulting behavior.

Isomorphic Representation. The mapping in cortical layers tends to map related events to nearby locations. If the relation of the events is spatial nearness on the sensory surface, the map is called topographic. If nearness corresponds to similarity in some other respect, the mapping is parametric.

Small number of Processing Steps. The elementary operations used by the brain (the "neural instruction set") are optimized in a way that allows for fast and reliable computations.

Of course, this list of differences between brains and computers is highly preliminary. It may, however, help to further an understanding of the neurobiology of information processing beyond the mere search for applications.

Acknowledgements. This work was supported by the Deutsche Forschungsgemeinschaft and by the Stiftung Volkswagenwerk. Prof. W. v. Seelen was in charge of the projects. We are grateful to Ms. R. Rehbinder who carefully prepared the line drawings.

6 References

Adelson EH, Bergen JR (1985) Spatiotemporal energy models for the perception of motion. *J Opt Soc Am* A2: 322-342

Barlow HB (1985) Cerebral cortex as a model builder. In: Rose D, Dobson VG (eds) *Models of the visual cortex.* John Wiley & Sons, Chichester, New York, pp 37-46

Barlow HB (1986) Why have multiple cortical areas? *Vision Res* 26: 81-90

Braccini C, Gambardella G, Sandini G (1981) A signal theory approach to the space and frequency variant filtering performed by the human visual system. *Signal Processing* 3: 231-240

Braitenberg V (1985) Charting the visual cortex. In: Peters A, Jones EG (eds) *Cerebral cortex, Vol 3: Visual cortex.* Plenum Press, New York, pp 379-410

Creutzfeldt OD, Sakmann B, Scheich H, Korn A (1970) Sensitivity distribution and spatial summation within receptive-field center of retinal on-center ganglion cells and transfer function of the retina. *J Neurophysiol* 33: 654-671

Dinse HRO, Best J, Krüger K, Mallot HA, Seelen W v (1988) The dynamics of receptive field organization in the cat visual cortex (areas 17, 18, 19, and PMLS): oscillations and non-separability. (in Preparation)

Enroth-Cugell C, Robson JG, Schweitzer-Tong DE, Watson AB (1983) Spatiotemporal interactions in the cat retinal ganglion cells showing linear spatial summation. *J Physiol (London)* 341: 279-307

Epstein LI (1984) An attempt to explain the differences between the upper and lower halves of the striate cortical map of the cat's field of view. *Biol Cybern* 49: 175-177

Fischer B (1973) Overlap of receptive field centers and representation of the visual field in the cat's optic tract. *Vision Res* 13: 2113-2120

Fischer B, May HU (1970) Invarianzen in der Katzenretina: Gesätzmäßige Beziehungen zwischen Empfindlichkeit, Größe und Lage receptiver Felder von Ganglienzellen. *Exp Brain Res* 11: 448-464

Götz KG (1988) Cortical templates for the self-organization of orientation-specific d- and l-hypercolumns in monkeys and cats. *Biol Cybern* 58: 213-223

Grimson WEL (1981) *From images to surfaces.* MIT Press, Cambridge, MA, London

Heeger DJ (1987) Optical flow from spatiotemporal filters. *Proc First Int Conf Comp Vision, London*, pp 181-190

Horn BKP (1986) *Robot vision.* MIT Press, Cambridge, MA, London

Jain R, Barlett SL, O'Brien N (1987) Motion stereo using ego-motion complex logarithmic mapping. *IEEE Trans Pattern Anal Machine Intel (PAMI)* 9: 356-369

Knudsen EI, du Lac S, Esterly SD (1987) Computational maps in the brain. *Ann Rev Neurosci* 10: 41-65

Koenderink JJ, Van Doorn AJ (1976) Local structure of movement parallax of the plane. *J Opt Soc Amer* 66: 717-723

Korn A, von Seelen W (1972) Dynamische Eigenschaften von Nervennetzen im visuellen System. *Kybernetik* 10: 64-77

Krone G, Mallot HA, Palm G, Schüz A (1986) Spatiotemporal receptive fields: a dynamical model derived from cortical architectonics. *Proc Roy Soc (Lond)* B 226: 421-444

Little JJ, Bülthoff HH, Poggio T (1987) Parallel optical flow computation. In: Baumann, L (ed) *Proc Image Understanding Workshop* Scientific Applications International Corp, Los Altos, CA, pp 915-920

Löwel S, Freeman B, Singer W (1987) Topographic organization of the orientation column system in large flat-mounts of the cat visual cortex: a 2-deoxyglucose study. *J Comp Neurol* 255: 401-415

Mallot HA (1985) An overall description of retinotopic mapping in the cat's visual cortex areas 17, 18, and 19. *Biol Cybern* 52: 45-51

Mallot HA (1987) Point images, receptive fields, and retinotopic mapping. *TINS* 10: 310-311

Marr D (1982) *Vision.* Freeman, San Francisco

Marroquin JL, Mitter S, Poggio T (1987) Probabilistic solution of ill-posed problems in computational vision. *J Amer Stat Assoc* 82: 76-89

Mitchison G, Durbin R (1986) Optimal numberings of an N x N array. *SIAM J Alg Disc Meth* 7: 571-582

Peters A, Jones EG (1984) *Cerebral cortex, Vol 1: Cellular components of the cerebral cortex.* Plenum Press, New York

Poggio T, Torre V, Koch C (1985) Computational vision and regularization theory. *Nature* 317: 314-319

Reitboeck HJ, Altmann J (1984) A model for size- and rotationinvariant pattern processing in the visual system. *Biol Cybern* 51: 113-121

Regan D, Beverly KI (1978) Looming detectors in the human visual pathway. *Vision Res* 18: 415-421

Sawchuk AA (1974) Space-variant image restoration by coordinate transforms. *J Opt Soc Amer* 64: 138-144

Schwartz EL (1980) Computational anatomy and functional architecture of striate cortex: a spatial mapping approach to perceptual coding. *Vision Res* 20: 645-669

Seelen W v, Mallot HA (1988) Parallelism and redundancy in neural networks. In: Eckmiller R, Malsburg C v d (eds) Neural computers. Springer-Verlag, Berlin Heidelberg New York, pp 51-60

Seelen W v, Mallot HA, Giannakopoulos F (1987) Characteristics of neuronal systems in the visual cortex. *Biol Cybern* 56: 37-49

Shen CW, Lee RCT, Chin YH (1987) A parallel nonlinear mapping algorithm. *Int J Pattern Recog Artif Intellig* 1: 53-69

Sherk H. (1986) Coincidence of patchy inputs from the lateral geniculate complex and area 17 to the cat's Clare-Bishop area. *J Comp Neurol* 253: 105-120

Swindale NV, Matsubara JA, Cynander MS (1987) Surface organization of orientation and direction selectiviy in cat area 18. *J Neurosci* 7: 1414-1427

Tusa RJ, Rosenquist AC, Palmer LA (1979) Retinotopic organization of areas 18 and 19 in the cat. *J Comp Neurol* 185: 657-678

Van Essen D (1985) Functional organization of primate visual cortex. In: Peters A, Jones EG (eds) *Cerebral cortex, Vol 3.* Plenum Press, New York, pp 259-329

Invariances in Pattern Recognition

HERBERT J. REITBOECK

Angewandte Physik und Biophysik, Philipps-Universität Marburg
D-3550 Marburg, FR Germany

Abstract. Humans and animals can recognize objects independent of object position in the visual field, largely independent of viewing angle and distance, and even independent of considerable variations in object shape. For object classification in the visual system, a comparison of sensory information with data in memory is required. Due to the large number of possible retinal pictures that can represent members of the same object class and even one individual object, such comparison is only feasible if an efficient object description can be generated that is object specific and invariant to changes in "non-essential" parameters. Of particular interest for models of invariance operations in the CNS are shift-invariant transforms. *Shift-invariance* is a primary invariance and other invariances can be derived from it. Size- and rotation-invariance, e.g., can be realized via a shift-invariance mechanism in combination with a logarithmic polar coordinate transform. With some approximations, the mapping of visual space to area 17 in the visual cortex of primates can be described by a logarithmic polar coordinate transform. Models of size-invariant processing in the CNS based on this mapping function have been proposed. In this paper, concepts for the generation of shift-, size-, and rotation-invariant pattern representations in the visual system are discussed, and a critical evaluation of their advantages, drawbacks, and neurophysiological implications will be given. Particularly, the paper will focus on two basic problems: (a) Neural networks are not well suited to perform exact arithmetic operations, as required in many models. A shift-invariant transform will be described that puts minimal demands on the neural transfer function. (b) Invariance operations generally require that the object has been separated from the background. Our model of region labeling via correlated neural activities preserves individual object contributions in composite pattern transforms, and is thus able to cope with the problem of figure/ground separation.

1 Introduction

Humans and animals can recognize objects independent of changes in object position within the visual field, largely independent of viewing angle and distance, and even independent of considerable distortions, i.e. limited variations in object shape, as long as essential object defining features are maintained. For any recognition task, objects or object attributes must be compared with data in memory. Because of the very large number of different pixel (picture element) configurations that may represent the same object, the recognition task would be impossible - in terms of both, memory size requirements and search time - if each pixel configuration were represented by a separate template in memory. The magnitude of this problem becomes evident from a simple example: If the retina had two receptors that respond to two illumination conditions only ("light" and "no light," with no grey shades inbetween), the number of possible pixel configurations is four (00, 01, 10, 11), and there are 16 ways (corresponding to the 16 Boolean functions of two variables) to separate these pixel configurations into two pattern classes. Generally, for n pixels there are $N=2^n$ binary images, and $K=2^N$ possibilities to divide them into two pattern classes. For a very coarse "retina" with 20 x 20 binary receptors, the number of possible binary patterns, already, is $N=2^{400}$ which is many orders of magnitude larger than the estimated number of elementary particles in the universe.

Before object classification can be attempted, the pictures therefore have to be pre-processed, and an object description must be generated that is object-specific, but independent of viewing conditions and independent of certain variations of the object shape. This is required in both, biological pattern recognition and in general purpose computer pattern recognition.

Shift-, size-, and rotation-invariance operations are of singular importance in pattern recognition, because these operations are not object-specific and they can be achieved via universal algorithms. Rotation invariance, here, only refers to object rotations around the viewing axis. The recognition of rigid 3-dimensional objects rotated around any arbitrary space axis poses no problems in computer pattern recognition, if the required object projectional views are known. It is, however, most unlikely that the visual system employs algorithms for this task that would be similar to those presently used in computer recognition of 3-dimensional objects. Other "higher" invariances as required, e.g., in order to recognize an animal independent of changes in posture, are object specific and cannot be realized via a general transform applicable to any object class.

The most basic invariance is *shift-invariance*. In a natural viewing situation, the objects to be recognized could be anywhere in the visual field. In addition, there is a permanent shift of the retinal image due to eye and body movements. To make the recognition task manageable, shape information and position information must be separated. Shift-invariance is of fundamental significance, since size- and rotation-invariance can be derived from it.

2 Shift-Invariance

Shift-invariant pattern recognition does not mean that the position of the pattern - particularly its position relative to other patterns - would be irrelevant. It only means that the task is split into two simpler tasks: The position-invariant pattern description facilitates comparison with data in memory for object identification. Positional information, generally is maintained and utilized at higher processing levels when the pattern is interpreted in the context of other patterns in the image. The same holds for other invariances, particularly for size- and rotation invariance.

In computer pattern recognition, a variety of techniques is used in order to generate a shift-invariant pattern description. Shift-invariance can be realized via Fourier amplitude- or power-spectra, Walsh power-spectra (Harmuth, 1972), moments (Hu 1962; Teague 1980), via certain nonlinear transforms (Reitboeck & Brody 1969), autocorrelation functions (Bracewell 1965), etc. In syntactic pattern recognition (Fu 1974; Fu & Rosenfeld 1976), specific shift-invariance operations are not required, since patterns are characterized by the structural relations of their components (sub-patterns or pattern primitives) whose absolute position can either be ignored, or retained to be used in a higher context. Although higher pattern recognition and scene classification operations - particularly in humans - might well be described by syntactic models, such models are unlikely to apply for the primary processing stages in the visual system, where essential processing mechanisms appear to be realized via specific network interconnections.

For models of primary invariance operations in the central nervous system, *shift-invariant transforms* are of particular interest, because such general operations, as mentioned above, are not object specific, i.e., the algorithm does not have to be matched to the object before it can be applied. (Note that shift-invariant transforms generally are non-linear. This means, that the pattern has to be separated from the background as well as from other objects, before the transform can be applied; more about this in Section 4.2). Shift-invariant transforms can be realized via a layered network of parallel computing elements. The structure of such networks could be realized, in principle, within parts of the CNS. A more serious question is, whether neurons are capable of executing with sufficient accuracy the arithmetic operations required in such transforms. We will come back to this question in Section 2.2.

2.1 Generation of Shift-Invariance via the Fourier Transform

2.1.1 Fourier Transform Technique

If a pattern f(x,y) has the Fourier transform F(v,w), then the shifted pattern f(x+a, y+b) has the Fourier transform

$$e^{i2\pi\,(av+bw)}F(v,w)$$

The Fourier transform of the original and the shiftet pattern only differ by a phase factor whose absolute magnitude is 1. Fourier amplitude- and power-spectra, therefore, are invariant to shifts in the object domain. Fourier transform techniques have been proven to be very useful in computer picture processing and pattern recognition, and they are widely used in applications, such as picture enhancement, digital filtering, holography, correlation analysis, etc.

Since the late sixtieth, the question whether signal processing in some parts of the visual system is in the spatial frequency domain has attracted considerable interest (for a review see Maffei 1978), and several models of information processing in the visual system based on Fourier domain operations have been proposed. These models are ranging from descriptions of visual receptive fields in terms of spatial frequency filters and Fourier domain invariance models, up to highly speculative models of "holographic" processes in the visual system.

On the other hand, there are also compelling arguments to dismiss Fourier domain concepts as unsuited descriptors of neural information processing: The Fourier transform is a tool in linear systems analysis; neural networks, however, are highly non-linear systems. The question of Fourier domain processing in the CNS and, particularly, spatial frequency domain processing in the visual system, therefore, should be addressed in more detail in the following Section.

2.1.2 Spatial Frequency Domain Processing in the Visual System

During the past 20 years, a large number of psychophysical results have been published that can be interpreted best in terms of spatial frequency domain parameters: Campbell & Robson (1968) reported results on the perception of square wave patterns that are in agreement with a model based on linear superposition of Fourier components. Blakemore & Campbell (1969) found that adaptation to a square-wave grating raises the threshold for the detection of the fundamental frequency and its third harmonic as would be predicted by a Fourier model. The McCollough-effect (McCollough 1965), a long lasting threshold after-effect, too, is spatial frequency specific (Stromeyer 1972). Additional evidence comes from a modified "Craik-Cornsweet illusion experiment" by Campbell et al. (1971): A square wave grating (of limited contrast, and spatial frequency lower than 1c/deg) whose fundamental frequency has been filtered out, still appears as if it had square wave luminance distribution. It seems as if the visual system had restituted the fundamental frequency.

The Fourier hypothesis was supported further by results from electrophysiological experiments with grating stimuli, which indicate that neurons at various processing stages in the visual system do respond selectively to specific ranges of spatial frequencies (Enroth-Cugell & Robson 1966; Campbell et al. 1969; Maffei & Fiorentini 1973; Schiller et al. 1976; DeValois et al. 1982).

These results have initiated arguments whether image processing in the visual system is in the object domain or in the spatial frequency domain (Macleod & Rosenfeld 1974; DeValois et al. 1978, 1979; Albrecht et al. 1981). Presently, there is some agreement that the two descriptions are complementary (Kulikowski & Bishop 1981; MacKay 1981; Daugman 1985; Kulikowski & Kranda 1986), and the question is rather which domain gives a more transparent description of specific effects. For the characterization of the receptive field properties of visual neurons at primary processing stages, both descriptions have their advantages. There do remain, however, psychophysical results that are not explainable by the transfer functions of receptive fields (e.g., the restitution of the fundamental frequency in the modified Craig-Cornsweet experiment (Campbell et al. 1971)). These results are more adequately described in terms of Fourier parameters.

Although the psychophysical and neurophysiological results indicate that certain properties of visual channels - in spite of the nonlinearity of neural networks - can be characterized very well in terms of Fourier parameters, there is no psychophysical or neurophysiological evidence that a *global Fourier transform* is executed anywhere in the visual system. Such global transform, however, would be required if shift-invariant pattern descriptions were realized via the Fourier transform.

2.2 R-Transforms as a Model of Shift-Invariance in the Visual System

R-transforms are a class of nonlinear irreversible transforms (Reitboeck & Brody 1969). Their common property is that they are invariant under cyclic permutations and thus generate a shift-invariant pattern description. Various members of these classes of transforms and their properties have been described since (Ulman 1970; Kunt 1975; Wagh 1975; Wagh & Kanetkar 1975a,b, 1977; Burkhardt & Mueller 1980; Rao & Ahmed 1980; Elliott & Rao 1982; Rao 1985).

R-transforms are of interest as possible models of information processing operations in the CNS for two reasons: (i) They only require two types of *commutative* operations that can be quite arbitrary in any other respect. Any type and combination of commutative arithmetic (or logic) operations will do. Particularly, any (even nonlinear) neural transfer function is acceptable, as long as the postsynaptic state is invariant to a permutation of the presynaptic inputs. (ii) R-transforms can be executed by a simple neural network of computing elements.

Network Model for Executing a 2-Dimensional R-Transform

The tree graph for the calculation of a 1-dimensional R-transform is shown in Fig. 1. The graph resembles the scheme for the execution of the Fast Fourier transform (FFT), except that the operations in the nodes indicated by "sigma" and "delta" (Fig. 1) are arbitrary commutative operations, such as the sum and absolute difference. For $P=2^P$ data samples, the span of the butterfly operations is P/2 in the first transform layer, and decreasing by a factor of 0.5 in each subsequent layer, until the span is 1. The signals thus arrive at the last layer after $p={}_2\log P$ transform steps. With some cautionary reservations, the span of these butterfly operations can be thought to model lateral interactions.

For execution on a digital computer, simple operations in the nodes of the graph in Fig. 1 can be chosen that need much less computation time than the summation of products of sine functions, as required in the nodes of the FFT. R-transforms, therefore, are much faster than the FFT. For the variables x and y, examples of suitable arithmetic operations are:

(x+y), (x*y), abs (x-y), etc.

For the Boolean variables X and Y, suitable logic operations would be:

(X AND Y), (X OR Y), (X NAND Y), (X EXOR Y), etc.

In a "hard-wired" special purpose processor or in a neural R-transform model, functions such as

(x^2+y^2), $(x+y)*x*y$, sqrt $[a*(x+y)+b*(x^2+y^2)]$, ln x + ln y, etc.

can be chosen, that more adequately describe the specific transfer function of the elements (neurons) at the nodal points. It should be kept in mind, however, that the operations in the nodal points only need to be specified in the (hardware or software)

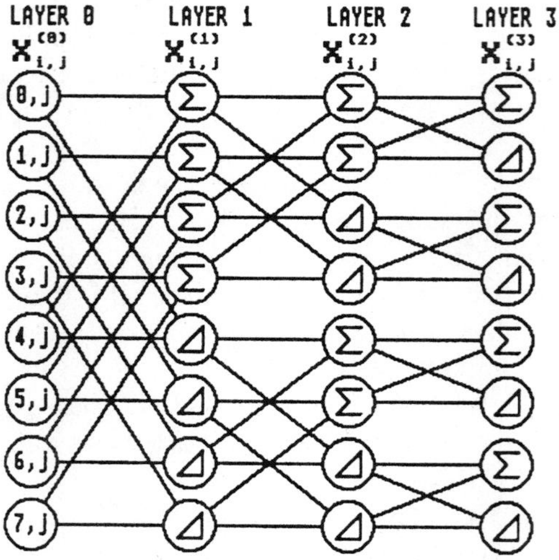

Figure 1. Tree graph for a 1-dimensional R-transform. Operations denoted by "sigma" and "delta" are arbitrary commutative operations. If a 2-dimensional transform is executed via subsequential vertical and horizontal transforms, this would correspond to the tree graph of column j of the vertical transform.

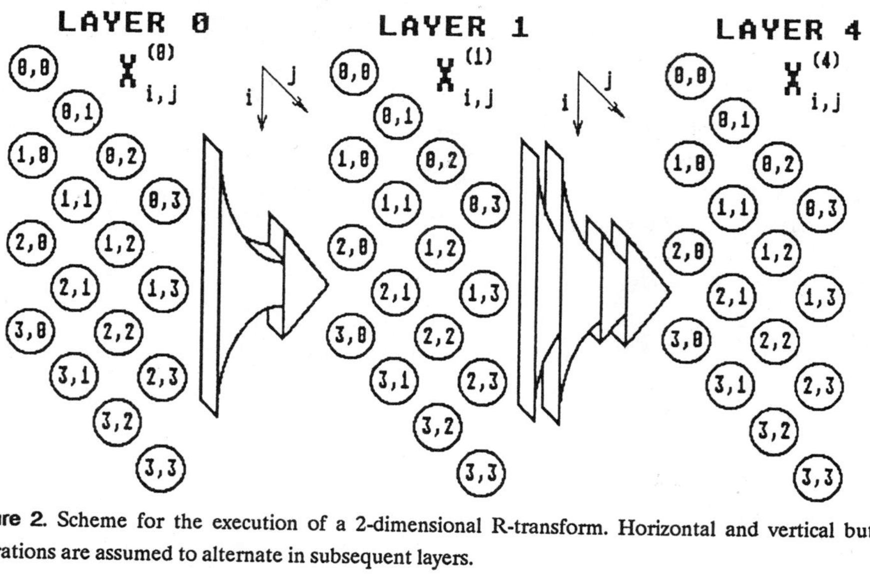

Figure 2. Scheme for the execution of a 2-dimensional R-transform. Horizontal and vertical butterfly operations are assumed to alternate in subsequent layers.

model. If the transform were to be realized by a neural network, the synaptic transfer functions would not need to have any analytic description; as long as two groups of synaptic inputs are commutative, the output will be invariant to shifts of the input pattern.

The execution of the R-transform of a function of two variables (i, j), such as an image of P*Q pixels ($0 \leq i \leq P-1$; $0 \leq j \leq Q-1$), requires ${}_2\log P + {}_2\log Q$ transform steps (= transform layers of computing elements). The data matrices for an array of 4*4 pixels are shown in Fig. 2. In this transform scheme, "horizontal" and "vertical" butterfly operations alternate in subsequent layers (Fig. 4). For a square matrix of neurons, the span of the butterfly operations (corresponding to the span width of lateral interactions) is the same in each pair of row- and column transform layers.

a

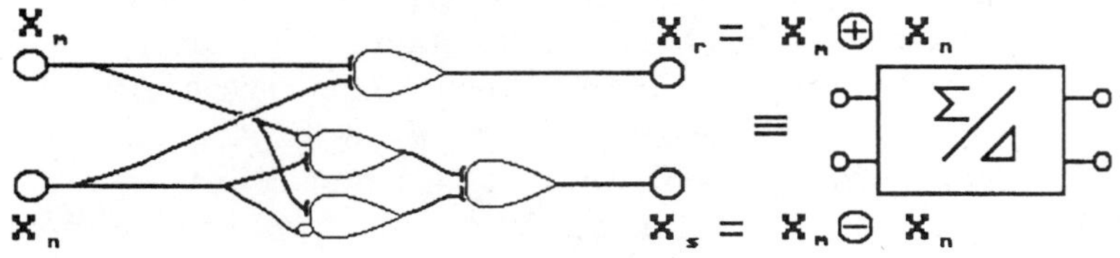

b

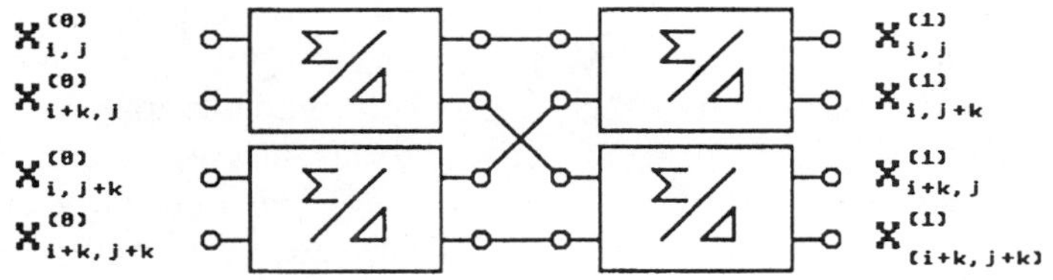

Figure 3. (a) Neural network whose output is invariant to commutations of the input signals ("sigma/delta"-block). (b) Realization of the butterfly operations in Fig. 3 via four "sigma/delta"-blocks.

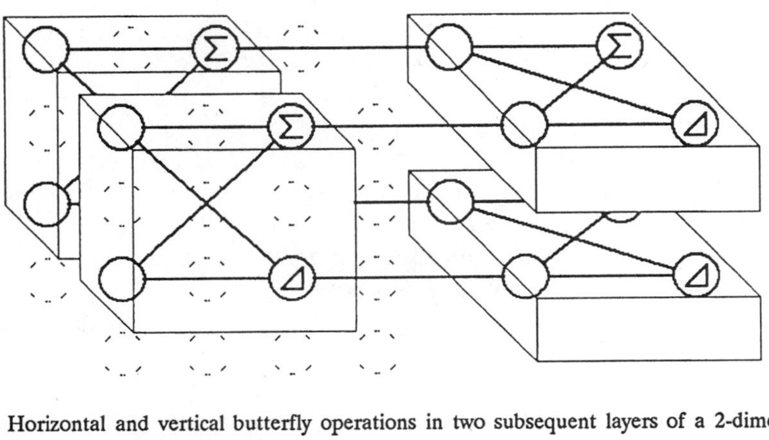

Figure 4. Horizontal and vertical butterfly operations in two subsequent layers of a 2-dimensional R-transform network.

A simple neuronal circuit that can execute the required commutative operations at the nodal points of the diagrams in Figs. 1 and 4 is shown in Fig. 3a. For clarity, the "axons" are drawn to terminate in both, excitatory and inhibitory synapses. In a more realistic scheme, interneurons could be added. For the 2-dimensional transform, such basic circuits (symbolized by the "sigma/delta" block in Fig. 3a) are combined to form one set of horizontal and vertical lateral interactions (see Fig. 3b).

3 Size- and Rotation-Invariance

Size- and (limited) rotation-invariance constitute the second group of basic invariances. In computer pattern recognition, scale-invariance can be achieved via the Mellin transform (see Casasent & Psaltis 1976, 1977; Huang et al. 1975; West & Reitboeck 1979). The Mellin transform can be realized via a logarithmic coordinate transform in combination with a Fourier transform (e.g., see Bracewell 1965). The logarithm converts size scalings (multiplication with a scale factor) into translations (addition to the log of the scale factor). These shifted images are logarithmically distorted. A subsequent shift-invariant transform results, thus, in a size-invariant pattern description. This method works only, if the original images are always presented in a specific position (e.g., by directing the gaze) or if an additional shift-invariant transform has been applied to the image before the logarithmic transform was done.

Within certain ranges of visual angle, the retinal projection to the striate cortex of some mammals can be described by a cortical magnification factor *CM* that is inversely proportional to retinal excentricity. (For cortical mapping see Mallot & v. Seelen, this volume.) The mapping of visual space to the cortical area V1 (Daniel & Whitteridge 1961; Cowey 1964; Allman & Kaas 1971) can thus be approximated by a logarithmic polar coordinate transform. (In primates, this holds for extrafoveal regions up to visual angles of about 20 degrees; the deviation at the fovea might be smaller than was thought previously (Dow et al. 1981); in cat the mapping function is more complex (Epstein 1984)). In combination with a shift-invariant transform (e.g., localized in non-visuotopic cortical areas), this logarithmic representation of visual space could provide a mechanism for a scale-invariant object description. Models of size-invariance in the visual system based on this concept have been proposed by Cavanagh and by Schwartz (see Cavanagh 1985; Schwartz 1985). Various problems associated with these models have been discussed by Cavanagh (1981, 1982), Schwartz (1981, 1983), and Reitboeck & Altmann (1984).

In a strict Mellin correlation model, a global Fourier transform would be required, which is an unlikely operation to be performed in the CNS, as mentioned before. Cavanagh (1985) suggested a model for realizing a global transform via the summation of local log polar frequency transforms. Due to the coarse spatial frequency resolution of cortical cells, this transform would probably not yield good shape discrimination. A scale-invariant correlation that does not involve a global Fourier transform has been defined by Altmann & Reitboeck (1984), and a model for size-invariance in the visual system based on the R-transform (that does not require neurons to perform exact arithmetic operations) has been proposed by Reitboeck & Altmann (1984).

An interesting aspect of Mellin-type invariance models is that there exists a corresponding model for auditory pattern recognition (Altes 1978). The logarithmic mapping of frequencies along the basilar membrane of the inner ear converts frequency scalings into shifts; a subsequent shift-invariant transform results in a pitch-invariant description of auditory signals. A further coincidence in the mapping functions of V1 (in the visual domain) and along the basilar membrane (in the auditory domain) is the deviation from the logarithmic map to an approximately linear representation. In the visual system this holds for the foveal region (high spatial frequencies), and in the auditory system for the basilar membrane region close to the helicotrema (below 500 Hz in humans). Although these deviations do not support the Mellin models, the coincidence might point to a common functional principle.

There remain other arguments against the hypothesis that the quasi-logarithmic mapping of visual space onto the striate cortex is of functional significance for size-invariant filtering. Ablation of V1 in cat does not seriously affect shape recognition (Sprague et al. 1977; Hughes & Sprague 1986). Although the cat has parallel inputs to areas V1, V2, and V3, one would expect shape recognition to be affected by ablation of V1 if an essential stage of the size-invariance mechanism were located there. Lesions in area V1 severely impair pattern vision in monkey (Mishkin & Ungerleider 1982). However, since the retino-geniculate inputs pass serially through V1 to higher regions in monkey, this is not an argument that the resulting loss in pattern vision is due to a damage to the size-invariance mechanism.

4 Invariance Generation in the Visual System via Global Transforms

4.1 Problems Due to the Cyclic Periodicity of the Transforms

The discrete Fourier transform of a function sampled at intervals T is periodic with the frequency 1/T, and the discrete representation of the spectrum (line spectrum), in turn, requires the assumption that the original function is periodic. A similar law holds for other discrete transforms, including the R-transforms and the discrete Mellin transform. (In the CNS, too, a visual stimulus is "sampled" by the receptor mosaic and at the level of single unit responses, the electrical activity evoked by the stimulus is represented in discrete channels). This cyclic periodicity results in a strange phenomenon: When parts of an object have been shifted beyond, say, the right border of the visual field, those parts must enter the visual field again at the left border to generate the same transform as an object that is totally within the visual field. Since this contradicts any psychophysical evidence, one could conclude, therefore, that discrete transforms with this property cannot be a model for the generation of invariant pattern descriptions in the visual system. This is, however, not the case; mathematical methods used in machine pattern recognition to overcome such problems of cyclic periodic transforms are finding some interesting analogies in the visual system: In Mellin-type transforms, the size invariant correlation can be improved if the logarithmically distorted picture is differentiated and smoothed before the shift-invariant transform is applied. The two operations can be realized via a convolution of the picture with a "Mexican hat" function (Reitboeck & Altmann 1984). The convolution with a filter function of constant size in the logarithmic domain approximately corresponds to a convolution with a space-variant filter in the object domain, whose size increases linearily with excentricity. Such filters, thus, closely resemble the receptive fields of retinal ganglion cells.

4.2 Object Separation

A common problem of all models of position-, size-, and rotation-invariant processing in the visual system via general transforms is that the matching of the transformed pattern with templates in memory gives a poor measure of similarity, as long as the object has not been separated from background structures and from other objects in the visual field.

We have developed a model for pattern separation in the visual system, where object regions are labeled in the time domain (Reitboeck 1983; Reitboeck et al. 1987a). A basic assumption in our model is that correlated neural activity plays a fundamental role in the function of the central nervous system (e.g., see v. d. Malsburg 1981; Reitboeck 1982; Johannesma et al. 1986). Our model is consistent with several operational principles of the visual system (Wilson & Singer 1981; Gray & Singer 1987; Eckhorn & Reitboeck 1988; Eckhorn et al. 1988); and its texture separation capacity is very well in agreement with results on the pre-attentive texture separation in humans (Reitboeck et al. 1987b).

In our model, texture regions are labeled, and linked to other regions that belong to a given object via correlated or quasi-synchronous neural activity. If the object is moving against the background, correlated activity is induced automatically in those neurons that are activated by the object. Object movement, therefore, is a powerful mechanism for figure/ground separation (Reichardt & Poggio 1979; Reichardt et al. 1983; see also Ewert, this volume). But even without a relative movement between object and background, a global correlation of neural activity is induced by the shifts of the retinal image during normal vision. Spatial differences, e.g., in local brightness, are thus transformed into (spatial and) temporal variations of neural activities. The discharge patterns of neurons that are activated by regions of similar texture are, therefore, similar, whereas neural activities evoked by discriminable textures are different. The difference is detectable by appropriate temporal sampling.

Texture discrimination experiments using only this initial processing stage of our model have been shown to be in agreement with psychophysical results on pre-attentive texture discrimination by humans (Reitboeck et al. 1987a,b). At higher visual processing levels (including the lateral geniculate nucleus) efferent signals from various visual areas are assumed in our model to selectively affect the temporal order in the neural assemblies that respond to specific object regions. Finally, interaction with memory is assumed to modify the temporal and spatial associations for object definition and object classification. An additional attractive property of our model is that neural assemblies in separate cortical areas can be linked via correlated activity without requiring dedicated lines. Object labeling via correlated neural activity can be easily preserved during the shift-, size-, and rotation-invariance transforms, and the contributions of a particular object to the transform of a composite pattern, thus, can be filtered out and operated on individually at all processing stages.

5 Conclusion

Although shift-, size-, and rotation-invariance transforms can "reduce the dimensionality of the recognition space" (i.e., reduce the number of pattern templates required in memory by many orders of magnitude), they alone could not explain the performance of the visual system in object recognition. In addition to position and size changes, 3-dimensional objects change appearance when seen from different directions, and living objects can assume a variety of shapes and postures. In a realistic situation, the position- and scale-invariant correlations still would have to be calculated for a very large number of templates.

Models of invariance generation in the visual system have many attractive features. However, at the present state of our understanding of pattern recognition mechanisms in the visual system, they should be considered to be preliminary hypotheses only, whose main function is to point out specific problems, to stimulate new concepts and to give a theoretical background for experiments.

6 References

Albrecht DG, De Valois RL, Thorell LG (1981) Visual cortical neurons: are bars or gratings the optimal stimuli? *Science* 207: 88-90

Allmann JH, Kaas JH (1971) Representation of the visual field in striate and adjoining cortex of the owl monkey (*Aotus trivirgatus*). *Brain Res* 35: 89-106

Altes RA (1978) The Fourier-Mellin transform and mammalian hearing. *J Acoust Soc Amer* 63: 174-183

Altmann J, Reitboeck HJ (1984) A fast correlation method for scale- and translation-invariant pattern recognition. *IEEE Transactions on Pattern Analysis and Machine Intelligence PAMI* 6: 46-57

Blakemore C, Campbell FW (1969) On the existence of neurons in the human visual system selectively sensitive to the orientation and size of retinal images. *J Physiol* 203: 237-260

Bracewell R (1965) *The Fourier transform and its applications.* McGraw-Hill Book Comp, New York

Burkhardt H, Mueller X (1980) On invariant sets of a certain class of fast translation-invariant transforms. *IEEE Trans ASSP* 28: 517-523

Campbell FW, Robson JG (1968) Application of Fourier analysis to the visiblity of gratings. *J Physiol* 197: 551-566

Campbell FW, Cooper FG, Enroth-Cugell C (1969) The spatial selectivity of the visual cells of the cat. *J Physiol* 203: 223-235

Campbell FW, Howell ER, Robson JG (1971) The appearance of gratings with and without the fundamental Fourier component. *J Physiol* 217: 17-18

Casasent D, Psaltis D (1976) Scale invariant optical correlation using Mellin transforms. *Opt Commun* 17: 59-63

Casasent D, Psaltis D (1977) New optical transforms for pattern recognition. *Proc IEEE* 65: 77-84

Cavanagh P (1981) Size invariance: reply to Schwartz. *Perception* 10: 469-474

Cavanagh P (1982) Functional size invariance is not provided by the cortical magnification factor. *Vision Res* 22: 1409-1412

Cavanagh P (1985) Local log polar frequency analysis in the striate cortex as a basis for size and orientation invariance. In: Rose D, Dobson VG (eds) *Models of the visual cortex.* John Wiley, Chichester, pp 85-95

Cowey A (1964) Projection of the retina onto striate and prestriate cortex in the squirrel monkey (*Saimiri sciureus*). *J Neurophysiol* 27: 366-393

Daniel PM, Whitteridge D (1961) The representation of the visual field on the cerebral cortex in monkeys. *J Physiol* 159: 203- 221

Daugman, JG (1985) Representational issues and local filter models of two-dimensional spatial visual encoding. In: Rose D, Dobson VG (eds) *Models of the visual cortex*. John Wiley, Chichester, pp 96-107

DeValois RL, Albrecht DG, Thorell LG (1978) Cortical cells: bar and edge detectors, or spatial frequency filters? In: Cool SJ, Smith EL (eds) *Frontiers of visual science*. Springer-Verlag, New York, pp 544-556

DeValois KK, DeValois RL, Yund EY (1979) Responses of striate cortex cells to grating and checkerboard patterns. *J Physiol* 291: 483-505

DeValois RL, Albrecht DG, Thorell LG (1982) Spatial frequency selectivity of cells in macaque visual cortex. *Vision Res* 22: 545-559

Dow BM, Snyder AZ, Vautin RG, Bauer R (1981) Magnification factor and receptive field size in foveal striate cortex of monkey. *Exp Brain Res* 44: 213-228

Eckhorn R, Reitboeck HJ (1988) Assessment of cooperative firing in groups of neurons: special concepts for multiunit recordings from the visual system. In: Başar E (ed) *Dynamics of sensory and cognitive processing*. Springer-Verlag, Berlin Heidelberg New York, pp 219-227

Eckhorn R, Bauer R, Reitboeck HJ (1988) Discontinuities in visual cortex and possible functional implications: relating cortical structure and function with multi-electrode/correlation techniques. In: Başar E (ed) *Springer series in brain dynamics, Vol 2* (in press)

Elliott DF, Rao KR (1982) *Fast transforms: algorithms, analyses, applications*. Academic Press, New York

Enroth-Cugell C, Robson JG (1966) The contrast sensitivity of retinal ganglion cells of the cat. *J Physiol* 187: 517-552

Epstein LI (1984) An attempt to explain the differences between the upper and lower halves of the striate cortical map of the cat's field of view. *Biol Cybern* 49: 175-177

Fu KS (1974) *Syntactic methods in pattern recognition*. Academic Press, London

Fu KS, Rosenfeld A (1976) Pattern recognition and image processing. *IEEE Trans Comput C* 25: 1336-1345

Gray CM, Singer W (1987) Stimulus-dependent neuronal oscillations in the cat visual cortex area 17. *Neurosci* 22: 1301

Harmuth HF (1972) *Transmission of information by othogonal functions*. Springer-Verlag, Berlin Heidelberg New York

Hu MK (1962) Visual pattern recognition by moment invarants. *IRE Trans Inform Theory* II-8: 179-187

Huang GC, Russell FD, Chen WH (1975) Pattern recognition by Mellin transform. *EIA/AIPR Symposium*, Univ of Maryland

Hughes HC, Sprague JM (1986) Cortical mechanisms for local and global analysis of visual space in the cat. *Exp Brain Res* 61: 332-354

Johannesma P, Aertsen A, van den Boogaard H, Eggermont J, Epping W (1986) From synchrony to harmony: ideas on the function of neural assemblies and on the interpretation of neural synchrony. In: Palm G, Aertsen A (eds) *Brain theory*. Springer-Verlag, Berlin Heidelberg New York, pp 25-47

Kulikowski JJ, Bishop PO (1981) Fourier analysis and spatial representation in the visual cortex. *Experientia* 37: 160-163

Kulikowski JJ, Kranda K (1986) Image analysis performed by the visual system: feature versus Fourier analyis and adaptable filtering. Pettigrew JD (ed) *Visual neuroscience.* Cambridge Univ Press, Cambridge Mass, pp 381-404

Kunt M (1975) On computation of the Hadamard transform and the R-transform in ordered form. *IEEE Comput Trans* C-24: 1120-1121

Macleod IDG, Rosenfeld A (1974) The visibility of gratings: spatial frequency channels or bar detecting units? *Vision Res* 14: 909-915

MacKay D (1981) Strife over visual cortical function. *Nature* 289: 117-118

Maffei L (1978) Spatial frequency channels: neural mechanisms. In: Held R, Leibowitz HW, Teuber HL (eds) *Handbook of sensory physiology, Vol 8.* Springer-Verlag, Berlin Heidelberg New York, pp 39-66

Maffei L, Fiorentini A (1973) The visual cortex as a spatial frequency analyzer. *Vision Res* 13: 1255-1267

Malsburg C v d (1981) The correlation theory of brain function. *Dept of Neurobiology Internal Report 81-2,* MPI Biophysical Chemistry Göttingen

McCollough C (1965) Color adaptation of edge-detectors in the human visual system. *Science* 149: 1115-1116

Mishkin M, Ungerleider LG (1982) Contribution of striate inputs to the visuospatial functions of parieto-preoccipital cortex in monkeys. *Behav Brain Res* 6: 57-77

Rao KR, Ahmed N (1980) A Class of discrete orthogonal transforms. *Comp and Electr Eng (USA) Co* 7: 79-87

Rao KR (1985) *Discrete transforms and their applications.* Van Nostrand Reinhold Comp, New York

Reichardt W, Poggio T (1979) Figure-ground discrimination by relative movement in the visual system of the fly. Part I: Experimental results. *Biol Cybern* 35: 81-100

Reichardt W, Poggio T, Hausen K (1983) Figure-ground discrimination by relative movement in the visual system of the fly. Part II: Towards the neural circuitry. *Biol Cybern* 46: 1-30

Reitboeck HJ (1982) in Altmann L (1982) *Psychophysische Untersuchungen zur Rolle der Synchronität bei der Mustererkennung mit Hilfe einer mikroprozessorgesteuerten LED-Matrix.* MS Thesis, Univ of Marburg

Reitboeck HJ (1983) A Multi-electrode matrix for studies of temporal signal correlations within neural assemblies. In: Başar E, Flohr H, Haken H, Mandell AJ (eds) *Synergetics of the brain.* Springer-Verlag, Berlin Heidelberg New York, pp 174-181

Reitboeck HJ, Altmann J (1984) A model for size- and rotation-invariant pattern processing in the visual system. *Biol Cybern* 51: 113-121

Reitboeck HJ, Brody TP (1969) A transformation with invariance under cyclic permutation for applications in pattern recognition. *Inf Control* 15: 130-154

Reitboeck HJ, Eckhorn R, Pabst M (1987a) A model of figure/ground separation based on correlated neural activity in the visual system. In: Haken H (ed) *Synergetics of the brain.* Springer-Verlag, Berlin Heidelberg New York, pp 44-54

Reitboeck HJ, Eckhorn R, Pabst M (1987b) Texture description in the time domain. In: Cotterill RMJ (ed) *Computer simulation in brain science.* Cambridge Univ Press, Cambridge Mass

Schiller PH, Finlay BL, Volman SF (1976) Quantitative studies of single-cell properties in monkey striate cortex. *J Neurophysiol* 39: 1288-1351

Schwartz EL (1981) Cortical anatomy, size invariance, and spatial frequency analysis. *Perception* 10: 455-468

Schwartz EL (1983) Cortical mapping and perceptual invariance: a reply to Cavanagh. *Vision Res* 23: 831-835

Schwartz EL (1985) Local and global functional architecture in primate striate cortex: outline of a spatial mapping doctrine for perception. In: Rose D, Dobson VG (eds) *Models of the visual cortex.* John Wiley, Chichester, pp 146-156

Sprague JM, Leby J, DiBerardino A, Berlucchi G (1977) Visual cortical areas mediating form discrimination in the cat. *J Comp Neurol* 172: 441-488

Stromeyer CF (1972) Edge-contingent color after effects: spatial frequency specificity. *Vision Res* 12: 717-733

Teague MR (1980) Image analysis via the general theory of moments. *J Opt Soc Amer* 70: 920-930

Ulman LJ (1970) Computation of the Hadamard transform and the R-transform in ordered form. *IEEE Comput Trans* C-19:359-360

Wagh MD (1975) R-Transform amplitude bounds and transform volume. *J Inst Electronics and Telecom Engrs* 21: 501-502

Wagh MD, Kanetkar SV (1975a) A multiplexing theorem and generalization of R-transform. *Intern J Comput* C-24: 1120-1121

Wagh MD, Kanetkar SV (1975b) A multiplexing theorem and generalization of R-transform. *Intern J Comput Math* 5: 163-171

Wagh MD, Kanetkar SV (1977) A class of translation invariant transforms. *IEEE Trans on Acoustics, Speech and Signal Processing, ASSP-25*: 203-205

West G, Reitboeck HJ (1979) Zur ähnlichkeitsinvarianten Mustererkennung mittels der Fourier-Mellin-Transformation. *Elektron Informationsverarb und Kybernetik* 15: 507-512

Wilson JTL, Singer W (1981) Simultaneous visual events show a long-range spatial interaction. *Perception and Psychophys* 30: 107-113

Perception by Sensorimotor Coordination in Sensory Substitution for the Blind

REINHARD KOY-OBERTHÜR

Institut für Elektrische Informationstechnik, Technische Universität Clausthal
D-3392 Clausthal-Zellerfeld, FR Germany

Abstract. A contour based technical aid for the blind is presented. This device is intended to enable the perception of object contours and their spatial relations, improving orientation and mobility for the blind. An optical contour sensor reduces environmental information to direction and position of object contours. These relevant contour informations are transmitted to the blind by the cutaneous system at the lower arm as substitution channel. The blind carries the contour sensor like a torch in his hand. He scans an object or a scene by contour tracking and saccades, depending on the context of actual acquired information and his experience knowledge. Recognition results from coordinating the sensory information and the motor reactions in connection with context. Based on *schema theory* a model of the perception process is proposed. The model is described by a formal grammar generating a symbolic description of the "world" by composing actual sensory informations with motor reactions, including the context of experience by an associator. This model can be extended to a hierarchical structure of concurrent models. Complex perceptual contents are produced stepwise by the elementary informations and reactions. The model is compaired with experimental data. Finally, schema theoretical implications are discussed from an information theoretical point of view.

1 Introduction

The problem of sensory substitution for the blind is not solved satisfactorily, although pattern recognition (computer vision) technics proceed, micro electronics allow developments of powerful portable micro computers, and neurophysiology and biological cybernetics provide us with advancing models concerning structure and function underlying information processing in the brain. The reason is threefold: (i) the insufficience of technical pattern recognition operating under the various conditions in a natural environment, (ii) the problem of limited information capacity of the available substitution channels, and (iii) our still limited understanding of perception processes.

A detailed survey of technical aids for the blind including perception, rehabilitation, and computer vision problems is given by Welsh & Blash (1980), Herzog & Kügle (1982), Warren & Strelow (1985). Most visual aid systems realized for the blind are working either with transmission of the entire picture with reduced pixle number to the skin (Collins & Bach-Y-Rita 1973; Bach-Y-Rita 1977; Collins 1978), or with pattern recognition in a technical system and appropriate transmission of the processed data to the blind (Collins & Deering 1981). Our approach compromizes between these ways. We reduce the pictorial information to relevant features, the contours in a fixed point, and use the human intelligence for recognition by sensorimotor coordination within the individual context of experience. Based on

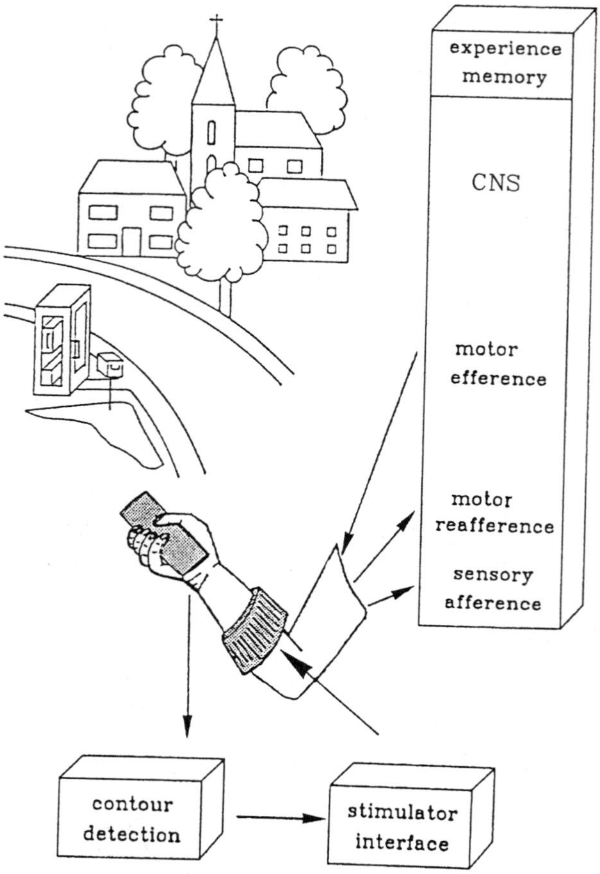

Figure 1. Illustration of perception with the KONWA-system. The technical system, consisting of contour sensor, contour detection unit, and stimulator interface, transmits the direction and the position of a fixed contour to the blind. Recognition results from coordinating direction information with scan movements in the context of experienced knowledge.

schema theory a model of the perception process is proposed. (For applications of schema theory see also Arbib, Arkin, and Cervantes-Pérez, this volume).

The relevance of contours in visual perception is shown by neurophysiological investigations (e.g., see Korn 1982; Frisby 1979) and clearly demonstrated by our daily experience. Thus, the restriction of transmitted information to contours is justified. It is quite obvious, too, that in this way only those objects can be recognized which are clearly defined by contours within the context.

Our generalized aim is "object recognition" in the broadest sense to improve independent orientation and mobility for the blind. Although at this time we cannot exactly describe all the possibilities and limitations of our method, first field trials in perception of undisturbed geometrical figures and combinations of those figures were successful.

2 General Description of the Contour-Perception System

The structure of the contour-perceiving KONWA-system is shown in Fig. 1. KONWA is derived from the German "*Kon*tur-*Wa*hrnehmung" (contour-perception). The technical system consists of an optical contour sensor, a contour detection unit, and a stimulator interface. (The contour sensor has been developed originally for robot vision application by Prof. E. Mühlenfeld and his group. An extended description of the contour sensor and the contour detection is given by Mühlenfeld & Therburg 1981, and the special contour sensor used in the KONWA-system is described by Koy-Oberthür 1987).

Reduction of environmental information to direction and position of a contour is achieved by the special structure of the contour sensor illustrated in Fig. 2. The sensor's circular target is composed of radial light sensitive stripes, averaging the intensity within every stripe by integration over the stripe area. The radial integration, i.e., the integration parallel to a possible contour, performs a suppression of contour disturbances and enables the detection of contours in spite of very low contrast, as has been shown by Mühlenfeld & Therburg (1981). The contour sensor is an incoherent optical correlator, and the radial stripes are correlation masks matched to contours with a curvature radius which is large against the sensor's radius. Contour direction and position are specified by the intersection points of the contour with the target circle. These intersections (see P_1 and P_2 in Fig. 2) are defined by the maximum or minimum gradients of the sensor signal as shown in Fig. 2. In analogy to the retinal filter operation, contour detection results from filtering the sensor signal by a band pass and computing the extreme values of the filtered signal. The angles of the extremes in reference to the sensor circle determine the searched intersection points, representing the detected contour.

These reduced contour informations are transmitted to the blind. His cutaneous system at the lower arm with electrical stimulation of the skin is used as the substitution channel. Sixteen stimulators, arranged in a circle around the lower arm, transmit the contour information. A pair of activated stimulators (see S_1 and S_2 in Fig. 3) according to the computed intersection points represents the fixed contour by the blind's imagination of the conjunction line between both stimulators. In this way,

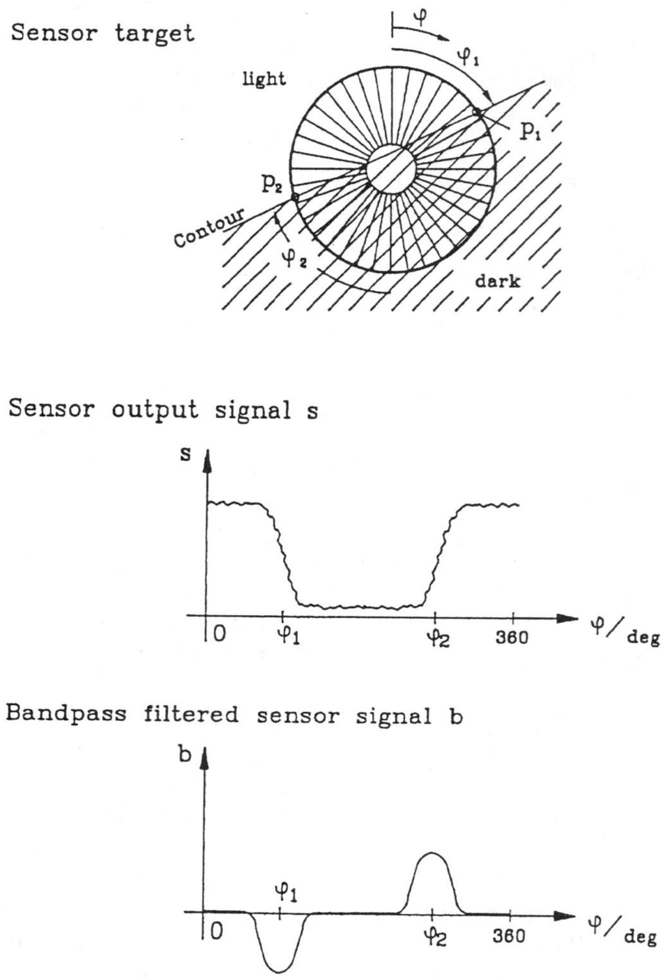

Figure 2. Contour detection with the contour sensor. A contour within the sensor target is defined by the intersection points P_1 and P_2. The appertaining angles are computed by looking for the extremes in the band pass filtered sensor signal b.

direction and position of the contour within the sensor circle are transmitted. The direct relation between the detected contour and the contour information received by the blind is shown in Fig. 3.

The sensor is carried by hand like an electric torch. If the direction of a contour is recognized, the blind is able to follow the contour and to scan the object with the sensor by movements with his hand and arm. The shape of an object can be recognized either by tracking the entire contour, or - if appropriate context information is available - by combining partial contour tracking with saccades to

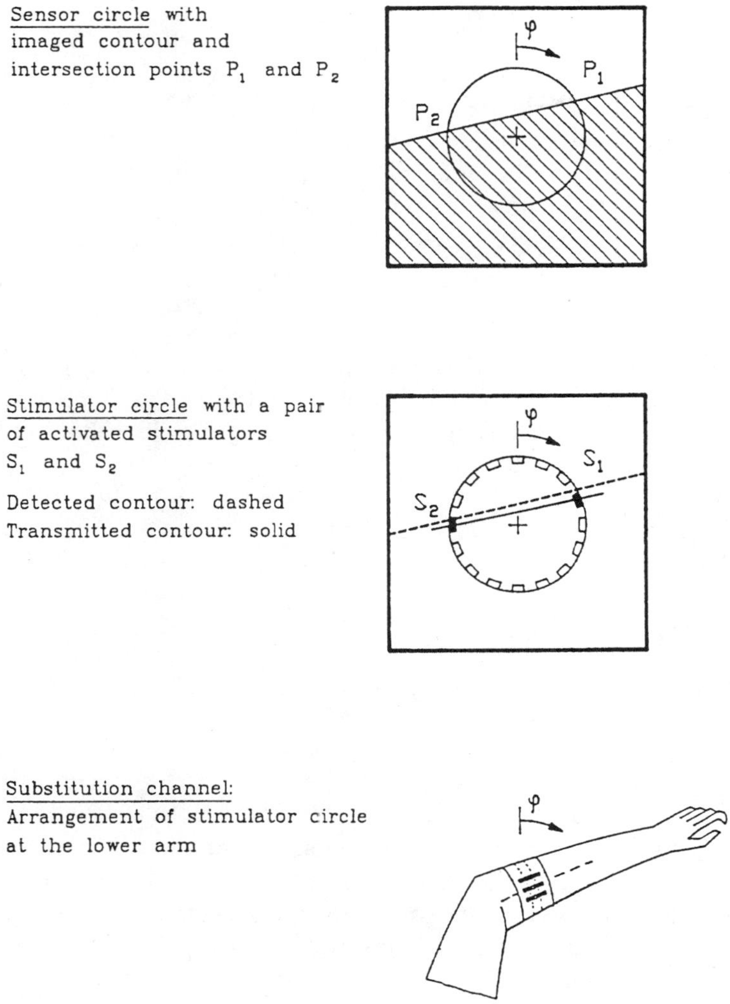

Figure 3. Transmission of the contour information by the lower arm as substitution channel.

expected contours. Object recognition, here, involves sensorimotor coordination in connection with available context information.

A seeing person's imagination of orientations in a scene scanned by means of an actual electric torch may illustrate the perception process of the blind provided with the KONWA-system. If absolute darkness without any scattered light is assumed, and if the searchlight beam of the torch is very small, the sight will be constrained to look for contours and to follow them in a manner described above. The meaning of a detected shape, however, essentially results from knowledge of the context.

3 The Sensory Substitution Channel

One main problem in sensory substitution is the channel that transmits the information from the technical system to the blind's sensory system. For this task, only the cutaneous somatosensory system or the auditory system is available. We have selected the cutaneous system, although audition has a much greater channel capacity. There are several psychological and information theoretical reasons for our choice: (i) Hearing is of such great importance in the blind's natural communication, so that it should not be disturbed by (other abstract) information which probably would involve much attention. (ii) To avoid interpretation problems of abstract sensory informations, that means to avoid semantical problems at the lowest level of the perception process, a well matching between the structure of the information to be transmitted and the structure of the receiving system must be demanded. In case of the KONWA-system, this matching is illustrated in Fig. 3 and it will be discussed at the end of this Section.

The design of the stimulators is based in principle on the investigations by Hoffmann (1981). Sixteen stimulators around the lower arm transmit the contour information by electrical stimulation of the skin. Note that electrical stimulation of the skin provides a better local resolution and a lower power consumption than mechanical stimulation (Hoffmann 1981). To avoid problems concerning discrimination and reproduction of intensity and frequency reception, only the cutaneous location works as information carrier. Every stimulator is built by a pair of rectangular, flexible surface electrodes (Fig. 4). Capacitive coupling of electrodes to pulsed current sources prevents direct current from flowing through the skin. Figure 4 shows the construction of the stimulators and the timing of the stimulation current.

According to the channel model of Shannon's information theory (Shannon & Weaver 1949), we investigated the transinformation and the channel capacity of the lower arm with stimulator circle. Our primary aim was to discover, which number of stimulators - i.e., the number of cutaneous locations - can be discriminated. Obviously, this number limits the direction resolution in contour detection. Following Shannon's channel model human senses have been investigated in a great number of experiments using absolut judgement tasks. (For experimental data obtained in visual, auditory, somatosensory, and gustatory domains see Miller 1956; MacRae 1970; Völz 1982; Geiser 1985). The general result was that only 7 to 9 stimulus classes can be discriminated - equivalent to transinformation values of 2.8 to 3.2 bit per transmitted symbol. In preliminary experiments with seeing probants, we have achieved an unexpected high transinformation value of 3.5 bit which is equivalent to 11 exactly discriminated stimulators (Koy-Oberthür 1987). Further experiments with blind probants showed a strong interindividual variability yielding transinformation values of 3.5 to 4.8 bit after 35 hours of learning. In summary, these results justify the number of 16 stimulators to transmit contour information. The theoretical direction resolution is 11.25 degrees.

The above statement deals with the quality of transmission. Investigating the required time, we found that transinformation is nearly independent of the time within 400 to 50 ms of stimulation time. (The transinformation decreased from 2.8 to

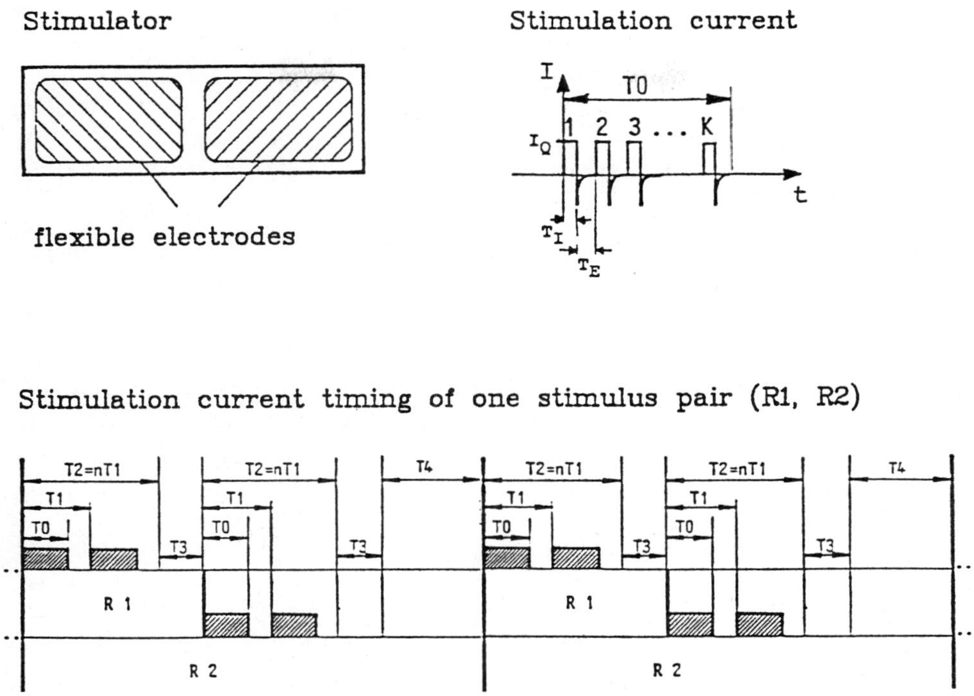

Figure 4. Stimulator design and timing of stimulation current. The area of each electrode is 8 mm x 18 mm. The stimulation current I_Q is adjustable for every stimulator from 1 to 12 mA. Pulse duration is $T_I = 100$ us and recharge duration is $T_E = 200$ μs. K determines the stimulus intensity (used values: K=1...3). A pulse train of K pulses is represented by a hatched rectangle in the lower diagram. The duration of activation of one stimulator is determined by the number n of pulse trains. The breaks are essential to avoid adaptation (with T1=5 ms after Hoffmann 1981); T2=25 ms, T3=T4=10 ms after Özdemir 1985).

2.6 bit in experiments with 8 stimulators). This result is important since it is related to the velocity of perception with the KONWA-system.

The described substitution channel enables a large amount of transinformation compared with the previous investigations. This fact can be explained by assuming wide equivalence between the channel structure and the structure of that part of the brain working as "receiver" for the substitution channel. In terms of cognitive psychology, we can assume the existence of a well matched schema receiving this specially structured information. The practical meaning of the expression *well matched schema* may be illustrated by the following simple idea: In our daily life the partition of a circle is frequently realized and can therefore easily be imagined, for example as a clock. The stimulator circle at the arm, being matched to the receiver, has also a direct spatial relation to the contour information to be transmitted.

Therefore, there exists practically no (semantical) problem of understanding the elementary information symbols transmitted by the substitution channel.

At the beginning of this Section, the auditory system was potentially considered as an alternative substitution channel. Now we learn that such a direct matching between contour information, channel, and receiver would hardly be possible with an auditory substitution channel. In summary, the channel investigations show that the stimulator circle at the lower arm is suitable to transmit contour information.

4 A Model of Sensorimotor Coordination

At the beginning it was mentioned that recognition of objects, here, takes advantage of sensorimotor coordination. Now the question has to be answered, how the blind may create complex perceptual contents, or how he can recognize objects from the elementary sensory information. More generally we ask: How does perception function? The answer will be given in the manner of a model of the perception process. We will start with definitions of the concepts "perception" and "schema":

Definition 1:
Perception is the active construction of a symbolic description of the world composed of the whole sensory information and the motor actions in coherence with experienced knowledge.
This definition is similar to a one given by Korn (1982). It is extended here to emphasize explicitly a symbolic information processing and an active role of the cognitive system. The process of perception - in the sense of the given definition: The process of active construction of a symbolic world model - is described by the psychological concept of "schema."

Definition 2:
Schema is a functional unit receiving special information. Anticipating a possible perceptual content, the schema looks for expected information. Anticipation is built by the actual sensory input in coherence with experienced knowledge. The schema is matching itself to the perceived information.
The concept of "schema," developed by the ethologists Jacob v. Uexküll and Konrad Lorenz (see Ewert, this volume), has been introduced as a basic concept of cognitive psychology by Jean Piaget. This concept is not generally and clearly defined. Although the definition problem will not be discussed in this paper, with the definition given here, schema serves as a foundation for our model. (For details on schema theory see Piaget 1954; Neisser 1976; Arbib & Hesse 1986; see also contributions by Arbib, Arkin, and Cervantes-Pérez, this volume).

The perception process is illustrated in Fig. 5 by the so-called "perception cycle" of Neisser (1976) describing perception as a cycle process of information acquisition, matching, and exploration for special information. This cycle process is controled by the schema. In search for special expected information, the respective anticipation implies the existence of a goal. Because the schema controls the perception process, the aimed model of perception must be also a model of the schema. Therefore, the model will furthermore be denoted as "SCHEMA."

The first task for the SCHEMA is the "construction of a symbolic description of the world" according to definition 1. This *symbolic description* consists of sensory informations and motor reactions. With respect to the assumed symbolic character of information processing, the SCHEMA is expressed in terms of formal languages. Thus, perception can be comprehended as a syntactic pattern recognition task. Using terms of pattern recognition, furthermore, information symbols are called *features* and motor reaction symbols are called *actions*. Features and actions are two kinds of terminals of a generative grammar describing the perception process or describing the SCHEMA. The perceptual content is symbolized by a string of features and actions. Recognition in the model is classification of a generated string to comprehend it as a superposed perceptual unit called "metasymbol." A metasymbol represents the result of the cognitive activity; it is the abstract of a string. In the case of perception by the KONWA-system, the schema at the lowest level of a hierarchical structure of schemata combines single contour directions (features) with sensor movements (actions) to contour segments (metasymbols, straight lines or circular arcs); the superposed schema comprehends segments in their relations to figures, and the next superimposed schema comprehends configurations in their relations to objects.

At this first state, the SCHEMA is structured in a grammar, generating feature-action-strings, and a classificator, generating a metasymbol as the abstract of the string. What does the SCHEMA perform at this state?

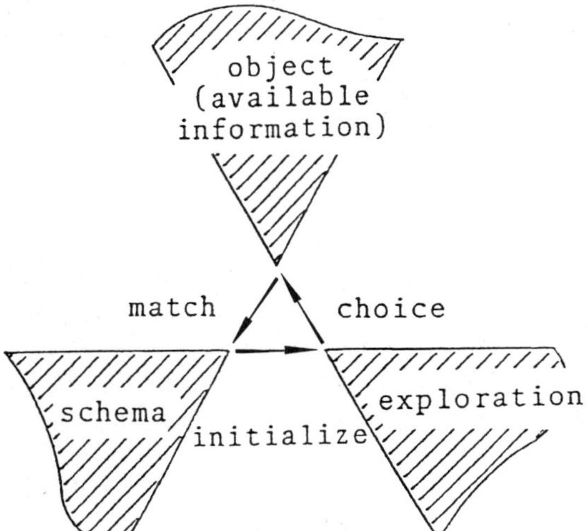

Figure 5. Illustration of perception as a cirular process of goal directed search for information, information acquisition, and matching to the available information, controled by the schema. (After Neisser 1979).

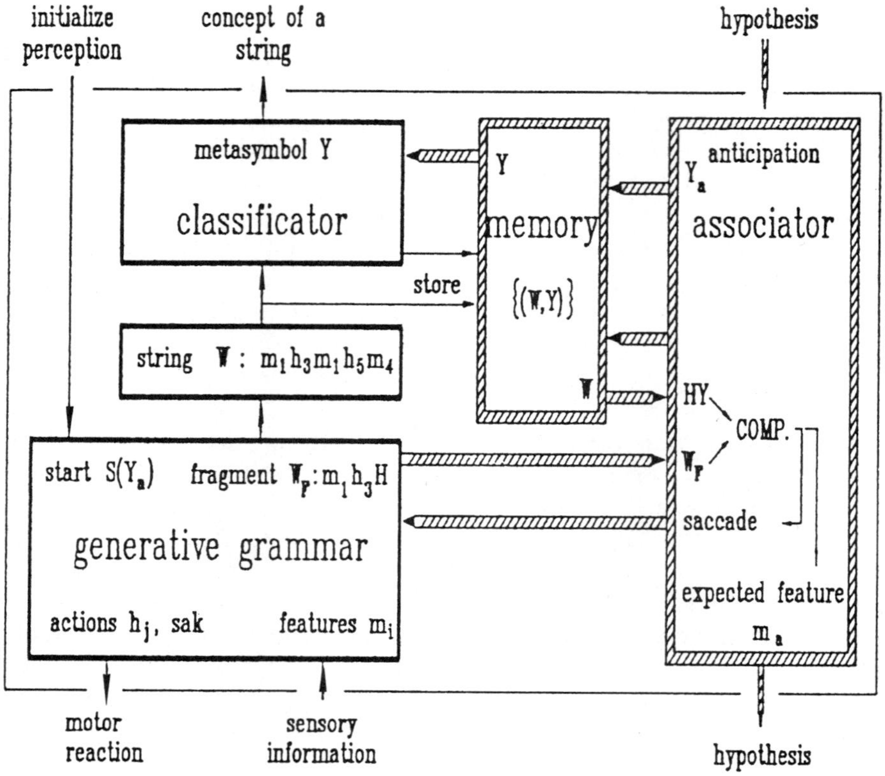

Figure 6. The SCHEMA: a model of perception by sensorimotor coordination based on schema theory. In the case of learning (solid framed blocks), the grammar generates a string W and the classificator finds the appertaining metasymbol Y; both, W and Y, are stored in the memory. In the case of available context information (hatched blocks) a fragmentary string W_F is compared with the memory content. A hypothesis HY is built and proved by the following action (saccade). If the hypothesis is confirmed, Y is taken from the memory.

The SCHEMA cannot yet deal with experienced knowledge. This situation is comparable with the case of learning a complete new perceptual content without any corresponding experience. Then the assignment of actions to actually acquired informations is exclusively determined by the rules of the generative grammar. Therefore, these rules determine the structure of perceptual contents, whichever can be constructed by the model.

Furthermore, following definition 2, the SCHEMA will be extended to create anticipations using the context of experience knowledge stored in memory containing strings together with their appertaining metasymbols. During generating a string by the grammar, the growing fragment of the string is currently compared with the memory content by a unit called "associator." If the fragment fits a stored string, that means if it is a part of a stored string, the associator starts a special action to prove relevant features of the hypothetic string. If the aquired information confirms the hypothesis, the appertaining metasymbol is taken from the memory without generating and classifying the complete string.

Using the context of experience, the associator drives the model to process disturbed information and to apply an efficient strategy in case of undisturbed information.

The SCHEMA is illustrated in Fig. 6. The symbolic denotations used in this figure are defined in the following section formalizing the model. Complex perceptual contents can be generated and described in a hierarchical structure of several SCHEMAs by using metasymbols of a lower level SCHEMA as features for the superposed one. A hypothesis in a superposed SCHEMA works as anticipation of an expected metasymbol in the lower one, and the action for proving the hypothesis in the superposed SCHEMA activates the lower one for information acquisition. Hence, the goal is determined by anticipation resulting from a superposed hypothesis. The model of perception by sensorimotor coordination shown in Fig. 6 may be built-up to a complex structure like the one shown in Fig. 7. In the subordinate levels several parallel SCHEMAs may be working in correspondence with the several sensory and motor channels. In the superposed steps information of all senses are comprehended to strings and metasymbols. SCHEMAs may also work in parallel at every level to represent the different perceptual qualities of reality and to fulfil different concurrent tasks, such as: recognition of three-dimensional objects, obstacles, and hollows in the sidewalk (Fig. 7). Thus, in principle it is possible to construct a complex symbolic description of the world by the developed model, independent of a special content.

A formalization of SCHEMA is described by Koy-Oberthür (1987). In the following, the principles are summarized in an abridged version. (For explanation of formal languages see Arbib 1968 and Herschel 1974, and for application of these in pattern recognition see Fu 1974 and Niemann 1983). The generating grammar G has the following generalized form:

$$G = \{ T, Z, S, R, B \}$$

set of terminals: $T = \{m_i(A), h_j(A)\}$

set of nonterminals: $Z = \{F(A), H(HY), C(A), S(Y_a)\}$

set of productions: $R = \{r_0, r_1, \dots, r_9\}$

set of conditions: $B = \{B1, B2, B3, B4\}$

The *terminals* $m_i(A)$ and $h_j(A)$ symbolize features and actions with an attributing vector A that is not specified at first. The *nonterminal* F points indicate that the growing string is a fragment; H means that a hypothesis HY is built and proved by a saccade (sak) as a special action; C symbolizes the search for confirmation the hypothesis; S is a starting symbole with a metasymbol Y_a as a possible anticipation

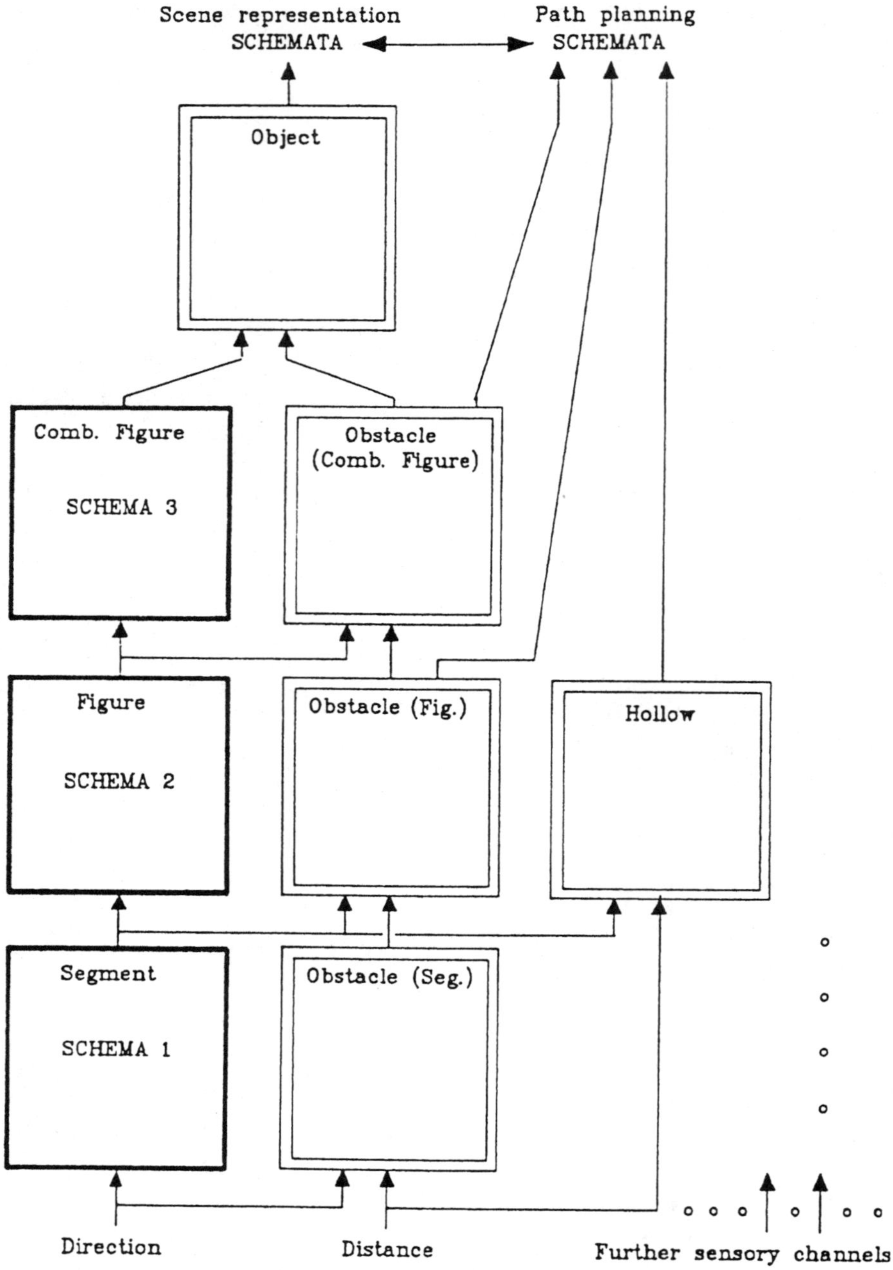

Figure 7. An example for a structure representing complex perceptual contents with hierarchical and parallel SCHEMAs. The solid framed blocks symbolize the perception of two-dimensional combined figures, realized in our current experiments. Distance information is required for the recognition of three-dimensional objects in space necessary for the blind's mobility to avoid obstacles (in different complexity) and hollows in the sidewalk.

by a superposed SCHEMA. Every *production* occurs as many times as combinations of different terminals are present. Therefore, for example r_0 would exactly be named $r_{0,n}$ with n=1...I·J. The abridged notation is intended to enable an easy synopsis of the syntactic description without respect to several special terminals. The *conditioned productions* can be expressed in an abridged manner where the indices and attributes are omitted to emphasize the principles:

$$
\begin{array}{lll}
r_0: & S \longrightarrow m\ h\ F & \\
r_1: & F \longrightarrow m & /\ B1 \\
r_2: & F \longrightarrow m\ h\ F & /\ \overline{B1}\ \&\ \overline{B2}
\end{array}
$$

$$
\begin{array}{lll}
r_3: & F \longrightarrow m\ h\ H & /\ B2 \\
r_4: & H \longrightarrow sak\ m & /\ B3 \\
r_5: & H \longrightarrow sak\ F & /\ B4 \\
r_6: & H \longrightarrow sak\ m\ C & /\ \overline{B3}\ \&\ \overline{B4}
\end{array}
$$

$$
\begin{array}{lll}
r_7: & C \longrightarrow h\ m & /\ B3 \\
r_8: & C \longrightarrow sak\ F & /\ B4 \\
r_9: & C \longrightarrow h\ m\ C & /\ \overline{B3}\ \&\ \overline{B4}
\end{array}
$$

B1 to B4 are conditions determining the application of the productions with the meaning:

B1: "a string is generated independently of experienced knowledge"
B2: "a hypothesis is built or an anticipation exists"
B3: "the hypothesis (anticipation) is confirmed"
B4: "the hypothesis (anticipation) is refuted"

The set of strings produced by this grammar gives the formal language L(G):

$$L(G) = \{W_k \ /\ G \longrightarrow W_k\};\ k=1\ \ldots\ K$$

The associator uses the experienced knowledge. This is symbolized by a memory M' containing all generated strings together with their appertaining metasymbols. The set of stored strings is called M and is a subset of the language L(G):

$$
\begin{array}{l}
M \in L(G) \\
M = \{W_n \ /\ G \longrightarrow W_n\};\ n=1\ \ldots\ N,\quad N<K \\
M' = \{\ (W_n, Y_n)\ \}
\end{array}
$$

After each production application of the grammar, the associator trys to build a hypothesis by comparing the fragment with the known strings. The strings W_n and the fragments W_{Fn} may be seen as arranged sets and HYP may be the set of possible hypotheses. With W_F as actually constructed fragment we can formulate the association of the hypothetic string HY:

$$HYP = \{W_n \mid (W_n \in M) \;\&\; (W_F \subset W_n)\},\ n=1 \ldots N$$
$$HY \in HYP$$

HY symbolizes the most likely element of HYP. Therefore, the string HY has to be verified by the subsequent actions of the model.

Now the conditions B2 to B4 can be expressed by the defined terms of the associator:

$$B2:\ \overline{(HYP = \emptyset)} \vee \overline{(Y_a = \emptyset)}$$
$$B3:\ HY \text{ ---> } W$$
$$B4:\ HY \text{ -+-> } W$$

The condition B1 always depends on the special perceptual content that shall be perceived by the model and, hence, it cannot be generally formalized. (For a detailed description of a general associator see Koy-Oberthür 1987; for a special associator working in a technical patttern recognition system see Mühlenfeld 1984).

The above described two cases of cognitive activity (without or with corresponding experienced knowledge) are explained in the grammar G by the productions r_0 to r_2 and r_3 to r_9, respectively. The productions r_0 to r_2 describe the generation of all possible strings of the form $W = m_i\, h_j\, m_i \ldots h_j\, m_i$ recursively and free of context. Therefore, classification of a string could in principle easily be performed by parsing with a machine. This fact, however, is of interest only from a theoretical point of view. It means that the "string problem" can be solved and that classification of the string is in principle always possible. The grammar used by Koy-Oberthür (1987) is extended to a fractionally fuzzy grammar (DePalma & Yau 1975) with respect to the fuzziness of human cognitive processes and in order to realize a simple "maximum membership classification." Classification is performed by computing the maximum value of the memberships of a string to the several classes representing several fuzzy subsets of the formal language L(G).

The productions r_3 to r_9 involve the context of experience in the cognitive process. However, they are in a formal way also free of context. This property becomes irrelevant at this place, since the final string and its metasymbol result from the memory M (M') by association. The context dependence is completely considered by the conditions B1 to B4, although strings are generated contextfree in the sense of formal languages.

A specialization of the model for the perception of figures with the aid of the KONWA-system is formulated in detail by Koy-Oberthür (1987). It consists of two hierarchical levels of SCHEMAs, one building contour segments from the elementary sensory information, and another combining segments to figures (configurations). In the first stage the contour informations at a fixed point are *features* and the movements to the next fixpoint are *actions*. This stage describes the construction of segments primarily by contour tracking. In the second stage, segments are features. The changes of the average direction of movement (or angles between neighbouring segments) and the saccades in response to expected segments are actions. The condition B1 is derived from the property of "closeness" of a closed geometrical figure or from missing a sensor signal in case of an open figure. Algorithms serve for computing actions from sensory and memory information. In case of unavailable information, searching SCHEMAs are working in a manner of "searching images." At this point the model is theoretically developed proposing a possible answer to the question of how the blind can create perception with the KONWA-system.

5 Preliminary Experimental Results

Characteristic examples for perception with the KONWA-system are shown in Figs. 8 and 9. These examples are taken from experiments with blind probants. The contour sensor, movable around two axes, was mounted on a stand to record the movements. We presented two-dimensional figures at a constant distance. After scanning figures with the contour sensor, these figures were also perceived by touching. In this way the new ("artificial") sensorimotor experience could be linked to the familiar ("natural") experience. As recognition with stimulator pairs had been acquired, our experiments proceeded with learning primary sensorimotor coordinations by complete scanning single contour segments (lines or arcs), simple geometrical figures (triangles, rectangles, circles), and combinations of two of those figures. The blind had the task to discriminate and to identify the presented figures by complete contour tracking. In terms of SCHEMA the blind learned this way certain defined contents of several memories M' according to the three stages of SCHEMAs in perceiving (i) contour segments, (ii) figures, and (iii) combined figures, respectively. Figure 8a demonstrates a characteristic example of a first trial of scanning a single line by sensorimotor coordination. This is an example for the recognition of a segment according to SCHEMA 1 shown in Fig. 7. The progress in learning to recognize a simple figure is illustrated in Fig. 8b, and scans of combined figures obtained in an advanced state of learning are depicted in Fig. 8c, according to SCHEMAs 2 and 3 of Fig. 7. The required recognition time in relation to the task serves as a measure of the progress in learning. The detection rate is not a useful measure, because it is not significantly less than 1 after a short time of learning. Recognition of the simple figures used here seems to be only a question of time. All these examples are related to the case of learning new perceptual contents, according to perception without using the context of experience. This process is described by the productions r_0 to r_2 in the grammar G or by the solid framed blocks of the model shown in Fig. 6.

In further experiments the task was changed by including the context information after a certain set of combined figures had been learned. Now, the figures should be recognized as fast as possible without scanning the entire contour. Figure 9 gives examples for recognizing combined figures corresponding to Fig. 8c, however, with the difference that only a part of the contour was scanned. Using the experienced knowledge, represented in a memory M', recognition is achived with much less actual information within much less time and without a significant decrease of the detection rate. This memory content consists of the learned set of combined figures illustrated in Fig. 9. The efficiency of the new scanning strategy becomes evident by comparing the required times in Figs. 8c and 9. The case of context-

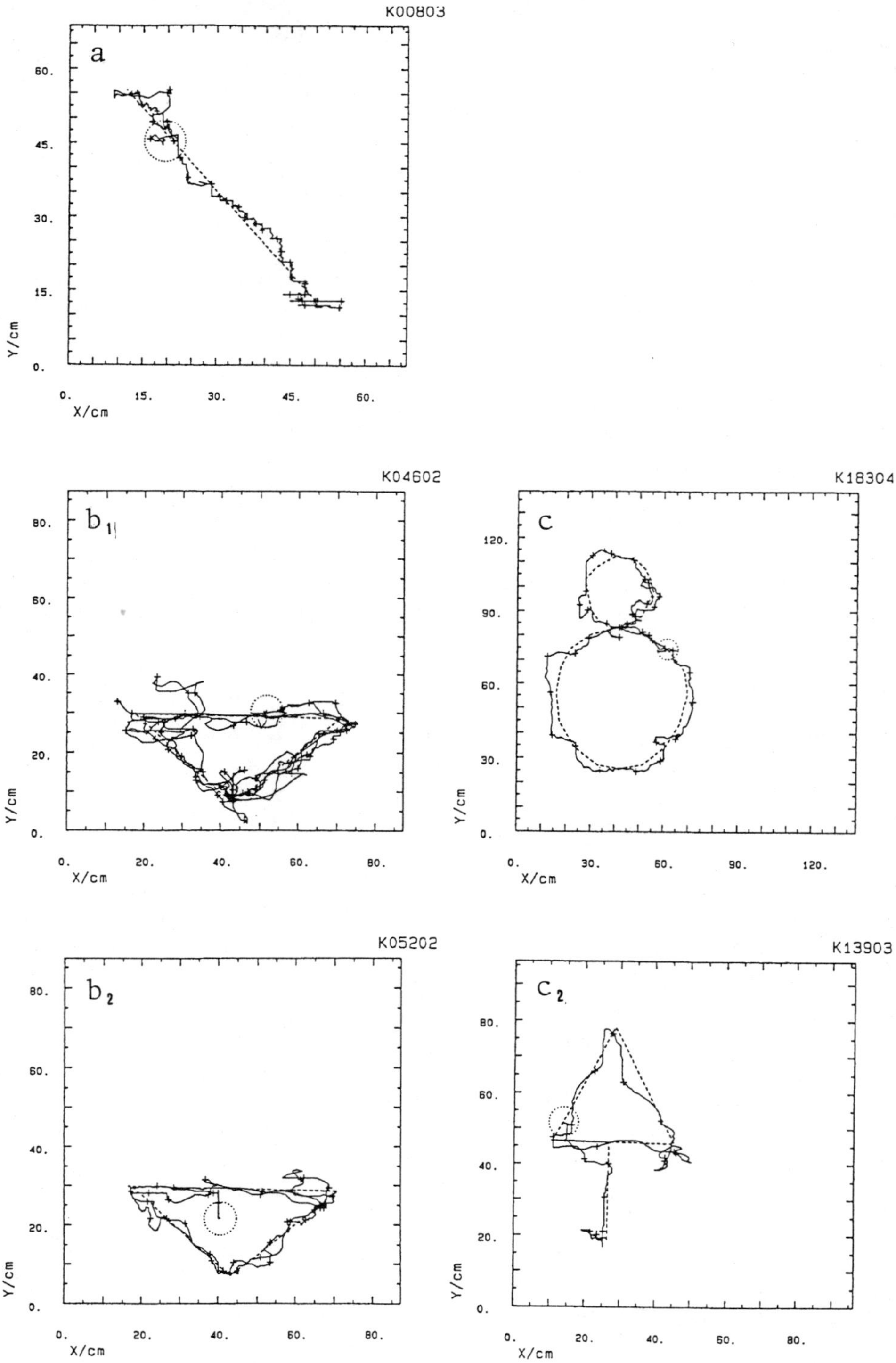
K00803
a
Y/cm
X/cm
K04602
b1
Y/cm
X/cm
K18304
c
Y/cm
X/cm
K05202
b2
Y/cm
X/cm
K13903
c2
Y/cm
X/cm

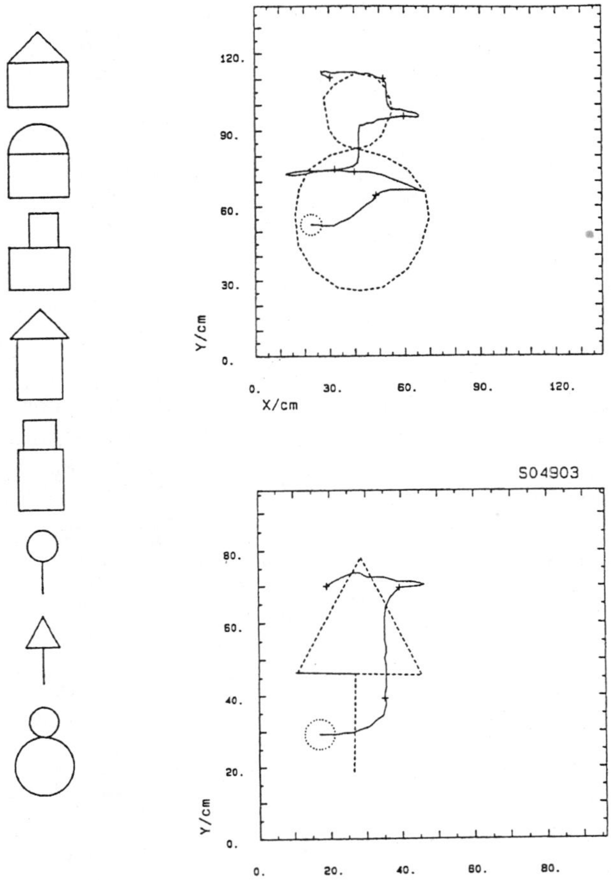

Figure 9. Recognition of the same combined figures as shown in Fig. 8c. Recognition is achieved without entire contour tracking, but using the context information according to the hatched blocks of the SCHEMA in Fig. 6. On the left the complete set of known figures (representing the context information) is depicted. The required recognition time increased essentially as shown in Fig. 8c (top: $T_r = 7$ s; bottom: $T_r = 3$ s).

Figure 8. Examples of movement records to recognize figures with the KONWA-system by contour tracking. In all following examples (a-c) the sensor target is drawn by dots, the dashed line indicates the presented figure, the solid line shows the blind's scan, and the crossings are drawn in constant time distances of 1 s. The distance between figure and object is 96 cm. The required time for recognition is T_r. (a) Example of a first trial in sensorimotor coordination, scanning a simple line; $T_r = 34$ s. (b) Example for a learning progress, scanning triangles; top: for the first trial in this figure $T_r = 63$ s; bottom: after nearly 60 min of learning $T_r = 36$ s. (c) Examples for scanning combined figures; top: $T_r = 36$ s; bottom: $T_r = 19$ s.

supported perception is described by the productions r_3 to r_9 of the grammar or by the hatched blocks of the model in Fig. 6. The assumptions in the model for involving the context of experience in the perception process are confirmed by the presented experimental results.

6 Discussion and Conclusions

6.1 What Has Been Shown in this Approach?

Firstly, the examples depcited in Figs. 8 and 9 demonstrate that the principles realized by the KONWA-system came into function and could be evaluated experimentally. Essentially this means for the blind that perception of the shape of objects by sensorimotor coordination is possible. Secondly, it is shown that the results can be interpreted in equivalence with the proposed model of the perception process. This *schema*-based model declares the different hierarchical steps of construction of perception from the actually acquired information in coherence with selecting actions and involving the context of experience. Schema theory performs the understanding of the perception process with the KONWA-system .

6.2 What Can Schema Theory Perform in Principle?

Toward an answer of this question we will discuss some theoretical implications of SCHEMA. The progress in developing the KONWA-system led in its theoretical part from Shannon's information theory to schema theory. Trying to transmit information from the natural environment to the blind the questions arose, what at all information is, and in our special case, what this information will be. In every case of dealing with information these questions must be answered. (For a detailed investigation see Koy-Oberthür 1987). Shannon gave a definition and a quantification of the information present in an ensemble of symbols in form of the entropy. Entropy is independent of a special content, but it can be applied only under specific conditions being mostly not fulfilled in the higher levels of biological information processing. Investigating a sensory channel in the sense of the channel model as described in Section 3, this problem must be solved by experimental design. In biological systems information implies, in addition to a quantitative aspect, also syntactical, semantical, and pragmatical aspects according to the structure, the meaning, and the goal-dependent use of the information. These aspects of information have to be considered by every model describing the perception of information, and they are also present in a SCHEMA: The grammar according to *structure*, the context-dependent actions derived from knowledge of the *meaning* of the acquired information, and the *goal* appearing explicitly in form of anticipation or hypothesis. (Note that the goal at the highest level of a SCHEMA structure does not grow out of the model, rather it has to be given externally to the model). Hence, SCHEMA contributes to an extended understanding of the concept "information" in those cases which are adequately described by the schema.

6.3 Situations Described by Schemas

With reference to the definition of *schema* obviously all sensorimotor coordination processes can be described by schemas. The following examples illustrate the spectrum of problems relating to schema theory:

(i) Based on data obtained in neuroethological studies Lara et al. (1982, 1984) have applied schema theory to develop a model of visuomotor coordination in toads (see also Arbib, this volume).
(ii) An example describing stimulus-specific habituation in toads with a schema-based model is presented by Lara (this volume).
(iii) Arbib (this volume) draws our attention to the link between technical sensorimotor problems and neural models of visuomotor coordination.
(iv) An application of schema theory to intelligent mobile robots is presented by Arkin (this volume).
(v) So-called "on-line specialists" mediating sensorimotor coordination in robots (see Torras, this volume) can be interpreted as motor schemas that have perceptual schemas embedded.
(vi) A pattern recognition system for robot vision is presented by Mühlenfeld (1984). Although not applying schema theory, the system is essentially based on associative sensorimotor coordination.
(vii) Cervantes-Pérez (this volume) proposes schema theory as a common language for Neuroethology, Brain Theory, Artificial Intelligence, Linguistics, and other disciplines concerned with the analysis of sensorimotor coordination processes.

From an information theoretical point of view, Cervantes-Pérez's proposal of schema theory as a "common language" is in agreement with the following statement: In biological systems acquisition of information is related to reaction, i.e., information without reaction is largely meaningless; relating to various behavioral alternatives, information gains meaning (semantic) in the system. A concept like SCHEMA, that integrates informations regarding *feature* and *action*, is an adequate tool of describing, understanding, and modeling higher level information processing, i.e., higher level cognitive processes, independent of a special scientific discipline.

If the described ideas were accepted, i.e., *schema* is accepted as the fundamental functional unit in biological information processing, then the question is addressed to Neuroethology, whether there is a structural equivalence for such a functional unit in the brain, and whether the neural code representing a "feature-action-string" can be discovered. The results reported by Ewert and other neuroscientists (this volume) provide us with evidences of parallel and hierarchically organized structures that underlie processing of sensory information and commanding of motor actions. Can neural structures be assigned to schemas, for example, like certain cortical "simple cells" are tuned to specific orientations of contrast borders? If schemas are really as meaningful for the brain as postulated, then corresponding neural equivalents (brain structures and codes) should be found. The neuroethological studies in toads, for example, indicate that a configural "prey schema" refers to stored information regarding an invariant structure of *action* relevant *feature* relationships provided by pretectal/tectal circuitry (continuum of functional units of connected cells); an output of this circuitry is mediated by configuration-selective T5.2 neurons that are acting as filters (Ewert, this volume). The neuroethological concept of "sensorimotor codes" may relate to some extent to components of "feature-action-string."

6.4 Further Investigations

Returning from speculation to the KONWA-system, the aim of stressing schema theory is first at all an understanding and modeling of perception. Although by this time the model is only a theoretical one, its implementation in simulating perceptual processes can subserve for parameter optimization of the system with less experimental expenditure. Another aim is the incorporation of learning, as a biological property, that will allow technical pattern recognition and robot vision to adapt their function to changing environmental conditions.

In our current experiments we are achieving the preliminary aim of shape recognition of two-dimensional figures with and without using context information, and our probants are learning suitable strategies for context included perception. However, recognition of objects as a precondition for a blind's orientation in a natural environment requires the consideration of spatial information. To make shape recognition feasible over variable and far distances, the essential advantage of the KONWA-system can only be used if the spatial depth relations will be integrated into the perception process.

Further system developments are required to provide measurements of the distance of a fixed contour and to convey the transmission of distance information to the blind. Correspondingly, our probants will learn to use this new information in further experiments. In terms of the model we can say that a new sensory input channel must be added and that new SCHEMAs have to combine shape and distance informations. As a result these SCHEMAs should exhibit a wide invariance of shape perception against changing spatial relations between an object and the blind, hence allowing the recognition of objects in their real spatial extension. Other important features for the blind concerned with the recognition of *obstacles* and *hollows* in the sidewalk (e.g., excavations, landings, edges) also require the integration of contour (shape) and distance information as illustrated in the model structure of Fig. 7.

In summary, the integration of distance information into the perception process is a nessesary condition to perform further investigations in natural environments to achieve the generalized aim formulated at the beginning: Improvement of independent orientation and mobility for the blind.

Acknowledgements. The author likes to thank all coworkers of the institute for their readiness to discuss problems and to prepare the illustrations. Particularly, I am indepted to Prof. W. Rath, Mrs. B. Bruch, Mr. U. Gerkens, and to our blind coworkers realizing the experiments at the University of Hamburg. These experiments are supported by the "Deutsche Forschungsgemeinschaft". I thank Prof. E. Mühlenfeld for his support and for his interest in my work.

7 References

Arbib MA (1968) *Algebraic theory of machines, languages, and semigroups.* Academic Press, New York [German edition (1973): *Algebraische Theorie abstrakter Automaten, formaler Sprachen und Halbgruppen.* Akademie-Verlag, Berlin]

Arbib MA, Hesse MB (1986) *The construction of reality.* Cambridge Univ Press, Cambridge MA

Bach-y-Rita P (1977) Hirnplastizität und sensorische Substitution. In: Boehringer Mannheim GmbH (ed) *Mannheimer Forum* 76/77. Mannheim, pp 89-144

Collins CC (1978) Visual substitution in blind mobility. *Proc VIIth Int Conf of Ergophthalmology.* Kagoya Japan

Collins CC, Bach-y-Rita P (1973) Transmission of pictorial information through the skin. In: Lawrence H, Hamilton JG (eds) *Advances in biological and medical physics Vol 14.* Academic Press, New York, pp 285-315

Collins CC, Deering M (1981) Real-time natural scene analysis for a blind prothesis. *Proc of the 7th Int Joint Conf on Artificial Intelligence,* Vancouver

Depalma GF, Yau SS (1975) Fractionally fuzzy grammars with application to pattern recognition. In: Zaheh LA, Fu KS, Tanaka K, Shimura M (eds) *Fuzzy sets and their applications to cognitive and decision processes.* Academic Press, New York, pp 329-351

Frisby P (1979) *Seeing, illusion, brain, and mind.* Oxford Univ Press, Oxford

Fu KS (1974) *Syntactic methods in pattern recognition.* Academic Press, New York

Geiser G (1985) Codierung optischer Information. In: Bodmann H-W (ed) *Aspekte der Informationsverarbeitung.* Springer-Verlag, Berlin Heidelberg New York, pp 221-253

Herschel R (1974) *Einführung in die Theorie der Automaten, Sprachen und Algorithmen.* Oldenbourg Verlag, München

Herzog HW, Kügle M (1982) *Untersuchung zu technischen Hilfsmitteln für Blinde.* Infratest Gesundheitsforschung, München

Hoffmann C (1981) *Eine tragbare Sinnesprothese für Gehörlose zur Übertragung von Sprachinformation durch elektrische Anregung der Haut.* PhD Dissertation, Inst Techn Univ Munich, München

Korn A (1982) *Bildverarbeitung durch das visuelle System.* Fachberichte Messen-Steuern-Regeln 8, Springer-Verlag, Berlin Heidelberg New York

Koy-Oberthür R (1987) *Ein Kontur-Wahrnehmungs-System für Blinde - Entwicklung des Systems, Optimierung der Informationsübertragung an den Blinden und Beschreibung der Wahrnehmung als syntaktischen Mustererkennungsprozeß.* VDI-Verlag, Düsseldorf

Lara R, Cervantes F, Arbib M (1982) Two-dimensional model of retinal-tectal-pretectal interactions for the control of prey-predator recognition and size preference in amphibia. In: Amari S, Arbib MA (eds) *Competition and cooperation in neural nets.* Springer-Verlag, Berlin Heidelberg New York, pp 371-393

Lara R, Carmona M, Daza F, Cruz A (1984) A global model of the neural mechanisms responsible for visuomotor coordination in toads. *J Theor Biol* 110: 587-618

Macrae AW (1970) Cannel capacity in absolut judgement tasks: an artifact of information bias? *Psychol Bull* 73: 112-121

Miller GA (1956) The magical number seven plus or minus two: some limits on our capacity for processing information. *Psychol Rev* 63: 81-97

Mühlenfeld E, Therburg RD (1981) Erkennung und Vermessung stark gestörter linienhafter Bildstrukturen. *Technisches Messen* 48(5): 179-186

Mühlenfeld E (1984) Robot vision by a contour sensor with associative memory. *Pattern Recognition* 17(1):

169-176

Neisser U (1976) *Cognition and reality.* Freeman, New York [German edition (1979): *Kognition und Wirklichkeit.* Klett-Cotta, Stuttgart]

Niemann H (1983) *Klassifikation von Mustern.* Springer-Verlag, Berlin Heidelberg New York

Özdemir M (1985) *Untersuchung zur Beschreibung der Informationsübertragung durch den Tastsinn des Menschen bei elektrischer Anregung der Haut.* Dipl Thesis, Institut Elek Informationstechnik Tech Univ Clausthal, Clausthal-Zellerfeld

Piaget J (1954) *The construction of reality in the child.* Basic Books, New York [German edition (1975): *Der Aufbau der Wirklichkeit beim Kinde, Band 2.* Ernst Klett Verlag, Stuttgart]

Shannon CE, Weaver WW (1949) *The mathematical theory of communication.* Univ of Illinois Press, Illinois [German edition (1976): *Mathematische Grundlagen der Informationstheorie.* Oldenbourg-Verlag, München]

Völz H (1982) Information I. Studie zur Vielfalt und Einheit der Information. Akademie-Verlag, Berlin

Warren DH, Strelow ER (eds) (1985) *Electronic spatial sensing for the blind: contributions from perception rehabilitation and computer vision.* Martinus Nijhoff Publishers, Dordrecht Boston Lancaster

Welsh RL, Blasch BB (eds) (1980) *Foundation of orientation and mobility for the blind.* American Foundation of the Blind, New York

PART V

The Visuomotor Interface

Schema Theory as a Common Language to Study Sensori-Motor Coordination

FRANCISCO CERVANTES-PEREZ

Departamento de Neurosciencias
Instituto de Fisiologia Celular
Universidad Nacional Autónoma de México
04510 México D.F., México

Abstract. Among sciences there is a tendency to generate data and knowledge in one discipline without making it available to other disciplines. We discuss how Arbib's *Schema Theory* represents an attempt to solve this *isolation problem* in Cognitive Science by providing us with a global language explaining cognitive processes at a level that can be used, and understood, within all disciplines of cognitive sciences. In addition, we show how Schema Theory adds a key methodology to the "top-down" approach which allows us to set the stage within the theory-experiment cycle in order to investigate the neural substrate of sensori-motor coordination. We use our analyses of visuomotor coordinations in toads and praying mantises as examples (i) to show the applicability of Schema Theory in the study of what processes should occur within an animal's brain in order to explain overall behaviors, and (ii) to point out how Schema Theory permits, in a very general way, to work top-down. The ultimate aim is to generate testable hypotheses about the neural mechanisms that underlie behavior.

1 Introduction

There are many disciplines sharing common concerns in their efforts towards the understanding of the brain and the mind. Most of the contributions in this book are related to the understanding of how the central nervous system (CNS) of animals, including humans, processes sensory information to control the animals' interactions with a complex environment in which they live.

Unfortunately, in this multidisciplinary approach, the disciplines have been developed in an independent manner, each one mostly isolated from the others. We call this the *isolation problem*. Until recently, there have been attempts to link some of these disciplines. Under the name of *Cybernetics*, the science of control and communication in animal and machine, Norbert Wiener (1948) gathered at least three different ideas about the functioning of the brain: first, Craik's view of the CNS as a sort of computer capable of creating an internal model of external events (Craik

1943); second, McCulloch & Pitts' (1943) formal model of a neuron as a threshold logic unit that layed the fundamentals for the "physiology of the computable," demonstrating that every Turing Machine program could also be programmed into a network of logical neurons; and third, Wiener's own work developed in collaboration with Rosenblueth and Bigelow in 1943 showing the existance of feedback within the human nervous system and laying the basis for analyzing it as a "purposive brain" working in a special framework with circular processes (Rosenblueth et al. 1943): The CNS, in response to sensory information, controls muscles whose actions produce changes in the environment that yield a new activation of the receptors which feed (back), this way, in the CNS.

More recently (Simon 1980; Arbib 1985; 1987a,b), various disciplines emerging originally from the field of Cybernetics (i.e., Cognitive Psychology, Brain Theory, and Artificial Intelligence) have been grouped under the umbrella of *Cognitive Science* together with other areas concerned with the study of information processing systems that are adaptive. Among these disciplines are Neuroscience, Computer Science, Robotics, Linguistics, Psychology, Sociology, and Philosophy. An important aspect of Cognitive Sciences is that it represents the recognition of a common set of concerns shared by all those disciplines that analyze the brain, the mind, and the machines under *serial* and *parallel* information processing paradigms. Furthermore, it calls for the need of a common theoretical framework global enough to permit an immediate transfer of new findings and concepts obtained within one of the disciplines to all the others.

In this paper, we discuss the *Schema Theory* offered by Arbib (1981, 1987a,b) as the interscientific bridge (see Section 3) that provides us, on the one hand, with a global language in which we can carry out high-level analyses of cognitive processes or other functions of interest to us in cognition and, on the other hand, with elements low-level enough to develop "top-down" analyses allowing us, in principle, to explain how schemas could be implemented by a neural network or a computer program. In addition, we use our analyses of visuomotor coordination in toads and praying mantises as examples to show the applicability of Schema Theory. This concerns the study of what processes should occur within the animal's brain to explain overall behaviors and the evaluation of how Schema Theory permits, in a very general way, to work top-down. In the latter case, we start with the description of a specific behavior as a "big-box" (big "black box") and continue by breaking it into smaller boxes, thus studying how that behavior can be provided by defined interactions among the functions carried out by the smaller boxes. At later stages in the study, these new boxes are also broken into much smaller pieces whose interactions must account for the functions of intermediate boxes. This process continues until we reach a stage where a neural network can be defined to implement those specific functions to be performed by the smallest boxes. The aim here is that of Neuroethology, namely to identify the neural mechanisms that underlie behavior.

2 The Cognitive Science Tri-Cycle

Within the disciplines working towards the understanding of the brain, the mind, and the design of intelligent machines, we find very diverse points of view, experimental and theoretical ones, and hence different levels of analysis. In the latter case, two extremes can be distinguished (see Arbib, this volume): *top-down* studies, trying to explain behavior in terms of coupling mechanisms among functional units that represent perceptions and actions of the animal, and *bottom-up* analyses whose aim is to work from neural networks, or computer programs, in order to establish correlations between them and specific overall behaviors that need to be explained in living organisms or to be performed by an intelligent machine.

In Cognitive Science, we are confronted with the *isolation problem*, the tendency to generate data and knowledge within one discipline without making it accessible to the other disciplines. Nevertheless, it is clear that the disciplines classified as Cognitive Sciences can make contact with each other through the *top-down* analysis where the main concern is the processes (algorithms) that have to be carried out by an organism, or a machine, to produce proper interactions with the environment and not the hardware that implements these processes.

For the purpose of our discussion, let us consider the following cognitive sciences: Brain Theory, Neuroethology, Philosophy, Robotics, Cognitive Psychology, Artificial Intelligence, and Linguistics (see Fig. 1):

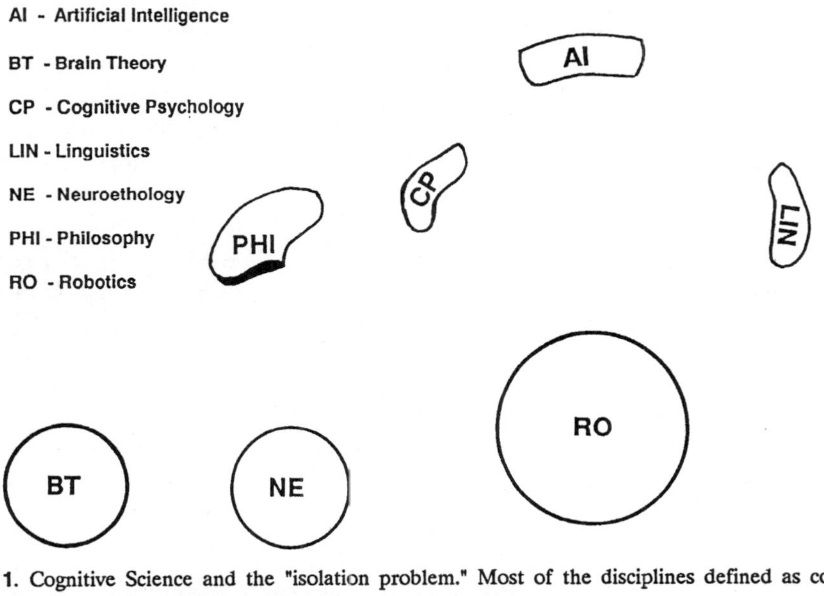

Figure 1. Cognitive Science and the "isolation problem." Most of the disciplines defined as cognitive sciences have been generating data and knowledge without making it available to the others.

(1) *Brain Theory* is concerned with the study of the CNS function with the aid of mathematical modeling and computer simulations; the aim is to explore, theoretically, what processes have to be carried out by the CNS in order to explain overall behavior, and how neural networks identified through anatomical and physiological studies may be the substrate of these processes (Arbib 1972, 1987a; Cervantes-Pérez 1985; see also Arbib, this volume).
(2) *Neuroethology* is devoted to the experimental analysis of physiologically and anatomically testable neural networks that underlie the releasing and control mechanisms of behavior patterns (Ewert 1980; Camhi 1984; see also Ewert, this volume).
(3) *Philosophy* is working towards the understanding of the human mind in relation to the construction of reality. It tries to identify the capabilities and processes (sensations, perceptions, etc.) used by the mind in order to acquire knowledge, or, as Isaiah Berlin puts it, philosophy is concerned with questions that bother, but we do not have any idea of how to search for the answer.
(4) *Robotics* studies the basic organization and operation of intelligent computer-based robots with sensorial capabilities which are reprogrammable, multi-functional manipulators designed to perform a variety of tasks within specific industrial environments (Lee et al. 1983; see also Arkin and Torras, this volume).
(5) *Cognitive Psychology* seeks to explain human cognition (i.e., perception, memory, problem-solving and thought) in terms of processes that transform, store, retrieve, and manipulate sensory information with no concern of what neural mechanisms might be involved (Piaget 1971; Neisser 1976).
(6) *Artificial Intelligence* tries to design computer programs that allow machines to exhibit some aspects of those behaviors considered to require intelligence as displayed by some animal, such as image understanding, cognition, learning, problem-solving, game theory, etc. (Minsky 1985; Arbib 1987a).
(7) *Linguistics* has followed two main approaches to the study of language. One is concerned with the specification of grammars that generate different natural languages, and the other with those processes that allow the understanding of a given sentence or the expression of an idea through an appropriate phrase (Arbib 1987a).

The selection of these disciplines is twofold; on the one hand, they represent different universes of knowledge (Biological and Exact Sciences, and Humanities) and, on the other hand, there have been attempts to establish cycles of collaboration among some of these disciplines. In the latter case, the purpose is to make the knowledge found in one area available to the others, hence expecting that once this knowledge is considered in a different scientific context, it may produce new ideas and hypotheses that lead to a better understanding of the phenomena under study, not only within those new areas but also in the discipline which originally generated them. The notion *recognition* is such an example of interscientific interest tackeled from molecular biological, neurophysiological, ethological, neuroethological, psychological, cybernetical, philosophical and theological points of view.

2.1 The Initial Cognitive Sciences

In the late seventieth, there was a strong interaction among Cognitive Psychology, Artificial Intelligence, and Linguistics which were the first disciplines that started to be known as *Cognitive Sciences* (see Fig. 2A). Workers in these areas have been interested in models of the processes that allow us to understand a given string of symbols, or to express with a proper sentence a given idea. At the beginning, these disciplines paid attention only to the *serial* information processing paradigm (Chomsky 1956; Minsky 1961), although recently many researchers in Cognitive Science have been concerned with the *parallel* information processing paradigm as well (Rumelhart & McClelland 1986; Arbib 1987a).

2.2 The Theory-Experiment Cycle

Based on experimental data, our group has developed a couple of family-models to study sensori-motor coordination, with which we try to establish a close interaction between Brain Theory and Neuroethology (see also Arbib, this volume). Drawing on the fact that animals' actions are the outcome of the processing of information carried out by neural circuits in the CNS, these disciplines address similar issues to neuroscience. They share the concern for the development of new experimental and theoretical methodologies to analyze the neural mechanisms subserving specific behaviors (Fig. 2B). In this approach, hypotheses generated by experimental data are integrated into a family-of-models which are investigated through computer simulations in order to define whether or not there is a parametric combination that yields results compatible with the available data. Based on the postulates embedded in the models, we then try to develop explicit hypotheses which may serve to stimulate new experiments whose results, in turn, could be used to refine and develop the models accordingly (Cervantes-Pérez 1985; Arbib 1989).

2.3 Intelligent Machines: Robotics and Artificial Intelligence

Modern industrial robots, although controlled by computers, only function as simple positional machines that execute a given task by playing back a set of pre-programmed (stored in the computer memory) sequences of motions. Therefore, they have to work in highly-structured environments; that is, once the robot has been programmed, all parts and tools have to be in place, otherwise the task cannot be executed. The current trend (see Torras and Arkin, this volume) is to provide robots with the capability of processing sensorial information (visual, tactile, etc.), so that they can perform a variety of tasks with no need of being completely re-programmed each time when they have to work in non-structured environments. The aim is to develop "intelligent robots" that, based on sensory information, may be capable of understanding specific environmental situations to define the best motor response that will lead to the completion of pre-established tasks (*goals*). This in particular has originated the interaction among the fields of Robotics and Artificial Intelligence (Fu 1971; Fikes et al. 1972) (Fig. 2C).

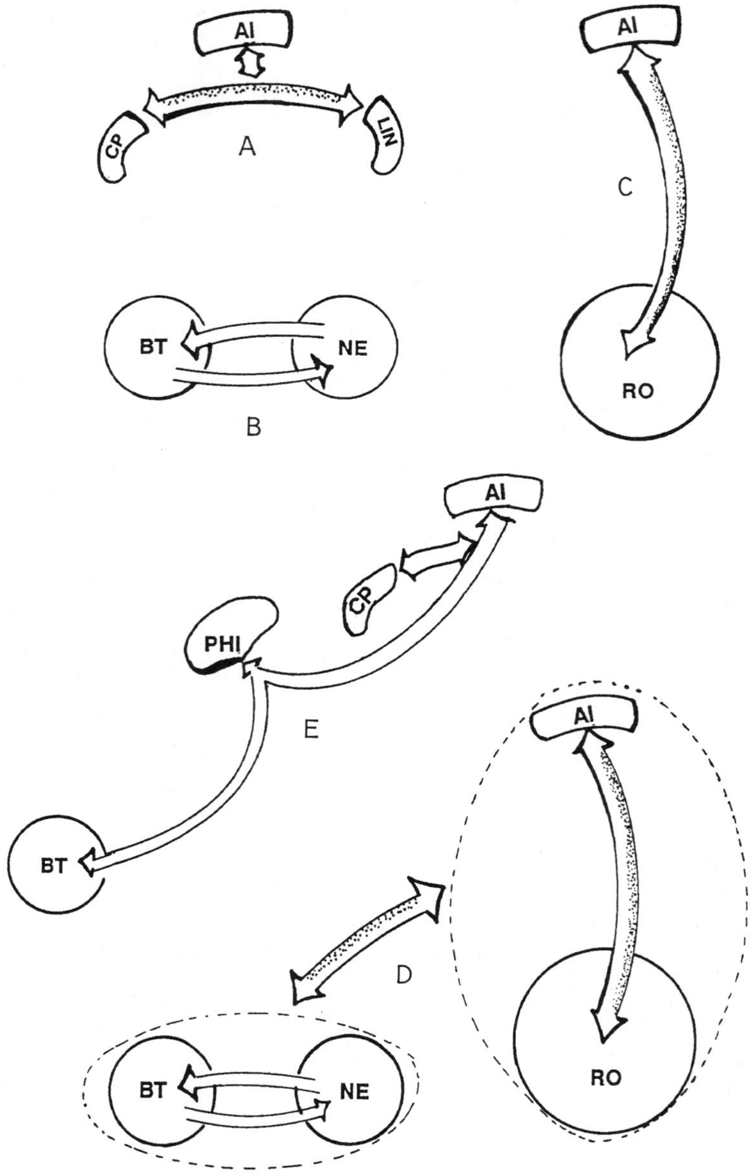

Figure 2. Cyclic studies in Cognitive Science. (A) The initial Cognitive Sciences. (B) The theory-experiment cycle between Brain Theory and Neuroethology. (C) Integration of Artificial Intelligence to the design of intelligent Robots. (D) Perceptual Robotics; integration of Neuroscience and Robotics. (E) Philosophy and Cybernetics represented by the study of the philosophical implications of concepts coming from disciplines that have originally emerged from Cybernetics (AI ,CP, and BT). For abbreviations see Fig. 1.

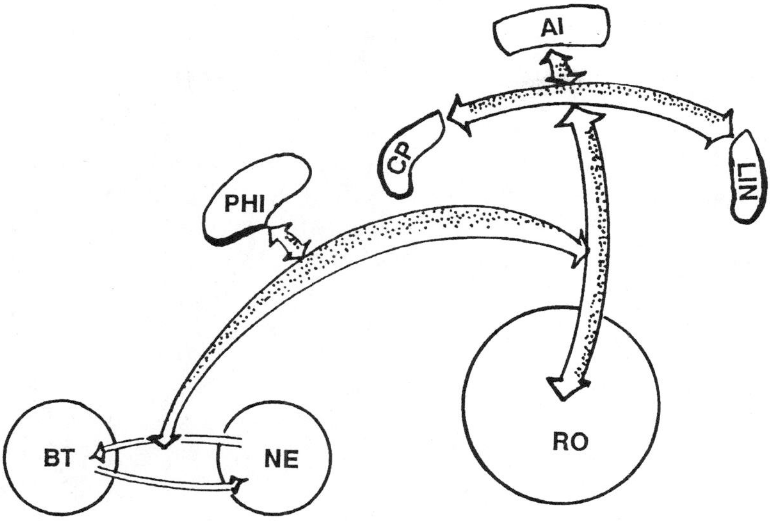

Figure 3. The Cognitive Science "tri-cycle." The solution for the *isolation problem* in Cognitive Science is represented by the tri-cycle body. It calls for a common framework to allow communication and transfer of data and knowledge among all disciplines defined as cognitive sciences. For explanations see text.

2.4 Perceptual Robotics

Iberall & Lyons (1984) defined *Perceptual Robotics* as the area of interaction among Neuroscience, Robotics, and Limb Prosthetics design (Fig. 2D). These interactions were defined based on the needs of each one of these disciplines: first, the design of sensor-based intelligent robots needs processing models for perception, feedback mechanisms, and distributed motor control; second, neuroscience needs theoretical models and artificial testbeds for understanding the viability of hypotheses generated by experimental data; and third, both of these disciplines will enhance the design of artificial limbs through a better understanding of how the CNS carries out sensori-motor coordination (see also Koy-Oberthür, this volume).

2.5 Philosophy and Cybernetics

If we consider the cybernetic origin of Brain Theory, Cognitive Psychology, and Artificial Intelligence (Arbib 1987a), then the connection between Philosophy and these disciplines (Fig. 2E), has been covered by Crosson & Sayre (1967) in their book *Philosophy and Cybernetics*. The papers in Crosson & Sayre's book explore the interaction between computer technology and the various philosophic conceptions about the nature of man. The participant Philosophers specialized in this matter obviously accepted - in general - that the possibilities of the mechanisms which are artificially intelligent will force basic changes on our concept of the human intelligent behavior.

2.6 A Multiple Interaction: The Tri-Cycle

If we overlap all those previous attempts to establish an interchange of knowledge between two or more disciplines belonging to Cognitive Science, discussed so far, we find that the cycles (Fig. 2) become a *tri-cycle* (see Fig. 3). The location of every discipline in the tri-cycle has a special meaning. The initial cognitive sciences (Cognitive Psychology, Artificial Intelligence, and Linguistics) were placed at the maneuver, since they showed the direction to be followed. Robotics is located at the big tire, because it represents the testbed were other disciplines - mainly Brain Theory and Neuroethology - may test how well the generated hypotheses account for the phenomena under study; that is, Robotics will keep on rolling other research areas by showing the gaps in the understanding of the phenomena and identifying the sort of elements that are missing in order to complete an adequate explanation. Among all cognitive sciences, the ones that generate knowledge - empirically and theoretically - about what possible mechanisms are used by the CNS to control behavior are the Neurosciences; therefore, we placed Brain Theory and Neuroethology as the small tires that give the tri-cycle support and stability. That is, the foundation of our understanding of cognition will become more sound as our neuroscientific knowledge becomes more complete. Finally, a common phrase with respect to Philosophy is that it is the mother of all sciences; thus, we have placed it as the seat of the tri-cycle to serve as support to the rest of the cognitive sciences.

The combination of those cyclic attempts also clears up what should be a common concern of all cognitive sciences, represented by the tri-cycle body in Fig. 3, which is the need of developing a theoretical framework to serve as the bridge that would allow each one of these disciplines to properly communicate with all the others. It must be noted that the development of such a theory for the tri-cycle frame will be the solution of the *isolation problem*.

3 Schema Theory as the Bridge to Study Sensori-Motor Coordination

Because of the nature of those disciplines in Cognitive Science, any theory developed to serve as a bridge to communicate between them has to go beyond the specific frames of those studies known as interdisciplinary. Rather, it has to be a theory based on an *interscientific* approach (Yankelevich & Méndez 1986), as it must explain phenomena occurring in a set of knowledge areas that belong to different scientific universes, whether to the Natural and Exact Sciences, or to the Humanities.

Within Neuroscience, studying perception and motor control and considering experimental and theoretical bases, Arbib (1981, 1985, 1987a) has developed a theoretical framework - he defined as *Schema Theory* - adequate to describe the relationship between sensory information and the control of motor response. Arbib's Schema Theory allows us to study sensory guided behaviors in the context of an *perception-action-cycle* (Neisser 1976; Arbib 1981). An organism perceives to act and may act to perceive under certain conditions; that is, new stimuli update the

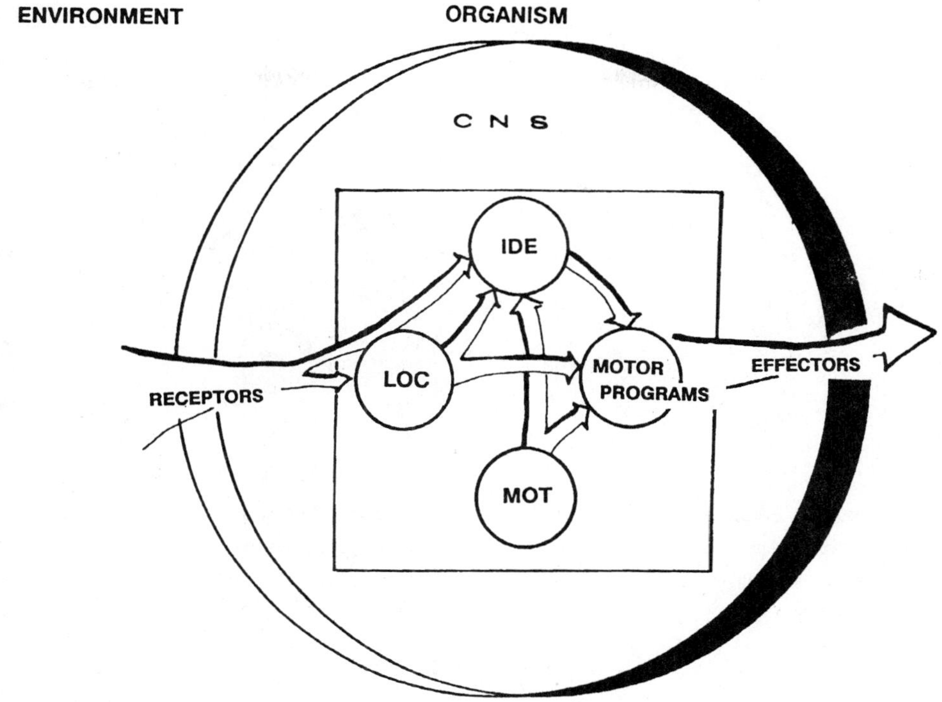

Figure 4. An organism interacting with a complex environment. Information processing steps in the central nervous system (CNS) of an organism leading to a behavioral action in response to an external situation. IDE, identification; LOC, localization; MOT, motivation. (Adapted after Ewert 1980).

organism's *internal model of the world*, hence modifying its perception of the spatio-temporal relation it keeps with domains of interaction in the environment. This directs the adequate (next best) action that would lead successfully to the consecution of its current *goal*.

To interact with complex environments, living organisms or certain machines must be able to *identify*, *locate*, and obtain a *parametric representation* of the different types of environmental stimuli. We refer to domains of interaction. This parametric representation is used by the "organism," a biological or technical one, to activate the appropriate *motor programs* to contend with external situations. In particular with animals, it has been found that the capabilities to yield specific behaviors are modulated by different factors that define the animal's motivational state (Ewert 1980) (Fig. 4). Formalizing experimental and theoretical postulates, Arbib's Schema Theory suggests that the processing of information, that has to be carried out by an animal's CNS - or by a computer in the robot's case - during sensori-motor coordination, can be usefully decomposed into three different types of entities (see Fig. 5):

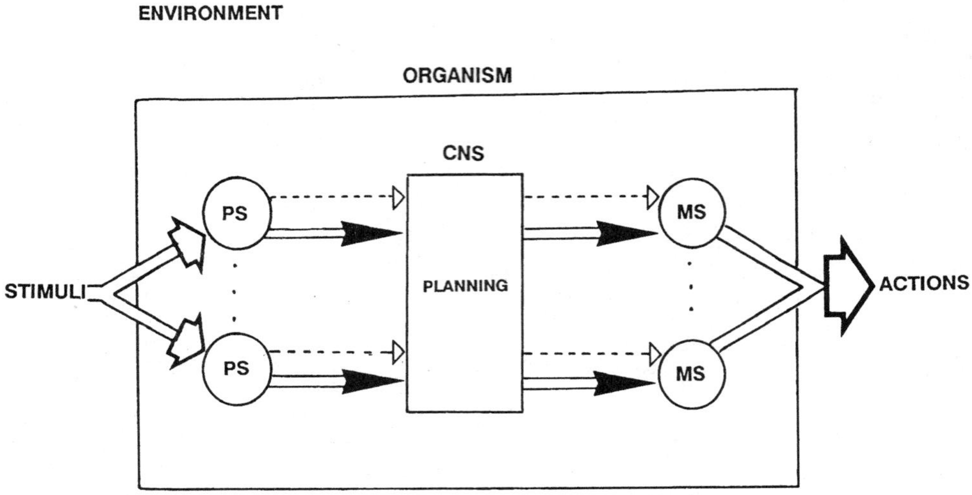

Figure 5. Sensori-motor coordination according to Schema Theory. The organism's internal model of the world is updated by new sensory information that activates Perceptual Schemas (PS's) which are used by the organism (planning) to decide what action (motor schema, MS) should be selected. This proceeds within the perception-action-cycle acccording to Arbib (1981) and Neisser (1976).

(1) *Perceptual Schemas* (PS) are defined as the units of knowledge that represent within the brain a domain of interaction; it is the environmental situation (for example, an object, or a detail of an object, or a social situation) the animal can interact with.

(2) *Motor Schemas* (MS) are defined as the units of motor control that use information obtained by one or more PSs to direct the adequate action of the animal to cope properly with the current situation.

(3) *Planning* is defined as the process that selects information from activated PSs in order to activate the proper MSs in accord with the animal's current goals.

Thus, PSs generate parametric representations of domains of interaction which are to be used by MSs, but need not to activate such MSs. Furthermore, there may be different PSs active at the same time, so that they may cooperate or complete with each other in giving the necessary conditions to control the activation of MSs.

Lara et al. (1984) developed an analysis of visuomotor coordination in toads (see Section 4) using a modified version of Arbib's Schema Theory. The main postulates, included in their adaptation, are:

(i) A schema is a unit of knowledge composed of: a *perceptual schema* that recognizes environmental stimuli that are important for the animal, a *motor schema* that represents a control system for simple motor behavior, a *goal state* that defines the state of the animal-environment relationship to which the schema directs, and a

program that relates perceptual processes to motor responses, coordinates spatio-temporally motor actions, and calls for other schemas.
(ii) Schemas are *goal-directed* units of information processing that link sensory stimuli with the best actions that may lead to reach the schema goal state.
(iii) The animal's CNS contains a group of *main schemas* whose function is to set the goal to be reached and which can activate other schemas, each one with its corresponding subgoal to reach the final goal. That is, there exists *hierarchical interactions* among schemas.
(iv) Schemas must compete with other schemas - at the same level of importance - in order to gain control to define the animal's appropriate overall behavior. That is, there also exist *heterarchical interactions* among schemas.

Lara et al. (1984) also developed graphic diagrams to represent schemas' interactions (Figs. 6-8). In these diagrams, main schemas are represented by vertical rectangles, perceptual schemas by triangles, motor schemas by squares, and the goal state by an hexagone. In addition, they indicated data flow with continuous lines and control flow with dotted lines, following Arbib's convention (Arbib 1981).

In this Section we have described the "bridge." A convincing example demonstrating how Schema Theory bridges (according to Fig. 2D) the "interscientific isolation problem" is provided by Arkin (this volume; see also the contribution by Koy-Oberthür).

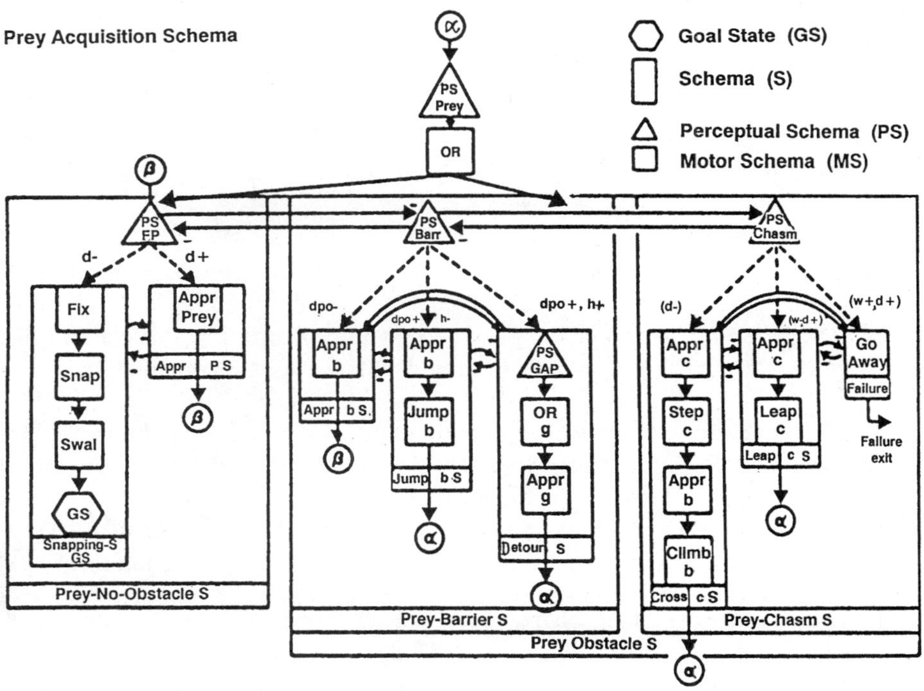

Figure 6. Four main schemas in the toad's visuomotor system. The competition among main schemas, represented by arrows, is based, first, by the stimulus movement, second, by its parametric specification and, third, by previous experiences of the animal. (From Lara et al. 1984).

4 Schema-Theoretic Models of Visuomotor Coordination

Through the integration of *top-down* and *bottom-up* approaches, our group has developed theoretical models of visuomotor coordination in amphibia and insects. In this section, we present examples of how Schema Theory can be used to study sensori-motor coordination and discuss some aspects of the schema-theoretic models we have developed for toads and mantises.

4.1 The Frogs and Toads Model

Lara et al. (1982, 1984) have analyzed visuomotor coordination in toads by means of a global model whose experimental bases are mainly neuroethological. In this model (Fig. 6), they considered four possible behavioral patterns represented by the so-called main schemas which compete to gain control of the animal's behavior:

(1) *Mating*. Toads respond with a specific repertoire in the presence of fellow mates: approaching, jumping, clasping (Heusser 1960; Kondrashev 1976, 1987). Further research is needed in this area, so that we can identify different schemas which might participate in the processing of the corresponding sensory information.

(2) *Prey Acquisition*. When confronted with a potential prey, and depending on the spatio-temporal relation between the toad and the stimulus, toads and frogs may display any of the motor responses that comprise the prey-catching behavior: orienting, detouring, jumping, approaching, binocular fixating, snapping, swallowing and wiping the mouth (Ewert 1976, 1984; Collett 1983; Ingle 1983a,b) (Fig. 7).

(3) *Predator Avoidance*. Toads respond to predators with one of various avoidance behaviors, depending on the parametric composition of the releasing stimulus. If the predator is airborne and close to the toad, the animal ducks to avoid it; if the predator is far afield, the toad searches for a hiding place: orienting, jumping, or running away. In the case of terrestrial predators, toads may puff-up, take a stiff-legged posture, orient towards the predator tilting the body and side-stepping, or they jump away from the direction where the predator is coming (Ingle 1976; Ewert & Traud 1979).

(4) *Response to Non-Moving Stimuli*. Toads respond to a situation with non-moving stimuli depending mostly on motivational factors, i.e., hunger, humidity, time of the day, etc.. Frogs in general seem to go to places related to blue colors, probably because this may direct to a pond where the animal can maintain its body humidity, find preys to eat and a place to hide (Muntz 1962a,b; Grüsser & Grüsser-Cornehls 1976).

The competition among these main schemas is mediated by the available sensory information through the activation of perceptual schemas (Fig. 7). The activation of the *Moving-Object* perceptual schema (*MO*) is postulated to exert an inhibition upon the *Non-Moving-Object* schema (*NMO*), and to bias the competition among the possible moving stimuli. The latter is considered in the model as a result of the animal's experience. This postulate is based on the fact that a prey-like stimulus,

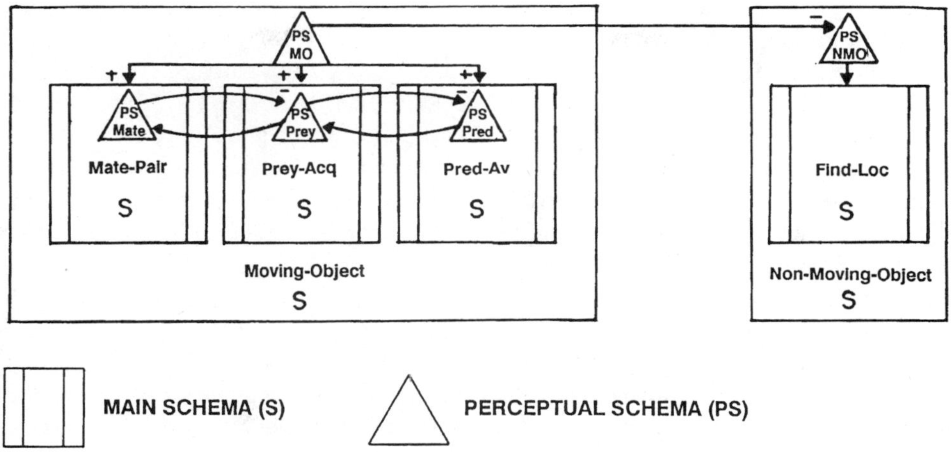

Figure 7. Prey-acquisition master schema. See text for explanation.

such as a bee, can be considered as enemy after knowing the noxious effect it produces when captured by the toad (Cott 1936; see also Lara, this volume). Similarly, the response to a predator-like stimulus can change after some training or its association with food (Brzoska & Schneider 1978; see also Finkenstädt, this volume). Lara & Arbib (1985) modeling habituation in toads postulated that the prey-predator discrimination system can be modulated by telencephalic regions as a result of experience, a function that corresponds to the one associated with the *MO* perceptual schema. (For details see Finkenstädt and Lara, this volume).

Main schemas may call for other schemas to continue the processing of sensory information for selecting the adequate action allowing the animal to cope appropriately with the external situation. Working top-down, Lara et al. (1984) analyzed a detailed expansion of the *Prey-Acquisition* schema (Fig. 6) which is activated by *MO* whenever there is a prey-like stimulus in the animal's visual field. The first motor response activated is an *orientation* towards the direction where the potential prey is located, to bring it into the binocular visual field. Then there may be competition between perceptual schemas to analyze whether or not there are obstacles in between.

If prey is directly available, the the *Free-Prey* (*FP*) schema is activated, and depending on the spatio-temporal relation *d* between the toad and its prey, it may *approach* the prey located far afield (d+) in the frontal visual field to place it in a zone where it can be reached (snapping zone); or *attack* the prey located close (d-) in the frontal visual field involving the sequential response of *fixating* the prey to estimate the distance where it is located, *snapping* at the prey to take it inside the mouth, and *swallowing* the prey. This will take the animal to reach the goal state of this main schema (Ewert 1976, 1980; Grüsser & Grüsser-Cornehls 1976; Ingle 1983b).

These animals may be confronted with different types of obstacles. In the paper by Lara et al. (1984) only barriers and chasms were considered (Fig. 6). When a prey is placed behind an obstacle, they "plan" a route to avoid the barrier, so that prey is reached by detour (Lock & Collett 1979; Collett 1983; Ingle 1983a). This situation activates the *Prey-Barrier* schema which is composed of three alternative behaviors. If prey is close to the barrier (dpo-), the toad *approaches* the barrier trying to get into a position where the prey can be snapped. When the prey is far afield from the barrier (dpo+), the activation of schemas depends on the barrier's hight *h*; if it is small (h-) the *Jump-Barrier* schema is activated and, after execution control is returned to the level of main schemas, the goal state may be reached. In case the hight is big (h+), the detour schema is activated. The model includes - as an important part of the *Detour* schema - a perceptual schema for detecting *gaps* in the environment. This allows the animal, if the gaps are large enough, to pass the gaps between barriers to capture the prey or, when gaps are to small, to walk around the barrier boundaries.

When prey is behind a chasm, the activation of three alternative schemas depends on the width *w* and the depth *d* of that chasm. If the chasm is shallow (d-), the *Cross-Chasm* schema is activated and, after execution, control is returned to the main schema's level, so that the goal state can be reached. For deep chasms (d+), *w* bias the schema activation. For a chasm deep and narrow (w-), the *Leap* schema gains control defining a sequence of motor actions to place the animal in front of the prey. In case the chasm is deep and long (w+), then the *Go-Away* schema is activated and represents a failure exit for the *Prey-Acquisition* schema (Fig. 6).

Drawing on the three-dimensional characteristics of a specific environmental situation, this model allowed Lara et al. (1984) to analyze how toads may plan routes in order to catch a prey in a complex environment (Collett 1983). They postulated that the equivalents of schemas in the toad's brain are "modular structures" (e.g., see *columns* defined anatomically by Lázár, this volume, or *functional units* and *goal-related neuronal assemblies* described by Ewert, this volume) that link perceptual information with motor responses to reach some pre-established goal. In their model they also considered that schemas have to be implemented in different brain regions, meaning that schema activation defines not only the sequence of actions to be executed but, in time, they will also be useful to point out those brain regions that must be active during specific situations (cf. *multiple instantiations of schemas* according to Arbib, this volume; *sensorimotor codes* after Ewert, this volume). Additionally, Lara et al. (1984) reproduced in a qualitative way Collett's (1983) results to explain detour behavior depending on the width, depth, and color of gaps, hence postulating the existence of a *Gap* perceptual schema, instead of an alternative proposed by Arbib & House (1985) where the detour path is a function of barrier and prey position with a gap determined by the relative position of barriers (see also Arbib, this volume). Through computer simulations with this model, these authors predicted that in case a toad confronts a prey that is behind a barrier with two adjacent gaps of different widths placed symmetrically with respect to the toad's position, the animal will choose to go through the wider gap. If also the depth of the gaps is changed, the model predicts that the toad will choose the deepest gap.

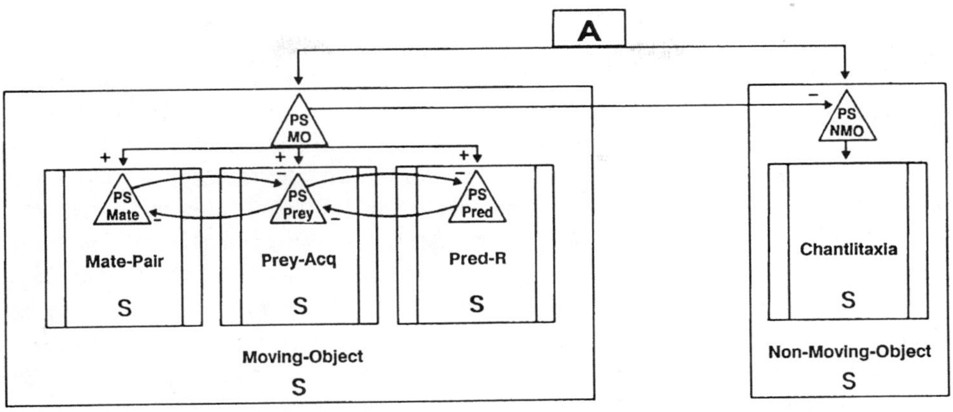

Figure 8. Four main schemas in the praying mantis visuomotor system.

4.2 The Praying Mantis Model

Schema Theory is intended as a general tool to be used in trying to identify the "*style of the brain*" (Arbib 1987a). Our group is also developing a schema-theoretic model of visuomotor coordination in a different biological system, the praying mantis (*Stagmomantis limbata*, Hahn). Insects have been used in particular to understand motor pattern generators (see Bässler, this volume), whereas studies in amphibia have more concern with understanding the processing of sensory information. By presenting the first stage of our model of visuomotor coordination in the praying mantis, we stress that at the level of analysis embedded in Schema Theory we may conduct in a general way comparative analyses asking how the CNSs of insects and amphibia guide (control) their motor actions based on the available sensory information about the three-dimensional configuration of the world. The aim is to see how much can be learned from what is known in one biological system, which may help us to understand similar phenomena occurring in an other system, hence taking advantage of the fact that in these animals scientists are working in different but complementary directions: tracing the information flow "down-stream" from receptors to effectors (main approach in amphibian neuroethology) or following it so-to-speak "up-stream" involving the analysis of processes controlling motor responses (main approach in insect neuroethology).

In Fig. 8 we show the global schema of the visually guided behavior in the praying mantis. In this case, we have only considered the capability of responding to moving and non-moving objects, comprised of four different behavioral patterns (main schemas) whose activation depends on the type of stimuli appearing in the animal's visual field: *Mate-Pair* displayed in the presence of a fellow mate of opposite sex; *Prey-Acquisition* activated in the presence of a potential prey; *Predator-Avoidance* exhibited when confronting a predator; *Chantlitaxia* or *Search-Of-A-Hiding-Place* displayed when the animal is placed in an open field. The first three are activated by moving objects, whereas the repertoire of the latter must allow the animal to interact with a variety of stationary objects.

The global schemas subserving toads (Fig. 6) and mantises (Fig. 8) to interact with visual stimuli in the three-dimensional world show remarkable similarities. That is, no matter whether it is an amphibian or an insect, to survive it seems that animals depending mainly on their visual system must have the ability of displaying comparable primary visually guided behaviors (main schemas) interacting properly with specific situations. Thus, working with schema-theoretic models of phenomena occurring in one animal, we may generate hypotheses about the existance of certain processes which are needed in order to explain similar phenomena occurring in different animals. This leads to the notion that our knowledge gained about processes and their interactions responsible for the control of specific sensori-motor behaviors can be made immediately available to other areas as well, like Robotics, for example.

5 Schema Theory for Top-Down Studies

The *isolation problem* not only occurs among disciplines from different universes of knowledge but also among areas belonging to the same scientific discipline. Taking Neuroscience from the Natural Sciences as an example, we see that Neuroanatomy has generated data about the structural organization of the brain, Neurophysiology has accumulated knowledge that corresponds to the functioning of certain morphological structures, while Ethology has explained the conditions for behavioral patterns. For a long time there was a lack of a "common language" for the development of integrative analyses to explain behavior in terms of neural network activity, until Neuroethology and Brain Theory emerged. Neuroethology (Ewert 1980), following a multimethodological approach, and Brain Theory (Arbib 1981), following an interscientific approach, develop appropriate concepts in order to identify and understant the neural mechanisms that underlie overall behaviors.

Within Brain Theory, Arbib's *Schema Theory* has been defined in such a way that it allows us to attack the isolation problem within Neuroscience following a *top-down* approach (Fig. 9): starting with the description of specific behaviors and then working towards the identification of the underlying neural substrate. Here, the CNS of the animal is considered as a "big-box" during the performance of a specific behavior. This box is then broken into smaller boxes to study how that behavior can be explained in terms of defined interactions among the functions carried out by the smaller boxes. This process continues by breaking these boxes into smaller pieces whose functions and interactions must account for the function of intermediate boxes, until we reach a stage where the function of a smallest box is executed by a well defined neural network.

Our group has developed so-called family-models to analyze the neural implementation of the *Prey-Predator-Discrimination* schema that controls activation of the prey-catching and predator-avoidance repertoire (Fig. 9). Cervantes-Pérez et al. (1985) presented a model that allows us to explain prey-predator discrimination in toads in terms of retino-tectal-pretectal interactions, hence testing Ewert's original hypothesis of pretectal inhibition upon tectal units responsible for prey-predator discrimination (Ewert 1968, 1984; see also Ingle 1983b; Grüsser & Grüsser-Cornehls

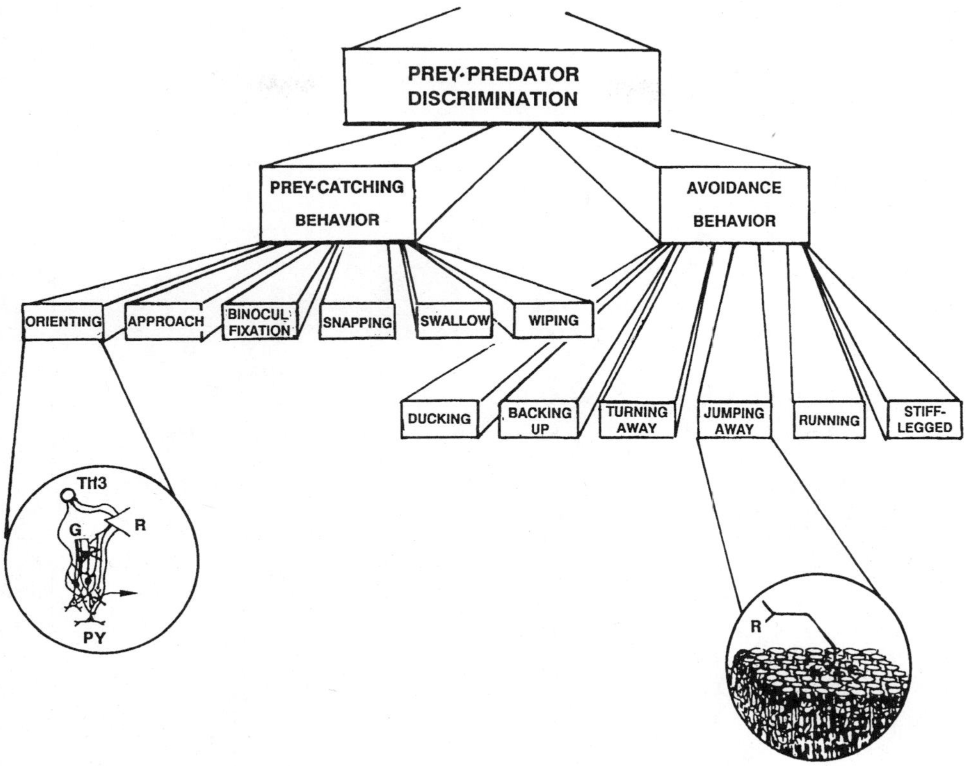

Figure 9. Schema theory and "top-down" analysis. Schema Theory allows us to work top-down, i.e., starting with one of the animal's pattern recognition behaviors (e.g., prey-predator discrimination) as a big-box, then breaking it into smaller boxes (prey-catching and avoidance behaviors) whose functions and interactions must account for the function of the bigger box; the analysis continues until the processing of information defined for the smallest boxes may be carried out by a neural network (cf. Fig. 11b,c).

1976; Grüsser-Cornehls 1984). In addition, our model explains movement direction invariance of prey-recognition as being a consequence of tectal architecture (cf. also Betts, this volume).

The model includes ethological data on fixed action patterns in toads and frogs released by relatively simple key-stimuli. Both the geometry of a moving stimulus with respect to the direction of its motion (Ewert 1976, 1984) and the absolute size of the stimulus (Ingle 1976; 1983b) play a prominent role in the toad's prey-catching behavior: Rectangular moving objects whose longest axes are oriented parallel to the direction of motion ("worm-like") are treated as potential preys, while if the long axis of the same object is oriented perpendicular to the direction of movement ("antiworm-like"), the animal does not respond with prey-catching behavior (Fig. 10A, left diagram) rather may assume a freezing posture or even show predator-avoidance. Ewert et al. (1979) showed that the "worm *vs.* antiworm" discrimination is invariant both under the direction of motion and the speed of the moving visual stimulus.

Ewert (1968, 1976) found that toads or frogs with thalamic pretectal lesions loose their ability of discriminating between prey and predator stimuli; these animals were responding with prey-catching behavior towards any moving object (Fig. 10A, right diagram).

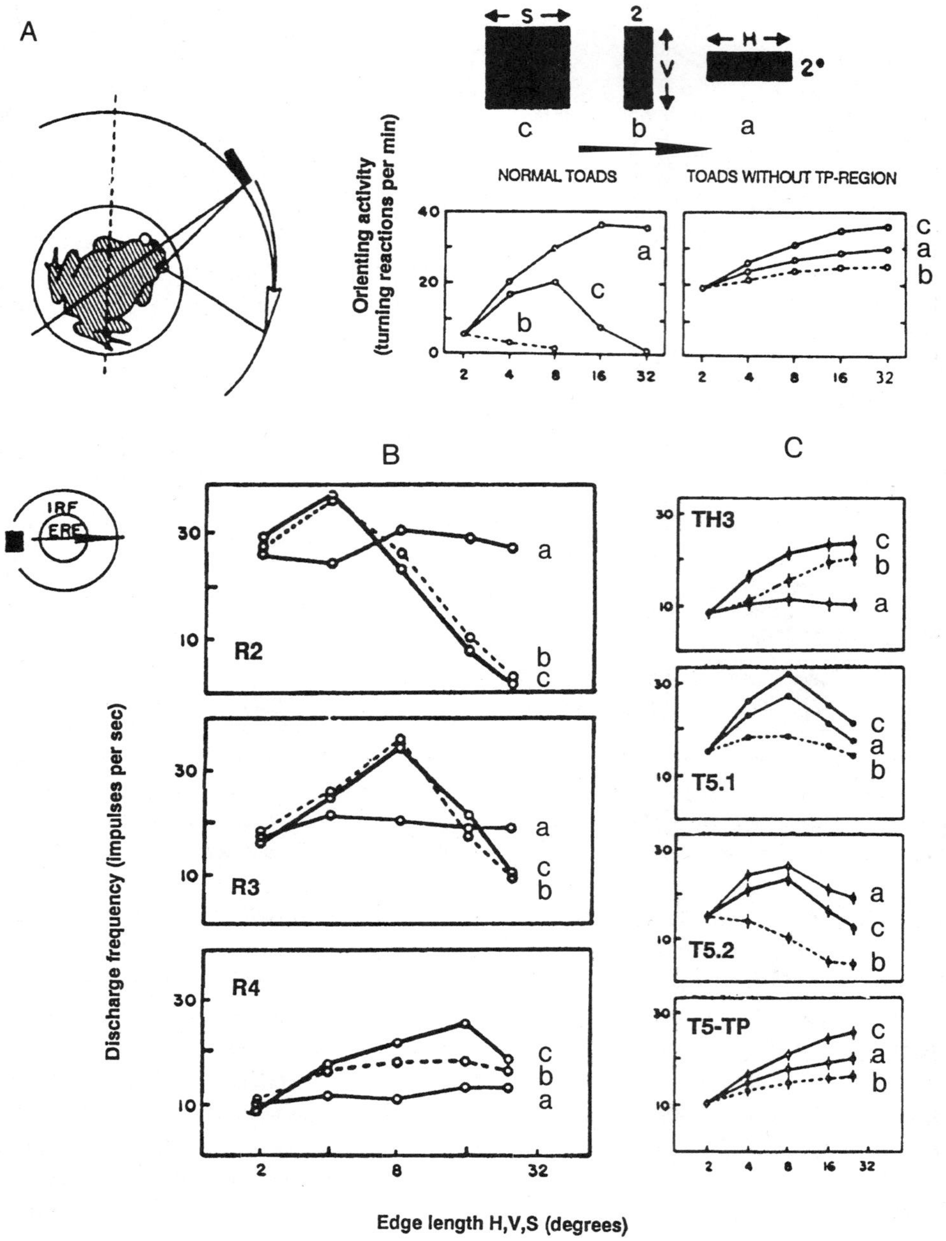

Our model also includes physiological data about firing rates of neurons from the retina, the optic tectum, and the pretectum. Trying to establish that various brain regions of the visuomotor system may play a role in the control of visually guided behaviors, physiological studies have been conducted to analyze retinal, pretectal, and tectal neural responses to changes in different parameters of visual stimuli (for reviews see Ewert 1984, 1987; Grüsser & Grüsser-Cornehls 1976; Grüsser-Cornehls 1984; see also contributions by Ewert and an der Heiden & Roth, this volume). It has been shown that retinal ganglion cells of the classes R2, R3, and R4 respond with almost the same intensity to "worm-like" moving stimuli of variable length (Fig. 11B), whereas to "antiworm-like" stimuli, or to a squares, the response rates of class R2 and R3 cells initially increase with increasing edge length up to the maximum sizes of the neurons' excitatory receptive fields (ERF) and then progressively decrease for larger objects. Class R4 ganglion cells increase their activity particularly in response to extension of the overall area of a stimulus, hence showing strongest activity to large square objects. In all classes of retinal ganglion cells, the discharge rates change with the angular velocity of a visual object (Ewert 1984; Grüsser-Cornehls 1984).

At the tectal level, it has been shown in toads (Ewert 1976, 1984; Roth & Jordan 1982) and in frogs (Schürg-Pfeiffer & Ewert 1981) that cells, classified by Ewert as T5.2, respond to moving configurational stimuli at firing rates which resemble the probability that the stimulus under investigation fits the prey category (cf. Fig. 10A, left diagram, and Fig. 10C, T5.2). After thalamic prectectal ablation, T5.2 cells' activity again resemble the animal's behavior toward different visual stimuli in that

Figure 10. Neuroethological studies on configural stimulus discrimination in toads. (A) Prey-catching orienting behavior. Prey-catching orienting activity in response to three different configurations of a moving visual stimulus. Horizontal ("worm": *type a*) and perpendicular ("antiworm": *type b*) rectangles, and squares (*type c*); the arrow indicates the direction of movement. Left diagram: normal animal's prey-catching orienting responses become more frequent if dimension H of a stimulus *type a* is larger, whereas response frequency rapidly drops to zero if dimension V of a *type b* stimulus is larger, and a sort of interaction of these two properties is obtained when both dimensions H and V in stimulus *type c* are increased. Right diagram: this discrimination is lost in toads with thalamic pretectal lesions. (From Ewert 1976). - (B) Retinal ganglion cells' responses. Class R2, R3, and R4 ganglion cells respond to different configurations (*types a, b* and *c*) of stimuli moving at a visual angular velocity of 7.6 deg/s. The receptive field of each neuron consists of a central excitatory (ERF) and a peripheral inhibitory (IRF) area. The three classes of neurons respond almost with the same intensity to change of stimulus *type a.* For *types b* and *c*, R2 and R3 cells increase their discharge rate up to their respective receptive field sizes and then drop it down, whereas R4 cells increase their rate as the stimulus size increases, giving the strongest response to stimuli of *type c.* (From Ewert 1976). - (C) Tectal and pretectal cell responses. Thalamic pretectal TH3 neurons are mostly sensitive to large (*type c*) and perpendicular (*type b*) stimuli. Tectal T5.1 neurons are more sensitive to stimulus *type c* than to *type a* or to *type b.* Tectal T5.2 neurons prefer stimulus *type a* to *type c* and give a very weak response to *type b.* This neuron's response resembles the animal's prey-catching behavior. After thalamic pretectal lesions (T5-TP), the discriminative abilities of T5.1 and T5.2 neurons are lost. (From Ewert 1976).

they respond indiscriminately to any moving object traversing their ERFs (cf. Fig. 10A, right diagram, and Fig. 10C, T5-TP) (see also single cell recordings in behaving toads by Schürg-Pfeiffer, this volume).

In our models, we have also taken into consideration anatomical data about the afferent and efferent projections of the optic tectum (Scalia & Fite 1974; Fite & Scalia 1976; Székely & Lázár 1976; Lázár 1984) as well as previous modeling efforts (Ewert & v. Seelen 1974; Didday 1976). In Fig. 11 we show the architecture of our two-dimensional model of the retino-tectal-pretectal interactions composed of an array of 8 by 8 tectal units ("columns") that have *focal* receptive fields equal to the maximum ERF of R2 ganglion cells (4 degrees), as well as a *non-focal* (8 degrees) overlap with its neighbors' receptive fields. This model builds on a previous model proposed by Lara et al. (1982), in which they have considered anatomical data from Székely & Lázár (1976) emphasizing on the vertical organization of the optic tectum with a local *vertical sample* defined as a *facilitation tectal column* (*FTC*). Each *FTC* is composed by one *thalamic-pretectal* cell (TH3) and six tectal elements: a *glomerulus* (GL), a *large pear* (LP) and a *small pear* (SP) shaped cell, a *stellate neuron* (SN), and a *pyramidal* cell (PY) which is the only efferent cell of this column (Fig. 13b-d). In the model, we hypothesized that PY cells correspond to Ewert's T5.2 neurons (for evidence see intracellular recording and labeling by Matsumoto, this volume).

Because of as yet limited data about the anatomy of the pretectal nuclei, TH3 cell responses were postulated to be the result of the integration of the information provided by R3 and R4 retinal ganglion cells. We also postulated TH3 cells to be responsible for modulating changes in the latency of PY cell response through a mechanism which is the outcome of a tonic inhibitory effect that depends on the motivational state of the animal. More specifically, we hypothesized that the responses of TH3 cells depend on two different mechanisms: one that varies with changes in the animal's motivational state for controlling tectal response latency, and another one which depends on retinal input to give tectal columns the capability of discriminating between preys and predators. In the latter case, every TH3 cell inhibits the activity of the LP, SP, and PY neurons of the *FTC* for corresponding loci within the central visual maps. (For a similar concept see Ewert and Betts, this volume; for a different opinion see an der Heiden & Roth, this volume).

Figure 11. Neural network interactions among retina, tectum, and pretectum. (a) The retina sends fibers retinotopically to both optic tectum (via classes R2, R3, and R4) and pretectum (via classes R3 and R4), and the latter structure maps upon the former. (b) TH3 neurons project retinotopically to the tectum; only the projections of three rows of TH3 cells mapping upon the tectal columns are shown; retinal inputs to tectal columns are not shown here. (c) Interactions among retinal, tectal, and prectectal cells within a tectal column. TH3 cell inhibits LP, SP, and PY cells of the tectal column corresponding to its retinotopic projection. (From Cervantes-Pérez et al. 1983; Cervantes-Pérez 1985). (d) Interactions of tectal columns within the tectal network. R, retinal input; G, glomerulus; LP, large pear-shaped cell; SP, small pear-shaped cell; SN, stellate neuron; PY, pyramidal cell corresponding to T5.2 neuron; TH3, thalamic pretectal input; arrows indicate excitatory influences and lines with bars inhibitory ones. (After Lara et al. 1982).

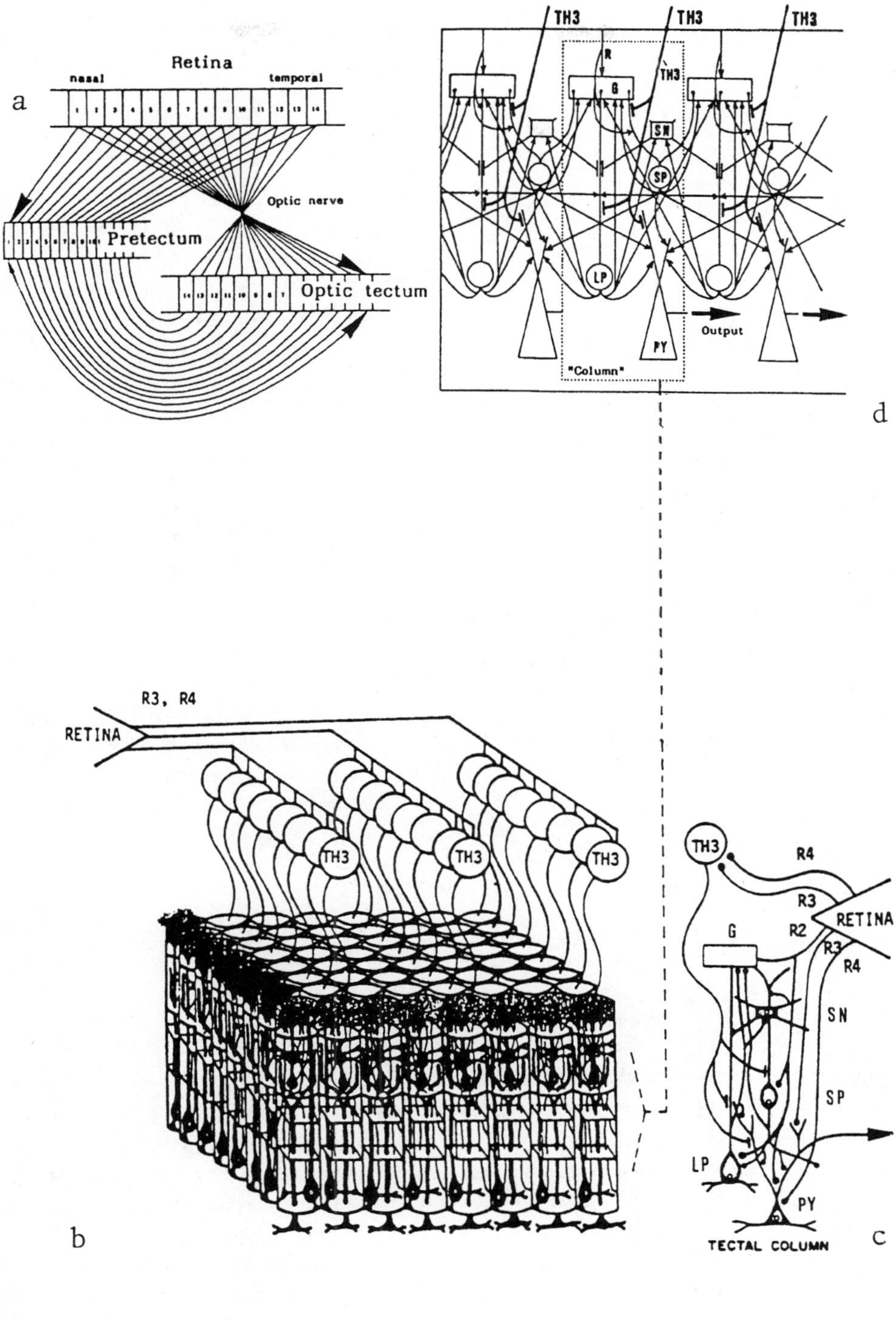

a
Retina
nasal
temporal
Optic nerve
Pretectum
Optic tectum
d
TH3
TH3
TH3
R
G
SN
SP
LP
PY
Output
"Column"
b
R3, R4
RETINA
TH3
TH3
TH3
c
TH3
R4
R3
RETINA
G
R2
R3
R4
SN
SP
LP
PY
TECTAL COLUMN

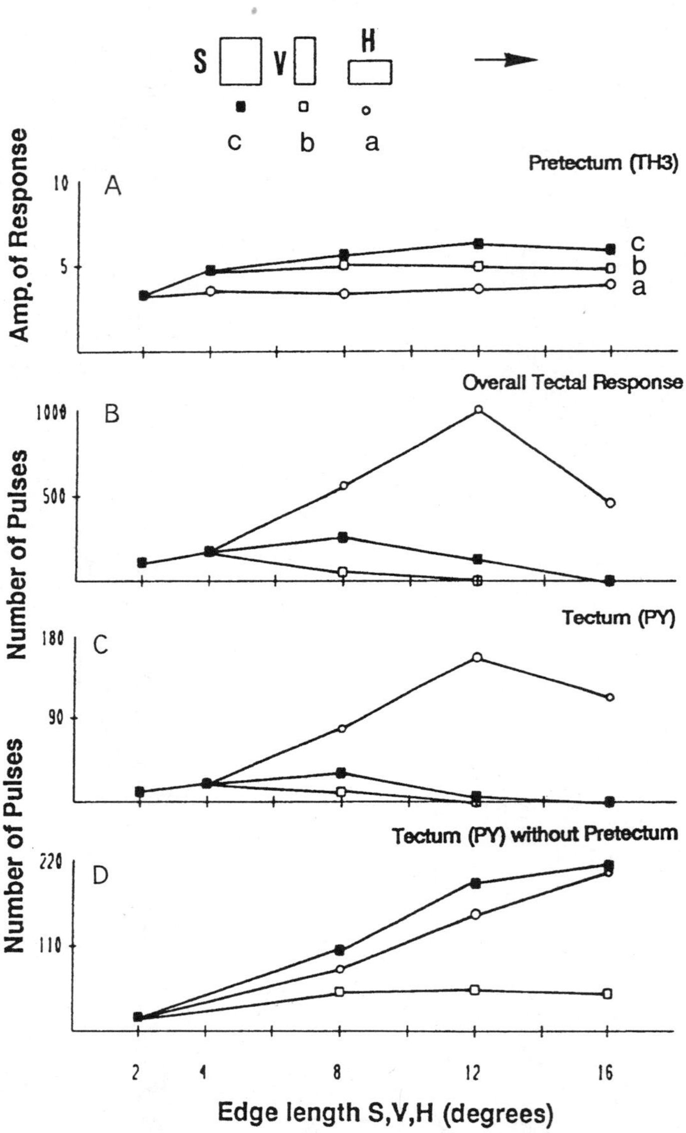

Figure 12. Computer simulation of the activities of pretectal and tectal cells (of the neuronal model network shown in Fig. 11b) in response to different configurations (*types a, b* and *c*) of moving stimuli (visual angular velocity is 8 degrees/s). (A) Pretectal TH3 cells are mostly sensitive to stimuli of *type c* and *b*. (B) Overall response of the tectum to the three types of stimulus (*a, b,* and *c*): tectum response is mostly sensitive to stimulus *type a* and is weakest to *type b*. (C) Response of a tectal PY cell (corresponding to T5.2 neuron) to the three different stimuli: it responds better to stimulus *type a* than to *type c,* and it gives a very weak response to *type b*. This curve corresponds to the one shown in B. (D) PY cell response when the pretectum is ablated: cells are then mostly sensitive to stimuli of *type c* or *a*. (From Cervantes-Pérez et al. 1983; Cervantes-Pérez 1985).

The activity of PY cells is suggested to define the spatial location of the stimulus, the direction of prey-catching orienting, and the prey-predator discriminative abilities. When several PY cells are active, the target location of a response is determined by the average of their target loci as weighted by their corresponding activities. It is important to point out that the neural organization proposed for pretectal-tectal interactions took into consideration the anatomical data available by the time. The purpose was to obtain the one that best reproduced the experimental results. A symmetric architecture was postulated for the lateral intertectal column connections (Fig. 11d) which explained the response invariance to the direction of stimulus movement, and could thus be interpreted as an immediate consequence of the structural organization of the tectum.

Computer simulations with PY cells of the model allowed us to reproduce the different physiological responses of T5.2 cells to moving configurational stimuli (Fig. 12) under conditions with and without pretectal inhibition, respectively. In Fig. 13 we observe that when a pretectal lesion is simulated, PY cells respond best to squares, but they hardly discriminate "worm-like" and "antiworm-like" stimulus configurations. In addition, these simulations show that the overall tectal response, disconnected from the pretectum, is wider spread, while the response to the "worm-like" stimulus, although it is strong too, is concentrated in a narrower area. Note that differences in response to worm- *vs.* antiworm-configurations (still present in a simulated *single* PY cell after pretectal ablation, Fig. 12D) are strongly reduced when the overall tectal response of several adjacent tectal columns (required for the release of a behavioral response) is considered (Fig. 13a,b).

Figure 14 shows the simulations of PY cells' responses to the same stimuli as in Fig. 13 when the pretectal inhibitory effect is present. Here, it can be observed that the response to squares, although wider spread, is weaker than that to the "worm-like" stimulus, whereas the weakest response is elicited by "antiworm-like" stimuli. Finally, we show a simple test of the direction invariance whereby a rectangle was moving in "worm"- (Fig. 15A) or in "antiworm"- (Fig. 15B) configuration in three different directions (Fig. 15ABa-c).

These modeling studies present only qualitative analyses of how the retino-tectal-pretectal interactions might be the neural implementation of the *Prey-Predator-Discrimination* schema. Nevertheless, they allow us to suggest that some brain regions in the amphibian visual system are organized in functional units consisting of interconnected neurons (the smallest pieces of the "big box" mentioned above, Fig. 9) with specific retinotopic mappings, where the functional interactions among elements belonging to different regions yield visual pattern recognition properties necessary for the activation of the proper motor response. According to our hypothesis, the *FTC* represents a recognition unit which is only an element integrated in a complex neural machinery (see "macro-network" concept discussed by Ewert, this volume). The interactions among different elements that integrate sensory information from different sources give the animal the ability of selecting the adequate motor behavior required to cope with a complex environmental situation.

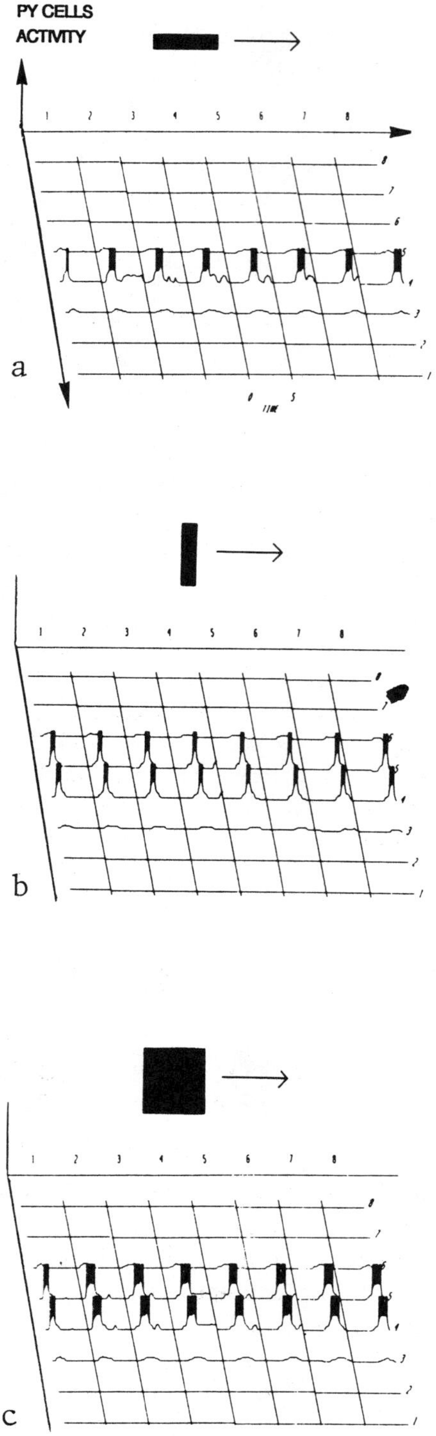

Figure 13. PY cell's responses of the neuronal network shown in Fig. 11b (consisting of 64 columns) when pretectal ablation is computer-simulated. Stimuli of the *types a, b,* and *c* (see Figs. 10 and 12) are moved at an angular velocity of 8 deg/s; the arrow indicates the direction of movement. In these graphs, the x and y axes represent the spatial localization of the (i*th*, j*th*) tectal column. The y axis of this plane is also used to show the time scale for the response of every column's PY cell, while the vertical axis (z axis) represents its local membrane potential. If the PY local potential is above the threshold, this is indicated by spikes. (a) Response to a stripe of 2°x 8° moving "worm-like" (*type a*). (b) Response to a same stripe moving "antiworm-like" (*type b*). (c) Response to a square stimulus of 8°x 8° (*type c*). At the level of one PY cell, the tectal response is stronger for stimulus *type c* than to *type a* and less to *type b*. Note that the overall response of the tectum is also stronger to *types c* and *b*, since it is wider spread than the respnse to *type a*. (From Cervantes-Pérez 1985).

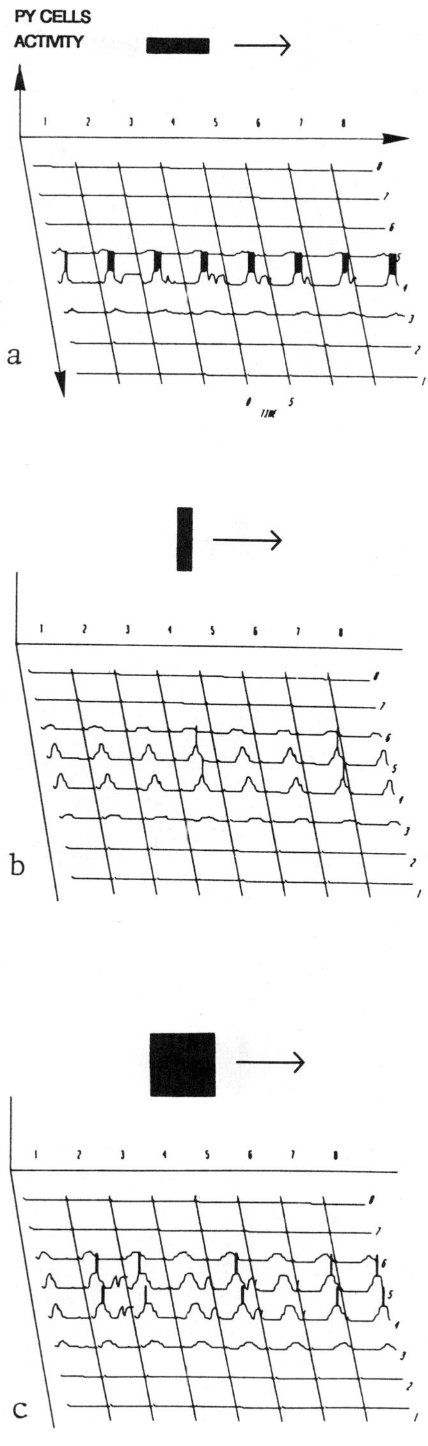

Figure 14. Computer stimulation of PY cells' responses when pretectal inhibition is present. Stimuli of the *types a, b* and *c* (cf. Fig. 13) are moved at 8 deg/s. (a) Response to a "worm-like" stimulus of 2°x 8° (*type a*). (b) Response to an "antiworm-like" stimulus of 8°x 2° (*type b*). (c) Response to a 8°x 8° square stimulus (*type c*). The response to *type c*, although wider spread, is weaker than that to *type a*; weakest response is to *type b*. PY cells are most sensitive to "worm-like" stimuli. (From Cervantes-Pérez 1985)

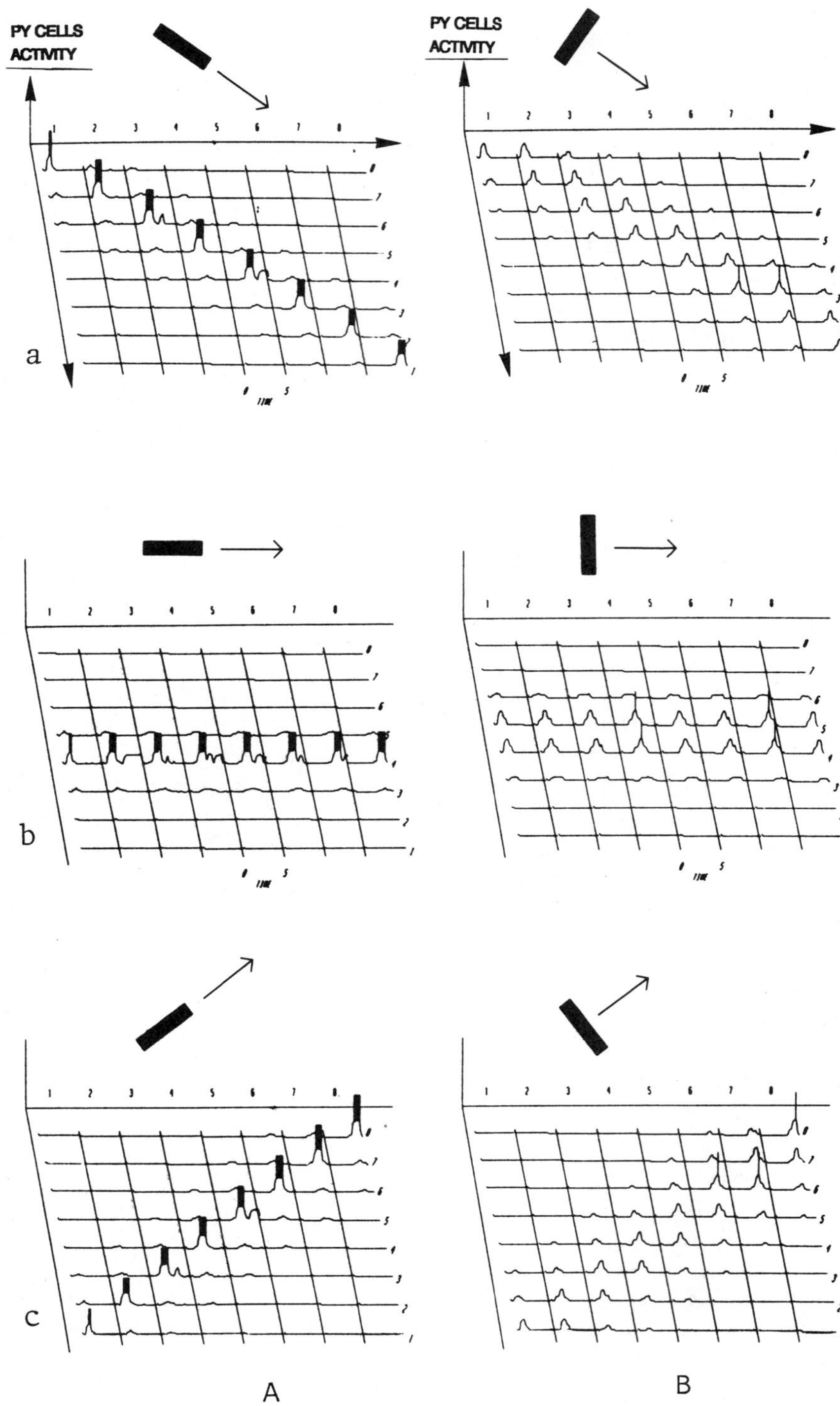
PY CELLS ACTIVITY
PY CELLS ACTIVITY
a
b
c
A
B

6 Conclusions

The definition of "interscientific umbrellas" to gather and to link different disciplines with common concerns is very important, but even more important is the proposition of theoretical frameworks that allow an intercommunication among those disciplines; so we can avoid having a set of areas studying similar phenomena but with different methodologies that isolate one from the others, eventhough they share common interests that are covered by the same "umbrella."

The *isolation problem* in Cognitive Science represents one of the main obstacles in the understanding of how the brain works. If we can solve this problem, then the communication of new findings and theories across all disciplines concerned with processes related to cognition will be greatly enhanced. Although there have been different cyclic attempts to solve this problem, they represent only partial solutions. We propose that in order to obtain a global solution, we rather need an integrative combination of what has been done with the cyclic attempts; that is, we need to follow what we defined as the tri-cyclic approach.

The cognitive *tri-cycle* shows the existence of common concerns among different disciplines from very diverse universes of knowledge, which try to understand what processes and what interactions among them have to be carried out by an animal's brain, or by a computer program in the case of robots, in order to explain overall behaviors during specific situations. In addition, it also points to the solution of the *isolation problem* representd by the frame of the tri-cycle; that is, we need to develop an *interscientific* bridge that offers, on the one hand, a global language in which high level analyses of phenomena in cognition of interest to us can be carried out and, on the another hand, theoretical elements low-level enough to allow us to work top-down trying to explain observations obtained in one discipline (e.g., ethology) in terms of mechanism studied in other disciplines (e.g., neuroanatomy and neurophysiology). In other words, we are trying to identify neural networks, or computer programs, that underlie behavior.

Arbib's *Schema Theory* has been discussed as a theoretical framework for the solution of the *isolation problem* in Cognitive Science, although we must say that it is still in the stage of development. Nevertheless, defined so far, Schema Theory has been already used with success at the higher levels of the top-down approach in the analysis of some aspects of sensori-motor coordination in biological organisms (see above examples in amphibians and insects; see also Arbib, this volume), machines (see Arkin and Torras, this volume), and "hybrid systems" (see Koy-Oberthür, this volume). Additionally, Schema Theory has been also used by our group to conduct

Figure 15. Computer simulation of the direction invariance of PY cells' responses. A stimulus of 2°x 8° is moved at 8 deg/s "worm-like" (A) or "antiworm-like" (B) in three different directions: moving from the upper left corner of the figure to the lower right corner (a), from left to right in the horizontal plane (b), or from the lower left to the upper right (c). (From Cervantes-Pérez 1985).

studies on the neural networks carrying out the processing of visual information with respect to the toad's prey-predator-discrimination schema. However, much more needs to be done, both theoretically and experimentally, to get Schema Theory to really become the interscientific bridge that will solve the isolation problem in Cognitive Science.

Acknowledgements. I want to thank Nidia Lara for her valuable comments on the manuscript, and I am also grateful to Carmen González and Silvia Coria for helping me with the drawings.

7 References

Arbib MA (1972) *The metaphorical brain: an introduction to cybernetics as artificial intelligence and brain theory.* Wiley Interscience, New York

Arbib MA (1981) Perceptual structures and distributed motor control. In: Brooks VB (ed) *Handbook of physiology - The nervous system II. Motor control.* American Physiological Society Bethesda MD, pp 1449-1480

Arbib MA (1985) *In search of the person.* Univ of Massachusetts Press, Amherst MA

Arbib MA (1987a) *Brains, machines and mathematics.* Springer-Verlag, Berlin Heidelberg New York

Arbib MA (1987b) *From schema theory to language.* Oxford Univ Press, Oxford New York

Arbib MA (1989) *The metaphorical brain 2: neural networks and beyond.* Wiley Interscience, New York (in press)

Arbib MA, House D (1985) Depth and detours: an essay on visually guided behavior. *COINS Tech Rep* 86-48, Univ of Mass, Amherst MA

Arbib MA, House D (1987) Depth and detours: an essay on visually guided behavior. In: Arbib MA, Hanson AR (eds) *Vision, brain, and cooperative compution.* A Bradford Book/MIT Press, Cambridge Mass, pp 129-163

Brzoska J, Schneider H (1978) Modification of prey-catching behavior by learning in the common toad (*Bufo bufo* L, Anura, Amphibia); changes in response to visual objects and effects of auditory stimuli. *Behav Processes* 3: 125-136

Camhi JM (1984) *Neuroethology.* Sinauer Ass Inc, Sunderland MA

Cervantes F, Lara R, Arbib M (1983) A neural model of interactions subserving prey-predator discrimination and size preference in anuran amphibia. In: Lara R, Arbib MA (eds) *Proceedings of the second workshop on visuomotor coordination in frog and toad: models and experiments.* COINS Tech Rep 83-19, Univ of Mass, Amherst MA

Cervantes-Pérez F (1985) Modelling and analysis of neural networks in the visuomotor system of anura amphibia. *COINS Tech Rep* 85-27, Univ of Mass at Amherst MA

Cervantes-Pérez F, Lara R, Arbib MA (1985) A neural model of interactions subserving prey-predator discrimination and size preference in anuran amphibia. *J Theor Biol* 113: 117-152

Chomsky N (1956) Three models of the description of language. *Proc Symp Inf Theory. IRE Transactions on Information Theory* IT-2(3): 113-124

Collett T (1983) Picking a route: do toads follow rules or make plans? In: Ewert J-P, Capranica RR, Ingle DJ (eds) *Advances in vertebrate neuroethology.* Plenum Press, London New York, pp 321-330

Cott HB (1936) The effectiveness of protective adaptations in the hive-bee, illustrated by experiments on the feeding reactions, habit formation and memory of the common toad (*Bufo bufo bufo*). *Proc Zool Soc* (London) 1: 113-133

Craik KJW (1943) *The nature of explanation.* Cambridge Univ Press, Cambridge MA

Crosson F, Sayre K (1967) *Philosophy and cybernetics.* Univ Notre Dame Press, Notre Dame, Indiana

Didday RL (1976) A model of visuomotor mechanisms in the frog optic tectum. *Math Biosci* 30: 169-180

Ewert J-P (1968) Der Einfluß von Zwischenhirndefekten auf die Visuomotorik im Beute- und Fluchtverhalten der Erdkröte (*Bufo bufo* L). *Z Vergl Physiol* 61: 41-70

Ewert J-P (1976) The visual system of the toad: behavioral and physiological studies on a pattern recognition system. In: Fite KV (ed) *The amphibian visual system: a multidisciplinary approach.* Academic Press, New York, pp 141-202

Ewert J-P (1980) *Neuroethology: an introduction to the neurophysiological fundamentals of behavior.* Springer-Verlag, Berlin Heidelberg New York

Ewert J-P (1984) Tectal mechanisms that underlie prey-catching and avoidance behaviors in toads. In: Vanegas H (ed) *Comparative neurology of the optic tectum.* Plenum Press, New York, pp 247-416

Ewert J-P (1987) Neuroethology of releasing mechanisms: prey-catching in toads. *Behav Brain Sci* 10: 337-405

Ewert J-P, Seelen W v (1974) Neurobiologie und Systemtheorie eines visuellen Mustererkennungs-mechanismus bei Kröten. *Kybernetik* 14: 167-183

Ewert J-P, Traud R (1979) Releasing stimuli for antipredator behaviour in the common toad *Bufo bufo* (L). *Behaviour* 68: 170-180

Ewert J-P, Arend B, Becker V, Borchers H-W (1979) Invariants in configurational prey selection by *Bufo bufo* (L). *Brain Behav Evol* 16: 38-51

Fikes R, Hart P, Nielsson N (1972) Some new directions in robot problem solving. In: Meltzer B, Michi D (eds) *Machine intelligence, Vol 7.* pp 405-430

Fite KV, Scalia F (1976) Central visual pathways in the frog. In: Fite KV (ed) *The amphibian visual system: a multidisciplinary approach.* Academic Press, New York, pp 87-118

Fu KS (1971) Learning control systems and intelligent control systems: an intersection of artificial intelligence and automatic control. *IEEE Transactions on Automatic Control* AC-16: 70-72

Grüsser O-J, Grüsser-Cornehls U (1976) Neurophyiology of the anuran visual system. In: Llinás R, Precht W (eds) *Frog neurobiology.* Springer-Verlag, Berlin Heidelberg New York, pp 298-385

Grüsser-Cornehls U (1984) The neurophysiology of the amphibian optic tectum. In: Vanegas H (ed) *Comparative neurology of the optic tectum.* Plenum Press, New York, pp 211-245

Heusser H (1960) Instinkterscheinungen an Erdkröten unter besonderer Berücksichtigung des Fortpflanzungsinstinktes der Erdkröte (*Bufo bufo* L). *Z Tierpsychol* 17: 67-81

Iberall T, Lyons D (1984) Towards perceptual robotics. *COINS Tech Rep* 84-17, Univ of Mass at Amherst MA

Ingle DJ (1976) Spatial vision in anurans. In: Fite KV (ed) *The amphibian visual system: a multidisciplinary approach.* Academic Press, New York, pp 119-140

Ingle DJ (1983a) Brain mechanisms of visual localization by frogs and toads. In: Ewert J-P, Capranica RR, Ingle DJ (eds) *Advances in vertebrate neuroethology.* Plenum Press, London New York, pp 177-226

Ingle DJ (1983b) Prey-selection in frogs and toads: a neuroethological model. In: Satinoff E, Teitelbaum P (eds) *Handbook of behavioral neurobiology; Vol 6 Motivation.* Plenum Press, New York, pp 235-261

Kondrashev SL (1976) Influence of the visual stimulus size on the breeding behavior of anuran males. *Akad Nank Zool J* 55: 1576-1579

Kondrashev SL (1987) Neuroethology and color vision in amphibians. A commentary. *Behav Brain Sci* 10:

385

Lara R, Arbib MA (1985) A model of the neural mechanisms responsible for pattern recognition and stimulus-specific habituation in toads. *Biol Cybern* 51: 223-237

Lara R, Cervantes F, Arbib MA (1982) Two-dimensional model of retinal-tectal-pretectal interactions for the control of prey-predator recognition and size preference in amphibia. In: Amari S, Arbib MA (eds) *Competition and cooperation in neuronal nets.* Springer-Verlag, Berlin Heidelberg New York, pp 371-393

Lara R, Carmona M, Daza F, Cruz A (1984) A global model of the neural mechanisms responsible for visuomotor coordination in toads. *J Theor Biol* 110: 587-618

Lázár G (1984) Structure and connections of the frog optic tectum. In: Vanegas H (ed) *Comparative neurology of the optic tectum.* Plenum Press, New York, pp 185-210

Lee C, González R, Fu KD (1983) *Tutorial on robotics.* IEEE Computer Society Press

Lock A, Collett T (1979) A toad's devious approach to its prey: a study of some complex uses of depth vision. *J Comp Physiol* 131: 179-189

McCulloch WS, Pitts WH (1943) A logical calculus of the ideas immanent in nervous activity. *Bull Math Biophys* 5: 115-133

Minsky M (1961) Steps toward artificial intelligence. *Proc IRE* 49: 8-30

Minsky M (1985) *Robotics.* Omni Publications International, New York

Muntz WRA (1962a) Microelectrode recordings from the diencephalon of the frog, *Rana pipiens,* and a blue sensitive system. *J Neurophysiol* 25: 699-711

Muntz WRA (1962b) Effectiveness of different colours of light in releasing the positive phototactive behavior of frogs, and a possible function of the retinal projection to the diencephalon. *J Neurophysiol* 25: 712-720

Neisser U (1976) *Cognition and reality: principles and implications of cognitive psychology.* WH Freeman, San Francisco CA

Piaget J (1971) *Biology and knowledge: an essay on the relations between organic and cognitive processes.* Edinburgh University Press, Edinburgh

Rosenblueth A, Wiener N, Bigelow J (1943) Behavior, purpose and teleology. *Philos Sci* 10: 18-24

Roth G, Jordan M (1982) Response characteristics and stratification of tectal neurons in the toad *Bufo bufo* (L). *Exp Brain Res* 45: 393-398

Rumelhart DE, McClelland JL (1986) (eds) *Parallel distributed processing. Explorations in the microstructures of cognition.* MIT Press, Cambridge MA

Scalia F, Fite KV (1974) A retinotopic analysis of the central connections of the optic nerve of the frog. *J Comp Neurol* 158: 455-478

Schürg-Pfeiffer E, Ewert J-P (1981) Investigations of neurons involved in the analysis of Gestalt prey features in the frog *Rana temporaria. J Comp Physiol* 141: 139-152

Simon HA (1980) Cognitive science: the newest science in the artificial. *Cognitive Sci* 4: 33-46

Székely G, Lázár G (1976) Cellular and synaptic architecture of the optic tectum. In: Llinás R, Precht W (eds) *Frog neurobiology.* Springer-Verlag, Berlin Heidelberg New York, pp 407-434

Wiener N (1948) *Cybernetics: or control and communication in the animal and the machine.* The Technology Press and Wiley, New York

Yankelevich G, Méndez A (1986) *Ensayos en interciencia.* UNAM Press, México

Behavior-Correlated Properties of Tectal Neurons in Freely Moving Toads

EVELYN SCHÜRG-PFEIFFER

Abteilung Neuroethologie, Fachbereich Biologie/Chemie, Universität Kassel
D-3500 Kassel, FR Germany

Abstract. Searching for the cellular fundamentals of visuomotor function, in the optic tectum of paralyzed toads and frogs various neurons can be classified with respect to their sensitivity or selectivity in response to moving configural visual stimuli. The tectal T5-type neurons were of particular interest, since these - as output elements of pretectal/tectal circuitry - are suggested to be involved in configural pattern recognition and in releasing various action patterns (concept of "command releasing system"). The present data obtained from T5.2 neurons in *freely moving* toads during prey-catching behavior support this concept: (i) These neurons are best activated by prey objects, e.g., worm-like stimuli; (ii) their discharge peak-frequencies depend on the animal's motivation and attention; (iii) discharges precede orienting and snapping toward prey; (iv) during orienting these neurons are typically active, but during snapping they are silent. The configuration-selective characteristics both of T5.2 neurons and prey-catching are abolished after lesioning caudal thalamic (pretectal) nuclei while the prey-catching system itself is functioning. Hence, T5.2 neurons display at least four behavior-correlated properties: recognition, pre-motor activity, motor-feedback, and state-dependent modulation.

1 Introduction

In the anuran optic tectum various types of neurons have been classified on the basis of angular sizes and arrangements of their visual receptive fields (Grüsser & Grüsser-Cornehls 1976) and their sensitivity or selectivity to configural features of moving visual objects (Ewert 1974, 1984; Ewert et al. 1983; Schürg-Pfeiffer & Ewert 1981). Of particular interest are monocular tectal small field neurons of the classes T5.1, T5.2, T5.3, and T5.4 which - partly projecting to bulbar/spinal systems (Satou & Ewert 1985) - are suggested to be involved in configural pattern recognition and to deliver inputs to motor pattern generating networks. In comparison to results obtained from behavioral experiments, extracellular recordings in *paralyzed* toads have shown that the discharge frequency of class T5.2 neurons is positively correlated with the probability that the configuration of a moving object fits the prey-category. This points to T5.2 neurons as essential components of command systems that, in combination with other neurons, release the various ballistic action patterns of prey-catching behavior. The concept of a "command releasing system, CRS" (Ewert, this volume) presumes that tectal T5.2 neurons are integrated in a neural macro-network whose prosencephalic (e.g., pretectal, hypothalamic) constituents determine sensory analytical and motivational properties of these neurons. This hypothesis, however, is hard to evaluate from recordings of neurons obtained in paralyzed animals alone. The problem is obvious: If a toad is motivated to catch prey, it responds to an adequate sign stimulus appropriately; but would an immobilized toad *want* to react correspondingly, if it could? In the present chapter the CRS hypothesis is investigated by means of extracellular recordings from single tectal neurons in the *freely moving* toad. This technique allows one to determine the relationship between (i) stimulus features, (ii) neuronal responses, and (iii) prey-catching action pattern and to study these (a) under different attentional states, and under (b) the influence of pretectal lesions.

2 Methodological Remarks

2.1 Terminology

About three decades ago the first extracellular recordings from the brains of conscious vertebrates were performed (Jasper et al. 1958; for reviews see Phillips 1973; Lemon 1984). Experimental animals were monkeys placed in a chair; the animal was either totally restrained (e.g., in order to study eye-movements) or had limited mobility (e.g., to investigate a stereotype action pattern, such as grasping toward a target) (Hubel 1957; Hubel 1959; Hubel 1960; Evarts 1968a,b). Mountcastle (1976) and Lynch et al. (1977) described a common recording method which is used today in many laboratories. For extracellular recordings from motor cortex or parietal cortex the animal's head is fixed to the electrophysiological set-up; the animal is trained to manipulate a key and to push a buttom with one hand while the other extremities are restrained. By means of such a procedure, Mountcastle et al. (1975) studied in monkey's parietal cortex command functions for selective attention. Evarts & Fromm (1977) investigated monkey's hand and finger movements in relation to discharges of neurons of the motor cortex (see also Fromm & Evarts 1977). Chronic unit recordings in behaving animals in the range over days required a different technique (Salcman & Bak 1973).

It is important to specify the terms "conscious," "behaving," and "freely moving" which are often used in the literature without referring properly to the conditions of the experimental animal. *Conscious* may refer to a preparation where the animal is awake, however, immobilized either by pharmacological blockage of neuromuscular transmission or by mechanical restrainment; in the latter case the animal is allowed to show a limited behavioral repertoire often restricted to the movements of the eyes, a limb, or the fingers. The terms *behaving* and *freely moving* are used differently in the literature, ranging from partly restrained situations, in which the animal fixed in an experimental set-up can only move an extremity, to telemetric methods that allow the animal to locomote and to display the entire behavioral repertoire. In the present study these terms are applied to investigations, in which fexible connections from the recording electrodes of the *freely moving preparation* to the electronic set-up exist. In this procedure toads show normal locomotion, avoidance, and prey-catching behaviors.

2.2 Requirements

For the extracellular recording experiments in freely moving toads, *Bufo bufo*, the following methodological problems must be solved:

(i) When the experimental animal moves, the tip of the recording electrode must be stable in the brain tissue. This requires the construction of a relatively small head-piece subserving as an electrode carrier firmely connected to a tiny socket (5 mm diam.) and fixed with special kinds of glue and dental cement onto a small area of the cranium near an exposed part of the brain. In our laboratory we have developed various kinds of head pieces with particular mechanical properties. The initial one consisted of a miniature drive by which the recording electrode was advanced vertically using a hydraulic device operating according to the "master-slave" principle (Ewert & Borchers 1974). Since the fluid-containing tubes can be very cumbersome to a behaving animal, in a further developement a mechanical microdrive was constructed involving the screw-axis technique (Borchers 1980, 1982b; Borchers & Pinkwart 1983). This head-piece is equipped with two stainless steel micro-electrodes whose depths in the brain tissue are adjustable in dorso-ventral direction by turning a screw. Recently, this technique was improved by adding the following features: (a) by means of three different electrodes it is possible to record successively or in parallel from three different sites of the brain, to subsequently stimulate or to lesion one of these, and with the aid of a fourth electrode to record neck muscle potentials which are good indices of the onset of animal's movements; (b) the microelectrodes can be positioned in vertical and in horizontal directions in the freely moving preparation; (c) a socket permanently fixed to the skull allows the experimenter to exchange the head-piece in the awake toad (for details see Schürg-Pfeiffer & Ewert 1988). Following an experiment, by passing an anodal direct current, the recording site can be marked for subsequent histological analysis.

(ii) When the animal moves, the recording electrode positioned in the brain also picks up muscle potentials and interferences that mask the neuronal response. Elimination of these disturbances requires differential recording techniques with impedance matching, in which the potential at the active electrode is led to the positive input and the potential of the indifferent electrode to the inverting input of a differential amplifier system with field effect transistor (FET) input stages. The miniature operational amplifiers (impedance converters) are attached directly to the head-piece (Ewert & Borchers 1974; Borchers 1980, 1982a-c). In a further elaboration of this technique Borchers & Pinkwart (1983) developed a miniature transmitting system for wireless single neuron recording and stimulating (see also Pinkwart & Borchers 1987; Borchers & Pinkwart, this volume). Although the mechanics (microdrive, electrodes, and socket) and the electric and electronic components for radio-transmission of neuronal signals (battery, resistors, capacitors, transistors, and coil antenna) are miniaturized, the mass of the fully-equipped head-piece of about 4.0 g sometimes disturbed the spontaneous behavior of our experimental toads of the species *Bufo bufo*. Another problem is an unpredictible proneness of the system for interferences. Nevertheless, this innovative technique is essential for the investigation of a variety of specific questions such as those related to behavioral situations, in which the animal must turn continuously (e.g., optokinetic nystagmus or habituation of orienting toward prey).

To reduce weight, we have recently improved the wirebound recording technique by setting the entire electronics outside the experimental arena (Fig. 1), so that the toad carries only the head-piece (0.7 g) consisting of socket, miniature microdrive, and electrodes (Fig. 2). Besides differential recording, bipolar techniques can also be applied which allows us to exactly localize the recording site for subsequent stimulation and marking (Schürg-Pfeiffer & Ewert 1988). The cylindrical arena, in which the neuro-behavioral experiments are conducted subserves as a Faraday cage. For extracellular recording we use stainless steel or tungsten microelectrodes of predictable shape (according to Gottschaldt's technique; see Gottschaldt & Vahle-Hinz 1979) and predictable electrical properties (according to

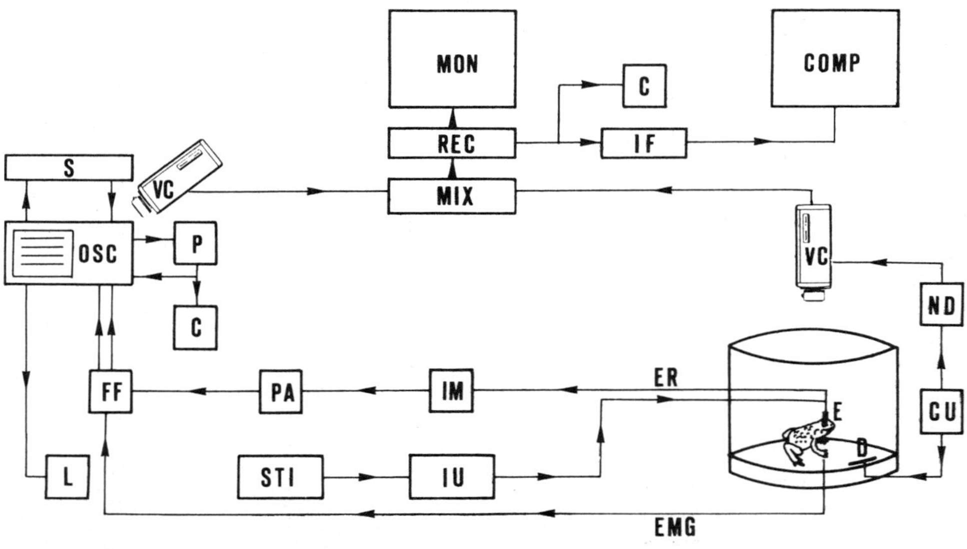

Figure 1. Neuroethological set-up for extracellular differential recordings of single neurons in the toad during prey-catching behavior and for data processing. Stainless steel microelectrodes (E) are fixed in a head piece mounted on the dorsal rostral cranium. The electrodes are connected to the preamplifier (PA) via very thin insulated wires. Both oscillogram (OSC) and animal's behavior are recorded with two videocameras (VC) brought together by means of a mixer (MIX); C, counter; COMP, computer; CU, control unit for constant velocity of the prey dummy; FF, frequency filter; IF, interface for pulse generation; IM, impedance matching; IU, insulation unit; L, loudspeaker; MON, monitor; ND, numerical display for different stimulus configurations; P, pulse sharper; S, stepper; STI, stimulation unit.

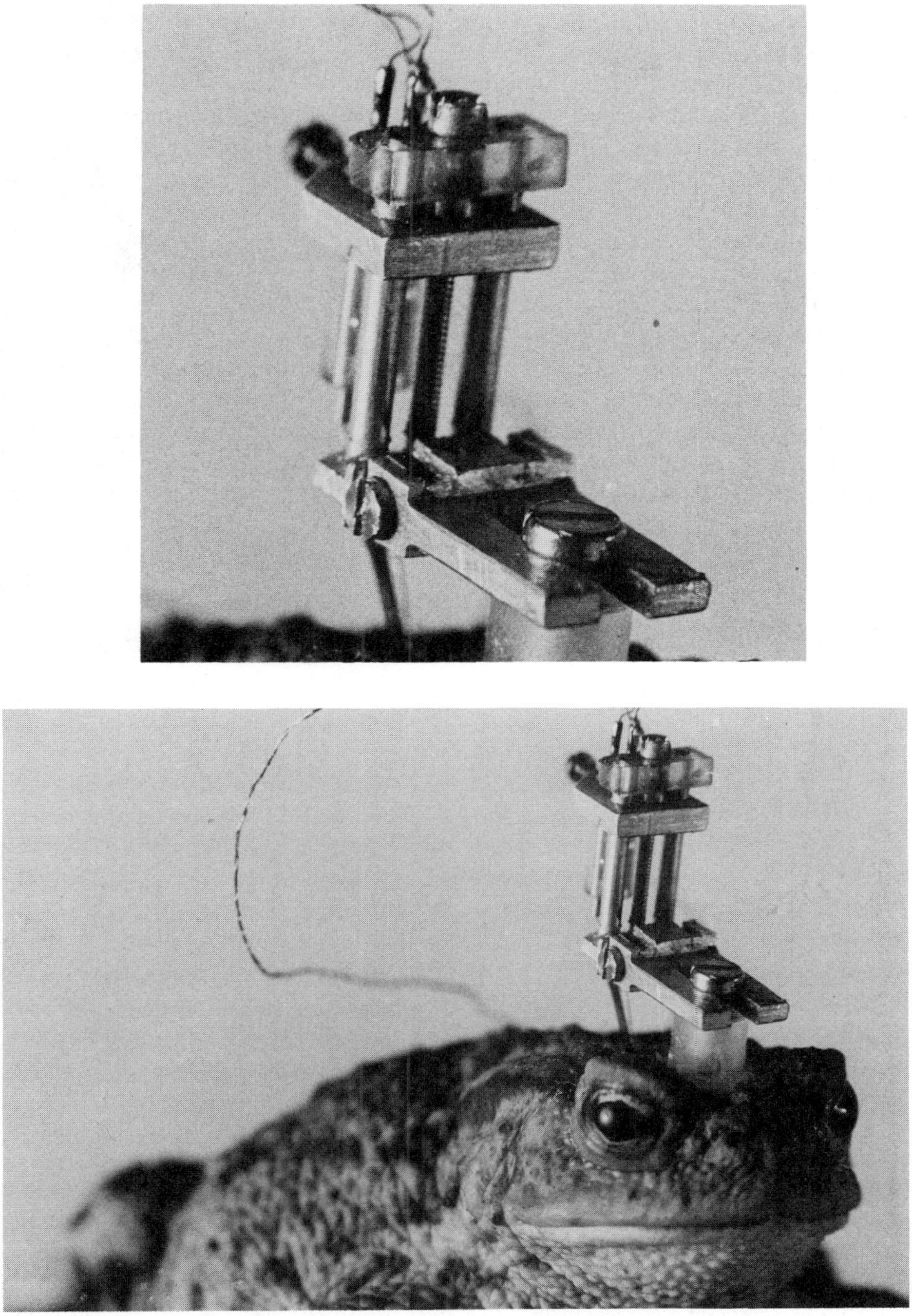

Figure 2. Freely moving preparation for wire-bound recording technique. Bottom: Toad *Bufo bufo spinosus* with a head-piece carrying the microelectrodes. The head-piece for screw-axis technique (top) is equipped with three stainless steel microelectrodes adjustable in dorsoventral and horizontal directions.

E. Schürg-Pfeiffer 1983). The connections between the electrodes and the 30 cm distant adapter at the top of the arena consist of flexible insulated micro-shielded silver wires of 0.1 mm in diameter. Within the adapter the silver wires are twisted into a common shielded cable leading to the electronic set-up (about 2 m from the arena). Besides reducing weight and elimination of interferences, the main advantage of this method lies in the flexibility of its application: All three recording electrodes are changeable from active to inactive states and each can be used for stimulation, lesioning, and marking in the respective brain area.

(iii) Analysis of behavior-correlated neuronal properties requires the monitoring of the visual stimulus (moving prey dummy), the neuronal activity, and the animal's behavior. In the initial studies (Borchers & Ewert 1978) the signal of the neuronal spike activity was mixed with a video signal of the visual stimulus and the behaving toad. The use of a video-scan converter permitted frame-by-frame analysis of all three events in space and time. In a recent elaboration of this method, video signals of stimulus and toad and of the spikes displayed on the storage oscilloscope are mixed and evaluated frame-by-frame. Simultaneously, with this image analysis the neuronal signals are stored on video tape and processed by a computer after converting the signals into transistor-transistor logic (TTL)-pulses. With appropriate software it is possible to detect the digitized spikes, to measure the interspike intervals of a spike train, and to calculate the interspike-frequency-time (ISFT) histogram. Due to a minimal resolution of 100 μs loss of information is neglectable.

(iv) Measurements of neuronal and behavioral responses, and their correlation, often require a technique with high temporal resolution. This problem cannot be solved adequately by means of frame-by-frame analysis alone with resolution of 20 or 40 ms. To determine the onset of motor activity, we simultaneously monitor the neck muscle EMG. A flexible thin insulated wire (0.1 mm diam.) inserted into the toad's neck muscle is used for recording. Experiments in which these wire electrodes were implanted in neck muscles and in various leg muscles have shown that the former were consistently activated before activation of the ones in the four extremities (Spreckelsen 1987). Neck muscle activity is thus indicative for the onset of the behavioral activities under investigation. The spatio-temporal pattern of the behavioral response (e.g., prey-catching, avoidance, locomotion) spatio-temporal pattern can be analyzed on video frame-by-frame.

(v) Presentation of various configural moving visual stimuli requires an appropriate experimental procedure (Ewert 1968; Burghagen & Ewert 1982) equipped for extracellular recording experiments in freely moving toads (Fig. 3). An electric device allows the experimenter to choose with the aid of motor driven steering mechanisms (concealed underneath the arena) one of six different black two-dimensional visual objects driven on the arena's bottom against a white background at a constant velocity. The exact position of the visual stimulus can be evaluated on video; its configuration was numerically faded into the video picture.

The neuroethological results reported here were obtained from *Bufo bufo spinosus* during observation times lasting up to 3 h. During a period of 20 min most of the animals were responsive to prey or to a prey dummy (e.g., a 2 mm x 20 mm stripe moving in "worm" configuration).

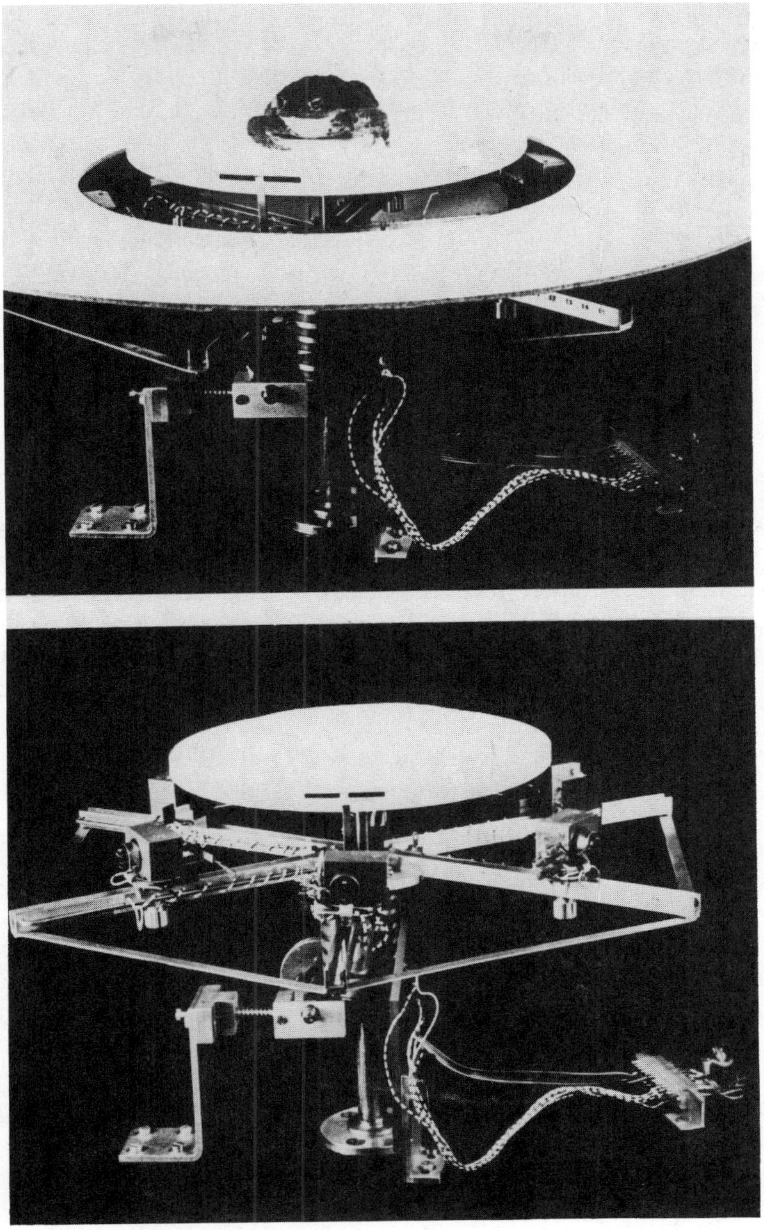

Figure 3. Experimental procedure for the investigation of toad's prey-catching and single cell activity in response to different moving stimuli (for purpose of illustration the arena wall is removed). Top: Set-up showing the arena bottom and the concentric slit in which a prey dummy can be moved around the toad; the slit is not seen by the animal. Bottom: Same set-up without peripheral arena bottom, showing electric motors, stimulus holders, and a worm dummy. (Photo H. Burghagen).

3 Responses of Retinal Ganglion Cells

Recordings from terminals of optic nerve fibers in the optic tectum of awake paralyzed frogs, *Rana pipiens, Rana esculenta, Rana temporaria*, have revealed at least four different classes of retinal ganglion cells: class R1, R2, R3, and R4 neurons (Barlow 1953; Lettvin et al. 1959; Grüsser et al. 1967, 1976; Schürg-Pfeiffer & Ewert 1981). Comparative recordings in paralyzed toads, *Bufo bufo*, showed that retinal ganglion cells display some sensitivity to the dynamic configuration of a moving visual stimulus (Ewert & Hock 1972): A stripe of 2° width traversing the excitatory receptive field (ERF) of a retinal ganglion cell activates maximally when the long axis is oriented perpendicular to the direction of movement and if its length equals the diameter of the ERF which is 4-6° in class R2 neurons and 8-10° in class R3 neurons. Recordings in freely moving toads toward configural moving visual stimuli answered the question whether retinal ganglion cells fulfill the criteria of prey detectors.

3.1 Are Barlow's "Fly Detectors" Correlated with Prey Capture?

Class R3 neurons are characterized by their response to a brisk change of diffuse illumination of the retina at *off* and *on*. Interestingly, in freely moving toads it could be shown that the *off* response fails to occur if the animal casts a shadow over the retina during complete eye closure (Ewert & Borchers 1974; Borchers & Ewert 1978), suggesting that *off* activation of these R3 neurons can be suppressed via "efferent commands" to the retina in order to focus visual information processing on changes of brightness that occur in the environment rather on those evoked by the animal itself.

Barlow has proposed that frog's R3 neurons are detectors for prey and thus responsible for the release of prey-catching: "*on-off* units seem to possess the whole of the discriminatory mechanism needed to account for this rather simple behaviour [prey capture], ... and it is difficult to avoid the conclusion that the *on-off* units are matched to this stimulus [image of a fly at 2 inch distance] and act as *fly detectors*" (Barlow 1953, p. 86).

In freely moving toads, we have recorded the responses of fiber endings of class R3 neurons whose ERF diameter ranged between 8° and 10°, independently of the distance of the test stimulus to the toad's eye. If during chronic recording in a prey-motivated toad a 6° x 2° (or 8° x 2°) black stripe traversed the ERF center at a visual angular velocitiy of 10°/s against a white background in a manner unlike a fly were walking, i.e. with its longer axis oriented perpendicular to the direction of movement (so-called "antiworm" configuration), the neuron was activated optimally, while prey-catching toward this stimulus failed to occur (Fig. 4b). The same stripe traversing the ERF at the same velocity fly-like in the direction of its longer axis (so-called "worm" configuration) reduced the neuronal discharge frequency, while the toad readily responded with prey orienting and snapping (Fig. 4a) (Schürg-Pfeiffer & Ewert 1988). These results ultimately show that retinal R3 neurons do not fulfill the criteria of a prey-detector matched to visual objects which are optimal for the release of prey-catching. Even if a 4° x 2° moving stripe was presented (which in worm configuration better fits the image of a walking fly), the R3 neuron's activity in response to the different dynamic stripe configurations did not change statistical significantly, whereas the release of prey-catching was positively correlated with the configuration resembling a walking fly (Ewert et al. 1978; Borchers & Ewert 1979).

3.2 Are Lettvin's "Bug Perceivers" Correlated with Prey Capture?

Retinal class R2 neurons are distinguished from R3 by the size of their ERF ranging in paralyzed and in freely moving toads between 4° and 6° diameter. These neurons show no *on-off* response to brisk change in diffuse illumination of the retina. Due to their small ERFs they are better tuned to smaller objects in comparison to R3. The preference of R2 neurons to convex shapes described by Lettvin et al. (1959) could not be confirmed by Grüsser et al. (1967) and Grüsser & Grüsser-Cornehls

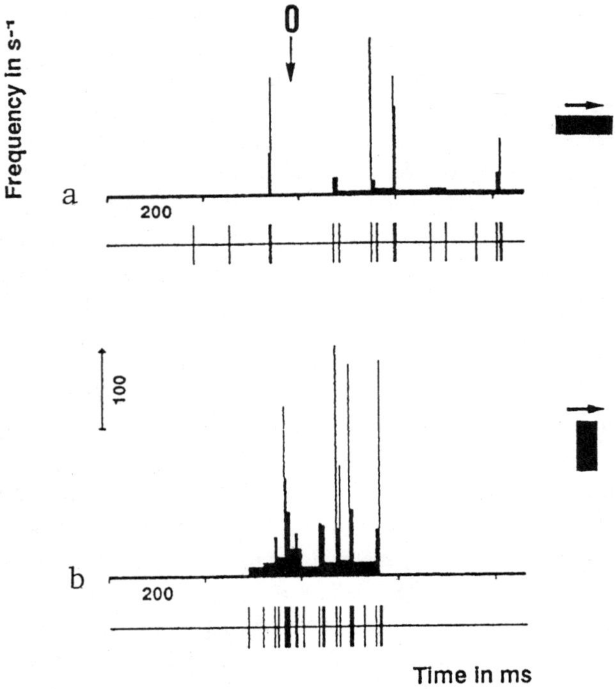

Figure 4. Mismatch between toad's retinal R3 neuronal activity and prey-selective behavior. Interspike-frequency-time (ISFT) histograms calculated from discharges of a R3 neuron to different configural moving stimuli in a prey-motivated toad. (a) To a 2° x 6° black stripe traversing the neuron's ERF center against a white background in the manner of a walking fly (at a visual angular velocity of 10°/s), the neuronal activity is relatively weak; the toad responds with prey-catching (O indicates the start of the orienting movement). (b) To the same stripe traversing the ERF center in "antifly" configuration (at the same velocity), the neuronal activity is stronger, while prey-catching fails to occur. (Note that resolving power of matrix printer is more than the 100 us underlying the computer calculation of the ISFT histograms; this must be considered when spike trains and ISFTs are compared in Figs. 4-8 and 11).

(1968). Lettvin and coworkers suggested R2 neurons as prey detectors: "We have been tempted, for example, to call the convexity detectors "bug perceiver" ... Could one better describe a system for detecting an accessible bug?" (Lettvin et al. 1959, pp. 1950-1951).

Bugs (beetles) have small elongated shapes, and these move "bug-like" in the direction of their longer body axes. If during chronic recording from a R2 neuron in a prey-motivated toad a 4° x 2° black shape traversed the ERF center at 10°/s against a white background "antibug-like" with the long axis oriented perpendicular to the direction of movement, the neuron was strongly activated, whereas no prey-catching occurred. The same stripe traversing the neuron's ERF at 10°/s bug-like in the direction of its long axis reduced the neuronal activity, while the toad readily responded with prey-catching (Borchers, unpubl.). The recordings from freely moving animals ultimately invalidate the retinal "bug detector" concept.

4 Responses of Feature Analyzing Tectal Cells

Behavioral experiments in toads have shown that monocular input is sufficient for configural prey-recognition (Ewert 1968, 1969). Therefore, we have focussed the recordings in freely moving toads on the various types of monocular driven tectal small-field neurons of the classes T5.1, T5.2, T5.3, and T5.4 that were optimally tuned to different sets of configural features of moving visual stimuli (Ewert & v. Wietersheim 1974a; Ewert et al. 1983; see also Ewert, this volume). These T5-type neurons received inputs from the contralateral eye; their ERFs ranged between 20° and 35° in diameter. The following report summarizes some essential data obtained by Schürg-Pfeiffer & Ewert (1988).

4.1 Properties of Feature-Sensitive T5.1, T5.3, and T5.4 Neurons

Class T5.1 neurons in freely moving as well as in paralyzed toads show highest discharge rates (R) to black moving square (S) objects, less to worm-like (W) stripes, and somewhat less to antiworm-like (A) stripes: $R_S > R_W > R_A$ provided that contrast and velocity parameters for S, W, and A are held constant. This relation, however, is statistically highly significant only for stimuli whose longest edge length is about 8° (S = 8° x 8°, W = 2° x 8°, and A = 8° x 2°). The neuronal response to a 2° x 20° stripe in W-configuration is moderate and almost equal to the releasing value of the same stripe in A-configuration, whereas the toad responds with prey-catching *only* to the W-configuration.

Class T5.3 neurons differ from T5.1 in that their activity to the A-configuration of a stripe is stronger than to W-configuration. The former stimulus elicits neither prey-catching nor escape, but sometimes avoidance.

Class T5.4 neurons are not sensitive to stripe stimuli presented either in W- or A-configuration. They show strongest activity in response to large compact square- or disc-shaped moving stimulus which elicited escape behavior.

Taking these results together, we can conclude that no type of tectal neuron investigated here in the freely moving toad is tuned to prey-catching. Class T5.4 neurons can be regarded as threat (predator)-sensitive neurons.

4.2 Properties of Prey-Selective T5.2 Neurons

Class T5.2 neurons in freely moving and paralyzed toads are distinguished from T5.1 by the relation $R_W > R_S > R_A$. An important property, not expressed by this relationship, refers to the progressive decrease of T5.2 neuron's activity with increasing edge length of a stripe perpendicular to the direction of movement. To a 2°x 20° (or 2 mm x 20 mm) stripe the neuron shows strong activity in response to the W-configuration, but very weak activity or no response to the A-configuration. This accords with the probability of prey-catching. Therefore, prey-selective T5.2 cells are investigated in more detail with respect to correlations between neuronal discharges, prey-catching action patterns, motivation and attention. The T5.2 neurons recorded from the border of tectal layers 6 and 7 had an ERF of about 35° in diameter.

4.2.1 Orienting Movement

The target oriented turning movement has the function to "catch" the visual image of an object into the frontal visual field. Akert (1949) has called this behavioral pattern the *visual grasp reflex*. In ethological terms we refer to *directed appetitive behavior* (Manning 1979) and in psychological terms to the directional (attentional) component of the *orientation reaction* (see Laming, this volume). Orienting toward prey in toads is a ballistic response that obtains no feedback from the target during the movement (Grobstein et al. 1983). Behavioral experiments have shown (Ewert 1968, 1969) that prey-recognition precedes orienting, so that the significance of visual features is expressed by the prey-catching activity (number of prey-catching responses per time interval). For a given stimulus position, the latency and the velocity of turning can vary depending on levels of motivation, attention, and releasing values of stimulus features.

As the worm-like prey dummy is moved at a distance of 5 cm from the toad's eyes at constant velocity through a T5.2 neuron's ERF (located more or less laterally in the visual field), the cell responds with a strong increase of the discharge frequency before the toad starts to orient (Fig. 5Aa). Increase of the neuronal activity in response to prey precedes and predicts the behavioral response. Figure 5a shows the ISFT histogram of a representative experiment with a prey-motivated, alert toad: The neuron discharges at a relatively low frequency as the 2 mm x 20 mm worm-like dummy enters the ERF; while the dummy traverses the ERF, the discharge rate increases in a manner of "warming up" reaching peak frequencies up to 700 spikes/s; orienting toward prey starts usually in the declining phase of the ISFT histogram. During the orienting movement the neuron keeps on firing although prey is outside the ERF; firing stops when turning is completed. The correlation between T5.2 neuron firing and orienting was consistent when the prey dummy repeatedly traversed the ERF in the fronto-lateral visual field (Fig. 6a). It is important to note that spontaneous turning movements, or orienting toward a prey object outside the ERF, were not introduced by activity of a T5.2 neuron under investigation.

To a stripe traversing the ERF in antiworm configuration neuronal activity was extremely weak and prey-catching failed to occur (Fig. 5Ab). The configural response characteristic was independent of ERF localization in the visual field. The latency between T5.2 neuronal activity and orienting response depended on the locus of the ERF in the visual field and the responsiveness of the toad.

4.2.2 Fixating and Snapping

Binocular fixation belongs to *directed appetitive behavior* (Manning 1979), whereas snapping is the *consummatory act* of prey-catching. Both the release of fixation and snapping, respectively, require prey recognition (Ewert, this volume). The time between fixation and snapping is variable depending on motivation, attention, and the releasing value of stimulus features.

Figure 7a shows the ISFT histogram of an experiment in which the worm-like prey dummy is moved at a distance of 3 cm from the toad's eyes at constant velocity through a T5.2 neuron's ERF located in the frontal visual field: When the dummy enters the ERF, the neuron warms up and discharges strongly before the toad starts to snap. Unlike to orienting the neuronal activity ceases during snapping, irrespective of whether toad's eyes are open or closed during this 200-300 ms period. Figure 6b depicts the neuronal activity during a behavioral sequence in response to prey at 5 cm distance. Before approaching, the neuron increases its firing frequency; approaching starts in the declining phase of the ISFT histogram; the activity lasts during the approaching movement; it stops for a moment when the toad sits still at a snapping distance, increases again before snapping, and ceases. After snapping, T5.2 neurons showed sometimes a weak (not stimulus related) rebound background activity (see also Fig. 7b).

4.2.3 Motivation and Attention

Motivational factors related to hunger, season, moulting etc. determine long-term adjustments of the prey-catching releasing mechanisms for days to months, whereas attentional alertness changes rapidly. It is possible to record neurons with T5.2 properties in satiated toads which are not motivated to catch prey. But their discharge frequencies to the 2 mm x 20 mm prey dummy (moving at 10°/s) are relatively low averaging around 10-30 spikes/s; compared with responsive toads, the pre-motor "warming up" is missing. Although these toads are not responsive to prey behaviorally, the T5.2 neurons by their discharges clearly display the sharp worm *vs.* antiworm preference, independent of the direction and the speed of stimulus movement.

Figure 5. Pre-motor and state-dependent properties of prey-selective tectal T5.2 neurons. ISFT histograms calculated from discharges of a representative tectal T5.2 neuron to different black configural moving stimuli at times the toad is responsive (A) and not responsive (B) to prey. Aa: Toward a worm-like 2 x 20 mm^2 stripe traversing the lateral ERF center at 10°/s, a strong increase of neuronal activity precedes the onset of orienting (indicated by O). Ab: To the same stripe in antiworm configuration the neuron is silent and along with this prey-catching fails to occur. Ba,b: Neuronal activity of the same T5.2 neuron to worm (a) and antiworm (b) during a phase in which the toad is not responsive to prey. All records are from the same T5.2 neuron. (See also technical comment of Fig. 4).

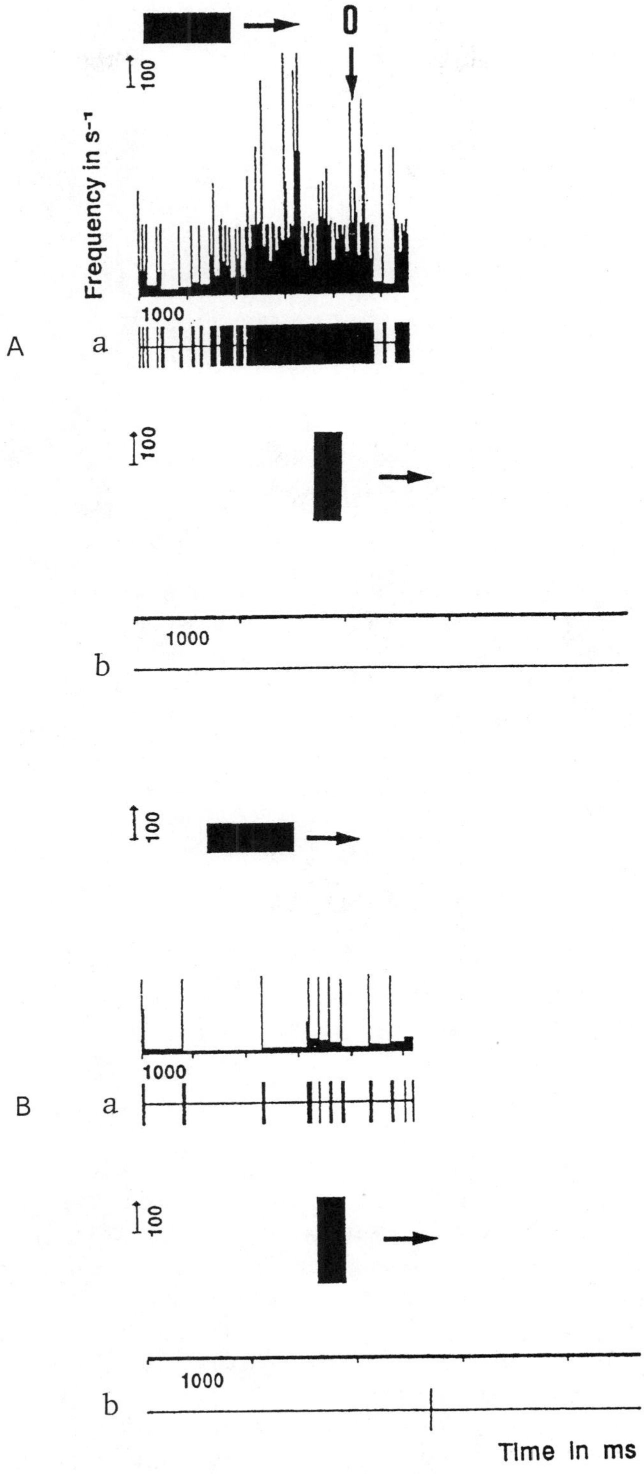
0
100
Frequency in s⁻¹
1000
A a
100
1000
b
100
1000
B a
100
1000
b
Time in ms

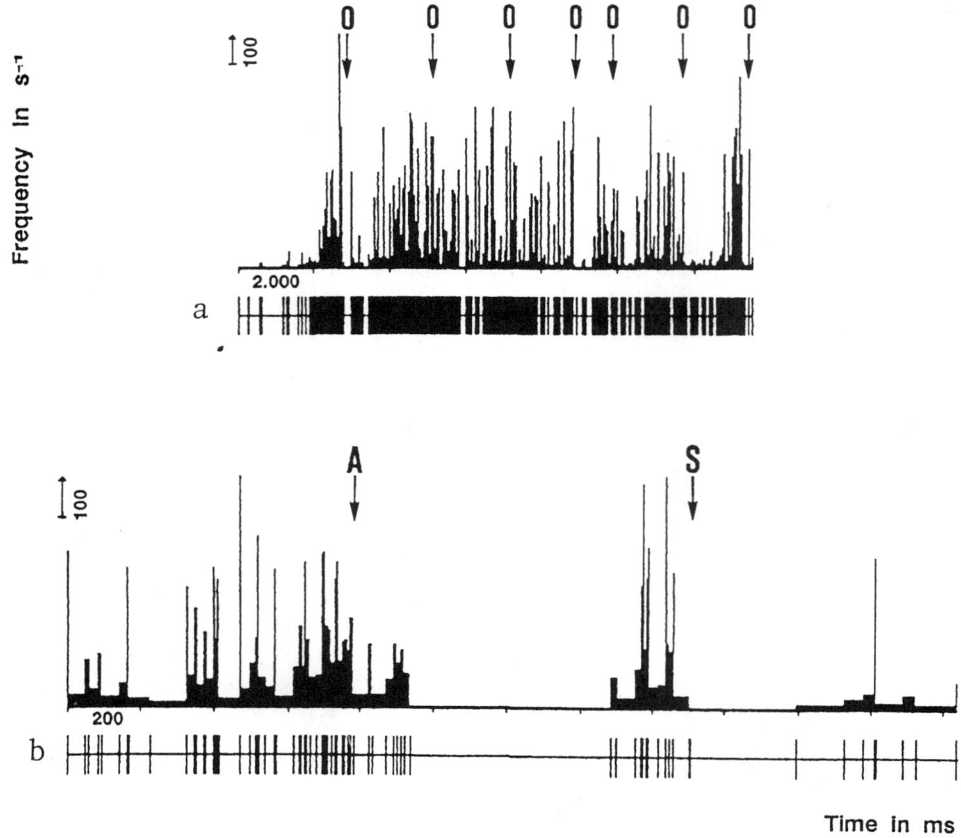

Figure 6. Activity of T5.2 neurons preceding orienting, approaching, and snapping. ISFT histograms calculated from (a) a T5.2 neuron with a frontolateral ERF in response to repetitive release of orienting with a black 2 x 20 mm^2 prey dummy and (b) a T5.2 neuron with a frontal ERF in response to the same prey dummy presented in the frontal visual field eliciting approaching and subsequently snapping. The onset of orienting O, approaching A, and snapping S, respectively, is indicated by arrows.

In prey-motivated, hungry toads prey-selective T5.2 cells can be recorded more frequently than in unmotivated animals. The neuronal discharge frequencies in response to the worm-like dummy averages between 100 and 150 spikes/s (Fig. 5Aa); that is more than four times the activity of T5.2 neurons recorded in unmotivated animals. Neuronal response to the antiworm configuration is zero and prey-catching fails to occur (Fig. 5Ab). During the observation period toad's prey-catching activity can suddenly alter from highly responsive to absolute indifference, due to short-term attentional shifts. This clear-cut change in the behavioral responsiveness is reflected by the discharge frequency of the same recorded T5.2 neuron. In an inattentive toad, prey-selective neurons are not silent, but their average discharge rates are 70-80%

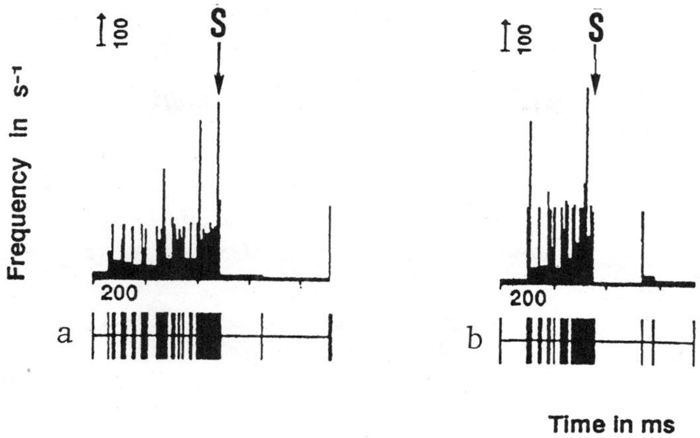

Figure 7. Discharge patterns of prey-selective neurons before and during snapping. ISFT histograms of two different T5.2 neurons (a and b) with frontal ERFs in response to a 2 x 20 mm^2 prey dummy presented at the snapping distance: Strong increase of neuronal activity precedes the onset of snapping; during snapping the neurons are silent.

less than in periods of alertness. Here, too, T5.2 neuron displays the worm *vs.* antiworm preference (cf. a and b in Fig. 5B).

Whereas chronic recordings from the same T5.2 cell in response to prey allowed one to calculate a certain level of neuronal activity that predicted a subsequent prey-catching response (see also Section 4.5), an absolute threshold could not be calculated for T5.2 neurons in different toads. The following properties characterize T5.2 neuronal activities in prey-responsive and unresponsive states:

Responsive toad ERF: ca 35°, firing rate: strong, ISFTH: "warming up"-like
Unresponsive toad ERF: < 35°, firing rate: weak, ISFTH: "plateau"-like

4.3 Properties of T5.2 Neurons Before and After a Pretectal Lesion

Behavioral experiments and single cell recording studies in paralyzed toads (and frogs) have shown that prey-catching and T5.2 neurons are disinhibited and loose their prey-selective properties if the lateral postero-dorsal (Lpd) and the lateral posterior (P) nucleus of the caudal thalamus in the pretectal region are lesioned (Ewert 1968, 1984; Ewert & v. Wietersheim 1974b; Ewert et al. 1974; Megela et al. 1983). The present experiments in freely moving toads allow us to study the response characteristic of the same T5.2 neuron *and* prey-catching before and after a pretectal lesion applied by electrocoagulation via a seperate electrode inserted in the Lpd/P region. In two animals we succeeded. These experiments are particulary convincing, since pre-lesion both toads were not prey-motivated: The average discharge rate of the recorded T5.2 neuron was relatively weak (ca. 10 spikes/s) in response to the black 2 mm x 20 mm worm-like stripe moving at 10°/s through the ERF of ca. 30° diameter; the neuron was silent to the same stripe offered in antiworm configuration (Fig. 8a_1), or to a black square of 6 mm x 6 mm or 60 mm x 60 mm in size (Fig 8b_1); the large moving square sometimes elicited avoidance behavior such as sidestepping. About 5 s following a pretectal lesion applied ipsilateral to the tectal recording position, a remarkable neuro-behavioral change occurs: Irrespective of stimulus configuration also the antiworm (Fig. 8a_1), or the large square (Fig. 8b_2), each one moving at 10°/s, elicits very strong activity in the same T5.2 neuron whose ERF had now doubled in size to about 70° diameter. The average discharge rates are between 120 and 160 spikes/s, reaching peak frequencies up to about 700 spikes/s, and this strong increase of activity always precedes and predicts subsequent prey-catching.

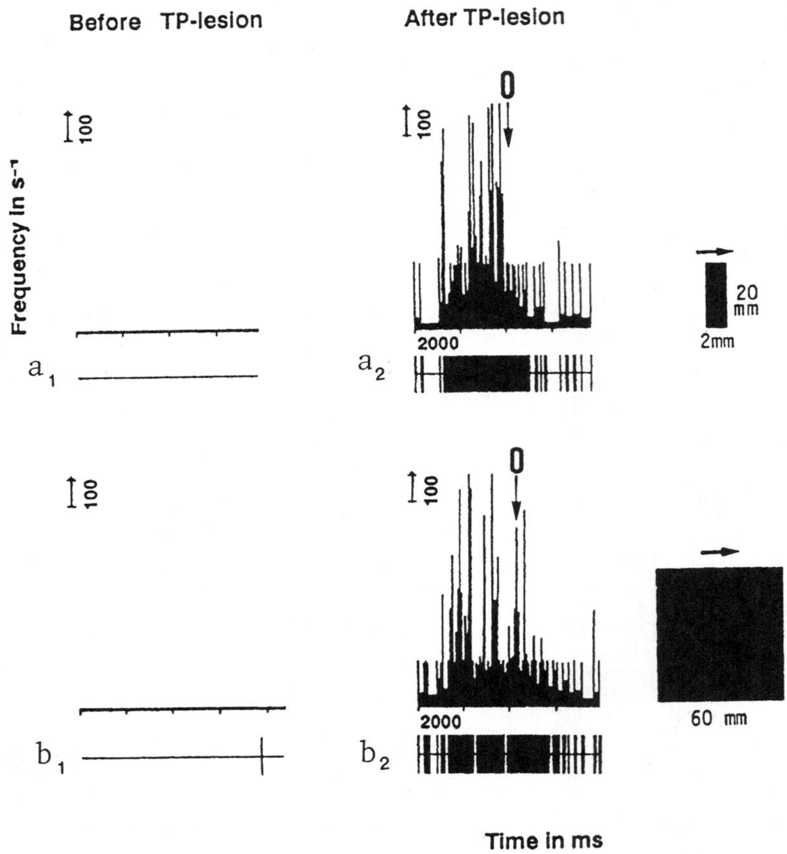

Figure 8. Response of a T5.2 neuron before (a_1, b_1) and after (a_2, b_2) a pretectal lesion applied with an electrode positioned in the ipsilateral caudal diencephalon. ISFT histograms of neuronal responses to a 20 x 2 mm^2 stripe traversing the ERF in antiworm configuration (a_1, a_2), and to a 60 x 60 mm^2 square (b_1, b_2). The onset of orienting is indicated by O.

Similar to intact prey-motivated toads, the discharges last during orienting and they occur during snapping. Since T5.2 neurons and prey-catching of both pretectally lesioned toads showed weak habituation, the neuro-behavioral correlations could be investigated in great detail. Sometimes, upon visual stimulation a T5.2 neuron could exhibit excessive self-sustained activities and after-discharges which resulted in corresponding hyperexcited prey-catching orienting. Similar results are reported from recordings in toads which received a micro-knife cut induced large pretectal lesion prior to recording (Megela et al. 1983).

Pretectally lesioned toads, too, were sometimes not responsive to visual stimuli. In these short periods the "disinhibited" T5.2 neuron showed a weaker "plateau"-like activity to a moving object, independent of its configuration; "warming up firing," indicative for subsequent prey-catching, failed to occur.

4.4 Evaluation of Temporal Discharge Patterns

Comparative investigations of the ISFT histogram of retinal ganglion cell fibers and tectal neurons show that certain kind of increase ("warming up") in the spike frequency is correlated with prey-catching. This is manifest in prey-selective T5.2 neurons of prey-motivated alert toads.

Another point of interest in the evaluation of the temporal discharge pattern of T5.2 cells refers to so-called pairwise discharges yielding "double-spikes" of about 6 ms interspike intervals. These intervals are remarkably constant during ERF traverse by a prey-dummy, and they can vary between 4 and 8 ms in different field traverses (see Fig. 9b,c). The occurrence of double-spikes probably depended on stimulus parameters and the toad's attentional state. An analysis of the interspike-interval (ISI) histogram for ERF traverses with the same prey dummy (2 mm x 20 mm worm-like stripe moving at 10°/s) exhibits a strong (n_{max} = 43) peak around 8 ms corresponding to 125 Hz. In unmotivated toads the peak is significantly weaker (n_{max} = 9); following a pretectal lesion, in the same T5.2 neuron a strong (n_{max} = 46) peak can be observed (Fig. 9c_1, c_2). In class R3 neurons there is a small peak (n_{max} = 3) around 2 ms (Fig. 9a).

4.5 Recording and Electrical Stimulation with Discharge Patterns

The above data pose the question of the significance of temporal spike discharge patterns currently investigated in our laboratory. Pilot studies indicate that electrical stimulation of the optic tectum with square wave impulse patterns derived from T5.2 spike patterns of prey-motivated toads, in contrast to trains of equidistant impulses at a comparable average pulse frequency, were more effective triggers of prey-catching in that they released finer tuned action patterns at a comparatively low stimulus current (less than 3 μA). Stimulation through a microelectrode close to a previously recorded T5.2 cell elicited an orienting movement that was correlated with the center of the neuron's ERF in the visual field. In response to stimulation at tectal electrode positions in the representation of the lateral extrafrontal visual field toads showed successive turning movements of approximately the same amplitude; upon an

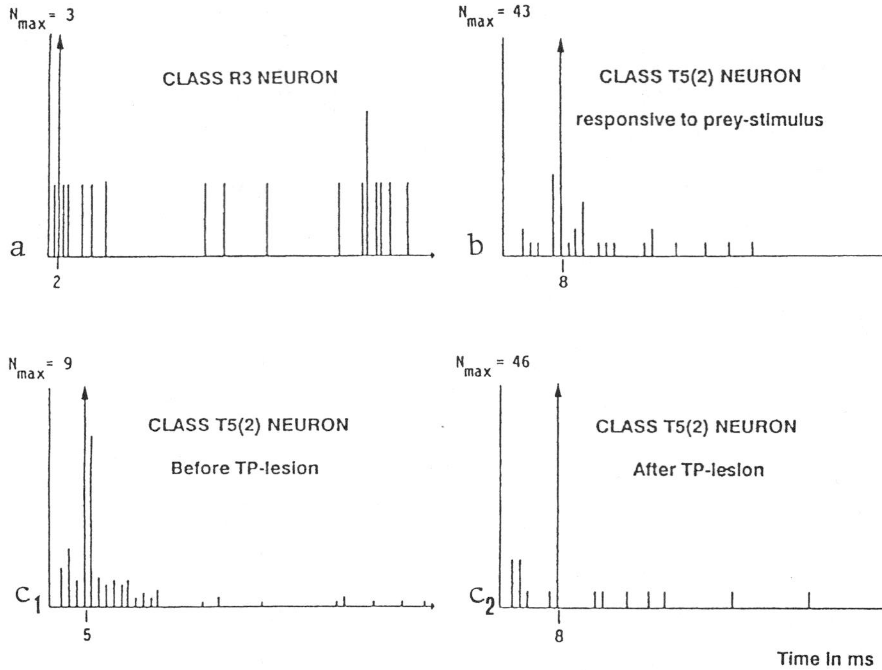

Figure 9. Interspike-interval (ISI) histogram of a retinal class R3 neuron (a), a class T5.2 neuron in a prey-motivated toad (b), a T5.2 neuron in a unmotivated toad before a pretectal lesion (c_1), and the same neuron after pretectal lesion (c_2); n_{max} refers to the most frequent interspike interval. Representative examples.

increase of stimulus current, or stimulus frequency, the response latency decreased while the turning frequency increased.

Although we doubt that tecto-bulbar/spinal projection neurons such as T5.2 use temporal spike patterns as stimulus-specific codes, the involvement of short bursts (e.g., "double spikes") - probably not restricted to T5.2 cells - may have something to do with "level setting" and "trigger efficacy". Note in Figs. 5, 6, 7, and 8 that toads always respond to a stimulus with prey-catching if "warming up" firing exceeds the inter-spike frequency of the double-spikes which itself is remarkably constant during an ERF traverse.

5 Spontaneously Active Tectal Neurons

Borchers and Ewert described in deep tectal and subtectal layers two types of spontaneously active neurons whose discharges were correlated with any motor activity, except respiratory movements (Ewert 1980a,b; Borchers 1980, 1982a,c; Borchers & Pinkwart 1983; Megela et al. 1983). In the following discussion, response properties of the so-called *movement neurons* will be reviewed to show similarities and differences to *prey-selective neurons*.

5.1 Class T8.1 Neurons

When the toad was sitting still, the weakly fluctuating discharge frequency of these neurons was around 5-6 spikes/s. An increase in the discharge rate over 40 spikes/s then literally predicted a subsequent spontaneous movement of the animal's head or leg, and even during walking each leg movement was preceded by a short burst of spikes (Fig. 10a) (Borchers 1980, 1982a,c). These T8.1 movement neurons had no visual ERF and could also not directly be activated by tactile or auditory stimuli. Movement neuron's spontaneous "pre-motor" activity was also recorded if the toad's environment was absolutely dark. In a bright environment, however, visually driven responses, such as prey-catching action patterns, were each introduced by a strong burst.

Hence, T8.1 neurons are different from T5.2 neurons in the absence of an ERF and the nonspecificity of neuro-behavioral correlations. Although an analysis of the ISFT histogram indicated stronger increases in spike frequency for fast movements, like snapping, than for slow movements, such as walking, stalking, and turning (Borchers 1980; Borchers & Pinkwart 1983, see also Megela et al. 1983), there is presently no evidence that a "fixed action pattern" is introduced by a T8.1 neuron's corresponding "fixed discharge pattern." The spontaneous discharges of T8.1 neurons seem to reflect the animal's readiness to move, and there are hints that in periods of increasing spike frequency visually controlled behavior can be elicited more frequently.

5.2 Class T8.2 Neurons

These neurons stopped their spontaneous discharges shortly before and during the toad (in bright environment or in absolute darkness) closed its eyes, or moved head or legs spontaneously, or (in brightness) snapped toward prey (Fig. 10b); while the animal was sitting still, the neuron regained its spontaneous activity, initially rebounding above the former level and then settling (Borchers 1980). The property of T8.2 neurons is different from T5.2 in that a remarkable decrese of spontaneous activity introduces motor behavior.

Here we are reminded by the suppression of *off* activity in retinal R3 neurons mentioned earlier in this chapter. These phenomena are to some extent comparable to "saccadic suppression" obtained among *visual fixation cells* recorded from speciosa monkey's parietal cortex (Mountcastle 1976): A powerful suppression of the neuron's spontaneous activity begins 60 ms before the onset of a visually evoked saccadic movement, reaching a maximum just at eye movement, persisting until the eye settled steadily in its new position, after which there was a rebound facilitation. These suppressive effects that occur during ballistic movements may contribute to "visual stability" (MacKay 1970, 1973; Brooks & Fuchs 1975; Mountcastle 1976).

6 Tectal Wide Field Neurons

Similar to paralyzed toads, in freely moving preparations the visual ERF of tectal wide field neurons encompassed the frontal (class T2.1) the inferior (class T2.2), or the entire (class T4) visual field of the contralateral eye. The visual responses of these neurons could be rather variable, as far as (i) sensitivities to moving configural stimuli, (ii) "sameness" and "newness" characteristics regarding weak or strong stimulus-specific habituation (Lettvin et al. 1961; Ewert 1984), and (iii) shifting areas of maximal visual responsiveness in the large ERFs are concerned.

Recordings from T2 or T4 neurons of a toad sitting in front of a mealworm have shown that discharges are related to the speed and to the changes of movement and dynamic shape of the prey object (Borchers 1980). In response to prey an increase of neuronal activity was not necessarily correlated with prey-catching (see also Megela et al. 1983). The neurons were silent toward a static retinal image. Borchers (1980) investigated the question as to whether shifts of such retinal images, caused by respiration-coupled eye movements (Schipperheyn 1973), are monitored by a T2.2 neuron. He has found that eye movements coupled with ventilatory movements of the buccal floor did not produce shifting retinal images sufficient to activate T2.2 neurons. However, the eye movements associated with deep breathing involving the lateral body flanks led to retinal shifts that released short bursts of neuronal activity (interburst intervals of 1-3 s and interspike frequency up to 70 spikes/s), each respiration cycle eliciting one burst. This information can be used to monitor the stationary visual environment.

In summary, we suggest that T5.2 neurons deliver commands related to *goal oriented prey-selective attention*, tectal wide field T2 and T4 type neurons (at least part of) have something to do with *general visual arousal*, while T8-type neurons are the equivalent to *general motor activity*. (For definitions of arousal and attention see Laming, this volume).

Figure 10. ISFT histograms of spontaneously active and visual widefield neurons recorded from the toad's optic tectum. (a) Frequency pattern of a class T8.1 neuron that increases its spontaneous activity shortly before the toad begins a spontaneous body movement: leg movement (1), head turning (2), and walking (3). Insert: During walking each leg movement (see arrows) is introduced by a short neuronal burst. (b) Activity of a T8.2 neuron that decreases its spontaneous activity shortly before and during orienting toward a mealworm (see O) and snapping (see S). (c) Class T2.2 widefield neuron of a toad sitting in homogeneous environment in front of a small stationary object (a) during deep breathing cycles that induce retinal static images to shift positions which release corresponding bursts, and (b) during buccal respiratory movements eliciting no visual response. (From Borchers 1980).

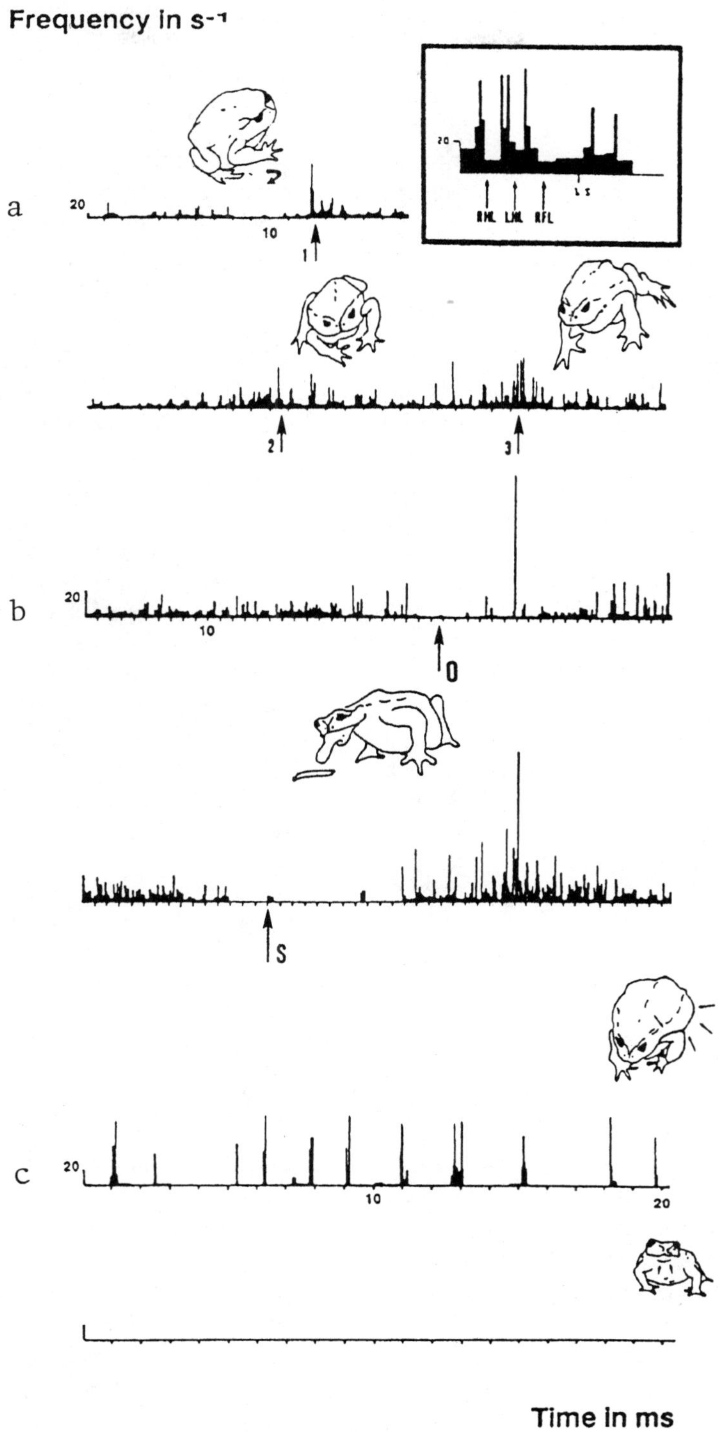
Frequency in s-1
a
20
10
1
RHL
LHL
RFL
2
3
b
20
10
O
S
c
20
10
20
Time in ms

7 Discussion and Conclusions

7.1 Tectal Command Functions

Electrical stimulation experiments in freely moving toads suggest that the different ballistic actions of prey-catching (orienting, approaching, binocular fixation, snapping) exist as prewired motor programs in the brain (Ewert 1967a,b): Focal activity in the optic tectum calls these programs into play, similar to a motor *command*. In visual stimulus situations, necessary condition for commanding prey-catching patterns is motivation and appropriate processing of the signals. The optic tectum is thus involved in sensory analysis integration of state-dependent inputs, and motor command functions for "visual grasping" (see Berkinblit, this volume, for comparable somatosensory functions in the spinal cord commanding "wiping" in response to cutaneous stimulation).

As described in detail by Ewert (1987, and this volume), converging lines of evidence derived from recording extracellular and intracellular potentials, lesioning brain areas by neurotoxins or direct current, retrograde cell labeling, antidromic cell activation, and functional staining with 2DG technique, support the hypothesis that certain excitatory and inhibitory interactions of tectal and thalamic/pretectal neurons are decisive for releasing prey-catching or avoidance behavior. Recalling the concept of "command releasing system" (CRS), certain combinations of different optimal coincidently activated classes of specialized neurons "command elements" cooperatively release the related behavioral response (Ewert 1980a,b, 1984; Schürg-Pfeiffer & Ewert 1981). Command elements are specified for stimulus feature analysis and localization, respectively, thus combining information from different sources about the same target; we refer to sensorimotor integration (see also Torras, this volume). Different combinations of command elements yield "sensorimotor codes" according to which motor pattern generating circuits for various action patterns are activated (see contributions by Ewert and Schwippert et al., this volume).

7.2 Command Elements for Toad's "Visual Grasp Reflexes"

The above concept suggests that class T5.2 neurons as command elements with configural prey recognition properties are shared by different CRSs addressed to the various action patterns of prey-catching. Single cell recordings in freely moving toads show that the activity of T5.2 neurons is positively correlated with orienting toward prey ("visual grasping"), approaching, and snapping ("tongue grasping") (Fig. 6a,b). That is, unlike retinal or tectal feature sensitive neurons, T5.2 prey-selective neurons are (optimally) matched to dynamic configural stimulus features that define a sign stimulus's prey-catching releasing value.

This does not deny that other classes of tectal [T5.1, T5.3, T5.4] and pretectal [TH3] neurons are involved in visual pattern recognition. These may fulfill various intrinsic functions within a global feature-analyzing network, and tecto-bulbar/spinal projective neurons of such types (Satou & Ewert 1985) could take part in CRSs for action pattern not related to prey-catching. For example, as shown in this study, it is unlikely that T5.3 neurons which prefer the antiworm (non-prey) to the worm (prey) configuration subserve as command element in a CRS for prey-catching.

The chronic recordings in behaving toads characterize T5.2 neurons by four important properties: Prey recognition, pre-motor activity, motor feedback, and state-dependent modulation:

Recognition properties are evidenced by the fact that in a prey-motivated toad the correlation between configural stimulus features and prey-catching is paralleled by the correlation between configural stimulus features and discharge frequency of T5.2 neurons. Pretectal lesion induced disinhibition yielding agnosia of these neurons is reflected in the toad's disinhibited prey-catching behavior. With reference to the CRS concept this means that the agnostic command elements are addressing prey-catching commands to inadequate targets. Probably, in intact toads, too, reduction of pretecto-tectal inhibitory input is required for appropriate activation of T5.2 cells, which is brought about optimally by a visual stimulus moving in worm configuration that maximally drives the tectal selfexcitatory networks in the presence of appropriate motivation.

Pre-motor properties become obvious in the ISFT histogram of T5.2 neurons. The onset of orienting or snapping toward prey is introduced by a "warming up" in the discharge frequency. We suggest that the pre-motor firing frequency must exceed the inter-spike frequency of intermingled "double-spikes" in order to elicit a prey-catching action pattern. Toads always start to move in the maximal or already declining phase of the IFST histogram. Spontaneous turning movements or orienting toward prey outside the ERF are not predicted by the recorded T5.2 neurons.

Motor feedback properties revealed by sustained discharges during orienting toward prey can be explained by positive feedback from the turn generating system which maintains the turning action. During snapping, T5.2 neurons are silent suggesting that neuronal activity is suppressed probably due to negative feedback from the snap generating system necessary for resetting. Hints for tecto-bulbar-tectal loops are provided by combined stimulation/recording studies (Ewert et al. 1988).

Modulatory properties are manifest in the correlation between the level of T5.2 neuron firing and prey-catching during short-term changes of the toad's responsiveness probably related to attentional shifts. Long-term state-dependent adjustments related to motivation, too, influence the level of firing and hence the pre-motor activity. It is interesting to note that configural selective properties of T5.2 cells are maintained in behaviorally unresponsive toads, but at a lower level of discharge frequency suggesting that "warming-up pre-motor activity" requires threshold operation. Presumably, certain input conditions (except excitatory inputs from R2/R3 and T5.1 cells) must be fulfilled to appropriately drive T5.2 neurons: (i) Reduction of inhibitory inputs from prosencephalic (telencephalic/ hypothalamic/pretectal) structures, and (ii) excitatory inputs from tegmental and bulbar reticular sources. A mechanism accounting for this modulation might be excitatory input from class T8.1 *movement cells* of deep tectal (and subtectal) layers, so that the response of T5.2 *prey-selective* cells of the tectal output layer would be enhanced if the activity in T8.1 overlapped in time with visual input. As to whether neuronal trigger efficacy takes advantage of temporal spike patterns (e.g., involving trains of "double-spikes") cannot as yet be answered.

From a broad view we suggest that T8 *movement cells* are generalized for motor activity ("motor arousal") in a comparable manner T2 and T4 *wide field neurons*, at least in part, are generalized for visual input ("visual arousal"), and that T5.2 *configural selective small field neurons* guide the toad's attention toward a specific goal, that is prey. (For mechanisms that underlie arousal see contributions by Laming, Finkenstädt, and Gonzalez-Lima, this volume). The question in which manner these neurons interact or cooperate, must be investigated by simultaneous recording from different types in the behaving toad. The present data also do not yet allow us to determine how prey-localization interferes, a topic tackeled by Grobstein (this volume) which also will receive support from parallel recordings of neurons in the freely moving animal.

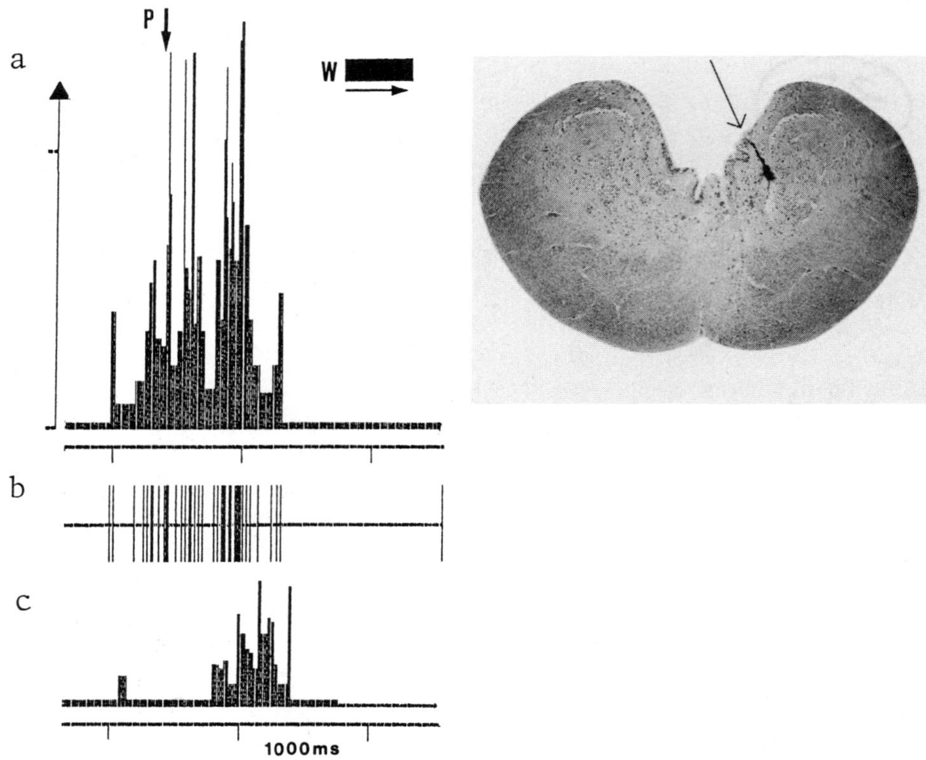

Figure 11. Correlation between extracellularly recorded activity of a single neuron in the toad's hypoglossal nucleus (a, b), tongue projection during snapping (P) as the tip of the tongue leaves the toad's mouth, and the summated electromyogram recorded from the neck musculature (c). In (a) the ISFT histogram and in (b) the neuron's spike train is shown. Snapping was elicited by means of a moving 2 x 20 mm^2 worm-like prey dummy. To fascilitate snapping to prey, the toad was pre-treated with apomorphin 10 min before recording. The photograph on the top depicts a histological transverse section of the hypoglossal nucleus showing the recording site of the stainless steel electrode marked (see dark label) by means of the Prussian blue reaction after passing an anodal direct current.

7.3 Command *vs*. Motor Functions

To study tectal *command* and bulbar *motor* activity in the snapping toad, we have recently also recorded the single cell responses from hypoglossal neurons. To fascilitate snapping toward prey, toads were pre-treated with a lymphatic injection of apomorphin (0.2 ml/40 g body weight). About ten minutes after this treatment, animals readily snapped at but did not orient toward (laterally placed) prey. (For details see Gladen 1984 and Quick 1984). Unlike following pretectal lesions, the visual prey-recognition in apomorphin-treated animals was quite normal. During snapping in response to a visual prey dummy, the recording electrode was advanced into the hypoglossal area (Fig. 11, top) until a single cell could be recorded whose activity correlated with tongue projection. Figure 11 shows the correlation between the activity of a single hypoglossal neuron (Fig. 11a,b), tongue projection (P) just as the tip of the tongue left the toad's mouth, and an electromyogram (Fig 11c) recorded from the neck musculature involved in the lunging movement. Similar to tectal T5.2 cells, strong hypoglossal cell activity always preceded snapping. However, while T5.2 cells were silent during snapping (Fig. 7), the hypoglossal cells discharged also during tongue projection. The neuronal events between tectal commands and snapping behavior are discussed by Weerasuriya (this volume).

7.4 Comparisons with Command Functions in Other Vertebrates

The concept of "command neuron," introduced to invertebrate neuroethology by Wiersma & Ikeda (1964) has undergone a kind of evolution in order to explain "command functions" for comparable tasks in vertebrates (e.g., see Eaton 1983; DiDomenico & Eaton 1987; Ewert, this volume). Common features of *command neuron* and *command releasing system* concepts are neurons (small population) that are target-related and goal-oriented.

In the superior colliculus of mammals neurons generate a relatively high-frequency spike burst before the onset of eye saccades. It is suggested that a specific subset of colliculus neurons must be activated in order to initiate a particular saccade. Data obtained by Ito (1986) in cats provide evidence of a sensory map in the superficial layer transformed into a motor map in the deep layers, in which a vector from an initial eye position to a goal eye position is represented. This is confirmed by McIlwain (1986) showing that directions and amplitudes of eye saccades are strongly influenced by the initial position of the eye in the orbit. In rhesus monkey, too, sensory signals in the superior colliculus seem to be organized in motor coordinates, so that maps of sensory space are dynamic, i.e., shifting with relative movements of eyes, head, and body (Sparks 1988). This hypothesis accords with results obtained from electric stimulation of the superior colliculus (Roucoux et al. 1980; Roucoux & Crommelinck 1987). In frogs, too, it is argued that head (and body) position signals must be integrated in the topographic correspondence between input and output (see Grobstein et al. 1983; and Grobstein, this volume).

Wurtz & Goldberg (1972) recording single unit activity from the superior colliculus of awake macaque monkey have found that *eye movement cells's* responses are always enhanced (warming and building up), when the animal uses a certain stimulus in the visual receptive field as a target for a saccade. Since this enhancement does not occur when the animal must saccade to another stimulus, the authors suggest that these cells are related to selective attention. Furthermore, this property is only displayed if the behavioral response is eye movement and not, e.g., hand movement. Wurtz (1979) suggests that eye movement cells are enhanced if activity in the *movement cells* of deep collicular layers overlaps in time with the responses from visual cells of superficial layers. We have adopted this hypothesis (Section 7.2) for putative connections between toad's tectal T5.1, T5.2, and T8.1 cells that may explain the pre-motor T5.2 warming-up. (Note that toads do not show voluntary tracking or overt eye movements).

Recording single unit activity from area 7 of the parietal lobe of speciosa monkey's association cortex Mountcastle et al. (1975) have described *visual tracking neurons* which selectively respond when the animal followed a rewarding food object with the eyes. The neuronal spike frequency increased almost to its maximal rate well before the first small on-to-track saccade and then decreased; a further tonic increase of activity led the eye movement. These are distinguished from *arm projection neurons* of area 5 that strongly discharge only before the monkey reaches forward with his arm to grasp a rewarding object. Here, too, neuronal activity begins to discharge before the arm starts to move, rises to a high level, then declines, but continues until the arm movement is completed. Motivation and attention play a prominent role in the activation of these neurons.

Similar to the toad's tectal T5.2 *prey-selective* neurons, monkey's parietal cortical *visual tracking* and *arm projection* neurons fulfill integrating tasks and - as components of circuits for directed visual attention (Mountcastle 1976) - they may exert command functions regarding "grasp reflexes." These commands are state-dependent and are thus formed by concurrent outputs from neuron populations of various sources in the brain that are not yet known in detail. In toads we suggest that the class T5.2 neurons and their inputs are necessary components of command releasing systems related to prey-catching.

Acknowledgements. I thank Mrs. G. Burggraf and Mrs. G. Kruk for technical assistance. The research was supported by the Deutsche Forschungsgemeinschaft Schu 526/1 - 1/2.

8 References

Akert K (1949) Der visuelle Greifreflex. *Helv Physiol Pharmacol Acta* 7: 112-134

Barlow H-B (1953) Summation and inhibition in the frog's retina. *J Physiol (Lond)* 119: 69-88

Borchers H-W (1980) *Untersuchungen zur Relation zwischen Verhalten und der Aktivität von Neuronen aus der retinotectalen Projektion freibeweglicher Erdkröten (Bufo bufo* L.*)*. Habil Thesis, Univ of Kassel

Borchers H-W (1982a) Correlation between behavior patterns and single unit responses from the optic tectum in freely moving toad (*Bufo bufo*). In: Trappel R, Ricciardi L, Pask G (eds) *Progress in Biocybernetics and Systems Res Vol 9*, Hemisphere Publication Corp, McGraw-Hill, Washington London, pp 109-117

Borchers H-W (1982b) Advances in the development of micro-recording systems and single unit analysis in toads (*Bufo bufo*). *J Theor Neurobiol* 1: 268-285

Borchers H-W (1982c) Behavior-correlated neurons in freely moving toads. *Neurosci Lett (Suppl)* 10: 84-85

Borchers H-W, Ewert J-P (1978) Eye closure in toads (*Bufo bufo* L) does not produce off responses in retinal on-off ganglion cells: a question of efferent commands. *J Comp Physiol* 125: 301-303

Borchers H-W, Ewert J-P (1979) Correlation between behavioral and neuronal activation of toads *Bufo bufo* (L) in response to moving configurational prey stimuli. *Behav Process* 4: 99-106

Borchers H-W, Pinkwart C (1983) A telemetry system for single unit recording in the freely moving toad (*Bufo bufo* L) In: Ewert J-P, Capranica RR, Ingle DJ (eds) *Advances in vertebrate neuroethology*. Plenum Press, New York, pp 603-612

Brooks BA, Fuchs AF (1975) Influence of stimulus parameters on visual sensitivity during saccadic eye movement. *Vision Res* 15: 1389-1398

Burghagen H, Ewert J-P (1982) Question of "head preference" in response to worm-like dummies during prey-capture of toads *Bufo bufo*. *Behav Process* 7: 295-306

DiDomenico R, Eaton RC (1987) Toward a reformulation of the command concept. *Behav Brain Sci* 10: 374-375

Eaton RC (1983) Is the Mauthner cell a vertebrate command neuron? A neuroethological perspective on an evolving concept. In: Ewert J-P, Capranica RR, Ingle DJ (eds) *Advances in vertebrate neuroethology*. Plenum Press, New York, pp 629-636

Evarts EV (1968a) Relation of pyramidal tract activity to force exerted during voluntary movement. *J Neurophysiol* 31: 14-27

Evarts EV (1968b) A technique for recording activity of subcortical neurons in moving animals. *Electroenceph Clin Neurophysiol* 24: 83-86

Evarts EV, Fromm C (1977) Sensory responses in motor cortex neurons during precise motor control. *Neurosci Lett* 5: 267-272

Ewert J-P (1967a) Elektrische Reizung des retinalen Projektionsfeldes im Mittelhirn der Erdkröte (*Bufo bufo* L). *Pflügers Arch* 295: 90-98

Ewert J-P (1967b) Aktivierung der Verhaltensfolge beim Beutefang der Erdkröte (*Bufo bufo* L) durch elektrische Mittelhirnreizung. *Z Vergl Physiol* 54: 455-481

Ewert J-P (1968) Der Einfluß von Zwischenhirndefekten auf die Visuomotorik im Beute- und Fluchtverhalten der Erdkröte (*Bufo bufo* L). *Z Vergl Physiol* 54: 455-481

Ewert J-P (1969) Quantitative Analyse von Reiz-Reaktionsbeziehungen bei visuellem Auslösen der Beutefang-Wendereaktion der Erdkröte (*Bufo bufo* L). *Pflügers Arch* 308: 225-243

Ewert J-P (1974) The neural basis of visually guided behavior. *Sci Amer* 230: 34-42

Ewert J-P (1980a) *Neuroethology: an introduction to the neurophysiological fundamentals of behavior.* Springer-Verlag, Berlin Heidelberg New York (German edition, Springer-Verlag, Berlin Heidelberg New York 1976; Japanese edition, Baifukan, Tokyo 1982; Chinese edition, Peking Science Press, Peking 1983)

Ewert J-P (1980b) Prey-catching sequence controlled by a multiple action system in toads. *Neurosci Lett (Suppl)* 5: 28

Ewert J-P (1984) Tectal mechanisms that underlie prey-catching and avoidance behaviors in toads. In:

Vanegas H (ed) *Comparative neurology of the optic tectum.* Plenum Press, London New York, pp 247-416

Ewert J-P (1987) Neuroethology of releasing mechanisms: prey-catching in toads. *Behav Brain Sci* 10: 337-405

Ewert J-P, Borchers H-W (1974) Antworten retinaler Ganglienzellen bei freibeweglichen Kröten. *J Comp Physiol* 92: 117-130

Ewert J-P, Hock FJ (1972) Movement sensitive neurones in the toad's retina. *Exp Brain Res* 16: 41-59

Ewert J-P, Wietersheim A v (1974a) Musterauswertung durch tectale und thalamus/praetectale Nervennetze im visuellen System der Kröte (*Bufo bufo* L). *J Comp Physiol* 92: 131-148

Ewert J-P, Wietersheim A v (1974b) Der Einfluß von Thalamus/Praetectum-Defekten auf die Antwort von Tectum-Neuronen gegenüber bewegten visuellen Mustern bei der Kröte (*Bufo bufo* L). *J Comp Physiol* 92: 149-160

Ewert J-P, Hock FJ, Wietersheim A v (1974) Thalamus/Praetectum/Tectum: Retinale Topographie und physiologische Interaktionen bei der Kröte (*Bufo bufo* L). *J Comp Physiol* 92: 343-356

Ewert J-P, Borchers H-W, Wietersheim A v (1978) Question of prey feature-detectors in the toad's *Bufo bufo* (L) visual system: a correlation analysis. *J Comp Physiol* 126: 43-47

Ewert J-P, Burghagen H, Schürg-Pfeiffer E (1983) Neuroethological analysis of the innate releasing mechanism for prey-catching behavior in toads. In: Ewert J-P, Capranica RR, Ingle DJ (eds) *Advances in vertebrate neuroethology.* Plenum Press, New York London, pp 413-475

Ewert J-P, Buxbaum-Conradi H, Framing EM, Schürg-Pfeiffer E, Weerasuriya A (1988) Extracellular activity of neurons of the toad's medulla oblongata in response to visual and tactile stimulation: a functional approach toward tectal/bulbar/spinal circuitry. (in prep)

Fromm C, Evarts EV (1977) Relation of motor cortex neurons to precisely controlled and ballistic movements. *Neurosci Lett* 5: 259-265

Gladen S (1984) *Der Einfluß der Apomorphindosis auf das Beutefangverhalten und auf die reizspezifische Gewöhnung bei Erdkröten.* Staatsexam Thesis, FB 19, Univ of Kassel

Gottschaldt K-M, Vahle-Hinz C (1979) A simplified method of making steel microelectrodes. *Pflügers Arch* 382: 52

Grobstein P, Comer C, Kostyk SK (1983) Frog prey-catching behavior: between sensory maps and directed motor output. In: Ewert J-P, Capranica RR, Ingle DJ (eds) *Advances in vertebrate neuroethology.* Plenum Press, New York London, pp 331-347

Grüsser O-J, Grüsser-Cornehls U (1968) Neurophysiologische Grundlagen visueller angeborener Auslösemechanismen beim Frosch. *Z Vergl Physiol* 59: 1-24

Grüsser O-J, Grüsser-Cornehls U (1976) Neurophysiology of the anuran visual system. In: Llinas R, Precht W (eds) *Frog neurobiology.* Springer-Verlag, Berlin Heidelberg New York, pp 297-385

Grüsser O-J, Grüsser-Cornehls U, Finkelstein D, Henn V, Patutschnik M, Butenandt E (1967) A quantitative analysis of movement detecting neurons in the frog retina. *Pflügers Arch* 293: 100-106

Hubel DH (1957) Single unit activity in visual cortex of the unanaesthetised cat. *Fed Proc* 16: 63

Hubel DH (1959) Single unit activity in striate cortex of unrestrained cats. *J Physiol (Lond)* 147: 226-238

Hubel DH (1960) Single unit activity in the lateral geniculate body and optic tract of unrestrained cats. *J Physiol (Lond)* 150: 91-104

Ito M (1986) Neural systems controlling movements. *TINS* 10: 515-518

Jasper H, Ricci GF, Doane B (1958) Patterns of cortical neuron discharge during conditioned responses in monkeys. In: Wolstenholme G, O'Connor C (eds) *Neurological basis of behaviour.* Little Brown, Boston

Lemon R, Prochazka A (1984) *Methods for neuronal recording in conscious animals.* Wiley, Chichester

New York Brisbane Toronto Singapore

Lettvin JY, Maturana HR, McCulloch WS, Pitts WH (1959) What the frog's eye tells the frog's brain. *Proc Inst Radio Eng NY* 47: 1940-1951

Lettvin JY, Maturana JY, Pitts WH, McCulloch WS (1961) Two remarks on the visual system of the frog. In: Rosenblith WA (ed) *Sensory communication.* MIT Press, Cambridge MA, pp 757-776

Lynch JC, Mountcastle YB, Talbot WH, Yin YCT (1977) Parietal lobe mechanisms for directed visual attention. *J Neurophysiol* 40: 362-389

MacKay DM (1970) Evaluation of visual threshold by displacement of retinal image. *Nature* 225: 90-92

MacKay DM (1973) Visual stability and voluntary eye movements. In: Jung R (ed) *Central processing of visual information. Handbook of sensory physiology,* Vol VII/3A. Springer-Verlag, Berlin Heidelberg New York, pp 307-332

Manning A (1979) *An introduction to animal behavior.* Springer-Verlag, Berlin Heidelberg New York

McIlwain JT (1986) Effects of eye position on saccades evoked electrically from superior colliculus of alert cats. *J Neurophysiol* 55: 97-112

Megela A, Borchers H-W, Ewert J-P (1983) Relation between activity of tectal neurons and prey-catching behavior in toads *Bufo bufo. Naturwissenschaften* 70: 100-101

Mountcastle VB (1975) The view from within: pathways to the study of perception. *The Johns Hopkins Med J* 136: 109-131

Mountcastle VB (1976) *The world around us: neural command functions for selective attention.* The FO Schmitt Lecture in Neuroscience 1975. Neurosci Res Progr Bull 14. MIT Press, Cambridge MA, pp 1-47

Mountcastle VB, Lynch JC, Georgpoulos A, Sakata H, Acuna C (1975) Posterior parietal association cortex of the monkey: command functions for operations within extrapersonal space. *J Neurophysiol* 38: 871-908

Phillips MI (1973) *Brain unit activity during behavior.* Charles C Thomas Press, Springfield Ill

Pinkwart C, Borchers H-W (1987) Miniature three-function transmitting system for single neuron recording, wireless brain stimulation, and marking. *J Neurosci Methods* 20: 341-352

Quick I (1984) *Auswirkungen von Apomorphin auf die visuelle Musterdiskriminationsfähigkeit und die Beutefangmotivation bei Erdkröten.* Staatsexam Thesis, FB 19, Univ of Kassel

Roucoux A, Guittan D, Crommelinck M (1980) Stimulation of the superior colliculus in the alert cat. II. Eye and head movements evoked when the head is unrestrained. *Exp Brain Res* 39: 75-85

Roucoux A, Crommelinck M (1987) Sensorimotor maps in the tectum. *Behav Brain Sci* 10: 386-388

Salcman M, Bak MJ (1973) Design, fabrication and *in vivo* chronic recording intracortical microelectrodes. *IEEE Trans Biomed Eng* 20: 253-260

Satou M, Ewert J-P (1985) The antidromic activation of tectal neurons by electrical stimuli applied to the caudal medulla oblongata in the toad, *Bufo bufo* (L). *J Comp Physiol* 157: 739-748

Schipperheyn JJ (1973) Respiratory eye movement and perception of stationary objects in the frog. *Acta Physiol Pharmacol Neerl* 12: 157-159

Schürg-Pfeiffer E (1983) *A simple method of testing tungsten microelectrodes by their galvanic properties.* Laboratory Report, Neuroethol Dept, FB 19, Univ of Kassel

Schürg-Pfeiffer E, Ewert J-P (1981) Investigation of neurons involved in the analysis of gestalt prey features in the frog *Rana temporaria. J Comp Physiol* 141: 139-152

Schürg-Pfeiffer E, Ewert J-P (1987) Correlation between responses of prey-selective T5.2 neurons and prey-catching in freely moving toads. In: Elsner N, Creutzfeldt O (eds) *New frontiers in brain research.* Thieme-Verlag, Stuttgart

Schürg-Pfeiffer E, Ewert J-P (1988) Correlation between tectal single neuron activity and prey-catching in freely moving toads (*in preparation*)

Schürg-Pfeiffer E, Spreckelsen C, Ewert J-P (1988) Correlation between behavior and neuronal activity in the medulla oblongata of toads. In: Elsner N, Creutzfeldt O (eds) *New frontiers in brain research.* Thieme-Verlag, Stuttgart

Sparks DL (1988) Neural cartography: sensory and motor maps in the superior colliculus. In: Finger TE (ed) *Neural cartography: how does the CNS use sensory maps?* Karger, Basel, pp 49-55

Spreckelsen C (1987) *Relation zwischen neuronalen Aktivitäten in der Medulla oblongata und Bewegungsmustern bei der Erdkröte.* Dipl Thesis, Univ of Marburg

Wiersma CAG, Ikeda K (1964) Interneurons commanding swimmeret movements in the crayfish *Procambarus clarkii* (Girard). *Comp Biochem Physiol* 12: 509-525

Wurtz RH (1979) Modulation of the primate visual system by attention and readiness to respond. In: Kandel ER, Krasne FB, Strumwasser F, Truman JW (eds) *Cellular mechanisms in the selection and modulation of behavior.* Neurosci Res Progr Bull 17. MIT Press, Cambridge MA, pp 562-576

Wurtz RH, Goldberg ME (1972) Activity of superior colliculus in behaving monkey. III: Cells discharging before eye movements. *J Neurophysiol* 35: 575-586

Visual Integration in Bulbar Structures of Toads: Intra/Extra-Cellular Recording and Labeling Studies

WOLFGANG W. SCHWIPPERT, THOMAS W. BENEKE, and EDDA M. FRAMING

Abteilung Neuroethologie, Fachbereich Biologie/Chemie, Universität Kassel
D-3500 Kassel, FR Germany

Abstract. The concept of coding by "command releasing systems" raises the question of how visual information is decoded in post-tectal structures of the medulla oblongata. In the present study, extra- and intracellular activities of neurons in the bulbar reticular formation (RetF) and in nuclei of the branchiomotor column (BMC) were analyzed in response to visual stimulation as well as to electrical stimulation of the optic tectum, the brain stem, or the spinal cord in slightly paralyzed toads. Similarities between tectal and bulbar visual neurons, based on receptive field sizes and responses to moving configural stimuli, suggest various kinds of integrating properties in the RetF and BMC. Besides configural selectivities, particular characteristics of bulbar neurons were tonic activity, or warming-up and burst activity in response to specific input. Transformation of a visual input into a standard pre-motor activity was shown by cells that displayed cyclic bursts as a post-stimulus event. Camera-lucida reconstructions of intracellularly labeled neurons revealed five morphological types distinguished by size, orientation, and extension of dendrites. Most labeled cells in the RetF and BMC showed dendritic trees with bewildering arborizations suitable to pick up inputs from different sites of the grey and white matter. Results obtained in amphibians and in mammals are discussed.

1 Introduction

In all vertebrates - from fish to primate - the optic tectum is the main projection field of retinal fibers. The tectal efferents are essential for the control of a variety of action patterns involving eye, head, and body movements. The question of the intervening processes between tectal output and motor activity has been tackled by studies applying anatomical (e.g., see Weerasuriya, this volume) and lesion techniques (see Grobstein, this volume). To understand the fundamental rules for the analysis of visual signals, the integration of sensory information as a basis for motor pattern generating processes, and the transformation into appropriate motor signals, electrophysiological investigations at the cellular level are indispensible. Extracellular recording and intracellular recording/labeling studies in toads and frogs have shown that tectal neurons are integrated in a complicated network extending across different cell layers and involving pre- and post-tectal structures (see Ewert, Manteuffel, Gruberg, Lázár, and Matsumoto, this volume). The tectal outflow to pre-motor/motor systems of the medulla oblongata and the spinal cord is mediated by classes of tectal neurons which have been studied anatomically by retrograde transport of HRP (Weerasuriya & Ewert 1981; Weerasuriya, this volume) and of Co^{3+}-lysine (Lázár et al. 1983; Tóth et al. 1985), by anterograde intracellular labeling with Co^{3+}-lysine (Matsumoto et al. 1985, 1986; Antal et al. 1986; Matsumoto, this volume), and physiologically by antidromic stimulation techniques (Satou & Ewert 1985). The results of injecting HRP or Co^{3+}-lysine or applying electric current to the medulla clearly show that at least three morphologically distinct types of tectal cells (ganglionic, pyramidal, and pear-shaped neurons) and at least eight classes of neurons with physiologically distinct properties (T1-4 and T5.1-4) project their axons from the tectal layers 6/7 to the medulla oblongata. Signals leaving the tectum via the tecto-bulbar/spinal tracts seem to express the result of information processing in functional units which have been described morphologically by Lázár (this volume) and physiologically by Ewert and Schürg-Pfeiffer (this volume). Ewert's concept of "command releasing systems" suggests that distinct combinations of tectal output neurons yield coded commands for motor pattern generating systems.

Which post-tectal brain stem structures receive monosynaptic descending tectal inputs? Of special interest here are two main descending tectal pathways, the tractus tecto-bulbaris et spinalis cruciatus (tr.tbs.c.) and the tractus tecto-bulbaris directus (tr.tb.d.).

The tr.tbs.c. innervates the nucleus interpeduncularis and the nucleus lateralis profundus at midbrain level (Rubinson 1968). The same tract probably feeds into other tegmental structures of which the anterodorsal, anteroventral, posteroventral nuclei, and the nucleus of the medial longitudinal fasciculus are discussed as (part of) a system responsible for visual space-coding concerning stimulus elevation, horizontal eccentricity, and depth (Grobstein, this volume). The tr.tbs.c., traveling around the periphery of the midbrain and crossing at its base (ansulate commissure), descends caudally in a ventromedial position of the brain stem. Many cells of the top of layer 6 of the lateral tectum arborize in toral and tegmental areas and project their axons within tr.tbs.c. toward the reticular formation (RetF) of the contralateral medulla oblongata (Fig. 1C) (Matsumoto et al. 1986; Schwippert et al. 1988) and to the medial portion of the ventral spinal grey up to the level of the brachial enlargement. According to the results from degeneration studies after unilateral tectal lesions (Rubinson 1968), the tecto-spinal tract tr.tbs.c. descends entirely contralateral and is confined to the medial ventral spinal grey, reaching the cervical enlargement. Rubinson (1968) and Ebbesson (1976) point out that these tecto-fugal projections probably provide the anatomical substrate for neck movements and upper limb control involved in visually guided orientation (see also Yerkes 1903), which has been confirmed by lesion experiments (Ingle 1983). The fact that prey-selective tectal T5.2 cells could be backfired in response to electrical stimulation of the contralateral ventromedial brain stem, caudal (Satou & Ewert 1985) or rostral (Fig. 2) to the obex, is evidence that tr.tbs.c. carries prey feature specific information.

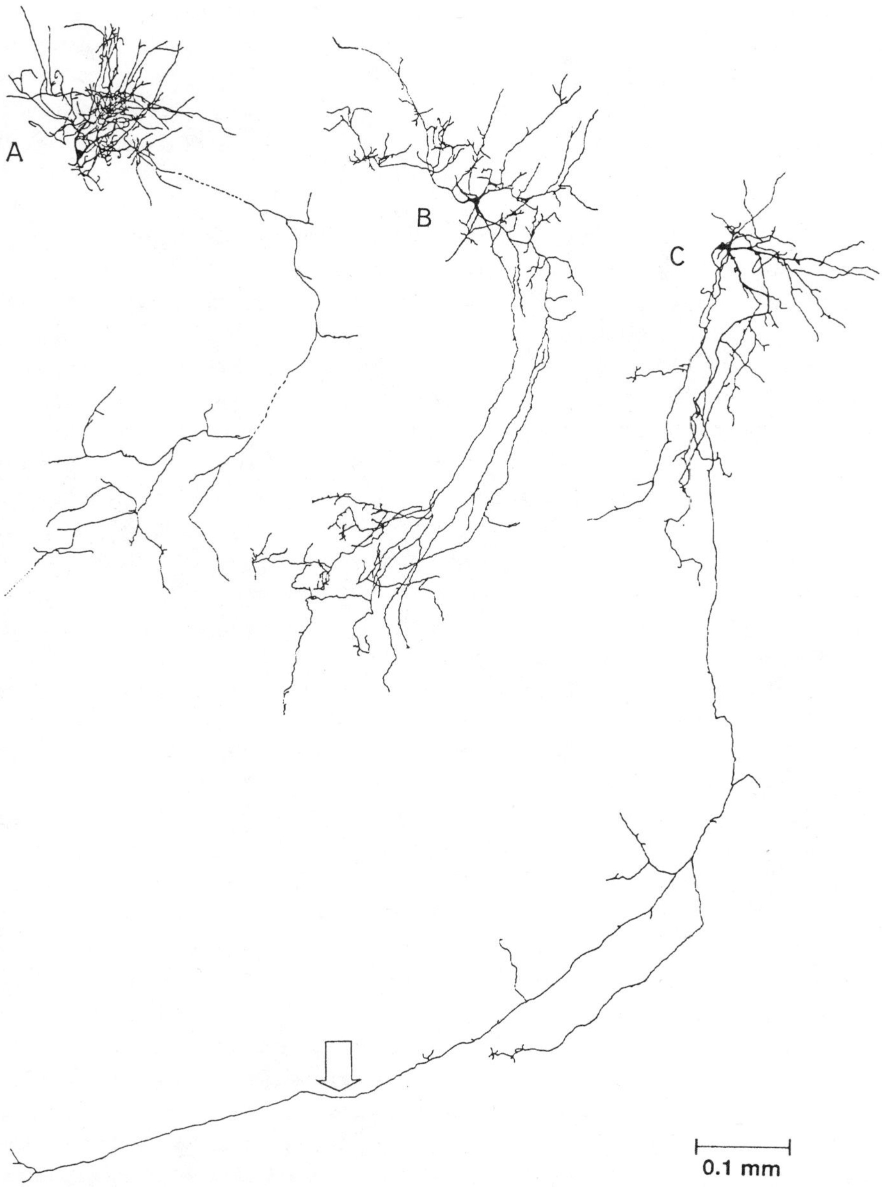

Figure 1. Three examples of projection neurons in the lateral optic tectum of the common toad *Bufo bufo spinosus.* These visual cells were intracellularly recorded, stained iontophoretically with cobalt-lysine, and reconstructed by means of the camera-lucida technique. The somata of two cells (A, B) are located in tectal layer 8; their dendritic trees extend into the superficial layer 9; axonal arborizations travel to the ipsilateral torus semicircularis or the nucleus isthmi. The long axon of another neuron (C), whose soma is located on the top of layer 6, projects towards the contralateral medulla oblongata (arrow indicates the ansulate commissure); some fiber branches terminate also in toral and in various tegmental structures. (From Schwippert et al. 1988).

The tr.tb.d., which is smaller than the tr.tbs.c., travels ipsilateral and sends fibers to the nucleus isthmi, the superior olivary nucleus, and the neuropil of the lateral rhombencephalic basal plate. In the medulla oblongata, this tract runs ventrolaterally and its fibers supply the ipsilateral and the contralateral medial RetF. (For anatomic evidence see Rubinson 1968; Lázár 1969; Ingle 1983; for a review see Nieuwenhuys & Opdam 1976). Our intracellular recording and labeling studies have shown that cells of layer 8 of the lateral tectum send axon collaterals to the nucleus isthmi, to toral, or to tegmental areas (Fig. 1A,B.).

The RetF - the main projection field of fibers running in tr.tbs.c. and tr.tb.d. - consists of lateral, medial, and median longitudinal zones (see Fig. 5a). The idea that the medial RetF is involved in motor pattern generation has a long tradition (Gaupp 1896; Edinger 1908; Röthig 1927; Herrick 1930; Tuge 1932). The medial RetF represents a continuous zone extending from the tegmentum through the medulla oblongata and including medium-sized and large cells located at the basal plate, the border between grey and white matter (see deep layer 3 in Fig. 5a). These cells have been termed "medial-ventrale Zellen" (medioventral cells) by Gaupp (1896) and "motorische Schaltzellen" (motor relay cells) by Röthig (1927). Edinger (1908) assumed that these cells send axons to bulbar and spinal motor nuclei suggesting

> "that these elements constitute the chief center of motor coordination and designated them together as the nucleus motorius tegmenti. It seems likely that it was on the basis of Edinger's ideas that Röthig characterized the large reticular units as motor relay cells. The term nucleus motorius tegmenti has been employed by Herrick (1930) in his description of the rhombencephalon of the urodele *Necturus*. ... He included in this nucleus ... also a considerable part of the small-celled periventricular grey of the basal plate. Ariens Kappers et al. (1936) concurred with Edinger (1908) in the opinion that the large reticular elements of the brain stem play an important role in the coordination of motor activity" (Nieuwenhuys & Opdam 1976, p.842).

Figure 2. Neurophysiological evidence of a tecto-bulbar/spinal class T5.2 projection neuron in the common toad. In the paralyzed animal, the neuron was extracellularly recorded from the top of layer 6 of the rostral optic tectum and physiologically determined by its response to configural moving visual stimuli. This neuron could be activated antidromically in response to a 0.1 ms electrical square wave impulse *e* (see vertical arrow) applied to the tractus tecto-bulbaris et spinalis cruciatus in the contralateral medulla oblongata at the level of the motor nucleus of the VII*th* cranial nerve. (a) Constant latency response of the antidromic spike *a* at 2.3 ms (several superimposed traces), two traces just below response threshold to determine the all-or-nothing reaction. (b) Collision test in which a visually elicited spike *v* triggered the electrical stimulus *e* that, following a stimulus artifact, evoked an antidromic spike *a* at a spike(*v*)-stimulus(*e*) delay of 9.5 ms (b_1); the antidromic spike was extinguished (b_2) at a critical spike(*v*)-stimulus(*e*) delay D = 3.3 ms that meets the collision criterion of Fuller & Schlag (1976), according to which the difference between the calculated delay C (= L + R) and the measured delay D must be less than 50 %: $[C - D]/C < 0.5$ (latency L = 2.3 ms; absolute refractory period R = 3.2 ms). (C) The refractory period was measured with electrical double stimuli e,e' of variable intervals. (J.-P. Ewert, unpublished).

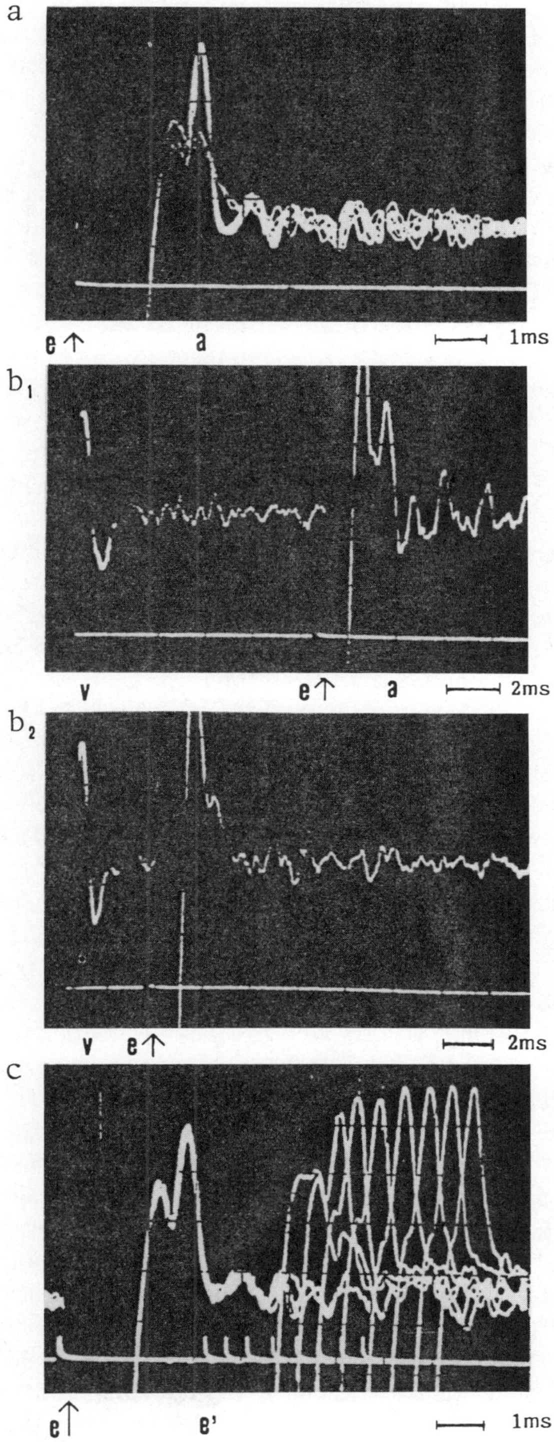
a
e↑
a
1ms
b1
v
e↑
a
2ms
b2
v
e↑
2ms
c
e↑
e'
1ms

Opdam et al. (1976) divide the medial RetF in three zones:
(a) The nucleus of the fasciculus longitudinalis medialis is situated in the more rostral part of the brain stem whose medium-sized cells are distributed in the most superficial part of the tegmental central grey.
(b) The nucleus reticularis isthmi is situated rostral and extends to the isthmic region and into the tegmentum mesencephali.
(c) The rhombencephalic medial RetF is a continuous column extending throughout almost the entire rhombencephalon. In sagittal direction, three nuclei are distinguished: nucleus reticularis *superior* (rostral), *medius* (at the level of the motor nucleus of the VIII*th* cranial nerve), and *inferior* (at the level of the motor nucleus of the X*th* nerve). Intracellular recordings from identified tongue muscle motoneurons of the toad's nucleus hypoglossus (XII*th*) in response to electrical stimulation of the optic tectum (Satou et al. 1984) have provided evidence of bilateral polysynaptic pathways, in which neurons of the medial RetF appear to serve as interneurons (for anatomic detail see Weerasuriya, this volume). Applying the same method, Manteuffel & Naujoks-Manteuffel (1987) suggest that fire salamender's neck muscle motoneurons, too, receive tectal input via at least one additional synapse while a few motoneurons seem to be directly coupled with visual centers. It is known that neurons of the rhombencephalic medial RetF receive inputs from the tectum (via tr.tbs.c. and tr.tb.d.), the pretectum, and the spinal cord, and there is also anatomic evidence of a reticulo-spinal system (Mensah 1974; Mensah et al. 1974; ten Donkelaar et al. 1981; Weerasuriya, this volume). Anatomical (Mensah 1974) and physiological (Cruce 1974) data indicate that cells of the medial RetF project to the cervical level of the spinal cord, cells of the lateral RetF principally to lumbar regions, while the magnocellular cell groups are projecting to the entire cord. Ebbesson (1976) argues that bilateral arrangement of ipsi- and contralateral tecto-reticular projections may allow a rather simple excitatory-inhibitory interaction required for head turning.

Whereas the function of the bulbar RetF is as yet largely speculative in amphibians, more results are available for mammals. The superior colliculus (homologous to the optic tectum) projects to different areas of the mesencephalic and pontine RetF, to the dorsolateral pons, and to the nucleus reticularis tegmenti pontis. Axons traveling via tr.tbs.c. to the pontine/bulbar medial RetF are monosynaptically connected to reticulo-spinal projection neurons (Peterson et al. 1975; 1978; Peterson & Fukushima 1982; Eckmiller 1987) which in turn are connected to spinal motoneurons (Anderson et al. 1971). Motoneurons innervating neck muscles receive at least one inhibitory and two excitatory inputs. Numerous investigations support the assumption that especially the more rostral bulbar reticular neurons "mediate" collicularly evoked orienting responses in terms of complex and coordinated eye-head movements (Peterson et al. 1978; Peterson & Fukushima 1982). In the sensorimotor coupling of cat's eye movements, bulbar bursting cells are of particular interest. To produce saccades, specific collicular output is fed into bulbar neurons that respond with burst activity if specific conditions of integration (spatiotemporal summation) are fulfilled. Certain bursting neurons have pre-motor properties and activate the corresponding motoneurons of the oculomotor nucleus (Robinson 1981). (See also Sections 4.3.2 and 4.3.3).

Which kind of information reaches single bulbar neurons? In toads, the concept of "command releasing system" suggests that internunclear (motor pattern generating) networks integrate inputs from various origins, such as *tectal* (e.g., locus-specific, feature-coded information), *tegmental* (e.g., space-coded information), and *vestibular* (proprioceptive information) (see contributions by Ewert, Finkenstädt, Grobstein, and Weerasuriya, this volume). In the present study, visual input to single neurons of the rostral RetF and the branchiomotor column (BMC) of the toad's medulla oblongata was investigated by applying extracellular and intracellular recordings in combination with stimulating and labeling techniques. In particular, the following questions arise: (1) How are receptive fields of visual neurons organized regarding size, shape, and localization? (2) To which extend are configural sensitivities of tectal neurons retained in bulbar neurons? (3) What are the intrinsic bulbar properties? (4) Are there interactions between optic tectum, medulla oblongata, and spinal cord? (5) Are there any relations between morphological and physiological properties of bulbar neurons? (6) Can comparisons be drawn between properties of bulbar neurons in amphibians and mammals?

2 Extracellular Approach toward Properties of Bulbar Neurons

By means of extracellular studies (Ewert et al. 1984, 1989; Framing 1986) single neurons were recorded mainly from the rostral bulbar medial RetF of slightly succinylcholine-paralyzed toads *Bufo bufo spinosus* in response to visual stimulation or to electrical stimuli applied to the optic tectum, to the medulla oblongata adjacent to the obex, or to the cervical spinal cord. In the following, the results are reviewed. (For anatomic investigations of bulbar structures see Weerasuriya, this volume).

To characterize bulbar neurons in response to visual input, six kinds of configural moving stimuli were tested: 2° x 2° square (S2), 8° x 8° square (S8), 16° x 16° square (S16), 32° x 32° square (S32), 2° x 16° worm-like moving stripe (W), 16° x 2° antiworm-like moving stripe (A). The black stimuli were moved at 10°/s against a white background in a perimetric hemisphere (for methodological details see Ewert & Borchers 1971; Ewert 1984). In addition, a neuron's response was tested toward a moving 15 cm x 25 cm background texture consisting of rectangular black and white patches in random distribution according to Julesz (1976); the minimal texture width was 5 mm. Standard electrical stimulation and recording techniques were used (see Ewert 1984; Satou & Ewert 1985).

During the intra- and extracellular recordings from the rostral medial and lateral RetF and the BMC different types of neurons could be distinguished whose response properties partly showed a surprising similarity with classes of tectal T1-9 neurons (Grüsser & Grüsser-Cornehls 1976; Ewert 1976; Ewert 1984), so that a corresponding characterization of medullary M-type neurons is reasonable (Figs. 3 and 4). However, we hasten to admit that this should not imply that the characteristics of M-neurons are necessarily determined by inputs of related T-neurons. Furthermore, the M-numbering system does not necessarily refer to separate classes of cells rather to properties displayed by these cells under the present criteria. (For a discussion on this topic see Section 4.2).

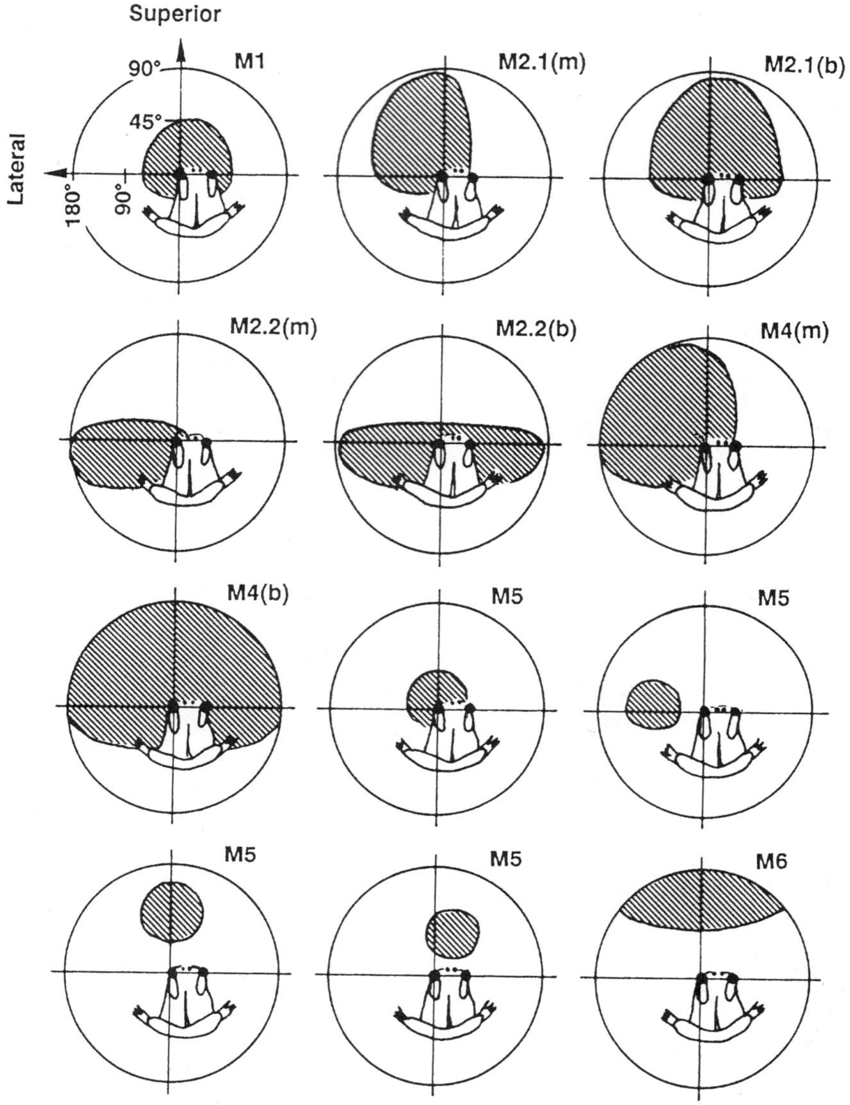

Figure 3. Arrangements and sizes of excitatory receptive fields, ERFs, of M-type neurons recorded extracellularly from the medial reticular formation of the common toad's medulla oblongata. The approximate ERFs are shown in schematic diagrams of the animal's visual field. M1: binocular smallfield neuron; M2.1: widefield neuron with a frontal monocular, m, or binocular, b, ERF; M2.2: widefield neuron with a horizontal to inferior monocular, m, or binocular, b, ERF; M4: monocular, m, or binocular, b, total field neuron; M5: different monocular, so-called smallfield neurons whose ERFs have different locations in the visual field; M6: widefield neuron with a large superior, mostly binocular, ERF. (From Ewert et al. 1989).

The histological evaluation of the recording sites (Fig. 5b) showed that most neurons with visual input were recorded from the ventral grey of the rostral (superior) rhombencephalic medial RetF which is the region of Röthig's (1927) "motorische Schaltzellen." Neurons were also recorded from the adjacent BMC. Caudally, there was a decrease in the frequency of recording visual sensitivity in favour of somato-sensory input, whereas in lateral positions auditory responses could be recorded.

2.1 "Monocular Small Field" Neurons

The T5 properties displaying M5 type neurons have a monocular excitatory receptive field (ERF) of 40 to 90° diameter which is relatively "small" for bulbar neurons (Fig. 3). The ERF of these neurons was typically located in the visual field of the eye ipsilateral (seldom contralateral) to the recording side. Although the ERFs of different neurons could be located in various parts of the visual field, these were found mostly in frontal regions. The ERF typically included a small area of 10 to 15° diameter in which the neuron was particularly responsive. According to their

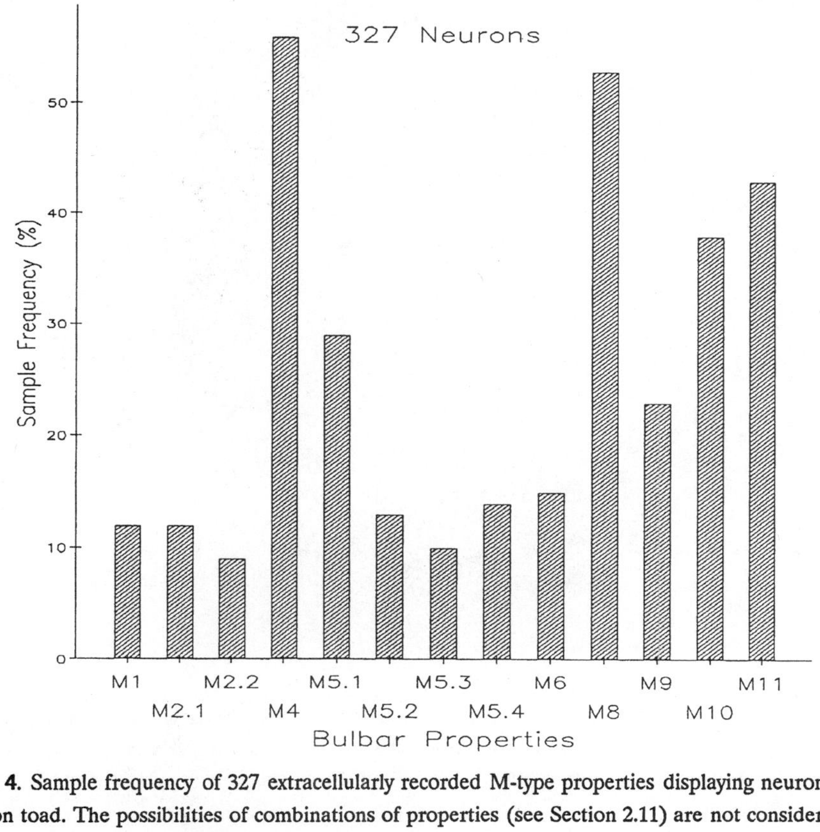

Figure 4. Sample frequency of 327 extracellularly recorded M-type properties displaying neurons in the common toad. The possibilities of combinations of properties (see Section 2.11) are not considered here. For explanations see text. (From Ewert et al. 1989).

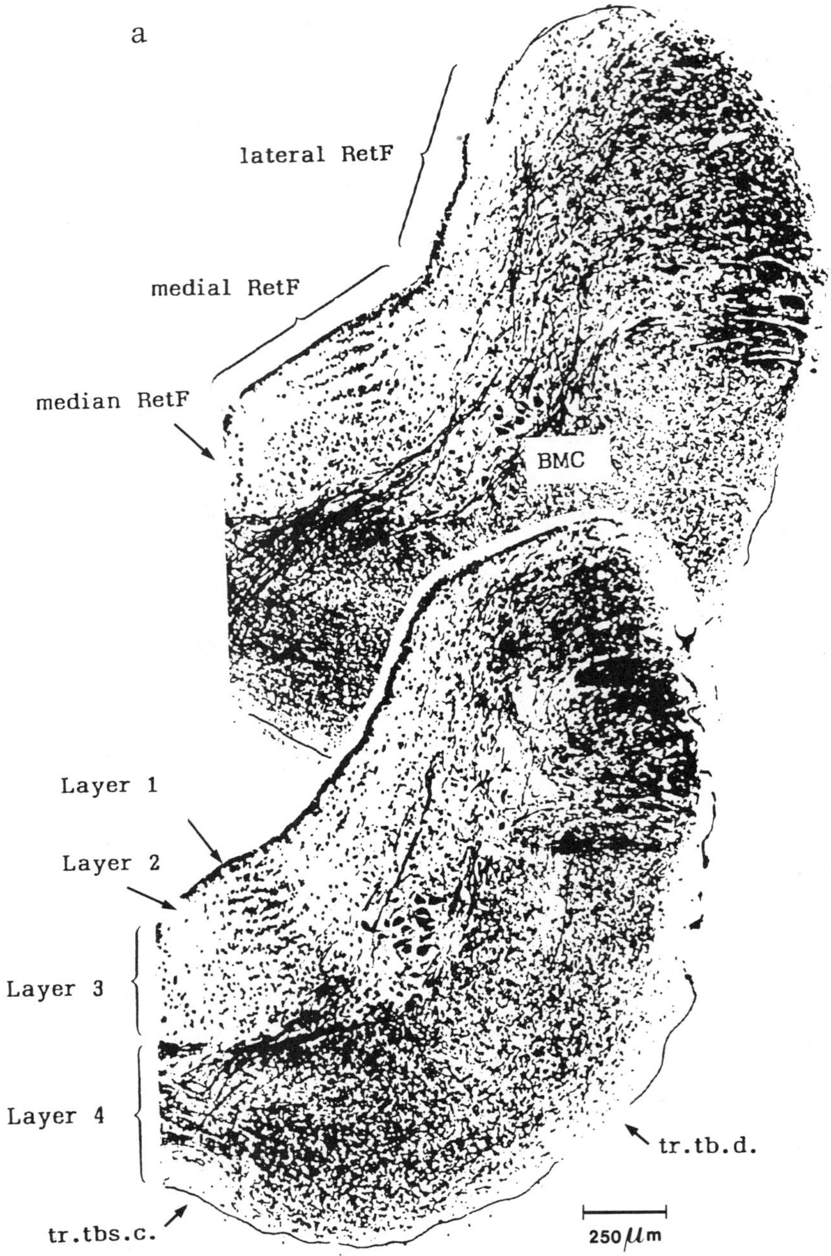

Figure 5. Localization of neurons recorded extracellularly from the medial reticular formation of the paralyzed common toad's medulla oblongata at levels of the motor nuclei of the V*th* (trigeminal) and the VII*th* (facial) cranial nerves. (a) Bulbar histology shown in half transverse sections stained with the Klüver-Barrera technique. (b) Examples of recording sites of various bulbar M-type neurons; the numbers refer to the response types M1-M11. The tips of the stainless steel recording electrodes were marked by means of

b

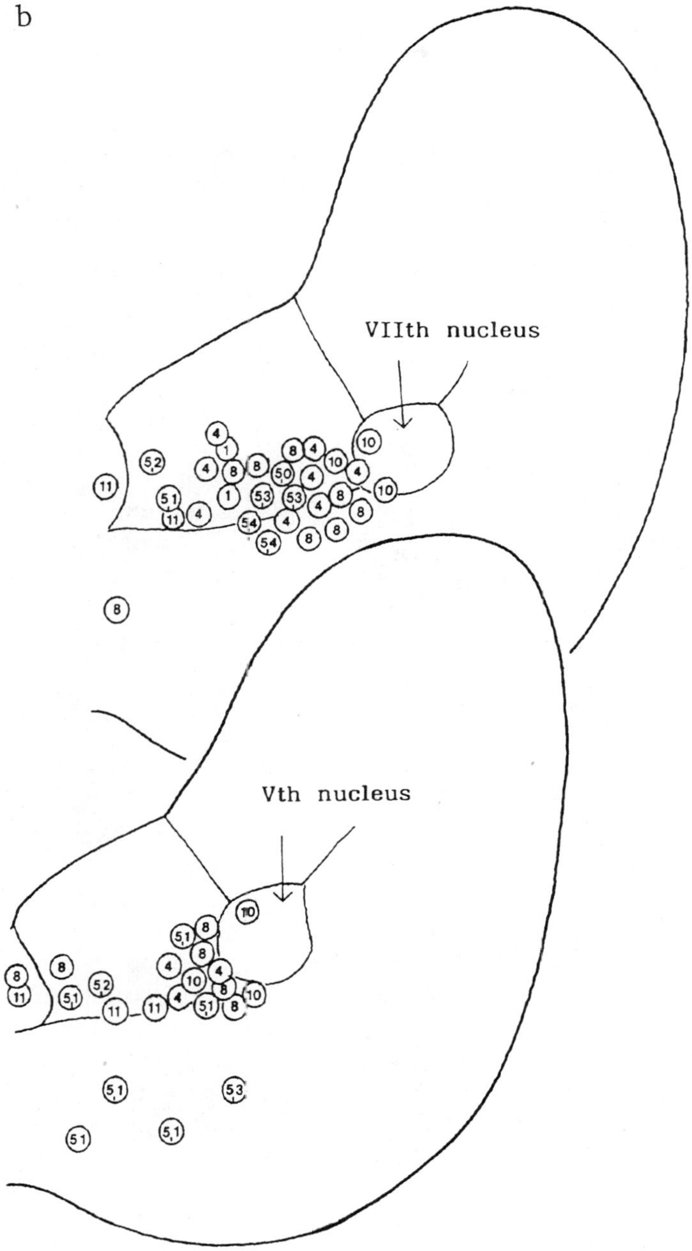

(Fig. 5, continued) the Prussian Blue reaction after passing an anodal direct current. Layer 1: stratum ependymale; layer 2: stratum subependymale; Layer 3: stratum griseum; layer 4: stratum album. The V*th* and VII*th* motor nuclei are located in the branchiomotor column, BMC; RetF, divisions of the bulbar reticular formation; tr.tb.d., tractus tecto-bulbaris directus; tr.tbs.c., tractus tecto-bulbaris et spinalis cruciatus. (From Framing 1986; Ewert et al. 1989).

different sensitivities to moving configural stimuli (S, W, A of comparable edge length), these neurons - in analogy to tectal T5-types - could be distinguished as

M5.1 $(R_S > R_W > R_A)$

M5.2 $(R_W > R_S > R_A)$

M5.3 $(R_S > R_A > R_W)$

M5.4 $(R_{S8} > R_{S4} > R_A = R_W = 0)$

in which R is the average discharge frequency. Whereas most of the M5.3 and M5.4 neurons were also activated by a moving background texture, the M5.1 and M5.2 neurons - similar to T5.1 and T5.2 (Tsai & Ewert 1988) - typically failed to respond. A case in which synaptic activities of a M5.2 neuron were recorded in response to electrical stimulation of the ipsilateral medulla (Fig. 6Bb,c), or to stimulation of the contralateral optic tectum, suggests that these cells integrate tectal and bulbar inputs. Antidromic stimulation/recording techniques have shown that T5-type neurons project to the medulla oblongata (cf. Fig. 2 for an example; see also Satou & Ewert 1985). It is possible that (some) M5.3 neurons resembling thalamic pretectal TH3 cells are driven by pretecto-bulbar projections described by Weerasuriya & Ewert (1981) and Ingle (1983) (see also Lázár and Matsumoto, this volume).

2.2 "Binocular Small Field" Neurons

M1 neurons have relatively small binocular frontal ERFs of 40 to 90° diameter (Fig. 3) and respond well to small moving objects (e.g., a S8-stimulus). The discharge rate to the A-stimulus was less than that to W or S8. By means of an eye cover procedure it could be shown that M1.1 neurons were activated by stimulation of either the ipsilateral or the contralateral eye (in a manner of an *OR*-gate) if a relatively small object was moved in a distant position. M1.2 neurons responded best to a moving object close to the toad. M1.3 neurons were activated only if corresponding areas of both retinae were stimulated simultaneously (*AND*-gate characteristic) by an object in a narrow distance to the eyes. The configural response property was similar to M1.1 and M1.2. After long-term monocular covering, the *AND* condition sometimes changed to *OR*. Taking together, these neurons strongly resembled the properties of tectal T1-type neurons.

2.3 "Frontal Wide Field" Neurons

M2.1 neurons typically show a monocular ERF with a frontal elevation up to 90° and a lateral extent of 90 to 120° (Fig. 3, M2.1[m]). The ERF of some neurons was located in the visual field of the eye ipsilaterally to the recording site and in others in the visual field of the contralateral eye. The responsiveness was strongest in a frontal section of about 40° diameter. Many neurons had binocular visual ERFs (Fig. 3), and they showed often a bilateral cutaneous somatosensory receptive field (in monocular neurons ipsilateral to the visual ERF) that mainly exhibited tactual sensitivity in the rostral part of the body skin. To moving visual stimuli, monocular M2.1 neurons usually displayed a configural sensitivity similar to that of T5.1. The neurons with binocular ERFs preferred large stimuli and responded well to a moving textured background. Many neurons were sensitive to repetitive ERF traverses with the same visual stimulus, hence displaying "sameness" properties; other neurons habituated to successive movement of the same stimulus but responded well to a new stimulus, hence showing "newness" properties, as has been described for tectal widefield neurons by Lettvin et al. (1959). Few cells were directionally selective to object motion from extrafrontal to frontal areas of the ERF. Taking together, the visual response properties of these bulbar neurons are similar to tectal T2.1 neurons (see Ewert 1984 for a review).

2.4 "Horizontal Wide Field" Neurons

M2.2 neurons have a monocular (ipsi- or contralateral) ERF of about 200° in horizontal rostro-caudal extension; its inferior extent measured from the horizon is about 40° (Fig. 3, M2.2[m]). Some neurons in addition received input from the body surface, ipsilateral to the visual ERF, particularly in the region of the hind extremities. There were also neurons whose ERFs had a corresponding bilateral extension (Fig. 3, M2.2[b]). All these neurons showed a configural sensitivity similar to T5.1. Few neurons responded directionally selective if an object was moved from the horizon to inferior field positions. The visual responses of these bulbar neurons resemble properties of tectal T2.2.

2.5 "Dorsal Wide Field" Neurons

M6 neurons are distinguished from M2.1 and M2.2 widefield neurons by a large ERF of about 120° diameter located in the dorsal visual field of the ipsilateral and/or the contralateral eye (Fig. 3). These neurons, resembling tectal T6 properties, were responsive only to large moving (looming) visual objects, like M5.4 cells.

2.6 "Total Field" Neurons

M4 neurons have a large monocular ERF encompassing the visual field of the ipsi- and/or contralateral eye (Fig. 3). Some of these neurons also obtained excitatory cutaneous input, particularly from the lateral body flanks, ipsilateral to the visual ERF. Various neurons displayed "newness" or "sameness" properties, respectively. Within the large visual ERFs there were small patches of particularly great sensitivity (about 20° in diameter); such an area could shift its position or suddenly expand with repetitive visual stimulation. Among M4 neurons, different sensitivities to moving configural stimuli were observed similar to the tectal types T5.1-4 or thalamic pretectal TH3. Of the M4 neurons investigated so far, most were synaptically activated (latency 4 to 7 ms) to ipsi- (or contra)lateral electrical stimulation of the rostral optic tectum. Since M4 neurons displayed properties like T4, stimulation and recording sites were interchanged to check for ortho/antidromic bulbo-tectal

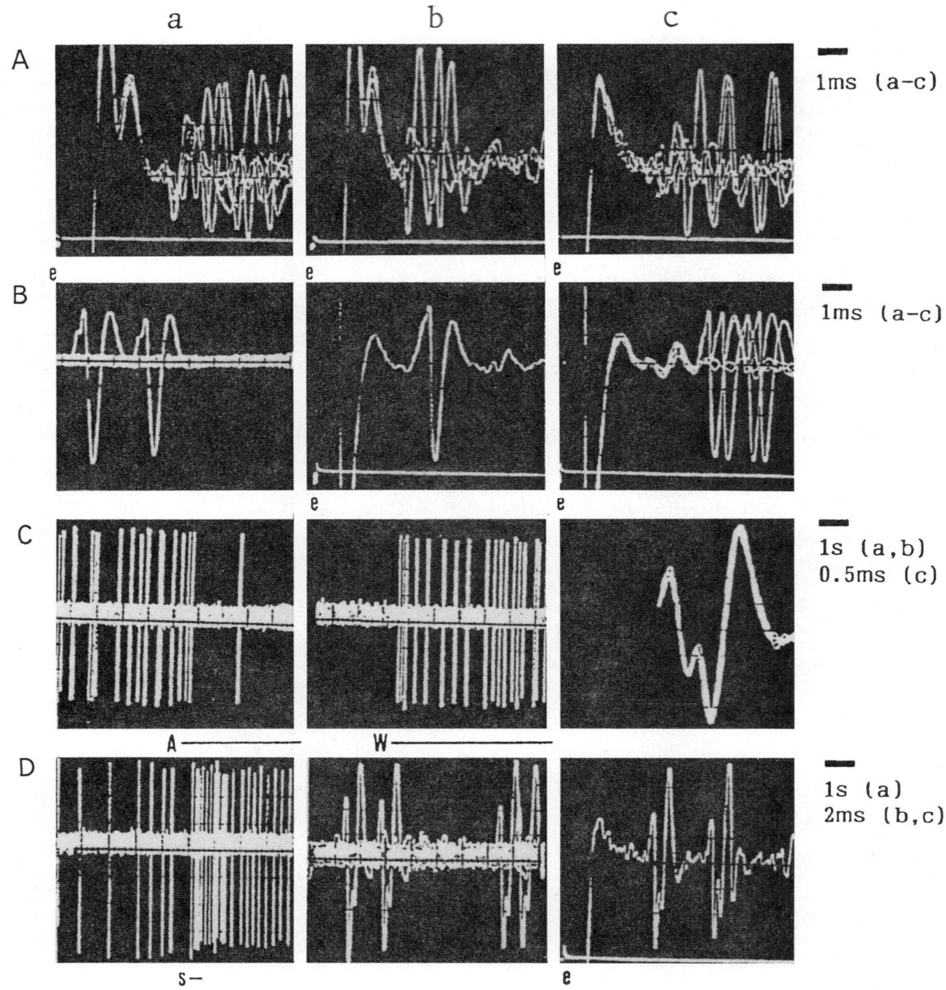

Figure 6. Extracellular responses of tectal and bulbar neurons in paralyzed common toads. (A) This tectal T4 neuron is synaptically activated (a-c) in response to a 0.1 ms electrical impulse applied to the contralateral rostral medulla oblongata (several superimposed traces). Note the initial constant-latency field potential in (a) and (b) followed by variable-latency spike responses. (B) Medullary M5.2 neuron of the medial reticular formation, RetF, responding to a 2°x 8° worm-like moving stipe (a) that traversed the ERF of the ipsilateral eye (note the inflection, probably due to IS-SD spike component); the cell was silent to the same stripe in antiworm configuration. The same neuron could be synaptically activated by electrical *e* stimulation of the ipsilateral branchiomotor column close to the nucleus of the solitary tract (b,c) or to the contralateral optic tectum (not shown here); note the variable latency in (c). (C) Combined property displaying M4-M8 neuron in the medial RetF whose spontaneous activity was inhibited during and shortly after presenting a 2°x 8° antiworm-like *A* stripe to the ipsilateral eye (a); the neuron could be re-activated by a worm-like *W* moving stripe (b). This neuron was synaptically activated by electrical stimulation of the contralateral optic tectum; (c) shows a superposition of several spontaneous, visually, and electrically

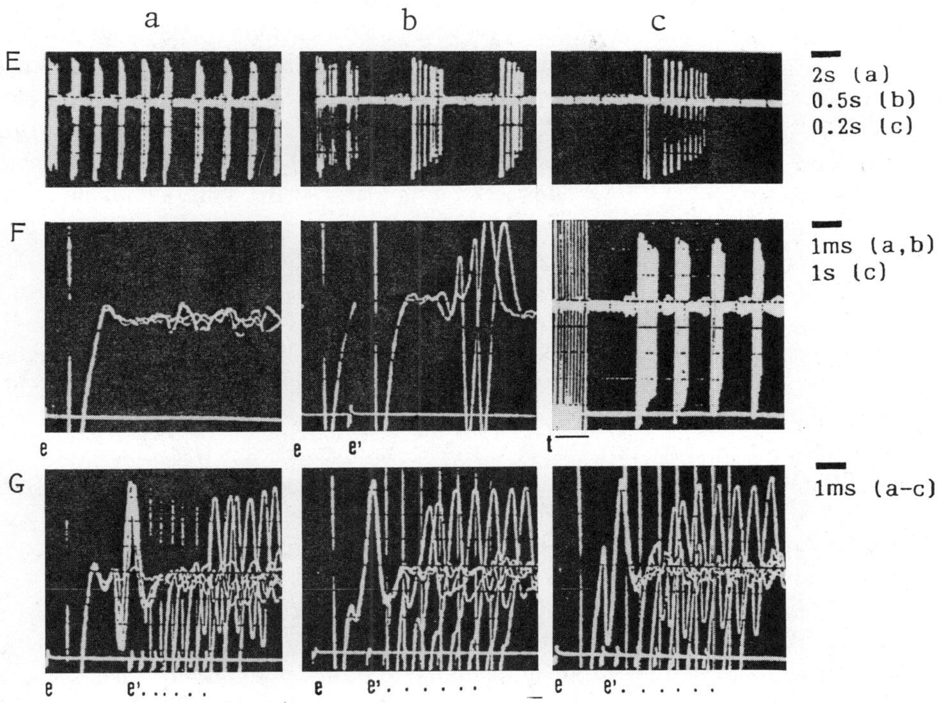

(Fig. 6, continued) elicited spikes; for the registration the oscilloscope sweeps were triggered by the spikes. (D) Multiple property displaying M4-M8-M9 neuron of the medial RetF whose spontaneous tonic activity was increased during and after moving a 8°x 8° square *S* in the ERF of the ipsilateral eye (a); the same neuron was synaptically activated by electrical stimulation *e* of the contralateral tectum (c); note the identical shapes of spontaneous (b) and electrically elicited (c) spikes. (E) Example of an M10 neuron, recorded from the medial RetF close to the VII*th* motor nucleus, which displays in response to a large looming stimulus a cyclic activity consisting of equidistant bursts of spikes as a post-stimulus event (a); in (b) three bursts and in (c) one burst is shown on larger time scales. (F) This M10 neuron in the branchiomotor column showed cyclic burst activity either spontaneously or in response to visual or to cutaneous stimulation. The neuron was synaptically activated in response to a short-interval electrical double stimulus *e-e'* applied to the ipsilateral medial RetF near the hypoglossal nucleus; see the two superimposed traces in (b); single electrical stimuli were not effective (a). A train *t* of several (e.g., ten) electrical double stimuli elicited cyclic bursting (c). (G) Examples of M11 neurons, recorded in the rostral third of the medial RetF, which occasionally discharged just one spike in response to visual or tactile stimulation. Such a neuron (a) could be activated antidromically by electrical stimulation of the contralateral hypoglossal nucleus. The latency varied between 2.5 and 2.8 ms. To double stimuli *e-e'* the neuron followed down to an interstimulus interval of 3.2 ms (absolute refractory period). Another neuron (b) showed a latency of 1.6 ms and an absolute refractory period of less than 1.9 ms. A neuron of the type M4-M11 (c) in the rostral median RetF readily responded to a moving visual stimulus or to cutaneous stimulation or, antidromically, to stimulation of the ipsilateral spinal cord at the level of the 3rd root; latency: 1.9 ms; absolute refractory period: less than 1.8 ms. (From Ewert et al. 1989).

influences. During the recordings close to a T4 neuron, an orthodromic constant-latency field potential in response to bulbar stimulation was often observed (Fig. 6Aa,b), and this was followed by synaptically elicited spikes (Fig. 6Aa-c) that were identical in amplitude and time course with the spikes elicited by visual stimulation from the registrated T4 neuron. In a few cases, also the tecto-bulbar projective character of a T4 neuron has been demonstrated by its antidromic activation (see Satou & Ewert 1985). The observation that not every T4 neuron could be activated antidromically by bulbar electrical (contra- or ipsilateral) stimulation - while many bulbar neurons resemble T4 features - suggests also the possibility of bulbo-tectal feedback mechanisms, besides intrinsic tectal properties responsible for T4 neurons. The same may be true for T2/M2 type neurons. Since not all M4 neurons could be activated by tectal electrical stimulation, corresponding relationships with other visual structures (e.g., regarding thalamic pretectal TH4 widefield neurons) are probable.

In response to spinal electrical (ipsi- or contralateral) stimulation, tactile sensitive M4 neurons showed synaptic activation by discharging spikes at a variable latency (6 to 9 ms). These extracellular records (Fig. 8b) often exhibited one or two constant-latency field responses whose orthodromic character was determined in single cell studies by the absence of collision in the collision test. Taking all results together, M4 neurons are suitable to integrate visual (tectal, pretectal) and somatosensory (spinal) activity, and in turn they may feed back to the sources of their inputs, which thus leads to an increase of the local excitability.

In a few cases of multiple unit recording, the extracellular response toward a large S32-object included spikes both from a M6 neuron (discharging "phasically" a few large spikes) and a M4 neuron (discharging "tonically" a number of small spikes). An increased activity precisely coincided in both types suggesting a sort of (bulbar) coupling or concurrent external inputs.

2.7 Spontaneously Active Neurons

M8 neurons show spontaneous, partly fluctuating discharge activity. Although these neurons do not necessarily have distinct visual ERFs, the discharge rate could be "modulated" by sensory stimulation, e.g., increased, decreased, or even inhibited by a visual moving stimulus. This effect could be selective for the configuration or the movement direction of the stimulus. For example, the spontaneous activity of such a combined M4-M8 neuron was suppressed in response to a stripe moving in A-configuration (Fig. 6Ca) and ceased for a while in the absence of this stimulus in terms of a "sample and hold" mechanism; the activity could then be regained by the same stripe moving in W-configuration (Fig. 6Cb). This neuron was synaptically activated by electrical stimulation of the contralateral optic tectum (Fig. 6Cc). All M8 neurons did not exhibit a correlation between visual stimulation and neuronal firing.

2.8 Reverberation Displaying Neurons

M9 neurons have large visual ERFs - similar to M4 - and they show reverberatory activity after visual stimulation. One type showed "newness" responses, since repeated traverses of the same path in the visual field with a moving stimulus led to habituation; but any change of the direction of movement - such as circular or zig-zag - strongly activated the neuron in the manner of "warming up" leading to a kind of self-excitatory state that persisted beyond the stimulation as a short-term reverberatory event. After warming-up, some neurons responded also to (relatively small visual) stimuli which were previously subliminal. Such a neuron with combined M4-M9 properties could be synaptically activated by electrical stimulation of the contralateral optic tectum, and there was an effect of temporal summation, since only electrical double impulses (interstimulus interval about 3 ms) were effective. Another M4-M9 type showing "sameness" properties always increased its firing rate in response to a moving stimulus, and this activity persisted beyond the stimulation for a couple of seconds as a reverberatory event (Fig. 6Da). These cells, too, could be synaptically activated by electrical stimuli applied to the contralateral tectum but, here, one impulse generally elicited two spikes (Fig. 6Dc).

2.9 Cyclic Bursting Neurons

M10 neurons are characterized by the property that they discharge a sequence of rhythmic bursts of spikes in response to visual (Fig. 6Ea), or to tactile, stimulation as post-stimulus events. Some neurons seemed to burst spontaneously. In a sequence of bursts, each burst displayed a relatively constant number (Fig. 6Eb) of rather equidistant (Fig. 6Ec) spikes, except the initial burst which was strongest (see also Fig. 6Fc). Among various neurons of this type, a burst consisted of 2 to 10 spikes at 15 to 20 Hz; the inter-burst intervals of such a sequence ranged between 1 and 2 s, and the number of bursts was 2 to 16. After discharging a sequence of bursts, there was a recovery period of 2-4 min. Neurons with this property have not been recorded from the optic tectum or the pretectum.

The M10 bursting characteristic was usually combined with other properties, mostly with those of M5.4, M4, M8, and M9 neurons. In such a M5.4-M4-M8-M9-M10 neuron, the response to a large looming stimulus resulted in a post-stimulus burst pattern, so that previously (M8-) tonic activity was converted to (M10-) periodicity: Weak tonic activity displaying neurons of this type first warmed up in response to visual stimulation by increasing the frequency of firing; suddenly they discharged a couple of bursts as a post-stimulus event. There was a correlation between an increase in tonic (acyclic) spontaneous activity and - from a certain level of frequency - the onset of cyclic bursting. Following the last burst, the activity became sometimes "free-running" in a tonic manner.

In visual sensitive M10 neurons, bursts could be elicited by a short train of electrical impulses applied to the optic tectum. Class M10 neurons of the BMC bursting in response to visual or tactile stimulation, could also be activated by electrical stimulation of the medial RetF: Whereas one electrical square wave impulse was ineffective (Fig. 6Fa), a double-impulse of a short (2 ms) interval elicited synaptically a spike from the M10 neuron (Fig. 6Fb); a short impulse train caused the neuron to discharge a series of bursts (Fig. 6Fc).

In a preliminary study (Framing & Ewert, 1989), the burst activities of two single M10-type neurons - located either unilaterally or bilaterally in the medulla oblongata - were recorded simultaneously with two separate micro-electrodes. In many

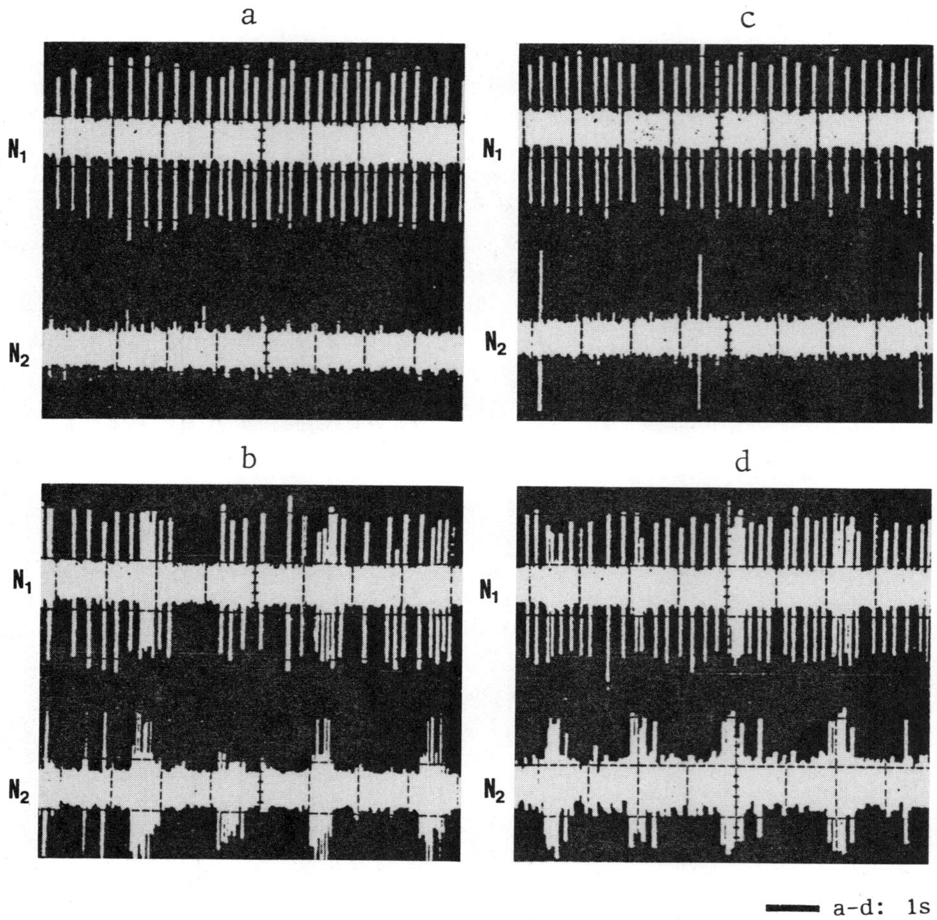

Figure 7. Examples of parallel extracellular recordings with two microelectrodes from each two medullary neurons N_1 and N_2 in the paralyzed common toad. Recording sites of the stainless steel electrodes - after passing an anodal direct current - were examined by means of the Prussian Blue reaction in Neutral Red counterstained transverse sections of the medulla. Time segments from long-term recordings are depicted. (a-d) Neurons N_1 and N_2 were recorded bilaterally from about opposite sites of the medial RetF adjacent to the motor nucleus of the V*th* cranial nerve. N_1 showed spontaneous tonic spike activity while N_2 was silent (a). In response to visual stimulation with a large looming square, N_2 discharged a series of bursts of spikes in M10-like manner and these coincided with a more or less increasing spike frequency in the M8-like tonic firing of N_1 (b). After several seconds, both neurons returned to their previous behavior (c). A few minutes later, the coincident cyclic burst activity in both neurons could be elicited again in a very similar manner (d). Moderate changes in the tonic firing of N_1 were not paralleled by any kind of activity change in N_2. - (e) Both neurons were recorded ipsilaterally adjacent to the VII*th* nucleus; the distance between the histologically verified recording sites was 70 µm. These neurons were silent without sensory stimulation. Cyclic bursting in response to visual stimulation, or sometimes occurring spontaneously, were strongly correlated in both neurons. The spike burst in N_1 was typically longer (about 250 ms) than the

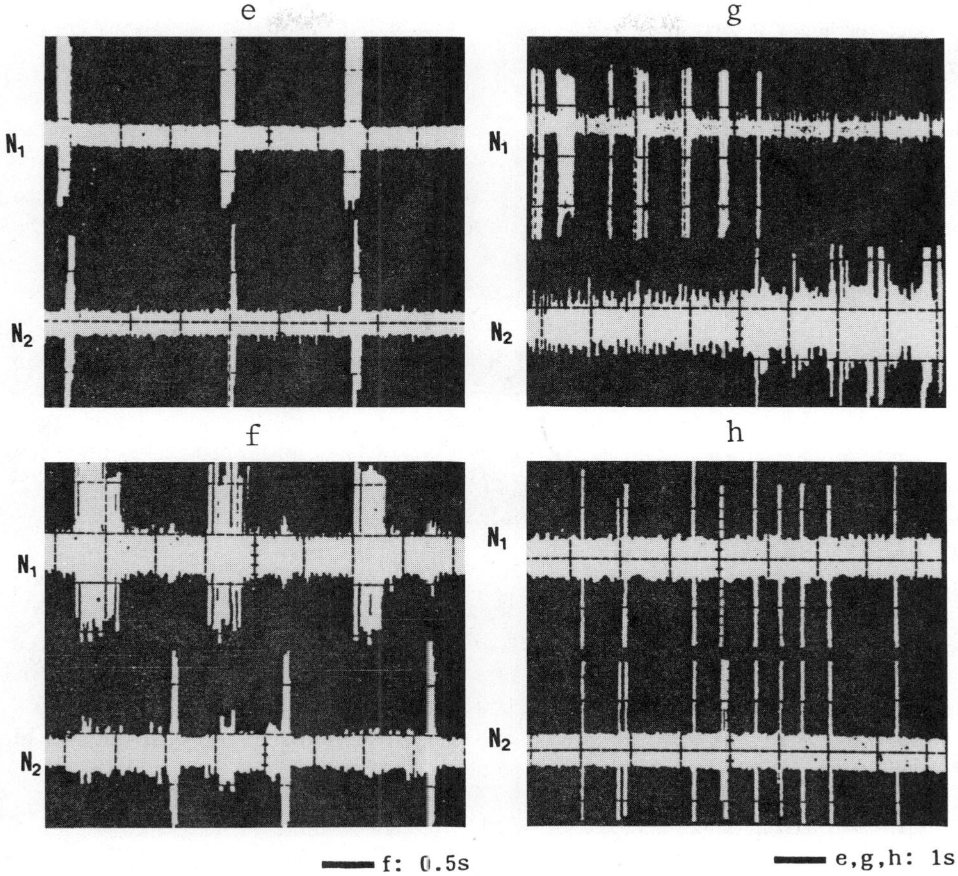

(Fig. 7, continued) one in N_2 (about 100 ms) and preceded the bursts in N_2 by 150 ms - (f) Bilateral recordings at the level of the VII*th* nucleus. Neuron N_1 was located in the lateral RetF very close to the VII*th* and N_2 on the opposite side within the VII*th*. In both neurons the spontaneous cyclic burst activities were phase-locked in such a manner that the burst of one cell occurred in the silent phase of the other cell. Note that some background burst activity in the record of N_1 coincided with the burst of N_2. - (g) Neurons N_1 and N_2 were recorded bilaterally from opposite sites of the medial RetF adjacent to the V*th* nucleus. Here a burst sequence of N_1 could be followed by a sequence of bursts in N_2; single bursts among both neurons were, as far as has been investigated, not phase-locked. These bursting neurons seemed to "avoid" to discharge a burst sequence in an activity period of the other neuron. - (h) Tonic firing neurons recorded ipsilaterally, N_1 from the medial RetF and N_2 from the lateral RetF, both adjacent to the V*th* nucleus. During long-term recording there were periods (as shown here) in which the spikes of both neurons obviously exhibited synchrony. - These preliminary investigations do not yet allow to suggest the means of coupling of neurons N_1 and N_2. (The time bases of the dualbeam oscilloscope were slightly delayed; in the present figures they are corrected, so that the beams but not the oscilloscope screens are precisely in register). (From Framing & Ewert 1989).

cases (Fig. 7), the burst activities seemed to be phase-locked or coincided in a manner of "absolute coordination" according to v. Holst (1939), which of course requires extensive mathematical examination (Székely 1965; Wendler 1969).

2.10 Projective Bulbar/Spinal Neurons

The property M11 refers to the projective character of neurons whose activities were recorded from the medial RetF in response to electrical stimulation of the caudal medulla (near the obex) or the spinal cord at cervical level. Some neurons had wide visual ERFs, like M4 "newness neurons," partly combined with wide cutaneous mechanoreceptive fields. Large looming objects were most effective. Many of these neurons did not respond to sensory stimulation, indicating their presence just by a rarely occurring spontaneous large spike.

From the recording sites in the rostral medial RetF, constant-latency spike activity of single neurons was recorded in response to ipsilateral cervical stimulation at latencies of $L_1 = 1.39 \pm 0.12$ SDM (36 %), $L_2 = 1.98 \pm 0.04$ SDM (25 %), or $L_3 = 2.97 \pm 0.19$ SDM (7 %). Variable-latency responses were found at $L_4 = 4.56 \pm 0.48$ SDM (14 %) and at $L_5 = 6.51$ ms $\pm$ 1.04 SDM (18 %). Some recordings of field potentials of the summated unit activity to electrical spinal stimulation showed various constant-latency responses in the same record (Fig. 8a,b) suggesting that different types of neurons are closely intermingled. The L_4- and L_5-latency responses have to be interpreted as postsynaptic; these spikes (Fig. 8b) could also be elicited by visual or tactile stimulation. The individual variability of the latency was one to several milliseconds. The L_1-, L_2-, and L_3-responses were non-synaptic; the individual

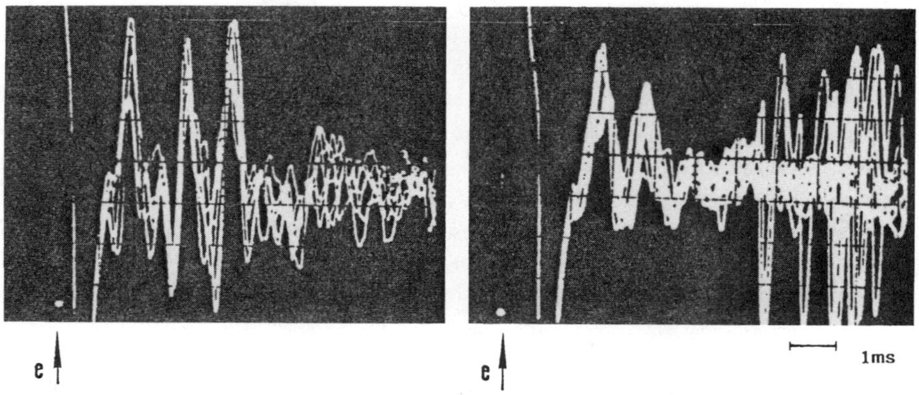

Figure 8. (a) Three constant-latency field potentials extracellularly recorded from the toad's medial RetF in response to electrical stimulation *e* of the ipsilateral spinal cord at the level of the 3rd route (five superimposed traces). (b) Similar experiment as (a); note two constant-latency field potentials followed by synaptic spikes with variable latencies (six superimposed traces); the spikes could be also activated by visual or cutaneous stimulation. (From Ewert et al. 1989).

variability of their latency was less than 0.2 ms. Whereas all these neurons satisfied the following-ability, only a few neurons fulfilled the criterion of the collision-test for antidromic activation (Lipski 1981). But there were other "indirect" indications accounting for the reticulo-spinal projective character of many of these neurons, such as responses to visual stimulation and the occurrence of soma injury spikes.

2.11 Multiple Property Displaying Neurons

In most cases, the described M-type property was not seen separately but in combination, such as :

M2.1 - M8	M4 - M8	M5.1 - M8	M6 - M4
M2.1 - M9	M4 - M9	M5.1 - M8 - M9	M6 - M9
M2.1 - M10	M4 - M10	M5.2 - M9 - M10	M8 - M10
M2.1 - M8 - M9	M4 - M11	M5.4 - M10- M11	M6 - M9 - M10
M2.1 - M8 - M9 - M10	M4 - M8 - M9	M5.4 - M4 - M8 - M9	
M5.4 - M8 - M9 - M10		M5.4 - M4 - M8 - M9 - M10	
M4 - M8 - M9 - M10		M5.4 - M4 - M8 - M9 - M10 - M11	

We hasten to add that neurons assigned to M-types may have additional properties that were not identified in the present study. Furthermore, the number of "multiple property" displaying neurons may be actually larger, so that the sample distribution shown in Fig. 4 must be evaluated with caution. This histogram only resembles the frequency of bulbar properties for the applied criteria obtained from extracellular recordings in the medial RetF of the immobilized toad. The M-classification is discussed in Section 4.2.

3 Description of Bulbar Neurons by Intracellular Recording and Labeling

Since it is accepted that the essential aspects of information processing are reflected in a "subthreshold activity," the intracellular recording technique is another approach to characterize neuronal activities. Applying the method of Antal and Matsumoto (Antal et al. 1986; Matsumoto et al. 1986), we have intracellularly recorded the membrane responses of rostral bulbar neurons by means of potassium-citrate/cobalt-lysine filled glasmicropipettes in toads *Bufo bufo spinosus* which were slightly paralyzed with succinylcholine (Schwippert & Ewert 1987; Schwippert et al. 1988, 1989). Visual stimuli were black squares (S4, S8, S16, S32) and stripes (W,A), brisk changes of the diffuse background illumination (*off-on*), and a moving background texture. The optic tectum was stimulated electrically in the ventrolateral region. The body surface was stimulated tactually with a bristle. Successfully penetrated cells were iontophoretically labeled and, after counterstaining, reconstructed with the camera-lucida technique, so that - in addition to physiological properties - the site of the soma, dendritic and axonal arborizations, and termination fields of axonal and dendritic projections could be evaluated.

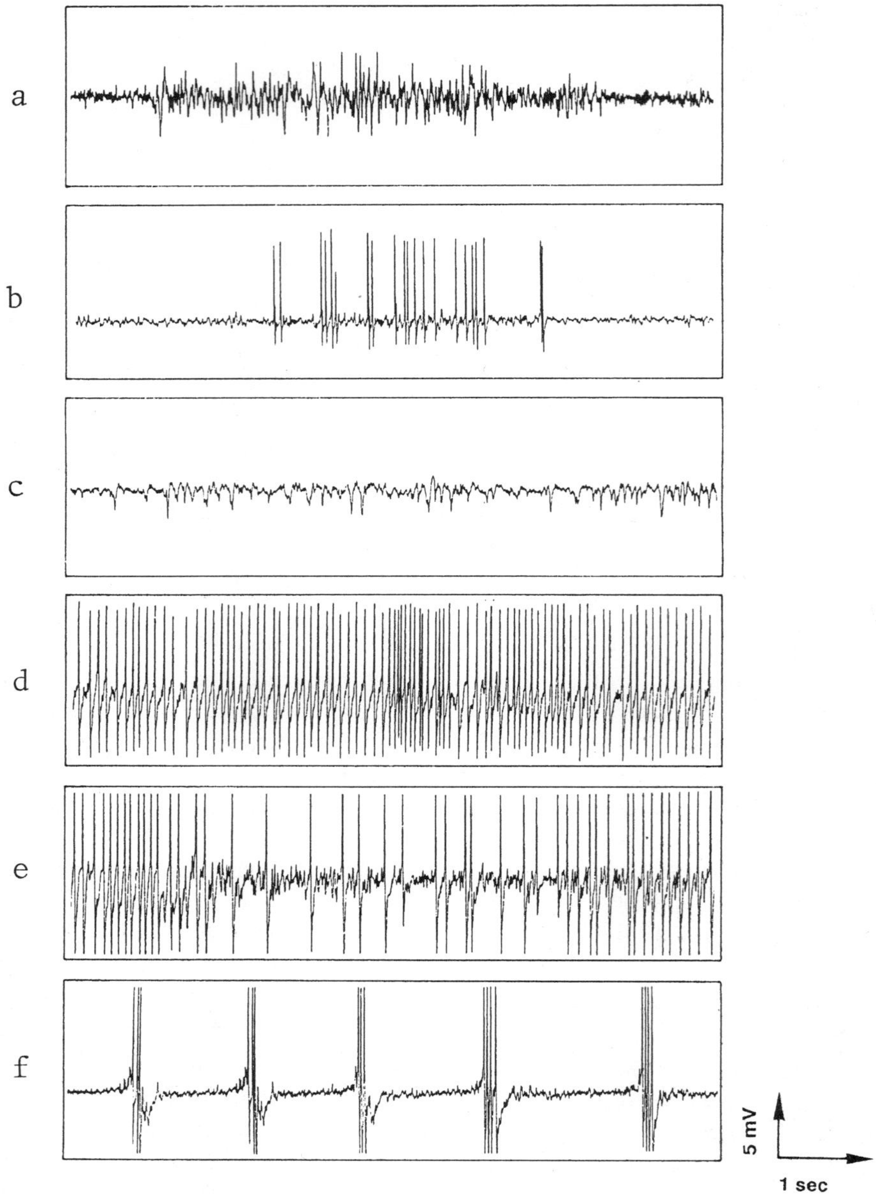

Figure 9. Examples of changes in membrane activities of intracellularly recorded bulbar neurons of paralyzed common toads in response to moving visual stimuli. (a) EPSPs and IPSPs; (b) EPSPs, spikes and IPSPs; (c) IPSPs mainly; (d) increase or (e) decrease of tonic spike activity; (f) post-stimulus cyclic burst activity, in which each burst of spikes is introduced by an EPSP and terminated by an IPSP. In some bulbar visual neurons, electrical stimulation of the optic tectum either inhibits ongoing tonic spike activity (g) (the arrow indicates the initial IPSP) or elicits spike activity (h); note that lasting trains of electrical impulses may lead to inhibitory (g) or excitatory (h) post-stimulus after-effects. (Schwippert et al. 1989).

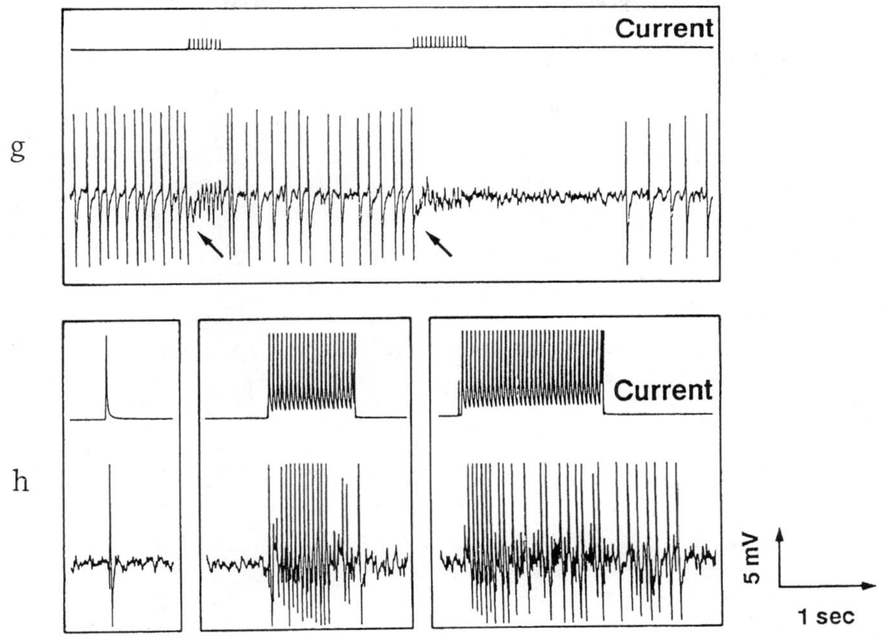

3.1 Survey of Intracellular Response Properties of Bulbar Neurons

In the rostral medulla oblongata, at its border between the cerebellum and the rostral end of the IX*th* nucleus, a total sample of 192 neurons with stable membrane potentials ranging between -45 and -70 mV were recorded intracellularly. The bulk of penetrated cells was located at the levels of the VI*th* and VII*th* nucleus. As in the extracellular studies, the somata of these neurons were located mainly in the central grey of the ventral RetF close to the border of the white matter and in the BMC. Response properties of visual M-type neurons (M1-11) obtained from extracellular recordings were detected and in addition emerged in postsynaptic activities. Of the sample of intracellularly recorded neurons, 82 % showed visual inputs as regards responses to S-, W-, and A- stimuli or to brisk change of diffuse illumination; the remaining 18 % failed to respond to any visual stimuli, while 5 % of these received somatosensory input particularly from the head region (snout and eye). Other sensory inputs were not tested in this study. Some neurons showed both visual and tactile inputs. In monocular neurons, the somatosensory RF/ERF was pronounced ipsilateral to the visual RF/ERF.

The question of preferences of bulbar neurons to moving configural (W-, A-, S-) stimuli was studied in order to compare these with the well-known tectal response properties and thus to trace information flow. In the sample of recorded neurons, 51 % were sensitive to moving stimuli, 80 % of these cells were most sensitive to large

S16 or S32 stimuli, hence resembling T5.3, TH3, or T5.4 properties. The receptive fields (RFs) of the recorded bulbar neurons were located mostly in the frontal visual field. Tectal input to these cells was confirmed by electrical stimulation (0.1 ms, 5-100 μA) in that area of the optic tectum where the frontal visual field is represented. In bulbar cells, the EPSPs and spikes (E-type responses) and IPSPs (I-type responses) elicited by electrical stimulation occurred in different combinations and sequences whose variety (EI, IE, EE, II, EII, EEI etc.) displayed a still higher diversity than in tectal cells to electrical stimulation of the optic tract (reported by Matsumoto, this volume; see also Matsumoto et al. 1986).

As regards the electrical membrane activities of different bulbar cells, there were either exclusively EPSP/IPSP activities (Fig. 9a) or EPSPs and IPSPs combined with spikes (Fig. 9b) in response to moving visual stimuli, or spontaneous IPSPs (Fig. 9c), or tonic spike activity which increased (Fig. 9d) or decreased (Fig. 9e) to visual stimulation, or cyclic patterns of spike bursts (Fig. 9f) that occurred either spontaneously or were elicited as a post-stimulus event. In bulbar cells, warming-up and reverberatory activities were often strongly pronounced. Sustained excitatory as well as inhibitory activities could also be evoked by electrical stimulation of the ventrolateral optic tectum (Fig. 9g,h); after-effects were only inducible if the optimal stimulation frequency of 10 to 30 Hz lasted longer than 500 ms, suggesting an activation of internal loops.

About 20 % of all recorded neurons showed M10-type burst characteristics (Fig. 9f). Depending on the individual cell, such repetitive burst-sequences could be "stimulus-specific," e.g. evoked only by presentation of S- or W-stimuli, respectively. Most M10 neurons responded specifically toward large looming squares irregularly moved in the toad's superior visual field. Other M10 neurons started to burst after mechanical stimulation of the head region with a bristle, or to vibration, or in response to brisk change of the background illumination.

3.2 Morphological Properties of Rostral Bulbar Neurons

Histological transverse sections of the toad's medulla oblongata - like in frogs (Nieuwenhuys & Opdam 1976) - show four layers in dorsoventral direction (Figs. 5a, 10): (i) Stratum ependymale (layer 1) consisting of a layer of columnar ependymal cells; (ii) stratum subependymale (layer 2) which is almost free of cells, located close underneath the stratum ependymale; (iii) stratum griseum (layer 3), also called the medullary central grey, characterized by cells of various sizes arranged in different cell clusters: the cranial nerve motor nuclei in the BMC, the median RetF, the medial RetF, and the lateral RetF; two zones are here distinguished, the tegmento-rhombencephalic periventricular layer (small-celled inner zone) and the central layer (large-celled zone including the cranial nerve nuclei); (iv) stratum album or the medullary white matter (layer 4) forming the ventral and peripheral half of the brain stem, containing the bulk of myelinated fibers and only a few scattered cells. The somata of the seventy successfully labeled neurons (Fig. 10) were found in the medial RetF (26 cells), the lateral RetF (10 cells), and in the V*th* (trigeminal), VI*th* (abducens), VII*th* (facial), and IX*th* (glossopharyngeal) cranial nerve nuclei of the BMC (34 cells). The proportion of recorded cells with visual inputs across medial

RetF, lateral RetF, and BMC was 18:8:29. Their somata were situated mainly in layer 3 at the border between grey and white matter and in the VII*th* nerve motor nucleus. These recording sites are approximately coincident with those of the extracellular studies (Fig. 5b). The soma diameter of large-celled neurons was between 15 and 30 µm and those of small-celled neurons below 10 µm. Taking into consideration the site of the cell body, the shape and extension of the dendritic trees, and the axonal projections, five morphological neuron types, distributed over the three brain stem areas, could be distinguished (Fig. 11):

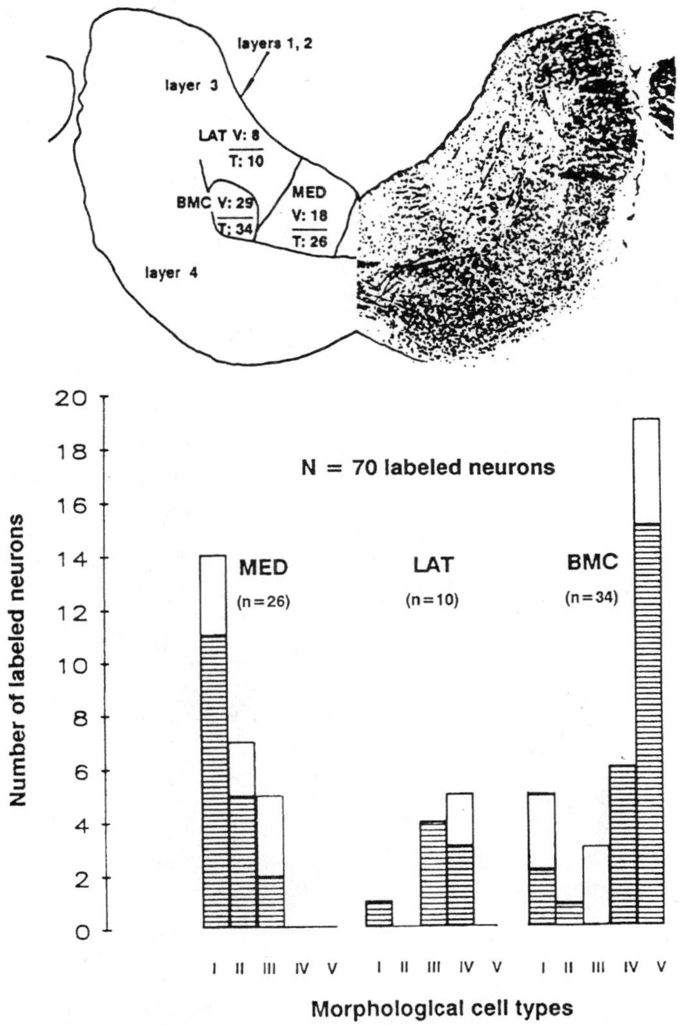

Figure 10. Sample distribution of intracellularly recorded and iontophoretically cobalt-lysine labeled neurons in the common toad's medial (*MED*) RetF, median (*MEN*) RetF, lateral (*LAT*) RetF, and the branchiomotor column (*BMC*); V, visual neurons; T, total neurons recorded in the respective area. The rastered parts in each column refer to the proportion of visual neurons. *I* to *V* refer to the different morphological *Types I-V* of bulbar neurons shown in Fig. 11.

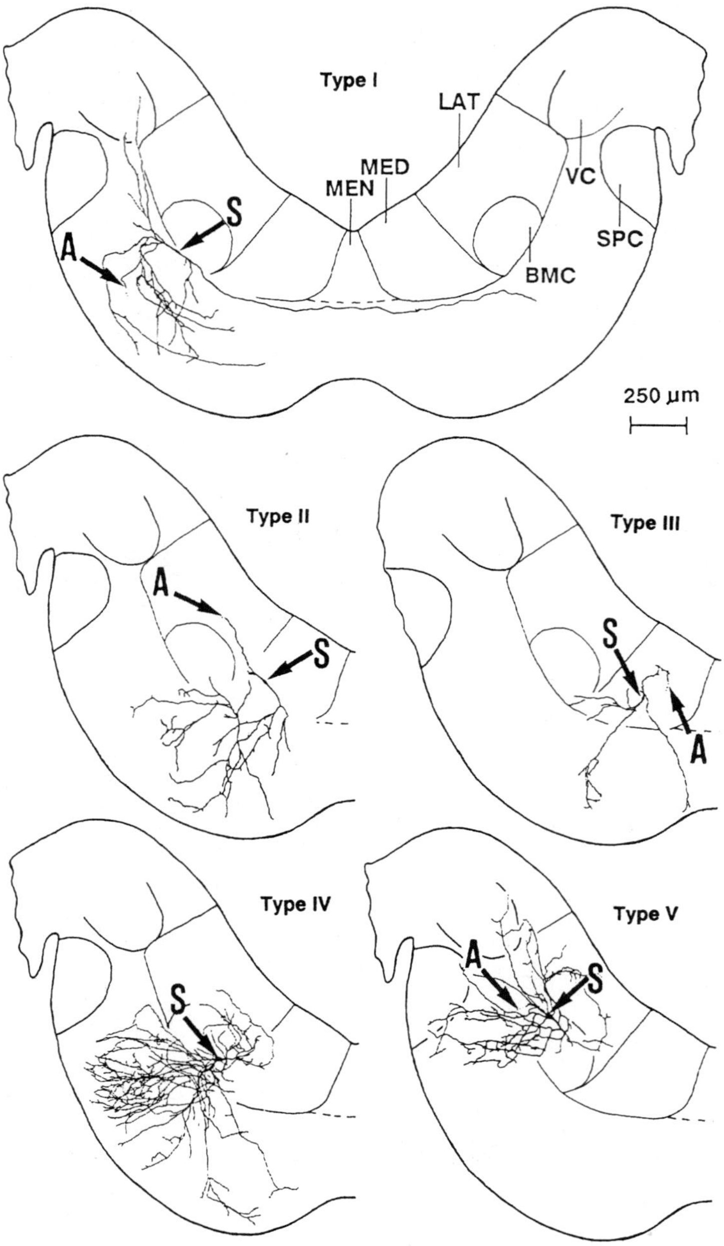

Figure 11. Examples of morphological distinguishable *Types I-V* of neurons in the common toad's medulla oblongata. Camera-lucida reconstructions of cobalt-lysine filled cells. The same method was applied to the examples shown in Figs. 12-17. SPC, spino-cerebellar complex; VC, vestibular complex; *A* indicates the axon and *S* the soma. For further abbreviations see Fig. 10. (From Schwippert et al. 1989).

Type I Neurons with Ventrolaterally Extending and Crossing Dendrites: Somata of these neurons with projections similar to bipolar cells are primarily located in the ventral central grey near the border between layers 3 and 4. Visual sensitive cells of this type (see Fig. 11) were most frequently recorded in the medial RetF, less in the lateral RetF, and seldom in the BMC. The bulk of the dendritic arborizations spread ventrally into the stratum album with termination fields close to the tecto-bulbar/spinal tracts, tr.tbs.c. and tr.tb.d. A second group of dendritic branches extended dorsolaterally along the borderline between the grey and the white matter toward the ipsilateral vestibular complex. In some cases, dendritic projections crossed the midline to reach the paths of the tr.tbs.c. and tr.tb.d. Dendrites literally appeared to avoid the grey matter. In cells located in the BMC, all somata were situated exactly close to the border with the white matter; the dendritic arborizations, especially the bidirectionally and laterally running branches, were longer than in neurons of the medial RetF, so that they also terminated in the vestibular complex of the contralateral side. Neurons of this type were described by Nieuwenhuys & Opdam in the frog's nucleus reticularis superior (cf. Fig. 2 on p. 816 in Nieuwenhuys & Opdam 1976).

Type II Neurons with Dendrites Extending Mainly Ventrally: The somata of these neurons are situated in the central grey of the medial RetF at the border between layers 3 and 4. The dendritic arborizations spread mainly ventrally toward the white matter reaching areas of the tecto-bulbar/spinal tracts, especially near the tr.tbs.c. Long ipsilateral dendritic projections typical for *Type I* were not observed. Unlike *Type I*, the *Type II* neurons showed a few axonal arborizations with terminal boutons in the grey matter. Multiple stainings due to transneuronal labeling were frequently found both in *Type I* and *II* neurons. Investigations in reptiles (garter snakes) suggest that bulbar reticular neurons comparable to the toad's *Types I* and *II* receive tectal information, since the bouton-rich axonal collaterals of tectal neurons form an overlapping meshwork within these neurons' dendritic fields (Dacey & Ulinski 1986a,b).

Type III Neurons with Tiny Somata and Small Dendritic Trees: The somata of these neurons, the smallest of all bulbar neurons recorded so far, were located in the ventral part of layer 3 of the medial or lateral RetF, and less in the BMC. The sparse dendritic ramifications spread in the central grey and, rarely, in the white matter. In a few cases, short axons and collaterals with terminal boutons could be observed in the grey matter. Neurons of *Type III* were described with the Golgi rapid technique by Nieuwenhuys & Opdam (1976) in the frog's nucleus reticularis medius and with the HRP technique by Kostyk & Grobstein (1987 a,b) in the frog's dorsolateral medulla oblongata close to the solitary tract. Cells of this morphological type were also described as preganglionic parasympathetic neurons by Oka et al. (1987).

Type IV Neurons with Ventrolaterally Extending Dendrites: The somata were located in the ventral part of layer 3 of the lateral RetF and in the BMC. These neurons showed two dendritic trees, a small one extending into the grey matter and a richly arborized one spreading ventrolaterally in the white matter. The former one, in particular, distinguishes these neurons from *Type I*. In one visual cell of the lateral RetF, the dendritic branches reached the grey matter of the cerebellum. In motoneurons of the BMC, the axon arising from the cell body or from a primary dendrite near the soma could often be followed into the ventral root.

Type V Neurons with Dorsolaterally Extending Dendrites: The somata of these motoneurons were found exclusively in the BMC. In contrast to *Type IV*, the dendritic tree extended in a dorsolateral direction with branches supplying the grey and the white matter. In some cells, the axon could be traced in the respective ventral root. The dendritic arborizations of *Type IV* and *V* motoneurons in various nuclei of the BMC were rather similar. Satou et al. (1985) in their intracellular recording and retrograde labeling studies of tongue-controlling motoneurons in the toad's hypoglossal XII*th* nucleus distinguished four subtypes characterized by (i) lateral or ventrolateral spread of the dendritic trees, (ii) medial orientation of dendrites, (iii) dendrites crossing the midline, and (iv) dendritic access to the solitary tract. A comparison with the morphology of neurons in the facial VII*th* nucleus of our study shows that (i) and (iv) are similar to the *Types IV* and *V* which were also described in the trigeminal and facial motor nuclei of frogs by Matesz & Székely (1978) and Székely & Matesz (1987).

In summary (Figs. 10 and 11), *Type I* and *Type II* neurons were more typical for the medial RetF, *Type V* was specific to BMC, *Type IV* was found in lateral RetF and BMC while *Type III* could be identified at a low distribution in all three bulbar structures. An interesting outcome of this study is that differences in neuronal morphological features were established mainly between BMC and medial RetF, and that BMC obviously does not only contain motoneurons. Dendrites crossing the midline, especially seen in *Typ I* cells, were sometimes also found in the remaining neurons of this sample. All these morphological observations are instructive because of hints and ideas about connectional and integrative capabilities. In general, for neurons of the BMC it is assumed that the ventrolateral dendritic field serves as a recipient of bulbar extrinsic input, whereas the dorsomedial portion seems to be responsible for intrinsic information processing (Matesz & Székely 1978). These studies underline the idea that brain stem structures across vertebrates display a surprising similarity both in cytology and cytoarchitecture (see Matesz & Székely 1978; Matesz 1979).

3.3 Intracellular Activities of Visual Sensitive Bulbar Neurons

The following M-type characteristics could be detected by intracellular recording where they are reflected by spike as well as subthreshold (EPSP/IPSP) activity. It was not possible to assign one of the aforementioned morphological types a physiological M-type property consistently.

M5.1 Characteristic

In response to moving configural S-, A-, and W-stimuli, the excitatory potentials (EPSPs and spikes) of these neurons resemble tectal T5.1 properties: $R_S > R_W > R_A$. Somata of bulbar neurons with M5.1 characteristic were located in the medial RetF and in the BMC (Fig. 12a). The diameters of the monocular receptive fields (RF) ranged between 30 and 90°, which suggests converging inputs of T5.1 cells whose RFs

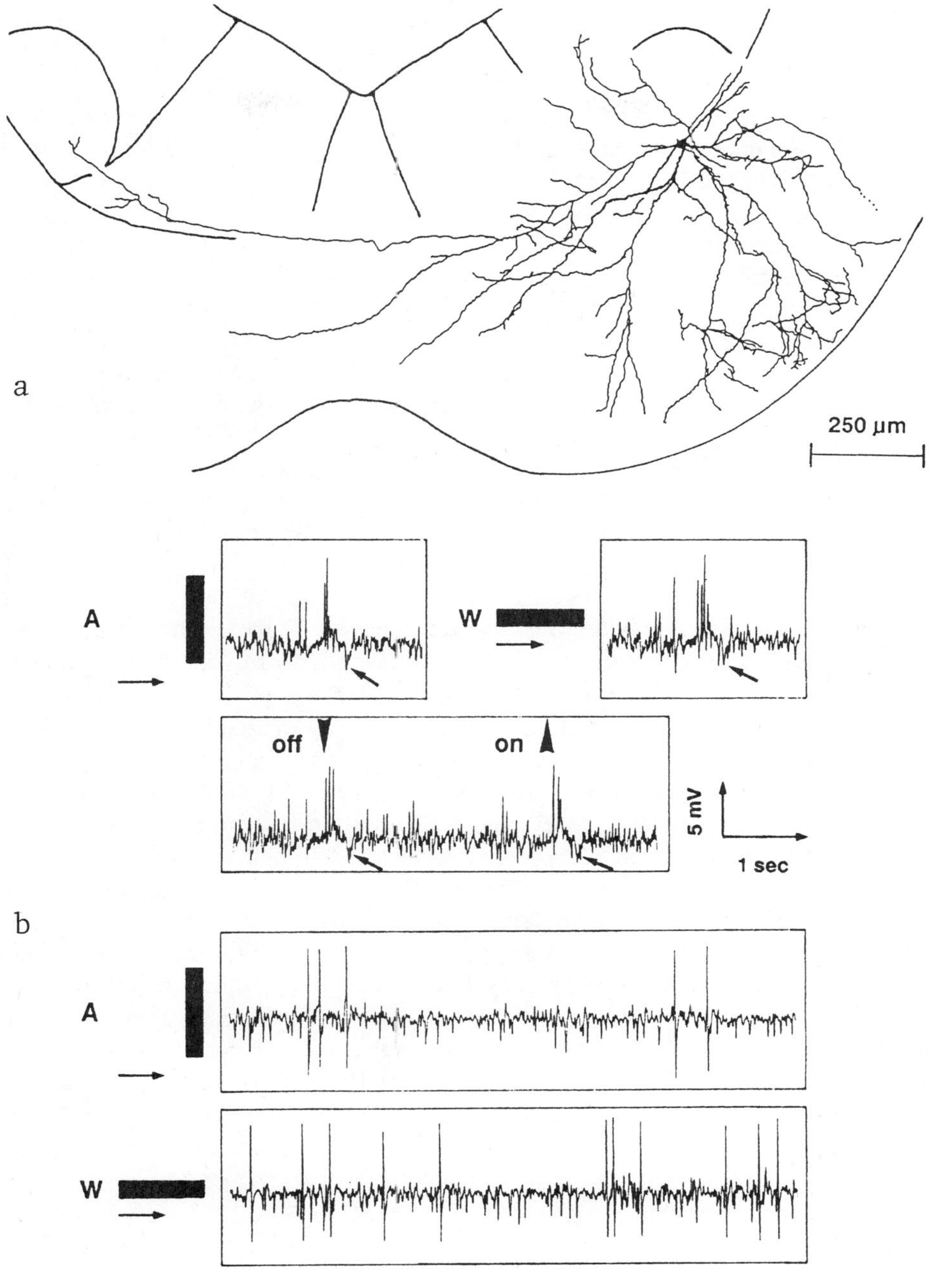

Figure 12. (a) Example of an intracellularly recorded and cobalt-lysine labeled M5.1 neuron of the morphological *Type II*; its soma was located in the VII*th* nucleus. This neuron showed spontaneous EPSPs and IPSPs of small amplitude. The worm-like *W* moving stripe was slightly more effective than the antiworm *A* one. To brisk change of diffuse light, EPSPs and IPSPs followed at *off* and *on*. Note that excitatory activity is always terminated by IPSPs (see arrows). (b) Example of another M5.1 neuron. RF diameter = 60° (a); = 90° (b). For further explanations see text. (From Schwippert et al. 1989).

were smaller (25-40° diameter; see Matsumoto et al. 1986). Tectal input to M5.1 was confirmed by electrical stimulation of the optic tectum.

Interestingly, EPSP and spike responses to a moving visual stimulus were in some cases terminated by small IPSPs and then followed by a strong IPSP (see arrows in the records of Fig. 12a). To brisk changes of diffuse illumination, the neuron exhibited the same response pattern at *off* and *on* (Fig. 12a). Both cases suggest an excitatory visual input and a feedback control indicated by the delayed inhibitory activity. Some neurons discharging spikes in response to moving visual stimuli in addition displayed a stimulus independent background activity of fast IPSPs (Fig. 12b). Similar to tectal T5.1 cells (Ewert et al. 1979; Matsumoto et al. 1986; Betts, this volume), the M5.1 configural sensitivity was sometimes not stable, so that occasionally $R_W = R_A$ or even $R_A > R_W$. Furthermore, within the RFs there were small areas of best excitability that could change in diameter and even shift their positions, which may be a hint for concurrent tectal inputs from several neurons of different classes, e.g., T4 and T5.1. A few M5.1 cells showed, ipsilateral to the visual field, cutaneous input in response to tactile stimulation, especially in the snout region.

M5.2 Characteristic

The spiking as well as the postsynaptic activities of these cells in response to moving configural stimuli was according to tectal T5.2 neurons: $R_W > R_S > R_A$ (Fig. 13a,b). The somata were located in the medial RetF or in the BMC. In different neurons, the RFs ranging between 30 and 90° in diameter could be located anywhere in the visual field, mainly in the frontal field. The large RFs indicate converging inputs from T5.2 cells whose RFs are significantly smaller: ERF=20 to 30° measured extracellularly by Ewert & v. Wietersheim (1974); RF=30 to 40° measured intracellularly by Matsumoto et al. (1986).

The facts that (i) M5.2 neurons' best excitatory activity in response to a moving object was restricted to a relatively small region in the center of the RF (Fig. 13a,b) and that (ii) the spiking activity in some cases was followed by IPSPs (see arrow in Fig. 13b) suggest additional inhibitory inputs and loops. For example, the "best activity" zone in the RF/ERF of a M5.2 neuron may be explained by lateral inhibition among adjacent T5.2 cells. In addition, we cannot exclude the possibility that the configural selectivity (W *vs.* A preference) of M5.2 neurons - which seems to be sharper than in T5.2 neurons - is increased by inhibitory inputs from neurons preferring the A- to the W-configuration, such as tectal T5.3 and/or pretectal TH3 cells.

M5.3 Characteristic

The excitatory activity of these cells showed configural response characteristics similar to T5.3 or TH3 neurons: $R_S \geq R_A > R_W \geq 0$. The somata of M5.3 neurons were located in the BMC or in the medial RetF close to the BMC. In M5.3 cells the A *vs.* W preference (Fig. 14) appeared to be sharper than in corresponding tectal or pretectal neurons suggesting that, here too, the discriminative acuity was increased. Also in these neurons, best activity zones in their RFs were observed during traverse of a visual object. Like TH3 cells (Tsai & Ewert 1988), these neurons were activated by a moving textured background.

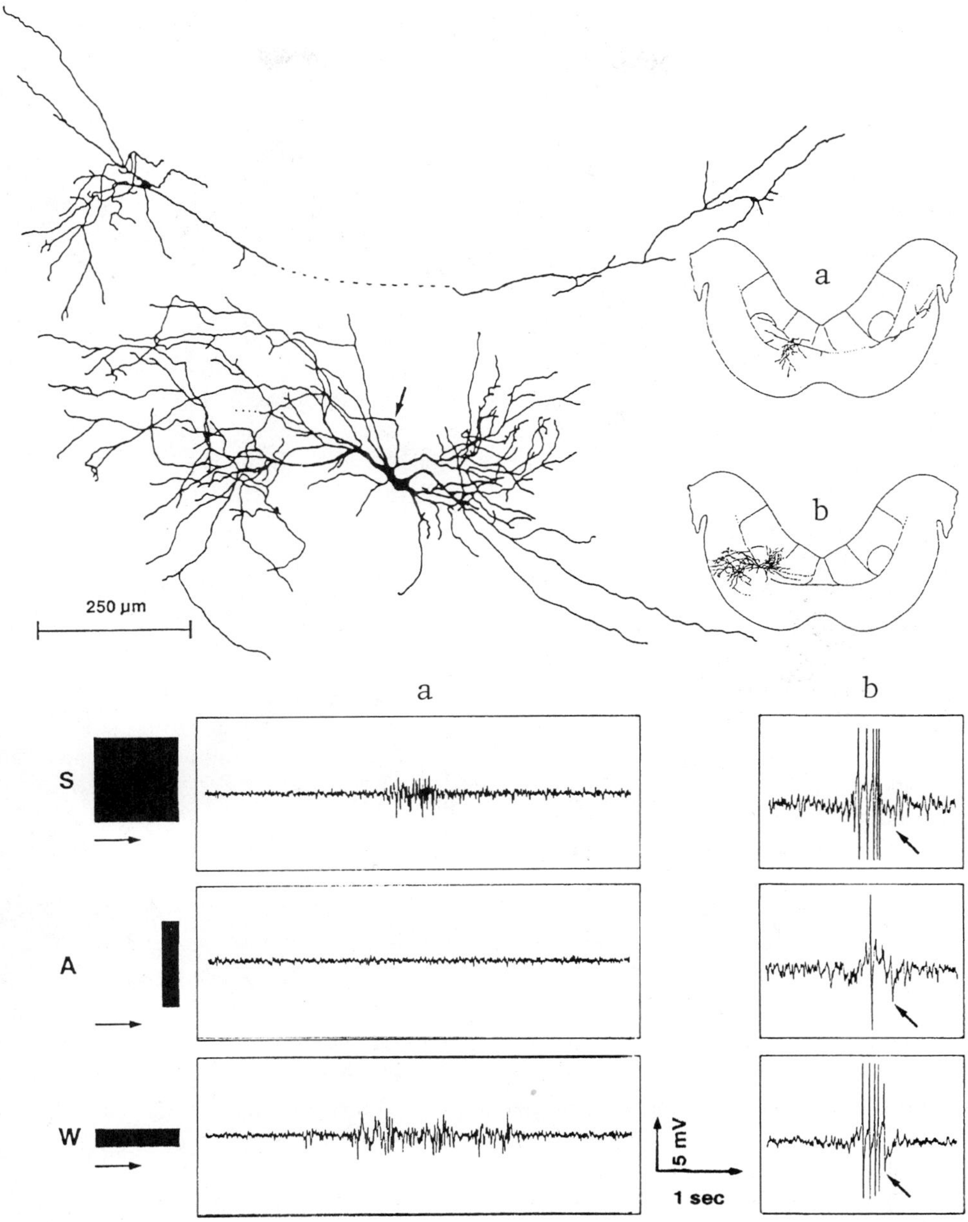

Figure 13. Examples of two intracellularly recorded and labeled M5.2 neurons, one (a) belonging to the morphological *Type III* whose soma was located in the medial RetF at the level of the *VIIth* motor nucleus, the other (b) belonging to *Type IV* whose soma was located in the VII*th* nucleus. The RF diameters were about 40° (a) and 90° (b). Note the intermingled EPSP/IPSP activity in (a); in (b) spikes followed an EPSP and were terminated by an IPSP (see arrows). In both neurons the excitatory responses to the worm-like *W* stimulus were remarkably stronger than to the antiworm *A* stimulus. (From Schwippert et al. 1989).

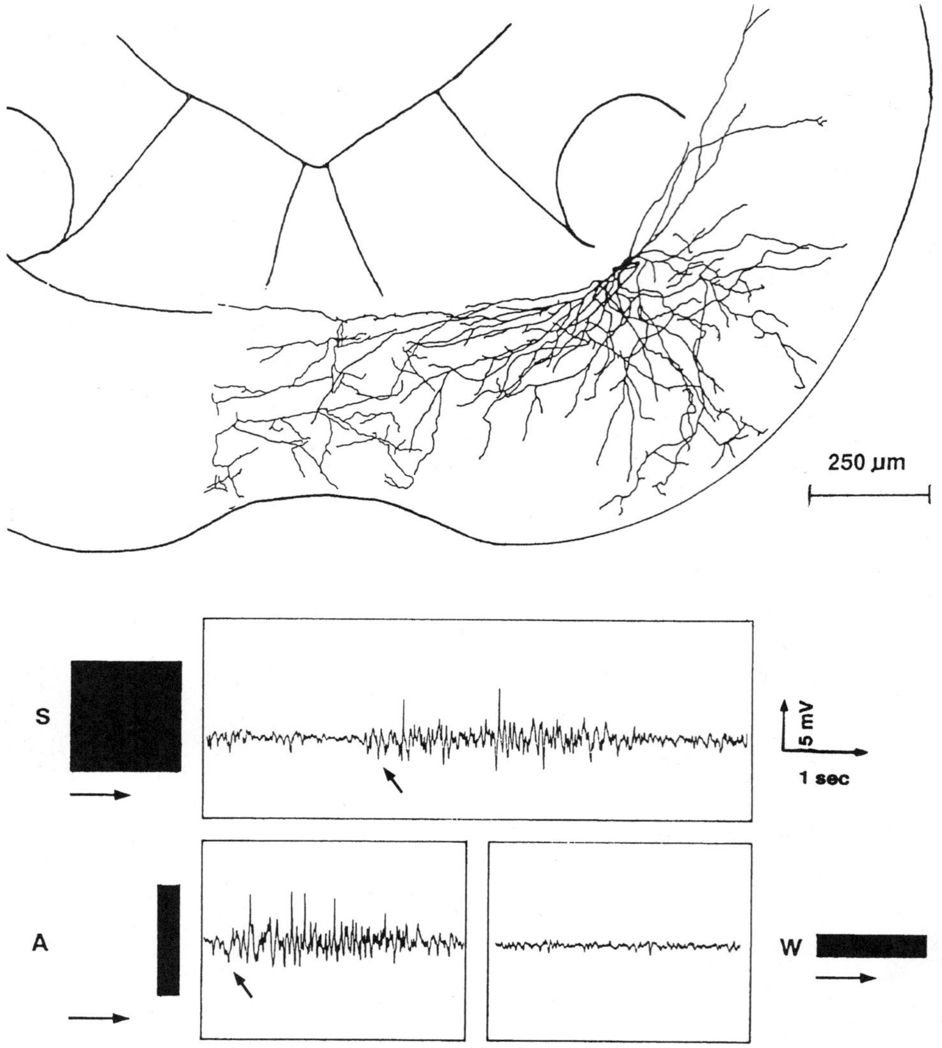

Figure 14. Example of an intracellularly recorded and labeled M5.3 neuron of *Type I* whose soma is located adjacent to the VII*th* nucleus. This neuron responded with EPSP/IPSP activity that was much stronger to the square *S* and to the antiworm *A* than the worm *W* stimulus. (From Schwippert et al. 1989).

The observation that in these bulbar cells moving visual objects elicited both EPSPs and IPSPs (Fig. 14) suggests excitatory and inhibitory inputs from different visual sources. Best activity zones were always smaller than the large RFs which sometimes reached 120° in diameter (compared to about 30° in T5.3 cells), and this again is an indication of concurrent inputs. These highly integrating cells characterized by extensive dendritic ramifications (Fig. 14) showed rather complex responses. For example, such a cell of the morphological *Type I* situated at the border between the grey and white matter near the VII*th* nucleus had an RF of about 120°, or even larger, which indicates additional T4 input. The excitatory activity in response to S or A was introduced by IPSPs (see arrows in Fig. 14) suggesting a pre-existing inhibitory loop. Change of the background illumination elicited only very small IPSPs at *off* and *on*. The activity during W-stimulation was not distinguishable from the control experiment without visual stimulation. Tectal input could be confirmed by electrical stimulation of the ventrolateral optic tectum. In this case, the response started with one or two spikes followed by sustained after-discharges in terms of IPSPs and EPSPs, for more than 10 s. Taking all these results together, it is likely that M5.3 are driven by excitatory *and* inhibitory visual inputs.

M5.4 Characteristic

These neurons responded T5.4-like with EPSPs (and/or spikes) to big compact moving visual stimuli, such as square and disc shaped objects, but not to small stripes: $R_{S32} > R_{S16} > R_A = R_W = 0$. They were most frequently recorded as morphological *Types I* and *V*. The RF diameters ranged between 40 and 100°. In response to brisk change of diffuse light, tonic-(phasic) *off* (and *on*) postsynaptic responses indicated inputs of tectal or pretectal cells driven by retinal R4 (and R3) ganglion cells. Compared with the control, a slightly enhanced inhibitory activity toward the W- and A-configuration was obtained, whereas the S-stimulus elicited spikes, EPSPs, and IPSPs (Fig. 15a,b). The "perceptual sharpening" between stripes and squares can be explained by a special computation of the stimulus area. Note that EPSP > IPSP for S-stimuli, whereas IPSP > EPSP for A- or W-stimuli (Fig. 15a,b). M5.4 neurons were strongly activated by moving textured backgrounds. Some cells showed also activation in response to cutaneous stimulation.

M5.0 Characteristic

In the medulla oblongata, there were a couple of neurons which did not show any clear discrimination between S-, W-, and A-stimuli. The RF diameters of these neurons were less than 100° and thus remarkable smaller than those of the M4 cells (> 180°) resembling the tectal T4 widefield neurons. These cells were called M5.0 in analogy to T5.0 neurons recorded from the optic tectum, predominantly after thalamic/pretectal lesions (Ewert 1984). In a few cases, the repetitive presentation of moving visual stimuli evoked in M5.0 cells a response shift to M5.1, M5.3, or M5.4 characteristics, in particular if these were traveling through the frontal visual field. Presumably, such neurons obtain concurrent inputs from a "modulated T5 base" (see Betts, this volume).

Summarizing the properties of M5-type neurons described from extra- and intracellular studies - in comparison to T5-types - four points are worth mentioning: (i) the acuity of configural selectivity seems to be stronger; (ii) the RFs/ERFs are larger; (iii) there are discrete zones of maximum activity within the RFs/ERFs; (iv) the received information is obviously richer, probably due to concurrent inputs from different tectal (and non-tectal) sources, for which the bewildering spread of dendritic trees seen in most bulbar cells provide a precondition. Although the large RFs/ERFs point to a non-topographic representation, just these neurons display a pronounced configural selectivity (e.g., responding to W- but not to A-stimuli or *vice versa*), and they are suitable to take yes/no decisions. Acuity of discrimination should

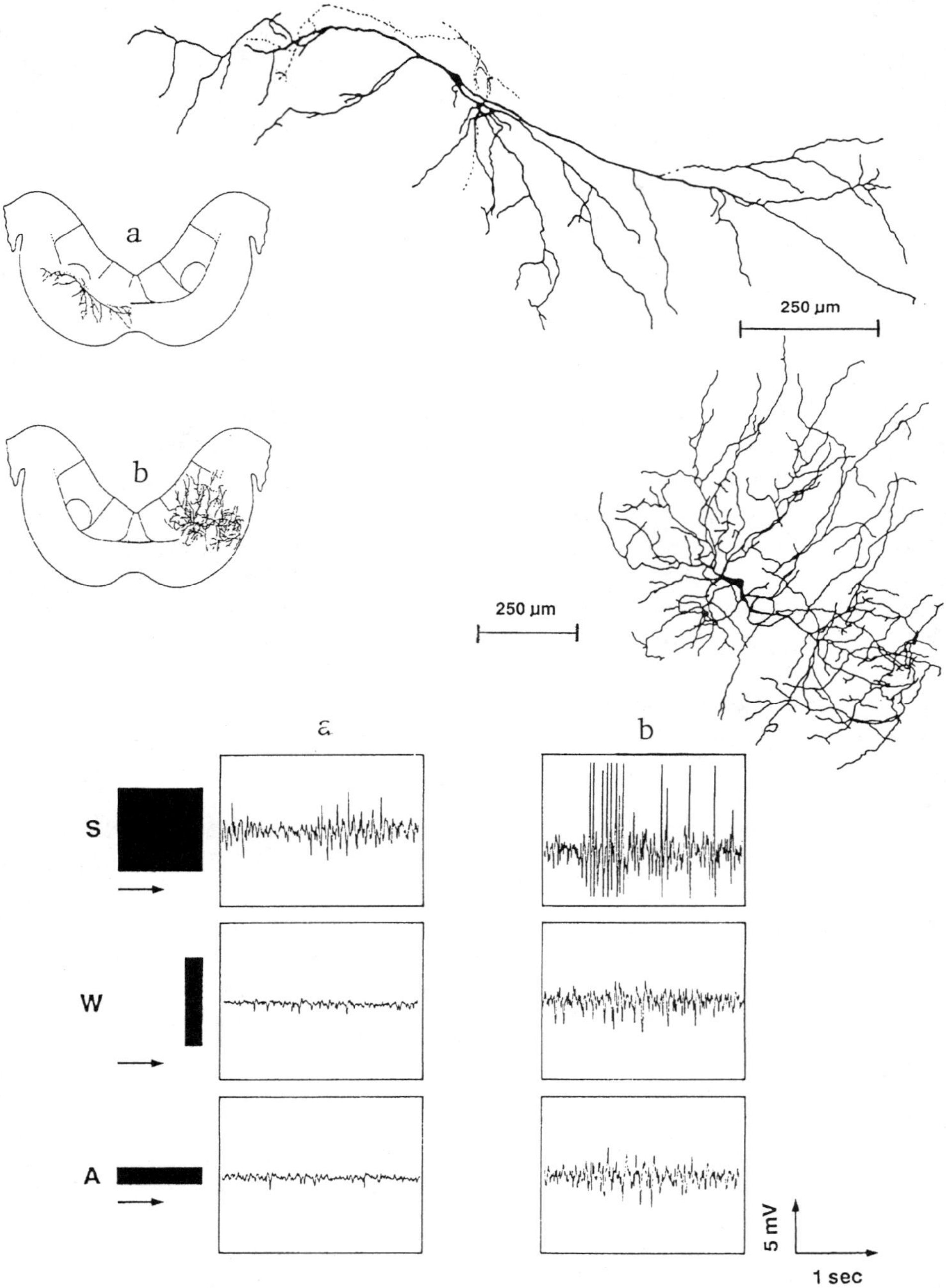

Figure 15. Examples of two intracellularly recorded M5.4 neurons, one (a) of the morphological *Type I*, the other (b) of *Type V*, the somata of both located in the VII*th* nucleus; the RF diameter was about 50°. Note that in both cases a square of 32° (or 16°, not shown here) strongly activated the cell, while worm-like *W* or antiworm-like *A* moving stripes did not elicit excitatory but more (b) or less (a) IPSP activity. (From Schwippert et al. 1989).

be one criterion for neurons that take part in pre-motor networks. Further criteria have to be investigated. Recalling the concept of "command releasing system," bulbar neurons at the pre-motor level indeed show combined tectal (and pretectal) properties, such as T4 and T5.2 or T4 and T5.4 or TH4 and T5.4, and there is indication that these properties are sharpened by lateral interactions among adjacent neurons of the same type.

M10 Characteristic

About 20 % of the bulbar neurons, so far investigated intracellularly, responded with cyclic bursting after visual or tactile stimulation (Figs. 9f, 16ab, and 17), comparable to the results obtained from extracellular recordings (Fig. 6E,F). From the ten successfully labeled M10 neurons, six cells belonged to the morphological *Type V* and one to *Type IV* all situated in the BMC, and three to *Type II* in the medial RetF (close to BMC). In particular, cyclic bursting as a post-stimulus event was frequently evoked by S32-stimuli, especially by circular, irregular movements (simulating features of a predator; Fig. 16a,b) or, in some cells, by a stripe in W-configuration (simulating features of prey; Fig. 17). In visual sensitive M10 neurons, electrical stimulation of the optic tectum with a train of negative square wave impulses of 10 to 30 Hz was most effective to trigger bursts (Fig. 16a, "current"). Among different cells, the average spike frequency within a burst ranged from 10 to 40 Hz and was mostly around 25 Hz. The firing rate was often higher at the beginning of a burst. An entire burst sequence consisted of 2 to 12 cycles. The inter-burst intervals were between 1 and 2 s and generally shorter at the beginning of a sequence. After the periodic activity, there was a refractory period of 1-5 min until the next burst sequence could be elicited.

The intracellular investigations allowed us to study membran activities of these neurons in more detail. In a spontaneously M8-like firing bulbar neuron the introducing burst in response to a visual S32-stimulus (Fig. 16a) was at first preceded by an increased postsynaptic activity followed by spikes of increasing frequency. The M9-like warming-up seemed to be the "generating signal" for the subsequent repetitive bursts. A sequence of bursts often began with a sort of "resetting" IPSPs. A typical example is shown in Fig. 16a (see arrow), in which a train of electrical impulses applied to the contralateral optic tectum elicited bursting. A burst was introduced by spike-level reaching EPSPs; each burst was "terminated" by a strong IPSPs which shifted toward the baseline (Figs. 9f and 18a). At the end, the cyclic burst activity was mostly finished by "free-running" tonic spike activity (e.g., see Figs. 16 and 17).

To electrical tectal stimulation, a single stimulus impulse typically elicited one or two spikes in the M10 neuron (Fig. 16b); bursting required a short impulse train (see also Fig. 6Fc). Burst sequences elicited by visual or electrical tectal stimulation were not necessarily identical; e.g., in the former case bursts could be phasic-tonic (Fig. 16a, see "S") and in the latter case phasic (Fig. 16a, see "current").

The releasing conditions, the temporal pattern, and the refractoriness of the cyclic bursts in toad's bulbar neurons resembled properties of vertebrate's (e.g., Matsuoka 1985; Willis 1986; Roberts 1987; Garcia-Rill & Skinner 1987) and invertebrate's (e.g., Bässler 1986; Gillette 1987) bursting neurons known to play the prominent role in pre-motor and motor activity. We have therefore studied in very

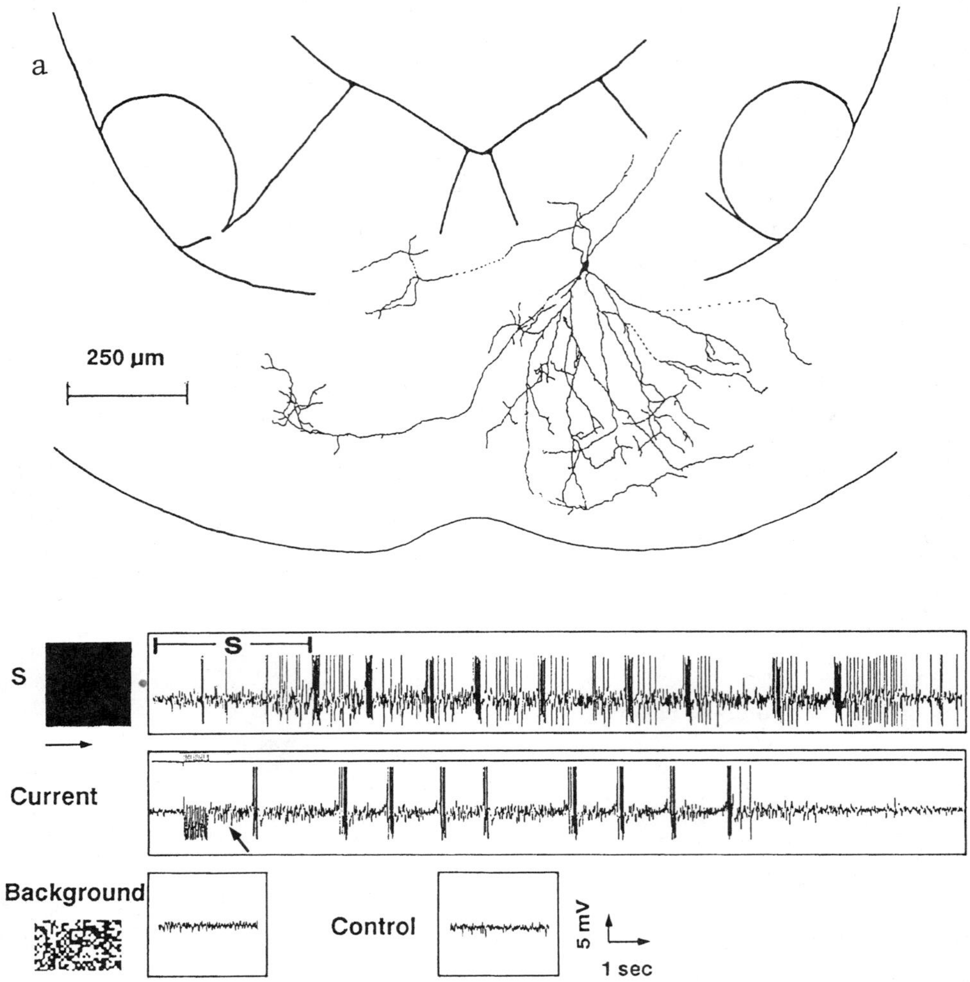

Figure 16. Examples of two intracellularly recorded and labeled M10 neurons, one (a, this page) of the *Type II* whose soma was located in the medial RetF at the level of the VII*th* nucleus, the other (b, opposite page) of *Type IV* whose soma was situated in the VII*th* nucleus. In both neurons a looming large stimulus S elicited spike activity which was "converted" into cyclic bursts of spikes as a post-stimulus event. Note in (a) that a similar burst activity can be elicited after stimulation of the contralateral rostrolateral optic tectum with a 1s lasting current consisting of a train of electrical impulses at 15 Hz; this burst sequence is preceded by small IPSPs (see arrow). In (b) single electrical impulses applied to the tectum initially elicit an IPSP followed by excitatory activity that depends on the strength of the current. (From Schwippert et al. 1989).

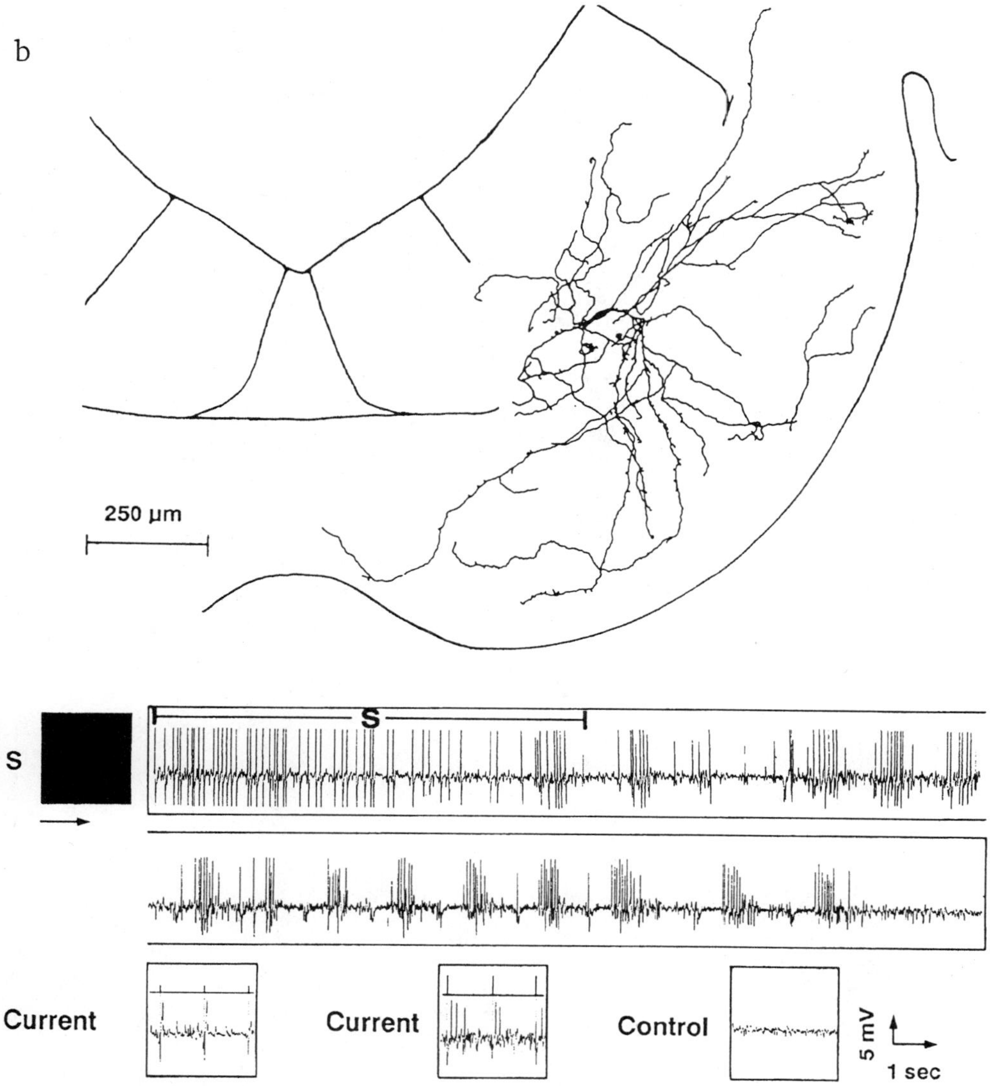

lightly paralyzed toads the mechanograms of neck muscles during intracellular recording. In a few successful cases, rhythmic contractions of eye and neck musculative accompanied by attempts to locomote were "introduced" by bulbar bursters (Fig. 18b). In another experiment, after bursting with the onset of a movement the electrode left the cell (Fig. 18c).

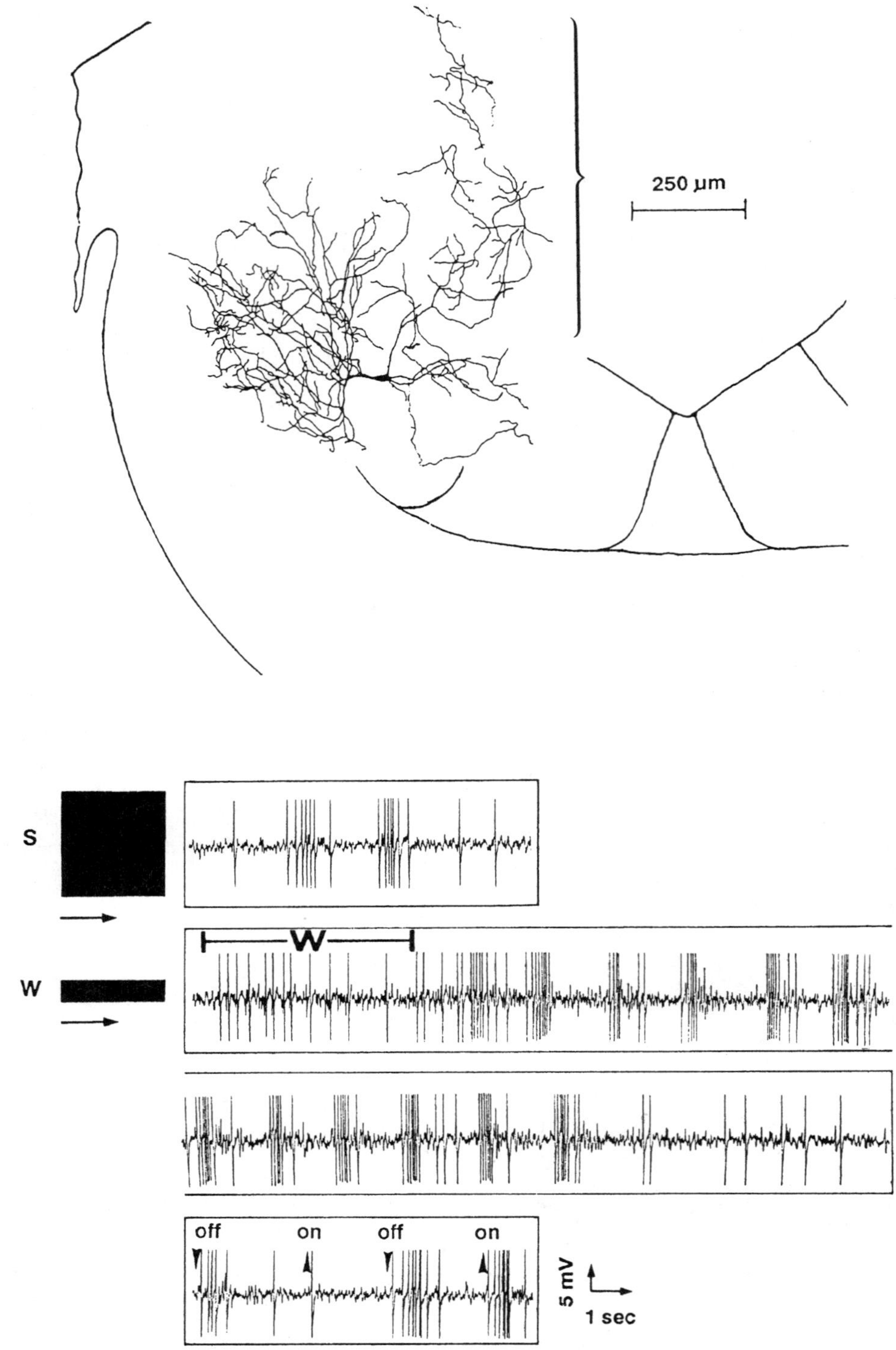
250 µm
S
W
W
off
on
off
on
5 mV
1 sec

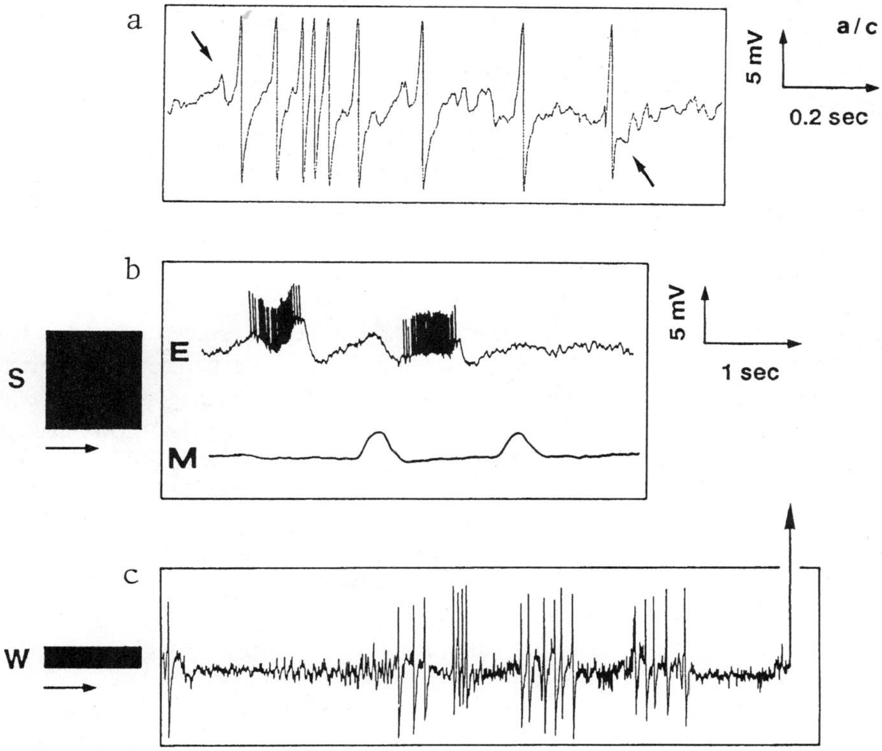

Figure 18. Some features of M10 neurons. (a) A burst of spikes - cut out of a burst sequence - is introduced by an EPSP and terminated by a strong IPSP (see arrows). (b) Correlation between bursting and neck muscle contraction in a very lightly succinylcholine-paralyzed toad; each intracellularly recorded burst event (E) is succeeded by neck muscle contraction (M) registrated by means of an electromechanic transducer. (c) In another slightly paralyzed toad, bursts elicited by a worm-like stimulus resulted in elevation of the animal's head, so that the intracellular recording electrode left the cell (see vertical arrow). (From Schwippert et al. 1989).

Figure 17. Example of an intracellularly recorded and labeled M10 neuron of the *Type V* whose soma is located in the V*th* motor nucleus. Unlike the neurons shown in Fig. 16a and b, this cell displayed cyclic burst activity specifically after presenting a worm-like *W* moving stripe. Note that the square *S* elicited spikes during stimulation but no post-stimulus bursts. To changes of diffuse light, the neuron showed brisk bursts of spikes at *off* and *on*, but no cyclic burst activity. The dendritic arborization of this neuron indicated by the large bracket extended towards the cerebellum. (From Schwippert et al. 1989).

4 Discussion

To investigate rules of morphological and physiological properties of integration by bulbar neurons, in the present extra/intracellular recording studies we have selected a segment of the medial/lateral RetF and its adjacent BMC - between superior and inferior divisions mainly at the level of the facial VII*th* motor nucleus - known to receive tectal inputs.

4.1 General Properties of Bulbar Neurons in Toads

The anuran medial RetF of the medulla oblongata is a projection field of tectal (Rubinson 1968; Lázár 1969), toral/tegmental, and thalamic/pretectal efferents (Weerasuriya & Ewert 1981; Ingle 1983; Tóth et al. 1985). The extra- and intracellular recordings from the RetF and the BMC, reviewed in this chapter, show that many medullary M-type cells display similarities to tectal T-type and thalamic-pretectal TH-type neurons regarding *receptive field arrangements* (M1- and T1-types; M2- and T2-types; M4-, T4-, and TH4-types), *configural sensitivities* (M5.1-4, T5.1-4, and TH3), *spontaneous activity* (M8, T8, and TH8), and *warming-up* and *reverberatory properties* (M9, T9, and TH9) (see Ewert 1984 for a review). The results from antidromic stimulation experiments (Satou & Ewert 1985) suggest that certain characteristics of bulbar neurons are determined by converging inputs of descending tectal cells.

4.1.1 Gaining and Filtering

The "perceptual sharpening," beginning in retinal layers (Lettvin et al. 1959) and established in pretectal/tectal networks (Ewert 1974, 1976), seems to be improved in bulbar reticular structures and thus occurs as a multi-level process that involves *gaining operations* (e.g., achieved by convergence) and *filtering operations* (e.g., mediated by lateral interactions). There are various hints suggesting that the M5-type characteristic is amplified via inputs of "command signal carrying" T5-type projection neurons: (i) the diameters of the RFs/ERFs of the bulbar neurons are about twice of the RFs/ERFs of the tectal T5-type equivalents; (ii) within the RFs/ERFs there are small areas of optimum excitability, and these areas may change their diameters or may even shift their positions; (iii) the configural tuning of M5-type neurons is sharper than in tectal T5-types; (iv) for certain stimulus configurations, IPSPs dominate EPSPs. The increase in the selective properties can be explained by threshold operations involving lateral inhibition between tecto-bulbar projecting T5-cells that belong to the same type and closely agree in threshold. Also the spatiotemporal overlap of receptive fields might play a role. In addition, there may be "heterogeneous convergence." For example, the T5.2-mediated M5.2 selectivity may be increased to some extent by inhibitory inputs from T5.3 and TH3 projection cells (for IPSPs in M5.2 cells see Fig. 13b). In M5.4 neurons, the "stripe neglect" (or even inhibition) could be explained by inhibitory inputs from T5.3 and T5.2 cells (see IPSPs in response to W- and A-stimuli in Fig. 15b). It seems that perceptual sharpening is reached at the expense of a topographical representation, which raises the question for the mechanisms responsible for stimulus localization (see Grobstein, this volume). An evaluation of M-neurons, based on receptive field sizes *alone*, would be an inappropriate approach to bulbar functional properties.

Another aspect of integration by bulbar structures is exemplified by multiple properties displaying neurons, which is in agreement with Ewert's notion of *sensorimotor codes* and the concept of "command releasing systems." The bewildering dendritic trees of most bulbar neurons provide a suitable substrate for such integrative and interactive processes.

4.1.2 Laterality, Binocularity, and Widefieldness

The finding that most of the monocular M5-type neurons were activated from the eye ipsilateral to the recording site (while only few neurons responded to stimulation of the contralateral eye) suggests main inputs to these cells from the contralateral tectum mediated by tr.tbs.c. (and less from the ipsilateral tectum via tr.tb.d.). The different proportion of tecto-bulbar/spinal mediated tectal efferents (tr.tbs.c. : tr.tb.d > 1) approximated from ipsi- and contralateral RF/ERF properties of M5-type neurons is consistent with results from antidromic stimulation/recording studies (Satou & Ewert 1985) and with the different morphological strength of both tracts (Ebbesson 1976). Binocularity in M-neurons at the bulbar level - in response to monocular inputs - could be achieved by convergence of tectal efferents traveling in tr.tbs.c. and tr.tb.d., by decussating fibers of tr.tb.d. in the rostral medulla, and by decussating dendrites (or axon collaterals) of M-neurons.

The extremely wide visual RFs/ERFs of M4 and M4-M9 neurons - which may have general arousing ("sameness" type), selectively sensitizing ("newness" type), or timing ("reverberatory" type) characteristics - allow different interpretations. The first which comes to mind is that tecto-bulbar descending T4 neurons provide a sort of an unspecific "hallo"-system that together with a certain type of T5 descending feature analyzing cells, releases a toad's goal-directed response. Since tectal T4 neurons are rarely antidromically activated by bulbar electrical stimulation, the M4 wide field properties may also arise from convergence of tectal T5-type neurons, or from bulbar intrinsic interactions, or from extratectal (e.g., thalamic pretectal) afferents, besides descending axonal T4 projections. Evidence of excitatory tectal inputs to M4 neurons is provided by their synaptic activation in response to tectal stimulation. Since T4 cells in turn can be activated by bulbar electrical stimulation, T4 properties could also be determined by bulbo-tectal feedback. These projections have been shown anatomically (Lázár, this volume). Such an ascending feedback-loop, for example, might influence the activity level of tectal feature-detecting T5 neurons. In this case, the widefield characteristics would not be a special tectal property.

Among bulbar visual widefield neurons (M2.1, M2.2, M4-types), many have large cutaneous mechanoreceptive fields. Here, too, the question arises whether this property is determined by inputs of similar bimodal thalamic neurons (Ewert 1971), subtectal neurons (Ewert & Borchers 1971), and/or by spino-bulbar connectivity. Results showing that visual/tactual sensitive bimodal M4 neurons can be synaptically activated by electrical stimulation of the spinal cord supports the latter possibility.

With respect to avoidance behavior, a fast visual "escape system" should be present whose neurons conducting information via a few synapses fire to threatening visual objects somewhere in the wide RF/ERF with an increasing rate depending on the stimulus size. Certain kinds of widefield neurons, e.g. T6/M6, may be involved in such a function that might exist besides a configural predator recognition system whose pretectal/tectal interactive computations take a longer time on account of more synapses. Another fast escape system could be also mediated by T5.4 neurons involving about 6 synapses, including the neuromuscular junction, from receptor to effector. Such a relatively fast system might be an ancestral one, although we must consider (see Ebbesson 1984) that a putative so-called simple system may have evolved from a complex structure.

4.1.3 Loop Interactions

Pretectal, tectal, bulbar reticular, and spinal structures in toads are functionally interconnected through loops. This has been shown for tectum and medulla, for medulla and spinal cord (Ewert et al. 1989), and for pretectum and tectum (Ewert 1968, 1974; Ewert et al. 1974; Brown & Ingle 1973). It is suggested that tectal intrinsic mutual excitatory loops between tectal units (Székely 1973) are selectivly controlled and goal-specified by pretecto/tectal inhibition for *visual pattern recognition* (Ewert 1968, 1987; Ewert & v. Seelen 1974; Lara et al. 1982). Pretectal/tectal excitatory loops may be related to *memory* (Ewert 1971; Ewert et al. 1974; see also contributions by Lara and Finkenstädt, this volume). It is reasonable to suggest that bulbar/tectal and bulbar/spinal excitatory loops (e.g., involving M11-type neurons) participate in *time programs* of motor pattern generating circuits. Schürg-Pfeiffer (this volume) shows that the activity of prey-selective tectal T5.2 neurons (descending in tr.tbs.c.) not only precedes toad's orienting toward prey in the manner of "warming-up," but also persists during the turning movement. It seems likely that T8, T9, M8, and M9 neurons on account of their tonic firing and warming-up properties are involved in circuits in which "movement predicting" T8 cells (Borchers & Pinkwart 1983; Ewert 1984) may be part of a general "go/nogo" system (for a further discussion see Schürg-Pfeiffer, this volume).

Both tectal and reticular structures are influenced (adjusted and modulated) by prosencephalic nuclei of the "limbic system" (e.g., medial pallium, nucleus amygdalae) and related structures (e.g., hypothalamus, preoptic area, anterior thalamus) (Wilczynski & Northcutt 1983a,b; Finkenstädt et al. 1985, 1986; Ewert & Finkenstädt 1987; Finkenstädt, this volume). These influences are partly mediated by the solitary system (Weerasuriya 1983; see also Weerasuriya, this volume).

4.1.4 The "Reticular" Network: An Interwoven Structure

The discovery of "multiple properties" displaying bulbar neurons suggests also intrinsic integration and interaction among bulbar cells. The fact that recording sites of different M- and combined M-type cells in the medial and lateral RetF are largely intermingled (Fig. 5b) is an indication for the interwoven nature of this structure in the sense of a "reticular" network that has access to motoneurons of the adjacent BMC as well as to caudal bulbar (e.g., hypoglossal) and to rostral spinal motoneurons, partly mediated by reticulo-spinal M11-projection cells. This is supported by neuroanatomic studies showing that iontophoretic application of HRP into the caudal medulla oblongata (Weerasuriya, this volume) or into spinal areas (ten Donkelaar et al. 1981) leads to backfilling of cells dispersed in the area of recording sites of M-type neurons in the ventral central grey of the medial RetF, mainly at the boundary between layers 3 and 4, coincident with Gaupp's (1896) "medioventral cells," Röthig's (1927) "motor relay cells," and Edinger's (1908) "premotor cells."

Although the present examination of M-neurons in paralyzed toads hardly gives information about motor pattern generating (MPG) circuits responsible for specific action patterns, the post-stimulus or spontaneous bursting M10 neurons provoke speculations. Toads possess various rhythmic action patterns (e.g., release calling, mating calling, snapping/gulping) whose generation involves bulbar central pattern generators. These might take advantage of (or are connected with) the breathing MPG that extends between the V*th* and the posterior edge of the XII*th* nucleus (e.g., see Viala 1986; Viala & Freton 1983; Euler 1983). Schmidt (1974a,b) suggests that the vocal phase of calling may have evolved from the expiratory-phase of breathing, and the inspiratory-phase of calling seems to be identical to the inspiratory-phase of breathing: A call-generating circuity, once triggered by an appropriate command system, thus would modify the breathing MPG to produce calling by appropriate coordination of the laryngeal muscles, hence changing the expiratory-phase of breathing into the vocal-phase of calling. (For a discussion on vocal cord mechanisms see also Schmidt 1971, 1973, 1976; for details on the origin and pattern control of breathing rhythmicity in vertebrates see Baumgarten & Kanzow 1958; Euler et al. 1973; Mitchel & Herbert 1974; Cohen & Feldman 1977; Merrill 1979, 1981; Lipski & Merrill 1980; Feldman et al. 1980; Madden & Remmers 1982; Rose et al. 1984).

With respect to the interwoven nature of the bulbar reticular structure, the question arises as to whether MPGs for different motor patterns are derived from a general bulbar pattern generating circuitry. Depending on specific inputs from different *command releasing systems* and on peripheral feedback this circuitry, for example, might develop motor patterns suitable to generate rhythmic vocal release croaks in response to mechanosensory input of a clasping rival male, or rhythmic tongue glicking as a threat noise in response to a visual predator (typical for the Blomberg toad), or swallowing prey, or eye-closing/jaw-grasping/swallowing of the slack body skin during moulting. Other MPGs for different action patterns - such as locomotion, or retreat from a predator, or turning toward prey - involve spinal circuitry that takes into account vestibular and proprioceptive inputs (Comer & Crain 1972; Pompeiano 1981; Ashton et al. 1984; Dieringer & Precht 1986).

The fact that bursting in many bulbar neurons can be triggered by a large looming stimulus (or by electrical tectal stimulation) suggests (i) links between tectal command releasing systems involving particularly class T4 and T5.4 neurons, (ii) integrative bulbar functions (M5.4-M4-M8-M9), and (iii) central pattern generation (M5.4-M4-M8-M9-M10-M11) for release of avoidance action patterns (e.g., retreat or hopping). In a few cases our intracellular recordings in slightly paralyzed toads have shown a correlation between visual threat stimulus, subsequent activation of a M5.4-M4-M8-M9-M10 system (warming-up in tonic firing and leading to bursting), and onset of eye/neck muscle contraction.

Surprisingly, in freely moving toads cyclic bursting neurons were seldom recorded. To study the influence of the succinylcholine-induced neuromuscular blockage, long-term recordings with the technique developed for freely moving preparations were performed. The results indicate a correlation between the fading paralyzing effect, an increased ability to contract eye and neck muscles, and a decrease in the number of repetitive bursts in the type of M10 neuron under investigation; as the toad is then able to move, head elevation is introduced by one burst whose spike frequency is correlated with the velocity of head movement (Schürg-Pfeiffer, in preparation.).

Our intracellular data from paralyzed toads suggest that there are bulbar burst rhythm generating circuits which - upon adequate trigger input ("sensorimotor code" according to Ewert) - drive M10 neurons to discharge sequential bursts of spikes. Taking into account the aforementioned data on bulbar neurons in freely moving animals, it seems that the cyclic burst activity continues in paralyzed toads partly due to a lack of feedback information, e.g. proprioceptive inputs (Takizawa 1981). The feedback (referring to the type of action pattern) then may decide as to how the cyclic activity either should persist, or should be modified, or should even be reduced to just one burst. The question whether succinylcholine induced blockage of neuromuscular transmission in addition influences toad's motivation, attention or arousal, has to be studied (Beneke & Schwippert, in preparation.).

4.2 Problems with the M-Classification

The distinction of M-type neurons is problematical and, as already mentioned in Section 2.11, only of heuristic value since we are dealing here with properties, their combinations and interactions. So, striktly speaking, for example, there may be no class M5.4 cells, but bulbar cells which in response to certain criteria display properties of T5.4 besides other properties not investigated here. The problems in the M-classification become particularly obvious with the M10 bursting neurons. If certain visual sensitive cells of the medial RetF integrate visual, proprioceptive, and state-dependent information, then it is reasonable to assume that bursting is only evoked if the combination of these inputs is appropriate. Neurons designated as M10 should then form the final pre-motor stage characterized by the ability to produce bursts. However, this does not exclude the possibility that other cells showing no burst activity in an ongoing experiment will not belong to M10. The absence of bursts, for example, can be due to a lack of state-dependent and subthreshold sensory inputs. Furthermore, it is probable that certain neurons are more or less "passively" driven by bursting cells. In summary, our proposed naming of neurons of the toad's medulla oblongata only accentuates properties under investigation. This is useful for the experimenter, from a conceptional point of view for the design of experiments. This does not exclude the existence of other important neuronal features that will assign certain cells described here to "other classes," which may be significant in another context.

4.3 Comparisons

There are several structural and functional homologies of colliculo(tecto)-bulbar structures among vertebrate species. Antidromic stimulation/recording techniques in mammals have revealed two-way connectivity between superior colliculus and bulbar

RetF (Peterson et al. 1974; Rose & Abrahams 1978; Peterson & Fukushima 1982). In amniote vertebrates it could be shown that locomotion involves ponto-medullary structures (Shik & Yagodnitsyn 1977; Kazennikov et al. 1980; Grillner et al. 1982; Steeves et al. 1986). Decerebrate geese or ducks, for example, locomote if certain areas of the ventromedial ponto-medullary RetF (gigantocellularis) or the more dorsal parvocellular RetF - which harbors reticulo-spinal projecting neurons - is stimulated electrically (Steeves et al. 1986). Electrical stimulation of reticular regions in the nucleus reticularis pontis caudalis (gigantocellularis, or ventralis) induces postsynaptic activities in motoneurons that are responsible for the activation and coordination of eye and neck muscles. The colliculo-reticulo-spinal pathway is necessary to activate eye and neck motoneurons for various orienting responses in mammals (Anderson et al. 1971). An intermediate influence of tegmental elements during a colliculo(tecto)-reticular signal transfer is probable (see also Grobstein, this volume). Single cell recording studies in mammals have shown that the colliculo-bulbar projections are not topographically organized (McIlwain 1986). Evidence of parallel processing is provided by data showing that cells triggering the horizontal saccades are localized in the paramedian RetF, whereas those triggering the vertical saccades are in the tegmental RetF (Ito 1986). The caudal pontine and bulbar RetF in cats exhibit zones of reticulo-spinal neurons with different functions regarding eye, neck, and back movements (Peterson et al. 1974; Peterson et al. 1978; Grantyn et al. 1977; Peterson & Fukushima 1982; Grantyn et al. 1987). In rats it has been shown that the nuclei gigantocellularis and paragigantocellularis of the medial and lateral RetF are essential for rhythmical jaw movements involved in mastication (Ito 1986; Moriyama 1987). Recording studies in the sheep suggest (Amri et al. 1984) that medullary swallowing neurons in the ventral RetF connect the medullary "swallowing center" to the V*th* motor nucleus; during swallowing, inputs originating from the dorsal region of the medullary center (interneurons programming the motor sequence) are probably relayed in the ventral RetF adjacent to the nucleus ambiguous before reaching the pontine swallowing neurons. Although amphibians do not show eye saccades or masticative movements comparable to mammals, their eye, neck, and jaw movements are integrated in a variety of action pattern.

4.3.1 General Properties of Bursting Neurons

Two major physiological types of cyclic bursting neurons are distinguished: (1) In one type a train of bursts is induced by synaptic interaction. (2) In another type bursts arise due to particular regenerative membrane properties involving special ionic mechanisms. Regarding (2) it is accepted that the depolarization plateau responsible for the repetitive spike-discharges is carried by inward sodium- and calcium-currents, and that the hyperpolarization which finishes a burst and which maintains the interburst interval, is caused by an outward potassium-current as an intrinsic membrane reaction. The "cardiac model" (Noble 1983) provides the classical example for cells whose pacemaker activity takes advantage of an intrinsic oscillating membrane mechanism. In neuronal networks (1) synaptic interactions are responsible for burst sequences broken by IPSPs via inhibitory interneuronal loops (Russel & Hartline 1978; Hartline & Gassie 1979; Selverston et al. 1983).

In general, four properties of neuronal loops are discussed (Bässler 1986; see also Hartline & Gassie 1979): (i) Endogeneously bursting cells, due to spontaneously oscillating membrane potentials, are built in single or combined cyclic loops that obey the classical positive and negative feedback mechanisms (Hartline 1967; Watanabe et al. 1967; Maynard 1972); (ii) networks with negative feedback properties consisting of recurrent cyclic inhibitory (Kling & Székely 1968; Friesen et al. 1976) or cyclic excitatory/inhibitory loops with an odd number of inhibitory connections; such networks oscillate if the open loop gain is kept within certain mathematical marginal conditions; (iii) networks with positive feedback properties - reverberatory circuits of mutually excitatory elements (Maynard 1955) - oscillate if a high-pass filter function is incorporated in the loop; precondition for bursting is that such systems have no (or an even number of) inhibitory connections; (iv) the loops themselves consist of several integrating sub-loops, so that the resulting network may reach a highly complicated organization level. Another feature of oscillating networks consists of neuronal adaptation of inhibitory elements (Wilson & Waldron 1968) of mutual inhibitory connectivity in conjunction with the presence of reverberatory loops (Matsuoka 1985).

4.3.2 "Eye Neck" Reticulo-Spinal Neurons in Mammals

Intracellular recording studies have shown that cat's eye-neck reticulo-spinal neurons relay collicular information to eye and neck motoneurons. These eye-neck cells work like integrators in a pre-motor network that produces the output signals for the eye position and the head movement (Willis 1986). Neurons situated in deep layers of the contralateral superior colliculus activate the reticulo-spinal cells in the ponto-medullary medial RetF at a latency of 2 to 10 ms (Peterson et al. 1974), suggesting both monosynaptic as well as polysynaptic pathways. The task of eye-neck cells is to control orientation behaviors involving cooperative movements of eyes and head. These neurons generate a special burst activity that influences the spontaneous discharges of motoneurons. As a result of spatial summation, the appropriate muscle system contracts. More specifically, eye-neck reticulo-spinal neurons generate a train of spike-bursts which is followed by a lasting declining firing towards a new steady-state that corresponds to the eye or head position during fixation. Bursting is correlated with the onset of muscle contraction at the beginning of saccadic and pursuit eye movements. The latencies between a burst and the onset of muscle contraction range between 100 and 150 ms, and the duration of a burst corresponds to the duration of a saccade (Grantyn & Berthoz 1987). Latencies of 3 ms and less between reticulo-spinal activity and neck musle motoneuron activation indicate monosynaptic reticulo-motoneuron connectivity (Peterson et al. 1974; Peterson & Fukushima 1982). The cell bodies of eye-neck neurons are distributed over various regions of the brain stem extending from the pontine to the bulbar RetF and branching into different motor nuclei such as the abducens (VI*th*), the facial (VII*th*), the vestibular nucleus (VIII*th*), and the nucleus prepositus hypoglossus (Grantyn & Berthoz 1987). Eye-neck neurons in mammals thus contribute to the reticular network responsible for orienting and fixating behaviors. Apart from superior collicular input, these cells integrate concurrent inputs from the mesencephalic RetF, the spinal cord, and from the cerebral cortex (Grantyn & Berthoz 1987).

4.3.3 The Mammalian Oculomotor Integrator

What are the cellular mechanisms that translate sensory information into bursting activity and the processes that synchronize bursting cells to control the appropriate target motoneurons? We choose the cat's (or monkey's) saccade producing oculomotor integrator as an example to show how different types of neurons - such as tonic, bursting, pause, and bursting-tonic, neurons - interact (Ito 1986; Willis 1986; for a review see Eckmiller 1987). To briefly repeat, bursting cells contact motoneurons directly, and in combination with tonically active neurons they create the decisive motor generating bursts. The pre-motor activity of bursting-tonic cells is closely correlated with the activity of the neck and eye muscle systems. The burst signal is proportional to the angular velocity of the saccadic eye movement, and the tonic signal is proportional to the steady-displacement of the eye. In the oculomotor adjustment and the coding of eye position, three types of other neurons contribute; these are pursuit, velocity, and intermediate phase neurons (Willis 1986; Eckmiller 1987).

In the oculomotor integrator of cats, three subcircuits are distinguished: (1) The excitatory circuit comprises bursting cells, tonic cells, bursting-tonic cells, and motoneurons; (2) an inhibitory loop enables the control of agonist-antagonist coordination; (3) a cerebellar circuit exerts an additional control of the oculomotor integration system by supplying the brain stem neurons with a standard activity below the threshold level. Figure 19 shows a principle that may underlie neuronal integration and interaction for saccadic generation (Willis 1986). We are ignoring here the contribution of cerebellar or muscular feedback systems. All cells of the

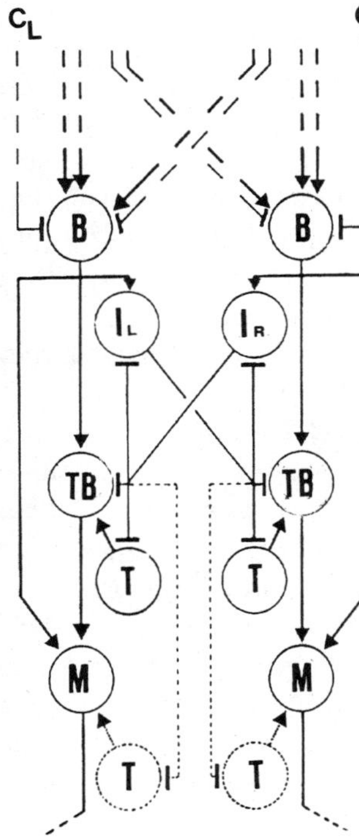

Figure 19. Proposed connectivity and interaction of neurons responsible for the translation of visual information into bursting activity controlling appropriate target motoneurons. Basic features of this network are derived from the mammalian oculomotor integrator (Willis 1986). The brackets in this legend refer to neurons recorded in toads, which may be involved in partly comparable functions. Left, C_L, and right, C_R, colliculo-bulbar/spinal output [tectal projection neurons]; bulbar bursting neurons, *B* [M10 neurons]; tonic neurons, *T* [M8 neurons]; tonic/bursting neurons, *TB* [M8-M10-M11 neurons], inhibitory interneurons, *I*; motoneurons, *M*. Arrows indicate excitatory and dashed lines inhibitory synapses. [Input from proprioceptors in the extrinsic ocular muscles to the vestibular nuclei in toads have been investigated by Ashton et al. 1984; for a consideration of eye position signals in frog's abducens motoneurons see Dieringer & Precht 1986]. For explanations see text.

oculomotor nuclei and the paramedian pontine RetF are represented symmetrically in each hemisphere. In the first step, information provided by superior collicular output cells (C_L and C_R in Fig. 19) is fed to bursting neurons (B) that respond with burst activity if certain conditions are fulfilled, such as spatiotemporal summation to exceed the "bursting level." The translation of a superior collicular trigger signal into bulbar standard burst signals is regarded as the proper interfacing process. These bursters (B) trigger another group of integrating neurons, the tonically active cells (TB) which have pre-motor properties and which activate the oculomotor neurons (M). The tonic activity of TB cells is maintained by pacemaker neurons (T). In a simpler scheme, B excites M driven by T, directly. Parallel output of B excites hypothetical inhibitory interneurons which, acting pairwise as bistable integrators (I_L,I_R), inhibit the corresponding cells, the TB cells, and the T cells in the contralateral hemisphere. It is suggested that a sufficient number of bursters must cooperate to generate the appropriate motoneuron activity (Grantyn et al. 1987).

4.3.4 Comparisons with Toads

Although toads do not show saccadic eye movements comparable to mammals, eye-neck coupled movements are integrated in various action patterns, such as orienting or avoidance turning. The question arises as to how the activation of toad's eye-neck system takes advantage of a bulbar network similar to the one of the mammalian oculomotor integrator. We find the following comparable data: (i) Monosynaptic tecto-bulbar connectivity provided by tectal T1, T2, T4, and T5-type cells whose properties might determine medullary M1, M2, M4 and M5-type neurons; (ii) tonic M8 neurons, (iii) bursting M10 neurons, (iv) tonic-bursting M8-M9-M10 neurons with warming-up properties, (v) reticulo-spinal M10-M11 bursting neurons. Furthermore, there is evidence that tectal output signals serve as a trigger to generate standard signals in terms of bursts. The intracellular records of toad's M10 neurons show spontaneous irregular EPSP and IPSP activity. The permament presence of subthreshold postsynaptic activities indicates concurrent inputs from several extrinsic and intrinsic sources. If the spatiotemporal summation reaches a defined level, then bursts will be evoked. Bursting is introduced by "warming-up activity." Long lasting repetitive bursts - in some neurons - may be due to the lack of feedback, e.g., from the inactivated muscle systems caused by pharmacological immobilization.

Following these considerations, there are several intermediate steps between sensory input and motor output: (i) In the optic tectum and pretectum visual information is processed preserving a topographic representation of the visual field; (ii) space- and feature related activity is transferred in the manner of "sensorimotor codes" to tegmental and bulbar structures: (iii) neurons of the tegmental and bulbar RetF combine space and feature coded signals into a frequency-modulated signal, in conjunction with proprioceptive inputs, recognizable on the discharges of pre-motor neurons. Parallel extra- and intracellular recordings from several neurons in paralyzed toads (e.g., see Fig. 7) and parallel extracellular recordings from different neurons in behaving toads are required to further elucidate the principles that underlie sensorimotor-tranformation and motor pattern generation for action patterns released by sign stimuli.

Acknowledgements: The authors thank Prof. Dr. J.-P. Ewert for detailed discussions. We are indebted to Mrs. G. Kruk for excellent technical assistance and to Gisela Burggraf for animal care. The investigations reviewed here were supported by grants of the Deutsche Forschungsgemeinschaft Schw 337/2-1 and Ew 7/9-1.

5 References

Amri M, Car A, Jean A (1984) Medullary control of the pontine swallowing neurons in sheep. *Exp Brain Res* 55: 105-110

Anderson ME, Yoshida M, Wilson VJ (1971) Influence of the superior colliculus on cat neck motoneurons. *J Neurophysiol* 34: 898-907

Antal M, Matsumoto N, Székely G (1986) Tectal neurons of the frog: intracellular recording and labeling with cobalt electrodes. *J Comp Neurol* 246: 238-253

Ariens Kappers CU, Huber CG, Crosby EC (1936) *The comparative anatomy of the nervous system of vertebrates, including man.* Vol.1, Macmillan, New York

Ashton JA, Boddy A, Donaldson IML (1984) Input from proprioceptors in the extrinsic ocular muscles to the vestibular nuclei in the giant toad, *Bufo marinus. Exp Brain Res* 53: 409-419

Bässler U. (1986) On the definition of central pattern generator and its sensory control. *Biol Cybern* 54: 65-69

Baumgarten R v, Kanzow E (1958) The interaction of two types of inspiratory neurons in the region of the tractus solitarius of the cat. *Arch Ital Biol* 96: 361-373

Borchers H-W, Pinkwart C (1983) A telemetry system for single unit recording in the freely moving toad (*Bufo bufo* L). In: Ewert J-P, Capranica R R, Ingle DJ (eds) *Advances in vertebrate neuroethology.* Plenum Press, New York

Boyle R, Pompeiano O (1981) Cerebellar influences on the response characteristics of vestibulospinal neurons to sinusoidal tilt. *Arch Ital Biol* 119: 208-225

Brown WT, Ingle D (1973) Receptive field changes produced in frog thalmic units by lesions of the optic tectum. *Brain Res* 59: 405-409

Bullock TH, Terzuolo CA (1957) Diverse forms of activity in the somata of spontaneous and integrating ganglion cells. *J Physiol* 138: 341-364

Cannon SC, Robinson DA (1985) An improved neural-network model for the neural integrator of the oculo-motor system: more realistic neuron behavior. *Biol Cybern* 53: 93-108

Cohen MI, Feldman JL (1977) Models of respiratory phase switching. *Fed Proc* 36: 2367-2374

Comer C, Schotland J, Grobstein P (1985) Short and longterm effects of unilateral vestibular lesions on posture and orieting movements in the frog. *Soc Neurosci Abstr* 11: 289

Corner MA, Crain SM (1972) Patterns of spontaneous bioelectric activity during maturation in culture of fetal rodent medulla and spinal cord tissues. *J Neurobiol* 3: 25-46

Cruce WLR (1974) A supraspinal monosynaptic input to hindlimb motoneurons in lumbar spinal cord of the frog *Rana catesbeiana. J Neurophysiol* 37: 691-704

Dacey DM, Ulinski PS (1986a) Optic tectum of the eastern garter snake *Thamnophis sirtalis.* I: Efferent pathways. *J Comp Neurol* 245: 1-28

Dacey DM, Ulinski PS (1986b) Optic tectum of the eastern garter snake *Thamnophis sirtalis.* V. Morphology of brainstem afferents and general discussion. *J Comp Neurol* 245: 423-453

Dieringer N, Precht W (1986) Functional organization of eye velocity and eye position signals in abducens motoneurons of the frog. *J Comp Physiol* 158: 179-194

Donkelaar, ten H-J, de Boer-van Juizen R, Schuoten FTM, Eggen SJH (1981) Cells of origin of descending pathways to the spinal cord in the clawed toad *Xenopus laevis. Neurosci* 16: 2297-2312

Ebbesson SOE (1976) Morphology of the spinal cord. In: Llinás L, Precht W (eds) *Frog neurobiology.* Springer-Verlag, Berlin Heidelberg New York, pp 679-706

Ebbesson SOE (1984) Evolution and ontogeny of neural circuits. *Behav Brain Sci* 7: 321-366

Eckmiller R (1987) Neural control of pursuit eye movements. *Physiol Rev* 67: 797-857

Edinger L (1908) *Vorlesungen über den Bau der nervösen Zentralorgane des Menschen und der Tiere, Vol.2.* Vogel, Leipzig

Euler C v (1983) On the origin and pattern control of breathing rhythmicity in mammals. In: Roberts A, Roberts BL (eds) *Neural origin of rhythmic movements.* Symp Sco Exp Biol. Cambridge Univ Press, Cambridge, pp 469-485

Euler C v, Hayward JN, Marttila I, Wyman RJ (1973) Respiratory neurones of the ventrolateral nucleus of the solitary tract of cat: vagal input, spinal connections and morphological identification. *Brain Res* 61: 1-22

Ewert J-P (1968) Der Einfluß von Zwischenhirndefekten auf die Visuomotorik im Beute- und Fluchtverhalten der Erdkröte (*Bufo bufo* L.). *Z Verg Physiol* 61: 41-70

Ewert J-P (1971) Single unit response of the toad (*Bufo americanus*) caudal thalamus to visual objects. *Z Verg Physiol* 74: 81-102

Ewert J-P (1974) The neural basis of visually guided behavior. *Sci Amer* 230: 34-42

Ewert J-P (1976) The visual system of the toad: behavioral and physiological studies on a pattern recognition system . In: Fite KV (ed) *The amphibian visual system: a multidisciplinary approach.* Academic Press, New York, pp 141-202

Ewert J-P (1984) Tectal mechanisms that underlie prey-catching and avoidance behaviors in toads. In: Vanegas H (ed) *Comparative neurology of the optic tectum.* Plenum Press, New York, pp 247-416

Ewert J-P (1987) Neuroethology of releasing mechanisms: prey-catching in toads. *Behav Brain Sci* 10: 337-405

Ewert J-P, Borchers H-W (1971) Reaktionscharakteristik von Neuronen aus dem Tectum opticum und Subtectum der Erdkröte *Bufo bufo* (L). *Z Vergl Physiol* 71: 165-189

Ewert J-P, von Seelen W. (1974) Neurobiologie und System-Theorie eines visuellen Muster- Erkennungs-mechanismus bei Kröten. *Kybernetik 14:* 167-83

Ewert J-P, Wietersheim A v (1974) Musterauswertung durch tectale und thalamus/praetectale Nervennetze im visuellen System der Kröte (*Bufo bufo* L). *J Comp Physiol* 92: 131-148

Ewert J-P, Finkenstädt T (1987) Modulation of tectal functions by prosencephalic loops in amphibians. A commentary. *Behav Brain Sci* 10: 122-123

Ewert J-P, Speckhardt I, Amelang W (1970) Visuelle Inhibition und Exzitation im Beutefangverhalten der Erdkröte *Bufo bufo* L. *Z Vergl Physiol* 68: 84-110

Ewert J-P, Hock FJ, Wietersheim A v (1974) Thalamus, Praetectum, Tectum: retinale Topographie und physiologische Interaktionen bei der Kröte *Bufo bufo* L. *J Comp Physiol* 92: 343-356

Ewert J-P, Borchers H-W, Wietersheim A v (1979) Directional sensitivity, invariance, and variability of tectal T5 neurons in response to moving configural stimuli in the toad *Bufo bufo* (L). *J Comp Physiol* 132. 191-201

Ewert J-P, Schürg-Pfeiffer E, Weerasuriya A (1984) Neurophysiological data regarding motor pattern generation in the medulla oblongata of toads. *Naturwissenschaften* 71: 590-591

Ewert J-P, Matsumoto N, Schwippert WW (1985) Morphological identification of prey-selective neurons in the grass frog's optic tectum. *Naturwissenschaften* 72: 661-662

Ewert J-P, Framing EM, Schürg-Pfeiffer E, Weerasuriya A (1989) Extracellular activity of neurons of the toad's medulla oblongata in response to visual stimulation: a functional approach toward tectal/bulbar/spinal circuitry (submitted)

Feldman JL, Sommer D, Cohen MI (1980) Short time scale correlations between discharges of medullary respiratory neurons. *J Neurophysiol* 43: 1284-1295

Finkenstädt T, Adler NT, Allen TO, Ebbesson SOE, Ewert J-P (1985) Mapping of brain activity in mesencephalic and diencephalic structures of toads during presentation of visual key stimuli: a computer assisted analysis of ^{14}C-2DG autoradiographs. *J Comp Physiol* 156: 433-445

Finkenstädt T, Adler NT, Allen TO, Ewert J-P (1986) Regional distribution of glucose utilization in the telencephalon of toads in response to configurational visual stimuli: a ^{14}C-2DG study. *J Comp Physiol* 158: 457-467

Framing EM (1986) *Extrazelluläre Ableitungen visuell sensitiver Neuronen in der Medulla oblongata der Erdkröte Bufo bufo spinosus* (L). Diploma Thesis, FB 19, Univ Kassel

Framing EM, Ewert J-P (1989) Parallel single cell recordings in the toad's medulla oblongata: correlation of spike activity (in preparation)

Friesen WO, Poon M, Stent G (1976) An oscillatory neural circuit generating a locomotory rythm. *PNAS* 73: 3734-3738

Fuller JH, Schlag JD (1976) Determination of antidromic excitation by the collision test: problems of interpretation. *Brain Res* 112: 283-298

Garcia-Rill E, Skinner RD (1987) The mesencephalic locomotor region I: Activation of a medullary projection side. *Brain Res* 411: 1-12

Gaupp E, (1896) *A.Ecker's und R.Wiedersheim's Anatomie des Frosches.* F Viehweg & Sohn, Braunschweig

Gillette R (1987) The role of neural command in fixed action patterns of behavior. In: Guthrie DM (ed) *Aims and methods in neuroethology.* Manchester Univ Press, Manchester, pp 46-79

Gillette R, Kovac MP, Davis WJ (1978) Command neurons in *Pleurobranchaea* receive synaptic feedback from the motor network that they excite. *Science* 199: 798-801

Görcs T, Antal M, Olah E, Székely G (1979) An improved cobalt technique with complex compounds. *Act Biol Acad Sci Hung* 30: 79-86

Grantyn A, Berthoz A (1987) Reticulo-spinal neurons participating in the control of synergic eye and head movements during orienting in the cat. II: Morphological properties as revealed by intra-axonal injections of horseradish peroxidase. *Exp Brain Res* 66: 339-354

Grantyn A, Grantyn R, Robiné (1977) Neuronal organization of the tecto-oculomotor pathways. In: Baker R, Berthoz A (eds) *Control of gaze by brain stem neurons.* Elsevier, Amsterdam. pp 197-206

Grantyn A, Ong-Meang Jacques V, Berthoz A (1987) Reticulo-spinal neurons participating in the control of synergic eye and head movements during orienting in the cat. *Exp Brain Res* 66: 355-377

Grillner S (1985) Neurobiological bases of rhythmic motor acts in vertebrates. *Science* 228: 143-149

Grillner S, McClellan A, Sigvardt K, Wallen P, Williams T (1982) On the neural generation of 'fictive locomotion' in lower vertebrate system, *in vitro.* In: Sjölund B, Björklund A (eds) *Brain stem control of spinal mechanisms.* Fernström Foundation Series, Elsevier Biomedical Press, Amsterdam New York Oxford, pp 273-295

Grillner S (1985) Neurobiological bases of rhythmic motor acts in vertebrates. *Science* 228: 143-149

Grossberg S (1987) Co-operative self-organisation of multiple neural systems during adaptive sensorimotor control. In: Guthrie DM (ed) *Aims and methods in neuroethology.* Manchester Univ Press, Manchester, pp 260-300

Grüsser O-J, Grüsser-Cornehls U (1976) Neurophysiology of the anuran visual system. In: Llinás L, Precht W (eds) *Frog neurobiology.* Springer-Verlag, Berlin Heidelberg New York, pp 297-385

Hartline DK (1967) *Integrative physiology of the lobster cardiac ganglion.* Thesis Harvard Univ Biology Department

Hartline DK, Gassie Jr DV (1979) Pattern generation in the lobster (*Panulirus*) stomatogastric ganglion. *Biol Cybern* 33:209-222

Herrick CJ (1930) The medulla oblongata of *Necturus. J Comp Neurol* 50: 1-96

Holst E v (1939) Die relative Koordination als Phänomen und als Methode zentralnervöser Funktionsanalyse. *Erg Physiol* 13: 228-306

Ingle DJ (1983) Brain mechanisms of visual localization by frogs and toads. In: Ewert J-P, Capranica RR, Ingle DJ (eds) *Advances in vertebrate neuroethology.* Plenum-Press, New York . pp 177-226

Ito M (1986) Neural systems controlling movements. *TINS* 9: 515-518

Julesz B (1976) Experiments in the visual perception of texture. In: Held R, Richards W (eds) *Recent progress in perception.* WH Freeman, San Francisco. pp 37-46

Kazennikov OV, Selionov VA, Shik ML, Yakokleva GV (1980) The rombencephalic 'locomotor region' in turtles. *Neurophysiol* 12: 251-257

Kling V, Székely G (1968) Simulation of rhythmic neurons activities. I: Functions of networks with cyclic inhibitions. *Kybernetik* 5: 89-103

Kostyk SK, Grobstein P (1987a) Neuronal organization underlying visually elicited prey orienting in the frog. - II: Anatomical studies on the laterality of central projections. *Neurosci* 21: 57-82

Kostyk SK, Grobstein P (1987b) Neuronal organization underlying visually elicited prey orienting in the frog. III: Evidence for the existence of an uncrossed descending tectofugal pathway. *Neurosci* 21: 83-96

Lara R, Cervantes F, Arbib MA (1982) Two-dimensional model of retinal-tectal-pretectal interactions for the control of prey-predator recognition and size preference in amphibia. In: Amari S, Arbib MA (eds) *Competition and cooperation in neural nets.* Springer-Verlag, Berlin Heidelberg New York, pp 371-393

Lázár G (1969) Efferent pathways of the optic tectum in frog. *Act Biol Acad Sci Hung* 20: 171-183

Lázár G, Tóth P, Csank G, Kicliter E (1983) Morphology and location of tectal projection neurons in frogs: a study with HRP and cobalt-filling. *J Comp Neurol* 215: 108-120

Lettvin JY, Maturana HR, McCulloch, Pitts WH (1959) What the frog's eye tells the frog's brain. *Proc IRE* 47: 1940-1951

Lipski J (1981) Antidromic activation of neurons as an analytic tool in the study of the central nervous system. *J Neurosci Methods* 4: 1-32

Lipski J, Merrill EG (1980) Electrophysiological demonstration of the projection from expiratory neurons in rostral medulla to contralateral dorsal respiratory group. *Brain Res* 197: 521-524

Lowe AA (1981) The neural regulation of tongue movements. *Prog Neurobiol* 15: 295-344

Madden KP, Remmers JE (1982) Short time scale correlations between spike activity of neighboring respiratory neurons of nucleus tractus solitarius. J Neurophysiol 48: 749-760

Manteufel G, Manteufel-Naujoks C (1987) Optic inputs to identified neck-muscle motoneurons of *Salamandra salamandra* (L) *J Hirnforsch* 28: 525-632

Matesz C (1979) Central projection of the VIII*th* cranial nerve in the frog. *Neurosci* 4: 2061-2071

Matesz C, Székely G (1978) The motor column and sensory projections of the branchial cranial nerves in the frog. *J Comp Neurol* 178: 157-176

Matsumoto N, Antal M (1984) Physiological and morphological study of prey and predator detecting neurons in the frog tectum. *Proc 6th Ann Meet Jpn Soc Gen Comp Physiol*, p 169

Matsumoto N, Bando T (1980) Excitatory synaptic potentials and morphological classification of tectal neurons of the frog. *Brain Res* 192: 39-48

Matsumoto N, Schwippert WW, Ewert J-P (1985) Receptive field and morphology of tectal neurons of the frog. *Proc 7th Ann Meet Jpn Soc Gen Comp Physiol*, p 138

Matsumoto N, Schwippert WW, Ewert J-P (1986) Intracellular activity of morphologically identified neurons of the grass frog's optic tectum in response to moving configurational visual stimuli. *J Comp Physiol A* 159: 721-739

Matsuoka K. (1985) Sustained oscillations generated by mutually inhibiting neurons with adaptation. *Biol Cybern* 52: 367-376

Matsushima T, Satou M, Ueda K (1984) Toad's snapping pathway: relationship to the excitatory pathways from glossopharyngeal nerve to tongue-muscle motoneurons. *Zool Sci* 1: 881

Maynard DM (1955) *Direct inhibition in the lobster cardiac ganglion.* Thesis Univ Calif, Los Angeles Dept Zoology

Maynard DM (1972) Simpler networks. *Ann NY Acad Sci* 193: 59-72

McIlwain JT (1986) Effects of eye position on saccades evoked electrically from superior colliculus of alert cats. *J Neurophysiol* 55: 97-112

Mensah PL (1974) *The course and distribution of descending fibers of the lateral funiculus of the amphibian spinal cord.* PhD Thesis Univ Calif Irvine

Mensah PL, Glanzman DL, Levy WB, Thompson RF (1974) The effects of 5,6-hydroxytryptamine in the amphibian spinal cord using silver staining techniques. *Brain Res* 78: 25-261.

Merrill EG (1979) Is there reciprocal inhibition between medullary inspiratory and expiratory neurones? In: Euler C v, Lagercrantz H (eds) *Central nervous control mechanisms of breathing.* Pergamon Press, Oxford, pp 239-254

Merrill EG (1981) Where are the *real* respiratory neurons? *Fed Proc* 40: 289-2394

Miles FA, Evarts EV (1979) Concepts of motor organization. *Ann Rev Psychol* 30: 327-362

Mitchel RA, Herbert DA (1974) Synchronized high frequency synaptic potentials in medullary respiratory neurons. *Brain Res* 75: 350-355

Moriyama Y (1987) Rhythmical jaw movements and lateral ponto-medullary reticular neurons in rats. *Comp Biochem Physiol A* 86: 7-14

Nieuwenhuys R, Opdam P (1976) Structure of the brain stem. In: Llinás L, Precht W (eds) *Frog neurobiology.* Springer-Verlag, Berlin Heidelberg New York, pp 811-855

Noble D (1983) Ionic mechanisms of rhythmic firing. In: Roberts A, Roberts BL (eds) *Neural origin of rhythmic movements.* Symp Soc Exp Biol, Cambridge Univ Press, Cambridge

Oka Y, Takeuchi H, Satou M, Ueda K (1987) Morphology and distribution of the preganglionic parasympathetic neurons of the facial, glossopharyngeal and vagus nerves in the Japanese toad: a cobaltic lysine study. *Brain Res* 400: 389-395

Opdam P, Kemali M, Nieuwenhuys R (1976) Topological analysis of the brain stem in the frogs *Rana esculenta* and *Rana catesbeiana. J Comp Neurol* 165: 307-331

Pearson K (1976) The control of walking. *Sci Amer* 235: 72-86

Peterson BW, Fukushima K (1982) The reticulospinal system and its role in generating vestibular and visuomotor reflexes. In: Sjölund B, Björklund A (eds) *Brain stem control of spinal mechanisms.* Fernström Foundation Series, Elsevier Biomedical Press, Amsterdam New York Oxford, pp 225-252

Peterson BW, Anderson ME, Filion M (1974) Responses of ponto-medullary reticular neurons to cortical, tectal and cutaneous stimuli. *Exp Brain Res* 21: 19-44

Peterson BW, Maunz RA, Pitts NG, Mackel RG (1975) Patterns of projecting and branching of reticulospinal neurons. *Exp Brain Res* 23: 333-351

Peterson BW, Pitts NG, Fukushima K, Mackel R (1978) Reticulospinal excitation and inhibition of neck motoneurons. *Exp Brain Res* 32: 471-489

Pompeiano O (1981) Sensory mechanisms and motor control. *Progress Rep Inst Human Physiol Univ Pisa*, pp 3-20

Roberts A (1987) Skin sensory modalities, free nerve endings and behaviour: a reappraisal based on studies of amphibian embryos. In: Guthrie DM (ed) *Aims and methods in neuroethology.* Manchester Univ Press, Manchester, pp 80-103

Roberts BL, Williamson RM (1983) Motor pattern formation in the dogfish spinal cord. In: Roberts A, Roberts BL (eds) *Neural origin of rhythmic movements.* Symp Soc Exp Biol. Cambridge Univ Press, Cambridge, pp 31-350

Robinson DA (1981) Control of eye movements. In: Brookhart JM, Mountcastle VB, Brooks VB, Geiger SR (eds) *Handbook of physiology, Section I: The nervous system, II: Motor control, Part 2,* American Physiological Society, Bethesda Maryland, pp 1275-1320

Rose D, Larnicol N, Duron B (1984) An HRP study of the cat's spinal respiratory motoneurones during postnatal development. *Exp Brain Res* 56: 458-467

Rose PK, Abrahams VC (1978) Tectospinal and tectoreticular cells: their distribution and afferent connections. *Can J Physiol Pharmacol* 56: 650-658

Röthig P (1927) Beiträge zum Studium des Zentralnervensystems der Wirbeltiere. 11. Über die Faserzüge im Mittelhirn, Kleinhirn und der Medulla oblongata der Urodelen und Anuren. *Z Mikr Anat Forsch* 10: 381-472

Rubinson E (1968) Projections of the tectum opticum of the frog. *Brain Behav Evol* 1: 529-561

Russell DF, Hartline DK (1978) Bursting neural networks: a reexamination. *Science* 200: 453-455

Satou M, Ewert J-P (1985) The antidromic activation of tectal neurons by electrical stimuli applied to the caudal medulla oblongata in the toad *Bufo bufo* L. *J Comp Physiol A* 157: 739-748

Satou M, Matsushima T, Ueda K (1984) Neuronal pathways from the tectal "snapping-evoking area" to the tongue muscle controlling motoneurons in the Japanese toad: evidence of the intervention of excitatory interneurons. *Zool Sci* 1: 829-832

Satou M, Matsushima T, Takeuchi H, Ueda K (1985) Tongue-muscle-controlling motoneurons in the Japanese toad: topography, morphology and neuronal pathways from the 'snapping-evoking area' in the optic tectum. *J Comp Physiol A* 157: 717-737

Schlag JD (1978) Electrophysiological mapping techniques. In: Robertson RT (ed) *Neuroanatomical research techniques.* Academic Press, New York, pp 385-404

Schmidt RS (1971) A model of the central mechanisms of male anuran acoustic behavior. *Behaviour* 39: 288-317

Schmidt RS (1973) Vocal cord mechanisms of release calling in Northern Leopard Frog. *Copeia* 1973(3): 624-627

Schmidt RS (1974a) Central mechanisms of frog calling. *Amer Zool* 13: 1169-1177

Schmidt RS (1974b) Neural correlates of frog calling: independence from peripheral feedback. *J Comp Physiol* 88: 321-333

Schmidt RS (1976) Neural correlates of frog calling. *J Comp Physiol* 108: 99-113

Schwippert WW, Ewert J-P (1987) Visual neurons in the medulla oblongata of common toads: intracellular recording and labeling. In: Elsner N, Creutzfeldt O (eds) *New frontiers in brain research.* Proc 15th Göttingen Neurobiol Conf. Thieme Verlag, Stuttgart, p 158

Schwippert WW, Beneke TW, Ewert J-P (1988) The processing of visual information in the rostral brain stem of toads. In: Rauschecker J, Mohn G (eds) *Visual processing of form and motion.* Europ Brain Behav Soc Workshop, p 77

Schwippert WW, Beneke TW, Ewert J-P (1989) Neurobiological studies of intracellularly recorded and labeled neurons in the toad's medulla oblongata (submitted)

Selverston AI (1980) Are central pattern generators understandable? *Behav Brain Sci* 3: 535-571

Selverston AI, Miller JP, Wadepuhl M (1983) Cooperative mechanisms for the production of rhythmic movements. In: Roberts A, Roberts BL (eds) *Neural origin of rhythmic movements.* Cambridge Univ Press, Cambridge, pp 55-87

Shik ML, Yagodnitsyn AS (1977) The pontobulbar 'locomotor strip'. *Neurophysiol* 9: 95-97

Steeves JD, Sholomenko GN, Webster DMS (1986) Stimulation of the pontomedullary reticular formation initiates locomotion in decerebrate birds. *Brain Res* 401: 205-212

Székely G (1965) Logical network for controlling limb movement in urodela. *Acta Physiol Hung* 27: 285-289

Székely G (1973) Anatomy and synaptology of the optic tectum. In: Autrum H, Jung R, Loewenstein WR, MacKay DM, Teuber HL (eds) *Handbook of sensory physiology VII/3, Central processing of visual information, Part B*, Springer-Verlag, Berlin Heidelberg New York, pp 1-26

Székely G, Antal M (1981) Significance of the dendritic pattern in the function of the neuron. In: Székely G, Labos E, Damjanovich S (eds) *Neuronal communication and control. Adv Physiol Sci 30,* Pergamon Press and Akademiai Kiad, Buapest, pp 171-182

Székely G, Matesz C (1987) Trigeminal motoneurons with disparate dendritic geometry innervate different muscle groups in the frog. *Neurosci Lett* 77: 161-165

Takizawa N (1981) Integral multiple interspike intervals of spontaneous activity in isolated medulla oblongata of frog. *Brain Res* 212: 466-469

Tóth P, Csank G, Lázár G (1985) Morphology of cells of origin of descending pathways to the spinal cord in *Rana esculenta.* A tracing study using cobalt-lysine complex. *J Hirnforsch 26:* 365-83

Tsai HJ, Ewert J-P (1988) Influence of stationary and moving background structures on the response of visual neurons in toads (*Bufo bufo*). *Brain Behav Evol* 32: 27-38

Tuge H (1932) Somatic motor mechanisms in the midbrain and medulla oblongata of *Chrysemys elegans* (Wied). *J Comp Neurol* 55: 185-271

Viala D (1986) Evidence for direct reciprocal interactions between the central rhythm generators for spinal "respiratory" and locomotor activities in the rabbit. *Exp Brain Res* 63: 225-232

Viala D, Freton E (1983) Evidence for respiratory and locomotor pattern generators in the rabbit cervico-thoracic cord and their interaction. *Exp Brain Res* 49: 247-256

Watanabe A, Obara S, Toyohiro A (1967) Pacemaker potentials for the periodic burst discharge in the heart ganglion of a stomatopod, *Squilla oratoria. J Gen Physiol* 50: 839-862

Weerasuriya A (1983) Snapping in toads: some aspects of sensorimotor interfacing and motor pattern generation. In: Ewert J-P, Capranica RR, Ingle DJ (eds) *Advances in vertebrate neuroethology.* Plenum-Press, New York, pp 613-628

Weerasuriya A, Ewert J-P (1981) Prey-selective neurons in the toad's optic tectum and sensorimotor interfacing: HRP studies and recording experiments. *J Comp Physiol* 144: 429-434

Wendler G (1969) Relative Koordination bei gekoppelten, rhythmisch tätigen Modellneuronen. *Zool Anz Suppl* 33: 477-482

Whittington DA, Lestienne F, Bizzi E (1984) Behavior of preoculomotor burst neurons during eye-head coordination. *Exp Brain Res* 215-222

Wilczynski W, Northcut RG (1983a) Connections of the bullfrog striatum: efferent projections. *J Comp Neurol* 214: 333-343

Wilczynski W, Northcut RG (1983b) Connections of the bullfrog striatum: afferent organization. *J Comp Neurol* 214: 321-333

Willis JB (1986) The modelling of neural circuitry. *Prog Neurobiol* 26: 93-118

Wilson DM, Waldron I (1968) Models for the generating of motor output pattern in flying locusts. *Proc*

IEEE 56: 1058-1064

Yerkes RM (1903) The instincts, habits, and reactions of the frog. *Harvard Psychol Studies* 1: 579-638

Organization in the Sensorimotor Interface: A Case Study with Increased Resolution

PAUL GROBSTEIN

Department of Biology, Bryn Mawr College
Bryn Mawr, Pennsylvania 19010, USA

Abstract. Recent studies on the neuronal organization underlying prey orienting behavior in the frog, *Rana pipiens*, are reviewed and discussed in the context of more general consideration of the information processing characterstics and nervous system structures involved in the production of directed movements. It is suggested that directed movements in general depend on "activity gated divergence," a property of some neuronal networks which allows them to associate with a given topographically specified sensory input any of a number of outputs, depending on the character of other relevant neuronal signals. It is further suggested that two important information processing steps occur between midbrain sensory maps involved in directed movement and the actual motor output pattern. The first of these involves the production of a signal related to stimulus location in an abstract coordinate frame, one closely related neither to sensory surfaces nor to movement parameters. The second involves the selection of the particular movement to be generated and the creation of the appropriate motoneuron discharge patterns. The first transformation seems to occur in the ventral midbrain, and the latter in the spinal cord. Both transformations are likely to involve an interaction between signals derived from the initial topographically specified sensory input and signals representing other relevant information, and hence both transformations probably contribute to the overall activity gated divergent character of the network involved in directed movement. The context-dependent nature of the neuronal signals, together with evidence that autonomous central organization is critical to their interpretation, raises new questions about how signals represented in different coordinate frames are combined, and suggests that reflex theories are ceasing to provide an adequate conceptual framework for the analysis of sensorimotor processing.

1 Introduction

The organization of the neuronal circuitry which underlies directed movement is a matter of active investigation in a variety of organisms. In anuran prey orienting, a paradigmatic example, an appropriate stimulus at a given location in space triggers a ballistic movement directed towards the stimulus. Until recently, the reliability and the map-like character of the behavioral input/output relation encouraged investigators to look for a relatively simple, comparably map-like organization in the underlying neuronal circuitry. The topographic retino-tectal projection, generally regarded as the first stage of the circuitry underlying visually triggered orienting (Ewert 1984, 1987; Grobstein et al 1983; Ingle 1983), has this character. So too does a more recently described topographic representation of the body surface in the midbrain (Comer & Grobstein 1981c), which almost certainly represents a comparable first stage of the circuitry underlying prey orienting movements made in response to tactile stimuli (Comer & Grobstein 1981a,b). Given the existence in the brain of such maps of relevant sensory surfaces, investigators were encouraged to look for a subsequent sensorimotor interface organization which associates with each point in the sensory maps a particular, distinctive, and approprate movement. Over the past five or ten years, however, it has became clear that the sophistication and complexity of both anuran orienting behavior and its associated neuronal circuitry are greater than had been previously thought. An important element in this altered perspective has been the realization that the map-like character of the behavioral input/output relation which constitutes anuran prey orienting could not, even in principle, be accounted for by a neuronal organization which extablishes a map-like relationship between a sensory map and a repertoire of directed movements. Correspondingly, recent studies have revealed that several unexpected processing steps and neural structures intervene between the sensory maps and the production of directed movements. In this chapter, I will discuss recent and ongoing studies from my laboratory aimed at providing new information about the neuronal organization underlying anuran orienting. Similar findings in a variety of organisms and behavioral situations suggest that what is emerging is a new general view of the nature of sensorimotor integration, involving a new class of conceptual models with important implications for both experimentation and theory.

2 Activity-Gated Divergence: A New Concept for the Sensorimotor Interface

A stimulus at a given location in visual space, or at a given location on the body surface, activates a population of receptor cells at a particular location within a two-dimensional array of receptors. Topographically organized sensory projections associate each such location within the receptor sheet with corresponding locations within two-dimensional sensory maps in the central nervous system, and hence preserve in a place-coded form information about simulus location. A number of years ago, Ewert showed that electrical stimulation of the anuran tectum causes orienting movements which vary with stimulation site in a fashion quite similar to the variation in retinal input (Ewert 1967, 1974). These observations led to the hypothesis of a quite simple interface organization underlying anuran orienting: arriving retinal input at a given tectal location activates tectal neurons at that location which in turn engage premotor circuitry in such a way as to cause the particular movement appropriate for that retinal input. The hypothesis is appealing, and motivated a number of subsequent quite significant observations with respect to the analysis of visual input for *pattern recognition* (Ewert 1987), including, for example, the demonstration that relevant tectal information is transferred directly to bulbar/spinal structures (Satou & Ewert 1985). The ultimate validity of the simple interface hypothesis for *stimulus localization* depends, however, on the validity of an important, implicit assumption: that the place-coded information about stimulus location available in sensory maps is adequate to specify the appropriate movement. This has turned out not to be the case for anuran prey orienting, triggered either visually or tactually.

A given retinal and associated tectal location represents the value of a two-dimensional variable and, accordingly, corresponds not to a stimulus location in three dimensions, but rather to a particular line of sight from the eye, an array of points at different distances all of which are imaged at the same location on the retina (cf. Grobstein et al. 1983, 1985). The directed movement of a frog in response to a visual stimulus, in contrast, reflects stimulus location in all three dimensions of space: The movement varies not only with the line of sight on which a stimulus falls but also with the stimulus distance along that line of sight (Ingle 1970; Comer & Grobstein 1981a; Grobstein et al. 1985; Grobstein 1988a). From this it follows straightforwardly that the place-coded information associated with activation of a particular location in the retino-tectal projection is insufficient to specify the appropriate movement. An additional signal, one related to stimulus distance, must also be involved. An alternate way of saying the same thing is that the neuronal organization underlying visually triggered movements in the frog must be such that each retinal and associated tectal locus can potentially activate any of a variety of movements, with the particular movement in a given situation depending on other neural activity related to stimulus distance. This form of organization I have called "activity-gated divergence" (Grobstein 1988a,b). Recent findings suggest that the movement associated with activation of a given retinal and associated tectal locus varies not only with stimulus distance but also with posture (Ingle 1970; Reyes & Grobstein 1984) and body size (Grobstein et al. 1985), implying that at least two additional signals are involved in gating the divergent circuitry intervening between the retino-tectal projection and the production of movement.

Several years ago, we suggested that tactually elicited orienting movements in frogs, like visually elicited orienting movements, depend on an activity-gated divergent network (Grobstein et al. 1983). Our evidence was indirect, and the suggestion seemed, on first consideration, to ignore the fact that tactually elicited orienting is in some respects simpler than visually elicited orienting: animals orient tactually only to stimuli which touch the body surface, making it seem more likely that an organization which associates a particular movement with a particular location in a sensory map would suffice in this case. Recently, we have acquired more direct evidence that tactually elicited orienting too depends on an activity-gated divergent network (Grobstein et al. 1988). As suspected based on other work (Fukson et al. 1980), we have found that the orienting movement associated with stimulation of a particular location on the skin depends on variables in addition to the particular location stimulated. As shown in Fig. 1, for hindlimb positions nearer the body, frogs respond to digit stimulation with a directed snapping movement, while for limb positions further from the body, the same cutaneous stimulation more frequently triggers a qualitatively different motor pattern: a turning movement lacking a snap.

The clear implication is that, as with visually triggered orienting, the place-coded information associated with stimulation of a particular region of the receptor sheet is inadequate to specify the appropriate movement; an additional signal, related to limb position in this case, must also be involved. Given observations discussed below, a further possible similarity between visual and tactile behavior is worth note. The additional signal involved in tactile behavior is presumably based on information about limb posture, derived either from afferent or corollary discharge signals. Given, however, that what appears to be critical is how far the digits are from the body, it seems possible that the postural information is subsequently synthesized into a signal representing digit distance relative to the body.

Visually and tactually elicited orienting in frogs are by no means the only situations in which it has recently been realized that the sensorimotor interface must have an activity-gated divergent character. Two other systems in which one might have presumed map-like relations between sensory input and motor output have both proven in the last few years to display a similar organization. Wiping movements in response to irritating stimuli in the spinal frog depend not only on the skin region activated but on limb position as well (Fukson et al. 1980), in much the same sense as do the quite different prey orienting movements in the intact frog. Visually evoked saccadic eye movements in primates, long thought to depend only on the retinal locus activated by a stimulus, have recently been shown to vary depending on an additional

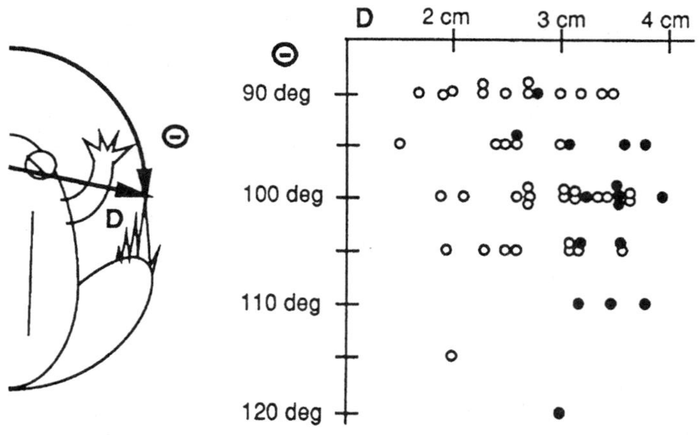

Figure 1. Evidence for activity-gated divergence in tactually elicited orienting behavior in the frog (Grobstein et al. 1988). Illustrated are orienting responses of an individual frog to gentle stroking of a single cutaneous site, the most rostral part of the foot of the right hindlimb. Normal variations in posture produce variations in the location of this site relative to the body of the animal. The location of the stimulus site relative to the body was recorded prior to each stimulus trial. Each dot in the graph to the right corresponds to one stimulus trial and indicates whether the response was a directed snap (open dots) or an orientation movement lacking tongue projection (filled dots). Dots are positioned so as to show the location of the tip of the forelimb relative to the body for that stimulus trial. The ordinate gives horizontal eccentricity and the abscissa distance, defined as shown to the left.

signal related to eye position (Mays & Sparks 1980; Sparks & Mays 1983). I have discussed elsewhere the theoretical criteria which must hold for a given behavior to allow one to continue to entertain the possibility of a map-like coupling between a receptor or sensory sheet and movement (Grobstein 1988a). The conclusion is that such hypotheses are tenable only in a quite restricted set of circumstances (cf. Finger 1988). For most examples of directed movement, some form of activity-gated divergence must exist.

Recognition of activity-gated divergence as a fundamental characteristic of sensorimotor organization leads both to an important general conclusion and to a host of new questions, of varying degrees of generality. The conclusion is that tracing anatomical connections from locations in a receptor or sensory sheet to motoneurons cannot provide an adequate explanation for behaviorally observed correspondences between stimulus location and movement direction. The correspondences depend on the interaction between such connectivity patterns and the contemporaneous pattern of electrical activity deriving from other sources (Grobstein 1988a,b; 1989a,b). From this follows two obvious general questions: What kinds of connectivity patterns support the phenomena of activity-gated divergence and how do the relevant patterns of electrical activity interact with such networks?

For the case of anuran orienting, these questions can be sharpened somewhat, with the longterm objective of providing concrete answers, by asking how many information processing steps and neural structures intervene between known sensory maps and motoneurons, and where in this sequence extrinsic signals are added. I will treat this matter at length in subsequent sections, but some preliminary remarks are relevant here. For visually triggered orienting, the first obvious processing step is between the superficial retino-tectal projection and the deeper lying neurons whose axons leave the tectum. One might in principle entertain the hypothesis that all necessary signals (including at least those related to distance, posture, and body size) are added during this step, so that what leaves tectum is a true motor command, a signal related in a one to one fashion to a particular movement pattern. This possibility would be consistent with electrical stimulation observations (Ewert 1967) if, as recently shown in primates (Sparks & Porter 1983), the effect of extrinsic signals was to produce an appropriate variation in the location of deep tectal activity associated with activity at a given superficial location. At the same time, as discussed below, there are reasons to doubt that all relevant signals are added prior to deep tectum, and the electrical stimulation observations in anurans leave the matter open. The key question is whether the movement triggered by stimulation at a given tectal locus does or does not vary with variation in the other relevant signals. If the movement were the same when triggered, for example, from different starting postures, the finding would imply that postural information must normally be added prior to deep tectum; if the movement varied when triggered from different starting postures, it would imply that postural information is added subsequently.

While the available results of electrical stimulation are consistent with either possibility, a number of other considerations imply that at least some addition of information occurs subsequent to tectal outflow. Among these is a curious lack of easily understandable order in tectal projections themselves. Prey items at different locations trigger quite different sorts of movements in the frog (Ingle 1970; Comer &

Grobstein 1981a; Grobstein et al. 1985). For example, for some locations head and tongue movements are primarily involved. For others, tongue movements are lacking entirely, and the predominant feature is stepping movements of the limbs. If the tectal outflow was actually a motor command, one might expect that different tectal regions would project to distinguishably different target structures. What one instead finds is that the projections of different tectal regions are substantially the same in this respect (Masino et al. 1984; Masino & Grobstein 1985, 1989c). This is perhaps most easy to understand if the final determinant of the movement pattern triggered is not the tectal outflow alone but the interaction of the tectal outflow with additional signals. It is worth noting that the anatomy of tectal projections in the frog is not unusual. Tectal projections from individual loci are proving to be remarkably widespread in virtually all organisms examined (see references in Grobstein 1988a; Masino & Grobstein 1989c). If properly understood in the frog, what this implies is that information addition subsequent to the tectum is probably a characteristic of tectally triggered directed movements in all vertebrates. For an explicit discussion of saccade control see Grobstein (1988a).

A second reason for thinking that there is probably some information addition subsequent to the tectum relates to an interesting geometrical problem associated with thinking of deep tectum as the source of motor commands, given the recognition that the movement repertoire varies in at least three dimensions. The results of electrical stimulation imply the same relation between deep tectum and movement as obtains between superficial tectum and retina: an orderly two-dimensional mapping apparently relating to elevation and horizontal eccentricity. It is not impossible that there is some as yet to be discovered form of deep tectal organization which relates to distance but evidence for this is currently lacking, and other findings raise the possibility that the distance signal may bypass the tectum entirely. The existence of a distance signal involved in visually triggered orienting does not depend on binocular vision (Ingle 1972; Grobstein et al. 1985), and it is likely that an adequate distance signal exists which is not of retinal origin at all (cf. Ingle 1972; Jordan et al. 1980; Collet & Harkness 1982; Grobstein et al. 1985; Douglas et al. 1986). In addition, recent findings from our laboratory (Schweichler & Grobstein, unpublished observations) indicate that an adequate distance signal survives complete unilateral tectal lobe removal. Finally, findings discussed further below indicate that a distance signal descends to the spinal cord independently of other signals involved in orienting, signals more obviously related to retinal input. Collectively, these observations make it likely that at least one of the signals that gates the divergent circuitry coupling the retino-tectal projection to movement, that related to distance, is added after the tectum.

A third line of evidence consistent with significant information addition after the tectum relates to a series of lesion observations, discussed more fully in a following section of this chapter. What these imply, at a minimum, is that the tectal outflow does not directly address motoneurons but rather is relayed at least twice, and undergoes at least two important transformations, prior to the production of motoneuron activity. The observations also suggest that, fairly early in this process but subsequent to the tectum, visual and tactile information may come to be represented in the same pathways. The character of the transformations observed, as well as the problems of representing visual and tactile information in a common coordinate frame, both suggest post-tectal information addition, as will be discussed further below.

The evidence for activity-gated divergence in the circuitry underlying both visually and tactually elicited orienting in the frog is compelling. The evidence that at least some of the relevant information addition occurs subsequent to the tectum, in the case of visually elicited behavior, is strong, if still somewhat indirect. This not only focuses attention on post-tectal organization as the key to progress in further understanding the neural basis of directed movement in the frog but has some more general implications as well. In particular, it suggests that the concept of a "motor map," frequently used to characterize deep tectum not only in the frog but in a variety of vertebrates, probably needs re-evaluation. The concept was useful so long as it appeared to be the case that there was a one-to-one relationship between the pattern of deep tectal activity and the resulting pattern of movement, but there is increasing evidence that this is not the case in the frog or any other vertebrate (Grobstein 1988a, 1989b). An obvious question at this point is how to characterize the signal leaving tectum, if not as a motor command? The findings discussed in the next sections suggest the tectal outflow might most appropriately be regarded as an input to circuitry establishing a representation of spatial location in an abstract, head- or body-centered coordinate frame.

3 Abstract Spatial Representation: A New Concept of Processing Within the Sensorimotor Interface

Over the past few years a number of lines of evidence (Comer & Grobstein 1978, 1981a,b; Weerasuriya & Ewert 1981; Weerasuriya 1983; Satou et al 1984) have led to a consensus among investigators that the tectal outflow signals involved in orienting are unlikely to engage involved motoneurons directly (Grobstein et al. 1983; Ewert 1987). Prior to motoneurons but subsequent to the tectum there must be intervening premotor circuitry (what I will call "pattern-generating circuitry") which elaborates the particular spatio/temporal motoneuron discharge patterns which constitute a given movement. Recently, we have acquired evidence for at least one additional kind of information processing in the sensorimotor interface, one distinguishable in its organization from both the two-dimensional tectal maps and conceivable forms of pattern-generating circuitry (Kostyk & Grobstein 1987a,b,c; Grobstein & Masino 1986; Masino & Grobstein 1986, 1989a,b). In this section, I want to briefly review this evidence (see also Grobstein 1988a, 1989b), as well as related findings in other organisms suggesting that the additional form of information-processing seen in the frog may be a general characteristic of sensorimotor processing.

Lesions of the optic tectum produce visual orienting deficits in the frog which reflect the organization of the retino-tectal projections. Of particular concern in the present context is that unilateral removal of one tectal lobe results in a scotoma in the contralateral monocular visual field. Lesioned frogs fail to respond to stimuli anywhere within this region. Accurately directed responses persist for stimuli in the ipsilateral monocular field as well as throughout the frontal binocular field, an area which includes locations up to about 45° to either side of the midsagittal plane (Kostyk & Grobstein 1982, 1987a). The deficit is straightforwardly understandable given that the visual field of each eye in the frog covers one hemifield and extends an additional 45° across the midsagittal plane in front (Grobstein et al. 1980), and that each retina projects in its entirety to the opposite tectal lobe (Fig. 2). Lesioned frogs respond for stimuli at all locations represented in the intact tectal lobe and fail to respond for stimuli at locations normally represented solely in the missing tectal lobe.

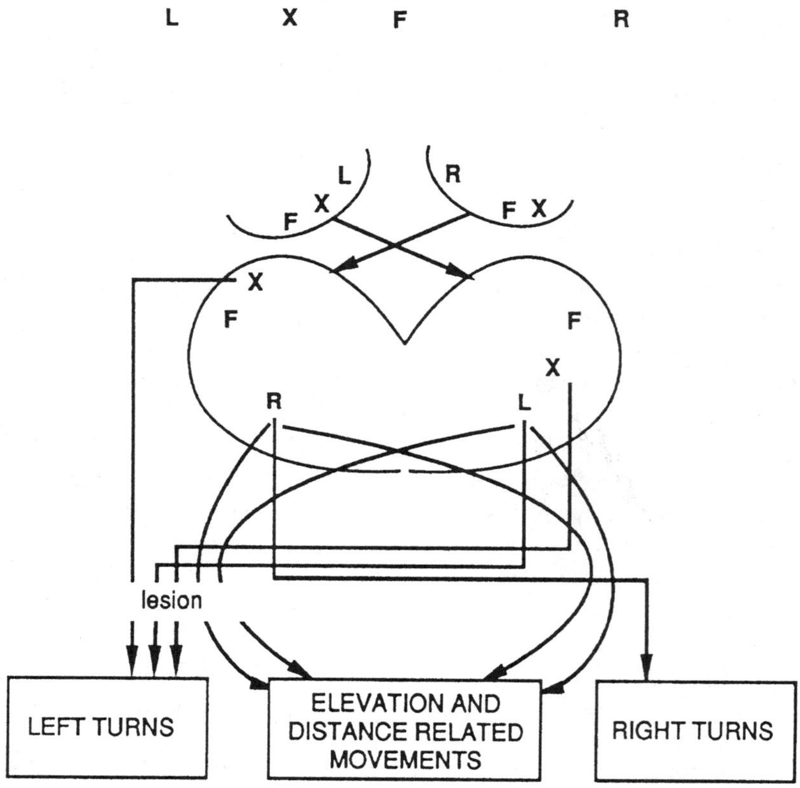

Figure 2. An early model of the organization of descending tectofugal pathways involved in visually elicited orienting behavior in the frog (modified from Kostyk & Grobstein 1982, 1987a,b,c). The visual field is represented above by four stimulus locations: F, a point directly in front of the frog; L and R, points sufficiently far to the left and right respectively so as to be imaged on only one eye; X, a point to the left of the midline but in binocular field, so that it is imaged on both eyes. The imaging of these visual field locations on the two retinas and their mapping on the two tectal lobes by way of the crossed retino-tectal projections is diagrammed in the middle part of the figure. Note particularly that location X is represented in both tectal lobes. Pathways linking locations in the two tectal lobes with three hypothetical sets of pattern-generating circuitry are shown to the bottom of the figure. For each, several sets of linking pathways are involved. To simplify the figure, the full set of pathways are shown only for points "R" and "L".

If the neural circuitry underlying visually triggered orienting consisted only of tectum, pattern-generating circuitry, and motor command pathways running from the former to the latter, then behavioral deficits resulting from damage to descending tectofugal projections should differ from those following tectal lesions, but should also be clearly characterizable in a motor co-ordinate frame, in the same sense that those resulting from tectal lesions are characterizable in a retinal coordinate frame. Unilateral lesions of a definded tract caudal of the tectum, in either rostral (Kostyk & Grobstein 1982, 1987a,b) or caudal medulla (Grobstein & Masino 1986; Masino & Grobstein 1986, 1989a), do in fact produce a visual orienting deficit which differs strikingly in two respects from that following unilateral tectal lesions. First, the deficit does not consist of a failure to respond to stimuli at any location in visual space, but rather of an abnormality in the movements which are triggered by stimuli at particular locations. Second, the region of visual space affected is not a restricted part of the contralateral visual field, but rather the entirety of the ipsilateral hemifield. Lesioned frogs respond with normally directed movements to stimuli in the contralateral visual

---

Figure 2 (continued). As discussed in the text, lesions of the descending pathways on the left side of the brain ("lesion") abolish orienting turns to stimuli in the left visual hemifield, while sparing turns toward stimuli in the right visual hemifield. Also spared, for locations throughout both hemifields, are movements which vary with stimulus elevation and distance. In this model, the forwardly directed movements for stimuli in the left visual field are accounted for by the assumption that each corresponding tectal location ("L" in the figure) projects by independent and differently lateralized paths to two sets of pattern-generating circuitry, one for left turns (the lower left box), and one for the elevation and distance related components of orienting movements (middle box). The link to turn-generating circuitry is by a crossed path, and hence interrupted by the lesion, while that to circuitry for elevation and distance related movements can be completed by an uncrossed path, and hence is spared.

The explanation provided by the model for the forwardly directed responses to stimuli in the left visual hemifield is incorrect, for reasons discussed in the text. Other aspects of the organization of descending projections, however, follow directly from observed persistances of behavior following complete unilateral interruption of all descending pathways, and are correct as diagrammed. The persistance throughout the visual field of movements which vary with stimulus elevation and elevation indicates that information relevant to such movements can descend by either a crossed or an uncrossed pathway (pairs of curved lines from "R" and "L"). The persistance of rightward turns indicates that information for contralateral turning can descend by a crossed pathway (straight lines from "R" and "L", as well as from "X" in the right tectal lobe). Interruption of such crossed pathways from the right tectal lobe cannot, however, fully account for the observed failure to turn toward leftward stimuli, since a significant part of the left visual hemifield ("X") is represented not only in the right but also in the left tectal lobe. To account for the hemifield character of the observed deficit, we hypothesized the existence of an uncrossed path from the left tectal lobe to turn generating circuitry (straight line from "X" in the left tectal lobe), a path which, like the crossed path from the right tectal lobe, would be interrupted by the lesion. This hypothesis, as discussed in the text, led to subsequent anatomical studies suggesting the existence of a tecto-tegmento-spinal pathway.

hemifield. For stimuli in the ipsilateral visual hemifield, lesioned frogs respond, regardless of stimulus eccentricity on the horizontal, with a forwardly directed movement. These movements, while not varying with the horizontal eccentricity of the stimulus, do vary with its elevation and distance.

These observations clearly indicate a form of representation of spatial information in the medulla quite different from that in the tectum. One difference is that information about the horizontal eccentricity of a stimulus can be disturbed independently of information about stimulus elevation and distance; such a separate representation of variables ("parcellation") is quite different from the joint representation of two variables in known tectal maps and their joint disturbance by tectal lesions. A second difference is that in the medulla information about horizontal eccentricity is oppositely and more completely lateralized. In the medulla, information about the eccentricity of stimuli to the left of the midsagittal plane is strictly dependent on the left side of the brain, rather than being bilaterally represented or dependent on the right as in the tectum. In these two respects, the medullary deficit not only differs from the tectal deficit but appears at first consideration straightforwardly interpretable in a hypothetical motor coordinate frame (Kostyk & Grobstein 1982). Indeed, a quite similar deficit is known in the case of saccade control and has been so interpreted (for a discussion see Grobstein 1988a). Borrowing and slightly modifying that interpretation, one could imagine that there exist distinct sets of pattern generating circuitry for movements related to stimulus elevation and distance and for movements related to horizontal eccentricity, with the latter subdivided into left-turn generating circuitry on the left side of the brain and right-turn generating circuitry on the right (Fig. 2). In this framework, the normal three-dimensional movement triggered by a stimulus at a given location would be imagined to result from the simultaneous activation of two functionally distinct pathways from a given tectal locus, one running to turn generating circuitry and the other to circuitry generating movements related to stimulus elevation and distance. Interruption of the pathways linking both tectal lobes with turn generating circuitry on one side of the brain and sparing of all other linking pathways would provide a reasonable explanation of the described behavior.

It turns out, however, that the deficit following medullary lesions cannot be so simply understood. There is an important, additional feature of the behavioral deficit, first noted by Kostyk & Grobstein (1987) and more extensively documented by Masino & Grobstein (1989a). As discussed earlier, the motor pattern used by frogs in visual orienting varies with stimulus distance. The variation is both quantitive and qualitative (Grobstein et al. 1985), with the latter involving motor pattern shifts much like those described above for tactile behavior. For nearer locations, the visually triggered motor pattern consists of a directed snap, including a characteristic lunging movement and projection of the tongue. For more distant locations, what is instead involved is a reorientation movement, a motor pattern which differs from snapping in a number of ways including the absence of lunging and tongue projection. In normal frogs, a transition between the two movement patterns occurs at a distance of about two body lengths for frontal stimuli and one body length for lateral stimuli (Ingle 1970; Grobstein et al. 1985; Kostyk & Grobstein 1987a; Grobstein 1988a; Grobstein & Masino 1989a). If the turning deficit in lesioned frogs reflected simply the absence of the turning component of the normal three-dimensional movement, these animals would be predicted to snap forward for lateral stimuli in the affected hemifield at distances out to one body length, and to hop forward for stimuli at greater distances. In fact, the transition distance for lesioned animals who respond to lateral stimuli with forwardly directed movements is not at a distance of one body length but rather at a distance of two (Kostyk & Grobstein 1987a; Masino & Grobstein 1989a), the same distance at which both lesioned and normal animals switch between motor patterns for frontal stimuli.

That lesioned animals snap rather than reorienting for distances at which normal animals reorient clearly indicates that the deficit is not due to the selective interruption of a signal activating circuitry responsible for patterning the motoneuron discharges underlying turns in one direction. The syndrome involves not only an absence of turning, but a correlated change from one kind of motor act to another. More generally, the lesioned animals, rather than displaying for lateral stimuli movements which are themselves abnormal, are instead displaying for such stimuli the normal movements appropriate for frontally placed stimuli. What this implies is that what has been directly disturbed by the lesion is not in any sense the motor act itself, but rather some necessarily prior process involved in determining the particular motor acts to be associated with stimuli at particular locations. Additional evidence for this point comes from observations on animals in which medullary lesions invade but do not totally destroy a critical region implicated in the hemifield deficit (see below and Fig. 3). Rather than failing entirely to turn toward lateral stimuli, these animals turn short of laterally placed targets. Correspondingly, the distance at which they switch from snapping to re-orienting is neither the normal distance for laterally placed stimuli nor that for frontally placed stimuli, but rather an intermediate distance comparable to that at which normal animals switch for stimuli at intermediate eccentricities (Masino & Grobstein 1989b).

In normal frogs, the motor act associated with a given stimulus location varies with both horizontal eccentricity and distance, indicating that some interaction of signals representing these two variables must occur prior to the selection of a particular motor act. The behavior observed in lesioned animals is difficult to account for in relation to a disturbed motor signal, but is readily understandable in relation to an interaction between eccentricity and distance signals. If a signal reflecting horizontal eccentricity relative to the mid-sagittal plane were eliminated prior to its interaction with a distance signal, one would expect to see exactly the behavior observed: forwardly directed movements together with a variation in motor pattern choices with stimulus distance which is appropriate for frontally placed stimuli. Similarly, a reduction in an eccentricity signal associated with partial lesions should, as observed, lead to reduction in turn amplitude and a correlated variation in motor pattern choices with stimulus distance appropriate for the observed turn amplitude.

What this indicates is that what one is dealing with in the medulla is a representation of information about stimulus location in a coordinate frame which, while clearly not that of the retina, is not one closely related to movement either (for a discussion of the possible forms of a movement coordinate frame see Grobstein 1989b). The lateralized signal disturbed by the medullary lesions is not directly related to any component of movement but appears instead to be related to stimulus location in a head or body-centered coordinate frame. More generally, the representation of spatial information in the medulla seems to involve a signal distributed over several distinct pathways, with the parcellation of the signal relating not to components of movement but to axes of some more or less Euclidean head or body-centered coordinate frame.

An additionally significant feature of the lateralized and parcellated form of spatial representation seen in the medulla is that it appears not to be an idiosyncracy specific to visually elicited orienting. We have recently been studying tactually elicited orienting behavior following medullary lesions (Grobstein et al. 1988). Consistent with preliminary observations (Kostyk & Grobstein 1987a,b,c), we have found that unilateral medullary lesions produce deficits in tactile orienting behavior strikingly similar to those seen in visual orienting behavior. An example is shown in Fig. 3. The deficit is a partial one, involving a reduction in turn amplitude rather than a complete absence of turns, but clearly relates relatively specifically to stimuli on the side of the body ipsilateral to the midsaggital plane and involves a disturbance in turning toward such stimuli, rather than a failure to respond to them. What is even more striking is that the deficit involves not only an altered turn amplitude but also an alteration in motor pattern, entirely analogous to that seen in visual behavior. Normal frogs almost invariably snap for rostral tactile stimuli and almost invariably re-orient without snapping for caudal stimuli (Comer & Grobstein 1981a). In the lesioned frog, there is an increase in the frequency with which more caudal stimuli ipsilateral to the lesion trigger snaps rather than re-orientations, an increase clearly evident in comparison with the responses of the lesioned animal to stimuli at comparable locations contralateral to the lesion (Fig. 3, locations B and C). The implication is that, as in visual behavior, what is being observed is not abnormal movements but normal movements for rostrally located stimuli, and hence that, in this case too, what has been disturbed is a horizontal eccentricity signal proir to a critical interaction with other signals which determines motor pattern choice. Our findings suggest not only that the medullary circuitry underlying tactually elicited orienting has the same organization as that underlying visual behavior, but that it may in fact be the same circuitry. While we have as yet studied a relatively small number of animals, the severity of the visual and tactile deficits observed in individual animals studied so far has been clearly correlated (Fig. 3).

There is reason to suspect that the lateralized and parcellated form of spatial representation observed in the medulla is not only a common characteristic of circuitry underlying both visually and tactually elicited orienting behavior in the frog, but may be a common characteristic of sensorimotor circuitry in vertebrates generally, and perhaps even of sensorimotor circuitry in all animals. Unilateral lesions produce unilateral orienting deficits in a variety of circumstances where the deficits cannot readily be accounted for in terms of disturbances of primary sensory maps. Parietal cortex lesions in humans, for example, can result in neglect of visual stimuli to one side of the body midline, regardless of eye position at the time the stimulus occurs (Critchley 1953; Hyvärinen 1982). Hemifield deficits following unilateral lesions are also seen in a number of situations where stimuli to one side of the body midline normally activate receptors on both sides, including mammalian sound localization (Masterton & Imig 1984), water wave localization in aquatic frogs

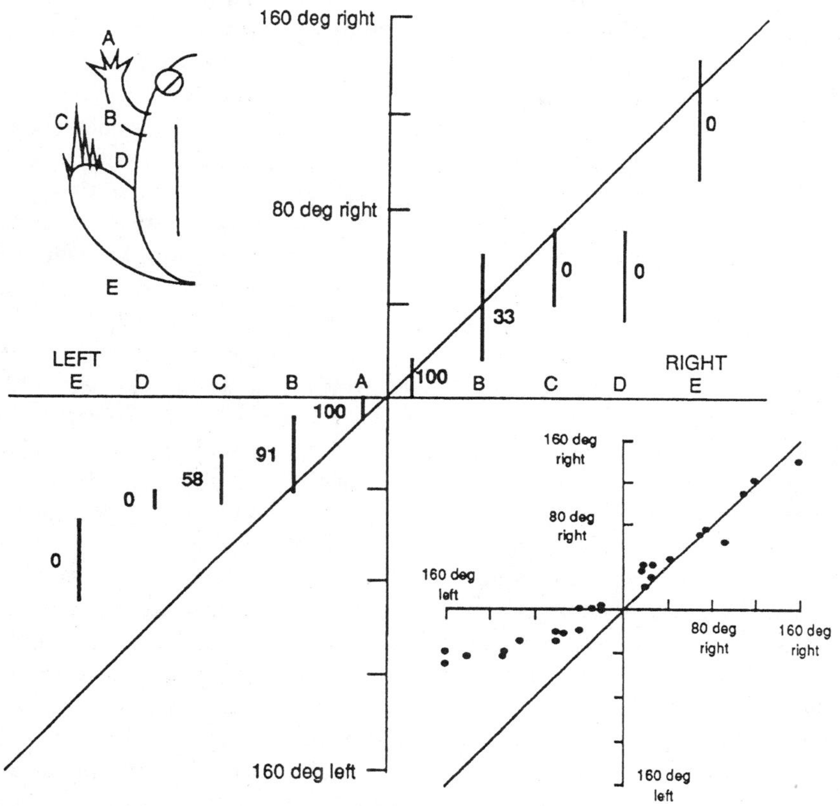

Figure 3. Evidence for an abstract spatial representation in the case of tactually elicited orienting behavior (Grobstein et al. 1988). Illustrated is the orienting behavior of an individual frog following a unilateral lesion on the left side of the brain at the junction between the mesencephalon and the medulla. The frog's visually elicited behavior was tested first and is shown to the lower right. Each dot corresponds to one trial. Horizontal stimulus eccentricity is shown on the abscissa; response angle, similarly defined, is shown on the ordinate. The deficit observed is characteristic of that seen following partial damage of a descending tract believed to be carrying information about horizontal stimulus eccentricity in the ipsilateral visual hemifield (Kostyk & Grobstein 1987b; Masino & Grobstein 1989); the animal undershoots in response to stimuli in the left visual hemifield. The frog was subsequently blinded and its tactually elicited orienting behavior studied. These observations are summarized in the larger graph. The abscissa indicates stimulus position, in terms of body locations identified in the upper left inset. The ordinate shows response angle in terms of horizontal eccentricity. Bars indicate the range of response angles observed in multiple trials. The angled line is provided as an aid to comparison of response to stimuli on the left and right sides of the body and is not intended as a summary of observations (responses to position "D" are characteristically not very different from those to position "C"). The frog displays undershooting for tactile stimuli on the left, much as it does for visual stimuli. Also noteworthy are the numbers associated with each bar in the display of tactile behavior, which indicate the percentage of stimulus trials at that location which elicited snaps as opposed to re-orientations. Note that correlated with the more forwardly directed movements triggered by stimuli at positions B and C on the left is an increased frequency of snapping.

(Elepfandt 1987a,b), and cockroach escape behavior in response to wind puffs (Comer & Dowd 1987).

Not only lateralization but the sidedness of the lateralization, as well as the parcellation of signals into separate components like those we have described in the frog, have parallels in other systems. The phenomenon of the elimination of horizontal components of movement with persistance of other components has been observed in the orienting movements of aquatic frogs (Elepfandt 1987a,b) as well as in cockroach escape behavior (Comer & Dowd 1987). As noted earlier, there are also marked similarities between deficits in the frog and certain deficits described in relation to primate saccade control: unilateral lesions in the pontine reticular formation abolish ipsilateral saccades while sparing contralateral as well as vertical saccades (Raphan & Cohen 1978). Still more striking are observations on human head control: lesions can produce a loss of ipsiversive head turns while contralateral and vertical movements are spared (Beevor 1909; Ianonne & Gerber 1982). Of particular interest in this case is that the deficit, as in frogs, seems to bear no simple relation to either muscles or motor synergies. What is disturbed is not an ability to activate muscles on, for example, the left side of the body, nor an ability to make leftward head movements: the patients described were unable to deviate the head to the left of the midsaggital plane, but were fully capable of leftward movement of the head from rightward deviated starting positions.

The problem of associating given locations in space with appropriately directed movements is one with which all animals have to deal for a variety of different behaviors, but one which might in principle be dealt with differently for each form of behavior (visually *vs*. tactually elicited orienting, for example) and by each animal. Against this background, the similarities in observations on two different behaviors in the frog, and between the frog and other verebrates, is striking. One cannot readily avoid the suggesion that the apparently arbitrary dissociation of information about stimulus location into a lateralized and parcellated form may be a rather general characteristic of the vertebrate sensorimotor interface. Indeed, given observations on invertebrates as well, it seems possible that this form of organization has some still more general significance, perhaps relating to the solution of computational demands intrinsic to the problem and invariant across organisms.

The discovery that there exists in the frog medulla a previously unexpected form of spatial representation raises not only interesting new questions about the computational problems underlying sensorimotor processing in general but also some new quite specific questions about sensorimotor circuitry in the frog. In particular, one would like to know the place of the medullary form of organization in the overall information processing sequence underlying directed movement. The observations discussed to this point indicate that it exists as a form distinct from that of sensory maps or immediately premotor circuitry. However, they leave open the question of whether the medullary organization is a processing step which intervenes between the two, or instead involves independent circuitry which modulates a more direct path linking sensory maps to premotor circuits. I will consider this question further in connection with recent anatomical observations in the following section.

4 Discovery of New Pathways and Structures in the Sensorimotor Interface

In the frog, as in many other vertebrates, each tectal lobe receives input from the contralateral eye and gives rise to a projection which recrosses the midline in the ventral midbrain and subsequently descends through the medulla, typically to the spinal cord, as a compact bundle. This crossed tecto-bulbar and tecto-spinal tract (CTBST) was until recently, presumed to be the key anatomical link between tectum and the premotor circuitry involved in orienting (see Ewert, this volume). Consistent with such a presumption was the finding that unilateral tectal lobe lesions in the newt abolish orienting toward contralateral visual stimuli, while more caudal lesions disturb orienting toward ipsilateral stimuli (Sperry 1948). Also consistent was evidence in several vertebrates, including the frog (Ingle 1983), that orienting movements to stimuli at locations throughout visual space are disturbed by ventral midbrain lesions made in the area through which tectal efferent fibers decussate (Sprague et al. 1963; Voneida 197; Raczkowski et al. 1976; Winterkorn & Meikle 1980).

New observations made over the last few years have substantially changed this picture. It is now clear that the crossed tectospinal and tecto-bulbar tract is not the only descending projection important for orienting in frogs. At least one, and more probably several, additional descending pathway is involved. Among these, recent findings suggest, is an indirect tectofugal pathway involving a relay in the midbrain tegmentum. The tecto-tegmento-spinal pathway is of particular interest in that it may turn out to be the pathway actually carrying important parts of the localization information previously suspected to travel on tecto-bulbar and tectospinal axons.

The optic tectum in frogs, as in other vertebrates, gives rise to a number of projections in addition to those generally associated with the CTSBT; among these are substantial projections into the ventral midbrain (Masino & Grobstein 1985; Masino & Grobstein 1989c). Evidence that projections other than the CTSBT play a role in orienting first emerged from independent but related studies in two different laboratories. Ingle (1983) reported that interruption of the crossed tecto-spinal tracts by ventral midbrain lesions does not abolish all orienting behavior; it instead produces a rather specific, more restricted deficit in turning. Lesioned frogs continue to respond to stimuli at all locations in visual space but do so with orienting movements which, regardless of stimulus eccentricity on the horizontal, tend to be forwardly directed (see also Kostyk & Grobsten 1987c). Kostyk & Grobstein (1982, 1987a) observed a similar selective deficit following a lesion like that reported by Sperry for the new, a hemisection immediately caudal to the tectum which interrupted not only the crossed tecto-spinal tract from the opposite tectal lobe but all connections between the midbrain and more caudal structures on one side of the brain. As described in the previous section, lesioned frogs turned normally to stimuli in the contralateral visual hemifield but responded with forwardly directed movements to stimuli in the ipsilateral visual hemifield.

Several conclusions emerged from these observations, of which perhaps the most generally significant and definitive is that each tectal lobe must be able to access pattern-generating circuitry for at least some aspects of orienting in some way other than via the CTSBT. Ingle (1982) speculated that the forwardly directed movements he observed in the absence of crossed tecto-bulbar and tecto-spinal projections might be triggered by uncrossed descending paths from each tectal lobe. Our own observations, in which not only the CTSBT but all crossed projections from one tectal lobe were interrupted and forwardly directed movements for stimuli at locations represented solely in that tectal lobe persisted, proved that such paths exist. What seemed to us equally noteworthy in our observations on unilaterally lesioned animals was that the misdirected movements to ipsilateral stimuli, while not varying with horizontal stimulus eccentricity, did vary with stimulus elevation and distance, and that the normally directed movements to contralateral stimuli did so as well (Kostyk & Grobstein 1982, 1987a). The former indicated not only that an uncrossed path was adequate to trigger forwardly directed movements but, additionally, that it carried adequate information to support variations in those movements with variations in stimulus location along two spatial axes. The latter, since it involves movements triggered from locations represented solely in the tectal lobe on the side of the lesion, (see Fig. 2) indicates that a crossed path is also adequate to support variations in movement with variations in stimulus location along the same two axes. Though incomplete, our findings suggest that this crossed path is distinct from the CTSBT: in a few cases our hemisection lesions invaded the remaining CTSBT without producing any obvious deficit in elevation or distance related responses. In sum, there is at least one descending pathway in addition to the CTSBT which is likely to be important for orienting behavior: an uncrossed pathway carrying information about stimulus elevation and distance. There is likely, as well, to be a third, crossed, pathway carrying similar information. Also noteworthy, for reasons discussed further in a later section, are some additional conclusions from these observations: forwardly directed movements can be triggered by descending signals restricted to one side of the brain, and information related to stimulus distance and eccentricity descends bilaterally.

That animals with a complete hemisection turn accurately toward stimuli contralateral to a lesion which interrupts all connections from one tectal lobe proves what had long been hypothesized: each tectal lobe gives rise to a crossed descending tectofugal path adequate to support turns in a direction opposite to that tectal lobe. Ironically, the same observations gave rise to subsequent studies which suggest that this crossed descending tectofugal path may not be identifiable with the CTSBT, and may instead involve a relay in the midbrain tegmentum. The hemifield turning deficit can be partially but not completely explained in terms of interruption of the CTBST from the opposite tectal lobe. As described earlier, each tectal lobe in the frog represents the entire visual field of the opposite eye, one visual hemifield together with an additional 45° across the midsagittal plane, and is alone capable of triggering turns toward stimuli at locations throughout this region. The hemifield deficit thus implies not simply interruption of the outflow from one tectal lobe but an additional quite specific disturbance in the output of the other, so as to produce a deficit with a border at the mid-sagittal plane rather than at 45° lateral (Fig. 2).

The interpretation of behavior following brain lesions needs to be approached with some care (Grobstein 1989a). Conclusions based on behavior which survives the lesion, of the sort drawn to this point, are straightforward; those based on behavioral deficits require more caution. In the present case, the failure to orient to stimuli in one hemifield clearly indicates a correlated disturbance in the outputs of both tectal lobes. One cannot, however, from these findings alone conclude that the disturbance is due to interruption of essential descending paths from the two tectal lobes; a variety of other possible explanations for the disturbance can be imagined (Grobstein 1989a). For this reason, we undertook to determine whether the hemifield deficit could be produced by more restricted damage to some tract which contains axons carrying descending signals from both tectal lobes (Kostyk & Grobstein 1987b). Large lesions at the junction between midbrain and medulla proved to be without effect unless they invaded a ventromedially located region of white matter. Small lesions restricted to this area, which includes the medial longitudinal fasciculus, produced a deficit indistinguishable from that resulting from complete hemisection. Retrograde labeling studies showed that the critical region contained axons of cells in the contralateral tectal lobe. It contains few or no axons from cells in the ipsilateral tectal lobe, but did prove to contain axons of cells from several ventral midbrain nuclei which themselves are in a position to receive such input. The findings are consistent with the hypothesis that the hemifield deficit results from interruption of descending paths from both tectal lobes, assuming at least one is a relayed projection. At the same time, the retrograde labeling studies also showed that the critical region at the midbrain/medulla junction contains as well ascending axons from cells in a number of medullary nuclei, making it possible that it was damage to these rather than to descending projections that produces the hemifield deficit.

More recently, we have found that a hemifield deficit in all important respects identical to that produced by unilateral lesions at the junction between midbrain and medulla can also be produced by unilateral lesions at the junction between medulla and spinal cord (Grobstein & Masino 1986; Masino & Grobstein 1986, 1989a), a location caudal to the nuclei giving rise to the previously observed ascending projections. Here, as rostrally, damage to a critical white matter region is necessary and sufficient to produce the deficit. The critical region defined was again medial but somewhat more dorsal than at the rostral site, and again includes the medial longitudinal fasciculus. This, and the similarity in the deficits, suggested that the same axons are being interrupted at both rostral and caudal locations. Retrograde labeling studies from the caudal site were consistent with this possibility but, significantly, did not show significant labeling of tectal cells in either tectal lobe (Masino & Grobstein 1986, 1989b). The only cells which were prominently labeled in both rostral and caudal studies were a population of large neurons on the lateral edge of the ventromedial tegmentum. Kostyk & Grobstein (1987b) identified this labeling with the nucleus of the medial longitudinal fasciculus. The observations of Masino & Grobstein (1986, 1989b) suggest that it is actually a longer continuous column, extending caudally through the anteroventral nucleus and into the posteroventral nucleus. Reconstructions of dendrites of labeled cells of this column show them to be in a position to receive tectal input; reconstructions of labeled axons showed directly that they pass through both the rostral and caudal critical regions (Masino & Grobstein 1989b).

These recent observations greatly strengthen the evidence that the hemifield deficit is due to interruption of descending axons. (For additional evidence on the existence of an uncrossed pathway adequate to support ipsilateral turns see Kostyk & Grobstein 1987c). Both rostral and caudal critical regions proved to contain axons likely to be carrying descending tectofugal signals. In addition, the caudal critical region is not only caudal to medullary nuclei projecting through the rostral critical region but it itself appears to contain almost exclusively descending axons, judging both from its location and a paucity of labeled neurons observed still more caudally in our retrograde studies. The recent observations also indicate that by the junction between medulla and spinal cord, information critical for ipsilateral turning is not being carried by tecto-spinal axons. Lesions which included this region did not produce a hemifield deficit, while lesions which did not did produce such a deficit. This in turn raised the obvious question of whether the interruption of tecto-bulbar and tecto-spinal axons, which clearly occurred in the case of the more rostral lesions, was actually necessary to produce the hemifield deficit in those cases. A simpler possibility, given the retrograde labeling studies and the similarity of the deficits at the two locations, is that information from both tectal lobes which is important for turning in one direction is transferred to the ventral tegmentum in the midbrain, and that interruption of tecto-spinal and tecto-bulbar axons subsequent to this transfer is irrelevant to the hemifield deficit. Consistent with this possibility is the finding that unilateral tegmental lesions, sparing the tecto-bulbar and tecto-spinal tracts, can produce a hemifield deficit (Masino & Grobstein 1989b). These studies also support the identification of the ventrolateral large cell column in the tegmentum as a critical source of descending signals. Large lesions sparing this column did not produce a hemifield deficit. Only when the lesions included this column was the hemifield deficit observed.

Our findings to date are consistent with an hypothesis that signals related to horizontal eccentricity are relayed in the ventral midbrain and descend to the spinal cord on tegmental axons. Additional work is, however, necessary before one can firmly conclude either that the particular pathway has been correctly identified or that it is the only relevant pathway. At the same time, recent findings in the rat (Dean et al. 1986) indicate that some persistance of orienting movements following interruption of crossed descending tecto-bulbar and tecto-spinal tracts is not a phenomenon unique to the frog. Similarly, our suggestion of a tegmental involvement in anuran orienting, while novel in the literature on this topic, is less so in the context of vertebrate sensorimotor processing generally. Tecto-tegmental relays are a prominent component of the neuronal organization underlying primate saccade control (Wurtz & Albano 1980; Sparks 1986), and tecto-tegmental projections are a significant component of tectal efferent projection patterns in all vertebrates, whether or not eye movements are a characteristic part of their orienting repertoire. It has also long been known that electrical or pharmacological activation of tegmental structures can produce lateralized movements, including ipsiversive ones, in a number of vertebrate species, and there is now increasing evidence of a variety of kinds that not only bulbar but midbrain tegmental neurons play an important role in both the vertical and lateral components of orienting head movements in mammals (Grantyn & Berthoz 1988).

While there remains some uncertainties about details of tectofugal organization (cf. Ingle 1983), several points seem firmly established by the evidence discussed in this section. It is quite clear that the crossed tecto-bulbar and tecto-spinal tract is not the sole pathway by which signals involved in orienting can reach pattern-generating circuitry. There exists, in addition, at least one uncrossed pathway, whose anatomy remains to be clarified, which carries information related to stimulus elevation and distance. Additional crossed and uncrossed paths are likely. What also seems reasonably certain is that different pathways carry different aspects of the descending signal involved in orienting: information relevant to horizontal turning can be disturbed independently of information relevant for distance and elevation related movements. Finally, there is an increasing body of evidence consistent with an hypothesis that at least one of the descending signals, that related to horizontal turning, descends in part or in whole via a relay in the ventral midbrain tegmentum.

5 A New Provisional Map of the Anatomical and Information Processing Organization of the Sensorimotor Interface

The findings reviewed in the previous section are of interest not only in their own right but also in relation to the discussions, in the two prior sections, of activity-gated divergence and of an abstract spatial representation. The finding that there are in fact at least two descending tectofugal pathways, and that locations where lesions reveal the existence of an abstract spatial representation turn out to contain axons probably carrying descending signals, makes it likely that such a representation is actually an intermediate stage in the tectofugal pathway, rather than a characteristic of some system which modulates direct connections between tectum and premotor circuitry. With regard to activity-gated divergence, the recent findings on the anatomical organization of tectofugal projections suggest some new loci at which information addition to the retino-tectal signal may be occurring. More generally, the findings discussed in the previous and prior two sections clearly indicate the need for some new models to guide future research on sensorimotor organization in the frog and, taken together, provide the basis for elaborating such a more general model. In this section, I want to outline a model of this kind. My objective is not to assert the reality of any of the elements in the model; the experimental evidence relevant to establishing their existence has been critically reviewed in prior sections. My objective instead is to show how these elements taken together suggest lines of new inquiry which are likely to productive. As I have argued elsewhere (Grobstein 1989a), the latter is frequently not only the best criterion for judging the scientific usefulness of hypotheses, but also provides the appropriate basis for their ultimate acceptance and rejection (cf. Grobstein et al. 1983; Grobstein 1988b).

A new provisional map of the sensorimotor circuitry underlying anuran orienting is illustrated in Fig. 4. Taken collectively, what the findings discussed in this chapter suggest is that at least two major transformations occur in going from the midbrain sensory maps involved in orienting to the production of appropriate motoneuron discharge patterns. The first of these appears to occur in going from dorsal midbrain to tegmentum, and involves a shift from initial two-dimensional signals place-coded in the coordinate frames of particular sensory modalities to a three-dimensional head or body centered signal whose components are carried by at least two distinct descending pathways. One is a lateralized pathway, in which information about horizontal stimulus eccentricity relative to the mid-sagittal plane descends on the side of the brain ipsilateral to the stimulus. Our findings suggest that, once elaborated in the ventral midbrain, this signal descends without interruption or further processing to the spinal cord, and it is so illustrated in Fig. 4. The other descending system carries information about stimulus elevation and distance. As illustrated, this information appears to descend simultaneously on both sides of the brain, but we can as yet say little else about the organization of this system, making it a subject of considerable interest for future investigation. It seems possible distinct tracts are involved, one related to elevation and the other to distance, but we do not yet have any evidence bearing on this point. Nor do we as yet have evidence as to whether these signals, like those related to horizontal eccentricity, run uninterrupted from midbrain to spinal cord. Both synaptic relays and intermediate processing steps are possible in this pathway or pathways. Finally, the source of the distance signal, as discussed, remains to be clarified.

Our findings further suggest that the lateralized and parcellated intermediate signal is not specifically a visual signal, but rather one common to at least the tactile modality as well, and this feature too is illustrated in Fig. 4. The evidence for convergence of visual and tactile signals at the tegmental level, while still indirect, is particulary significant given prior evidence for lack of a critical convergence of these signals in the tectum (Comer & Grobstein 1978, 1981b). Here, too, the model poses some interesting new issues for exploration. What our findings imply is that visual and tactile signals related to horizontal eccentricity are recoded into a similar horizontal eccentricity signal in a coordinate frame common to the two. The computational problems of doing so are probably non-trivial, and the step may well require some information addition to assure alignment of body and retinally derived signals. The horizontal components of these signals are, however, at least similarly place-coded in the initial sensory maps. What our observations raise as an additonal possibility is that visual and tactile signals related to stimulus distance are also recoded into similar signals in a common coordinate frame. These signals may be initially coded in quite different ways, with some combination of binocular parallax (Collett et al. 1987), monitoring of accommodative state (Douglas et al. 1986), and probably other cues for the former (Grobstein et al. 1985; Collett & Harkness 1982), and proprioceptive input and/or efference copy signals for the latter. How these might be recoded into a common signal would seem to be a problem offering major challenges both for neurobiological analysis and computational theory.

An additional point which is worth making in the present context is that the nature of information coding in the descending pathways should be regarded as an entirely open question. Prior work has presumed at least some degree of place-coding, that information about stimulus location is coded in terms of the subset of descending axons active for a given stimulus. The primary reason for so believing has been the presumption that these pathways derive fairly directly from sensory maps. With the recognition that an important information processing step intervenes between the sensory maps and the descending pathways, this basis for expectation has been removed. In addition, as suggested, important parts of the information input to the descending pathways may in fact be represented in non-place coded forms. A signal related to limb position, or to accommodative state, for example, may well be frequency coded rather than place coded. The finding that partial damage to the descending pathways produces similar behavioral effects apparently irrespective of

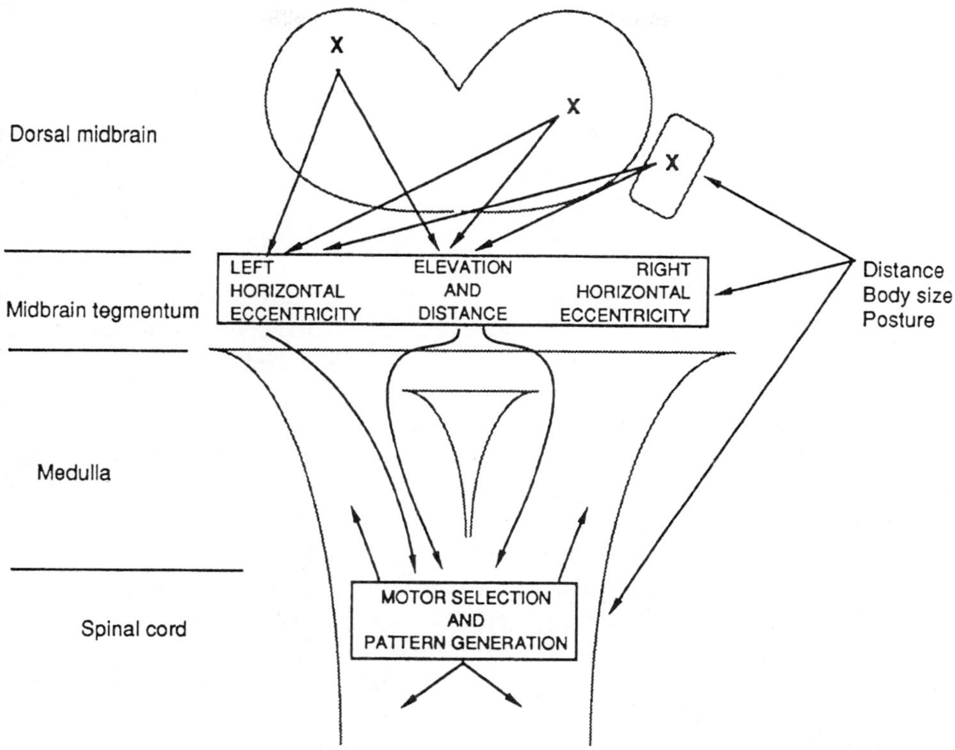

Figure 4. A new provisional map of the sensorimotor interface in the frog. Lighter outlines schematize the optic tectum (above); the right lateral torus semicircularus (right rectangle), a dorsal midbrain structure which receives cutaneous input from the left side of the body (Comer & Grobstein 1981c); and the medulla and spinal cord (below). As in Fig. 2, "X" corresponds to a stimulus location a short distance to the left of the midsagittal plane, one which is represented, as indicated, in both tectal lobes, as well as in the torus (whether this representation too is bilateral is uncertain and so only the crossed representation is shown). The dorsal midbrain maps project to the midbrain tegmentum, where there is created (long middle rectangle) a lateralized and parcellated representation of stimulus location in a head or body centered coordinate frame, as discussed in the text. This signal is carried by descending projections (curved arrows) to the spinal cord where a second transformation (lower rectangle) determines the motor act to be generated and creates the appropriate patterns of motoneuron discharges. The horizontal eccentricity component of the descending signal descends ipsilaterally and other components descend bilaterally, as illustrated. This part of the provisional map is organized in a fashion similar to that of the descending projections illustrated in Fig. 2, and is similarly motivated. Following motor selection and pattern generation, signals must subsequently be distributed bilaterally to both more rostral and more caudal locations (lower straight arrows) since relevant motoneurons are located in both the medulla and the spinal cord on both sides of the nervous system. The overall function of the schematized organization depends on the addition of information over and above that represented in the initial dorsal midbrain sensory maps. Such information may be added in the tectum, the midbrain tegmentum, and the spinal cord, as indicated to the right.

the particular subset of axons damaged (Kostyk & Grobstein 1987a; Masino & Grobstein 1989a) is consistent with some non-placed coded form of representation in the descending pathways. Electrophysiological recordings in these pathways would be useful in clarifying their form of coding, as well as to verify the presumed convergence of visual and tactile information.

The model of Fig. 4 motivates intriguing questions not only about the organization of the descending pathways but also about how this organization is produced. Among these is the obvious issue of how the topographically coded signals of the initial sensory maps are transformed into the parcellated, lateralized, and body-centered signals evident in what we presume to be the descending pathways. Our current findings provide no information as to how separate signals related to elevation and horizontal eccentricity are extracted from the two-dimensional signals of the topographic maps, but do provide some information about the lateralization of the horizontal eccentrity signal. Perhaps the simplest interpretation of this aspect of the initial transformation is that there is a topographic partial decussation in the descending tectofugal projections, with tectal loci representing visual field loci to the left of the midsaggital plane projecting to the left side of the brain and those representing right visual field loci to the right. We have looked for but been unable to find such a pattern of projections (Masino et al. 1984; Masino & Grobstein 1985; Kostyk & Grobstein 1987b; Masino & Grobstein 1989c), indicating that the lateralization apparently results from some interaction between tectal projections and post-tectal circuits. This conclusion is of course of particular interest in connection with our observations indicating that a synaptic relay in the tegmentum probably intervenes between tectal outflow and the descending lateralized signal, and makes an exploration of tecto-tegmental interactions of particular interest for future research. In considering the nature of this interaction, some additional observations are relevant. As discussed previously, these indicate that the creation of an appropriately lateralized signal on either side of the brain can occur in the absence of one tectal lobe, and, further, that the readout of a lateralized signal on one side of the brain does not depend on the existence of a pathway carrying related signals on the other. At the same time, certain midbrain lesions result in apparent abnormalities in the sidedness of the descending signal (Kostyk & Grobstein 1987c), suggesting that its elaboration may involve some side to side comparison at the tegmental level.

The conclusion that lateralization reflects an interaction between tectal projections and post-tectal circuits is also of interest in connection with likely locations at which non-retinal signals gate the divergent circuitry linking the retino-tectal projection with movement. We do not yet know whether the lateralization of the descending signal relates to the midsagittal plane of the head or that of the body. Were the latter the case, it would imply that postural information must be added to retinal information somewhere in the midbrain, either in the tectum or in the tegmentum. The problems of creating appropriate convergence of visual and tactile signals in descending circuitry also suggests important information addition in the tegmentum. The likelihood of information addition at this level is indicated in the model of Fig. 4.

The second major transformation illustrated in the model of Fig. 4 involves the selection of an appropriate motor pattern, and requires at a minimum that signals related to horizontal eccentricity and distance be combined. Since the behavior resulting from lesions in caudal midbrain and caudal medulla both indicate disturbances prior to this transformation, our findings imply that the selection of an appropriate motor program must involve processes occuring in the spinal cord, and this too is illustrated in Fig. 4. The implied sophistication in spinal cord information processing is somewhat greater than might have been expected based on the orienting literature, but is not inconsistent with evidence of other sorts (cf. Fukson et al. 1980; Berkinblit et al. 1986a,b; Berkinblit et al., this volume).

How neuronal circuitry is organized so as to combine signals related to horizontal eccentricity with those related to distance is an open question, one again posing interesting computational problems. At the moment, we have only two observations bearing on this point. One is that the computation seems to regard as equivalent and sufficient either of two distance signals descending on opposite sides of the brain. The other is that interruption of a path presumed to be carrying the horizontal eccentricity signal does not disturb the computation, being instead treated as a signal indicating zero eccentricity in one direction. Of more general interest is an open question having to do with the relation between the processing involved in selecting a motor program and that involved in patterning the motoneuron discharges for that program. Anuran orienting movements in general must involve elaborate, specific, and correlated discharges in motoneurons at a large number of locations in both the spinal cord and the brainstem. These, in turn, presumably depend on a number of premotor circuits patterning, for example, limb movements, head movements, and tongue movements. It is possible that the process of motor pattern selection is distinct from and prior to these pattern-generating circuits. It is, however, also possible that the descending signals are distributed independently and directly to the various pattern-generating circuits. In the latter case, interconnections between the pattern-generating circuits would have to be of such a kind as to yield not only coordination among the various output patterns but also to account for the act of motor program selection itself in terms of some kind of distributed processing. Noteworthy is that, in either case, having traced descending signals to the spinal cord does not complete an analysis of the central pathways involved in orienting. There must be subsequent ascending signals, not entirely of sensory origin, which assure coordinated activation of motoneurons residing in the brainstem. Such ascending pathways are also shown in Fig. 4. These, as well as necessary descending pathways also shown, are presumably bilateral, since orienting movements in general require coordinated activity in both bulbar and spinal motoneurons on both sides of the body.

The likely complexity and interconnectedness of the premotor circuitry involved in orienting has recently been stressed by Ewert (1987), who also reviews recent electrophysiological findings from his laboratory on medullary organization. Given our evidence that one of the descending signals important for orienting apparently descends to the spinal cord without interruption in the medulla, the evidence from Ewert and his colleagues that there is substantial medullary activity related to orienting is particularly significant (see also Schwippert et al., this volume). In the context of the model offered here, such activity might indicate the existence of synaptic relays in the descending pathways related to elevation and distance. Of perhaps still greater interest is the alternative or additional possibility that observed activity relates to the ascending pathways illustrated in Fig. 4, and hence that studies of medullary organization might help to resolve the question of whether motor pattern selection is a unitary or a distributed information processing event. Finally, as discussed earlier, it remains possible that information about stimulus eccentricity descends not only via a tegmento-spinal projection but along other pathways as well, and that the observed medullary activity will ultimately prove to establish the existence of such additional descending pathways.

In addition to tectum and tegmentum, our observations suggest that the level of motor selection and patterning is a third possible location of information input, and this is also indicated in the model of Fig. 4. While we have at the moment no direct evidence for information addition at this level, the possibility is worth entertaining for conceptual reasons alone. A key feature of the model of Fig. 4 is the existence, between sensory input and motor output, of a signal related to stimulus location in some coordinate frame related to the body. Since parts of the body can move relative to one another and vary in size, the generation of such a signal requires that some information addition occurs rostrally. The interesting conceptual point is that, given a signal related to stimulus location in some generalized body coordinate frame, a second information addition may well be necessary as part of the process of motor pattern selection and generation. Just as the creation of a signal representing stimulus location in some generalized coordinate frame depends on knowing the position of parts of the body relative to one another, so too may the creation of a motor pattern appropriate for a stimulus location specified in the generalized coordinate frame. This would clearly be the case if, as in saccade control (Mays & Sparks 1980; Sparks & Porter 1983), appropriately directed orienting movements can be produced by a frog even when postural shifts have occurred between the disappearance of a stimulus and the production of the movement. An additional reason to suspect information addition at the level of motor pattern selection and generation is the increasing evidence indicating, in a variety of systems, that not only the latter but the former as well may make use of reafferent information to deal with unanticipated variations in the environment or in the body periphery (cf. Gallistel 1980; Berkinblit et al. 1986a,b; Grillner 1986; Bässler 1986; see also Bässler, this volume). It has long been known, for example, that afferent signals can affect not only limb movement amplitudes in locomotion but overall gait patterns as well. Some preliminary observations in our lab are consistent with such an interaction between central and peripheral signals in motor pattern choice in frog orienting. Also relevant in this regard is the suggestion of von Holst & Mittelstaedt (1950) that directed movements in general may involve some comparison between a central signal and a reafferent vestibular signal (see also Mittelstaedt & Eggert, this volume), and recent findings indicating that frog orienting turns are in fact altered by vestibular damage (Comer et al. 1985).

6 General Concepts of Sensorimotor Processing: What's New?

I have remarked elsewhere on the important, if often underappreciated, role that conceptual frameworks play in science (Grobstein 1988b, 1989a), both in defining questions for inquiry and in determining what inquiries and observations are regarded as meaningful. I have also at several points in this chapter characterized recent findings on the frog as new in the sense of unexpected, and hence with the implication that there has existed a conceptual framework for approaching problems of sensorimotor processing which requires alteration. This I think is true, if for no further reason than the new questions to which the model discussed in the previous section calls attention. At the same time, the model itself reflects broader changes in perspective which seem to me to be taking place. In this final section, I want to try and make clear the nature of this more general change in the asking and answering of questions not only about sensorimotor processing but about neuronal information processing in general.

With some noteable exceptions, western neurobiology has, for an extended period of time, been dominated by the notion of "reflexes" as the basis of an explanatory framework for nervous system function. With regard to the problem of orienting behaviors, and directed movements generally, what this translated into is the presumption that the underlying neuronal circuitry would consist of a parallel array of reflex-like pathways, each associating a particular stimulus location with a particular and appropriate motor pattern. There are at least three important characteristics inherent in such a picture of neuronal organization and function. First, the behavioral significance of activity at a given location in one of the pathways is relatively independent of that at other locations either in that pathway or in other pathways. Second, except for an ultimate convergence onto some common final motor path, different behaviors depend on different pathways constructed specifically for those behaviors. Finally, the information processing business of the nervous system is largely restricted to the task of extablishing parallel stimulus-response linkages. Investigators were certainly aware that behavior is a good deal less machine-like than implied by this picture, but assumed that given a description of the basic parallel reflex array it would be subsequently possible to identify and characterize additional circuitry which acts to modulate the basic neuronal organization. The emerging picture of the neuronal organization underlying sensorimotor integration in the frog, and indeed of the neuronal organization underlying a variety of behaviors in a variety of organisms (Grobstein 1988b, 1989a), is different, so much so as to suggest that the notion of reflexes might best be discarded from the neurobiological vocabulary. In general, neuronal organization seems to be related most closely to those behavioral characteristics which neurobiologists regarded as secondary to basic neuronal function, and least closely to those which neurobiologists regarded as basic (see also Berkinblit et al. 1986a,b; Berkinblit et al., this volume).

Not even at the sensory receptor sheets is it the case that the behavioral significance of the location of activity is independent of activity at other locations in the nervous system. Our finding that the initial sensory signals are quickly transformed into a three-dimensional coordinate frame, whose axes differ from that of the sensory sheets, implies that the context-dependence of the behavioral significance of neuronal activity is almost certainly a primary factor in the basic organization of the nervous system, rather than one dealt with by modulating circuitry which can be subsequently explored as a secondary problem. The descending signals are not only fundamentally three-dimensional rather than two-dimensional with some subsequent modification, but their creation may well involve addition of information related to body posture and other characteristics as well. It is not only the early inclusion of additional information which indicates the importance of context-dependence in neuronal processing, but the subsequent nature of signaling as well. Our observations imply that the behavioral significance of the signals in an individual descending pathway is not invariant but rather depends, at a minimum, on contemporaneous signals in other descending pathways. This is certainly true at the motoneuron level as well (cf. Feldman 1986; Berkinblit et al. 1986a,b; see also Berkinblit, this volume) and hence likely to be true throughout the entire neuronal circuitry underlying orienting. Finally, it is not only the evident importance of context dependence which distinguishes the observed organization from reflex models, but the likely character of the added information. While some of this information may come in on reafferent and other sensory pathways, it seems likely that some of the information is provided by corollary discharge signals. What this implies is that even if one defined sensory stimulus in the broadest possible sense, as all arriving sensory information, one would still not be able to predict the resulting output. Not only a context dependent handling of information but some degree of autonomous function are fundamental characteristics of neuronal organization.

Our observations on the organization of the descending paths involved in anuran orienting also conflict with the second aspect of the reflex concept of neuronal function, the notion that, with the exception of final common paths, different behaviors depend on different sets of circuitry organized in specific relation to those behaviors. Our finding suggesting that visual and tactile signals converge not at the level of a final common motor path but rather at an intermediate level represented by the descending pathways implies that an understanding of the organization of this level probably cannot be achieved from considerations of visually elicited behavior alone. Indeed, the existence of a processing level in which stimulus location is represented in some non-retinal, non-motor coordinate frame was not even considered in early models of the circuitry underlying either anuran orienting, or the related phenomena of primate foveation, and now needs to be entertained in both (see Grobstein 1988a, 1989b). More generally, there is, as discussed, reason to suspect that the same processing level may be involved in a large variety of behaviors in a wide array of vertebrates, and that something similar to it may be a characteristic of invertebrates as well. What this suggests is that the organization of this level may relate not to the characteristics of any particular sensory channel or form of motor output but instead to problems of sensorimotor processing which are more general than those which would be imagined when thinking of the nervous system in terms of an array of reflex pathways.

As discussed elsewhere (Grobstein 1989b), there is for any given example of orienting behavior no obvious reason why the nervous system needs to create a signal which varies in precisely three dimensions. There are almost certainly more than three independent variables in the motor control circuitry itself, and probably more than three in the initial input pathways. At the same time, a signal based on three orthogonal axes is somehow not surprising; it has a curious air of obviousness to it, as if one would have expected it all along. Humans are of course accustomed to thinking of space in terms of location relative to three orthogonal axes; that space can be described in other ways is a rather late mathematical discovery. One cannot help but suspect that the three-dimensional signal in the frog brain is not more surprising than it is precisely because it is a form of representation common to all vertebrates, including man. This, however, begs the question of why not only a three-dimensional representation, but a particular form of a three-dimensional representation, one involving for example, among other characteristics, ipsilateral representation of horizontal eccentricity, should be so widespread.

I have suggested elsewhere (Grobstein 1989b) that the commonness of the observed organization may be a clue to the explanation for its existence: that in general nervous systems have a problem of relating a variety of forms of more specific information represented in a variety of different coordinate frames so as to yield a coordinated relation of the organism to the external world, and that the observed form of spatial representation may be in some sense the mathematically simplest way to do this. The lateralization phenomenon, for example, which seems to have no obvious rationale in terms of visual processing, may have some explanation in terms of a coding scheme applicable to a variety of kinds of information. In making the suggestion that one needed to worry about the handling of several sorts of information, I was unaware that it had earlier been made and indeed somewhat more explicitly developed by Mittelstaedt (1975, 1983, 1986). Further work along these lines (cf. also Pellinonisz 1983; Grossberg & Kuperstein 1986; Bullock & Grossberg 1987) seems to me to have a high probability of significant outcome, with regards to both problems of sensorimotor integration and problems of computational theory generally. Regardless, it seems fairly clear that the nervous system has an internal form of organization, one unlikely to have become apparent from the study of isolated reflexes, which strongly influences both the handling of sensory input and the production of motor output but is not equatable with either one.

Autonomous central nervous system processing principles and functions, together with a pronounced context dependence, are, it might be argued, no more than some additional complexities which can be accommodated within a basic reflex picture of nervous system function. One might add, furthermore, that they are not new ideas. The emerging picture of the neuronal circuitry underlying orienting in the frog bears some interesting similarities to general models of sensorimotor processing proposed by von Holst & Mittelstaedt (1950) over thirty years ago. Their remark that "new results resist any description using reflex terminology" is strikingly similar to a comment of Lashley (1951) who wrote, also about sensorimotor processing, that "Explanation of these activities ... in terms of simple isolated interconnections of all the sensory and motor elements involved seems quite improbable on anatomical grounds and is ruled out by results of our experiments." Both papers contain clear descriptions of autonomous central nervous system processing as well as of the importance of context dependence in the interpretation of neural signals.

Similar descriptions of central autonomy and context-dependence are available in the Soviet literature on sensorimotor processing (cf. Gelfand et al. 1971; Luria 1980), a literature which, at least formally, is even more firmly based in reflex theories of brain function than that in the west. If these ideas are both old and can arise within a reflex framework, then is there really anything new? At a pragmatic level, the answer is clearly yes. Until recently, neither von Holst and Mittelstaedt, nor Lashley, nor the Soviet literature have had much impact on the way most western neurobiologists have gone about exploring nervous system organization. As I have argued elsewhere (Grobstein 1988a,b; 1989a), we are emerging from a period dominated by the microelectrode and modern tracing techniques, a period in which the availability of new technology which facilitated the description of interconnections between individual neurons interacted with reflex theories of neuronal function in such a way as to encourage the belief that descriptions of the connectivity pattern of the nervous system would suffice to account for its information processing characteristics and functions. During this period, both observations and questions relevant to principles of neuronal organization at levels above the neuron were largely ignored in the west, though, significantly, not in the Soviet Union. At the neuronal level, matters of context dependence and central autonomy were evident primarily as a disturbing complexity in neuronal circuitry, and the limitations of a fixation of this level have been increasingly recognized (Grobstein 1989a). At a minimum then, what is new is a renewed interest in, and respect for, studies of neuronal information processing at a variety of levels of complexity (cf. Ewert 1987; Arbib 1987), and a consequent improvement in our understanding of sensorimotor integration, among other aspects of neuronal function.

My suspicion is that these new movements will carry us beyond reflex models of neuronal information processing, and perhaps of information processing in general, in a more fundamental sense as well. It is increasingly clear that the nervous system, even at its most basic levels, is not organized so as to yield particular responses for particular stimuli but instead organized so as to accomplish particular objectives. For a given organism, the latter not only reflects the existence in the nervous system of an information pool which is totally independent of any environmental experience of that organism, but also depends on a capacity to vary input/output relations over a wide range depending on past, recent, and ongoing experience with the environment. At its most rudimentary levels, the nervous system is proving to be not a machine but a constantly changing information-gathering device (Grobstein 1988b). Should it turn out, as seems likely, that the nervous system is not only continually gathering information but continually creating it as well, the usefulness of the reflex metaphor will disappear entirely.

Acknowledgements. The author's work is currently supported by NIH Grant 1 R15 NS 24968. This chapter reflects long and productive associations with a number of student and faculty colleagues, including Christopher Comer, Sandra Kostyk, Paul Patton, and Philip Ulinski. I am particularly grateful for recent collaborations and associated discussions with Tom Masino, out of which emerged not only many of the observations mentioned in this article but also a number of the ideas put forth. For more recent valuable collaborations and discussions, I also want to thank Cathy Crowley, John Spiro, Maria Schweichler, and Cynthia Smeraski.

7 References

Arbib MA (1987) *Brains, machines, and mathematics,* 2nd edn SpringerVerlag, New York

Bässler U (1986) On the definition of central pattern generator and its sensory control. *Biol Cybern* 54: 65-69

Berkinblit MB, Feldman AG, Fukson OI (1986a) Adaptability of innate motor patterns and motor control mechanisms. *Behav Brain Sci* 9: 585-638

Berkinblit MB, Gelfand IM, Feldman AG (1986b) A model for the aiming phase of the wiping reflex. In: Grillner S, Stein PSG, Stuart DG, Forssberg H, Herman RM (eds) *Neurobiology of vertebrate locomotion.* Macmillan Press, London, pp 217-227

Beevor CE (1909) Remarks on paralysis of the movement of the trunk in hemiplegia. *Brit Med J 1*: 881-885

Bullock D, Grossberg S (1987) Neural dynamics of planned arm movements: emergent invariants and speed-accuracy properties during trajectory formation. *Psych Rev* (in press).

Collett T, Harkness L (1982) Depth vision in animals. Ingle D, Goodale R, Mansfield R (eds) *Analysis of visual behavior.* MIT Press, Cambridge, pp 111-176

Collett TS, Udin SB, Finch DJ (1987) A possible mechanism for binocular depth judgements in anurans. *J Comp Physiol* 66: 35-40

Comer C, Dowd JP (1987) Escape turning behavior of the cockroach. *J Comp Physiol* 160: 571-583

Comer C, Grobstein P (1978) Prey acquisition in atectal frogs. *Brain Res* 153: 217-221

Comer C, Grobstein P (1981a) Tactually elicited prey acquisition behavior in the frog, *Rana pipiens,* and a comparison with visually elicited behavior. *J Comp Physiol* 142: 141-150

Comer C, Grobstein P (1981b) Involvement of midbrain structures in tactually and visually elicited prey acquisition behavior in the frog, *Rana pipiens. J Comp Physiol* 142: 151-160

Comer C, Grobstein P (1981c) Organization of sensory inputs to the midbrain of the frog, *Rana pipiens. J Comp Physiol* 1432: 161-168

Comer C, Schotland J, Grobstein P (1985) Short and long term effects of unilateral vestibular lesions on posture and orienting movements in the frog. *Soc Neurosci Abstr* 11: 289

Critchley M (1953) *The parietal lobes.* Hafner, New York

Dean P, Redgrave P, Sahibzada N, Tsuji K (1986) Head and body movements poduced by electrical stimulation of superior colliculus in rats: effects of interruption of crossed tectoreticulospinal pathway. *Neurosci* 19: 367-380

Elepfandt AA (1987) Central organization of wave localization in the clawed frog, *Xenopus leavis.* I. Involvement and bilateral organization of the midbrain. *Brain Behav Evol* 31: 347-357

Elepfandt AA (1987) Central organization of wave localization in the clawed frog, *Xenopus leavis.* II. Midbrain topology for wave directions. *Brain Behav Evol* 31: 358-368

Ewert J-P (1967) Aktivierung der Verhaltensfolge beim Beutefang der Erdkröte (*Bufo bufo* L) *Z Vergl Physiol* 54: 455-481

Ewert J-P (1974) The neural basis of visually guided behavior. *Sci Amer* 230: 34-42

Ewert J-P (1984) Tectal mechanisms that underlie prey-catching and avoidance behaviors in toads. In: Vanegas H (ed) *Comparative neurology of the optic tectum.* Plenum Press, New York, pp 247-416

Ewert J-P (1987) Neuroethology of releasing mechanisms: prey catching in toads. *Behav Brain Sci* 10: 337-405

Feldman AG (1986) Once more on the equilibrium-point hypothesis for motor control. *J Motor Behav* 18: 17-54

Finger TE (1988) Sensorimotor mapping and oropharyngeal reflexes in goldfish, *Carassius auratus. Brain Behav Evol* 31: 17-24

Fukson OI, Berkinblit MB, Feldman AG (1980) The spinal frog takes into account the scheme of its body during the wiping reflex. *Science* 209: 1261-1263

Gallistell CR (1980) *The organization of action.* Plenum Press, New York

Gelfand IM, Gurfinkel VS, Tsetlin ML, Shik ML (1971) Some problems in the analysis of movements. In: Gelfand IM, Grufinkel VS, Fomin SV, Tsetlin ML (eds) *Models of the structural-functional organization of certain biological systems.* MIT Press, Cambridge, pp 329-345

Grantyn A, Berthoz A (1988) The role of the tectoreticular system in the control of head movement. In: Peterson BW, Richmond FJ (eds) *Control of head movement.* Oxford University Press, New York, pp 224-244

Grillner S (1986) Interaction between sensory signals and the central networks controlling locomotion in lamprey, dogfish, and cat. In: Grillner S, Stein PSG, Stuart DG, Forssberg H, Herman RM (eds) *Neurobiology of vertebrate locomotion.* Macmillan Press, London, pp 505-512

Grobstein P (1988a) Between the retino-tectal prejection and directed movement: topography of a sensorimotor interface. Brain, *Behav Evol* 31: 34-48

Grobstein P (1988b) On beyond neuronal specificity: problems in going from cells to networks and from networks to behavior. In: Shinkman PG (ed) *Advances in neural and behavioral development, Vol III.* Ablex Publishing Company, New Jersey, pp 1-58

Grobstein P (1989a) Strategies for analyzing complex organization in the nervous system. I. Lesion experiments, the old rediscovered. In: Schwartz E (ed) *Systems development foundation benchmark series in computational neuroscience.* MIT Press, Cambridge (in press)

Grobstein P (1989b) Strategies for analyzing complex organization in the nervous system. II. A case study: directed movement and spatial representation in the frog. In: Schwartz E (ed) *Systems development foundation benchmark series in computational neuroscience.* MIT Press, Cambridge (in press)

Grobstein P, Masino T (1986) Sensorimotor circuitry underlying directed movement in the frog: evidence for an intermediate representation of space in the tectofugal pathways. *Soc Neurosci Abstr* 12: 684

Grobstein P, Comer C, Kostyk SK (1980) The potential binocular field and its tectal representation in *Rana pipiens. J Comp Neurol* 190: 175-185

Grobstein P, Comer C, Kostyk SK (1983) Frog prey capture behavior: between sensory maps and directed motor output. In Ewert J-P, Capranica RR, Ingle DJ (eds) *Advances in vertebrate neuroethology.* Plenum Press, New york, pp 331-347

Grobstein P, Reyes A, Zwanziger L, Kostyk SK (1985) Prey orienting in the frog: accounting for variations in output with stimulus distance. *J Comp Physiol* 156: 775-785

Grobstein P, Crowley K, Spiro J (1988) Neuronal organization underlying directed movement in the frog: similarities between visually and tactually elicited orienting. *Soc Neurosci Abstr* 14 (in press)

Grossberg S, Kuperstein M (1986) *Neural dynamics of adaptive sensory-motor control: ballistic eye movements.* Elsevier/North Holland, Amsterdam

Holst E v, Mittelstaedt H (1950) Das Reafferenzprinzip. *Naturwissenschaften* 37: 464-476. (English translation in: *The behavioral physiology of animals and man,* Vol I. Holst E v, University of Miami Press, Coral Gables, pp 139-173

Hyvärinen H (1982) Posterior parietal lobe of the primate brain. *Physiol Rev* 62: 1060-1129

Ianonne AM, Gerber AM (1982) Brown-Sequard syndrome with paralysis of head turning. *Ann Neurol* 12: 116

Ingle D (1970) Visuomotor functions of the frog optic tectum. *Brain Behav Evol* 3: 57-71

Ingle D (1972) Depth vision in monocular frogs. *Psychonom Sci* 29: 37-38

Ingle D (1983) Brain mechamisms of visual localization by frogs and toads. In: Ewert J-P, Capranica RR, Ingle DJ (eds) *Advances in vertebrate neuroethology.* Plenum Press, New York, pp 177-226

Jordan M, Luthardt G, Meyer-Naujoks C, Roth G (198) The role of eye accommodation in the depth perception of common toads. *Z Naturforsch* 35: 851-852

Kostyk SK, Grobstein P (1982) Visual orienting deficits in frogs with various unilateral lesions. *Behav Brain Res* 6: 379-388

Kostyk SK, Grobstein P (1987a) Neuronal organization underlying visually elicited prey orienting in the frog. I. Effects of varius unilateral lesions. *Neurosci* 21: 41-55

Kostyk SK, Grobstein P (1987b) Neuronal organization underlying visually elicited prey orienting in the frog. II. Anatomical studies on the laterality of central projections. *Neurosci* 21: 57-82

Kostyk SK, Grobstein P (1987c) Neuronal organization underlying visually elicited prey orienting in the frog. III. Evidence for the existence of an uncrossed descending tectofugal projection. *Neurosci* 21: 83-96

Lashley KS (1951) The problem of serial order in behavior. In: Jeffries, LA (ed) *Cerebral mechanisms in behavior.* John Wiley and Sons, New York, pp 112-136

Luria AR (198) *Higher cortical functions in man.* Basic Books, New York

Masino T, Grobstein P (1985) The organization of tectal projections to the ventral midbrain in *Rana pipiens. Soc Neurosci Abstr* 11: 289

Masino T, Grobstein P (1986) Sensorimotor circuitry underlying directed movement in the frog: organization of tectofugal pathways and likely involvement of the nMLF. *Soc Neurosci Abstr* 12: 684

Masino T, Grobstein P (1989a) The organization of descending tectofugal pathways underlying orienting in the frog, *Rana pipiens.* I. Lateralization, parcellation, and an intermediate spatial representation (submitted)

Masino T, Grobstein P (1989b) The organization of descending tectofugal pathways underlying orienting in the frog, *Rana pipiens.* II. Evidence for the involvement of a tecto-tegmento-spinal pathway (submitted)

Masino T, Grobstein P (1989c) Tectal connectivity in the frog, *Rana pipiens*: tectotegmental projections and an analysis of topographic organization generally (submitted)

Masino T, Kostyk SK, Grobstein P (1984) Laterality of tectal efferent projections in *Rana pipiens. Soc Neurosci Abstr* 10: 60

Masterton RB, Imig TJ (1984) Neural mechanisms for sound localization. *Ann Rev Physiol* 46: 275-287

Mays LE, Sparks DL (198) Saccades are spatially, not retinocentrically, coded. *Science* 208: 1163-1165

Mittelstaedt H (1975) On the processing of visual information. *Fortschr Zool* 23: 128-141

Mittelstaedt H (1983) Introduction into cybernetics of orientation behavior. In: Hoppe W, Lohmann W, Markl H, Ziegler H (eds) *Biophysics.* Springer-Verlag, Berlin, pp 794-801

Mittelstaedt H (1986) The subjective vertical as a function of visual and extraretinal cues. *Acta Psychol* 63: 6-85

Pellionisz A (1983) Brain theory: connecting neurobiology to robotics. Tensor analysis: natural coordinates to describe, understand and engineer functional geometries of intelligent organisms. *J Theoret Biol* 2: 185-211

Raphan T, Cohen B (1978) Brainstem mechamisms for rapid and slow eye movements. *Ann Rev Physiol* 4: 527-552

Raczkowski K, Casagrande VA, Diamond IT (1976) Visual neglect in the tree shrew after interruption of the descending projections of the deep superior colliculus. *Exp Neurol* 5: 14-29

Reyes A, Grobstein P (1984) Frog prey orienting: effects of cerebellectomy. *Soc Neurosci Abstr* 10: 61

Satou M, Ewert J-P (1985) The antidromic activation of tectal neurons by electrical stimuli applied to the caudal medulla oblongata in the toad *Bufo bufo. J Comp Physiol* 157: 739-748

Satou M, Matsushima T, Ueda K (1984) Neuronal pathways from the tectal "snapping-evoking area" to the tongue-muscle-controlling motoneurons in the Japanese toad: evidence of the intervention of excitatory interneurons. *Zool Sci* 1: 829-832

Sparks DL, Mays LE (1983) Spatial localization of saccade targets. I. Compensation for stimulation-induced perturbations in eye position. *J Neurophysiol* 49: 45-63

Sparks DL, Porter JD (1983) Spatial localization of saccade targets. II. Activity of superior colliculus neurons preceding compensatory saccades. J Neurophysiol 49: 64-74

Sperry RW (1948) Orderly patterning of synaptic associations in regeneration of intracentral fiber tracts mediating visuomotor coordination. *Anat Rec* 102: 63-75

Sprague JM, Levitt M, Robson K, Liu CN, Stellar E, Chambers WW (1963) A neuroanatomical and behavioral analysis of the syndromes resulting from midbrain lemniscal and reticular lesions in the cat. *Arch Ital Biol* 11: 225-295

Voneida TJ (1970) Behavioral changes following midline section of the mesencephalic tegmentum in the cat and monkey. *Brain Behav Evol* 3: 241-26

Weerasuriya A (1983) Snapping in toads: some aspects of sensorimotor interfacing and motor pattern generation. In: Ewert J-P, Capranica RR, Ingle DJ (eds) *Advances in vertebrate neuroethology.* Plenum Press, New York, pp 613-628

Weerasuriya A, Ewert J-P (1981) Prey-selective neurons in the toad's optic tectum and sensorimotor interfacing: HRP studies and recording experiments. *J Comp Physiol* 144: 429-434

Winterkorn JMS, Meikle TH (1980) Lesions of the tectospinal tract do not produce compulsive circling. *Brain Res* 19: 597-600

How to Transform Topographically Ordered Spatial Information into Motor Commands

HORST MITTELSTAEDT and THOMAS EGGERT

Max-Planck-Institut für Verhaltensphysiologie Seewiesen
D-8130 Seewiesen/Starnberg, FR Germany

Abstract. The question of how topographically ordered sensory information is transformed into motor commands is largely unsolved. Our theoretical approach to the positive phototaxis of an "idealized planaria" leads to the concept of *map-weighting* provided by an intermediate information processing structure between eyes and motor apparatus. Map-weighting and subsequent summation generate the components which, reciprocally facilitated and inhibited in the CNS, yield appropriate commands. It is suggested that map-weighting may be adjusted to produce or restore an optimal match with the external luminance distribution. The matching may be achieved by comparing the reafference of the animal's movements with an efference copy of the motor commands. The concept of map-weighting, in principle, may be applied to other sensori-motor functions as well, such as the transformation of the locus of prey in the toad's tectum into an appropriate motor output.

1 Introduction

On the retina, due to the construction of the eye, the spatial distribution of external luminance is represented as a two-dimensional array. This order is largely preserved in the first stages of central nervous processing. And clearly, according to present knowledge, this topographical representation of stimulus location is to the advantage of, or even necessary for, the subsequent pattern recognition processes. Eventually, however, in order to control goal directed movements, the map-ordered information must be transformed into a motor innervation pattern (see also Grobstein, this volume). The question is, how this may be performed.

Let us first try to understand what is supposed to happen if that performance is to succeed. This may help us to formulate the problem more clearly, and subsequently put us into the position of suggesting reasonable hypotheses of how it may be solved.

In order to grasp at an object by way of an open-loop hand movement, the motor apparatus needs information about the location of the target with respect to the shoulder. If the target will always appear at the same distance, the problem reduces to processing a two-dimensional vector. Now, a motor apparatus such as the human arm and hand certainly presents several anatomical constraints, but they cannot fully specify the directional vector and hereby reduce the information necessary to define the movement. Hence, on the motor side, the vector is largely represented by neuronal activity, rather than by anatomical arrangement. On the retina and on the retinotopical maps in the CNS, however, the information about the direction of the target is entirely represented by the anatomical location of the neurons which are stimulated by the target. In this case, therefore, spatial information which is purely defined by anatomy must be transformed into information which is mainly defined by neuronal activity. On the other hand, we may envisage cases, where this transformation is not necessary, that is, where the anatomical arrangement of receptors and effectors obviates it, or cases where both types of representations are mixed.

Ideally, in trying to tackle such problems one should want to separate the deductive fractions of possible solutions from those which require experimentation. In order to achieve this, it seems mandatory to know, firstly, the *necessary* constituents of the performance under study and, secondly, all combinations of causal pathways which are *sufficient* to produce it. Unfortunately, many of the former are likely to be trivial or uninformative, and the latter too numerous and too intricate. However, theoretically guided modeling and experimentation does neither require a list of all necessary constituents nor of all sufficient combinations of causal pathways, but rather the logical tree of *basic alternative classes* of sufficient solutions to the problem at hand (Mittelstaedt 1960, 1985, 1988b; Rose & Dobson 1985). The analysis may then "climb the logical tree" by experimentally deciding at every node into which class the given case belongs.

The present essay does not aspire to deduce a logical tree for our problem. But it will at least try to lay the ground for such a venture.

2 Fundamentals of Orientation

Let us start with the idea of Jacques Loeb (1890, 1918) to explain phototaxis. We have a bilateral pair of sense organs, each oriented such as to signal the luminance within its own lateral hemisphere, and a bilateral pair of motor apparatus', each oriented such as to produce a contralateral torque. All we need to achieve positive phototaxis is to connect the left eye with the right motor apparatus and vice versa. Another case in point would be a vertebrate where a pair of motor apparatus' were oriented such as to exclusively produce torques in the planes of the corresponding semicircular canals.

These two cases belong to a class of solutions to our problem where the spatial arrangements of both, the sensory and the motor system, are matched in such a way that no transformation is required. Hence, the spatial information which controls the performance is not, and need not be, explicitly represented in neuronal terms. The neuronal signals represent an ambiguous combination of stimulus direction and intensity. Indeed, if a neuronal representation were necessary for other purposes than to produce positive phototaxis, it had to be derived by neuronal interaction from the two afferent signals.

2.1 Understanding Loeb's Phototaxis

It is commendable to understand why and under which conditions such an information processing structure performs so well. As a model we will envision a somewhat idealized planaria, a translucent flatworm equipped with two primitive eyes. They are assumed to consist of a light-sensitive eyespot partially and incompletely shielded by a pigmented cup whose opening is directed at an angle ψ_i of $\pm$ 45 degrees with respect to the animals median plane (Fig. 1). We shall adopt an

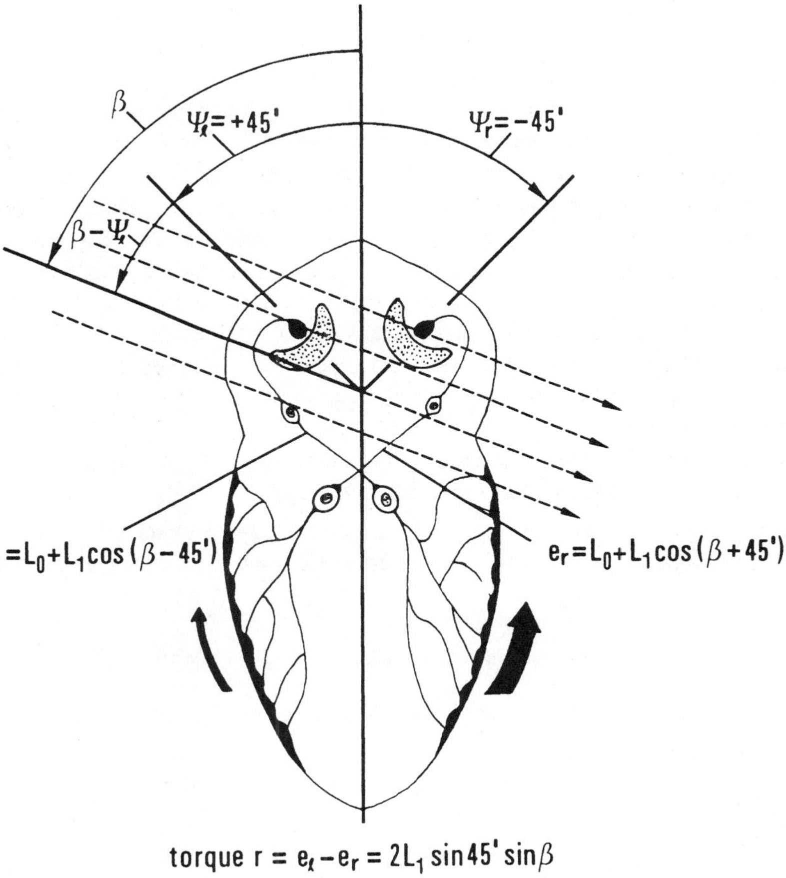

Figure 1. Idealized planaria with ocelli which respond to a point source of light presented at a deviation β from the animal's median plane such that the impulse rate $e_i = t(\beta-\psi_i) = T_0 + T_1 \cos(\beta-\psi_i)$. If the external light is more broadly distributed but still symmetrical, the discharge rate e_i will still be a cosine function, namely $e_i = L_0 + L_1 \cos(\beta-\psi_i)$, where L_0 is the DC component and L_1 the modulation amplitude. If the ocelli are directed at $\psi_{l/r} = +/- 45$ degrees and each output is transferred to the respective contralateral motor apparatus, the resulting torque $r = e_l - e_r = 2L_1 \sin 45° \sin\beta$ will turn the animal until $\beta = 0$, that is until it faces the point of maximal luminance density.

animal-fixed right handed coordinate system and further simplify matters by considering just one plane. The intensity tuning function $t(\beta-\psi_i)$ of such an eye shall be defined as the firing rate e_i of the afferent neuron in dependence of the angle $(\beta-\psi_i)$ between the angular deviation β of a point source of ligth and the cup's axis. It shall be assumed that this tuning function turns out to be

$$e_l = t_l = T_0+T_1\cos(\beta-\psi_l) = T_0+T_1\cos(\beta-45^\circ) \tag{1a}$$

for the left eye and

$$e_r = t_r = T_0+T_1\cos(\beta-\psi_r) = T_0+T_1\cos(\beta+45^\circ) \tag{1b}$$

for the right eye, where T_0 is the DC component rate and T_1 the modulation amplitude of the response to the point source.

Such a tuning function demands nothing of the optical apparatus but an appropriate arrangement of the cup's pigments. Yet for the purpose of orientation it provides a perfectly sufficient filter under any conditions: however the light may be distributed around the animal, the output e will be exclusively dependent on the first circular harmonics of the light distribution function. This function shall be characterized by the luminance density q in dependence of the head-fixed variable C (cf. also Fig. 2):

$$q = Q_0+\sum_{n=1}^{\infty} Q_{sn}\sin[n(\epsilon-\beta)]+\sum_{n=1}^{\infty} Q_{cn}\cos[n(\epsilon-\beta)] \tag{2}$$

The animal will turn towards the brightest spot, if the luminance density is symmetrical about $\epsilon = \beta$, that is, if the sine components $Q_{sn}\sin[n\,(\epsilon-\beta)]$ of its circular Fourier expansion are zero. For the bilaterally symmetrical motor apparatus will then produce a torque r which is proportional to the difference between the impulse rates:

$$\begin{aligned} r = e_l-e_r &= L_0+L_1(\cos\beta\cos45^\circ+\sin\beta\sin45^\circ) \\ &\quad -[L_0+L_1(\cos\beta\cos45^\circ-\sin\beta\sin45^\circ)] \end{aligned} \tag{3a}$$

hence

$$r = 2L_1\sin45^\circ\sin\beta \overset{!}{=} 0. \tag{3b}$$

with L_0 and L_1 to be presently determined.

The reason why these primitive eyes provide the brain with just the right information is that, mathematically, their output is the convolution of the external luminance distribution q with their tuning functions t: hence, in the circular frequency domain, the complex amplitudes $\underline{L}_n$ of the circular Fourier components of their outputs e_i are the *products* of the components of their tuning functions with the respective components of the external light distribution. Because (*vide* Eqs. 1a, 1b) the eye's tuning function contains only first harmonics, apart from the DC component $L_0 = 2\pi Q_0 T_0$, nothing but the first harmonics of the external light distribution can pass the filter, hence, with q symmetrical, $L_1 = \pi Q_{c1} T_1$.

We shall come back to these relationships, but first turn to discussing the functional merits and limitations of such an information processing structure (IPS).

2.2 Performance

In addition to positive phototaxis, this IPS will allow orientation at an angle to the light source ("menotaxis"), either by changing the DC component L_0 of one eye with respect to that of the other ($L_{0l} \neq L_{0r}$) or by changing the modulation amplitude of one with respect to that of the other ($L_{1l} \neq L_{1r}$). Both modes allow what amounts to a change of reference but, again, without an explicit neuronal representation of a "set point" ("Sollwert") and an explicit "comparator" producing the difference between set point and afferent input. However, both modes permit only a limited range of reference change ($\pm$ 45 degrees in the latter case) and cause a reduction of efficiency (of the slope of the torque function at the reference angle) with increasing reference deviation.

Consequently, orientation at any angle and hence all higher orientation performances, such as navigation, demand a structural improvement. The minimal requirement is an additional connection of the two eyes with the motor apparatus' of their own side. But this entails, for the first time in our deduction, an intermediate neuronal instance which controls the crossed pathways in a mode of reciprocal inhibition and facilitation. Such a mechanism could provide the sums and the differences of the output of the eyes, namely

$$e_l - e_r = 2L_1 \sin 45^\circ \sin\beta \tag{4a}$$

$$e_r - e_l = -2L_1 \sin 45^\circ \sin\beta \tag{4b}$$

$$e_l + e_r = 2L_0 + 2L_1 \cos 45^\circ \cos\beta \tag{4c}$$

$$-(e_r + e_l) = -2L_0 - 2L_1 \cos 45^\circ \cos\beta \tag{4d}$$

From these four any reference angle could be attained, provided the DC components in Eqs. 4c,d could be eliminated. Because, with $L_0 = 2\pi Q_0 T_0$, they depend on the DC component Q_0 of the light distribution, a third eye appears to be needed, for instance an ocellus pointed upwards, aligned with the axis of rotation and hence yielding a measure of intensity, and thus of Q_0, independent of the animal's

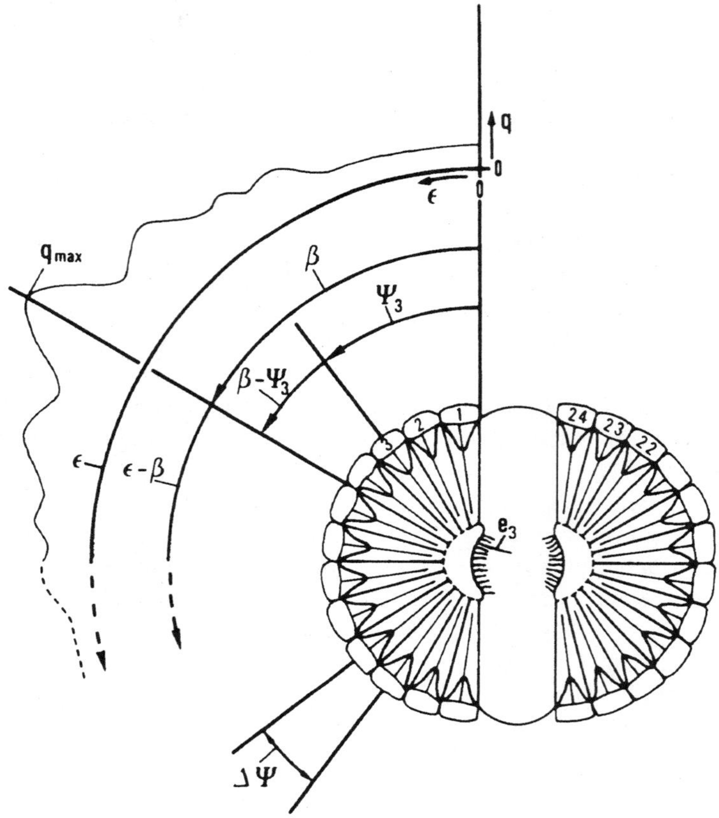

LUMINANCE DENSITY $q(\epsilon-\beta)$

$$q = Q_0 + \sum_{n=1}^{\infty} Q_{cn} \cos n(\epsilon-\beta)$$

TUNING FUNCTION $t(\beta-\Psi_i)$

$$t = T_0 + \sum_{n=1}^{u} T_n \cos n(\beta-\Psi_i) \qquad i=1,2,\ldots k \quad k=24$$

AFFERENT OUTPUTS $e_i(\beta-\Psi_i)$

$$e_i = L_0 + \sum_{n=1}^{u} L_n \cos n(\beta-\Psi_i)$$

MOTOR COMMANDS

$$s_1 = \sum_{i=1}^{k} e_i \sin\Psi_i = \frac{1}{2} k L_1 \sin\beta$$

if: $k > u+1$

$$c_1 = \sum_{i=1}^{k} e_i \cos\Psi_i = \frac{1}{2} k L_1 \cos\beta$$

Figure 2. Idealized compound eye consisting of 24 equally spaced ommatidia. Their tuning functions $t(\beta-\psi_i)$ contain higher circular harmonics up to n = u. Convolution with the external luminance density $q(\epsilon-\beta)$ yields their outputs $e_i(\beta-\psi_i)$. By weighting them with $\sin\psi_i$ and $\cos\psi_i$, respectively, and subsequent summation, the components s_i and c_i are gained. If $k > u+1$, they are proportional to $(1/2)kL_1 \sin\beta$ and $(1/2)kL_1 \cos\beta$, respectively, and hence may serve as motor commands.

yaw orientation. Alternatively, an additional eye might be directed backwards (at ψ = 180 degrees) or, if that is not possible for anatomical reasons, two additional eyes directed at $\psi = \pm 135$ degrees.

But such arrangements may waste information and still not lead to an optimal phototactic performance. Moreover, an increase in number of eyes will create both, new functional opportunities and new problems of information processing, among them those from which we started and which we try to clarify. In order to attain generally valid solutions, we shall at once proceed to eyes with many directionally sensitive and more broadly tuned receptors.

3 Generation of Motor Commands by Map-Weighting

As a model we shall choose an idealized compound eye consisting of k=24 ommatidia with equal optical, photoreceptive and neurophysiological characteristics, equidistant axes and perfect bilateral symmetry (Fig. 2). Their intensity tuning functions, as defined above, measured at their 24 outputs e_i (for instance at the output of the lamina cartridges) are assumed to be symmetrical about $\beta - \psi_i$, that is, any optical or neurophysiological cross-talk between them up to that point is assumed to be also equal and symmetrical. Thus

$$e_i = t(\beta-\psi_i) = T_0 + \sum_{n=1}^{u} T_n \cos[n(\beta-\psi_i)] \tag{5a}$$

hence with any symmetrical luminance density q

$$e_i = L_0 + \sum_{n=1}^{u} L_n \cos[n(\beta-\psi_i)] \tag{5b}$$

where u is the highest circular frequency which the tuning function contains and hence also the highest circular frequency of the external luminance distribution which can pass this filter. Clearly, from optical reasons alone it cannot just contain the first circular harmonics. Moreover, if the potential for pattern discrimination which such an eye offers is to be exploited, the tuning function must needs include higher harmonics. How then can the first harmonics, which are necessary for higher orientation performances, be extracted from such an array of outputs?

3.1 Weighting and Necessary Conditions

A hypothesis suggested earlier by one of us (Mittelstaedt 1975, 1983) lets each output e_i be used twice, on the one hand weighted with a constant "built-in" (but see Section 5) factor proportional to the sine of ψ_i (of the angle by which its respective optical axis deviates from the animal's median), on the other hand with a constant factor proportional to the cosine of ψ_i. The sine and cosine fractions are then summed to yield the required first harmonic components s_1 and c_1, namely

$$s_1 = \sum_{n=1}^{k} e_i \sin\psi_i = \frac{1}{2}\, k\, L_1 \sin\beta \tag{6a}$$

and

$$c_1 = \sum_{n=1}^{k} e_i \cos\psi_i = \frac{1}{2}\, k\, L_1 \cos\beta \tag{6b}$$

provided that $k > u + 1$. This procedure, we shall name it "map-weighting" for short, yields the information sufficient for light orientation and yet permits a reasonably good angular resolution, resulting from the comparatively broad tuning, up to $u = 22$ in our case, of the 24 units.

With the procedure described by Eqs. 6a,b, we have arrived at an answer to the question of our title, namely of how topographically ordered spatial information may be transformed into motor commands: map-weighting and subsequent summation generates the components s_1 and c_1, which are then reciprocally facilitated and inhibited by the CNS to yield the appropriate commands for a set of antagonistic motor apparatus'. By means of such a bi-component modulation (for details see Mittelstaedt 1962, 1964a) the animal may set itself at, and maintain, any angle with respect to the light source.

The question arises whether the exclusive extraction of first harmonics is not only sufficient but also necessary, that is, the *only* way to yield orientation performances which demand the two first harmonic components, e.g., the sun-compass (Mittelstaedt 1962), homing by path integration (Mittelstaedt & Mittelstaedt 1982; Mittelstaedt 1985) or head-to-trunk coordinate transformation (Mittelstaedt 1964b). In most of these cases, the external light distribution is generally symmetrical about $\epsilon = \beta$. Hence, also the higher harmonics contain information about the direction β of the light source. This information is left unused in the suggested weighting procedure, but it need not be. For, in general, the necessary and sufficient condition for the extraction of a circular harmonic n reads

$$k > u+n \tag{7}$$

Consequently, the highest harmonic which can be extracted is $n = u$ and hence $k > 2u$.

In order to exploit the usable higher harmonics, the IPS might operate according to an extension of Eqs. 6a,b, namely by also weighting the outputs e_i with $\sin(n\,\psi_i)$ and $\cos(n\,\psi_i)$ to obtain the components s_n and c_n:

$$s_n = \sum_{n=1}^{k} e_i \sin(n\,\psi_i) = \frac{1}{2}\, k\, L_n \sin(n\,\beta) \tag{8a}$$

$$c_n = \sum_{n=1}^{k} e_i \cos(n\,\psi_i) = \frac{1}{2}\,k\,L_n \cos(n\,\beta) \tag{8b}$$

A motor command r which would turn the animal towards the point of symmetry of the external light distribution, may then be generated by summation of all s_n, namely

$$r = \sum_{n=1}^{u} s_n = \sum_{n=1}^{u} \frac{1}{2}\,k\,L_n \sin(n\,\beta) \tag{9}$$

Instead of by summation, the components s_n and c_n may be processed separately, by cross-multiplication with a set of centrally generated standard components $\sin(n\,\varphi)$ and $\cos(n\,\varphi)$ of an "internal estimate" φ of the light direction β. The sum of the cross-products acts as the "turning tendency" f of an internal feedback loop which changes the internal estimate φ until

$$\begin{aligned} f &= \sum_{n=1}^{u} \frac{1}{2}\,k\,L_n [\sin(n\,\beta)\cos(n\,\varphi) - \cos(n\,\beta)\sin(n\,\varphi)] \\ f &= \sum_{n=1}^{u} \frac{1}{2}\,k\,L_n \sin[n(\beta-\varphi)] \overset{!}{=} 0. \end{aligned} \tag{10}$$

Examples of why this may be expedient or even necessary and of how it may be executed are worked out for spider navigation (Mittelstaedt 1978, 1983) and for the effect of visual information on the apparent vertical in man (Mittelstaedt 1986, 1988a).

3.2 Resolving *vs.* Tuning Power

In our case, then, if all available information for orientation is to be exhausted, the tuning function must be cut off above n = u = 11, that is, at half the limit considered above. Therefore, if orientation in specific external conditions demands a narrow tuning to higher circular harmonics while pattern discrimination demands narrow angular visual fields, we meet with another case of the classical trade-off, here between power of angular resolution and circular frequency tuning. Our idealized planaria presents one extreme: its eye is tuned to just one circular frequency, the one which is wanted, but its "visual field" comprises the entire range of 360 degrees and hence cannot resolve anything within it. Towards the other extreme, with k constant, the angular visual field might become ever narrower, enhancing resolution. But then the circular frequency tuning would comprise an ever increasing range of frequencies, and hence the extraction of the first (or any other) harmonic became less and less effective. An organism interested in both, good angular resolution and selective circular frequency tuning, must therefore seek a viable compromise. For the mathematically minded and for those intent on a deeper understanding, we shall add an appendix (Section 4), in which these relationships and their optimization will be further developed. Readers not familiar with the mathematical methods applied will find a verbal formulation of the results in Section 5.

The minimal condition for extraction of both first harmonics in our metaphorical planaria becomes clear right away though: with u=1 and n=1, it follows that k > 2, and hence the number of ocelli must at least be 3, symmetrically distributed at an equal angular distance of 120 degrees, e.g., directed at $\psi = 0$, and $\pm$ 120 degrees, their outputs weighted with sin ψ_i and cos ψ_i respectively. Real planarias do not comply with this, apparently, and hence cannot navigate, whereas agelenid spiders, who can, with their 6 secondary eyes and k at least 12, may extract up to the 5th harmonic indeed.

The principle of map-weighting may be readily extended to three-dimensional maps and allows to combine mapped with scalar information. Beyond the visual and vestibular information processing for which it was originally proposed (Mittelstaedt 1975, 1978, 1983, 1985, 1986), it may be applicable to other sensory modalities (e.g. see Baldi & Heiligenberg 1988) and to higher central nervous functions such as, for instance, the transformation of the location of prey in the toad's tectum (cf. Ewert, this volume) into an appropriate motor command - lastly, in reverse as it were, to the distribution of the motor commands to a topographically ordered array of muscles (Mittelstaedt 1983, p 798/799; Haustein 1988, 1989 submitted).

4 Mathematical Appendix

In this Section we shall try to determine how accurately the direction β of a 2π periodic luminance density function $q(\epsilon-\beta)$ may be gained from the output magnitudes of k equidistant and uniformly distributed receptors. To that end we define the following periodic functions and their Fourier expansions. The maximally attainable accuracy may then be expressed in terms of the system's spectral properties.

4.1 Definitions

(1) The angular dependency of the luminance density q, (abbreviated LD):

$$q(\epsilon) = \sum_{n=-\infty}^{+\infty} \underline{Q}_n \exp(jn\epsilon) \tag{11}$$

(2) The intensity tuning function of a noise free receptor looking into the direction ψ, its axis direction, in dependence of the direction β of a point source:

$$t(\beta-\psi) = \sum_{n=-\infty}^{+\infty} \underline{T}_n \exp[jn(\beta-\psi)] \tag{12}$$

(3) The output $b(\psi)$ of this receptor in dependence of the axis direction ψ and the luminance density $q(\epsilon-\beta)$:

$$b(\psi) = \int_0^{2\pi} t(\epsilon-\psi)q(\epsilon-\beta)\, d\epsilon = \sum_{n=-\infty}^{+\infty} \underline{L}_n \exp[jn(\psi-\beta)] \tag{13}$$

This function is here defined as a continuous cyclic cross-correlation between the tuning function t and the LD function q. It describes the linear transformation, the LD signal would undergo if the angular distance between the receptors were infinitely small. Actually, this continuous function is sampled at equidistant angular intervals of $\Delta\psi = 2\pi/k$ (cf. Fig. 2). Hence the receptor output vector consists of k equidistant samples of the function $b(\psi)$:

$$b_i = b(i \cdot \Delta\psi) \quad , \quad 1 \leq i \leq k \tag{14}$$

From these, by the weighting given in Eqs. 8a,b, a discrete but now periodic spectrum may be computed by means of Fourier transformation. Its spectral amplitudes $\underline{\tilde{L}}_n$ are related to the harmonics of the continuous functions as follows:

$$\underline{\tilde{L}}_n = \frac{1}{k} \sum_{i=1}^{k} b_i \exp(-j2\pi \cdot i\frac{n}{k}) \tag{15a}$$

$$\underline{\tilde{L}}_n = \sum_{m=-\infty}^{+\infty} \underline{L}_{n-mk} \exp[-j(n-km)\beta] \tag{15b}$$

The linear transformation of the b_i is invertible by:

$$b_i = \sum_{i=1}^{k} \underline{\tilde{L}}_n \exp(j2\pi \cdot n\frac{i}{k}) \tag{15c}$$

The relation between the spectral amplitudes $\underline{L}_n$ of the continuous signal b and the harmonics $\underline{\tilde{L}}_n$ gained by the weighting is shown in Fig. 3. The Fourier coefficients $\underline{L}_n$ and the expansions of t and q are further related as

$$\underline{L}_n = 2\pi\, \underline{Q}_n \underline{T}_{-n} \;. \tag{16}$$

If t or q are band limited, the spectrum of b contains likewise merely harmonics up to a limit u:

$$\underline{L}_n = 0 \text{ if } |n| > u \tag{17}$$

Fig. 3 makes clear, then, that at $k > |n| + u$ the spectral amplitudes $\tilde{\underline{L}}_n$ are identical with the harmonics $\underline{L}_n$ of the continuous function b.

(4) Furthermore, we define the differential quotient of b. The function b' expresses the sensitivity of a noise-free receptor for β at the angle ψ:

$$b'(\psi) = \frac{\delta}{\delta\beta}[b(\psi)] = \sum_{n=-\infty}^{+\infty} \underline{L}'_n \exp[jn(\psi - \beta)] \quad \text{with} \qquad (18)$$

$$\underline{L}'_n = -jn\underline{L}_n$$

The function b' (ψ), just as b (ψ), is a continuous extension of the actual sensitivities

$$b'_i = \frac{\delta}{\delta\beta} b_i = b'(i \cdot \Delta\psi) \ . \qquad (19)$$

Hence, analogously, we define a discrete periodic spectrum:

$$\tilde{\underline{L}}'_n = \frac{1}{k} \sum_{i=1}^{k} b'_i \exp(-j2\pi \cdot i\frac{n}{k}) \qquad (20a)$$

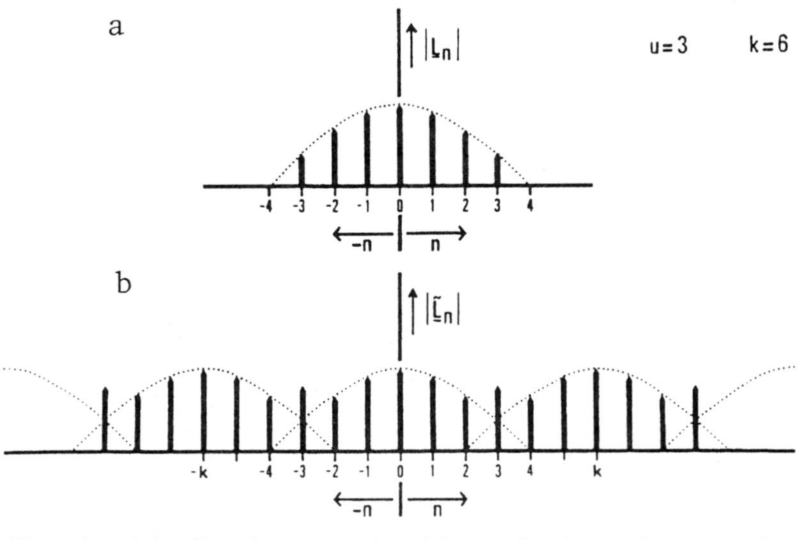

Figure 3. Illustration of the effect of a transgression of the sampling theorem $k > u+n$, where k is the number of equidistant ommatidia, u the highest harmonic in their tuning function $t(\beta-\psi)$ and n the highest harmonic to be extracted by map-weighting. (a) Discrete spectrum of the periodic continuous function $b(\psi)$ (see Eq. 13). (b) Periodic spectrum of the samples b_i (Eqs. 14, 15a,b). With $k = 6$ and $u = 3$, as in this example, it follows that, since $n < k-u$, n must not be larger than 2. In fact, due to overlap of the spectra, the amplitude (as well as the phase) of the third harmonic is contaminated!

$$b'_i = \sum_{i=1}^{k} \tilde{\underline{L}}'_n \exp(j2\pi \cdot n\frac{i}{k}) \tag{20b}$$

As shown for $\underline{L}_n$ and $\tilde{\underline{L}}_n$, the periodic spectrum gained solely from the sampled values generate a periodic extension of the spectrum of the continuous function:

$$\tilde{\underline{L}}'_n = \sum_{m=-\infty}^{+\infty} \underline{L}'_{n-km} \exp[-j(n-km)\beta] \tag{20c}$$

(5) Lastly, we assume that the output magnitudes e_i of the k receptors are additively contaminated by noise. The noise R is assumed to be Gaussian and its variance σ_R^2 independent of b_i. The mean of the distribution of R be zero.

$$e_i = b_i + R \quad , \; 1 \leqslant i \leqslant k$$
$$\tilde{\underline{E}}_n = \frac{1}{k}\sum_{i=1}^{k} e_i \exp(-j2\pi \cdot i\frac{n}{k}) \tag{21}$$

4.2 Comparison of the Result of "Map-Weighting" with the Statistical Optimum

We shall now compare the estimate of β which can be gained by the suggested IPS of map-weighting with the statistical estimate $\hat{\beta}$ of β which may be determined by means of the maximum likelihood method. An estimate of $\hat{\beta}$ demands certain assumptions about the functions q, t, and b. We shall begin with the ideal case that the coefficients of the functions $q(\epsilon)$ and $t(\beta-\psi)$ hence also those of $b(\psi)$ are given and constant. We shall see later under which conditions this restriction may be unnecessary.

According to the maximum likelihood method (Fisher 1925) the optimal estimate $\hat{\beta}$ of β is given by

$$\sum_{i=1}^{k} b'_i(e_i - b_i) \overset{!}{=} 0 \tag{22}$$

where β is to be substituted by $\hat{\beta}$ in the functions b_i and b_i'. Since they depend non-linearly on $\hat{\beta}$, the latter can only be determined by iteration. As a simple procedure we may compute a "turning tendency" f_{ML} (analogous to the turning tendency f in Eq. 10) which is proportional to successive changes of $\hat{\beta}$:

$$f_{ML} = \sum_{i=1}^{k} b'_i \cdot e_i - \sum_{i=1}^{k} b'_i \cdot b_i \tag{23}$$

By recourse to the theorem of PARSEVAL for finite signals (Marko 1982) the second summand may also be expressed by the discrete periodic spectra:

$$\sum_{i=1}^{k} b'_i \cdot b_i = k \sum_{n=1}^{k} \tilde{\underline{L}}'_n \cdot \tilde{\underline{L}}_{-n} \tag{24}$$

If k > 2u, and hence no "aliasing" occurs, the sum will become

$$k \sum_{n=1}^{k} \tilde{\underline{L}}'_n \cdot \tilde{\underline{L}}_{-n} = k \sum_{n=-u}^{u} -jn|\underline{L}_n|^2 = 0 \; . \tag{25}$$

Hereby we have also used the symmetry property of the spectra of real signals: $\underline{L}_{-n} = \underline{L}_n{}^*$.

Only if k > 2u, therefore, the maximum likelihood formula (Eq. 22) reduces to a cross-correlation with a subsequent detection of the maximum:

$$f_{ML} = \sum_{i=1}^{k} b'_i \cdot e_i = \frac{\delta}{\delta\hat{\beta}} \left(\sum_{i=1}^{k} b_i \cdot e_i \right) \stackrel{!}{=} 0 \tag{26}$$

Again by recourse to the PARSEVAL theorem this may be expressed in dependence of the spectral amplitudes:

$$f_{ML} = k \cdot \sum_{n=-u}^{u} -jn\underline{L}_n \exp(-jn\hat{\beta}) \cdot \tilde{\underline{E}}_{-n} \tag{27}$$

Taking account of the symmetry properties $\underline{L}_{-n} = \underline{L}_n{}^*$ and $\tilde{\underline{E}}_{-n} = \tilde{\underline{E}}_n{}^*$ we obtain

$$f_{ML} = 2k \sum_{n=1}^{u} n \cdot \mathrm{Im}\{\underline{L}_n \exp(-jn\hat{\beta}) \cdot \tilde{\underline{E}}_n^*\} \stackrel{!}{=} 0 \; . \tag{28}$$

Where Im {...} denotes the imaginary part of the term set in braces.
In the noise-free case $\sigma_R^2 = 0$, because of k > 2u and

$$\tilde{\underline{E}}_n = \underline{L}_n \exp(-jn\beta) \tag{29}$$

this becomes

$$f_{ML} = 2k \sum_{n=1}^{k} n|\underline{L}_n|^2 \sin[n(\beta-\hat{\beta})] \overset{!}{=} 0 \ . \tag{30}$$

In this form the "turning tendency" f_{ML} of the maximum likelihood procedure is identical, but for a factor of the summands, with the turning tendency of Eq. 10. The latter, as we have seen, is assumed to drive an internal feedback loop until the internal estimate φ of β is identical with β.

The variance σ_β^2 of the optimal estimate of Eq. 22 is given (Fisher 1925), independently of the sampling, by

$$\sigma_\beta^2 = \frac{\sigma_R^2}{\sum_{i=1}^{k} b_i'^2} \tag{31}$$

It yields a measure of the accuracy of the maximum likelihood estimate, and may also be expressed, again under recourse to the PARSEVAL theorem, in terms of the spectral amplitudes $\tilde{\underline{L}}_n'$:

$$\sigma_\beta^{-2} = k \cdot \sigma_R^{-2} \sum_{n=1}^{k} |\tilde{\underline{L}}_n'|^2 \tag{32}$$

If the sampling theorem $k > 2u$ is met, the variance may be also expressed in terms of the spectral amplitudes of t and q:

$$\sigma_\beta^{-2} = k \cdot \sigma_R^{-2} \sum_{n=-u}^{u} n^2 \cdot 4\pi^2 \cdot |\underline{T}_n|^2 \cdot |\underline{Q}_n|^2 \tag{33}$$

5 Conclusion

The mathematical analysis has revealed two fundamental traits of the envisaged map-weighting:

(1) Maximal accuray at any orientation β is only attainable if the sampling theorem $k > 2u$ is met. Consequently, compliance with the sampling theorem is indeed a necessary condition for an optimal determination of the animal's orientation to the light distribution. But even if the sampling theorem is actually transgressed, β may still be rather well estimated by means of map-weighting, provided the amplitudes of the harmonics higher than (1/2)k are small, as they are in a realistic tuning function.

(2) The preceding mathematical analysis has been carried out under the condition that the amplitudes of the Fourier coefficients of the external luminance function as well as the tuning function and hence also of their convolution are given and constant. But even if the external luminance distribution $q(\epsilon)$ is not constant at the time and place when orientation must be optimal, the tuning function or the weighting may be adjusted to produce or restore an optimal match to $q(\epsilon)$. Since the tuning function must serve also other functions than orientation, e.g., pattern discrimination, it better not be tampered with. For anatomical, optical and neurophysiological reasons that may not be possible anyway. It may well be feasible, however, to change the weighting until the match to $q(\epsilon)$ is perfect. This may be executed by modulating the k outputs e_i, or, if k is very large, more economically by modulating the factors $\sin(n\varphi)$ and $\cos(n\varphi)$ in Eq. 10 (analogous to the factors nL_n in Eq. 27). This way the IPS may also be able to adjust to non-symmetrical light distributions, even if all harmonics have different phases. The matching may be achieved by comparing the reafference of the animal's movements with an efference copy of the motor commands.

Acknowledgements. We are grateful to Romi Weber for drawing the figures and her tolerance of petty cavils, we thank Anke Bock for transforming an illegible handwriting into a flawless computer print-out, and express appreciation and praise to Peter Ewert for consummating this essay with an abstract of his own.

Note. In Eqs. (15c) and (20b) $i=1$ must be replaced by $n=1$.

6 References

Baldi P & Heiligenberg W (1988) How sensory maps could enhance resolution through ordered arrangements of broadly tuned receivers. *Biol Cybern* 59: 313-318

Fisher R A (1925) Theory of statistical estimation. *Proc Cambridge Philos Soc* 22: 700

Haustein W (1988) *Die Steuerung des Auges unter Listings Gesetz: Ein Matrizen-Modell der Okulomotorik und die Rolle der visuellen Reafferenz.* PhD Dissertation Tech Univ München

Haustein W (1989) Considerations on Listing's Law and the primary position by means of a matrix description of eye position control. *Biol Cybern* (submitted)

Loeb J (1890) *Der Heliotropismus der Thiere und seine Übereinstimmung mit dem Heliotropismus der Pflanzen.* Verlag Georg Hertz, Würzburg

Loeb J (1918) *Forced movements, tropisms and animal conduct.* JB Lippincott Company, Philadelphia London

Marko H (1982) *Methoden der Systemtheorie.* Springer-Verlag, Berlin Heidelberg New York

Mittelstaedt H (1960) The analysis of behavior in terms of control systems. In: Schaffner B (ed) *Group processes, transactions of the fifth conference,* Oct 1958. The Josiah Macy, Jr Foundation, New York, pp 45-84

Mittelstaedt H (1962) Control systems of orientation in insects. *Ann Rev Entomol* 7: 177-198

Mittelstaedt H (1964a) Basic solutions to a problem of angular orientation. In: Reiss RF (ed) *Neural theory and modeling*. Stanford University Press, pp 259-272

Mittelstaedt H (1964b) Basic control patterns of orientational homeostasis. *Symp Soc Exp Biol* 18: 365-385

Mittelstaedt H (1975) On the processing of postural information. *Fortschr Zool* 23: 128-141

Mittelstaedt H (1978) Kybernetische Analyse von Orientierungsleistungen. In: Hauske G, Butenandt E (eds) *Kybernetik 1977*. R Oldenbourg-Verlag, München, pp 144-195

Mittelstaedt H (1983) Introduction into cybernetics of orientation behavior. In: Hoppe W, Lohmann W, Markl H, Ziegler H (eds) *Biophysics*. Springer-Verlag, Berlin Heidelberg New York, pp 794-801

Mittelstaedt H (1985) Analytical cybernetics of spider navigation. In: Barth FG (ed) *Neurobiology of arachnids*. Springer-Verlag, Berlin Heidelberg New York, pp 298-316

Mittelstaedt H (1986) The subjective vertical as a function of visual and extraretinal cues. *Acta Psychologica* 63: 63-85

Mittelstaedt H (1988a) The information processing structure of the subjective vertical. A cybernetic bridge between its psychophysics and its neurobiology. In: Marko H, Hauske G, Struppler A (eds) *Processing structures for perception and action*. Final report of the "Sonderforschungsbereich Kybernetik." Verlag Chemie, Weinheim

Mittelstaedt H (1988b) Basic solutions to the problem of head-centric visual localization. In: Warren R, Wertheim AH (eds) *Perception and control of self-motion*. Erlbaum, Hillsdale, New Jersey (in press)

Mittelstaedt H, Mittelstaedt M-L (1982) Homing by path integration. In: Papi F, Wallraff HG (eds) *Avian navigation*. Springer-Verlag, Berlin Heidelberg New York, pp 290-297

Rose D, Dobson VG (1985) *Models of the visual cortex*. Wiley, New York

PART VI

Motor Control

In Search of the Motor Pattern Generator for Snapping in Toads

ANANDA WEERASURIYA

Department of Physiology, Faculty of Medicine, University of Colombo
Colombo-8, Sri Lanka

Abstract. The hypothesis that a neuronal network interposed between the output elements of the optic tectum, and hypoglossal and other cranial motoneurons subserves ballistic snapping in anurans is critically examined. Morphological and electrophysiological data have not revealed a direct monosynaptic connection between the optic tectum and hypoglossal motoneurons. On the contrary, the available evidence supports the presence of interneurons mediating impulse traffic from the tectum to cranial motoneurons. It is argued that these interneurons are part of a motor pattern generator (MPG). The afferent connections of the bulbar reticular formation (RetF) and the hypoglossal nucleus were studied using the horseradish peroxidase method. The results substantiate the hypothesis that a vital component of the neural substrate for snapping generation is located in the medial RetF (MRF), has access to the motoneurons relevant for snapping, and receives inputs from the tectum and the torus semicircularis. Bilateral tectal and toral inputs to MRF suggest that the MPG consists of two unilateral subunits each having independent access to the relevant motoneurons. This is consistent with the observations that toads with unilateral tectal, or medullary lesions are still capable of snapping. Connections between the nucleus of the solitary tract and the hypoglossal nucleus may be important for elementary bulbar reflexes and manipulation of prey within the mouth. Interconnections between hypoglossal nucleus and other craniomotor nuclei might play a role in the coordination of tongue and jaw musculature necessary for precise timing of the ballistic snapping movement. From a review of the available evidence, a more detailed scheme of the MPG is proposed.

1 Introduction

Snapping in anurans, the consummatory act of the prey-catching sequence (orienting, approaching, fixating, snapping), is a stereotyped ballistic action pattern which, when elicited by an appropriate sign stimulus, proceeds to completion without peripheral feedback from the target (Hinsche 1935; Gans 1961; Grobstein et al. 1983). The release of snapping requires particular visual or tactile stimuli. An adequate stimulus in each of these sensory modalities is sufficient by itself to release prey-catching behavior and, additionally, a cooperative interaction among different sensory modalities (including olfaction) has been demonstrated (for a review see Ewert 1984). The critical features of the visual key stimulus have been investigated in great detail, and corresponding neurophysiological studies have shown that the sensory analysis of visual stimuli is performed in tectal, pre-tectal, sub-tectal, and post-tectal areas (see contributions by Ewert, Manteuffel, Grobstein, Schwippert et al., and Gruberg, this volume). The essential characteristics of the tactile prey sign stimulus have not yet been described. Lesion studies point to subtectal (toral) areas in the mesencephalon as being involved in the analysis of cutaneous stimuli as well as being necessary for tactually elicited prey-catching (Comer & Grobstein 1981a,b). (For an analysis of cutaneous "local signs" responsible for the release of the various wiping movement patterns see Berkinblit et al., this volume). Less is known about the neural substrates for processing chemical stimuli in general and learned prey-related odors in particular (Ewert 1968; Shinn & Dole 1978).

In order for an appropriate stimulus to release snapping, it was postulated (a) that the outcome of the sensory analysis from the three sensory modalities converges on an internuncial neuronal network which has direct access to the motoneuron pools, (b) that activation of this network, the motor pattern generating circuitry (MPG), leads to a relatively stereotypical pattern of activation and inhibition of motoneurons in the medulla and the rostral spinal cord, and (c) that these motoneurons in turn are responsible for elaborating the patterned, coordinated sequence of muscle activation which forms the basis of snapping (Weerasuriya 1983). Hence, snapping is effected through a basically unidirectional flow of coded neural information from sense organs to sensory analyzers, motor pattern generating network (including internal feedback loops), motoneurons, and finally the respective muscles.

The analysis of neural systems subserving motor control proceeds at various levels of organization. As pointed out by Ito (1986) this includes:
(1) behavioral characterization,
(2) identification of neural and contractile components,
(3) elucidation of neural circuitry, and
(4) modeling of the entire system.

The behavioral characterization of snapping by anurans towards prey has been described in detail (Gans 1961; Burghagen 1979; Comer & Grobstein 1981a) and analyzed by means of kinesiological and electromyographic methods (Gans & Gorniak 1982a,b). The experiments presented and discussed in this chapter are related to the second and third levels, i.e., identification of neural components, their interconnections, and characterization of information flow through this network. At the fourth level, this will eventually lead to the construction of models of snapping which may not only yield empirically verifiable predictions about the neural organization of snapping, but also provide us with concepts and hints about the general organization of neural systems responsible for ballistic, stereotyped action patterns in vertebrates. Furthermore, it is not inconceivable that such models, when developed in adequate quantitative detail, will be useful in terms of their structures and algorithms for the design of biomedical prosthetic devices (see Koy-Oberthür,

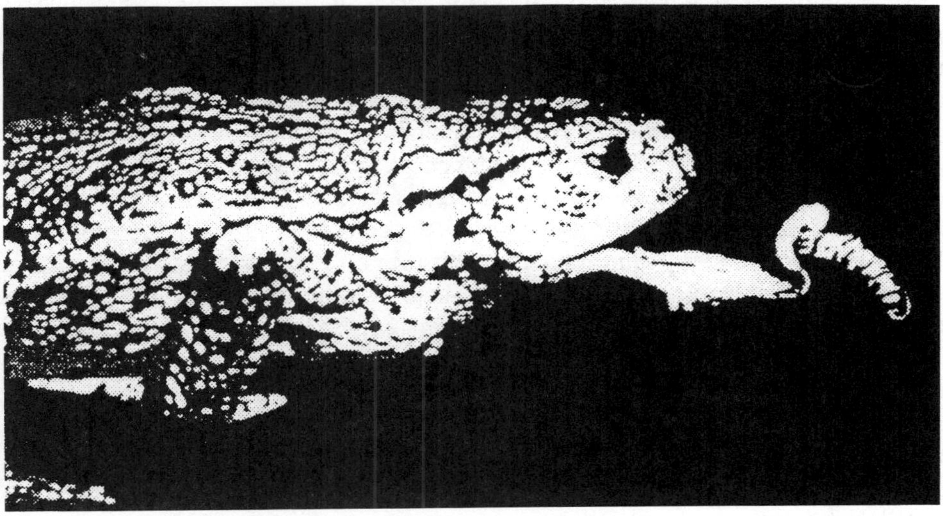

Figure 1. Common toad, *Bufo bufo*, snapping at prey (photo by H.-G. Meyer).

this volume) and components of industrial robots solving comparable sensorimotor tasks (see contributions by Arkin and Torras, this volume).

What do we know already about the neural aspects of the MPG for snapping? Various electrophysiological (Maeda et al. 1977; Satou et al. 1985) and morphological investigations (Lázár 1969; Rubinson 1968) suggest that the output elements of the optic tectum, the sensory analyzer for visually guided snapping, are not in direct monosynaptic contact with motoneurons of the hypoglossal nucleus and other motor nuclei of the medulla oblongata and spinal cord - the Sherringtonian final common pathway for tongue and jaw movements. Therefore, interneurons interposed between the tectal efferent elements and the hypoglossal motoneurons are suitable candidates for the constituent elements of the internuncial network of the MPG. The medullary (bulbar) medial reticular formation (MRF) is involved in generating the motor patterns of various oscillatory movements such as locomotion (Grillner 1981), respiration (Feldman 1986), frog calling (Schmidt 1976) as well as numerous bulbar reflexes such as swallowing (Doty 1976; Jean 1984), eye movement (Kaneko et al. 1981), and mastication (Nozaki et al. 1983). It was proposed that (part of) the MPG for snapping, too, is located in the MRF (Weerasuriya 1983). In order to subject this proposal to experimental investigations and to delineate the pathways of information flow from the tectum to the medullary (bulbar) motoneurons, the afferents of the MRF and the hypoglossal nucleus were examined using the HRP method.

2 Why Analyse Snapping?

The experimental analysis of neural networks underlying the control of a behavior pattern has its advantages and disadvantages, in so far as (i) sensory stimulus pattern recognizing and (ii) motor pattern generating systems are concerned.

With respect to (i), it is common to open the "black box" from the sensory side and follow the impulse traffic *downstream* through various levels and stages of information processing in parallel/hierarchically organized neuronal networks. Hence, if the animal's motivation is appropriate, it is feasible to evaluate the behavior releasing stimulus features (sign stimuli) by measuring the probability of occurrence of the respective action pattern and to test stimulus-response relationships of neurons of the peripheral and central sensory systems accordingly. Problems often arise with the question of sensorimotor linkage (see also contributions by Grobstein and Mittelstaedt, this volume). One reason for this is that the effectors, such as neck and leg muscles and the respective motoneuronal pools, are involved in many different behavioral patterns; furthermore, an action pattern (e.g., orienting) can be incorporated in various behavioral functions (e.g., prey-catching, mating, hiding) depending on the releasing sign stimuli and the animal's motivation.

With respect to (ii), the "effector specificity" of the common toad's tongue for snapping provides an advantage to open the "black box" from the motor side in tracing behavior-specific information flow so-to-speak *upstream*: firstly, by identifying the relevant tongue motoneurons and, subsequently, by investigating their afferents.

Snapping in anurans is related to feeding. There are two kinds of snapping patterns: biting prey with the jaws, or acquiring prey with the tongue (Fig. 1). Whereas jaw grasping in common toads, *Bufo bufo*, is suitable for relatively large prey objects, tongue flipping is adequate for small ones. Hence, there are two "motor schemas" according to the terminology of Arbib (this volume). Tongue flipping is the dominant pattern, whereas jaw grasping is used mostly after unsuccessfull tongue projections (Eikmanns 1955), which means that the "coordinating control programs" (Arbib 1987) of the latter motor schema takes advantage of experience gained from trial and error. While the jaw muscles are involved in various other behaviors, such as yawning and moulting, the tongue in *Bufo bufo* is specific to feeding (in *Bufo blombergi* it is also used for clicking threat noises). Tongue snapping consists of a sequence of ballistic acts: (1) opening the mouth, (2) projecting the tongue toward prey, (3) grasping the prey with the tongue, (4) retracting the tongue, (5) closing the mouth, and (6) gulping. The sequence from 1 to 5 takes about 200 ms (Burghagen 1979; for details in *Bufo marinus* see Gans & Gorniak 1982a,b).

3 Topography of Motoneurons

The initial step in the analysis of snapping MPG is to determine the topography of motoneurons that innervate the toad's tongue and jaw muscles.

3.1 Tongue Muscle Motoneurons

The tongue flip is controlled by two extrinsic tongue muscles, the musculus genioglossus (protractor) and m. hyoglossus (retractor). If horseradish peroxidase (HRP) was injected into one of the identified extrinsic tongue muscles, while transecting the nerves supplying the others muscles, the corresponding motoneurons were labeled (backfilled) ipsilaterally in the caudal medulla oblongata around the obex (Fig. 2A-D) (Weerasuriya & Ewert 1981). Within the hypoglossal nucleus, the motor nucleus of the XII*th* cranial nerve (Nieuwenhuys & Opdam 1976), there is a definite topographical cell distribution: the genioglossal motoneurons are located caudally (Fig. 2A, right), whereas the hyoglossal motoneurons are more rostral (Fig. 2D, left). These data have been confirmed anatomically in frogs by soaking transected single hypoglossal nerve branches in HRP (Stuesse et al. 1983), and physiologically in toads by intracellularly recording the antidromic activity of tongue muscle motoneurons in response to electrical stimuli applied to identified branches of the hypoglossal nerve (Satou et al. 1985). After intracellular iontophoretic application of HRP into a tongue muscle motoneuron, extensive spread of branching of dendritic trees becomes visible, mainly in the lateral and dorsolateral directions, even extending into the nucleus of the solitary tract; some dendrites crossed the midline to the contralateral hypoglossal nucleus (see Satou et al. 1985). Studies using other morphological tracing techniques in frogs have also revealed extensive dendritic arborizations of hypoglossal motoneurons (Matesz & Székely 1977).

Interestingly, a comparable separation between tongue protractor and retractor motoneurons has been demonstrated in mammals by Kramer et al. (1979) and Uemura-Sumi et al. (1981), although here a more marked mediolateral separation of motoneuron pools was observed. For a discussion on controversies of the homologies between anurans and mammals in this context see Stuesse et al. (1983).

3.2 Jaw Muscle Motoneurons

The jaw is lowered by the musculus depressor mandibuli and elevated by the m. temporalis, m. pterygoideus, and m. masseter. We injected HRP into the respective muscles and traced the backfilled neurons (Fig. 2E,F). As reported for other anuran species (Gaupp 1899; Nieuwenhuys & Opdam 1976), in toads too, it can be shown that the m. depressor muscle is innervated by the motoneurons of the VII*th* (facial motor) nerve nucleus (Fig. 2F, right), and the elevator muscles by motoneurons of the V*th* (trigeminal motor) nerve nucleus (Fig. 2F, left). The rostral pole of the former nucleus is about 250 µm caudad of the caudal pole of the latter, and both are part of the branchiomotor column. There is a gap of about 1200 µm between the rostral pole of the hypoglossal nucleus and the caudal pole of the facial motor nerve nucleus. For topography of motor nerve nuclei see also Fig. 4.

4 Tecto-Bulbar Outflow

To obtain a gross characterization of the tectal neurons projecting toward bulbar/spinal motor systems, HRP was deposited unilaterally into the caudal medulla oblongata within and rostral to the hypoglossal nucleus (Weerasuriya & Ewert 1981). In these studies the injection sites were too large either to determine the topography of the descending pathways or to differentiate among afferents to various medullary nuclei. Furthermore, all retrogradely labeled cells are not necessarily medullary afferents since HRP uptake by fibers of passage cannot be excluded.

These unilateral HRP injections produced retrograde bilateral labeling of cells in the optic tectum. HRP backfilled cells were also seen in the sub-tectal region (principal and laminar nuclei of the torus semicircularis), the ventral tegmentum (nucleus anteroventralis and the interstitial nucleus of the fasciculus longitudinalis medialis), and the caudal dorsal thalamus (posterocentral and lateral posterodorsal nuclei). For comparable results in frogs using HRP labeling see Ingle (1983) and using cobalt labeling see Lázár et al. (1983) (cf. also contributions by Lázár and Matsumoto, this volume).

More specifically, about 80% of the labeled tectal cells were in the lateral half of the tectal lobe (Fig. 3a-c), and there were more cells in the contralateral tectum than in the ipsilateral one. The stained cells were situated at the top of layer 6 and within layer 7; they were identified as either pyramidal, ganglionic, or piriform cells (Fig. 3, top) (For nomenclature of tectal layers and morphological cell types see Székely & Lázár 1976). The axons of some of these cells could be traced in layer 7 (see arrows in Fig. 3, top), and fibers of the contralateral projection were observed to cross the midline in the ansulate commissure (see arrow in Fig. 3b). The results are in agreement with previous suggestions from silver impregnation studies (Rubinson 1968; Lázár 1969) showing that the tectum has a bilateral projection to bulbar/spinal nuclei mediated by the tractus tecto-bulbaris directus (t.tb.d.) and the tractus tecto-spinalis cruciatus (t.ts.c.). The present study in toads characterizes tectal cell types and their locations. Furthermore, since Ingle (1983) has shown that cutting the t.ts.c. at its tegmental decussation (ansulate commissure) abolishes prey-orienting but not snapping, it can be concluded that visual information mediated by the ipsilaterally projecting t.tb.d. is sufficient for triggering snapping.

The dominance of backfilled tectal cells in the ventrolateral tectal lobe in its rostrocaudal extent (see Fig. 3b,c) strongly coincides with (i) the map of the snapping releasing visual field (Ingle 1976), (ii) the "snapping evoking area" determined by electrical brain stimulation of the optic tectum (Ewert 1967), and (iii) the topography of regional ^{14}C-2-deoxyglucose uptake in the tectum during visually elicited snapping (Finkenstädt et al. 1985). These results suggest that tectal neurons, located in layers 6/7 with their axons projecting toward the medulla oblongata, play a pivotal role in the sensorimotor interfacing of visually guided prey snapping by toads. Thus, they appear to represent a vital link in the correlation between afferent messages and efferent motor commands, i.e., between behaviorally relevant stimuli and the

Figure 2. Topography of motoneurons involved in snapping. Dorsal view of the common toad's brain. T, telencephalon; OT, optic tectum; C, cerebellum; M, medulla oblongata; O, obex; double arrow: levels of transverse sections (segments) shown in A-D, running from caudal (A) to rostral (D). Calibration bar: 1000 μm. (A-D) Topographical representation of tongue muscle motoneurons in transverse sections through the hypoglossal nucleus. HRP was injected into the left hyoglossus and right genioglossus muscles, respectively. A: 300 μm caudad of the obex; B: 200 μm rostrad of the obex; C: 500 μm rostrad of the obex; D: 800 μm rostrad of the obex. (From Weerasuriya & Ewert 1981). (E,F) Horizontal sections through the medulla oblongata; E: bilateral labeling of hypoglossal motoneurons following injection of HRP into tongue muscles of both sides. F: Labeling of motoneurons in the V*th* (trigeminal motor) nerve nucleus and the VII*th* (facial motor) nerve nucleus following injection of HRP into the muscles that elevate and depress the jaw. In both figures (E, F), rostral is to the right. Calibration bar: 100 μm, common to both. (From Weerasuriya 1983).

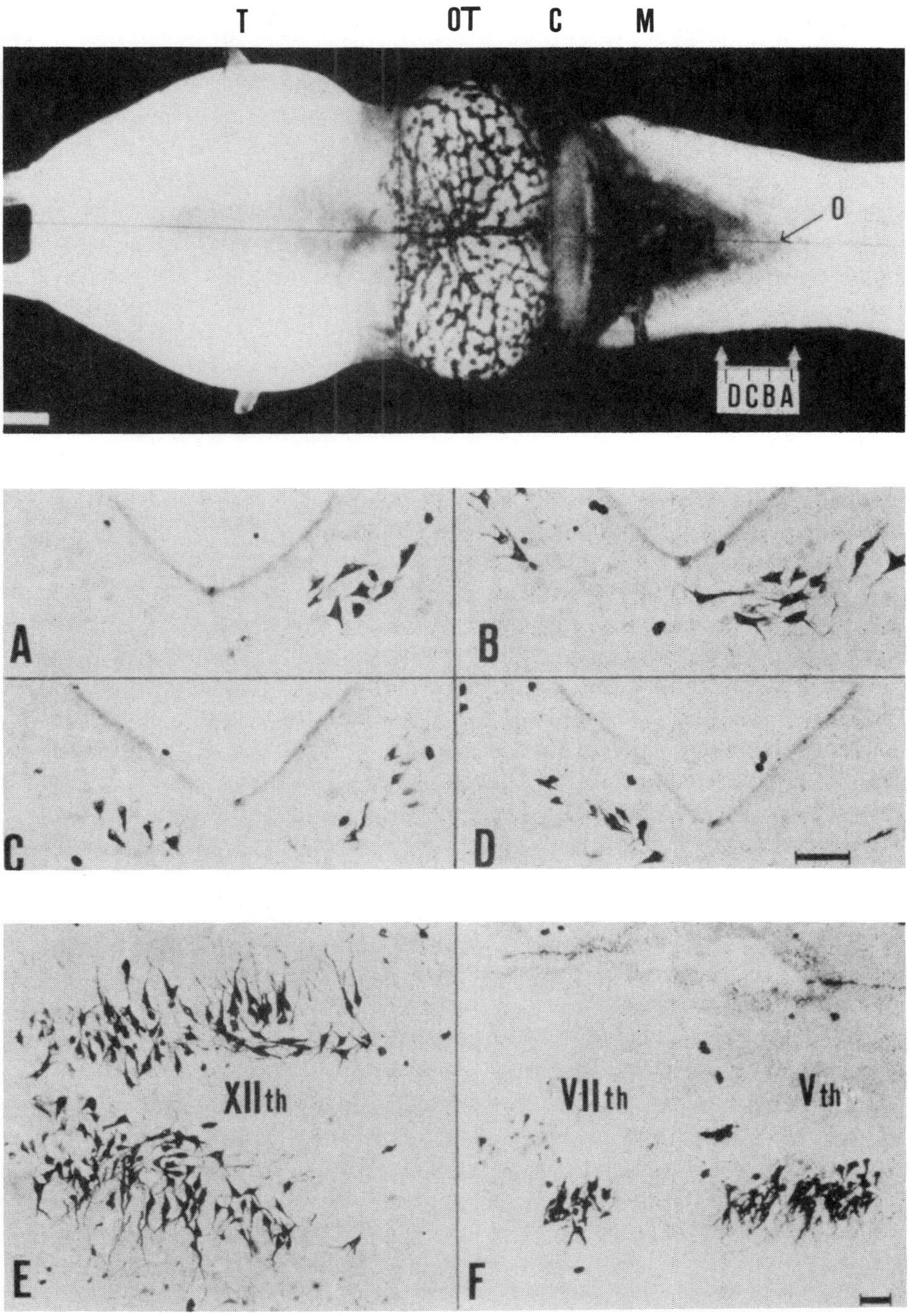
T
OT
C
M
O
DCBA
A
B
C
D
E
F
XIIth
VIIth
Vth

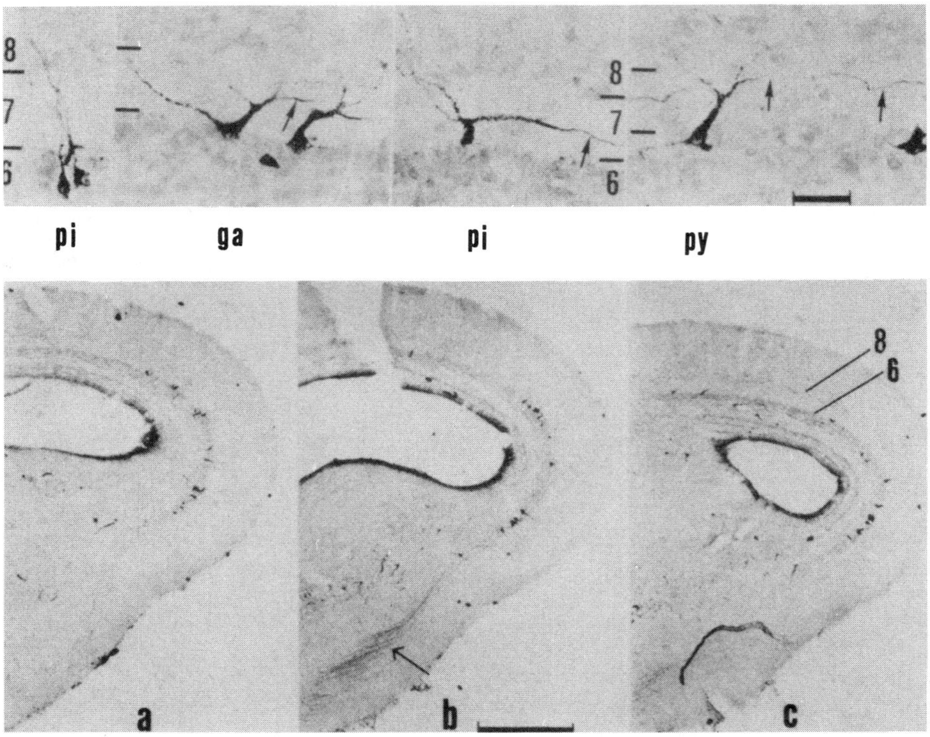

Figure 3. Labeling of tecto-bulbar/spinal projecting neurons. Bottom: Transverse hemi-sections (a-c) through the optic tectum of the common toad illustrating the pattern of cell labeling following injection of HRP into the contralateral caudal medulla oblongata. The crossing fibers of the contralateral projection are indicated by arrow. Sections are running from caudal to rostral. Calibration bar: 500 µm. (From Weerasuriya 1983). Top: Morphological details and locations of tectal output cells with descending projections labeld after HRP injection into the caudal medulla. The axons of some of the cells are indicated by arrows. Calibration bar: 50 µm, common to all cells; ga, ganglionic cell; pi, piriform cell; py, pyramidal cell; 6-8, tectal layers. (From Weerasuriya & Ewert 1981).

appropriate motor behavior released by such stimuli. Antidromic stimulation/recording techniques by Satou & Ewert (1985) indeed have shown that prey-selective T5.2 neurons, for example, are output elements of the tectum with their axons reaching the medulla oblongata (see also Matsumoto, this volume). These neurons come as close as possible to fulfilling the role of "command elements" in a vertebrate (see also Schürg-Pfeiffer, this volume). Although it is extremely unlikely that the criteria for "command neurons" as defined by Kupfermann & Weiss (1978) can be applied for anuran prey-catching, it can be suggested that T5.2 neurons cooperating with other (tectal) neurons are *command elements* (i.e., components) in visual command releasing systems, CRS (see Ewert, this volume). In this respect it is interesting to note the near indistinguishability (from videotape analysis) between the snapping responses elicited by visual or (in blind frogs) by tactile stimuli (Comer & Grobstein 1981a,b). Since snapping can also be elicited by touch in frogs after tectal ablation (Grobstein et al. 1983), in addition to visual CRS, tactual CRS must exist whose command elements are located in subtectal (toral) regions and which have access to the same snapping generating circuitry.

5 Toward the Internuncial Circuitry

Since neuroanatomical (Cajal 1909) and neurophysiological (Porter 1965; Satou et al. 1985) studies failed to detect axon collaterals in hypoglossal motoneurons, this would favour the suggestion that these motoneurons, when driven by a pattern generating internuncial neuronal networks, participate in a stereotyped, ballistic fixed action pattern.

5.1 Spatial Summation

Evidence for interneurons has been provided by intracellular recordings of the membrane potential of hypoglossal motoneurons (Satou et al. 1985) during electrical stimulation of tectal output fibers in the "snapping evoking area" (defined by Ewert 1967). It could be shown that activation of hypoglossal motoneurons from either tectal hemisphere was mediated by interneurons operating in polysynaptic pathways. Furthermore, the amplitude of excitatory postsynaptic potentials recorded either from tongue protractor or retractor motoneurons increased if both tectal lobes were stimulated simultaneously. Hence, activation of hypoglossal motoneurons is subject to spatial facilitation, a phenomenon also known from snapping behavior: The time interval between onset of binocular fixation and snapping at prey could be relatively long; this period was further prolonged in the monocular animal (see Schneider 1954), however, it was remarkably reduced, if the visually "deprived" tectal lobe was simultaneously electrically stimulated at subliminal intensities through an electrode implanted in the snapping evoking area (Ewert 1967). There are many possibilities for convergence of information leaving both eyes, at levels downstream from mesencephalic structures, e.g., (i) in tegmental nuclei via axon collaterals of tectofugal fibers in the ansulate commissure, (ii) in the medial reticular formation (MRF) integrating inputs from the ipsilaterally and the contralaterally projecting tecto-bulbar and tecto-spinal tracts, respectively (Rubinson 1968; Lázár 1969), (iii) in the rostral MRF integrating inputs from commissural fibers of the tecto-bulbar tract (Ingle 1983), and (iv) in the hypoglossal nucleus linking tongue muscle motoneurons through their dendritic commissural fibers (Satou et al. 1985).

5.2 Specification of Afferents to Bulbar Nuclei

5.2.1 Methodological Comments

To determine the afferents of bulbar nuclei, micro-injections of HRP were necessary. Some methodological comments are in order. The experiments were carried out on anesthetized common toads, *Bufo bufo*. After exposing the medulla oblongata and gently moving aside the choroid plexus, an iontophoretic deposit of

HRP solution was delivered through a glass micropipette (10-20 μm tip diam.) into the medial reticular formation or the hypoglossal nucleus at an anodal current of 1.5-2.5 μA during a 10 s on and 10 s off cycle for a total injection period of 3-5 min. After a survival period of 4-5 days, the deeply anesthetized toads were transcardially perfused with phosphate-buffered (pH 7.2-7.4) Ringer solution and then with 50-80 ml of a fixative containing 1.5% glutaraldehyde, 4.0% sucrose, and 1.0% dimethylsulfoxide in 0.05 M phosphate buffer (pH 7.3-7.4). The dissected brain and rostral spinal cord was stored in the above fixative for 6-10 h at 4°C and then transferred to a 30% sucrose solution in phosphate buffer at 4°C (for about 12 h), or until the tissue sank. The brain was mounted in 1% gelatin and sectioned transversely at 40 μm in a cryostat at -10°C. Sections were transferred directly onto gelatin-coated slides, air dried, and then kept at 4°C (Ebbesson et al. 1981). Every section was collected. Within eight hours of being cut the sections were reacted with substrate and H_2O_2. The DeOlmos-Heimer (1977) method modified by Ebbesson et al. (1981), using benzidine dihydrochloride as the chromogen, was applied.

The iontophoretic micro-applications of HRP were as small as possible (Fig. 4). An example is shown in Fig. 5. The unilateral injection sites were in the hypoglossal nucleus (rostral to the obex, at the obex, or caudal to it) or in the medial reticular formation (Fig. 4). For hypoglossal injections the patterns of retrograde cell labeling were strikingly similar, so that the following description applies to all three injection sites.

The usefulness of an injection site was judged in terms of its size, discreteness and position in relation to the intended target. The success rate was about 50% for

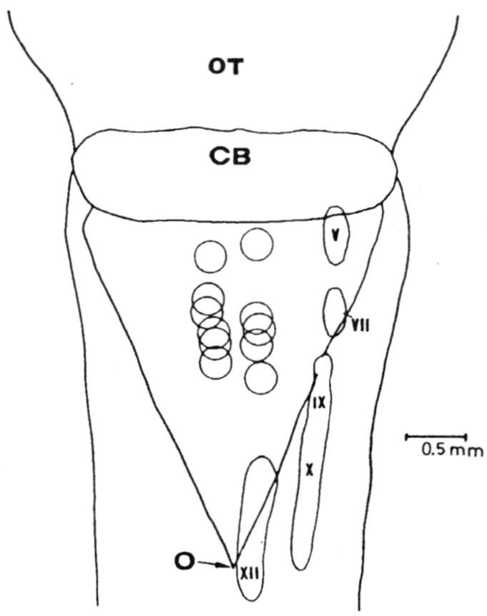

Figure 4. The iontophoretic deposits of HRP in the medial reticular formation are mapped as circles onto a dorsal view of the medulla oblongata. OT, optic tectum; CB, cerebellum; O, obex. The roman numerals indicate the location of the cranial motor nuclei.

the MRF injections and about 30% for the hypoglossal nucleus (a much smaller target).

5.2.2 Afferents of the Hypoglossal Nucleus

The overall pattern of retrograde cell labeling (backfilling), obtained from seven toads following unilateral HRP micro-injection into the hypoglossal nucleus, is shown in a series of transverse sections through the medulla oblongata in Fig. 6 (left). If the injection sites were confined to the hypoglossal nucleus, labeled cells were not found rostral to the rhombencephalon. No cells were backfilled in the optic tectum. Heavy cell labeling was seen bilaterally in the MRF up to the plane of the VIII*th* cranial nerve. According to the nomenclature of Opdam et al. (1976), this encompasses also the nucleus reticularis inferior and n. reticularis medius. Rarely, an HRP-stained cell was observed in the n. reticularis superior. Medium sized cells as well as the larger multipolar elements of the medial reticular formations were labeled (Fig. 7A,B). Cells were mostly located in the ventral zone of the grey matter (central grey) adjacent to the white matter (stratum album). Two main types of neurons were distinguished, those with laterally and ventrally spreading dendrites (Fig. 7A) and others with dendrites extending preferentially towards the subjacent white matter (Fig. 7B). A few cells in the *median* reticular formation (the area of the raphe nucleus) were labeled, too; most of these were at a level corresponding to the rostral end of the solitary tract.

Consistent retrograde cell labeling was seen bilaterally in the lateral and ventral portions of the nucleus of the solitary tract (NST) with ipsilateral labeling being heavier. Stained cells in NST were distributed both rostral and caudal to the injection

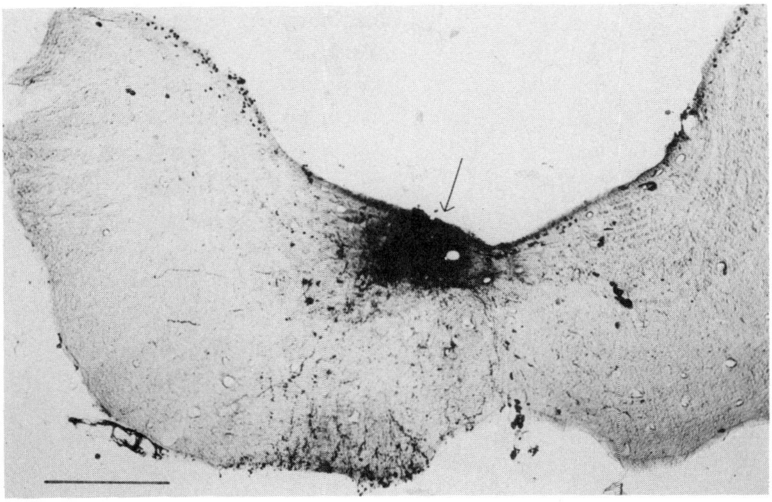

Figure 5. Example of an HRP micro-injection site in the medulla oblongata limited to the grey matter and not spreading into the subjacent white matter. Counter staining: neutral red. Calibration bar: 200 μm.

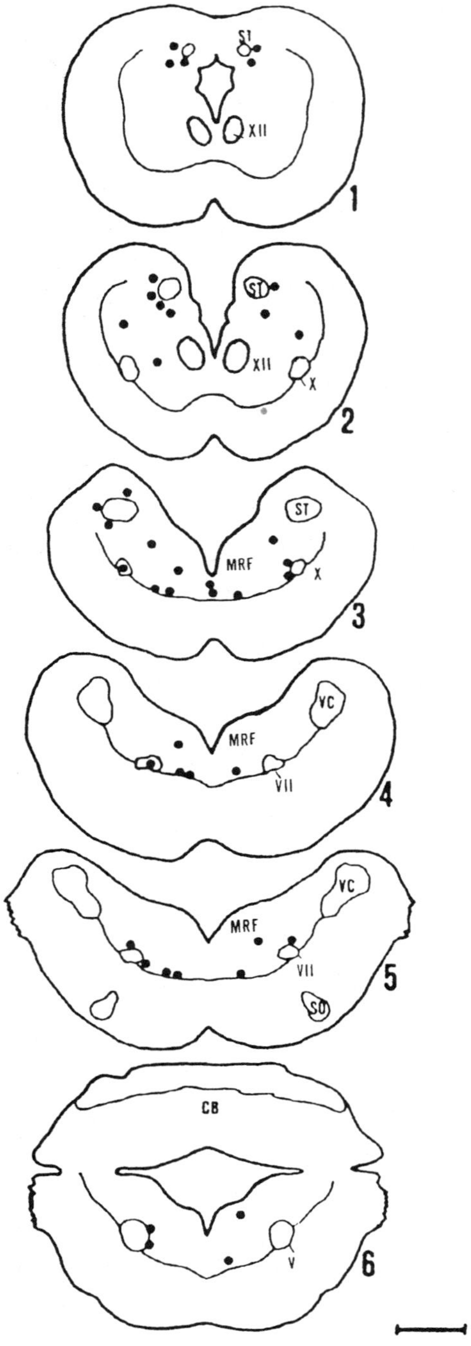

Figure 6. Retrograde cell labeling in the medulla oblongata following micro-injection of HRP into the left hypoglossal nucleus (see Figure on this page). Various levels (1-6) of transverse sections through the medulla oblongata running from caudal (1) to rostral (6). Pattern of retrograde cell labeling is indicated by dots. Opposite page: Sections stained with the Klüver-Barrera method, running from caudal (top) to rostral (bottom). V*th*, trigeminal motor nucleus; VII*th*, facial motor nucleus; X*th*, vagal motor nucleus; XII*th*, hypoglossal nucleus; CB, cerebellum; LRF, lateral reticular formation; MRF, medial reticular fromation; mRF, median reticular formation; SO, superior olive; ST, solitary tract; SPC, spino-cerebellar complex; SPM, spinal motor nucleus; TB, tractus tecto-bulbaris directus, tr.tb.d.; TS, tractus tecto-bulbaris et spinalis cruciatus, tr.tbs.c; TV+VII, descending tracts of V*th* and VII*th* motor nuclei; VC, vestibular complex. Calibration bar: 500 µm. (From Weerasuriya & Ewert 1988).

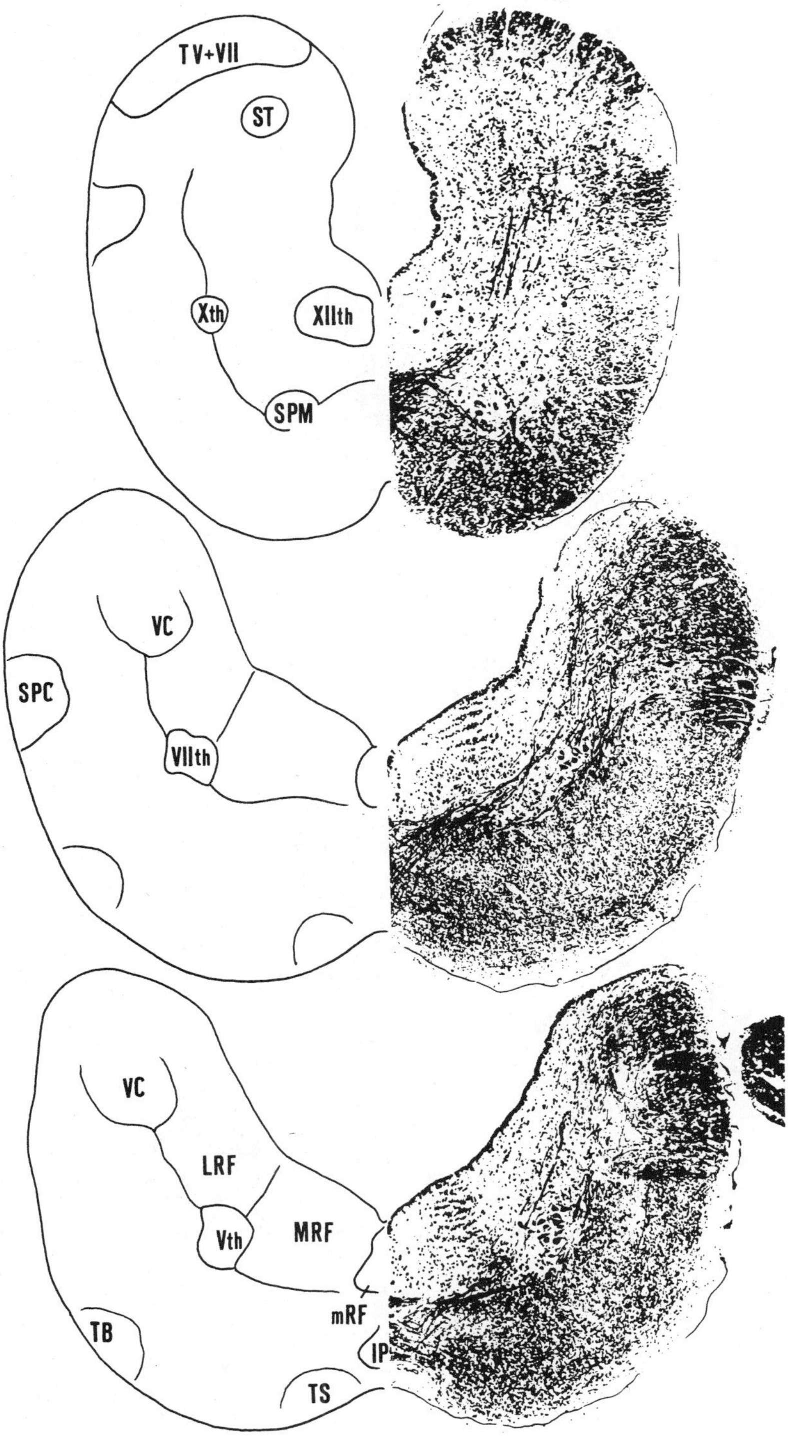
TV+VII
ST
Xth
XIIth
SPM
VC
SPC
VIIth
VC
LRF
Vth
MRF
mRF
TB
IP
TS

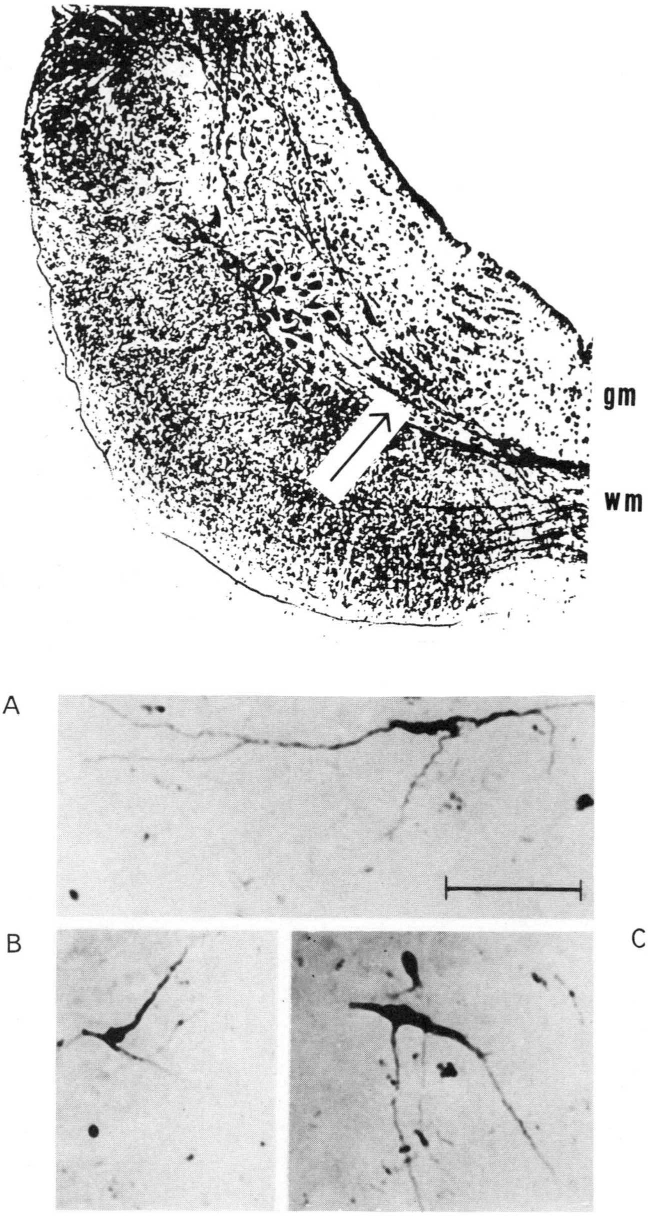

Figure 7. Examples of retrogradely labeled large multipolar cells in the medial reticular formation after micro-injection of HRP into the ipsilateral hypoglossal nucleus. (A) The dendrites of this cell are at the border between grey matter, gm, and white matter, wm (cf. arrow in stained histological section on top), extending partly into the lateral branchiomotor column, the subjacent white matter, and the opposite side of the medulla oblongata. (B) Cells located at the border between grey and white matters. In contrast to the cell shown in A, dendrites are branching mostly towards the white matter and less in lateral and medial directions. Calibration bar: 100 μm. (From Weerasuriya & Ewert 1988).

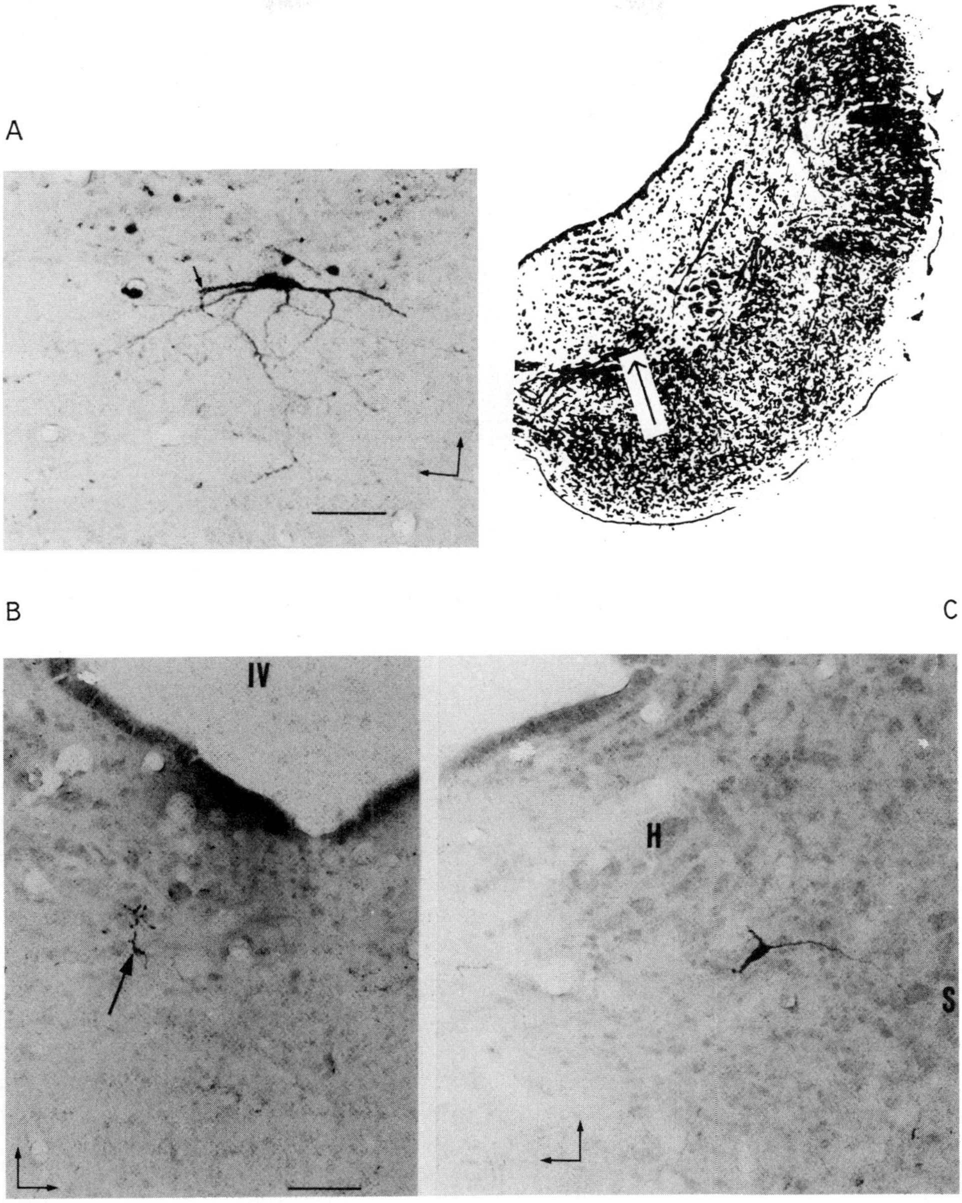

Figure 8. Examples of retrograde labeling of bulbar neurons after unilateral micro-injection of HRP into the medial reticular formation (see Fig. 5). (A) Large cell situated at the border between grey matter and white matter (see arrow in Klüver-Barrera stained histological section on top, right); note the extensive dendritic spread towards the white matter. (B) Terminal labeling of axonal buds on hypoglossal motoneurons. (C) A labeled cell between hypoglossal nucleus (*H*) and rostral end of the spinal motoneuron column; a dendrite of this cell extends into the spinal motor column (*S*). Calibration bar: 100 μm; the horizontal arrow points to the midline.

site, but the majority was rostral. A few cells situated ventral and lateral to NST were labeled bilaterally. Backfilled cells were found bilaterally within and near the branchiomotor column. The peribranchiomotor labeling was associated with the motor nuclei of the V*th*, VII*th*, IX*th*, and X*th* cranial nerves.

5.2.3 Afferents of the Medial Reticular Formation

The pattern of retrograde cell labeling described here is based on the results from eleven toads in which the injection sites were restricted to the MRF. The location of the successful injection sites are plotted onto a dorsal view of the toad's medulla oblongata (Fig. 4), and one of these is depicted in a transverse section in Fig. 5. Heaviest labeling was seen in the adjacent MRF. The stained cells were distributed bilaterally and quite often were the magnocellular elements at the border between the grey and white matters (Fig. 8A). These cells had extensive dendritic trees with branches extending into the surrounding white matter. Smaller cells located in the middle of the grey matter were also labeled and they had smaller dendrites confined to the grey matter. Backfilled cells were also seen bilaterally in and near the branchiomotor column. In the rostral spinal cord, some cells distributed bilaterally and near the spinal motor column were labeled (Fig. 8C). A few cells near the ipsilateral NST and the ipsilateral lateral reticular formation were labeled, too.

In the mesencephalon, backfilling was limited to the optic tectum, the torus semicircularis, and the ventral tegmental nuclei. Labeled tectal cells were distributed bilaterally (contralateral > ipsilateral) in the ventrolateral and lateral regions, heaviest at the border between layers 6 and 7. In the torus, bilateral labeling was seen in the principal and magnocellular nuclei. The ipsilateral anterior and posterior ventral tegmental nuclei and the ipsilateral posterior nucleus of the thalamus contained a few stained cells.

Terminal labeling, indicative of anterograde transport of HRP, was seen bilaterally in the MRF and ipsilaterally in the hypoglossal nucleus (Fig. 8B).

6 Discussion, Conclusions, and Speculations

The preponderance of tectal efferent cells in the central and ventrolateral part of the tectal layers confirms earlier morphological (Weerasuriya & Ewert 1981) and physiological (Satou & Ewert 1985) reports of a similar nonuniform distribution of efferent elements in the toad's tectum. A recent morphological study, using HRP in the frog, reported comparable results (Kostyk & Grobstein 1987). Thus, motor commands related to prey stimuli in the snapping-evoking visual field are relayed to brainstem motor circuits (e.g., in the medulla oblongata and the spinal cord) via output elements in the rostrolateral tectum and a few efferent cells are probably adequate for this purpose. For an involvement of mesencephalic, tegmental reticular structures in motor pattern generation of prey-catching orienting see Grobstein (this volume).

The concept of a motor pattern generating circuitry (MPG) for snapping was first postulated (Weerasuriya & Ewert 1981) and then discussed in some detail (Weerasuriya 1983; Grobstein et al. 1983) more than five years ago. Since then numerous reports (Ewert et al. 1984a; Satou et al. 1985; Satou & Ewert 1985;

Weerasuriya & Ewert 1984, 1988; Weerasuriya 1988; Matsushima et al. 1986; Schwippert et al., this volume) have provided evidence that has substantially confirmed this concept permitting experimental verification of it through more detailed and specific testable predictions. The data and the model of the MPG are still at a rather qualitative level and do not allow a quantitative analysis such as has been performed for sensory tectal/pretectal functions (see contributions by Arbib, Betts, and Cervantez-Pérez, this volume). In order to approach some properties of the snapping MPG, we shall tackle the following issues:

6.1 What is the Evidence for a Snapping MPG?

Earlier studies with silver degeneration techniques (Rubinson 1968; Lázár 1969) and more recent investigations with HRP (Kostyk & Grobstein 1987) have shown that neither the ipsilateral nor the contralateral tectofugal projections terminate in motor nuclei. HRP studies of the afferents to the hypoglossal nucleus did not label any cells in the tectum (Weerasuriya & Ewert 1988). Electrophysiological studies (Satou et al. 1985) demonstrated that focal electrical stimulation of the tectum produces in both the tongue retractor and protractor motoneurons postsynaptic potentials whose latencies suggest a polysynaptic pathway. These authors found no evidence for a monosynaptic tecto-hypoglossal pathway. In contrast, neck and forelimb motoneurons receive a disynaptic input from the tectum (Maeda et al. 1977). Therefore, the available evidence consistently indicates that the information relayed by tectal efferent neurons reaches hypoglossal (and probably other cranial) motoneurons only through a "chain" (network) of interneurons.

6.2 How Many Interneurons are Traversed from Tectal Efferents to Motoneurons?

The mean latency of the tecto-hypoglossal inhibitory postsynaptic potential, IPSP, was around 9 ms (Satou et al. 1985). Allowing 5 ms for conduction time (at 1 m/s for 5 mm) and assuming a synaptic delay of 1 ms (Brookhart & Fadiga 1960), at least 3 interneurons seem to be interposed between the tectal efferents and the hypoglossal motoneurons. The excitatory postsynaptic potential, EPSP, had a longer latency suggesting that there are more interneurons in this pathway. If either conduction distances are shorter or conduction velocities are faster (the longest antidromic latencies measured by Satou & Ewert, 1985, were about 3 ms), then a larger proportion of the latency has to be accounted for by synaptic delays and hence more interneurons may be involved. In contrast, tectofugal impulses cross only one or two interneurons to reach neck and forelimb motoneurons in the brachial enlargement of the spinal cord (Maeda et al. 1977) and glossopharyngeal afferents have monosynaptic access to hypoglossal motoneurons (Matsushima et al. 1986). Thus, the longer latencies of tectally evoked postsynaptic potentials in hypoglossal motoneurons seem best accounted for in terms of the interneurons which are part of the MPG. Furthermore, it is predicted that tectally evoked potentials in branchial motoneurons would also have similar long latencies, and that subtectal stimulation (corresponding to tactually elicited snapping) would also give rise to synaptic potentials in hypoglossal and branchial motoneurons with comparable long latencies.

The above electrophysiological evaluation indicates how fast stimulus-responses for snapping behavior could be. Schürg-Pfeiffer's (this volume) recordings from freely moving toads show that the actual delays between the onset of visually evoked activation in tectal (or hypoglossal) neurons and the beginning of snapping may take up to 400 ms which suggests intervening integrative (threshold operational) and interactive (excitatory/inhibitory) processes (see also Schwippert et al.; Grobstein, and Mittelstaedt & Eggert, this volume).

6.3 Which Kinds of Cells are Probably Involved in Snapping?

The hypoglossal nucleus receives afferent connections from the MRF, the NST, and from neurons in and around other craniomotor nuclei (Weerasuriya & Ewert 1984, 1988). In this context it is interesting to note that MRF cells projecting to the hypoglossal nucleus (Fig. 6 left) are, with respect to their location and dendritic trees (Fig. 7A,B), similar to intracellularly recorded cobalt-stained neurons responding to visual stimuli (Schwippert et al., this volume). In addition, some of these neurons displayed cyclic burst activities in response to visual stimulation. The connections between hypoglossal nucleus and NST are probably important for the elementary bulbar reflexes and manipulation of prey within the mouth. Interconnections between the hypoglossal nucleus and other craniomotor nuclei are likely to play a role in the coordination of tongue and jaw musculature necessary for the precise timing of the ballistic snapping movement (Weerasuriya 1983). Similar interconnections among the III*rd*, IV*th*, and VI*th* cranial motor nuclei in mammals significantly contribute to the coordination of the extra-ocular muscles in executing various saccadic eye movements related to pursuit and vestibular-ocular reflexes (Robinson 1981).

6.4 Functional Aspects of Snapping Pattern Generation

The high degree of stereotypy, the extremely rapid execution of snapping, and the inability to rectify or terminate the movement midstream point to a ballistic fixed action pattern similar to swallowing, vomitting, and other bulbar motor synergies. All these motor patterns require a highly synchronized and coordinated sequence of activation and inhibition of many motoneurons and the impulses reaching the respective motoneurons to enact these synergies are believed to originate in a network of premotoneurons. From the available evidence it is possible to infer some aspects of the functional organization of the underlying MPG and to predict certain other features of this internuncial network. In terms of the direction of impulse traffic, MPG is defined as the chain of interneurons beginning with those that receive direct input from efferent tectal neurons and other sensory analyzers and continuing upto those which exert monosynaptic control over bulbar motoneurons. Although it appears that upon release the performance of the MPG is uninfluenced by external peripheral feedback loops, it is unlikely that the various components of the snap proceed in a completely "open loop" fashion. Internal feedback loops within the internuncial network probably serve to coordinate the precise timing between the jaw and tongue movements.

Since bilateral lesions of the MRF abolish snapping, whereas extensive ipsilateral lesions do not disrupt this action pattern (Weerasuriya 1988), we suggest that the MPG is bilaterally represented. Each of the two parts should have independent access to the motoneurons on both sides and also receive independent polysynaptic input from both tectal hemispheres (Weerasuriya & Ewert 1988). The nearly identical latencies of postsynaptic potentials of hypoglossal motoneurons elicited by ipsi- and contralateral tectal stimulation (Satou et al. 1985) suggest that bilateral tectal efferents converge on MPG interneurons. From there onwards the impulse traffic from both hemispheres probably follows a common pathway to the hypoglossal motoneurons. If the impulse traffic leaving both hemispheres were to converge only on or immediately before the motoneurons, then a wider disparity between the ipsi- and contralateral latencies would have been observed. Such a bilateral organization of the MPG is in contrast to what is known about the MPG for swallowing in mammals (Doty et al. 1967) which is organized as two half-centers with each half-center controlling only the bulbar motoneurons of that side. This dual representation of the MPG, apart from conferring the advantage of a back-up system, is probably related to the fact that snapping is elicited by stimuli in the frontal visual field and *optimal* activation of the motor pattern requires simultaneous impulse traffic from both tectal hemispheres via the MPG to the relevant motoneuron pools. In support of this, Satou et al. (1985) demonstrated a range of temporal and spatial facilitations between the tectal hemispheres in evoking postsynaptic effects on hypoglossal motoneurons. This is in accordance with behavioral data indicating that the occurrence and accuracy of snapping in monocular frogs is less than that in binocular animals, and that the latency of snapping is also increased in the monocular animals (Schneider 1954; Ewert 1967). Translated into neuronal events, the activation arriving from only one tectal hemisphere (in monocular toads) takes much longer to reach the threshold for triggering the MPG.

The lowest threshold as well as shortest latency effect of tectal stimulation on hypoglossal motoneurons was an IPSP (Satou et al. 1985). During swallowing in mammals, too, hypoglossal motoneurons exhibit a brief hyperpolarization before the onset of action potentials that activate the tongue muscles (Sumi 1969). Presumably, these IPSPs ensure that ongoing subliminal activity in hypoglossal (and probably other bulbar) motoneurons is terminated, hence "resetting" the system before the initiation of the ballistic motor pattern (see also inhibitory "pause-neurons" responsible for the coordination of saccadic eye movements discussed by Fuchs & Kaneko 1985, and for spinal wiping reflexes suggested by Berkinblit et al., this volume). The tectofugal volleys that elicit the higher threshold and longer latency EPSPs probably travel along the neuronal pathways taken by tectofugal impulses when they elicit snapping under normal conditions.

6.5 Predicted Locations and Interconnections of the Interneurons

The MRF is involved in the generation of the snapping response (Weerasuriya 1983; 1988). The preponderance of cells with tectal input in the rostral MRF (Ewert et al. 1984a; Schwippert et al., this volume) suggests that the tecto-receptive (and probably the major) part of the MPG is located between the level of the nucleus isthmi and the rostral pole of the vagal branchiomotor column. The parts of the MPG that have monosynaptic access to the motoneurons are probably located near the respective motor nuclei (Fig. 9). Intracellular recording and cobalt-staining studies have shown that stimulus-event sustaining cyclic bursting cells are mostly situated close to branchiomotor nuclei (Schwippert et al., this volume). In mammals, too, it has been suggested that integrative motor control functions are performed by interneurons situated immediately adjacent to bulbar motoneurons (Landgren et al. 1986). Interneurons in such perimotor nuclear locations are considered to play an integrative role in recruiting their neighbouring motoneurons into various motor synergies (see also Olsson et al. 1986). Interconnections among such interneurons, associated with the various motor nuclei (Weerasuriya & Ewert 1984, 1988; Landgren et al. 1986), are probably the neural substrate for coordinating the various subsynergies of a motor pattern. When the hypoglossal nerves are transected bilaterally, toads lunge toward prey objects without any accompanying tongue movements and, interestingly, also without opening their jaws (Weerasuriya 1988). The absence of tongue movement is a direct consequence of the hypoglossal nerve transections, but the absence of jaw movement (the motoneurons responsible for elevating and lowering the jaw are intact) is likely to be due to an inhibition arising from the absence of any tone in the lingual musculature detected by glossopharyngeal (Matsushima et al. 1986) or hypoglossal (Stuesse et al. 1983) afferents. Furthermore, this inhibition is likely to be mediated by the interneurons situated adjacent to motor nuclei. The above experiment also demonstrates that certain components of the snap can be uncoupled (lunge *vs*. tongue and jaw movements), whereas others are more tightly coupled (tongue, jaw, and gulping movements) suggesting a hierarchical

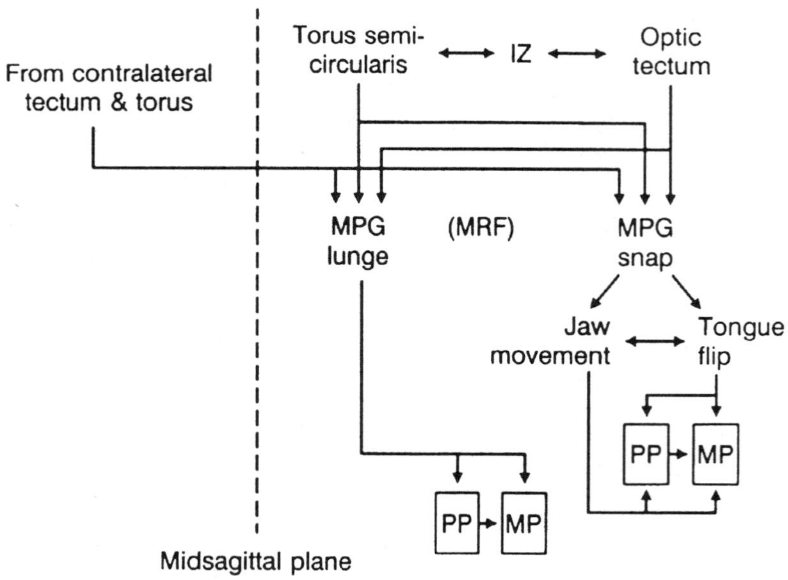

Figure 9. Proposed functional organization of the motor pattern generator for snapping in anurans. The arrowheads indicate the direction of information flow. IZ, intermediate zone wedged between the ventrolateral tectum and torus semicircularis; MP, motoneuron pool; PP, premotoneuron pool.

organization among the respective MPGs. As already mentioned, internal feedback loops within local circuits of the MPG may serve to coordinate the precise timing of tongue and jaw. Since hypoglossal transected toads do not open their jaws, it is probable that activation of the tongue muscle motoneurons is tightly synchronized with activation of jaw muscle motoneurons. Furthermore, it seems likely that (bilateral) co-activation of homonymous motoneurons as well as short delays in sequential actions of effectors are responsible for spatiotemporal aspects of the snapping pattern (Westerfield & Frank 1982; Brenowitz et al. 1983). In this context, variations in dendritic architecture and topography of synapses of the different motoneurons may be important for the generation of the precise spatiotemporal sequence of motoneuron activation and inhibition. Since neuroanatomical as well as physiological studies have failed to reveal axon collaterals in hypoglossal motoneurons (Cajal 1909; Porter 1965; Satou et al. 1985), this would favor the suggestion that central programs which are addressed to these motoneurons do not depend on feedback loops from motoneurons to interneurons.

6.6 Complexity and Variability of Snapping

First of all we recall that snapping can be elicited by stimuli form different sensory modalities (Kicliter 1973; Shinn & Dole 1978, 1979; Comer & Grobstein 1981a,b). The ability of atectal (sightless) frogs to respond to tactile stimuli, and of those with toral damage to respond to visual stimuli, suggests that somatosensory and visual systems at mesencephalic levels have separate access to the motor programs governing prey acquistion (Comer & Grobstein 1981a,b). Shinn & Dole (1978) demonstrated that frogs with bilateral enucleations will migrate toward a discrete source of mealworm odor and sometimes even respond with snapping. Information from the olfactory bulb cannot reach the hindbrain directly and has to cross at least two synapses before reaching mesen- or rhombencephalic levels (Northcutt & Kicliter 1980). Whereas the innate basis of visual releasers of anuran prey-catching is well established (see Ewert, this volume), it is yet to be demonstrated, though likely, that tactile releasers too are innately determined. However, the olfactory releasers of prey-capture have to be learnt by toads (Dole et al. 1981). Facilitation of visual prey-catching in the presence of known prey odor has been reported by Ewert (1968).

Secondly, some comments on the stereotypy of snapping are in order. Although the degree of stereotypy is suspected to be rather high, it is not clear whether variations in the snapping distance, estimated at 1.7-3.0 cm for the toad, are surmounted by changing the amplitude of the lunge, the degree of extension of the tongue, or both. It is also uncertain whether anurans can flip their tongue sideways. More precise behavioral and electromyographic data would certainly help to answer these questions. Experimental lesions of the motor apparatus and variations of the imposed load would aid in revealing the degree of plasticity (short- and long-term), if any, of the snapping response and this in turn would reflect its level of stereotypy.

Jaw and tongue grasping are examples for two different patterns of the consummatory act. Within the stereotyped snapping response there is certainly allowance for elementary bulbar reflexes, e.g., in the case of a toad trying to eat an earthworm: After the initial snap which brings a part of the earthworm into the toad's mouth, subsequent attempts to ingest the rest of the earthworm - supplied by alternative use of the forelimbs - are guided probably by sensory feedback from the trigeminal-facial-glossopharyngeal-vagal afferent systems. In this context the integration of snapping into the rest of the prey-catching motor pattern is of interest, too. Swallowing (or gulping) follows the strike with a relatively constant latency (Ewert 1967). Presumably, striking and swallowing are centrally coordinated in such a way that they are separated by a predetermined constant latency or, alternatively, swallowing is triggered by the presence of prey in the mouth. It is difficult to decide between these possibilities, since contact between the tip of the returning tongue and the pharyngeal surface, following an unsuccessful strike, might be an adequate stimulus to initiate swallowing.

A third important point concerns the "state of the organism" affected by level of satiation, season, time of day, presence of predators, etc., and defined by parameters operating at different levels (Hobson & Scheibel 1980). Drawing on the available anuran neuroanatomical data, Ewert (this volume) has suggested pathways through which seasonal changes and levels of satiation may influence sensory analyzers directly. Given the presence of both catechol- and indoleaminergic cells in the anuran rhombencephalon (Parent 1973) and the postulated role of monoaminergic cells in behavioral state control (Bloom 1979), it is also possible that the MPG is a recipient of the outputs of "level setting interneurons." Thus, the greatly reduced level of prey-capture observed in toads during winter might be related to changes in the sensory analyzers, the MPG, and even the motoneurons (direct serotonergic influences on anuran spinal motoneurons have been described by Soller 1977; Cardona & Rudomin 1983). An example of such macro-network influences may be the raphe nucleus located in the bulbar median reticular formation harboring serotonin containing cells projecting to the intermediate tectal layers (Braak 1970; Parent 1973) and to the hypoglossal nucleus (Weerasuriya & Ewert 1988). Catecholamines have been identified in pear-shaped tectal neurons of layer 6 (Bieger & Neuman 1984); piriform cells of this layer project to the posterior thalamic neuropil (Ewert et al. 1984b). Interactions of the hypothalamus and the reticular formation with the motor systems of prey-capture - taking into account visceral sensory information - are mediated by the nucleus of the solitary tract. Székely et al. (1983) from electron microscopic studies in frogs imply the presence of various neurotransmitters in axon terminals of the latter nucleus (catecholamines, somatostatin, and enkephalin). Certain motor functions appear to involve dopaminergic transmission, since application of apomorphine (a dopamine agonist) causes a "sort of disinhibition" of snapping, while simultaneously inhibiting prey-orienting (Gladen 1984; Quick 1984). In contrast to the "disinhibition phenomenon" caused by pretectal lesions, this syndrome is not paralleled by changes in prey-selective properties of tectal neurons.

6.7 Are Interneurons of the MPG "Monopolized" for Snapping?

As discussed earlier (Weerasuriya 1983), the interneurons of the MPG must not function exclusively for the generation of snapping, but may also take part in other motor synergies. For example, it is conceivable that some components of the MPG are in the reticular formation immediately adjacent to cranial motor nerve nuclei, and in turn are accessed by another component receiving direct input from the sensory analyzers. The former components may be utilized in other bulbar motor synergies such as the rejection of an unpalatable object from the mouth. Furthermore, the identity and the characterization of a MPG rests not only on certain interneurons and their interconnections but also on the patterns of impulse traffic in this network during the generation of the sequence of motoneuronal activation and inhibition that leads to a snap. In addition, this MPG is triggered by a specific spatiotemporal pattern of impulses reaching it along pre-determined channels - the command elements of a command releasing system (see also contributions by Ewert and Schürg-Pfeiffer, this volume). A heuristic scheme for the MPG for snapping as discussed above is presented in Figure 9.

Acknowledgements. The helpful discussions and gracious hospitality of Prof. Dr. J.-P. Ewert and his co-workers, and text processing by Mrs. Ursula Reichert are gratefully acknowledged. The research conducted at the Department of Neuroethology of the University of Kassel was supported by the Deutscher Akademischer Austauschdienst (DAAD). The author was a Visiting Professor at the University of Kassel during the preparation of this manuscript.

7 References

Arbib MA (1988) Levels of modelling of mechanisms of visually guided behavior. *Behav Brain Sci* 10 (in press)

Bieger D, Neuman RS (1984) Selective accumulation of hydroxytroptamines by frog tectal neurons. *Neurosci* 12: 1167-1177

Bloom FE (1979) Chemical integrative processes in the central nervous system. In: Schmidt FO, Worden FG (eds) *The neurosciences: fourth study program.* MIT Press, Cambridge Mass

Braak H (1970) Biogene Amine im Gehirn vom Frosch (*Rana esculenta*). *Z Zellforsch* 106: 269-308

Brenowitz GL, Collins WF Jr, Erulkar SD (1983) Dye and electrical coupling between frog motoneurons. *Brain Res* 274: 371-375

Brookhart JM, Fadiga E (1960) Potential fields initiated during monosynaptic activation of frog motoneurones. *J Physiol (Lond)* 150: 633-655

Burghagen H (1979) *Der Einfluß von figuralen visuellen Mustern auf das Beutefangverhalten verschiedener Anuren.* Ph.D. Thesis, Univ Kassel

Cajal SR y (1909) *Histologie du systeme nerveux de l'homme et des vertebres.* Maloine, Paris

Cardona A, Rudomin P (1983) Activation of brain stem serotonergic pathways decreases homosynaptic depression of monosynaptic responses of frog spinal motoneurons. *Brain Res* 280: 373-378

Comer C, Grobstein P (1981a) Tactually elicited prey acquisition behavior in the frog, *Rana pipiens*, and a comparison with visually elicited behavior. *J Comp Physiol* 142: 141-150

Comer C, Grobstein P (1981b) Involvement of midbrain structure in tactually and visually elicited prey acquisition behavior in the frog, *Rana pipiens. J Comp Physiol* 142: 151-160

DeOlmos J, Heimer L (1977) Mapping of collateral projections with the HRP method. *Neurosci Lett* 6: 107-114

Dole JW, Rose BB, Tachiki KH (1981) Western toads (*Bufo boreas*) learn odor of prey insects. *Herpetologica* 37: 63-68

Doty RW, Richmond WH, Storey AT (1967) Effect of medulla lesions on coordination of deglutition. *Exp Neurol* 17: 91-106

Ebbesson SOE, Hansel M, Scheich H (1981) An 'on the slide' modification of the DeOlmos-Heimer HRP method. *Neurosci Lett* 22: 1-4

Eikmanns K-H (1955) Verhaltensphysiologische Untersuchungen über den Beutefang und das Bewegungssehen der Erdkröte (*Bufo bufo* L). *Z Tierpsychol* 12: 229-253

Ewert J-P (1967) Aktivierung der Verhaltensfolge beim Beutefang der Erdkröte (*Bufo bufo* L) durch elektrische Mittelhirnreizung. *Z Vergl Physiol* 54: 455-481

Ewert J-P (1968) Der Einfluß von Zwischenhirndefekten auf die Visuomotorik im Beute- und Fluchtverhalten der Erdkröte (*Bufo bufo*). *Z Vergl Physiol* 61: 41-70

Ewert J-P (1984) Tectal mechanisms that underlie prey-catching and avoidance behaviors in toads. In: Vanegas H (ed) *Comparative neurology of the optic tectum.* Plenum Press, New York, pp 247-416

Ewert J-P (1987) Neuroethology of releasing mechanisms: prey-catching in toads. *Behav Brain Sci* 10: 337-405

Ewert J-P, Schürg-Pfeiffer E, Weerasuriya A (1984a) Neurophysiological data regarding motor pattern generation in the medulla oblongata of toads. *Naturwissenschaften* 71: 590-591

Ewert J-P, Finkenstädt T, Weerasuriya A (1984b) Concepts for neuronal correlates of Gestalt perception: visual prey recognition in toads. In: Aoki K, Ishii S, Morita H (eds) *Animal behavior, neurophysiological and ethological approaches.* Japan Scientific Soc Press, Tokyo, pp 137-159

Feldman JL (1986) Neurophysiology of respiration in mammals. In: Bloom FE (ed) *Handbook of physiology. The nervous system. Intrinsic regulatory systems of the brain.* American Physiological Society, Bethesda pp 463-524

Finkenstädt T, Adler NT, Allen TO, Ebbesson SOE, Ewert J-P (1985) Mapping of brain activity in mesencephalic and diencephalic structures of toads during presentation of visual key stimuli: a computer assisted analysis of ^{14}C-2DG autoradiographs. *J Comp Physiol* 156: 433-445

Fuchs AF, Kaneko CRS (1985) A brain stem generator for saccadic eye movements. In: Evarts EV, Wise SP, Bousfield D (eds) *The motor system in neurobiology.* Elsevier Biomedical Press, Amsterdam New York Oxford, pp 126-132

Gans C (1961) The bullfrog and its prey. *Nat Hist* 70: 26-37

Gans C, Gorniak GC (1982a) Functional morphology of lingual protrusion in marine toads (*Bufo marinus*). *Amer J Anat* 163: 195-222

Gans C, Gorniak GC (1982b) How does the toad flip its tongue? Test of two hypotheses. *Science* 216: 1335-1337

Gaupp E (1899) *A Ecker's und R Wiedersheim's Anatomie des Frosches.* Vieweg & Sohn, Braunschweig

Gladen S (1984) *Der Einfluß der Apomorphindosis auf das Beutefangverhalten und auf die reizspezifische Gewöhnung bei Erdkröten.* Staatsexam Thesis, Univ Kassel

Grillner S (1981) Control of locomotion in bipeds, tetrapods, and fish. In: Brooks VB (ed) *Handbook of physiology. The nervous system. Motor control.* American Physiological Society, Bethesda, pp 1179-1236

Grobstein P, Comer C, Kostyk SK (1983) Frog prey capture: between sensory maps and directed motor output. In: Ewert J-P, Capranica RR, Ingle DJ (eds) *Advances in vertebrate neuroethology.* Plenum Press, New York, pp 331-347

Hinsche G (1935) Ein Schnappreflex nach "Nichts" bei Anuren. *Zool Anz* 111: 113-122

Hobson JA, Scheibel AB (eds) (1980) *The brain stem core: sensorimotor integration and behavioral state control. Neuroscience research program bulletin Vol 18.* MIT Press, Cambridge Mass

Ingle DJ (1983) Brain mechanisms of visual localization by frogs and toads. In: Ewert J-P, Capranica RR, Ingle DJ (eds) *Advances in vertebrate neuroethology.* Plenum Press, New York, pp 177-226

Ito M (1986) Neural systems controlling movement. *TINS* 9: 515-518

Jean A (1984) Brainstem organization of the swallowing network. *Brain Behav Evol* 25: 109-116

Kaneko CRS, Ewinger C, Fuchs AF (1981) Role of cat pontine burst neurons in generating of saccadic eye movements. *J Neurophysiol* 46: 387-408

Kicliter E (1973) Flux wavelength and movement discrimination in frogs: forebrain and midbrain contributions. *Brain Behav Evol* 8: 340-365

Kostyk SK, Grobstein P (1987) Neuronal organization underlying visually elicited prey orienting in the frog. II: Anatomical studies on the laterality of central projections. *Neurosci* 21: 57-82

Kramer EB, Rath T, Lischka MF (1979) Somatotopic organization of the hypoglossal nucleus: HRP study in the rat. *Brain Res* 170: 533-537

Kupfermann I, Weiss KR (1978) The command neuron concept. *Behav Brain Sci* 1: 3-39

Landgren S, Olsson KA, Westberg KG (1986) Bulbar neurones with axonal projections to the trigeminal motor nucleus in the cat. *Exp Brain Res* 65: 98-111

Lázár G (1969) Efferent pathways of the optic tectum in the frog. *Acta Biol Acad Sci Hung* 20: 171-183

Lázár G, Toth P, Csank G, Kicliter E (1983) Morphology and location of tectal projection neurons in frogs: a study with HRP and cobalt-filling. *J Comp Neurol* 215: 108-120

Maeda M. Magherini PC, Precht W (1977) Functional organization of vestibular and visual inputs to neck and forelimb motoneurons in the frog. *J Neurophysiol* 40: 225-243

Matesz C, Székely G (1977) The dorsomedial nuclear group of cranial nerves in the frog. *Acta Biol Acad Sci Hung* 28: 461-474

Matsushima T, Satou M, Ueda K (1986) Glossopharyngeal and tectal influences on tongue muscle motoneurons in the Japanese toad. *Brain Res* 265: 198-203

Nieuwenhuys R, Opdam P (1976) Structure of the brain stem. In: Llinás R, Precht W (eds) *Frog neurobiology.* Springer-Verlag, Berlin Heidelberg New York, pp 811-855

Northcutt RG, Kicliter E (1980) Organization of the amphibian telencephalon. In: Ebbesson SOE (ed) *Comparative neurology of the telencephalon.* Plenum Press, New York, pp 203-255

Nozaki S, Enomoto S, Nakamura Y (1983) Identification and input-output properties of bulbar reticular neurons involved in the cerebral cortical control of trigeminal motoneurons in the cat. *Exp Brain Res* 49: 363-372

Olsson KA, Landgren S, Westberg KG (1986) Location of and peripheral convergence on the interneurone in the disynaptic path from the coronal gyrus of the cerebral cortex to the trigeminal motoneurones in the cat. *Exp Brain Res* 65: 83-97

Opdam P, Kemali M, Nieuwenhuys R (1976) Topological analysis of the brain stem of the frogs *Rana esculenta* and *Rana catesbeiana. J Comp Neurol* 165: 307-331

Parent A (1973) Distribution of monamine-containing neurons in the brain stem of the frog, *Rana temporaria. J Morph* 139: 67-78

Porter R (1965) Synaptic potentials in hypoglossal motoneurones. *J Physiol (London)* 180: 209-244

Quick I (1984) *Auswirkungen von Apomorphin auf die visuelle Musterdiskriminationsfähigkeit und die Beutefangmotivation bei Erdkröten.* Staatsexam Thesis, Univ Kassel

Robinson DA (1981) Control of eye movements. In: Books VB (ed) *Handbook of Physiology. The nervous systems. Motor control.* American Physiological Society, Bethesda, pp 1275-1320

Rubinson K (1968) Projections of the tectum opticum of the frog. *Brain Behav Evol* 1: 529-561

Satou M, Ewert J-P (1985) The antidromic activation of tectal neurons by electrical stimuli applied to the caudal medulla oblongata in the toad *Bufo bufo* (L). *J Comp Physiol* 157: 739-748

Satou M, Matsushima T, Takeuchi H, Ueda K (1985) Tongue-muscle controlling motoneurons in the Japanes toad: topography, morphology, and neuronal pathways from the 'snapping evoking area' in the optic tectum. *J Comp Physiol* 157: 717-737

Schmidt RS (1976) Neural correlates of frog calling. Isolated brainstem. *J Comp Physiol* 108: 99-114

Schmidt RS (1984) Neural correlates of frog calling: preoptic area trigger of 'mating calling'. *J Comp Physiol* 154: 847-853

Schneider D (1954) Das Gesichtsfeld und der Fixiervorgang bei einheimischen Anuren. *Z Vergl Physiol* 36: 147-164

Shinn EA, Dole JW (1978) Evidence for a role for olfactory cues in the feeding response of leopard frogs, *Rana pipiens. Herpetologica* 34: 167-172

Soller RW (1977) Monoaminergic inputs to frog motoneurons: an anatomical study using fluorescence histochemical and silver degeneration techniques. *Brain Res* 122: 445-458

Stuesse SL, Cruce WLR, Powell KS (1983) Afferent and efferent components of the hypoglossal nerve in the grass frog. *J Comp Neurol* 217: 432-439

Sumi T (1969) Synaptic potentials of hypoglossal motoneurons and their relation to reflex deglutition. *Jap J Physiol* 19: 68-79

Székely G, Lázár G (1976) Cellular and synaptic architecture of the optic tectum. In: Llinás R, Precht W (eds) *Frog neurobiology*. Springer-Verlag, Berlin Heidelberg New York, pp 407-434

Székely G, Levai G, Matesz K (1983) Primary afferent terminals in the nucleus of the solitary tract of the frog: an electron microsopic study. Exp Brain Res 53: 109-117

Uemura-Sumi M, Mizuno N, Iwahori N, Tackeuchi Y, Matsushima R (1981) Topographical representation of the hypoglossal nerve branches and tongue muscles in the hypoglossal nucleus of macaque monkeys. *Neurosci Lett* 22: 31-35

Weerasuriya A (1983) Snapping in toads: some aspects of sensorimotor interfacing and motor pattern generation. In: Ewert J-P, Capranica RR, Ingle DJ (eds) *Advances in vertebrate neuroethology.* Plenum Press, New York, pp 613-627

Weerasuriya A (1988) Effect of medullary lesions and hypoglossal nerve transections on snapping in toads (in preparation)

Weerasuriya A, Ewert J-P (1981) Prey-selective neurons in the toads optic tectum and sensori-motor interfacing: HRP studies and recording experiments. *J Comp Physiol* 144: 429-434

Weerasuriya A, Ewert J-P (1984) Afferents of the hypoglossal nucleus in the common European toad, *Bufo bufo. Amer Ass Anat Abstr* 1984: 192A

Weerasuriya A, Ewert J-P (1988) Afferents of the hypoglossal nucleus in the toad, *Bufo bufo* (submitted)

Westerfield M, Frank E (1982) Specificity of electrical coupling among neurons innervating forelimb muscles of the adult bullfrog. *J Neurophysiol* 48: 904-913

Wiping Reflex in the Frog: Movement Patterns, Receptive Fields, and Blends

M.B. BERKINBLIT, A.G. FELDMAN, and O.I. FUKSON

Institute for Information Transmission Problems
Academy of Sciences and Moscow State University
Moscow 101447, USSR

Abstract. Chemical stimulation of a site on the skin of the back anterior to the iliosacral joint in high spinal frogs usually elicits a rostral wiping reflex, in which the tip of the hindlimb rubs against the site; stimulation of a site near the anus elicits a caudal wiping reflex in which the heel or side of the foot rubs against the site. The spinal frog also generates (i) a sequence of rostral-caudal wiping reflexes in response to stimulation of a site in the transition zone located between the rostral wiping receptive field (RF) and the caudal RF and (ii) a lateral wiping reflex to stimulation of a site on a side of the body. Rostral, caudal ipsilateral, and caudal contralateral wiping forms can be executed in a variable sequence to stimulation of a site in the transition zone. The map of the RFs of various wiping forms depends on the level of spinal cord transection. At a low level of transection the rostral RF can disappear, whereas the transition zone between the two RFs becomes a pure RF for a caudal wiping form. In rare cases, both intact and high spinal frogs generate an ineffective *hybrid* wiping reflex to stimulation of a site in the transition zone: the hindlimb then takes a conflict configuration in which the distal joints behave as in a rostral wiping, whereas the proximal joints do as in a caudal wiping. This conflict is solved by the spinal cord: after a time delay, it re-organizes the hindlimb and performs a pure wiping reflex. Thus, the spinal cord has the adaptive ability to correct occasional errors without descending neural inputs from supraspinal structures. Various aspects of the sensorimotor parallel-hierarchical organization of the wiping reflex are discussed.

1 Introduction

The present paper addresses the problem of sensorimotor coordination or, more specifically, the question of how information about the site of a stimulus on the body surface is transformed into motor patterns associated with the site. There are features bringing together this kind of motor task with that of visuomotor coordination. For example, the "*visual grasping reflex*" (Akert 1949) such as orienting toward prey in amphibians (Ewert 1987; see also Arbib 1972) has much in common with the *wiping reflex* generated in response to stimulation of the skin of the body: both ballistic movements are goal-directed and adaptively adjusted as a function of the position of the stimulus. In both motor tasks, an intrinsic sensory neural map is likely used: in the visuomotor task, a *retinotopic* map representing the animal's external environment in the activity of optic tectal neurons and, in the somatosensorimotor task, a *somatotopic* map representing the animal's body surface in the activity of dorsal horn neurons. Both behaviors are complex although elicited by relatively simple sign stimuli (releasers) and make, as a rule, the use of innate releasing mechanisms (see also contributions by Arbib, Ewert, Grobstein, and Schürg-Pfeiffer, this volume). Although there are obvious distinctions between the two behavioral patterns, their similarities justify the consideration of the wiping reflex in the present volume that chiefly deals with visuomotor coordination problems.

A spinal frog with a complete transection of the spinal cord at the level of the obex generates a site-specific wiping reflex of the ipsilateral hindlimb in response to tactile or chemical stimulation of the skin of the back (Baglioni 1913; Stein 1983; Berkinblit et al. 1986a). The spinal frog has the ability to select various inter-joint coordinations (synergies) and trajectories of movement to a given site (Berkinblit et al. 1986a). In addition, the frog can use each of several movement strategies or forms to remove a stimulus from a site of the skin. In particular, as is shown in the present study, the spinal frog generates a rostral wiping form (R form) in response to stimulation of a site of the skin of the back anterior to the iliosacral joint. The frog removes the stimulus with the tips of the digits and throws it away from the body. The spinal frog also generates a caudal wiping form (C form) to stimulation of a site on the skin near the anus. In this case, the heel or side of the foot of the ipsilateral or contralateral hindlimb wipes the site. The reflex forms were observed in a fictive wiping (Simpson 1976).

Mortin et al (1985), Robertson et al. (1985), and Stein et al. (1986) have observed various forms of the scratch reflex in the spinal turtle: a pure rostral and a pure caudal form and their blends. The blends could be performed in response to stimulation of a site in a transition zone between two receptive fields. A blend can be either a "switch" or a "hybrid." In a switch, the hindlimb produces a rostral scratch in one cycle and a caudal scratch in the following cycle. In a hybrid, the hindlimb rubs twice against the site in a continuous movement sequence. These findings show that the spinal turtle has the ability to use various movement strategies and their blends to perform a given motor task. In the present study, we analyzed similar properties of sensorimotor coordination during the performance of the wiping reflex in spinal frogs.

2 Methodological Comments

Grass frogs, *Rana temporaria*, were spinalized under ether anesthesia by a complete transection at the level of the obex (high spinal preparations) or at various levels between *3rd-7th* spinal segments (low spinal preparations; Fig. 1). The level of spinal cord transection was verified visually through a laminotomy made after the end of experiments. Each of the fifty investigated frogs was tested to determine if it could

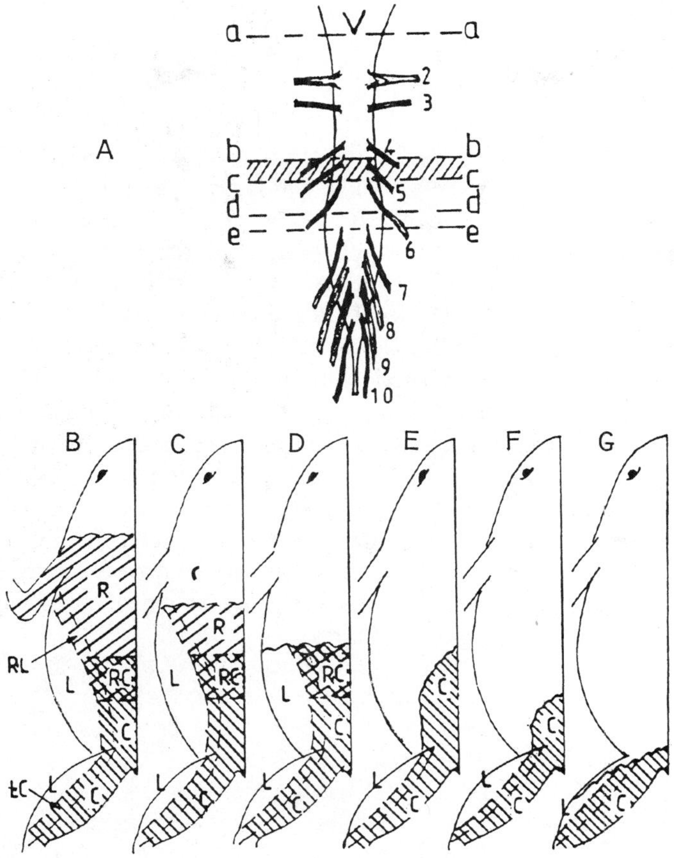

Figure 1. Levels of spinal cord transection (A) and receptive fields, RFs, of wiping reflex forms (B to G). A: aa is a transection level for high spinal preparations with RFs as in B; transections between aa and bb yield preparations with RFs as in C; transections between bb and cc (hatched strip) lead to preparations with a partical (D) or a complete (E) loss of the RF for a rostral *R* wiping form; transections at dd or ee give rise to a further decrease of RFs (F,G). B: receptive fields for a rostral *R*, caudal *C*, and lateral *L* wiping forms and transition receptive zones (*R-C*, *R-L*, and *L-C*).

produce site-specific wiping reflexes in response to chemical stimulation (a small filter paper of 4-9 mm^2 soaked in a weak solution of 2-5% sulphuric acid) applied to a site on the dorsolateral surface of the body (the back, sides, femurs, and the area near the anus). A minimal strength (2% acid) was firstly tested. If it was ineffective, we used a higher strength of stimulation.

The movements were filmed (48 frames/s) with a 16 mm camera situated at a distance of 0.5 m from the frog supported against the background by a measuring grid. We determined the coordinates of the sites of stimulation and the time course of joint angles in the hip, knee, ankle, foot, and digits.

3 The Wiping Reflex

A high spinal frog generates a sequence of motor patterns in response to stimulation of a site of the back. The frog initially contracts muscles under the site and pulls locally up the skin for about 1 s. In an intact frog this local reaction ("skin vibration") is combined with locomotion, i.e., jumping away. Under natural conditions, the two reactions can be sufficient if it is necessary, for example, to drive an insect off the skin, but they are ineffective in the case of chemical stimulation used in our experiments, so that the frog applies an additional strategy: it wipes the stimulus with a hindlimb using one or another form of the wiping reflex. The latency of a wiping reflex after the end of the local skin vibration depends on the strength of stimulation (typically, about 1 s for 4-5% acid and about 5 s for 2% acid). In addition, stimulation at a low intensity can be subliminal for eliciting a wiping reflex. However, two subthreshold stimuli applied simultaneously to two distant sites could elicit a wiping reflex directed to one of the sites in one cycle and then to the other site in the following cycle.

Thus, the high spinal frog can respond by various movement strategies in response to stimulation of a site on the skin. The large latencies of motor responses allow us to assume the existence of a neuronal structure that performs a spatiotemporal integration of sensory signals from the skin receptors and accumulates its own activity before eliciting a motor response.

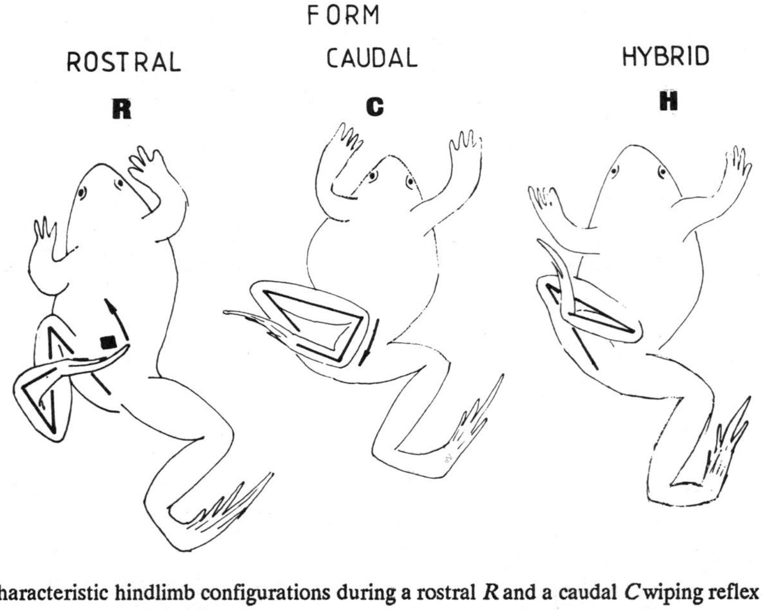

Figure 2. Characteristic hindlimb configurations during a rostral *R* and a caudal *C* wiping reflex generated in response to stimulation of a site on the *R-C* transition zone in a high spinal frog. The configurations are set just before the onset of removing the stimulus (filled square) in the direction shown by arrows. In *H*, an ineffective hybrid hindlimb configuration is shown.

4 Wiping Forms and Their Receptive Fields

A sketch of wiping receptive fields (RFs) in a high spinal preparation is shown in Fig. 1B. As a rule, a pure rostral wiping (R form) can be elicited in response to stimulation of a site on a widespread surface of the back anterior to the level of the iliosacral joint and posterior to the shoulders. The stimulus is removed with the tips of the digits of the ipsilateral hindlimb.

Stimulation of a site on rostral and caudal surfaces of the sides of the body elicits a site-specific wiping reflex which is similar to R form wiping, however, modified in such a way that the frog does not place the foot of the hindlimb above the femur as in R form wiping, but lifts the femur and moves the foot under it and then wipes the stimulated site. We utilize the term L form to movement patterns specific for the lateral surface of the body. The RF of the L form wiping extends to lateral surfaces of the femurs (Fig. 1B).

Stimulation of a site near the anus or medial surfaces of the femurs elicits a pure caudal form of the wiping reflex (C form). Its RF occupies a part of the skin of the back. The movements are usually performed with the two hindlimbs, either simultaneously or successively, or one hindlimb can repeat a wiping reflex in isolation (see below).

There is a transition zone posterior to the iliosacral joint, in which stimulation of a site elicits either an R wiping form or a C form, or a sequence of the two forms (Figs. 1B and 2). We call the transition zone the R-C zone. There are additional transition zones for the L, C, and R forms which were not analyzed here in detail (Fig. 1B). Figure 3 depicts in normalized coordinates, how often an R or a C wiping form can be elicited in response to stimulation of a site in the R-C transition zone. Data were obtained from eight high spinal frogs about (50 x 8 trials) having close maps of RFs. The results show that the probability to observe an R form in response to stimulation of a site in the R-C transition zone gradually increases when the stimulus is shifted to a rostral position, whereas the probability to observe a C form decreases correspondingly. Thus, for the transition zone there is a rostrocaudal gradient of the probability to observe one or the other form.

A spinal frog can sometimes generate, by mistake, an ineffective C wiping form in response to stimulation of a site on the R zone, far from the R-C transition zone, and an ineffective R form in response to stimulation of a site of the C zone. These errors show that, strictly speaking, there is no pure RF for a given wiping form: the probability to observe a given wiping form for some RFs can be close to 1 but, in fact, it is always slightly less than 1.

A high spinal frog usually performs, especially in response to strong stimulation of a site, a sequence of 3-5 wiping reflexes. If the site is on the R-C transition zone, the frog can randomly switch one wiping form to the other in successive cycles. A probability matrix for such switches for the R-C zon is shown in Fig. 4. The symbols R, C, and cC in a vertical column indicate wiping forms performed in a cycle and those in a horizontal row indicate wiping forms performed in the following cycle. Caudal ipsilateral and caudal contralateral wiping forms are marked by the symbols C and cC, correspondingly. The figures in squares show how often (in %) a given

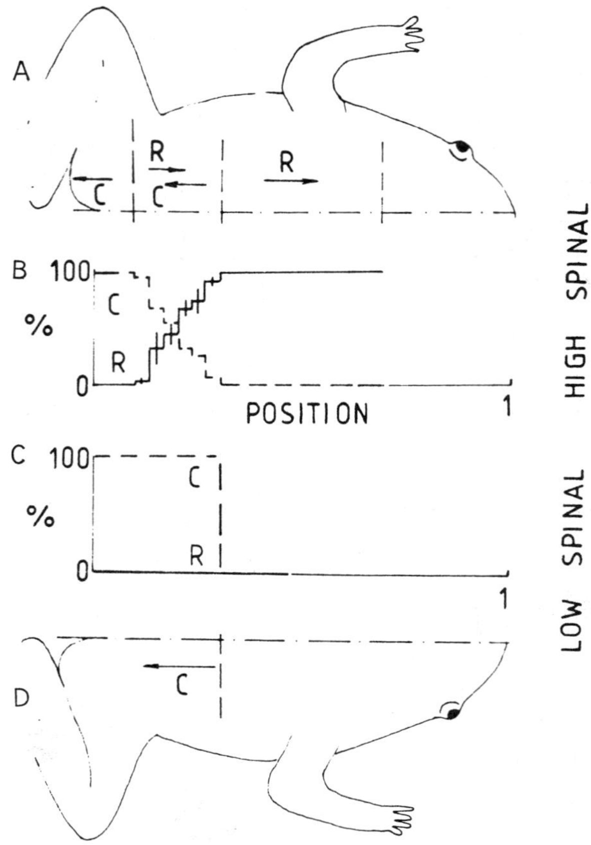

Figure 3. The *R-C* transition receptive field, RF, for rostral *R* and caudal *C* wiping forms in high spinal frogs and its transformation in a pure RF for a *C* wiping form (C,D), as the result of a spinal cord transection inside the hatched strip shown in Fig. 1A. Arrows in (A) show the direction of removing the stimulus for corresponding wiping reflex forms and vertical dashed lines indicate the borders of receptive zones. (B,C) The percent of *R* and *C* wiping reflexes observed as a function of the position of the site of skin stimulation; vertical bars for *R* indicate standard deviations for the number of *R* forms.

sequence of two wiping forms occurs. A given wiping form can be repeated in the following cycle or replaced by another form. Such random sequences of wiping forms can sometimes cooperate in the execution of the motor task. For example, a spinal frog using C form wiping can shift the stimulus to a rostral site of the skin and thus facilitates removing the stimulus in the following cycle by R form wiping.

Figure 1C-G shows maps of the RFs of wiping reflex forms in preparations with spinal cord transections at various low levels. The maps remain, in the main, the same (just the anterior border of the RF of the R wiping form shifts in a posterior direction), if the level of transection is progressively replaced from the obex to an anterior edge of the entry zone of the dorsal roots of *5th* segment (Fig. 1C,D).

There is a crucial level of spinalization between dorsal roots of segments 5 and 6, after which either an essential part of R form's RF or both R and L form's RFs disappear, and, as a consequence, the transition zone becomes a pure RF of C form wiping (Fig. 1E). Following spinalization posterior to the entry zone of the dorsal roots of segment 6, the caudal RF gets narrow and finally only a small zone near the anus and medial surfaces of the femur remain effective for eliciting C form wiping (Fig. 1F,G).

PRECEDENT FORM → SUBSEQUENT	R	C	cC	N
R	44	48	8	25
C	57	36	7	28
cC	12.5	50	37.5	8

Figure 4. Matrix representation of sequences of rostral *R*, ipsilateral caudal *C* and contralateral caudal *cC* wiping reflex forms generated in response to stimulation of a site near the middle of the *R-C* transition zone in 6 high spinal frogs. The figures in squares show how often, in % of N, one wiping form follows another; N is the number of coupled reflex forms of which the first is fixed.

5 Hindlimb Movement Patterns

The kinematics of R form wiping has been described in detail (Berkinblit et al. 1986a). Briefly, there are 5 phases of the reflex: flexion, placement, aiming, whisking, and extention. All the phases are well expressed in high spinal preparations. One or two phases (e.g., flexion and extension) can be missed, which is regularly observed in intact frogs. In the flexion phase, the frog executes a deep flexion in all joints, the digits of the foot are clenched and then the frog raises its foot above the femur. In the following, placing phase, the digits strighten and are placed on the dorsal surface of the back, behind the stimulus, whereas the knee joint is extended to a position value near 50 deg and the ankle joint is maximally flexed to a value near of 30 deg. In the aiming phase, knee and tarsometatarsal extensor movements are performed and the tip of the foot moves more to the stimulus. In the wisking phase, the frog removes the stimulus by a rapid extensor movement in the tarsometatarsal joint and digits, then costs it away from the body by means of a ballistic extensor movement of the whole hindlimb (extensor phase).

In C form wiping, there is a flexor phase of the movement of the whole limb, but after that a knee movement to a hyperflexor position near -20 to -30 deg follows, such that the side of the heel moves in a rostral direction along the skin near the anus and back surface. In this phase, the heel can hit and shift the stimulus to a more rostral site. After a small delay (20-30 ms), ankle and tarsometatarsal extensor movements follow and almost the same angle of about 80 deg is established in the two joints. This time a hyperflexor knee position remains, such that the heel of the limb is situated above a caudal surface of the body. Then a rapid knee extensor movement follows and, as a result, the heel wipes the caudal surface of the body and removes the stimulus. There is no stereotyped pattern of hip joint movements during the execution of a C form. In particular, the hip can be motionless all the time of C form wiping. Thus, not only are the limb sites with which the frog removes the stimulus different for the two wiping forms, but so are the inter-joint coordinations inherent in them. Figure 5 shows the most characteristic difference between the two forms in the aiming phase: in R form wiping, a knee extensor movement and an ankle flexor movement

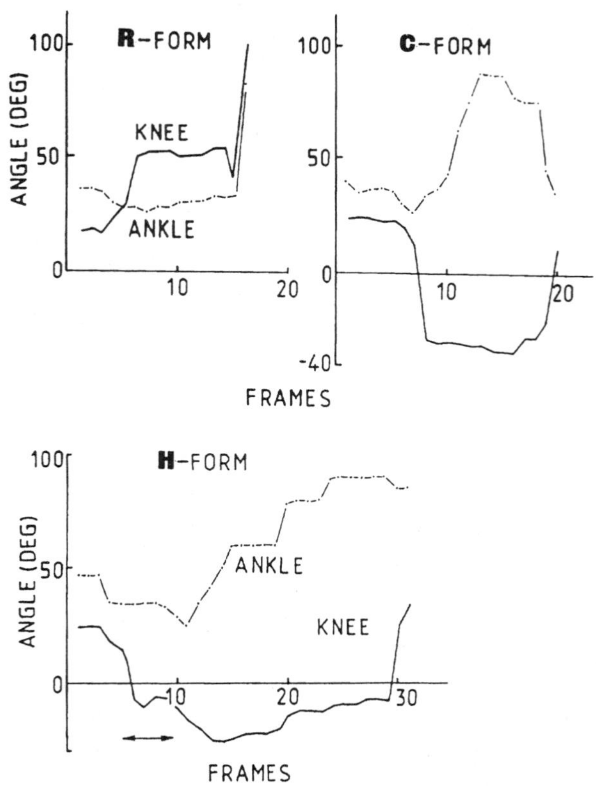

Figure 5. Examples of knee and ankle movement patterns in a rostral *R*, a caudal *C*, and a hybrid *H* wiping reflex in a high spinal frog. Arrows in *H* show the time interval when a hybrid limb configuration was observed. This hybrid was transformed in a pure *C* wiping pattern. Movie speed 48 frames/s.

are simultaneously executed, whereas in a C form a hyperflexor knee movement is associated with an extensor ankle movement. In addition, in R form wiping, the stimulus is usually removed by extensor movements in the ankle and digits, whereas the same goal is achieved by means of a knee extensor movement in a C form.

6 Hybrid Form of Wiping

In rare trials (about 1 of 300) in spinal preparations and intact frogs, a mixture of R and C form wiping movements was observed in response to stimulation of a site in the R-C transition zone (Figs. 2 and 5). In a hybrid, a hyperflexor knee movement to a negative position value as is the case in C form wiping is combined with a lift of the distal part of the limb above the femur as is typical for R form. This presumably creates inappropriate mechanical tensions in the limb, which prevents its further movement. The conflict situation is solved by the spinal frog by one way or another after a time delay: either a knee extensor movement begins and then a pure R form wiping is executed or an ankle extensor movement is performed and then a pure C form follows.

It should be recalled that spinal turtles have the ability to generate a hybrid scratch reflex in response to stimulation of a site in a transition zone between two RFs (Stein et al. 1986); here, the hybrid is effective: the hindlimb rubs twice the stimulated site in a continuous movement sequence. In contrast, the hybrid wiping form in frogs is ineffective. However, its occurrence tells us that the spinal cord has the ability to correct occasional errors and, in addition, to combine segments of different programs to form new movement patterns.

7 Discussion and Conclusions

7.1 Movement Strategies

A motor task can usually be solved in different ways (Bernstein 1967; Mortin et al. 1985; Robertson et al. 1985; Berkinblit et al. 1986a; Stein et al. 1986), and the present data also show that the frog's spinal cord has the ability to use various movement strategies to respond to skin stimulation: the local skin reaction, a number of wiping forms and their blends ("motor schemas" according to Arbib, this volume). In addition, the spinal frog can vary the trajectory of movement to a skin site though the form of wiping reflex can remain the same (Berkinblit et al. 1986a,b). An intact frog can use an additional strategy, locomotion, to respond to skin stimulation. The motor patterns are performed in a certain sequence. In spinal frogs, one or another form to wiping is executed after the local skin reaction. Thus, the spinal frog has the ability to make use of a set of possible solutions ("motor schemas") of the motor task, selected by "coordinated control programs" (see Arbib, this volume).

7.2 Signal Integration

A motor response to stimulation of a site in a spinal frog arises after a long delay even if the strength of stimulation is high. In addition, two stimuli subliminal in isolation can become suprathreshold when applied simultaneously to two sites. These properties of sensory processing can be associated with a neuronal system that integrates signals from skin receptors and accumulates its own activity before eliciting a motor response. This integration is performed in such a way that the specific information about the sites of stimulation is derived from the afferent signals in a pure form. As a consequence, the frog can wipe every stimulated site in isolation and does not wipe an intermediate site. It is interesting to note that an intact frog subjected to visual stimulation by two flies acts less successfully: it jumps to a site between the flies (Ingle 1968; for a discussion of this "average fly phenomenon" see Arbib 1972).

7.3 Specification of Wiping by Command Neurons

The present study has also shown that every site of the skin in high spinal frog is associated with a probability to observe a wiping form in response to stimulation of the site. This probability varies as a function of the position of the site. There are relatively "pure" receptive zones, each one associated chiefly with a wiping form. A wiping form arising in response to stimulation of a site on the transition zone located between the rostral and caudal RFs can be repeated or alternate with the other wiping form in the following cycle (Fig. 4). To discuss this phenomenon, we shall assume first of all that the form of a wiping reflex is specified by corresponding command neurons - in the sense of cooperatively working command elements - as has been suggested for release of other motor behaviors (Davis 1977; Kupfermann & Weiss 1978; Ewert, this volume). For example, a population of R command neurons is associated with R form wiping, and appropriate C command neurons are associated with the C form. In addition, we assume that the map of RFs for various wiping forms is essentially determined by the projections of skin afferents on such command neurons. For example, it is possible that the afferents from rostral areas of the frog's skin give more projections to R command neurons and do less projections to C command neurons. In contrast, the connection strength of the afferents from the caudal skin fields can be more effective for C command neurons than for R command neurons (Fig. 6). The R and C command neurons could receive comparable inputs from the same sites of the R-C transition zone, which could give rise to some kind of instability, so that the probability to observe an R or a C wiping form in response to stimulation of a site on this zone could be comparable.

This hypothesis is partly supported by the data on a change in the maps of the RFs as a function of the level of spinalization (Fig. 1). Actually, it is known that afferents from the skin near the anus and from the hindlimbs give projections to the caudal spinal segments 7-10, whereas afferents from the skin of the back and sides give projections to the more rostral segments (Davis & Constantine-Paton 1983). As a result, spinal cord transections at progressively lower levels could first interrupt the

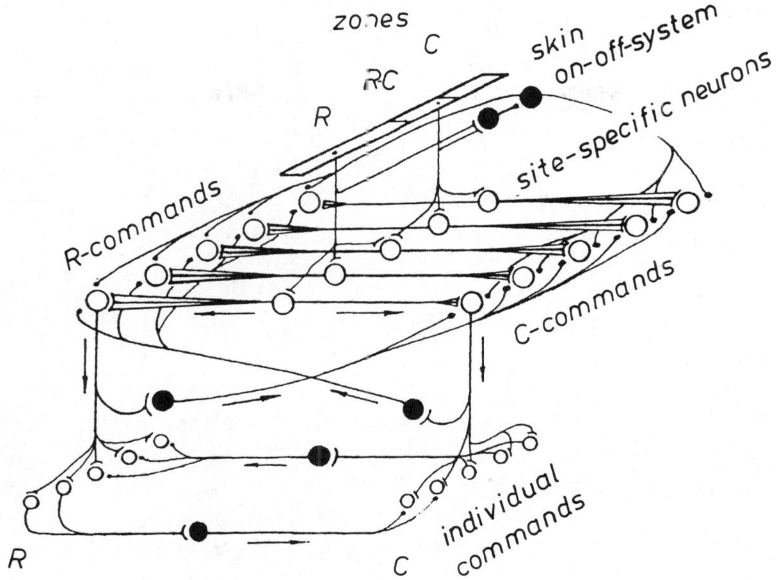

Figure 6. Hypothetical representation of the parallel-hierarchical organization of the wiping reflex. Only a rostrocaudal coordinate of the stimulus position on the skin is considered. The magnitude of axonal terminations of site-specific neurons symbolizes the strength of their connections with commmand neurons. Excitatory and inhibitory neurons are marked by open and filled circles, respectively.

afferent inputs to the R command neurons and thus the R-C transition zone could become a pure RF for the C wiping form as was observed (Fig. 1E). Another explanation of this effect is that a spinal cord transection at a low level could interrupt the connections of the R command neurons with their addresses and exclude, in such a way, the possibility to elicit R form wiping.

7.4 The Problem of Choice

When a site of the R-C transition zone is stimulated, R and C wiping forms are usually performed in isolation, i.e., separately. How does the spinal cord avoid their simultaneous using (although not always as the occurrence of a hybrid form show)? The problem of choice of a motor reaction of a set of possible responses to a given stimulus is a typical behavioral problem in animals (Arbib 1972). For example, a medicinal leech in response to tactile stimulation can perform either a withdrawal reflex or swimming. The two reactions are incompatible: the withdrawal reflex is performed by a rapid simultaneous contraction of all the muscles of the body, whereas swimming is executed by reciprocal activation of dorsal and ventral longitudinal muscles (Friesen & Stent 1978). It has been shown that simultaneous execution of the two movements is excluded due to inhibitory connections between corresponding sets of command neurons. We can assume that an inhibitory interaction between the R and C command neurons in the frogs also excludes their simultaneous activation. This organization may preserve the possibility of using the two wiping forms in a sequence: both afferent inputs from the transition zone to the command neurons and their interaction can essentially change during the performance of a wiping form, so that any wiping form becomes possible in the following cycle.

7.5 Hybrid Wiping

It is necessary to modify the above scheme in order to explain the possibility of hybrid wiping. A hybrid is a complex combination of R and C wiping forms: the distal joints take the configuration as in the case of R form wiping, whereas the proximal joints take the configuration typical for a C form. In other words, the proximal and distal joints participate in the execution of different motor patterns. To explain this phenomenon, we assume that for every joint there are two sets of command neurons, one for R form wiping and the other for C form wiping, and the two sets compete with each other in the control over the joint. A hybrid occurs when the command neuronal sets subserving the distal joints and associated with R form wiping dominate the command neuronal sets associated with C form wiping, whereas for the proximal joints the dominance is reverse.

It is interesting that, in a hybrid, both programs continue performing until an inconvenient configuration of the limb is reached. It compels us to think that central control processes, but not afferent signals from the muscles and joints, here play a predominant role. In this connection, it should be recalled that blends of scratch reflex forms were observed both in intact and immobilized turtles, in which afferent signals from the limb were essentially suppressed (Robertson et al. 1985). The inadequate limb configuration in a hybrid is transformed by the frog spinal cord into an effective wiping pattern. An intact frog can also display a similar hybrid, so that the hybrid is not the result of artificial conditions, for example, due to the absence of descending control in spinal preparations.

7.6 Sets of Collectively Working Control Systems

The hybrid form of the wiping reflex corroborates the hypothesis that motor behaviors are organized of compartments or segments that can be combined in different ways to form new movement patterns (e.g., Berkinblit et al. 1986a; Fentress 1986; Viviani & Terzuolo 1982), which is consistent with the concept of *motor schema* (Arbib, this volume). In addition, we can hypothesize that every limb joint is subserved by a set of individual control systems which interact in the process of solving a common motor task. Similar organizations probably exists in locomotion: it has been argued that limb joints in cats are subserved by a set of individual generators which interact with each other and, as a result, produce an overall locomotor pattern (Grillner 1981; see also Bässler, this volume). A comparable idea has been realized in a model for the control of a multi-joint limb in a reaching task similar to that performed during a wiping reflex in a spinal frog (Berkinblit et al. 1986a,b; cf. also Hinton 1984; McClelland et al. 1986). In this model, a control system of the whole limb consists of a set of individual control systems, each one associated with a single joint. They act in parallel and interact with each other in order to solve a common task (reaching a target in space with the limb's tip). Every control system produces an individual movement in the corresponding joint based on the information on the position of the target and on the result of collective work of all the limb's joints, in particular, the knowledge of the position of the limb's tip relatively to the target.

7.7 Levels of Sensorimotor Processing

Figure 6 summarizes our views on the organization of wiping reflex in the spinal frog. We assume that the wiping reflex is performed by a number of levels of sensorimotor processing. One level is the above-mentioned on-off system that integrates signals from the skin and prevents the execution of a wiping reflex if the strength of stimulation did not reach a threshold, or if a precedent motor reaction (e.g., the local skin reaction) has not been completed. Another level is a set of neurons of short-term memory for the coordinates of all stimulated sites of the skin. These site-specific neurons are related to set of command neurons for various wiping forms. A small population of command neurons is switched on, if an excitatory inflow to it from the corresponding site-specific neurons is associated with a cessation of inhibition from on-off neurons. An occasional execution of a wiping reflex is excluded by this system. Every command neuron population is sensitive to a certain pattern of excitation of the site-specific neurons, so that simultaneous stimulation of two distant sites could give rise to excitation of two command neuron populations, each of which could provide a movement to the corresponding site, but their parallel action is excluded due to their antagonistic interaction. For a given wiping form there are individual neuronal groups, each of which generates corresponding movement pattern in a single joint. The groups subserving a given joint, but eliciting different wiping forms, are related by inhibitory connections. The connection strength can be variable and not always symmetric for the two wiping forms, and dependent on the joint, so that some joints can participate in the execution of R form wiping, whereas the rest joints can do a C form. As a consequence, a hybrid configuration of the limb can occur.

The scheme presented in Fig. 6 needs an experimental testing and a further theoretical elaboration. However, the hypothetical sensorimotor organization resembles to that of saccadic eye movements in cats (Fuchs & Kaneko 1985). It has been shown that there are brain stem inhibitory neurons that interrupt their activity for the time of saccadic movements (pause-neurons) and release the saccadic generator from their inhibitory control (for inhibitory "reset functions" in the toad's snapping pattern generator see also Weerasuriya, this volume). The neurons do not specify the size of saccadic movements. There are other neurons of the brain stem that modify their tonic activity as a function of the saccadic size. They also give projections to the saccadic generator and, as a result, its action is controlled by the pause-neurons and the size-specific neurons. Thus, the comparison between the two seemingly different motor tasks suggests common principles of their organization.

Acknowledgement. We thank Tanya Tkach for her technical assistance.

8 References

Akert K (1949) Der visuelle Greifreflex. *Helv Physiol Pharmacol Acta* 7: 112-134

Arbib MA (1972) *The methaphorical brain. An introduction to cybernetics as artificial intelligence and brain theory.* Wiley-Interscience, New York London Toronto

Baglioni S (1913) Die Hautreflexe der Amphibien (Frosch und Kröte). *Erg Physiol* 13: 454-546

Berkinblit MB, Feldman AG, Fukson OI (1986a) Adaptability of innate motor patterns and motor control mechanisms. *Behav Brain Sci* 9: 585-599

Berkinblit MB, Gelfand IM, Feldman AG (1986b) A model for the aiming phase of the wiping reflex. In: Grillner S, Stein PSG, Stuart H, Forssberg H, Herman RM (eds) *Neurobiology of vertebrate locomotion.* Wener-Gren International Symposium Series Vol 45. MacMillan Press, Houndsmills Basingstoke Hampshire London, pp 217-227

Bernstein NA (1967) *The co-ordination and regulation of movements.* Pergamon Press, Oxford

Davis MR, Constantine-Paton M (1983) Hyperplasia in the spinal sensory system of the frog. II: Central and peripheral connectivity patterns. *J Comp Neurol* 221: 453-465

Davis WJ (1977) The command neuron. In: Hoyle G (ed) *Identified neurons and arthropod behavior.* Plenum Press, New York, pp 293-305

Ewert J-P (1987) Neuroethology of releasing mechanisms: prey-catching in toads. *Behav Brain Sci* 10: 337-405

Fentress JC (1987) Compartments and cohesions in adaptive behavior. *J Comp Psychol* 101:254-258

Friesen WO, Stent GS (1978) Neuronal circuits for generating rhythmic movements. *Ann Rev Biophys Bioengin* 7: 37-61

Fuchs AF, Kaneko C (1985) A brain stem generator for saccadic eye movements. In: Evarts EV, Wise SP, Bousfield D (eds) *The motor system in neurobiology.* Elsevier Biomedial Press, Amsterdam New York Oxford, pp 126-132

Grillner S (1981) Control of locomotion in bipeds, tetrapods, and fish. In: Brooks V (ed) *Handbook of physiology, Section 1, The nervous system, Part 2.* American Physiol Society, Bethesda, pp 1179-1236

Hinton GE (1984) Parallel computations for controlling an arm. *J Mot Behav* 16: 171-194

Ingle D (1968) Visual releasers of prey-catching behavior in frogs and toads. *Brain Behav Evol* 1: 500-518

Kupfermann J, Weiss RW (1978) The command neuron concept. *Behav Brain Sci* 1: 3-39

McClelland JL, Rumelhart DE, Hinton GE (1986) The appeal of parallel distributed processing. In: Rumelhart DE, McClelland JL (eds) *Parallel distributed processing: explorations in the microstructure of cognition, Vol 2.* MIT Press, Cambridge Mass, London, pp 3-44

Mortin LJ, Keifer J, Stein PSG (1985) Three forms of the scratch reflex in the spinal turtle: Movement analyses. *J Neurophysiol* 53: 1501-1516

Robertson GA, Mortin LJ, Keifer J, Stein PSG (1985) Three forms of the scratch reflex in the spinal turtle: Central generation of motor pattern. *J Neurophysiol* 53: 1517-1534

Simpson JJ (1976) Functional synaptology of the spinal cord. In: Llinas R, Precht W (eds) *Frog neurobiology.* Springer-Verlag, Berlin Heidelberg New York, pp 728-749

Stein PSG (1983) The vertebrate scratch reflex. *Symp Soc Exp Biol* 37: 383-403

Stein PSG, Camp AW, Robertson GA, Mortin LI (1986) Blends of rostral and caudal scratch reflex motor patterns elicited by stimulation of two sites in the spinal turtle. *J Neurosci* 6: 2259-2266

Viviani P, Terzuolo C (1982) Trajectory determines movement dynamics. *Neurosci* 7: 431-437

Pattern Generation for Walking Movements

ULRICH BÄSSLER

Fachbereich Biologie, Universität Kaiserslautern
D-6750 Kaiserslautern, FR Germany

Abstract. Walking is a very variable kind of movement. The analysis of the neural network producing such a variable motor output is shown for the stick insect, but there is some probability that many of the statements are also valid for walking in general. The experimental results allow us to make the following hypotheses. Each leg has its own pattern generator for walking movements. Higher centers recommend walking direction and gait by opening or closing coordinating pathways between these pattern generators, but the actual decision is made under the influence of sense organs. Each pattern generator is built-up of subunits (approximately one for each joint). The coordination of the subunits may vary, and the subunits can also be used for the control of other active movements. Each subunit contains sense organs as integral parts. In the denervated preparation at least some of the subunits are able to produce an irregular alternation, but there is no coordination between them.

1 Introduction

Walking is the most common mode of locomotion for terrestrial animals. It is highly developed in two phyletic groups, arthropods and four-legged vertebrates. In each of these animals walking movements can be performed in several different ways depending on the actual requirement of the animal. This flexibility is expressed on three levels:

(i) Most animals can adopt different walking gaits like the trot or gallop. That is, they can change the temporal and spatial relationships between the movements of particular legs mainly depending on the walking speed.

(ii) All animals can walk forwards and backwards and in many cases sideways as well. When a leg changes its walking direction, the coordination of the movements of its joints also changes. As an example: leg lifting and leg protraction occur in synchrony during forward walking, but during backward walking, leg lifting is associated with leg retraction.

(iii) Each part of a step has to be adapted to irregularities of the ground; otherwise the animal is unable to walk over uneven terrain.

This shows that flexibility is present in the movements of a single joint, in the coordination of the joints of a leg, and in the coordination of all the legs. The neural system producing the motor output pattern for walking movements must take into account these requirements, i.e., it must be very flexible.

In this paper I want to describe the construction principles of such a neural system. I use the stick insect as an example. As far as we know, many of the statements are valid for walking in general, so that the stick insect is used here as a model for all walking animals including vertebrates. The considerations are restricted to the control of timing, and most of the influences on the shape of the motor output are not mentioned here. The variability of the movements of single joints is mainly caused by shaping influences and, therefore, not discussed here.

Insects are ideal for studying principles of motor control because their walking movements are performed in much the same way as the corresponding movements of vertebrates, but are easier to investigate. In particular, the number of their neurons is much smaller than in vertebrates, so that the analysis can use identified neurons. Furthermore, many of their proprioceptors lie outside the musculature and thus can easily be stimulated mechanically.

Stick insects or phasmids form a group of approximately 2500 species, including the largest insect species. We work mainly with a rather small animal called *Carausius morosus* (body length 7 cm), but we sometimes also use larger species. Stick insects are slowly climbing animals and are mainly active during the night. Their habitat is normally the crowns of trees and bushes in tropical and subtropical regions.

2 Movement of All Legs

To find out to what extent the movements of an individual leg can be independent from the movements of other legs, a stick insect was fastened above a light treadwheel (Foth & Bässler 1985). The wheel consisted of two independent halves which the animal could turn with five of its legs. The left hindleg rested on a motor-driven conveyor belt (Fig. 1a). When the animal walks with its five legs on the wheels with a rather constant velocity, one can make the belt velocity either similar to or faster, or slower, than the wheel speed. When the belt was moved slower than the wheel, the leg on the belt took fewer steps than the legs on the wheel. When the belt speed was higher than the wheel speed, the leg on the belt took more steps than those on the wheels. In all these cases the belt leg remained coordinated with the other legs (Fig. 1b and c). Electrophysiological records showed that the leg on the belt actually walked: its retractor motoneurons were strongly excited during the time the leg was moved backwards passively (Fig. 2). The modulation of their firing rate in the stepping frequency of the other legs is apparently the expression of coordinating influences (Bässler et al. 1987).

When different legs can operate at different frequencies, there cannot be a single "master center" which determines the movements of all legs. Instead, each leg must have its own control network for walking movements. The concept of a control network is understood here to encompass all the sensory and central nervous structures responsible for timing the motor output (see also "coordinated control programs" discussed by Arbib, this volume). The fact that the step rhythm was influenced by the belt speed, shows that such a control network can be influenced by changing external conditions, i.e., it must include sense organs in addition to its

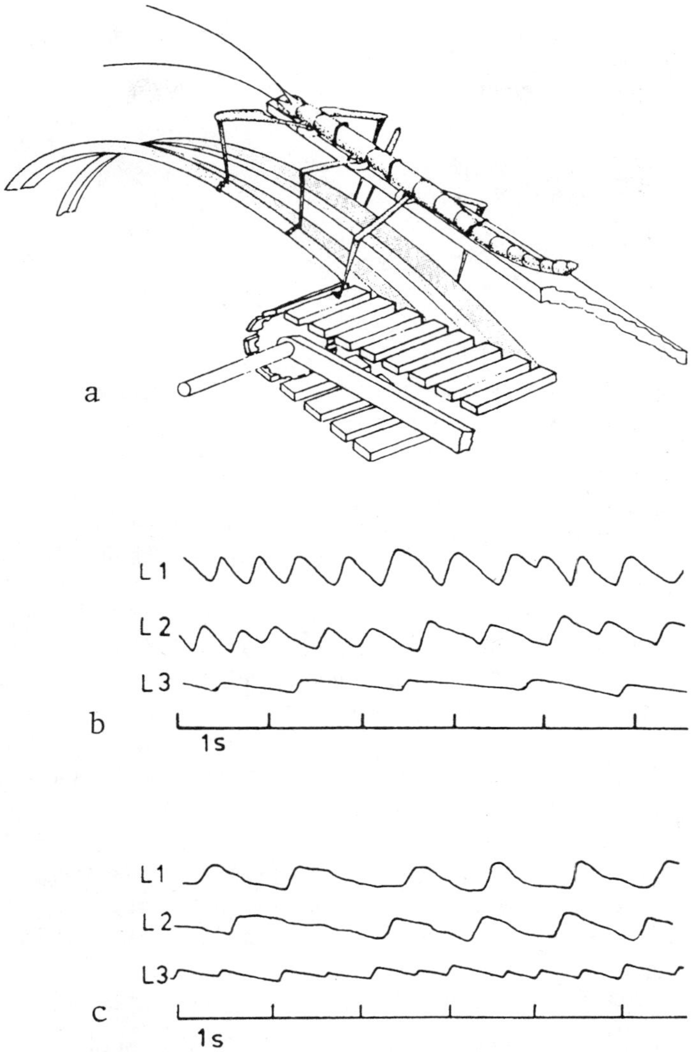

Figure 1. A stick insect is fastened above a double treadwheel. The left hindleg is moved by a motor-driven belt. (a) Experimental situation. (b) Movements of the left foreleg (L1), middle leg (L2), and hindleg (L3), when the belt speed is slower than the wheel speed. Up represents forward movement (swing) of each leg. (c) Like b, but belt speed is faster than wheel speed. (From Foth & Bässler 1985).

central nervous parts. In this paper the control network for the walking movements of a single leg is called a walking-pattern generator, although this term is equivocal to some extent (see Section 5).

The stick insect possesses two walking gaits, tetrapod and tripod gait (Graham 1972), which are very similar to walk and trot in quadruped animals. These gaits are likely to be produced by specific interactions between the six walking-pattern

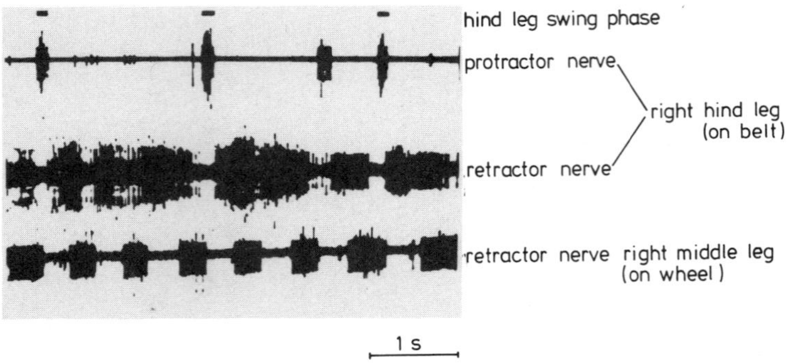

Figure 2. Records from the protractor and retractor coxae motoneurons of the right hindleg (on the slowly moving belt) and from retractor coxae motoneurons of the ipsilateral middle leg (gives the stepping frequency of the legs on the wheel). A bar in the top trace shows the occurrence of the hind leg swing phase.(From Foth & Bässler 1985).

generators of the animal (for details see Bässler 1983; Cruse et al. 1987). As insects have six legs, there is some probability that their coordinating mechanisms differ from those of vertebrates. Therefore, the investigations of the coordinating mechanisms of the stick insect are not reviewed here.

Where is a particular walking-pattern generator situated? This question will now be answered for the foreleg. A stick insect possessing only forelegs can walk quite regularly when the prothoracic ganglion (which innervates the forelegs) is either connected to the intact suboesophageal ganglion (a head ganglion) or to the rest of the thoracic and abdominal nervous system. In the latter case, walking movements only occur when the abdomen is mechanically stimulated. Apparently, the prothorax contains all neuronal and sensory structures for the performance of a step, but this walking-pattern generator only functions either when it is connected to the suboesophageal ganglion, or when the rest of the ventral nerve cord is strongly excited.

What kind of movement is produced by a single pattern generator? To answer this question we amputated all legs except for the two legs of one segment and fixed the animal above a walking wheel (Fig. 3). The whole CNS was intact. Forelegs walked only forwards under these conditions. Hindlegs walked mainly backwards, and middle legs walked forwards and backwards. Apparently, the inherent walking direction of the foreleg pattern generator is forwards and that of the hindleg pattern generator is backwards. "Inherent direction" refers here to the direction in which the leg pulls the body when the generation of the walking pattern for that leg is free of external coordinating influences. A single intact foreleg is sufficient to cause the hindlegs to walk forwards. The foreleg must only build-up force and needs not be walking in order to produce forward walking of hindlegs.

The inherent walking directions of fore- and hindlegs were also expressed in the following experiment. An animal was fastened above a glassplate coated with silicon

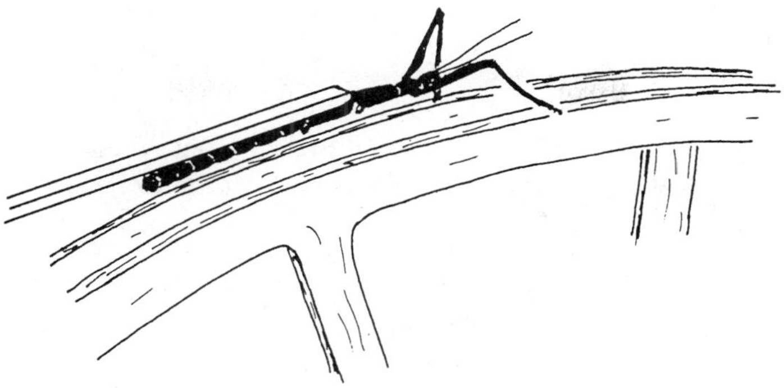

Figure 3. An animal possessing only forelegs fixed above a double treadwheel.

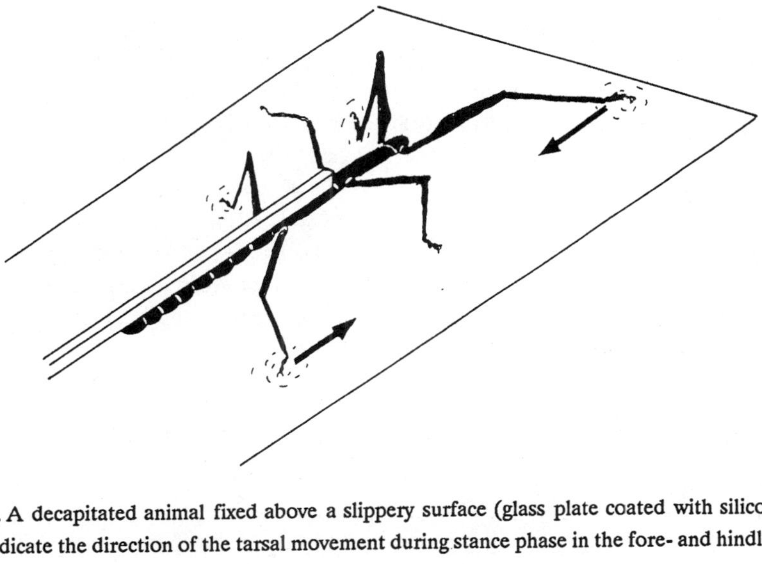

Figure 4. A decapitated animal fixed above a slippery surface (glass plate coated with silicone oil). The arrows indicate the direction of the tarsal movement during stance phase in the fore- and hindlegs.

oil. Stick insects walk quite regularly under these conditions, although there is no mechanical coupling between the legs as there ordinarily would be. Removal of the brain (supraoesophageal ganglion) did not change anything. However, if also the suboesophageal ganglion was removed, the inherent walking direction was expressed, i.e., the forelegs walked forwards and the hindlegs backwards (Fig. 4). On a solid ground, with mechanical coupling of the legs, the animal is unable to walk any longer (Bässler et al. 1985).

These simple experiments allow the following conclusions:

(a) There is a coordinating pathway from sense organs of the foreleg to the hindleg pattern generators which is able to influence the walking direction of the hindlegs.

This pathway only functions when the suboesophageal ganglion is intact.
(b) The walking direction of a hindleg is apparently not commanded by higher centers. Higher centers may submit a recommendation (Kien 1983), however, the decision about the walking direction is produced under the influence of sense organs in the legs. This is not a hierarchical command structure but a democratic decision-making process on a rather low level. We will meet other democratic processes later on.

Perhaps the recommendation to walk forwards consists of opening the coordinating pathways from fore- to hindlegs by signals from the suboesophageal ganglion. Then it is conceivable that a recommendation to walk backwards would consist of opening coordinating pathways in the opposite direction. Perhaps higher centers (especially in the suboesophageal ganglion) are able to open and close other coordinating pathways between the six pattern generators as well. Then, the kind of functioning pathways would not only recommend walking direction but also gait. As the actual decisions are taken under the influence of sense organs, the system can adapt to new situations like that of Fig. 1a, or can change the temporal relationships between the remaining legs when some legs have been amputated. The flexibility in the coordination between legs would then be based on a change in the combination of open pathways in conjunction with strong influences from sense organs.

3 Control of a Single Leg

What are the functional properties of a single walking-pattern generator? The first step in answering such a question is normally to denervate the central nervous system and to find out whether the isolated CNS is able to produce some kind of rhythmic motor output. We have also done this kind of experiment (Bässler & Wegner 1983). The whole thoracic CNS was denervated by cutting all the nerves, but the CNS in the head and abdomen remained intact. The activity of the protractor and retractor motoneurons was recorded in right middle and hindlegs from the nerve stumps. The animals were continuously stimulated mechanically on the abdomen to keep them active. Figure 5 shows the most regular sequence of nearly two hours of recording. Like in all other parts of the records, there was a reciprocal activity in protractor and retractor motoneurons. The burst durations varied (in the other parts of the record more strongly), however, only within certain limits (Fig. 6). There was no obvious coordination between motoneurons of other joints of the same leg (not shown here), or between corresponding motoneurons of other legs (Fig. 5). Apparently, the isolated thoracic CNS is able to produce only some features of the walking motor output (alternation of antagonists, termination of bursts after some time), but rather important features are not present (i.e., regularity of alternations, coordination of the joints of one leg and of several legs).

On the other hand, when one leg stood on a motor-driven belt and the other five legs on a self-propelled wheel (Fig. 1a), it was often possible to produce regular stepping of the belt-leg when the wheel-legs did not walk. This often occurred at the end of a longer walking sequence and the wheel-legs stopped walking a long time before the belt-leg also stopped.

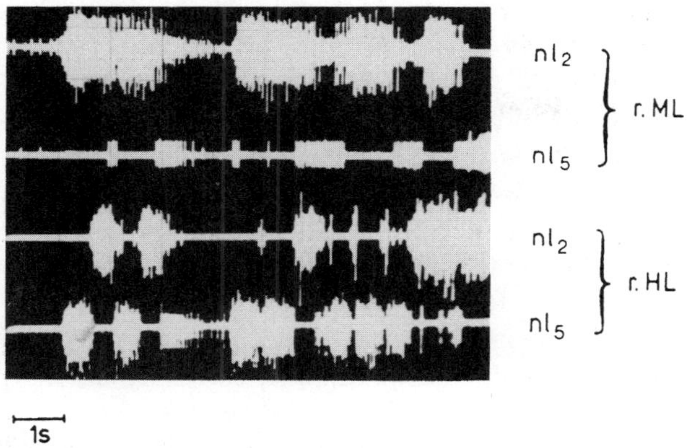

Figure 5. Records from the protractor (nl2) and retractor (nl5) coxae motoneurons of right middle (r. ML), and hind legs (r. HL). Thoracic ventral nerve cord denervated. This was the most regular sequence recorded. (From Bässler & Wegner 1983).

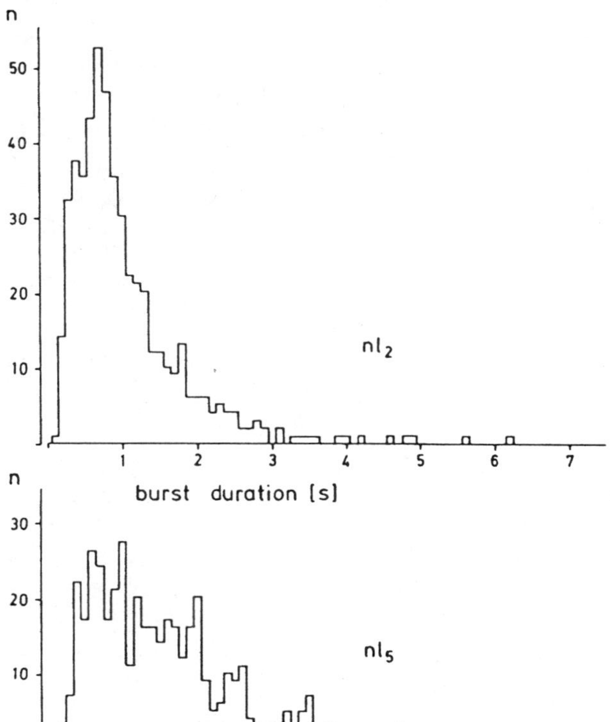

Figure 6. Histograms of the durations of protractor (nl2) and retractor (nl5) bursts in hindleg records in the denervated thoracic ventral nerve cord. (From Bässler & Wegner 1983).

These and a large number of other experiments (summarized by Bässler 1983) demonstrated that the timing of the walking movements of an individual leg is mainly, if not exclusively, controled by sense organs. Central elements are only of minor importance. Not all animal species are unable to produce a walking motor output in the denervated preparation. As an example, the cat is able to produce a detailed pattern without feedback (i.e., Grillner & Zangger 1984). It may well be that the stick insect represents one extreme in a continuum of walking-pattern generators which reaches from mainly centrally driven networks to systems that only function properly when feedback is present.

What kind of feedback is incorporated in the pattern generator? I shall try to answer this question for the stance-swing transition. A step consists of two parts: stance phase (for forward walking: retraction of the leg with the tarsus on ground) and swing phase (for forward walking: protraction of the leg through the air). As all legs that are in stance phase at the same time are mechanically coupled via the substrate, and the swing phase is normally performed with constant velocity, the main working points for timing influences are the transitions between the two phases. The transition from stance to swing phase is influenced by position measuring sense organs and by campaniform sensilla that measure load. This has been shown several times (e.g., Cruse 1985). The first indications came from the behavior of animals with surgically manipulated sense organs (Bässler 1977). It was possible to surgically manipulate sense organs that measure the position of the subcoxal joint (the basal leg joint) in such a way that they continuously signaled a protracted leg position, no matter where the leg actually was. Such an operated leg moved backwards during

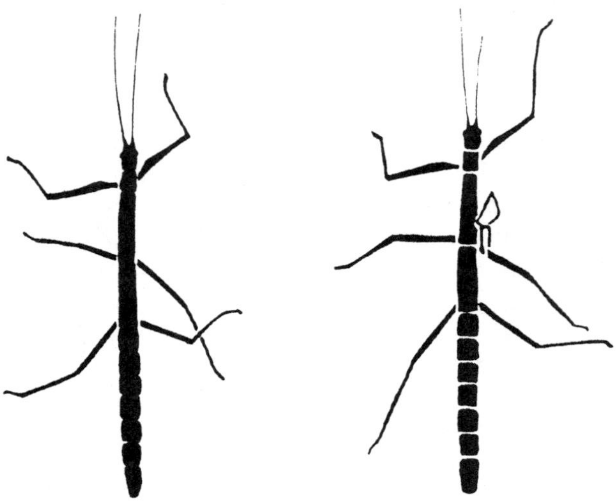

a b

Figure 7. Walking stick insects with manipulated sense organs of the right middle leg. (a) Sense organs at the subcoxal joint (the basal leg joint) continuously signaled a protracted leg position. (b) Campaniform sensilla continuously stimulated. (From Bässler 1977).

stance phase as far as anatomically possible (Fig. 7a). Often it did not release the ground after it had reached its posterior extreme position. As stick insects grasp the ground during stance phase, this indicates that stance phase could not be terminated. In this case the leg was pulled off the ground with a distinct jerk by the activity of the other legs. Apparently, the stance phase is only terminated when sense organs signal that the normal posterior extreme position has been reached.

The second type of sense organs are the campaniform sensilla which measure load. When the campaniform sensilla on the base of a leg were continuously stimulated by attaching a metal clamp, such a leg performed a normal stance phase. The tarsus released the ground at the end of the stance phase, but the leg was not swung forward, i.e., the swing phase did not begin (Fig. 7b) and the leg was dragged over the ground while the other legs walked.

Apparently, the stance-swing transition in one leg is influenced by position sense organs and campaniform sensilla of that leg and by coordinating influences from other legs. Each of these influences can trigger a stance-swing transition, provided the transition is not inhibited by the two other kinds of influences. The probability that a sudden decrease of load triggered a swing phase increased the more posterior the leg position was (Cruse 1985). This shows that the transition from stance to swing phase is determined by another democratic decision-making process and not by a hierarchical command structure.

The transition from swing to stance phase is not so well understood. Only one mechanism was studied in detail, the so-called targeting response. Free walking stick insects set the hindleg tarsus down at the end of swing phase right behind the tarsus of the ipsilateral middle leg. This occurs irrespective of where the middle leg tarsus is (Fig. 8). The adaptive purpose of this response is obvious for a climbing animal. Where the middle leg has found a solid foothold, the hindleg will with a high probability also find one just behind (Cruse 1979). A similar response was found between fore- and middle leg.

The targeting response only functions during forward walking. When a standing animal first lifts and then sets down a hindleg, the tarsal position of the leg is not affected by the position of the middle leg. In animals walking backwards, the middle leg is set down at the end of its swing phase (which is now a retraction) close to the actual position of the hindleg tarsus. In this case, the targeting response is reversed, i.e., it works from hind- to middle leg (Dean & Wendler 1983). Thus, the targeting response loop only functions in a certain behavioral context.

4 The "Active Reaction"

From all these experiments it was obvious where the pattern generators are located and possible to determine at least some of the sense organs involved. Therefore, the next step was to study the neural basis of components of a pattern generator. This is only possible with intracellular methods, and for these recordings a preparation would be helpful which allows the release of the walking motor output, or at least part of it, in a totally restrained animal. As the isolated thoracic CNS produces no walking motor output, we looked for a reaction to a sensory stimulus which corresponds to a part of the walking motor output. We found it in the so-called active reaction of the femur-tibia joint of the foreleg. This is a reaction to a position and velocity signal.

To understand the active reaction, some anatomical remarks are necessary. The femur-tibia joint is a hinge joint which can only be moved in one plane. Its position is measured by the femoral chordotonal organ (Fig. 9), a sense organ with approximately 500 sense cells which measure position, velocity or acceleration, or a combination of any two of these parameters (Hofmann et al. 1985; Hofmann & Koch 1985). The organ lies at the proximal end of the femur and is connected to the tibia by a long apodeme. This receptor apodeme travels outside the musculature and ends dorsal to the axis of rotation of the femur-tibia joint. Flexion of the joint, therefore, stretches the chordotonal organ and extension releases it. The femur-tibia joint is moved by two muscles, the extensor and the flexor tibiae muscles (for details see Bässler 1983).

There exists a very strong resistance reflex when the animal is inactive, i.e., mechanically stretching the chordotonal organ (which signals a flexion movement) strongly excites the motoneurons of the extensor muscle (Bässler 1983; see also Fig. 12). The same stretch stimulus releases the active reaction when the animal is active.

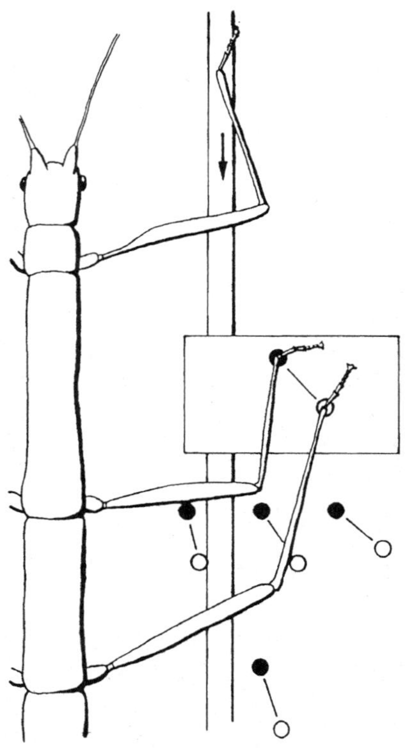

Figure 8. Targeting response. The animal walked on a wheel. The right middle leg stood on a stationary platform. The black dots indicate the position of the middle leg tibia in different experiments, the circles the mean position of the swing phase endpoint of the right hindleg. The leg positions are shown for one experiment. (Modified from Cruse 1979).

In stick insects the inactive and the active states are well defined and easily recognized. The transition from the inactive to the active state is abrupt. When the animal is active first and then becomes inactive, the transition is sometimes also abrupt, but sometimes an intermediate state can be observed. In this state, the animal does not actively move, however, its responses are like those in the active animal. The intermediate state is normally rather short in intact animals, but when the brain has been removed (decerebrate animal), it lasts much longer (Bässler 1983). This intermediate state is ideal for the study of responses of the active animal. Therefore, many studies on the active reaction have been performed in decerebrate animals.

The active reaction was elicited in the following way (Fig. 10a): The animal, which had only forelegs left, was restrained. The right femur was opened and the activity of two excitatory motoneurons of the extensor tibiae muscle was recorded extracellularly. The flexor possesses 14 excitatory motoneurons which cannot be individually recognized (Debrodt & Bässler, in preparation). Therefore, the force of the whole muscle was measured by a transducer attached to the tibia. Additionally, the force produced by the retractor unguis muscle was measured. The retractor unguis muscle flexes the claw which is responsible for grasping the ground during the stance phase of walking. The femoral chordotonal organ was stimulated mechanically in such a way that it signaled a flexion movement of the joint from approximately 140° to 60° (Bässler 1986a).

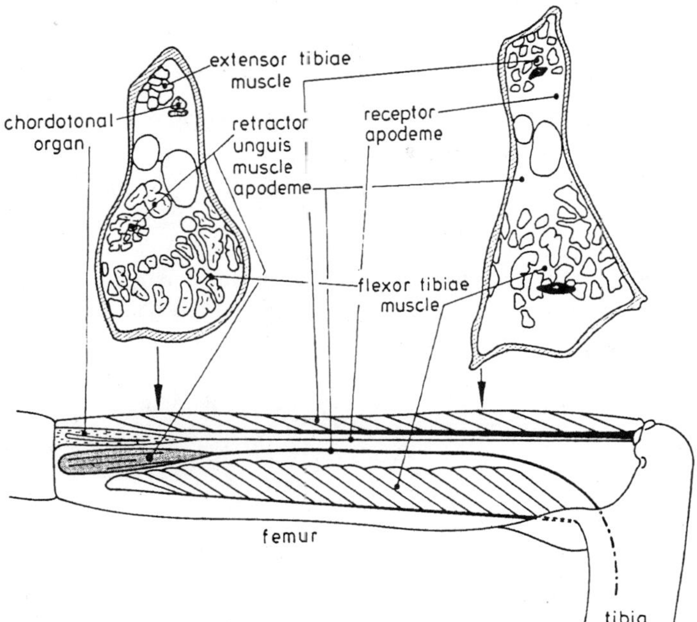

Figure 9. Schematic diagram of the femur of a stick insect with two representative cross-sections. For details see text.

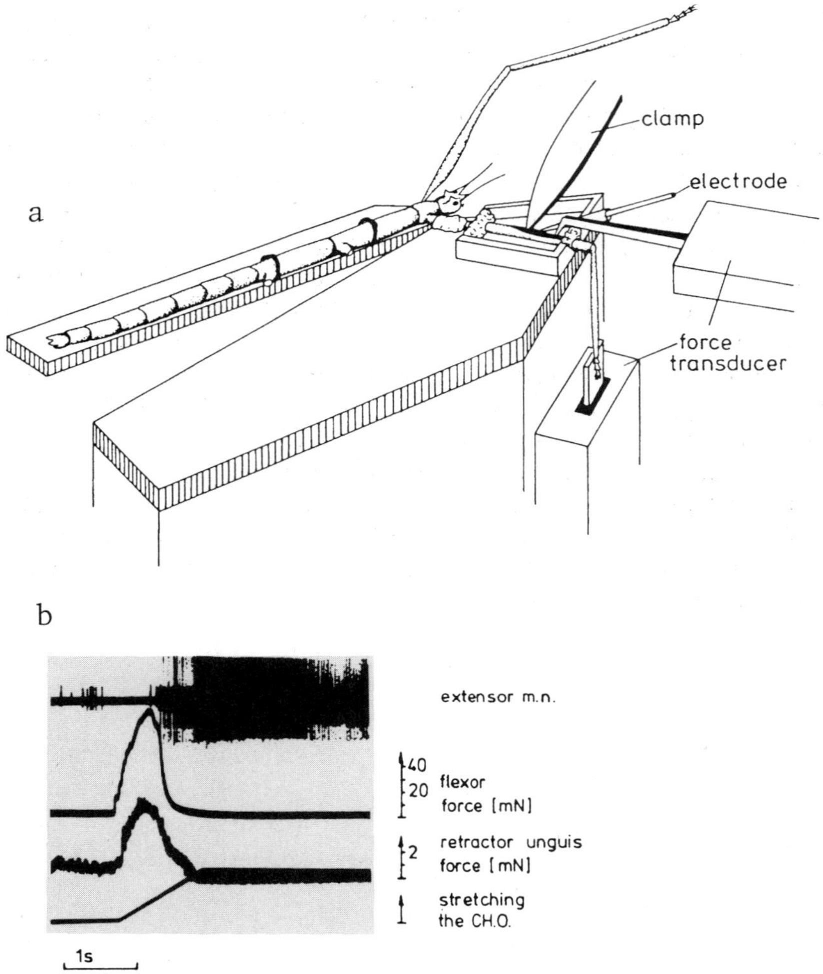

Figure 10. Experimental set-up (a) and typical result (b) for an active reaction. Recordings of the activity of the extensor motoneurons (by suction electrode), the force of the flexor tibiae muscle (force transducer attached to the tibia), and the force of the femoral part of the retractor unguis muscle (other force transducer) in response to a stimulation of the femoral chordotonal organ by a clamp attached to the receptor apodeme corresponding to a flexion movement from 140° to 60° of the femur-tibia joint. (From Bässler 1986a).

In the inactive animal this stimulus excites the extensor motoneurons, inhibits flexor motoneurons (resistance reflex), and has no influence on the retractor unguis. In an animal that is in the active or intermediate state, the same stimulus releases an active reaction. The active reaction consists of two parts: (i) At first the stimulus strongly excites the flexor muscle. Since the stimulus signals a flexion movement, this is a positive feedback. At the same time the retractor unguis muscle also contracts.

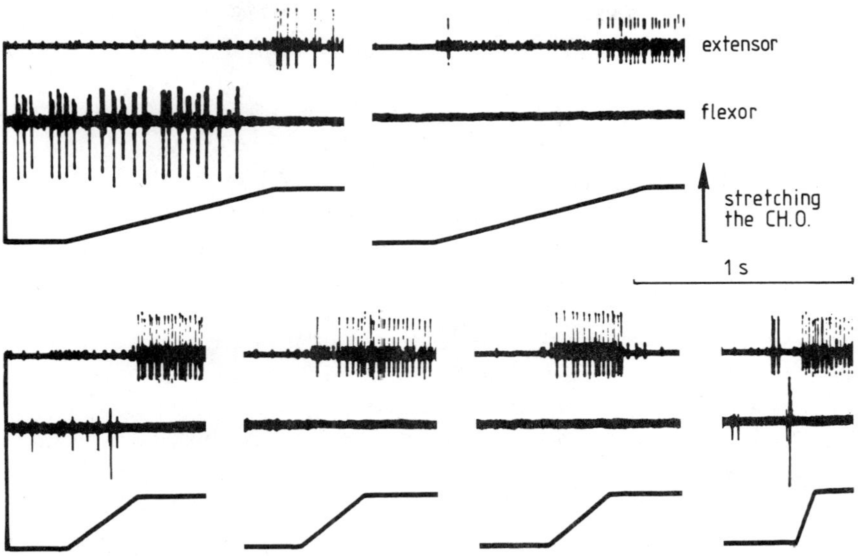

Figure 11. Recordings from the extensor nerve and the flexor muscle in an active animal during stimulation of the femoral chordotonal organ with stretches of different velocities.

(ii) When the chordotonal organ signals a fairly flexed joint position, the force of the flexor and retractor unguis muscles declines and the extensor motoneurons strongly fire (Bässler 1986a). In the foreleg, the flexor and retractor unguis muscles contract during the stance phase of forward walking and the extensor muscle during the swing phase. When an active reaction is released in a hindleg, the retractor unguis muscle is excited together with the flexor when middle and forelegs have been removed. This is the coordination during the stance phase of backward walking, and freely moving animals possessing only hindlegs normally walk backwards. When all the legs are present during the release of an active reaction in a hindleg, the retractor unguis neurons fire together with the extensor motoneurons. This is the coordination during a stance phase of forward walking, which corresponds to the fact that hindlegs of intact animals preferentially walk forwards (Nothof & Bässler, unpublished observation). These and some other evidences mentioned below suggest that the first part of the active reaction in the foreleg corresponds to the stance phase and the second part to the swing phase of forward walking.

The active reaction, elicited as described in Fig. 10, only occurs with slow ramp speeds. With high ramp speeds the extensor motoneurons are excited at the beginning of the ramp. This is a negative feedback. With middle ramp velocities both kinds of reactions are possible (Fig. 11). As mentioned earlier, the first part of an active reaction (occurring at slow ramp speeds) forms a positive feedback. In the closed loop system (when the chordotonal organ is stimulated by the movement of the joint) the positive feedback would accelerate any slow flexion movement. On the other hand, fast flexion movements are braked by the negative feedback occurring in response to fast velocities. A system which accelerates slow movements and brakes fast movements is a kind of velocity control system. We can therefore state that the first part of the active reaction produces a velocity control for flexion movements (Bässler 1988).

Whenever an active reaction occurred, the transition from part 1 (flexor excitation) to part 2 (extensor excitation) was independent of ramp velocity always occurring at the same position of the receptor apodeme (Bässler 1986a).

What is the behavior of other leg muscles when an active reaction is released in the tibial flexor and extensor motoneurons? It was mentioned above that the retractor unguis muscle of the foreleg always contracts together with the flexor, however, in the hindleg it contracts together with the extensor when the animal "wants" to walk forwards and together with the flexor when it "wants" to walk backwards. Thus, this muscle always behaved as it did during a normal step.

The muscles moving other joints of the same leg were more loosely coupled to the extensor-flexor system. That means for the foreleg that muscles being normally active during a stance phase of straight, forward walking tended to be active together with the flexor, and muscles normally active during swing phase tended to be active together with the extensor. But the coupling was far from being absolute and other coordinations between the activity of several joints of a leg were also possible (Bässler 1986a). Apparently, a walking-pattern generator for a single leg consists of subunits which, in the case of the hindleg retractor unguis, can change the kind of coordination, or which are only loosely coupled to other subunits (in the case of the other subunits). The loose coupling between the subunits might be the basis for the flexibility in the coordination of leg joints during walking in different directions. The kind of coupling between the different subunits during an active reaction in forelegs corresponds to the behavior of that leg during walking. Forelegs are often used to explore the surroundings and, therefore, often change walking direction, i.e., the coordination between femur-tibia joint and subcoxal joint. But in all these cases retractor unguis and flexor tibiae muscles contract at the same time.

Is the active reaction also involved in the control of active movements other than walking? To answer this question the femur of a leg was restrained, so that the animal could only move the femur-tibia joint. When the animal performed active movements with this joint (seeking movements), it was possible to show that the position information from the femoral chordotonal organ was used to control the endpoints especially of flexion movements. During the middle part of the movement, velocity was the controlled variable. Additionally, the flexion movements were performed with a velocity which corresponds to that ramp velocity during an active reaction at which the transition from positive to negative feedback took place in the active reaction (Weiland & Koch 1987). Thus, the joint flexed as if its movement had been controled by the active reaction.

If the network producing the active reaction is a subunit of the walking-pattern generator and if it is also involved in the control of seeking movements, it is possible to conclude that the subunits forming the walking-pattern generator are also involved in other pattern generators. In this context, the experiments described by Figs. 5 and 6 can be interpreted in a new way: apparently each subunit can produce a rather irregular rhythm in the denervated preparation, but coordination between the subunits and between different legs is lacking.

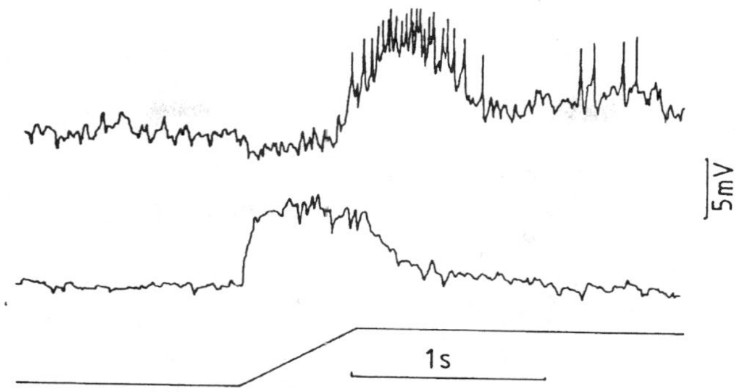

Figure 12. Two intracellular records from a fast extensor tibiae motoneuron during two separate stimulations of the femoral chordotonal organ. Top trace: active reaction; middle trace: resistance reflex in the inactive animal; bottom trace (valid for both records): stimulus.

According to this hypothesis there are several subunits (approximately one for each joint). These subunits can be combined and coordinated in different ways to produce the different kinds of active movements (like forwards, backwards, and sideways walking, seeking movements, rocking etc.). The pattern generator for a certain kind of movement (e.g., forward walking) would then be put together by opening certain coordinating pathways between the subunits (and closing others). There is some probability that the coordinating pathways between the subunits forming a particular walking-pattern generator are only functioning when sensory feedback is present. (The subunits are not coordinated in the denervated preparation.) If this could be verified, it would mean that the walking-pattern generator is a functional unit only if sensory feedback is present. In other words, a walking motor output could only be produced when feedback signaled that walking was possible. This would mean that walking could not be commanded by higher centers. It would only be recommended (Kien 1983) and the decision to walk would be produced on a rather low level under the influence of sense organs.

The next step was to evaluate the neural basis of the active reaction which we now consider to represent only one subunit. At first we recorded intracellularly from motoneurons. Figure 12 shows two different intracellular records from the fast extensor tibiae motoneuron (FETi), one of the two excitatory motoneurons of the extensor tibiae muscle. In the inactive animal a stretch stimulus excites FETi (resistance reflex), whereas in the active animal FETi is hyperpolarized during the first part of the stimulus (active reaction). The latency between stimulus onset and depolarization was 8 to 15 ms, whereas the latency of the hyperpolarization ranged from 15 to 30 ms (Bässler 1988). This shows that there are different pathways for the resistance reflex in the inactive animal and the active reaction. As there is no hint of a depolarization just after stimulus onset in the active reaction one must conclude that the fast loop of the resistance reflex (at least its fastest components) is switched off during an active reaction. In the resistance reflex, there is also no hint of a superimposed active reaction. Apparently, the active reaction does not function in the inactive animal. So, the behavioral state determines which one of the two pathways is used.

There are two possible explanations for this result: (i) The network generating the active reaction is only functioning in the active animal; (ii) the network generates the active reaction in the inactive animal as well, but it is connected to the motoneurons only in the active animal. At the moment it is not possible to distinguish between these alternatives, although in both cases the subunit responsible for the active reaction only seems to be able to influence the motoneurons when the animal is active. Apparently, the subunits only function in the active animal.

Thus, we can summarize the hypotheses: The subunits forming a pattern generator of a single leg function only when the animal is active (or in the intermediate state). However, they are properly coordinated only when some sensory feedback and coordinating influences from other legs are present. The coordination may change according to recommendations from higher centers.

5 Discussion

In the preceding description I have tried to avoid many of the terms and concepts normally used to describe motor pattern generation, because I feel that they often have a meaning that is too narrow or imply too many interpretations and hypotheses.

This is most obvious with the term central pattern generator (CPG). At least some of the subunits responsible for the movements of a single joint can produce an irregular rhythm (Fig. 5). According to the general understanding of the term, they could be called CPG provided the definition does not require that the rhythm be regular. But it is not possible to term a subunit "CPG for walking movements" because its output is only relevant for walking when this subunit is connected in a certain way to other subunits. Isolated subunits produce a motor output with no special meaning. Only if they are connected in certain ways are they able to generate meaningful movements.

Similar objections exist to the term "pattern generator" in the sense used in this article (for a delimitation from CPG see Bässler 1986b). The term walking-pattern generator is used here for the network (including sense organs) generating the timing cues for the movements of a single leg. The reason is that this is the smallest unit which is able to produce a motor output for walking movements. However, when a coordination between the six pattern generators does not exist (e.g., after ablation of the suboesophageal ganglion), the animal is unable to walk any longer on a solid ground (see Fig. 4). Thus, if the term walking-pattern generator is used for the smallest unit which is able to generate walking, one could also argue that this term can only be used for the whole system controling all the legs.

The system generating walking movements includes a large amount of democracy. But this does not exclude the possibility that higher centers (e.g., in the suboesophageal ganglion) are themselves hierarchically organized and that, under natural circumstances, they really command walking. Higher centers can only submit a recommendation to the system generating the walking movements. However, the system generating the walking movements can only resist the recommendation when extreme situations are present, such as denervation or removal of most legs.

Apparently, it is dangerous to apply concepts derived from other relatively simple systems to the complicated and flexible system controling walking movements. But terms are necessary to describe the phenomena. Therefore, one should always be aware that most terms include interpretations and hypotheses which means that deductions based on these terms are inadmissible.

6 References

Bässler U (1977) Sensory control of leg movement in the stick insect *Carausius morosus. Biol Cybern* 25: 61-72

Bässler U (1983) Neural basis of elementary behavior in stick insects. In: Braitenberg V (ed) *Studies in brain function.* Springer-Verlag, Berlin Heidelberg New York

Bässler U (1986a) Afferent control of walking movements in the stick insect *Cuniculina impigra.* II: Reflex reversal and the release of the swing phase in the restrained foreleg. *J Comp Physiol* 158: 351-361

Bässler U (1986b) On the definition of central pattern generator and its sensory control. *Biol Cybern* 54: 65-69

Bässler U (1988) Functional principles of pattern generation for walking movements of stick insect forelegs: the role of the femoral chordotonal organ afferences. *J Exp Biol* 136: 125-147

Bässler U, Dübner C, Fahrig T (1987) Motor output oscillations in denervated thoracic ganglia of walking stick insects. *Zool Jb Physiol* 91: 393-401

Bässler U, Foth E, Breutel G (1985) The inherent walking direction differs for the prothoracic and metathoracic legs of stick insects. *J Exp Biol* 116: 301-311

Bässler U, Wegner U (1983) Motor output of the denervated thoracic ventral nerve cord in the stick insect *Carausius morosus. J Exp Biol* 105: 127-145

Cruse H (1979) The control of the anterior extreme position of the hindleg of a walking insect, *Carausius morosus. Physiol Entomol* 4: 121-124

Cruse H (1985) Which parameters control the leg movement of a walking insect? II: The start of the swing phase. *J Exp Biol* 116: 357-362

Cruse H, Knauth A, Schwarze W (1987) Coordination of middle and hind legs of walking stick insects. In: Mazzini M, Scali V (eds) *Stick insects, phylogeny and reproduction.* Univ Siena and Univ Bologna Italy, pp 87-96

Dean J, Wendler G (1983) Stick insect locomotion on a walking wheel: interleg coordination of leg position. *J Exp Biol* 103: 75-94

Foth E, Bässler U (1985) Leg movements of stick insects walking with five legs on a treadwheel and with one leg on a motor-driven belt. I: General results and 1:1 coordination. *Biol Cybern* 51: 313-318

Graham D (1972) A behavioural analysis of the temporal organisation of walking movements in the first instar and adult stick insects (*Carausius morosus*). *J Comp Physiol* 81: 23-52

Grillner S, Zangger P (1984) The effect of dorsal root transection on the efferent motor pattern in the cat's hindlimb during locomotion. *Acta Physiol Scand* 120: 393-405

Hofmann T, Koch UT (1985) Acceleration receptors in the femoral chordotonal organ of the stick insect, *Cuniculina impigra. J Exp Biol* 114: 225-237

Hofmann T, Koch UT, Bässler U (1985) Physiology of the femoral chordotonal organ in the stick insect, *Cuniculina impigra*. *J Exp Biol* 114: 207-223

Kien J (1983) The initiation and maintenance of walking in the locust: an alternative to the command concept. *Proc R Soc (Lond)* B 219: 137-174

Weiland G, Koch UT (1987) Sensory feedback during active movements of stick insects. *J Exp Biol* 133: 137-156

Neuroscience in Motion: The Application of Schema Theory to Mobile Robotics

RONALD C. ARKIN

Department of Information and Computer Science
Georgia Institute of Technology
Atlanta, Georgia 30332, USA

Abstract. Theories for path planning behavior in animals can be of great value in providing significant insights into the design of functioning mobile robot systems. A mobile robot path execution system has been developed that strongly correlates with a model for detour behavior in the frog. The robot's motor schema based navigation system draws on potential field methodology to produce "intelligent" behavior based on environmental perception. Both simulation and actual experimental results are presented.

1 Introduction

Research in mobile robot navigation is currently heavily funded and is proceeding in many laboratories with varying degrees of success. Unfortunately, in most cases, little attention has been given to those systems that have already successfully tackled the problem of complex navigation in the real world, namely biological systems. Artificial intelligence has consistently drawn on cognitive psychology and its models of human thought processes to deal with the problems of machine learning and reasoning. Robotics researchers would do well to give heed to the theories of navigation and path planning in living organisms.

This paper will present a mobile robot navigation system that bears a strong resemblance to a model by Arbib & House (1985) for detour behavior in the frog. Section 2 will describe the intention for this schema-based path execution system. Section 3 will discuss the operation of the path execution component of the system in the overall mobile robot architecture. The potential field methodology used to implement the system will then be presented in Section 4 along with simulation results. Section 5 will discuss the behaviors exhibited by an actual mobile robot using this system.

2 Intention

2.1 Path Planning and Execution

Path execution at its lowest level can be viewed as the dynamic modification of higher-level plans based on environmental sensing. Goal-achieving behavior is not necessarily the primary determinant in such a system. Barto (1983) describes models and simulates the behavior of an organism attracted by the presence of food. Its motor behavior is predominantly random with the frequency of changes in direction increasing as the organism moves farther and farther from its goal.

Motivated by insect behavior, Rod Brooks and his colleagues at MIT are building working mobile robots (Brooks 1986a). A subsumption architecture is forwarded (Brooks 1986b) that enables the layering of multiple behaviors in a convenient yet somewhat rigid manner. Neuroscientific models of insect navigation are not presented, however, to support this architecture as the means by which insects actually accomplish this task. Mittelstaedt (1985) presents path planning models for spiders and millipedes that warrant such an investigation (see also Mittelstaedt et al. 1979).

Arbib & House (1985) have developed a behavioral model that forwards a mechanism for path planning in amphibians. It is this model which bears a strong resemblance to the path execution mechanisms employed in the Autonomous Robot Architecture (AuRA) developed for use with our mobile robot HARV (Arkin

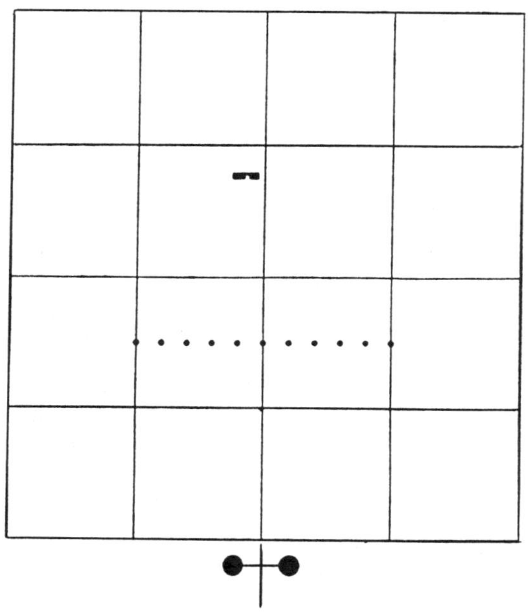

Figure 1. A depiction of a frog prey-selection scenario. The two large blackened circles at the bottom of the figure denote the frog's eyes, the smaller circles are fence-posts, and the darkened rectangle a supply (small cup) of worms. (From Arbib & House 1985).

1987c). The Arbib & House model specifically describes prey-acquisition behavior in an obstacle-strewn environment. This environment consists of the frog (or toad) itself, a small cup of worms, and a series of spaced fenceposts (Fig. 1), according to the experimental paradigm investigated by Collett (1983; see also Lock & Collett 1979). No direct claim is made that this model actually reflects the way these animals truly conduct path planning, but rather that it "will serve to stimulate the design of new behavioral and physiological experiments" (Arbib & House 1985). For us it has gone further, stimulating new experiments in the domains of artificial intelligence and robotics. The basic operation of the Arbib & House model can be summarized as follows: A vector determining the motion of the frog is associated with the frog's

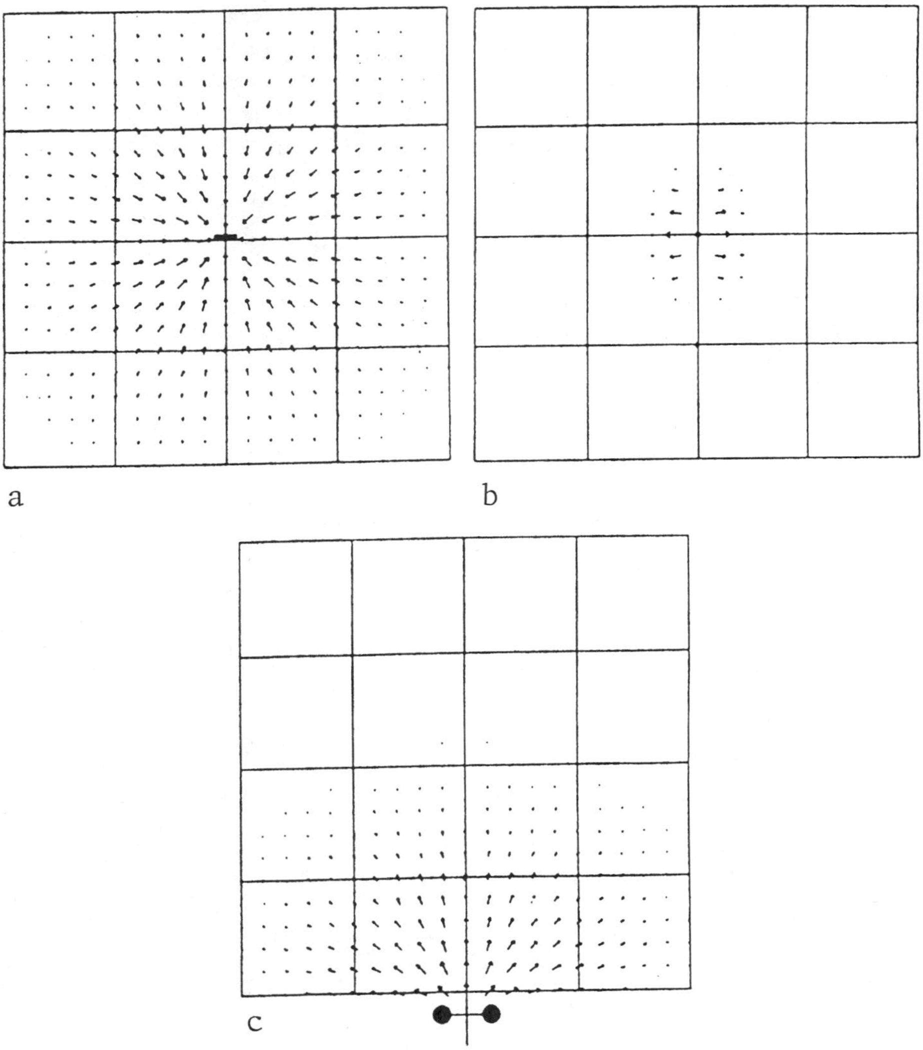

Figure 2. Primitive vector fields associated with Fig. 1. (a) Prey-attractant field. (b) Barrier repellent field. (c) Frog representation field. (From Arbib & House 1985).

position in the environment. These path-determining vectors are computed from the combination of vectors that result from the frog's perception of its environment. Figure 2 illustrates three of these primitive vector fields: prey-attractant, repellant, and a field representing the animal itself. These individual vector components are then summed, with their values arising from the currently perceived environmental conditions, to produce more complex fields (Fig. 3).

It is stated that both ballistic and tracking behaviors can be represented in this manner. These individual fields can be computed in a distributed and parallel manner. No complete path ("continuous trajectory") is computed at this level, but the model "rather serves to generate a map of motor targets" (Arbib & House 1985). The divergence operator is forwarded as the means by which these maps are computed (Fig. 3b). The animal does not make choices, but rather reacts to its environment based on its current perceptions. Vector fields for other motor actions are envisioned, including side-stepping, turning, and snapping. A body-centered frame of reference coordinate system is advocated. Arbib & House (1985) state that "these preliminary results are in strong agreement with the behavioral data." They propose that it is possible that this form of processing can be conducted by neural structures and as new neuroscientific data arrives, refinement of the model will occur.

2.2 Schemas

Schema theory has a wider literature than does path planning models for the frog. Ethologists, psychologists, neuroscientists, computer scientists, and roboticists all have interpretations (see contributions by Ewert, Arbib, Cervantes-Pérez, and

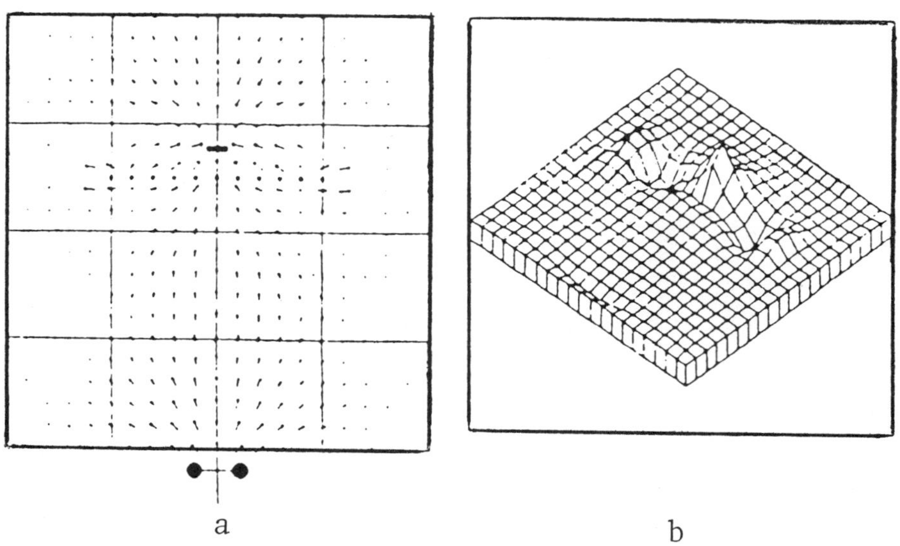

Figure 3. Resultant frog-prey selection vector field. (a) Result produced by vector field. (b) Associated divergence field. (From Arbib & House 1985).

Koy-Oberthür, this volume). The following list presents a subset of schema definitions; most of these are compatible with the concept of schema as used within AuRA.

A schema is:

- an internal small-scale model of external reality (Craik 1943);
- a perceptual whole corresponding to a mental entity (Piaget 1971);
- a pattern of action as well as a pattern for action (Neisser 1976);
- an adaptive controller that uses an identification procedure to update its representation of the object being controled (Arbib 1981);
- a mental codification of experience that includes a particular organized way of perceiving cognitively and responding to a complex situation or set of stimuli (Webster 1984);
- a control system which continually monitors feedback from the system it controls to determine the appropriate pattern of action for achieving the motor schema's goals (Overton 1984);
- a generic specification of a computing agent (Lyons 1986);
- a functional unit receiving special information, anticipating a possible perceptual content, matching itself to the perceived information (Koy-Oberthür, this volume).

The key features for our concept of schema in the mobile robotics domain are:

(i) they act as individual yet communicating computing agents (amenable to distributing processing);
(ii) they contain the ability to access environmental information (perceptual schemas) or effect motor action (motor schemas);
(iii) each represents a basic behavior which when combined with other schemas yields more complex behavior.

The mobile robot schemas employed within AuRA (Arkin 1987a) do not necessarily represent the perception and motor schemas forwarded by other researchers in other fields. Our schema-based navigation system is predominantly concerned with producing reliable and effective navigation in a dynamically changing world in real-time. The mechanisms by which our mobile robot navigates are not proposed to be the way in which animals accomplish this task. Nonetheless, it is a goal of this paper to show that theories for path planning behavior in animals can be of great value in providing significant insights into the design of functioning mobile robot systems.

3 Methodology of Schema-Based Navigation within AuRA

Motor schemas, when instantiated, must drive the robot to interact with its environment. On the highest level, this will be to satisfy a goal developed within the planning system, and on the lowest level, to produce specific translations and rotations of the robot vehicle.

AuRA's pilot is charged with implementing leg-by-leg the piecewise linear path developed by the navigator (Arkin 1986). To do so, the pilot chooses from a repertoire of available sensing strategies and motor behaviors (schemas) and passes them to the motor schema manager for instantiation. Distributed control and low-level planning occur within the confines of the motor schema manager during its attempt to satisfy the navigational requirements. Multiple concurrent behaviors (schemas) may be present during any leg, for example:

- *Stay-On-Path* (a sidewalk or a hall)
- *Avoid-Static-Obstacles* (parked cars, trees, etc.)
- *Avoid-Moving-Obstacles* (people, moving vehicles, etc.)
- *Find-Intersection* (to determine end of path)
- *Find-Landmark* (e.g., building for localization)

The first three are examples of motor schemas, the last two perceptual schemas. To provide the correct behavior, a subset of perceptual schemas must be associated with each motor schema. For example, in order to stay on the sidewalk, a "*Find-Path-Edge* (sidewalk) perceptual schema" can be instantiated to provide the necessary data for the "*Stay-On-Path* motor schema" to operate. If the uncertainty in the actual location of the sidewalk can be determined, the SI's (schema instantiations) associated velocity field, applying pressure to remain on the sidewalk, will reflect this uncertainty measure. The same holds for obstacle avoidance: if a perceptual schema for obstacle detection returns the position of a suspected obstacle and the relative certainty of its existence, the actual avoidance maneuvering will depend not only on whether an obstacle is detected, but also on how certain we are that it exists. Differing strategies within each SI will determine how to manage the perceptual uncertainty. If an event is potentially fatal, even large amounts of perceptual uncertainty would produce motor behavior, but erring in the direction of safety.

An example illustrating the relationship between motor schemas and perceptual certainty follows. The robot is moving across a field in a particular direction (*Move-Ahead* schema). The *Find-Obstacle* schema is constantly on the look-out for possible obstacles within a subwindow of the video image (windowed by the direction and velocity of the robot). When an event occurs, (e.g., a region segmentation algorithm detects an area that is distinct from the surrounding backdrop or an interest operator locates a high-interest point in the direction of the robot's motion), the *Find-Obstacle* schema spawns off an associated perceptual schema (*Static-Obstacle* SI) for that portion of the image. It is now the *Static-Obstacle* SI's responsibility to continuously monitor that region. Any other events that occur elsewhere in the image spawn off separate *Static-Obstacle* SIs. Additionally, an "*Avoid-Static-Obstacle* SI motor schema" is created for each detected potential obstacle.

The motor schema SI pauses waiting for notification that the perceptual schema is sufficiently confident in the obstacle's existence to warrant motor action. If the perceptual schema proves to be a phantom (e.g., a shadow) and not an obstacle at all, both the perceptual and related motor SIs are deinstantiated before producing any motor action. On the other hand, if the perceptual SI's confidence (activation level) exceeds the motor SI's threshold for action, the motor schema starts producing a repulsive field surrounding the obstacle. The sphere of influence (spatial extent of repulsive forces) and the intensity of repulsion of the obstacle are affected by the distance from the robot and the obstacle's perceptual certainty. Eventually, when the robot moves beyond the perceptual range of the obstacle, both the motor and perceptual SIs are deinstantiated. In summary, when obstacles are detected with sufficient certainty, the motor schema associated with a particular obstacle (its SI) starts to produce a force tending to move the robot away from the object. Figure 4a shows a typical repulsive field for an *Avoid-Static-Obstacle* SI. The control of the priorities of the behaviors (e.g., when is it more important to follow the sidewalk than to avoid in certain but possible obstacles) is partially dependent on the uncertainty associated with the obstacle's representation. Other isolated motor schema velocity fields are shown in Fig. 4b-d. A typical combination of motor schemas is illustrated in Fig. 5. Note the fact that although the entire field is expensive to compute, each active motor SI need only determine the velocity vector at the robot's current location relative to the environmental object, making the computation very rapid. Furthermore, as the SIs are activated in parallel, even better performance is attainable.

The similarity between the repellant field of the Arbib & House model for detour behavior in the frog and the *Avoid-Static-Obstacle* SI should be apparent upon inspection of the figures pertaining to both (Figs. 2b and 4a). Note, however, the asymmetry in the Arbib & House fencepost representation has been removed. The relationship between the prey-attractant field and the *Move-To-Goal* SI (Figs. 2a and 4b) is also readily apparent. Less obvious is the relationship between the animal field and the *Move-Ahead* SI (Figs. 2c and 4c). The formulation for our velocity fields are different in most cases than the Arbib & House approach. There is nothing intrinsic in our representation that makes them superior, other than the fact they are simple and they work for our mobile robot. Other formulations might provide even better results, especially for higher-velocity robots.

In AuRA, a combination mechanism must be applied to all the SI-produced vectors. The result is then used to provide the necessary velocity changes to the robot. The simplest approach is vector addition. By having each motor SI create a normalized velocity vector, a single *Move-Robot* schema monitors the posted data for each SI, adds them together, makes certain that it is within acceptable bounds and then transmits it to the low-level robot control system. In essence, the specific velocity and direction for the robot can be determined at any point in time by summing the output vectors of all the active individual SIs. As each motor SI is a distributed computing agent, preferably operating on separate processors on a parallel machine, with each needing only to compute the velocity at the point the robot is currently located and a few points in its projected track (not the entire velocity field), real-time operation is within reach.

Note that in AURA's method the divergence field is not computed, in contrast to the Arbib & House model, so that no local map is generated. Instead the vector addition approach computes instantaneous velocities for our robot. The control issue now is to match the current state of the robot to the velocity determined by the interaction of all the active motor schemas.

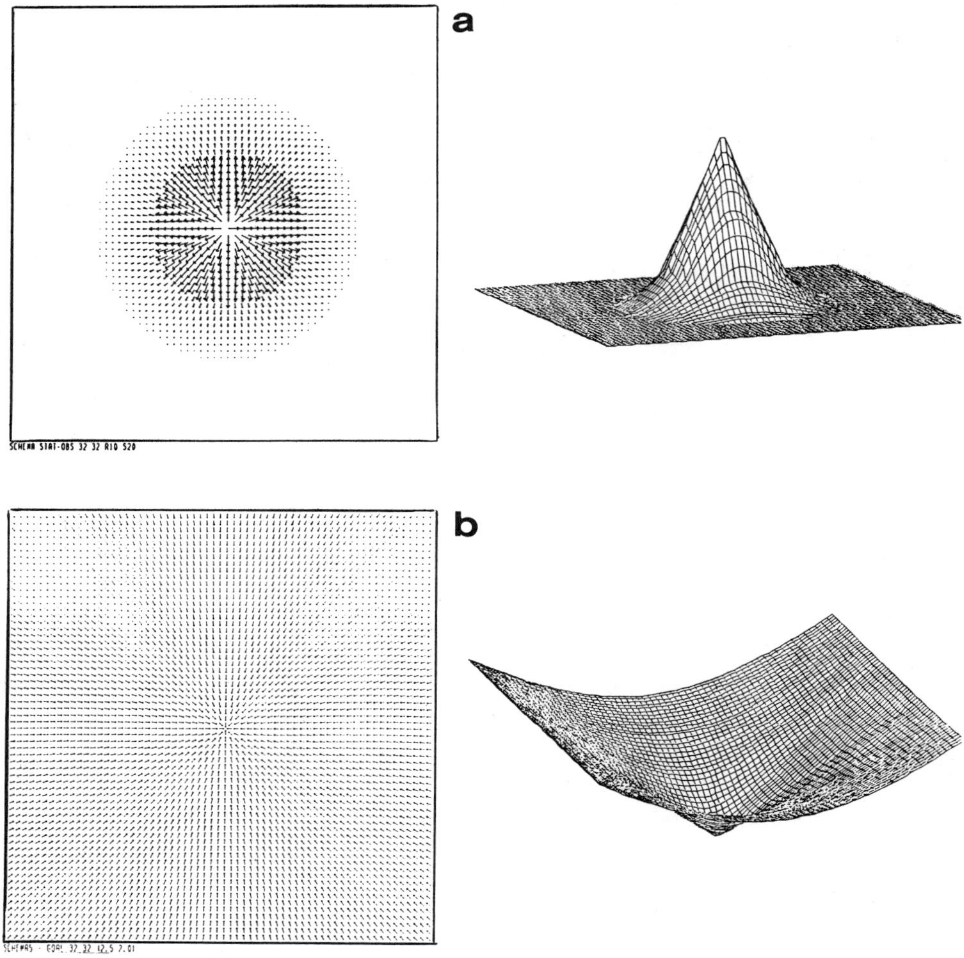

Figure 4. AuRA's isolated motor schema SI vector fields. The figures to the left are the 2-D field representations, while the figures to the right are the 3-D analogs of the velocity fields. (a) *Avoid-Static-Obstacle.* (b) *Move-To-Goal.* (c) *Move-Ahead.* (d) *Stay-On-Path.* (Fig. 4 cont'd on opposite page).

Aside from the obvious differences in the Arbib & House model and AuRA's approach, there still exists a strong parallel. Both account for observed motor behavior; in the first case for a frog, in the second case for a robot.

It is entirely possible that the trajectory of the robot can be computed for a small distance rather than just its instantaneous velocity at the immediate location. Each motor schema would have to interact with the *Move-Robot* SI, using the vector summation output to determine the position of the robot relative to its perceptions for the next time step. This is of particular significance if the sensor sampling rates are low. Trajectories can be determined that reflect the robot's perceptions at a given

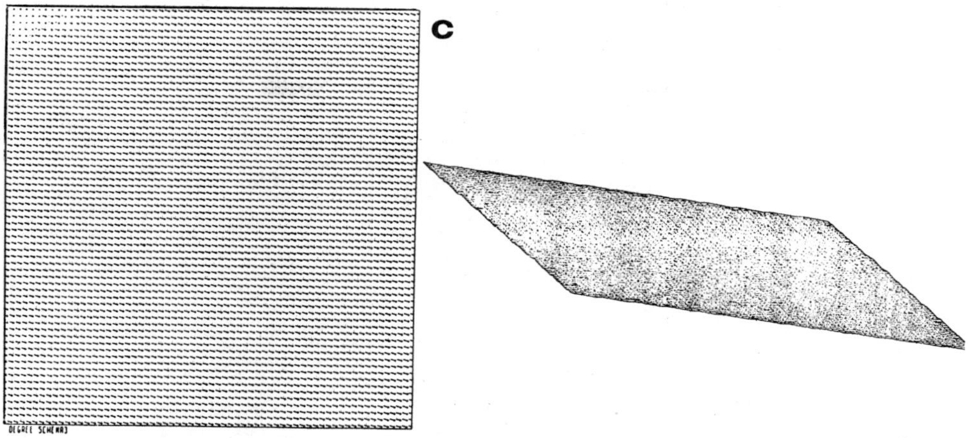

c

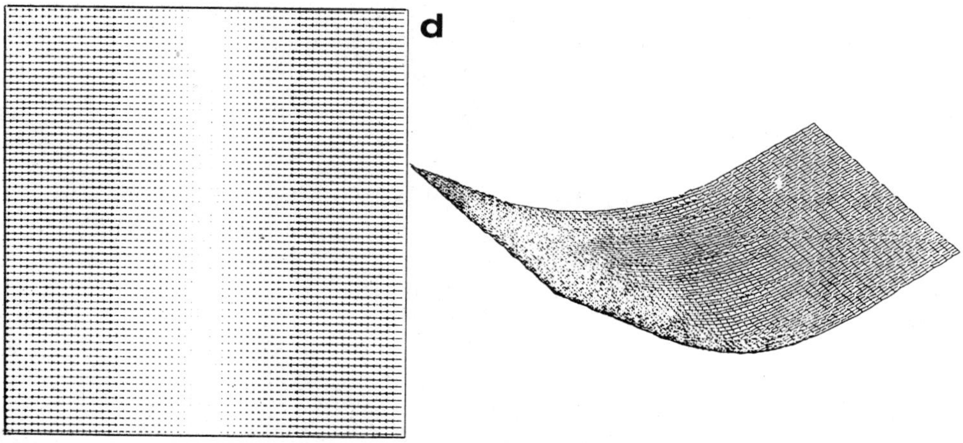

d

point in time, rather than just reacting to current sensing. This is of value in determining when to activate other schemas in anticipation of special problems or needs. Care must be taken in highly dynamic environments (e.g., moving objects), so that the plans developed do not ignore changes in the world that are evidenced only through perception.

Another approach explored is the addition of a background stochastic noise schema. This SI produces a low-magnitude random direction velocity vector that changes at random time intervals, but persists sufficiently long to produce a change in the robot's position if the robot's velocity was otherwise zero. Its role is to perturb the velocity of the robot slightly, removing it from undesirable equilibrium points which arise when the active motor SIs counterbalance each other. This schema serves to remove the robot from any potential field plateaus or ridges upon which it becomes perched. Other traps common to potential field approaches (e.g., box canyons) can be handled by establishing hard time deadlines for goal attainment.

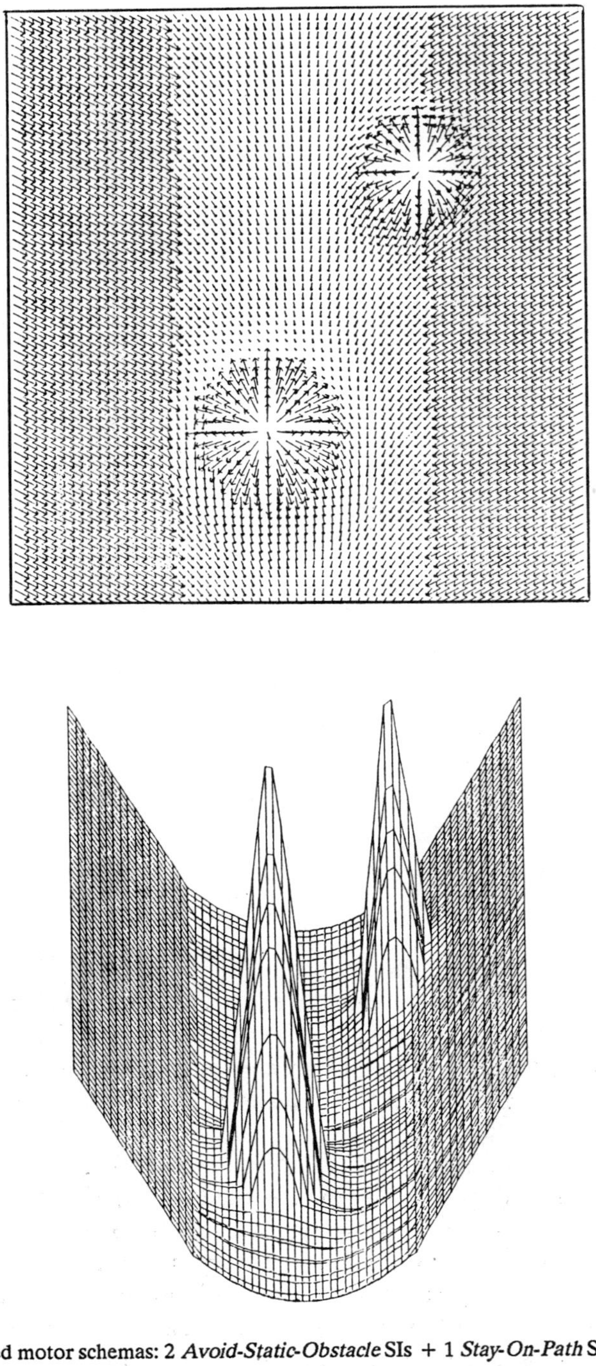

Figure 5. Combined motor schemas: 2 *Avoid-Static-Obstacle* SIs + 1 *Stay-On-Path* SI + 1 *Move-To-Goal* SI. The top figure is the 2-D representation of the field, the bottom is the 3-D analog.

4 Simulation

Simulations were run on a VAX 750 computer using the following motor schemas: *Stay-On-Path*, *Move-Ahead*, *Move-To-Goal*, *Avoid-Static-Obstacle*. A simulation run (Fig. 6) shows the sequence of resultant overall force fields based on perceived entities. These entities include path borders, goals, and obstacles.

The field equations for both the *Avoid-Static-Obstacle* and *Stay-On-Path* schemas are linear. An example showing the velocity produced by an obstacle is given below:

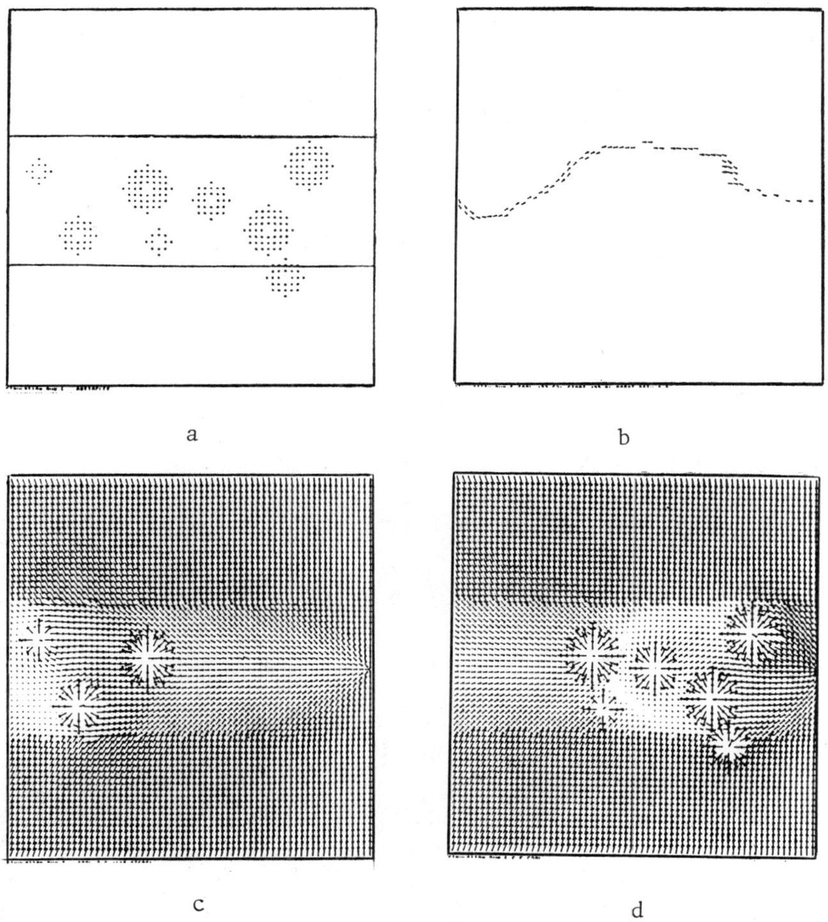

Figure 6. Schema simulation run. This simulation shows 8 *Avoid-Static-Obstacle* SIs, a *Move-To-Goal* SI, and a *Stay-On-Path* SI. (a) Shows the layout of the obstacle ridden course. (b) Simulated robot path through course. (c-d) With the robot starting at the far left, the robot's progress through the course can be observed. Note that the obstacles are added as they are perceived by the sensory system. No *a priori* knowledge of their location is assumed.

Avoid-Obstacle

$V_{magnitude}$ = 0 for d > S
{[(S-d)/(S-R)]}G for R < d ≤ S
infinity for d ≤ R

where: S = sphere of influence (radial extent of force from the center
of the obstacle); R = radius of obstacle; G = gain; d = distance of robot to center of obstacle

$V_{direction}$ = along a line from robot to center of obstacle moving away from obstacle

Stay-On-Path

$V_{magnitude}$ = P for d > (W/2)
{d/(W/2)}G for d ≤ (W/2)

where: W = width of path; P = off path gain; G = on path gain; d = distance of robot to center of path

$V_{direction}$ = along a line from robot to center of path heading toward centerline

Move-Ahead

$V_{magnitude}$ = fixed gain value

$V_{direction}$ = in specified compass direction

Move-To-Goal

$V_{magnitude}$ = fixed gain value

$V_{direction}$ = in direction towards perceived goal

More complex equations could be used (e.g., cubic as in Krogh & Thorpe 1986) but were deemed unnecessary for the low velocities employed.

Figure 6a illustrates the robot's course on a sidewalk moving towards a goal on a course studded with 8 obstacles. Note how the vector fields change as the robot encounters more obstacles along the way (Figs. 6c-d). When it has successfully navigated obstacles and they have moved out of range, their representation is dropped from short-term memory and the associated motor schema is deinstantiated (Fig. 6d). The robot stays on the path for the complete course, successfully achieving its goal while avoiding each obstacle.

5 Experimental Results

A real-time schema experimentation/demonstration system was developed based on sensing using ultrasonic and encoder data. Five different motor schemas have been implemented: *Move-Ahead* (encoder based), *Move-To-Goal* (encoder based), *Avoid-Static-Obstacle* (ultrasonic based), *Noise* (sensor independent), and *Follow-The-Leader* (ultrasonic based). The user is able to select the collection of motor schemas to use and associate a perceptual schema with each. After schema selection is complete, robot motion is initiated. The vehicle then behaves in an intelligent manner in response to its environmental stimuli. Several of the more interesting behaviors are described below. It should be noted that the schemas for this system emulate distributed processing, but actually are evaluated sequentially.

5.1 Avoidance

By instantiating the *Avoid-Static-Obstacle* schema with ultrasonic perception, the robot manifests avoidance behavior. The schema instantiation can be controlled by altering both the gain, which affects the velocity of the vehicle, or the sphere of influence of detected obstacles, which increases its sensitivity to the environment. When activated, the robot seeks out a potential field minimum and remains there unmoving (or slightly oscillating in cluttered environments due to the limited sampling rates and noise in the sensor data).

If a change in the environment occurs, (e.g., a person approaches), the robot is repulsed and seeks out a new potential minimum. The robot can be "herded" by following behind it, forcing it to move to a desired location. It avoids obstacles during its journey and then settles into its new location when the environment stabilizes. Figure 7 illustrates this process. The behavior superficially resembles a frog hopping away when it is approached.

5.2 Exploration

By combining the noise schema with the *Avoid-Static-Obstacle* schema, exploration behavior can be observed. The noise schema's gain (strength) and persistence (how frequently the direction changes) can be set at start-up by the experimenter. The robot meanders about the laboratory exploring different regions while avoiding collision with the obstacles.

The robot reacts quite well in response to changes in its environment, as when people surround it during a demonstration. The biggest problem HARV faces is the slow sensor sampling which makes its reflexes quite slow. The vehicle can also be herded when running in this behavioral mode, but it is not quite as obedient, moving in the general direction forced upon it but occasionally making some slight sidesteps due to the presence of noise. Actually, this form of herding is more reminiscent of an animal's behavior, and HARV has been likened to a sheep by some observers when it's running in this mode.

a

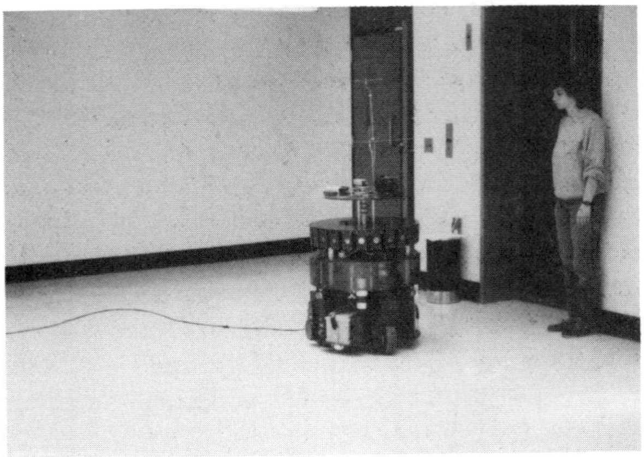

b

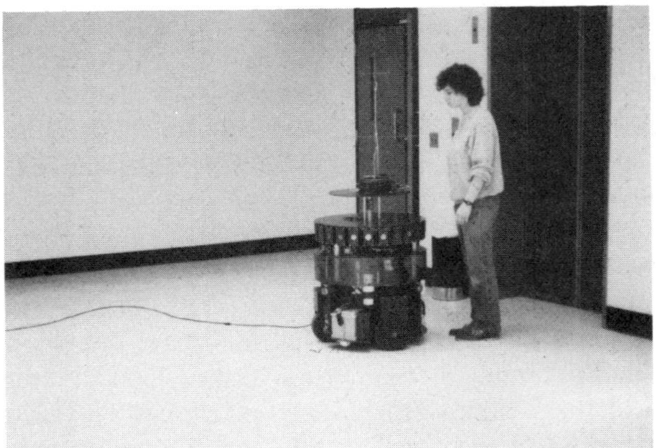

c

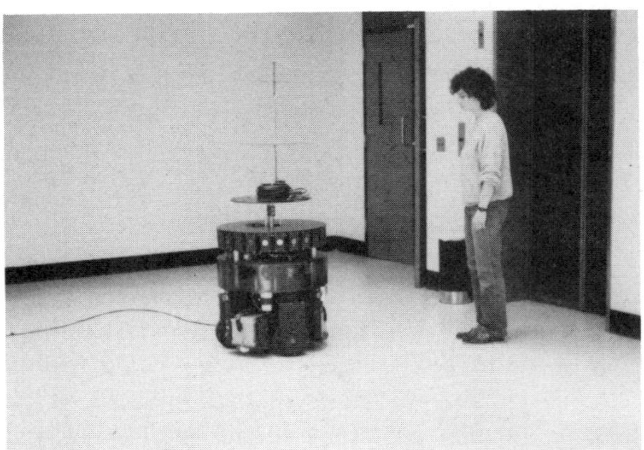

Figure 7. Avoidance behavior. In this sequence, the robot is initially stationary (a). When the robot is approached (b), it is repulsed and moves to its new position and then stops (c). This behavior is produced by the instantiation of an *Avoid-Static-Obstacle* schema.

5.3 Hall Following

By instantiating the *Move-Ahead* schema with an *Avoid-Static-Obstacle* schema, the robot is able to safely travel down a hall. Although there is no model of the hall (at the schema level), the robot senses the two sides as obstacles and positions itself near the center. The *Move-Ahead* schema drives it forward.

HARV responds to changes in the hall structure itself such as pillars, seeking out the middle of the hall. Small obstacles placed near the wall's sides are interpreted as part of the wall by the ultrasonic sensors. Open doorways pose no problem as long as the *Move-Ahead* direction is roughly parallel to the hall. If this direction points steeply into the wall, the robot might pass through an open door (see door entry below), but since the angle into the wall typically must be at least 45 degrees or more for normal doorway entry, this allows a latitude of about 90 degrees ($\pm$ 45) of error in the move-ahead direction for successful navigation of halls using this method.

5.4 Navigation in the Presence of Obstacles

The same combination of schemas allows for navigation in cluttered hallways or outdoor situations. The *Move-To-Goal* schema can be substituted freely for the *Move-Ahead* schema if desired. Figure 8 shows the robot's course through a series of cluttered obstacles. The *Move-To-Goal* schema relies on the shaft encoders to move the robot to a particular point. This is accomplished by specifying the distance to and direction towards the goal and then monitoring the encoder data to provide the input to the *Move-To-Goal* SI.

5.5 Single Wall Following

A "drunken sailor" walk can be produced by directing a *Move-Ahead* schema into a wall with the obstacle avoidance behavior active (Fig. 9). The robot slides along a repulsive field a specified distance from the wall, reacting to obstacles as it moves. This also allows the robot to enter doorways if the vector pointing into the wall is sufficiently large in magnitude and its angle is sufficiently steep.

5.6 Impatient Waiting

The potential fields methodology is strongly in evidence when the robot moves into a box canyon. The robot ends up in a potential well and is not able to make meaningful progress. Normally, after a time-out occurs based on a hard real-time deadline, the pilot and/or navigator would be reinvoked to compute an alternate path. Suppose, however, the blockage is temporary, due to closed doors at the end of a hall or an elevator. The robot can wait, rocking around in its potential well, until the obstruction is removed. This behavior is similar to what might be evidenced by a fly at a window. When the path becomes unblocked, the robot continues ahead, trying to satisfy its initial goals. Figure 10 depicts this process.

a b c

Figure 8. Navigation. A *Move-To-Goal* SI (shaft-encoder based) was used along with an *Avoid-Obstacle* SI, the goal being a location near the distant cone that does not have a ball on it. The robot's course winds through the obstacles while making its way towards the goal.

d

e

f

a

b

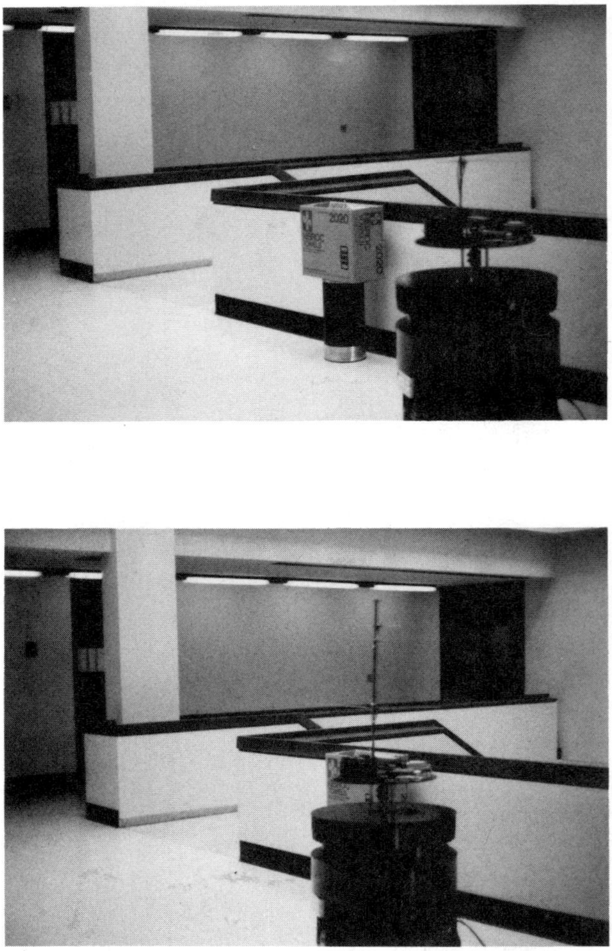

Figure 9. Wall following ("drunken sailor"). This behavior is produced by the combination of a *Move-Ahead* SI directed obliquely into the wall while keeping the obstacle avoidance behavior active. In the sequence shown, the robot follows the single wall, moving around the obstacle, and then staggers towards the staircase. The behavior appears as if HARV was leaning against the wall while moving forward, conforming to its contours. HARV would have fallen down the stairs (to the right), if it was not shut off by our technician (classic "drunken sailor" behavior).

c

d

e

a

b

c

Figure 10. Impatient waiting ("fly at a window"). This sequence shows what can occur when the robot enters a potential well. A *Move-Ahead* SI directs the robot to the end of the hall, while the *Avoid-Static-Obstacle* behavior keeps it from colliding with the door. The robot roves impatiently at the end of the corridor until the door is opened. At that point, it moves through the door and continues on its way.

d

e

f

5.7 Follow-The-Leader

HARV can track a moving object (with additional obstacle avoidance behavior to aid maneuvering in tight situations if desired). The appearance is similar to walking the robot on an invisible leash. It turns and tracks the nearest object within a given ultrasonic sensor spread, moving more rapidly as the distance between robot and object increases, but avoiding contact with the tracked object. The potential well (the ideal distance separating the tracked object and the robot) is user specifiable as is the separation distance at which the robot abandons following.

5.8 Motor Schema Experiment Summary

The motor schema experimentation/demonstration system enables visualization of motor schema-based potential field navigation. The greatest drawback lies in the relatively slow sampling rates for processing the ultrasonic data on a shared VAX-750 (typically 2-3 s for 24 sensor readings) causing minor oscillations of the vehicle in highly dynamic or extremely cluttered environments when continuous motion is allowed, since the sensor data becomes old relative to the robot's position.

6 Summary and Conclusions

A working mobile robot system capable of interacting effectively with a dynamic world has been constructed. Multiple active and concurrent schemas contribute to the determination of the overall path of the vehicle. Perceptual schemas are associated with specific goal-achieving motor behaviors to provide the essential data for the task a hand (cf. also "on-line specialists" in robotic sensorimotor coordination; Torras, this volume). The AuRA architecture has been profoundly influenced by cybernetic considerations. This is perhaps nowhere more apparent than at the motor schema navigational level. The Arbib & House model of detour behavior of the frog provides a strong correlate between the approach used within our mobile robot system and actual animal behavior.

Future work includes development of the AuRA architecture to encompass navigation in a three dimensional world (Arkin 1987b). Studies of the behavior of fish and flying insects may add important strategies to employ along these lines.

One thing that cannot be overemphasized is the need for the robotics researcher to take advantage of models created by neuroscientists. Although it is generally inadvisable (and impractical) to map these models unchanged onto robotic systems due to the obvious differences in the underlying hardware, much can be gained by looking to this work for insights and inspiration.

Acknowledgement. This research was supported in part by the U.S. Army under ETL grant DACA76-85-C-008 and by the NSF under grant DCR-8318776 while the author was employed at the University of Massachusetts, Amherst.

7 References

Arbib M (1981) Perceptual structures and distributed motor control. In: Brooks R (ed) *Handbook of physiology. The nervous system II.* American Physiological Society, Bethesda MD, pp 1449-1465

Arbib M, House D (1985) Depth and detours: an essay on visually guided behavior. *COINS Tech Rep* 85-20

Arkin R (1986) Path planning for a vision-based mobile robot. *COINS Tech Rep* 86-48

Arkin R (1987a) Motor schema based navigation for a mobile robot: an approach to programming by behavior. *Proc IEEE Int Conf on Robotics and Automation*, pp 264-271

Arkin R (1987b) Reactive/reflexive navigation for an autonomous vehicle. *Proc of the American Institute for Aeronautics and Astronautics Computers in Aerospace VI Conference*, Wakefield MA, pp 298-306

Arkin R (1987c) *Towards cosmopolitan robots: intelligent navigation in extended man-made environments.* PhD Dissertation, Dept of Computer and Information Science, Univ of Mass, Amherst (see also *COINS Tech Rep* 87: 80)

Barto A, Anderson C, Sutton R (1982) Synthesis of nonlinear control surfaces by a layered associative search network. *Biol Cybern* 43: 175-185

Brooks R (1986a) Achieving artificial intelligence through building robots. *MIT A I Lab Memo* 899

Brooks R (1986b) A robust layered control system for a mobile robot. *IEEE Journal of Robotics and Automation* 2(1): 14-23

Collett T (1983) Picking a route: do toads follow rules or make plans? In: Ewert J-P, Capranica RR, Ingle DJ (eds) *Advances in vertebrate neuroethology.* Plenum Press, New York, pp 321-330

Craik KJW (1943) *The nature of explanation.* Cambridge Univ Press, London

Krogh B, Thorpe C (1986) Integrated path planning and dynamic steering control for autonomous vehicles. *Proc IEEE Int Conf on Robotics and Automation*, pp 1664-1669

Lock A, Collett T (1979) A toad's devious approach to its prey: a study of some complex uses of depth vision. *J Comp Physiol* 131: 179-189

Lyons D (1986) *RS: a formal model of distributed computation for sensory-based robot control.* PhD Dissertation, Dept of Computer and Information Science, Univ of Mass, Amherst (see also *COINS Tech Rep* 86-43)

Mittelstaedt H (1985) Analytical cybernetics of spider navigation. In: Barth FG (ed) *Neurobiology of arachnids.* Springer-Verlag, Berlin Heidelberg New York, pp 298-316

Mittelstaedt M, Mittelstaedt H, Mohren W (1979) Interaction of gravity and idiothetic course control in millipedes. *J Comp Physiol* 133: 267-281

Neisser U (1976) *Cognition and reality: principles and implications of cognitive psychology.* Freeman, San Francisco

Overton K (1984) *The acquisition, processing, and use of tactile sensor data in robot control.* PhD Dissertation, Dept of Computer and Information Science, Univ of Mass, Amherst (see also *COINS Tech Rep* 84-08)

Piaget J (1971) *Biology of knowledge.* Edinburgh Univ Pess, Edinburgh

Webster M (1984) *Webster's ninth new collegiate dictionary.* Merriam-Webster, Springfield, MA

Sensorimotor Integration in Robots

CARME TORRAS

Institut de Cibernètica (CSIC-UPC)
Diagonal 647, 08028-Barcelona, Spain

Abstract. The chapter starts with a survey of the different types of sensors used in robotics, to next introduce the degrees of sensorimotor integration required to accomplish some generic tasks. Three such degrees are discussed: Two-stage processing, sensation-action interleaving, and simultaneous operation of sensors and actuators. For the purpose of illustration, a robotic assembly system currently under development at the Institut de Cibernètica is presented, and the way in which the system handles some sample tasks involving various degrees of sensorimotor integration is described. A selected set of general references about the different topics is provided.

1 Introduction

Sensorimotor integration has been the subject of numerous neuroethological and neurophysiological studies, but surprisingly such term is almost entirely missing in the robotic literature. However, many robots have been endowed with sensorimotor capabilities, since these are required to accomplish interesting tasks when there are uncertainties in either the environment or the functioning of the robot itself. Thus, although the robotics community has not yet reached the stage at which it would be possible to formulate a unified theory of sensorimotor integration in robots, it is certainly possible to point out successful ways in which the sensorimotor interface has been implemented (see also contributions by Arkin and Koy-Oberthür, this volume). The aim of this chapter is to provide the reader with enough background and references to construct his/her own picture of the state-of-the-art in the field.

The chapter is structured as follows. In Section 2, the most common types of sensors used in robotics are surveyed: vision, range, proximity, force and torque, and tactile sensors. Section 3 introduces sensorimotor integration as falling somewhere in between multisensory integration and motor coordination. Section 4 describes a robotic assembly project that is now under development at the Institut de Cibernètica, and provides an overview of the various modules designed to first plan and then actually carry out sensorimotor integration. In Section 5, three degrees of sensorimotor integration are distinguished, and the factors that determine the degree actually implemented in a robotic system are sketched. Some sample tasks requiring each of the three degrees of sensorimotor integration are described in Section 6. Finally, some points for discussion are raised in Section 7.

2 Types of Sensors Used in Robotics

Robots, like living beings, have the possibility to sense both their posture (i.e., the position of their articulations) and the environment around them. In this chapter, we consider only the latter sensory capability, and thus by sensors we refer only to "external" sensors. According to this meaning, the most common types of sensors with which robots have been equipped are: vision, range, proximity, force and torque, and tactile sensors.

Visual sensors (i.e., linear or array cameras) provide information about the intensity of light at different points in the scene. This information is usually digitized and processed to extract a symbolic representation of the scene in terms of edges, surfaces and spatial relationships between objects (Hanson & Riseman 1978; Barrow & Tenenbaum 1981; Ballard & Brown 1982; Brady 1982; Marr 1982; Binford 1982; Torras 1984). Although for some constrained tasks (e.g., recognition of flat objects on a table) a two-dimensional (2-D) representation suffices, the majority of real-world tasks require a 3-D representation (Besl & Jain 1985). While a sweep with a linear camera or a single shot of an array camera can provide sensory data needed for 2-D vision, additional information is required to obtain 3-D structure.

Several techniques have been used to gather this information, some inspired by biological vision and others exploiting available scientific and technological capabilities (Amat et al. 1983; Kak 1985). Among the first, stereopsis and depth from motion require matching corresponding patterns in separate image frames (Ullman 1979; Barnard & Fischler 1982; Anandan 1985). The techniques driven by technological developments amount to augmenting the 2-D information supplied by a camera with a depth measure obtained by means of a *range finding* device (Jarvis 1983). These devices calculate the time-of-flight of either a laser or an ultrasound signal from emission to receival after reflection by an object. Other techniques exploit knowledge of physical constraints of either the environment or the image acquisition process; this is the case in the procedures that extract shape from shading, occlusion and texture information from a single image (Barrow & Tenenbaum 1978; Ikeuchi 1980; Ikeuchi & Horn 1981; Witkin 1981), and those that derive depth from either the deformation suffered by a striped pattern of light projected onto the scene (VanderBrug et al. 1979; Hall et al. 1982; Sato et al. 1982) or the difference between the original image and the one obtained by projecting a graded light pattern onto the scene (Schwartz 1983; Bastuscheck 1984).

Proximity sensors are also used to measure distances, but within a short range (Espiau & Catros 1980; Balek & Kelley 1985; Thomas 1987). They can serve other purposes too, like estimating surface orientation or edge tracking. Although the most common of these sensors consist of pairs of emitters and receivers of infrared light (see Fig. 5), other configurations are also possible (Masuda 1986). Distance information is obtained from either the intensity of the reflected light, or the phase shift between the signals emitted and received in the case that two phase-shifted emitters are used.

	VISION	RANGE	PROXIMITY	FORCE	TOUCH
CONTACT				▨	▨
NONCONTACT	▨	▨	▨		
POINT		▨	▨	▨	
MAP	▨	□			▨
MOUNTED	□		▨	▨	▨
EXTERNAL	▨	▨		□	□
STATIC	▨	□		□	□
DYNAMIC	▨	▨	▨	▨	▨

Figure 1. Characterization of the five types of sensors most widely used in the robotics domain.

Force and torque sensors provide the 3 force and 3 torque components referred to a coordinate frame (Nevins & Whitney 1977). These sensors are used to implement compliance strategies for carrying out fine motions while maintaining contact with an object, such as those required in tight insertions and edge polishing. Although the most widely used configuration of a force sensor is to have it mounted on the wrist of a robot arm (see Fig. 3), other possibilities are available, such as incorporating it in the table that serves as support for the assembly task or installing it in the fingertips of the gripper.

Tactile sensors are matrices of elements sensitive to force (Overton 1984; Kak & Albus 1985). The most common among them are based on anisotropically conductive materials whose resistivity varies under mechanical pressure; others rely on optical techniques (see Fig. 4). These sensors are usually mounted on the fingertips of a gripper.

In Fig. 1 the five sensor types described above are characterized according to four dichotomic attributes: contact/noncontact, point/map, mounted/external and static/dynamic. Only force and torque sensors and tactile sensors operate by making contact with an object. Vision and touch provide a map of values, while range, proximity and force and torque sensors supply measurements at only one point. A depth map can be obtained by scanning a scene with a range finding device, this being a widely used technique in 3-D vision applications. Visual and range sensors are used mainly as external devices when working with robot arms, although a camera mounted on the wrist can be useful for some applications (Hill & Park 1979).

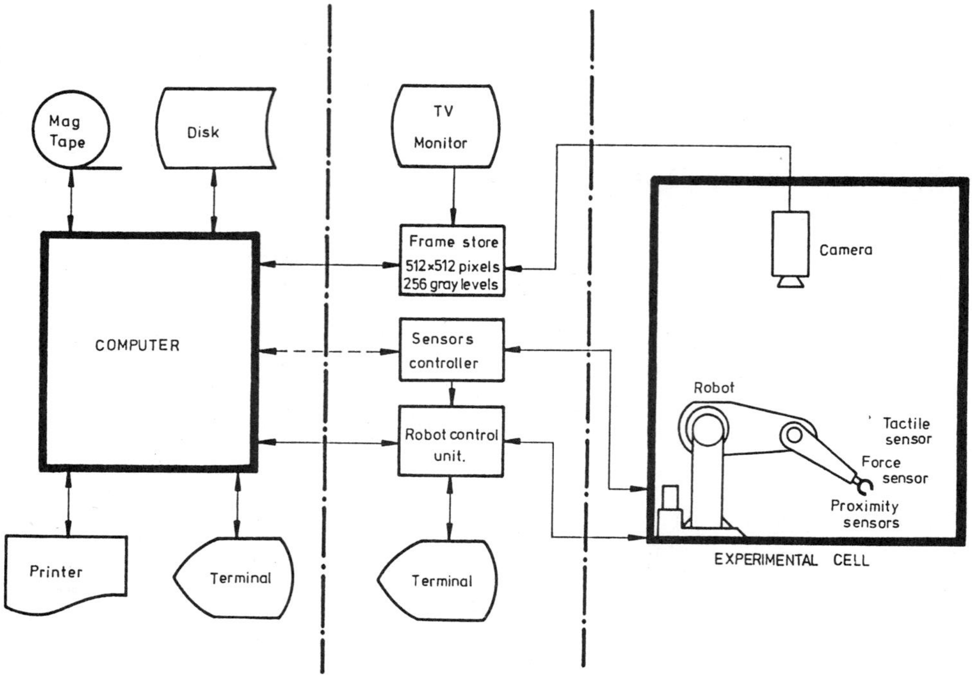

Figure 2. Current set-up of the Robotics Laboratory at the Institut de Cibernètica.

Obviously, mobile robots must carry these sensors in order to make use of their capabilities for navigation purposes. Proximity sensors are always used mounted on the robot, while force and torque sensors, although preferentially used mounted on the robot arm, can also work as external devices. Vision is evenly used in both a static and a dynamic fashion, while all the remaining sensor types are preferentially used in a dynamic way, that is, the acquired information changes with time due to the motion of either the sensor or the object. Other sensor taxonomies somewhat related to the one presented here have been proposed by Rosen & Nitzan (1977), Nitzan (1981), and Overton (1984). Pugh (1986) provides an extensive description of sensors that have been used in robotic applications.

3 Placing Sensorimotor Integration in Context

We can think of sensorimotor integration as falling somewhere in between *multisensory integration* and *motor coordination*. Integration implies combining information from different sources about the same target, while coordination implies combining actions by different agents to attain a common goal. Therefore, in sensory systems we speak of integration, whereas in motor systems we use the term coordination (see Bässler, Berkinblit, Ewert, Grobstein, Mittelstaedt, Schwippert et al., and Weerasuriya, this volume). In my opinion, this is the reason why when referring to the sensorimotor interface, both words are often not distinguished.

Throughout this chapter, I have preferred to use the term integration to make clear that sensory information is "integrated" into a motor pattern, much in the same sense as perceptual schemas are embedded into motor schemas (see contributions by Arbib, Arkin, Cervantes-Pérez, and Koy-Oberthür, this volume). In this way, I emphasize that sensory and motor information are deeply intervowen to come up with a unique pattern of motions, and it is not just that sensory and motor actions are synchronized to attain a unique goal.

Multisensory integration is now a very important research topic within robotics, also referred to as "sensor fusion." Several attempts have been made to integrate the information provided by different sensors into a unique object representation (Allen 1984; Bajcsy & Allen 1984; Magee et al. 1985). These range from combining qualitatively similar signals (e.g., gripper equipped with several proximity sensors, stereovision) to integrating data from different sensory modalities (e.g., range and vision to obtain 3-D models, vision and touch to disambiguate between several objects or between different poses of a known object, vision and proximity sensors to first coarsely locate an object and then accurately locate a suitable grasping point on it).

At least four instances of pure motor coordination have been studied within the robotics domain, namely those involving two or more robot arms (Freund & Hoyer 1986), a robot arm and its gripper (Arbib et al. 1983), two or more robot carts (Schwartz & Sharir 1983; Hopcroft et al. 1984; Yap 1984), and two or more robot legs (McGhee 1977; Raibert & Sutherland 1983). To these, we should add all developments concerning the coordination of an independently moving sensory device with a motor device.

4 An Illustrative Project

Figure 2 shows the set-up we currently have at the Robotics Laboratory of the Institut de Cibernètica. The host computer is a VAX 11/785 from DEC, the robot is a Puma 560 from Unimation, the force and torque sensor is a FS6 from Barry Wright Co. (Fig. 3), and the tactile sensor is a LTS 200 from Lord Co. (Fig. 4). The gripper has been especially designed to incorporate 6 proximity sensors, consisting of one emitter and one receiver of infrared light each (Fig. 5).

A project that is now under development based on the set-up described above will serve to illustrate specific needs for sensorimotor integration in robots, as well as details of the implementation of this capacity in a working system. The project, entitled "Implementing sensor-based assembly robot systems with visual instruction", is financed by the US-Spain Joint Committee for Scientific and Technological Cooperation and is being developed in cooperation with the Robotics and Automation Laboratories of the Rensselaer Polytechnic Institute in Troy (New York).

The goal of the project is the design and implementation of a robotic assembly system which, when given a very high-level specification of an assembly task (roughly in terms of only what pieces have to be inserted in what other pieces), is able to physically carry out the assembly. This involves automatically planning the detailed

a

b

Figure 3. Example of a wrist force and torque sensor. (a) Outside view. (b) Inside view, showing the preprocessing unit included in the sensor that allows high-speed transmission of already preprocessed signals.

sequence of assembly operations to be performed, converting the plan into a program that can be executed by the available robot and sensors, and monitoring task execution, so that the system can recover from errors. A general description of the envisaged system can be found in Kelley & Bonner (1985) and Basañez (1986), and an overview of the accomplishments until the end of 1986 is provided by Torras (1986).

a

b

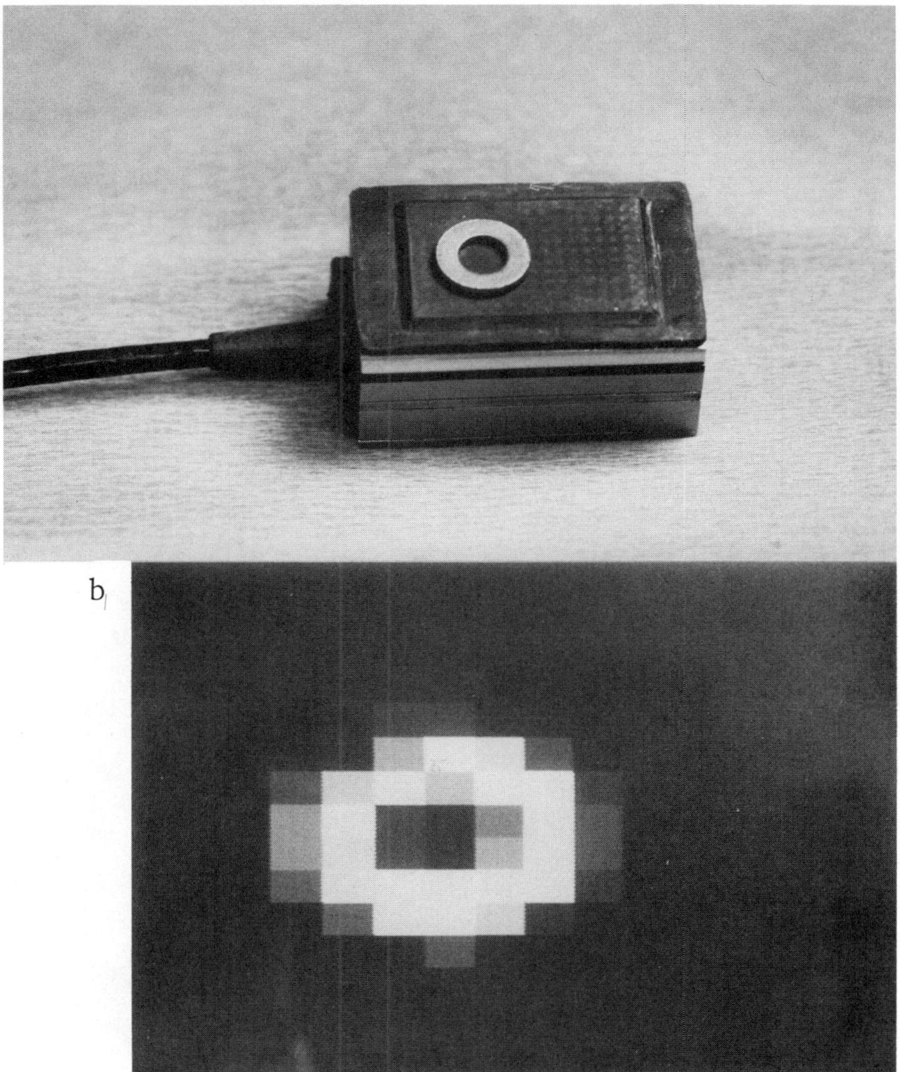

Figure 4. (a) Tactile sensor consisting of a matrix of 10 x 16 elements spaced every 1.8 mm. (b) Tactile image obtained when pressing upon the washer placed on the tactile sensor in (a).

a

b

Figure 5. (a) Parallel jaw gripper equipped with proximity sensors. (b) Detail of one of the fingers of the gripper in (a), showing the three pairs of proximity sensors that it incorporates.

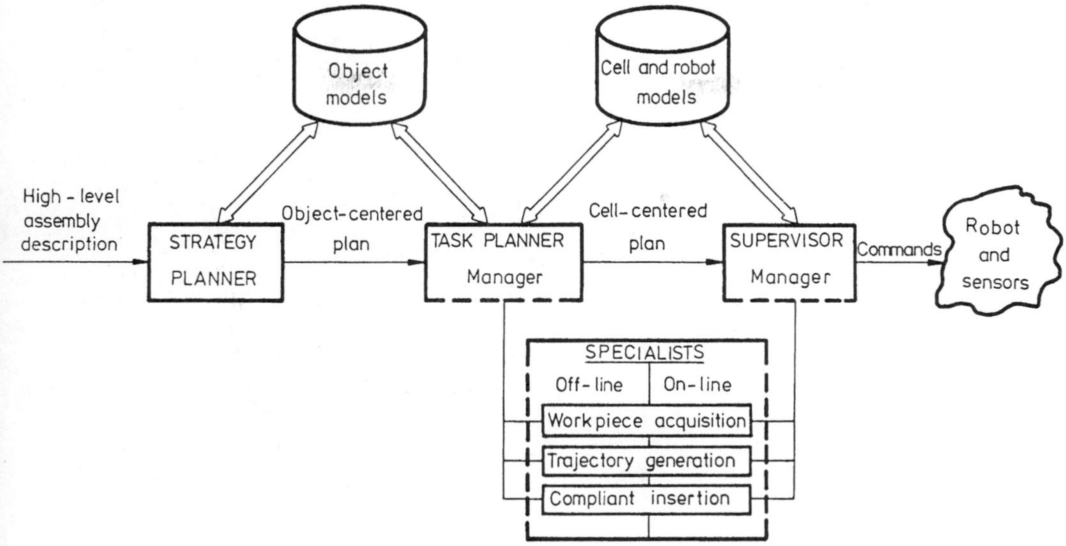

Figure 6. General structure of the robotic assembly system being developed (see text for details).

The general structure of the assembly planning-and-monitoring system under development is shown in Fig. 6. It consists of three main modules: strategy planner, task planner, and supervisor (For a related hierarchical decomposition of human motion planning into three levels - object, joint, and muscle - see Hollerbach 1982). From the initial assembly description, the strategy planner will generate an *object-centered plan*, consisting of the set of workpiece spatial relationships that characterize the final assembled product, and a sequence of assembly instructions, specified in terms of the relative positions and orientations of the workpieces, without any reference to robot movements.

Taking the above object-centered plan as input, the task planner will synthesize a *cell-centered plan*, consisting of a sequence of robot movements and commands to the different sensors. This planner has access to the robot kinematic model and is organized as a manager plus three off-line specialists - in workpiece acquisition, collision-free trajectory generation, and compliant insertion, respectively.

Finally, the supervisor controls the execution of the cell-centered plan, sending appropriate commands to the different devices and monitoring their execution, in order to detect errors and ask for replanning if necessary. It is also organized as a manager plus the on-line counterparts of the three specialists in the task planner.

Sensorimotor integration is planned by the off-line specialists and is accomplished by their on-line counterparts. To explain how this is done, the two workpiece acquisition specialists are described next (see Fig. 7). From a description of the assembly operation to be performed, the off-line grasping specialist generates a list of

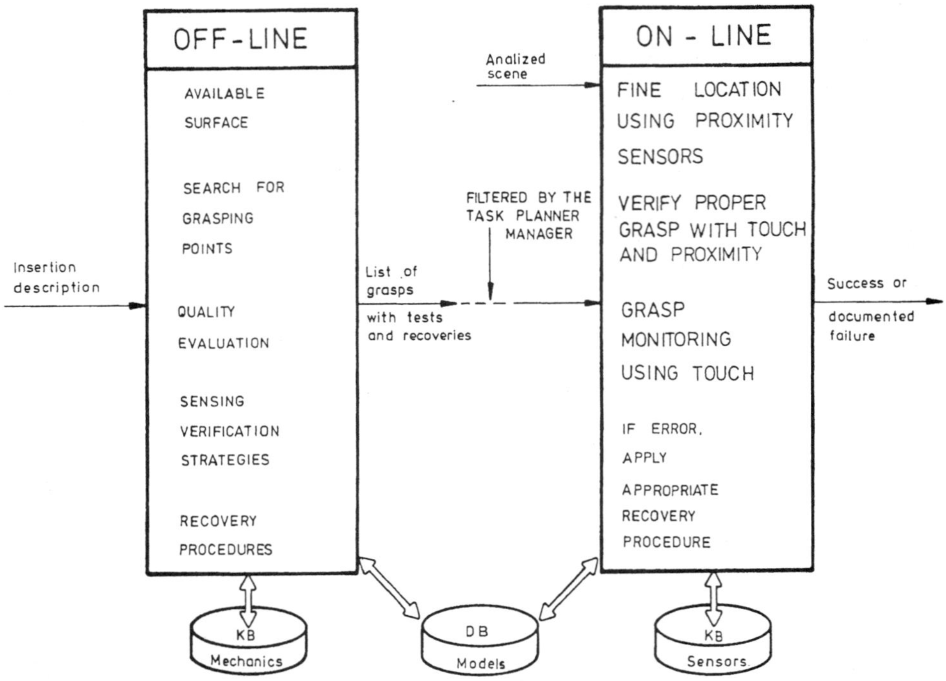

Figure 7. Schematic view of both the off-line and on-line workpiece acquisition specialists.

grasping sites of the workpiece to be assembled, ordered according to a measure of their qualities (stability, tolerance to uncertainties, etc.). The generation process starts by determining the portion of the workpiece surface that will remain accessible after the assembly operation has been completed; this is done by performing some specific subtraction operations on the boundary-based representation of objects provided by the CAD database. Next, a search for certain types of grasping sites (two parallel faces, face-edge, face-vertex, etc.) is carried out on this surface, guided by an evaluation function reflecting the *a priori* quality of each site. Finally, sensory strategies to assure the correct execution of each grasp, together with recovery procedures to apply in case of slippage, rotation of the workpiece within the gripper and other anomalous situations such as these, are included for each site.

The included sensory strategies are of three types: The first implies using proximity sensors to precisely locate the grasping sites, the second involves combining information supplied by both proximity and tactile sensors to verify that the grasping operation has been successfully completed, and the third consists of monitoring tactile information to detect movement of the workpiece in the gripper. Note that for the last two strategies to be useful in devising error recovery procedures, it is necessary to have models of what sensory information will look like in both the correct and the unwanted situations.

The list of grasping sites produced by the off-line workpiece acquisition specialist is filtered by the task planner manager, to meet the requirements of the other two off-line specialists. The filtered list is what is fed onto the on-line workpiece acquisition specialist, together with a description of the current state of the robot workspace in a neighbourhood of the workpiece to be acquired. On the basis of this state, the on-line specialist selects the best *a posteriori* grasping site, and controls the actual execution of the selected grasping by scheduling the operations to be performed by the different devices and then sending appropriate orders to their controllers. The on-line specialist encompasses the parametric code for the sensory strategies available to its corresponding off-line specialist as primitives upon which to build its plans (this is true for the three pairs of specialists). Some of these sensory strategies, for instance those that make use of proximity sensors to accurately locate grasping sites, are actually sensorimotor strategies, since they involve appropriately moving the gripper to collect further information on the basis of previously acquired information. In sum, on-line specialists can be seen as complex motor schemas that have perceptual schemas embedded into them (cf. contributions by Arbib, Arkin, Cervantes-Pérez, and Koy-Oberthür, this volume).

So far, we have not considered the case in which no grasping site in the list is accessible - due to the way the workpiece is lying on the table - and it is necessary to go through an intermediate stage to place the workpiece in a proper position for the gripper to be able to grasp it in a way suitable for carrying out the desired insertion. However, some work has already been done on the problem of freeing grasping sites obstructed by other workpieces (Sanfeliu & Torras 1984; Ruiz & Torras 1988).

The outcome of the on-line specialist can be either a successful workpiece acquisition operation or a documented failure. In both cases, control is subsequently transferred to the supervisor manager either to proceed with the assembly task by performing a workpiece transfer operation, under the control of the on-line trajectory generation specialist, or to stop the system and call for human intervention.

A preliminary description of the workpiece acquisition specialists, which are at present being implemented, is provided in Avila & Torras (1988). The competence of the off-line specialist could be enlarged by taking into account, in the *a priori* evaluation of the quality of each grasping site, the degree of difficulty that the application of sensory strategies to this site would entail.

5 Degrees of Sensorimotor Integration

In general, the degree of sensorimotor integration attained between sensory data and motor skills in robotic systems is higher as the amount of information provided by the sensors is lower. Thus, it has been possible to dynamically control robots by making use of force and torque information, while the real-time feedback control through vision has up to now proved to be computationally unfeasible, except for very simple tasks involving little information processing (e.g., line following). This has triggered considerable research on synchronous parallel architectures (Duff et al. 1985; Lougheed 1985).

Sensorimotor integration can either take the form of a *two-stage process*, be accomplished by *interleaving sensation and action*, or arise from the truly *simultaneous operation* of sensors and actuators. Besides technological limitations, like those pointed out above, task requirements and economical considerations determine the degree of integration implemented in a robotic system for a given application. Fig. 8 summarizes the degree of sensorimotor integration required by some generic tasks, as well as the sensory modalities involved in particular instances of these tasks. Some of these instances will be described in the next section.

6 Some Sample Tasks

A task for which a *two-stage process* generally suffices is the recognition, localization, and acquisition of randomly oriented objects, by making use of vision and/or range data; Pugh (1983) provides several examples of this approach. This obviously requires a previous camera-robot calibration (Rembold & Blume 1985).

Another such task is object recognition by combining either vision and touch or vision and range data. The idea here is to make some initial hypotheses about the contents of a scene, on the basis of only vision information, to subsequently use these hypotheses to guide the acquisition of tactile or range information. This newly acquired information can serve three purposes: to verify the initial hypotheses in the case that there are uncertainties in the environment, to disambiguate between several objects, and to disambiguate between several possible poses (position plus orientation) of a known object. Some research has been done on optimizing the acquisition of further sensory information on the basis of that previously acquired (Ellis 1986; Grimson 1986).

The already mentioned task of first coarsely locating an object through vision, to subsequently refine this initial location estimate by using proximity sensors, is also an example of a two-stage sensorimotor integration process.

Sensation-action interleaving is required in those recognition tasks in which, because of sensor limitations, the robot has to cooperate in the recognition process by appropriately moving the object in front of the sensor. An example of one such task is the recognition of overlapping, adjacent or partially occluded planar workpieces lying on a flat surface, when there is only one camera available. The approach we have followed to accomplish this task in our system has been to define a heap-crumbling operator and a workpiece-sliding operator whose parameters are adequately set in each situation so as to minimize the entropy introduced (direction of motion opposite both to where other workpieces are lying and to the limits of the vision field, intermediate storage of already recognized workpieces, etc.). Details are described in Sanfeliu & Torras (1984) and Sanfeliu et al. (1985).

Another task that requires vision-robot cooperation is the recognition, using only one camera, of objects whose identity is ambiguous from a single view. The robot has to find suitable grasping points on the object and turn it around to obtain several views of it in its different stable poses.

Sensation-action interleaving is also used to accomplish tasks whose very high computational requirements forbid their implementation as truly simultaneous sensor-motor processes. For instance, this is often the way in which mobile robots use dynamic vision or range sensors to avoid obstacles (see Arkin, this volume, for a discussion of the state-of-the-art in this research field).

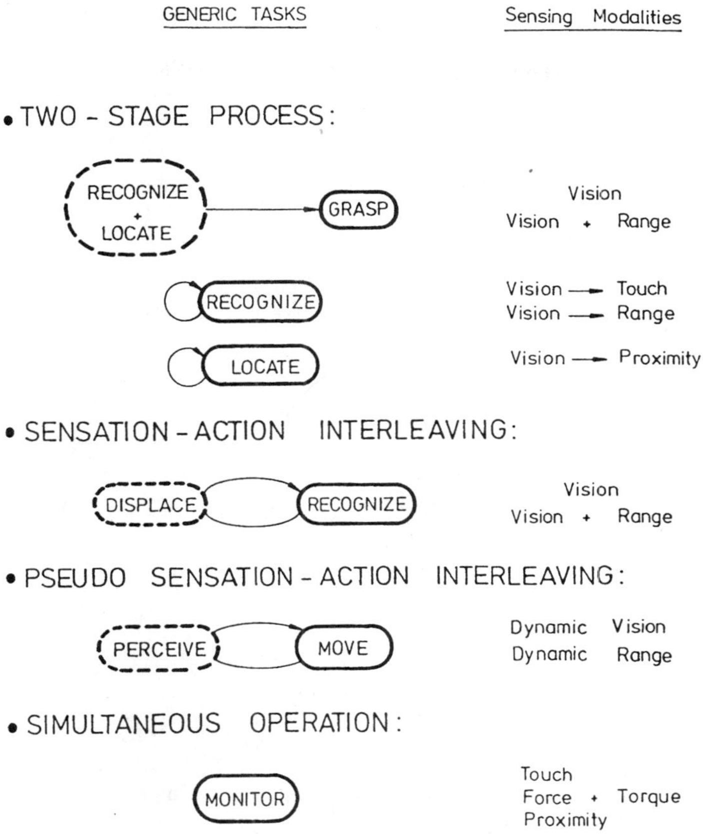

Figure 8. Examples of tasks requiring various degrees of sensorimotor integration. The objective of each task is shown encircled with a continuous line, while intermediate steps in the process are shown encircled with a dashed line. The sensory modalities involved in particular instances of these tasks are listed in the righthand side of the figure.

Simultaneous sensor-motor operation is a requirement of monitoring tasks, often implemented by means of contact sensors. An example is that of monitoring the correct subjection of an object into a gripper by means of a tactile sensor, to prevent slippage during a transfer operation. Other examples are compliant insertions and the manipulation of fragile objects by controlling the applied forces and torques (This is called active compliance to distinguish it from passive compliance, which is based on elastic arm-gripper links that try to emulate the elasticity of animal muscles). Proximity sensors can also be used for monitoring purposes, for instance to detect collisions before they occur and take the appropriate actions.

This is, of course, a somewhat simplified picture of the wide spectrum of robotic tasks that require various degrees of sensorimotor integration. Furthermore, even simple tasks such as bin picking (Kelley et al. 1981) involve combining a two-stage process (localization of grasping sites through vision and then grasping) with a simultaneous sensor-motor operation (monitoring with proximity sensors to prevent collisions).

7 Discussion and Conclusions

Since I am not able to trigger the really useful discussion, that which should ideally take place between *Rana* (or *Bufo*) and *robot*, I will suggest some points for a substitute discussion between *Rana* observers, *Rana* modelers, and roboticists.

My first point concerns vision. In this field, a very fruitful interaction between neurophysiologists and computer scientists already occurs (e.g., see Ewert and Arbib, this volume), resulting in a reasonably good understanding of relatively low-level visual processes both in animals and machines (see Arkin, this volume). This understanding is at present being extended in two directions: first, towards higher-level processes involving knowledge-based interpretation and, second, in the design of neuron-based massively-parallel architectures for pattern recognition and constraint satisfaction (Eckmiller & von der Malsburg 1988; see also von der Malsburg, this volume). These architectures are aimed at exploiting the parallelism inherent in some tasks to gain speed. In this regard, it is worth citing Overton (1984): "The present use of vision by robots assumes the form of take a picture, analyze, plan, and then blindly act with no regard for changes in the environment."

In my opinion, work on the synthesis of optimal sensing strategies (e.g., visually-guided acquisition of tactile, range, and proximity information) that has arisen within the field of robotics could be used by neurophysiologists and neuroethologists to investigate similar processes in animals and humans. Conversely, studies on how these strategies are progressively improved with experience in living beings could be profitably exploited by roboticists.

Hollerbach (1982) claims that work on biological motor control has focussed on free arm motions as opposed to contact movements. On the other hand, he hypothesizes that much of the biological control apparatus is concerned with the control of compliance rather than with the control of free movements (cf. also Berkinblit, this volume). Therefore, research on this apparatus should take advantage of the quite large body of knowledge there is about the control of compliance in robotics.

Finally, dexterous manipulation has been quite thoroughly studied in humans, while within robotics this is a research topic that is just beginning to develop. As a matter of fact, some dexterous robotic hands with about ten degrees-of-freedom are already available, and distributed control algorithms for them have also been developed, the open question now being the integration of sensory information in a way amenable to real-time dexterous manipulation.

To close up, biological systems provide a "proof of existence" and an endless source of inspiration for engineers designing autonomous robots able to accomplish useful tasks in continuously changing environments while, on the other hand, robotic systems constitute a useful test-bed to assess the biological plausibility of some hypotheses by making explicit the computational resources needed to achieve the desired level of performance.

Acknowledgements. The photographs in this chapter were supplied by Federico Thomas. Preparation of the manuscript was supported in part by the US-Spain Joint Committee for Scientific and Technological Research through grant CA83-188.

8 References

Allen P (1984) Surface descriptions from vision and touch. *Proc Int Robotics Conf*, Atlanta, pp 394-397

Amat J, Basañez L, Torras C (1983) Sistemas de visión tridimensional en los robots industriales. *Mundo Electrónico* 130: 71-78

Anandan P (1985) A review of motion and stereopsis research. *COINS Tech Rep* 85-52 Univ Massachusetts, Amherst

Arbib MA, Iberall T, Lyons D (1983) Coordinated control programs for movements of the hand. *COINS Tech Rep* 83-25 Univ Massachusetts, Amherst

Avila F, Torras C (1988) Propuesta de un sistema de planificación y monitorización de la aprehensión de piezas por un robot. *1er Congreso Iberoamericano de Inteligencia Artificial (IBERAMIA'88)*, Barcelona, pp 405-407

Bajcsy R, Allen P (1984) Converging disparate sensory data. In: Hanafusa H, Inoue H (eds) *Proceedings of the 2nd International Symposium on Robotics Research*, Kyoto. The MIT Press, Cambridge MA, pp 81-86

Balek DJ, Kelley RB (1985) Using gripper mounted infrared proximity sensors for robot feedback control. *Proc IEEE Int Conf on Robotics and Automation*, St. Louis, pp 282-287

Ballard D, Brown C (1982) *Computer vision.* Prentice-Hall, Englewood Cliffs, NJ

Barnard ST, Fischler MA (1982) Computational stereo. *Computing Surveys* 14(4): 553-572

Barrow HG, Tenenbaum JM (1978) Recovering intrinsic scene characteristics from images. In: Hanson AR, Riseman ER (eds) *Computer vision systems.* Academic Press, London New York, pp 3-26

Barrow HG, Tenenbaum JM (1981) Computational vision. *Proc IEEE* 69(5): 572-595

Basañez L (1985) Robots inteligentes. *Revista del CEAM* 193: 11-17

Bastuscheck CM (1984) Look-up table computation for ratio image depth sensor. *NYU-Courant Inst Robotics Lab Rep* 34

Besl P, Jain R (1985) An overview of three-dimensional object recognition. *Computing Surveys* 17(1): 75-145

Binford TO (1982) Survey of model-based image analysis systems. *Int J of Robotics Res* 1(1): 18-64

Brady M (1982) Computational approaches to image understanding. *Computing Surveys* 14(1): 3-71

Duff MJB, Levialdi S, Preston K, Uhr L (1985) *Computing structures and image processing.* Academic Press, New York

Eckmiller R, von der Malsburg C (eds) (1988) *Neural computers.* NATO ASI Series F Vol 41, Springer-Verlag, Berlin Heidelberg New York

Ellis RE (1986) A tactile recognition strategy for planar objects. *COINS Tech Rep* 86-32 Univ Massachusettts, Amherst

Espiau B, Catros JY (1980) Use of optical reflectance sensors in robotic applications. *IEEE Trans on Systems, Man, and Cybernetics* 10(12): 903-912

Freund E, Hoyer H (1986) Pathfinding in multi-robot systems: solution and applications. *Proc IEEE Int Conf on Robotics and Automation, San Francisco CA*, pp 103-111

Grimson WEL (1986) Sensing strategies for disambiguating among multiple objects in known poses. *IEEE Journal of Robotics and Automation* 2(4): 196-213

Hall EL, Tio JBK, McPherson ChA, Sadjadi FA (1982) Measuring curved surfaces for robot vision. *Computer* 15(12), 42-54

Hanson AR, Riseman ER (eds) (1978) *Computer vision systems.* Academic Press, New York

Hill J, Part WT (1979) Real-time control of a robot with a mobile camera. *Proc 9th Int Symp on Industrial Robots,* Washington DC, pp 233-246

Hollerbach JM (1982) Computers, brains, and the control of movement. *MIT AI Lab Memo* 686

Hopcroft JE, Schwartz JT, Sharir M (1984) On the complexity of motion planning for multiple independent objects; PSPACE hardness of the warehouseman's problem. *NYU-Courant Inst Robotics Lab Rep* 14

Ikeuchi K (1980) Shape from regular patterns. *Proc 5th Int Conf on Pattern Recognition,* Florida, pp 1032-1039

Ikeuchi K, Horn BKP (1981) Numerical shape from shading and occluding boundaries. *Artificial Intelligence* 17: 141-185

Jarvis RA (1983) A perspective on range finding techniques for computer vision. *IEEE Trans on Pattern Analysis and Machine Intelligence* 5(2): 122-139

Kak AC (1985) Depth perception for robots. In: Nof SY (ed) *Handbook of industrial robotics.* Wiley & Sons, New York, pp 272-319

Kak AC, Albus JS (1985) Sensors for intelligent robots. In: Nof SY (ed) *Handbook of industrial robotics.* Wiley & Sons, New York, pp 214-230

Kelley R, Birk J, Dessimoz J-D, Martins H, Tella R (1981) Acquiring connecting rod castings using a robot with vision and sensors. *Proc 1st Int Conf on Robot Vision and Sensory Controls,* Stratford-upon-Avon, UK, pp 169-178

Kelley RB, Bonner S (1985) Understanding, uncertainty and robot task execution. *1st IFAC Symp on Robot Control (SYROCO),* Barcelona, pp 43-47

Lougheed RM (1985) Multi-processor architectures for machine vision and image analysis. *Proc Int Conf on Parallel Processing*

Magee MJ, Boyter BA, Chien C-H, Aggarwal JK (1985) Experiments in intensity-guided range sensing recognition of three-dimensional objects. *IEEE Trans on Pattern Analysis and Machine Intelligence* 7(6): 629-636

Marr D (1982) *Vision.* Freeman, San Francisco

Masuda R (1986) Multifunctional optical proximity sensor using phase modulation. *J Robotic Systems* 3(2) 137-147

McGhee RB (1977) Control of legged locomotion systems. *Proc 18th Jt Automatic Control Conf,* pp 567-573

Nevins JL, Whitney DE (1977) Research on advanced assembly automation. *Computer* 10(12): 24-38

Nitzan D (1981) Assessment of robotic sensors. *Proc 1st Int Conf on Robot Vision and Sensory Controls,* Stratford-upon-Avon, UK, pp 1-11

Overton KJ (1984) The acquisition, processing, and use of tactile sensor data in robot control. *COINS Tech Rep* 84-08 Univ Massachusetts, Amherst

Pugh A (1983) *Robot vision.* Springer-Verlag, New York Berlin Heidelberg

Pugh A (1986) *Robot Sensors.* IFS Publications Ltd, UK and Springer-Verlag, Berlin Heidelberg New York

Raibert MH, Sutherland IE (1983) Machines that walk. *Sci Amer* 248(1): 44-53

Rembold U, Blume C (1985) Interfacing a vision system with a robot. In: Nof SY (ed) *Handbook of industrial robotics.* Wiley & Sons, New York, pp 262-271

Rosen CA, Nitzan D (1977) Use of sensors in programmable automation. *Computer* 10(22): 12-23

Ruiz J, Torras C (1988) Extensión del algoritmo CF para que contemple ciclos en un cierto tipo de grafos Y/O. *1er Congreso Iberoamericano de Inteligencia Artificial (IBERAMIA'88)*, Barcelona, pp 69-88

Sanfeliu A, Torras C (1984) Cooperación visión-robot en tareas de ensamblaje. *II Simposio Nacional de Automática en la Industria, Zaragoza*, pp 147-154

Sanfeliu A, Torras C, Font J, Ruiz J (1985) Active-recognition system for the acquisition of overlapping or partially-hidden workpieces. *1st IFAC Symposium on Robot Control*, Barcelona, pp 355-359

Sato Y, Kitagawa H, Fujita H (1982) Shape measurement of curved objects using multiple slit-ray projections. *IEEE Trans on Pattern Analysis and Machine Intelligence* 4(6): 641-646

Schwartz JT (1983) Structured light sensors for 3-D robot vision. *NYU-Courant Inst Robotics Lab Rep* 8

Schwartz JP, Sharir M (1983) On the piano mover's problem. III Coordinating the motion of several independent bodies: the special case of circular bodies moving amidst polygonal barriers. *Int J Robotics Res* 2(3): 46-75

Thomas F (1987) Sensores táctiles y de proximidad para robots. *Revista Española de Electrónica* 395: 100-105

Torras C (1984) Concepto y perspectivas de la visión por computador. In: Valle R, Barberá J, Ros F (eds) *Inteligencia artificial: introducción y situación en España*. Colección Informes de Fundesco, Madrid, pp 85-103

Torras C (1986) Automatic planning of assembly robot tasks. In: Banatre M (ed) *2émes Journées INRIA - Universitat Politècnica de Catalunya sur Informatique et Robotique, Rapport INRIA*, pp 107-118

Ullman S (1979) *The interpretation of visual motion*. MIT Press, Cambridge, MA

VanderBrug GJ, Albus JS, Barkmeyer E (1979) A vision system for real time control of robots. *Proc 9th Int Symp on Industrial Robots, Washington DC*, pp 213-231

Witkin AP (1981) Recovering surface shape and orientation from texture. *Artificial Intelligence* 17: 17-45

Yap CK (1984) Coordinating the motion of several discs. *NYU-Courant Inst Robotics Lab Tech Rep* 16

PART VII

Arousal, Habituation, Conditioning

Central Representation of Arousal

PETER R. LAMING

Department of Biology, Queen's University of Belfast
Belfast BT7 1NN, Northern Ireland

Abstract. There are two types of mechanism by which the brain modulates its own responsiveness, those of *arousal* and *attention*. Comparative studies suggest that in fish and amphibia the nondirectional *arousal* component of the orientation reaction (OR) predominates over the directed *attention* component, the latter of which is more advanced in mammals. Fish and amphibia are therefore useful subjects for the study of arousal. Associated with behavioral arousal in these animals is an increase in amplitude and apparent synchrony of high frequency waveforms in the electroencephalogram (EEG). A model is presented of how such waveforms may sensitize neurons and thus increase responsiveness to subsequent stimuli. Also associated with neuronal responses to sensory experience are sustained potential shifts (SPSs) of probably glial origin. These SPSs may represent a change in the environment of neurons deeper in the sensory processing pathway which would increase the probability of these neurons responding to activation of that pathway. Hyperactivity of the brain, associated with seizures is also correlated with high levels of measures of arousal and clinically with a gliosis. Fundamental studies on the neuronal sensitization which occurs during arousal may also reveal the causes of neuronal hyperactivity in clinical disorders like epilepsy.

1 The Orientation Reaction: Arousal and Attention

When vertebrates are presented with a novel stimulus, or one which has inherent or learnt significance, they exhibit an *orientation reaction* (OR). During the OR, posture may change, ongoing activity may briefly cease, and the external sensory appendages like pinnae, eyes, whiskers, or nostrils may move, apparently in an effort to locate the stimulus source (Pavlov 1927; Sokolov 1960). Observations on mammals indicate that there are two components to the OR (Pribram & McGuinness 1975; Turpin 1983). The initial "involuntary" (Maltzman 1979) or "stimulus registration" response (Barry 1984) is non-directional and seems to correspond to recognition by the animal that an environmental change has occurred. Here, it will be referred to as *arousal*. Subsequently, the animal orients towards the stimulus source; this "voluntary orienting" (Maltzman 1979), or "specific registration" response (Barry 1984), is here referred to as *attention* (see also Rizzolatti 1983).

In mammals there appears to be a later ontogenetic development of the attentional component of the OR than that of arousal. At birth, young mammals have been reported to exhibit non-directional arousal (Bridger & Reiser 1959; Vinogradova 1961) though orientation and attention develops later; after 24-45 days in dogs and rabbits (Obraztsova et al. 1958) or 5-10 days in humans (Degtyar 1963).

Less phylogenetically advanced vertebrates than mammals often exhibit less complex forms of behaviors, yet with similar underlying mechanisms. This appears to be the case with the OR of fish and amphibia in which, as in neonate mammals, the arousal component predominates. Thus, several fish species have been shown to respond to a novel stimulus with stabilizing movements of the pectoral, dorsal and caudal fins which do not change the animal's position in the water or orientate it towards the stimulus. This behavior is not species-specific as it has been reported for the goldfish *Carassius auratus* (Savage 1971; Laming & Savage 1980), the roach *Rutilus rutilus* (Laming & Hornby 1981), the chub *Leuciscus cephalus* (Laming & Savage 1978), the slippery dick *Halichoeres bivittatus* (Laming & Ebbesson 1984), and the squirrel fish *Holocentrus rufus* (Laming 1987). A similar non-directed arousal response, involving elevation of the head and extension of the forelimbs has been reported by Laming (1982) in response to novel, poorly-localizable stimuli in the common toad *Bufo bufo*. (For "attentional" orienting responses in terms of "directed appetitive behavior" see contributions by Ewert and Finkenstädt, this volume; for comparative data obtained in rats see also Gonzalez-Lima, this volume).

2 Peripheral Physiological Correlates

The OR of mammals, unless it is divided into two components, is associated with equivocal changes in peripheral physiology (Barry 1984). Thus, either or both cardiac or ventilatory accelerations or decelerations can occur (Sokolov 1963a,b). Separation of the arousal and attentional components of the OR suggest that cardiac deceleration is associated with arousal (Barry 1984; Maltzman 1979), acceleration with attention (Eason & Dudley 1971; Maltzman 1979; Barry 1984). This accords well with the interpretation of cardiac and ventilatory responses found in fish, in which both show a deceleration for 4-10 s on presentation of a novel, arousing stimulus (Fig. 1) (Laming & Savage 1980). The magnitude of the cardiac response has been measured by the ratio of the longest interbeat interval in the electrocardiogram (ECG) or the interbreath (ventilations) interval in the ten seconds after a stimulus to that in the preceding 10 s (Laming & McKee 1981). With this measure it has been found that the magnitude of response to a novel stimulus (the onset of illumination) is directly correlated with the intensity of that stimulus, as is the number of stimulus presentations for the response to habituate (Rooney & Laming 1986).

These results have shown that in fish and amphibians the arousal component of the OR can be behaviorally described and in fish at least, it can be quantitatively assessed by changes in peripheral physiology. The apparently simpler form of the behavioral response and the unequivocal nature of its peripheral physiological correlates in these animals makes them useful models for the examination of the function of, and mechanisms that underlie arousal and attention.

3 Adaptive Nature of Responses to Novelty

Animals are not in a constant state of alertness. There are periods when they are asleep and relatively unresponsive, and within the waking period both the general level of arousal varies (Schürg-Pfeiffer, this volume), and there is selection of sensitivity to incoming stimuli on the basis of their likely relevance to the animal (Finkenstädt, this volume). Thus, a novel stimulus has two components, one is its *novelty* regardless of its modality, and the other is its *specific features*. Conceptually,

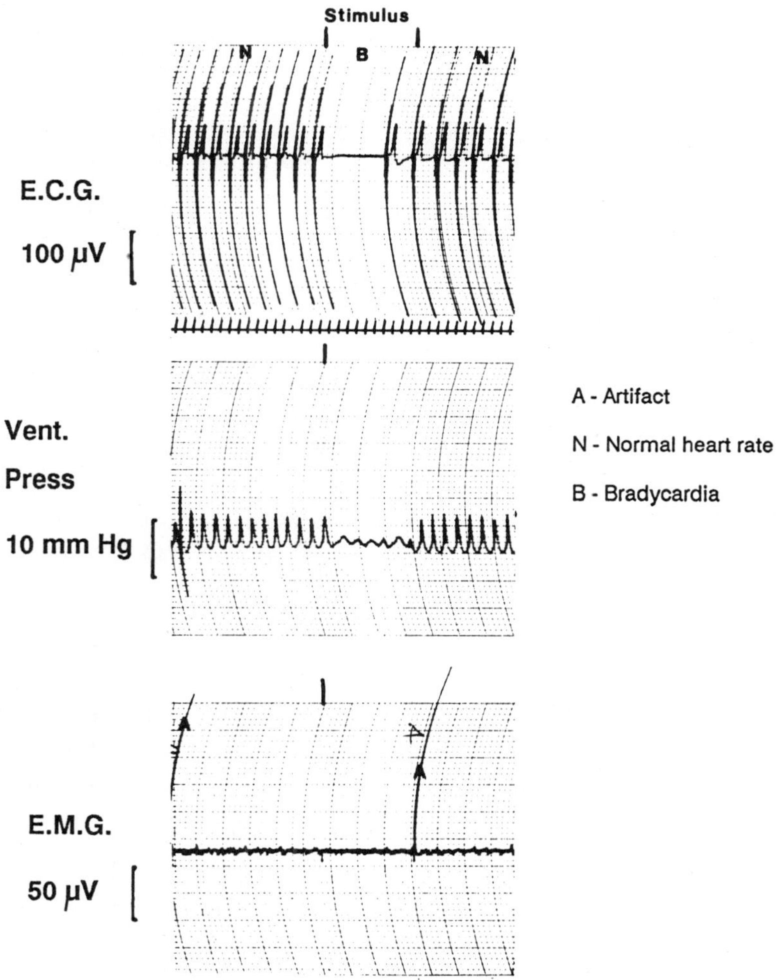

Figure 1. Peripheral physiological changes during behavioral arousal. A reduction in heart rate (top) and ventilatory rate and amplitude (center) accompany behavioral arousal during which there is no evident change in the electromyogram (bottom). Time marking in seconds.

the novelty is considered to induce an adaptive sensitization of neuronal elements involved in all sensory modalities and possibly motor functions as well. After identification of the source and modality(ies) of the stimulus, sensitization is considered to be restricted to the relevant modality(ies) during attention. Sensitization may also be restricted to particular neuronal elements involved in the motor output. Thus, the OR is a response whose two components enable animals to expend energy selectively, and only when triggered by an environmental change. The response involves changes in the gain of the amplifiers of incoming signals and possibly the associated motor outputs. Evidence for such changes in sensitivity during the OR comes primarily from studies in mammals.

4 Changes in Sensory and Motor Performance During Arousal and Attention

London (1954) reviewed the Russian literature in this area which consistently illustrated that the "generalized" OR of mammals (arousal?) was elicited by novel, low intensity, accessory stimuli and was associated with decreased sensory thresholds. Similar studies have been sparse in the West and are reviewed extensively by Spinks & Siddle (1983). Symons (1963) showed enhanced visual sensitivity after presentation of a, presumably arousing, non-visual stimulus. Similarly, auditory sensitivity has been shown to be enhanced by prior experience of a visual stimulus (Thompson et al. 1965) and visual sensitivity by a prior auditory stimulus (Watkins & Feehre 1965). A measure of stimulus detection and response capability that can be psychophysically determined in primates, is reaction time. In monkeys, reaction time to a visual stimulus was decreased by "prearousing" with an auditory stimulus (Lansing et al. 1959). Comparably, aroused humans respond faster than if non-aroused, and if made to attend, respond faster still to visual stimuli (Eason et al. 1969). Similar increases in responsivity seem to occur with lower vertebrates; toads respond better to equilibratory stimuli when illumination is increased (Birukow 1951). In contrast to arousal, "attention" defines a state where stimuli not relevant to the object being attended to are ignored. It is considered by many to involve inhibitory as well as facilitatory influences (Posner & Snyder 1975).

More direct evidence of increased sensitivity during arousal and attention has come from recording sensory unit activity or averaged evoked responses in the brain during behavioral responses to novelty or during arousal induced by electrical stimulation of the midbrain. Thus, visual evoked responses recorded from the lateral geniculate of cats have been shown to be enhanced during behavioral arousal (Livingstone & Hubel 1981).

Studies utilizing electrical stimulation of the midbrain to elicit generalized arousal (Moruzzi & Magoun 1949) have shown facilitation of evoked potentials from sensory cortex (Jasper 1960). Specifically, visual evoked and unit responses of the cortex are enhanced by electrical stimulation of the reticular formation combined with electrical or photic stimulation of the visual pathway (Creutzfeldt et al. 1974; Singer 1973, 1979; Suzuki & Jacobson 1971). The recently developed, powerful tool of ^{14}C-2-deoxyglucose autoradiography (see contribution by Finkenstädt, Gonzalez-Lima, and Horwitz, this volume) has also illustrated increased activation of the visual and auditory system's responses to visual objects or to sound during electrical stimulation of the reticular formation (Finkenstädt & Ewert 1985; Gonzalez-Lima & Scheich 1984). These studies suggest that arousal, associated with midbrain reticular formation activation, increases sensitivity of most sensory inputs to the brain. In toads Finkenstädt (1987; and this volume) proposes a neural circuit involving the medial pallium, the preoptic-hypothalamic area, and the tegmental reticular formation that affects the retino-pretectal/tectal stimulus-response pathway depending on arousal and past experience.

For attention, there appears to be inhibition of sensory inputs not associated with the modality to which attention is directed. Thus, in cats, auditory responses are inhibited during attention to a visual stimulus (Oatman 1971). Comparable results have been described by Garcia-Austt et al. (1964), Jane et al. (1962), and Eason et al. (1969).

Recently, attempts to detect changes in sensitivity during arousal and attention have been performed on less phylogenetically advanced vertebrates. In teleosts, different modalities of superimposed stimuli interact in different ways to either enhance or inhibit auditory responses (Laming & Brooks 1985). Thus, chemical stimuli tend to enhance, and pressure wave or tactile stimuli inhibit, auditory evoked responses. Experiments relating these changes in auditory evoked responses to ongoing activity in the electroencephalogram (EEG) support the contention that general EEG activation (arousal) is associated with facilitation of sensory input (Laming & Bullock, unpublished). In fish, inhibition of auditory input can occur to electrical stimulation of the midbrain (Piddington 1971a,b). Cerebellar stimulation can either enhance visually evoked tectal responses, or inhibit auditory evoked responses recorded in the fish octavolateralis region (Crispino 1983). Similar, though more complex, response changes have been observed in rats (Crispino & Bullock 1984). Descending control systems on audition have similarly been described in guinea pigs (Buno et al. 1966). It is these descending inhibitory processes which seem to "screen out" unwanted inputs during attention (Eason et al. 1969).

These studies have demonstrated that during arousal there appears to be general sensitization and during attention, selective inhibition of sensory input. Caution is necessary for the unequivocal acceptance of these data, however, as few controls have been made to determine if sensitivity has been enhanced by improved directionality of secondary sensory appendages like ear pinnae or eyes. Unfortunately, few workers have recorded simultaneously from regions involved in several sensory modalities during supposed arousal to investigate if all had enhanced responses. Also recordings have rarely been made from different points of the same sensory processing pathway during attention, to test if responses were consistently enhanced when the modality was the one to which attention was directed, or if all were inhibited when it was of another modality.

Experiments performed on restrained fish (Laming & Brooks 1985) or those which were immobilized (Laming & Bullock, unpublished) suggest that selectively inhibitory and possibly attentional influences on auditory responses occur, even in these animals, and that the inhibition may occur at the level of either the medulla or the midbrain. Similarly, facilitation of auditory responses could occur at either region. An important result was that a flashing light inhibited the medullary auditory evoked response near the auditory input, though higher up the auditory processing pathway, at the level of the tectum, the auditory response was enhanced compared to a control situation with no flashing light (Laming & Brooks 1985). Thus, even in fish which normally show little indication of directed attentional responses, there is an evident ability to inhibit or facilitate sensory input. Although much more research is necessary to clearly define the changes in sensitivity which occur during arousal and attention, it is apparent that significant facilitatory and inhibitory influences are involved. In order to determine what brain mechanisms might be involved, it is important to define the changes in general brain activity associated with non-specific arousal.

5 Changes in EEG During Behavioral Arousal

In all species of fish studied so far, environmental stimuli evoke increases in the frequency of maximum power of the EEG (Laming 1981). In the resting cod *Gadus callarias*, the telencephalon, mesencephalon, and medulla have most prominent frequencies of 4-6, 8-13, and 14-32 Hz, respectively; on photic stimulation this latter high frequency became predominant over the whole brain (Enger 1957). In the resting salmon, Hara et al. (1965) found dominant frequencies of 8-10 Hz in both the telencephalon and midbrain, and Schade & Weiler (1959) found a similar (8-14 Hz) activity in all brain regions of goldfish at rest. The dominant frequency increased to 18-24 Hz when the animal was presented with a presumably arousing visual stimulus. Schade (1959) reported that this faster activity showed bilateral synchrony which could be eliminated by midline section of the thalamus. In the author's studies on goldfish the predominant, resting, regular frequency of all brain regions was 6-9 Hz which was largely replaced by regular, 16-25 Hz activity when the fish showed behavioral arousal, associated with peripheral physiological changes (Laming 1980; Laming & Savage 1981). Similarly, Mori et al. (1981) have associated 14-25 Hz activity with arousal in goldfish. Other studies by Gilbert et al. (1964) with sharks and Godet et al. (1964) with the lungfish *Protopterus annectens* have shown that arousal in these fish is associated with an increase in amplitude, especially of high frequency components of the EEG. Power-spectral EEG analysis by the author in the laboratory of Theodore H. Bullock have confirmed that higher frequencies (12-36 Hz) are enhanced most during arousing stimulus presentation, though the power of lower frequencies also increases. None of the fish studies has specifically examined attentional behavioral responses, and the stimuli used were all of a diffuse nature, so it is considered probable that the EEG results reported here were all associated with arousal, rather than attention.

Studies of EEG responses in amphibia have revealed a similar type of response during arousal or on novel stimulus presentation. In the mudpuppy *Necturus maculosus*, moderate voltage regular activity (2-6 Hz) and high voltage fast activity (20 Hz) both become more prevalent during active behavior (Goodman & Weinberger 1969) supporting earlier work on urodeles (Peters & Vonderahe 1954). The more extensive studies of anuran amphibian EEGs also demonstrate increases in the amplitude of the higher ranges of frequency present in the animals during increased active behavior or arousal. Segura & DeJuan (1966) found rhythmic (7-11 Hz) frequency in the olfactory bulb of the toad *Bufo arenarum*. This frequency was of greatest amplitude in active animals examined in summer. Hobson (1967) obtained similar results in the bullfrog *Rana catesbeiana* and found that 5-12 Hz regular activity increased in conditions of behavioral arousal. In a study to relate EEG responses specifically to behavioral arousal in the toad *Bufo regularis* and the frog *Rana temporaria*, Laming (1982) found that in the midbrain and forebrain of both species a regular 14-18 Hz activity increased in amplitude by 50% either during presentation of a novel moving shadow stimulus with no behavioral response or during orienting to a prey object (Fig. 2). The lower frequencies prominent in the resting animals, of from 4-8Hz, became less evident during this behavior. Thus, in amphibia the amplitude of high frequency waveforms seems to increase during arousal.

The EEG correlates of aroused behavior in reptiles also show that, higher frequency waveforms predominate. In the lizard *Ctenosaurus pectinata*, Tauber et al. (1968) found that the dominant frequency increased from 8-15 to 22 Hz with increased voltage during behavioral arousal. Karmanova et al. (1971) recorded rhythmic patterns of 3-7 Hz during sleep and 8-16 Hz during arousal in the monitor *Varanus griseus*. Klein (1963) in a comparative study of sleep in birds and reptiles

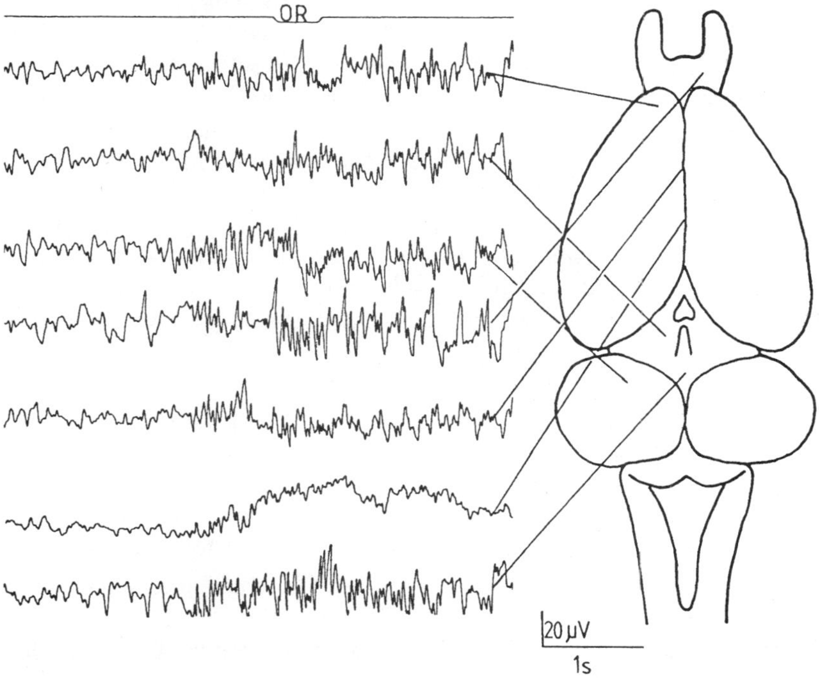

Figure 2. EEG responses to the presentation of a mealworm to common toads *Bufo bufo*. Toads presented with a mealworm show increases in amplitude of the EEG high frequency waveforms prior to and during orienting movements of the head (OR). (After Laming 1982).

showed that the EEG of these classes dropped, from fast to slow activity, as sleep ensued. The higher frequency patterns associated with arousal can also be induced by electrical stimulation of the midbrain, as demonstrated by Segura & Kacelnik (1977) in tegu lizards.

The combined evidence from studies on fish, amphibia, and reptiles suggests that in these vertebrates behavioral arousal, and in reptiles the transition from sleep to wakefulness, are associated with an increase in frequency and amplitude of the EEG. The possible source of this regular high frequency activity should be considered in an attempt to suggest how it might contribute to heightened sensitivity of the central nervous system (CNS) during arousal.

6 Oscillating Membrane Potentials: The Basis of the Synchronized EEG

6.1 Origin of EEG Waveforms

Grossly recorded regular EEGs are correlated with larger changes in intracellularly recorded potentials (Elul 1972). Thus, the EEG is generally considered to consist of field potentials generated by excitatory or inhibitory postsynaptic potentials (Creutzfeldt & Houchin 1974; Hubbard et al. 1969). Synchronous bursts of discharges in neurons are considered to generate dendritic postsynaptic field potentials and give rise to regular waveforms in the grossly recorded EEG. These bursts of discharges are believed to be caused by pacemakers (Elul 1972) which may be formed by thalamo-cortical loops (Anderson & Anderson 1968) or intrathalamic loops (Verzeano 1972). Two pieces of evidence from work on non-mammalian vertebrates suggest that the nature of pacemakers for synchronized waveforms does not depend on these specific loops. Firstly, a cortex does not exist in vertebrates less phylogenetically advanced than reptiles, yet both fish and amphibia show regular EEG waves. Secondly, separation of the telencephalon from the rest of the brain of toads does not remove regular EEG waveforms from either region (Laming et al. 1984b) suggesting that both the isolated telencephalon and the rest of the brain are intrinsically capable of synchronizing membrane potential oscillations. These results suggest that the "pacemaker" may have a much simpler basis than previously conceived. Much of the evidence for a regular postsynaptic potential pacemaker activity comes from data relating neuronal spike activity to the phase of EEG oscillations. The interpretation of these results has been that the EEG waveform is a result of post synaptic potentials generated by these bursts. An alternative possibility, however, consistent with sensitization and large, rapid EEG waveforms during arousal is that the membrane potential oscillations make spike activity more likely during particular phases of the EEG. Evidence for this has come from indications that averaged phase shifts precede unit activity (Frost & Gol 1966). Another objection to a complex neuronal circuitry or synaptic activity for generation of synchronous gross waves is the close relationship between action potentials and gross waves in fiber tracts (Laufer & Verzeano 1967).

Although both postsynaptic field potentials due to spiking activity and feedback circuits between neurons will inevitably contribute to the grossly recorded EEG, a simpler oscillator, the neuronal membrane itself, may form the basis of the field potentials recorded as the EEG.

Individual neurons maintain electrical potential across their membranes primarily due to their restricted permeability to Na^+. Their relatively greater permeability to K^+ makes this ion's distribution the primary determinant of the resting potential. In a constant environment and with perfect permeability to only this ion type, the resting membrane potential might be stable. This does not occur, since neurons live in an environment constantly responding to activity of other cellular elements, and are themselves being directly bombarded by subthreshold transmitter release and electrical activity at synapses. Indeed, oscillations in membrane potential in individual neurons may be a result of subthreshold stimulation of neurons (Jakobsson & Guttmann 1981). Added to these destabilizing influences are the leaky nature of the membrane to ions other than K^+, like Na^+ (Hodgkin & Keynes 1955). In order to keep the potential as constant as possible, an energetic Na^+/K^+ pump is sporadically active. This may be the cause of endogenous slow oscillations in membrane potential of *Aplysia* neurons (Chalazonitis 1978). Alternatively, slow oscillations studied in invertebrates may derive from an activation of delayed K^+ conduction, induced after slow depolarization due to Na^+ and Ca^{2+} leakage (Gola 1978). Faster metabolic or ionic events may also be involved in regulation of the transmembrane ionic potential (Tasaki 1978).

The resting individual neuron is thus dependent of its relatively maintained potential equilibrium on intrinsic servo mechanisms. Such mechanisms are, by their reliance on feedback, inherently oscillating. If this argument is accepted, then a neuronal population may be considered to consist of a number of membrane potential oscillators. If the nature of the unit oscillators, and their servo mechanisms, are similar in many neurons, it is not inconceivable that these may have similar oscillatory patterns.

This simplistic hypothesis of the nature and origin of regular EEG waveforms, implies that the neuronal membrane itself is the oscillator, and does not localize the potentials recorded with respect to synapses though it is here that changes in potential will have their greatest effects. It is apparent from this hypothesis that any change in either the environment or the electrochemical stability of the membrane of our "unitary generator," the neuron, will increase the activity of the servo-mechanism and, hence, the oscillatory amplitude. Frequency of waveforms may be affected by changing the response time characteristics of the servo mechanism, like tightening the strings of a violin. If enough neurons have oscillations in membrane potentials of similar frequency, and if these are synchronized, then the resultant field potentials should be recordable. It is interesting to note that parallel extracellular recordings from the toad's bulbar medial reticular formation have shown spontaneously active neurons whose discharges during certain periods exhibit synchrony (Schwippert et al., this volume).

What then is the relevance of the EEG changes on arousal and how does our view of arousal, as an "activation" of the brain, coincide with a view of synchronous oscillations in membrane potential as the basis of the EEG? Here, I reintroduce the view of EEG arousal described earlier, namely that in fish, amphibia, and reptiles activation is associated *with regular high amplitude, usually fast activity*. What might this represent? From the present discussion it is apparent that either increased synchronization of neuronal membrane potential oscillations or increased amplitude of oscillations in a relatively few neurons may have the effect of increasing power of a high frequency waveband. If the overall frequency of oscillations increased, then frequency, amplitude, and synchronization could each change the response properties of individual neurons to synaptically conducted information, or that of groups of neurons. These changes may then affect the characteristics of the population's output. It is worth mentioning that in frogs Matsumoto et al. (1986) have intracellularly recorded spontaneous fluctuations of excitatory and/or inhibitory postsynaptic potentials in the deep tectal layers (Matsumoto, this volume) from which class T8 "movement cells," extracellularly recorded in the freely moving toad, predicted animal's spontaneous movements (Borchers 1982; see also Schürg-Pfeiffer, this volume).

6.2 Frequency Increase in Membrane Potential Oscillations

Acting on an individual neuron, which has increased the frequency of its membrane potential oscillation, an electrochemical event would be more likely to drive that neuron to firing threshold, as its incidence is more likely to temporally

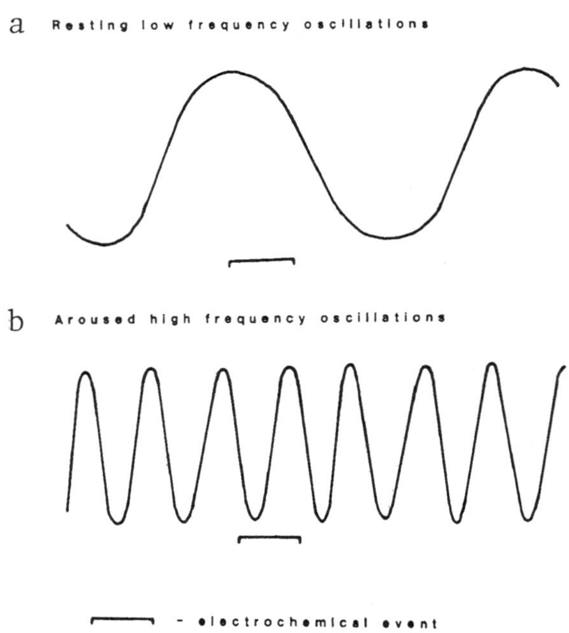

Figure 3. Effects of changes in frequency of membrane potential oscillations on the coincidence of membrane potential polarity changes and brief synaptic electrochemical events. The suggested effect is that the change from the EEG low frequency of an animal at rest (a) to the high frequency (b) when it is aroused, will increase the probability that depolarizing or hyperpolarizing synaptic events will coincide with similar endogenous directions of change in membrane potential.

coincide with a *depolarizing* phase of oscillations. The short event may be a brief transmitter release, or brief electrical depolarization. The same temporal relationship would hold for *hyperpolarizing* influences (Fig. 3). In any case brief influences, of either a hyperpolarizing or depolarizing nature, will be more likely to coincide with a similar direction of change in the normal membrane potential oscillation, and thus have a summative effect. Total cancellation of such an incoming effect would be impossible, if its duration exceeded half the wavelength of the EEG oscillation, and so briefer events would always be facilitated during faster aroused activity, rather than be swamped half the time during rest. This increase in effectiveness of brief input signals is due to more rapid "time sampling." In brain regions where the majority of neurons are excitatory, the tendency would be for an increase in excitatory (depolarizing) neuronal activity though the ratio of enhancement of excitatory and inhibitory activity would be maintained.

6.3 Amplitude Increases in Oscillations

Increases in the amplitude of the membrane potential oscillations in an individual neuron represent larger partial de- and hyper-polarizations. This would increase the summative effect of depolarizing or hyperpolarizing synaptic influences temporally coincident with the oscillation. Thus, a depolarizing input would be more likely to achieve threshold during coincidence with a large depolarizing wave than in the resting condition where such a wave had a lesser amplitude. The effect is of decreasing the thresholds either for depolarization or hyperpolarization. Again, if there were a predominance of excitatory influences, these would be more facilitated.

6.4 Changes in Synchrony of Oscillations

Frequency and amplitude characteristics in membrane potential oscillation are properties of individual neurons, but increase in synchrony, which may be reflected in the amplitude of the grossly recorded EEG, are properties of neuronal populations. Here, we have to consider the effects of input in several, often converging, pathways. Thus, as we have seen, membrane potential oscillations will affect the probability that an individual neuron will respond at a given time. If several neurons in the same synchronized oscillating population converge synaptically, then their effects are more likely to summate because they are time-locked together.

Therefore, it is hypothesized that increased frequency, amplitude, and synchrony of neuronal membrane potential oscillations during arousal responses will have several effects on incoming signals each of which is potentially sensitizing: The thresholds for their successful transmission will change more often, to a greater degree, and will thus have a greater summative effect. If a local population is primarily of neurons excitatory (depolarizing) in their transmission function, the greatest effect will be one of excitation or decreased threshold for activity.

This hypothesis predicts that spiking will be associated with EEG high voltage synchronous activity and that spike frequency will be highest during negative-going EEG waves. This relationship was tested (Laming & Ewert 1984) in common toads, *Bufo bufo*, in which recordings were made from tectal class T5.2 neurons known to be

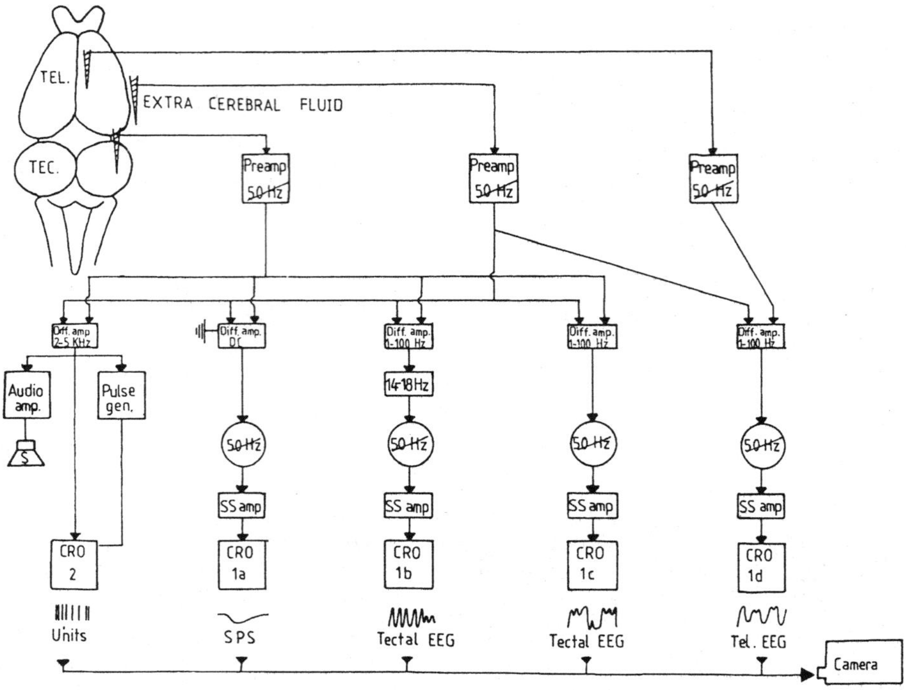

Figure 4. Circuit for recording a unit's spike activity, SPSs, and tectal EEG from the same electrode with simultaneous telencephalic EEG recording. Preamp: preamplifier with a gain of 10; Diff.amp.: differential amplifier; ss amp: single sided amplifier; audio amp: audio amplifier; s: loudspeaker; pulse gen: pulse generator; CRO: oscilloscope, /50 Hz: active filtration of this frequency; 1-100 Hz, 2-5 kHz: band pass filter between these frequencies. (After Laming & Ewert 1984).

best activated on presentation of a prey or a prey dummy stimulus, a situation which also evokes behavioral arousal prior to orientation (Laming 1982). Briefly, the methodology of the experiments involved recording a tectal unit, the EEG, and the sustained potential shifts (SPSs) by selective filtering of the signal derived from a microelectrode (Fig. 4). The toad was immobilized with 0.2 ml of 2% succinyl choline injected into the caudal lymph sacs. Recorded tectal units were identified on the basis of the size of their excitatory receptive field (ERF), spike activity in response to moving visual stimuli, and to the onset or offset of diffuse illumination. The most effective moving stimulus to evoke responses was that used during the experiments. The frequency and amplitude of unit responses were each calculated for each of four phases of the waveform of the 14-18 Hz EEG tuned to the frequency with highest power in the brain of individual toads presented with arousing stimuli prior to the experiment. The results of these experiments can be summarized in three important points. Firstly, the presentation of a stimulus was associated with unit activity and an increase in power and synchrony in the EEG (Fig. 5); indeed the EEG waveform showed some of the characteristics of a relaxation oscillator. Secondly, the highest number of spikes ($P < 0.05$) occurred during the quarter phase of the EEG leading to maximum negativity (see Table 1). Thirdly, well isolated spikes had their lowest amplitude during this quarter phase of the EEG ($P<0.05$) (Table 2). Finally, associated with the presentation of the visual stimulus was a large (55-90 µV) and sustained (>2 s) potential shift (SPS), a result to be discussed more fully later.

The first three points derived from these results support the hypothesis presented earlier, in that they confirm that EEG amplitude increases are associated with increased unit activity and that this unit activity is time-locked to the EEG phase. The lower amplitude of spikes during the negative-going phase also suggests that the EEG oscillations do represent changes in membrane potential and the lowered membrane potential was responsible for the decrease in spike amplitude and, therefore, also the increase in spike frequency. Although this is the most likely explanation of the results, caution must be taken as it is possible that these results may reflect an increased frequency of firing of a smaller (more distant) spike in recordings from more than one neuron.

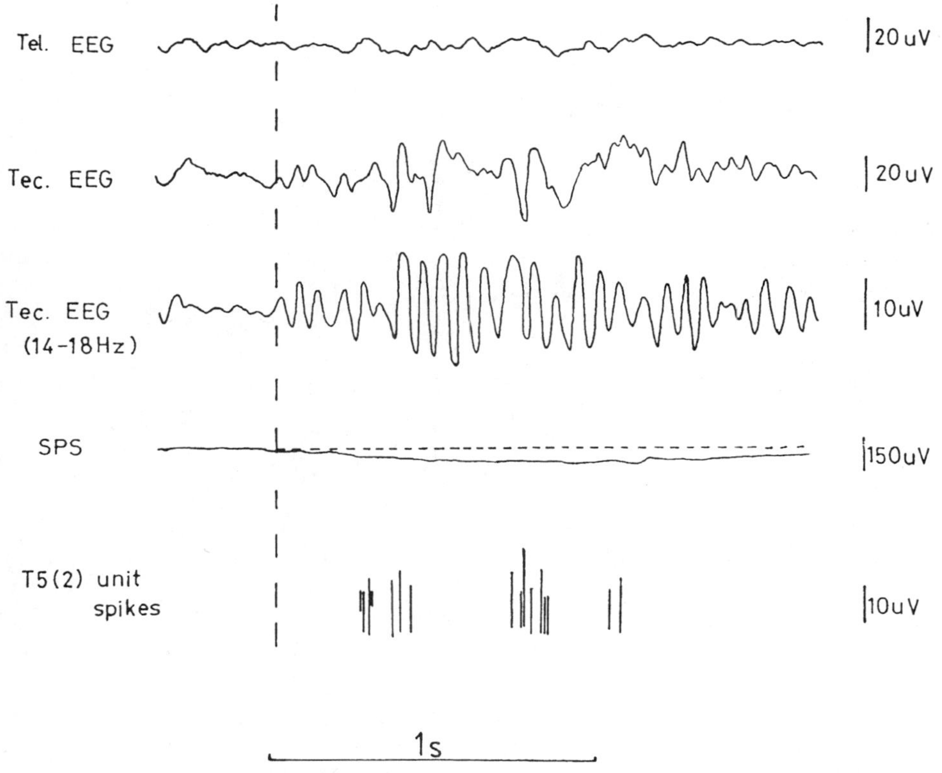

Figure 5. Responses of toad's tectal class T5.2 units, SPS, and tectal and telencephalic EEGs in response to presentation of a worm-like stimulus traversing the neuron's excitatory receptive field at a visual angular velocity of 22°/s. The vertical dotted line represents the deviation of the SPS from the baseline level.

Table 1. Spike frequencies of retinal ganglion cell fibers and of tectal T5-type neurons in response to a visual stimulus in relation to the phase of the 14-18 Hz electroencephalogram (mean ± SE of average number of spikes/phase for 4 stimuli). For nomenclature of tectal neurons see Ewert, this volume. (After Laming & Ewert 1984).

Unit type	n	Positive-going EEG		Negative-going EEG	
		Initial rise	Near peak	Initial fall	Near through
Retinal	5	9.40	10.70	12.80	18.80
		±2.16	±2.16	±2.16	±2.16
T5.1	4	2.62	2.87	3.63	4.81
		±0.62	±0.62	±0.62	±0.62
T5.2	17	5.19	5.80	6.67	9.61
		±0,56	±0.56	±0.56	±0.56
T5.3	3	3.92	4.58	3.65	5.58
		±1.09	±1.09	±1.09	±1.09
T4	6	4.04	4.58	4.63	6.46
		±0.40	±0.40	±0.40	±0.40

Table 2. Spike amplitudes of retinal ganglion cell fibers and of tectal T5-type-neurons in response to a visual stimulus in relation to the phase of the 14-18 Hz electroencephalogram (mean ± SE of average amplitudes in μV of spikes for 4 stimuli). (After Laming & Ewert 1984).

Unit type	n	Positive going EEG		Negative-going EEG	
		Initial rise	Near peak	Initial fall	Near trough
Retinal	5	16.84	15.87	15.87	16.62
		±0.45	±0.45	±0.45	±0.45
T5.1	4	17.38	17.59	17.68	16.58
		±0.91	±0.91	±0.91	±0.91
T5.2	16	12.95	12.68	12.54	11.95
		±0.27	±0.27	±0.27	±0.27
T5.3	3	13.99	14.61	14.24	13.37
		±1.00	±1.00	±1.00	±1.00
T4	6	17.41	16.63	16.44	16.69
		±0.41	±0.41	±0.41	±0.41

The general concept of a dynamic, synchronously oscillating, population of neurons has led to interest in how this might affect information processing and the transmission of neuronal spike activity in mammals. Researchers using either two closely placed electrodes, or a single electrode with an extracerebral reference, each with differential filtering to obtain unit and EEG activity, have examined therefore relationships between unit firing and the phase of regular EEG waveforms. Phase relationships have been found in the thalamus of mammals (Verzeano et al. 1970; Verzeano 1972) and the cortex (Fox & Norman 1968; Frost & Gol 1966). The EEG frequency ranges involved varied from 4-30 Hz though most were in the alpha (8-13 Hz) range. Slower potential changes recorded in the extracellular environment have also excited attention. The negative SPS found by Laming & Ewert (1984) for instance, suggests another potentially sensitizing mechanism, assuming it was derived from an increase in negativity and, therefore, a potentially depolarizing event in the extracellular environment. The nature and origin of such SPS responses is thus of interest.

7 Nature and Origin of the SPS

7.1 The SPS as a Feature of Brain Activation

In vertebrates, during presentation of a novel sensory stimulus or electrical stimulation to elicit behavioral arousal, a sustained steady shift in brain potential (SPS) of up to 10 s in duration has been found. This has been described for mammals (O'Leary & Goldring 1964; Loveless 1979), birds (Durkovic & Cohen 1966, 1968), and amphibia (Laming 1983; Laming & Ewert 1983, 1984; Laming et al. 1984a,b). Early beliefs that these SPSs were solely cortical in origin, cannot be sustained by these studies on lower vertebrates lacking a neocortex. The association between SPS responses and behavioral arousal has been demonstrated by the effects of novel stimuli in the human (Kohler & Wegener 1955), the cat (Gumnit 1961), and toads (Laming et al. 1984a). As with other correlates of arousal, the shift magnitude and duration habituates with repetitive stimulation (Rowland & Goldstone 1963; Laming & Ewert 1984). The SPS is most evident in the area of the brain concerned with the particular stimulus modality used. This is true for recordings made from cat sensory cortex (Lickey & Fox 1966; Kelly & van Essen 1974) or toad optic tectum (Laming & Ewert 1984). In these areas the SPS is negative with respect to a neutral reference. Indeed, negativity may occur in any part of the brain maintained in a state of readiness (Pribram & McGuinness 1975). Though the SPS is commonly negative, some areas may show a positive shift and locally a delayed positive "rebound" can occur (Gumnit 1961).

The SPS is thus a feature of activation of the brain and when elicited by a novel stimulus is called the reactive response (Rowland, pers. communication). It can be enhanced by the level of motivation of an animal, if the stimulus is food, water, or sexual, and the animal has been previously deprived of such stimuli (Rowland et al. 1967; Rowland 1968). Furthermore, anticipatory, proactive SPSs can be obtained when animals anticipate a rewarding stimulus (Rowland & Goldstone 1963; Morrell 1960). This is similar to contingent negative variations (CNVs) which may be related to the magnitude of heart rate changes (Lacey & Lacey 1974). CNVs also occur in the optic tectum to stimuli omitted during habituation of visual responses to stimuli otherwise regularly presented to toads (Laming, unpublished).

These studies suggest that arousal is associated with SPSs, and that both previous experience and motivation can influence the occurrence and amplitude of the SPS. Negative SPSs may also occur, spontaneously, and cognitive task performance in humans during these periods is enhanced (Bauer & Nirnberger 1981) compared to when the shifts are positive, indicating that they represent periods when brain operations are more "efficient."

7.2 Involvement of Neuroglia

The origin of SPSs has long remained a mystery because their time-course, in the order of seconds or more, did not correspond to any known neuronal events. It was with the description of neuroglial potential shifts, associated with local neuronal events and with similar time courses to SPS (Kuffler et al. 1966; Orkand et al. 1966), that glia became popular candidates for SPS generation. Such a glial origin for the major part of the SPS waveform has recently been confirmed by Roitback et al. (1982), though early components may include field potentials from active neurons.

SPSs of glial origin have been generally considered to be passive responses to enhanced extracellular potassium concentration resulting from neuronal activity (Ransom & Goldring 1973a,b,c). The highly selective permeability of glial cells to potassium (Futamachi & Pedley 1976) would then allow them to disperse the consequently absorbed potassium away from the site of neuronal activity (Kuffler 1967). This ability would be supported by the presence of low resistance potassium-permeable junctions between glial cells (Cohen 1970; Kuffler et al. 1966), now also demonstrated for astrocytes in guinea pig neocortex by dye coupling (Gutnick et al. 1981). This passive, spatial buffering, or redistribution of locally acquired potassium released into extracellular space is a probable function of glia (Gardner-Medwin 1983), a function also found in the electrically coupled glia in the retina of the drone, *Apis mellifera* (Coles & Orkand 1983). These results and interpretations imply that glial cells may just form a homeostatic sink for potassium, to regulate the extracellular environment of neurons. This is in line with the traditional metabolically supportive role for glia, especially vertebrate astrocytes. This passive role for glia would indicate that SPSs would always bear a direct relationship to neuronal activity (which is not obtained during proactive, anticipatory SPSs in the cat; see Shaefor & Rowland 1974).

Furthermore, a purely passive action of glia in dispersing neuronally derived potassium does not explain much of the data obtained for SPS responses by Laming & Ewert (1984). SPS responses in that work were recorded from the tectal surface (in the region of fiber terminals of retinal ganglion cells) and from deeper tectal regions associated with T5-type and T4 neurons. When comparisons between the latency of the activity of the first spike and the latency of the SPS are made (Fig. 6), it is apparent that the SPS in deeper regions of the tectum *precedes* the activity of the local unit. It must therefore be a response to neuronal activity earlier in the stimulus processing pathway, and implies a faster glial than neuronal conduction pathway. This would be possible due to the non-synaptic connectivity of glia which forms a functional syncitium for potassium conductance. These SPSs therefore occurred in regions not subject to local neuronal activity, but by spatially buffering K^+ would partially depolarize those local neurons. This suggests that glial cells are not only concerned with homeostatically removing extracellular K^+ from regions of active neurons, but are also involved in sensitizing neurons. Although these two properties seem mutually exclusive, I should like to propose that neuroglia, especially astrocytes, may have both these functions. The concept of conduction through neuroglial/neuronal pathways of K^+ exchange was proposed by Hertz (1965). It now seems more likely that glial cells act as *modulators* of excitability in the neuronal conduction routes carrying specific information.

7.3 Implication of the Cytoarchitecture

A clue to the modulatory role of neuroglia lies in the ontogeny of the brain and in the way in which neuronal activity and SPSs are conducted. During early development of the brain of vertebrates there is a radial proliferation of glia, forming long chains of glial "guides" projecting from germinal (periventricular) regions towards the surface (Rakic et al. 1974). Neuroblasts migrate along these glial guides, those to the cortical periphery migrating first and differentiating *in situ*, to be

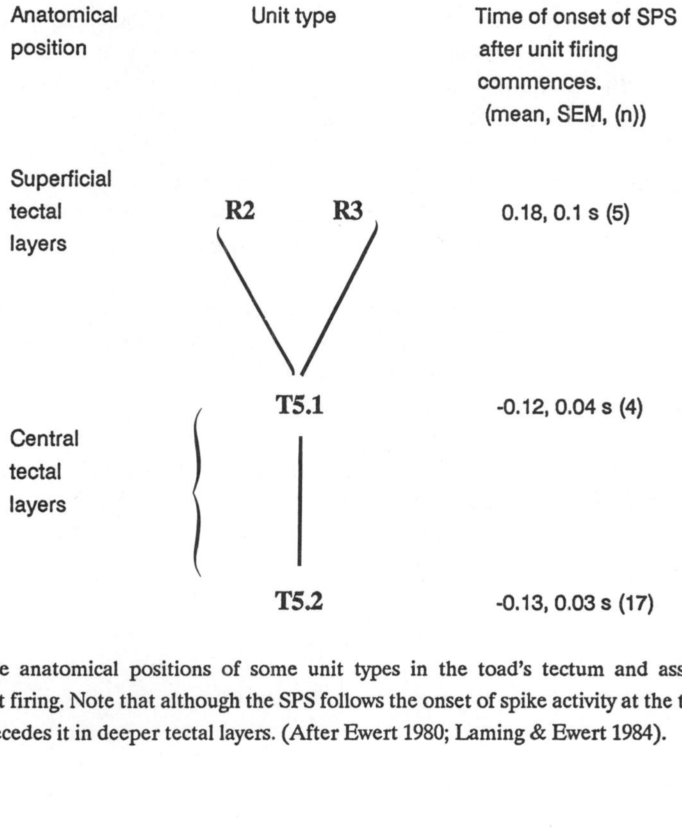

Figure 6. Relative anatomical positions of some unit types in the toad's tectum and associated SPS latencies from unit firing. Note that although the SPS follows the onset of spike activity at the tectal surface (the "input") it precedes it in deeper tectal layers. (After Ewert 1980; Laming & Ewert 1984).

followed by others which halt earlier in their migration, and make contact with the processes of their predecessors. Often the progressive migrations of neurons cease at similar distances from the core, presumably in response to biochemical and physical conditions in their environment induced by their neuronal predecessors and the local glia. Whatever causes cessation of migration and subsequent differentiation, many brain structures develop a laminated appearance as a result. In most vertebrates studied, neuroglia, like the neurons, retain their radial organization (King 1966). This is especially true of fish and amphibia, in which radial ependymal cells and astrocytes are evident (Rakic 1984).

The laminated tectum of the toad and the pre-emptive SPS, which occurs in deeper tectal layers, indicate that glia (i) are providing a more rapidly conducting pathway than spiking neurons, (ii) have a radial organization, and (iii) would sensitize neurons in a sensory processing pathway prior to their receipt of specific neuronal activation. Therefore, it is proposed that glia removes potassium from the space around active neurons associated with the retinal input, and by spatial buffering "prime" related neurons later in the sensory processing pathway (Fig. 7). This sensitization, by changing extracellular potassium, is mainly localized by the radial organization of neurons and glia, and takes place in relevant areas of the deep visual projection, though adjacent, associated regions would also receive some sensitization. This would promote recognition of stimuli which moved in the visual environment of the animal (Fig. 7). Thus, when neuron A1 fires, neurons of deeper areas A2 and A3 would be strongly sensitized, neurons in retinotopically adjacent areas, B1, B2, and B3 less so. Natural objects, moving from visual field areas represented by A to those represented by B, would be more likely to evoke a response in B due to its prior sensitization. One important feature of this mechanism is that although the neuronal pathway is unidirectional, transmission of glial depolarizations may also occur in the reverse direction, due to electrotonic coupling. Hence, when a signal from A1 arrives at A3, the glia will buffer in the opposite direction and sensitize the input. Incidentally, this could be a mechanism by which reticular formation activation may sensitize the brain, and its input: not by an action on neurons alone, but (also) by depolarizing glial populations with origins in deep "cortical" (or tectal) regions. Astrocytes are the most likely candidates for this feedback and SPS generation, because of their preponderance in the grey matter of the brain (Pope 1978). The current hypothesis explains how glia and the passively induced SPS, associated with them, may play a role in not only removing excess K^+ as a result of local neural activity, but may also have an adaptive function in relocating this excess, so that the specific neuronal processes will become more sensitive. Experimental injections of other cations (Na^+, Li^+) has the effect of depolarizing glia and also causing intense firing of local neurons (Grossman 1978).

7.4 Correlation with Sensory Experience

Anatomically, the current hypothesis has assumed only the maintenance of glial ontogenetic cytoarchitecture, and physiologically, their totally passive cellular role in response to neural activity. There is, however, evidence that SPSs are not simply the result of recordable neural activity, as they change with experience of a stimulus.

Thus, in the toad tectum, pre-emptive SPSs in the region of prey-selective T5.2 units became better correlated in amplitude with the local neuron's frequency during habituation, suggesting that a reduction in spatial buffering, possibly involving sensitization, had occurred due to experience (Laming & Ewert 1984). In other

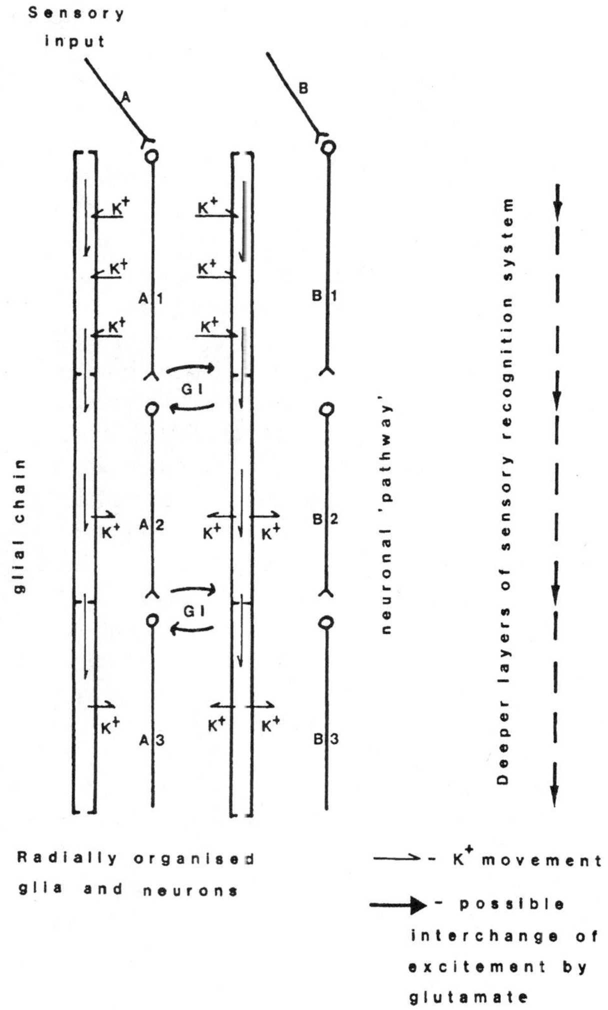

Figure 7. Proposed mechanism for sensitization of neurons by glia in a laminated sensory system. On arrival of the sensory input, active neurons (e.g., A1) release K^+ ions into extracellular space which is rapidly translocated by "spatial buffering" to the regions of A2 and A3 which are, therefore, partially depolarized prior to the specific neuronal input being conducted through the A1/A2 synapse. This sensitization mechanism is passive but would be more dramatic if glial cells were actively depolarized, e.g., by release of glutamate, or by affecting glutamate interchange between glia and neurons. In sensory systems which are organized topographically to represent spatial sensory experience like the visual system, neurons in areas close to the part of the sensory map being stimulated would also be sensitized (B1, 2, 3).

experiments SPSs have been obtained to missed stimuli during habituation, suggesting that SPSs can result from no local neural activity (Laming, Ewert, Rooney, unpublished). Similar anticipatory SPSs, not associated with recordable neuronal activity have been found in learning situations in mammals by Rowland & Goldstone (1963). These changes in SPS as a result of experience would suggest that glial cells are not totally passive in their role as extracellular potassium redistributors. Indeed, they themselves may become more easily depolarized during learning, with the presumed effect of highly sensitizing local neurons previously associated with a particular sensory experience.

Recent evidence has shown that astrocytes (in primary culture) not only depolarize passively, but also actively, by an Na^+ dependent mechanism, in response to the excitatory amino acids, glutamate and aspartate (Bowman & Kimelberg 1984). This may suggest another feedback between neurons and glia, not necessarily directly-related to ionic distribution. It is known that glia "mop up" glutamate, produced during neuronal synaptic transmission, by a carrier mediated process (Schousboe 1978; Reynolds & Herschkowitz 1986) and convert it to glutamine (Hamberger et al. 1978; Quastel 1978). They then return it to neurons for re-conversion. Alternatively, astrocytic glia may be directly involved in the synthesis of glutamate itself as glutaminase has been found to be present (Schousboe 1981). Astrocytes are thus in an ideal position to regulate the extracellular concentration of glutamate, a powerful excitatory amino acid. For glutamate studies in the frog's tectum see Sandoval et al. (this volume). If both neurons and glia can be actively depolarized by glutamate *in vivo*, then a potentially non-ionic (positive) feedback may thus obtain. Although the adaptive function of passive directional "spatial buffering" of K^+ by glia has been stressed, interchange of other ions, like Ca^{2+} (Heinemann et al. 1981) or Cl^- between neurons, glial cells and the extracellular environment may also represent important means by which glia can alter neuronal excitability.

8 Abnormal Sensitization, Arousal, and Glia

Central associations between arousal, sensitization, synchronized EEGs, and SPSs of glial origin led to exploration of data relating sensitization, arousal, and glial pathology. The most evident hypersensitization occurring in the vertebrate brain is that which in clinical, or experimentally induced, epilepsy result in seizures.

8.1 Epileptogenesis

Human epilepsy is often the result of brain damage or tumours, though genetic factors may contribute in the causation of low seizure thresholds in some individuals (Taylor & Bower 1971). Some animals, like the gerbil, exhibit seizures spontaneously (Marston & Chang 1965), especially if selectively bred for seizure susceptibility (Kaplan 1981). In general, however, animal models of epilepsy have utilized chemical treatments for seizure induction. Topical application of aminopyridine (Mihaly et al. 1983), aluminium hydroxide (Hatton & Ellisman 1984; Strain et al. 1979; Soper et al. 1978) and penicillin (Koroleva & Bures 1983), among others, have epileptogenic effects in mammals. Penicillin also induces seizures if injected intramuscularly

(Quesney 1984; Taylor-Courval & Gloor 1984; Chen et al. 1986) or applied to the bathing fluid of brain slices (Swann & Brady 1983). Epileptiform activity may also be induced by elevation of the extracellular concentration of potassium (Rutecki et al. 1985). Seizures may not only be induced by chemical treatment but also by repeated electrical stimulation of the brain or "kindling." Kindling effects may outlast the inducing stimulation by one month or more (McIntyre & Chew 1986).

Although most researchers have used mammals as models for induction of seizures, treatments like topical application of penicillin also induces seizures in fish and amphibia (Servit 1977). In fish topical application of aluminium hydroxide or intramuscular injection of penicillin also induce seizures (Laming et al. 1987). These results suggest that a cerebral cortex is not necessary for epileptogenesis.

8.2 Arousal and Seizure Incidence

Several authors have described links between arousal and seizure incidence in humans. In primary generalized epilepsy, seizure incidence is associated with arousal from light sleep or during the awake state (Niedermeyer 1982). In Lennox-Gastaut epileptic disorders, seizure activity is most prevalent during wakefulness and least during sleep or vigilance, drowsy states having intermediate seizure incidence (Papini et al. 1984). Thus, vigilance or attention reduce seizure incidence presumably by inhibition of brain regions not involved in the modality used for attending. A link between increases in arousal and seizure activity is often found (Niedermeyer 1972); however, there is more convincing evidence that epilepsy may be a dysfunction of the arousal system. This comes from studies of focal epilepsy where the focus is in the temporal lobes (Rogozea et al. 1983). Patients with temporal lobe tumours gave more elevated measures of the OR to a novel, acoustic, stimulus than those with tumours elsewhere. The measures used were finger vasoconstriction, alpha blocking in the EEG, electromyogram (EMG), and auditory evoked potential amplitude increases. In all the patients examined, elevation of these responses was related to tumour size, regardless of whether the tumour was epileptogenic. The authors stress that habituation was prolonged though examination of the slope of the habituation curve shows that it was equivalent in normal and epileptic individuals. This strongly suggests that it was the initial response level or degree of arousal which was elevated in these patients. A close correlation was found not just between tumour size and the OR magnitude, but also between tumour size and seizure incidence and severity. As the tumours were glial, these results indicate correlations between glial population, arousal, and the incidence and severity of seizures.

An indicant of sensitization, and possibly arousal, in epileptics who are photosensitive is that visual evoked potentials are shorter in latency than normal in these patients (Faught & Lee 1984). Also, audiomotor reaction time is shortened prior to absence attacks (Lehmann 1963), and cortical neurons are more responsive to thalamic stimulation prior to seizures (Kostopoulos & Avoli 1983).

Links between arousal and seizures have also been obtained from fish with aluminium hydroxide topically applied to the brain or with intramuscular injections of sodium penicillin (Laming et al. 1987). The roach *Rutilus rutilus* with these treatments consistently took more presentations of tap or "light on" stimuli for habituation of behavioral arousal responses than controls. Other individuals showed increases in EEG amplitude from a maximum of 70 μV to over a millivolt. The goldfish *Carassius auratus* with aluminium hydroxide applied topically to the brain showed elevation of cardiac arousal responses to nearly twice the normal value (Fig. 8) (Laming et al. 1987).

Studies on mammals have also shown an association of seizures with arousal. In susceptible gerbils, seizures are initiated in novel situations (Thiessen et al. 1968). In a study with gerbils, using ambulation as a measure of arousal (Thiessen et al. 1968) and directional "rearing" as a measure of attention, it has been found that seizure

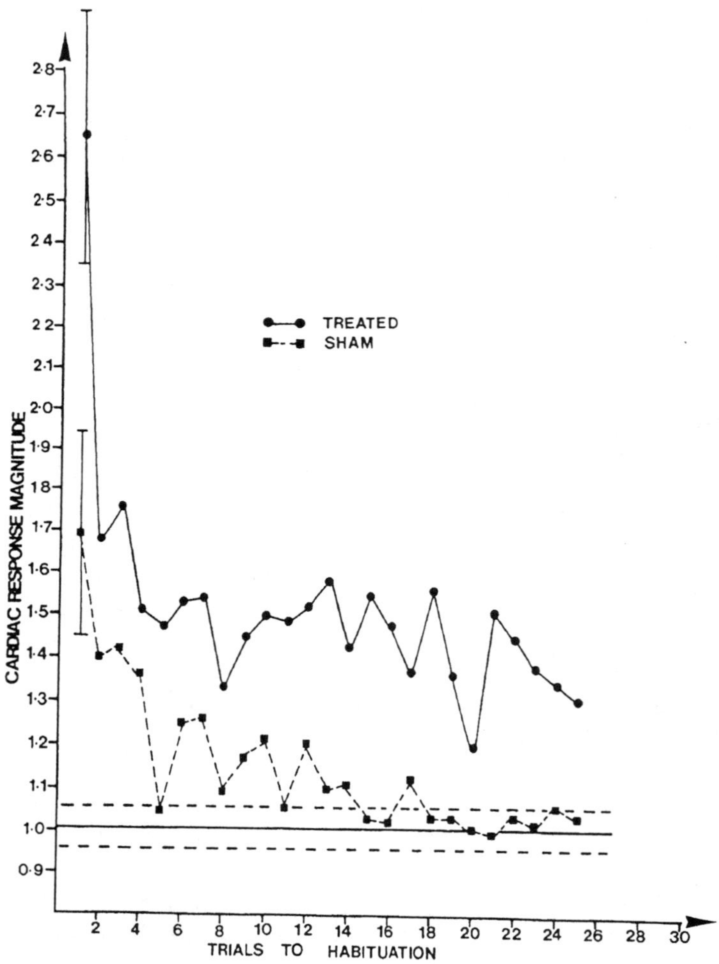

Figure 8. Effect of topically applied aluminium hydroxide on cardiac arousal responses and their habituation in goldfish. Elevated initial responses (mean ± SE) were accompanied by decreased habituation. (After Laming et al. 1987).

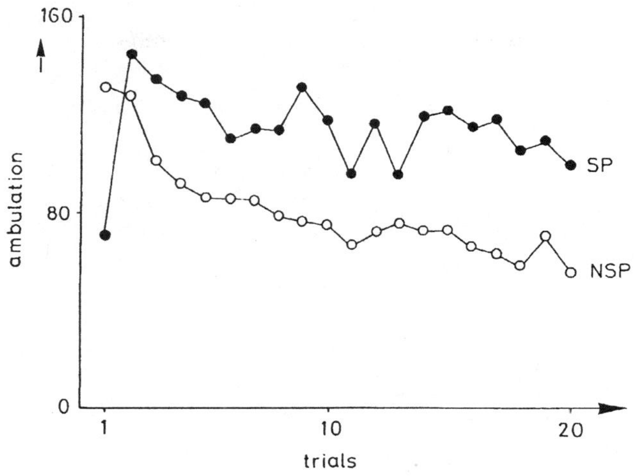

Figure 9. Ambulation, measured as the total number of lines crossed by seizure-prone (SP) and non-seizure prone (NSP) gerbils is shown for each trial. (After Laming et al. 1988).

severity is directly related to arousal and inversely related to attention (Laming et al. 1988). Gerbils previously categorized as seizure prone (SP) or non-seizure prone (NSP) by the method of Kaplan (1981) were tested by gently placing them in the center of a matt black 1 x 1 x 0.35 m high arena for 3 min at 4 min intervals. The results showed that apart from trial 1, when SP animals exhibited seizures, ambulation was thereafter more pronounced in SP animals (Fig. 9). Rearing activity was higher for NSP animals both on trial 1 and for 8 to 9 trials thereafter (Fig. 10). Subsequently, SP animals showed little decline in this activity which habituated in NSP animals. Principal component analysis of the behavioral data obtained in these experiments classified ambulation and rearing as two different classes of behavior suggesting differing neuronal mechanisms for their evocation. These results suggest that if ambulation and rearing are considered measures of arousal and attention, respectively, SP animals are hyperaroused and less able to attend. Inhibitory processes previously discussed as being components of attention would explain negative correlations between attention and seizure genesis.

8.3 Histopathology

Apart from these links between arousal and seizure proneness there are also consistent links between seizures and histopathology. Gliosis (usually astrocytosis) is the most common histopathological feature of epileptic foci, whether they be caused by injuries or tumours (Picker et al. 1981; Rassmussen 1983). Brain damage is often associated with glial proliferation, as part of the natural repair process. As glial organization within the brain of normal animals is presumably under genetic control,this may explain why genetic factors contribute to epilepsy in gerbils (Kaplan 1981) and humans (Taylor & Bower 1971).

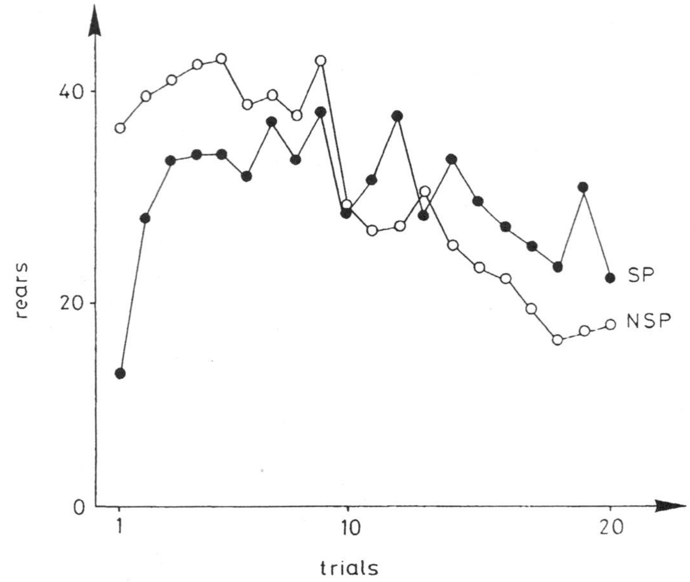

Figure 10. Rearing, measured as the total number of times a bipedal stance was adopted by seizure-prone (SP) and non-seizure prone (NSP) gerbils is shown for each trial. (After Laming et al. 1988).

In many cases the histopathology of chemically induced epilepsies is similar to that found in normal human epileptic foci, i.e., there is an astrocytic gliosis. Astrocytes in these conditions have been shown to swell (Mihaly et al. 1983) and to develop an increase in fibrils especially in the end feet (Yasuzumi et al. 1983) which increase in number and are often associated with neuronal processes (Hatton & Ellisman 1984). During "kindling," too, astrocytes swell (Wolff & Guldner 1978). If this is associated with a gliosis, then it may explain why pathways of conduction are facilitated by repeated seizure propagation (Petsche et al. 1974).

Although the most evident histopathological correlate of epileptic foci is the preponderance of glia, neurons are also affected. Neuron loss in the brain of epileptics (Mouritzen-Dam 1982) is presumed by many to be primarily due to ischaema caused by the effects of convulsive activity in CNS blood vessels.

8.4 Seizure Electrophysiology

During seizures there is a massive increase in CNS activation, whether this be measured by metabolic utilization of glucose (Susuki et al. 1983; Chapman 1985), or by analysis of spike and wave activity throughout the brain (Petsche et al. 1984), or by neuromagnetic recording techniques (Barth et al. 1984). This activity can be maintained in isolated hippocampal slices (Johnston & Brown 1981) treated with penicillin, or in pieces of temporal cortex excized from epileptic patients (Schwartzkroin & Knowles 1984). These pieces of tissue can be as small as 0.5 cm^3.

The latter reports suggest that a highly synchronized paroxysmal depolarizing shift (PDS) of 50-100 ms in duration (Ayala 1983) occurs. This is considered to be a result of a giant synaptic excitatory postsynaptic potential, with a duration of about 100 ms and a frequency according to Schwartzkroin & Knowles (1984) of about 2 Hz. These neuronal intracellular shifts are of shorter duration than glial depolarizations (1-2 s) and are of a frequency similar to that of the hyper-synchronized wave forms of the EEG of epilepsy. Such spike and wave activity at about 2-4 Hz (Petsche et al. 1984) and associated with behavioral arrest (Maxson et al. 1983) may be the result of these giant excitatory postsynaptic potentials (EPSPs), though higher frequency EEG activity (about 10 Hz) is also present in the initial tonic phase. Both types of pattern are locally highly coherent, suggesting a large degree of synchrony of neuronal populations (Rappelsberger et al. 1979), however this may be induced. The amplitude of these potentials recorded extracellularly is of the order of from 1 to 3 mV (Petsche et al. 1984), the slower wave activity which may correspond to the PDS being at the higher end of this voltage range. The highly synchronized nature of the PDS suggests that many neurons are receiving regular facilitation from a common source, even though this may have to derive from the small slices of tissue used. A neuronal feedback mechanism for such a source is normally assumed though glia may sensitize that source. Alternatively, they may have a direct effect via their active depolarization (Bowman & Kimelberg 1984) being transmitted through the glial syncytium (Gutnick et al. 1981) to the abnormally closely apposed neuronal/glial membranes in tumours (Harris 1975). As well as hyper-synchronized waveforms being present in the EEG, seizures are associated with a massive SPS of the order of 5 mV (Swann & Brady 1983) to 10 mV (Koroleva & Bures 1983). The latter authors describe this SPS as spreading depression (lasting for about 1 min) triggered by spikes or, occasionally, occurring spontaneously. Other workers have also described SPSs which may precede seizures (Caspers & Speckmann 1969). The ability of these depolarizing shifts to occur spontaneously may implicate active depolarization of astrocytes in their generation and spread. Shifts of this magnitude would have a profound effect on neuronal sensitivity.

8.5 Changes in Potassium Dynamics

The association of gliosis with epilepsy and the ability of increased extracellular K^+ to induce seizures (Rutecki et al. 1985) suggest that the increased number of glia or their cytological differentiation may affect the potassium dynamics of the neuronal environment. This may result from astrocytic swelling and consequent decreased extracellular space (Ward 1978). Glia in gliotic tissue have a normal (Heinemann & Dietzel 1984) or enhanced ability to remove K^+ (Ward 1983) though not necessarily to translocate it. The reduced extracellular space associated with the borders of gliotic tissue and the site of greatest extracellular concentration changes in K^+ (Heinemann & Dietzel 1984; Prince et al. 1978) may suggest a change in the K^+ dynamics in focal tissue. Rises in extracellular K^+ concentration are a feature of seizure activity (Greenwood et al. 1981; Swann et al. 1986). Although the mechanisms by which sensitization of neurons in epileptic foci are not fully understood, it is likely that the increase in numbers and/or changes in glial morphology associated with foci affect the dynamics of the ionic balance of the extracellular environment (Ward 1983; Laming, unpublished).

Although the role of K^+ in neuronal sensitization has been most considered here, changes in Na^+, Ca^{2+}, or Cl^- dynamics, and the interchange of neurotransmitters between glia and neurons may also be affected by increases in the relative numbers and abnormal morphology of reactive glia. In whatever manner glial cells are ultimately shown to interact with neurons, it is extremely unlikely that the consistent correlation between gliosis and focal seizures is coincidental.

The association of arousability, gliosis, massive SPSs, a hypersynchronized EEG, and epilepsy may therefore be further evidence for a glial role in sensitization during arousal in vertebrates in general.

9 Summary and Conclusions

Studies on fish and amphibia have revealed two central correlates of arousal which may have sensitizing functions. One is an increase in amplitude and apparent synchrony of high frequency wave forms in the EEG, the other a SPS which is negative with respect to an extracerebral reference. Biological mechanisms by which sensitization may derive from these changes have been proposed with some supporting evidence. Much more research is necessary to establish precisely how the brain modulates its own responsiveness, which it evidently does. Two types of modulatory mechanism appear to regulate responsiveness, those of arousal and attention.

Neuronal responsiveness appears to obtain general sensitization in *arousal*. Many, if not all input (sensory) pathways may incur lowered thresholds and the same may obtain for motor pathways. The non-specificity of the sensitization suggest a wide-ranging mechanism possibly mediated through the environment of the neurons themselves, with glia as important regulators of that environment. The effect is that of an amplifier of all inputs and possibly outputs. As novelty and "biological relevance" of inputs are still recognized, some filtering of irrelevant information must still occur, possibly in the sensory pathways themselves. A change in signal/noise ratio may therefore not be a necessary consequence of the increased gain.

When *attention* is directed towards particular features of the environment, as during the OR, inhibition of some sensory responses occurs. Unlike arousal which may have non-neural modulation, these selectively inhibitory (and excitatory) influences are most likely to be a result of neuronal activity (e.g, see Rizzolatti 1983). They can operate across and between modalities and are apparently of later ontogenetic and phylogenetic origin. The selectivity of the input inhibition must in part be a result of experience.

It is probable that the changes in responsiveness shown by animals are never purely those of arousal or attention. The degree to which general sensitization occurs compared to selective sensitization and inhibition may simply reflect the balance between the operational significance of two sets of mechanisms. Other selective influences will also participate separately or through selective facilitation or inhibition. These are those which influence responsiveness to particular stimuli and are concerned with motivational states like hunger, thirst, or sexual appetite. It is clear that any adequate model of a neuroethological system must include *modulation* of the sensitivity of that system. Similarily, any device which attempts to respond optimally and appropriately to a wide variety of potential signals in the environment must be able to select the signals and responses that are appropriate and energetically conservative. The mechanisms by which animals achieve this end, are a subject of continuing research using the simplest examples which illustrate the phenomenon, fish and amphibia.

10 References

Anderson P, Anderson SA (1968) *Physiological basis of the alpha rhythm.* Meredith Corporation, New York

Ayala GF (1983) The paroxysmal depolarising shift. *Prog Clin Biol Res* 124: 15-21

Barry RJ (1984) Preliminary processes in OR elicitation. *Acta Physiol* 55: 109-142

Barth DS, Engle WSJ, Beatty J (1984) Neuromagnetic evidence of spatially distributed sources underlying epileptiform spikes in the human brain. *Science* 223: 293-296

Bauer H, Nirnberger G (1981) Concept identification as a function of preceding negative or positive spontaneous shifts in slow brain potentials. *Psychophysiol* 18(4): 466-469

Birukow G (1951) Ermüdung und Umstimmung bei Gleichgewichtsreaktionen der Amphibien. *Verhandl Deutsch Zool Ges* 16:144-150

Borchers H-W (1982) Correlation between behavior patterns and single unit responses from the optic tectum in the freely moving toad (*Bufo bufo*). In: Trappel R, Ricciardi L, Pask G (eds) *Progress in biocybernetics research, Vol. 9.* McGraw-Hill, London, pp 109-117

Bowman CL, Kimelberg HK (1984) Excitatory amino acids directly depolarise rat brain astrocytes in primary culture. *Nature* 311: 656-659

Bridger WH, Reiser MF (1959) Psychophysiological studies of the neonate: an approach toward the methodological and theoretical problems involved. *Psychosomatic Med* 21: 165-176

Buno W, Velluti R, Handler P, Garcia-Austt E (1966) Neural control of the cochlear input in the wakeful-free guinea pig. *Physiol Behav* 1: 23-35

Caspers H, Speckmann EJ (1969) DC potential shifts in paroxysmal states. In: Jaspers HH, Ward AA, Pope A (eds) *Basic mechanisms of the epilepsies.* Little Brown & Co, Boston, pp 375-388

Chalazonitis N (1978) Some intrinsic and synaptic properties of abnormal oscillators. In: Chalazonitis N, Boisson M (eds) *Abnormal neuronal discharges.* Raven Press, New York, pp 115-132

Chapman AA (1985) Cerebral energy metabolism and seizures. In: Pedley TA, Meldrum BS (eds) *Recent advances in epilepsy II.* Churchill Livingstone, Edinburgh London New York, pp 19-63

Chen RC, Huang YH, How SW (1986) Systemic penicillin as an experimental model of epilepsy. *Exp Neurol* 93: 533-540

Cohen MW (1970) The contribution of glial cells to surface recordings from the optic nerve of an amphibian. *J Physiol* 210: 565-580

Coles JA, Orkand RK (1983) Modification of potassium movement through the retina of the drone (*Apis mellifera*) by glial uptake. *J Physiol* 340: 157-174

Creutzfeldt OD, Houchin J (1974) Neuronal basis of EEG waves. In: Remond A (ed) *Handbook of electroencephalography and clinical neurophysiology, Vol. 2.* Elsevier, Amsterdam, pp 5-55

Creutzfeldt OD, Kuhn V, Benevento LA (1974) An intracellular analysis of visual cortical neurones to moving stimuli: responses in a cooperative neuronal network. *Exp Brain Res* 21: 251-274

Crispino L (1983) Modification of responses from specific sensory systems in midbrain by cerebellar stimulation: experiments in a teleost fish. *J Neurophysiol* 49(1): 3-15

Crispino L, Bullock TH (1984) Cerebellum mediates modality-specific modulation of sensory responses of midbrain and forebrain in rat. *Proc Natl Acad Sci USA* 81: 2917-1920

Degtyar EN (1963) Conditions required for the formation of a CR system at various functional levels of children's nervous activity. *Zhur Vyssei Nerv Deiatel* 13: 631-637

Durkovic RG, Cohen DH (1966) DC potential activity in a nervous system lacking neocortex: the pigeon telencephalon. *Anat Record* 154: 341

Durkovic RG, Cohen DH (1968) Spontaneous, evoked and defensively conditioned steady potential changes in the pigeon telencephalon. *Electroenceph Clin Neurophysiol* 24: 474-481

Eason RG, Dudley LM (1971) Physiological and behavioral indicants of activation. *Psychophysiol* 7(2): 223-232

Eason RG, Harter MR, White CT (1969) Effects of attention and arousal on visually evoked cortical potentials and reaction time in man. *Physiol Behav* 4: 283-289

Elul R (1972) The genesis of the EEG. In: Pfeiffer CC, Smythies JR (eds) *International review of neurobiology, Vol. 15.* Academic Press, New York London, pp 228-272.

Enger PS (1957) The electroencephalogram of the codfish. *Acta Physiol Scand* 39: 55-72

Ewert J-P (1980) *Neuroethology: an introduction to the neurophysiological fundamentals of behavior.* Springer-Verlag, Berlin Heidelberg New York

Ewert J-P (1984) Tectal mechanisms that underlie prey-catching and avoidance behaviors in toads. In: Vanegas H (ed) *Comparative neurology of the optic tectum.* Plenum Press, New York London, pp 247-416

Faught E, Lee SI (1984) Pattern-reversal visual evoked potentials in photosensitive epilepsy. *Electroenceph Clin Neurophysiol* 59: 125-133

Finkenstädt T (1987) Verschaltung, Interaktion und Funktion visuell beeinflußbarer Hirngebiete bei Amphibien. Habilitation Thesis, Univ of Kassel

Finkenstädt T, Ewert J-P (1985) Glucose utilization in the toad's brain during anesthesia and stimulation of the ascending reticular arousal system: a ^{14}C-2-deoxyglucose study. *Naturwissenschaften* 72: 161-162

Fox SS, Norman RJ (1968) Functional congruence: an index of neural homogeneity and a new measure of brain activity. *Science* 159: 1257-1259

Frost JD, Gol A (1966) Computer determination of relationships between EEG activity and single unit discharges in isolated cerebral cortex. *Exp Neurol* 14: 506-519

Futamachi KJ, Pedley TA (1976) Glial cells and extracellular potassium: their relationship in mammalian cortex. *Brain Res* 109: 311-322

Garcia-Austt E, Bogacz J, Venzulli A (1964) Effects of attention and inattention upon visual evoked response. *Electroencephal Clin Neurophysiol* 17: 136-143

Gardner-Medwin AR (1983) Analysis of potassium dynamics in mammalian brain tissue. *J Physiol* 335: 393-426

Gilbert PW, Hodgson ES, Mathewson RF (1964) Electroencephalogram of sharks. *Science* 145: 949-951

Godet R, Bert J, Collomb H (1964) Apparition de la reaction d'eveil telencephalique chez *Protopterus annectens* et cycle biologique. *Comptes Rendus des Seances de la Societe de Biologie* 158: 146-149

Gola M (1978) A model for the production of slow potential waves and associated spiking in molluscan neurons. In: Chalazonitis N, Boisson M (eds) *Abnormal neuronal discharges.* Raven Press, New York, pp 243-262

Gonzalez-Lima F, Scheich H (1984) Functional activation of the auditory system of the rat produced by arousing reticular stimulation: a 2-deoxyglucose study. *Brain Res* 299: 201-214

Goodman DA, Weinberger NM (1969) An electroencephalographic study of *Necturus maculosus* (mud puppy). *Physiol Zool* 42: 398-410

Greenwood RS, Takato M, Goldring S (1981) Potassium activity and changes in glial and neuronal membrane potentials during initiation and spread of after discharges in cerebral cortex of cat. *Brain Res* 218: 279-298

Grossman RG (1978) Glial-neural interaction studies with intracellular injection of ions into cortical glia. In: Schoffeniels E, Franck G, Hertz L, Tower DB (eds) *Dynamic properties of glial cells.* Pergamon Press, New York, pp.105-113

Gumnit RJ (1961) The distribution of direct current responses evoked by sounds in the auditory cortex of the cat. *Electroenceph Clin Neurophysiol* 13: 889-895

Gutnick MJ, Connors BW, Ransom BR (1981) Dye coupling between glial cells in the guinea pig neocortical slice. *Brain Res* 213: 486-492

Hamberger A, Cotman CW, Sellstrom A, Weiler CT (1978) Glutamine, glial cells and their relationship to transmitter glutamate. In: Schoffeniels E, Franck G, Hertz L, Tower DB (eds) *Dynamic properties of glial cells.* Pergamon Press, Oxford New York, pp 163-172

Hara TJ, Ileda K, Gorbrian A (1965) Electroencephalographic studies of homing salmon. *Science* 149: 884-885

Harris AB (1975) Cortical neuroglia in experimental epilepsy. *Exp Neurol* 49: 691-715

Hatton JD, Ellisman MH (1984) Orthoganal arrays are redistributed in the membranes of astroglia from alumina-induced epileptic foci. *Epilepsia* 25(2): 145-151

Heinemann U, Dietzel I (1984) Extracellular potassium concentration in chronic alumina cream foci of cats. *J Neurophysiol* 52: 421-434.

Heinemann U, Konnerth A, Lux HD (1981) Stimulation induced changes in extracellular free calcium in normal cortex and chronic alumina cream foci of cats. *Brain Res* 213: 246-250

Hertz L (1965) Possible role of neuroglia: a potassium-mediated neuronal-neuroglial-neuronal impulse transmission system. *Nature* 4989: 1091-1094

Hobson JA (1967) Respiration and EEG synchronisation in the frog. *Nature* 213: 988-989

Hodgkin AL, Keynes RD (1955) Active transport of cations in giant axons from *Sepia* and *Loligo. J Physiology (London)* 128: 28-60

Hubbard JI, Llinas R, Quastel DMJ (1969) Electrophysiological analysis of synaptic transmission. *Monographs of the physiological society.* Edward Arnold, London, p 372

Jakobsson E, Guttmann R (1981) Continuous stimulation and threshold of axons. In: Adelman WJ, Goldman DE (eds) *The biophysical approach to excitable systems.* Plenum-Press, New York London, pp 197-213

Jane JA, Smirnov GD, Jasper HH (1962) Effects of distraction upon simultaneous auditory and visual evoked potentials. *Electroenceph Clin Neurophysiol* 14: 344-357

Jasper HH (1960) Unspecific thalamocortical relations. In: Field JJ, Magoun HW, Hall VE (eds) *Handbook of physiology, section I. neurophysiology, Vol II.* American Physiological Society, Washington, pp 1307-1321

Johnston D, Brown TH (1981) Giant synaptic potential hypothesis for epileptiform activity. *Science* 211: 294-297

Kaplan H (1981) Effects of fostering on seizure activity in the Mongolian gerbil. *Developmental Psychobiol* 14(6): 565-570

Karmanova IG, Belekhova MG, Tchurnosov, EU (1971) Specifics of behavioral and electrographic patterns of sleep and wakefulness in reptiles. *Sechenov Physiol J USSR* 57: 504-511

Kelly JP, Van Essen DC (1974) Cell structures and function in the visual cortex of the cat. *J Physiol* 238: 515-547

King JS (1966) A comparative investigation of neuroglia in representative vertebrates. *J Morphol* 119: 435-466

Klein M (1963) *Etude polygraphique et phylogenique des etats de sommeil.* Bosc, Lyons France

Kohler W, Wegener J (1955) Currents of the human auditory cortex. *J Cell Comp Physiol Suppl* 1: 25-54

Koroleva VI, Bures J (1983) Cortical penicillin focus as a generator of repetitive spike-triggered waves of spreading depression in rats. *Exp Brain Res* 51: 291-297

Kostopoulos G, Avoli M (1983) Enhanced response of cortical neurons to thalamic stimuli precedes the appearance of spike and wave discharges in feline generalized penicillin epilepsy. *Brain Res* 278: 207-217

Kuffler SW (1967) Neuroglial cells: physiological properties and a potassium-mediated effect of neuronal activity on the glial membrane potential. *Proc Royal Soc B* 168: 1-21

Kuffler SW, Nicholls JG, Orkand RK (1966) Physiological properties of glial cells in the central nervous system of amphibia. *J Neurophysiol* 29: 768-787

Lacey JI, Lacey BC (1974) On heart rate responses and behavior: a reply to Elliot. *J Personal Soc Psychol* 30: 1-18

Laming PR (1980) Electroencephalographic studies on arousal in the goldfish (*Carassius auratus*). *J Comp Physiol Psychol* 94: 238-254

Laming PR (1981) The physiological basis of alert behaviour in fish. In: Laming PR (ed) *Brain mechanisms of behaviour in lower vertebrates.* Cambridge Univ Press, Cambridge England, pp 203-224

Laming PR (1982) Electroencephalographic correlates of behavior in the anurans *Bufo regularis* and *Rana temporaria. Behav Neural Biol* 34: 296-306

Laming PR (1983) Relationship between the responses of visual units, EEGs, and slow potential shifts in the tectum of the toad, *Bufo bufo.* In: Ewert J-P, Capranica RR, Ingle DJ (eds) *Advances in vertebrate neuroethology.* Plenum-Press, New York London, pp 595-603

Laming PR (1987) Behavioural arousal and its habituation in the squirrel fish (*Holocentrus rufus*); the role of the telencephalon. *Behav Neural Biol* 47: 80-104

Laming PR, Brooks M (1985) Effects of visual, chemical, and tactile stimuli on the auditory evoked response of the teleost *Rutilus rutilus. Comp Biochem Physiol* 82A(3): 667-673

Laming PR, Ebbesson SOE (1984) Arousal and fright responses and their habituation in the slippery dick, *Halichoeres bivittatus. Experientia* 40: 767-769

Laming PR, Ewert J-P (1983) The effects of pretectal lesions on neuronal, sustained potential shift, and electroencephalographic responses of the toad tectum to presentation of a visual stimulus. *Comp Biochem Physiol* 76A(2): 247-252

Laming PR, Ewert J-P (1984) Visual unit, EEG, and sustained potential shift responses to biologically significant stimuli in the brain of toads (*Bufo bufo*). *J Comp Physiol* 154: 89-101

Laming PR, Hornby P (1981) The effect of unilateral telencephalic lesions on behavioral arousal and its habituation in the roach, *Rutilus rutilus. Behav Neural Biol* 33: 59-65

Laming PR, McKee M (1981) Deficits in habituation of cardiac arousal responses incurred by telencephalic ablation in goldfish (*Carassius auratus*) and their relation to other telencephalic functions. *J Comp Physiol Psychol* 95(3): 460-467

Laming PR, Savage GE (1978) Flow changes in visceral blood vessels of the chub (*Leuciscus cephalus*) during behavioural arousal. *Comp Biochem Physiol A* 59(3): 291-293

Laming PR, Savage GE (1980) Physiological changes observed in the goldfish (*Carassius auratus*) during behavioural arousal and fright. *Behav Neural Biol* 29: 255-275

Laming PR, Savage GE (1981) Seasonal differences in brain activity and responsiveness shown by the goldfish (*Carassius auratus*). *Behav Neural Biol* 32: 386-389

Laming PR, Borchers HW, Ewert J-P (1984a) Visual unit, EEG, and sustained potential shift responses in the brains of toads (*Bufo bufo*) during alert and defensive behavior. *Physiol Behav* 32: 463-468

Laming PR, Ewert J-P, Borchers HW (1984b) The effects of telencephalic ablation on unit, EEG, and sustained potential shift responses of the toad tectum to a visual stimulus. *Behav Neurosci* 98(1): 118-124

Laming PR, Rooney DJ, Ferguson J (1987) Epileptogenesis is associated with heightened arousal responses in fish. *Physiol Behav* 40: 617-624

Laming PR, Elwood RW, Best PM (1988) Arousal, attention, and epilepsy in the gerbil. *Behav Neural Biol* (in press)

Lansing RW, Lindsley DB, Schwartz E (1959) Reaction time and EEG activation under alerted and non-alerted conditions. *J Exp Psychol* 58: 1-7

Laufer M, Verzeano M (1967) Periodic activity in the visual system of the cat. *Vision Res* 7: 215-219

Lehmann HJ (1963) Praparoxymale Weckreaktionen bei pyknoleptischen Absenzen. *Arch Psychiat Nervenkr* 204: 417-426

Lickey ME, Fox SS (1966) Localisation and habituation of sensory evoked DC responses in cat cortex. *Exp Neurol* 15: 437-454

Livingstone MS, Hubel DH (1981) Effects of sleep and arousal on the processing of visual information in the cat. *Nature* 291: 554-561

London ID (1954) Research on sensory interaction in the Soviet Union. Psychol Bull 51: 531-568

Loveless NE (1979) Event-related slow potentials of the brain as expressions of orienting function. In: Kimmel HD, von Olst EM, Orlebeke JF (eds) *The orienting reflex in humans.* Erlbaum, Hillsdale, pp 77-100

Maltzman I (1979) Orienting reflexes and significance: a reply to O'Gorman. *Psychophysiol* 16: 274-282

Marston JH, Chang MC (1965) The breeding management and reproductive physiology of the Mongolian gerbil (*Meriones unguiculatus*). *Lab Anim Care* 15: 34-48

Maxson SC, Fine MD, Ginsburg BE, Konieck DL (1983) A mutant for spontaneous seizures in C57BL/10Bg mice. *Epilepsia* 24: 15-24

McIntyre DC, Chew GL (1986) Power-spectral analysis of electroencephalographic activity in kindled rats. *Exp Neurol* 92: 261-266

Mihaly A, Joo F, Szente M (1983) Neuropathological alterations in the neocortex of rats subject to focal aminopyridene seizures. *Acta Neuropathol* 61: 85-94

Mori S, Mitarai G, Takagi S, Usui S (1981) Electroencephalographic analysis of activities in the optic tectum of unrestrained carp. *Behav Brain Res* 2: 335-346

Morrell F (1960) Micro-electrode and steady potential studies suggesting a dendritic locus of closure. *Electroenceph Clin Neurophysiol* 13: 1553-1593

Moruzzi G, Magoun HW (1949) Brain stem reticular formation and activation of the EEG. *Electroenceph Clin Neurophysiol* 1: 455-473

Mouritzen-Dam A (1982) Hippocampal neuron loss in epilepsy and after experimental seizures. *Acta Neurol Scand* 66: 601-642

Niedermeyer E (1972) *The generalized epilepsies.* Thomas, Springfield

Niedermeyer E (1982) Epileptic seizure disorders. In: Niedermeyer E, Lopes da Silva F (eds) *Electroencephalography.* Urban & Schwarzenberg, Baltimore Munich, pp. 339-428

Oatman LC (1971) Role of visual attention on auditory evoked potential in unanaesthetised cats. *Exp Neurol* 32: 341-356

Obraztsova LF, Pomazanskaya LN, Stelmakh VA, Troshikhin VA (1958) On the orientation reaction to neutral and signal stimuli in dogs and rabbits in ontogenesis. In: Lynn R (ed) *Attention, arousal and the orientation reaction.* Pergamon Press, Oxford 1966, p 118

Orkand RK, Nicholls JG, Kuffler SW (1966) The effect of nerve impulses on the membrane potential of glial cells in the central nervous system of amphibia. *J Neurophysiol* 29: 788-806

Papini M, Pasquinelli A, Armellini M, Orlandi D (1984) Alertness and incidence of seizures in patients with Lennox-Gastaut syndrome. *Epilepsia* 25(2): 161-167

Pavlov IP (1927) *Conditioned reflexes.* Oxford Univ Press, London

Peters TJ, Vonderahe AR (1954) Electroencephalographic studies, induced seizures, and their modification by phenobarbitol, dilantin, and phenurone in the salamanders *Triturus viridescens* and *Ambystoma tigrinum. Electroenceph Clin Neurophysiol* 6: 253-260

Petsche H, Pockberger H, Rappelsberger P (1984) On the search for the sources of the electroencephalogram. *Neurosci* 11(1): 1-27

Petsche H, Prohaska O, Rappelsberger P, Vollmer R, Kaiser A (1974). Cortical seizure patterns in multidimensional view. The information content of equipotential maps. *Epilepsia* 15: 439-463

Picker S, Pieper CF, Goldring S (1981) Glial membrane potentials and their relationship to $[K^+]_o$ in man and guinea pig. *J Neurosurg* 55: 347-363

Piddington RW (1971a) Central control of auditory input in the goldfish. I: Effects of shocks to the midbrain. *J Exp Biol* 55: 569-584

Piddington RW (1971b) Central control of auditory input in the goldfish. II: Evidence of action in the free swimming animal. *J Exp Biol* 55: 585-610

Pope A (1978) Neuroglia: quantitative aspects. In: Schoffeniels E, Franck G, Hertz L, Tower DB (eds) *Dynamic properties of glial cells.* Pergamon Press, Oxford New York, pp 13-20

Posner MI, Snyder CRR (1975) Facilitation and inhibition in the processing of signals. In: Rabbitt PMA, Dornic S (eds) *Attention and performance V.* Academic Press, London New York, pp 669-682

Pribram KH, McGuinness D (1975) Arousal activation and effort in the control of attention. *Psychol Review* 82: 116-149

Quastel JH (1978) Cerebral glutamate-glutamine interrelations *in vivo* and *in vitro.* In: Schoffeniels E, Franck G, Hertz L, Tower DB (eds) *Dynamic properties of glial cells.* Pergamon Press, New York, pp 153-162

Quesney LF (1984) Pathophysiology of generalised photosensitive epilepsy in the cat. *Epilepsia* 25(1): 61-69

Rakic P (1984) Emergence of neuronal and glial cell lineages in primate brain. In: Black IB (ed) *Cellular and molecular biology of neuronal development.* Plenum Press, New York, pp 29-50

Rakic P, Stensaas LJ, Sayre EP, Sidman RL (1974) Computer aided three-dimensional reconstruction and quantitative analysis of cells from serial electron microscope montages of fetal monkey brain. *Nature* 250: 31-34

Ransom BR, Goldring S (1973a) Ionic determinants of membrane potential of cells presumed to be glia in cerebral cortex of cat. *J Neurophysiol* 36: 855-868

Ransom BR, Goldring S (1973b) Slow depolarisation in cells presumed to be glia in cerebral cortex of cat. *J Neurophysiol* 36: 869-878

Ransom BR, Goldring S (1973c) Slow hyperpolarisation in cells presumed to be glia in cerebral cortex of cat. *J Neurophysiol* 36: 879-892

Rappelsberger P, Petsche H, Vollmer R, Lapins R (1979) Rhythmicity in seizure patterns; intracortical aspects. In: Speckmann EJ, Caspers H (eds) *Origin of cerebral field potentials.* Georg Thieme, Stuttgart, pp 80-97

Rassmussen T (1983) Characteristics of a pure culture of frontal lobe epilepsy. *Epilepsia* 24: 482-493

Reynolds R, Herschkowitz N (1986) Selective uptake of neuroactive amino acids by both oligodendrocytes and astrocytes in primary dissociated culture: a possible role for oligodendrocytes in neurotransmitter metabolism. *Brain Res* 371: 253-266

Rizzolatti G (1983) Mechanisms of selective attention in mammals. In: Ewert J-P, Capranica RR, Ingle DJ (eds) *Advances in vertebrate neuroethology.* Plenum Press, New York, pp 261-297

Rogozea R, Florea Ciocoiu V, Constantinovici A (1983) Habituation of the orienting reaction in patients with epileptogenic cerebral tumours. *Biol Psychol* 16: 65-84

Roitback AI, Fanardjhyan VV, Melkonyan DS, Melkonyan AA (1982) Glial origin of slow negative potential of the cortical direct response. *Neirofiziologiia* 14(1): 76-84

Rooney D, Laming PR (1986) Cardiac and ventilatory arousal responses and their habituation in goldfish: effects of intensity of the eliciting stimulus. *Physiol Behav* 37: 11-14

Rowland V (1968) Cortical steady potential (direct current potential) in reinforcement and learning. In: Stellar E, Sprague JM (eds) *Progress in physiological psychology, Vol 2.* Academic Press, New York, pp 2-77

Rowland V, Goldstone M (1963) Appetitively conditioned and drive-related bioelectric baseline shift in cat cortex. *Electroenceph Clin Neurophysiol* 15: 474-485

Rowland V, Bradley H, School P, Deutschman D (1967) Cortical steady potential shifts in conditioning. *Cond Reflex* 2: 3-22

Rutecki DA, Lebeda FJ, Johnston D (1985) Epileptiform activity induced by changes in extracellular potassium in hippocampus. *J Neurophysiol* 54(5): 1363-1374

Savage GE 1971. Behavioural effects of electrical stimulation of the telencephalon of the goldfish (*Carassius auratus*). *Anim Behav* 19: 661-8.

Schade JP (1959) Bilateral synchrony and arousal in EEG of fish. *Electroenceph Clin Neurophysiol* 11: 613-614

Schade JP, Weiler IJ (1959) EEG patterns of the goldfish. *J Exp Biol* 36: 435-52

Schousboe A (1978) Glutamate, GABA, and taurine in cultured, normal glia cells. In: Schoffeniels E, Franck G, Hertz L, Tower DB (eds) *Dynamic properties of glia cells.* Pergamon Press, Oxford New York, pp 173-183

Schousboe A (1981) Transport and metabolism of glutamate and GABA in neurons and glial cells. *Intern Rev Neurobiol* 22: 1-45

Schwartzkroin PA, Knowles WD (1984) Intracellular study of human epileptic cortex; in vitro maintenance of epileptiform activity? *Science* 223: 709-712

Segura ET, DeJuan A (1966) Electroencephalographic studies in toads. *Electroenceph Clin Neurophysiol* 21: 373-380

Segura ET, Kacelnik A (1977) Cardiorespiratory and electroencephalographic responses to stimulation of the mesencephalic tegmentum in toads, lizards, and rats. *Exp Neurol* 57: 363-373

Servit Z (1977) Phylogenetic aspects of synchronisation in the electrogenesis of epileptic phenomena. Thalamo-cortical mechanisms in lower vertebrates. In: Petsche H, Brazier MA (eds) *Epilepsies.* Springer-Verlag, New York Berlin Heidelberg, pp 291-306

Shaefor PJ, Rowland V (1974) Dissociation of cortical steady potential shifts from mass action potentials in cats awaiting food rewards. *Physiol Psychol* 2(4): 471-480

Singer W (1973) The effect of mesencephalic reticular stimulation on intracellular potentials of cat lateral geniculate neurons. *Brain Res* 61: 35-54

Singer W (1979) Central core control of visual-cortex functions. In: Schmitt FO, Worden FG (eds) *The neurosciences fourth study program.* MIT Press, Cambridge, pp 1093-1110

Sokolov EN (1960) Neuronal models and the orienting reflex. In: Brazier MA (ed) *The central nervous system and behavior.* Macey, New York, pp 187-276

Sokolov EN (1963a) Higher nervous functions: the OR. *Ann Rev Physiol* 25: 545-580

Sokolov EN (1963b) *Perception and the conditioned reflex.* Pergamon Press, Oxford

Soper H v, Strain GM, Babb TL, Lieb JP, Crandall PH (1978) Chronic alumina temporal lobe seizures in monkeys. *Exp Neurol* 62: 99-121

Spinks JA, Siddle D (1983) The functional significance of the orienting response. In: Siddle D (ed) *Orienting and habituation: perspectives in human research.* John Wiley, New York, pp 237-314

Strain GM, Babb TL, Soper HV, Perryman KM, Lieb JP, Crandall PH (1979) Effects of chronic cerebellar stimulation on chronic limbic seizures in monkeys. *Epilepsia* 20: 651-664

Susuki J, Nakamoto Y, Shinkawa Y (1983) Local cerebral glucose utilisation in epileptic seizures of the mutant El mouse. *Brain Res* 266: 359-363

Suzuki TA, Jacobson JH (1971) Reticular stimulation and the light-adapted discharges of the visually-evoked cortical response in cat. *Tohoku J Exp Med* 103: 269-283

Swann JW, Brady RJ (1983) Penicillin-induced epileptogenesis in immature rat CA3 hippocampal pyramidal cells. *Develop Brain Res* 12: 243-254

Swann JW, Smith KL, Brady RJ (1986) Extracellular K^+ accumulation during penicillin induced epileptogenesis in the CA_3 region of immature rat hippocampus. *Develop Brain Res* 30: 243-255

Symons JR (1963) The effects of various heteromodal stimuli on visual sensitivity. *Quart J Exp Psychol* 15: 243-251

Tasaki I (1978) Chemical stimulants and real-time spectrum analyzer used for studying properties of membrane excitable sites. In: Chalazonitis N, Boisson M (eds) *Abnormal neuronal discharges.* Raven Press, New York, pp 233-242

Tauber ES, Rojas-Ramirez J, Hernandez-Peon R (1968) Electrophysiological and behavioural correlates of wakefulness and sleep in the lizard *Ctenosaura pectinata. Electroenceph Clin Neurophysiol* 24: 424-433

Taylor DC, Bower BD (1971) Prevention in epileptic disorders. *The Lancet* 2: 1136-1138

Taylor-Courval D, Gloor P (1984) Behavioural alterations associated with generalised spike and wave discharges in the EEG of the cat. *Exp Neurol* 83: 167-186.

Thiessen DD, Lindzey G, Friend HC (1968) Spontaneous seizures in the Mongolian gerbil. *Psychon Sci* 11: 227-228.

Thompson RF, Voss JF, Brogden WJ (1965) Effects of brightness of simultaneous visual stimulation on absolute auditory sensitivity. *J Exp Psychol* 55: 45-50

Turpin G (1983) Unconditioned reflexes and the autonomic nervous system. In: Siddle D (ed) *Orienting and habituation: perspectives in human research.* John Wiley, New York Toronto, pp 1-71

Verzeano M (1972) Pacemakers, synchronisation, and epilepsy. In: Petsche H, Brazier MA (eds) *Synchronisation of EEG activity in epilepsies.* Springer-Verlag, Wien New York, pp 154-188

Verzeano M, Laufer M, Spear P, McDonald S (1970) The activity of neuronal networks in the thalamus of the monkey. In: Pribram KH, Broadbent DE (eds) *Biology of memory.* Academic Press, New York, pp 239-271

Vinogradova OS (1961) *The orientation reaction and its neurophysiological mechanisms.* Acad Ped Sci RSFSR, Moscow

Ward AA (1978) Glia and epilepsy. In: Schoffeniels E, Franck G, Hertz L, Tower DB (eds) *Dynamic properties of glial cells.* Pergamon Press, New York, pp 413-427

Ward AA (1983) Physiological basis of chronic epilepsy and mechanisms of spread. In: Delgado Escueta AV, Wasterlain CG, Treiman DM, Porter RJ (eds) *Advances in neurology, 34: status epilepticus.*

Raven Press, New York, pp 189-197

Watkins WH, Feehrer CE (1965) Acoustic facilitation of visual detection. *J Exp Psychol* 70: 332-333

Wolff JR, Guldner FH (1978) Perisynaptic astroglial reactions to neural activity. In: Schoffeniels E, Franck G, Hertz L, Tower DB (eds) *Dynamic properties of glial cells.* Pergamon Press, New York, pp 115-118

Yasuzumi G, Aoyama N, Yabimoto N (1983) Ultrastructural changes of basal laminae and protoplasmic astrocytes in craniostenosis with epilepsy. *J Submicrosc Cytol* 15(2): 583-592

Functional Brain Circuitry Related to Arousal and Learning in Rats

FRANCISCO GONZALEZ-LIMA

Department of Anatomy, College of Medicine, Texas A&M University, College Station, TX 77843, USA

Abstract. The findings reviewed provide the first demonstration of how arousing and learning experiences modify the metabolic activity of the rat brain, and point to discrete functional pathways that seem to play key roles in the brain circuitry related to arousal and learning in mammals. Autoradiographic 2-deoxyglucose (2DG) techniques were used to map the functional brain circuitry influenced by *arousing* electrical stimulation of the midbrain reticular formation (RET) in behaving rats. RET stimulation produced selective patterns of metabolic activation and suppression in discrete regions. The findings point to an integrative role of ascending and descending RET pathways in alerting and defensive reactions. During *classical conditioning*, involving an auditory stimulus and RET activation, metabolic changes were clearly different from those seen after RET stimulation alone. For example, opposite effects of learning and arousing conditions on the hippocampus were clearly discriminated. It is suggested that the structures affected by RET form part of the ascending reticular activating system and that some of these structures are also engaged in classical conditioning. Besides subserving arousal, the structures metabolically activated during conditioning may play different roles in signal learning, such as: Selective perception (auditory system), directed attention (prefrontal and posterior parietal cortex), motivation (thalamic and limbic structures), and temporal association of the conditioned and unconditioned *CS-US* stimulus pair (auditory nuclei, hippocampal formation). Finally, there are descending systems mediating the conditioned response made of mixed sensorimotor (flocculus, vestibular nuclei, central grey, spinal cord) and autonomic (medullary nuclei) components involved in the defensive "freezing" reaction to RET stimuli.

1 Introduction

Early anatomical studies suggested that the reticular formation is an integrative network involved in the processing of information from virtually all levels of the neuraxis (Cajal 1909; Kohnstamm & Quensel 1908). In particular, the mesencephalic reticular formation (RET) when stimulated electrically produces generalized electrocortical desynchronization and behavioral arousal (Moruzzi & Magoun 1949). The finding of powerful effects of RET stimulation on thalamo-cortical activity led to the concept of an ascending reticular activating system.

Multiple circuits influenced by RET stimulation are thought to mediate the arousing and defensive reactions elicited by stimulation of this region. Therefore, identification of the structures influenced during arousing RET stimulation may be used to delineate the functional circuitry related to arousal in behaving animals. The ^{14}C-2-deoxyglucose (2DG) autoradiographic method was used to map the functional effects of arousing electrical stimulation of RET in the brain of behaving rats (Gonzalez-Lima & Scheich 1985). This method was applied because the functional activity of all brain structures can be mapped simultaneously. The present chapter reviews the first demonstration of brain structures with localized metabolic alterations related to arousal and learning in mammals. Table 1 lists the structures that show significant changes in 2DG uptake during the defensive "freezing" reaction evoked by *arousing* RET stimulation in rats. The major finding was that RET stimulation produces selective patterns of metabolic activation and suppression in discrete neural regions. Rostral to the stimulation site, the regions activated were limited to the so-called "nonspecific" thalamic nuclei and to the auditory system. In contrast, significant decreases in the "spontaneous" 2DG labeling characteristic of unstimulated control brains were observed in many extrathalamic regions of stimulated rats. A survey of ascending subcortical pathways from structures showing metabolic increases or decreases following RET stimulation is illustrated in Fig. 1C. Table 2 summarizes the effects of different stimulus conditions on metabolic changes in major forebrain structures during *classical conditioning.*

The first part of this chapter deals with a discussion of the structures summarized in Table 1 and their possible roles in *arousal.* The second part considers the structures summarized in Table 2 and their proposed roles in *learning* functions related to classical conditioning. Most of the discussion is based on results obtained with the 2DG method in rats under various stimulus situations.

2 Methodological Comments

Fully alert and unrestrained male albino rats (200 g) were investigated in all of our 2DG studies reviewed here. All rats underwent surgery and electrode implantation under pentobarbital anesthesia. Experiments started about one week after surgery. The surgical and 2DG procedures followed have been recently described in detail (Gonzalez-Lima 1987). 2DG was injected into rats subjected to RET stimulation, into control animals without stimulation, and into control rats stimulated in the superior colliculus dorsal to the RET site. In every RET-stimulated rat, histological verification of the electrode placements showed that the tips of the electrodes were within the deep mesencephalic nucleus of the dorsal midbrain reticular formation (referred to as RET).

The RET stimulation produced a region of high 2DG uptake of about one millimeter in diameter surrounding the electrode tips (Fig. 1 A and B). All RET-stimulated rats showed a short defensive "freezing" reaction (2-3 s duration) evoked by each 0.5 s train of pulses. This reaction is commonly seen in rats when confronted

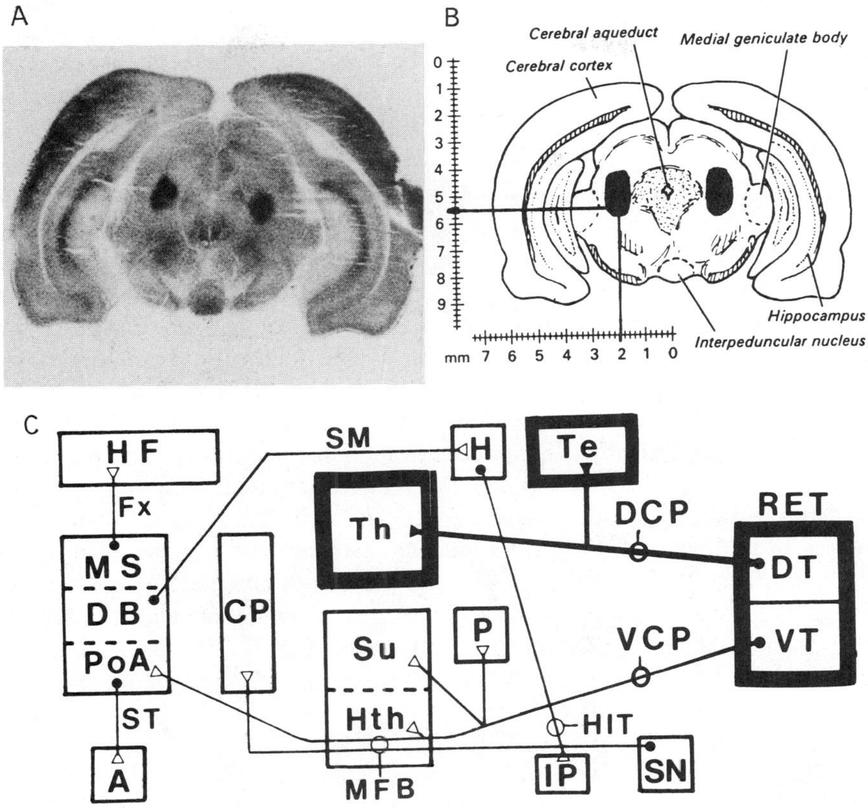

Figure 1. (A) 2DG autoradiograph of a brain section from a stimulated rat showing the activation produced by a stimulation site in the midbrain reticular formation. (B) Transverse section reproduced from the atlas of Hart (1969), demonstrating the extension of the electrically stimulated area in all experimental rats. The black space in the diagram represents a composite of the reticular area of the autoradiographs which displayed a significantly increased optical density following electrical stimulation. The exact localization of the electrode tips and adjacent structures were determined from the corresponding Nissl-stained sections. The stereotaxic coordinates used to place the electrodes are also indicated. The anteroposterior plane was 6 mm posterior to bregma. (C) Schematic block diagram of the subcortical regions affected by RET stimulation and their major cholinergic connections. Structures which showed metabolic activation are indicated by thick heavy lines, and structures which showed metabolic suppression are indicated by thin lines. The dorsal cholinergic pathway, DCP, mainly projects to midline regions of the thalamus, Th, and to the tectum, Te. The ventral cholinergic pathway, VCP, projects to the pretectal area, P, the subthalamus, Su, the hypothalamus, Hth, and the preoptic area, PoA. Higher-order projections of the VCP terminate in the amygdala, A, the hippocampal formation, HF, and the habenula, H. Nigrostriatal fibers, SN-CP, are also indicated. For further abbreviations see also Section 12. (From Gonzalez-Lima & Scheich 1985).

suddenly with a mild aversive stimulus such as touching with the hands of the experimenter or an air puff. The RET stimulus was repeated every 8.5 s on average (range 5-10 s) using a random interstimulus interval schedule for 90 min after 2DG injection. In most rats the "freezing" reaction was accompanied by bradycardia (decrease of heart rate of 1-2 beats/s) which lasted for 2-3 s. This reflex decrease in heart rate has also been found by others during RET stimulation of awake and urethane-anesthetized rats (Kawasaki et al. 1980;. Malmo & Mundl 1983; Versteeg et al. 1982). The specific effects of the RET-evoked bradycardia will be discussed separately in a comparison of rats with and without bradycardiac responses (Gonzalez-Lima 1988).

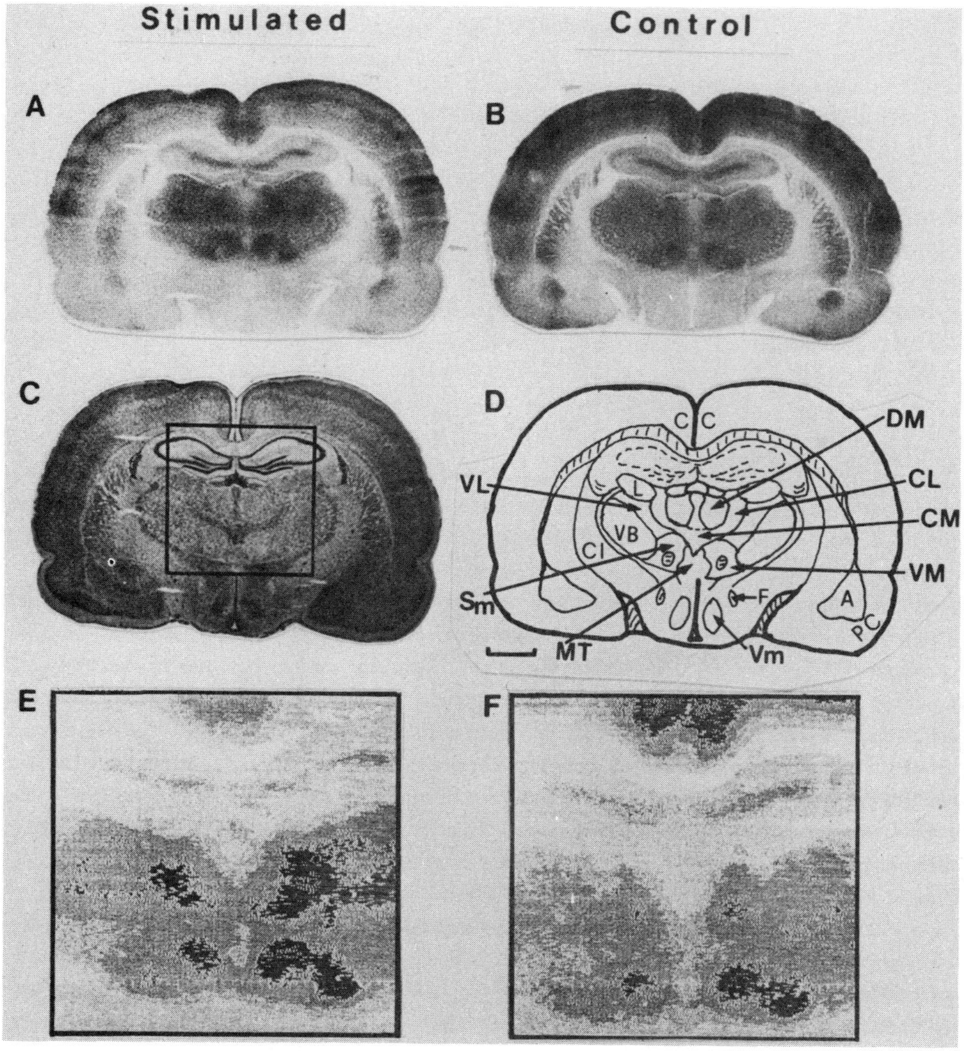

The incorporation of 2DG in brain tissue visualizes the activity of the sodium-potassium pump (Sokoloff 1984). Because of the highest sodium load in excited areas with a large fraction of terminal, dendritic, and synaptic membranes, dominant 2DG uptake commonly is in synaptic neuropil rather than in perikarya (see also Horwitz, this volume). Incorporation of 2DG has been well-documented in neuropil as well as in perikarya of brain cells (Ryan & Sharp 1982). Increased labeling appears to be correlated with synaptic as well as postsynaptic activity of cells (Nudo & Masterton 1986; Theurich et al. 1984), and there is a strong correlation between neural activity and stimulus-evoked 2DG labeling (Juliano & Whitsel 1987). Since the 2DG method maps electrical activity changes throughout the processes and body of brain cells, it is ideal for developing comprehensive functional maps of regions affected by a particular stimulation. The electrophysiological effects of electrical stimulation in a given brain area usually lead to antidromic activation of input structures and orthodromic activation of output structures. Studies with 2DG in which these effects could be clearly discriminated showed no significant elevation of 2DG uptake in antidromically stimulated central structures and a clear increase in 2DG uptake in the orthodromic projection fields (Gonzalez-Lima 1986; Nudo & Masterton 1986; Sharp & Ryan 1984). Similarly, in the RET stimulation findings described here, the structures known to relay only input fibers to RET were not significantly affected in terms of 2DG labeling. This is well in line with the reasoning that the changes in 2DG uptake described in Table 1 are related to neural activity evoked by output fibers projecting from RET to target structures rather than to the antidromic stimulation of afferent axons from neurons in the affected structures. Furthermore, monosynaptic as well as polysynaptic pathways affected by RET stimulation are possible to map using the 2DG method.

Figure 2. (A)-(F) Incorporation of 2DG in the medial thalamus (2 mm posterior to bregma). The first column shows the effect of electrical reticular stimulation, and the second column shows the spontaneous uptake of a control without stimulation. The first row shows autoradiographs of stimulated (A) and control brains (B). The Nissl-stained section in (C) was used to produce the autoradiograph in (A) and the diagram in (D). The schematic diagram contains initials of brain areas according to the abbreviations, and a scale bar equal to 2 mm. The third row contains computer-generated densitometric matrices of the autoradiographs divided into five relative intensity classes. The boundary of each class division is indicated by a contour line of white dots. The area used to generate the matrix in (E) and (F) is indicated by the square drawn in (C). A clear pattern of activation can be readily observed in the stimulated brain. The region of higher density forms an "X" pattern in the medial thalamus which includes the dorsomedial, DM, central lateral, CL, central medial, CM, and midline thalamic, MT, nuclei. The strongest uptake of 2DG corresponds to a region extending throughout both sides of the cytoarchitectonic boundary between DM and CL. Note in contrast the relative decrease in labeling of the cortex, dorsal hippocampus, and amygdala, A, of the stimulated brain. Other abbreviations: CC, cingulate cortex; CI, internal capsule; F, fornix; L, lateral thalamic nucleus; PC, periamygdaloid cortex; Sm, submedial thalamic nucleus (gelatinosus); VB, ventral basal nucleus; VL, ventral lateral nucleus; VM, ventral medial nucleus; Vm, ventromedial hypothalamic nucleus. (Adapted from Gonzalez-Lima & Scheich 1985).

Table 1. Summary of rat neural structures showing significant increases or decreases in 2DG uptake during defensive "freezing" behavior produced by midbrain reticular stimulation. Data based on the comparison of grey/white matter ratios expressed as the mean percent difference in 2DG uptake between corresponding structures in reticular-stimulated and unstimulated control brains. Summarized from Gonzalez-Lima & Scheich (1985) and Gonzalez-Lima (1986, 1987, 1988).

Activated Structures	% Difference	Proposed Role in Reticular Arousal
AUDITORY SYSTEM		Reticular-evoked sensitization of auditory system serves to enhance responsiveness to sound stimuli in the alert animal (auditory arousal).
Cochlear n.	26	
Trapezoid body	14	
Inferior colliculus	69	
Medial geniculate n.	79	
Auditory cortex	42	
THALAMUS (nonspecific)		Enhanced medial thalamic neuronal activity serves to discriminate the negative affective component produced by aversive reticular stimulation (negative affect coding).
Intralaminar n.	115	
Dorsomedial n.	45	
Anterior n.	78	
Midline n.	64	
Anterior reticular n.	109	
MIDBRAIN		Origin of ascending fibers with aversive information from reticular neurons and origin of descending pain-inhibitory fibers from the central gray (defensive reflex initiation).
Deep mesencephalic n.	37	
Central gray (dorsolateral)	27	
CEREBELLUM		Purkinje cell activation to control vestibulo-cerebellar motor output during "freezing" behavior (immobilization reflex).
Flocculus	63	
MEDULLA		Parasympathetic autonomic activation of vagal preganglionic neurons for heart rate deceleration (bradycardiac response).
Caudal ambiguus n.	25	
Vagus dorsal motor n.	19	
Solitary tract n.	14	
SPINAL CORD		Descending reticular inhibition of spinal nociceptive transmission (somatosensory analgesia).
Substantia gelatinosa	31	

3 Thalamo-Cortical Involvement in Arousal

The large activation produced by RET in the medial thalamus is illustrated in Fig. 2. Strong increases of 2DG uptake were present in the intralaminar central medial (200%) and central lateral (110%) nuclei. The dense region of activation in these nuclei formed an "X" pattern within the thalamus (Fig. 2, E *vs*. F). The band of higher density observed in the medial thalamus corresponds best to the region called

Table 1 (continued)

Suppressed Structures	% Difference	Proposed Role in Reticular Arousal
CORTEX		Cortical metabolic suppression with reticular-evoked decrease in high-frequency repetitive discharges of cortical interneurons (electrocortical response during "freezing" behavior).
Frontolateral (area 4)	-71	
Pyriform	-66	
Anterior parietal	-70	
Insular	-61	
Occipital	-52	
LIMBIC STRUCTURES		Reticular-evoked metabolic suppression associated with decreased neuronal transmission inversely related to arousal level and to unpredictable stimulation (limbic arousal coding).
Cingulate gyrus	-71	
Hippocampal formation	-69	
Basolateral amygdala	-76	
Lateral habenula	-42	
Interpeduncular n.	-66	
Mamillary body	-79	
BASAL FOREBRAIN		Basal forebrain neuronal activity is suppressed by aversive reticular stimulation. In contrast, stimulation of basal forebrain produces rewarding and sedative effects ("positive" affect coding pattern contrary to thalamus).
Medial septum	-80	
Diagonal band of Broca	-83	
Lateral preoptic area	-73	
Lateral hypothalamus	-78	
SENSORIMOTOR STRUCTURES		Sensorimotor suppression for tonic phase of immobilization reflex and suppression of vestibulo-cerebellar feedback during reticular stimulation (defensive motor paralysis).
Caudate-putamen	-59	
Substantia nigra compacta	-56	
Subthalamus	-71	
Pretectal n.	-82	
Deep cerebellar n.	-24	
Vestibular n.	-34	

nucleus medialis thalami pars lateralis in the rat (König & Klippel 1963). Midline thalamic nuclei showed uneven increases in 2DG uptake. A very small increase occurred in the reuniens-rhomboidal nuclei (33%); however, at more rostral levels there was a large focal increase (94%) in the paratenial nucleus (Fig. 3). These increases of 2DG uptake in the thalamus are compatible with the activation of direct ascending projections from RET. The major thalamic targets of RET projections are in the intralaminar, midline and medial nuclei according to studies using axonal degeneration (Nauta & Kuypers 1958), antidromic invasion (Ropert & Steriade 1981; Steriade et al. 1980), and autoradiography (Edwards & de Olmos 1976; Eberhart et al. 1985).

Most neurons in the reticular thalamic nucleus of the rat show decreased firing lasting seconds after RET stimulation (Keene 1978). In contrast, some neurons in the anterior reticular thalamus around the anterior ventral lateral nuclei show a large

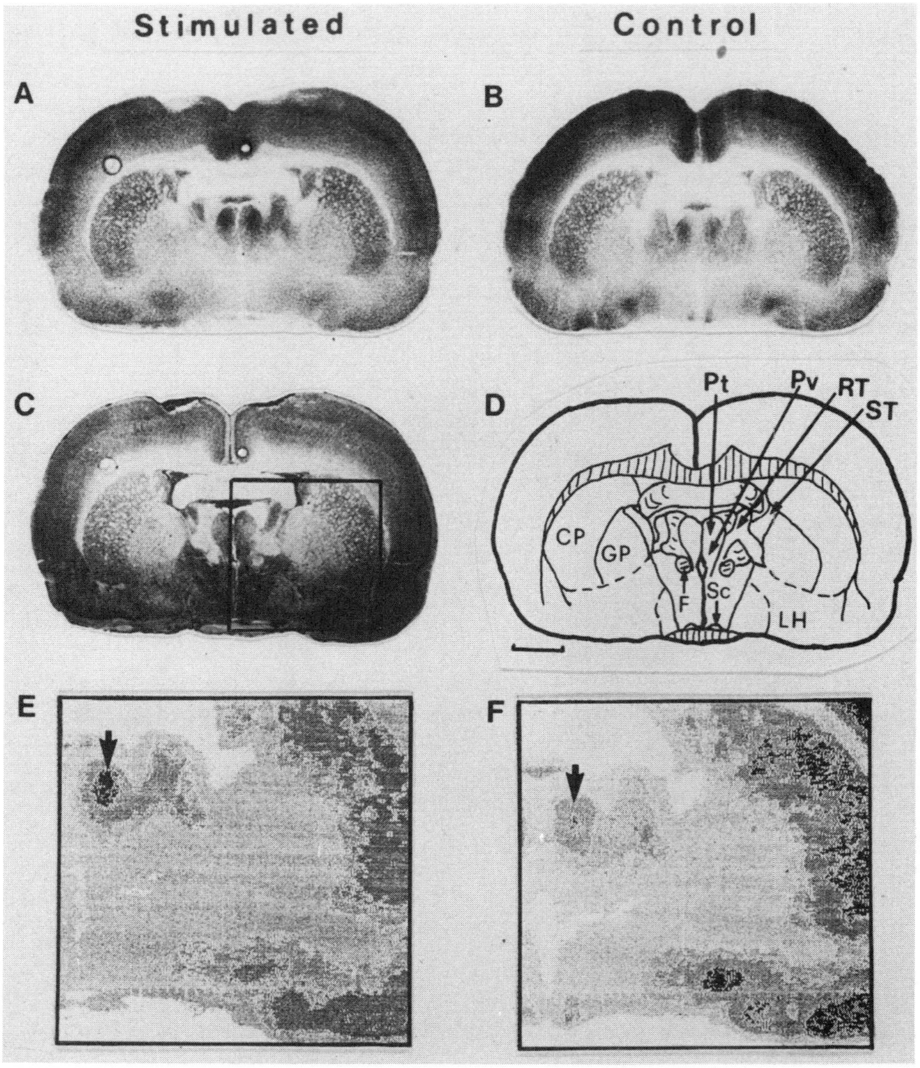

Figure 3. (A)-(F) Incorporation of 2DG in the paratenial nucleus, Pt, and rostral tip of the reticular nucleus, RT, of the anterior thalamus (1.0 mm posterior to bregma). The organization of the figure and scale bar in (D) are the same as in Fig. 2. The densitometric matrices of (E) and (F) correspond to the area inside the square in (C). Uptake of 2DG was increased in the Pt (arrows), and RT of stimulated rats. In contrast, the lateral hypothalamus, LH, caudate-putamen, CP, and the cortex illustrated, showed a decrease in labeling in the stimulated as compared to the control rat. Other abbreviations: GP, globus pallidus; Pv, paraventricular nucleus; Sc, suprachiasmatic nucleus; ST, stria terminalis. (Adapted from Gonzalez-Lima & Scheich 1985).

increase in firing rate during stimulation of RET in cats (Sasaki et al. 1976). These biphasic (inhibitory-excitatory) effects of RET on unit activity may explain the finding that 2DG uptake increased only in the most anterior portion of the reticular thalamic nucleus (Fig. 3).

Increases of 2DG uptake were also observed in the anterior ventral (97%) and anterior medial (58%) thalamic nuclei, for which no definitive evidence of direct RET projections has been reported. This suggests that RET stimulation activated anterior thalamic nuclei polysynaptically. Sasaki et al. (1976) reported responses to RET stimulation in thalamo-cortical neurons of the anterior ventral nucleus in the cat. These neurons displayed an initial decrease in firing rate during and, often for a short time, after RET stimulation, but they showed later tonic excitatory discharges for many seconds or minutes. In their studies, the anterior thalamic neurons exhibited sustained firing during repetitive stimulation of the centromedian-parafascicular complex of the intralaminar thalamus. These findings suggest that intralaminar neurons may be involved in the activation of the anterior thalamus during RET stimulation.

The largest activating effect of RET stimulation in cerebral uptake of 2DG was observed in the intralaminar thalamus (Table 1). This remarkable 2DG activation is consistent with unit activity data showing excitatory responses lasting many seconds in intralaminar neurons after RET stimulation in rats (Keene 1973). Interestingly, intralaminar neurons displaying prolonged excitatory responses after aversive RET stimulation, show long-lasting inhibitory responses following rewarding stimulation (Keene 1973). Therefore, intralaminar neurons appear to discriminate the negative from the positive affective component of stimuli by opposite responses. This has led Keene (1978) to the suggestion that intralaminar neurons are involved in "negative affect coding," i.e., the discrimination of the aversive component that may be associated with noxious arousing stimuli leading to defensive reactions.

The metabolic suppression observed in various cortical regions may be mediated by RET effects on the medial thalamus. A brain mechanism of electrocortical desynchronization ("EEG arousal") may involve reticular activation of nonspecific thalamo-cortical neurons which, in turn, have widespread inhibitory influences in cortical interneurons. There is convincing electrophysiological evidence that a monosynaptic relay in the intralaminar thalamus conveys RET efferents to many separate cortical areas and to the caudate nucleus (Glenn & Steriade 1982; Steriade 1980). Arousal is associated with a decrease in high-frequency repetitive discharges of cortical interneurons. These interneurons "strikingly decrease rates of spontaneous firing and probability of synaptically elicited responses during reticular-induced arousal or natural waking state" (Steriade 1980). In contrast to interneurons, cortical output cells do not become suppressed by RET stimulation.

The conclusion that the 2DG effects observed in the cortex are mediated by the activated nonspecific thalamic nuclei is further supported by the finding of largest metabolic suppression in cortical layers I and VI of RET-stimulated animals (Gonzalez-Lima & Scheich 1985). The classical "dual concept" of thalamo-cortical systems of Lorente de No (1938) shows that nonspecific thalamic nuclei project to cortical layers I and VI, whereas specific thalamic nuclei terminate in midlayers.

Further, electrophysiological evidence indicates that arousal-related stimuli selectively affect cortical neurons at the junction of layers I-II and in layer VI (Hyvärinen et al. 1980). These findings support the conclusion that arousal is linked to a suppression of the activity of cortical interneurons through the influence of intralaminar thalamo-cortical neurons activated by ascending RET fibers.

4 Arousal Effects on Limbic System

Stimulation of RET was correlated with a metabolic suppression of structures that form part of the limbic system (Table 1). The spontaneously displayed pattern of 2DG uptake in limbic structures was reduced in particular regions such as the posterior cingulate cortex, septal end of hippocampal formation, basolateral amygdala, lateral habenula, interpeduncular nucleus, and mammillary body. The observed decreases in 2DG uptake are correlated with the available electrophysiological data on prolonged unit responses to RET stimulation in rats (Keene 1978). There is also evidence that neuronal transmission through hippocampal circuits is inversely related to arousal level. For example, stimulation of the perforant pathway in rats evokes different responses in the hippocampal formation during sleep and waking states. Winson & Abzug (1978) showed that the lowest evoked responses in the hippocampal formation were produced during a waking state of "still alertness." This behavior occurs typically in the rat when the animal is startled by a sudden stimulus. This situation is very similar to the "freezing" reaction evoked by RET stimulation in our 2DG studies. These findings are compatible with the observed decrease of 2DG uptake in the hippocampal formation of RET-stimulated rats (Fig. 6, panel 2).

The decrease of "spontaneous" 2DG uptake produced by RET stimulation in limbic structures is compatible with data indicating that 2DG uptake in hippocampus and amygdala is relatively decreased during active wakefulness characterized by cortical asynchrony and high-amplitude burst of muscle activity. In contrast, sleep produces increases of 2DG uptake in hippocampus and amygdala that are opposite to the suppressive effects obtained during RET stimulation or in the active waking state (Ramm & Frost 1983).

The decrease in 2DG uptake observed in the lateral habenula of RET-stimulated rats may also be arousal-dependent. The lateral habenula shows a suppression of metabolic activity during rewarding medial forebrain bundle (MFB) stimulation (Gallistel et al. 1985). Since aversive RET stimulation also decreases 2DG uptake in the lateral habenula, this structure is not likely to discriminate the motivationally opposite (i.e., reward *vs*. punishment) effects of MFB and RET stimuli. Unit activity studies in the rat indicate that the lateral habenula contains neurons that show similar decreases in firing rates following either MFB or RET stimulation (Keene 1973). Moreover, the simultaneous administration of MFB plus RET stimuli leads to further reduction of habenular unit activity, suggesting a summation of similar effects as would be the case of units responding to the common arousal component evoked by motivationally opposite stimuli. Therefore, the uptake of 2DG in the lateral habenula seems to be related to arousal functions evoked by both positive and negative reinforcers. The 2DG uptake of the lateral habenula is decreased by arousing MFB and RET stimuli (independently of their opposite positive and negative reinforcing effects) while it is increased by sedation produced with neuroleptics such as pimozide (Gomita & Gallistel 1982) and haloperidol (Pizzolato et al. 1985), and by depressed behavior in rats (Caldecott-Hazard et al. 1988).

5 Basal Forebrain Involvement in Arousal

Basal forebrain regions showed reductions in 2DG uptake greater than those seen in the limbic structures (Table 1). The largest suppression evoked by RET stimulation was observed in the diagonal band of Broca and the medial septum which are components of the basal nucleus of Meynert. A significant decrease of labeling was found in the lateral preoptic area and lateral hypothalamus. It is noteworthy that reduced 2DG uptake was focused to the region occupied by the MFB.

Neurons showing strong long-lasting inhibition after RET stimulation have been described in a region of the ventral pallidum that forms part of the basal forebrain area which showed a large metabolic suppression after RET stimulation. Interestingly, units recorded from this region in rats show responses opposite to those of intralaminar units, i.e., reticular-evoked inhibition and MFB-evoked excitation lasting seconds (Keene 1975). These electrophysiological findings are fully in agreement with the observation of opposite metabolic effects of RET stimulation on intralaminar and basal forebrain regions. In addition, the basal forebrain regions activated by MFB stimulation in the 2DG study of Gallistel et al. (1985) correspond to regions strongly suppressed in terms of 2DG uptake during RET stimulation. The possibility that these differential effects of MFB and RET stimulation are related to the discrimination of positive and negative reinforcement components of the MFB and RET stimuli, respectively, requires further investigation.

Experiments based on electrical stimulation of basal forebrain regions in cats (Clemente 1968) indicate that activation of the basal forebrain is correlated with a decrease in arousal and induction of sleep. This finding is compatible with the observation of RET-evoked suppression of the basal forebrain. Indeed, Clemente (1968) has proposed the idea that the basal forebrain represents an "inhibitory system" that acts generally in opposition to the ascending reticular activating system. The 2DG studies reviewed here are compatible with this view. They suggest that two major neural systems, showing opposite effects during RET stimulation, may be involved in the *modulation* of overt behavioral patterns dependent on the arousal state of the organism.

Ascending RET fibers diverge and follow two separate courses as described by Edwards & de Olmos (1976). One division of fibers spreads around the dorsomedial thalamus and moves forward through the intralaminar thalamus. This division of fibers, together with fibers continuing through the midline thalamic nuclei, encapsulates the medial thalamus. The regions innervated by this ascending division correspond to the thalamic nuclei with increased 2DG uptake during RET stimulation, with the exception of the anterior nuclei that were probably influenced polysynaptically. On the other hand, a second division of ascending RET fibers is known to continue in the sub- and hypothalamus. These fibers pass into the zona incerta and then turn and ascend to distribute to the lateral preoptic area and the basal nucleus of Meynert. This second ascending division of fibers innervates regions that have all decreased 2DG uptake during RET stimulation. These two hierarchical levels, defined by two major parallel ascending divisions of RET fibers, appear to form two separate anatomical and functional systems. These systems could mediate, through the interplay of opposite patterns of activity, the modulation of increases and decreases in the arousal state of behaving rats.

There are also remarkable similarities betwen the two functional pathways affected differentially by RET stimulation and the dorsal and ventral tegmental cholinergic pathways which originate from cells of the midbrain tegmental area stimulated in the 2DG studies of RET stimulation (Fig. 1). Structures which showed metabolic activation are part of the dorsal cholinergic pathway, whereas the structures suppressed are part of the ventral cholinergic pathway and its higher-order cholinergic projections (Shute 1973).

It is of interest in this context that injection of cholinergic agonists in the RET produces a "hyper-arousal" response (Hobson et al. 1983), whereas injections of cholinergic agonists in the preoptic area/hypothalamic region suppress general activity in rats (Brudzynski & Mogenson 1986). These results support the hypothesis that cholinergic fibers may form most of the two separate functional pathways responsible for the ascending arousing reticular effects.

Finkenstädt & Ewert (1985) found that patterns of 2DG uptake during stimulation of the reticular formation in toads are comparable to those found in homologous brain structures of rats. For example, toads showed significant decreases in limbic structures such as the lateral amygdala, habenula, ventral medial pallium ("primordium hippocampi") and the hypothalamic/preoptic region. On the other hand, toads stimulated in the reticular formation showed increases in 2DG uptake in the tectum and rostral thalamus in agreement with the increases in 2DG uptake of RET-stimulated rats (Fig. 1C). It is of interest that anesthesia and behavioral habituation result in 2DG effects in toads that are opposite to those produced by reticular stimulation (Finkenstädt, this volume). A review of the 2DG effects of anesthesia and sleep in rats is also compatible with the findings of opposite effects to those produced by RET stimulation in rats (Gonzalez-Lima & Scheich 1985; Ramm & Frost 1983). The agreement between the data obtained in rats and toads during reticular stimulation and other arousal-related conditions supports the idea that there is a basic organizational plan for the structures related to arousal in various vertebrate groups (see also the results in fish reported by Laming, this volume).

6 Motor and Autonomic Correlates of Arousal

Caudal to the RET stimulation site there were a number of brainstem structures showing 2DG changes that may be involved with descending output systems mediating the defensive "freezing" response evoked by RET stimulation. Activated structures included the flocculus of the cerebellum, central gray matter, and medullary autonomic nuclei (Gonzalez-Lima 1988). In addition, a mapping of the descending effects of RET stimulation on the rat spinal cord showed a highly localized increase in 2DG uptake in the substantia gelatinosa (Rexed laminae 2-3) (Gonzalez-Lima 1986). Descending suppressive effects were limited to the deep cerebellar and vestibular nuclei (Gonzalez-Lima 1987).

The suppressive 2DG changes in sensorimotor structures (Table 1) are consistent with the immobilization reflex and the suppression of vestibulocerebellar feedback during RET stimulations. The defensive motor paralysis or "freezing" momentarily evoked by RET stimulation is also compatible with an activation of the cerebellar flocculus. Tract-tracing studies have shown direct RET projections to all the vestibular nuclei (Fukushima et al. 1982). Thus, the action of RET stimulation on vestibular nuclei can be accounted for by their monosynaptic connections. Our 2DG studies (Gonzalez-Lima 1987) favor the possibility that the vestibular nuclei mediate the RET-evoked metabolic activation of the flocculus. First, no changes in 2DG uptake were observed in brainstem reticular nuclei other than the stimulated midbrain RET region. Second, the largest suppression produced by RET stimulation in the brainstem was in the vestibular nuclei which together with the perihypoglossal nuclei and inferior olive send the largest number of projections to the flocculus. Third, given the inhibitory effects of vestibular inputs on the flocculus (Ito 1984) it may be concluded that a supression of these inhibitory inputs may result in activation of the flocculus, as occurred during RET stimulation. Further, the output from the flocculus is only mediated by the Purkinje cells and is inhibitory on neurons of both vestibular and deep cerebellar nuclei (Ito 1984). In this way a positive feedback circuit may operate in vestibular-floccular-vestibular inhibitory interactions and result in a long-lasting inhibition of the vestibular nuclei triggered by the RET effect, i.e., the suppression of vestibular nuclei disinhibits the flocculus which in turn inhibits back the vestibular nuclei (Gonzalez-Lima 1987). These inhibitory vestibulo-cerebellar interactions are compatible with the immobilization response observed in RET-stimulated rats.

In addition, RET stimulation may also produce a reflex decrease in heart rate. This bradycardiac response is correlated with the "freezing" behavior of rats. A 2DG study in RET-stimulated rats showing bradycardia and control rats (Gonzalez-Lima 1988) revealed that descending pathways from RET and the midbrain central grey may activate cardioinhibitory neurons responsible for bradycardia (Fig. 4). In particular, the largest increase in 2DG uptake was observed in the caudal nucleus ambiguus. Smaller increases in 2DG uptake were also found in the dorsal motor nucleus of vagus, the nucleus of the solitary tract, and the caudal inferior olive (Table 1). It was concluded that bradycardia induced centrally by RET stimulation may be mediated by well-defined brainstem descending pathways between the activated regions of the midbrain and the various autonomic medullary nuclei known to induce bradycardia upon electrical stimulation.

Finally, the descending effects of RET stimulation on the spinal cord may represent the activation of pathways which terminate in neurons of the substantia gelatinosa involved in the control of pain transmission (Gonzalez-Lima 1986). This possibility suggests an integrative role of descending RET effects in defensive reactions such as "freezing" behavior caused by aversive stimuli. Since 2DG techniques appear to map preferentially active synaptic terminals rather than cell bodies, it is likely that the RET effects are mediated by active synapses of fibers terminating in the spinal cord. Indirect pathways via RET relays in bulbar reticular nuclei, locus coeruleus, central grey or raphe neurons may be involved. But of the latter structures, only the central gray showed an increase in 2DG uptake in RET-stimulated rats. A descending pathway from the central grey to the raphe magnus, with serotonergic fibers terminating on interneurons of the substantia gelatinosa, is thought to induce analgesic effects (Basbaum & Fields 1984). However, RET stimulation produces descending inhibition of spinal nociceptive transmission by non-serotonergic mechanisms (Carstens et al. 1981). These results support the conclusion that the substantia gelatinosa may be an important site of action for arousal-related descending RET information at the spinal cord level.

7 Arousal Effects on the Auditory System

Little is known about the anatomic-physiologic relationship between sensory perception and reticular-evoked arousal. However, most observations suggest the existence of reticular mechanisms for sensory modulation (Hernandez-Peon 1961).

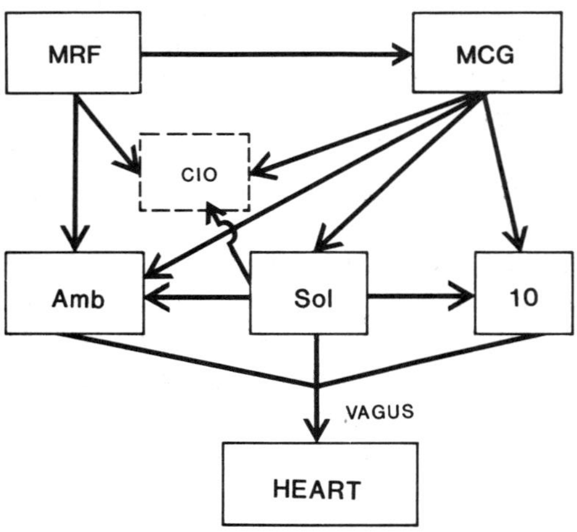

Figure 4. Schematic block diagram of the rat brain stem regions activated during RET-evoked bradycardia and their major descending connections reported in the literature. The midbrain RET, MRF, projects directly to nucleus ambiguus, Amb, caudal inferior olive, CIO, and midbrain central gray, MCG. In turn, MCG projects to CIO, Amb, solitary nucleus, Sol, and dorsal motor nucleus of vagus, 10. Cardioinhibitory neurons in the latter 3 nuclei send their axons to the heart via the vagus nerve to produce bradycardia. (From Gonzalez-Lima 1988).

The 2DG method was used to map functional responses in the auditory system during electrical RET stimulation (Gonzalez-Lima & Scheich 1984a).

Uptake of 2DG during RET stimulation was compared to the effect of a frequency modulated tone (4-5 kHz, 60 dB SPL) and to controls without stimulation. The major finding was a specific pattern of increased metabolic activation throughout the auditory pathway evoked during RET stimulation. The observed increases in 2DG uptake were always greater in RET-stimulated rats as compared to sound-stimulated or control rats. The dorsal cochlear nucleus showed the largest incorporation of 2DG among the auditory nuclei of the brainstem in RET-stimulated rats. In the central nucleus of the inferior colliculus, a layered pattern made of three discrete bands of high 2DG uptake was visible in the RET-stimulated rats (Fig. 5, row 2). The medial geniculate and the auditory cortex also showed highly significant increases in 2DG uptake induced by RET stimulation. The method provided correlations between classical morphological schemes of parcellation on nuclei and functionally defined areas of increased 2DG uptake. These observations were the first anatomical demonstration of the activating effects of RET stimulation in a sensory system, and they support the early concepts of arousing reticular mechanisms for sensory control proposed by Hernandez-Peon (1961).

Based on the new findings with the 2DG method it is possible to clarify significant problems found unresolved in the literature. For example, there are specific questions concerning inputs to the cochlear nucleus (CN), namely about the functional significance of axon collaterals of RET terminating in the CN (Scheibel & Scheibel 1958) and of the efferent fibers from CN to RET (Brodal 1969). The 2DG results may resolve this issue since RET stimulation produced a large increase in the metabolic activity of the CN. It is well established that RET stimulation produces behavioral arousal (Segundo et al. 1955) and lowers the perceptual threshold for auditory stimuli (Chin et al. 1965). Together with the fact that auditory and reticular systems are linked by reciprocal connections at the level of the CN, the 2DG results indicate that mechanisms of behavioral arousal and lowering of perceptual threshold interfere with the auditory system as peripheral as at the level of CN.

There are other reasons to favor the possibility that 2DG uptake increased in the CN due to activation of functionally relevant connections with RET: First, RET stimulation selectively produced a distinctive pattern of activation in the intermediate layers of the dorsal cochlear nucleus (DCN). Osen & Mugnaini (1981) have described excitatory mossy fibers of unknown origin terminating in the granular layer of the DCN. Some of these fibers were thought to be collaterals of olivo-cochlear fibers. But stimulation of the olivocochlear bundle has no effect on auditory responses that can be altered by attentive tasks (Glenn & Oatman 1980). Therefore, some of these mossy fibers may be of RET origin as suggested by the pattern of 2DG labeling concentrated in the region of termination of these fibers, a region which also corresponds to the terminal zone of fibers in the Golgi study of Scheibel & Scheibel (1958). If this speculation about the origin of input fibers is correct, one would expect differential responsiveness between the dorsal and ventral CN as a function of arousal level and consequently also of anesthetic state. Indeed, Evans & Nelson (1973) showed that barbiturate and halothane anesthesia reduced the inhibitory responses of neurons in the dorsal but not in the ventral division of the CN. Similarly, according to Young & Voigt (1981) "one striking feature of DCN responses is the extent to which they are dominated by inhibition which is weakened or abolished by anesthesia," and they present evidence that some of this inhibition is generated by inhibitory interneurons within the DCN. The mossy fibers entering the DCN terminate at inhibitory Golgi cells and at granular cells related to inhibitory interneurons called stellate cells.

Second, there are questions related to inferior colliculus (IC) functions, for instance why high intensities of wide band noise (WBN) produce a 2DG pattern of layering in the IC and low intensities of WBN do not (Sharp et al. 1981). The 2DG layering in IC of RET stimulated rats (Fig. 5, row 2) suggest an alternative answer for the functional significance of the banding pattern observed in the IC of rats exposed to very loud noise. It is well established that noise above 75 dB SPL causes a sympathicotonic activation of the autonomic nervous system (Kryter 1979). If these functional bands are related to the arousal components of the stimuli, one would be able to explain why low intensity WBN (65 dB or less) failed to produce this pattern of activity. The effects of RET stimulation suggest that the functional bands of the IC are associated with the processing of arousal-related information. The arousal necessary to produce a layered pattern in the IC may come from intense acoustic stimuli or RET activation elicited by non-acoustic stimuli.

Other questions are why the rat IC shows the highest level of spontaneous activity in silence within the auditory pathway (Woolf et al. 1983) and why this metabolic activity can be restored after some time when normal inputs from the cochlea are removed (Taniguchi 1980). The above line of reasoning may also help to explain the large level of "spontaneous" activity in the IC in silence. In alert, awake animals activity in the IC may to some extent be a reflection of the general level of arousal, which can also contribute to the large variability of spontaneous labeling observed in IC activity from animal to animal. Spontaneously emitted vocalizations and movement-induced noise may also play important roles. But Taniguchi's (1980) findings support the idea that some IC spontaneous activity is independent of acoustic stimulation. After unilateral cochlear destruction, the IC contralateral to the lesion exhibited markedly less labeling than the ipsilateral side. However, after a few days the activity of the contralateral IC started to increase and by the 20th postoperative day the IC showed an equal 2DG uptake on both sides. Furthermore, additional destruction of the remaining cochlea did not produce any remarkable decrease in 2DG uptake of the contralateral IC. Even if acoustic inputs are removed by bilateral cochlear destruction, the IC can restore its metabolic activity.

Birt et al. (1978) showed that bilateral ablation of the auditory cortex results in increased spontaneous unit activity in the rat IC, and they concluded that a non-specific activation of the IC from a non-cortical structure is selectively antagonized by cortical efferents. This non-cortical structure may involve the reticular cells and fibers stimulated in our experiments. These data taken together favor the idea that the reticular system influences the "spontaneous" activity of the IC.

A third group of questions relates to the medial geniculate body (MG) and the auditory cortex (AC). General anesthetics such as barbiturates have been shown to block ascending transmission in the reticular system while leaving relatively unchanged the transmission in "specific" sensory pathways (Brodal 1969). The fact that in anesthetized animals the MG and AC showed the largest decrease in 2DG utilization among the brain structures analyzed by Sokoloff et al. (1977) argues in favor of a significant reticular involvement in the spontaneous activity of these auditory structures. Further, the 2DG studies of Ramm & Frost (1983, 1986) in animals awake or asleep, indicate that functional activity in higher sensory regions does not depend primarily on external stimulation. Even if an animal is exposed to high levels of auditory stimulation during sleep, the auditory forebrain will show lower relative activity as compared with the waking state under the same constant stimulus level. Ramm & Frost (1986) concluded that the activity "of the sensory forebrain is set by an internal state-dependent influence arising from the reticular formation."

The large increase in metabolic activity in MG and AC during RET stimulation demonstrates that activity at higher levels of the auditory system is highly dependent on arousing RET effects. This conclusion is further substantiated by a number of electrophysiological studies in awake animals from which it is evident that the behavioral or arousal state of the animal can influence neural activity in the AC (Miller et al. 1982). By the same token there is a behavioral dependency of evoked potentials and single unit activity in the AC related to the level of arousal or "attentiveness" of the animal (Molnar et al. 1982).

8 Learning Effects on the Auditory System

Various aspects of associative learning were investigated in the auditory system using a classical conditioning paradigm. This approach allows the study of the interaction of two stimuli in paired and unpaired presentations. Therefore, the learned motivational value of a stimulus (conditioned stimulus, *CS*) can be discriminated from its physical parameters. Arousal-related *vs.* learning-related processes can also be discriminated by the use of sensitization groups receiving the same amount of stimulation but not developing conditioned responses to the *CS*.

Studies of unit activity during learning indicate that auditory responses of awake, behaving animals to acoustic stimuli are dependent not only on the physical properties of a stimulus, but also on its learned motivational significance (Birt & Olds 1982; Weinberger & Diamond 1987). Therefore, the 2DG method was used to analyze the effects of an aversive learning experience on the response of the auditory pathway to an acoustic stimulus. The experiment involved freely behaving rats in a Pavlovian conditioning paradigm in which a sound *CS* (4-5 kHz frequency modulated tone) was temporally paired with aversive RET stimulation (unconditioned stimulus, *US*). The unconditioned response was a rapid decrease in heart rate evoked by the *US*. The conditioned response (*CR*) was a bradycardiac response (criterion: 1 beat/s reduction of heart rate) following onset of the sound *CS*. Eight groups of rats were subjected to: (1) the tone *CS* without conditioning; (2) the aversive *US* alone; (3) the paired *CS-US* (acquisition); (4) the tone *CS* after conditioning (extinction); (5) the *US* prior to the *CS* (sensitization); (6) the unpaired *CS-US* (pseudoconditioning); (7) the *CS* after pseudoconditioning, and (8) no stimulation (Gonzalez-Lima & Scheich 1984b,c; 1986a).

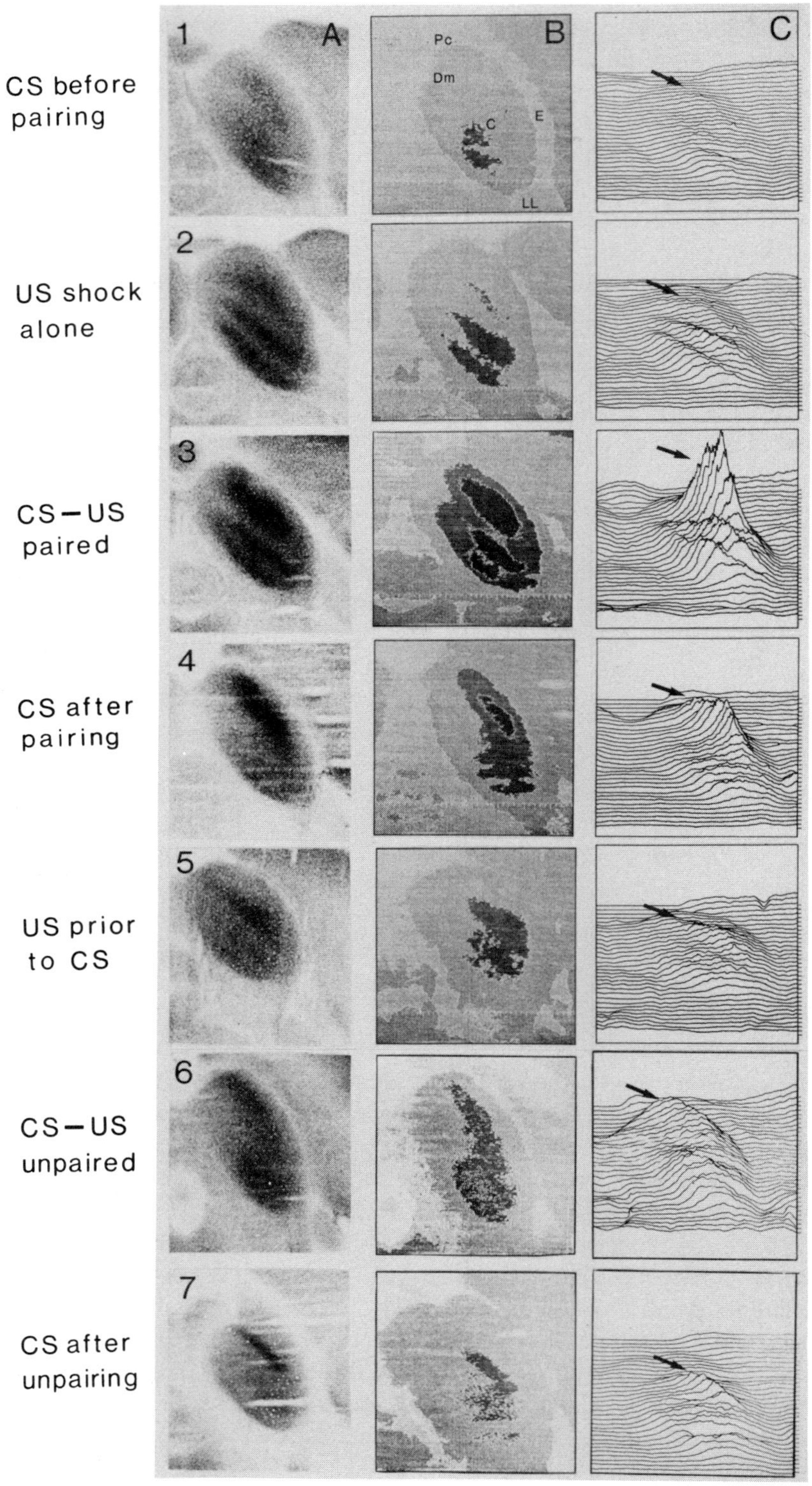
1
A
B
C
Pc
Dm
C
E
LL
CS before pairing
2
US shock alone
3
CS–US paired
4
CS after pairing
5
US prior to CS
6
CS–US unpaired
7
CS after unpairing

The major finding was the differential effect produced by the same tone before (Fig. 5, row 1) and after (Fig. 5, row 4) conditioning. The results also showed that reticular mechanisms interact with incoming acoustic stimuli and modulate the response of auditory structures (Fig. 5; compare row 1 with the sound alone with rows 5 and 6 with the sound plus reticular stimulation). Within each auditory structure, from cochlear nucleus to auditory cortex, the regions of overlap of the spatial representations of *CS* and *US* developed an enhanced metabolic response during conditioning (Fig. 5; indicated by arrow in row 3). The learned motivational value of the tone *CS* was represented in its tonotopically defined neuronal maps throughout the auditory pathway with the exclusion of thalamo-cortical levels. At these higher levels, the learned value of the tone *CS* after conditioning appears to be represented outside of its tonotopically defined space. The changes revealed by the 2DG method in conjunction with a classical conditioning paradigm were the first anatomical demonstration of the effects of sensitization and learning on a sensory system (Gonzalez-Lima & Scheich 1984b). The observations support the idea that auditory responses are dependent both on the *physical* and the *motivational* parameters of a stimulus.

It is not certain whether functional changes in the forebrain are essential for the maintenance of the simple conditioned bradycardiac response (*CR*) used in our experiment. Decorticated and decerebrated animals can learn simple conditioned reflexes under certain conditions (Russell 1980). Lesions of the auditory cortex do not affect heart rate responses conditioned to pure tones in the rat, whereas lesions of medial geniculate and inferior colliculus eliminate autonomic responses conditioned to tones (LeDoux et al. 1984). It is reasonable to expect that the large metabolic changes observed in subcortical auditory nuclei during conditoning (e.g., in the

Figure. 5. Incorporation of 2DG in the inferior colliculus of rats from groups 1-7 of Gonzalez-Lima & Scheich (1984c). The effects of the stimulus regimes are compared in 3 ways. (A) 2DG autoradiographs from comparable anteroposterior locations. (B) Computer-generated densitometric plots of the autoradiographs divided into 5 relative intensity classes; the boundary of each class division is indicated by a contour line of white dots. (C) Densitometric profiles corresponding to the same areas shown in column B. The 4-5 kHz FM tones produced a light band of high 2DG uptake in the dorsal region of the central nucleus (row 1, see arrow). RET stimulation produced 3 bands of high 2DG uptake (row 2). The paired *CS-US* produced an increase in the 3 bands of high 2DG uptake evoked by the US alone, with a very clear selective enhancement of the dorsal region (row 3). The tone *CS* presented alone after training produced a large increase in 2DG uptake in the dorsal region (row 4). The region of highest 2DG uptake corresponded to the site of overlap between the independent effects of the *CS* and *US* (see arrows). Presentation of the tone immediately after RET stimulation resulted in a selective (dorsal band) as well as a general increase in labeling (row 5); note that the 3-band pattern of RET activation is not well defined in row 5. Note the smaller changes in 2DG uptake produced by the *CS-US* unpaired (row 6) and after unpairing (row 7). Abbreviations: Pc, paracentral cortex; E, external nucleus; C, central nucleus; Dm, dorsomedial division of C; LL, lateral lemniscus.

inferior colliculus; Fig. 5, row 3) may be sufficient to sustain the necessary processes involved in tone-conditioned bradycardia. A recent study of 2DG uptake during tone-signaled classical conditioning of the rabbit nictitating membrane found an enhancement of activity in the cochlear nucleus in agreement with our results in rats (Winsky et al. 1986). Together these findings point to a critical role for the peripheral auditory nuclei in simple conditioned reflexes signaled by acoustic stimuli.

But to define the "minimal" neural substrate capable of sustaining a particular conditioned reflex is not sufficient to understand the functional organization of auditory learning in the mammalian brain. Instead, emphasis should be placed on the identification of all neural structures normally involved in the formation of a learned response to complex sound signals such as those used for communication by animals and humans.

For example, lesions of the auditory cortex (AC) disrupt behavioral responses conditioned to the emotional meaning of complex sounds such as frequency modulated tones (Kelly 1974). In conditioning 2DG studies with 4-5 kHz frequency modulated tones (Gonzalez-Lima & Scheich 1986a) most regions of the AC appear to become involved in the representation of the arousal parameters of the *CS*, while the discrete region of the *CS-US* convergence which represents the frequency parameters of the *CS* is preferentially activated. Even though the resolution of the 2DG technique is limited, it appears that the cortical region immediately surrounding the tonotopic space between 4 and 5 kHz is also enhanced during and after conditioning but to a lesser degree.

The results suggest that during Pavlovian conditioning, 3 types of mechanisms appear to interact with the incoming auditory *CS* and modify the activity of the auditory system. The first mechanism may relate to Hebb's ideas (1949) on associative learning. According to this view, the effects of two stimuli must converge in space and time at some point in the stimulated pathways in order for an association to take place. Consequently, temporal pairing of acoustic and reticular stimuli would result in a learning-related metabolic response in the region of spatial convergence of these stimuli within the auditory system. The effect observed in the 4-5 kHz tonotopically defined neuronal space during conditioning may involve a Hebb-type associative mechanism, as indicated by a significant enhancement of *CS*-evoked activity in the region of convergence of *CS-US* effects in the paired but not the unpaired condition. Within each major auditory structure the specific region of overlap of the spatial representations of *CS* and *US* developed an increased metabolic response as a result of conditioning (Fig. 5, row 3), but not as a result of pseudoconditioning (Fig. 5, row 6) or sensitization (Fig. 5, row 5). These findings support Hebb's ideas on the spatial-temporal convergence of two stimuli at some point of the stimulated pathways in order for an association to occur. During conditioning, the concurrent and repeated activation of convergent synapses which represent the *US* and *CS* leads to an enhanced metabolic response to the *CS*.

The second effect observed in the AC "extratonotopic" space of the *CS* after conditioning, but not after pseudoconditioning, is suggestive of a different mechanism related to learning, however, not to sensitization. This is particularly interesting in view of the activation pattern in the superficial cortical layers of the auditory cortex. A cortical laminar analysis of neurons related to the detection of relevant and irrelevant stimuli indicated that attention selectively affected cells at the junction of layer I-II and in layer VI of the somatosensory cortex (Hyvärinen et al. 1980). According to Lorente de No's dual concept of thalamo-cortical systems, most projections from specific nuclei terminate in midlayers and most axons from non-specific nuclei terminate in superficial and deep layers. Neurons from intralaminar nuclei respond to many sensory stimuli, including auditory, and they show a large increase in 2DG uptake (Fig. 2) and prolonged increases (up to 10 s) in firing activity following RET stimulation in the rat (Keene 1978). Therefore, the possibility that some of the effects seen in AC after conditioning may be mediated by the so-called non-specific thalamic nuclei implicated in motivational processes deserves further investigation.

The third effect observed was characterized by a widespread increase in 2DG uptake in all auditory nuclei, the AC and remaining portions of the temporal cortex. Most likely this third effect reflects a reticular arousal mechanism because it was common to all rats subjected to RET stimulation (e.g., see groups 2, 3, 5, and 6 in Fig. 5). These data are compatible with a great body of evidence suggesting that the auditory system, and particularly the AC, are highly responsive to arousal changes (Chin et al. 1965; Buchwald et al. 1966; Miller et al. 1982).

The effects on AC and on auditory subcortical nuclei taken together suggest that the learned signal properties of the auditory *CS* are differentially represented at various levels of the auditory pathway. First, the magnitude of the changes in the region of spatial *CS-US* convergence in the inferior colliculus is several times greater than the increase observed in AC. For example, 2DG uptake densitometry revealed an increase in the inferior colliculus during conditioning that was 2.5 times greater than the algebraic sum of its optical density in the *CS* and *US* groups, whereas the increase in AC did not exceed this sum. The inferior colliculus always showed the greatest activity change in the auditory pathway during and after conditioning (Gonzalez-Lima & Scheich 1984c). Second, the thalamocortical components of the auditory system (medial geniculate and AC) showed a metabolic change to the auditory *CS* after conditioning (extinction) different from the one observed at the more peripheral levels. Indeed, during extinction the conditioned signal properties of the *CS* were still represented (in terms of greater metabolic activity) within its tonotopically defined neuronal maps throughout the auditory pathway with the exclusion of thalamo-cortical levels. At these higher levels, presentation of the *CS* after conditioning failed to produce significant peak increases in tonotopic labeling and it produced instead a different pattern of labeling outside of the *CS* tonotopic bands. Therefore, at thalamocortical levels the uptake of 2DG in the tonotopic neuronal space defined by the *CS* paralleled the behavioral processes of *CR* acquisition and extinction, whereas at the peripheral subcortical levels the conditioning procedure resulted in a long-lasting metabolic enhancement in the *CS* tonotopic maps which persisted after *CR* extinction. A similar long-lasting effect of conditioning on 2DG uptake in another peripheral sensory structure has been reported in the olfactory bulbs of rats (Coppersmith et al. 1986).

9 Nonauditory Forebrain Structures Involved in Associative Learning

The 2DG findings in the auditory system lead to the hypothesis that associative learning represents a *distributed property* of neuronal maps, which may be demonstrated in any structure which meets the appropriate requirements of time and space convergence of *CS* and *US* inputs. The question remained of how classical conditioning would modify neural activity in nonauditory forebrain structures. This question is particularly relevant in view of the extensive literature available on the role of some cortical and limbic structures in mammalian learning (Rosenzweig & Bennett 1976). In an attempt to answer this question, the uptake of 2DG during and after auditory Pavlovian conditioning was examined in all telencephalic and diencephalic structures of the rat brain (Gonzalez-Lima & Scheich 1986b).

The most important outcome of this study was the identification of forebrain structures which showed significant metabolic changes during conditioning, namely prefrontal cortex, posterior parietal cortex, medial thalamus, lateral habenula, caudal striatum, and hippocampal formation. The prefrontal cortex showed discrete regions with enhanced 2DG uptake during conditioning and pseudoconditioning. The medial

Figure 6. Incorporation of 2DG in the hippocampal formation of rats from groups 1-7 of Gonzalez-Lima & Scheich (1986b). Autoradiographs (A) and computer-generated densitometric plots of the autoradiographs (B). Maximal labeling was concentrated in a central band located along the sides of the hippocampal fissure. The band of high 2DG uptake corresponds to the stratum lacunosum-moleculare, L-M, of the hippocampus proper and the underlying outer stratum moleculare, m, of the inner blade of the dentate gyrus. Uptake of 2DG in this band tended to decrease in RET-stimulated rats (row 2). The hippocampal sector CA1 outside of the central band region also showed a tendency to decrease 2DG uptake in RET-stimulated rats. The central band of high 2DG uptake increased significantly during conditioning (row 3) and in rats subjected to the tone after conditioning (row 4). Labeling was also increased in sectors CA2 and CA3 of groups 3 and 4. But labeling of sector CA1 (excluding central band region) in these groups remained within or below control levels. The dentate gyrus and sector CA4 showed a tendency to increase 2DG uptake in group 4 as compared with other groups. A generally lower overall uptake of 2DG can be seen in groups 6 and 7. Panel B5 shows a densitometric plot of an image made by super-imposition of the autoradiographic image upon the histological image stained with cresyl violet. Using this approach, the activity seen in the autoradiograph can be accurately assigned to morphologically defined hippocampal lamina. Observe how the central band of high 2DG uptake, L-M and m, lies outside the pyramidal, P, and granular cell layers, IG and OG. The dashed line running in the middle of this central band indicates the position of the hippocampal fissure. The region of highest 2DG uptake corresponds to the region of dendritic trees of pyramidal and granular cells together with the terminal fields of entorhinal and septohippocampal fibers. The arrow in panel B5 points to a narrow band of decreased labeling adjacent to the lower border of the central band of high 2DG uptake. Abbreviations: O, stratum oriens; P, pyramidal cell layer; R, stratum radiatum; L-M, stratum lacunosum-moleculare; m, stratum moleculare of dentate gyrus; S, supragranular band; IG, granular cell layer of inner blade; H, hilus of fascia dentata; OG, granular cell layer of outer blade.

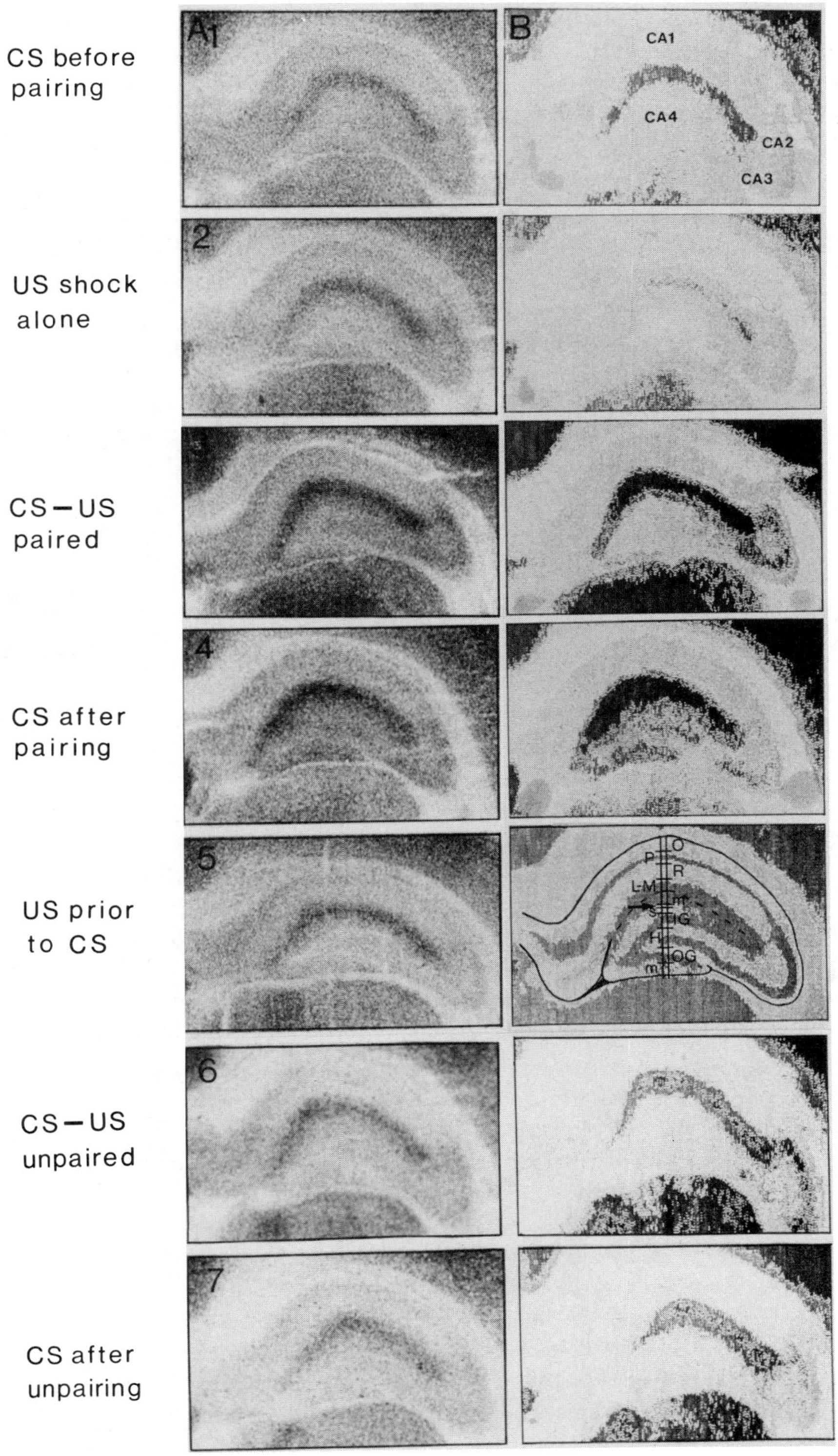
A
B
1
CS before pairing
CA1
CA4
CA2
CA3
2
US shock alone
CS–US paired
4
CS after pairing
5
US prior to CS
O
P
R
L-M
S
IG
H
OG
m
6
CS–US unpaired
7
CS after unpairing

thalamus was greatly activated in all groups subjected to reticular stimulation. The dorsomedial nucleus showed its largest activation during conditioning. The caudal portion of the caudate-putamen and the fundus striati, corresponding to a projection field from the medial geniculate, showed an overall increase in 2DG uptake during conditioning. A columnar organization was well-defined in the posterior parietal cortex of rats subjected to *CS-US* pairing. Significant 2DG changes in parietal cortex and lateral habenula were found mainly in the right hemisphere. These changes revealed by 2DG were the first demonstration of forebrain structures with localized metabolic alterations related to learning and reticular sensitization. Similar 2DG results have been found recently in monkeys by Friedman et al. (1987).

In addition, of the above regions activated during conditioning, only the hippocampal formation showed persistent changes in response to the acoustic *CS* presented alone after conditioning (*CR* extinction). The mapping of this effect in the hippocampal formation provides an excellent example of the high level of resolution that can be achieved by 2DG autoradiography (Fig. 6). A detailed laminar densitometric analysis of the hippocampal formation showed that 2DG uptake was concentrated in a central band along the sides of the hippocampal fissure. This band corresponded to the inner dentate and hippocampal molecular layers. Only this neuropil band of greater metabolic activity showed the learning-related changes. The increased 2DG uptake in this neuropil band during (Fig. 6, row 3) and after (Fig. 6, row 4) conditioning may visualize exceedingly strong activity of input terminals and local potentials in dendrites of pyramidal and granular neurons, i.e., in the vicinity of modifiable synapses. These synapses have been proposed as a prototype for the "Hebb synapses" that could subserve memory functions (Levy & Steward 1983). Since classical conditioning may involve an increase in the excitability of axodendritic synapses in the molecular layer, these results suggest that the 2DG method may be suitable to detect such changes. The learning-related changes in the hippocampal formation have been explained by long-term potentiation of neuronal activity (Voronin 1983). The postulated mechanism of long-term potentiation during conditioning seems to be revealed by the 2DG method in a central hippocampal band of high input convergency. The change visualized with 2DG labeling was the first anatomical identification of a hippocampal substrate with localized and long-lasting metabolic alterations as a result of learning (Gonzalez-Lima & Scheich 1986b).

It is likely that nonsensory structures metabolically activated by presentation of a *CS* after Pavlovian conditioning form part of a distributed neural system involved in the memory representation of that particular *CS* in mammalian brains. The idea of a distributed representation is further substantiated by the 2DG study of John et al. (1986). They showed that all the nonauditory structures activated during and after Pavlovian conditioning in our experiments were involved also in a visually guided task performed by operantly conditioned split-brain cats. However, they found a considerably larger number of structures involved in operant performance of a runaway task for food reward than we found using a simple classical conditioning of heart rate. The combined results of 2DG studies of auditory conditioning (Gonzalez-Lima & Scheich 1984a-c, 1985, 1986a,b) and recent findings with visual conditioning (Gonzalez-Lima et al. 1986c, 1987b) support the hypothesis that sensory learning resulting from Pavlovian conditioning constitutes an anatomically *distributed property* of neuronal maps. This property is functionally represented in the brain throughout the specific sensory maps of the auditory or visual stimuli used as *CS* as well as in integrative structures outside of the conventional sensory systems. Relevant discrete regions with high metabolic activity during classical conditioning are part of association cortex, thalamus, striatum, and limbic structures. These structures may be part of subsystems serving various different functions involved in classical conditioning.

10 Implications for the Neuroanatomical Organization of Auditory Learning

The 2DG studies of learning allowed us to draw several general conclusions regarding brain functional organization. Table 2 summarizes the major structures showing metabolic changes under the different stimulus regimes and their proposed functional roles. Although these conclusions are generally in agreement with the results of most other studies, they should be taken as highlights for further research rather than definitive answers. Neural structures involved in auditory learning processes may be organized along the following lines:

(1) The auditory structures may be specifically involved in sensory plasticity to the acoustic *CS*. The data explained above support the interpretation that the learned behavioral value of the tone *CS* is represented in its tonotopically defined neuronal maps throughout the auditory pathway with the exclusion of thalamo-cortical levels. At these higher levels, the learned value of the tone *CS* after conditioning appears to be represented outside of its tonotopically defined space (Weinberger & Diamond 1987).

(2) The medial thalamus was activated in all groups subjected to RET stimulation. The dorsomedial nucleus was activated during presentation of the *US* alone and even more during *CS-US* pairing. Unit activity studies in rats are in agreement with our 2DG studies because they provide evidence of learning-related increases in activity in the dorsomedial nucleus during tone-signaled conditioning (Delacour 1984) as well as responses in the intralaminar nuclei related to aversive stimulation (Keene 1978). Thus, the greater metabolic enhancement found in the dorsomedial nucleus of rats during aversive conditoning may be related to the acquired aversive component conditioned to the tone *CS*. The reticular *US* produced an aversive response by itself. This aversive response becomes enhanced during acquisition. The most simple explanation may be that the metabolic activity of medial thalamic structures is a reflection of the emotional (aversive) components of the *US* and *CS* rather than the result of memory functions (Markowitsch 1982). Therefore, these metabolic changes may indicate a differential thalamic involvement in the aversive unconditioned and conditioned properties of the stimulus situation.

(3) The prefrontal cortex was similarly activated during paired and unpaired *CS-US* presentations. These findings can be explained in terms of the sensitization effects of the combined stimuli. They are compatible with an involvement of the prefrontal cortex in emotional processes (Markowitsch 1982). But the possibilities of

Table 2. Simplified summary of the effects of the different stimulus conditions (groups 1 to 7 of Gonzalez-Lima & Scheich 1986a,b) on major forebrain structures which showed significant metabolic changes during Pavlovian conditioning. The number of plus and minus signs is an approximate indication of the proportional level of 2DG uptake change of a given structure relative to each stimulus condition.

GROUP					
1(*CS* alone) Auditory	2(*US* alone) Aversive	3(*CS-US* paired) Acquisition	4(*CS* after pairing) Extinction	5-7(*US, CS* unpaired) Sensitization	Proposed Role in Learning
Auditory system(+)	Auditory system(+)	Auditory system(++++)	Auditory system(+++)	Auditory system(++)	Perceptual *CS* component: For stimulus pattern recognition and selective perception of signal *CS* (sensory learning).
	Intralaminar thalamus (+)	Intralaminar thalamus (+)		Intralaminar thalamus (+)	*US* motivational component: For discrimination of aversive component of *US* (affect coding)
	Dorsomedial thalamus (+)	Dorsomedial thalamus (++)		Dorsomedial thalamus (+)	*CS* motivational component: For additional discrimination of acquired aversive component of *CS* (combined *CS* and *US* aversive effects).
		Prefrontal cortex (+)		Prefrontal cortex (+)	Interaction component: For temporal integration of the two stimuli effects in any combination of aversive *US* with auditory *CS* (sensitization).
		Posterior parietal cortex (+) and Lateral habenula (+)			Alerting *CS* Component: For attention to reinforced *CS* during *CS-US* pairings (directed attention).
	Neostriatum (-) and basolateral amygdala (-)	Auditory projection field in caudal neo-striatum (+) and central amygdala (+)			Motor and autonomic *CR* components: Activated for interphasing defensive conditioned responses to the *CS* during *CS-US* pairings (adaptive response learning).
	Hippocampal formation(-)	Hippocampal formation(+)	Hippocampal formation(+)	Hippocampal formation(-)	Predictive *CS* component: Activated by *CS* for representation of predicted *CS-US* temporal sequence (assoc. repres.).

prefrontal cortex involvement in emotional processes and short-term memory (such as in delayed-response tests) are not mutually exclusive. Prefrontal neurons become selectively active during the time between presentation of an emotionally relevant signal stimulus and the anticipated response (Fuster 1982). Prefrontal cortex activity was linked to the various paired and unpaired stimulus combinations of *CS-US* but not to their independent auditory or aversive effects. This interaction effect is consistent with the idea that prefrontal cortex activity reflects both a summation effect and the temporal integration between *CS* and *US* presentations, rather than any particular stimulus sequence or any independent stimulus effect.

(4) Parietal cortex and lateral habenula were activated only during *CS-US* pairing. This activation cannot be explained in terms of sensitization or in terms of the effects of the two stimuli alone before and after conditioning. These structures were probably involved in learning-related functions that may be described as directed attention to a reinforced stimulus (Mesulam 1983; Gomita & Gallistel 1982). Sensory-evoked neuronal responses in the posterior parietal cortex show a behavioral enhancement related to directed attention or to the motivational relevance of the stimuli (Hyvärinen 1982).

(5) A caudal portion of the neostriatum and the underlying amygdalostriatal transition zone showed a metabolic increase during *CS-US* pairing. This region corresponds to a site of convergence of thalamo-striate pathways with information from both *CS* and *US*. As in the case of the auditory system, the field of *CS-US* convergence developed enhanced metabolic responses. Lesions to auditory nuclei or its projection field to the striatum eliminate tone-conditioned emotional responses (LeDoux et al. 1984), suggesting a role for this striatal field in the mediation of auditory conditioned responses (Iwata et al. 1986).

The amygdalostriatal transition zone described by de Olmos et al. (1985) corresponds with the putamen area of the amygdala defined by Turner & Zimmer (1984) and the fundus striati of König & Klippel's atlas of the rat brain (1963). This transition zone has been regarded as either part of the neostriatum or the central amygdala. The central amygdala seems to be important for the autonomic output component of conditioned bradycardia (Kapp et al. 1984). But it appears that the increase in 2DG uptake during conditioning in our experiments does not involve much of the amygdaloid complex and it is limited to a caudal neostriatal field and the underlying amygdalostriatal transition zone. This region receives auditory *CS* information via projections from the medial geniculate (LeDoux et al. 1984). This neostriatal field may be particularly well suited to mediate the motor components involved in defensive responses conditioned to auditory stimuli. On the contrary, other adjacent regions such as the basolateral amygdala showed reductions in 2DG uptake during arousing RET stimulation. But this suppressive effect of RET was not observed during conditioning, suggesting that basolateral amygdala activity returned to levels similar to those observed in control rats without stimulation. This suggests that arousal-related reductions in 2DG uptake in the basolateral amygdala are antagonized during conditioning resulting in a net metabolic effect suggestive of a decrease in some arousal effects during associative learning (For a similar finding in amphibians see Finkenstädt, this volume).

(6) The hippocampal formation was the only nonauditory forebrain structure activated during and after conditioning. The finding that the hippocampal formation is metabolically activated in response to the *CS* after conditioning supports the idea that this long-lasting activation may be related to associative memory functions (Thompson et al. 1982; Laroche et al. 1987). Furthermore, *CS-US* associative processes may be selectively represented at the site of modifiable synapses in the molecular layer.

The 2DG method revealed that the hippocampal formation is clearly involved in the representation of the learned value of an auditory *CS*. A similar involvement has been observed using visual *CS*s in rats (Gonzalez-Lima et al. 1986) and toads (the "primordium hippocampi" described by Finkenstädt, this volume). Finkenstädt's findings in amphibians are of high significance for the formulation of a comparative neuroanatomy of vertebrate learning. The 2DG findings in rats and toads together with the well-known amnesic effects of hippocampal damage in humans and monkeys support the notion that the hippocampal formation is involved in associative functions both in higher and lower vertebrates. Functionally homologous structures in the various vertebrate groups such as the hippocampal formation may be particularly interesting for developing neural network models of learning (see contributions by Arbib and Lara, this volume).

However, the hippocampal formation may not be the only structure subserving memory of a *CS* in the forebrain. Structures which failed to reveal net metabolic changes after conditioning cannot be excluded from participation in memory processes. Memory-related changes may not be correlated with a net change in 2DG uptake in the case of heterogeneous neuronal populations showing both increased and decreased unit activity as a function of learning. For example, in the medial division of the medial geniculate nucleus, two neuronal populations which showed opposite firing responses have been observed (Weinberger & Diamond 1987). One group of neurons developed increases and the other decreases in activity during conditioning. Thus, learning-related changes in such a heterogeneous neuronal space may not be correlated with a regional net change (i.e., increase *vs.* decrease) in metabolic activity. The potential limitation of a metabolic mapping approach to exclude "negative findings" should be taken into account in the interpretation of 2DG studies of learning (for a discussion see also Horwitz, this volume).

Another consideration is that while all the structures described above were involved in auditory classical conditioning using frequency modulated tones (natural-like stimuli) in normal rats, in the case of brain-lesioned rats the organization of learning is forcibly restricted to the structures and circuitry available after brain injury. In this abnormal condition, animals with complete decerebration or decortication (Russell 1980) are still able to learn simple conditioned responses to pure tones (which are acoustic stimuli that do not exist in natural situations). Therefore, in brain-lesioned animals brainstem structures are sufficient for the reorganization of learning involving simple stimuli and conditioned responses. This suggests that the large metabolic changes observed in auditory brainstem nuclei in the 2DG studies of conditioning (Gonzalez-Lima & Scheich 1984b,c) may represent sufficient substrates to sustain the minimal processes needed for simple tone-conditioned responses. The structures and circuitry destroyed in decerebrated animals become critical under conditions involving more complex stimuli and learning tasks (Kelly 1974; Olton et al. 1987).

It is concluded that the structures showing significant metabolic changes during classical conditioning form part of subsystems, serving other functions besides general arousal, that may be involved in auditory signal learning. In brief, the proposed learning-related functions and key structures involved are: (i) selective perception of signal stimulus by auditory structures; (ii) discrimination of aversive motivational component by medial thalamic structures; (iii) emotional sensitization and directed attention to reinforced stimulus by prefrontal and posterior parietal cortex; (iv) adaptive response formation by striatal structures relevant to motor (neostriatum) and autonomic (amygdalostriatum) components of defensive reactions; and (v) long-lasting associative representation of the *CS-US* stimulus pair by the hippocampal formation. Finally, there are output systems mediating particular aspects of the defensive conditioned response made of mixed *sensorimotor* (flocculus, vestibular and deep cerebellar nuclei, central gray, spinal cord) and *autonomic* (medullary nuclei) components.

The described behavioral experiments with the 2DG method may well provide a useful approach to the determination of the functional brain organization of various learning processes in two ways. First, by the identification of the structures (or subsystems) within the nervous system that participate in the organization of a learned response. Second, by the identification of the cellular substrates (e.g., axodendritic synaptic substrate) underlying plasticity in suitable structures with clearly defined cytoarchitecture and connections (e.g., hippocampus). Once the substrates involved in integrative arousal and learning processes are localized at cellular and subsystem levels, then key processes of plasticity and neural organization of behavior can be analyzed.

11 Summary and Conclusions

Autoradiographic 2-deoxyglucose (2DG) techniques were used to map the functional brain circuitry influenced by *arousing* electrical stimulation of the midbrain reticular formation (RET) in behaving rats. RET stimulation produced selective patterns of metabolic activation and suppression in discrete regions. Ascending pathways activated by RET were limited to specific intralaminar, medial and anterior thalamic nuclei, and to the entire auditory system. Ascending pathways suppressed by RET were distributed widely in cortex (area 4, pyriform, anterior parietal, insular and occipital) as well as the basal forebrain (lateral hypothalamus, preoptic area, nucleus basalis, etc.) and limbic structures (hippocampal formation, cingulate gyrus, amygdala, lateral habenula). Descending pathways activated by RET included the flocculus of the cerebellum, central gray matter, medullary autonomic nuclei (particularly nucleus ambiguus), and the substantia gelatinosa of the dorsal horn of the spinal cord. Descending suppressive effects were limited to the deep cerebellar and vestibular nuclei.

The ascending effects may be related to the reticular activating system for arousal and appear localized to major cholinergic pathways. Anatomically defined ascending RET fibers diverge and follow two separate courses as they approach the ventral thalamus. These ascending fiber systems appear to form two separate anatomical substrates which could mediate the metabolic activation and suppression effects produced by RET. The descending effects may be correlates of the autonomic (bradycardia) and defensive "freezing" response to RET stimuli. For example, the nucleus ambiguus showed the largest metabolic increase in the medulla and this nucleus contains most of the cell bodies of preganglionic vagal neurons innervating the heart. This finding is consistent with the RET-evoked decrease in heart rate. The activating effect of RET stimulation on the substantia gelatinosa may represent the activation of descending RET pathways terminating in dorsal horn neurons involved in the control of pain transmission. Together these findings point to an integrative role of ascending and descending RET pathways in alerting and defensive reactions.

Using a *classical conditioning* paradigm the effects of a learned cardiac response to a previously indifferent auditory stimulus were investigated with 2DG. A trial was made of 4-5 kHz FM tones (conditioned stimulus or *CS*) paired with RET (unconditioned stimulus or *US*). The unconditioned response was the RET-evoked bradycardia. The effects were compared between groups subjected to acquisition (*CS-US* pairing), extinction (*CS* alone after pairing), sensitization (*US* prior to *CS*), pseudoconditioning (*CS-US* unpaired), and post- pseudoconditioning (*CS* after unpairing) trainings. Metabolic changes during conditioning were clearly different from those seen after RET alone. Two types of modulatory actions appeared to interact with the acoustic *CS* effects on the auditory system. The first may involve associative mechanisms as indicated by a facilitation of evoked activity in the specific region of convergence of *CS-US* effects. The second may reflect reticular sensitization or arousal effects because it was common to all groups subjected to RET stimulation. Cortical changes were highly localized. The prefrontal cortex showed discrete regions with enhanced 2DG uptake during conditioning, sensitization and pseudoconditioning. A columnar organization was well-defined in the posterior parietal cortex only during conditioning. Subcortical structures showing 2DG increases during conditioning included the lateral habenula and a caudal portion of the neostriatum. The anterior, intralaminar, and dorsomedial thalamic nuclei were greatly activated in all groups subjected to RET, although the dorsomedial nucleus showed its largest activation during conditioning.

The hippocampal formation showed a laminar pattern of activation during acquisition and extinction. 2DG uptake was concentrated in the molecular layers of neuropil along the sides of the hippocampal fissure. The hippocampal formation was the only nonauditory structure which clearly responded to the signal value of the acoustic *CS* during extinction. The changes in the hippocampal formation provide a first anatomical identification of a subcellular (axodendritic synaptic) substrate with long-lasting metabolic changes related to learning. Contrary to the activating effects of learning on the hippocampal formation, the arousing RET stimulus alone or combined in the sensitization groups lead to decrease in 2DG uptake in the hippocampal formation. Thus, the opposite effects of learning and arousing conditions on the hippocampus were clearly discriminated.

It is suggested that the structures affected by RET form part of the ascending reticular activating system and that some of these structures are also engaged in classical conditioning. Besides subserving arousal, the structures metabolically activated during conditioning may play different roles in signal learning, such as: Selective perception (auditory system), directed attention (prefrontal and posterior parietal cortex), motivation (thalamic and limbic structures), and temporal association of the conditioned and unconditioned *CS-US* stimulus pair (auditory nuclei, hippocampal formation). Finally, there are descending systems mediating the conditioned response made of mixed sensorimotor (flocculus, vestibular nuclei, central gray, spinal cord) and autonomic (medullary nuclei) components involved in the defensive "freezing" reaction to RET stimuli. The findings point to discrete functional pathways that seem to play key roles in the brain circuitry related to arousal and learning in mammals.

12 Abbreviations of Brain Structures

A, amygdala
AC, auditory cortex
C, central nucleus
CC, cingulate cortex
CI, internal capsule
CIO, caudal inferior olive
CL, central lateral thalamic nucleus
CM, central medial thalamic nucleus
CN, cochlear nucleus
CP, caudate-putamen
DCN, dorsal cochlear nucleus
DCP, dordal cholinergic pathway
DM, dorsomedial thalamic nucleus
Dm, dorsomedial division of C
E, external nucleus
F, fornix
GP, globus pallidus
H, habenula
HF, hippocampal formation
Hth, hypothalamus
IC, inferior colliculus
IP, interpeduncular nucleus
L, lateral thalamic nucleus
LH, lateral hypothalamus
LL, lateral lemiscus
MCG, midbrain central grey
MFB, medial forebrain bundle
MG, medial geniculate body
MRF, medial reticular formation
MS, medial septum
MT, midline thalamic nucleus
P, pretectal area
PC, periamygdaloid cortex

Pc, paracentral cortex
PoA, preoptic area
Pt, paratenial nucleus
Pv, paraventricular nucleus
RET, midbrain reticular formation
RT, rostral tip of reticular nucleus of the thalamus
SN, substantia nigra
SN-CP, nigro-striatal fibers
ST, stria terminalis
Sc, supra chiasmatic nucleus
Sm, submedial thalamic nucleus (gelatinosus)
Sol, solitary nucleus
Su, subthalamus
Te, optic tectum
Th, thalamus
VB, ventral basal nucleus
VCP, ventral cholinergic pathway
VL, ventral lateral nucleus
VM, ventral medial nucleus
Vm, ventromedial hypothalamic nucleus

Acknowledgements. The research reported here was supported in part by the Alexander-von-Humboldt Foundation, the DFG, SFB 45, the NIH-MBRS grant RR08067, and the NIH-BRS grant RR05814. I express my profound gratitude to my colleague H. Scheich who collaborated with me. I am also grateful to E. Gonzalez-Lima, I. Röder, M. Hansel, M. Camargo, and J. Natowitz for their assistance.

13 References

Basbaum AL, Fields HL (1984) Endogenous pain control systems: brainstem spinal pathways and endorphin circuitry. *Ann Rev Neurosci* 7: 309-338

Birt D, Nienhuis R, Old J (1978) Effects of bilateral auditory cortex ablation on behavior and unit activity in rat inferior colliculus during differential conditioning. *J Neurophysiol* 41: 705-715

Birt D, Olds ME (1982) Auditory response enhancement during differential conditioning in behaving rats. In: Woody CD (ed) *Conditioning: representation of involved neural functions.* Plenum Press, New York, pp 483-502

Brodal A (1969) *Neurological anatomy: in relation to clinical medicine.* Oxford University Press, London

Brudzynski SM, Mogenson GJ (1986) Decrease of locomotor activity by injections of carbachol into the anterior hypothalamic/preoptic area of the rat. *Brain Res* 376: 38-46

Buchwald JS, Halas ES, Schramm S (1966) Changes in cortical and subcortical unit activity during behavioral conditioning. *Physiol Behav* 1: 11-12

Cajal S R y (1909) *Histologie du systeme nerveux de l'homme et des vertebres.* Consejo superior de investigaciones cientificas. Instituto Ramon y Cajal, Madrid

Caldecott-Hazard S, Mazziota J, Phelps M (1988) Cerebral correlates of depressed behavior in rats, visualized using ^{14}C-2-deoxyglucose autoradiography. *J Neurosci* 8: 1951-1961

Carstens E, Fraunhoffer M, Zimmermann M (1981) Serotonergic mediation of descending inhibition from midbrain periaqueductal gray, but not reticular formation, of spinal nociceptive transmission in the cat. *Pain* 10: 149-167

Chin JH, Killam EK, Killam KF (1965) Factors affecting sensory input in the cat: modification of evoked auditory potentials by reticular formation. *Electroenceph Clin Neurophysiol* 18: 567-574

Clemente CD (1968) Forebrain mechanisms related to internal inhibition and sleep. *Cond Reflex* 3: 145-174

Coopersmith R, Lee S, Leon M (1986) Olfactory bulb responses after odor aversion learning by young rats. *Develop Brain Res* 24: 271-277

Delacour J (1984) Two neuronal systems are involved in a classical conditioning in the rat. *Neurosci* 13: 705-715

De Olmos J, Alheid GF, Beltramino CA (1985) Amygdala. In: Paxinos G (ed) *The rat nervous system.* Academic Press, Sydney, pp 223-334

Eberhart JA, Morrell JI, Krieger MS, Pfaff DW (1985) An autoradiographic study of projections ascending from the midbrain central gray, and from the region lateral to it, in the rat. *J Comp Neurol* 241: 285-310

Edwards SB, De Olmos JS (1976) Autoradiographic studies of the projections of the midbrain reticular formation: ascending projections of nucleus cuneiformis. *J Comp Neurol* 165: 417-432

Evans, EF, Nelson PG (1973) The responses of single neurones in the cochlear nucleus of the cat as a function of their location and the anaesthetic state. *Exp Brain Res* 17: 402-427

Finkenstädt T, Ewert J-P (1985) Glucose utilization in the toad's brain during anesthesia and stimulation of the ascending reticular arousal system. *Naturwissenschaften* 72: 161-162

Friedman HR, Janas J, Goldman-Rakic PS (1987) Metabolic activity in the thalamus and mammillary bodies of the monkey during spatial memory performance. *Soc Neurosci Abstr* 13: 207

Fukushima K, Ohno M, Takahashi K, Kato M (1982) Location and vestibular responses of interstitial and midbrain reticular neurons that project to the vestibular nuclei in the cat. *Exp Brain Res* 45: 303-12

Fuster JM (1982) Cortical neuron activity in the temporal organization of behavior. In: Woody CD (ed) *Conditioning: representation of involved neural functions.* Plenum Press, New York, pp 293-306

Gallistel CR, Gomita Y, Yadin E, Campbell KA (1985) Forebrain origins and terminations of the medial forebrain bundle metabolically activated by rewarding stimulation or by reward-blocking doses of pimozide. *J Neurosci* 5: 1246-1261

Glenn JF, Oatman LC (1980) Stimulation studies in the descending auditory pathway. *Brain Res* 196: 258-261

Glenn LL, Steriade M (1982) Discharge rate and excitability of cortically projecting intralaminar thalamic neurons during waking and sleep states. *J Neurosci* 2: 1387-1404

Gomita Y, Gallistel CR (1982) Effects of reinforcement-blocking doses of pimozide on neural systems driven by rewarding stimulation of the MFB: A ^{14}C-2-deoxyglucose analysis. *Pharmacol Biochem Behav* 17: 841-845

Gonzalez-Lima F (1986) Activation of substantia gelatinosa by midbrain reticular stimulation demonstrated with 2-deoxyglucose in the rat spinal cord. *Neurosci Lett* 65: 326-330

Gonzalez-Lima F (1987) Midbrain reticular stimulation produces patterns of metabolic activation and suppression in the cerebellum and vestibular nuclei: a 2-deoxyglucose study. *Brain Res* 412: 275-284

Gonzalez-Lima F (1988) Functional mapping of the brainstem during centrally evoked bradycardia: a 2-deoxyglucose study. *Behav Brain Res* 28: 325-336

Gonzalez-Lima F, Scheich H (1984a) Functional activation in the auditory system of the rat produced by arousing reticular stimulation: a 2-deoxyglucose study. *Brain Res* 299: 201-214

Gonzalez-Lima F, Scheich H (1984b) Classical conditioning enhances auditory 2-deoxyglucose patterns in the inferior colliculus. *Neurosci Lett* 51: 79-85

Gonzalez-Lima F, Scheich H (1984c) Neural substrates for tone-conditioned bradycardia demonstrated with 2-deoxyglucose. I: Activation of auditory nuclei. *Behav Brain Res* 14: 213-233

Gonzalez-Lima F, Scheich H (1985) Ascending reticular activating system in the rat: a 2-deoxyglucose study. *Brain Res* 344: 70-88

Gonzalez-Lima F, Scheich H (1986a) Neural substrates for tone-conditioned bradycardia demonstrated with 2-deoxyglucose. II: Auditory cortex plasticity. *Behav Brain Res* 20: 281-293

Gonzalez-Lima F, Scheich H (1986b) Classical conditioning of tone-signaled bradycardia modifies 2-deoxyglucose uptake patterns in cortex, thalamus, habenula, caudate-putamen, and hippocampal formation. *Brain Res* 363: 239-356

Gonzalez-Lima F, Russell IS, Gonzalez-Lima EM (1986c) Representation of visual memory demonstrated with 2-deoxyglucose: lateralization of learning effects in split-chiasm rats. *Neurosci Lett Supp* 26: S620

Gonzalez-Lima F, Russell IS, Gonzalez-Lima E (1987) Localization of neural substrates for visual memory demonstrated with 2-deoxyglucose. *Neurosci Supp* 22: S423

Hart BL (1969) *Experimental neuropsychology: a laboratory manual.* Freeman, San Francisco

Hebb DO (1949) *The organization of behavior: a neuropsychological theory.* Wiley, New York

Hernandez-Peon R (1961) Reticular mechanisms of sensory control. In: Rosenblith WA (ed) *Sensory communication.* Wiley, New York, pp 497-520

Hobson JA, Goldberg M, Vivaldi E, Riew D (1983) Enhancement of desynchronized sleep signs after pontine microinjection of the muscarinic agonist bethanechol. *Brain Res* 275: 127-136

Hyvärinen J, Poranen A, Jokinen Y (1980) Influence of attentive behavior on neuronal responses to vibration in primary somatosensory cortex of the monkey. *J Neurophysiol* 43: 870-882

Ito M (1984) *The cerebellum and neural control.* Raven Press, New York

Iwata J, LeDoux JE, Meeley MP, Arneric S, Reis DJ (1986) Intrinsic neurons in the amygdaloid field projected to by the medial geniculate body mediate emotional responses conditioned to acoustic stimuli. *Brain Res* 383: 195-214

John ER, Tang Y, Brill AB, Young R, Ono K (1986) Double-labeled metabolic maps of memory. *Science* 233: 1167-1175

Juliano SL, Whitsel BL (1987) A combined 2-deoxyglucose and neurophysiological study of primate somatosensory cortex. *J Comp Neurol* 263: 514-525

Kapp BS, Pascoe JP, Bixler MA (1984) The amygdala: a neuroanatomical systems approach to its contributions to aversive conditioning. In: Butters N, Squire LR (eds) *The neuropsychology of memory.* The Guilford Press, New York

Kawasaki H, Watanabe S, Ueki S (1980) Potentiation of pressor and behavioral responses to brain stimulation following bilateral olfactory bulbectomy in freely moving rats. *Brain Res Bull* 5: 711-718

Keene JJ (1973) Reward-associated inhibition and pain-associated excitation lasting seconds in single intralaminar thalamic units. *Brain Res* 64: 211-224

Keene JJ (1975) Reward-associated excitation and pain-associated inhibition lasting seconds in rat medial pallidal units. *Exp Neurol* 49: 97-114

Keene JJ (1978) Affect-related unit activity in forebrain. *Fed Proc* 37: 2246-2250

Kelly JB (1974) Polysensory cortical lesions and auditory temporal pattern discrimination in the cat. *Brain Res* 80: 317-327

Kohnstamm O, Quensel F (1908) Centrum receptorium der Formatio reticularis und gekreutzt aufsteigenden Bahnen. *Dtsch Z Nervenhk* 36: 182-188

König JFR, Klippel RA (1963) *The rat brain: a stereotaxic atlas.* Williams & Wilkins, Baltimore

Kryter KD (1979) Extraauditory effects of noise. In: Henderson D, Hamernik RP, Dosanjh DS, Mills JH (eds) *Effects of noise on hearing.* Raven Press, New York, pp 531-546

Laroche S, Neuenschwander-El Massioui N, Edeline JM, Dutrieux G (1987) Hippocampal associative cellular responses: dissociation with behavioral responses revealed by a transfer-of-control technique. *Behav Neural Biol* 47: 356-368

LeDoux JE, Sakaguchi A, Reis DJ (1984) Subcortical efferent projections of the medial geniculate nucleus mediate emotional responses conditioned to acoustic stimuli. *J Neurosci* 4: 683-698

Levy WB, Steward O (1983) Temporal contiguity requirements for long-term associate potentiation/ depression in the hippocampus. *Neurosci* 8: 791-797

Lorente de No R (1949) The structure of the cerebral cortex. In: Fulton JF (ed) *Physiology of the nervous system.* Oxford Univ Press, New York, pp 288-330

Malmo RB, Mundl WL (1983) Cardiovascular and respiratory responses to electrical stimulation of the midbrain in the rat. *Int J Psychophysiol* 1: 75-81

Markowitsch HJ (1982) Thalamic mediodorsal nucleus and memory: a critical evaluation of studies in animals and man. *Neurosci Biobehav Rev* 6: 351-380

Mesulam MM (1983) The functional anatomy and hemispheric specialization for directed attention. *TINS* 6: 384-387

Miller JM, Pfingst BE, Ryan AF (1982) Behavioral modification of response characteristics of cells in the auditory system. In: Woody CD (ed) *Conditioning: representation of involved neural functions.* Plenum Press, New York, pp 345-362

Molnar M, Karmos G, Csepe V (1982) Intracortical auditory evoked potentials in the sleep-waking cycle of the cat. In: Sinz R, Rosenzweig MR (eds) *Psychophysiology 1980.* VEB Gustav Fischer-Verlag, Jena, and Elsevier, Amsterdam, pp 549-555

Moruzzi G, Magoun HW (1949) Brain stem reticular formation and activation of the EEG. *Electroenceph Clin Neurophysiol* 1: 455-473

Nauta WJH, Kuypers H (1958) Some ascending pathways in the brain stem reticular formation. In: Jasper HH, Protor L (eds) *Reticular formation of the brain.* Little-Brown, Boston, pp 3-30

Nudo R, Masterton RB (1986) Stimulation-induced [^{14}C]2-deoxyglucose labeling of synaptic activity in the central auditory system. *J Comp Neurol* 245: 553-565

Olton DS, Meck WH, Church RM (1987) Separation of hippocampal and amygdaloid involvement in temporal memory dysfunctions. *Brain Res* 404: 180-188

Osen KK, Mugnaini E (1981) Neuronal circuits in the dorsal cochlear nucleus. In: Syka J, Aitkin L (eds) *Neuronal mechanisms of hearing.* Plenum Press, New York, pp 119-125

Pizzolato G, Soncrant TT, Holloway HW, Rapoport SI (1985) Reduced metabolic response of the aged rat brain to haloperidol. *J Neurosci* 11: 2831-2838

Ramm P, Frost BJ (1983) Regional metabolic activity in the rat brain during sleep-wake activity. *Sleep* 6: 196-216

Ramm P, Frost BJ (1986) Cerebral and local cerebral metabolism in the cat during slow wave and REM sleep. *Brain Res* 365: 112-124

Ropert N, Steriade M (1981) Input-output organization of the midbrain reticular core. *J Neurophysiol* 46: 17-31

Rosenzweig MR, Bennett EL (1976) *Neural mechanisms of learning and memory.* MIT Press, Cambridge

Russell IS (1980) Encephalization and neural mechanisms of learning. In: Jeeves M (ed) *Psychology survey No. 3.* Allen & Unwin, London, pp 92-114

Ryan AF, Sharp FR (1982) Localization of (^{3}H)2-deoxyglucose at the cellular level using freeze dried tissue and dry-looped emulsion. *Brain Res* 252: 177-180

Sasaki K, Shimono T, Oka H, Yamamoto T, Matsuda Y (1976) Effects of stimulation of the midbrain reticular formation upon thalamocortical neurones responsible for cortical recruiting responses. *Exp Brain Res* 26: 261-273

Scheibel, ME, Scheibel AB (1958) Structural substrates for integrative patterns in the brain stem reticular core. In: Jasper H, Proctor LD, Knighton RS, Noshay WS, Costello RT (eds) *Reticular formation of the brain.* Little Brown, Boston, pp 31-55

Segundo JP, Arana R, French JD (1955) Behavioral arousal by stimulation of the brain in the monkey. *J Neurosurg* 12: 601-613

Sharp FR, Ryan AF (1984) Regional (^{14}C)2-deoxyglucose uptake during forelimb movements evoked by rat motor cortex stimulation: pons, cerebellum, medulla, spinal cord, and muscle. *J Comp Neurol* 224: 286-306

Sharp FR, Ryan AF, Goodwin P, Woolf NK (1981) Increasing intensities of wide band noise increase (^{14}C)-2-deoxyglucose uptake in gerbil central auditory structures. *Brain Res* 230: 87-96

Shute CCD (1973) Cholinergic pathways in the brain. In: Laitinen LV, Livingston KE (eds) *Surgical approaches in psychiatry.* Univ Park Press, Baltimore

Sokoloff L (1984) Modeling metabolic processes in the brain in vivo. *Ann Neurol Suppl* 15: S1-S11

Sokoloff L, Reivich M, Kennedy C, DesRosiers MH, Patlak CS, Pettigrew KD, Sakurada O, Shinohara M (1977) The (^{14}C)deoxyglucose method for the measurement of local cerebral glucose utilization: theory, procedure, and normal values in the conscious and anesthetized albino rat. *J Neurochem* 28: 897-916

Steriade M (1980) State-dependent changes in the activity of rostral reticular and thalamocortical elements. In: Hobson JA, Scheibel AB (eds) *The brainstem core: sensorimotor integration and behavioral state control.* Neurosci Res Progr Bull 18. MIT Press, Cambridge, pp 83-91

Steriade M, Ropert N, Kitsikis A, Oakson G (1980) Ascending activating neuronal networks in midbrain reticular core and related rostral systems. In: Hobson JA, Brazier MAB (eds) *The reticular formation revisited.* Raven Press, New York, pp 125-167

Taniguchi I (1980) Changes in metabolic activity in the inferior colliculus following removal of auditory input. *Biomed Res* 1: 510-516

Theurich M, Müller CM, Scheich H (1984) 2-deoxyglucose accumulation parallels extracellularly recorded spike activity in the avian auditory neostriatum. *Brain Res* 322: 157-161

Thompson RF, Berger TW, Berry SD, Clark GA, Kettner RN, Lavond DG, Mauk MD, McCormick DA, Solomon PR, Weisz DJ (1982) Neuronal substrates of learning and memory: hippocampus and other structures. In: Woody CD (ed) *Conditioning: representation of involved neural functions.* Plenum Press, New York, pp 115-130

Turner BH, Zimmer J (1984) The architecture and some of the interconnections of the rat's amygdala and lateral periallocortex. *J Comp Neurol* 227: 540-557

Versteeg CAM, Bohus B, De Jong W (1982) Attenuation by arginine and desglycinamide-lysine-vasopressin of a centrally evoked pressor response. *J Aut Nerv Syst* 6: 253-262

Voronin LL (1983) Long-term potentiation in the hippocampus. *Neurosci* 10: 1051-1069

Weinberger NM, Diamond DM (1987) Physiological plasticity in auditory cortex: rapid induction by learning. *Prog in Neurobiol* 29: 1-55

Winsky L, Schindler CW, McMaster SE, Welsh JP, Harvey JA (1986) Enhanced uptake of ^{14}C-2-deoxy-D-glucose (2DG) in the dorsal cochlear nucleus during Pavlovian conditioning. *Soc Neurosci Abstr* 12: 181

Winson J, Abzug C (1978) Neuronal transmission through hippocampal pathways dependent on behavior. *J Neurophysiol* 41: 716-732

Woolf NK, Sharp FR, Davidson TM, Ryan AF (1983) Cochlear and middle ear effects on metabolism in the central auditory pathway during silence: a 2-deoxyglucose study. *Brain Res* 274: 119-127

Young ED, Voigt HF (1981) The internal organization of the dorsal cochlear nucleus. In: Syka J, Aitkin L (eds) *Neuronal mechanisms of hearing.* Plenum Press, New York, pp 127-133

Stimulus-Specific Habituation in Toads: 2DG Studies and Lesion Experiments

THOMAS FINKENSTÄDT

Abteilung Neuroethologie, Fachbereich Biologie/Chemie, Universität Kassel
D-3500 Kassel, FR Germany

Abstract. The innate releasing mechanisms of prey-catching in toads can be modified by experience. The present study sheds light on neural substrates involved in stimulus-specific long-term habituation (non-associative learning). Applying the ^{14}C-2-deoxyglucose (2DG) technique, the regional distribution of cerebral glucose utilization was compared between (i) toads after stimulus-specific long-term habituation of orienting toward a repeatedly presented prey dummy ("habituation group") and (ii) toads readily orienting toward the same prey stimulus ("naive group"). In the habituation group, the caudal ventral medial pallium (vMP) (so-called primordium hippocampi), a portion of the preoptic area (PO), and the dorsal hypothalamus (dHYP) showed statistically significant increases in 2DG-uptake; decreases were observed in the anterior thalamic (A) nucleus, the medioventral layers of the optic tectum (vOT), a portion of the tegmental reticular formation (RET), the striatum (STR), and the ventral cerebellum (vCB). The results suggest that stimulus-specific long-term habituation of prey-catching involves structures of the stimulus-response mediating circuit (e.g., vOT), extrinsic structures belonging to a modulatory circuit (e.g., vMP, RET), and relays connecting both circuits (e.g., PO/dHYP, A). According to Sokolov's hypothesis it is assumed that habituation leads to stimulus-specific after-effects in the modulatory system which accumulate with repetitive presentation of prey and suppress the response toward prey with reference to specific cues. Bilateral vMP-lesions strongly delay habituation without impairing innate prey recognition. Various brain structures (e.g., vMP, vOT, RET) showed opposite changes in 2DG-uptake during *habituation* and *arousal*. The underlying brain circuitries are discussed.

1 Introduction

To survive, organisms must control the interactions between their external and internal worlds and accordingly adapt their behavioral responses to environmental situations and to the internal states. Prey-catching behavior in toads is mediated by releasing mechanisms whose recognition properties are to a considerable extent innate. Adaptation of prey-catching to external and/or internal signals requires processes that modulate or modify the innate mechanisms. Suppose two toads in the same experimental paradigm, one of them frequently snapping and orienting toward a prey object, the other not responding at all. Various explanations are possible: Motivation is different, attention is different, or previous learning experiences with the prey object are different. Regarding the latter possibility, for example, the unresponsive animal may be habituated to certain cues of prey in response to repetitive presentation of the same object, called "non-associative" learning. It is also possible that the animal has aquired negative experience (e.g., bitter taste) in association with certain prey cues, called "associative" learning.

In the present study, we are especially interested in habituation defined as a decrement in response activity with repeated stimulation (Thompson & Spencer 1966; Groves & Thompson 1970). Habituation of orienting toward prey in toads can be obtained following long-term exposure of the animal to the same moving visual prey stimulus within a stimulus series (Eikmanns 1955; Ewert 1967a,b). If the stimulus is withheld following habituation, spontaneous recovery of orienting to the same stimulus occurs. In a sequence of stimulus series, the time course of habituation is strongly dependent on the inter-series intervals (Ewert 1967a). If the same prey object is presented in a sequence of stimulus series for several hours, orienting toward this object is abolished for more than 24 hours; the animal is then *long-term* habituated. Recovery is much faster if prey is withheld after the initial occurrence of response decrement in one stimulus series, called *short-term* habituation (Ewert 1967a; for a review see Ewert 1984). Habituation is specific (selective) to stimulus features (Birukow & Meng 1955) and to the locus of the stimulus in the visual field (Ewert & Ingle 1971) suggesting that motor pattern generating systems are not involved in processes that underlie habituation. Dishabituating stimuli (e.g., tactile or visual) are mainly effective in the phase of short-term habituation (Ewert 1967b). Habituation is influenced by the motivational state of the animal (Ewert 1984).

The neural mechanisms that underlie habituation have been investigated in great detail both in vertebrates and invertebrates (Bruner & Tauc 1966; Kandel 1976). The functional concepts emerging from studies in invertebrates, however, cannot necessarily be conferred to stimulus-specific long-term habituation phenomena in vertebrates (e.g., see Sokolov 1960; Ewert 1967a; Lara & Arbib 1985). Little is known about habituation relevant brain structures in lower vertebrates such as amphibians. The present investigation was conducted to determine the functional brain circuitry for long-term habituation of prey-catching in toads with consideration of the animal's arousal state (see also the contributions by Laming, Lara, and Gonzalez-Lima, this volume).

Since it is generally assumed that learning affects various structures throughout the brain, the ^{14}C-2-deoxyglucose (2DG) autoradiographic technique (Sokoloff et al. 1977) provides a suitable method to map experience-dependent regional metabolic changes of brain activity. Incorporation of 2DG, reflecting the activity of the sodium-potassium pump, is dominant in synaptic neuropil (Ryan & Sharp 1982; Sokoloff 1984). Enhanced 2DG-uptake is thus strongly correlated with both synaptic and postsynaptic neuronal activities (Theurich et al. 1984; Juliano & Whitsel 1987). Since this technique determines the sites and areas of brain activity, it can be regarded as a "functional stain" (see also Horwitz, this volume) that provides us with important "landmarks" for subsequent recording, stimulation, and lesion studies.

2 Methodological Comments

Experimental subjects, common toads *Bufo bufo spinosus*, were selected for comparable body weight and comparable prey motivation. The number of prey-catching orienting reactions in response to a black 4 mm x 28 mm worm-like prey dummy moving at 10°/s during a 1-min lasting presentation was taken as a measure of the toad's prey responsiveness (Ewert 1969). The habituation experiments were conducted with the procedure of Ewert & Kehl (1978) (Fig. 1a).

To map energy metabolism in the brain, ^{14}C-2-deoxyglucose (2DG) (15 µCi/ 100 g body weight) was injected into the animal's caudal lymphatic sacs. Autoradiographic x-ray film images of brain transverse sections were developed. Optical density was measured by means of a computer-assisted image analyzing system (Quantimet 920). For habituated (H) and "naive" nonhabituated (N) toads the concentration values $[^{14}C](H)$ and $[^{14}C](N)$ - taken via ^{14}C-calibrated methacrylate standards from optical density measurements - of single measurements of a brain structure were expressed in relation to the concentration value $[^{14}C]_{wh}$ of a "white matter" (wh) reference structure, yielding $X(N) = [^{14}C](N)/[^{14}C]_{wh}$ and $X(H) = [^{14}C](H)/[^{14}C]_{wh}$, respectively. The reference structure, an area below the nucleus ventralis nervi octavi, remained unaffected with regard to our experimental parameters, shown by means of the one-way analysis of variance *ANOVA*-test (for details see Gallistel et al. 1982; Finkenstädt et al. 1985, 1986). The X(N,H)-values were calculated for a number of *i* histological transverse sections on which the brain structure appeared along its sagittal extension. For a brain structure under investigation the bilateral *i* X(N)-values taken from a number of *j* "naive animals" (altogether $n_N = ij$ samples) were averaged, yielding X'(N), and in a comparable manner the X(H)-values from the "habituated toads" (n_H samples) were averaged to X'(H).

Comparison between the respective X'(N)- and X'(H)-values for a given brain structure was performed by means of the Kolmorov-Smirnov-Omnibus *(KSO)*-test. This non-parametric statistical test is suitable for comparison between two *independent* samples of equal or unequal sizes. The test of significance was two-tailed. To evaluate whether the samples obtained were independent, we applied the one-way *ANOVA*-test for unequal group sizes on randomly chosen structures. This test showed that the variances were about the same, both *between* and *within* the treatment groups. Reference of the measured values to $[^{14}C]_{wh}$ within each brain kept the experimentally unavoidable individual variabilities between brains below 0.10. Such variabilities may be due to film quality, differences in film treatment, differences in glucose blood level of the toads before the experiment, etc.

3 The Experimental Paradigm: "Habituation Group" and "Naive Group"

In the "habituation group," orienting behavior of toads in response to a prey dummy was habituated on two consecutive days with two sessions each. The prey dummy, a 4 x 28 mm^2 piece of black rectangular cardboard was moved in a worm-like fashion against a white background at a constant angular velocity of 10°/s and a constant distance to the cylindrical glass vessel in which the animal was placed (Ewert & Kehl 1978) (Fig. 1a). The dummy was presented until the toad failed to respond. Interestingly, the response started to fade in the lateral visual field, so that the "response field" progressively shrinked toward the frontal field. Following this initial response decrement (short-term habituation), the movement direction of the prey object was immediately reversed, and the response to the stimulus was then habituated again. After 90 min of the above procedure, a pause of 15 min was given, and the previous stimulation modus was repeated once, so that the net stimulation period for two successive days resulted in six hours. The pauses also gave us the opportunity to check for stimulus-specificity (Fig. 1b) and hence to verify that a lack in motivation was not the reason for response decrement. In the long-term habituated toads spontaneous recovery to the same prey object following a short pause of a few minutes could not be observed. "Dishabituation" due to stimulus-specificity was eliminated in the present study by using the same 4 x 28 mm^2 worm-like dummy moving in nasotemporal and temporonasal directions; "dishabituation" related to locus-specificity was avoided by stimulating a relatively wide area of frontal/lateral retinal positions of both eyes.

Toads of the "naive group" received no such pretreatment, except for a test on their prey responsiveness. During the six hours, they remained in the same experimental arena, but - in order to prevent habituation phenomena - they were not exposed to the moving prey stimulus.

Immediately following 2DG injection into the dorsal lymphatic sacs, toads of the *habituation group* and the *naive group*, respectively, were stimulated with the 4 x 28 mm^2 worm-like prey dummy in the same experimental environment for 60 min: The movement direction was altered every full rotation with 1 min lasting pauses every 3 min. During the entire 60 min period, animals of the "habituation group" did not respond. At the end of the 60 min session toads of the "naive group" still showed orienting responses.

At the beginning of the 2DG treatment, toads of both experimental groups differed in previous experience with the stimulus (stimulation-related after-effects). The 60 minutes lasting test thus allowed us to compare 2DG-uptake during the

Figure 1. a) Experimental procedure for the quantitative investigation of stimulus-specific habituation of prey-catching-orienting in toads. Above: Procedure. Below: Flow chart of an automatic device for the change of stimulus objects (prey dummies); a, arena base; c_{1-4}, electrical contacts; e_A, electric motor for presentation of the stimulus A; e_B, motor for presentation of a different stimulus B; h, stimulus holder; s, sliding contacts. (From Ewert & Kehl 1978). - b) Test of stimulus-specificity between a 4 mm x 28 mm rectangular black stripe and a black isosceles triangle of 28 mm length and 4 mm height. Each stimulus was moved continuously at 10°/s against a white background in horizontal direction around the experimental animal. Habituation of the toad's prey-catching orienting response first to the black stripe and, immediately thereafter, test of response activity (see vertical arrow) to the triangle. Abscissa: Stimulus presentation time t (min). Ordinate: Successive number of orienting responses/min (orienting activity, R). Each data point represents average values of experiments with 15 toads (after Ewert & Kehl 1978).

animal's response to the worm-like prey stimulus depending on the history of exposure to this stimulus in the two experimental groups. During this treatment, toads of both experimental groups differed in the way they expressed the *newness* and *sameness* effects of the stimulus, namely by prey-catching ("naive group") or not responding ("habituation group"), respectively. It must be mentioned that it was technically not possible to keep behavior-correlated variables constant in both

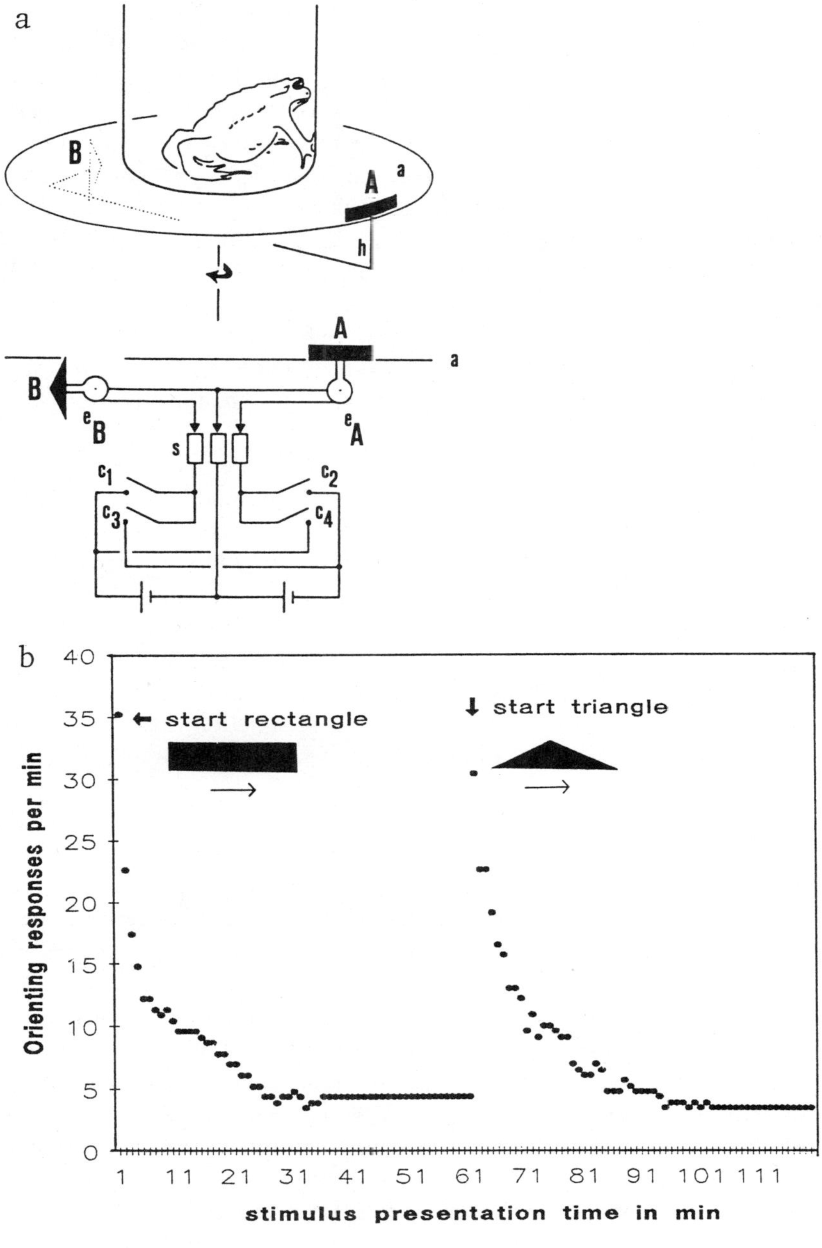

experimental groups. This "draw-back" - which could not be avoided by other kinds of control experiments without creating different serious problems - will be considered in the discussion of the data.

4 Comparison of 2DG-Uptake Across Brain Structures

Table 1 and Fig. 2 give a comparative survey of glucose utilization in toads of the habituation and naive group. In the *mesencephalon* 2DG-uptake in the "habituation group" *vs.* the "naive group" was significantly decreased in the medioventral tectal layers 2-7, vOT (Fig. 3c,d) (nomenclature see Székely & Lázár 1976). No significant metabolic differences between both experimental groups were seen in the afferent dorsal tectal layers 8-9 (dOT). In the tegmental reticular formation (RET), 2DG-uptake in the habituation group was significantly lower than in the naive group. RET refers to a tegmental portion of the medial reticular region with its caudal extent slightly posterior to the cerebellum (Nieuwenhuys & Opdam 1976). Its rostral boundary is approximately at the level of the nucleus isthmi (functionally determined by 2DG studies in toads aroused by novel visual stimuli; Finkenstädt et al. 1985). In the ventral tegmentum (vTEG, anteroventral tegmental nucleus of Potter 1972),

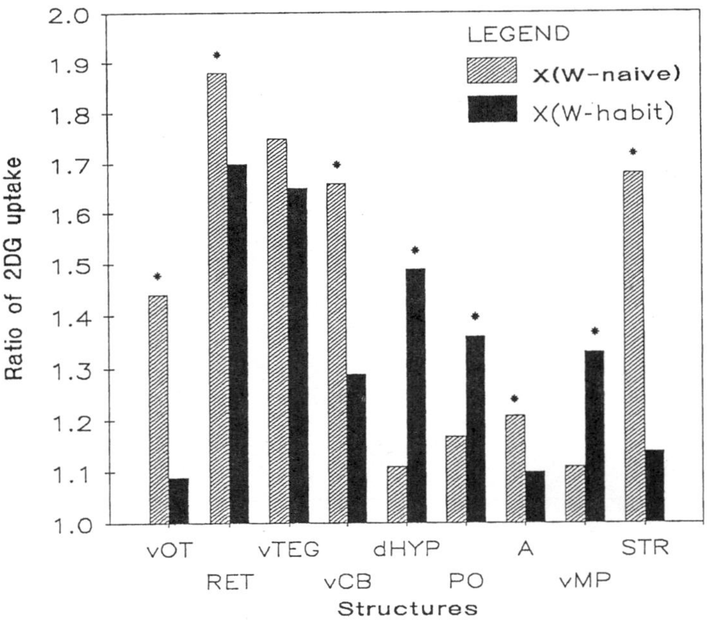

Figure 2. Uptake of 2DG in various brain areas following presentation of the 4 x 28 mm^2 worm-like prey dummy in naive (n=6) and long-term habituated (n=9) toads. X(W-naive) indicates the ratio of 2DG-uptake for animals of the "naive group", X'(N), and X(W-habit) for the "habituation group" X'(H) in comparison with a "white matter" reference structure *wh.* For a ratio of 1.0, glucose utilization in a structure is not different from the one measured in *wh.* Asterisks indicate statistical significance on the $P<0.05$ level. For details see text, for abbreviations see Table 1. (From Finkenstädt & Ewert 1988b).

Table 1. Comparison of 2DG-uptake measured by X'(H,N) ± SE in brain structures of toads of the "habituation group" (H) and the "naive group" (N) during presentation of a 4 x 28 mm^2 worm-like prey dummy. For explanations see text. Abbreviations of brain structures: A, anterior dorsal thalamic nucleus; AL, amygdala, lateral portion; dCB, dorsal half of the cerebellum; vCB, ventral half of the cerebellum; DP, dorsal pallium; dHYP, dorsal hypothalamus; pLP, posterior half of the lateral pallium; Lpd, lateral posterodorsal thalamus; Lpv, lateral posteroventral thalamus; aMP, anterior third of the medial pallium; dMP, dorsal portion of the posterior medial pallium; vMP, ventral two-third of the posterior medial pallium; MS, medial septum; dOT, dorsal layers of the optic tectum; vOT, ventral layers of the optic tectum; P, posterior thalamic nucleus; PO, preoptic area; RET, tegmental portion of the medial reticular formation; STR, striatum, dorsal and ventral portion; vTEG, ventral tegmentum. For abbreviations of brain structures see also Section 10. (From Finkenstädt & Ewert 1988b).

Structure	X'(N)	n_N	X'(H)	n_H	*KSO*-test
AL	1.01 ± 0.11	18	1.04 ± 0.13	27	---
DP	1.12 ± 0.15	21	1.08 ± 0.22	39	---
aMP	1.35 ± 0.13	18	1.31 ± 0.15	27	---
dMP	1.47 ± 0.13	20	1.39 ± 0.17	45	---
vMP	1.11 ± 0.13	35	1.33 ± 0.17	52	$P<0.002$
MS	1.65 ± 0.16	21	1.58 ± 0.15	45	---
pLP	1.30 ± 0.13	31	1.24 ± 0.18	54	---
STR	1.68 ± 0.11	35	1.14 ± 0.09	43	$P<0.001$
A	1.21 ± 0.12	19	1.10 ± 0.11	30	($P<0.05$)
dHYP	1.11 ± 0.10	19	1.49 ± 0.08	33	$P<0.001$
PO	1.17 ± 0.12	26	1.36 ± 0.14	28	$P<0.01$
Lpd	1.29 ± 0.18	19	1.25 ± 0.15	28	---
Lpv	1.51 ± 0.31	18	1.43 ± 0.30	27	---
P	1.38 ± 0.20	21	1.41 ± 0.18	31	---
dOT	1.48 ± 0.25	21	1.46 ± 0.13	38	---
vOT	1.44 ± 0.13	20	1.09 ± 0.10	32	$P<0.001$
vTEG	1.75 ± 0.17	18	1.65 ± 0.18	32	---
RET	1.88 ± 0.13	29	1.70 ± 0.16	41	$P<0.05$
dCB	1.29 ± 0.20	18	1.34 ± 0.17	27	---
vCB	1.66 ± 0.17	20	1.29 ± 0.11	28	$P<0.01$

2DG-uptake between habituation and naive groups, however, was not statistically significant. In the ventral *cerebellum* (vCB) (stratum granulare cerebelli), 2DG-uptake was significantly lower in the habituation group than in the naive group, whereas no significant differences were seen in the dorsal portion (dCB).

In the *diencephalon* we found significantly enhanced glucose utilization in the dorsal hypothalamus (dHYP) of the habituation group. Similar increases were seen in a portion of the preoptic (PO) area (Fig. 4c) which can be assigned broadly to the suprachiasmatic nucleus as delineated by Neary & Northcutt (1983), but with its anterior boundary extending more rostrally. Measurements of 2DG-uptake in the anterior (A) thalamic nucleus yielded a decrease of activity in the habituation group, however, the difference just reached the *P* (5%) level. In other visual sensitive diencephalic nuclei, such as the lateral posterodorsal (Lpd), the lateral posteroventral (Lpv), and the posterior (P) thalamic nuclei, statistical significant differences in 2DG-uptake could not be detected. The observed small differences in 2DG-uptake among both experimental groups were below 4% and thus within the resolution limits of the 2DG evaluation technique.

In the *telencephalon* habituation strongly enhanced 2DG-uptake in the posterior half of the ventral two-thirds of the medial pallium (vMP) compared with the naive group (Fig. 5b,c). In the striatum (STR) the opposite was the case. No significant differences were found in the posterior dorsal portion of the medial pallium (dMP),

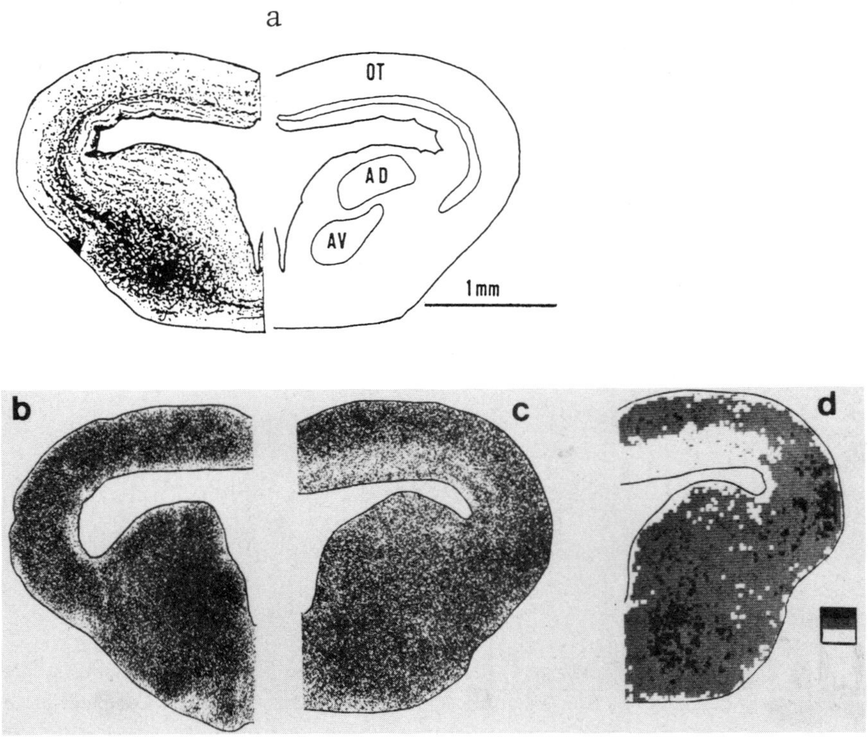

Figure 3. 2DG-uptake in the optic tectum. (a) Transverse half-section through the toad's mesencephalon at a medial level; left: high contrast photograph of a Klüver-Barrera stain; right: line drawing. (b),(c) Electronically unprocessed 2DG autoradiographic images taken from comparable brain sections in toads of the "naive" (b) and the "habituation" (c) group (representative examples). (d) This autoradiographic image is computer-processed by means of an image analyzing system. The gray scale reflects the relative amount of ^{14}C-2DG-uptake in the tissue with increasing values from light to dark. Note the lower activity in the medioventral layers of the optic tectum in c and d. OT, optic tectum; AD, anterodorsal tegmental nucleus; AV, anteroventral tegmental nucleus. (From Finkenstädt & Ewert 1988b).

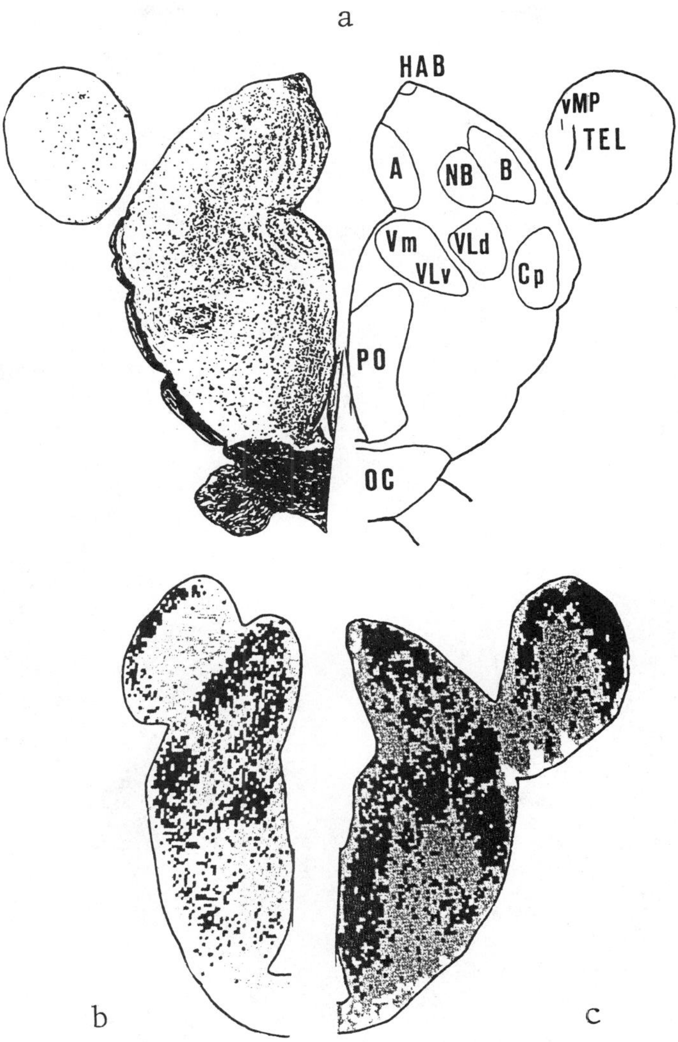

Figure 4. 2DG-uptake in the diencephalon. (a) Transverse half-section through the anterior diencephalon; left: high contrast photograph of a Klüver-Barrera stain; right: line drawing. (b),(c) Computer-processed 2DG autoradiographic images taken from comparabel brain sections in toads of the "naive" (b) and the "habituation" (c) group (representative examples). Note in c the increased 2DG-uptake in the preoptic area (PO) and the decrease in the anterior thalamic (A) nucleus. A, anterior thalamic nucleus; B, neuropil of Bellonci; Cp, corpus geniculatum thalamicum; HAB, habenular nucleus; NB, nucleus of Bellonci; OC, optic chiasm; TEL, telencephalon; Vm, ventromedial thalamic nucleus; VLd, dorsal part of the ventrolateral thalamic nucleus; VLv, ventral part of the ventrolateral thalamic nucleus; vMP, ventral posterior medial pallium (Finkenstädt & Ewert 1988b).

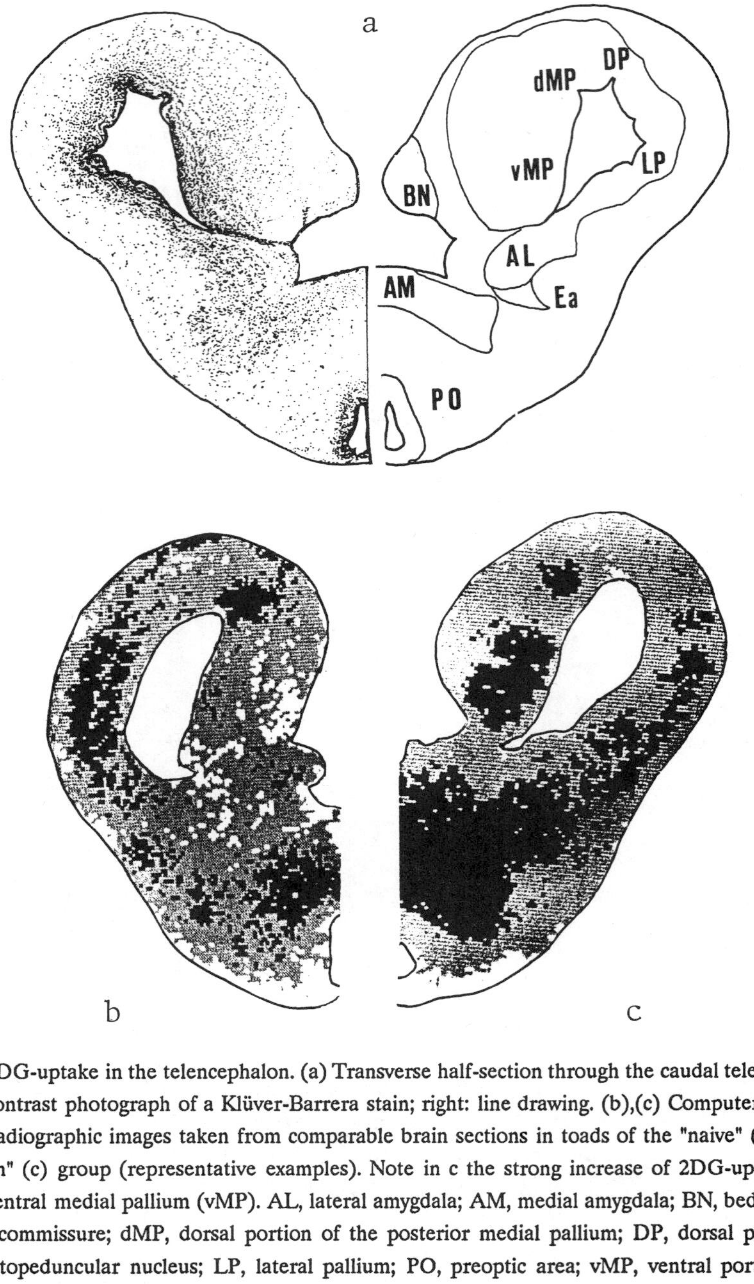

Figure 5. 2DG-uptake in the telencephalon. (a) Transverse half-section through the caudal telencephalon; left: high contrast photograph of a Klüver-Barrera stain; right: line drawing. (b),(c) Computer processed 2DG autoradiographic images taken from comparable brain sections in toads of the "naive" (b) and the "habituation" (c) group (representative examples). Note in c the strong increase of 2DG-uptake in the posterior ventral medial pallium (vMP). AL, lateral amygdala; AM, medial amygdala; BN, bed nucleus of the pallial commissure; dMP, dorsal portion of the posterior medial pallium; DP, dorsal pallium; Ea, anterior entopeduncular nucleus; LP, lateral pallium; PO, preoptic area; vMP, ventral portion of the medial pallium. (From Finkenstädt & Ewert 1988b).

the anterior medial pallium (aMP), the dorsal pallium (DP), the posterior half of the lateral pallium (pLP), the medial septum (MS), and the lateral amygdala (AL).

5 Brain Lesions

An important result of the 2DG studies was the strong increase of energy metabolism in vMP of the habituation group *vs.* naive group, which is opposite to the effects obtained in vOT. The relevance of vMP for habituation in prey-catching

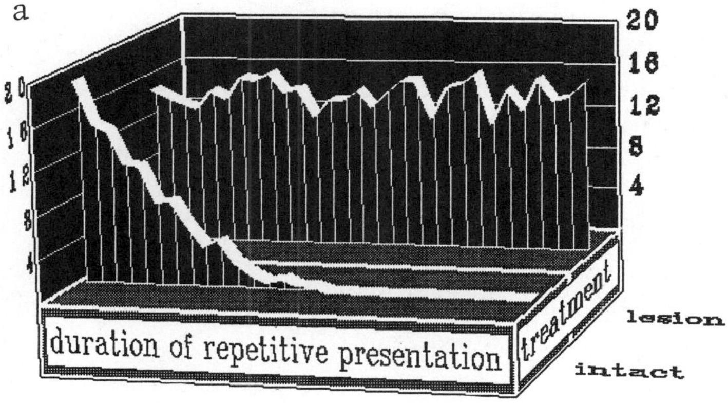

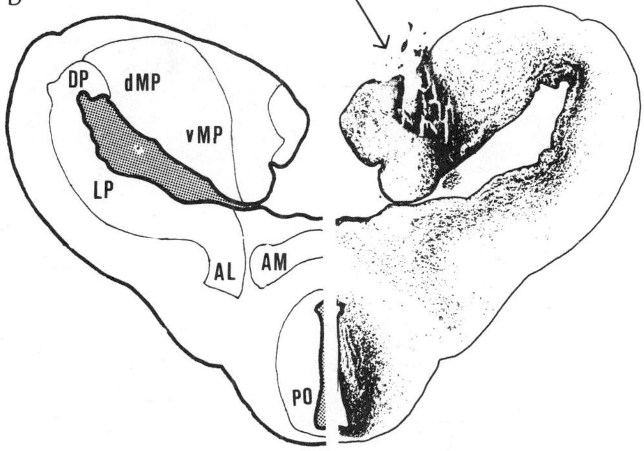

Figure 6. (a) Effects of medial pallial lesions on habituation. Prey-catching orienting activity (ordinate: orienting movements per min) in response to the continuously presented 4 x 28 mm^2 worm-like prey dummy measured in the procedure shown in Fig. 1. Average responses of naive intact toads (n = 9, intact) and after bilateral lesions in the ventral medial pallium (vMP) (n = 3, lesion). On the abscissa every 2.5 minutes one data point is depicted. (b) Representative example of a lesion in the ventral portion of the posterior medial pallium of toads. For morphology see Fig. 5a. (After Finkenstädt & Ewert 1988b).

behavior could be further tested by means of lesion experiments (Fig. 6b). All lesions were applied by passing an anodal direct current of 50-60 µA for 8-10 s through an insulated stainless steel electrode with an exposed tip of 10-20 µm, or aspiration suction (for technical details see Finkenstädt & Ewert 1983). In sham-operated controls the skull was opened and, sparing brain tissue, the brain meninges from the posterior telencephalon were removed.

In the intact naive toads (pre-lesion), habituation of orienting toward prey was completed after 32 $\pm$ 4.7 min (Fig. 6a, intact). Twenty-four hours after bilateral lesions in the posterior half of vMP, these animals showed accurate prey-catching, and the number of orienting responses in the initial 1-min interval was not much less than in naive animals pre-lesion. However, upon repeated stimulation the vMP-lesioned toads showed strongly delayed habituation (Fig. 6a, lesion); i.e., there was no progressive decrease of the orienting activity as observed in intact animals. But the vMP-lesioned toad's innate ability to distinguish between different configural (worm-like, antiworm-like, and square) stimuli was maintained. In sham-operated controls or in toads with bilateral lesions in DP or dMP, the time course of habituation was normal.

Following unilateral LP-lesions, prey-catching failed to occur in the hemifield of the eye contralateral to the lesioned hemisphere. After bilateral LP-lesions, visually guided prey-catching was totally abolished, although visually guided predator avoidance remained intact.

Unilateral lesions in the thalamic A-nucleus affected configural pattern discrimination in the frontal visual field of both eyes (about 70-90° diameter). Independently of size and configuration every moving visual stimulus was immediately responded with prey-catching. After bilateral A-lesions no habituation in response to repetitive presentation of prey occurred, similar to vMP-lesioned toads. (For comparable results in salamanders see Finkenstädt & Ewert 1983). The difference between vMP- and A-lesioned toads differed in configural pattern discrimination which was abolished in the latter.

6 Electrical Stimulation and 2DG-Uptake

Electrical stimulation of freely moving toads in the posterior MP elicited various motor responses such as jaw opening, gulping, snout wiping or puffing up to assume a threat posture, but no prey-catching (mode of stimulation: monopolar, cathodic square wave impulses of 100 cps and 5 ms pulse duration). If the vMP of toads was electrically stimulated during orienting toward prey, the turning movement was interrupted. A few seconds, sometimes even several minutes, elapsed before the toad's interest in prey returned. Stimulation of STR led to varying results: Some animals showed the aforementioned threat posture, stretching their forelimbs, some were responding with locomotion, others showed no response at all. Stimulation of the thalamic A-nucleus elicited predator avoidance responses and locomotion, similar to stimulation of the thalamic pretectal Lpd/P nuclei. Tectal stimulation was mainly responded with prey-catching orienting and/or snapping (see Ewert, this volume). Stimulation of the tegmental reticular formation (RET) led to arousal behavior, such as watchful head elevation, freezing, or unrest. Occasionally, trunk elevation and a predator-avoidance stifflegged posture could be observed.

Electrical stimulation of RET in immobilized toads (paralyzed with succinylcholin) strongly enhanced 2DG-uptake (Fig. 7) in dorsal and ventral layers of both dorsal tectal lobes (v,dOT), in vHYP (Fig. 8b), in the A-nucleus (Fig. 9a), the STR, and the cerebellum. Reduced 2DG-uptake was found in vMP and in the PO/dHYP region.

7 Narcosis and 2DG-Uptake

In mammals it has been observed that certain anesthetics block the midbrain ascending activating reticular system (Brodal 1969; reported in Gonzalez-Lima & Scheich 1985). There is evidence for comparable effects in amphibians. If toads were narcosized by an injection of MS222 (50 mg/100 g body weight) into the dorsal lymphatic sac (sufficient to keep animals anesthetized for about 60 min) and 2DG was injected with the onset of narcosis, the following brain structures displayed a specific activity pattern (Fig. 7): The vMP in its entire extent, the dorsal portion of MS (Fig. 10a), the aLP, the PO/dHYP region (Fig. 10b), and the dCB exhibited strong 2DG-uptake; within the optic tectum, only the dorsal layers showed 2DG-

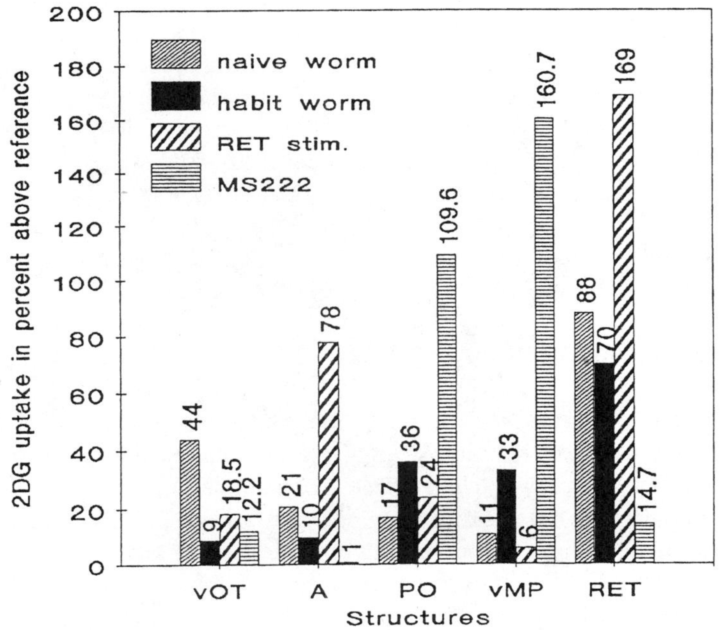

Figure 7. Influence of electrical stimulation of the tegmental reticular formation (RET stim; n = 3), MS222 anesthesia (MS222; n = 4), worm dummy stimulation (naive worm; n = 9), and habituation to a worm dummy (habit worm; n = 6) on 2DG-uptake (in percent above reference "white matter" structure) in the ventral portion of the optic tectum (vOT), the anterior dorsal thalamic nucleus (A), the preoptic region (PO), the ventral medial pallium (vMP), and in the tegmental reticular formation (RET).

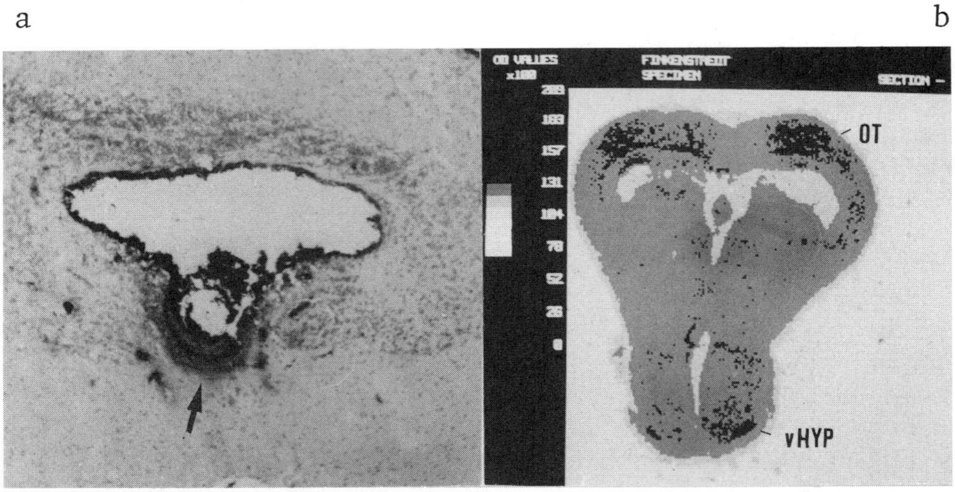

Figure 8. (a) Neutral red stained section at the very caudal end of the mesencephalon. The electrolytic lesion (arrow) indicates the position of the stimulation electrode in the RET. (b) Computer-processed representative autoradiographic image of the same toad receiving electrical stimulation in the RET. Note the strongly labeled tectal layers (OT) and the ventral hypothalamic activity (vHYP).

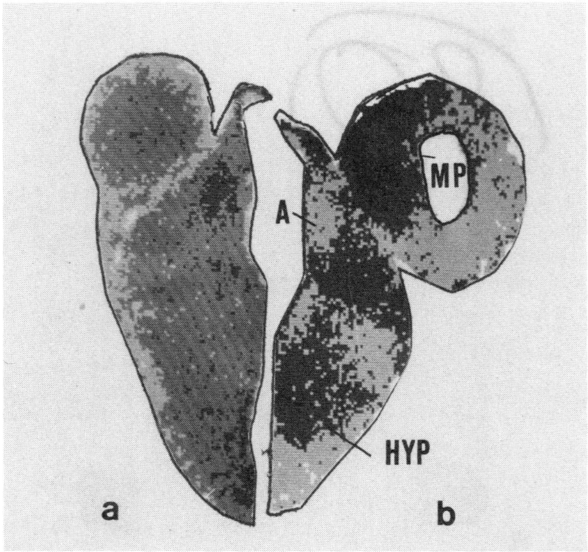

Figure 9. Computer-processed half sections of representative autoradiographs at rostral diencephalic levels. (a) Depicts strongly increased 2DG-uptake in the anterior thalamic nucleus (A) following electrical RET stimulation. (b) Depicts 2DG-uptake in a MS222 anesthetized toad; note high activity in the PO/dHYP-region and the ventral medial pallium (vMP) of the very caudal telencephalon, as well as very low label in the anterior dorsal thalamic nucleus (A). For morphology see Fig. 4a.

uptake; extremely weak metabolic activity was found in RET (Fig. 10c,d) and A-nucleus (Fig. 9b). The increase of metabolic activity in the vMP and PO/dHYP of MS222-narcosized toads could be reduced during electrical RET stimulation (Finkenstädt & Ewert 1985; Finkenstädt 1987). We conclude that RET stimulation in MS222-treated toads to some extent compensates the metabolic effects of narcosis.

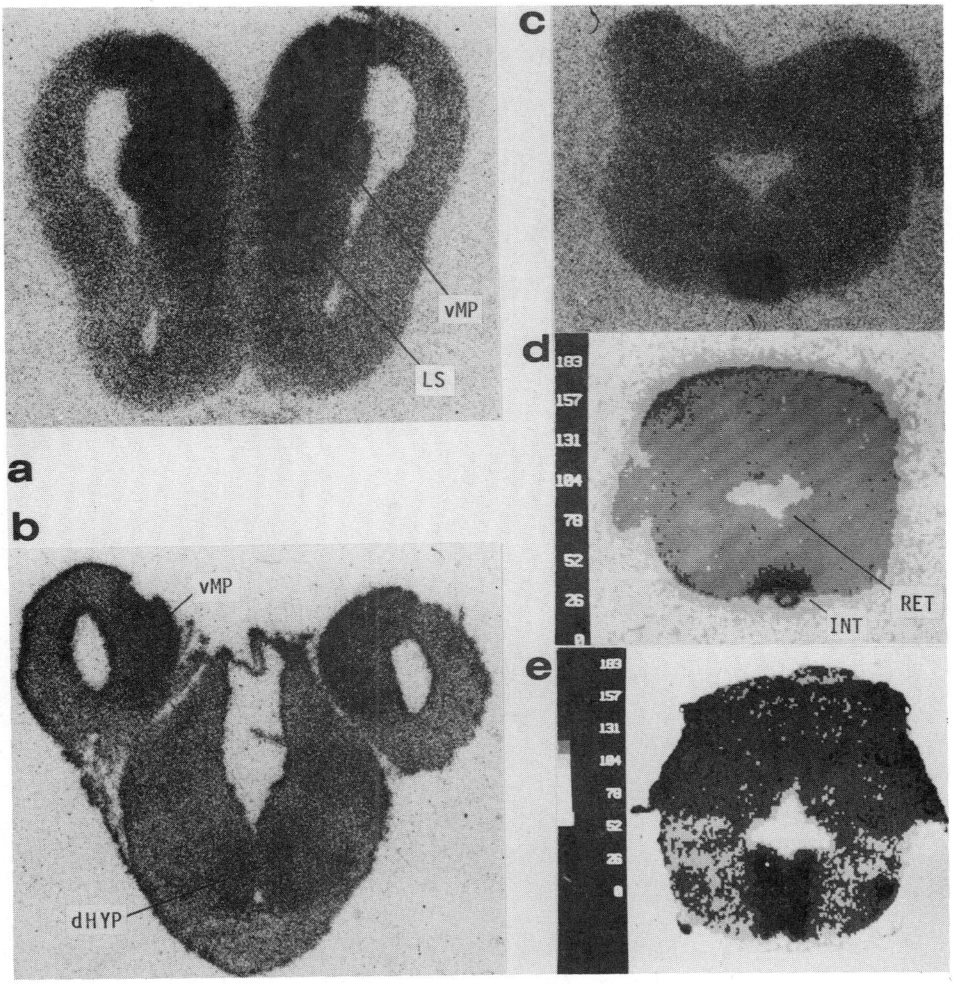

Figure 10. Photo taken from the x-ray film, depicting representative brain sections of a MS222 anesthetized 2DG treated toad. (a) Mid-telencephalic level. Note the strong accumulation of 2DG in the medial pallium (vMP) and medial septum (MS). (b) Rostral diencephalic level. Note the strong activity in the PO/dHYP-region and in the vMP. For computer-processed data according to (a) and (b) see Fig. 11. (c) Cerebellar level. Note the strong metabolic activity in the nucleus interpeduncularis (INT), but extremely low 2DG-uptake in the RET. (d) Similar to c, but computer-processed. For comparison, photo (e) depicts enhanced 2DG accumulation in the RET of a fully awake toad (at about the same level as c and d) frequently orienting toward the prey dummy. (For morphology and abbreviations see Figs. 4 and 5).

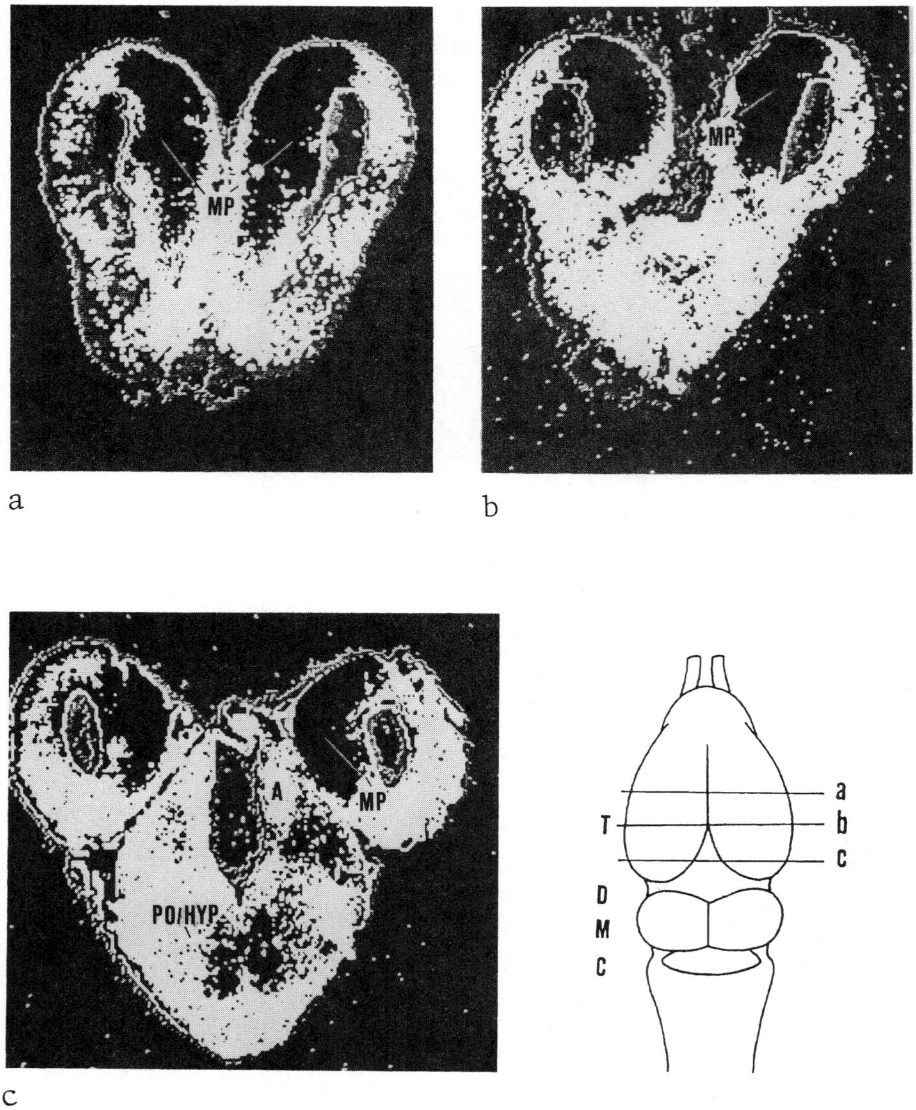

Figure 11. 2DG-uptake in MS 222 narcosized toads. Computer-processed, binary coded autoradiographic images from three transverse sections through the toad's telencephalon running from rostral (a) to caudal (b). Representative examples as shown in Fig. 10a and b. Note the strong 2DG-uptake in the ventral medial pallium (vMP) (a-c) and in the preoptic (PO)/hypothalamic area (c), while the anterior thalamic nucleus (A) is not activated. C, cerebellum; D, diencephalon; M, mesencephalon; T, telencephalon. For anatomy see Figs. 4 and 5.

8 Discussion

Before discussing neural structures and their functions for habituation of visually guided prey-catching in toads, we must take notice of the following facts: (1) Since the subject matter refers to *long-term* habituation, the underlying lasting storage processes cannot be explained by a self-generated suppression of synaptic sensitivity in a serially organized information processing system (as assumed by Horn 1967). (2) Habituation is stimulus-specific, suggesting that neither (a) neuronal adaptation of retinal ganglion cells (claimed by Grüsser & Grüsser-Cornehls 1976) nor (b) prey-catching motor pattern generating systems are involved in the mechanism of long-term habituation. (3) Habituation is locus-specific, presuming a representation of spatial information. (For a review see Ewert 1984).

8.1 Concepts of Stimulus-Response Mediating and Modulating Systems

According to Krasne et al. (1979) we distinguish between a stimulus-response (*S-R*) mediating circuit and an extrinsic circuit modulating *S-R* (Ewert 1987a). The results of the present study indicate that stimulus-specific long-term habituation of orienting toward a visual prey object affects structures of the *S-R* circuitry (e.g., vOT) and involves structures extrinsic to it (e.g., vMP). The latter belong to polysynaptic medialpallio-tectal pathways which could modulate *S-R* mediation via relay structures (e.g., PO/dHYP, A-nucleus) (Ewert & Finkenstädt 1987; Finkenstädt & Ewert 1988b; for anatomic connections see Kicliter & Ebbesson 1976; Northcutt & Kicliter 1980; Wilczynski & Northcutt 1977, 1983a,b; Neary & Northcutt 1983). With reference to this concept, prey-catching is released via a retino-pretectal/tectal pattern recognition system (mediating system), and this is controlled via an OT-A-vMP-PO/dHYP-OT loop (modulating system). Based on results of 2DG-, lesion-, and electrical brain stimulation studies we suggest that the modulatory circuit influences *S-R* mediation and thus the behavioral response, but itself does not elicit prey-catching.

What are the putative tasks of both circuits? The *S-R* mediating system must decide whether a present visual stimulus fits the prey category; the modulatory system must evaluate whether the stimulus has been previously encountered (evaluation of "sameness") or is novel (evaluation of "newness"). All this requires (i) recognition of prey, i.e., comparison of a present stimulus with an innate prey schema that has emerged during evolution, (ii) memorizing the neural image, within the prey-schema, related to stimulus-specific cues, during repetitive occurrence of prey, and (iii) comparison between information related to the cues of the present stimulus and information related to the neural representation of its image. While steps (ii) and (iii) can be assigned to short-term processes, long-term ones may in addition include (iv) stimulus-specific storage of the established representation of the image (Sokolov 1975; see also Lara & Arbib 1985). Comparison (iii) might - without analysis of the concrete properties of the signal - be acting on a binary principle of match-mismatch evaluation (Vinogradova 1970). We suggest that the net outcome of the respective processes following long-term habituation eventually inhibits orienting toward prey.

Innate visual prey recognition is mediated by pretectal/tectal structures whose interactions determine prey-selective T5.2 neurons (Ewert 1974, 1987b) which are known to project to bulbar/spinal motor systems (Satou & Ewert 1985). If the pretectal Lpd/P nuclei of the *S-R* mediating system are lesioned, prey-catching can be readily elicited by visual stimuli, however, the prey-selective property of T5.2 neurons and - along with this - the prey recognition of toads are impaired. (For a review see Ewert 1987b; Ewert, this volume). If the telencephalic medial pallium vMP of the putative *S-R* controlling system is lesioned, prey recognition remains intact, but stimulus-specific experience cannot be acquired, so that habituation fails to occur.

8.2 Sokolov's Hypothesis of Habituation

Sokolov (1960, 1975) was among the first suggesting that habituation in mammals is the result of an extrinsic control of a *S-R* mediating circuit by a modulating circuit. He assumed that an afferent signal is conveyed to the sensory cortex and, through axon collaterals, to the reticular formation. Access of the signal to the reticular formation is controlled by fibers originating in the cortex. The cortex stores neuronal representations of stimuli which have been presented in the past. If this representation matches the incoming signal, then impulses descend from the cortex to prevent the signals in the afferent collaterals from activating the reticular formation. If the stimulus is novel, it will have no neuronal representation, and the descending cortical fibers gate (or even facilitate) signals from the afferent collaterals to activate the reticular formation. Thus, response decrement to a repeatedly applied stimulus would be the result of an active gating process controlled by the output of a matching system.

Horn (1967) has criticized Sokolov's "active gating" concept as unnecessary to explain habituation. He proposed that the gradual waning of the response of a group of neurons to a repeated stimulus is caused by a self-generated depression of synaptic sensitivity at one or more places within the neuronal group. However, Horn explicitly emphasized that his arguments against Sokolov's hypothesis only apply for short-term effects, and not for long-term habituation processes. Our experimental results, in concert with those reported by Vinogradova (1970, 1975), support Sokolov's view (see also Foreman & Stevens 1987; Ewert & Finkenstädt 1987; Finkenstädt & Ewert 1988b).

8.3 The Lara & Arbib-Model of Prey Habituation in Toads

Based on computer simulations, Lara & Arbib (1985) have proposed a neural mechanism for stimulus-specific habituation in toads. Adopting Sokolov's concept (1960, 1975) they postulate that stimulus-specific habituation requires at some stage the "creation" of a model of the stimulus and its "comparison" with the ongoing stimulus (see also Lara, this volume). They argue that stimulus-specificity of prey-habituation involves the forebrain in connection with tectal columns (functional units of interconnected neurons) whose output cells (T5.2 neurons) act as complex feature extractors for the spatiotemporal properties of the stimulus (see Ewert, this volume). The model of the stimulus is created in an active way as a change in the average frequency and duration of response of a given neuron (or a set of neurons), and comparison is performed in an active way through the activity of another neuron (or a set of neurons) whose activity is correlated with match-mismatch between model and stimulus. The authors interpret retrieval of information in this system as an increase or decrease of the rate of firing of a given neuron, depending on whether the consequence of the prey stimulus was "positive" (i.e., fitting the stored model) or "negative" (i.e., being novel). Lara & Arbib (1985) refer here to short-term processes and they predict that the related match-mismatch operations involve thalamic pretectal Lpd/P and telencephalic structures. Stimulus-specific habituation is interpreted as a build-up of inhibition from pretectum to tectum and controlled through telencephalic modulation.

8.4 Brain Structures Involved in Prey Habituation

The model of Lara & Arbib (1985) predicts that forebrain lesions disrupt stimulus-specific habituation. This proved to be true so far MP is concerned: Bilateral vMP lesions in habituated toads strongly delayed habituation, while the innate configural pretectal/tectal prey recognition and localization systems remained functioning, which does not speak against an extrinsically controlled build-up of inhibition to the tectum. Patton & Grobstein (1986), too, did not find an impairment of prey-catching behavior in frogs following extended lesions in the medial pallium, which is consistent with our results. Lesions to dMP or DP had no comparable effect on the time course of habituation. The fact that the prey-catching activity progressively decreases with the mass of bilaterally ablated telencephalic tissue (Ewert 1967a) can be explained with the involvement of lateral pallial (LP) and striatal (STR) structures whose localized interruption abolishes prey-catching behavior.

In long-term habituated toads, we found that 2DG-uptake in vMP, PO, and dHYP is elevated, reduced in vOT, RET, A, STR, and vCB, and not remarkably changed in Lpd/P. We suggest a descending medial pallial inhibitory modulation of tectal activity via a preoptic/dorsohypothalamic relay. This assumption is backed by results from vMP lesions and by the observation that electrical stimulation of vMP interrupts the ongoing orienting toward prey. Moreover, it can be shown in long-term habituated toads, subsequently lesioned in vMP, that presentation of the same prey dummy leads to no comparable increases of 2DG-uptake in PO/dHYP, while RET exhibits an increase in glucose utilization indicating that this structure is released from suppression (Finkenstädt & Ewert 1985). We hence regard the tectum as one of the structures, where the *S-R* mediating and modulatory systems converge. The decrease of 2DG-uptake in vCB and STR of the habituation group is probably correlated with the different motor activities between both experimental groups (Section 8.10; see also Finkenstädt et al. 1985, 1986; Finkenstädt & Ewert 1985).
Since MP receives retinal and tectal information via the A-nucleus, the question arises through which pathways tectal neurons may be modulated according to the activity of certain vMP cells. There are two suitable, anatomically distinct pathways: Pathway (a) projects from MP via the A-nucleus (and P-nucleus) to OT (cf. "loop 2a", Ewert, this volume); pathway (b) relays MP efferents via PO (and HYP) to OT (cf. "loop 2b", Ewert, this volume; for anatomical details see Northcutt & Kicliter 1980; Wilczynski & Northcutt 1977, 1983a,b; Neary & Northcutt 1983). RET activity might be controlled by fibers originating in the vMP, monosynaptically or via PO/dHYP.

Table 2. Schematic survey of metabolic activity patterns across various brain structures in prey-habituated toads, in toads aroused either by electrical stimulation of the tegmental reticular formation (RET) or by visual stimulation with prey, and in MS222-narcosized toads; (+) refers to an increase, (-) to a decrease, and (o) to no significant change of 2DG-uptake, compared with controls. For abbreviations see Table 1. (From Finkenstädt et al. 1985, 1986; Finkenstädt & Ewert 1985, 1988b).

	2DG-Uptake			
Brain Structure	Habituation to prey	Arousal RET stim.	Arousal vis. stim.	Narcosis MS222
vMP	+	-	-	+
PO/dHYP	+	-	-	+
A	-	(o)	(o)	-
vOT	-	+	+	-
RET	-	+	+	-

Monosynaptic pathways between MP and RET, or between PO/dHYP and RET, however, have not yet been identified neuroanatomically.

Following Nauta (1963), we must also consider endocrine preoptic/hypothalamic effects. In this context a quantitative study by Horn et al. (1979) showed that administration of the pituitary hormone ACTH in toads significantly delays the course of recovery after stimulus-specific habituation of prey-catching. As a result, long-term habituation in a sequence of stimulus series is increased in ACTH-treated animals compared with controls. The aforementioned term "preoptic/hypothalamic relay" must be evaluated correspondingly.

From the survey given in Table 2 we are tempted to suggest reciprocal activity patterns for vMP, A, and PO/dHYP, on the one hand, and for vOT and RET on the other hand, which are opposite in habituated and aroused toads. The parallel activity patterns of energy metabolism related to habituation and narcosis suggest that both involve active processes. The latter effects of narcosis have not always been sufficiently considered in the literature reporting data on stimulation and recording studies in anesthetized animals.

8.5 Hypothesis for Processes Underlying Prey Habituation

Having Sokolov's hypothesis in mind, we infer that during stimulus-specific long-term habituation of prey-catching in toads the following processes take place: The visual pattern recognition system sends its processed information regarding stimulus cues via the A-nucleus to vMP. After repetitive presentation of a prey stimulus, appropriate information (a "model") is stored in connection with and through interaction of vMP interneurons, and this is compared with the information related to the present stimulus. If the prey object has been previously encountered, i.e., there is matching between the perceived visual signal and the stored representation of its image, the vMP output-neurons are activated and, in connection with PO/dHYP structures, inhibit tectal neurons. As long as these vMP neurons are active, they may suppress A-cells (by recurrent inhibition), which would explain the reduced 2DG-uptake in the A-nucleus of the habituation group. In parallel, RET may be inhibited as well, hence reducing its sensitizing effect on the visual pathways. If the encountered signal is novel, i.e. mismatch is detected, then vMP neurons are inhibited and the PO/dHYP descending pathway will not affect tectal neurons. Unretarded RET activity sensitizes the visual pathway; the reticular information feeds back to MP via an ascending tecto-anteriorthalamic pathway.

This speculation is supported by results from recording and lesion studies: Spontaneously active visual sensitive MP1 neurons strongly increase, MP3 neurons decrease (Fig. 12), and MP2 neurons do not alter their discharge rates in response to 30 min lasting repetitive stimulation with a visual moving object traversing their excitatory receptive fields which encompass the entire field of vision (Finkenstädt 1987; Böker, Finkenstädt, Ewert, unpubl.). Evidence that MP neurons can be inhibitory modulated by visual input has been provided by Trepakov (1974) recording inhibitory postsynaptic potentials from MP in response to electrical stimulation of the frog's optic nerve. In a recent electrophysiological study (Merkel-Harff, Finkenstädt

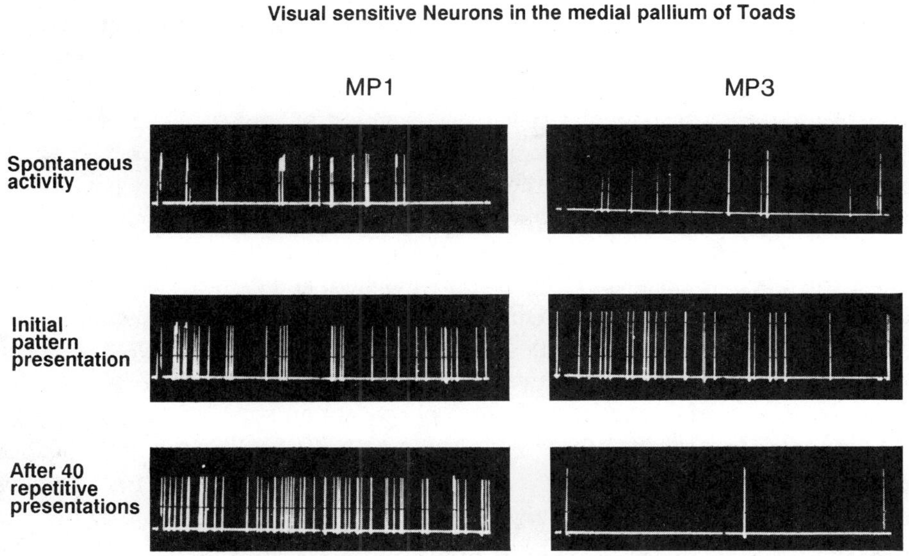

Figure 12. Representative extracellular single cell recording from visual sensitive ventral medial pallial neurons of the common toad, types MP1 und MP3. First row depicts spontaneous activity which is about the same in both neurons. After initial presentation of the dummy, both neurons enhance discharges at about equal amounts. Following 40 repetitive presentations of the same visual stimulus without pauses, type MP1 strongly increases, and type MP3 strongly decreases its activity even below spontaneous level.

and Ewert, in prep.) it was shown that olfactory stimulation in naive toads led to a dramatic decrease of spontaneous neuronal activity within the vMP.

It is worth remembering that among class T4 visual sensitive wide field neurons in deep tectal layers, the so-called "newness" and "sameness" cells exist (Lettvin et al. 1961; see also reviews by Grüsser & Grüsser-Cornehls 1976; Ewert 1984): "Newness" neurons show pronounced adaptation in response to repetitive traverse of their receptive fields with a moving stimulus; adaptation is locus-specific within the excitatory receptive field. "Sameness" neurons are activated by almost any moving object, independently of the frequency of its presentation. No results are presently available to decide whether T4- and T5-type neurons are modulated polysynaptically by medial pallial activity.

Accounting on the aforementioned results alone, the possibility of habituation-related medial pallial influences on pretectal Lpd and P nuclei via the thalamic A-nucleus (see "loop 2a", Ewert, this volume), in place of (or additional to) PO/dHYP cannot be excluded, although this seems to be unlikely as far as *long-term* habituation is concerned. It is possible that pretectal nuclei are involved in *short-term* habituation (Lara & Arbib 1985) which may explain the loss of habituation properties in prey-selective T5.2 neurons following pretectal lesions (Ewert, this volume). In this context, it is important to note that toads with lesions in the pretectal Lpd/P region as part of the pretectal/tectal pattern recognition system (Ewert 1967a, 1974), and/or in the anterior thalamic A-nucleus (Finkenstädt 1987), may show stong deficits of habituation of visually released prey-catching. Contrary to vMP lesions, however, prey-recognition too is strongly impaired. The response type of pretectally lesioned toads in which a strong delay of long-term habituation was observed (see response type I in Ewert 1967a), can be explained by accidental interruption of ascending tecto-anteriorthalamic pathways, which technically is difficult to avoid.

8.6 Habituation of Other Behaviors in Toads

Habituation of avoidance behavior in response to a repetitively presented large looming object is less lasting than habituation to prey (Ewert & Rehn 1969; Finkenstädt, unpublished results). Interestingly, predator habituation too is strongly delayed in toads with vMP lesions, suggesting that this structure has a general function for habituation. Repeated presentation of a predator dummy in vMP-lesioned toads may even lead to a build-up of avoidance resulting in a lasting "freezing posture." Our pilot studies with mechanically and acoustically released behavior patterns suggest that vMP is critically involved also in habituation of other types of behavior in response to different stimulus modalities. For example, habituation of the male toad's release call - elicited by bilateral mechanical pressure applied to both body flanks - also leads to a strong increase of 2DG-uptake in vMP (Finkenstädt & Ewert, unpublished).

8.7 Comparisons with Other Vertebrates

In lower vertebrates, comparable data are scarce. In teleost fish, damage of the telencephalon causes deficits in habituation of arousal responses (Laming & McKee

1981; Laming & Ennis 1982; Laming, this volume). In the urodele, *Salamandra salamandra*, MP-lesions lead to "disinhibited" visual prey-catching behavior characterized by hyperexcitation, impairment of configural pattern discrimination, and deficits in habituation (Finkenstädt & Ewert 1983). Urodeles possess a monosynaptic telencephalo-tectal connection which is missing in anurans (Finkenstädt et al. 1983). This fact might explain the partly different effects of vMP lesions between toads and salamanders (Finkenstädt 1987).

Drawing on its intrinsic interhemispheric connections and on histochemical investigations, Northcutt & Kicliter (1980) envision the medial pallium homologue to the hippocampus of mammals hence supporting Herrick's (1933) concept of "primordium hippocampi." Following hippocampal lesions in rats, repetitive non-habituating sniffing was found (Jarrard 1973), environmental cues remained novel for a longer time (Altman et al. 1973), and behavior was "hyperaroused" (Campbell et al. 1971). The latter authors speculate that the hippocampus is affecting "generalized arousal reactions" in an inhibitory manner. Douglas & Pribram (1966) mentioned in their prominent theory of hippocampal function in mammals that electrical stimulation of hippocampus produces inhibition of the so-called "orienting reflex," and that hippocampal lesions greatly attenuate any blocking effects on this behavior. In line with Sokolov's (1960) hypothesis, Foreman & Stevens (1987) conclude that the hippocampus may be envisioned as an interface that compares current sensory events with previous experience and elicits orientational responses when novelty ("mismatch") is detected. Removal of the hippocampus thus causes release of the appropriate motor system from inhibitory influences. The view that the hippocampus in mammals plays a primary role in so-called "reflexive orientation" has a long tradition (for review see Foreman & Stevens 1987). In accordance with Sokolov (1960, 1975), Vinogradova (1970, 1975) assumed that one of the hippocampal functions is to detect novelty and, as appropriate, to block or release the arousal system, and thus to alter motor behavior accordingly - perhaps via inhibitory control of brainstem reticular nuclei. Inhibition of RET by hippocampal inputs, via relay in the preoptic area, has been frequently proposed for mammals, but not yet been proven. During habituation of the rabbit's "orienting reflex," Vinogradova (1970, 1975) recorded in the hippocampus neurons that progressively enhanced their activity in the course of prolonged exposure to the same stimulus, which is similar to our findings in toad's MP1 neurons, and others that decrease activity, which match our class MP3 neurons.

8.8 Habituation and Arousal

Based on the present results, it appears that stimulus-specific long-term habituation in toads involves both, active modulatory processes on premotor structures and changes in the arousal state of the animal. In toads, arousing visual stimulation and visual habituation produce opposite patterns of 2DG-uptake in vMP, PO, dHYP, A, RET, and vOT (see Table 2) (Finkenstädt & Ewert 1985; Finkenstädt 1987). 2DG-uptake in vMP is strongly reduced after arousing electrical RET stimulation or arousing prey presentation. Comparable effects on RET were confirmed in a 2DG study of acoustically stimulated frogs (Zimmermann & Rahmann 1987).

Relationships between arousal and habituation have been frequently discussed in the mammalian literature (e.g., see Moruzzi & Magoun 1949; Segundo et al. 1955; Groves & Thompson 1970). RET stimulation studies in rats show remarkable similarities in changes of 2DG-uptake of homologous brain structures, such as the septal end of the hippocampal formation and anterior thalamic nuclei (Gonzalez-Lima & Scheich 1985; Gonzalez-Lima, this volume). Although electrical stimulation leads both to orthodromic and antidromic effects, it could be shown that the evoked metabolic activation results from orthodromic activation of RET-descending pathways projecting mono- or polysynaptically to their target structures (Sharp & Ryan 1984; Gonzalez-Lima 1986).

Between toads and rats there are also striking similarities in the effect of MS222 narcosis and behavioral habituation on the 2DG-uptake in structures of the "limbic" system and their associates, and these effects are opposite to the ones obtained after RET stimulation or behavioral arousal. Opposite effects on the hippocampus were also observed during wakefulness (Gonzalez-Lima & Scheich 1985; Gonzalez-Lima this volume) and sleep (Ramm & Frost 1983). MS222 seems to block ascending transmission in the reticular system of toads. Whether the transmission in the visual pathways remained largely unaffected, as described by Brodal (1969) in anesthetized rats, can hardly be answered for toads, since we did not explicitly stimulate the MS222 treated animals visually. However, our unpublished recording studies have shown that MS222 narcosis is paralleled with extremely weak visual responses of tectal neurons, compared to experiments in which toads were paralyzed with succinylcholine.

Comparing 2DG studies in toads and in rats, it is reasonable to assume a common basic functional organization of habituation and arousal in vertebrates (Hernandez-Peon 1961; Chin et al. 1965; Gonzalez-Lima & Scheich 1985; Finkenstädt & Ewert 1985, 1988b).

8.9 Consideration of Other Brain Structures

In the ventral and dorsal striatum (v,dSTR), glucose utilization in prey-habituated toads was significantly less than in naive controls. This might be due to differences in the motor activities between the two groups during the 60 min treatment, since previous experiments have shown an increased 2DG-uptake in the vSTR during visual release of prey-catching or predator avoidance behaviors (Finkenstädt et al. 1986). Striatal lesions abolish visually guided prey-catching (Finkenstädt 1987), which is in agreement with earlier findings obtained following large telencephalic ablations (Diebschlag 1935; Ewert 1967a). Lesions encompassing caudal STR and the most rostral portion of the lateral amygdala produced increases in response latency and decreases in response frequency as well (Patton & Grobstein 1986). The latter authors suggest that the observed deficit results from disturbance of modulatory processes rather than from impairment of the circuitry generating visual orienting, which accords well with our results (Finkenstädt et al. 1986; see also Ewert, this volume). The anuran paleo-striatum is homologous to a portion of the amniote basal ganglia (Wilczynski & Northcutt 1983b) which in mammals are suggested to be involved in initiating or controlling orientational and attentional motor programs

(Hore & Vilis 1980). Chevalier et al. (1984) conclude from physiological studies in rats that basal ganglia via a disynaptic pathway are acting on the tectum to switch on an appropriate sensorimotor link by which a significant stimulus can be taken as a target for an orienting movement. Similar to mammals, anurans possess a system of descending fibers originating in the striatum. This descending system is potentially capable of influencing tectal structures involved in orienting through various pathways, including two major disynaptic routes via the pretectal Lpd/(P) nuclei and the ventral tegmentum vTEG (Wilczynski & Northcutt 1977, 1983a,b). The one via Lpd/(P) is suggested to be disinhibitory (Ewert 1987a; Ewert, this volume). Since neither Lpd nor P or vTEG displays significant changes in 2DG-uptake between toads of habituation and naive groups, this technique gives no clear hints of their function for habituation. As already mentioned, unilateral thalamic lesions in Lpd/P disinhibit prey-catching and impair configural discrimination of objects moving in the contralateral visual field, suggesting Lpd/P to be involved in pattern recognition (Ewert 1967a, 1984), while unilateral vTEG ablation abolishes prey-orienting in the hemifield, suggesting vTEG to be taking part in prey localization (Masino & Grobstein 1986; see also Grobstein, this volume).

In the ventral half of the cerebellum (vCB), a significant decrease of 2DG-uptake was found in the "habituation group." Since this structure is mainly involved in posture and locomotion control, we assume that the reduced metabolic activity is correlated with reduced motor activity during the 60 min treatment (Finkenstädt et al. 1985). The vestibular system provides a major input to the anuran's cerebellum (Llinás & Precht 1969). Visual units were recorded in the cerebellum as well (Grüsser & Grüsser-Cornehls 1976).

8.10 Critical Evaluation of 2DG-Uptake in "Habituation" and "Naive" Groups

Finally, some comments on the different motor activities between the two experimental groups of this study are in order. The increase of 2DG-uptake in vCB and STR in toads of naive group *vs.* habituation group is probably primarily correlated with differences in motor activity and less specific to prey-catching, since increase of metabolic activity in these structures was also found during release of other motor behaviors (e.g., escape, locomotion). The changes of 2DG-uptake in vMP and PO/dHYP are not necessarily related to different motor activities between both experimental groups, since metabolic activity in nonmoving toads could be increased by MS222 and decreased in MS222 treated toads by arousing electrical RET stimulation (Finkenstädt & Ewert 1985). The same is true for RET whose metabolic activity in response to sensory stimuli can be altered independent of motor activity. The changes in vMP and PO/dHYP are not restricted to habituation and arousal specific to prey-catching, because both phenomena are correlated with comparable changes of 2DG-uptake in these structures for other stimulus-responses, such as predator-elicited avoiding and clasping-elicited release calling. The changes of 2DG-uptake in vOT between both experimental groups cannot be explained alone by the difference in prey-catching activity, since glucose utilization in the habituation group was *reduced* compared to naive non-responding toads which were stimulated for a comparable period with an antiworm-like stimulus (Finkenstädt et al. 1985).

9 Conclusions

Stimulus-specific habituation of prey-catching in toads involves "distributed properties" of brain function. The results of the present study indicate that stimulus-specific long-term habituation of orienting toward a visual prey object affects structures of the *S-R* mediating circuit (e.g., OT) and involves other structures extrinsic to it (e.g., vMP). The latter belong to a polysynaptic tectal-medialpallio-tectal system which - involving diencephalic relays (A, PO/dHYP) - during repetitive stimulus presentation modulates *S-R* mediation, probably by inhibiting tectal neurons and suppressing sensitizing (arousing) effects of ascending RET activity to tectal cells (Ewert 1987a,b; Finkenstädt & Ewert 1988b). Based on the results of 2DG-, lesion-, and electrical stimulation, we suggest that the modulatory system influences the response to prey, but itself does not elicit prey-catching.

This accords well with Sokolov's (1960, 1975) hypothesis of habituation as an active gating process extrinsic to the *S-R* pathway (see also Ewert & Birukow 1965). The concept is generally in agreement with the prey habituation model by Lara & Arbib (1985; see also Lara, this volume) and with results obtained in mammals (e.g., see Vinogradova 1970, 1975; Foreman & Stevens 1987). It is not in contradiction to the comments of Horn (1967).

The function of vMP in toads is not restricted to habituation and arousal of prey-catching. MP is involved in the control of comparable processes of *non-associative learning* in other *S-R* mediating systems (e.g., predator-elicited escape; clasping-elicited release call) and in *associative learning* (classical conditioning) (Finkenstädt & Ewert 1988a; Finkenstädt this volume).

Comparing the data obtained for the function of homologue brain structures in amphibians and mammals, we are tempted to assume a common fundamental organization for *selective attention* (disynaptic striato-tectal circuits), its modulation in relation to *habituation* (polysynaptic "limbic"-tectal circuits), and *arousal* (involving reticular structures). Presumably, evolution is much more conservative than once believed (see also Reiner et al. 1984).

10 Abbreviations of Brain Structures

A, anterior dorsal thalamic nucleus
ABO, accessory olfactory bulb
AC, nucleus accumbens
AL, nucleus amygdalae, lateral portion
AM, nucleus amygdalae, medial portion
B, neuropil of Belloncі
BN, bed nucleus of the pallial commissure
dBO, dorsal (granule cell layer) portion of the olfactory bulb
mBO, medial (mitral cell layer) of olfactory bulb
vBO, ventral (glomerula layer) of olfactory bulb
Cp, corpus geniculatum thalamicum
dCB, dorsal half of the cerebellum
vCB, ventral half of the cerebellum

DP, dorsal pallium
Ea, entopeduncular nucleus
HAB, nucleus habenulae
dHYP, dorsal hypothalamic area
vHYP, ventral hypothalamic area
IST, nucleus isthmi
INT, nucleus interpeduncularis
La, lateral anterior thalamic nucleus
LPp, anterior half of the lateral pallium, periventricular
aLP, anterior lateral pallium
pLP, posterior half of the lateral pallium
LPR, lateral prominence
LS, lateral septum
Lpd, lateral posterodorsal thalamus
Lpv, lateral posteroventral thalamus
aMP, anterior third of the medial pallium
dMP, dorsal portion of the posterior medial pallium
vMP, ventral two-third of the posterior medial pallium
MS, medial septum
NB, nucleus of Bellonci
OC, optic chiasm
dOT, dorsal layers of the optic tectum
sOT, snapping evoking area of the optic tectum
vOT, medio-ventral layers of the optic tectum
P, posterior thalamic nucleus
PO, preoptic region
RET, tegmental portion of the reticular formation
d,vSTR, striatum, dorsal (d) or ventral (v) portion
vTEG, ventral tegmentum
Vld, ventrolateral dorsal thalamus
VLd,v, dorsal (d) or ventral (v) part of the ventrolateral thalamic nucleus
Vm, ventromedial thalamic nucleus

Acknowledgement. I thank Bert Klöppel and Torsten Siebert for computer assistance, Mrs. Burggraf for attentive animal care, and Mrs. Gerda Kruk for expert histology and general assistance. This study was supported by the Deutsche Forschungsgemeinschaft (Az.: Fi 315/3-2,3).

11 References

Altman J, Brunner RL, Bayer SA (1973) The hippocampus and behavioural maturation. *Behav Biol* 8: 557-596

Birukow G, Meng M (1955) Eine neue Methode zur Prüfung des Gesichtssinnes bei Amphibien. *Naturwissenschaften* 49: 652-653

Brodal A (1969) *Neurological anatomy in relation to clinical medicine.* Oxford Univ Press, London

Bruner J, Tauc L (1966) Habituation at the synaptic level in *Aplysia. Nature* 210: 37-39

Campbell BA, Ballantine P, Lynch G (1971) Hippocampal control of behavioral arousal: duration of lesion effects and possible interactions with recovery after frontal cortical damage. *Exp Neurol* 33(1): 159-170

Chevalier G, Vacher S, Deniau JM (1984) Inhibitory nigral influence on tectospinal neurons, a possible implication of basal ganglia in orienting behavior. *Exp Brain Res* 53: 320-326

Chin JH, Killam EK, Killam KF (1965) Factors affecting sensory input in the cat: modification of evoked auditory potentials by reticular formation. *Electroenceph Clin Neurophysiol* 18: 567-574

Diebschlag E (1935) Zur Kenntnis der Großhirnfunktionen einiger Urodelen und Anuren. *Z Vergl Physiol* 21: 343-394

Douglas RJ, Pribram KH (1966) Learning and limbic lesions. *Neuropsychologia* 5: 197-220

Eikmanns KH (1955) Verhaltensphysiologische Untersuchungen über den Beutefang und das Bewegungssehen der Erdkröte (*Bufo bufo* L). *Z Tierpsychol* 12: 229-253

Ewert J-P (1967a) Untersuchungen über die Anteile zentralnervöser Aktionen an der taxispezifischen Ermüdung beim Beutefang der Erdkröte (*Bufo bufo* L). *Z Vergl Physiol* 57: 263-298

Ewert J-P (1967b) Der Einfluß von Störreizen auf die Antwortbereitschaft bei der Richtbewegung der Erdkröte (*Bufo bufo* L). *Z Tierpsychol* 24: 208-312

Ewert J-P (1969) Quantitative Analyse von Reiz-Reaktions-Beziehungen bei visuellem Auslösen der Beutefang-Wendereaktion der Erdkröte (*Bufo bufo* L). *Pflügers Arch* 308: 225-243

Ewert J-P (1974) The neural basis of visually guided behavior. *Sci Amer* 230: 34-42

Ewert J-P (1980) *Neuroethology. An introduction to the neurophysiological fundamentals of behavior.* Springer-Verlag, Berlin Heidelberg New York

Ewert J-P (1984) Tectal mechanisms that underlie prey-catching and avoidance behaviors in toads. In: Vanegas H (ed) *Comparative neurology of the optic tectum.* Plenum Press, New York, pp 247-416

Ewert J-P (1987a) Neuroethology: toward a functional analysis of stimulus-response mediating and modulating neural circuitries. In: Ellen P, Thinus- Blanc C (eds) *Cognitive processes and spatial orientation in animal and man, Vol I.* Martinus Nijhoff, Dordrecht, pp 177-200

Ewert J-P (1987b) Neuroethology of releasing mechanisms: prey-catching in toads. *Behav Brain Sci* 10: 337-405

Ewert J-P, Birukow G (1965) Über den Einfluß des Zentralnervensystems auf die Ermüdbarkeit der Richtbewegung im Beuteschema der Erdkröte *Bufo bufo* (L). *Naturwissenschaften* 52: 68-69

Ewert J-P, Finkenstädt T (1987) Modulation of tectal functions by prosencephalic loops in amphibians. A commentary. *Behav Brain Sci* 10: 122-123

Ewert J-P, Ingle DJ (1971) Excitatory effects following habituation of prey-catching activity in frogs and toads. *J Comp Physiol Psychol* 77: 369-374

Ewert J-P, Kehl W (1978) Configurational prey selection by individual experience in the toad *Bufo bufo. J Comp Physiol* 126: 105-114

Ewert J-P, Rehn B (1969) Quantitative Analyse der Reiz-Reaktionsbeziehungen bei visuellem Auslösen des Fluchtverhaltens der Wechselkröte (*Bufo viridis* L). *Behaviour* 35: 212-234

Finkenstädt T (1987) *Verschaltung, Interaktion und Funktion visuell beeinflußbarer Hirngebiete bei Amphibien.* Habil Thesis, Univ of Kassel

Finkenstädt T, Ewert J-P (1983) Visual pattern discrimination through interactions of neural networks: a combined electrical brain stimulation, brain lesion, and extracellular recording study in *Salamandra salamandra. J Comp Physiol* 153: 99-110

Finkenstädt T, Ewert J-P (1985) Glucose utilization in the toad's brain during anesthesia and stimulation of the ascending reticular arousal system. *Naturwissenschaften* 72: 161-162

Finkenstädt T, Ewert J-P (1988a) Effects of visual associative conditioning on behavior and cerebral metabolic activity in toads. *Naturwissenschaften* 75: 95-97

Finkenstädt T, Ewert J-P (1988b) Stimulus-specific long-term habituation of visually guided orienting behavior toward prey in toads: a ^{14}C-2DG study. *J Comp Physiol* 163:1-11

Finkenstädt T, Ebbesson SOE, Ewert J-P (1983) Projections to the midbrain tectum in *Salamandra salamandra* L. *Cell Tiss Res* 234: 39-55

Finkenstädt T, Adler NT, Allen TO, Ebbesson SOE, Ewert J-P (1985) Mapping of brain activity in mesencephalic and diencephalic structures of toads during presentation of visual key stimuli: A computer assisted analysis of (^{14}C)2DG autoradiographs. *J Comp Physiol* 156: 433-445

Finkenstädt T, Adler NT, Allen TO, Ewert J-P (1986) Regional distribution of glucose utilization in the telencephalon of toads in response to configurational visual stimuli: a ^{14}C-2DG study. *J Comp Physiol* 158: 457-467

Foreman N, Stevens R (1987) Relationships between the superior colliculus and hippocampus: neural and behavioural considerations. *Behav Brain Sci* 10: 101-152

Gallistel CR, Piner CT, Allen TO, Adler NT, Yadin E, Negin M (1982) Computer assisted analysis of 2-DG autoradiographs. *Neurosci Biobehav Rev* 6:409-422

Gonzalez-Lima F (1986) Activation of substantia gelatinosa by midbrain reticular stimulation demonstrated with 2-deoxyglucose in the rat spinal cord. *Neurosci Lett* 65: 326-330

Gonzalez-Lima F, Scheich H (1985) Ascending reticular activating system in the rat: a 2-deoxyglucose study. *Brain Res* 344: 70-88

Groves PM, Thompson RF (1970) Habituation: a dual process theory. *Psych Rev* 77: 419-450

Grüsser O-J, Grüsser-Cornehls U (1976) Neurophysiology of the anuran visual system. In: Llinás R, Precht W (eds) *Frog neurobiology.* Springer-Verlag, Berlin Heidelberg New York, pp 298-385

Hernandez-Peon R (1961) Reticular mechanisms of sensory control. In: Rosenblith WA (ed) *Sensory communication.* Wiley, New York, pp 497-520

Herrick CJ (1933) The amphibian forebrain. VIII: Cerebral hemispheres and pallial primordia. *J Comp Neurol* 58: 737-759

Hore J, Vilis T (1980) Arm movement performance during reversible basal ganglia lesions in the monkey. *Exp Brain Res* 39: 217-228

Horn G (1967) Neuronal mechanisms of habituation. *Nature* 215: 707-711

Horn E, Greiner B, Horn I (1979) The effect of ACTH on habituation of the turning reaction in the toad *Bufo bufo L.*. *J Comp Physiol* 131:129-135

Jarrard LE (1973) The hippocampus and motivation. *Psychol Bull* 79: 1-11

Juliano SL, Whitsel BL (1987) A combined 2-deoxyglucose and neurophysiological study of primate somatosensory cortex. *J Comp Neurol* 263: 514-525

Kandel E (1976) *Cellular basis of behavior: an introduction to behavioral neurobiology.* Freeman, New York

Krasne FB, Kandel ER, Truman JW (1979) Simple systems revisited. *Neurosci Res Prog Bull* 17(4): 529-538

Laming PR, McKee M (1981) Deficits in habituation of cardiac arousal responses incurred by telencephalic ablation in Goldfish (*Carassius auratus*) and their relation to other telencephalic functions. *J Comp Physiol Psychol* 95 3: 460-467

Laming PR, Ennis P (1982) Habituation of fright and arousal responses in the teleosts *Carassius auratus* and *Rutilus rutilus. J Comp Physiol Psychol* 96 3: 460-466

Lara R, Arbib MA (1985) A model of the neural mechanisms responsible for pattern recognition and stimulus specific habituation in toads. *Biol Cybern* 51: 223-237

Lettvin JY, Maturana HR, Pitts WH, McCulloch WS (1961) Two remarks on the visual system of the frog. In: Rosenblith WA (ed) *Sensory communication.* MIT Press, Cambridge MA, pp 757-776

Llinás R, Precht W (1969) Inhibitory vestibular efferents system and its relation to the cerebellum in the frog. *Exp Brain Res* 9: 19-29

Masino T, Grobstein P (1986) The organization of tectal projections to the ventral midbrain in *Rana pipiens. Soc Neurosci Abstr* 84: 11

Moruzzi G, Magoun HW (1949) Brain stem reticular formation and activation of the EEG. *Electroenceph Clin Neurophysiol* 1: 455-473

Nauta WJH (1963) Central nervous organization and the endocrine motor system. In: Nalbandov AV (ed) *Advances in neuroendocrinology.* Univ Illinois Press, Urbano, pp 5-21

Neary TJ, Northcutt RG (1983) Nuclear organization of the bullfrog diencephalon. *J Comp Neurol* 213: 262-278

Nieuwenhuys R, Opdam P (1976) Structure of the brain stem. In: Llinás R, Precht W (eds) *Frog neurobiology.* Springer-Verlag, Berlin Heidelberg New York, pp 811-855

Northcutt RG, Kicliter E (1980) Organization of the amphibian telencephalon. In: Ebbesson SOE (ed) *Comparative neurology of the telencephalon.* Plenum Press, New York, pp 203-255

Patton P, Grobstein P (1986) Possible striatal involvement in prey orienting behavior in the frog. *Soc Neurosci Abstr* 10(I): 61

Potter HD (1972) Terminal arborizations of retinotectal axons in the bullfrog. *J Comp Neurol* 144: 269-283

Ramm P, Frost BJ (1983) Regional metabolic activity in the rat brain during sleep-wake activity. *Sleep* 6: 196-216

Reiner A, Brauth SE, Karten HJ (1984) Evolution of the amniote basal ganglia. *TINS* 9: 320-325

Ryan AF, Sharp FR (1982) Localization of (^{3}H)2-deoxyglucose at the cellular level using freeze-dried tissue and dry-looped emulsion. *Brain Res* 252: 177-180

Sachs L (1978) *Statistische Methoden.* Springer-Verlag, Berlin Heidelberg New York

Satou M, Ewert J-P (1985) The antidromic activation of tectal neurons by electrical stimuli applied to the caudal medulla oblongata in the toad *Bufo bufo* (L). *J Comp Physiol* 157: 739-748

Segundo JP, Arana R, French JD (1955) Behavioral arousal by stimulation of the brain in the monkey. *J Neurosurg* 12: 601-613

Sharp FR, Ryan AF (1984) Regional (^{14}C)2-deoxyglucose uptake during forelimb movements evoked by rat motor cortex stimulation: pons, cerebellum, medulla, spinal cord, and muscle. *J Comp Neurol* 224: 286-306

Sokoloff L (1984) Modeling metabolic processes in the brain in vivo. *Ann Neurol Suppl* 15: 1-11

Sokoloff L, Reivich M, Kennedy C, DesRosiers MH, Patlak CS, Pettigrew KD, Sakurada O, Shinohara M (1977) The (^{14}C)-deoxyglucose method for the measurement of local cerebral glucose utilization: theory, procedure and normal values in the conscious and anesthetized albino rat. *J Neurochem* 28: 897-916

Sokolov EN (1960) In: Brazier MAB (ed) *The central nervous system and behaviour.* Josiah Macy Jun. Foundation, New York

Sokolov EN (1975) Neuronal mechanisms of the orienting reflex. In: Sokolov EN, Vinogradova O (eds) *Neuronal mechanisms of the orienting reflex.* Lawrence Erlbaum/Hillsdale, New York, pp 217-235

Székely G, Lázár G (1976) Cellular and synaptic architecture of the optic tectum. In: Llinás R, Precht W (eds) Frog neurobiology. Springer-Verlag, Berlin Heidelberg New York, pp 407-434

Theurich M, Müller CM, Scheich H (1984) 2-deoxyglucose accumulation parallels extracellularly recorded spike activity in the avian auditory neostriatrum. *Brain Res* 322: 157-161

Thompson RF, Spencer WA (1966) Habituation: a model phenomenon for the study of neural substrates of behavior. *Psych Rev* 73: 16-43

Trepakov VV (1974) Postsynaptic inhibition in the frog's primordial hippocampus. *Neurofiziologia* 5: 583-592

Vinogradova O (1970) Registration of information in the limbic system. In: Horn G, Hinde RA (eds) *Short term changes in neural activity and behavior.* Cambridge University Press, Cambridge, pp 95-140

Vinogradova O (1975) Hippocampus and the orienting reflex. In: Sokolov EN, Vinogradova O (eds) *Neuronal mechanisms of the orienting reflex.* Lawrence Erlbaum/Hillsdale, New York, pp 128-154

Wilczynski W, Northcutt RG (1977) Afferents to the optic tectum of the leopard frog: an HRP study. *J Comp Neurol* 173: 219-229

Wilczynski W, Northcutt RG (1983a) Connections of the bullfrog striatum: afferent organization. *J Comp Neurol* 214: 321-332

Wilczynski W, Northcutt RG (1983b) Connections of the bullfrog striatum: efferent projections. *J Comp Neurol* 214: 333-342

Zimmermann E, Rahmann H (1987) Acoustic communication in the poison-arrow frog *Phyllobates tricolor.* advertisement calls and their effects on behavior and metabolic brain activity of recipients. *J Comp Physiol* 160: 693-702

Visual Associative Learning: Searching for Behaviorally Relevant Brain Structures in Toads

THOMAS FINKENSTÄDT

Abteilung Neuroethologie, Fachbereich Biologie/Chemie, Universität Kassel
D-3500 Kassel, FR Germany

Abstract. Prey-recognition in toads can be modified by classical conditioning. When toads are fed several times on mealworms (unconditioned stimulus, *US*) presented with the experimenter's hand covered by a black glove (conditioned stimulus, *CS*), they finally respond to the *CS* alone with the conditioned response *CR*, prey-catching. Toads were trained to respond to a large (80 x 80 mm^2) black moving square (subserving as a generalized *CS*) with prey-catching. Following 2DG injection, the local cerebral metabolic effects of training, were quantitatively compared to naive animals. Our major finding was the differential metabolic effect produced by the same square shaped stimulus before and after conditioning. Associative learning not only changes metabolism in various cerebral structures of the visual pathway, but also in certain non-visual areas of the forebrain. Results of lesions in the medial pallium (MP) and the lateral amygdala (AL) further support the idea of the importance of these "limbic" structures in memory representation: Toads with bilateral ablation of the ventral MP prior to or post conditioning proved to be unable to show the *CR*. These findings are in line with the hypothesis that associative learning represents a spatially distributed property of neuronal maps, which may be demonstrated in any structure that meets the appropriate requirements of time and space convergence of *CS* and *US* inputs. The learning-related metabolic changes in the visual pathway let one assume that visual circuits cannot simply be regarded as mere input channels for the *CS*. Non-visual cerebral structures, changing metabolism in this context, may be regarded as part of a neural system critically involved in establishing the memory trace of the particular *CS* in the amphibian brain.

1 Introduction

Thorndike's (1932) "continuum hypothesis" infers that learning capacity across species is not different in general, but rather a matter of degree regarding speed, quality, quantity, and persistence of association, hence suggesting that animal's minds are similar to human minds, but only simpler. Today, this view appears to be oversimplified. On the one hand adaptive specializations were ignored (Rozin &

Kalat 1971), on the other hand general principles of learning were overestimated (Seligman 1970). Nevertheless, basic mechanisms of associative learning seem to exist across species both in vertebrates and invertebrates. Research in the marine molluscs *Aplysia* and *Hermissenda*, for example, have pushed forward our general understanding of learning and memory processes on neuronal and molecular levels (Kandel 1979).

From a neuroethological point of view, it is of particular interest to investigate how innate releasing mechanisms can be modified by learning (Baerends 1987; Fentress 1987). Prey-catching and predator avoidance behaviors of toads *Bufo bufo*, largely mediated by innate releasing mechanisms, can be modified by individual experience (for a critical evaluation of the term "innate" see Ewert, this volume). It becomes more and more evident that amphibians are not the poor learners they were suggested to be (Thorpe 1963). For example, Cott (1936) reported that toads after just one (or a couple of) negative experience(s) with hive bees associate the visual appearance of the bee with the painful stimulus, which leads to "bee avoidance." Brower & Brower (1962) have observed in toads that avoidance of honey bees is also "conferred" to their dronefly mimics. Boice (1970) speculated that mainly active hunters such as *Bufo woodhousii* display avoidance learning, in contrast to anuran species which are waiting for prey, such as the leopard frog *Rana pipiens*. Ewert (1968) has demonstrated that toads associate olfactory and visual cues of prey, which - in the presence of learned odor - leads to a decrease in the acuity of configural pattern discrimination, resulting in an enlargement of the prey category. Brzoska & Schneider (1978) discovered a modification of toad's prey-recognition in the course of hand-feeding as a result of visual conditioning. Elepfandt et al. (1985) showed that the clawed frog *Xenopus laevis* is able to discriminate water wave frequencies by so-called go/no-go conditioning. (For a review on visual learning in amphibians see Ewert 1984).

Nothing is as yet known about the underlying neural substrates of learning in amphibians, while in mammals and invertebrates a vast literature exists. This is surprising, since amphibians as the "lowest" terrestrial vertebrates are particularly suitable experimental subjects for various reasons: The basic "Bauplan" of the amphibian brain is comparable to mammals, although its structural organization is "simpler", i.e., less differentiated (Fite & Scalia 1976; Llinás 1976; Ebbesson 1980); the connectivity of brain structures is well understood and homologies with other vertebrates exist (e.g., Northcutt 1974; Kicliter & Ebbesson 1976; Wilczynski & Northcutt 1977, 1983a,b; Northcutt & Kicliter 1980; Neary & Northcutt 1983; Reiner et al. 1984); visually guided behaviors and the underlying neural substrates are well investigated (Ewert 1974, 1984, 1987; Ingle 1983).

To approach neural mechanisms of learning and memory, it is first necessary to map brain structures that seem to be involved in learning (Thompson et al. 1984). A suitable method applicable to behaving animals is the ^{14}C-2-deoxyglucose (2DG) technique which, combined with recording, stimulation, and lesion experiments, has already proved its usefulness for the investigation of non-associative learning in toads (Finkenstädt, this volume). As an example of associative learning, in the present study we have examined the aforementioned "hand-feeding" phenomenon.

2 The Training Paradigm: "Training Groups" and "Naive Groups"

2.1 Training Groups

Configural pattern discrimination in naive toads *Bufo bufo spinosus* was measured by their prey-catching responses to moving worm-like (W), antiworm-like (A), and square (S) objects of various sizes (Fig. 1a and Table 1) presented according to the procedure of Ewert (1969; see also this volume). Toads were then trained to respond to a moving predator dummy with prey-catching by pairing a prey object with a large object. In the course of "hand-feeding," a mealworm (*Tenebrio molitor*) when presented with the experimenter's hand covered with a black glove served as unconditioned stimulus, *US*. Initially, the moving hand released avoidance behavior. In the course of 2-4 weeks daily binocular training (hand-feeding), the hand in the black glove became the conditioned stimulus, *CS*, and released prey-catching as the conditioned response *CR* in the absence of *US*, as described by Brzoska & Schneider (1978).

As a result of this kind of visual/visual conditioning, toads show long-lasting modifications of configural pattern discrimination (Fig. 1b, Table 1), indicated by their prey-selecting characteristics (Burghagen 1979). Conditioning during hand-feeding is thus not specific to the experimenter's hand; rather the response spectrum to W-, S-, and A-stimuli is shifted towards larger sizes in conditioned animals, suggesting a generalization along the parameter "area." More specifically, whereas in unconditioned "naive" toads, prey-catching is optimally tuned to W- and S-stimuli in a certain (for W- and S-stimuli different) range of object size, in conditioned animals this range includes larger sizes, which reduces the preference of large W- over large S-stimuli (Fig. 1, cf. a and b). Furthermore, it is important to note that the prey-catching activity of conditioned toads shows no progressive decrease with edge elongation of A-stimuli (Fig. 1b).

Avoidance behavior in response to a black predator dummy (e.g., a 80 x 80 mm^2 square) moving against a white background is abolished in conditioned toads, if the object is presented as a ground-borne enemy. However, moving over-head as an airborne enemy, some avoidance responses can still be elicited. Toads appear thus to be conditioned mainly for those parts of the visual field, in which *US* and *CS* were presented during hand-feeding, especially involving the "snapping zone." To allow quantitative measurements, the 80 x 80 mm^2 square was used as a "generalized" *CS*.

Monocular training allowed us to pursuit the question of interhemispherical transfer in toads. Toads with the right eye firmly covered by a black sticky cap during training received a conditioning procedure that was otherwise identical to the one in the binocular group. During monocular rehearsal with the right or "naive" eye occluded, similar prey-catching characteristics were observed in both groups. (Covering the left eye led to comparable results). Immediately after the daily training sessions, the eye cap was removed. However, when the untrained or trained eye was covered, no interocular transfer could be observed in the lateral monocular "naive" visual field. When the "trained eye" was covered and the naive exposed to the *CS*, "interocular transfer" could be observed only in the frontal visual field: if a large

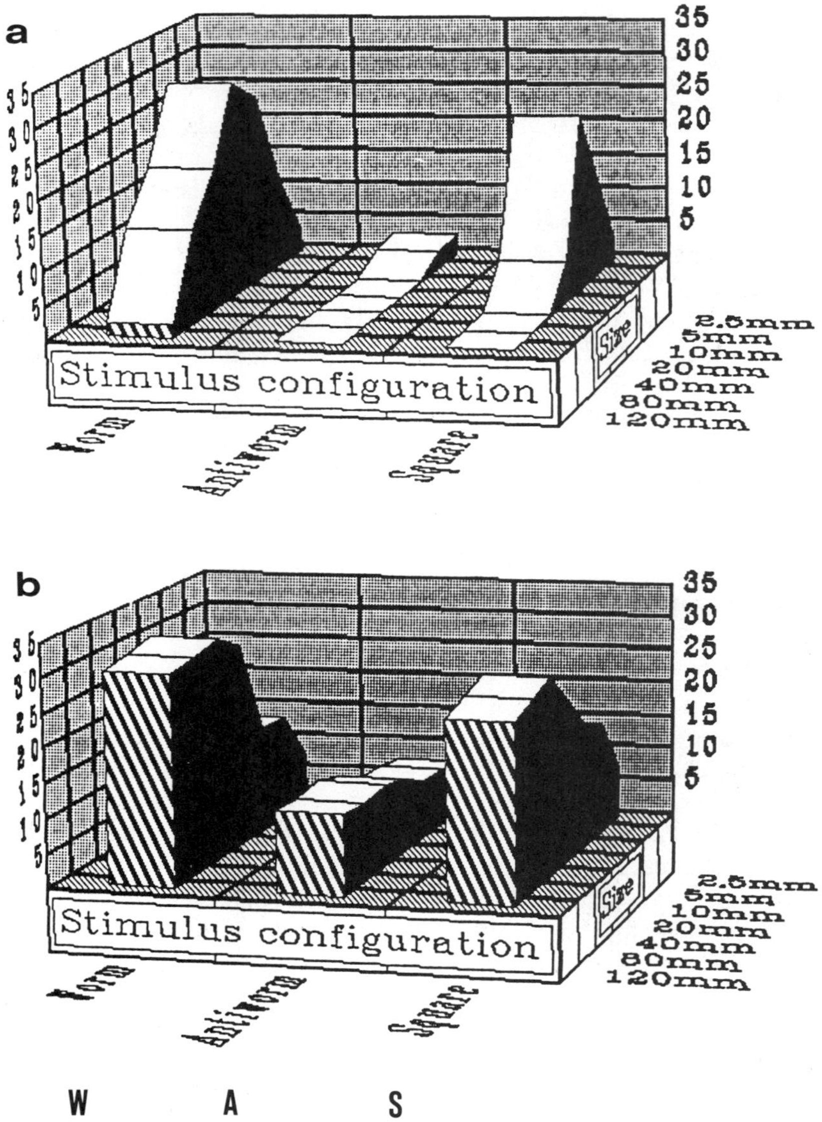

Figure 1. Influence of visual conditioning during hand-feeding on the discrimination of visual configural stimuli. Average orienting activity (orienting responses per 36 s) of nine toads (*Bufo bufo spinosus*) in response to worm-like (W), antiworm-like (A), and square (S) black stimuli of various edge lengths, before (a) and after (b) conditioning. See also Table 1. Each black stimulus was moved at v = 10°/s against a white background around the toad in the procedure of Ewert (1969). (After Finkenstädt & Ewert 1988a).

Table 1. Configural pattern discrimination in naive and trained (hand-fed) toads *Bufo bufo spinosus*. Average prey-catching orienting activity R ± SE (responses per 36 s) of nine binocularly trained toads. Responses to black worm, antiworm, and square stimuli of various sizes (moving at v = 10°/s against a white background) prior to and after training.

Stimulus	Edge length [mm]	Prey-Catching Orienting Activity Before Training	Prey-Catching Orienting Activity After Training
Square (S)	2.5	3.2 ± 1.9	7.5 ± 2.0
	5	14.1 ± 1.9	16.3 ± 3.1
	10	24.8 ± 2.4	17.9 ± 3.8
	20	5.9 ± 1.8	23.4 ± 2.9
	40	0.7 ± 0.6	28.2 ± 3.4
	80	0.0 ± 0.0	27.0 ± 4.5
	120	0.0 ± 0.0	25.7 ± 3.6
Worm (W)	5	12.8 ± 2.3	15.4 ± 3.9
	10	26.7 ± 3.0	15.5 ± 4.1
	20	30.4 ± 3.2	29.3 ± 4.7
	40	20.3 ± 2.8	32.0 ± 3.1
	80	13.2 ± 4.1	31.1 ± 3.5
	120	1.9 ± 1.8	30.4 ± 4.4
Antiworm (A)	5	2.5 ± 1.7	8.3 ± 2.2
	10	0.4 ± 0.5	8.4 ± 2.3
	20	0.0 ± 0.0	10.6 ± 2.8
	40	0.0 ± 0.0	11.8 ± 3.2
	80	0.0 ± 0.0	11.2 ± 3.4
	120	0.0 ± 0.0	11.7 ± 4.2

object was quickly moved in lateral or dorsal visual field sections of the "naive eye," the toad displayed avoidance behaviors.

2.2 Naive Group

Control animals were exposed to the mealworm and to the experimenter's hand, however, no temporal and spatial pairing was allowed, i.e., both stimuli were presented independently in space and time. Although control animals received about the same amount of stimulus presentations, no associative learning occurred. In this context, these toads are regarded as "naive."

3 Mapping of Brain Activity with 2DG

3.1 Technical Comments: "Interhemispherical" and "Brain-to-Brain" Comparisons

To investigate training-correlated cerebral glucose metabolism, the 2DG technique was applied in toads of (i) the monocular training group, (ii) the binocular training group, and in (iii) the naive group. In the "2DG experiment," ^{14}C-2-deoxyglucose (15µCi/100g body weight) was injected into the toad's caudal lymphatic sacs. Subsequently, according to a particular procedure, a black 80 x 80 mm^2 square, subserving as a generalized *CS*, was presented for 45 min. The stimulus was mechanically moved at constant speed (10°/s) around the toad sitting in a glass vessel. The procedure assured that individuals of all three groups (i)-(iii) received about the same amount of stimulations (five rounds to the left, one minute break, five rounds to the right, one minute break, etc.). By means of this procedure, stimulus habituation within the 45min was prevented. In monocularly as well as in binocularly trained toads, both eyes were exposed to the stimulus during the 2DG experiment. After processing 2DG, the optical density of brain structures in the autoradiographic images of brain transverse sections was measured by means of an image analyzing system (Quantimet 920, Cambridge Instr.; for details see Finkenstädt et al. 1985, 1986). The local cerebral 2DG-uptake measured in µCi of [^{14}C]-activity (Sokoloff et al. 1977) was quantitatively compared between animals of the three groups. In monocularly trained (M) toads, the percentage of 2DG-uptake between structures of the contralateral (*c*) and the ipsilateral (*i*) brain hemisphere was compared, 100 x ([^{14}C]M_c-[^{14}C]M_i)/[^{14}C]M_i; the interhemispheric comparison was evaluated with the *t*-test. In the binocular group, the percentage of 2DG-uptake between the brains of binocularly trained (B) and naive (N) animals was compared, 100 x ([^{14}C]'B-[^{14}C]'N)/[^{14}C]'N; the brain-to-brain comparison of bilaterally averaged [^{14}C]' values, calculated in relation to a "white matter" reference structure, was evaluated by means of the Kolmogorov-Smirnoff-Omnibus-(*KSO*)-test. (For details see Finkenstädt et al. 1985, 1986).

3.2 Survey of Metabolic Activity Across Brain Structures

Statistical significant differences of 2DG-uptake both in "brain-to-brain" (Table 2 and Fig. 2) and "interhemispheric" (Table 2; Fig. 3) comparisons were found in structures of the visual pathway as well as in forebrain areas not primarily visual. Regarding the latter, the caudal half of the ventral medial pallium, vMP (Fig. 4b), the lateral region of the nucleus amygdalae, AL (Fig. 4b), a portion of the periventricular layer of the anterior lateral pallium, LPp (Fig. 4a), and the preoptic/dorso-hypothalamic region PO/dHYP showed increases in 2DG-uptake (see also Table 2 and Fig. 2). In the central visual pathway, the medioventral tectal layers (vOT), the pretectal lateral posteroventral thalamic nucleus (Lpv), the pretectal posterior thalamic nucleus (P), the anterior division of the lateral thalamic nucleus (La), and the "neuropil" of Bellonci (NB) together with the dorsal portion of the ventrolateral thalamic nucleus (Vld), treated as "NB/Vld region," displayed increased 2DG-uptake. The pretectal lateral posterodorsal thalamic nucleus (Lpd), on the other hand, as the only

Table 2. Mean percentage of 2DG-uptake for various cerebral structures compared between contralateral and ipsilateral hemisphere of four monocularly trained toads ("interhemispheric comparison") and between nine binocularly trained and nine naive toads ("brain-to-brain comparison"). Training refers to hand-feeding (conditioning). During the 2DG experiment either eye of the previously monocularly trained, binocularly trained, and naive toads was stimulated with the generalized *CS*, a 80 x 80 mm^2 black square moving at 10°/s around the animal in variable horizontal direction. 2DG-uptake was measured in µCi of $[^{14}C]$-activity. For "interhemispheric comparison," in monocularly trained (M) toads the differences in 2DG-uptake across structures of the contralateral (c) and the ipsilateral (i) hemisphere (evaluated with the *t*-test) are expressed here by $XM_{c/i}[\%] = 100 \times ([^{14}C]M_c - [^{14}C]M_i)/[^{14}C]M_i$. For "brain-to-brain compari-son," the differences in 2DG-uptake between structures of binocularly trained (B) and naive (N) toads (evaluated with the *KSO*-test) are expressed here by $X'B/N[\%] = 100 \times ([^{14}C]'B - [^{14}C]'N)/[^{14}C]'N$. Note that the statistical tests are not based on the transformed %-values. For abbreviations of brain structures see Fig. 2 and Section 8.

	Interhemispheric Comparison		Brain-to-Brain Comparison	
Brain Structure	2DG-Uptake $XM_{c/i}$ (%)	*t*-test	2DG-Uptake X'B/N (%)	*KSO*-test
AM	1.9	-	2.3	-
AL	8.5	$P<0.05$	17.9	$P<0.05$
DP	1.7	-	4.0	-
LPp	16.9	$P<0.02$	15.5	$P<0.05$
aLP	2.8	-	3.0	-
dMP	2.1	-	3.5	-
aMP	-0.7	-	4.2	-
vMP	11.7	$P<0.05$	17.5	$P<0.02$
MS	3.6	-	3.3	-
STR	4.8	-	2.3	-
Lpd	-8.9	$P<0.05$	-6.8	$P<0.05$
Lpv	9.0	$P<0.05$	16.7	$P<0.02$
La	7.5	$P<0.05$	11.0	$P<0.05$
P	4.7	-	22.2	$P<0.01$
NB/Vld	8.0	$P<0.05$	18.6	$P<0.02$
A	6.0	-	4.9	-
PO/dHYP	0.0	-	17.2	$P<0.05$
dOT	3.3	-	1.6	-
vOT	6.7	$P<0.05$	15.7	$P<0.02$
sOT	9.5	$P<0.05$	38.8	$P<0.02$
vTEG	2.1	-	18.7	$P<0.02$
RET	-1.9	-	3.0	-
IST	11.4	$P<0.02$	31.7	$P<0.05$
dCB	-1.7	-	17.1	$P<0.02$
vCB	-0.82	-	-5.2	-

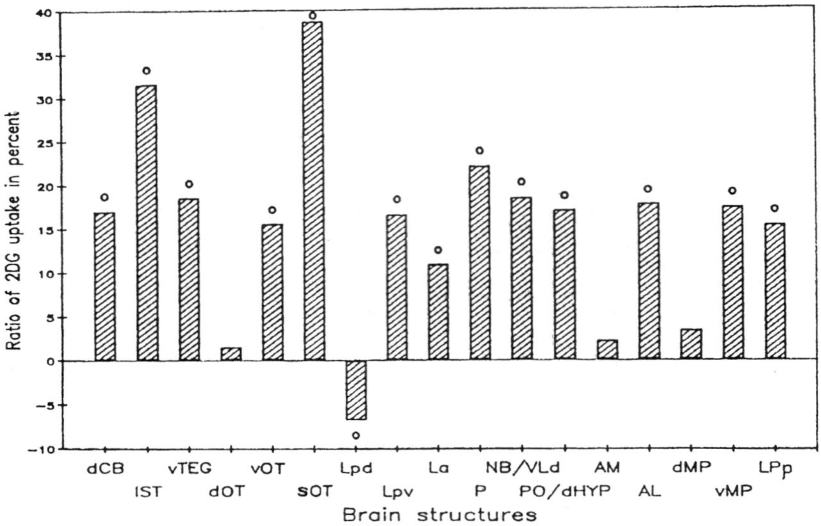

Figure 2. Average percentage of differences in 2DG-uptake across various brain structures of nine binocularly trained (hand-fed) toads in comparison to nine untrained naive controls ("brain-to-brain comparison"). Circles indicate statistical significance ($P < 0.05$). During 2DG application, the toads were stimulated with a "generalized *CS*" consisting of a 80 x 80 mm^2 black object (presented to either eye) moving at 10°/s against a white background in the procedure of Ewert (1969). Toads responded with prey-catching. For statistical evaluations see Table 2. Abbreviations of brain structures (see also Table 2): A, anterior dorsal thalamic nucleus; AM, nucleus amygdalae, medial portion; AL, nucleus amygdalae, lateral portion; dCB, dorsal half of the cerebellum; vCB, ventral half of the cerebellum; DP, dorsal pallium; IST, isthmic nucleus; aLP, anterior half of the lateral pallium; LPp, periventricular layers of aLP; Lpd, lateral posterodorsal thalamus; Lpv, lateral posteroventral thalamus; aMP, anterior third of the medial pallium; dMP, dorsal portion of the posterior medial pallium; vMP, ventral two-third of the posterior medial pallium; MS, medial septum; NB/Vld, "neuropil" (nucleus) of Bellonci/ventrolateral dorsal thalamus; dOT, dorsal layers of the optic tectum; sOT, snapping evoking area of the optic tectum; vOT, medio-ventral layers of the optic tectum; P, posterior thalamic nucleus; PO/dHYP, preoptic/dorsal hypothalamic area; RET, tegmental portion of the reticular formation; STR, striatum, dorsal and ventral portion; vTEG, ventral tegmentum. (From Finkenstädt & Ewert 1988c).

exception so far investigated showed a reduction in glucose utilization (Table 2, Figs. 2 and 3; for abbreviations of brain structures see Section 8).

Since 2DG-uptake was measured both for "interhemispheric" and "brain-to-brain" comparisons, correlations with differences in an animal's motor activity can be evaluated best for cases in which $[^{14}C]M_c = [^{14}C]M_i > [^{14}C]N$. This is the case, e.g., for the dorsal cerebellum (dCB), the ventral tegmentum (vTEG), and the striatum (STR). Previous studies have shown that 2DG-uptake in these structures in naive toads indeed is correlated with visually elicited motor activity (Finkenstädt et al. 1985, 1986).

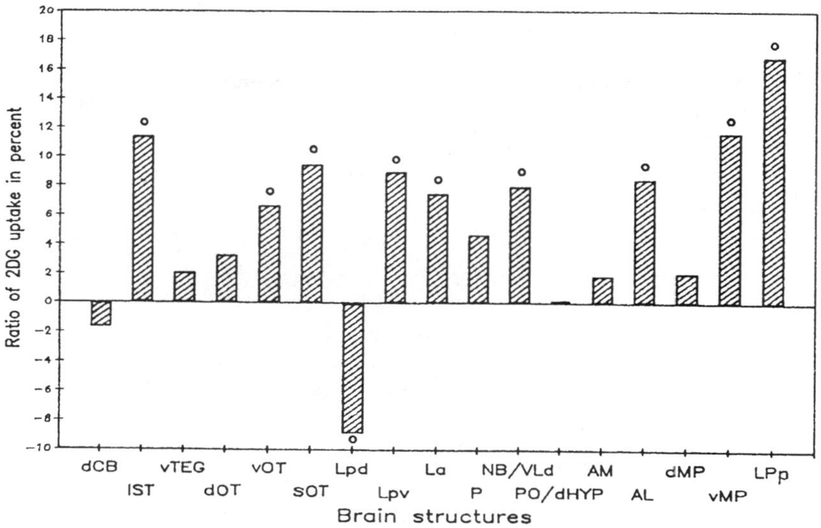

Figure 3. Average percentage of differences in 2DG-uptake across various brain structures of four monocularly trained (hand-fed) toads in comparison between contralateral and ipsilateral hemisphere ("interhemispheric comparison"). Circles indicate statistical significance ($P<0.05$). During the 2DG experiment, both eyes were uncovered and either eye was exposed to the "generalized *CS*", a 80 x 80 mm^2 black object moving at 10°/s against a white background. For statistical evaluations see Table 2 and for abbreviations see legend to Fig. 2. (From Finkenstädt & Ewert 1988c).

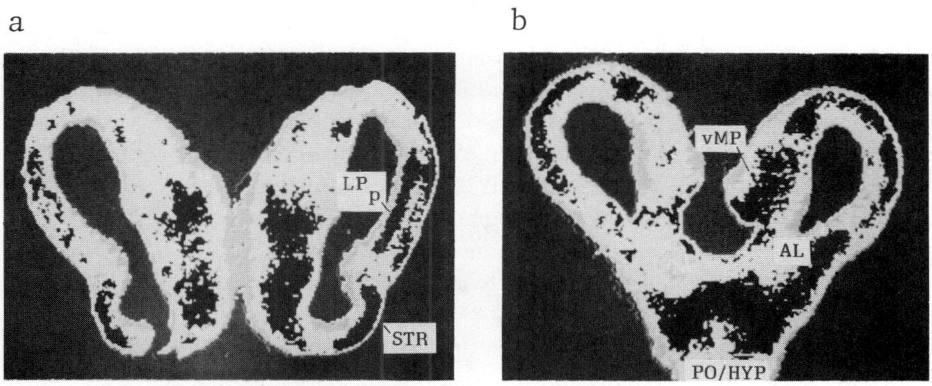

Figure 4. Interhemispheric comparison of 2DG-uptake in monocularly trained (hand-fed) toads. Representative computer-processed 2DG autoradiographs, binary coded and converted to grey levels (black indicating strongest 2DG-uptake); the right brain half here is contralateral to the "trained (left) eye." (a) Transverse section through the second third of the telencephalon; in the periventricular layer of the lateral pallium (LPp) the 2DG-uptake is considerably enhanced in the contralateral hemisphere. (b) Transverse section through the caudal one-third of the telencephalon; the posterior ventral medial pallium (vMP) and the lateral amygdala (AL) accumulate significantly more 2DG in the contralateral hemisphere than in the ipsilateral one. The PO/HYP area is bilaterally activated. (From Finkenstädt & Ewert 1988c).

4 Brain Lesions

The results from 2DG studies provide us with important cerebral "land marks" for subsequent experimental tests. In this context, the vMP, LP, AL, and Lpd nuclei are of particular interest. To further investigate the role of these structures for toad's associative learning and memory-related processes, the stimulus/response relationships for prey-catching of trained toads toward W-, S-, and A-stimuli of different sizes were determined before (Fig. 5a) and after lesions applied to different brain areas: vMP, aMP, dMP, AL, LP, STR, and Lpd/P.

Bilateral MP-lesions were applied by passing an anodal direct current of 50-60μA for 8-10 s through an insulated steel electrode (Fig. 6d), or by suction aspiration with a vacuum pump, or by mechanical impairment with a micro-scalpel. About 48 hours after surgery these animals were tested. The bilaterally vMP-lesioned toads identified and localized prey accurately, however, the training-induced modifications of configural pattern discrimination were partially abolished (Fig. 5 and Table 3). In addition to that, no obvious change in the general activity level like hyperactivity or increased spontaneous locomotion could be observed. The animals appeared to behave rather normal. Pilot studies revealed that in naive toads impairments of MP strongly delay or even prevent associative learning.

Following unilateral lesions in the lateral pallium, LP (Fig. 6a), prey-catching failed to occur toward prey objects presented in the ipsilateral visual field. After bilateral LP-lesions, visually guided prey-catching was abolished, while visually guided predator avoidance remained intact.

Unilateral lesions applied to the striatum, STR (Fig. 6b), impaired prey-catching toward prey objects moving in the contralateral visual field, both in conditioned and naive toads. After bilateral lesions, prey-catching failed to occur in the visual field of both eyes. Comparable results are reported by Patton & Grobstein (1986). Similar prey-neglect was obtained after total ablation of one or both telencephalic hemispheres, respectively (Ewert 1967).

After lesions in the lateral amygdala, AL (Fig. 6c), spontaneous activity in these animals was considerably increased. These toads appeared to be continuously searching for something, displaying frequent locomotion. The motor hyperactivity rendered testing the lesion-induced learning-related deficits more difficult. However, it appeared that the *CS* in AL-lesioned toads is less frequently responded than pre-lesion.

Unilateral or bilateral lesions in the pretectal Lpd/P region of naive toads (see Ewert, this volume) led to changes of configural prey-selection similar to those obtained after visual conditioning of prey-catching to predator objects. These animals largely disregarded differences in configural and size parameters. Bilateral pretectal lesions in conditioned toads did not change significantly the response to squares of 80 x 80 mm^2 or smaller stimuli.

Following sham-operations as well as control lesions in the anterior medial pallium (aMP), the dorsal medial pallium (dMP), or the dorsal pallium (DP) comparable behavioral effects were not observed.

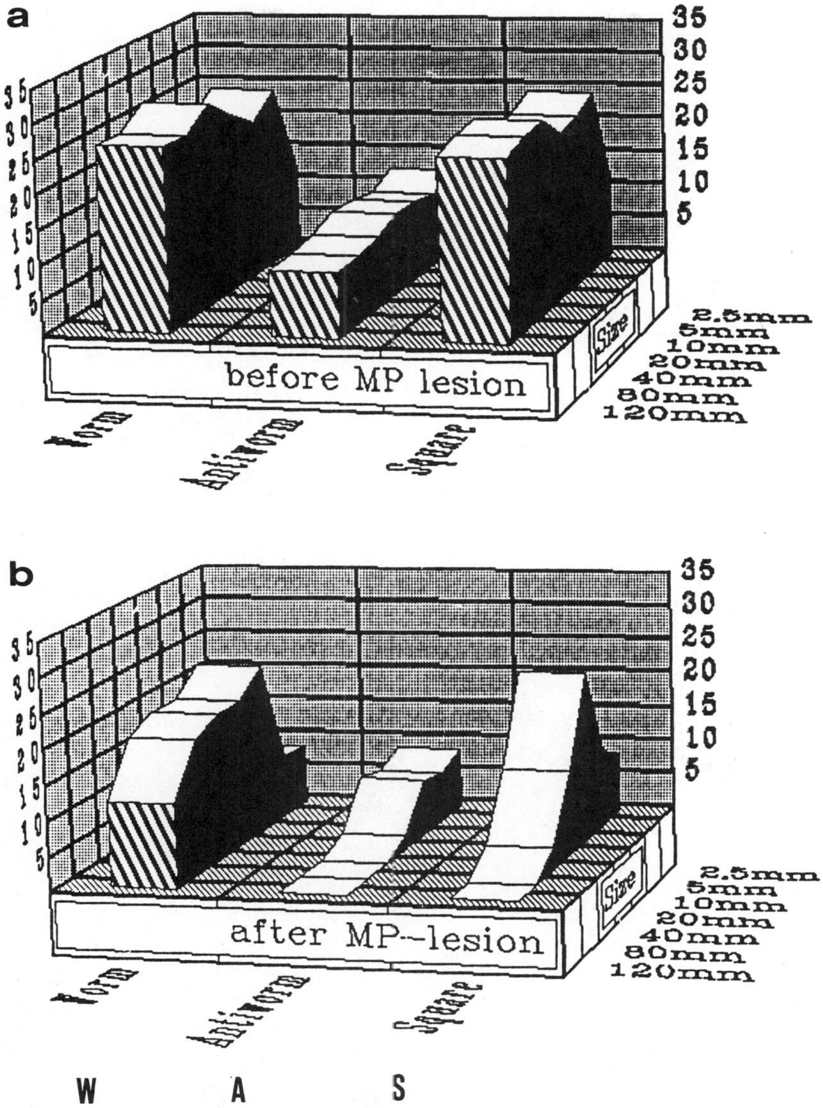

Figure 5. Effect of lesions in the medial pallium (MP) on visual configural pattern discrimination in conditioned (hand-fed) toads. Average orienting activity (orienting responses per 36 s) of four conditioned toads in response to black worm-like (W), antiworm-like (A), and square (S) stimuli of various edge lengths moving at v = 10°/s, before (a) and after (b) bilateral MP-lesions. Note that this lesion abolishes the effects of conditioning (cf. Figs. 1a and b). See also Table 3. An example of a MP-lesion is shown in Fig. 6d. (After Finkenstädt & Ewert 1988a).

Table 3. Effects of bilateral lesions in the medial pallium (MP) on configural pattern discrimination in trained (hand-fed) toads *Bufo bufo spinosus.* Average prey-catching orienting activity R $\pm$ SE (responses per 36 s) of four binocularly trained toads. Responses to black worm, antiworm, and square stimuli of various sizes (moving at v = 10°/s against a white background) prior to and after bilateral MP-lesions.

Stimulus	Edge length [mm]	Prey-Catching Orienting Activity	
		Pre-Lesion	Post-Lesion
Square (S)	2.5	12.5 ± 7.4	8.5 + 4.7
	5	26.0 ± 11.8	11.3 + 2.9
	10	24.3 ± 4.7	24.5 + 6.5
	20	24.0 ± 8.3	12.3 + 4.2
	40	27.3 ± 9.1	3.0 + 1.9
	80	27.0 ± 7.6	0.0 + 0.0
	120	25.8 ± 7.5	0.0 + 0.0
Worm (W)	5	25.8 ± 9.8	9.5 ± 3.2
	10	22.5 ± 6.5	24.0 ± 7.8
	20	25.3 ± 3.3	21.3 ± 3.3
	40	28.8 ± 8.1	21.0 ± 5.3
	80	26.3 ± 5.3	19.7 ± 2.1
	120	26.3 ± 3.9	12.3 ± 2.5
Antiworm (A)	5	10.5 ± 3.8	7.3 ± 3.6
	10	11.3 ± 4.7	8.3 ± 4.6
	20	11.5 ± 3.8	2.3 ± 1.3
	40	10.0 ± 3.2	0.0 ± 0.0
	80	9.5 ± 2.5	0.0 ± 0.0
	120	9.3 ± 3.2	0.0 ± 0.0

Figure 6. Survey of various forebrain lesions. (a) Electrolytic lesion (arrow) in the lateral pallium (LP) of a conditioned toad; representative example leading to failure of visual prey-catching behavior; the left side depicts the approximate borders of the nuclei. Lesions in dMP/DP have neither an effect on conditioning (hand-feeding) nor on pattern recognition in prey-catching. (b) Electrolytic lesion (arrow) in the striatum (STR) of a conditioned toad; representative example leading to failure of prey-catching behavior. (c) Electrolytic lesion (arrow) in the lateral amygdala (AL) of the caudal telencephalon in a trained toad; representative example leading to a neglect of the "generalized *CS*." (d) Right telencephalic hemisphere of a conditioned toad showing an electrolytic lesion (see arrow) in the posterior ventral medial pallium (vMP); representative example abolishing the effects of conditioning on visual configural pattern discrimination (see Fig. 5 and Table 3). AC, nucleus accumbens; AL, amygdala pars lateralis; AM, amygdala pars medialis; DP, dorsal pallium; LP, lateral pallium; LS, lateral septum; MS, medial septum; PO, preoptic area; dMP, dorsal medial pallium; vMP, ventral medial pallium; STR, striatum.

5 Olfactory Learning in Toads

In a recent study, Claudia Merkel-Harff in collaboration with Finkenstädt and Ewert from the neuroethological lab in Kassel has examined the distribution of 2DG accumulation in the brain of toads, which were classical conditioned to associate the behaviorally irrelevant odor "Cineol" with prey. Only a few training sessions were required to acquire the visual/olfactory association. Cineol smells very strong. Naive toads do not seem to alter their behavior during Cineol stimulation. In conditioned

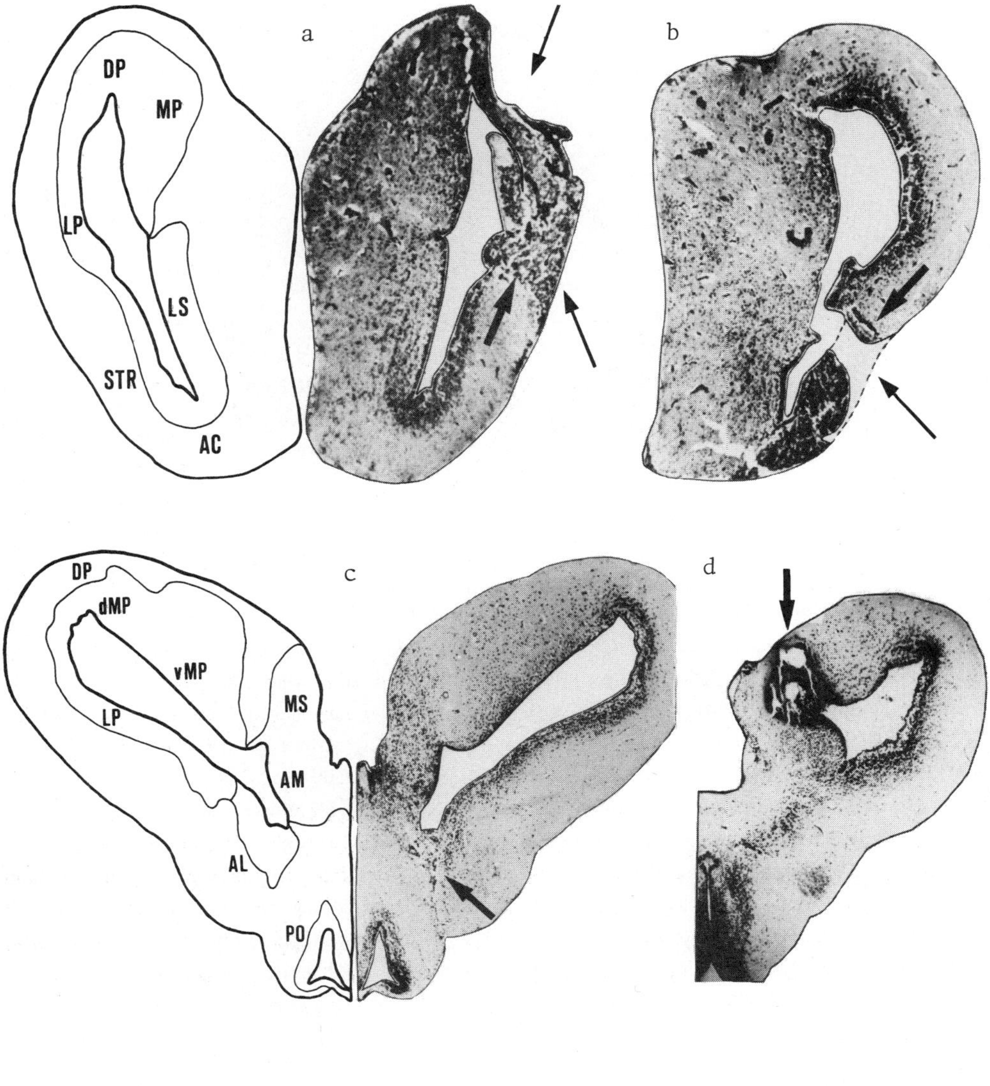

toads, Cineol stimulation alone led to increased motor activity. Prey-catching motivation was strongly enhanced. In the presence of Cineol, any moving visual object, even a conspecific, was responded with orienting and snapping. Success of conditioning was determined by presenting the conditioned stimulus Cineol together with a predator dummy (80 mm x 80 mm square black cardboard moved against white background). Only toads responding to this visual object with prey-catching in the presence of Cineol were further considered. Following successful conditioning, the animals received binocular enucleation and were compared to binocular enucleated "naive" toads. The latter received the same amount of olfactory stimulation and prey presentation before enucleation as the experimental group, however, both stimuli were given temporally and spatially unpaired. One week after recovery, 2DG was injected and Cineol was applied to animals of both groups in a closed arena. Odor was blown into the container for 60 s. Immediately thereafter, air was exchanged to

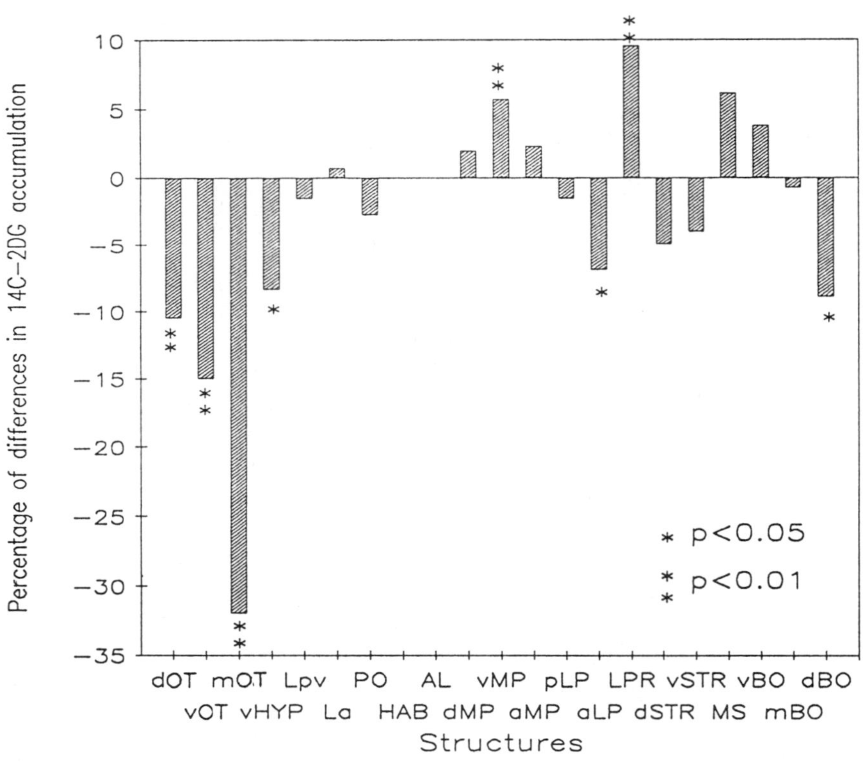

Figure 7. Average percentage of differences in 2DG-uptake across various brain structures of nine bilaterally olfactory (Cineol) conditioned toads in comparison to four "naive" controls ("brain-to-brain" comparison). Asterisks indicate statistical significance (*KSO*-test). Partly in addition to Fig. 2, the following structures were investigated: ABO, accessory olfactory bulb; vBO, ventral (glomerula layer); mBO, medial (mitral cell layer); dBO, dorsal (granule cell layer) portion of the bulbus olfactorius; HAB, nucleus habenulae; aLP, anterior lateral pallium; MS, medial septum; LPR, lateral prominence. For further abbreviations see Fig. 2. (Merkel-Harff, Finkenstädt & Ewert, in preparation).

prevent habituation to the Cineol odor. Following 2 min of stimulation-free intervals, the same procedure was repeated for a total of 90 min.

Since the animals of both groups were blinded, no prey-catching behavior occurred. Thus, secondary effects caused by different motor activity were prevented. The further processing and evaluation procedure was identical to the visual conditioning experiments.

In olfactory conditioned toads, compared to "odor naive" control animals, 2DG uptake was significantly reduced in the following structures (Fig. 7): anterior portion of the lateral pallium (aLP), dorsal portion of the bulbus olfactorius (dBO, granule cell layer), optic tectum (OT) (Fig. 8a) and the ventral hypothalamus (vHYP). On the other hand, significantly enhanced 2DG uptake was found in the ventral medial pallium (vMP) (Fig. 9b) and the lateral prominence (LPR) (Fig. 7). Other structures that were found to increase metabolism following presentation of odor in comparison to unstimulated controls, such as the lateral amygdala (AL), medial septum (MS), posterior portion of the lateral pallium (pLP), ventral (vBO, glomerula layer) and medial (mBO, mitral cell layer) bulbus olfactorius, habenular (HAB) and preoptic region (PO), do not significantly change glucose metabolism in conditioned toads in the "brain-to-brain comparison" paradigm (Fig. 7) (Merkel-Harff et al. 1988).

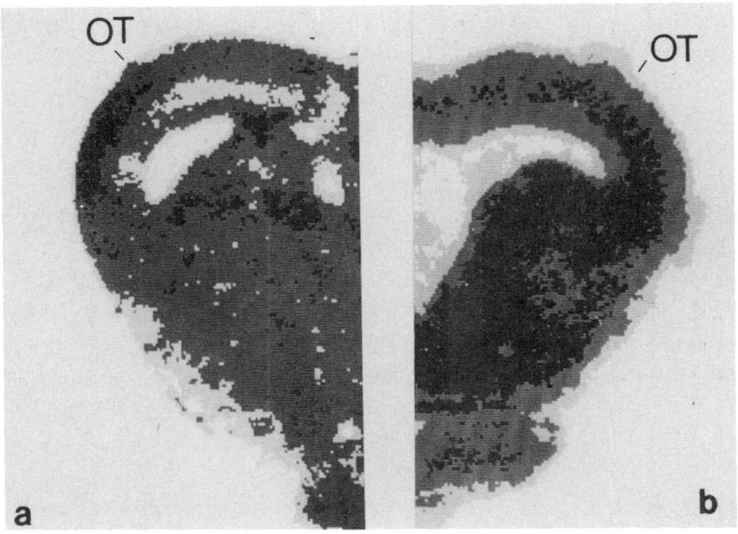

Figure 8. Representative computer-processed autoradiographic images of half sections at midbrain level. OT, optic tectum. (a) The olfactory conditioned toad was blinded and bilaterally stimulated with the odor Cineol. Note the weak 2DG uptake in the central tectal layers 6/7. (b) A blinded olfactory naive toad received Cineol stimulation. Note the heavy 2DG-uptake in the more superficial tectal layers 8/9. For further details see text.

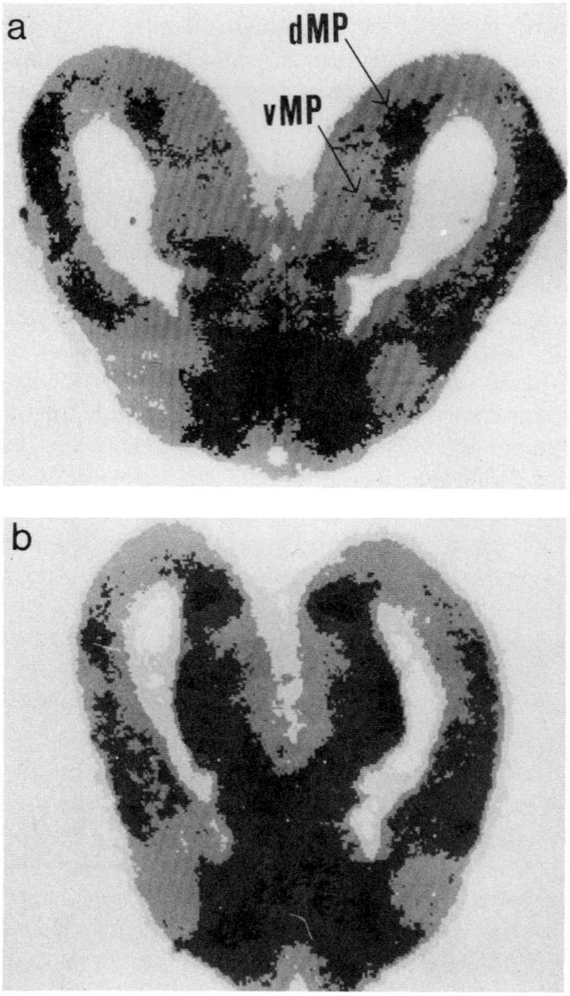

Figure 9. Representative computer-processed autoradiographic images at a posterior telencephalic level. (a) Cineol stimulated, blinded "naive" toad. (b) Cineol stimulated, blinded olfactory conditioned toad. Note the differences in 2DG-uptake in the ventral medial pallium (vMP) between (a) and (b). The dosal medial pallium (dMP) displayed strong 2DG-uptake in both experimental groups. (See Fig. 6c for histology).

6 Discussion

6.1 Associative Learning

The discussions in the subsequent Sections 6.1 to 6.5 primarily focus on the results obtained from visual/visual associative learning ("hand feeding" phenomenon). For visual/olfactory conditioning we refer to Section 6.6.

It has been shown that the behavioral response to *CS* is not only dependent on the physical properties of the stimulus, but also on its learned motivational significance (see Birt & Olds 1982; Weinberger & Diamond 1987).

A visual prey stimulus thus has two values, a key effect activating the stimulus-response (*S-R*) mediating system that analyzes the stimulus, deciding whether to release the behavior and to generate the motor pattern, and a modulating effect that influences the animal's behavioral "state," which has control over the mediational circuitry (Krasne et al. 1979). Depending on the physical properties of the stimulus, the *CS* appears to have a direct mediating action and, due to its learned motivational significance may also provoke a change in the status of the modulatory system.

These suggestions are supported by similar studies in rats in which auditory responses are dependent on the physical as well as on the motivational parameters of a stimulus (see Gonzalez-Lima, this volume). Thus, it may be concluded that during classical visual conditioning in toads both Hebb's (1949) suggested mechanisms of associative learning and the modulatory influences due to the motivational parameters of the *CS* interact with its perceived physical properties.

The training related behavioral changes in the conditioned toads become most obvious (Fig. 1; Table 1) in the prey-catching response toward previous predator dummies, in the shift of the optimal releasing size of square objects from about 10 mm to 80-120 mm edge lengths (cf. S in Fig. 1a and b), as well as in the enhanced response to edge extension of the antiworm stimulus (see A in Fig. 1a and b). Since the animals were trained to respond to the experimenter's hand, the latter two observations may be explained with response generalization along the parameter stimulus "area." Response generalization can be defined as a process, in which the organism performs the conditioned response towards a novel stimulus that is to a certain degree different from the original *CS* (see also Woody 1982): the bigger the difference, the weaker the response. This principle can be observed in the present behavioral data as well. However, at equal area size, worm-like objects were always preferred to antiworm configurations, and the releasing value of large worm patterns always exceeded that of square objects of comparable edge lengths. It can be concluded that conditioning does not substitute the innate prey recognition system, but rather modifies it. Since the *S-R* relationships of conditioned toads are to some extent similar to those obtained following pretectal lesions in naive toads, the possibility of an involvement of pretectal Lpd/P nuclei (connected with telencephalic structures) in associative learning cannot be overlooked.

In conditioned toads, escape behavior can still be elicited by large objects simulating airborne predators. An explanation for this observation may be that the superior visual field was spared during training, hence suggesting that visual conditioning is to some extent locus-specific. There might be a parallel to the results from monocular training, showing that there is no "interocular transfer" for the lateral visual field of the "naive eye."

6.2 General Comments on Learning-Correlated Activity in Brain Metabolism

Conceptionally, we must distinguish between the learned-response-circuit, that contains all the essential structures conveying the *CS* from the receptor to the *CR* effectors, involving the *S-R* system, and structures as sites of neuronal plasticity that reflects the memory trace itself. The latter may be considered as part of the former (see also Thompson et al. 1984). Whether the observed metabolic changes in the respective structures are primary consequences of learning and memory, or are due to the attentional, arousal, or motivational relevance of the stimulus remains uncertain. Therefore, further comparisons with data obtained from other techniques and paradigms are required.

A drawback of the present method lies in the different motor activities of trained toads (responding to the large generalized *CS* with prey-catching) and naive toads (showing no response or escape to the moving large object). 2DG-uptake must therefore also be compared with results from previous studies (Tables 4 and 5) in which glucose utilization was measured in unconditioned (naive) toads toward various configural (W, A, and S) stimuli that elicited different motor behaviors (Finkenstädt et al. 1985, 1986; see also Finkenstädt, this volume). It is obvious that not all possible "non-specific" influences, which may be affecting an experimental group, can be considered in control experiments. Interhemispherical comparison (in monocularly trained toads), additional to brain-to-brain analysis, was utilized in the present study to minimize interindividual variability, and to evaluate differences in motor activity. However, since no split-brain preparations were used, one has to consider the possibility of "bilateral transfer" effects in learning and storage.

Failure to detect metabolic changes after conditioning does not necessarily mean that the respective structure under investigation is not involved in learning-related processes. Several classes of neurons may form intermingled neuron populations, enhancing or reducing each others neuronal discharge. Thus, structures that do not reveal a net metabolic learning-related effect are not discussed in the present study. (See the critical evaluation of the 2DG method in the contribution by Gonzalez-Lima, this volume; cf. also the 3rd Session of the introductory discussion to this book).

6.3 Changes of Metabolic Activity in the Central Visual Pathway

Both by means of "interhemispherical" and "brain-to-brain" comparisons it was shown that the central and deep layers of the optic tectum (vOT), the snapping evoking area of the tectum (sOT), the nucleus isthmi (IST), the lateral anterior (La) nucleus, the lateral posteroventral (Lpv) thalamic nucleus, and the posterior (P) thalamic nucleus exhibited a training-correlated increase of glucose utilization in response to the *CS* (Figs. 2 and 3; see also Table 2).

The increased 2DG-uptake in sOT is obviously correlated with snapping activity. It has been shown that this tectal region represents a vital link between afferent messages and efferent motor commands (Ewert 1974; 1984; Ingle 1983; Weerasuriya, this volume). Changes in IST, La, and Lpv may be due to respective activity in the tectum with which these structures are directly connected. IST receives ipsilateral tectal projections from layer 6 (and deeper) and projects back to layers 8 and 9 of both tectal hemispheres (see also Gruberg, this volume). La is regarded as the main tectal relay point with the striatum (Lázár 1969; Wilczynski & Northcutt 1979). Similar to La, the P and Lpv nuclei project to the striatum and in turn receive

Table 4. Changes of 2DG-uptake in tectal (vOT and sOT) and in pretectal thalamic structures (Lpd, Lpv, and P) during release of prey-catching with a 4 x 28 mm^2 worm-like (W) stripe, predator avoidance with a 80 x 80 mm^2 square (S), and no response toward a 4 x 28 mm^2 antiworm (A) stripe in naive (*nav*) toads. For comparison: release of prey-catching with the S-stimulus in conditioned (hand-fed) toads (*cond*), and no response to W in toads habituated (*hab*) to the W-stimulus. The black stimuli were moving at 10°/s against a white background. Increase, decrease, and no change in 2DG-uptake, compared to reference structures and controls, are indicated by +, -, and o, respectively; in conditioned (*cond*) and habituated (*hab*) animals, 2DG-uptake was compared with naive controls. Qualitative description derived from quantitative data reported by Finkenstädt et al. 1985; Finkenstädt & Ewert 1988c. For abbreviations see Fig. 2 and Section 8.

Experimental Animal	Stimulus	Behavior	2DG-Uptake in Lpd	Lpv	P	vOT/sOT
Nav	W	Prey-Catching	(-)	(+)	++	++
Nav	A	No Response	+	(+)	o	+
Nav	S	Avoidance	++	++	-	+
Hab	W	No Response	o	o	o	-
Cond	S	Prey-Catching	-	++	++	++

tectal input (see also Lázár, this volume). Neary & Northcutt (1983) suggest that the main function of the medial thalamic region consists of relaying visual, auditory, and somatosensory information to the striatal complex.

The metabolic activity patterns in the three pretectal nuclei is not uniform, neither in naive nor in conditioned toads (Table 4). Furthermore, contrary to non-associative learning (Finkenstädt & Ewert 1988b; Finkenstädt, this volume), the conditioned toads showed learning-related differences of metabolic activity in the pretectal thalamic Lpd, Lpv, and P nuclei. This is of particular interest, since Lpd and lateral P are suggested to take part in innate visual pattern recognition in which pretecto-tectal inhibitory connections play an important role (see Ewert, this volume). Previous experiments in naive toads have shown that 2DG-uptake by Lpd in response to W, A, and S

$$[^{14}C]_S > [^{14}C]_A > [^{14}C]_W$$

is inversely correlated with 2DG-uptake in the optic tectum that in turn is directly correlated with the probability of prey-catching (Finkenstädt et al. 1985). Among all brain structures investigated, only Lpd showed training-correlated reduced glucose utilization in response to the generalized *CS*, the moving large object, which suggests training-dependent suppression. It is interesting to note that Lpd-lesioned naive toads display similar changes in configural pattern discrimination like conditioned intact animals. Lpd activity is suppressed in response to the *CS* following conditioning, hence suggesting an impairment of configural prey selectivity of tectal T5.2 neurons. In other words, decrease of Lpd (metabolic) activity in response to a moving visual stimulus appears to require that this stimulus is interpreted as prey. The Figure 25A shown in Ewert's contribution (this volume) suggests a possible neural framework that may underlie this sort of "associative disinhibition." For a corresponding computer simulation see Lara (this volume).

A role of pretectal nuclei in memory related processes is supported by studies recording TH9 "memory neurons" that are characterized by long-lasting after-discharges (Ewert 1971). The concept of pretectal involvement in visual conditioning is also supported by findings in pigeons (Cohen 1984), cats (Sprague et al. 1970), and rats (Gonzalez-Lima et al. 1986; Gonzalez-Lima, this volume).

Lpv shows strong glucose utilization in response to large moving objects, independently of the type of elicited behavior (prey-catching in conditioned toads or escape in naive toads) (Table 4). This suggests that the net metabolic activity in Lpv is correlated with stimulus size, while the one in P corresponds with prey-catching, particularly for snapping. 2DG-uptake in the P-nucleus of naive toads is significantly enhanced during snapping and less pronounced if only orienting toward prey is elicited (Finkenstädt et al. 1985). In binocularly trained toads frequently orienting and snapping toward *CS*, the P-nucleus considerably enhanced its metabolic activity compared to monocularly trained animals which were snapping rarely while orienting activity was about the same as in binocular subjects. Indication of a possible function of certain sets of neurons of the P-nucleus for snapping, suggested from 2DG studies, must be investigated in future experiments by applying recording, stimulation, and lesion techniques.

The fact that the three pretectal nuclei Lpd, Lpv, and P display different patterns of metabolic activity in response to various configural moving stimuli in naive, habituated, and conditioned toads (Table 4) accords well with suggestions on pretectal functions (see contributions by Ewert, Finkenstädt, and Manteuffel, this volume). Whereas neurons in Lpd and lateral P monitor moving (textured) large surfaces and determine configural stimulus selectivity by inhibitory connections to tectal neurons (Ewert 1987), certain sets of neurons of the P nucleus seem to play a role for the detection of stationary objects (Ingle 1983) and others for prey-catching, in particular snapping (Finkenstädt et al. 1985). The observation that following lesions to Lpd and lateral P some toads show disinhibited orienting but not snapping, could be explained by accidental lesions in central P, a possibility which should be carefully evaluated in future experiments.

The anterior thalamic NB/Vld region considerably increased 2DG-uptake in trained toads during prey-catching toward the large generalized *CS*. In naive animals, this region accumulated 2DG only during prey-catching in response to prey, but not during escape toward a predator object. Thus, this region may not be restricted to sensory processing of the perceived signal alone, it might be involved in processing the learned motivational significance of a stimulus and its attentional or arousal properties as well. The NB/Vld region relays retinal inputs to the P-nucleus and is reciprocally connected with Lpd which - like P - is reciprocally connected with OT (Neary & Wilczynski 1980; Neary & Northcutt 1983). Intrinsic Lpd-P connections have been shown by Golgi studies (Lázár 1979).

6.4 Changes of Metabolic Activity in Non-Visual Forebrain Areas

An important result of this study is the identification of non-visual forebrain structures that are receiving no primary visual information but show remarkable training-correlated increases in glucose utilization: The telencephalic posterior ventral medial pallium (vMP), lateral pallium (LP), lateral amygdala (AL), and the diencephalic preoptic/hypothalamic region (PO/dHYP). In amphibians, this is the first indication of involvement of so-called "limbic" forebrain structures in associative learning (see Figs. 2, 3, and Table 2).

6.4.1 Medial Pallium

The observed enhanced metabolic activity in vMP in response to the *CS* appears to be related to associative memory functions, a suggestion which is in line with similar results obtained in rats (Gonzalez-Lima et al. 1986; Gonzalez-Lima, this volume) (see also Thompson et al 1982; Laroche et al. 1987).

The lesion studies clearly indicate that MP in toads is essential for learning/memory related processes: Toads with bilateral ablation of the ventral medial pallium prior to or post conditioning proved to be unable to show the proper conditioned response, *CR*. Sham operations did not impair conditioning. In naive animals, MP lesions did not have an influence on configural prey-selection, in contrast to pretectal or anterior thalamic lesions (summarized in Ewert 1984). The behavioral results post-lesion prove, that the vMP-lesion effect is not necessarily a deficit in visual information processing; rather it can be interpreted as a memory dysfunction. The observations indicate that vMP in toads is not necessary for primary processing of visual information, which is also confirmed by lesion studies of Patton & Grobstein (1986) who did not find an effect on prey-catching in frogs following medial pallial lesions. However, intact MP appears to be indispensable for learning and memory processes, with regard to the paradigm used in the present study.

In the toad's vMP, metabolic enhancement after conditioning was mainly restricted to the region with low cell density, sparing both, the cell dense periventricular region as well as a dorsal portion of the medial pallium (dMP) (compare left and right in Fig. 4b; cf. also a and b in Fig. 9). The latter clearly does not overlap with the dorsal pallium (DP), as delineated by Kicliter & Ebbesson (1976) and Northcutt & Kicliter (1980). In the present as well as in a previous study (Finkenstädt et al. 1986) dMP is strongly activated apparently independently of configural, memory (cf. a and b in Fig. 9), or habituation related visual influences.

Similarly, Gonzalez-Lima & Scheich (1986) found learning-related metabolic changes only in a neuropil band of the hippocampus, reflecting enhanced activity of input terminals and local potentials in dendrites of pyramidal and granular neurons, i.e. in the vicinity of modifiable synapses. Gonzalez-Lima (this volume), too,

Table 5. Changes of 2DG-uptake in tectal (vOT) and telencephalic structures (vMP, dMP, LPp, aLP, AL, STR) during release of prey-catching with a worm-like (W) 4 x 28 mm^2 stripe, predator avoidance with a 80 x 80 mm^2 square (S), and no response toward a 4 x 28 mm^2 antiworm (A) stripe in naive toads (*nav*). For comparison: release of prey-catching with the S-stimulus in conditioned (hand-fed) toads (*cond*), and no response to W in toads habituated (*hab*) to the W-stimulus. The black stimuli were moved at 10°/s against a white background. Increase, decrease, and no change in 2DG-uptake, compared to reference structures and controls, are indicated by +, -, and o, respectively; in conditioned (*cond*) and habituated (*hab*) animals, 2DG-uptake was compared with naive controls. Qualitative description derived from quantitative data obtained by Finkenstädt et al. 1986; Finkenstädt & Ewert 1988b,c. For abbreviations see Fig. 2 and Section 8.

Experimental Animal	Stimulus	Behavior	2DG-Uptake in: vMP	dMP	LPp	aLP	AL	STR	vOT
Nav	W	Prey-Catching	-	++	-	+	o	+	++
Nav	A	No Response	o	++	++	+	o	o	(+)
Nav	S	Avoidance	-	++	+	+	o	+	+
Hab	W	No Response	++	++	+	+	o	-	-
Cond	S	Prey-Catching	++	++	++	+	++	+	++

concludes that the 2DG method is suitable to detect learning-related changes, such as increases in the excitability of axodendritic synapses in the molecular layer.

Although the term medial pallium is used here, this region has also been labeled "archipallium" and "primordium hippocampi" (Herrick 1948; summarized in Kicliter & Ebbesson 1976). The anuran medial pallium is thought to be homologous to the mammalian hippocampus based on its location in the dorsomedial wall of the hemisphere, in a position similar to that of the mammalian hippocampal region, as well on its histochemical similarities to the mammalian hippocampal-parahippocampal fields, and its intrinsic telencephalic connections (Northcutt 1974).

Hoffman (1963) divided the medial pallium into dorsal and dorsomedial divisions. These divisions could be confirmed in Golgi preparations (Kicliter & Ebbesson 1976). The dorsomedial division, encompassing the portion with migrated, cell poor regions, contains as the principal cell type pyramidal neurons. As stated by Hoffmann (1963), the dendrites of these pyramidal cells arborize medially and laterally and are especially plentiful on the medial side (Kicliter & Ebbesson 1976). Learning-related changes in the hippocampal formation have been explained by long-term potentiation of neuronal activity, preferentially in "Hebb"-type (modifiable) synapses, as found in pyramidal and granular cells of the hippocampal molecular layer of mammals (Voronin 1983; Levy & Steward 1983).

Based on extensive neuroanatomical studies on afferents and efferents of the medial pallium, Northcutt & Kicliter (1980) conclude that the medial pallium "is not a simple olfactory precursor of the mammalian hippocampus; instead, in amphibians it receives direct thalamic projections carrying visual and somatic information after relay in the dorsal thalamus. A picture seems to be emerging in which the amphibian medial pallium is a major multimodal target of ascending information, as well as being the sole output of the pallium, with the exception of the pars lateralis of the amygdala" (p. 245). The multimodal connectivity of MP in amphibians is a prerequisite for its proposed relation to associative memory functions, which is presumably independent of the sensory category of the *CS*. The 2DG studies in our lab (Merkel-Harff, Finkenstädt & Ewert, in preparation; see also Section 5) indeed prove that vMP plays a prominent role in olfactory as well as in visual associative learning.

It has been previously reported that damage to the hippocampus and its associated pathways in man (Milner 1972) results in anterograde amnesia evident as a major deficit in learning to recognize novel stimuli. Other reports in monkeys suggest alternatively that the combined impairment of both hippocampus and amygdala is crucial for anterograde amnesia (e.g., Mishkin 1982). However, lesions in monkey's in the hippocampo-fornical system produce learning deficits on tasks which require memory not only for *which* object was shown but *where* it was shown, thus information about position in space (Gaffan & Saunders 1985). The observation that hippocampal damage may impair learning, but not necessarily retention of the memory tasks, leads Rolls (1987) to conclude that the hippocampus may be involved in the storage process, but may not itself be the location where the information is stored. This accords well with the conclusions about the learning-related role of the medial pallium of toads (Finkenstädt & Ewert 1988a).

Previous studies in toads indicate that vMP plays a role in habituation processes, associated with detecting novelty and influencing attentive behavioral states (Finkenstädt & Ewert 1988b; Finkenstädt, this volume; see also Section 6.6 and Table 5), conclusions that are well in line with results on the hippocampus in mammals. Rats, hippocampectomized in adulthood, cannot inhibit movement (Teitelbaum & Milner 1963), which may be explained if the motor system is released from inhibitory influence (Foreman & Stevens 1987). In accordance with the latter is the observation that electrical stimulation of the hippocampus causes behavioral arrest (McLean 1957). Several authors have previously commented that the increased activity in hippocampus, when it occurs, appears to be a consequence of delayed or absent inter- and intrasession habituation (Foreman & Stevens 1987).

To bridge the gap between the apparent two major tasks of the hippocampal formation, learning/memory and inhibitory premotor functions related to shifts in attention, both in mammals and amphibians, the view of Lindsley & Wilson (1975) may be shared that "orienting, scanning and other evidences of the shifting of attention during the search for visual cues are necessary to obtain information which leads to learning and storage of information, whether in the hippocampus or elsewhere" (p. 274).

6.4.2 Nucleus Amygdalae

Learning-related increase of 2DG-uptake was present in the lateral amygdala (AL), restricted to its lateral portion also encompassing the lateral forebrain bundle and projection fields of the accessory olfactory tract (anatomically described by Northcutt & Royce 1975). Lesions to the AL region impaired learning/memory, independently of whether the AL nucleus (sparing lateral forebrain bundles) or the lateral forebrain bundles alone were destroyed. It appears that these lesions hamper learning-related performance and produce increased spontaneous activity.

The functional significance of AL in amphibians is largely unknown. It is also uncertain whether the pars lateralis is continuous with the dorsal striatum or with the lateral pallium (Kicliter & Ebbesson 1976). In concert with previous 2DG studies (Finkenstädt et al. 1986), the present results support the idea (Northcutt 1974) that amygdala pars lateralis is a distinct division. AL receives afferents from the accessory olfactory bulb (Northcutt & Kicliter 1980) and projections from medullary neurons via the lateral forebrain bundles (Kicliter 1979), also containing trigeminal, and statoacoustical afferents (Fuller & Ebbesson 1973) and ascending visual projections (Gruberg & Ambros 1974). Moreover, preoptic, hypothalamic, and ventrothalamic afferents reach AL. The efferents of AL terminate in hypothalamus (Kokoros 1973; Scalia 1976), ventral thalamus, posterocentral (P), and lateral posterodorsal (Lpd) thalamic nucleus (Northcutt & Kicliter 1980). Thus, AL may exert influences on the optic tectum via hypothalamus, Lpd- and/or P-nucleus.

McGaugh et al. (1984) suggested that the nucleus amygdalae of mammals is part of a memory modulating system. In a 2DG study on acoustic classical conditioning in rats, a caudal portion of the neostriatum and the underlying AL/STR transition zone showed increase of metabolic activity during *CS-US* pairing (Gonzalez-Lima, this volume). This transition zone has been regarded as either part (precursor) of the neostriatum or of the central amygdala. An adjacent region, the basolateral amygdala, displayed reductions during arousing electrical RET stimulation which was antagonized during conditioning (Gonzalez-Lima, this volume). In primates, the nucleus amygdalae appears to be responsible for cross-modality association of complex stimulus patterns and for stimulus-reinforcement associations and motivation (Spiegler & Mishkin 1981; Rolls 1987).

6.4.3 Lateral Pallium

2DG-uptake was considerably increased in a periventricular portion of the lateral pallium (LPp) in conditioned toads during prey-catching in response to the generalized *CS* (Table 2). Its position is at the rostral end of the posterior LP, as functionally determined by a previous 2DG study in naive, visually stimulated toads (Finkenstädt et al. 1986). The anterior one third of the LP (aLP) does not indicate training dependent differences in 2DG accumulation. This can be regarded as an indication that LPp is involved in processes related to associative learning, to which extent, however, remains to be elucidated. The results of LP-lesions only show that the lateral pallium influences prey-catching behavior. There are observations that the posterior LP may be involved in the integration of visual and olfactory modalities (see also Section 5). Recent 2DG studies demonstrate that the posterior two third of the

LP does not only process olfactory information, but is visual sensitive as well (Finkenstädt et al. 1986).

Based on histochemical studies in frogs, Northcutt (1974) suggested that the dorsal LP corresponds to a part of the mammalian isocortex (primary visual cortex) while the ventral portion is homologous to mammal's piriform cortex (olfactory cortex) (see also Kicliter & Ebbesson 1976). LP is the major target for secondary olfactory fibers. The 2DG studies in odor stimulated naive toads showed increased uptake in the LP as compared to non-stimulated animals (Merkel-Harff et al. 1988; see also Sections 5 and 6.6). LP projects to the rostral third of the striatum (Paul Desan, pers. communication) and to the MP. There is no LP projection known to descend further caudally. Thus, modulation of non-telencephalic structures is only possible via septum, medial pallium, and striatum.

6.4.4 Striatum

Metabolic differences in the ventral and dorsal striatum between trained and untrained controls could not be detected. This may be explained by the assumed function of the striatum in amphibians which appears to be homologous to a portion of the amniote basal ganglia. This structure is putatively involved in initiating or controlling orientation and attentional motor programs. Wilczynski & Northcutt (1983b) speculated that part of this major output of the amphibian forebrain may be involved in the organization of multimodal sensory information into meaningful attentional behavior (Ewert, this volume) and the modulation of the appropriate orientation to this information. In naive toads, glucose uptake was increased at about equal amounts during orienting towards prey and during predator avoidance behavior. While the naive untrained controls display predator avoidance, the trained toads respond with prey-catching towards the same stimulus.

6.4.5 Hypothalamus

Similar to vMP, the PO/dHYP-region increases 2DG-uptake during associative and non-associative learning. Both structures are also involved in arousal and habituation (Finkenstädt & Ewert 1988b; Finkenstädt, this volume). The PO/dHYP-region appears to play an essential role in conveying descending medial pallial activity to the optic tectum and to the tegmental reticular formation (RET). Since glucose utilization in PO/dHYP - in contrast to vMP and vOT - was symmetric across both hemispheres in monocularly trained toads (see Fig. 4b), its function cannot be restricted to relay information between vMP and vOT.

Nauta (1963) suggested that mammal's hypothalamus links "limbic forebrain" areas, in which he included the hippocampus, amygdala, and septum with "limbic" midbrain structures. According to the results of recording studies by Rolls (1987), the autonomic/endocrine hypothalamic system takes part in reward-related coding and thus plays a role for reinforcement. It is known that learning and memory can be influenced by a variety of hormones. Recent findings suggest that hormones released during learning, exert modulatory function in the endogeneous regulation of storage processes. Hormones do not appear to be essential for learning, rather they appear to have a modulatory role (summarized in McGaugh et al. 1984).

6.5 Changes of Metabolic Activity in the Cerebellum

The cerebellum is mainly involved in the control of body posture and locomotion, which requires integration of visual, proprioceptive and vestibular inputs. Visual afferents are conveyed via the optic tectum (OT) and the basal optic nucleus (BON). Visual input has been found bilaterally in the dorsal cerebellum (dCB), whereas vestibular signals enter mainly the ipsilateral ventral vCB (Llinás 1976). Multimodal cerebellar neurons integrating visual and vestibular signals have been recorded too (Llinás & Precht 1969; Dieringer 1974). Glucose utilization in dCB, however, cannot simply be interpreted as a result of "learning-related" enhanced activity in vOT, if the results from "interhemispheric" and "brain-to-brain" investigations are compared (Table 2; Figs. 2 and 3). The fact that in habituated, non-responsive toads the metabolic activity in vCB was significantly less than in responding animals, while glucose utilization in dCB was independent of prey-catching activity, suggests for the former structure a correlation with motor activity (Finkenstädt, this volume).

In mammals, the cerebellum is reported to take part in operant conditioning, i.e. learning and memory of motor behavior (Thompson et al. 1982). It could also be shown that the ipsilateral cerebellum (superior cerebellar peduncle) of the rabbit is essential for acquisition, retention, and relearning of the classical conditioned nictitating membrane reflex (McCormick et al. 1982).

6.6 Olfactory Learning

Our preliminary results of visual/visual and visual/olfactory conditioning cannot be easily compared, since in the latter differences in elicited motor behavior between the experimental and the control groups were avoided. In the former these differences were accepted as part of the memory retrieval process.

The main result of the olfactory conditioning experiments is the strongly enhanced 2DG uptake in the vMP (Fig. 9b). This result confirms the hypothesis, derived from the 2DG experiments on visually conditioned toads, that vMP is generally involved in associative learning, independent of the sensory modality involved.

According to Northcutt & Royce (1975) and Scalia (1976), the medial pallium receives olfactory information from the olfactory bulb via the medial olfactory tract. However, in naive toads olfactory stimulation does not increase 2DG uptake in vMP; compared to unstimulated animals, metabolism is even decreased (Merkel-Harff et al. 1988). This observation is in accordance with results from visual stimulation (Finkenstädt et al. 1985) showing that 2DG uptake is reduced following presentation of novel sign stimuli. These signals strongly decrease spontaneous neuronal activity in the vMP, as revealed by extracellular single cell recordings, while previously experienced signals, e.g. following habituation, strongly enhance discharge frequency in certain neuronal types (Finkenstädt, this volume). Similar results were found in the vMP of olfactory stimulated naive toads (Fig. 8a). Mealworm dust stimulation strongly depressed neuronal discharge below spontaneous activity level. In the recovery pause in odor-free air, neuronal activity regained quickly (Fig. 10a-f).

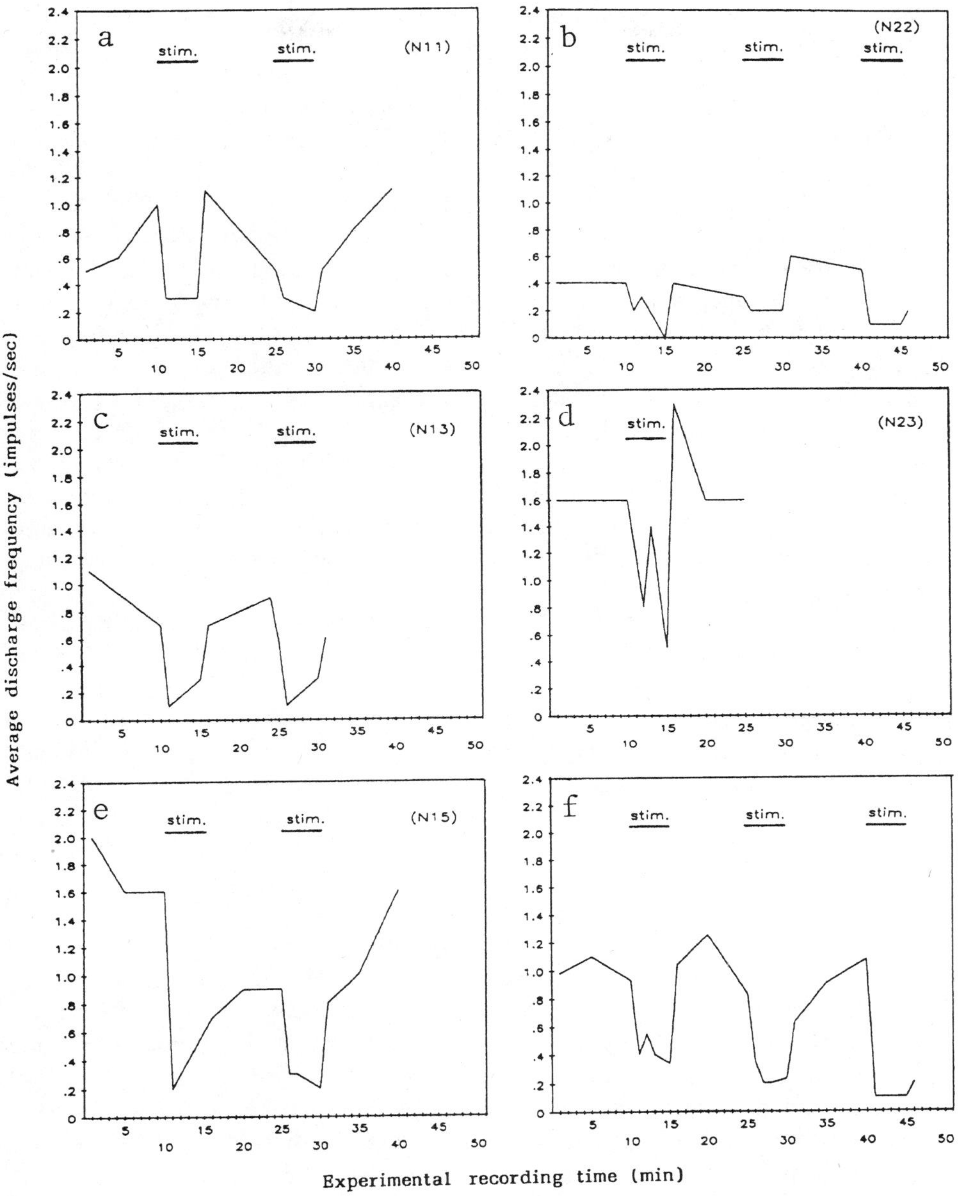

Figure 10. Discharge rate in spikes/s from five (a-e) single neurons, long-time extracellularly recorded in the ventral medial pallium (vMP) of toads *Bufo bufo* over a period of maximal 45min. The olfactory stimulus, given in intervals as indicated by the bars (stim.), was mealworm dust. The naive animals were not conditioned to this odor. The last plot (f) depicts the average discharge rate of the five (a-e) single neurons. (Merkel-Harff, Finkenstädt & Ewert, in preparation).

The preliminary results from single cell recordings (Fig. 10) are in accordance with investigations in the hippocampus of mammals (Vinogradova 1970, 1975; Douglas & Pribram 1966; Foreman & Stevens 1987). One of the possible functions of the medial pallium as a "primordium hippocampi" is detection of novelty of signals. Previously experienced signals, which includes habituation and associative learning, enhance metabolism (see also Section 6.4.1), while novel signals depress 2DG uptake in the vMP (see also Finkenstädt, this volume). The latter result is consistent with the single cell recordings shown in Fig. 10.

In the lateral portion of the amygdala (AL), olfactory stimulation strongly increases 2DG uptake in naive toads, as compared to unstimulated controls. This might be due to the fact that the accessory olfactory tract relays olfactory information from the accessory olfactory bulb (ABO) to the ipsilateral AL. However, this structure does not alter metabolism during olfactory conditioning. AL plays a prominent role during visual associative conditioning, which might point towards sensory specificity. Another explanation can be derived from AL lesions. In contrast to the visual training in the olfactory conditioning experiment no difference in motor behavior was observed between the experimental and the "naive" group. Lesions to the AL of visually conditioned toads produced hyperactivity and deficits in acquired performance. Thus, AL might be unspecifically involved in eliciting prey-catching behavior as well.

The effect on metabolism in the optic tectum in olfactory and visually trained toads appears to be different. We might conclude that the observed metabolic effect in OT is rather a secondary than a primary effect of conditioning. However, as already emphasized, the limited comparability between the two studies due to differences in the evoked motor response makes interpretation presently difficult, especially for the optic tectum.

7 Conclusions

The innate visual pattern recognition system in toads can be modified by training. In most of the visual pathways where temporal and spatial convergence between *CS* and *US* occurred, specific functional plasticity can be assumed, as indicated by changes in 2DG uptake. Thus, structures of the visual pathway cannot simply be regarded as mere input channels for the *CS*.

Information processing underlying associative learning (classical conditioning) in toads involves also non-visual forebrain structures outside the *S-R* system. Although it is difficult to decide for any investigated brain structure, whether it participates "primarily" or "secondarily" or "not at all" in conditioning, various lines of evidence suggest a prominent role at least for vMP, AL, LPp, PO/dHYP, Lpd/P, and OT in the spatiotemporal pairing of *US* and *CS* to acquire the conditioned response *CR*: Prey-catching toward a previously aversive or indifferent object. The resulting associative memory related information after telencephalic (vMP, APL, LPp) processing, relayed via diencephalic structures (e.g., P, PO/dHYP), influences prey-selective tecto-bulbar/spinal neurons (e.g., T5.2 cells) directly and/or alters the pretectal-tectal (Lpd/P, OT) configural pattern recognition system by inhibiting neurons (e.g., TH3 cells) of the pretectal Lpd nucleus. For a computer simulation see Lara (this volume).

8 Abbreviations of Brain Structures

A, anterior dorsal thalamic nucleus
ABO, accessory olfactory bulb
AC, nucleus accumbens
AL, nucleus amygdalae, lateral portion
AM, nucleus amygdalae, medial portion
dBO, dorsal (granule cell layer) portion of the olfactory bulb
mBO, medial (mitral cell layer) of olfactory bulb
vBO, ventral (glomerula layer) of olfactory bulb
dCB, dorsal half of the cerebellum
vCB, ventral half of the cerebellum
DP, dorsal pallium
HAB, nucleus habenulae
dHYP, dorsal hypothalamic area
vHYP, ventral hypothalamic area
IST, nucleus isthmi
La, lateral anterior thalamic nucleus
LPp, anterior half of the lateral pallium, periventricular
aLP, anterior lateral pallium
pLP, posterior half of the lateral pallium
LPR, lateral prominence
LS, lateral septum
Lpd, lateral posterodorsal thalamus
Lpv, lateral posteroventral thalamus
aMP, anterior third of the medial pallium
dMP, dorsal portion of the posterior medial pallium
vMP, ventral two-third of the posterior medial pallium
MS, medial septum
NB/Vld, "neuropil" (nucleus) of Bellonci/ventrolateral dorsal thalamus
dOT, dorsal layers of the optic tectum
mOT, medial layers of the optic tectum
sOT, snapping evoking area of the optic tectum
vOT, medio-ventral layers of the optic tectum
P, posterior thalamic nucleus
PO, preoptic region
RET, tegmental portion of the reticular formation
d,vSTR, striatum, dorsal (d) or ventral (v) portion
vTEG, ventral tegmentum

Acknowledgements. I thank Bert Klöppel, Torsten Siebert and Jochen Ruhland for computer assistence, Mrs. Gerda Kruk for histological preparation, Mrs. Gisela Burggraf for animal care, and Mrs. Ursula Reichert for text processing. This work was supported by the Deutsche Forschungsgemeinschaft (Fi 315/3-3).

9 References

Baerends GP (1987) Ethology and physiology: a happy marriage. A commentary. *Behav Brain Sci* 10: 369-370

Birt D, Olds ME (1982) Auditory response enhancement during differential conditioning in behaving rats. In: Woody CD (ed) *Conditioning: representation of involved neural functions.* Plenum Press, New York, pp 483-502

Boice R (1970) Avoidance learning in active and passive frogs and toads. *J Comp Physiol Psychol* 70(1): 154-156

Brower JVZ, Brower LP (1962) Experimental studies of mimicry. 6: The reaction of toads (*Bufo terrrestris*) to honey bees (*Apis mellifera*) and their dronefly mimics (*Eristalis vinetorum*). *Amer Naturalist* 96: 297-307

Brzoska J, Schneider H (1978) Modification of prey-catching behavior by learning in the common toad (*Bufo b bufo* L, Anura, Amphibia): changes in response to visual objects and effects of auditory stimuli. *Behav Processes* 3: 125-136

Burghagen H (1979) *Der Einfluß von figuralen, visuellen Mustern auf das Beutefangverhalten verschiedener Anuren.* PhD Thesis, Univ of Kassel

Cohen DH (1984) Identification of vertebrate neurons modified during learning: analysis of sensory pathways. In: Alkon DL, Farley J (eds) *Primary neural substrates of learning and behavioral change.* Cambridge Univ Press, Cambridge Mass, pp 129-154

Cott HB (1936) The effectiveness of protective adaptations in the hive-bee, illustrated by experiments on the feeding reactions, habit formation and memory of the common toad (*Bufo bufo bufo*). *Proc Zool Soc (London)* 1: 113-133

Dieringer N (1974) Responses of Purkinje cells in the cerebellum of the grassfrog (*Rana temporaria*) to somatic and visual stimuli. *J Comp Physiol* 90: 409-436

Douglas RJ, Pribram KH (1966) Learning and limbic lesions. *Neuropsychologia* 5: 197-220

Ebbesson SOE (1980) The parcellation theory and its relation to interspecific variability in brain organization, evolutionary and ontogenetic development , and neuronal plasticity. *Cell Tiss Res* 213: 179-212

Elepfandt A, Seiler B, Aicher B (1985) Water wave frequency discriminationin in the clawed frog, *Xenopus laevis. J Comp Physiol* 157: 255-261

Ewert J-P (1967) Untersuchungen über die Anteile zentralnervöser Aktionen an der taxisspezifischen Ermüdung der Erdkröte (*Bufo bufo* L). *Z Vergl Physiol* 57: 263-298

Ewert J-P (1968) Der Einfluß von Zwischenhirndefekten auf die Visuomotorik im Beute- und Fluchtverhalten der Erdkröte (*Bufo bufo* L). *Z Vergl Physiol* 61: 41-70

Ewert J-P (1969) Quantitative Analyse von Reiz-Reaktions-Beziehungen bei visuellem Auslösen der Beutefang-Wendereaktion der Erdkröte (*Bufo bufo* L). *Pflügers Arch* 308: 225-243

Ewert J-P (1971) Single unit responses of the toad (*Bufo americanus*) caudal thalamus to visual objects. *Z Vergl Physiol* 74: 81-102

Ewert JP (1974) The neural basis of visually guided behavior. Sci Amer 230: 34-42

Ewert J-P (1984) Tectal mechanisms that underlie prey-catching and avoidance behaviors in toads. In: Vanegas H (ed) *Comparative neurology of the optic tectum.* Plenum Press, New York, pp 247-416

Ewert J-P (1987) Neuroethology of releasing mechanisms: prey-catching in toads. *Behav Brain Sci* 10: 337-405

Fentress JC (1987) Ethological invariants: boxes, rubber bands, and biological processes. A commentary. *Behav Brain Sci* 10: 377-378

Finkenstädt T, Ewert J-P (1988a) Effects of visual associative conditioning on behavior and cerebral metabolic activity in toads. *Naturwissenschaften* 75: 95-97

Finkenstädt T, Ewert J-P (1988b) Stimulus-specific long-term habituation of visually guided orienting behavior toward prey in toads: a ^{14}C-2DG study. *J Comp Physiol* 163: 1-11

Finkenstädt T, Ewert J-P (1988c) Visual associative learning in toads involves a distributed property in brain function (in preparation)

Finkenstädt T, Adler NT, Allen TO, Ebbesson SOE, Ewert J-P (1985) Mapping of brain activity in mesencephalic and diencephalic structures of toads during presentation of visual key stimuli: a computer assisted analysis of ^{14}C-2DG autoradiographs. *J Comp Physiol* 156: 433-445

Finkenstädt T, Adler NT, Allen TO, Ewert J-P (1986) Regional distribution of glucose utilization in the telencephalon of toads in response to moving configurational visual stimuli: a C-2DG study. *J Comp Physiol* 158: 457-467

Fite KV, Scalia F (1976) Central visual pathways in the frog. In: Fite K (ed) *The amphibian visual system: a multidisciplinary approach.* Academic Press, pp 87-118

Foreman N, Stevens R (1987) Relationships between the superior colliculus and hippocampus: neural and behavioral considerations. *Behav Brain Sci* 10: 101-152

Fuller PM, Ebbesson SOE (1973) Central projections of the trigeminal nerve in the bull frog (*Rana catesbeiana*). *J Comp Neurol* 152: 193-200

Gaffan D, Saunders RC (1985) Running recognition of configural stimuli by fornix transected monkeys. *Q J Exp Psychol* 37B: 61-71

Gonzalez-Lima F, Scheich H (1986) Classical conditioning of tone-signaled bradycardia modifies 2-deoxyglucose uptake patterns in cortex, thalamus, habenula, caudate-putamen and hippocampal formation. *Brain Res* 363: 239-256

Gonzalez-Lima F, Russell IS, Gonzalez-Lima EM (1986) Representation of visual memory demonstrated with 2-deoxyglucose: lateralization of learning effects in split-chiasm rats. *Neurosci Lett Suppl* 26: S620

Gruberg ER, Ambros VR (1974) A forebrain visual projection in the frog (*Rana pipiens*). *Exp Neurol* 44: 187-197

Hebb DO (1949) *The organization of behavior: a neuropsychological theory.* Wiley, New York

Herrick CJ (1948) *The brain of the tiger salamander.* Chicago, Univ of Chicago Press

Hinton GE, Anderson JA (1981) *Parallel models of associative memory.* Erlbaum, New Jersey

Hoffman HH (1963) The olfactory bulb, accessory olfactory bulb, and hemisphere of some anurans. *J Comp Neurol* 120: 317-368

Ingle D (1976) Spatial vision in anurans. In: Fite KV (ed) *The amphibian visual system: a multidisciplinary approach.* Academic Press, New York, pp 119-140

Ingle D (1983) Brain mechanisms of visual localization by frogs and toads. In: Ewert J-P, Capranica RR, Ingle DJ (eds) *Advances in vertebrate neuroethology.* Plenum Press, pp 177-226

Ito M (1984) *The cerebellum and neural control.* Raven Press, New York

Kandel ER (1979) *Behavioral biology of Aplysia: a contribution to the comparative study of opisthobranch molluscs.* WH Freeman Co, San Francisco

Kicliter E (1979) Some telencephalic connections in the frog *Rana pipiens. J Comp Neurol* 185: 75-86

Kicliter E, Ebbesson SOE (1976) Organization of the 'non olfactory' telencephalon. In: Llinás R, Precht W (eds) *Frog neurobiology*. Springer-Verlag, Berlin Heidelberg New York, pp 946-972

Kohonen T, Oja E, Lethio P (1981) Storage and processing of information in distributed associative memory systems. In: Hinton GE, Anderson JA (eds) *Parallel models of associative memory*. Erlbaum, New Jersey, pp 105-143

Kokoros JJ (1973) *Efferent connections of the telencephalon in the toads Bufo marinus, and the tiger salamander, Ambystoma tigrinum.* PhD Thesis, Case Western Reserve University

Krasne FB, Kandel ER, Truman JW (1979) Simple systems revisited. *Neurosci Res Prog Bull* 17(4): 529-538

Laroche S, Neuenschwander EL, Massioui N, Edeline JM, Dutrieux G (1987) Hippocampal associative cellular responses: dissociation with behavioral responses revealed by a transfer-of-control technique. *Behav Neural Biol* 47: 356-368

Lázár G (1969) Efferent pathways of the optic tectum in the frog. *Acta Biol Acad Sci Hung* 20: 171-183

Lázár G (1979) Organization of the frog visual system. In: Lissak R (ed) *Recent development of Biology in Hungary Vol 8.* Akademiai Kiado (Budapest)

Levy WB, Steward O (1983) Temporal contiguity requirements for long-term associate potentiation/ depression in the hippocampus. *Neurosci* 8: 791-797

Lindsley DB, Wilson CL (1975) Brainstem-hypothalamic systems influencing hippocampal activity and behavior. In: Isaacson RL, Pribram KH (eds) *The hippocampus Vol 2.* Plenum Press, New York

Llinás R (1976) Cerebellar physiology. In: Llinás R, Precht W (eds) *Frog neurobiology*. Springer-Verlag, Berlin Heidelberg New York, pp 892-923

Llinás R, Precht W (1969) Inhibitory vestibular efferents system and its relation to the cerebellum in the frog. *Exp Brain Res* 9: 19-29

McCormick DA, Guyer PE, Thompson RF (1982) Superior cerebellar peduncle lesions selectively abolish the ipsilateral classically conditioned nictitating membrane/eyelid response of the rabbit. *Brain Res* 244: 347-350

McGaugh JL, Liang KC, Bennett C, Sternberg DB (1984) Adrenergic influences on memory storage: interaction of peripheral and central systems. In: Lynch G, McGaugh JL, Weinberger NM (eds) *Neurobiology of learning and memory*. The Guilford Press, New York, London, pp 313-367

McLean PD (1957) Chemical and electrical stimulation of the hippocampus in unrestrained animals. *Arch Neurol Psychiat* 78: 113-142

Merkel-Harff C, Finkenstädt T, Ewert J-P (1988) Odor-sensitive structures in the telencephalon of common toads. In: Elsner N, Barth F (eds) *Sense organs between environment and behavior.* Thieme Verlag, Stuttgart New York, p 75

Milner B (1972) Disorders of learning and memory after temporal lobe lesions in man. *Clin Neurosurg* 19: 421-446

Mishkin M (1982) A memory system in the monkey. *Phil T Roy Soc B* 298: 85-95

Nauta (1963) Central nervous organization and the endocrine motor system. In: Nalbandov AV (ed) *Advances in neuroendocrinology* 5-21. Univ of Illinois Press Urbano

Neary TJ, Northcutt RG (1983) Nuclear organization of the bullfrog diencephalon. *J Comp Neurol* 213: 262-278

Neary TJ, Wilczynski W (1980) Descending inputs to the optic tectum in ranid frogs. *Soc Neurosci Abstr* 6: 629

Northcutt RG (1974) Some histochemical observations on the telencephalon of the bullfrog *Rana catesbeiana* (Shaw). *J Comp Neurol* 157: 379-390

Northcutt RG, Kicliter E (1980) Organization of the amphibian telencephalon. In: Ebbesson SOE (ed) *Comparative neurology of the telencephalon.* Plenum Press, New York, pp 203-255

Northcutt RG, Royce GJ (1975) Olfactory bulb projections in the bullfrog *Rana catesbeiana. J Morphol* 145: 51-268

Patton P, Grobstein P (1986) Possible striatal involvement in prey orienting behavior in the frog. *Soc Neurosci Abstr* 10: 61

Reiner A, Brauth SE, Karten HJ (1984) Evolution of the amniote basal ganglia. *TINS* 7:320-325

Rolls ET (1987) Information representation, processing, and storage in the brain: analysis at the single neuron level. In: Changeux J-P, Konishi M (eds) *The neural and molecular bases of learning.* John Wiley & Sons Limited, pp 503-540

Rozin P, Kalat JW (1971) Specific hungers and poison avoidance as adaptive specializations of learning. *Psychol Rev* 78: 458-486

Scalia F (1976) Structure of the olfactory and accessory olfactory systems. In: Llinás R, Precht W (eds) *Frog neurobiology.* Springer-Verlag, Berlin Heidelberg New York, pp 213-233

Seligman MEP (1970) On the generality of the laws of learning. *Psychol Rev* 77: 406-418

Spiegler BJ, Mishkin M (1981) Evidence for the sequential participation of inferior temporal cortex and amygdala in the acquisition of the stimulus-reward associations. *Behav Brain Res* 3: 303-317

Sprague JM, Berlucchi G, DiBerardino A (1970) The superior colliculus and pretectum in visually guided behavior and visual discrimination in the cat. *Brain Behav Evol* 3: 285-294

Sokoloff L, Reivich M, Kennedy C, DesRosiers MH, Patlak CS, Pettigrew KD, Sakurada O, Shinohara M (1977) The (^{14}C)-deoxyglucose method for the measurement of local cerebral glucose utilization: theory, procedure and normal values in the conscious and anesthetized albino rat. *J Neurochem* 28: 897-916

Teitelbaum H, Milner P (1963) Activity changes following partial hippocampal lesions in rats. *J Comp Physiol Psychol* 56: 284-289

Thompson RF, Berger TW, Berry SD, Clark GA, Kettner RN, Lavond DG, Mauk MD, McCormick DA, Solomon DA, Weisz PR, Weisz DJ (1982) Neuronal substrates of learning and memory: hippocampus and other structures. In: Woody CD (ed) *Conditioning: representation of involved neural functions.* Plenum Press, New York, pp 115-130

Thompson RF, Clark GA, Donegan NH, Lavond DG, Lincoln JS, Madden IV J, Mammounas LA, Mauk MD, McCormick DA, Thompson JK (1984) Neuronal substrates of learning and memory: a "multiple-trace" view. In: Lynch G, McGaugh JL, Weinberger NM (eds) *Neurobiology of learning and memory.* The Guilford Press, New York, London, pp 137-180

Thorndike EL (1932) *Fundamentals of learning.* New York. Teachers College.

Thorpe L (1963) *Learning and instinct in animals.* Harvard Press Cambridge Mass

Vinogradova O (1970) Registration of information in the limbic system. In: Horn G, Hinde RA (eds) *Short term changes in neural activity and behavior.* Cambridge Univ Press, Cambridge, pp 95-140

Vinogradova O (1975) Hippocampus and the orienting reflex. In: Sokolov EN, Vinogradova O (eds) *Neuronal mechanisms of the orienting reflex.* Lawrence Erlbaum/Hillsdale, New York, pp 128-154

Voronin LL (1983) Long-term potentiation in the hippocampus. *Neurosci* 10: 1051-1069

Weinberger NM, Diamond DM (1987) Physiological plasticity in auditory cortex: rapid induction by learning. *Prog Neurobiol* 29: 1-55

Wilczynski W, Northcutt RG (1977) Afferents to the optic tectum of the leopard frog: an HRP study. *J Comp Neurol* 173: 219-229

Wilczynski W, Northcutt RG (1979) Striatal efferents in the bullfrog, *Rana catesbeiana. Soc Neurosci Abstr* 5: 147

Wilczynski W, Northcutt RG (1983a) Connections of the bullfrog striatum: afferent organization. *J Comp Neurol* 214: 321-332

Wilczynski W, Northcutt RG (1983b) Connections of the bullfrog striatum: efferent projections. *J Comp Neurol* 214. 333-342

Woody CD (1982) *Memory, learning, and higher function: a cellular view.* Springer-Verlag, New York Heidelberg Berlin

Learning and Memory in the Toad's Prey/Predator Recognition System: A Neural Model

ROLANDO LARA *)

Centro de Investigaciones en Fisiologia Celular
Universidad Nacional Autonoma de México
04510 México D.F., México

Abstract. A neural model of the mechanisms responsible for learning in the prey/predator recognition system of the toad is proposed. The prey/predator recognition system consists of retina-tectum-thalamus-pretectum interactions. This system is modulated by a group of (telencephalic and thalamic) neurons which can change the inhibitory effect of pretectal neurons on tectal cells as a result of experience. The model can simulate the following phenomena: (1) Long-term stimulus-specific habituation; (2) learning to avoid bees; (3) learning not to avoid, and even to snap at, predator-like stimuli; (4) learning to discriminate between palatable and unpalatable stimuli based on color cues. All these changes are produced not as a consequence of change in the transmission of information across subsequently connected neurons, but as a change in the dynamics of a whole network. Learning in our model is exhibited as the capacity of the system to choose for a specific situation the proper motor pattern rather than as the mere storage of information. Learning in this sense can be regarded as the dynamic changes of the system that allow the animal to shift from one to another motor schema or to assign different perceptual schemas to the same motor schema.

1 Introduction

Neuroethological studies have shown (Ewert 1968, 1980) that prey-catching behavior in toads is strongly released by small stimuli which are elongated along the direction of motion (worm-like stimulus, W), while this behavior is greatly inhibited if the stimulus's longest axis is elongated perpendicular to the direction of motion (antiworm-like stimulus, A). There is evidence that the optic tectum controls the commands for prey-orienting and snapping; but the capacity to discriminate between

---

*) This paper was completed by the late Rolando Lara in August 1983. We publish it here in a copy-edited form. The Editors.

prey and non-prey stimuli seems to be the result of the *joint function* of pretectum and tectum, because animals without pretectum fail to discriminate the nature of the stimulus. It has also been shown that avoidance behavior can be elicited by electrical stimulation of the thalamic-pretectal region, and that this behavior is no longer present after thalamic-pretectal lesions; but due to the fact that tectal lesions also disrupt predator avoidance, it seems that it is the *joint activity* of pretectum and tectum that controls the release of this response (Ewert 1968, 1976).

Anatomical (Scalia 1976; Székely & Lázár 1976) and physiological (Ewert 1976; Grüsser & Grüsser-Cornehls 1976) studies of the prey/predator recognition system have shown that the retina sends in a retinotopic way information to both tectum and thalamus pretectum on the contrast, velocity, and area of moving objects (giving less information on object elongation in the direction of motion than to the elongation perpendicular to the direction of movement). Cells have been recorded in the tectum that are sensitive to several properties of the stimulus, such as W-like, A-like features, depth, novelty, etc. (Ewert 1970, 1984; Schürg-Pfeiffer & Ewert 1981). The cell type that is mostly sensitive to W-stimuli (class T5.2 of Ewert) responds toward different types of configural stimuli in a way similar to the overall behavior of the animal, i.e., they increase their response to W-stimuli and reduce their response to (or are inhibited by) A-stimuli, and this discrimination is disrupted after pretectal lesions. Another type of cell (class T5.3 of Ewert), on the other hand, responds stronger to A-stimuli and less to W-stimuli. Studies on the physiological responses of the thalamic pretectal cells (Ewert 1971) reveal a great variety of properties, but most of them display sensitivity to predator-like stimuli, i.e., they have wide receptive fields and are mostly sensitive to large objects and to A-stimuli. One type of these cells (class TH3 of Ewert) has a receptive field of about 46° diameter and might project in a retinotopic way to tectal neurons. The response of TH3 neurons is thus very similar to the response of T5.3 neurons.

Toad's prey/predator recognition can be modified as a result of experience in the following cases: (1) Prey-orienting behavior habituates specifically to a given stimulus in a short- and a long-term manner after the repetitive presentation of the stimulus (Ewert & Kehl 1978; Lara & Arbib 1985; see also contributions by Finkenstädt, this volume); (2) predator avoidance behavior also habituates when the predator stimulus is repetitively presented; predator habituation lasts for a shorter period than prey habituation (Ewert 1984); (3) learning not to attack harmful stimuli, e.g., toads learn to not attack bees in one to seven trials (Brower & Brower 1960; 1962; Cott 1936); (4) learning not to avoid predator-like objects, e.g., when being fed by the experimenter's hand (Brzoska & Schneider 1978; Ewert 1982), which however, is non-specific, because there is a generalization for large moving objects; (5) learning to discriminate palatable from unpalatable stimuli based on the prey's color (Mikulka et al. 1980); (6) learning to discriminate palatable from unpalatable stimuli based on the background color (Mikulka et al. 1980); (7) changes in prey-predator recognition based on olfactory cues (Ewert 1965, 1968).

Asking for neural functions involved in learning, it has been shown that habituation of prey orienting is disrupted after pretectal lesions, while with a partial or total telencephalic ablation, the number of responses is reduced and habituation is faster and stronger (Ewert 1965, 1984). Pretectal lesions also disrupt the toad's learned behavior to not attack bees; however, it is not known whether the prectectal region stores the learned information or whether the discriminative capacity of the animal is impaired. Short-term changes are known to occur in the response of retinal, tectal, and thalamic neurons. Although some of these changes represent the general properties observed in habituation, all of these have a shorter time course than those observed in stimulus-specific habituation of the animal's behavior. Preliminary studies on telencephalic functions indicate that some cells may be showing long lasting changes, but these have not been closely related to processes of habituation or learning. With respect to the physiological evidence related to processes of learning and memory in frogs and toads, correlations between the changes in the overall behavior and changes in neuronal response have yet to be investigated systematically. (For recent 2DG studies on brain structures involved in associative and non-associative learning see Finkenstädt, this volume).

In the present paper, we propose a model of the neural changes that may be occurring in the toad's prey-predator recognition system during processes related to learning and memory. The model is a continuation of previously proposed models of prey/predator recognition (see Arbib, Cervantes-Pérez, and Betts, this volume) and stimulus-specific habituation of prey orienting (Lara & Arbib 1985).

2 Basic Structure of the Model

Based on the aforementioned evidence, and taking into consideration the anatomical studies on the tectum (Lázár, this volume), it has been suggested (Ewert & v. Seelen 1974; Lara et al. 1982) that the prey recognition system involves the interaction among retinal ganglion cells, pretectum, and tectum (Figs. 1 and 2). The retinal (R) ganglion cells send their fibers to both pretectum (classes R3 and R4) and tectum (classes R2, R3, and R4). It is hypothesized that thalamic pretectal TH3 cells activated by R3 and R4 inhibit, in a retinotopic way, the activity of tectal neurons. The activity of tectal T5.2 cells is the result of the global activity of several tectal neurons, constituting the tectal column, and the inhibitory effect of TH3 neurons (Ewert, this volume). The tectal column (Lara et al. 1982) consists of a glomerulus (*GL*), large (*LP*) and small pear shaped (*SP*) neurons (probably T5.1-type cells), a stellate neuron (*SN*), and a pyramidal (*PY*) neuron which is a T5.2 cell. (For identification of physiologically classified tectal neurons see also Matsumoto, this volume; italics refer to the anatomical classification of tectal neurons by Székely & Lázár 1976.) The glomerulus receives optic input mediated by R2 fibers and the recurrent axons of both *LP* and *SP*. The *LP* neurons receive the excitatory effect of both R2, *GL*, and *SP*, and this is inhibited by *SN* and TH3. *SN*, the inhibitory neuron of the tectal column, is excited by *LP*. Finally, the T5.2 neurons (*PY* neurons) receive the excitatory inputs of R2, R3, and R4, *LP* and *SP*, and are inhibited by TH3, thus integrating the activity of tectal and pretectal cells.

Predator recognition, on the other hand, has been postulated to be the result of the joint activity of TH3 and T5.3 neurons (large ganglionic *LG* neurons) (Ewert 1976; Schürg-Pfeiffer & Ewert 1981). Class T5.3 cells are excited by R2, R3, and R4 and inhibited by *SN* of its corresponding recognition column (Figs. 1 and 2).

With respect to habituation, we have postulated that short-term stimulus specific habituation of prey orienting occurs as a result of a build-up of inhibition from the thalamus-pretectum to the tectum (Lara & Arbib 1985). This inhibition is the result of various processes: (i) the creation of a model of the tectal spatiotemporal pattern generated by the stimulus, (ii) its comparison with the actual pattern, and (iii) building-up of inhibition from pretectum to tectum if process (ii) shows matching; if (ii) yields mismatch, dishabituation occurs. All these processes, we postulate, involve thalamic and telencephalic structures (see contribution by Finkenstädt, this volume).

As described above, we have suggested that prey/predator recognition is the result of the joint retino-pretectum-tectum interactions (Figs. 1 and 2). We postulate that these can be externally modulated by telencephalic structures through a system that increases or reduces the level of excitation (or inhibition) of a hypothetical cell, H, depending on the association of a given stimulus to the consequence of its past motor responses. This excitation (inhibition) can regulate the prey/predator recognition system in order to change the motor response to a given stimulus as a consequence of experience, e.g., learning to avoid a prey-like stimulus or learning to snap at a predator-like stimulus. We have postulated that there are several types of H neurons, each regulating a specific feature of the external world, such as prey, predators, background color, odor, etc. (Fig. 2). In case of prey, because of its stimulus specificity, we have to postulate that H has a way to store and code its response to each stimulus as a result of experience. We have simulated this function simply as a storage of a matrix that has the expected consequence of each stimulus in it. This value is then retrieved in a context-addressable manner through a postulated neural code which defines the proper place to retrieve the expected value (see Section 4). The specificity of the stimulus is simulated through a system that creates a model (MO) of the stimulus and compares (CO) it to the present object. If model and stimulus are similar, H retrieves the expected response and changes the inhibitory effect from TH3 to the tectum; but if MO is different from the stimulus, H is reset and the new expected value of the stimulus is then retrieved and a new inhibition change occurs.

In the case of predators, on the other hand, because of less specificity, once a predator-like stimulus is associated with a specific consequence, H is changed for all predator-like stimuli. We have shown in Figs. 1 and 2 that the group of H cells regulates the inhibitory effects from pretectal TH3 cells to the tectum, but it may be also possible that H itself regulates the activity of tectal neurons. In the former case, H will excite TH3 while in the latter H will inhibit the tectal neurons.

This model allows us to consider stimulus-specific habituation as a special case of a more general function that associates a specific stimulus to an expected response, suggesting that after the repetitive presentation of the stimulus, a negative value is associated to that stimulus which is then translated into a build-up of inhibition from pretectum to tectum, as described by Lara & Arbib (1985).

In Fig. 2 we show how the basic structure of the model can be expanded to 8 functional units (columns). The way the units interact have been described in detail elsewhere (Arbib & Lara 1982; Lara et al. 1982) and is shown in the Appendix (Section 7) where the mathematical model is presented.

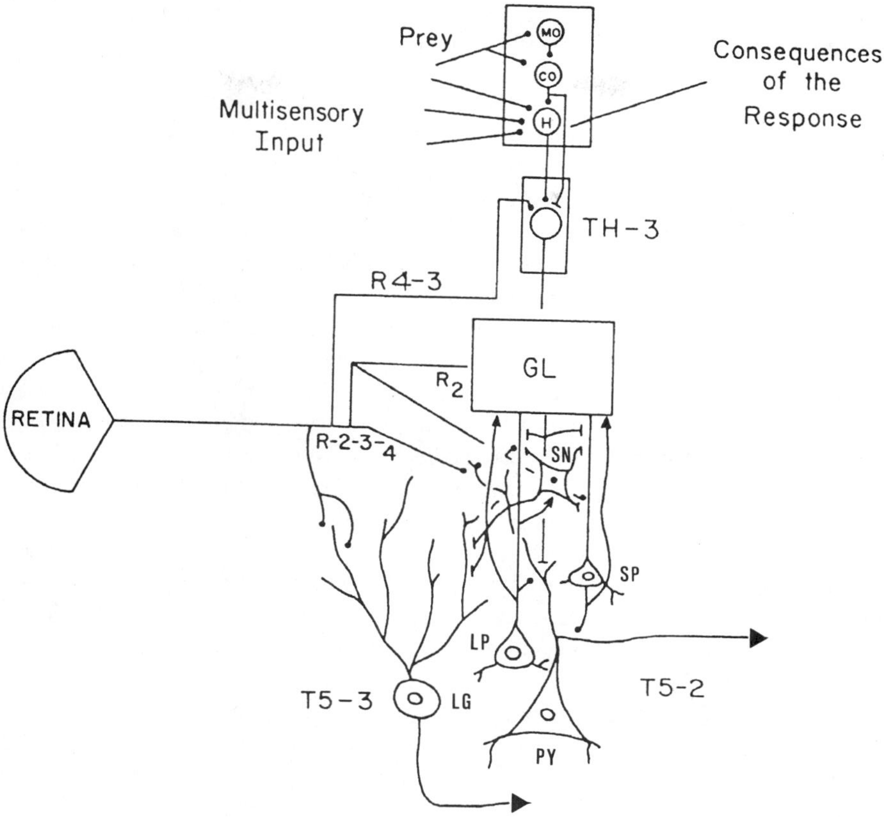

Figure 1. Basic model for the modulation of the prey/predator recognition system in toads during processes related to learning and memory. The prey/predator recognition system involves retinal ganglion cells R2, R3, and R4, thalamic pretectal neurons TH3, the tectal column, and tectal T5.3 neurons. The tectal column consists of the glomerulus (*GL*), large (*LP*), and small (*SP*) pearshaped neurons (T5.1-type neurons), the stellate neurons (*SN*), and pyramidal (*PY*) neurons (= T5.2 neurons) (italics refer to anatomic nomenclature of tectal neurons by Székely & Lázár 1976). Prey-catching response is controlled by T5.2 cells through the joint activity of retinal (R2, R3, R4), tectal (*LP, SP*), and pretectal (TH3) neurons. Avoidance behavior is controlled by the joint activity of TH3 and T5.3 cells . T5.3 cells (large ganglionic *LG* neurons) are activated by retinal ganglion cells (R2, R3, R4) and are inhibited by the *SN* cells of its corresponding tectal column. The modulatory effect from telencephalic structures to this prey/predator recognition system is exerted over the inhibitory effect from the pretectal neurons (TH3) over the tectal cells (*LP, SP*, and T5.2), indirectly controlling the response of T5.3 cells. The telencephalic modulation is constituted by several cells coming from different brain regions and sensing specific properties of the external world. H is the hypothetic learning neuron that stores the expected consequence of the response to a specific stimulus, coded as an increase or a decrease of excitation to TH3. For prey, we have also considered a short-term memory system which creates a model of the stimulus (MO) and compares it (CO) with the actual one, and, if they are different, resetting of H occurs and the new expected value is retrieved, while if they are similar, then the effect of H is continued over TH3.

3 General Behavior of the Model

The basic assumption of this model is that the inhibitory effect from pretectal neurons can control the prey/predator recognition system and allows the animal to change the innate response to a specific stimulus through learning. The prey/predator recognition system simply states that prey-catching behavior occurs if T5.2 cells are active, because of a low level of inhibition from TH3 neurons; while predator avoidance is the result of the joint activity of T5.3 and TH3 neurons. Modulation of this system is executed in the following way: Whenever a visual stimulus is presented to the animal, prey-catching or predator-avoidance behavior is elicited. The telencephalic cells H first store the consequence of the response, simulated as a positive or negative constant, depending on whether the consequence was positive or negative (see Section 4). The next time the same stimulus is presented, H retrieves the corresponding expected value and changes the inhibitory effect from TH3 (or by itself) to the tectal neurons: If the expected consequence was positive, the inhibition is reduced proportional to the retrieved value, while if it was negative the inhibition is increased. In the first case, the activity of the tectal column is increased and if the inhibition is greatly reduced, then it can start to respond to stimuli to which originally it did not respond, accompanied by an inhibition of tectal T5.3 neurons. In the second case of an increased inhibition, the activity of the tectal column is inhibited up to a point that it may disinhibit the response of T5.3 neurons. The joint activity of TH3 cells and T5.3 cells would allow the animal to avoid a prey-like stimulus.

Figure 2. Extended model of the modulation of the prey-predator recognition system with eight units. Note that there are different H for each feature of the world, such as prey, predator, color, olfaction, etc., coming from different brain regions (TH = thalamus, TPH = telencephalon, TEC = Tectum). For each of these stimuli, H associates a consequence of the response, possibly modulated by the hypothalamus. All H's can modulate the inhibitory effect of TH3 over the tectal column. Abbreviations of tectal neurons: GL, glomerulus; LG, large ganglionic cell (= T5.3 neuron); LP, large pear-shaped cell (= T5.1 neuron); SP, small pear shaped cell (= T5.1 neuron); SN, stellate neuron; PY, pyramidal cell (= T5.2 neuron).

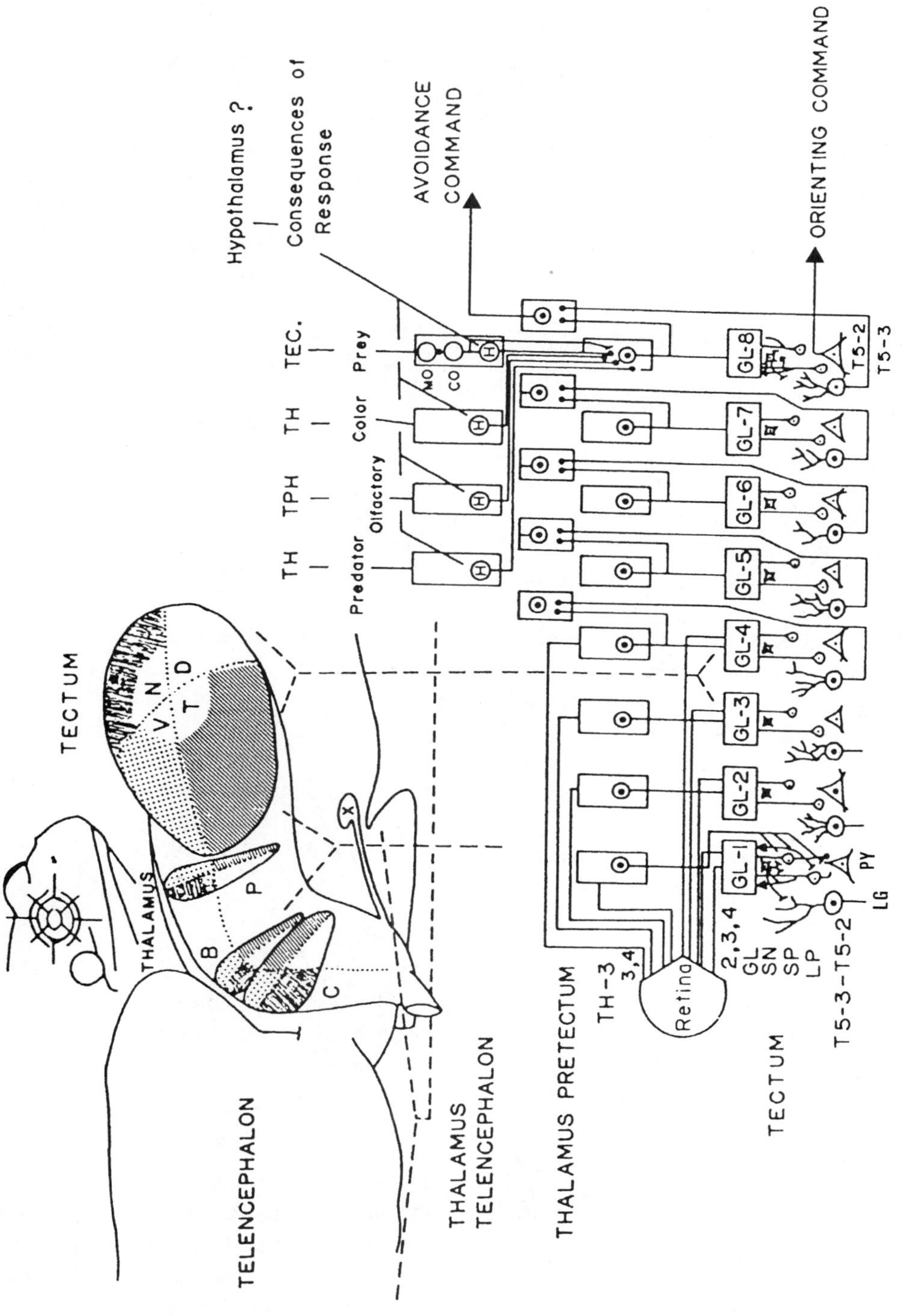
TECTUM
THALAMUS
TELENCEPHALON
THALAMUS
TELENCEPHALON
THALAMUS PRETECTUM
TECTUM
Hypothalamus ?
Consequences of Response
AVOIDANCE COMMAND
ORIENTING COMMAND
Predator
Olfactory
Color
Prey
Retina
GL-1
GL-2
GL-3
GL-4
GL-5
GL-6
GL-7
GL-8
T5-3-T5-2
T5-2
T5-3
LG
PY

4 Mathematical Definition of the Model

We have simulated the response of the different neurons using the simplest model of the electrical properties of the membrane (capacitance and resistance) and a threshold function to simulate the generation of spikes. The basic definition of this model is the following:

$$dm/dt = -m + U_i$$

where U_i is the sum of the excitatory and inhibitory inputs and m is the membrane potential. We have considered two types of threshold function:

$$f(m) = \begin{cases} 1 \text{ if } m > \text{threshold} \\ 0 \text{ else} \end{cases}$$

and for inhibitory neurons:

$$h(w) = \begin{cases} w \text{ if } w > \text{threshold} \\ 0 \text{ else} \end{cases}$$

The mathematical definition of the constituents of the prey/predator recognition system has been defined in detail elsewhere for the tectal column (Lara et al. 1982), retino-pretectal-tectal interactions for prey/predator recognition (Lara et al. 1982), and stimulus-specific habituation (Lara & Arbib 1985).

The only addition to these models to simulate learning and memory is in the definition of the equation defining H, which modulates the inhibitory effect of TH3 over the tectal neurons as a result of experience. H is defined as follows:

$$dH/dt = (H_0 - H)k_3 + EV(D,F)$$

where H_0 is the tonic effect of H over a TH3 cell; k_3 is a constant, and EV(D,F) represents the association of a given stimulus U(D,F) of duration D and frequency F, considered as a possible simplified code that the nervous system may use to define different stimuli, with a positive or negative consequence of the response, and is defined in the following way:

$$dEV(D,F)/dt = -EV(D,F)*k_4 + k_5$$

where k_4 and k_5 are constants. The sign of k_5 indicates whether the stimulus U(D,F) was associated with a positive or a negative consequence; k_4 regulates the rate of forgetting of EV(D,F). The memory of the system to several stimuli is simulated as a matrix whose values are the EV of each stimulus U(D,F) and are retrieved as EV(D,F) in a content-adressable way.

Note that we have defined only a group of H neurons for the control of the multiple pretectal and tectal units. We have postulated elsewhere that short-term habituation is a local process and for each tectal column we propose one habituation column. It seems, however, that for long-term processes related to memory and learning the stimulus has to be recognized independently of the site of stimulation. For this reason we propose two, of many, ways in which the toad's nervous system could achieve this property: Through a generalization of the stored information mediated by neurohormonal factors for each habituation column, or through a different system that controls the storage and retrieval of information independently of the stimulated site. At the moment, we do not know of any evidence that favours either of these hypotheses.

5 Computer Simulation

The computer simulation of the present model was done in a microcomputer and the graphic system of Cromemco. The model is constituted of retinal ganglion cells R2, R3, and R4, eight tectal columns, eight thalamic TH3 units, and the learning cells H, each associated with a specific feature of the external world (Fig. 2).

The simulation of the retinal response was simply done taking the experimentally obtained rate of firing to a specific size and configuration of the stimulus used. If not indicated elsewhere, we have used the values obtained for 2° x 12° stripes either moved with the longest axis along (W-configuration) or perpendicular to (A-configuration) the direction of motion. We have chosen these stimuli because there is a clear-cut discrimination both behaviorally and cellularly. The movement of the stimulus was simulated by presenting it to each column for a specific period of time considering that each column covers approximately 12 degrees of the visual field.

We illustrate the results in two ways (Figs. 4 to 8), showing the response, represented as bars, of the eight tectal T5.2 neurons and of pretectal avoidance (AV = TH3 + T5.3) neurons to the presentation of the stimulus, or showing the response of the cells within one unit (R2, T5.2, TH3, T5.3, and AV) to see the way the inhibitory effect changes in each case.

We have simulated the following phenomena: (1) Long-term habituation and its dependence on inter-stimulus intervals; (2) learning to avoid bees; (3) learning to attack a predator-like stimulus; (4) learning to discriminate between palatable and unpalatable stimuli based on prey's color; and (5) learning to discriminate palatable from unpalatable stimuli based on background color.

5.1 Long-Term Habituation

It has been shown (Ewert 1965, 1984) that after several series of habituation trials, habituation is stronger and faster and dishabituation (recovery) lasts longer. The reason for this is that recovery after habituation in a stimulus series has (at least) two time constants, an initial fast one, for 50% of recovery, and the subsequent one slow. To simulate this, we simply assign to EV(D,F) a negative value, thus increasing the inhibitory effect from TH3 to the tectal column, through excitation of H to TH3 or inhibition of H to the tectal columns directly. At the same time, we have shown that long-term habituation can be simulated by a change of k_3 in the following equation:

$$dH/dt = k_3(H_o - H)/x(D,F) + EV(D,F)$$

where

$$x(D,F)(t+1) = \begin{cases} x(D,F)(t) + 1 & \text{if the animal habituated to the stimulus } U(D,F) \text{ and } H>0.5*H_o \\ 1 & \text{else.} \end{cases}$$

In this way, H will be more sensitive to EV(D,F) and will recover to H_o in a longer period depending on the number of times the animal has habituated to that stimulus. The condition $H>0.5*H_o$ simulates the fact that recovery has two time constants: k_3 if $H>0.5+H_o$ and $k_3/x(D,F)$ if $H>0.5*H_o$. Because of the nature of the equation for H, for different interstimulus intervals, habituation is faster for short ones than for large ones. In Fig. 3 we show the computer simulation of long-term habituation using

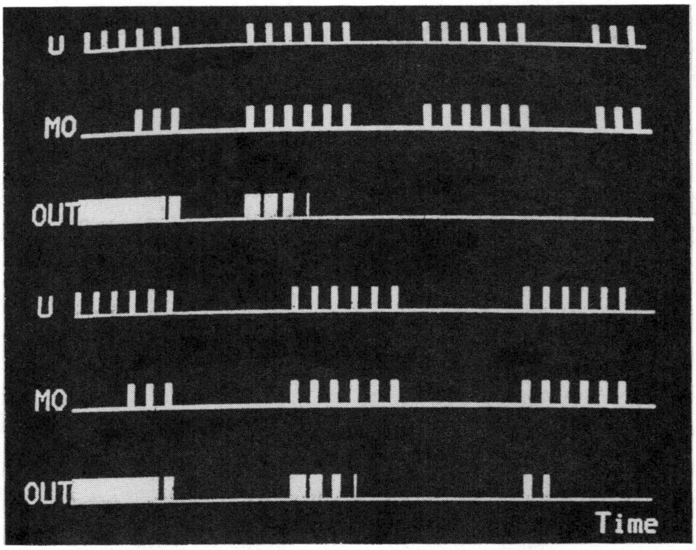

Figure 3. Computer simulation of long-term habituation. Each presentation of the stimulus is represented by a group of bars. U represents the stimulus, MO the creation of the model of the stimulus, and OUT is the response of T5.2 cells. In the upper graph the stimulus U is presented 6 times and the response starts to decrease when the creation of the model of the stimulus has been obtained. In the second series habituation is faster, and no response is observed in the third one, because the interseries interval is short. In the lower graph we show that if the interseries interval is longer, then at the third series we can get a response.

two different inter-stimulus intervals. U represents the optic input, MO is the model of the stimulus, and OUT is the response of T5.2 cells, which simulates prey-orienting response. In the first series notice that T5.2 response is reduced; while in the second series it is completely suppressed; in the third presentation of the stimulus, no response is obtained because of the short time for recovery. In the lower graph, the inter-series interval is longer, thus in the third presentation of the stimulus there is some response.

5.2 Learning to Avoid Bees

Several authors have shown that toads can learn to avoid harmful prey-like stimuli, such as bees or similar insects (Cott 1936; Brower et al. 1960; Brower & Brower 1962). To simulate this phenomenon, we presented a prey-like stimulus and we gave a negative value for the consequence of the prey-catching response (EV(D,F) = -k). Figure 4 depicts the response of the eight T5.2 neurons and the eight AV neurons to the presentation of the stimulus. Figure 4A shows that when the stimulus was presented for the first time, the tectal response was very strong thus the animal snapped at the stimulus. In the second presentation of the stimulus, the tectal response was greatly inhibited and the thalamic neurons started to be activated, thus inhibiting the snapping response and even producing avoidance. The responses to the third and fourth presentation of the stimulus make evident that the animal has learned to avoid the stimulus.

5.3 Learning to Attack a Predator-Like Stimulus

Toads learn to eat from the experimenter's hand and after several trials they even orient to and snap at the moving hand itself (Brzoska & Schneider 1978; Ewert 1982). This effect was generalized to other predator-like stimuli (see Finkenstädt, this volume). Similarly, it has been found that toads can attack a 30° x 2° antiworm stripe that has been associated with food (however, the same stripe in worm configuration still is the stronger prey-catching releaser).

To simulate these effects we associated a predator-like stimulus with a positive consequence, EV = k (notice it is non-specific). We have chosen the constants to get learning very fast in order to show the graphic results in Fig. 5. The first time the stimulus was presented, the thalamic cells were greatly activated, but in the second presentation of the stimulus the activity of thalamic neurons and T5.3 tectal cells was greatly inhibited (see Fig. 5B), and as a consequence tectal T5.2 cells were activated. Further presentations of the stimulus facilitates the tectal response.

5.4 Learning to Discriminate Between Palatable and Unpalatable Stimuli

Toads learn to discriminate between palatable and unpalatable stimuli depending on the color of the stimulus (Mikulka et al. 1980). We simulate the association to each stimulus, represented by a different code U_1(D1,F1), U_2(D2,F2), of a positive and negative consequence, respectively. At the beginning, the tectal response to both stimuli is very strong; but in the second trial, the tectal response to

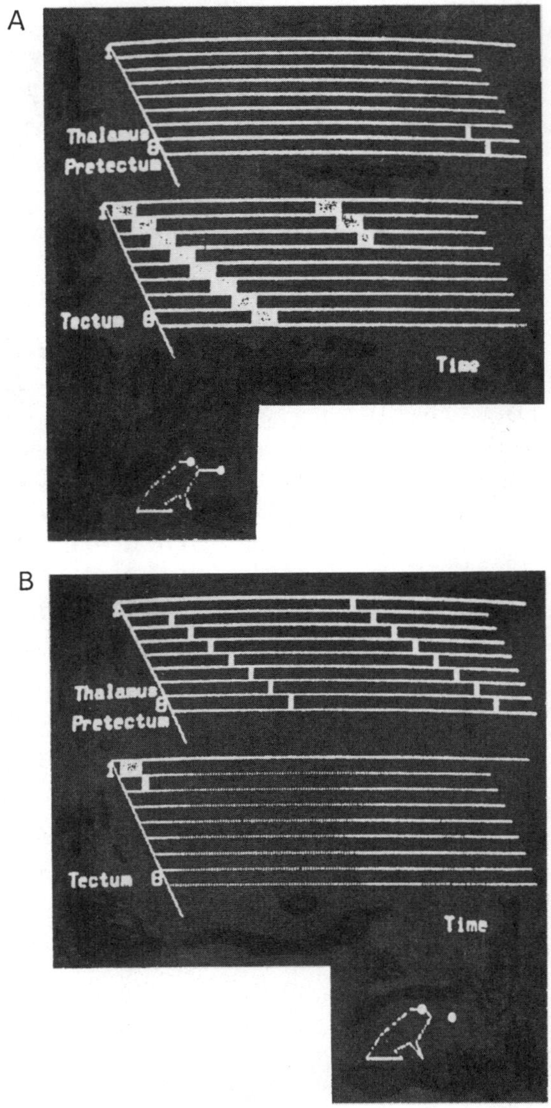

Figure 4. Computer simulation of the response of eight tectal T5.2 and eight pretectal AV=TH3+T5.3 neurons when a prey-like stimulus is associated with a harmful effect, i.e., learning to avoid bees. The stimulus is presented from column 1 to column 8 and the response is represented as bars. (A) When the stimulus is presented the first time, T5.2 cells of the eight tectal columns respond strongly, thus attacking (snapping at) the prey stimulus as is shown in the lower part of the graph; but at the second presentation of the stimulus, the tectal activity is greatly reduced and a weak response is observed in the thalamic units. (B) Further presentation of the stimulus yields a continuous inhibition of tectal activity with a concomitant increased thalamic response that allows the animal not to attack (see lower part of the graph), and even to avoid the stimulus.

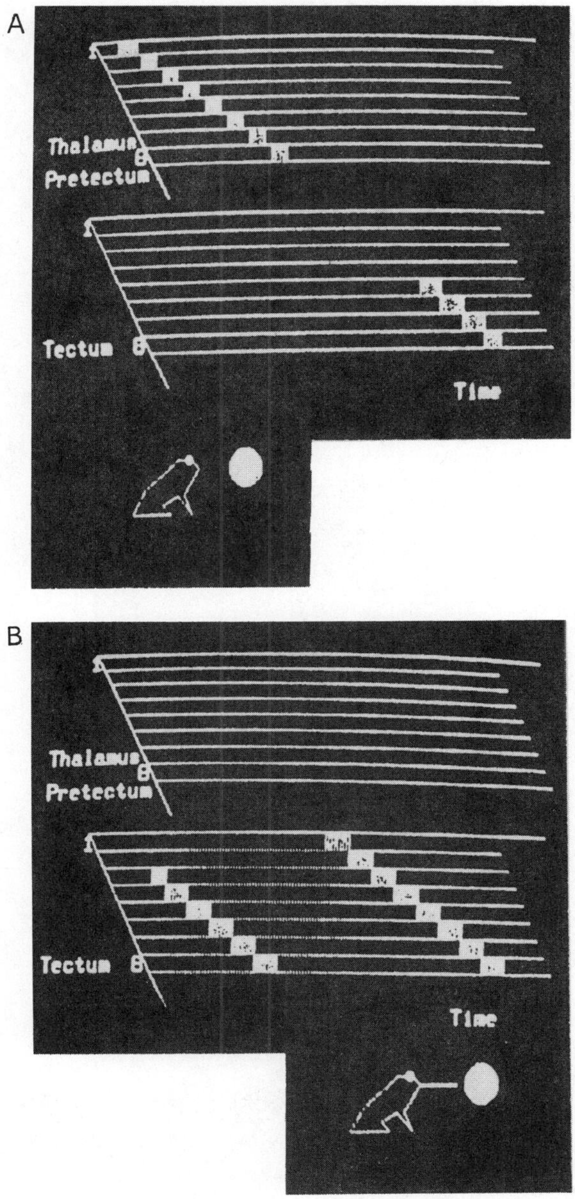

Figure 5. Computer simulation of pretectal and tectal response when a predator-like stimulus is associated with a positive consequence, i.e., learning to be fed from the experimenter's hand. (A) The first time the hand is presented, there is a strong pretectal activity, thus the animal does not attack but avoids the hand (see lower part of the figure). (B) Further presentation of the hand stimulus inhibits the thalamic response, continuously disinhibiting tectal activity that, in the fourth presentation of the stimulus gives a strong response, with the animal attacking, i.e. orienting toward and snapping at, the hand (see lower part of the figure).

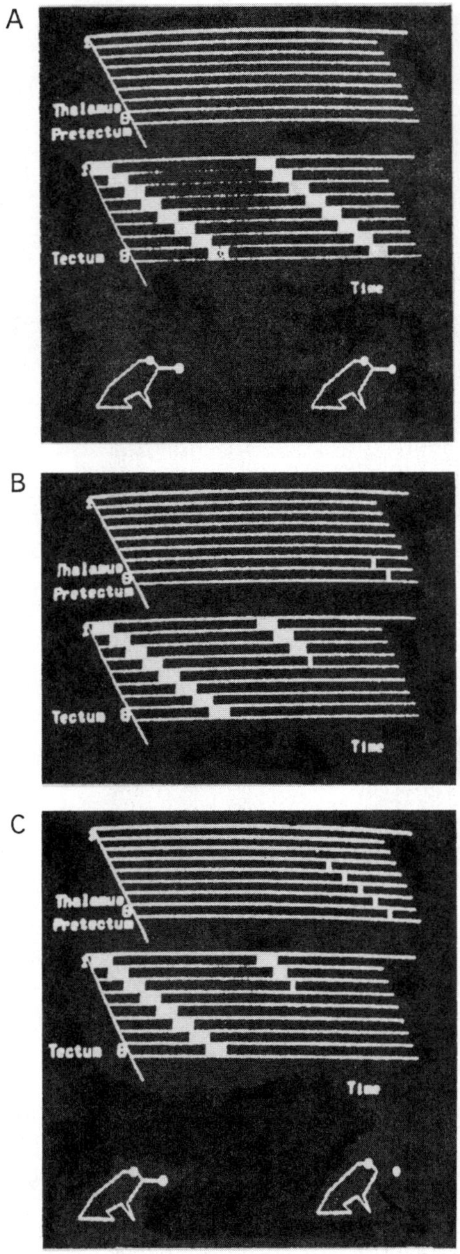

Figure 6. Computer simulation of the response of pretectal and tectal cells when the animal learns to discriminate palatable from unpalatable stimuli based on the color of the prey stimulus. We present the two stimuli with different color alternatively, first the palatable one and then the unpalatable. (A) At the beginning the tectal response to both stimuli is strong, producing a snapping response; but in the second (B) and third (C) trials the tectal response to the second stimuli is greatly reduced while the thalamic activity builds up, thus the animal does not snap at the second stimulus (see lower part of the figure).

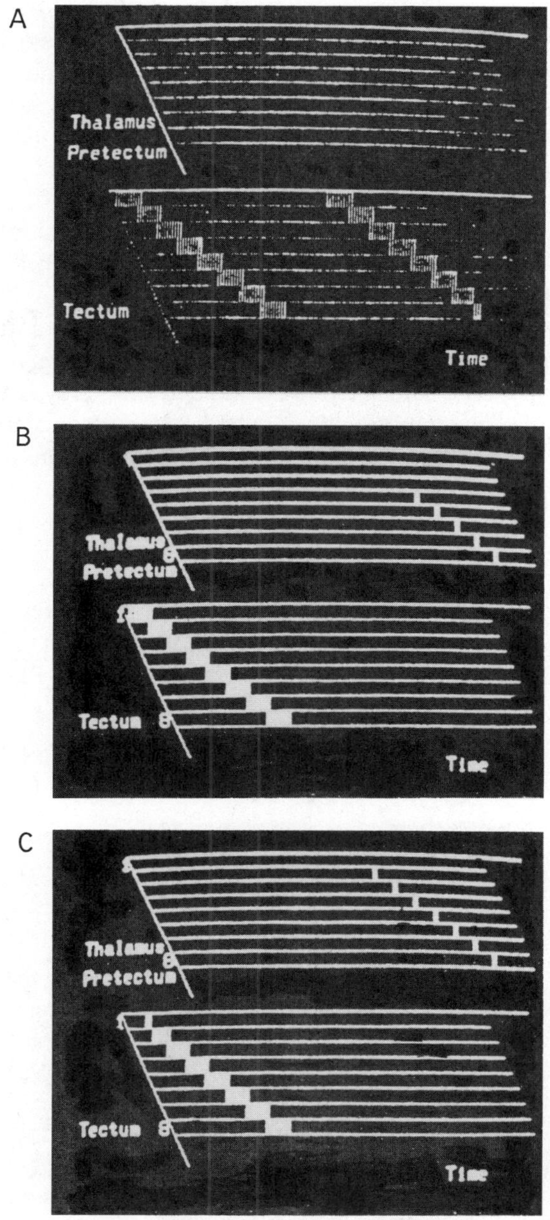

Figure 7. Computer simulation of tectal and pretectal cells when the animal learns to discriminate palatable from unpalatable stimuli based on the background color. In this case, we alternate the background color for the same stimulus, but the first is palatable, while the second is unpalatable. At the beginning (A) the tectal response to both stimuli is strong; but in the second (B) and third (C) trials there is no tectal response to the stimulus associated with the second background color and the thalamic cells are activated. Note that in this case, because the background color is changed before the presentation of the stimulus, the inhibitory effect is stronger than in that shown in Figure 6.

the first stimulus remains the same, while that to the second one is strongly reduced. This inhibitory effect builds up in subsequent presentations (see Fig. 6C), thus inhibiting completely the tectal response and even activating the avoidance system.

Background color can also be used by toads to discriminate between palatable and unpalatable stimuli (Mikulka et al. 1980). We simulated this phenomenon assigning a positive and a negative consequence not to the prey stimulus, but to the cell sensing the background color (see Fig. 2), possibly coded by thalamic neurons (Kondrashev 1987). Whenever this color is present, the animal will associate a positive or a negative consequence. Figure 7 shows that when we present the stimuli the first time, the tectal response is very strong; but in the next trial, the tectum does not respond to the stimulus that was associated with the second color. Note that the effect in this experiment is stronger than in the one presented above, because we have changed the background colors before the presentation of the prey stimulus; thus the expected response acts for a longer period over the prey/predator recognition system (see Fig. 8C to obtain the alternate effects of the two stimuli on the inhibitory effect of pretectal neurons).

6 Discussion

In the present paper, we have presented a model to show how excitatory and inhibitory modulation from telencephalic structures may change an innate discriminatory system for prey/predator recognition as a result of experience. This change is produced not as a consequence of a change in the transmission of information among subsequently connected neurons, but as a change in the dynamics of a whole network through a change in the inhibitory effect from pretectal to tectal

Figure 8. Computer simulation of the response of tectal T5.2, T5.3 and pretectal TH3, AV neurons to the presentation of the stimulus during (A) learning to avoid bees, (B) learning to attack a predator-like stimulus, and (C) learning to distinguish between palatable and unpalatable stimuli based on color cues. In (A) we can see that the first time the stimulus U is presented the T5.2 cell gives a response; but at the second presentation the inhibitory effects of TH3 neurons start to increase, thus inhibiting T5.2 response. The inhibition of T5.2 cells disinhibits and at its turn, the activity of T5.3 cells together with TH3 neurons makes AV respond. In (B) we can see that in the first presentation of the stimulus, TH3, T5.3 and AV cells are activated; but in the second presentation of the stimulus, the inhibitory effect from TH.3 cells is reduced down to a point that T5.2 neurons are now activated. Note, however, that T5.3 neurons are still activated, although the lower the level of inhibition of TH3 cells, the stronger the activity of T5.2 cells and the stronger the inhibition over T5.3 neurons. In (C) we can observe that in the first presentation of the two stimuli, the tectal response is the same, and observe how the inhibitory effect of TH3 changes because of the consequences of the response. In the second trial, the presentation of the first stimulus produces a stronger tectal response, because the thalamic inhibition is greatly reduced, while when we present the second stimulus the second time, the inhibitory effect of TH3 is greatly increased, thus there is no tectal T5.2 response and T5.3 is activated causing AV to fire. Note the alternation of increase and decrease of the inhibitory effect of TH3 to tectal cells depending on the association of the stimulus.

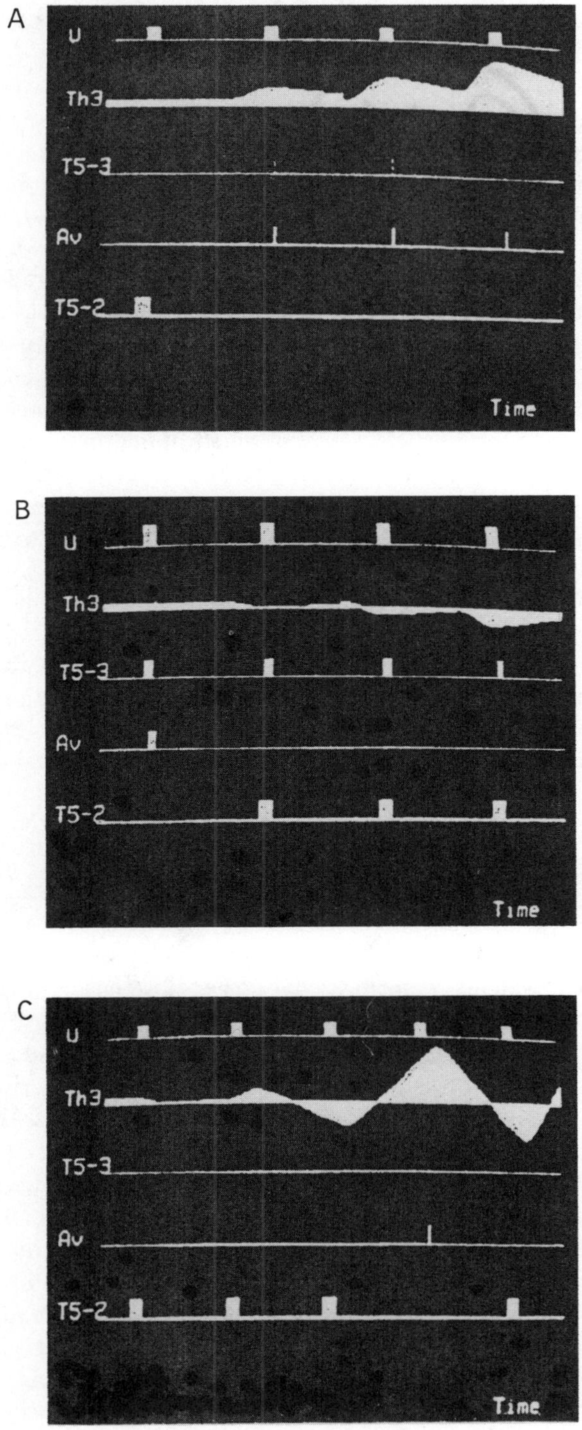
A
U
Th3
T5-3
Av
T5-2
Time
B
U
Th3
T5-3
Av
T5-2
Time
C
U
Th3
T5-3
Av
T5-2
Time

neurons. Moreover, we have elsewhere proposed that the toad's brain can be considered as a superstructure constituted by perceptual and motor schemas and by a command and information system (see also contributions by Arbib, Cervantes-Pérez, and Ewert, this volume). The perceptual schemas are systems that recognize relevant stimuli of the external world, such as prey, predators, barriers, etc. The motor schemas are innate simple motor patterns, such as walking, snapping, orienting, avoiding, etc., which can be activated depending on the specific global configuration of the external world, i.e., the relation between perceptual schemas. These schemas are then activated and directed depending on the information of the command system. Learning in this model is the association of a given configuration of the world to a given motor pattern, and not the storage of a given engram. Thus learning can be considered as the capacity of the system to choose for a specific situation the proper motor pattern rather than the mere storage of information. In the case of toads, the model suggests that this ability to change motor schemas is an innate one and can be considered as a general property of the system, which, at the same time, sets the limits of the learning capacity of the animal. Learning in this sense can be considered as comprising the dynamic changes of the system that allow the animal to shift from one to another motor schema or to assign different perceptual schemas to the same motor schema, respectively.

We have, however, postulated that there is stored information of a given stimulus with a specific consequence of the response. Note again, in the model we are not storing the visual image of the stimulus but the association of the stimulus to a given expected consequence whose retrieval produces not a specific message, but a neural action (excitation or inhibition) that would modify the dynamics of the system.

We have postulated in our model that the storage of information related to a specific feature of the world (such as prey, predator, background, etc.) may be distributed to cells of various nuclei of the telencephalon, the thalamus, and the pretectum (see Finkenstädt, this volume) and that the generalization of the learned process may be obtained through neurohormonal factors. A second possibility could be that there is a single system which associates a specific stimulus, independent of the site of stimulation, to an expected response and can control the general behavior of the animal. Initial evidence indicates that some substances such as ACTH can alter the storage and retrieval of information (e.g., see Horn et al. 1979), because it can modify the dynamics of dishabituation in toads. Other studies in mammals (Lashley 1929) also suggest that memory is not located in a single place but distributed in (almost) the whole brain. For this reason, we think that the first of these hypotheses is more likely to occur, although further research should clarify this point.

With respect to the claim that H could act independently of TH3 or associated with it, it is interesting to note that telencephalic (lateral pallial or striatal) lesions produce a reduction of prey orienting activity, possibly because of an increase in the inhibitory effect from pretectum to the tectum. Therefore, it seems likely that certain telencephalic structures are inhibiting the thalamic inhibition to the tectum, so that after telencephalic lesions this inhibition is increased (see Ewert, this volume).

We have also shown that the model can integrate habituation as a particular case of a more general property of the nervous system to anticipate expected responses. In this sense, habituation in toads cannot be considered simply as a reduction of synaptic efficiency, but as a complex process of storage-retrieval and control of motor responses (see also Finkenstädt, this volume).

The nature of the present model allows us to predict the changes in the neural response of the different cells involved in processes related to learning and memory. It is interesting to note that in our model most of the proposed changes are only manifest when the stimulus is presented, and these are of a short-term and functional nature. That is, the consolidation of learning is seen as the strength and speed of a specific response to a specific stimulus, i.e., how fast and strong is the inhibitory effect associated with a specific stimulus, and not as a permanent change in the rate of firing of a single neuron (see also Doty 1987; Ewert, this volume). Because most of the observed changes in tectal and pretectal cells in processes related to habituation are short-term, our model could be an explanation for this "paradoxical" result.

Finally, we stress that we have been developing our models in a continuous and hierarchical order to integrate new phenomena not by rejection of past models but by the integration of new properties that allow us to interpret new results (Arbib, this volume). The initial models were based, as much as possible, on anatomical and physiological results. The recent models we have developed go beyond this data and postulate specific hypotheses through the creation of new concepts that could allow us to guide experiments and to better understand the complex processing of information related to visuomotor coordination in frogs and toads.

7 Appendix

7.1 Description of the Basic Structure of the Model

(1) $\tau_{gl}\dot{gl}_i(t) = -k_1 \cdot gl_i(t) + R2_i(t) + I_i(t)$

where

$$I_i(t) = w_{gl.p}(SP_i(t) + SP_{i+1}(t) + w_{gl.lp}\,[LP_{i-1}(t) + LP_i(t) + LP_{i+1}(t)]$$

(2) $\tau_{sn}\dot{sn}_i(t) = -k_2 \cdot sn_i(t) + w_{sn.lp}\,[LP_i(t) + LP_{i+1}(t)]$

(3) $\tau_{lp}\dot{lp}_i(t) = -lp_i(t) + w_{lp.sp}\,[SP_{i-1}(t) + SP_i(t)] + gl_i(t)$

$$+\, w_{lp.R2}R2_i - w_{lp.sn}\,[SN_{i-1}(t) + SN_i(t)]$$

(4) $\tau_{sp}\dot{sp}_i(t) = -sp_i(t) + gl_i(t) + gl_{i+1}(t) - w_{sp.sn}\,[SN_i(t) + SN_{i+1}(t)]$

$$+\, w_{sp.R2}R2_i(t) - w_{sp.th3}TH3_i(t)$$

(5) $\tau_{py}\dot{py}_i(t) = -py_i(t) + w_{py.sp}SP_i(t) + w_{py.lp}\,[LP_i(t) + LP_{i+1}(t)]$

$$+\, w_{py.R2}R2(t) + w_{py.R3}R3(t) + w_{py.R4}R4(t) - w_{py.th3}TH3(t)$$

(6) $\tau_{t53}\dot{t53}_1(t) = -t53_1(t) + w_{t53.R2}R2(t) + w_{t53.R3}R3(t) + w_{t53R4}R4(t)$
$-w_{t53.sn}SN_1(t)$

(7) $\tau_{th3}\dot{th3}_1(t) = -th3_1(t) + w_{th3.R3}R3(t) + w_{th3.R4}R4(t) + w_{th3.h}H(t)$

(8) $\tau_{av}\dot{av}_1(t) = -av_1(t) + w_{av.th3}h[TH3(t) - 33.5] + w_{av.t53}T53_1(t)$

(9) $\dot{h}(t) = [h_0-h(t)]k_3 + EV(D,F)$

(10) $d\dot{EV}(D,F,t)/dt = -EV(D,F,t)k_4 + k_5$

(11) $MO(t) = \sum_{n=0}^{Dmo-1} d(t-nFmo)$

(12) $CO = ABS\ [u(t)-MO(t)]$

7.2 Threshold Functions

```
LP=f(lp-1.0)
SP=f(sp-3.15)
T5.2=PY=f(py-23)
T5.3=f(t53-15)
AV=f(av-1.2)
TH3=TH=h(th)
SN=h(sn-0.2)
```

where

$$f(x)= \begin{cases} \text{if } x>\text{threshold} \\ 0 \text{ else} \end{cases}$$

and

$$h(x)= \begin{cases} w \text{ if } w>\text{threshold} \\ 0 \text{ else} \end{cases}$$

(Note that t53 refers to T5.3, etc.)

7.3 Membrane Constants

$\tau_{gl}=1.0$	$k_1=0.5$
$\tau_{sn}=0.65$	$k_2=0.5$
$\tau_{lp}=0.3$	
$\tau_{sp}=0.9$	
$\tau_{py}=0.12$	
$\tau_{av}=0.25$	
$\tau_{t53}=0.25$	
$\tau_{th3}=1.0$	

7.4 Weights

$w_{gl.lp}=1.5$	LP to GL	$w_{py.th}=0.4$	TH3 to T5.2
$w_{gl.sp}=0.1$	SP to GL	$w_{th.h}=3.0$	H to TH3
$w_{lp.sp}=0.8$	SP to LP	$w_{py.r2}=4.5$	R2 to T5.2
$w_{lp.sn}=8.0$	SN to LP	$w_{py.r3}=0.3$	R3 to T5.2
$w_{lp.th}=0.2$	TH3 to LP	$w_{py.r4}=6.0$	R4 to T5.2
$w_{lp.r2}=1.0$	R2 to LP	$w_{th.r3}=0.3$	R3 to TH3
$w_{sp.sn}=1.0$	SN to SP	$w_{th.r4}=5.0$	R4 to TH3
$w_{sp.th}=0.2$	TH3 to SP	$w_{t53.r2}=0.4$	R2 to T5.3
$w_{sp.lp}=1.0$	LP to SP	$w_{t53.r3}=0.4$	R3 to T5.3
$w_{sp.r2}=0.5$	R2 to SP	$w_{t53.r4}=11.5$	R4 to T5.3
$w_{py.lp}=0.8$	LP to T5.2	$w_{av.t53}=1.0$	T5.3 to AV
$w_{py.sp}=1.0$	SP to T5.2	$w_{av.th}=1.0$	TH3 to AV

7.5 Constants

$k_3=0.5$
$k_4=0.001$
$k_5=0.1$
$k_0=10$

8 References

Arbib MA, Lara R (1982) A neural model of the role of the tectum in prey-catching behavior. *Biol Cybern* 44: 185-196

Brower JVZ, Brower LP (1962) Experimental studies of mimicry 6: the reaction of toads (*Bufo terrestris*) to honey bees (*Apis mellifera*) and their dragonfly mimics (*Eristalis vinetorum*). *Amer Naturalist* 96: 297-307

Brower LP, Brower JVZ, Westcott PW (1960) Experimental studies of mimicry 5: the reaction of toads (*Bufo terrestris*) to bumble bees (*Bombus americanorum)* and their robberfly mimics (*Mallophora bomboides*), with discussion of aggressive mimicry. *Amer Naturalist* 94: 343-356

Brzoska J, Schneider H (1978) Modifications of prey catching behavior by learning in the common toad (*Bufo b bufo* L, Anura, Amphibia): Changes in response to visual objects and effects of auditory stimuli. *Behav Process* 3: 125-136

Cott HB (1936) The effectiveness of protective adaptation in the hive bee illustrated by experiments on the feeding reactions, habit formation and memory of the common toad (*Bufo bufo* L). *Proc Zool Soc Lond* 1: 113-133

Doty RW (1987) Has the greedy toad lost its soul; and if so, what was it? A commentary. *Behav Brain Sci* 10(3): 375

Ewert J-P (1965) Der Einfluß peripherer Sinnesorgane und des Zentralnervensystems auf die Anwortbereitschaft bei der Richtbewegung der Erdkröte (*Bufo bufo* L). PhD Thesis, Univ of Göttingen

Ewert J-P, (1968) Der Einfluß von Zwischenhirndefekten auf die Visuomotorik im Beute- und Fluchtverhalten der Erdkröte (*Bufo bufo* L). *Z Vergl Physiol* 61: 41-70

Ewert J-P (1970) Neural mechanisms of prey-catching behavior and avoidance in the toad (*Bufo bufo* L). *Brain Behav Evol* 3: 35-56

Ewert J-P (1971) Single unit responses of the toadss caudal thalamus to visual objects. *Vergl Physiol* 74: 81-102

Ewert J-P (1976) The visual system of the toad: behavioral and physiological studies on a pattern recognition system. In Fite KV (ed) *The amphibian visual system: a multidisciplinary approach.* Academic Press, New York, pp 141-202

Ewert J-P (1980) *Neuroethology. An introduction to the neurophysiological fundamentals of behavior.* Springer-Verlag, Berlin Heidelberg New York

Ewert J-P (1982) Configurational prey-selection by individual experience in the toad. Univ of Mass, Amherst, *COINS Tech Rep* 82-16

Ewert J-P (1984) Tectal mechanisms that underlie prey-catching and avoidance behaviors in the toad. In: Vanegas H (ed) *Neurology of the optic tectum.* Plenum Press, New York, pp 247-416

Ewert J-P, Kehl W (1978) Configurational prey selection by individual experience in the toad *Bufo bufo. J Comp Physiol* 126: 105-114

Ewert J-P, Seelen W v (1974) Neurobiologie und System-Theorie eines visuellen Muster-Erkennungs-mechanismus bei Kröten. *Kybernetik* 14: 167-183

Ewert J-P, Borchers HW, Wietersheim A v (1979) Directional sensitivity, invariance and variability of tectal T5 neurons in response to moving configurational stimuli in the toad *Bufo bufo. J Comp Physiol* 132: 191-201

Grüsser O-J, Grüsser-Cornehls U (1976) Neurophysiology of the anuran visual system. In: Llinás R, Precht W (eds) *Frog neurobiology.* Springer-Verlag, Berlin Heidelberg New York, pp 297-385

Horn R, Greiner B, Horn I (1979) The effect of ACTH on habitutation of the turning reaction in the toad *Bufo bufo* L. *J Comp Physiol* 131: 129-135

Ingle D (1973) Disinhibition of tectal neurons by pretectal lesions in the frog. *Science* 180: 422-424

Kondrashev SL (1987) Neuroethology and color vision in amphibians. A commentary. *Behav Brain Sci* 10(3): 385

Lara R, Arbib MA (1985) A model of the neural mechanisms responsible for pattern recognition and stimulus-specific habituation in toads. *Biol Cybern* 51: 223-237

Lara R, Arbib MA, Cromarty A (1982) The role of the tectal column in facilitation of amphibian prey-catching behavior: a neural model. *J Neurosci* 2: 90-100

Lara R, Cervantes F, Arbib MA (1982) Two-dimensional model of retinal-tectal-pretectal interactions for the control of prey-predator recognition and size preference in amphibia. In: Amari S, Arbib MA (eds) *Competition and cooperation in neural nets.* Springer-Verlag, Berlin Heidelberg New York, pp 371-393

Lashley KS (1929) *Brain mechanisms and intelligence: quantitative studies on injuries to the brain.* Chicago Univ Press, Chicago

Mikulka P, Hughes J, Aggerup G (1980) The effect of pretraining procedures and discriminative stimuli on the development of prey selection behaviors in the toad (*Bufo terrestris*). *Behav Neurol Biol* 29: 52-62

Scalia F (1976) The optic pathways of the frog: nuclear organization and connections. In: Llinás R, Precht W (eds) *Frog Neurobiology.* Springer-Verlag, Berlin Heidelberg New York, pp 386-406

Schürg-Pfeiffer E, Ewert J-P (1981) Investigation of neurons involved in the analysis of gestalt prey features in the frog *Rana temporaria. J Comp Physiol* 141: 139-152

Székely G, Lázár G (1976) Cellular and synaptic architecture of the optic tectum. In: Llinás R, Precht W (eds) *Frog neurobiology.* Springer-Verlag, Berlin Heidelberg New York, pp 407-434

Telemetric Transmission System for Single Cell Studies in Behaving Toads

HANS-WILHELM BORCHERS*) and CHRISTOPH PINKWART

Abteilung Neuroethologie, Fachbereich Biologie/Chemie, Universität Kassel
D-3500 Kassel, FR Germany

Abstract. Comparison between neuronal and behavioral activities in response to sign stimuli requires recordings from single cells during the animal's behavior. For studies of successive action patterns, such as habituation of orienting in response to repetitively presented prey, *wireless transmission* of electrical signals is required. In this paper, we present a miniature three-function system developed for toads that allows us, by means of a telemetric procedure, to record single neurons from the brain and to stimulate and to subsequently lesion the recording site through the recording electrode. Spikes are transmitted by using frequency modulation for distances up to 10 m. Electrical brain stimulation through the recording electrode is performed with the aid of an inductively powered circuit which is combined with the transmitter (mass of the system 2.38 g). Biphasic current pulses can be applied. Pulse frequency and duration are wireless remotely controlled.

1 Introduction

To evaluate "the role of single neurons in the psychology of perception" (Barlow 1985) single cell studies in behaving animals are indispensable (for investigations in toads see Ewert & Borchers 1974; Borchers & Ewert 1978; Borchers 1982a,b; Borchers & Pinkwart 1983). For correlations between a sign stimulus, the neuronal activity, and the onset of an action pattern, wire-bound recording techniques are adequate (see Schürg-Pfeiffer, this volume). However, if the analysis of behavioral sequences is required in which twisting of the signal-transmitting wires must be avoided, or arousal is studied in which any disturbances to the animal must be

---

*) The paper presented here in an edited form was prepared in 1985. Parts were published by C. Pinkwart and the late H.-W. Borchers in the *Journal of Neuroscience Methods* 20: 341-352 (1987). The Editors.

prevented, wire-less recording is necessary. In the present paper, we describe a telemetric three-function transmitting system. The three functions are:
(1) Wireless transmission of extracellularly recorded action potentials from single neurons in the toad's brain does not disturb the animal;
(2) wireless remotely controlled electrical micro-stimulation of the brain through the recording electrode allows us to check the "behavioral relevance" of neurons close to the recorded brain site;
(3) telemarking the recording (and stimulation) site by means of electrolytic micro-lesions is prerequisite for the histological evaluation.

All these functions can be performed with the aid of a radio transmitter, VHF receiver, inductive energizer, telestimulator, and a telemarker. The principle of this three-function system is shown in Fig. 1. Since the animal has to carry the miniature equipment (transmitter, stimulator-marker) on its head during the experiment, a minimum of weight and size as well as lowpower consumption are of great importance. The radio transmitter and the stimulator-marker are composed of miniature chip components that are soldered directly together. The transmitter is embedded in the surface of a small chip of glue, and the stimulator is attached to the bobbin of the coil. Both units are fixed to a microdrive which carries the electrodes. By means of dental cement, the entire head piece is mounted onto the toad's skull (Borchers 1982b). The flow-chart shown in Fig. 2 gives an overview of the methodology and the facilities for data collecting and processing (Glaser & Ruchkin 1976).

2 Telemetric Recording System

2.1 The Radio Transmitter

In radio telemetry, the information is transmitted by means of high frequency (RF) carrier waves (Demling & Bachmann 1970; Fryer 1970; Ko 1970; Mackay 1970; Kimmich & Vos 1972; McElligot 1973; Phillips 1973; Jeutter & Fromm 1980). In the construction of such a system, three points of special importance generally have to be considered:
(1) The choice of the carrier frequency influences the band-width of the transmitted signal, the length of the antenna, and the propagation conditions. The frequency is mainly determined by the tele-communication authorities of each country.
(2) There are different kinds of modulation, for example, pulse amplitude modulation, pulse duration modulation, pulse position modulation, and pulse code modulation. In those methods that use sampling technique, multichannel processing is possible by time-division-multiplexing. Sample systems yielding high precision in the transmission require more elements and are more expensive and bigger. In biotelemetric systems, frequency modulation (FM) is mostly applied. The carrier frequency is modulated according to the amplitude of the information signal. Simultaneous transmission of several channels is possible by means of "frequency division multiplexing."
(3) The power which the transmitter radiates to the receiver depends on the required transmitting range as well as the maximum size, weight and volume of the device, the power source, and the length of the operating period.

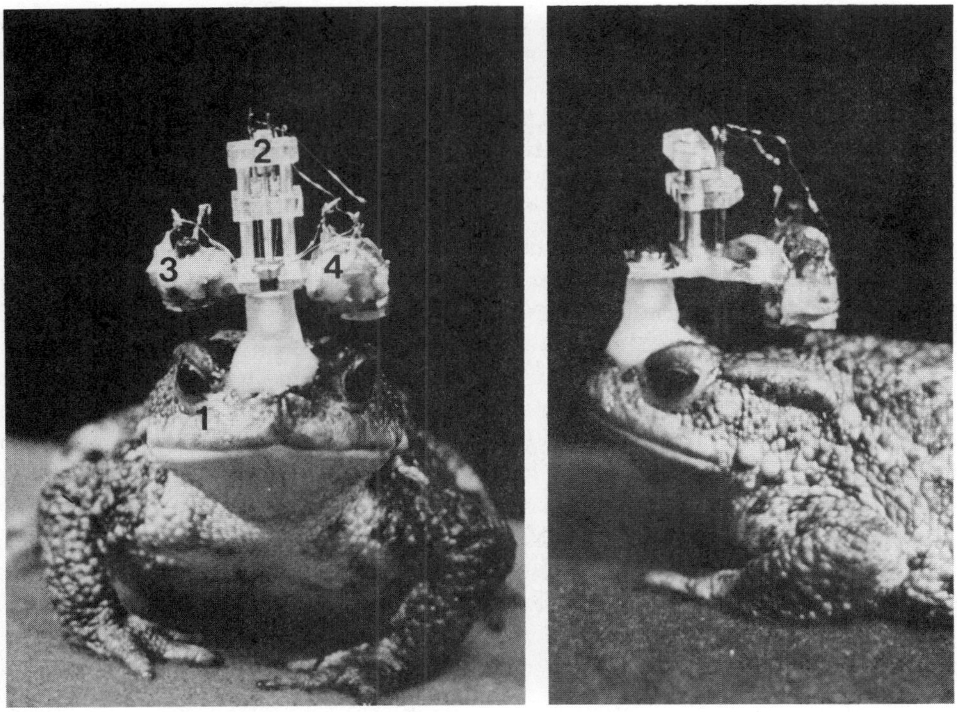

A

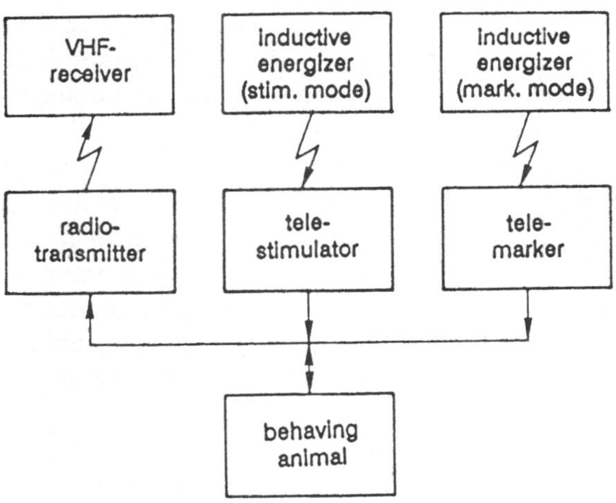

B

Figure 1. (A) Miniature three-function transmitter. 1, toad with head piece; 2, microdrive (width =8 mm); 3, radio transmitter; 4, telestimulator with telemarker. (B) Principle of the three-function system with wireless transmission for single neuron recording, wireless brain stimulation and marking at the same site via the same electrode. (From Pinkwart & Borchers 1987).

Since miniaturization and power consumption are strongly emphasized, the transmitter is designed to operate on frequency modulation at distances up to 10 m. A carrier frequency of 150 MHz is licensed by the German Postal Administration (telecommunication authorities). Depending on the requirements, the transmission frequency can be tuned in the range from 90 to 160 MHz. The transmitter is powered by a small 1.5 V silver oxide cell (Varta V 364) with a resulting operating time of about 50 h.

Figure 3 shows the circuit of the three-function transmitter. The radio-transmitter (Fig. 3A) is designed to work with an electrode pair consisting of an active and an indifferent stainless steel or tungsten electrode (Borchers 1982a). The electrical insulation from ground of the single ended radio transmitter yields a quasi-differential input stage. This permits us to reject various spurious muscle potentials and reduces background noise. The recorded biopotential is amplified and used to modulate the frequency of a VHF transmitter stage whose tuned circuit radiates energy to the VHF receiver. In the input stage, the potential is fed by the capacitor C1 to the gate of the field effect transistor T1 which works in a source circuit (Fig. 3A). T1 provides a high input impedance and an amplification of approximately 7. The high input impedance and amplification free of feedback in the first stage yield a good signal-to-noise ratio (SNR) of 5 and a common mode rejection ratio (CMR) of

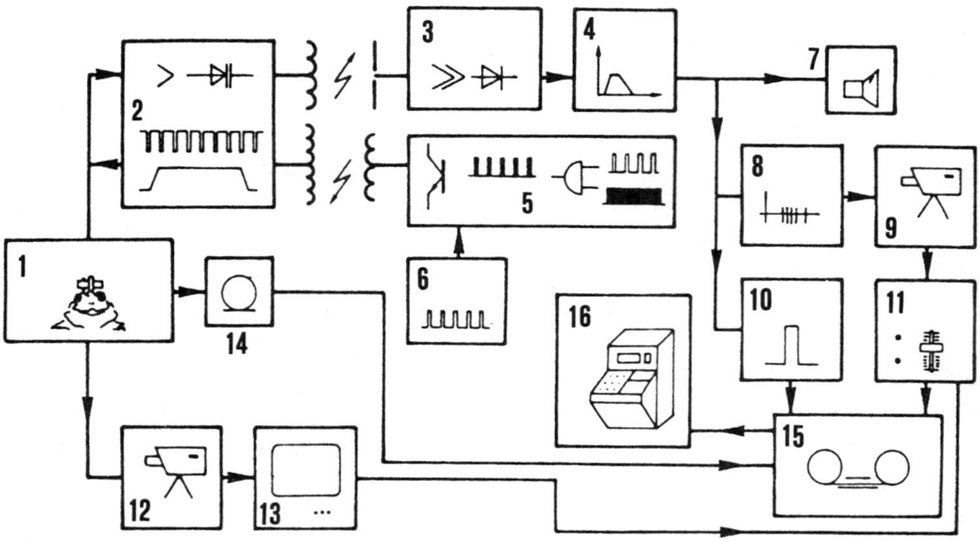

Figure 2. Flow-chart illustrating the procedure for collecting data during telemetric single cell recording, telestimulation, and telemarking in the freely moving animal. (1) Experimental animal, *Bufo bufo spinosus*; (2) miniature transmitter for teletransmission of action potentials; (3) receiver; (4) bandwidth selector; (5) inductive energizer; (6) external pulse generator; (7) audiomonitor; (8) oscilloscope; (9) video camera; (10) pulse former, window discriminator; (11) scan converter; (12) video camera; (13) video monitor; (14) microphone; (15) videotape; (16) computer. (From Pinkwart 1983).

100 dB for the transmitter-receiver system. After current amplification in the second stage, which is realized by means of a low noise npn-transistor T2 in a common collector circuit, the signal modulates the radio frequency of the Colpitts oscillator stage T3 (Tietze & Schenk 1970). Its resonant frequency is mainly determined by the inductance of the coil L1 and the capacitances C5 and C6. The inductor L1 serves also as the transmitting antenna. Frequency modulation is accomplished by the collector capacitance of T3 which is linearly related to the base voltage (Fryer 1970). The transmitter is powered by a single 1.5 V miniature silver oxide cell (Varta V 364). Figure 4A-F illustrates the different steps in the construction of the miniature transmitting device.

In summary, the *VHF radio transmitter system* has the following specifications: Carrier frequency = 150 MHz, tunable ± 1 MHz; modulation sensitivity = 42.3 kHz/mV; signal strength at 1 m from the transmitter = 50 µV; broadcast distance = 10 m in air; input impedance = 20 MΩ; bandwidth = 40 Hz to 20 kHz (-3 dB); noise level (40 Hz to 20 kHz, -3 dB) measured at receiver output = 30 µV peak to peak with 1MΩ source resistance, related to transmitter input = 12 µV peak to peak with both inputs grounded; common mode rejection ratio CMR = 100 dB; input signal range = 0 to 1 mV; DC power drain = 0.3 mA at 1.5 V; battery life = up to 50 h; mass = 0.73 g (VHF transmitter), = 0.35 g (battery); volume = 11 x 7 x 7 mm^3 (incl. battery).

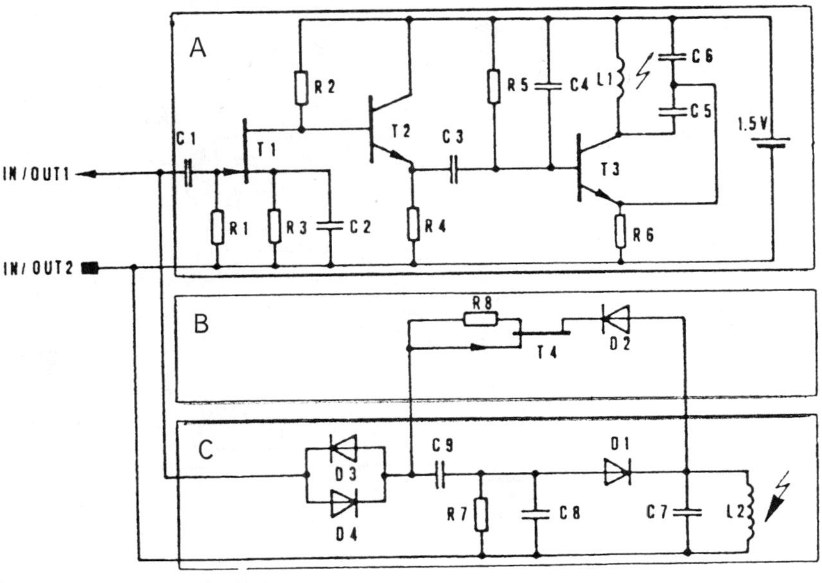

Figure 3. Circuit of the miniature three-function transmitter. (A) Radio-transmitter; (B) telemarker; (C) telestimulator. T1, T4 = BFT 46; T2 = BCW 32; T3 = BFR 93; D1 + D2, D3 + D4 = BAV 99; R1 = 20M; R2, R3, R5, R8 = 100k; R4 = 10k, R6 = 2,5k; C1, C9 = 10n; C2, C3 = 220n; C4, C8 = 1n, C5 = 5p, C6 = 20p, C7 = 330p. BFT 46, BCW 32, BFR 93, and BAV 99 are product numbers of the transistors and diodes. (From Pinkwart & Borchers 1987).

Figure 4. Steps in the construction of the miniature radio transmitter. (A) Components on an assembly plate; (B) plastic-embedded components; (C) soldered components; (D) transmitter in test status; (E) transmitter piece united; (F) end product of the radio transmitter including transmitting coil and battery embedded in plastic. For size comparison cf. distance of holes (4mm) in the plate. (From Pinkwart 1983).

2.2 The VHF Receiver

The flow-chart of the receiver, a modified VHF receiver MIHO 1980, is shown in Fig. 5A. Its main components are: (1) The preamplifier realized with a field-effect-transistor (BF 900); (2) the mixer (SO 42P) that converts the 150 MHz frequency to the intermediate frequency of 10.7 MHz; (3) the oscillator 139.3 MHz; and (4) the demodulator (TBA 120S) which restores the original wave form of the biopotential (Fig. 5B; 1-4). The receiver is equipped with an automatic frequency control, AFC,

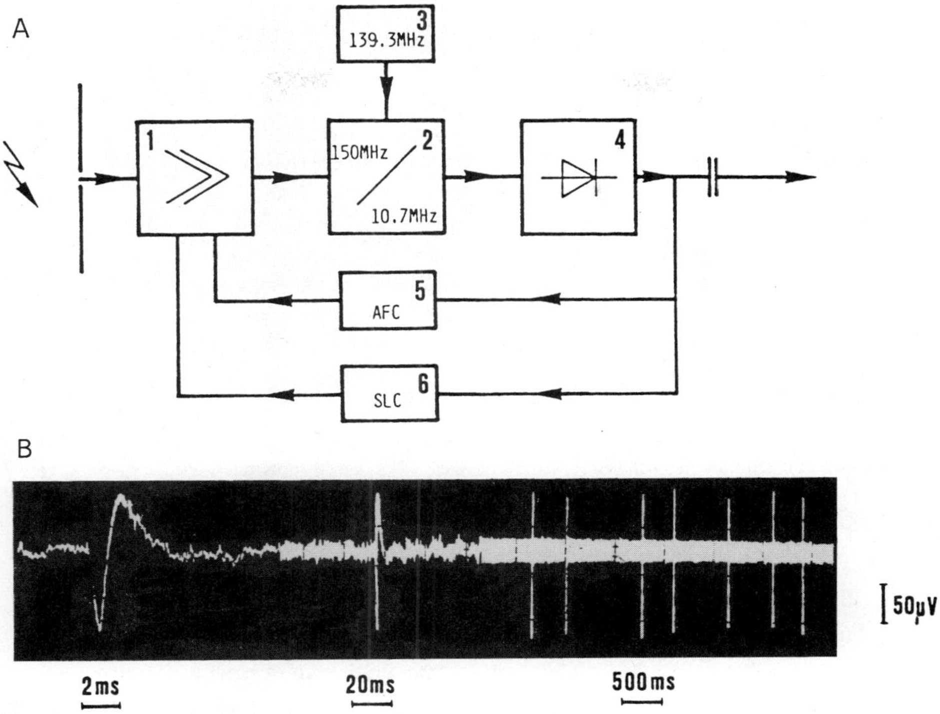

Figure 5. (A) Flow-chart of the VHF receiver. 1, Preamplifier; 2, mixer; 3, oscillator; 4, demodulator; 5, automatic frequency control, AFC, circuit; 6, servo locked circuit, SLC. (B) Example of action potentials telemetrically recorded from a cell in the optic tectum of the behaving toad in response to a moving visual stimulus. Note the different time scales, 2 ms, 20 ms, and 500 ms, of the spike traces. (From Pinkwart 1983).

that increases the tuning band and a servo locked circuit, SLC, which maintains the "optimal tuning point" irrespective of transmitter frequency shifts. The AFC - and SLC-circuits are depicted in Fig. 6. The VHF receiver is driven by the transmitter output via a folded dipole antenna. In order to record and process the received and demodulated signal (Fig. 5B), it can be filtered and amplified.

3 Telestimulation System

3.1 The Inductive Energizer

During a stimulation experiment the animal sits within a plastic cylinder of 28 cm diameter and 20 cm height with an energizing coil wound around it. To get the magnetic field more homogeneous without increasing the inductance, the wire is wound around the cylinder 4 times in parallel. Each coil carries 3 turns with a

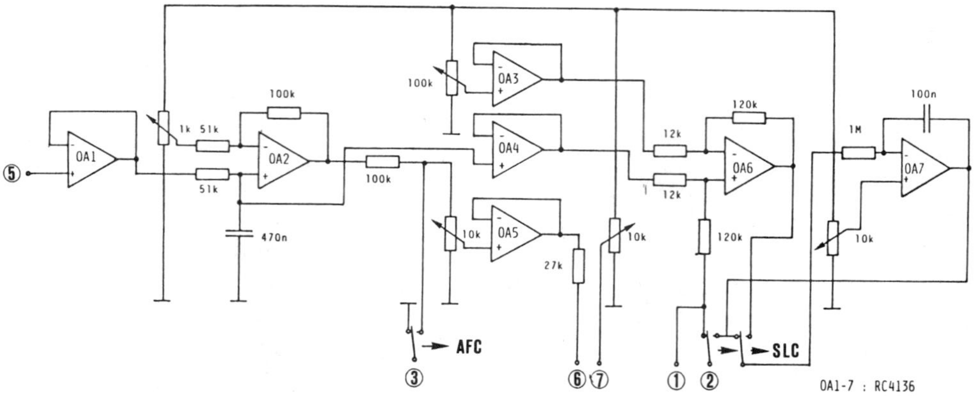

Figure 6. AFC- and SLC-circuits of the 150 MHz- receiver. 1,2, SLC-connection; 3, AFC-connection; 5, AFC-SLC-input; 6,7, adjustment display. (From Pinkwart 1983).

gradient of 1 cm/turn. In this way the magnetic field intensity and the pulse amplitude are independent of the height at which the experiments are performed.

The energizing coil is fed by the inductive energizer. This is built up with integrated circuits for pulse generation and with discrete transistors for power amplification. The flow-chart and its circuit are shown in Fig. 7A and B. By means of an internal pulse generator which is built with two timers NE555, the pulse width, frequency, and train duration of the stimulus are determined. The pulses and a 500 kHz AC voltage from the oscillator ICL8038 are fed to an AND-gate ITT7409 whose output shows a pulsed 500 kHz signal. This signal is preamplified in an npn transistor Darlington stage (2 x 2N1613) and coupled inductively to the transmitting stage. The power transistor T2 (2N3055) works in the Class C-Mode. It yields an output signal with a peak-to-peak voltage swing across the resonated coil L3 and capacitor C3 of nearly twice the supply voltage (Fryer et al. 1978). The amplitude is determined by the emitter follower stage T1 (2N3055).

Because the maximum cylinder diameter depends on the driving voltage of transistor T2 of the inductive energizer, cylinder diameter for use with larger animals, e.g. rats, needs a final stage with a high driving voltage and a high voltage transistor (e.g., MOS-FET).

3.2 The Telestimulator

The telestimulator is designed to stimulate brain tissue electrically by means of biphasic pulses delivered extracellularly via the recording electrode. The circuit of the telestimulator is illustrated in Fig. 3C. To minimize the weight and volume of the telestimulator, it is powered inductively with an energizing frequency of 500 kHz. In the mode "stimulation," this frequency is pulsed according to a chosen

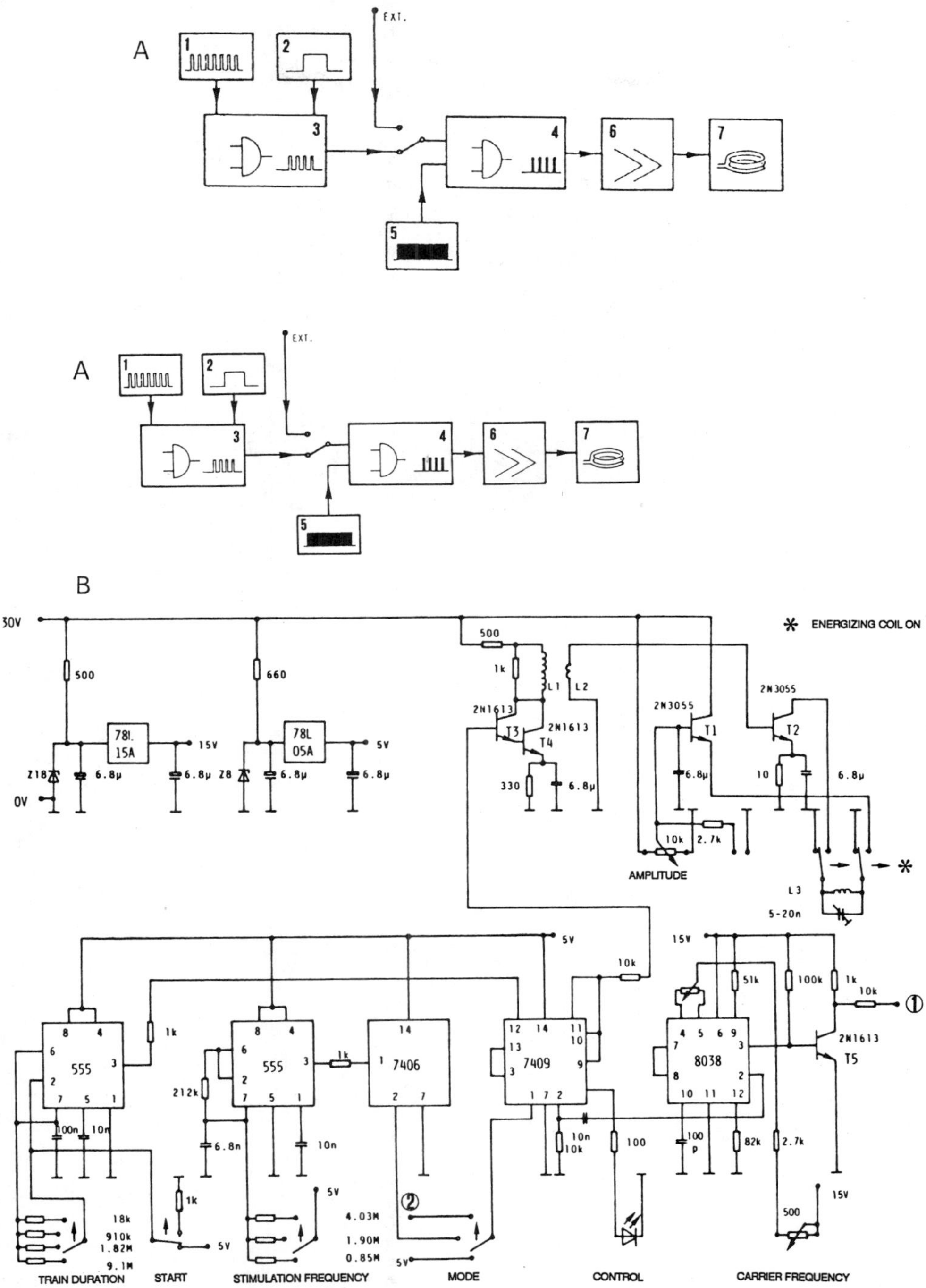

Figure 7. (A) Flow-chart of the inductive energizer. (1) Pulse generator; (2) train duration; (3), (4) AND-gate; (5) 500 kHz-oscillator; (6) amplifier; (7) energizing coil. - (B) Electronic circuit. 1, connection to frequency counter; 2, connection to external input. For details see text. (From Pinkwart 1983).

stimulus pattern, e.g., 1-ms pulses with a pulse frequency of 100 Hz within a train duration of 200 ms. The pick-up coil L2 of the telestimulator (Fig. 3C) consists of approximately 100 turns (insulated copper wire; diameter = 0.1 mm) wound around a ferrite core (diameter = 2 mm, length = 10 mm). This coil is resonated at the

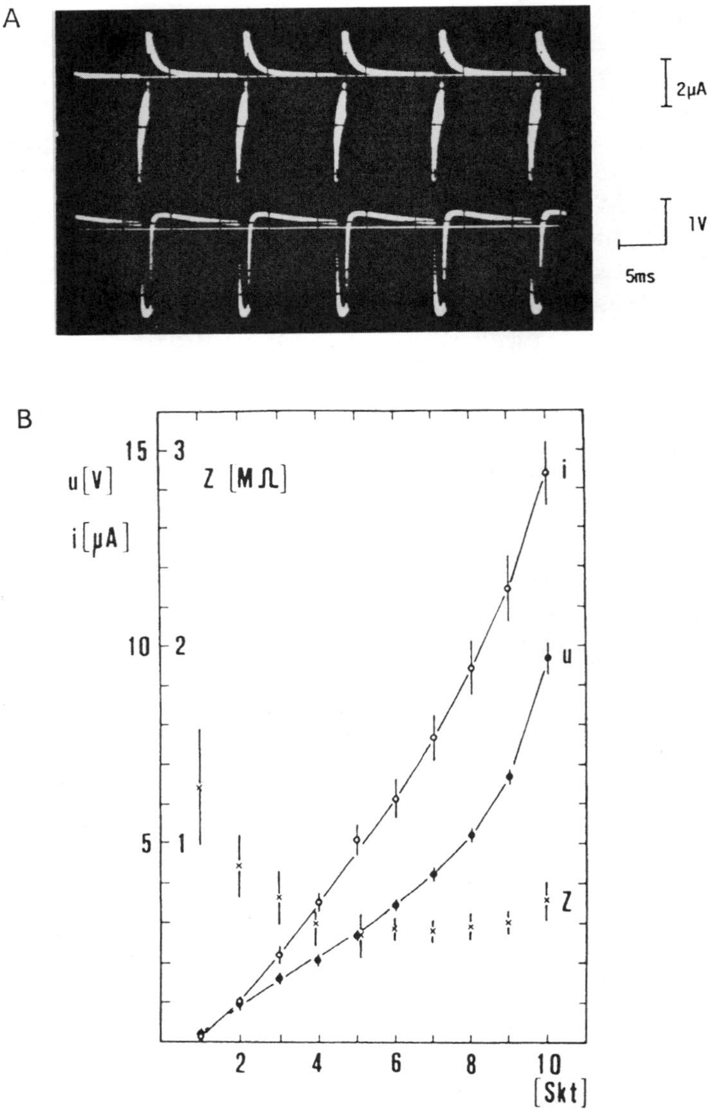

Figure 8. (A) Telestimulator-transmitted current *i* (cf. Aa_1) and assigned voltage *u* (cf. Aa_2) depending on time (abscissa). - (B) Mean value and standard deviation of 10 arrangements (n = 10) of current *i*, voltage *u*, and electrode impedance *z* (using stainless steel electrodes) in dependence of the impulse generator's display (at DC, 1 MHz, -3 dB) in units of scales, Skt (abscissa). (From Pinkwart 1983).

energizing frequency with the HF ceramic chip capacitor C7. During pulses the AC voltage is rectified by the diode D1 to a negative voltage which is collected by the capacitor C8. Simultaneously, the capacitor C8 will be discharged via R7 and via capacitor C9, the electrodes and the brain tissue. Capacitor C9 subserves in combination with the electrode impedance as the differentiator and produces in the mode "stimulation" a biphasic pulse with a net flow of current kept to zero (Fig. 8A) (Seligman 1982). The stimulus current is determined by this impedance and the output voltage of the energizer power stage (Fig. 8B). Therefore, the stimulus current can be remotely controlled (0-15 μA in cylinder center, 0-20 μA near cylinder wall).

In summary, the *telestimulator system* has the following specifications: Energizing frequency = 500 kHz, tuned to the pick up LC circuit, L2, C7, in Fig. 3C; energizing coil size = 280 mm diam, 200 mm height; standard stimulus train durations = 100 ms, 200 ms, 1 s; standard pulse frequencies within train = 50 Hz, 100 Hz, 200 Hz; standard pulse width = 1 ms; pulse shape = biphasic, sharp negative pulse, smooth charge compensation; maximal pulse amplitude = 15 μA, 10 V in cylinder center; external input = from external generator.

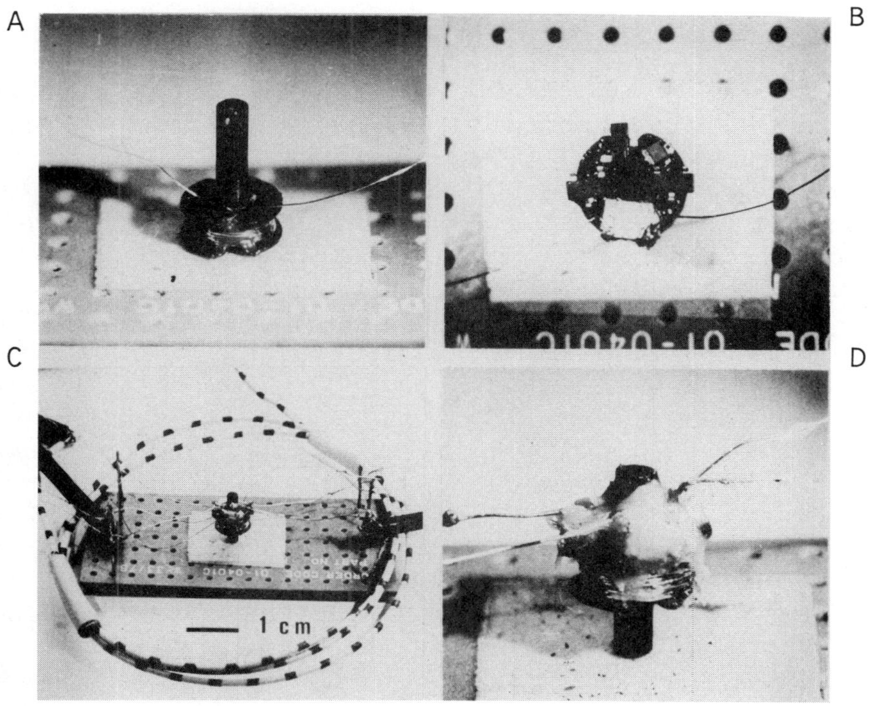

Figure 9. Steps in the construction of telestimulator and telemarker. (A) Transmitting coil on an assembly plate; (B) coil with chips; (C) connected components of the telestimulator in test position; (D) completed telestimulator embedded in plastic. (From Pinkwart 1983).

4 Telemarking System

The telemarking system is designed to mark the recording/stimulation site for subsequent histological evaluation. The telemarker (Fig. 3B) delivers a constant anodal direct current through the recording steel electrode. The current produces an iron deposit at the tip of the electrode that can be identified later in a histological section by means of the Prussian Blue staining procedure (Lee et al. 1969). The circuit is a supplement to the telestimulator and consists of only 3 components: the diode D2, the resistor R8, and the transistor T4 (Fig. 3B). It is powered inductively via the energizer coil L3 and the resonance circuit L2-C7 in the same way as the telestimulator. The AC voltage is rectified by the diode D2 to a positive voltage that supplies the marking circuit. In the "marking" mode the energizing voltage works continuously at the highest level and the constant-current source R8-T4 feeds the current through the antiparallel diodes D3 and D4 into the active electrode. By resistor R8 (Fig. 3B) the marking current is limited to 2.5 µA. Figure 9 illustrates various steps in the construction of the telestimulator and telemarker.

In summary, the *telemarker* has the following specifications: Constant direct current = 2.5 µA $\pm$ 10%; maximal magnetic field intensity = 0.65 mT; mass = 0.67 g (telestimulator plus marker); volume = 10 x 7 x 7 mm^3 (telestimulator plus marker).

5 Subsystems in Independent Operation

With regard to the subsequently described conditions, the VHF radio transmitter, the telestimulator, and the telemarker - all three connected with the same active and indifferent electrodes - operate independently (Fig 3A-C): (i) In the mode "recording" the low impedance of the telestimulator and telemarker is blocked by the antiparallel diodes D3 and D4: at less than the forward voltage they show an extremely high resistance for small signals such as the recorded action potentials. (ii) If the unit works in the mode "stimulation," the high input resistance R1 of the radio transmitter prevents a drop of current. The telemarker subtracts only a very small part of current during pulses. The difference is taken into consideration by means of the calibration of the stimulation current (Fig. 8B). (iii) Since the constant current does not pass the input capacitor C1 in the mode "marking," the marking current is independent of the VHF transmitter. The marking current is independent of the telestimulator if the primary voltage of the energizing coil is increased slowly. Then the negative current is blocked by capacitor C9.

In case the amplitude of the stimulus pulse current remains small ($I < 3.5$ µA), the time of saturation remains shorter than 10 ms. This is accomplished by the input resistor R1 of the radio transmitter. Thus, in the case of small stimulus pulse current it is possible to record and stimulate simultaneously.

6 Conclusions

The three-function system, with emphasis on extremely small size and low weight (Eichenbaum et al. 1977), allows the experimenter to transmit single neuron activity, to stimulate the recorded brain tissue, and to mark the recorded site. The three modes are usually carried out sequentially in an experiment. Since the stimulation and recording sites are identical, it is possible to get an immediate feedback of the neuronal activity in the stimulated region. This permits one to stimulate with very small current intensities (Hambrecht 1978) certain groups of neurons previously identified by their response characteristics. The change-over from one mode to the other (recording-stimulation-marking) can be carried out by remote control, thus the whole experiment can be performed without the need to handle the animal, so that shifts of the electrode tip can be avoided.

Our experiments in freely moving toads of the subspecies *Bufo bufo spinosus* (L.) have demonstrated that the three-function transmitting system works very well for the correlation of neuronal pulse trains and corresponding motor patterns (Borchers 1982b; Borchers & Pinkwart 1983; Pinkwart 1983). Furthermore, information on interconnections between brain regions can be obtained by stimulating parts of the brain different from that where the neuronal activity is being recorded (for a review see Ewert 1984). Pulse frequency, pulse width, stimulation current intensity or train duration can be changed, remotely controlled from the pulse generator. The stimulation current is calibrated in relation to the position of the pick up coil of the telestimulator at the animal's head. The telestimulator demodulates the signal of the pulse generator and produces biphasic stimulation pulses. The curve shape of the pulses is similar to the shape of extracellularly recorded action potentials.

Finally, we select two examples illustrating the advantage of this recording technique. Figure 10 shows that we are able to electrically stimulate and to record

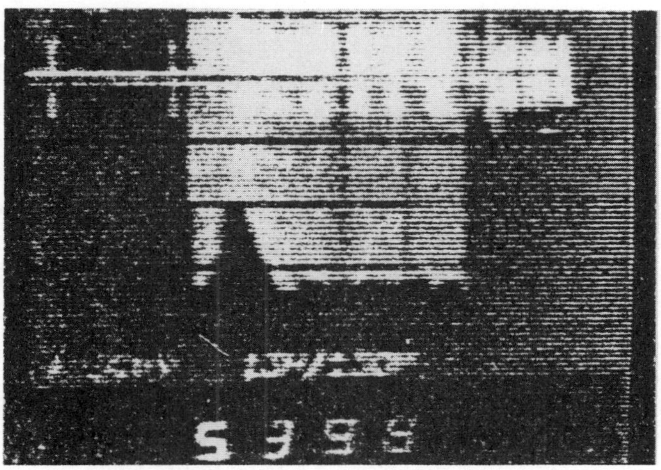

Figure 10. Example of simultaneous telestimulation and recording. Oscillogram (2s/sweep) of a teletransmitted spike train of a neuron recorded in the toad's subtectum, superimposed by an electrical impulse train (current: 3μA) delivered by the telestimulator via the same microelectrode. Note a weak background activity of the visual-tactile-vibratory sensitive neuron prior to electrical brain stimulation; the neuronal activity is strongly increased during brain stimulation and persists for a while after stimulation. The toad responded with avoidance behavior, such as fearful ducking. (From Pinkwart 1983).

single unit activity simultaneously. Figure 11 depicts single cell recording and toad's behavior video-monitored frame-by-frame. For a review on behavior-correlated neuronal activity in common toads see Schürg-Pfeiffer (this volume).

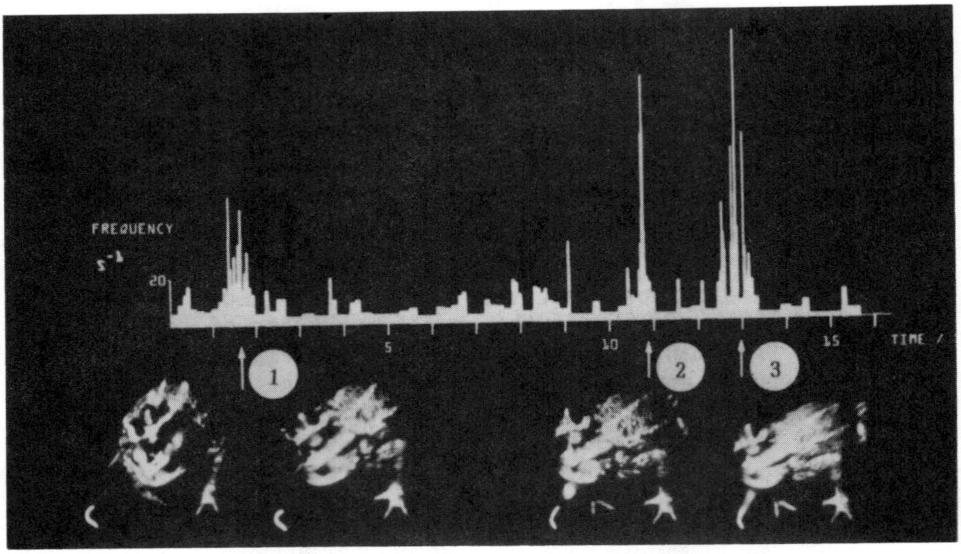

Figure 11. Example of tele-recording a single cell and parallel video-monitoring toad's behavior in response to a meal worm. The interspike-frequency-time-histogram shows that the response of a tectal T8 "movement cell" precedes any kind of behavioral movement, here, during toad's prey-catching: (1) orienting, (2) successless snapping, (3) successful snapping; selected video frames are shown. Unlike T5.2 cells, the T8 cell's activity was also correlated with movements (e.g., locomotion) not related to prey-catching, not shown here. (From Borchers & Pinkwart 1983).

Acknowledgement. This research was supported by the Studienstiftung des Deutschen Volkes.

7 References

Barlow HB (1985) The role of single neurons in the psychology of perception. The Twelfth Bartlett Memorial Lecture. *Quart J Exp Biol* 37: 121-145

Borchers H-W (1982a) Advances in the development of micro-recording systems and single unit analysis in toads (*Bufo bufo* L). *J Theor Neurobiol* 1: 268-285

Borchers H-W (1982b) Correlation between behavior patterns and single unit responses from the optic tectum in the freely moving toad (*Bufo bufo* L). In: Trappl R, Pask G, Ricciardi L (eds) *Progress in cybernetics and systems research, Vol 9.* Hemisphere Publishing Corporation, Washington, pp 109-117

Borchers H-W, Pinkwart C (1983) A telemetry system for single unit recording in the freely moving toad (*Bufo bufo* L). In: Ewert J-P, Capranica RR, Ingle DJ (eds) *Advances in vertebrate neuroethology.* Plenum Press, New York, pp 603-612

Borchers H-W, Ewert J-P (1978) Eye closure in toads (*Bufo bufo* L) does not produce *off*-responses in retinal *on-off* ganglion cells: a question of efferent commands. *J Comp Physiol* 125: 301-303

Demling L, Bachmann K (eds) (1970) *Biotelemetric.* Thieme-Verlag, Stuttgart

Eichenbaum H, Pettijohn D, Deluva AM, Chorover S (1977) Compact miniature microelectrode system. *Physiol Behav* 18: 1175-1178

Ewert J-P (1984) Tectal mechanisms that underlie prey-catching and avoidance behaviors in toads. In: Vanegas H (ed) *Comparative neurology of the optic tectum.* Plenum Press, New York, pp 247-416

Ewert J-P, Borchers H-W (1974) Antworten retinaler Ganglienzellen bei freibeweglichen Kröten. *J Comp Physiol* 92: 117-130

Fryer TB (1970) Implantable biotelemetry system. *Report NASA SP* 5094, USA

Fryer TB, Gordon FL, Bill AW (1978) An inductively powered telemetry system for temperature, EEG, and activity monitoring. *Biotelemetry Patient Monitoring* 5: 53-76

Glaser EM, Ruchkin DS (1976) *Principles of neurobiological signal analysis.* Academic Press, New York

Hambrecht FT (1978) Functional electrical stimulation. In: Cobb WA, Van Duijn H (eds) *Contemporary clinical neurophysiology* (EEG Suppl No 34). Elsevier, Amsterdam, pp 369-372

Jeutter DC, Fromm E (1980) A modular expandable implantable temperature biotelemeter. *IEEE Trans Biomed Engineer* 27: 5

Kimmich HP, Vos JA (1972) *Biotelemetry.* Meander, Leiden

Ko WH (1970) Biotelemetry. In: Cynes M, Milsum JH (eds) *Biomedical engineering systems.* Inter-University Electronics Series, McGraw-Hill, New York

Lee BB, Mandl G, Stean JPB (1969) Micro-electrode tip position marking in nervous tissue: a new dye method. *J Electroenceph Clin Neurophysiol* 27: 610-613

Mackay RS (1970) *Biomedical telemetry.* Wiley & Sons Inc, New York London Sidney Toronto

McElligott JG (1973) A telemetry system for the transmission of single and multiple channel data from individual neurons in the brain. In: Phillips MJ (ed) *Brain unit activity during behavior.* CC Thomas, Springfield, IL, pp 53-66

MIHO Electronic Corporation (1980) *Rundfunkempfänger 7805 and 2m Konverter 7712V.* Kassel FR Germany

Phillips MI (ed) (1973) *Brain unit activity during behavior.* CC Thomas, Springfield Ill

Pinkwart C (1983) *Entwicklung eines Systems zur radiotelemetrischen Einzelzellableitung mit der Möglichkeit der Telestimulation und -Markierung im Mittelhirn freibeweglicher Erdkröten* (*Bufo bufo L).* PhD Thesis, Fachbereich 19, Univ Kassel

Pinkwart C, Borchers H-W (1987) Miniature three-function transmitting system for single neuron recording, wireless brain stimulation and marking. *J Neurosci Methods* 20: 241-352

Seligman LJ (1982) Physiological stimulators: from electric fish to programmable implants, IEEE *Trans Biomed Eng* 29 (4): 270-284

Tietze U, Schenk C (1970) *Halbleiterschaltungstechnik.* Springer-Verlag, Berlin Heidelberg New York

Functional Neural Systems Analyzed by Use of Interregional Correlations of Glucose Metabolism

BARRY HORWITZ

Laboratory of Neurosciences, National Institute on Aging
National Institutes of Health, Bldg. 10, RM. 6C-103
Bethesda, MD 20892, USA

Abstract. In order to learn which brain regions are functionally associated with one another in a specific group of subjects during a particular experimental paradigm, a computer-assisted quantitative method was developed for analyzing regional rates of glucose uptake. For each pair of brain regions, the partial correlation coefficient, controlling for whole brain glucose utilization, is calculated between the regional cerebral metabolic rates for glucose for all the subjects. A mathematically derived correlation matrix is constructed in which the statistically significant relationships are displayed. Applications of this novel approach to both human and animal studies are discussed. Similar techniques using electrical data (obtained from both micro-electrodes and scalp electrodes) are reviewed, and the strengths and weaknesses of these approaches are examined.

1 Introduction

Many of the papers included in this volume attempt to elucidate the neural mechanisms involved in sensorimotor behavior in amphibians. This elucidation requires that we determine the neural substrates mediating the *functional activity* (e.g., which neural components are involved when a toad sees a worm, and attacks it) and show how these neural elements implement the behavior under examination (e.g., see Betts, this volume). In essence, this is an example of what neuroscientists do for the many other behaviors and organisms that they study. The general problem then in defining *functional systems* in the brain is first determining what neural elements comprise the system (or behavior) of interest, and then ascertaining how the functioning of these elements results in the behavior under study (i.e., what algorithms are "computed" by the neural structures).

Grobstein (this volume) provides an interesting illustration of the complexity involved in understanding functional systems. He points out that the same object, a worm, can evoke at least one of two different prey-catching action patterns: if the worm is sufficiently close to the frog, orienting and snapping are released. However, if the frog is not close enough to reach the worm with his tongue, he jumps toward the prey object, and then attempts to catch it. Thus, diverse motor outputs can be produced in response to visual inputs that share prey features, but are distinct by the local sign of the stimulus (see concepts of "sensorimotor code" and "command releasing system" discussed by Ewert, this volume).

Given that, at least in the adult organism, the anatomical relations among brain elements are relatively fixed, a key problem in understanding a set of behaviors involves determining what changes take place in the *functional relations* among neural elements as the different behaviors are performed. To do this, it is necessary to obtain and interrelate data from many brain structures at the same time. Although there have been several efforts in the last 20 years to do this, this approach is essentially still in its infancy. In this chapter, I shall review our efforts at attempting to interrelate data from many brain regions (Horwitz et al. 1984, 1986, 1987, 1988; Soncrant et at. 1986a,b). We have used metabolic measures of *functional activity* as the baseline data to interrelate. Others have used electrical activity (Gerstein 1970, 1987; Gerstein et al. 1978, 1985; Gevins et al. 1985, 1987; Livanov et al. 1973) which will be reviewed briefly.

2 Simultaneous Recordings of Electrical Activity

2.1 Multiunit Microelectrode Recording

One of the first significant efforts at trying to develop techniques for displaying and understanding data obtained from simultaneous microelectrode recordings was that of George Gerstein and his collaborators. The initial approaches, devised to assess whether two neurons had covarying patterns of impulse activity (and if such patterns were related to the times of stimulation of the neural system under study), employed such techniques as the cross-interval histogram, the cross-correlation histogram, and the joint peri-stimulus scatter diagram (reviewed in Gerstein 1970). More recently, an ingenious method has been developed for investigating the *cooperative firing pattern* of many neurons recorded simultaneously (Gerstein et al. 1985; Gerstein & Aersten 1985). The key ingredient involves representing each neuron as a particle in a multidimensional space (one dimension for each recorded neuron). Assigned to each particle is a "charge" that is related to the spike train of the corresponding neuron. The resulting forces on the particles lead to an aggregation of those particles whose corresponding neuronal firings are related in time to one another.

The use of such techniques has shown that cooperative firing of neuronal assemblies can occur over short, medium, and long time scales (Gerstein 1987), and that temporal coherence in neuronal firing can be a dynamic, time varying property of the neuronal ensemble (Gerstein 1974).

2.2 Optical Recording of Voltage Sensitive Dyes

There have been a number of attempts to obtain many *in vivo* simultaneous microelectrode recordings (e.g., Kruger 1982; Reitboeck 1983; see also Reitboeck, this volume). However, the difficulty of this procedure has led others to try indirect approaches. One technique that is starting to provide interesting results involves the use of voltage sensitive dyes that fluoresce when the transmembrane voltage changes (Cohen et al. 1978, Grinvald 1985). Figure 1 illustrates one example from the work of Grinvald et al. (1984). Shown is the real time mapping of visually evoked responses (an average of 40 trials) in the intact frog tectum. Two loci of activity, corresponding to two different light spots presented sequentially to the contralateral eye, are apparent. Another example comes from the work of London et al. (1985), in which the firing pattern of 46 neurons in the *Aplysia* buccal ganglion during repetitive stimulation of the cerebral buccal connective was recorded.

The technical development of these optical methods is continuing, and data soon will be available that can be analyzed by the procedures worked out by Gerstein and his collaborators (Gerstein 1987).

2.3 Scalp Recorded Electrical Activity

At the more macroscopic level, scalp recorded evoked electrical activity has been used as the database from which to infer *functional relations*. The most elaborate effort of this sort has been undertaken by Alan Gevins and his collaborators (Gevins et al. 1985, 1987). In one experiment (Gevins et al. 1987), the subject was to respond to a visual stimulus by pressing a button with either his left or right hand, depending on the nature of the visual stimulus. Strong correlations in the shape of the evoked potential during a 187 ms time interval centered about the finger response were found between central supplementary motor cortex and regions in the expected hemisphere (i.e., left, if the right finger was used). This pattern of correlation was different during other temporal segments of the evoked response (e.g., during the stimulus to response interval). These results suggest the presence of shifting functional relations over different portions of the behavior.

3 Metabolic Measures of Functional Activity

3.1 Principles and Techniques

We have used the *correlation method* with data obtained by measuring regional cerebral metabolism (e.g., the 'deoxyglucose method'). The idea here is that the more active a nerve cell is, the greater its oxidative metabolism (Roy & Sherrington 1890; Sokoloff 1981). Thus, at least for normal subjects, regional glucose metabolism is a measure of a brain region's functional activity (see also contributions by Gonzalez-Lima and Finkenstädt, this volume).

The method most commonly used to examine cerebral glucose metabolism was developed by Sokoloff et al. (1977). Reivich et al. (1979) modified this method to make it applicable for use in humans utilizing *positron emission tomography* (PET). The radiolabeled tracer employed in the Sokoloff techniques is 2-deoxyglucose (2DG), which differs from glucose in having a hydrogen atom, instead of an OH group, attached to the second carbon. Like glucose, 2DG crosses the blood-brain barrier, enters the cell, and is phosphorylated by hexokinase into deoxyglucose-6-

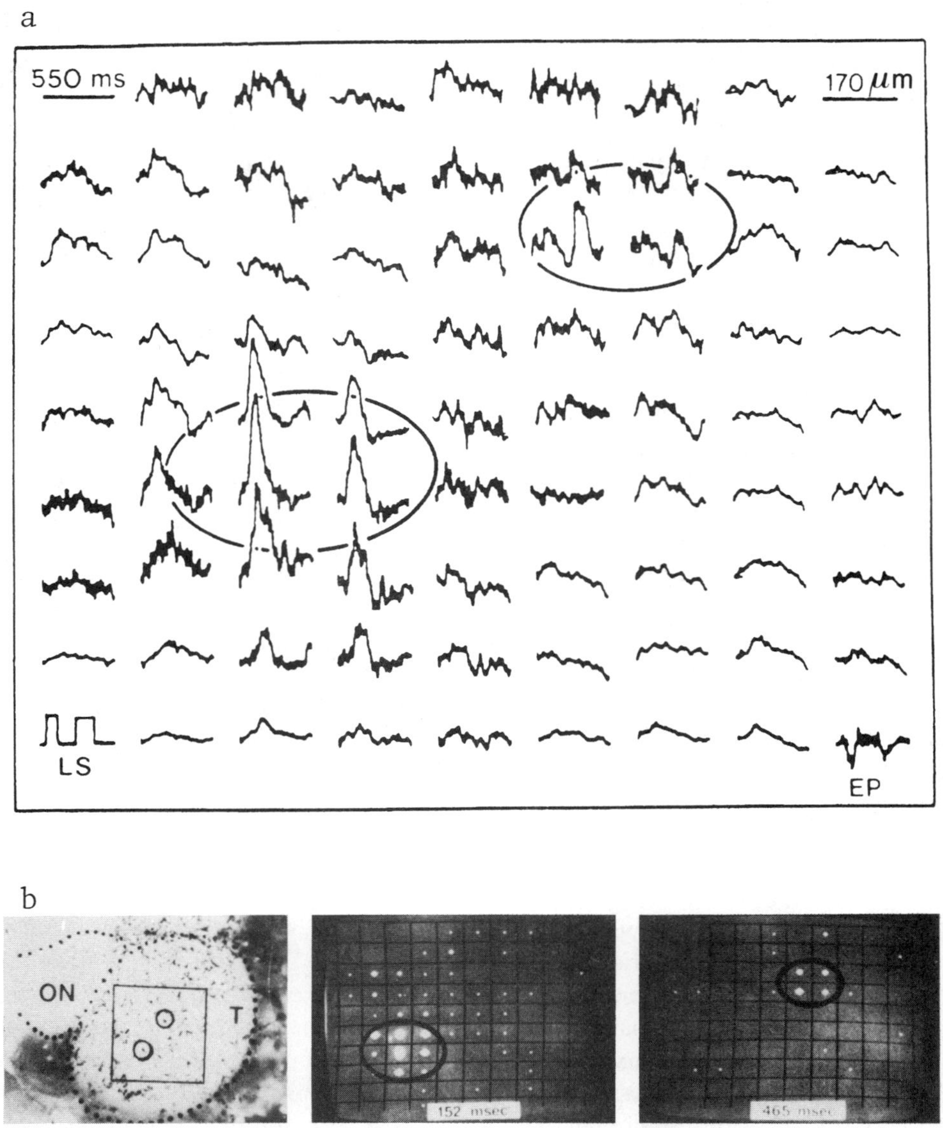

Figure 1. Real time mapping of visually evoked responses in the intact frog tectum using a voltage sensitive dye. (a) Displayed at the top is the fluorescent response recorded from 9x9 photodectors positioned over the tectum image. The average of 40 trials following the presentation of two different light spots given one right after the other to the contralateral eye is shown; two loci of activity are observed. (b) On the bottom right is the visualization of the spatial patterns of activity, at two different times, depicting the two loci of activity; ON, optic nerve; T, tectum. The superimposed image of the tectum is shown at the bottom left. (From Grinvald et al. 1984; reprinted by permission from Nature, Vol. 308, p.849, copyright 1984 Macmillan Magazines Ltd.).

phosphate (D6P). However, the absence of the OH group on the second carbon prevents the labeled substance from proceeding further down the glycolytic pathway, essentially trapping D6P in the cell. Because of the absence of branch points prior to this step, the cerebral metabolic rate for glucose (CMRglc) can be measured, at steady-state, by the rate of accumulation of D6P.

The Sokoloff method (Sokoloff et al. 1977) entails a bolus injection of labeled 2DG, waiting about 45 min for D6P to accumulate, and then using quantitative autoradiography to determine the distribution of label in the brain. If one also measures the time course of plasma radioactivity, then an operational equation derived by Sokoloff enables one to calculate regional CMRglc (rCMRglc). The advantage of the 2DG method is that it allows one to look at the functional activity in the entire brain, although what one is seeing is the integrated activity of the functional activity over the 45 min time interval of tracer accumulation.

For animal studies, ^{14}C-2DG is generally employed, whereas for human studies with PET, either ^{11}C-2DG or ^{18}F-fluorodeoxyglucose (FDG) is used. PET is based on the principle of positron-electron annihilation. When a positron (positive electron) is emitted from a radionuclide, it soon encounters an electron, and the two mutually annihilate and produce two 511 KeV gamma rays that move away from one another at an angle equal to approximately 180°. Detecting these two gamma rays in coincidence determines the line along which the radioactive decay occurred. Thus, combining the techniques of computed tomography with coincidence detection of annihilation gamma rays allows one to reconstruct an image of the distribution of radioactivity in a subject's brain following the administration of a positron emitting radionuclide (for a comprehensive review of PET, see Phelps et al. 1986).

3.2 Problems in Interpretation

As with any technique, there are problems in interpreting *metabolic maps*. One problem has to do with the fact that both excitatory and inhibitory synaptic potentials may entail increases in cerebral metabolism. This was probably best demonstrated in a study by Ackermann et al. (1984) who produced long-duration recurrent inhibition of hippocampal pyramidal cells by means of low frequency stimulation of the fornix. During the stimulation period, intravenously injected 2DG was taken up. The resulting autoradiograms showed that suppression of pyramidal cell firing was accompanied by hippocampal 2DG uptake. This result is reasonable because the highest neuronal metabolism is thought to occur in the neuropil, where axonal terminals make large numbers of synapses. Thus, rather than associating regions of hypometabolism with inhibition, Ackermann et al. (1984) concluded that disfacilitation is the more likely mechanism. Nevertheless, what determines whether a region is hyper- or hypometabolic is the balance between excitatory and inhibitory activity, which can depend very much on the particular geometrical configuration within a neural structure.

A second problem has to do with how one should interpret changes in regional metabolism. Functional activation of the brain can be produced by sensory stimulation, cognitive activation, or drug action, for example. The usual way in which these experiments are done is as follows: one has a control state (e.g., the resting state) in which rCMRglc is determined; then the particular activation paradigm is used, and rCMRglc again determined. The interpretation is straight-forward enough if a region increases its metabolism, and this region is thought to be involved in the paradigm.

This is illustrated in Fig. 2A. Given the results of Ackermann et al. (1984), we may get the same result whether region 1 excites or inhibits region 2. For the remainder of his paper we shall assume for simplicity that both excitation and inhibition lead to metabolic increases.

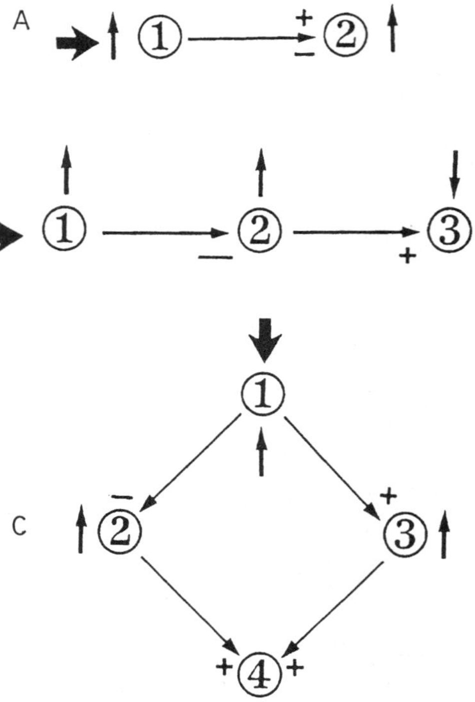

Figure 2. (A) Brain region 1 projects to region 2 where it makes either excitatory (+) or inhibitory (-) synapses. Under the experimental conditions being studied, region 1 is activated (thick arrow). Glucose metabolism in both regions 1 and 2 increases (thin, upward-pointing arrows). (B) Brain region 2 normally excites region 3. During activation, region 1 inhibits region 2, thus reducing the excitation of region 3. Glucose metabolism increases in 1 and 2, and decreases in 3. (C) Prior to activation, region 2 excites region 4. During stimulation, region 1 is activated, inhibits region 2, and excites region 3; although the excitation from 2 to 4 is reduced, the activity of region 4 remains high because 3 now excites it. Although region 4 is part of the functional system in both the activated and non-activated states, there is no change in its glucose utilization during activation.

What do we infer if we get a decrease in some region's metabolism following stimulation? One interpretation is shown in Fig. 2B. Region 1 inhibits region 2 (and hence rCMRglc in both 1 and 2 increases), but since 2 is now inhibited, it no longer activates region 3, as it had done in the baseline state. Thus, rCMRglc in region 3 goes down (note that 2 could have been inhibiting 3 in the baseline state, as well as exciting it as shown in Fig. 2B). To make this inference, however, one must know something about the functional neuroanatomy of the regions to start with. Moreover, in a typical PET or 2DG study, one has not three, but many regions to take into account.

The most difficult situation to deal with is the case where no change in rCMRglc is detected following stimulation. The correct statistical interpretation is the following: No inference can be made about the effect of the stimulus on that region's metabolism, although the result is not inconsistent with the hypothesis that the stimulus does not affect the region's functional activity. However, the notion often developes that if no change is found, the stimulus has no effect. As illustrated by Fig. 2C, no change in rCMRglc in region 4 can be consistent with the stimulus having a significant effect on region 4. Suppose prior to stimulus onset, region 2 excited 4. As region 1 is stimulated, it inhibits 2 and excites 3, and now 3 excites 4. Although rCMRglc in region 4 is unchanged, it is clear that 4 participates in the functional neuroanatomy associated with this stimulus (as do regions 1, 2, and 3), as it did in the baseline paradigm. However, looking at the absolute values of rCMRglc one might suggest that 4 does not participate in the stimulus paradigm.

We conclude from these illustrations that not only must we determine whether a stimulus paradigm causes a region's metabolism to increase or decline, but to understand the *functional interactions* of the system, one must as well determine how a brain region's relation to other regions is altered by the paradigm, and this cannot be determined just by looking at the change in value of the imaged parameter. This point is particularly important when considering studies of cognitive functions, where the stimulus paradigm is much more complicated than for simple sensory or motor behaviors, and the functional neuroanatomy is even less well understood.

4 Correlation Matrix Method

4.1 Correlation Method

In order to determine which brain regions are functionally associated for a specific paradigm, we have developed (Horwitz et al. 1984) a correlation method [similar in spirit to that used by Gevins in his electrical studies (Gevins et al. 1985)], whose fundamental hypothesis is shown in Fig. 3. If two brain areas, such as region 1 and region 2, are *functionally coupled* during the experimental conditions under study, so that activity in one region is a function of the activity in the second, then if we plot the rCMRglc in region 1 against that in 2 (each data point represents the values obtained from a single subject), one will find a statistically significant correlation coefficient. A larger correlation coefficient implies a stronger functional association between the two regions. Likewise, if two regions, such as 1 and 3 are not functionally

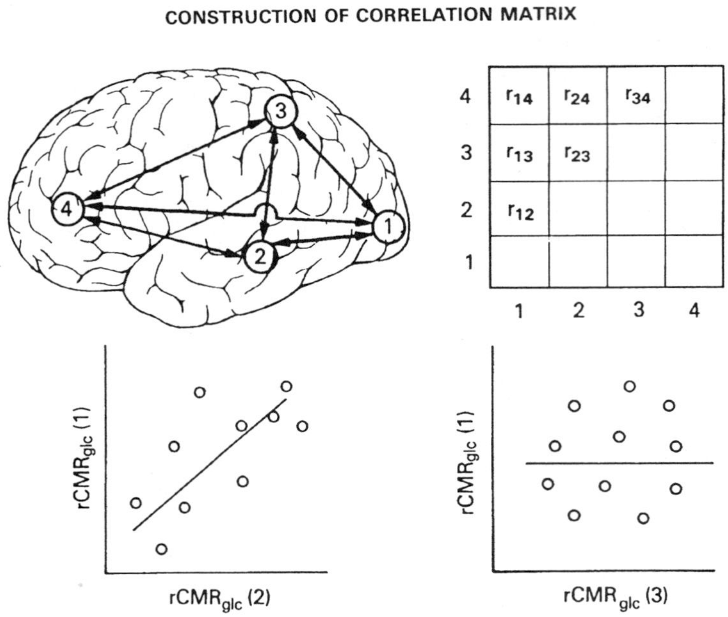

Figure 3. Drawing illustrating the fundamental hypothesis of the correlation method. The correlation matrix is shown in the upper right. r_{ij} is the correlation coefficient for the ij*th* region pair. Because $r_{ij} = r_{ji}$, and $r_{ii} = 1$, only the coefficients above the diagonal are indicated. For the specific case shown, represented by the plots in the lower half of the figure, r_{12} would have a large value, whereas r_{13} would have a value close to zero. (From Horwitz 1987).

coupled under the conditions studied, the correlation coefficient will have a small value.

This notion is turned around and it is hypothesized that if we find a large correlation coefficient between the glucose metabolism in two regions, we infer that these two regions are functionally coupled in the particular experimental state being investigated. Of course, there also can be artifactual correlations, and care must be taken to minimize these.

The way in which this correlation approach is implemented also is illustrated in Fig. 3. We evaluate the correlation coefficient between every pair of regions of interest. For example, we use the rCMRglc values in regions 1 and 2 in each subject to evaluate r_{12}. We repeat this for each pair of regions (e.g., r_{12}, r_{23}, etc.). The

---

Figure 4. The 59 brain regions from which glucose metabolic data was obtained in resting humans by PET. The names of the regions are given in Horwitz et al. (1984). (From Horwitz et al. 1984).

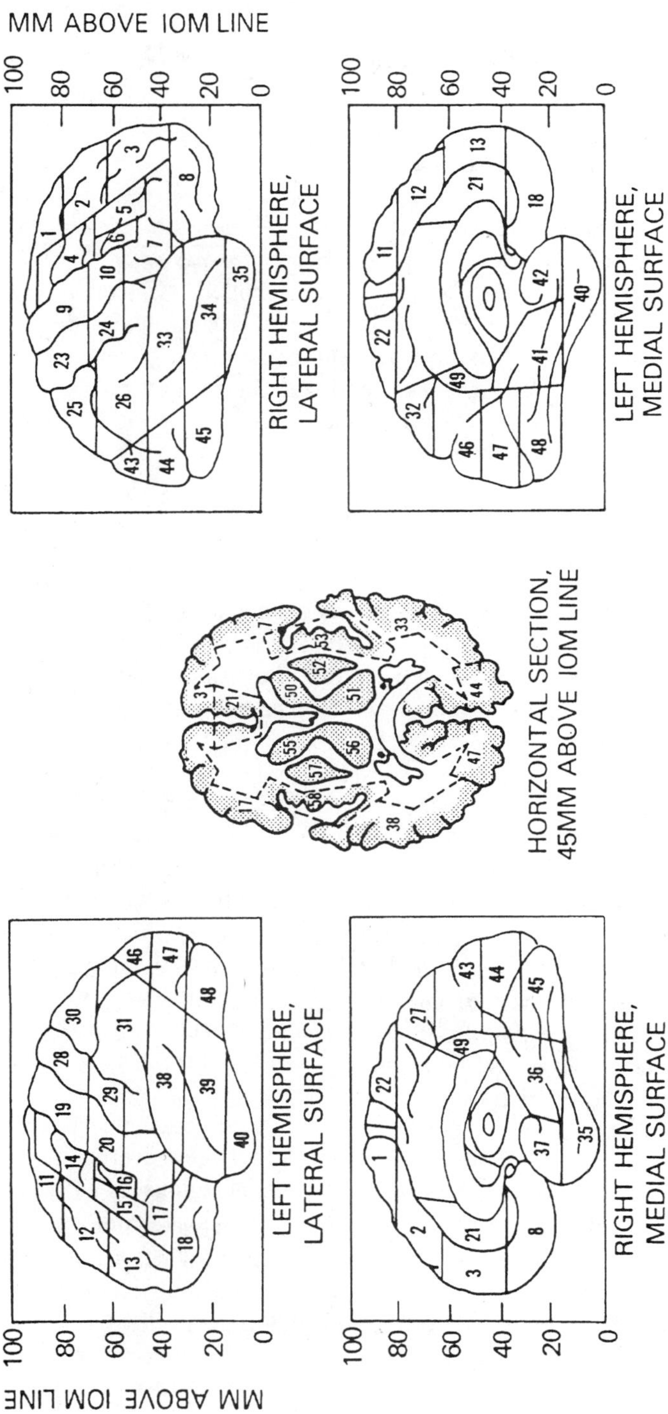
MM ABOVE IOM LINE
100
80
60
40
20
0
RIGHT HEMISPHERE, LATERAL SURFACE
LEFT HEMISPHERE, MEDIAL SURFACE
HORIZONTAL SECTION, 45MM ABOVE IOM LINE
LEFT HEMISPHERE, LATERAL SURFACE
RIGHT HEMISPHERE, MEDIAL SURFACE
MM ABOVE IOM LINE

correlation matrix, using all these values, is thus assembled. Only one half of the matrix is needed, since the diagonal elements are always equal to 1 ($r_{ii}=1$), and the matrix is symmetric (i.e., $r_{ij}=r_{ji}$).

This method has been used in both human and rat studies (Horwitz et al. 1984, 1986, 1987, 1988; Metter et al. 1984a,b; Clark et al. 1984; Soncrant et al. 1986a,b). For our human studies, rCMRglc is determined in 59 regions of interest (ROIs): 28 bilateral and 3 midline regions (see Fig. 4; the names of the regions are given in Horwitz et al. 1984). Having 59 ROIs results in a correlation matrix with 1711 entries. Because intraindividual differences in cerebral glucose metabolism are small compared to interindividual differences, *partial correlation coefficients* (Thorndike 1978), in which whole brain metabolism is the controlled variable, are used.

The power of the correlation approach resides in being able to look at all correlations at once, so that patterns of relationships can be discerned. In order to present the entire matrix, a symbol is used to represent a correlation coefficient that meets some criterion of statistical acceptability. If the correlation does not meet the criterion, it is left blank in the matrix.

Our first use of the correlation method was to see if the distribution of correlation coefficients was non-random (Horwitz et al. 1984). PET data corresponding to 40 very healthy males aged 21-83 yrs were used (Duara et al. 1984). Scanning was done on an ECAT II tomograph (Phelps et al. 1978) with a spatial resolution of 1.7 cm in both the image and axial planes. The subjects were in the resting state during the 45 min uptake period for the radiolable ^{18}FDG: lying supine, awake, blindfolded, ears plugged with cotton. Thus, the dominant sensory input was somatosensory, but there was no specific activation, no patterned input.

Figure 5 shows the correlation matrix obtained. Displayed are those correlations statistically significant at the $P < 0.01$ level. Positive correlations are denoted by "+", negative ones by "0". A negative correlation means that an increase in metabolism in one region is related to a decrease in the second. The brain regions have been grouped by lobe (frontal, F; parietal, P; temporal, T; occipital, 0); some noncortical regions, such as the cerebellum and several of the basal ganglia, have been collected together, as shown. For each large grouping, regions in the right (R) hemisphere are listed first, followed by regions in the left (L). Dashed lines are used to separate the hemispheric groups; the dashed line separating the left frontal lobe from the right parietal lobe distinguishes two midline regions. Similarly, the dashed line separating the left occipital lobe from the right noncortical areas distinguishes a midline occipital region. The numbers along the axes correspond to the region numbers of Fig. 4.

The major pattern of the correlation matrix shows many interactions within and between frontal and parietal regions, within and between temporal and occipital regions, but little coupling between these two large domains. An important finding is that all bilateral regions have statistically significant correlations with their homologues on the opposite side of the brain, and generally, these correlations are among the largest in value.

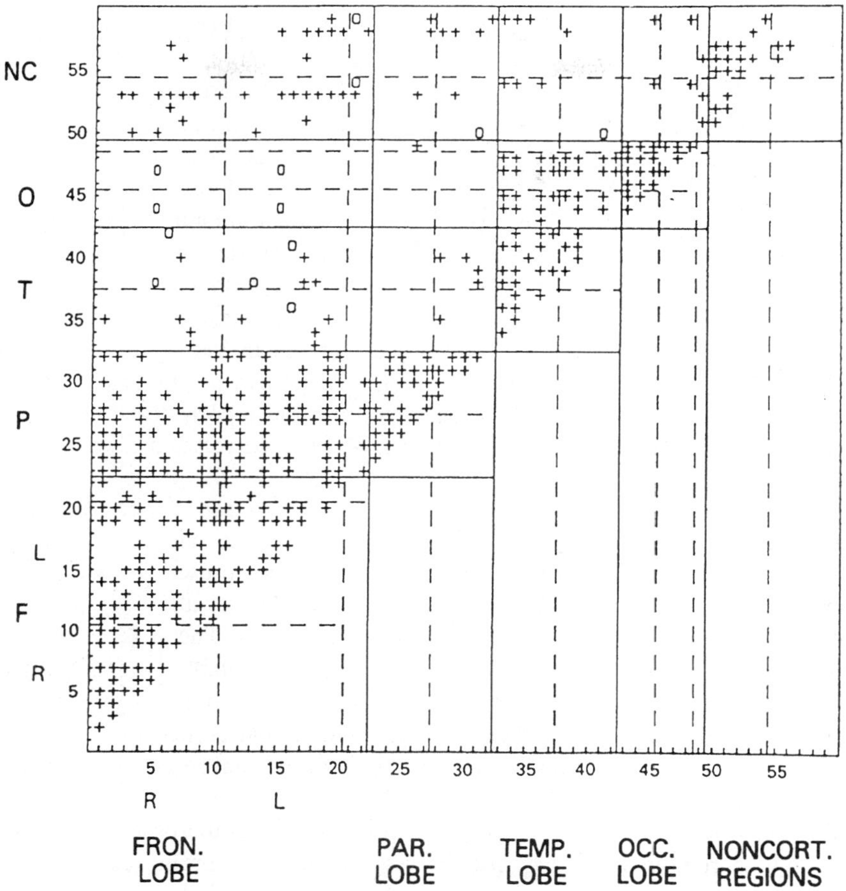

Figure 5. Partial correlation matrix for the 40 healthy men at rest. Displayed are the statistically significant partial correlations ($P<0.01$) which are positive (+) or negative (0). The numbers along the axes correspond to the region numbers of Fig.4. F, frontal; P, parietal; T, temporal; O, occipital; NC, noncortical; R, right hemisphere; L, left hemisphere. (From Horwitz et al. 1984).

4.2 Application to Alzheimer's Disease

The correlation matrix method has been used by our laboratory to investigate differences in functional associations between a variety of subject groups. Differences were found between healthy young and elderly men (Horwitz et al. 1986), and between autistic men and age- and sex-matched controls (Horwitz et al.1988). Differences also were obtained between 21 Alzheimer disease (AD) patients and 21 healthy age-matched control subjects (Horwitz et al. 1987). The AD group, as shown by their scores on the Folstein Mini-Mental State exam (Folstein et al. 1975), were

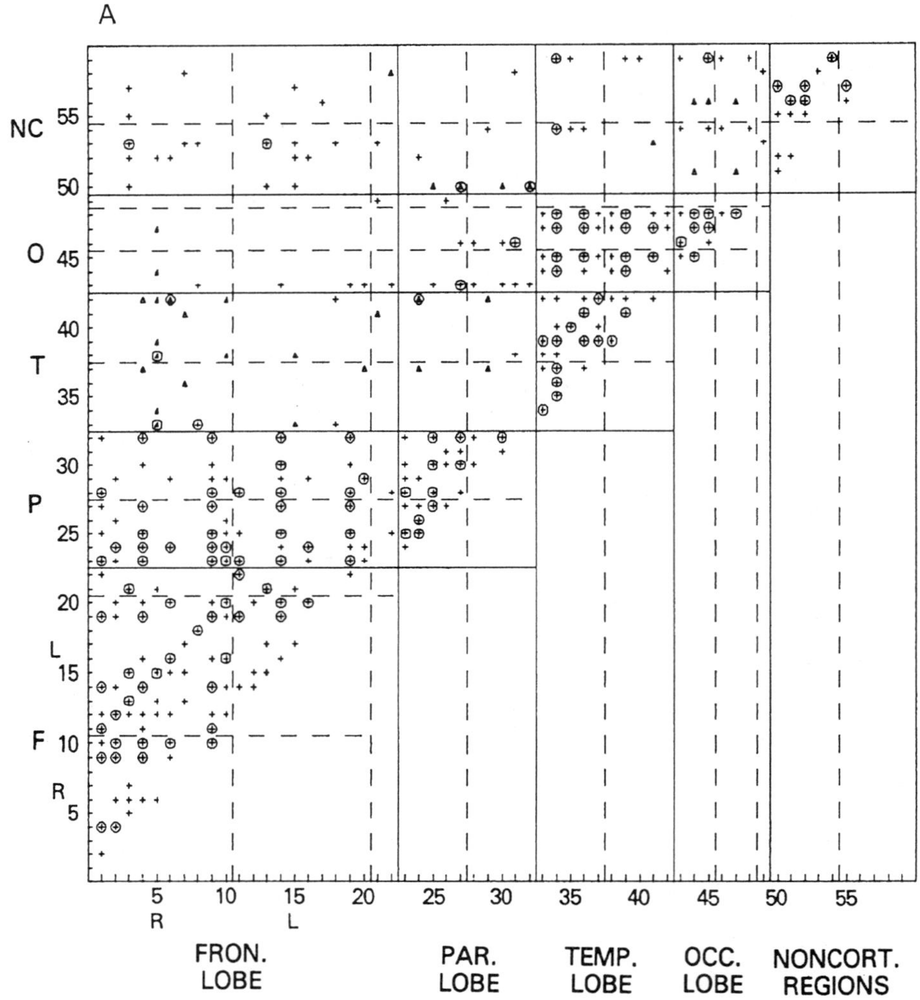

Figure 6. (A) Partial correlation matrix for 21 healthy age-matched control men. Shown are the correlations whose absolute magnitudes exceeded 0.5. Positive correlations are denoted by "+", negative ones by small triangles. Those which are circled meet the reliability criteria (all jackknifes > 0.5; 19 of 20 bootstraps > 0.5). (B, opposite page) Partial correlation matrix for 21 Alzheimer's disease patients. Notation same as in (A). Abbreviations are the same as in Fig.5.

predominantly mildly-to-moderately demented. Subjects were in the same resting state described earlier.

The correlation matrices for both groups are shown in Fig. 6. Displayed are those correlations whose absolute value was greater than 0.5 (significant at $P < 0.025$). Positive correlations are denoted by "+", negative correlations by the small triangles. Because correlation coefficients can be influenced by a few values when the

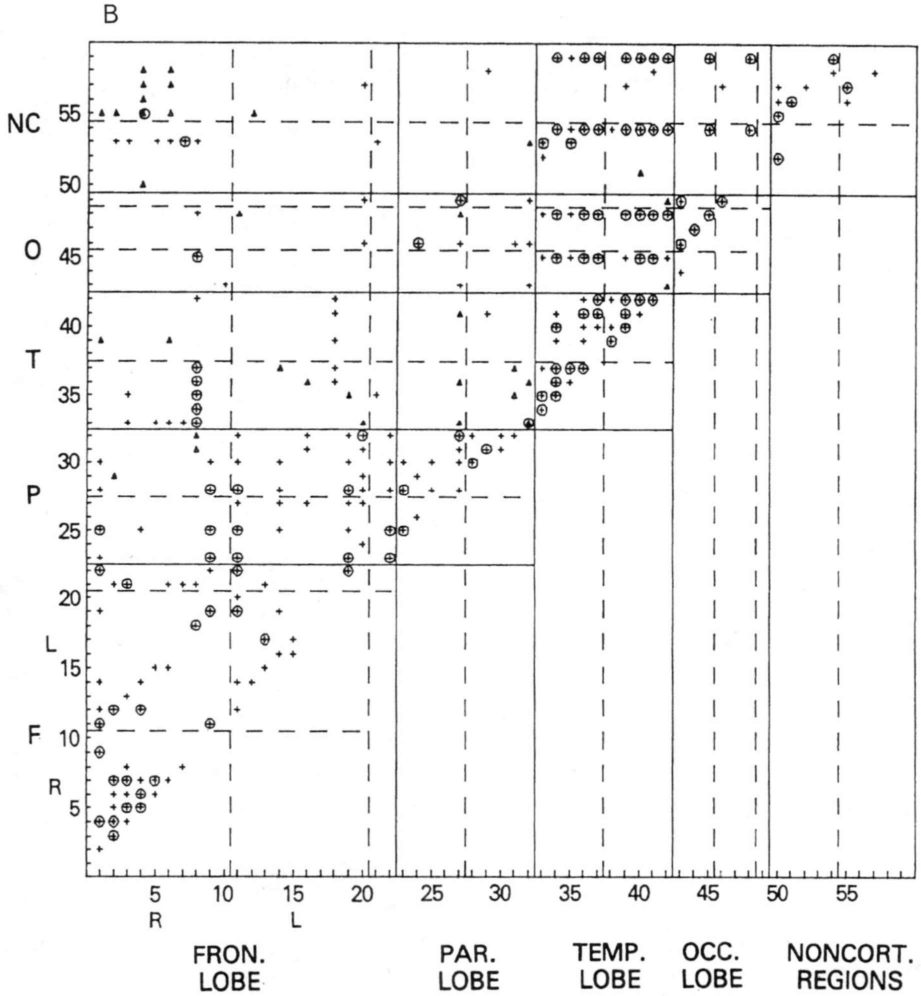

number of data points is small, several procedures were used to make the criteria for statistical acceptability more strict. One procedure, the "jackknife" (Efron 1981), is used to eliminate large correlations due to a single outlier. In this technique, one subject is eliminated from the sample, and the partial correlation coefficient is recomputed for the remaining 20 subjects. This is repeated 20 times, eliminating each subject once. Any correlation that has a large value due to one outlier will show a large drop in value when that subject is eliminated. A "bootstrap" procedure (Efron 1981) is employed to eliminate correlation coefficients with large variances. In the bootstrap, 21 subjects are randomly selected, with replacement, from the sample of 21, and the partial correlation coefficient recomputed. This is repeated 20 times. The bootstrap provides a measure of the variance of each correlation, thus estimating its statistical accuracy. Therefore, we could define a set of "reliable" partial correlation coefficients which not only were statistically significant ($P<0.025$), but also met certain jackknife and bootstrap criteria (all 20 jackknifed correlations > 0.5; at least 19 of 20 bootstrapped correlations > 0.5). "Reliable" correlations are circled in Fig. 6.

There is no statistically significant difference in the total number of reliable correlations between the whole brain matrices for the AD patients and the controls. The matrix for the AD group has fewer reliable correlations between regions in the frontal lobe with regions in the parietal lobe than does the matrix for the control group. This result is consistent with neuropathological findings (Pearson et al. 1985; Rogers & Morrison 1985) that suggest a loss of cortico-cortical fibers between the frontal and parietal lobes in AD, which in turn implies a breakdown in functional communication between these regions (Rapoport et al. 1986).

A finding of great interest is that while most (82%) of the control group's homologous correlations (those between right-left homologous brain regions) are reliable, 50% of those for the AD group are not reliable. This loss of reliable homologous correlations in the AD group corresponds to an increased asymmetry in metabolism between the two hemispheres in this disease. Haxby et al. (1985) in our laboratory have demonstrated that this metabolic asymmetry correlates with an asymmetry in neuropsychological perfomance. For example, AD patients with greater language rather than visuo-constructive deficits have decreased left-sided *vs.* right-sided glucose utilization. Similarly, AD patients with greater visuo-constructive *vs.* language difficulties have a greater right-sided hypometabolism. Our finding of a loss of large correlations between homologous brain regions in AD is not inconsistent with the idea that homologous correlations arise primarily from a mutual activation of homologous brain regions by lower brain regions, such as the locus coeruleus and the basal forebrain, whose neurons project widely to cortical areas, and that in AD asymmetric cell loss in these lower brain areas results in asymmetric deafferentation of their target sites, thus leading to asymmetric metabolism. Recently, Arendt et al. (1985) found hemispheric asymmetry in the amount of cell loss in the basal forebrain in AD. The corpus callosum may also mediate, in part, the large homologous correlations.

4.3 Application to Deoxyglucose Data from Rats

Correlational analysis also was performed on deoxyglucose data from twenty-four 3-month old Fischer-344 rats studied in the resting state (Soncrant et al. 1986a). Large positive correlations between nearby regions, both in cortical and noncortical areas of the brain, and large negative correlations between distant pairs of regions were found. Left-right homologous regions generally were among the most highly correlated region pairs. These results are somewhat similar to those found in humans, studied at rest by PET (Horwitz et al. 1984).

As stated above, the strong functional couplings between left-right homologues may be mediated in part by the corpus callosum. To test this, the correlation matrix for 16 rats that had undergone corpus callosotomy was examined and compared to the correlation matrix for 15 sham-operated controls (Soncrant et al. 1986b). The following questions were asked: (1) How do correlations between cortical regions in opposite hemispheres change following transection of the corpus callosum? (2) How do correlations between cortical regions in the same hemisphere change following corpus callosotomy? (3) In particular, what is the effect of corpus callosotomy on correlations between right-left homologous cortical regions?

Sectioning of the corpus callosum was performed via a paramedian incision and verified later histologically. Because axon degeneration is fairly complete by 7 days, the deoxyglucose procedure was undertaken 10 days post-surgery. All rats were studied in the resting state: awake, hindlimb restraint, no specific sensory stimulation. Following the injection of ^{14}C-2DG, a 45 min time interval was allowed before the animals were sacrified. Since we are interested only in relative rates of glucose consumption (one region relative to another), brain radioactivity, measured by the optical density of autoradiographs, was used as the indicator of relative regional glucose uptake.

No major focal differences in brain radioactivity between groups were evident from qualitative examination of the autoradiographs. However, transection of the corpus callosum reduced the values of partial correlations between rCMRglc for regions in opposite hemispheres. Furthermore, cutting the corpus callosum reduced the values of *partial correlation coefficients* between rCMRglc for pairs of cortical regions in opposite hemispheres, but increased them for cortical region pairs within the same hemisphere. Finally, correlations between homologous cortical regions were reduced in value following corpus callosotomy, although generally, the size of the reduction was small.

These results show that transection of the corpus callosum changes not only the *functional interactions* between regions in opposite hemispheres, but it also alters intrahemispheric relations.

5 Discussion and Conclusions

It is important to keep in mind that the *correlation methods* we have described are exploratory data analysis techniques. In this sense, the statistical techniques employed are being used descriptively in an attempt to assess the pattern of correlations so that *functional associations* may be inferred. Obviously, the results suggested by these methods must be compared with information obtained by other physiological and anatomical techniques.

However, it is equally important to realize that correlation-like methods are likely to be the only basis for defining and understanding *functional systems* in the brain, at least in higher vertebrates. With their limited repertoire of behaviors, and their relatively small number of neurons, invertebrates can be studied essentially by numerous single unit recordings (or, at least by recording from only a few neurons simultaneously), from which data for determining functional relationships can be obtained (e.g., see Getting & Dekin 1985). Although many papers in this volume demonstrate that such techniques can be used for some vertebrate studies, nevertheless, it seems clear that understanding the neurobiology of higher vertebrate behaviors will require the analysis of *simultaneous multiunit data*, obtained during the performance of the behavior.

Each of the methods we have reviewed that attempt to ascertain functional associations among brain regions (or neurons), as with all scientific methods, have individual strengths and liminations. For *multiunit neuronal recording*, the major strength lies in the rich detail of temporal activity one obtains, and in the rather precise spatial localization of this activity. Specific limitations include the fact that very few of the neurons involved in responding to the particular experimental conditions under study can be recorded from, and the anatomical connectivity of the recorded units often is unknown.

In principle, the use of *voltage sensitive dyes* permits one to observe a much larger array of neuronal activity (although usually only in two dimensions), and at the same time to maintain the high spatial and temporal resolution that comes from working at the cellular level. However, formidable technical problems limit the quality of information that can be obtained using this technique (Grinvald 1985). Besides problems of dye toxicity, and finding the appropriate dye for a given experimental preparation, two interrelated problems make it difficult to evaluate quantitatively the neuronal responses in a precise manner: (1) The size of the fluorescent signal is hard to calibrate, being a function, as it is, of membrane area, dye binding, and dye sensitivity; thus, the relative sizes of the optical signal from different parts of the same preparation, or from different preparations, may not be directly related to differences in transmembrane voltage; (2) dye bleaching during the course of an experiment decreases the size of the optical signal, making it difficult to interpret differences over time. Nevertheless, determining functional relations over a limited time course is possible using the presence or absense of a spike as the significant datum.

The use of *macroscopic measures* of functional activity (i.e., scalp recorded evoked potentials, glucose metabolism or blood flow), although yielding a more comprehensive collection of data, clearly suffers from poorer spatial resolution. Like microelectrode data, scalp recorded evoked potentials have excellent temporal resolution, and thus, functional relations between brain regions that change over time intervals on the order of a few hundred milliseconds can be studied. Moreover, because multiple trials within a single subject are performed, and what is beeing correlated are the shapes of the potential waveforms, it is possible to ascertain how functional relationships change within a single individual during different tasks.

Two major limitations of *scalp recorded evoked potentials*, however, are apparent. First, the spatial resolution of the scalp recordings is quite poor, even compared to other macroscopic techniques such as PET. Furthermore, interpretation of the recorded potentials with respect to spatial localization of functional activity can be confused because of volume conduction effects. That is, the response recorded at one electrode is not due necessarily to electrical activity near that electrode. The second major problem is the fact that the sources of event related potentials are not known (Gevins & Cutillo 1986). Consequently, the exact physiological meaning of significant correlations in unclear.

Using metabolic data as the basis to infer functional relations has a number of significant advantages. First, spatial resolution, while coarse, is fairly good (for animal studies, the *deoxyglucose technique* using quantitative autoradiography has a resolution of about 200 μm for ^{14}C-2DG, and on the order of 20 μm for ^{3}H-2DG (Smith 1983); for human studies employing PET, current tomographs provide a spatial resolution of about 5 mm). Moreover, activity in both cortical and subcortical structures can be obtained.

As we have discussed, one key limitation in the use of metabolic data is that it is hard to differentiate changes in metabolism due to excitation from those due to inhibition. The temporal resolution of these techniques is also quite poor, ranging as it does from about 1-2 min for blood flow measurements using PET (Herscovitch et al. 1983), to 30-45 min for the deoxyglucose method (Sokoloff et al. 1977). Therefore, what is determined by these procedures is related to the time integrated functional activity over the measurement period. Although this may be the quantity of interest, it is important that the subjects be maintained in a steady-state during the measurement period.

Perhaps the most limiting factor associated with employing metabolic data to ascertain functional relations is that the number of data values used in evaluating a correlation coefficient is relatively small. Obviously, a separate animal is required to obtain one set of values in the deoxyglucose method. Although repeat studies can be performed with PET (e.g., Chang et al. 1987; Fox et al. 1984), the number of repetitions is limited by the amount of radioactivity that can be given. Moreover, because PET is a relatively invasive and complex procedure, it is difficult to do studies with large samples.

Certain problems are common to all the methods that attempt to interrelate data from multiple units. Uncertainty exists as to the linearity of the relation between functional activity and the measured parameter. Another problem concerns the use of traditional statistics on data that comes from nonlinearly-interacting elements. Many brain regions that project to an area receive projections back from that area, and thus it will be important to understand how these feedback interactions are reflected in terms of functional associations.

Because better techniques are now available to acquire and analyze large quantitaties of data, these *correlation-type methods* are starting to be more widely used. Understanding sensory, motor, and cognitive behavior requires that the *interrelational aspects* of functional brain activity be ascertained. These relations are not apparent from examining data from a single recording site. Consequently, the use of the correlation-type technqiues reviewed here will be needed to probe these associational features of brain function.

Acknowledgements. Some of the work reviewed here was carried out in collaboration with, among others, R. Duara, T. Soncrant, C. Grady, J. Haxby, and S.I. Rapoport. I wish to thank these colleagues for useful discussions, and T. Soncrant for a careful reading of the manuscript. I am grateful to A. Grinvald for kindly providing me with Figure 1.

6 References

Ackermann RF, Finch DM, Babb TL, Engel J (1984) Increased glucose metabolism during long-duration recurrent inhibition of hippocampal pyramidal cells. *J Neurosci* 4: 251-264

Arendt T, Bigl V, Tennstedt A, Arendt A (1985) Neuronal loss in different parts of the nucleus basalis is relatad to neuritic plaque formation in cortical target areas in Alzheimer's disease. *Neurosci* 14: 1-14

Chang JY, Duara R, Barker W, Apicella A, Finn R (1987) Two behavioral states studied in a single PET/FDG procedure: theory, method, and preliminary results. *J Nucl Med* 28: 852-860

Clark CM, Kessler R, Buchsbaum MS, Margolin RA, Holcomb HH (1984) Correlational methods for determining regional coupling of cerebral glucose metabolism. A pilot study. *Biol Psychiat* 19: 663-678

Cohen LB, Salzberg BM, Grinvald A (1978) Optical methods of monitoring neuron activity. *Ann Rev Neurosci* 1: 171-182

Duara R, Grady C, Haxby J, Ingvar D, Sokoloff L, Margolin RA, Manning RG, Cutler NR, Rapoport SI (1984) Human brain glucose utilization and cognitive function in relation to age. *Ann Neurol* 16: 702-713

Efron B (1981) Nonparametric estimates of standard error: the jackknife, the bootstrap, and other methods. *Biometrika* 68: 589-599

Folstein MF, Folstein SE, McHugh PR (1975) "Mini-Mental State" - a practical method for grading the cognitive state of patients for the clinician. *J Psychiat Res* 12: 189-198

Fox PT, Mintun MA, Raichle ME, Herscovitch P (1984) A non-invasive approach to quantitative functional brain mapping with $H_2{}^{15}O$ and positron emission tomography. *J Cereb Blood Flow Metabol* 4: 329-333

Gerstein GL (1970) Functional association of neurons: detection and interpretations. In: Schmitt FO, Quarton GC, Melnechuk T, Adelman G (eds) *The neurosciences: second study program.* Rockefeller University Press, New York, pp 648-661

Gerstein GL (1974) Plasticity in small neuronal assemblies. *Acta Neurobiol Exp* 34: 81-92

Gerstein GL (1987) Information flow and state in cortical neural networks: interpreting multi-neuron experiments. In: Seelen W v, Shaw G, Leinhos UM (eds) *Organization of neural networks: structures and models.* Verlag Chemie Verlagsgesellschaft, Weinheim, FR of Germany (in press)

Gerstein GL, Aersten AMHJ (1985) Representation of cooperative firing activity among simultaneously recorded neurons. *J Neurophysiol* 54: 1513-1528

Gerstein GL, Perkel DH, Subramanian KN (1978) Identification of functionally related neural assemblies. *Brain Res* 140: 43-62

Gerstein GL, Perkel DH, Dayhoff JE (1985) Cooperative firing activity in simultaneously recorded populations of neurons: detection and measurement. *J Neurosci* 5: 881-889

Getting PA, Dekin MS (1985) *Tritonia* swimming: a model system for integration within rhythmic motor systems. In: Selverston AI (ed) *Model neural networks and behavior.* Plenum Press, New York, pp 3-20

Gevins AS, Cutillo BA (1986) Signals of cognition. In: Lopes da Silva FA, van Leeuwen WS, Remond A (eds) *Handbook of electroencephalography and clinical neurophysiology, Vol 2.* Elsevier, Amsterdam, pp 335-381

Gevins AS, Doyle JC, Cutillo BA, Schaffer RE, Tannehill RS, Bressler SL (1985) Neurocognitive pattern analysis of a visuospatial task: rapidly-shifting foci of evoked correlations between electrodes. *Psychophysiol* 22: 32-43

Gevins AS, Morgan NH, Bressler SL, Cutillo BA, White RM, Illes J, Greer DS, Doyle JC, Zeitlin GM (1987) Human neuroelectric patterns predict performance accuracy. *Science* 235: 580-585

Grinvald A (1985) Real-time optical mapping of neuronal activity: from single growth cones to the intact mammalian brain. *Ann Rev Neurosci* 8: 263-305

Grinvald A, Anglister L, Freeman JA, Hildesheim R, Manker A (1984) Real-time optical imaging of naturally evoked electrical activity in intact frog brain. *Nature* 308: 848-850

Haxby JV, Duara R, Grady CL, Cutler NR, Rapoport SI (1985) Relations between neuropsychological and cerebral metabolic asymmetries in early Alzheimer's disease. *J Cereb Blood Flow Metabol* 5: 193-200

Herscovitch P, Markham J, Raichle ME (1983) Brain blood flow measured with intravenous $H_2{}^{15}0$. Theory and error analysis. *J Nucl Med* 24:782-789

Horwitz B (1987) Brain metabolism and blood flow during aging. In: Ellington RJ, Murray NMF, Halliday AM (eds) *The London symposium (EEG Suppl 39).* Elsevier Science Publishers, Amsterdam, pp 396-402

Horwitz B, Duara R, Rapoport SI (1984) Intercorrelations of glucose metabolic rates between brain regions: application to healthy males in a state of reduced sensory input. *J Cereb Blood Flow Metabol* 4: 484-499

Horwitz B, Duara R, Rapoport SI (1986) Age differences in intercorrelations between regional cerebral metabolic rates for glucose. *Ann Neurol* 19: 60-67

Horwitz B, Grady CL, Schlageter NL, Duara R, Rapoport SI (1987) Intercorrelations of regional cerebral glucose metabolic rates in Alzheimer's disease. *Brain Res* 407: 294-306

Horwitz B, Rumsey J. Grady C, Rapoport SI (1988) The cerebral metabolic landscape in autism: intercorrelations of regional glucose utilization. *Arch Neurol* 45: 749-755

Kruger JA (1982) A 12-fold microelectrode for recording from vertically aligned cortical neurons. *J Neurosci Methods* 6: 347-350

Livanov MN, Gavrilova NA, Aslanov AS (1973) Correlation of biopotentials in the frontal parts of the human brain. In: Pribram K, Luria A (eds). *Psychophysiology of the frontal lobes.* Academic Press, New York, pp 91-107

London JA, Zecevic D, Cohen LB (1986) Simultaneous monitoring of activity of many neurons from invertebrate ganglia using a multielement detecting system. In: De Weer P, Salzberg BM (eds) *Optical methods in cell physiology.* Wiley, New York, pp 115-132

Metter EJ, Riege WH, Kameyama M, Kuhl DE, Phelps ME (1984a) Cerebral metabolic relationships for selected brain regions in Alzheimer's, Huntington's, and Parkinson's diseases. *J Cereb Blood Flow Metabol* 4: 500-506

Metter EJ, Riege WH, Kuhl DE, Phelps ME (1984b) Cerebral metabolic relationships for selected brain regions in healthy adults. *J Cereb Blood Flow Metabol* 4: 1-7

Pearson RCA, Esiri MM, Hiorns RW, Wilcock GK, Powell TPS (1985) Anatomical correlates of the distribution of the pathological changes in the neocortex in Alzheimer's disease. *Proc Natl Acad Sci USA* 82: 4531-4534

Phelps ME, Hoffman EJ, Huang SC, Kuhl DE (1978): A new computerized tomographic imaging system for positron-emitting radiopharmaceuticals. *J Nucl Med* 19: 635-647

Phelps ME, Mazziotta J, Schelbert H (eds) (1986) *Positron emission tomography and autoradiography.* Raven Press, New York

Rapoport SI, Horwitz B, Haxby JV, Grady CL (1986) Alzheimer's disease: metabolic uncoupling of associative brain regions. *Canadian J Neurol Sci* 13: 540-545

Reitboeck HJA (1983) A 19-channel matrix drive with individually controllable fiber microelectrodes for neurophysiological applications. *IEEE Trans Syst Man Cybern SMC* 13: 676-682

Reivich M, Kuhl D, Wolf A, Greenberg J, Phelps M, Ido T, Casella V, Fowler J, Hoffman E, Alavi A, Som P, Sokoloff L (1979) The (18-F)fluorodeoxyglucose method for the measurement of local cerebral glucose utilization in man. *Circ Res* 44: 127-137

Rogers J, Morrison JH (1985) Quantitative morphology and regional and laminar distribution of senile plaques in Alzheimer's disease. *J Neurosci* 5: 2801-2808

Roy CS, Sherrington CS (1890) On the regulation of the blood supply of the brain. *J Physiol (Lond)* 11: 85-108

Smith CB (1983) Localization of activity-associated changes in metabolism of the central nervous system with the deoxyglucose method: prospects for cellular resolution. In: Barker JL, McKelvy JF (eds) *Current methods in cellular neurobiology Vol. I.* Wiley, New York, pp 269-317

Sokoloff L (1981) The relationship between function and energy metabolism: its use in the localization of functional activity in the nervous system. *Neurosci Res Prog Bull* 19: 159-210.

Sokoloff L, Reivich M, Kennedy C, Des Rosiers MH, Patlak CS; Pettiigrew KD, Sakurada O, Shinohara M (1977) The (^{14}C)-deoxyglucose method for the measurement of local cerebral glucose utilization: theory, procedure, and normal values in the conscious and anesthetized albino rat. *J Neurochem* 28: 897-916

Soncrant TT, Horwitz B, Holloway HH, Rapoport SI (1986a) The pattern of functional coupling of brain regions in the awake rat. *Brain Res* 369: 1-11

Soncrant TT, Horwitz B, Sato S, Holloway HH, Rapoport SI (1986b) Left-right regional brain functional interactions are disrupted by corpus callosotomy in the rat. *Soc Neurosci Abstr* 12: 177

Thorndike RM (1978) *Correlational procedures for research.* Gardner Press, New York

Neural Models, Rana and Robots

CHRISTOPH VON DER MALSBURG

Max-Planck-Institut für Biophysikalische Chemie
D-3400 Göttingen, FR Germany

Abstract. The technological analog of the model "*Rana computatrix*" - the construction of flexible robots - constitutes a renewed source of interest in neuroscience. This chapter gives a short overview of goals, problems, and perspectives involved in modeling a relatively simple behavior with respect to the various levels of data structure.

1 Introduction

Neurobiological work on frog and other "small systems" is largely motivated by the prospect of gaining a comprehensive view of the control of behavior. An essential pacemaker in this venture is the modeling effort (e.g., see Arbib 1987a and this volume). The back-engineering involved in it is a research tool in itself (see Arkin, this volume). For certain important aspects of the function of the nervous system this research tool may prove decisive. In the past, modeling has been very much hampered by the quantitative limitations of simulation tools. With the development of massively parallel computers we are now entering a new era (Eckmiller & v. d. Malsburg 1988). Within a few years it may become feasible to construct an actual "*Rana computatrix*" (Arbib 1982). Present efforts by engineers to achieve flexible control of robots have renewed their interest in the neurosciences. I will attempt here to give a short overview of the goals and problems involved in producing an integral model of a simple behaving animal.

2 Goals and Problems

The central issue is the representation of the scene relevant for the animal. "Scene" is meant here in a broad sense, including the layout of the perceived environment, the identity and significance as well as coordinates and states of movement of objects, the postural, dynamical, and behavioral state of the animal itself, and its intended movements. The scene is to be represented in terms of a data structure which can serve as a natural basis for various relevant processes of organization. This data structure must constitute a "working memory" (the actual scene itself) and a "knowledge base." Certain parts of the data structure must be able to exert global control on others, for instance, implementing different behavioral states.

Organization acts on both levels of the data structure and is responsible for the establishment of behavioral states, for structuring perception, and for generating motor patterns. The modeling of perception still poses major problems, if one wants to go beyond classification of stimuli under a small number of categories (Piaget 1971; Arbib & Hesse 1986). The problem of reconstructing object data from image data (e.g., 3-D shape from stereo and movement information and from graded gray levels) is intensely persued at the moment (Ballard et al. 1983; Lehky & Sejnowski 1988) and there seems to be rapid progress. The next step will be the integration of different "specialists" into a coherent system. Segmentation of sensory fields into sub-fields which are likely to correspond to independent objects or processes is an important perceptual problem (Creutzfeldt 1983). It has long been neglected but may be nearing plausible model solutions now (e.g., see v. d. Malsburg & Schneider 1986). A number of solutions has been proposed to the problem of invariance, i.e., of object representation and recognition independent of stimulus position, size, deformation, illumination etc (e.g., Bienenstock & v. d. Malsburg 1987; v. d. Malsburg 1988b). It remains to be seen whether these solutions stand up to tests under realistic natural conditions. Entirely unsolved is the problem of composing a complex scene out of individually recognized objects and their relationships. One may generally say that, whereas there is much progress now with low level perception, high level perception is still in a very crude state. The main reason for this seems to be inadequacy of the data structures employed.

Another field which will need rapid progress in the coming years is the modeling of motor control. As with high level perception, the main difficulty here seems to be inadequate data structure. It is, for instance, not clear how to represent high-dimensional coordinate vectors in visual or in limb-position space, and thus how to represent object position and state of movement. The engineer's solution, strings of numbers, does not lend itself to adaptation and self-organization as required for the nervous system. The same applies to the problem of representing limb trajectories. Well understood are simple reflex arcs controling a few degrees of freedom by feedback. Sensorimotor tasks like navigation in a complex environment or manipulation of an object by a hand under visual control are way beyond neural models of today. Again, much progress can be expected once the appropriate data structure is found. It may then even be feasible to implement and control a repertoire of different modes of behavior in the same system.

The classical neural data structure is based on "units" (which may directly correspond to neurons), each of which is an elementary symbol (e.g., for "bug, detected in position x of retina"). Each symbol or unit can be active and relevant to the present scene, or inactive and irrelevant. The working memory for a whole scene is the set of currently active units. The knowledge base of a neural system is represented by excitatory and inhibitory connections between units (Hebb 1949; Eccles et al. 1967; Creutzfeldt 1983). In the simplest case, the dynamics of the network are designed such that it tends to stabilize certain activity patterns. Organization of the working memory takes the form of a relaxation of activity towards stable states. The knowledge base is organized with the help of synaptic plasticity which stabilizes the current activity state. The adaptation of this type of neural architecture to the needs of a specific modeling task usually takes the form of an assignment of specific meaning to units (by connecting them appropriately to inputs or outputs) and of introducing an appropriate initial pattern of connections. It may, however, turn out that the simple neural architecture described is not appropriate and will have to be modified in fundamental aspects. One proposal stipulates the introduction of a syntactical structure of dynamic links which allow the system to represent hierarchically structured objects and scenes. This may not be necessary in simple animals with stereotyped behavior, but is probably an absolute requirement for flexible behavior in complex scenes. For a more extensive discussion and references regarding the neural architecture the reader is referred to v. d. Malsburg (1981, 1988a).

3 Perspective

Neural modeling is currently profiting intensely from computer science and electrical engineering, not only by exploiting computers and parallel machines for simulations, but also from learning about the practical problems encountered in processing actual signals derived from a camera or a microphone. Engineers trying to construct flexible robots are actually realizing that the know-how on parallel processing and self-organization implicit in neural models (Rumelhart & McClelland 1986) may be a valuable return (Arbib 1987b).

4 References

Arbib MA (1982) Rana computatrix: An evolving model of visuomotor coordination in frog and toad: In: Hayes JE, Michie D, Pao YH (eds) *Machine intelligence.* Ellis Horwood, Chichester, pp. 501-517

Arbib MA (1987a) Levels of modelling of mechanisms of visually guided behavior. *Behav Brain Sci* 10 (in press)

Arbib MA (1987b) *Brains, machines, and mathematics.* Springer-Verlag, New York Berlin Heidelberg

Arbib MA, Hesse MB (1986) *The construction of reality.* Cambridge Univ Press, Cambridge Mass

Ballard DH, Hinton GE, Sejnowski TJ (1983) Parallel visual computation. *Nature* 306: 21-26

Bienenstock E, Malsburg C vd (1987) A neural network for invariant pattern recognition. *Europhys Lett* 4(1): 121-126

Creutzfeldt, OD (1983) Cortex Cerebri. Leistung, strukturelle und funktionelle Organisation der Hirnrinde. Springer-Verlag, Berlin Heidelberg New York

Eccles JC, Ito M, Szentagothai J (1967) *The cerebellum as a neural machine.* Springer-Verlag, Berlin Heidelberg New York

Eckmiller R, Malsburg C v d (1988) *Neural computers.* Springer-Verlag, Berlin Heidelberg New York

Hebb DO (1949) *The organization of behavior.* John Wiley, New York

Lehky SR, Sejnowski TJ (1988) Neural network model for the cortical representation of surface curvature from images of shaded surfaces. In: Lund JS (ed) *Sensory processing.* Oxford Univ Press, Oxford

Malsburg C vd (1981) The correlation theory of brain function. *Internal Rep 81-2 MPI Biophysical Chemistry,* 3400 Göttingen, FR Germany

Malsburg C vd (1988a) A neural architecture for the representation of scenes. In: McGaugh JL, Weinberger NM, Lynch G (eds) *Brain organization and memory: cells, systems, and circuits.* Oxford Univ Press, New York

Malsburg C vd (1988b) Pattern recognition by labeled graph matching. *Neural Networks* (in press)

Malsburg C vd, Schneider W (1986) A neural cocktail-party processor. *Biol Cybern* 54: 29-40

Piaget J (1971) *Biology and knowledge.* Edinburgh Univ Press, Edinburgh

Rumelhart DE, McClelland JL (1986) *Parallel distributed processing: explorations in the microstructure of cognition.* MIT Press, Cambridge London

Subject Index

A-configuration; *see Antiworm-*
Abdominal nervous system 634
Absolute
  coordination 500
  objective size; *see Size*
Abstract spatial representation 543, 549, 555
Accessory
  olfactory tract 822, 826
  optic system 16, 194, 317
Accommodation; *see also Depth*
  cues 148, 149, 151, 152
  field 148, 149
  mechanism 93
  state 556
Acetylcholine 79, 293, 345, 350; *see also Cholinergic*
  esterase 345
  receptors 351
  transferase 350
Acquired information; *see also Associative, Conditioning, Learning*
  auditory 745, 749, 758
  olfactory 55, 99, 100, 609, 811-813, 824
  robots 165, 683, 684
  sensory substitution 407
  visual 54, 55, 62, 99, 100, 107, 768-851
ACTH-treated animals 786, 850
Action
  patterns 43, 167, 405, 410, 415
  perception cycle 41, 136
Active gating process of habituation 784, 792; *see also Habituation*
Activity gated divergence xiii, 157, 538, 539-541, 543, 555; *see also Localization*
Acuity of selectivity 49-52, 74-78, 271, 363, 514, 577; *see also Antiworm, Discrimination, Stimulus*
Adaptive
  adjustments 616, 636
  command systems 103
  database techniques 167
  modification of prey-selection 297; *see also Acquired, Modulation, Modulatory*
  network mechanism 100, 124, 839
  programming 167
  properties 69; *see also Acquired*
  response formation 695, 757, 768
  sensitization 696
Adequate model 5; *see also Model*
Afferents of
  bulbar nuclei 194, 195, 487, 527, 557, 598, 605; *see also Medulla*
  hypoglossal nucleus 599, 600
  nucleus isthmi 35, 181, 194, 343
  optic tectum 83, 101, 176-181, 187, 193-195, 351
  pretectum 101, 106, 189-191, 194, 195
  tegmentum 101, 194, 195, 557
Affine transformations 366
After-; *see also Post-*
  discharges 30, 82, 467, 496, 497, 514, 515
  nystagmus 325, 328
Agnosia 66, 67, 473
Agnostic
  command elements 473
  tectal neurons 82, 466
  visual field 67, 79
Alertness 462, 738; *see also Attention, Arousal, Wakefulness*
Algorithms xii, 14, 15, 84, 411, 654, 873
  configural stimulus selection 60, 78, 85, 89
  control 686
  iterative 363
  levels 107, 357, 379
  line analysis 24
  motor 328
  network 32
  path planning 157
  specification 136
  recursive 364
  universal 384
All-or-none modes 56, 328, 484, 514
Aluminum hydroxide 712-714
Alytes obstetricans 54
Alzheimer's disease 883-886
Amine neurotransmitters 224, 230, 236; *see also Neurotransmitters*
Amino acid neurotransmitters 224, 229; *see also Neurotransmitters*
Aminopyridine 712
Amnesia 756, 821
Amphibians 16, 84-86, 106-108, 243, 248, 324, 694, 698, 700, 710, 800; *see also Anuran, Bufo, Frogs, Salamanders, Rana, Toads*
AND-gate characteristic 95, 328, 387, 492
Amygdala; *see Nucleus*
Amygdaloid complex 755
Amygdalo-striatal
  complex 757
  transition zone 755
An der Heiden's model of retino-tectal function 248
Analgesic effects 742
Analyzer made of neurons; *see Assembly, Bullock, Ensemble*
Anesthesia and brain activity 740, 744, 745, 779, 790
Anisotropy
  computation 25, 91, 108
  computational maps 91, 378

configuration discriminators 25, 91
cortex-like structure 91, 176
direction-invariance 53, 58, 78, 91, 283, 298, 446
feature maps 91, 372
histological structure 91, 358
implicit vs. explicit computing 91
orientation columns 91, 358-361
spatio-temporal stimulus patterns 91, 281
Ansulate commissure 594, 598
Anterior
reticular thalamus 735
thalamus 66, 67, 69, 101, 523, 734, 736, 772, 774
Antidromic
collision test 93, 484, 485
latencies 485, 495, 500, 606
stimulation/recording 93, 218, 492, 597, 733
Antiworm configuration; *see also Configuration, Discrimination, Stimulus, Worm*
2DG uptake in the brain 63, 817, 820
category formation 61
comparison to "hawk configuration"
continuum 60
definition 50
discrimination value 51, 272, 277
movement direction 53, 78, 283
neural filters 81, 129
neuronal responses 70-92, 459-467, 492, 509-512
pretectal lesions 67, 77, 82-84, 145, 442, 444, 466
response modulation 69, 100, 271-290, 802
responses of conditioned toads 802
simulated neuronal responses 144, 145, 251-264, 279-290, 442-446
threat signal 61
velocity 53
Anuran
amphibians 53, 85, 106, 107, 539, 541, 562; *see also Amphibians, Bufo, Frogs, Rana, Toads*
species 84, 85, 248, 295
Aperture problem 371
Apis 708
Aplysia 23, 125, 800, 875
Apomorphin 475, 611
Apparent movement 335; *see also Background, Object*
Appetitive orientation
toward mate 99, 271, 296
toward prey 40, 43, 99, 271, 296
Approach-vector-selector schema 161
Approachable object selector 271, 296
Approximateness vs. exactness 61, 108, 377
Arbib-House's model of path planning 133, 650
Arbib-Lara's model of tectal column 137
Archipallium 820; *see also Pallium*
Areas V1-V3; *see Cortical*
Areas 17-19; *see Cortical*
Arithmetic operations 387; *see also Algorithms, Computer, Model*
Arm projection
neurons 476
schemas 123, 161, 162
Arousal xiv, 28, 92, 107, 237, 525, 693, 696, 697, 700, 707, 708, 712, 713, 729, 730, 734, 737, 739, 742, 759, 789, 792, 816, 819
and habituation 789, 790
autonomic correlates 740
brain structures involved 734-745, 757, 789
central representation 694-719
definition 693
EEG 698
effects 718, 738, 757, 792
epilepsy 713
hippocampal formation 28, 738
neuroglia 712
related information 742, 744
reticular formation 730, 778, 789
stimuli 704, 778, 780, 781, 786
ventral medial pallium 28, 778, 779, 789
Array of tectal columns 81, 140, 141, 142, 144, 441, 444, 838, 844; *see also Column*
Arthropods 631
Artificial; *see also Computer, Machine, Technological, Robot*
environment 13, 46
grammars 15, 405-407
image processing 336, 399, 678, 679
intelligence 6, 15, 16, 33, 41, 165, 167, 422, 424, 425, 649, 651
stimuli 46, 49, 244; *see also Sign stimulus, Stimulus*
Ascending; *see also Afferent, Axonal*
reticular activating system 730, 739, 779, 792
subcortical pathways 730
Aspartic acid 238, 712
efflux 235
release 235
Assembly
of connected cells 81, 88, 100, 129, 138, 142, 177, 205, 263, 276, 360, 392, 441, 839; *see also Column, Ensemble, Functional unit*
operations 677, 679, 681; *see also Robot*
planning and monitoring 681; *see also Robot*
Assignment problem
credit 33
structural 17, 27, 33, 94, 218; *see also Dendrites, Morphological*
Association cortex 476, 753
Associative; *see also Adaptive, Acquired, Conditioning, Learning*
conceptual models 29, 33, 100, 753, 758, 814-826, 848-851
disinhibition 100, 101, 818
learning 33, 40, 42, 55, 99, 100, 107, 745, 750, 799, 800, 808, 814, 821, 822, 826, 835
hand-feeding phenomenon 55, 801, 843
memory 33, 406, 758, 759, 814-826, 848-851; *see also Memory*
odor learning 55, 811
prey-recognition 55, 801
sensorimotor coordination 406, 415; *see also Sensory substitution*
Associator 406, 409
Astrocytic; *see also Glial*
glia 712
gliosis 716
Asymmetric
directional sensitivity 78, 275, 285, 299, 319
processors 25, 91, 108, 359
Atectal; *see also Brain, Optic tectum*
frogs 610
toads 65
Attention 98, 99, 462, 474, 525,

693, 694, 696, 697, 713, 715, 718, 745, 816, 819; *see also Alertness, Wakefulness*
definition 98, 693
motor programs 790, 823
orienting 462, 694
shifts 99, 462, 464
Auditory 40
cortex 743, 744
learning 753, 755
nuclei 743, 757
system 742, 745
AURA; *see Robot*
Autonomic functions 523, 759
Autoradiography 64, 66, 227, 730- 832, 875, 888; *see also Deoxyglucose*
Average
anatomy 358, 359, 360
fly phenomenon 152, 624
pulse analogue correlation method 213, 214, 220
Aversive
motivation 757
reticular stimulation 737, 738
Avoidance behavior 65, 74, 193, 209, 482, 522, 524, 661, 662, 788, 801, 834; *see also Defensive, Detour, Escape*
barriers 68, 131-133, 136, 152-159, 193, 244, 376, 654
brain structures involved 65, 68, 97, 129, 344
computer model 839
door frame 134, 668
global models 66, 96
maps 65
predator 58-61, 63, 65, 68, 129, 344, 432, 443, 521, 843
relevant neurons 74, 97
robot 654-662
Awake state 713; *see also Wakefulness*
Axo-dendritic circuits 219
Axonal projections; *see also Loops, Network, Neuro-*
isthmo-tectal 350, 351
pallio-hypothalamic 101, 107, 783
pretecto-bulbar/spinal 213
pretecto-tectal 66, 67, 81, 83, 193, 194, 212
striato-pretectal 66, 101, 106, 107, 195
tectal 73, 81, 179, 205
tecto-bulbar/spinal 80, 93, 94, 96, 106, 181, 184, 210, 483, 485, 593, 596
tecto-isthmic 181
tecto-tegmental 554
tecto-thalamic 183, 210

Back-up systems 103, 608
Background 13, 24, 41, 46, 52, 67, 392; *see also Environment, Object, Textured*
color 836, 841, 848
motion compensation 314, 334
motion detectors 313, 320
texture 46, 47, 67, 69, 76, 392, 492; *see also Textured*
Ballistic behavior 43, 56, 87, 123, 538, 589, 592, 607, 616
Barbiturates 743, 744
Barlow's "fly detector concept" 86, 458
Barr schema 134, 135
Barrier; *see also Avoidance, Detour*
avoidance 68, 128, 131, 134
negotiation models 153
Basal
forebrain 106, 195, 739, 886
ganglia 106, 195, 790, 823; *see also Striatum*
optic nucleus 189, 194, 317
Bauplan 800; *see also Homologies*
Bee avoidance 55, 800, 843, 844
Behavior correlated
neuronal activity 857, 869, 870
neuronal properties 456, 461
Behavioral
flexibility 270; *see also Acquired, Associative, Learning*
input/output relation 538; *see also Black box*
relevant brain structures 63, 66, 77, 162, 313, 332, 343, 734, 735, 754, 773, 779, 805-807, 812, 875,
sequences 41, 44, 123, 134, 162
specificity 22
state 611, 815; *see also State*
Bett's model of tectal modulation 271
Bi-component modulation
of map weighted information 576
pretecto-tectal inhibition 80, 84, 264, 440
Bifurcating axons 209, 212, 218
Big box 422, 436, 437, 443
Bilateral
comparison of 2DG uptake 805, 807
neuronal projections 212
transfer in learning 816
Binocular
bulbar neurons 521
disparity 147, 336
gain 321
small field neurons; *see Small field neurons*
training 801, 805, 806, 818
wide field neurons; *see Wide field neurons*
Biochemical approaches 106, 107, 193, 191, 193-195, 223-238, 350-353, 610; *see also Neuro-transmitters*
Biocybernetics 15; *see also Cybernetics*
Biological
information processing 9, 14, 16, 21, 358, 384, 415, 649, 686
thinking 9
Birds 59, 106, 699
Black box studies 43, 168, 277, 592
Blackbird 59
Blind 397, 398, 399
probants 411
Body centered coordinate frame 475, 548, 557, 560; *see also Coordinates*
Bombina variegata 54
Boolean variables 387
Bootstrap procedure 885
Bottom-up studies 122, 423
Box-models 6, 44, 66, 129, 313, 332
Bradycardia 741, 742, 758
Brain
and computer 14, 121; *see also Artificial*
evolution 54, 61, 86, 107, 108
lesions 65-68, 77, 83, 145, 264, 317, 808; *see also Brain lesions*
functional activity 63, 707, 873-889; *see also Deoxyglucose*
slices 224, 227, 713, 716
style 121, 126, 435
theory 167, 422, 424
Brain lesions
auditory cortex 748
anterior thalamus 67, 778
basal optic nucleus 317
hippocampus 821
isthmic nucleus 343, 346
medial pallium 777, 808-811
medulla oblongata 546, 547, 606
optic tectum 65, 542, 543, 546, 598, 610; *see also Atectal*
pontine reticular formation 550
pretectal thalamus 66, 77, 83, 91, 145, 297, 442, 444, 466, 804
spinal cord 617

striatum 808
tecto-bulbar/spinal tracts 45, 94, 544, 545, 551, 552,
tegmentum 553
telencephalon 68
Brain stimulation
medial pallium 778
mesencephalic reticular formation 737-744, 779-781
optic tectum 64, 65
pretectal thalamus 65
striatum 778
Brainstem; *see also Medulla*
motor circuits 525-528, 538-564, 591-611
reticular nuclei 484, 486-528, 550, 604-610, 741, 789
Branchiomotor column 490, 491, 494, 495, 505-507, 521, 593, 603-606
Breathing rhythmicity 524
Breeding behavior
explosive 295
object discrimination 97, 99, 295
prolonged 295
Bridging gapes xii
cognitive and neuro-sciences 126, 423
cognitive science "tricycle" 427, 428
isolation problem 423
morphology/physiology 16, 192; *see also Dendrites, Morphological*
schema theory 428
Bufo; see also Toads
arenarum 698
asper 54
bufo 40, 44, 54, 84, 85, 94, 127, 176, 245, 247, 248, 269, 432, 438, 457, 471, 487, 591, 592, 694, 769, 800, 801, 833, 844, 859
marinus 592
regularis 699
viridis 54
woodhousii 800
Building block approaches 16, 21, 122, 537-564; *see also Model*
Bulbar; *see also Medulla*
bursting cells 486, 495, 497-500, 502, 515-519, 525-528
functional properties 520-529, 546-550, 604-611
reflexes 610
Bulbar/spinal; *see also Medulla*
circuits 92, 539, 784; *see also Brain stem*
excitatory loops 523
projection neurons 500
Bulbo-tectal feedback 493, 496, 522, 523
Bulbus olfactorius 813
Bullock's concept of multiple input metastable feedback loop 106
Bursting cells 495, 497-499, 525, 634, 637; *see also Bulbar*
Butterfly operations 389

C language 4; *see also Computer*
Calcium-dependent transmitter release 225-238
Call-generating circuitry 524
Camera-lucida reconstructions 184, 210, 483, 506
Campaniform sensilla 638, 639
Carassius
auratus 694, 714
morosus 632
Cardiac
arousal 714
model 526
responses 694; *see also Bradycardia*
Catecholamines 224, 611
Category; *see also Stimulus*
definition 57
formation 61
mathematical description 89
stimulus continuum concept 49-51, 57-59
Cats 220, 371, 527, 627, 696, 697, 707, 708, 737
Caudal thalamic lesions 67; *see also Pretectum, Thalamus*
Caudate nucleus 737, 752
Cell-centered plan 681
Cellular architecture 22; *see also Network, Neuro-*
medulla oblongata 506-519
optic tectum 176-187, 192, 193, 205-210, 211-213, 218, 219, 483
pretectum 187-191, 193-195, 322, 323
Central
autonomy 564
fovea 69, 103
lateral inhibition 66, 81, 261, 263, 576
oscillator 700
pattern generator 22, 638, 646; *see also Motor*
Cerebellum 740, 806, 824
Cerebral cortex 527; *see also Cortical*
Cerebral energy metabolism 29-31, 733; *see also Deoxyglucose*
brain to brain comparison 769
interhemispheric comparison 804
interregional correlations 879
Changes of diffuse light
eye closure inhibition of off response 470
on/off responses of neurons 71, 72, 74, 75, 79, 204, 220, 469, 501; *see also Luminance*
Changing neuron types 91, 103, 270-300, 513
Chasm schema 134, 135; *see also Schema*
Chemical markers 7
Chemoarchitecture 79, 176, 223-238; *see also Biochemical, Neurotransmitters*
Cholinergic; *see also Acetylcholine*
agonists 740
fibers 350
pathways 740, 757
transmission 79
Chordotonal organ 642-645
Chromatic directional sensitivity 299; *see also Color*
Chronic unit recordings in behaving toads 72, 452-476, 857-871
Cineol 811
Cingulate cortex 738, 757
Circadian determinants 88
Circuitry; *see also Cellular, Model, Network, Neuro-*
stimulus-response compensating 313, 327
stimulus-response mediating 66, 81, 142, 162, 406, 429-431, 575, 608, 625, 676, 681
stimulus-response modulating 42, 66, 98-107, 271-300, 351-355, 429, 610, 696, 712, 742, 783; *see also Modulation, Modulatory*
Circular
Fourier expansion 572
frequency domain 572
frequency tuning 577
receptive fields 71, 74, 91
Clasping behavior 270; *see also Breeding*
Classical conditioning 730, 745, 753, 758, 822, 826; *see also Acquired, Associative, Conditioning, Learning*
Classification of neurons
medial pallial 787
medullary 487-489, 501, 506, 524
retinal 70-72, 86

tectal 72-76, 85, 87, 91, 92, 179, 184, 205, 247, 248, 271, 442, 469, 470
thalamic-pretectal 72-76, 187, 442
Clinical disorders xiv
Alzheimer's disease 883
epilepsy 712
Closed loop system 361, 643; *see also Feedback, Loops, Reafference*
Co-activation 610; *see also Combination, Concurrent*
Co-existing circuits 103
Cobalt labeling 177-191, 205-217, 504-519
Cochlear nucleus 743
Cockroach 550
Coding; *see also Information*
color 97
eye position 528
head/body position 475, 541
map-weighted commands 576
place information 95, 540, 555, 556, 558
problems 27
reward 823
sensorimotor 95, 107
stimulus significance 29
temporal spike pattern 467, 524
Cognition 40, 107, 108
Cognitive xiii
activity 405, 410
functions 879
instruments 107
processes 410
psychology 33, 422, 424, 649
science "tri-cycle" 423, 427, 447
sciences 57, 122, 422, 425, 426
Coherent movements 47, 336; *see also Background, Optokinetic*
Coincidence; *see Combination, Concurrent, Cooperation*
Colliculus; *see also Optic tectum*
bulbar projections 525, 526
inferior 747
neurons 475
reticulo-spinal pathway 526
superior 730
Collision test; *see Antidromic*
Color 97, 841
Column
architecture 10, 88, 137-146, 177
concept 80, 137, 177
cortical 359, 360
model 137-146, 168, 202, 835
organizations 137, 177, 192, 331, 359, 377, 378, 379, 752, 758, 836
tectal 81, 137-146, 177, 440-446, 835-848
tectal vs. cortical 90, 91, 177
Combination
bulbar properties 501
command elements 95-97, 104
feature action string 405, 416
of actions 43, 123, 626, 676
of information 95, 359, 472, 497, 676, 682; *see also Associative, Learning*
of motor schemas 123, 128, 134, 152, 159, 654, 658
of motor segments to movements 626
of open pathways 636
of space related signals 558, 559
of stimulus features to configurations 50, 51, 56-61
of stimulus segments to figures 410, 411
sensorimotor codes 95, 104
space and feature signals 95, 482
Command xiv, 104, 168, 193, 209, 472, 645, 681
elements 95-97, 102, 296, 472, 597, 612, 624
feature coded 95, 102
functions 104, 475, 476
map-weighting 575
neuron populations 95, 627
neurons 102, 475, 597, 624, 626, 627
pathways 545
place coded 541; *see also Coding*
prey-catching 95
releasing system 92, 95-97, 102, 103, 107, 452, 472, 475, 482, 487, 516, 522, 524, 597, 612, 874
space coded 95; *see also Code, Localization, Prey*
system 92, 95, 102, 452, 475, 850
vs. motor functions 475
walking 646
wiping 624
Common
language in neurosciences 415, 436
programming language 4
tools for modeling 4, 168
Commutative operations 387
Comparators 328, 573, 835-839
Comparisons 6, 89, 106, 436, 524, 525, 482, 784; *see also Homologies*
across species 54, 84, 85, 90, 244, 248
definition of schema 121, 404, 430, 653
models of prey-recognition 86-91, 262, 297
preying toads, praying mantises 436
romantic vs. responsible modelers 6, 7, 12
Compartmental analysis 18
Compensating
head movements 324, 325, 328
image shifts 312
Competitive interactions; *see also Interactions, Lateral, Mutual*
features 47, 48, 69, 82
interocular 47
intraocular 47
motor schema 123, 148
perceptual schema 123, 148
Complex logarithmic mapping; *see Mapping*
Complexity 9, 10, 16, 68, 104, 195, 219, 609
Compound
eye 574, 575
motor coordinations 43, 98; *see also Motor*
Computational; *see also Computer, Machine, Robot*
frog 125; *see also Rana computatrix*
maps 23, 90, 378
neuroscience 121, 122, 125
physiology 422
problems 550, 556, 559
strategies 125
theory of vision 357
time 363
Computer; *see also Machine, Technological*
6th generation 35, 121, 167
and brain 14, 121
analogies 26
assisted image analyzer 769
C language 4
CRAY 16, 168
evolving creatures 125
image processing 13, 14, 24, 384, 385; *see also Robot*
LISP 168
models 5; *see also Model*
NSL language 168
parallel 14, 15, 32, 33, 122
Quantimet 769
reconstruction of dendrites 18
science 167, 422, 686, 895
serial 14, 15

software explosion 15
tomography 877
VAX 659, 677
vision 14, 125, 165, 397; *see also Image, Machine*
processing
VISIONS
VLSI 122
volume 15
Computer simulations of; *see also Concepts, Model*
cortical information processing 358-379
habituation 841
learning 843
map-weighting 575
modulation of selectivity 271-289
path planning 152-159, 650-670
pretecto-tectal interaction 89, 90, 142
prey/barrier perception 128-136, 150
prey-recognition 89, 140, 263, 436-446
retinal network 251
robot navigation 650-670
sensory substitution 411-414
tectal column 137-146
tectal network 137, 257
Concepts of; *see also Dual, Model*
activity gated divergence 538
apparent anisotropy 91, 108
cognition 107, 423
cognitive "tricycle" 423
column 137, 177, 358, 440
command system 95, 102, 624
computational maps 90, 378
coordinated control programs 122, 123, 162
distributed properties; *see Distributed, Parallel*
fixed action pattern 40, 62
innate releasing mechanism 40, 62, 95
instinct 40, 107
interacting configural cues 47, 48, 69, 82, 88
interface 92, 97, 107, 538, 560, 597
intermediate processing 165, 538, 597
iterative tuning 363
lateral central inhibition 66, 81, 261
map weighting 575
multiple input metastable feedback loop 106
pattern generator 523, 543, 604, 646
parallel processing; *see Distributed, Parallel*
representational maps 90, 365
schema 121, 126, 134, 404, 414
self motion computer 329
serial processing; *see Hierarchical, Serial*
sensorimotor code 95
sign stimulus 41, 45, 56-62
stimulus response modulating circuitry 42, 783
structural anisotropy 358
TS base modulator 271
Conceptual
boxes 6, 66, 123, 354, 436
building blocks 16, 21, 122, 537-564
models 538; *see also Model*
Concurrent; *see also Combination, Cooperation*
activities 32, 95, 103, 123
inputs 44, 45, 496, 513, 514
Conditioned
bradycardia 747, 755
go/no-go 523, 800
prey-catching 55, 100, 801, 809, 843
reflexes 747, 748, 750, 757, 810
Conditioning 29, 55, 99, 105, 747, 748, 750, 757, 810, 843; *see also Acquired, Associative, Learning; Model*
Configuration; *see also Combination; Sign stimulus, Stimulus*
feature maps 90
features 46-51; 56-61
interaction concept; *see Concepts*
discrimination 49, 52, 81, 129, 205, 272, 789, 803, 809, 810
discriminators 90
dynamic 50, 57-59
hybrid form of wiping 627
modulation of selectivity 100, 271
movement direction invariance 53, 78
moving configural stimuli 46-51, 77, 129, 144, 205, 244, 247, 248, 252-263, 261, 272, 463, 511, 771, 802, 832
neuronal selectivity 76, 85, 89-91, 205
neuronal tuning 77
predator selectivity 58, 60, 61, 523
prey selectivity 51, 60
retina-pretectum-tectum models 81, 89, 129, 137-146, 440-447
retina-tectum model 248-264
sensitive model neurons 144, 145, 256-263, 442-447
sensitive pretectal neurons 72-78
sensitive retinal ganglion cells 70-72
sensitive tectal neurons 72-78, 245-247
species differences 54, 84, 85, 90, 244, 245, 248
static 57-59
velocity dependence 247
velocity invariances 53, 78, 90, 246-248
Connectionist studies 42, 81, 167, 358; *see also Cellular*
Connectivity patterns 42, 81, 96, 100, 139, 142, 205, 210, 263, 276, 353, 441, 527, 564, 837, 839; *see also Axonal*
Consolidation of learning 851
Constant latency responses 484, 494, 500; *see also Antidromic*
Consummatory act 40, 43, 462, 589, 610
Context
addressing 836, 840
dependent actions 414
dependent neuronal processing 561-564
ensemble coding 108
experience 398, 406, 411
grammars 15, 405-407
information 164, 385, 400
schematic description 153, 164
Continuum 296, 416
hypothesis 799
model 361
network 88, 361
of columns 142, 361, 441, 486
of functional units 88, 416, 441
of stimulus features 22, 60, 271, 296
of walking-pattern generators 638
Contour 57
information 399, 404
perception system 399
sensor 399, 411
tracking 400, 411, 412
Control systems 123, 627, 632, 636, 644; *see also Command, Coordination*
Convergence 521, 522, 608; *see also Concurrent, Hierarchical*
command elements 96, 103
neuronal network 42

visual and tactile signals 556
Cooperation; *see also Combination, Concurrent*
AND gate; *see AND*
command elements 95, 103, 624-627
computation 125, 126, 134, 167
depth perception 147
mechanisms 103, 333, 472, 590
neuronal firing 498, 499, 874
neuronal properties 501
neuroscience and computer science 4, 166, 167; *see also Computer, Model*
sensorimotor function 97, 627; *see also Sensorimotor*
Coordinates
body centered 475, 548, 555, 557, 560
Cartesian 153
common frame 556
eye centered 369
head centered 543
head-to-trunk 576
prey in space 43-45, 92, 95
retinal 545
transforms 366-368, 374
world 369
Coordination 676; *see also Cooperation, Coupling*
control programs 97, 122, 123, 135, 136, 162, 592, 623, 632
influences 632, 634, 639, 646
motor schemas 133
of leg 631, 636
of leg joint 616, 622, 631, 644
of motoneurons 559
pathways 635, 636, 645
patterns of activity 23, 559
subunits 644
Corollary discharge signals 540, 562; *see also Efference copy, Loops, Feedback, Reafference*
Corpus callosum 886, 887
Correlation
dendritic morphology and function 17, 18, 94, 202, 205, 210, 218, 501
matrix 879-886
metabolic and electrical activity 30
methods 874, 879, 880, 889
neuronal and behavioral selectivity 76, 458-467
pulse analogue correlation 213
retinal and tectal responses 216
Cortex 361, 707, 823
comparisons with tectum 91, 177
equivalent structures 176
Cortical
average anatomy 358, 359
columnar organization 88, 358-361
intrinsic connections 359, 363
iterative algorithm 363
networks 357, 379
organization 358-360, 376
patchy connectivity 359
processing 358-379
retinotopic mapping 358, 366, 367, 374
spatiotemporal activity 361
structural principles 358, 359
visual field maps 366, 367, 374
Cortical areas
area 5 162
area 6 162
area 7 161, 476
areas 8,9 161
area 17 91, 371
area 18 365
areas 17-19 366, 376
areas V1-V3 391
Cortico-
cortical connectivity 363, 886
limbo-cortical circuits 107
CRAY computer 16, 168; *see also Computer*
Coupling of cells 359, 497-500; *see also Co-activation, Combination, Cooperation*
Craig-Cornsweet experiment 386
Craniomotor nuclei 482, 490, 607; *see also Branchiomotor, Medulla*
Credit assignment problem 33
Crickets 40
Cross-correlations 89, 214-217, 874; *see also Correlation*
Cross-disciplinary studies xi, 4, 166, 421; *see also Inter-, Multidisciplinary*
Crossed and uncrossed tectal descending pathways 93, 94, 184, 193, 210, 482, 490, 521, 551, 593, 604; *see also Optic tectum*
Ctenosaurus pectinata 699
Cue Interaction model 147, 148, 151
Cutaneous
inputs 399, 492, 493, 557, 590
mechanoreceptive fields 500, 619
released orienting 540
released wiping 618
Cybernetics xi, 4, 421, 427; *see also Biocybernetics*
Cyclic
burst activity 495, 497, 499, 502, 516, 525, 634; *see also Bursting*
excitatory/inhibitory loops 527; *see also Feedback, Loop*
periodic transforms 391
Cytoarchitecture 69, 175-195, 508, 709, 757; *see also Axonal, Cellular, Network, Neuro-*
medullary neurons 491, 504-519
nucleus isthmi 342
optic tectum 73, 175, 179, 205, 210
pretectal region 73, 189, 209-213

Data structure xv, 893, 894
Decerebration 617, 641, 756
Decision making; *see also AND, OR, Prey, Recognition*
democratic 41, 633, 636, 639
hierarchical 41
Decoding premotor circuitry 104, 107, 520, 524; *see also Coding*
Decortication 756
Dedicated functions 19
Defensive responses 730, 734, 740, 755, 757, 758, 759; *see also Avoidance*
Democratic decision-making; *see Decision*
Dendrites
analysis 17
and axonal arborizations 205, 359, 507, 513
architecture 610
column 179
commissural fibers 598
computer simulation 17, 18
connections 219
correlation with behavior 19, 218
correlation with electrophysiology 19
correlation with receptive fields 19, 218
geometry 358-361
model studies 20
morphology and function 17, 19, 20, 94, 202, 205, 210, 218, 501
problems of functional assignments 17, 27, 94, 218
trees 17, 19, 20, 184, 205, 513, 514, 522
variability 18, 19
Deoxyglucose (2DG) auto-

radiography 63, 64, 66, 696, 730, 731, 734, 737, 767, 774, 789, 800, 812, 817, 820, 875, 888
advantage 32, 730, 768, 769
anesthesia 779, 786
arousal 734-745, 757, 786
associative learning 745-759, 801-814
cells involved 28, 29
configural stimuli 63, 64, 817
correlation with electrophysiological activity 29
critical evaluation 29-31, 742, 756, 791, 816, 877, 889
depolarization vs. hyperpolarization 30, 31
habituation 769-777, 786
modeling 32
neuroglia 28
neuropil 31
pre- vs. postsynaptic inhibition 31
predator avoidance 63, 64
prey-catching 63, 64, 66
sodium/potassium pump 31
spatial resolution 29
temporal dimension 29

Depth; *see also Accommodation, Stereopsis*
cue sources 151
map 675
perception 34, 90, 92, 146, 149, 151, 334, 335, 336; *see also Coordinates, Localization, Prey*
perception model 146-149

Descending tectofugal pathways; *see Crossed, Optic tectum*

Detaillism 26; *see also Model*

Detector concept 22, 41; *see also Barlow, Lettvin*

Detour behavior 103, 128, 133, 157, 649, 670; *see also Avoidance, Barrier, Path*
comparisons 128-133
models 128-136, 149-159
pretectal stimulation 65
robots 654
schemas 128

Dev's model of stereopsis 146

Dexterous manipulation; *see also Arm*
control schemas 159-163
robot 686

Didday's model of prey-selection 125

Digital machine analogies 26; *see also Computer, Machine, Robot, Technological*

Directed
appetitive behavior 43, 461, 694
attention 99, 470, 757, 759
movements 43, 537, 539, 541, 545, 546, 550, 552, 561; *see also Arm, Dexterous, Orienting*

Directional
invariance 25, 53, 78, 91, 146, 276, 283, 298, 384, 437, 446
selectivity 319, 321, 331, 493
sensitivity 271, 298
tuning curves 78, 283-300, 319
vector 570
weights 300

Discrimination; *see also Prey, Recognition*
configural moving objects; *see Configuration*
directional dependence; *see Directional*
directional invariance; *see Directional*
filters 67, 77, 81, 89, 129, 142, 263, 276, 360, 372, 441, 837, 839
function 51, 89, 137-146, 261, 281-290
object vs. background 47, 163, 193, 312, 334, 392
values 52, 71, 76, 78, 272, 273, 511

Dishabituation 770, 791; *see also Habituation*

Disinhibition
associative 55, 101, 801-803
disynaptic 101, 106
polysynaptic 101
pretectal lesion induced 67, 82, 83, 444, 466
prey-catching 67
tectal and behavioral 465-467
tectal neurons 77, 82, 83, 193, 467

Disparity; *see also Depth, Stereopsis*
field 149
matching 92, 148, 149

Disproportionate
distributions of specialized neurons 103
visual maps 366

Distance 44, 45, 92, 492, 542, 559; *see also Depth*
and elevation 555
signals 542, 547, 556, 674

Distortions in
directional sensitivities 78, 284, 319
retinal maps 366-367, 372, 374

Distributed; *see also Parallel*
control 654, 686
feedback 363, 364
maps 62, 63, 142, 366-375
motor control 123, 578
processing emulation 661
properties of brain function 27, 28, 33, 34, 41, 42, 104, 105, 559, 560, 652, 750, 752, 753, 792, 850

Divergence
maps 62, 63, 378
network properties 42, 104, 105; *see also Bifurcating*
operator 156, 652

Dogs 694

Domestic fowls 60

Dopaminergic transmission 236, 611; *see also Neurotransmitters*

Dorsal
cholinergic pathway 740; *see also Cholinergic*
pallium 777, 808; *see also Pallium*
thalamus 279; *see also Thalamus*

Dorsomedial thalamic nuclei 739, 758; *see also Thalamus*

Downstream analysis 592

Dual concept of
pattern generators 608
pretecto-tectal inhibition 80, 84, 440
striato-tectal influences 106
thalamo-cortical systems 737, 749

D(W,A) values 71; *see also Discrimination*

Dyes
cobalt lysine 180, 182, 205, 483, 511, 512
ethidium bromide 215
Nissl 73, 178
toxicity 888
voltage sensitivity 874

Dynamic; *see also Modulation, Modulatory, Neuro-, Plastic*
assembly of cells 100, 300
configuration 57, 59, 459
control of robots 683
dishabituation 850
Gestalt 50
membrane potentials 700, 709
metastable feed back loop 106
model of T5 neuron function 269
modulation 42, 66, 98-107, 271-289, 351-355, 429, 610, 696, 709, 712, 742, 783
networks 7, 100-102, 126, 276, 894
oscillating neurons 707

time constant 324

Eccentricity signal 546, 547; *see also Coding, Localization, Prey*
Edge
detectors 362
orientation detectors 24, 90, 91, 362
positioning properties 72, 92
preference 53, 72, 255, 294
tracking 674
EEG
during arousal 698, 737
origin 700, 701
responses 699, 700, 705, 707
Effector specificity 592
Efference copy 47, 313, 314, 556, 584; *see also Corollary, Feedback, Loops, Reafference*
Efferent
channels 94
commands 458; *see also Command*
Electric fish 40
Electrical brain stimulation; *see Brain*
Electrocortical desynchronization 730, 737
Eleutherodactylus coqui 54
Elevation and distance information 546, 552; *see also Coding, Localization, Prey*
Emotional processes 753, 755, 757; *see also Motivation, State*
Enclose schema 161, 163; *see also Schema*
Engineering 15, 41, 108, 336, 686, 895; *see also Robot, Technological*
Engram 850; *see also Memory*
Enhancement of excitation 469, 703; *see also Facilitation, Sensitization*
Enkephalin; *see also Neuropeptides*
isthmic nucleus 183
nucleus of the solitary tract 610
optic tectum 107, 183, 191, 293
pretectal nuclei 80, 106, 107, 194
lateral forebrain bundle 195
Ensemble coding 22, 41, 81, 95, 263, 501; *see also Assembly, Coding, Functional unit*
Entropy 414, 684
Enucleation experiments 228, 233, 350, 607; *see also Monocular*
Environment 57, 125, 311, 397, 399, 421, 425, 893; *see also Background, Object, Scene, Textured*
change of 661, 686, 696
inhibitory surround 47, 69, 82
masking 47, 48, 69, 82
uncertainties 684
Epilepsy xiv, 712, 716-718
disorders 713
foci 716
EPSP/IPSP activities 31, 79, 202-220, 322, 501-528
Epstein's model of detour behavior 131, 152; *see also Barrier, Detour*
Escape system 65, 68, 74, 522; *see also Avoidance*
Essential neurons 91, 96
Estimating absolute size 50, 103; *see also Depth, Size, Stimulus*
Ethidium bromide 215
Ethological analysis 5, 45-61
Euclidian norm 364
Evolutionary aspects 15, 57, 107, 108; *see also Brain evolution, Homologies, Phylogenetic*
Ewert-v.Seelen's filter interaction model 89, 129, 137, 263
Exact sciences 12, 424
Excitatory receptive fields; *see Receptive*
Experience; *see Acquired, Associative, Conditioning, Habituation, Learning*
Experimentation; *see also Methods, Model*
and theory 3-36, 168, 538
black box studies; *see Blackbox*
down-stream analysis 592
inferential level 20
multidisciplinary approach 42, 421
top-down studies 42, 122, 422, 423, 436
upstream analysis 592
Explicit
neuronal representation 570, 573
vs. implicit processing 90, 91, 108
Exploration behavior 661
Extinction 745, 749, 758; *see also Memory*
Extracellular recordings
freely moving animals 458-476, 858-870
paralyzed animals 70-94, 247, 252, 273, 274, 319-321, 485-501, 634-637, 643, 705, 787
Extrastriate cortex 336; *see also Cortex, Cortical*
Extratectal
influences 297
inputs 522
modulation 294; *see also Modulation, Modulatory*
Eye; *see also Coordinates*
and head movements 482, 486
and neck motor units 524-527
centered coordinate 369
closure phenomenon 458
enucleation 228, 233, 350
movements 168, 314, 470, 591
movement cells 475; *see also Saccades*
neck cells 527
neck movements 482, 518
position signal 541

Facial motor nerve nucleus 593
Facilitation effects 137, 573; *see also Enhancement, Sensitization*
Family-of-models 125, 152, 425, 436; *see also Model*
Feature; *see also Combination, Configuration, Sign stimulus, Stimulus*
action string 405, 415, 416
adaptive properties 100
analysis 22, 49, 51, 58, 60, 89, 95, 129, 142, 271, 329
analyzing model networks 142, 146, 150, 263, 276, 361, 372, 441, 837, 839
analyzing network 42, 79, 81, 88, 91, 472
analyzing neurons 41, 69, 70, 72, 78, 81, 88, 460
detector concept 41, 42, 86; *see also Barlow, Lettvin*
extraction 22, 77, 89, 90, 92, 100, 336, 362, 372, 521
maps 23, 90, 353, 354
relationships 60, 61, 90, 247
Feedback 482, 638; *see also Loops*
control in robots 683
distributed 363
guidance 43
information 525
iterative algorithm 363
loops 101, 104, 195, 522, 610
metastable loop 106
negative 644; *see also Efference copy, Reafferent*
positive 137, 219, 298, 363-365, 527, 642, 643; *see also Reverberatory*

Femoral chordotonal organ 640
Figure/ground separation 25, 53, 392; *see also Background, Object, Textured*
Filtering operations; *see Configuration, Discrimination, Feature, Stimulus*
Find-Landmark schema 159; *see also Schema*
Fire salamander 67, 85, 244, 248, 311, 313; *see also Salamanders*
Fish 670, 694, 697, 700, 710, 713, 740
Fixed
  action pattern 40; *see also Action*
  discharge pattern 469
  neural ciruitry 125
Flexibility; *see also Acquired, Assosicative, Dynamic, Modulation*
  coordination 636
  neuronal tuning 98, 100, 270-300
  robotics 893, 895
Force and torque sensors 675, 676; *see also Robot*
Form 57; *see also Configuration, Conture, Feature, Gestalt, Shape*
Formal
  languages 405
  model of a neuron 422; *see also Model*
Fourier
  domain invariance models 385
  domain operations 385, 386
  transform 90, 385-387
Free-Prey schema 134; *see also Schema*
Freely moving preparation 453, 455, 859; *see also Wire-*
Freezing behavior 730, 738, 741, 742, 779, 788; *see also Avoidance, Defensive*
Frequency
  coding 467, 556; *see also Coding*
  division multiplexing 858
Frogs 125, 175, 223, 238, 248, 351, 540, 544, 615, 875; *see also Rana*
Functional
  associations 879, 887, 889; *see also Acquired, Adaptive, Associative*
  brain activity 66, 730, 768, 800, 873, 889, *see also Deoxyglucose*
  brain circuitries 28, 61, 142, 873, 887; *see also Cellular, Circuitry, Neuro-*
  brain organization 68, 142, 195, 757; *see also Brain, Deoxyglucose*
  columns 137-146, 331; *see also Column, Model*
  context 98, 164, 406; *see also Context*
  homologies 106, 107, 525; *see also Homologies*
  interactions 47, 88, 129, 142, 147, 879, 887, 879; *see also Axonal, Brain, Combination, Concepts, Interaction*
  mapping 29, 63, 176; *see also Deoxyglucose, Map*
  neuroanatomy 879; *see also Cellular, Deoxyglucose*
  recovery from brain lesions 84
  relations 874, 875, 889; *see also Combination, Correlation, Feature*
  stains 769, 874; *see also Deoxyglucose, Dyes*
  units 69, 80, 81, 88, 177, 192, 423, 434, 645, 646, 836, 837; *see also Assembly, Ensemble*
Functionality problems 10

GABA
  efflux 233, 234
  ergic transmission 79, 80, 194, 223, 224, 229, 236, 238, 293
  pretectal nuclei 194
  release 233
Gabor function 362
Gadus callarias 698
Gaining and filtering 520
Game playing machines 33; *see also Artificial, Machine*
Gamma-aminobutyric acid, *see GABA*
Ganglionic neurons
  basal 106
  retinal 70, 86, 141, 244, 248, 361
  tectal 73, 80, 179, 184, 220
Gap recognition 134, 135
Gaussian
  functions 109, 258, 362, 363
  rate 250
Gaze stabilization; *see also Background, Optokinetic*
  reflex 312, 313, 314, 317, 326, 330, 333
Generalizations 23, 470, 522, 696, 789, 815, 841, 850
Genetic programming 10, 26, 56, 62; *see also Brain, Evolutionary, Homologies, Phylogenetic*
Gerbils 712, 714
Gestalt 50, 57; *see also Combination, Configuration, Feature*
Glial cells 28, 236, 237, 708, 713
  cytoarchitecture 710
  depolarization 710, 717
  syncytium 717
  transmitter 236, 237, 710, 711
Gliosis 236, 715, 717, 718
Global
  architecture 358; *see also Dendrites, Neuroarchitecture*
  language 422, 447; *see also Grammar*
  models 6, 8, 9, 129; *see also Model*
  schemas 436; *see also Schema*
  transforms 391
Glomerulus 139, 140, 440
Glossopharyngeal afferents 606
Glucose utilization; *see Deoxyglucose*
Glutamic acid 223, 227, 228, 231, 236, 237, 711, 712
  decarboxylase 193, 194
  dehydrogenase 228
  efflux 232
  glial cells 237, 711
  release 230
Glycine 223, 224, 229, 234-237
Go/nogo
  conditioning 800
  system 523
Goal xv, 4, 414, 425
  achieving behaviors 43-45, 405, 650, 670
  directed movements 522, 569, 616
  directed search for information 405
  eye position 475
  oriented selective attention 106, 470, 523
  related neuronal assemblies 80, 88, 431, 434
  state 430; *see also State*
Goldfish 238, 351, 698, 714
Goose/hawk discrimination 59
Grammar 15, 405-407, 411, 414
Grandmother cell 41
Grasping 682; *see also Arm, Dexterous*
  tactual 616-627
  visual 6, 15, 45, 65, 472, 476, 592
Grey matter 507, 600
Guidance of fibers 7, 709; *see also Glial*
Guinea pigs 697

Gulping 43, 523, 611, 778; *see also Swallowing*

Habenular nucleus 738, 740, 755, 757, 775
Habituation xiv, 82, 99, 107, 125, 707, 711-713, 768, 773, 784, 792, 826, 841
  2DG uptake 769-792
  arousal 714, 786, 789
  avoidance 788
  column 841
  hippocampal formation 28, 107, 789
  hypotheses 786-789, 821
  long-term 28, 55, 768, 771, 788, 841
  memory 28, 789
  models 784, 841
  prey-catching 55, 771
  short-term 28, 768, 788
  stimulus-specificity 55, 771
  ventral medial pallium 28, 777
Half-centers 608
Halichoeres bivittatus 694
Hall following 663; *see also Robot*
Hallo-system 522
Haloperidol 739
Halothane 743
Hand
  feeding phenomenon 99, 297, 800, 801-803, 810, 843, 845
  movement 569; *see also Arm, Dexterous*
Hardware; *see also Computer*
  adaptation 358
  level 357
  neurons 220
HARV; *see Robot*
Head; *see also Coding, Coordinate, Localization*
  centered coordinate 543
  fixed variable 572
  movements 314, 559
  position signals 92, 475
  trunk coordinate transformation 576
Hebb-type
  associative mechanism 29, 748, 815
  modifiable synapses 752, 821
Heeger's model 362
Hemifield visual deficit 547, 553; *see also Localization*
Hemispheric asymmetry 886
Hermissenda 800
Heterarchical interactions 431; *see also Interaction*
Heterogeneous convergence 521; *see also Concurrent, Convergence*
Hierarchical; *see also Information, Serial*
  and parallel processing 41, 88, 104, 408, 431, 624, 625; *see also Distributed, Parallel*
  command structure 636, 639
  levels 104, 740
  robotics 14
  schemas 405, 407, 408, 431; *see also Schema*
  structures 41, 609, 851, 895
  systems 10, 14, 40, 41
High-affinity transmitter
  transport 224, 236
  uptake 236
High-pass filter 250, 253, 526
Hippocampus 823, 826
  2DG uptake 738, 750, 877
  arousal 738
  fissure 752, 758
  formation 28, 30, 31, 107, 738, 750, 752, 756, 758
  fornical system 821
  functions 789
  habituation 107
  learning 756
  lesions 821
  molecular layers 752
  primordium hippocampi 107, 740, 756, 758, 789, 819-821
  slices 716
Histopathology 715, 716
Hollow perception 407, 416
Holocentrus rufus 694
Homeostasis 98; *see also Motivation, State*
Homologies xi, xiv
  arousal 28, 29, 694, 696, 707, 740
  bulbar reticular formation 524-528
  cognition 107
  command functions 102, 475
  cortical structures 91, 108, 358, 887
  habituation 28, 29, 107, 789
  isocortex 823
  isthmic nucleus 342
  limbic structures 107, 723, 738, 756, 790, 792, 823
  medial pallio-tectal modulation 107
  nucleus amygdala 822
  piriform cortex 823
  premotor suppression 469
  pretectal structures 84-86, 106, 107, 335, 818
  prey selection 54, 90, 244
  primordium hippocampi 107, 740, 756, 758, 789, 819-821
  schema theory as common language 421, 422, 428-431
  sign stimuli 57-62
  striato-tectal modulation 107
  striatum 106, 823
  swallowing 607
  tecto-tegmental control of orienting 554; *see also Localization, Prey*
Horizontal
  movements 320, 555
  OKR 317, 321
  saccades 526
  stimulus eccentricity 542, 546-549, 554-559
  turning 49, 127, 555
Hormones 88, 823
Horseradish peroxidase (HRP)
  studies 179, 191, 322, 323, 347-350, 593-605
Humanities 12, 424
Humans 61, 108, 385, 392, 548, 577, 686, 694, 696, 707, 708, 713, 756, 875, 877, 880, 882, 886, 888
Hunting season 98; *see also Breeding*
Hybrid
  forms of wiping 623, 626, 627
  sensorimotor systems 398-416
Hydromantes italicus 84, 245, 248, 315
Hyla
  arborea 54
  cinera 54
Hyper arousal 740, 789
Hypercolumns 378
Hyperexcitatory effects 67; *see also Disinhibition, Enhancement, Sensitization*
Hyperexcited prey-catching 467; *see also Disinhibition, Prey*
Hypermetabolism 877
Hypersensitization 712
Hypoglossal motoneurons 475, 495, 523, 591-593, 598, 600, 604-608
  facilitation 608
  hyperpolarization 608
  recording during snapping 474
  snapping generation 605
  spatiotemporal activation 608
  topography 592, 593, 595, 598, 601
Hypometabolism 877, 886
Hypothalamic/preoptic region; *see Preoptic*
Hypothalamus 99, 101, 523, 739, 740, 757, 773, 780, 783, 785, 788, 791, 813, 823

Ibotenic acid 67, 345
Identification of
  neurons 16-20, 23, 102, 205, 632
  neurotransmitters 223, 225
Image acquisition 674; *see also Acquired, Associative, Habituation, Learning, Robot*
Immunocytology 80, 106, 176, 191, 193-195; *see also Enkephalin, Neuropeptides*
Implementation
  algorithm 84
  neural circuitry 136
  schema 124
Implicit vs. explicit computing 25, 90, 108
Individual experience 54, 297; *see also Acquired, Associative, Conditioning, Habituation, Learning*
Indoleaminergic cells 224, 611; *see also Neurotransmitters*
Inductive energizer 858, 863-865
Industrial robots 425; *see also Robot*
Inferential level of analysis 20; *see also Experimentation*
Inferior
  colliculus 744, 747, 748
  olive 741
Information
  bearing parameters 59, 60, 84; *see also Sign stimulus, Stimulus*
  coding 95, 540, 556; *see also Coding*
  symbols 405
  theory 27, 402
Information processing; *see also Brain, Concepts, Concurrent, Decision making, Democratic, Integration, Interaction, Model, Network, Neuro-, Recognition*
  columns 137-146, 359, 440-447, 835, 838
  feature detector 41, 86
  maps 365-379
  models 89, 126-166, 248-264, 271-300, 358-379, 440, 835
  multilevel 104, 520, 528; *see also Multi-*
  network analyzer 41, 42, 81, 96, 573
  pallio-tectal 107, 783
  parallel/hierarchical 41, 88, 104, 408, 431, 624, 625; *see also Distributed, Parallel*
  pretectal 72-90, 317-332
  retinal 70, 86, 248
  retino-pretecto-tectal 88, 129, 142, 440
  sensory substitution 399-416
  strategy 358
  striato-tectal 106, 791
  tectal 72-90, 137-146, 275-264
  tecto-bulbar 94, 520
  tecto-tegmental 554
  wetware 358
Inhibition; *see also Interaction, Lateral, Mutual*
  build-up 101, 785, 836
  connectivity 321, 627
  lateral 66, 81, 261, 263, 576
  pretecto-tectal, *see Pretecto-*
  systems 739
Inhibitory
  interaction 47, 69, 82, 88, 129, 142, 261, 263; *see also Pretecto-*
  interneuron 81, 191, 285, 286
  large pear cells 286
  loops 101, 514, 526
  neurotransmitter 191; *see also GABA*
  postsynaptic potentials; *see EPSP/IPSP*
  receptive field 71; *see also Receptive*
  surround hypothesis 47; *see also Interaction*
Innate xii; *see also Phylogenetic*
  concept 40, 62, 800
  releasing mechanism 40, 95, 616, 800
  releasing schema 40
Insects 670
Instantiation of schemas 123, 125, 126, 134, 135, 654; *see also Robot, Schema*
Instinct 40, 107
Integration 42, 66, 194; *see also AND, Concurrent, Cooperation, Information*
  concurrent inputs 44, 45, 496, 513, 514, 527
  medulla oblongata 93, 194, 520-528, 593, 599, 604, 606
  neurons 205, 481, 508, 513, 521
  pretectum 104, 106, 194
  properties 88
  retino-tectal inputs 80
  spinal cord 624
  striatum 106, 823
  tectum 81, 104-107, 142, 194, 276, 837
  tegmentum 194, 482, 554
Intelligence 108; *see also Artificial*
Intelligent
  interfaces 167; *see also Artificial*
  machines 167, 423, 425, 661; *see also Artificial, Computer, Robot*
Intensity tuning function 572, 575
Inter area
  connectivity 377
  correlation 879-887
  matrix 141
Interaction 42, 194, 522; *see also Axonal, Connectivity, Information, Integration, Intrinsic, Lateral, Mutual*
  antagonists 627
  command neurons; *see Command*
  computing 124
  concept 47, 55, 68, 80-82, 88; *see also Inhibitory*
  graphics 167
  schemas 135, 430, 431; *see also Schema*
  tectum and medulla 94, 194, 487, 520
  tectum and pretectum 66, 79, 81, 142, 194, 212
Interdisciplinary research xii, 35; *see also Cross-, Multi-*
Interface; *see also Intermediate, Sensorimotor*
  language 95, 168
  sensorimotor 106, 538, 543, 551, 555, 560
  organization 95, 481, 538
  sensory/associative/motivational 106
Interhemispherical transfer 801
Intermediate xiv
  database 165
  internuncial network 598, 590, 591, 607
  levels 8, 22, 23, 104, 124, 168, 551
  processing steps 33, 551, 555, 556, 562, 573
  state 641
  storage 684
Internal
  and external signals 89
  estimate 577
  feedback loops 100, 607, 610; *see also Intrinsic*
  patterns 26; *see also Central*
  state 88; *see also Motivation, State*
Interneuronal communication 42, 223, 607; *see also Information, Interaction, Inter-*

nal, Intrinsic
Interocular
  competition 47
  transfer 801, 815; *see also Associative*
Interpeduncular nucleus 738
Interpretation 165; *see also Prey, Recognition*
Interregional correlation xv, 873
Interrelational aspects 889
Interscientific xiii
  bridge 422, 447; *see also Bridging, Schema theory*
  umbrella 447
Intervening processes 481, 543, 598, 607; *see also Integration, Intermediate*
Intracellular recording 79, 80, 201-220, 528, 639
Intrinsic
  bulbar properties 487
  cortical feedback 363-365
  hemispheric relations 887
  periodic organization 377; *see also Column, Functional*
  self-motion computer 312, 330
  tectal connectivity 84, 137, 151, 213
  tectal lateral inhibition 261, 334; *see also Lateral*
  thalamic loops 700
Invariance 57, 383, 416, 443
  directional 52, 53, 91, 272, 283, 298, 446
  operations 59, 384, 391
  rotation 390
  shift 384-390
  size 390
  velocity 52, 53, 78, 90, 244, 246-248
Inverse perspective mapping 369, 370, 376; *see also Map*
Invertebrates 550, 562
Iontophoretic applications; *see also Intracellular*
  cobalt-lysine 177-191, 205-217, 504-519
  horseradishperoxidase 179, 191, 322, 323, 347-350, 593-605
Ipsilateral turning 65, 553, 554
Isocortex 823
Isolated nerve endings 224
Isolation problem 421, 423, 427, 428, 436, 447; *see also Bridging*
Isomorphisms xii, 17, 22; *see also Dendritic, Morphological*
Isotropic histological structure 25, 91
Isthmic nucleus; *see Nucleus*
Isthmo-
  tectal fibers 342
  tectal neurotransmitter 350
Iterative algorithm 363, 365

Jacobian matrix 370
Jaw
  motoneurons 593
  muscles 592
  opening 592; *see also snapping*
Joint representation 546
Jumping 65, 618; *see also Avoidance, Escape*

Kainic acid 67, 82, 83
Kernel function 372
Kindling 713, 716
Knowledge 15, 16, 40, 57, 108, 123, 163, 398, 406; *see also Acquired, Associative, Conditioning, Learning, Memory*
  based interpretation 686, 893, 894
  individual 54, 57, 107
  innate 40, 57, 61, 62
KONWA-system (contour perception) 398, 399, 405

Labeling of cells; *see Intracellular, Iontophoretic*
Laminated structures 72, 73, 90, 176, 178, 358
  cortical 359, 360
  tectal 73, 176, 178, 205
Language 4, 5, 57, 61, 108, 168, 415, 422, 428; *see also Computer, Grammar*
Lara's model of learning 835
Lara-Arbib's model of prey habituation 784
Large field neurons; *see Wide field*
Large ganglionic neurons 80, 179, 181, 183, 191, 192, 206, 209, 215-220, 294
Large pear shaped cells 80, 192, 277, 440, 835
Lateral; *see also Mutual*
  excitation 69, 82, 89
  inhibition 66, 68, 82, 89, 219, 261-264, 331, 334, 378, 511, 521
  interactions 387, 516, 521
Lateral amygdala 100, 740, 819, 822, 826; *see also Nucleus*
  anterior thalamic nucleus 101, 804, 816
  bulbar reticular formation 490, 505, 506, 521
  forebrain bundle 101, 107, 187, 195, 822
  geniculate body 342, 365, 376, 392
  habenula 738, 750, 755; *see also Habenula*
  pallium 777, 808, 813, 819, 822
  posterodorsal thalamic nucleus 67, 774, 806, 807, 815, 817, 818; *see also Pretectum*
  posteroventral thalamic nucleus 63, 773, 774, 805, 816, 817; *see also Pretectum*
Lateralization 522, 550, 554, 555, 558, 563
Learning xiv, 33, 42, 59, 69, 92, 98, 107, 729, 730, 757, 759, 833, 834, 850; *see also Acquired, Associative, Conditioning, Habituation, Memory, Model*
  cardiac response 745, 758
  definition 98
  game playing machines 33
  models 33, 833-851
  motivational significance 819
  synaptic matrix 33
  to attack predators 55, 801, 841, 843; *see also Handfeeding*
  to avoid bees 55, 841, 843, 844
  to distinguish palatable and unpalatable objects 841, 843, 846, 847
  related metabolic changes 748, 804-826
  related processes 752-759, 815-826; *see also Hebb*
  reorganiation 756
Leech 625
Lens accommodation
  mechanism 93, 148; *see also Accommodation, Depth*
Lesion experiments, *see Brain*
Lettvin's "bug perceiver" concept 86, 459
Leuciscus cephalus 694
Level setting 468
Levels of
  complexity 564
  information processing 104-107, 520, 557
  modeling 122-167
  sensorimotor processing 104, 520, 528, 557, 627
Limb
  control 482
  coordination 616, 622, 631, 636, 644
  movements 559
  position 539, 540
  prosthetics design 427
  walking 631-647
  wiping 540, 618-628

Limbic system 107, 523, 723, 738, 740, 750, 753, 756, 790, 792, 823; *see also Homologies*
Linguistics 108, 422, 424; *see also Language, Grammar*
LISP; *see Computer*
Local
  circuits xii, 80, 93, 179, 183, 191, 372, 610; *see also Interactions, Intrinsic*
  neural activity 712
  skin reaction 623
Localization; *see also Coordinates, Model, Prey*
  command releasing systems 92, 95, 96
  depth 146
  information 22, 43, 311, 551
  maps 365, 367, 555
  releasing mechanisms 43
  resolution 22, 23, 520
  robot 684, 685
  schema-based navigation 653
  schemas 124, 149
  sensorimotor processes 538, 543, 551, 555, 560
  spatial representation 543
  systems 312, 313, 331
Locomotion 526, 591, 618, 626, 631, 778; *see also Limb*
  motor pattern 626
  routines 103
  walking 631-647
Locus coeruleus 742, 886
Locus-specific habituation 55, 770
Loeb's concept of phototaxis 571
Logical
  entities 10
  operations 108; *see also AND, OR*
Long-term
  changes in tectal neurons 287
  habituation 99, 768, 783, 784, 788, 792, 841, 842
  neural adjustments 80, 98
  potentiation 752
  recordings 293
Looming; *see also Avoidance, Escape, Threat*
  detectors 362
  objects 74, 493, 517
Loops 104, 195, 522, 610; *see also Efference copy, Feedback, Reafference*
  arousing 101
  closed 361, 643
  extending 101, 107
  gating 101, 106
  interactions 523
  modifying 101, 107
  open 361
  priming 101
  screening 101; *see also Habituation*
Low-pass
  characteristic of OKR 314
  filter 251, 365, 378
Luminance
  density function 578, 579
  distribution 572, 573, 584
  sensitive neurons 74; *see also Changes*
Lunging movement 546

M-type classification; *see Medulla*
Machine; *see also Artificial, Computer, Robot, Technological*
  community 14
  grammars 15, 405-407, 411, 414
  learning 649
  pattern recognition 391
  vision xii, xiii, 15, 121
  vocabulary 15
Macro
  network xii, 68, 104, 105, 219, 271, 443, 452, 611
  schema 167
Mammals 106, 107, 342, 554, 607, 608, 693, 694, 707, 712, 714, 759, 779, 784, 789, 822, 823, 826
Mammillary body 738
Man; see Humans
Map 62, 63, 365, 441, 482, 521, 538, 558, 569; *see also Mapped, Mapping*
  coupling 541
  feature 74, 90
  interacting 62, 63
  ordered information 74, 142, 569
  organization 331, 538
  sensorimotor 555
  weighting xiii, 575, 576, 578, 580, 581, 583
Mapped; *see also Mapping, Maps*
  complex logarithmic 368, 371
  filters 372, 373
  function 365
  information processing 555
  inverse perspective 369, 370, 376
  velocity field 368
Mapping; *see also Coordinates, Mapped*
  computational 90, 378
  extratonotopic space 748
  functional brain activity 30, 804; *see also Deoxyglucose*
  non-topographic 331, 377
  parallel 62, 63, 372-377
  parametric 90, 331
  perspective 371
  receptotopic 365
  representational 22, 29, 90
  rules 7
Mastication 526, 591; *see also Jaw*
Match-mismatch evaluation 783, 785; *see also Information*
Mate recognizers 97, 296; *see also Breeding*
Mathematical models; *see Computer simulations, Models*
Mating; *see also Breeding*
  motivation 99, 296
  period 99
  season 99, 296
McCollough effect 386
Medial
  bulbar reticular formation 490, 494, 495, 497, 504, 505, 507, 517, 521, 591, 599, 603, 604, 607; *see also Bulbar, Medulla*
  forebrain bundle 101, 738, 739
  geniculate body 744, 755
Medial pallium 68, 100, 101, 107, 523, 697, 774, 777, 784, 785, 788-789, 804, 808-810, 813, 819, 820, 821, 823-826
  MP-type neurons 787
Medial thalamus 737, 739, 750, 753, 757
Medialpallio-
  hypothalamo-tectal pathways 107, 783; *see also Circuitry*
  tectal pathways 783
Median bulbar reticular formation 505; *see also Raphe nucleus*
Medioventral bulbar cells 524
Medulla oblongata 187, 192, 212, 244, 482, 483, 494, 504, 521, 546, 550, 553, 557, 591, 600, 601, 697
  autonomic nuclei 740, 757
  binocularity 521
  comparisons 524
  cyclic bursting neurons 497, 515, 525
  extracellular recording 488-501
  function 520-529
  gaining and filtering 520
  intracellular recording/

labeling 501-519
laterality 521
lesions 546-548
loops 522
M1-M11 type neurons 488-519
M5.1-4 type neurons 489-492, 508-515
morphological classification of neurons 504-508
multiple properties 501
organization 560
parallel recording 497-499
physiological classification of neurons 524
pretectal afferents 94, 194, 213
reticular formation 354, 484, 742; *see also Bulbar*
reticular network 522
snapping neurons 592, 593, 599, 605
snapping pattern generation 606-611
tectal afferents 93, 94, 194, 210, 593-597
tegmental afferents 194, 484, 557, 558
wide-field neurons 492-496, 521
swallowing neurons 526

Mellin
correlation model 390
invariance 391

Membrane
activities 502
potential 138, 302, 840; *see also EPSP/IPSP*
potential simulation 138, 282, 302, 840
oscillation 700, 701, 702

Memory 27, 108, 523, 833; *see also Acquired, Associative, Knowledge, Storage*
functions 752, 893, 894
KONWA system 398
loops 522
modulating system 822; *see also Modulation, Modulatory*
neurons 74, 497, 818, 836
processes 756, 800, 808
representation 752

Menotaxis 573

Mesencephalic reticular formation 527, 729-745, 757, 772, 773, 779-781, 786, 790, 791

Metabolic; *see also Deoxyglucose*
mapping 28-31, 730, 756, 816
measures of functional activity 733, 769, 804, 874, 875
suppression 737

Metasymbol 405-407; *see also Contour perception*

Methods; *see also Experimentation, Model, Modeling*
antidromic stimulation/recording 93, 218, 485, 492, 597, 733
autoradiography 64, 66, 227, 70-832, 875, 888
camera lucida reconstructions of neurons 184, 205, 210, 323, 483, 506-519
Cobalt-lysine labeling 178, 180, 186, 506-519
collision test of antidromic activation 485
computational frog "Rana computatrix" 125, 168
computer assisted analysis of autoradiographic images 732, 746, 774
computer simulated cell architectures 137, 276, 441, 837
computer simulated membran potentials 138, 260, 282, 301, 846, 851
cooperation in brain research 4, 167
deoxyglucose (2DG) 733, 763, 804, 875
electrical stimulation; *see Brain*
electroencephalogram (EEG) analysis 698, 875
experimentation and modeling 3
extracellular recording in paralyzed animals 70, 247, 494
feature discrimination value 52, 71, 78, 277
fluorodeoxyglucose (FDG) 877
global model approach 66, 81, 100, 129
horseradishperoxidase (HRP) labeling 187, 348, 597, 595
Ibotenic acid 345
immunocytochemistry 106, 193, 352
interdisciplinary approach; *see Inter-*
interregional correlation of glucose metabolism 879
interspike-frequency-time (ISFT) histogram 456-463
intracellular recording/labeling 203-217, 502, 506, 509-519
Kainic acid 83
mathematical analyses of information processing 89, 243, 357, 383, 571
lesioning; *see Brain*
model-family approach 125, 152, 425, 436
motor coordination 517, 621, 633
multidisciplinary approach; *see Multi-*
multiunit recording 874
neuroethological analysis 43, 452, 538, 591, 632, 801
neuron/behavior correlation 76
neurotransmitter identification 225, 350
Nissl stains 73, 178
optical recording of voltage sensitive dyes 874
parallel recording 498
positron emission tomography 875
pulse-analogue correlation 213-216
schema-based models 121, 134, 167, 404, 422, 431, 653
schema-based robots 653, 661
sensorimotor robots 676
sign stimulus analysis 45-54, 127, 457
single cell wire-bound recording in freely moving animals 452
stimulus continuum analysis 51, 60, 77, 85
systems-theoretical analysis 89
wire-less recording/stimulation/marking in freely moving animals 858

Midbrain/medulla junction 553

Millipedes 650

Miniature telemetry; *see also Telemetric recording*
radio transmitter 862
three-function transmitter 859, 861

Minimal neural substrate 748; *see also Functional unit*

Misdirected movements 552

Mobile robot 14, 159, 649-670, 676, 684; *see also Robot*

Model of; *see also Computer, Concepts*
associative learning 29, 33, 100, 834-851
ballistic movements 43, 123, 608, 624
column 137
command function 95-97, 102,

569-584, 625
configural discrimination 80-91, 261-264, 440-447
control of limbs 625, 627
cortical intrinsic tuning 363
cortical organization 360
cue interaction 81, 129, 147
depth perception 146, 651
detour behavior 131, 152
dynamic selectivity 100, 276-300
epilepsy 712
filter interaction 89, 90, 129
gaze stabilization 327
grasping 123, 159-163
habituation 784, 786, 841
invariance generation 384-393
learning 100; *see also associative*
locomotion 635-647
map-weighting 575
mapped cortical filters 372
modulation of prey recognition 100, 837, 838
non-topographic maps 377
object localization 149-152, 544, 557, 569-584
oculomotor integrator 527
path planning 152-159, 650-670
pretecto-tectal interaction 66, 89, 90, 140-145
prey/barrier perception 128-136, 149-152, 433
prey selection 80-91, 125, 127, 145, 440-447
retinal network 248-257
retinotopic mapping 365
self motion computation 333-336
sensorimotor codes 44, 95-97
sensory substitution 399-416
schemas for vision 163
tectal columns 137
tectal network 81, 257-264

Modelers
romantic 6, 7, 10, 13
responsible 6, 10, 12

Modeling xi
adequate 5
aim 3-16, 24, 122, 153, 167, 169, 670
and engineering 14-16, 108, 124; *see also technical*
and experimentation 3-36, 570
approaches 121-169, 248-264, 269-305, 358-379, 384-393, 436-447, 769-784, 833-853
bottom up 122, 423
complexity 16, 26
concepts 6; *see also Concepts*
constitution of cell types 257
credit assignment problem 33
design of experiments 153, 168; *see also Experimentation*
detailism 5, 6, 9, 26
dynamics 7, 21, 27
evolving complexity 167
families 125, 152, 425, 436
global box 6, 8, 44, 66, 129
heuristic value 8, 13
inferential level 20, 22
intermediate steps 8, 22, 124
metabolism 31, 32
network 81, 89, 100, 139, 142, 248, 257, 276, 360, 441, 837, 839
parallel paradigm 15, 32, 33
physical vs. biological thinking 9-11
processing schemes 153
raw data 4, 32
reality problem 9, 10, 124
regularity problem 10
schema-theoretic approach 122, 404, 422, 430, 649
simple principle 10, 25, 26
social structures 4, 11, 166
structural assignment problem 17-20, 33
structure/function 9, 17-20, 26
technical applications 108, 122, 165, 379, 397-416, 649-670, 673-686
top down 122, 423, 436

Modifiable synapses 29, 752, 756, 821; *see also Associative, Hebb*

Modification by learning; *see Acquired, Associative, Dynamic, Learning, Modulation, Modulatory*

Modular architectures 377, 434; *see also Assembly, Column, Ensemble, Functional unit*

Modulation of 27, 55, 98, 271, 473, 709, 718, 719, 739, 742, 758, 783, 792, 815, 823, 850; *see also Cellular, Circuitry, Modulatory, Network, Neuro-*
and EEG; *see EEG*
arousal; *see Arousal*
attention 106; *see also Attention*
configural discrimination 59, 69, 89, 91, 98-103, 271, 800-826
neuropeptides 80, 106, 107, 176, 191, 193-195, 293, 610; *see also Enkephalin*
pretecto-tectal inhibition 68, 69, 84, 99, 100, 816-818, 834-851
prey-selection 55, 59, 68, 69, 98, 100, 800-826
reticular formation; *see Arousal, Mesencephalic*
sensorimotor performance 106, 107, 696
snapping 609

Modulatory
circuits 5, 42, 98, 101, 103, 107, 276, 297, 473, 561, 783; *see also Circuitry, Modulation*
command elements 100, 103
dynamics 21, 27, 276-300
enhancement; *see Enhancement, Sensitization*
excitability 709; *see also Enhancement, Sensitization*
glial function 711
isthmic function 353
loops 101, 106, 107
network structure 100, 101, 276
pallio-hypothalamo-tectal influences 101, 107, 522-783
prosencephalic influences 68, 89, 90, 101, 106, 107, 747, 783, 836, 848
reticular formation; *see Mesencephalic*
sensitization; *see Sensitization*
striato-pretecto-tectal influences 101, 106, 195
synaptic weights 126; *see also Weights*

Monkeys 368, 371, 391, 469, 475, 696, 756

Monoaminergic cells 611; *see also Amine*

Monocular
depth cues 148; *see also Accommodation, Depth*
feature analyzing neurons 74-78
OKR-gain 321
training 801, 805, 807
vs. binocular inputs 73, 521

Morphological features; *see also Dendrites, Network, Neuro-*
and physiological properties 17-20, 33, 94, 218
bulbar neurons 506-519
cortical neurons 360
pretectal neurons 187, 212, 213, 322, 323
tectal neurons 179, 184, 205, 210

Motion
analysis 362, 363; *see also Background*
parallax 335

perception 362
Motivation 59, 69, 92, 98, 105, 107, 462, 472, 525, 592, 707, 759, 768, 812, 816, 822; *see also Reinforcement, State*
aversive 757
command releasing systems 98
functional contexts; *see Context*
hibernation 98
modulation; *see Modulation, Modulatory*
parameters 738, 747
processes 98, 749
significance 98, 745, 755, 815
state 429, 440; *see also State*
Motoneurons 508, 528, 543, 636
snapping 592-594, 608
walking 634, 637, 642, 643
wiping 617
Motor
arousal 474
choice 43-45, 625; *see also Schema*
commands 44, 94-96, 313, 541-543, 569, 574-578, 584, 594, 605, 624-627, 646; *see also Command*
contribution of compartments 626
control 632, 636; *see also Control*
coordinate frame 475, 545-547; *see also Coordinates*
coordination 484, 500, 607, 632, 634, 635, 676, 677
cortex 875
democracy 633
feedback 473
foreleg pattern generator 634
free arm motions 686; *see also Arm*
generating bursts 519, 527, 528, 634, 637
hindleg pattern generator 634
hindlimb configurations 618
innervation pattern 569
map 543, 555
movement strategies 616, 618, 623
pattern alteration 548
pattern generating circuitry 91, 95, 96, 102, 523, 524, 546, 589, 590, 605, 626, 646
pattern generator for snapping 592-611
pattern generator for walking 632-647
pattern selection 43-45, 547, 548, 559, 560, 625
pattern shifts 546
predicting neurons 468, 469, 473, 476, 523
recommendation 636, 645-647
relay cells 524
schema 97, 123, 133, 136, 156, 430, 592, 623, 626, 653, 654, 670, 677, 683; *see also Schema*
skills in robotic systems 683; *see also Robot*
Move Arm schema 161, 162; *see also Schema*
Move-Ahead schema 159
Move-To-Goal schema 159
Movement-direction invariance 53, 78, 108, 273, 283, 284; *see also Directional, Invariance*
Moving
background patterns 47, 321; *see also Background, Object, Textured*
configural stimuli 47-52, 58, 60, 63, 77, 85, 129, 144, 205, 206, 244, 246, 247, 252, 263, 270, 272, 443-446, 492, 509, 802, 809
MS 222 narcosis 779, 782; *see also Anesthesia*
Multi-
channel processing 32, 858; *see also Distributed, Parallel*
component models 7
disciplinary approach 42, 421
dimensional feature vectors 377
dimensional space 874
layer adaptive networks 33
level processes 104, 520, 521, 627
modal connectivity 821
plexing 27
sensory integration 676, 677
unit recording 498, 874, 888, 889
Multiple
depth cues 149
input metastable feedback loop 106; *see also Feedback, Functional unit, Loops*
instantiations of schemas 21, 98, 104, 434; *see also Schema*
property displaying neurons 495, 501, 523
representation of the visual field 62, 63, 366; *see also Map*
unit recording 496
Muscular feedback 481
Mutual; *see also Lateral*
excitation and inhibition 81, 88, 441
excitatory loops 81, 257, 264, 523; *see also Reverberatory*
inhibition 66, 81, 261
M-type classification; *see Medulla*

Narcosis 781, 786, 790; *see also Anesthesia*
Naso-temporal image shift 312
Natural scene 13, 363; *see also Background, Object, Textured*
Navigation 148, 664, 894
in cluttered hallways 663
purposes 676
requirements 654
robot 650; *see also Robot*
toad 43, 152; *see also Detour*
Necessary
conditions 102, 575
constituents 102, 570
Neck muscle systems 482, 486, 528
Necturus maculosus 698
Negative
affect coding 737
feedback 527, 644; *see also Feedback, Loops, Reafference*
Neonates 694; *see also Ontogenetic, Phylogenetic*
Neostriatum 755, 757, 758, 822; *see also Striatum*
Network; *see also Cellular, Circuitry, Model*
analyzer hypothesis 41
compensation 91
continuum 81, 91, 361, 441
interconnections 80-82, 202, 257, 261, 385, 441
logical neurons 422
model 81, 142, 261, 360, 387, 441, 837; *see also Model*
properties 16, 41, 91, 363, 522, 523
Neuro-
anatomy 27, 32, 176
architecture 73, 81, 139, 142, 176-220, 895; *see also Cellular, Network*
behavioral correlations 76, 467, 469, 858
biology 15, 41
computers; *see also Computer, Robot, Transputer*
ethology xii, 6, 21, 22, 40, 41, 63, 122, 152, 169, 415, 424, 439, 454, 686, 800
glia; *see Glial*
hormonal factors 841, 850; *see also ACTH*
leptics 739
modulation; *see Dynamic,*

Modulation, Modulatory, Neurotransmitters
muscular blockage 524
pathology 716-718, 886
peptides 80, 106, 107, 176, 191, 193-195, 293, 610; *see also Enkephalin*
physiology 686
sciences 422
transmitter identification 225
Neuronal
assemblies; *see Assembly, Ensemble, Functional*
blueprint 167
building blocks 21; *see also Building block*
circuits; *see Circuitry, Models*
computing 121, 167; *see also Computer, Model, Simulation*
ensembles; *see Ensemble*
hardware 21, 26, 220, 357, 358
loops; *see Feedback, Loops*
models; *see Models*
networks 42, 80, 124, 142, 276, 357, 360, 379, 422, 442, 837, 839; *see also Cellular, Models, Network*
parsimony 104
population 80
reality 124; *see also Model*
synaptic weights 143; *see also Weights*
thresholding functions 140; *Threshold*
tuning 77, 81, 270, 363, 577
units; *see Assembly, Ensemble, Functional*
Neurotransmitters xii
acetylcholine; *see Acetylcholine, Cholinergic, Nicotinic*
aspartic acid 235, 238, 712
dopamin 236, 611
enkephalin; *see Enkephalin*
GABA; *see GABA*
glutamic acid; *see Glutamic acid*
glycine 223, 224, 229, 234-237
norepinephrin 236
peptides; *see Neuropeptides*
serotonin 224, 230, 236, 610
Newness
neurons 470, 493, 496, 500
properties 783, 788
Nicotinic cholinergic 79
Nociceptive transmission 742; *see also Pain*
Non-
associative learning 99, 768, 792; *see also Habituation*
directional arousal 694
linearities 328, 329
placed coded form of representation 556, 558; *see also Coding, Localization, Map*
specific arousal 698
specific thalamic nuclei 730, 737, 749
topographic mapping 377, 514, 526
verbal species 108
visual forebrain areas 819, 826
Norepinephrine 236
Novelty 695, 789, 821
detection 107
sign stimuli 824
situations 714
Nucleus amygdalae 107, 523, 738, 757, 777, 790, 804, 808, 813, 822, 823, 826
Nucleus isthmi xiii, 34, 68, 149, 151, 179, 181, 341-353, 448,
anatomy 347
controversial results 34, 151, 346
depth perception 34, 151, 346
equivalence to parabigeminal nucleus 342
interactions 342, 346, 348, 351, 353
lesions 343-345
neurotransmitter 345, 350
receptive fields 35, 343
tectal binocularity 346
weighting function 34
Nucleus lentiformis mesencephali 187, 189
Nucleus motorius tegmenti 484
Nucleus of Meynert 739, 740
Nucleus of the longitudinal fasciculus 486, 553, 594
Nucleus of the solitary tract 494, 600, 607, 611, 741
Nucleus reticularis isthmi 486
Nystagmus 317; *see also Optokinetic*

Object
background discrimination 46, 47, 335, 336, 392; *see also Background, Textured*
centered plan 681
classification 45, 56-62, 384, 392
localization; *see Localization*
motion; *see Motion*
recognition; *see Prey, Recognition*
Obstacle; *see also Avoidance, Barrier, Detour, Model, Path*
avoidance 219, 371, 376, 407, 416, 663, 666, 670
detection 661, 818
environment 651
Ocularity stripes 26, 359, 377
Oculomotor integrator 527
Odor; *see Olfactory*
Off-line; *see also Robot*
grasping specialist 681, 683
workpiece acquisition 682, 683
Olfactory
and visual cues 788, 800, 812, 823, 825, 826, 834, 836
bulb 824; *see also Accessory*
conditioning 55, 100, 811-814, 824, 826
On- and off
bursts to stripe edges 72, 254, 255; *see also Edge*
channels 249, 250, 275
responses 70, 79, 204, 250, 509, 627
zones in the receptive field 249
On-line; *see also Robot*
grasping specialist 415, 670, 682, 683
workpiece acquisition 683
Ontogenetic aspects 43, 54, 85, 86, 297
Open loop 607; *see also Feedback, Loop*
Operant conditioning 824; *see also Associative, Conditioning*
Operational principles 13, 106
Optic tectum 63, 65, 67, 72, 101, 175, 176, 209, 212, 218, 219, 223, 236, 243, 244, 269, 330, 335, 342, 482, 543, 605, 707, 708, 774, 813, 826; *see also Tectal neurons*
2DG uptake 63, 64, 774
and avoidance 65, 68
and prey-catching 65, 66, 68, 76, 137, 248, 436, 544, 557; *see also Prey*
antidromic activation 93, 485
Cobalt-labeling 177, 205, 483
columns 137, 139, 144, 177, 183, 192, 297, 298, 331, 440, 784, 835, 838, 840, 842
commands 94, 95, 472
disinhibition 77, 82, 83, 145, 442, 444, 446
efferents 93, 94, 210, 485, 542, 543, 558, 593, 605
functional units 81, 137, 276, 837; *see also columns*
HRP labeling 596
integration 69, 81, 101, 106, 107, 258

intrinsic excitatory processes 48, 69, 81, 82, 137, 257, 264, 298, 440
intrinsic inhibition 261, 334
intrinsic interaction 469, 541
lesions; *see Brain*
models; *see Model*
modulatory afferents 98, 100, 101, 106, 107, 271, 353, 783
networks 73, 81, 142, 257, 441, 839
neurons; *see Tectal*
neuroarchitecture 176-220
neurotransmitters 224-238
premotor function 461
pretectal inhibition 69, 80-86, 129, 142, 441, 835, 848
specifying afferents 83, 84, 106, 466
stimulation; *see Brain*

Optic
accessory system 16, 194, 317
tract 202

Optical
correlator 399
flow analysis 368, 369, 376
match 584

Optoelectronic devices 14, 15, 121; *see also Computer, Machine, Technological*

Optokinetic
afternystagmus 325
nystagmus 68, 192, 194, 209, 314
pretectal cells 324
reflex 314
turntable 324

OR-gate 97, 387, 492

Orient-Trunk schema 161, 162

Orientation
behavior 106, 575, 576
columns 91, 108, 360
detectors 90
iso-slabs 10
model 133, 152
of dendrites 508; *see also Dendrites*
reaction 461, 693
schema 156
tuned cells 377

Orienting behavior 44, 45, 51, 94, 461, 482, 527, 538, 551, 554, 555, 694, 789
localization properties; *see Prey, Localization*
path planning; *see Detour, Path*
pretectal influences; *see Brain, Pretectal, Thalamic*
recognition properties; *see Prey*
tectal mechanisms; *see Brain, Model, Optic tectum*

Oscillating
membrane potentials 700
networks 527; *see also Bursting, Medulla*

Overt behavior 739

P nucleus; *see Pretectal*

Pacemakers 700

Pain 741; *see also Nociceptive*

Paleo-striatum 101, 790; *see also Striatum*

Pallium
lesions 777, 811
medial pallium; *see Medial pallium*

Pancreatic polypeptide substance 106; *see also Neuropeptides*

Panoramic vision 317

Parabigeminal nucleus 342; *see also Nucleus isthmi*

Parallel; *see also Distributed*
actions 627
and hierarchical information processing 39, 42, 88, 104, 415, 592, 625; *see also Hierarchical*
architectures 377, 683, 686
ascending reticular formation 740
channels 32
computer architectures 15, 32, 121, 893
distributed processing 14, 23, 25, 32, 33, 68, 90, 91, 104, 335
extracellular recordings 498
processing 108, 385, 425, 895
reflex array 561
schemas 407
vs. serial processing 41

Parallelism 16, 122, 377, 686

Paralyzed toads 518, 525

Paramedian pontine reticular formation 481

Parametric; *see also Coordinates, Map*
code 683
maps 23, 90, 331, 377
representations 429, 430

Parcellated
body-centered signals 558
nuclei 743
posterior dorsal thalamus 86, 297
representation of space information 546-548, 550, 556

Parcellation theory 86

Parietal cortex 476, 548, 750, 755, 757; *see also Cortex*

Parseval theorem 582

Past experience 697; *see also Acquired, Experience*

Patchy cortical connectivity 359, 377; *see also Column*

Path
behavior 649, 650, 653
determining vectors 652
gerbil's detour behavior 131
models 133, 152, 650, 652
planning 130, 131, 133, 152-154, 430
robot's obstacle behavior 661-670
toad's prey/barrier behavior 130

Pattern
generating circuitry 22, 23, 144, 543-545, 551, 554, 559; *see also Motor*
generator 631, 636, 646
recognition 14, 25, 57, 86-90, 129, 137-159, 263, 383, 384, 538, 569, 686; *see also Prey, Recognition*

Pause-neurons 528, 608, 627

Pavlovian conditioning 745, 748, 750, 753, 754; *see also Associative, Conditioning*

Pear-shaped neurons 73, 80, 179, 206, 209, 218, 286, 294, 611

Penicillin 712

Peptidergic neuroregulators 80, 177, 191, 293; *see also Enkephalin, Neuropeptides*

Perception 23, 108, 397, 404, 416
action cycles 41, 405, 428
cycles 404
definition 404
of depth 133, 146
of figures 56-61, 410; *see also Configuration, Prey*
with KONWA-system 399-416

Perceptual
contents 405, 411
motor schemas 122, 123, 430
operations 41, 44, 80-91, 126-166, 248-264, 271, 317, 359, 399, 431, 441, 538, 575
robotic systems 121, 159, 167, 427, 650-670, 674-686
schemas 97, 98, 135, 136, 159, 430, 654, 670, 677, 683, 850; *see also Schema*
sharpening 54, 514, 521
threshold 743
uncertainty 654

Perihypoglossal nuclei 741

Periodic
activity 516; *see also Bursting, EEG*

organization 359
Peripheral feedback 524, 607; *see also Efference, Feedback, Reafference*
Perspective mapping 369; *see also Map*
Pharmacology; *see Neurotransmitters*
Phase-locked activities 499, 500
Philosophy 422, 424, 427
Phototaxis 570
Phylogenetic points of view 26, 40, 54, 57, 61, 62, 84-86, 90, 106-108; *see also Evolutionary, Homologies*
Physical
continuum 51, 58
environments 57
sciences 11
thinking 9, 10
Picrotoxin 79
Pigeon 220, 236, 238, 818
Piriform
cells 179, 189, 594, 611; *see also Pear-shaped*
cortex 823
Pituitary hormone 786
Place-coding 538-540, 556; *see also Localization, Prey*
Planaria 571, 578
Planning 430; *see also Path, Programs*
Plasticity 105, 757; *see also Dynamic, Modulation, Modulatory*
Pontine
medullary reticular formation 486, 526
reticular formation 526, 550
swallowing neurons 526
Population coding 22, 82, 102, 300; *see also Assembly, Coding, Ensemble*
Position-coded signal 328; *see also Coding, Coordinates, Localization, Prey*
Positive
and negative reinforcement 739; *see also Motivation*
feedback; *see Feedback, Iterative, Reverberatory*
phototaxis 570, 573
Positron emission tomography 875, 889
Post-stimulus events 497, 503, 516, 517; *see also After-, Reverberatory, Warming-up*
Post-tectal structures 354, 482, 590, 543, 558; *see also Bulbar, Medulla, Tegmentum*
Posterior thalamic nucleus 67, 73, 86, 189, 211, 317, 350, 816; *see also Pretectum, Thalamus*
Posteroventral medial pallium; *see Medial pallium*
Power gate 355
Praying mantis 435
Pre-motor
activity 462, 469, 473, 525, 528
cortex 162
network 91, 516, 527, 538, 550, 551, 555, 559
neurons 461, 469, 482, 607
properties 461, 462, 469, 473, 481, 486, 519, 523, 525, 821
Predator; *see also Avoidance, Escape*
avoidance 58, 60, 66, 68, 70, 97, 126, 134, 778, 779, 790, 823
habituation 788, 834
recognition schema 127
Prefrontal cortex 750, 753, 755
Preoptic area 101, 523, 739, 740, 757, 773, 813
Preoptic-hypothalamic
area 697, 740, 785, 804, 823; *see also Hypothalamus*
Preshape schema 161-163
Pretectal/tectal; *see also Pretecto-, Pretectum*
circuitry 66, 80-84, 88, 194, 195, 416, 521
excitatory loops 523
interactive computations 65, 88, 90, 334-336, 443, 523, 783, 834
Pretecto-tectal inhibition 68, 69, 80-86, 90, 92, 129, 142, 144, 194, 195, 264, 294, 440, 523, 817, 836
functional recovery 84, 91, 297
ontogenetic studies 85, 86
projections 67, 106, 187-191, 193-195, 209-213
stimulus-specific tuning 83, 84, 271, 292-294, 297
transmission 194, 195
two-stage mechanisms 84, 85
urodeles vs. anurans 86
Pretectum 16, 106, 126, 151, 157, 175, 176, 220, 245, 334-336, 482; *see also Thalamic*
2DG uptake 63, 64, 773, 805-807, 817
and learning 816-819, 834-851
and prey recognition 65-90, 129, 142, 441, 442
comparison across vertebrates 86, 106
immunocytochemical studies 106, 194, 195
lesions 67, 77, 83, 141, 145, 193, 270, 297, 320, 334, 443, 444, 465, 466, 788, 808, 815
Lpd/P region 67, 72, 101, 788, 791, 804, 808, 827
modulation of tectal cells 82, 84, 100, 104, 107, 271, 276, 331, 783, 837
neurons 74-77, 209, 321, 322, 324, 328; *see also Thalamic*
nuclei 67, 68, 189, 191, 193, 293, 317, 321, 440, 516, 523, 590, 774, 817, 818, 835, 836, 848, 850
optokinetic nystagmus 317-336
projection neurons 189, 191, 193, 194, 212, 213, 323, 329, 492
stimulation 65
tuning of tectal cells 69, 77, 84, 106, 271, 276, 293, 440, 442, 837
Prey; *see also Stimulus*
acquisition schema 40, 57, 88, 133, 134, 415, 433, 783
and background 47, 48, 53, 55
and barrier 135, 136
barrier schema 434
category 46, 56-61, 69, 89, 783
configuration 45, 47, 49; *see also Configuration*
dummies 45-54, 58-61, 244
edge preference 53; *see also Edge*
feature maps 23, 90
filters 89, 90
habituation 55, 785, 834
localization 44, 45, 92, 149, 151, 152, 538-564, 791
motivation 40, 54, 98, 473
natural objects 46
odor 55, 610; *see also Olfactory*
selective attention 99, 470
selective model pyramidal neurons 142-145, 276, 439-447, 835-851
selective T5.2 neurons 25, 76-91, 100, 129, 205, 206, 245-248, 263, 461, 462, 468, 476, 523, 597; *see also Tectal*
vs. predator-discrimination 50, 61, 69, 97, 220, 436, 833, 837, 840, 848
Prey-catching 41, 45, 55, 56, 64-66, 68, 78, 79, 126, 193, 209, 245, 270, 296, 465, 651, 767, 778, 783, 790, 801, 810, 823
2DG-mapping 63, 64, 66, 817, 820; *see also Deoxyglucose*
action patterns 43, 45, 56, 431

commands 95, 98, 472, 475, 541
conditions of release 45, 95
measures 47, 52-56
modification 55
modulation 55, 59, 98, 783
ontogeny 43
optic tectum; *see Optic tectum*
orienting 44, 49, 51, 461, 540
pre-motor responses 461-468
releasing mechanisms 40, 44, 45, 95
sequential release 44, 45
snap evoking area 64
snapping 44, 51, 64, 462, 591
thalamic pretectal structures; *see Pretectum, Thalamic*
Prey-localization; *see also Stimulus*
command elements 95, 96
Grobstein's concept of activity gated divergence 539-541
locus dependend action 43-45, 95
models 149, 544, 557
patterns 43-45
specialized neurons 92, 93
Prey-recognition 45, 56, 57, 69, 82, 86-89, 125, 128, 137, 152, 473, 784, 801, 815; *see also Configuration, Discrimination, Recognition, Stimulus*
Arbib & Lara's column model 137, 140
an der Heiden & Roth's retino-tectal model 248-263
Barlow's retinal detector concept 86, 458
Betts's pretecto-tectal dynamic model 271-297
Cervantes, Lara & Arbib's pretecto-tectal column interaction model 142, 440
Didday's tectal competition model 125
Ewert & v. Seelen's pretectal/tectal filter interaction model 80, 88, 89, 128, 129
Grüsser's motor concept 45, 87
history 86-88
innateness 62
Lettvin's retinal detector concept 86, 459
neuronal tuning 77, 84, 100, 271
species differences in toads 54, 84, 85, 90, 244, 248
Primary visual cortex; *see Cortex*
Primates 61, 540, 541, 554, 562; *see also Humans*
Primordium hippocampi 107, 740, 756, 789, 820, 826; *see also Homologies, Medial pallium*
Procion Yellow 202
Programming language; *see also Computer, Language*
Programs 43
ballistic action patterns 43, 56
coordinated control; *see Coordination, Schema*
prey-catching 43
route planning 43
Projection neurons 73, 93, 94, 179, 181, 183, 189, 191, 194, 205, 210, 212, 213, 218, 219, 360, 483, 485; *see also Axonal*
Proprioceptive information 159, 481, 482, 525, 556, 632
Prothoracic ganglion 634
Protopterus annectens 698
Pseudoconditioning 745, 748, 750, 758; *see also Conditioning*
Psychology 422
Pulse-
analogue correlation 213, 214, 220
triggered averaging 213, 220
Pyramidal
cortical neurons 360, 752, 820, 877
tectal model neurons 137-145, 276-290, 294, 440-446, 835-851
tectal neurons 73, 80, 179, 192, 205-207, 594, 596

Quadratic regularization 363

R-type neuron classification; *see Retinal*
Rabbits 694, 748, 824
Radially symmetric structure 108; *see also Circular*
Radio telemetry 858
Rana 40, 94, 893; *see also Frogs*
berlandieri forreri 227
catesbeiana 699
computatrix xiv, 23, 34, 125, 126, 140, 147, 159, 168, 220, 248, 334, 893
esculenta 54, 176
pipiens 176, 237, 342, 800
temporaria 54, 84, 203, 205, 248, 252, 270, 616, 699
Raphe nucleus 600, 611, 742
Rats 554, 694, 697, 729, 731, 734, 737, 740, 757, 789, 815, 818, 819, 882
Reaching 123, 627, 682; *see also Arm, Dexterous, Grasp*
Readiness 469, 707
Reafference: *see also Corollary, Efference, Feedback, Loops*
information 314, 560, 562, 584
principle 47
vestibular signal 560
Receptive field 82, 202, 218
arrangements 75, 249, 521
excitatory 71, 249, 258, 259
inhibitory 70, 71, 249
medullary neurons 488
retinal neurons 70, 71, 244, 249, 259, 458
small; *see Small field*
tectal neurons 74-76, 92, 93, 202, 259, 460
thalamic pretectal neurons 74-76, 209
wide; *see Wide field*
zones of maximum activity 514
Reciprocal inhibition 573; *see also Antagonist, Interaction, Lateral, Mutual*
Recognition 57, 89, 140, 152, 424, 833, 894; *see also Configuration, Discrimination, Prey, Stimulus*
algorithms 24
invariance 52, 384-392
prey; *see Prey*
syntactic 385
workpieces 684; *see also Robot*
Recommendation to walk 636, 645, 646
Recording
and iontophoretic labeling 70, 202-220, 322, 501, 519
EEG; *see EEG*
extracellular 70-94, 246, 317, 487, 634
intracellular 202-220, 322, 501
multiunit 874
parallel 498, 499
scalp 875
telemetric in behaving animals 858-870
voltage sensitive dyes 874
wirebound in behaving animals 452-476
Recurrent; *see also Feedback, Lateral, Mutual*
excitation 298; *see also Reverberatory*
inhibition 261
Redundance 13
Reflex
concept 562
models 564
Regional metabolism 877; *see also Correlation, Deoxy-*

glucose
Regularity
mechanism 21
problem 10
Reinforcement 737, 739, 822; *see also Associative, Conditioning, Motivation*
Reinstallation of function 21, 23; *see also Functional, Plasticity*
Release call 788
Release of behavior
conditions 44, 45, 516
current concepts 40, 95, 107
innateness 40, 62
mechanisms 40, 43-45, 92, 97
schemas 40, 43, 57, 95, 97, 98; *see also Schema*
systems 95; *see also Command*
Reproductive behavior 295; *see also Breeding*
Reptiles 106, 699, 700
Reset functions 469, 607, 608, 627
Resolving vs. tuning 22, 520, 577; *see also Localization, Prey*
Respiration 591
Retention 108
Reticular; *see also Mesencephalic*
activating system 30, 523, 757
descending pathways 790
evoked arousal 737, 742, 749
evoked bradycardia 732
evoked inhibition 739
formation 696, 729, 745, 784
mechanisms 742, 747
sensitization 752, 758
stimulation 737, 740, 752
thalamic nucleus 735
Reticular formations
medullary (bulbar); *see Bulbar, Medial, Medulla*
pontine ; *see Pontine*
rhombencephalic 486
tegmental; *see Mesencephalic, Tegmental*
Reticulo-spinal
activity 527
bursting neurons 482; *see also Bursting*
projection cells 193, 486, 501, 523, 526, 527
Retinal
detector concepts; *see Prey*
ganglion cells 71, 72, 176, 201, 244, 248, 252, 255-257, 279, 392, 458, 459, 835, 841
R1-R4 types 70-72, 80-82, 88, 126, 244, 276, 301, 458
R1, R2 type 86, 244, 249, 252, 297, 294, 459
R3 type 86, 279, 294, 458
R4 type 244, 277
information processing 70, 86, 248-257
maps 7, 74, 168, 354, 358, 365, 366, 570, 616; *see also Map*
prey recognition concept; *see Prey*
projection fields; *see Map*
Retrieval of information 108 841, 850
Retrograde cell labeling 599-604
Reverberatory activities 79, 137, 140, 219, 496, 497, 521, 522, 527; *see also Interaction, Lateral, Mutual, Recurrent*
Reward related coding 823; *see also Reinforcement*
Rewarding stimulation 737
Rhythmic; *see also Cyclic, Bursting*
action patterns 524, 526
motor output 634, 636
Robot xii, 14, 121, 124, 651, 664, 666, 668, 673, 686, 893; *see also Machine, Technological*
actuators 167
applications 676
architecture 649, 650
assembly system 14, 677, 681
AURA 650, 159
dexterous manipulation 686
grasping specialists 415, 670, 618-683
grippers 677, 680
HARV 650
intelligent systems 425, 427, 649; *see also Artificial*
mobile 415
perception 122
Puma 677
schema-based navigation 159, 649, 650-670
sensorimotor coordination 670, 676, 685
sensors 674-676, 680-685
vision 379, 415
work piece acquisition 681-684
Robotics science 6, 15, 415, 422, 424-427, 649, 675, 676
Roof schema 165
Rostral thalamus 66, 67, 740; *see also Anterior, Thalamic*
Rotation invariance 384, 390; *see also Invariance*
Rutilus rutilus 694, 713
Saccade
control 550, 554, 560
generator 627
movements 469, 527, 528, 540, 607, 608, 627
oculomotor integrator 528
suppression 469
Salamanders 244-248, 311, 313, 329, 778; *see also Hydromantes*
Salamandra salamandra 84, 85, 86, 176, 245, 248, 313
Saliency map 353, 354; *see also Map*
Sameness properties 470, 493, 497, 522, 783, 788; *cf. also Newness*
Sample
and hold mechanism 496
tasks 684
Sampling theorem 580, 583
Scalar information 578
Scale invariance 390; *see also Invariance*
Scalp recorded electrical activity 875, 888; *see also Recording*
Scene 385, 893, 894; *see also Background, Environment*
Schema xii, 4, 40, 43, 57, 61, 121, 123, 404, 415
assemblage 136
based models 123, 124, 128, 133, 159, 163, 414, 415, 431, 432, 653; *see also Model*
change 850
combination 658, 670
coordinated control programs 122, 123, *see also Coordination*
definitions 57, 121, 123, 404, 415, 430, 653
deinstantiation 655
goal directed 431
heterarchical interactions 431
hierarchical/parallel 405, 407
innate releasing 40
instantiations 98, 123, 124, 126, 434, 654, 661
interactions 405
master 433
motor 133, 430, 592
perceptual 122, 430
plasticity 61
subschema 123, 124, 159
superposition 405
terminology 97, 98; *see also definitions*
theory 122, 399, 406, 414, 415, 421, 422, 428, 430, 432, 436, 447, 652
tuning 124

well matched 403
Schema based model of
contour perception in sensory substitution 398, 399
detour behavior 128
finger adjustment 159
hand control 159
path planning 152
prey-acquisition 133, 134, 431, 432
prey recognition 126
robot navigation 653-670
vision 163
Schemas (robot's navigation)
Avoid-Moving-Obstacle 654
Avoid-Static-Obstacle 654-656, 659, 661, 662, 668
Find-Intersection 654
Find-Landmark 654
Find-Obstacle 654
Find-Path-Edge 654
Follow-The-Leader 661, 670
Impatient Waiting 663, 668
Move-Ahead 654-656, 659, 661, 663, 666, 668
Move-Robot 655
Move-To-Goal 655, 656, 659, 663, 664
Static-Obstacle 654
Stay-On-Path 654, 656, 659
Schemas (toad's prey-catching and detour)
Barr 134
Chasm 134
Cross-Chasm 434
Free Prey 134, 433
Jump-Barrier 434
Moving Object 432
Non-Moving Object 432
Prey 134
Prey-Barrier 135, 434
Prey-Chasm 135
Prey-No-Obstacle 135
Scotoma 344, 543
Searching image 411
Season 611; *see also Breeding*
Seeing machines 379, 415; *see also Computer, Machine*
Seizures 712, 713
Selective
attention 476, 792; *see also Attention*
inhibition of sensory input 697
perception 757, 759
Self-
computation 314, 329, 331, 332
excitatory networks 473, 497; *see also Reverberatory*
filter 312
induced moving images 47, 67, 69
motion 312, 313, 327, 330, 332
organization 91, 894, 895
sustained activities 467; *see also Reverberatory*
Sensitization 92, 237, 697, 700, 711, 712, 718, 745-748, 753, 758, 792; *see also Enhancement, Facilitation*
mechanism 707, 711
of neurons 709, 710, 717
Sensors; *see Robot*
fusion 677
limitations 684
Sensorimotor xii, xiii
circuitry 96, 522, 527, 548, 550, 555, 741; *see also Intermediate*
codes 23, 95-98, 102, 107, 416, 434, 472, 522, 482, 525, 874
coordination 97, 399, 401, 404, 421, 430, 575, 615, 634
integration 97, 472, 538, 563, 564, 624, 673, 676, 677, 683-686, 759
interface 42, 92, 95, 97, 537, 538, 540, 543, 550, 551, 555, 557, 594, 597, 604-611, 624, 673, 677
levels 104, 520, 528
link 592, 791
processing levels 543, 550, 555, 561, 627, 741
strategies 44, 45, 682, 683
tasks 894
terminology 97
transformation 97, 123
Sensory
associative/motivational interface 107
cortex 385, 696, 707; *see also Cortex*
experience 55, 99, 710, 758, 768, 783, 800, 801, 815, 834
feedback 611, 645; *see also Feedback*
maps 29, 63, 365, 406, 548, 556; *see also Map*
modulation; *see Modulatory*
plasticity 753; *see also Dynamic, Plasticity*
substitution 87, 397, 402
Septum 739, 777, 813, 823
Sequential
actions 44, 45, 95, 123, 134, 431, 610
machines 32; *see also Computer, Machine*
Serial; *see also Hierarchical*
computers 14, 121, 122
information processing 32, 41, 42, 422, 425, 783
Serotonin 224, 230, 236, 610, 611, 742
Sex hormones 99
Shannon's channel model; *see Coding, Information*
Shape 57; *see also Form*
Sharks 698
Shift-invariance 384-387, 390; *see also Invariance*
Short-term
attentional shifts 99, 462
habituation 768, 788, 841
memory 137, 165, 627, 660, 755
modulation 80, 84, 293
Sidestepping 65, 133, 652, 661; *see also Barrier, Detour, Obstacle, Path*
Sidewalk 407
Sign stimulus 39, 41, 56-61, 592, 616
Signal/noise separation 41; *see also Background, Object, Textured*
Significant coincidence 29, 45, 95; *see also AND, Configuration, Combination, Concurrent*
Simulation studies 9, *see also Computer*
Simultaneous; *see also Combination, Concurrent, Cooperation*
multiunit data 498, 887
operation of sensors and actuators 683
sensor-motor operation 685
telestimulation and recording 869
Simple principles 26; *see also Model*
Site-specific
habituation 55
neurons 627
wiping 619
Size
constancy 50, 90, 92
invariance 390
Skin vibration 618
Sleep 700, 713, 738, 739, 740, 790
Small
field neurons 74, 77, 202, 206, 208, 247, 259, 271, 460, 489, 492, 508
pear shaped neurons 73, 440, 835; *see also Pear-shaped*
Snapping
behavior 45, 462, 589, 591, 592, 598, 607-610

cells involved 605, 607, 818
evoking tectal area 64, 594, 597, 598, 816
evoking visual field 605, 801
generating circuitry 598, 605-608, 627
interneurons 605, 608
premotor tectal cells 465
releasing stimuli 51, 56
tectal efferents 45, 94, 551
variability 609
Sodium
dependent uptake 236
potassium pump 733
Software explosion 15; *see also Computer*
Sokoloff's 2DG technique 733, 769, 804, 875, 877; *see also Deoxyglucose*
Sokolov's hypothesis of habituation 784, 786, 792
Solitary system 507, 508, 523, 600; *see also Nucleus*
Somatosensory
cortex 749; *see also Cortex*
information 610, 616
Somatostatin 611; *see also Neuropeptides*
Somatotopic map 616; *see also Map*
Songbirds 61
Sound localization 548
Space; *see also Coordinates, Localization, Place, Prey, Spatial*
coding 92, 95, 97, 482, 540, 555, 556, 558; *see also Coding*
invariance 375, 379
monitoring neurons 92
variance 373, 375, 377
variant filtering 375, 376
Spatial; *see also Space*
facilitation 598
frequency filters 250, 364-386
information 570, 576
representation 543, 549, 550, 555; *see also Map*
temporal convergence 748; *see also Associative*
Spatiotemporal; *see also Spatial*
filtering 362, 363, 372
frequency channel 362
patterns 68, 121, 141
properties 47, 91
Specialist
for work piece acquisition 681, 682; *see also Robot*
neurons 23, 61, 85, 92, 93, 95, 205, 472, 485
purpose processor 167, 358
Species-specific; *see also Evolutionary, Homologies, Phylogenetic*
cognition 107
differences 54, 84-86, 90, 244-248
Spiders 577, 650
Spinal
cord 523, 527, 557
frogs 618, 619
motor column 604, 605
motor neurons 193
Spino-bulbar connectivity 522
Split-brain 752
Spontaneously active neurons 468, 496, 521, 527, 744, 745, 788; *see also Tonic*
Spreading depression 717
State-dependent; *see also Modulation, Modulatory, Motivation*
alertness 695
information 9, 102, 472, 525, 611, 745
modulation 41, 54, 59, 61, 69, 80, 84, 88, 89, 91, 98-101, 103, 473, 610
properties of neurons 271, 295, 297, 462
Static configuration 59; *see also Configuration, Gestalt*
Stationary obstacles; *see Avoidance, Barrier, Detour, Obstacle, Path*
Stellate neurons 73, 137-143, 276, 277, 440, 441, 835, 837, 851
Step rhythm 632
Stepping movements 542; *see also Motor, Walking*
Stereopsis 34, 146, 148; *see also Depth, Localization, Prey*
Stereoscopic disparity 34
Stick insects 632
Stimulus; *see also Combination, Configuration, Form, Prey, Shape*
angular velocity 53, 78, 90, 244-248, 257, 316, 319, 321
category 56
concurrent cues 44, 45, 95
concurrent events 55, 99, 100
configuration vs. area 50, 52, 281
configuration vs. shape 50, 57
continuum 41, 50, 51, 58, 59-61, 69, 77, 84, 85, 88, 89
discrimination; *see Discrimination*
distance 539, 546, 547, 555, 556; *see also Depth, Stereopsis*
elevation 555, 556
habituation; *see Habituation*
localization 521, 539; *see also Localization, Prey*
movement direction 53, 78, 91, 273, 283, 318-322, 327; *see also Directional*
specific codes 468; *see also Coding, Sensorimotor*
successive stimuli; *see Habituation*
Stimulus-response; *see Circuitry, Network, Neuro-*
Storage 783, 821, 836, 850; *see also Learning, Memory*
retrieval 851
Strategy planner 681; *see also Robot*
Striatal efferents 106
Striatum 68, 101, 195, 244, 750, 753, 757, 774, 785, 790, 791, 806, 808, 816, 817, 823
String
of features 405
of symbols 425
Structural; *see also Dendrites, Model, Morphological*
assignment problem 17-20, 33, 94, 218
principles 358
Structure; *see Background, Texture*
Strychnine 79, 237
Suboesophageal ganglion 634, 636
Substantia gelatinosa 741, 742, 757
Substitution
command elements 103
sensory channel 399
Subtectal neurons 522, 590
Subtractive
binocular pretectal cell 321
pretecto-tectal interactions 25, 65, 80-86, 88-91, 129, 142, 440-444
Sufficiency criterion 91, 570
Superstructure 378, 850; *see also Macro-network*
Supervisor manager 681, 683; *see also Robot*
Supra-
chiasmatic nucleus 773
oesophageal ganglion 635
spinal control system 195
sylvian cortex 335; *see also Cortex*
Surround inhibition 47, 55, 69, 72, 82; *see also Background, Object, Subtractive*
Sustained potential shifts 704,

707, 712, 718
Swallowing 43, 591, 608, 611; *see also Gulping*
Swarm phenomenon 47
Symbolic; *see also Schema*
description 57, 405, 407, 425
picture 57, 167
structure 108
Symmetric directional sensitivity 78, 275, 280, 285, 299; *see also Directional*
Synaptic 6, 16, 192
connections 203, 220
matrix 33
plasticity 894
tectal organization 176-183, 192, 202, 223-238
transmission; *see Neuropeptides, Neurotransmitters*
weights 142, 302
Synaptosomes 223
Synchronization
bursts of discharges 497-500, 700
cell activities 499, 528, 610
EEG 700, 712, 718
membrane potentials 702
oscillating populations 703
wave forms 700, 717
Syntactic pattern recognition 13, 385; *see also Recognition*
Systems theoretical model 108, 128

T-type classification; see *Tectal*
Tactile; *see also Cutaneous*
information 497, 511, 538, 542, 682, 697
releasers 493, 610
sensors 675, 679, 685; *see also Robot*
Tactually elicited
orienting 539, 540, 548-550
snapping 97, 606
Task planner manager 681, 683; *see also Robot*
Technological applications 24, 125, 159, 357, 674; *see also Computer, Machine, Robot*
configural analysis 14, 108
general considerations xii, 14-16, 122, 379, 427
image analysis 14, 165, 336, 379
path planning 650-670
sensory substitution 399-416
work piece acquisition 676
Tectal neuron classes
T1-T7 types 74-97, 179, 201, 202, 245
T1.1-T1.3 types 74, 75, 92-95
T2.1 type 74, 75, 91-93, 95, 219, 470, 474
T2.2 type 74, 75, 91, 92, 219, 470, 474
T3 type 74, 75, 93, 95
T4 type 74, 75, 92-95, 219, 470, 474, 496, 520, 708, 788
T5 base modulator hypothesis 271, 276
T5 types 74-97, 492, 521, 522, 706, 708, 788
T5.0 type 270, 513
T5.1 large ganglionic neuron 80, 205, 207, 219
T5.1 pear-shaped neurons 80, 205, 219, 837
T5.1 type 74-97, 99, 202, 205, 206, 218-220, 246-248, 264, 270, 460, 835
T5.2 pyramidal neurons 80, 145, 205, 206, 442-446
T5.3 large ganglionic neurons 80, 219, 837
T5.3 type 74-97, 202, 207, 218, 220, 246, 247, 270, 460, 504, 511, 834, 838, 843
T5.4 type 74-97, 202, 219, 460, 504
T7 type 74, 92, 93
T8.1 type 469, 702, 870
T8.2 type 469
T9 type 520
T5.2 type configuration selective neurons 74-99, 202, 205, 206, 209, 218-220, 246-248, 263, 264, 270, 287, 294, 440, 443, 461, 470, 473, 476, 482, 703, 705, 711, 783, 784, 788, 818, 834, 835, 838, 843, 844,
command elements 95, 472
comparison with cortical orientation detectors 25, 90, 91, 108
configuration selectivity 82-91, 76-78, 463, 473; *see also Configuration*
dynamic model 271, 276
enhancement effect 473, 469, 476
equivalent bulbar properties 492, 501, 510, 511, 520
essential vs. sufficient properties 91
implicit vs. explicit processing 90
learning models 100, 835-851
loss in selectivity after pretectal lesion 83, 84, 465, 466, 473
models 69, 129, 142, 144, 261, 440-446; *see also Concepts, Model*
motor feedback 473
pre-motor properties 461, 473
prey-catching correlation 76, 463, 473
prey recognition 88-90, 463, 473; *see also Prey*
projections 93, 485
pyramidal neurons 73, 80, 142, 144, 205, 206, 276, 440-446, 837
species differences 84-86, 90, 248
state dependence 98-100, 462-465, 473
tuning ranges 77
velocity invariance of worm vs. antiworm-preference 78, 90, 246
Tecto-
bulbar/spinal connections 80, 88, 92-94, 181, 184, 193, 454, 472, 473, 482, 483, 485, 496, 498, 507, 521, 522, 551, 554, 593, 596-598, 826
hypoglossal pathway 101, 606
isthmic connections 181, 184, 195, 347, 484
pretectal connections 195, 210, 522
reticular connections 184, 482, 486
tegmental connections 195, 482, 483, 545, 551, 554, 557, 558
thalamic connections 80, 183, 210
Tegmentum 16, 187, 195, 244, 350, 354, 551-558, 594, 791, 806
afferents/efferents 194
axons 554
cell groups 101, 554
cholinergic pathways 740
nuclei 68, 92, 191, 482, 526, 556, 605
reticular formation 526, 605, 697, 772, 778, 779, 786; *see also Mesencephalon*
stimulus localization 554, 557; *see also Localization, Prey*
Telemetric
marking 858, 868
recording 858, 860
stimulation 858, 863, 864, 866, 867
three-function transmitting system 857, 858
Telencephalon 69, 101, 106, 107; *see also Cortex*
amygdala; *see Nucleus*
medial pallium; *see Medial*
pallial structures 101, 773, 805-

808
septum; *see Septum*
striatum; *see Striatum*
Teleost fish 324, 697, 788
Textured background 25, 46, 47, 55, 219, 392, 493, 511, 514, 674; *see also Background, Object*
Thalamic pretectal neuron classes
TH1-TH10 types 73-77, 201, 209, 440, 522
TH3 type 74-88, 100, 141, 219, 220, 270, 287, 321, 504, 511, 834, 835, 838, 840, 841, 850
TH4 type 74, 92, 97, 520
TH9 type 818
Thalamic pretectal nuclei 63, 65, 84-86, 187, 244, 269, 270, 297, 707, 737, 785, 834, 850; *see also Pretectum*
immunocytological studies 80, 106, 107, 194
lesions 65-70, 83-86, 514, 834
phylogenetic aspects 106, 107
Thalamo-
tectal connections *see Brain, Pretectum*
tectal inhibition; *see Brain, Pretectum*
cortical connections 700, 730, 734, 737, 738, 749, 753
Theory and experiment 3, 108, 153, 425; *see also Model*
Three-dimensional; *see also Coding, Coordinates, Localization*
coordinate frame 539, 561
head/body centered signal 555
maps 578
movement 546, 547
world 336
Threshold; *see also Weighting*
function 138, 140, 303, 305, 840, 852
operations 88, 89, 253, 282, 521
weigths 141-143, 305, 853
Time-division-multiplexing 858
Toads 39, 40, 125, 269, 415, 589, 696, 740, 756, 767, 778, 858; *see also Alytes, Bufo*
Tongue; *see also Snapping*
grasping 472, 592
motoneurons 508, 592, 593, 595
movements 475, 542, 546, 559, 592
muscles 592, 593
snapping 44, 95, 472, 591
Tonic firing neurons 74, 481, 482, 469, 496, 502, 523, 527, 528; *see also Spontaneous*
Tonotopic
labeling 749
maps 747
Top-down studies 42, 122, 422, 423, 436; *cf. also Bottom-up*
Topography; *see Map*
Torus semicircularis 483, 521, 557, 593, 604, 605, 608
TP; *see also Pretectal, Thalamic-, Thalamo-*
modulated T5-base 264, 297
phenomenon 67, 83, 263, 264
Tractus tecto-bulbaris/spinalis 52, 93, 94, 184, 193, 210, 482, 484, 490, 522, 523, 551, 593, 594, 604; *see also Bulbar, Medulla*
Transputer 14; *see also Computer*
Triangulation 149; *see also Depth*
Trigeminal
motor nerve nucleus 593
facial-glossopharyngeal-vagal afference 611
Trigger efficacy 468
Tuning
function 572, 575, 583, 584
model parameters 144
ranges of visual neurons 77
Turing Machine 32, 422; *see also Computer*
Turning; *see also Localization, Orienting, Prey*
amplitude 548
generating circuitry 546
measures 56
movements 551, 552, 652, 778
procedures 49, 127, 457
releasers 45-54, 60
tendency 581, 583
Turtles 616, 623, 626
Two-stage; *see also Levels*
pretecto-tectal inhibitory inputs 80, 85, 440
sensorimotor integration 104, 520, 683, 684; *see also Sensorimotor*

Ultrasonic perception 661, 663, 670; *see also Robot*
Uncinate field 187
Upstream analysis 592
Urodeles 53, 86, 698, 789; *see also Hydromantes, Salamandra*

Varanus griseus 699
Velocity 481; *see also Stimulus*
coded neuronal signal 328
dependent activities 53, 57, 246, 247, 257
invariance in configural perception 53, 78, 90, 246, 247, 273
storage integrator 328
vectors 53, 78, 273, 377, 378
Ventral
cholinergic pathway 740
medial pallium 28, 740; *see also Medial pallium*
pallidum 739
Veratrine 228, 238
Vertebrates 548, 554, 631, 788; *see also Homologies*
Vertical
OKR 317; *see also Optokinetic*
saccades 526
Vestibular
background motion systems 329
collic reflex 314, 324
compensation movement 325
gaze stabilization reflex 314, 327
nuclei 741
ocular reflexes 195, 314, 607
postrotatory nystagmus 325
proprioceptive information 524, 578
semicircular canal system 314, 327
VISIONS machine 165
Visual
and olfactory association 55, 811, 824
arousal 474; *see also Arousal*
attention 99, 462, 476, 694
channels 62, 63, 94
conditioning 55, 801
fixation cells 469
grasping 65, 461, 472, 615
maps; *see Map*
pattern recognition; *see Prey, Recognition*
pecularities in amphibians 56
tracking neurons 476
vestibular interactions 325, 328; *see also Vestibular*
Visuomotor coordination; *see Motor, Sensorimotor*
Voltage sensitive dyes 875, 876, 888

W-configuration; *see Worm-*
Wakefulness 700, 713, 737, 738, 790; *see also Arousal, Attention*
Walking
decision-making 645

gaits 631, 633
generator 633-636, 644
motor output 638, 639, 645
movements 631, 635, 638, 646
wheel 634
Walsh power-spectra 385
Warming-up properties 104, 461-468, 473, 476, 482, 497, 520-524
Water wave localization 548
Weighting; *see also Threshold*
function 34, 363, 575, 576
interaction 90
map ordered information 569-584
patterns 140
Wetware 358
White matter 507, 553, 600
Widefield neurons 74, 75, 92, 97, 157, 202, 470, 474, 488, 492-496, 521, 522
Wiping 608, 615-619, 624, 625 778
command functions 625, 626
forms 616, 619-621, 623, 625-627
functional compartments 626
matrix representation 621
problem of choice 623-625
receptive fields 617, 619
signal integration 624
Wire-
bound recording 452-456
less recording 858-870
Working memory 893, 894; *see also Memory, Storage*
Workpiece acquisition operations 681, 683; *see also Robot, Specialist*
Worm; *see also Antiworm, Configuration, Prey*
and antiworm continua 51, 60, 77, 271
configuration 50, 59,-61, 72, 128, 144, 146, 205, 206, 219, 245, 272, 278, 461, 815, 833
filter 89
information-bearing parameters 61
vs. antiworm discrimination 50-91, 263, 437, 511

X*enopus laevis* 800